Physical Constants

Standard acceleration of terrestrial gravity	$g = 9.80665$ m s^{-2} (exactly)
Avogadro's number	$N_0 = 6.022137 \times 10^{23}$
Bohr radius	$a_0 = 0.52917725$ Å $= 5.2917725 \times 10^{-11}$ m
Electron charge	$e = 1.6021773 \times 10^{-19}$ C
Faraday constant	$\mathscr{F} = 96{,}485.31$ C mol^{-1}
Universal gas constant	$R = 8.31451$ J mol^{-1} K^{-1}
	$= 0.0820578$ L atm mol^{-1} K^{-1}

Masses of fundamental particles:

Electron	$m_e = 9.109390 \times 10^{-31}$ kg
Proton	$m_p = 1.672623 \times 10^{-27}$ kg
Neutron	$m_n = 1.674929 \times 10^{-27}$ kg
Ratio of proton mass to electron mass	$m_p/m_e = 1836.15270$
Planck's constant	$h = 6.626076 \times 10^{-34}$ J s
Speed of light in a vacuum	$c = 2.99792458 \times 10^8$ m s^{-1} (exactly)

Values are taken from "Quantities, Units and Symbols in Physical Chemistry," International Union of Pure and Applied Chemistry, Blackwell Scientific Publications, 1988.

Conversion Factors

Standard atmosphere	1 atm $= 1.01325 \times 10^5$ Pa $= 1.01325 \times 10^5$ kg m^{-1} s^{-2} (exactly)
Atomic mass unit	1 u $= 1.660540 \times 10^{-27}$ kg
	1 u $= 1.492419 \times 10^{-10}$ J $= 931.4943$ MeV
	(energy equivalent from $E = mc^2$)
Calorie	1 cal $= 4.184$ J (exactly)
Electron volt	1 eV $= 1.6021773 \times 10^{-19}$ J $= 96.48531$ kJ mol^{-1}
Foot	1 ft $= 12$ in $= 0.3048$ m (exactly)
Gallon (U.S.)	1 gallon $= 4$ quarts $= 3.78541$ L (exactly)
Liter-atmosphere	1 L atm $= 101.325$ J (exactly)
Metric ton	1 metric ton $= 1000$ kg (exactly)
Pound	1 lb $= 16$ oz $= 0.45359237$ kg (exactly)

CHEMISTRY

science of change

third edition

David W. Oxtoby
The University of Chicago

Wade A. Freeman
University of Illinois at Chicago

Toby F. Block
Georgia Institute of Technology

Saunders Golden Sunburst Series

SAUNDERS COLLEGE PUBLISHING
Harcourt Brace College Publishers

PHILADELPHIA FORT WORTH CHICAGO SAN FRANCISCO
MONTREAL TORONTO LONDON SYDNEY TOKYO

Requests for permission to make copies of any part of the work should be mailed to: Permissions Department, Harcourt Brace & Company, 6277 Sea Harbor Drive, Orlando, Florida 32887-6777.

Publisher: Emily Barrosse
Publisher: John Vondeling
Acquisitions/Development Editor: Kent Porter Hamann
Product Manager: Angus McDonald
Project Editor: Frank Messina
Art Director: Caroline McGowan
Text Designer: Chazz Bjanes
Cover Designer: Kathleen Flanagan
Production Manager: Charlene Catlett Squibb

Cover Credit: Charles D. Winters

Printed in the United States of America

Chemistry: Science of Change, Third Edition

ISBN: 0-03-020088-1

Library of Congress Catalog Card Number: 97-67375

7890123456   048   10 987654321

In Memory of Norman H. Nachtrieb

Norman H. Nachtrieb, our colleague, friend, and co-author, died in September 1991. His careful examination of what actually works in teaching chemistry and his years of successful experience as a teacher and researcher provided the original impetus for this book. His spirit continues, guiding us in everything from the proper use of language to a more enduring appreciation of the beauty of the chemistry that is described. We dedicate this third edition of Chemistry: Science of Change *to his memory.*

David W. Oxtoby

Wade A. Freeman

Toby F. Block

Preface

Chemistry: Science of Change, Third Edition, is intended for use as the text in introductory chemistry courses taken by students of biology, chemistry, geology, physics, engineering, and related subjects. The text builds on students' knowledge of algebra but does not require calculus. Although some background in high-school science is assumed, no specific knowledge of topics in chemistry is presupposed; this book is a self-contained presentation of the fundamentals of chemistry. Its aim is to convey to students the dynamic and changing aspects of chemistry in the modern world.

The fundamental organization of the third edition of *Chemistry: Science of Change* is the same as that of the first two editions. Chemistry, as an experimental science, is introduced to students in terms of macroscopic concepts and principles that have their origins in the laboratory. The more abstract, theoretical interpretative aspects of the subject then follow. The overall goals are to persuade students that both theory and experiment are indispensable to the progress of science and to help them appreciate how each aspect informs and guides the other.

Changes in This Edition

A number of changes and additions have been made to improve the text within the "macro-micro" framework just described, and several chapters have been extensively rewritten:

■ The introduction to the constituents of the atoms (electrons, protons, neutrons) and the masses and charges in Chapter 1 has been expanded, so that the subsequent discussion of the periodic table and Lewis theory in Chapter 3 has a more solid foundation. This material also connects closely to the general theme of Chapter 1: the analysis of what makes up matter.

■ Chapter 3 now presents the prediction of molecular shapes (using valence shell electron-pair repulsion theory) because it is a natural extension of Lewis theory. The revised discussion of Lewis theory now uses the electronegativity concept explicitly.

■ Chapter 4 has been completely rewritten. The emphasis is on using patterns of chemical reactivity to make reasonable predictions of the outcome of chemical reactions. Although the focus of the revised Chapter 4 remains inorganic chemistry, organic compounds and reactions now appear as well. The discussion of acid–base reactions in this chapter uses the Arrhenius model exclusively. This simplifies and shortens the discussion. However, there are numerous lead-ins for the immediate introduction of Brønsted-Lowry and Lewis acid–base theories, which now appear in Chapter 8. Solution stoichiometry appears in the last section of Chapter 4, where

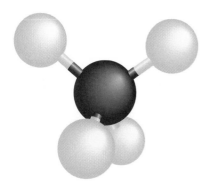

it provides required background for early laboratory work that includes standardizations and titrations. The first four chapters now constitute a solid practical introduction to aqueous reaction chemistry.

■ The discussion of phase equilibria now appears in Chapter 6 alongside the discussion of colligative properties instead of in a separate chapter. Also, the discussion of intermolecular forces in this chapter has been made less abstract, and the presentation of colloidal dispersions has been substantially rewritten and expanded because of the importance of this material in biochemistry.

■ In Chapter 10, the discussion of the definition and use of thermodynamic standard states, a topic students often find difficult, has been expanded and improved.

■ Chapter 15, which addresses nuclear chemistry, has been extensively rewritten and updated.

■ In Chapter 18, the discussion of molecular orbitals has been revised to emphasize the importance of the relative phase of the atomic orbitals that are being combined.

Some general changes were made as well. A new series of essays entitled "Chemistry in Your Life" has been introduced. These boxed features discuss the chemistry behind matters of public or student concern, usually in medical or environmental areas. Also, the art program has been enhanced, with the consistent integration of new, more vivid representations of molecular structure. Throughout the book, numerous small changes with the aim of clarifying and expanding explanations appear in both the figures and the text.

Organization

The underlying premise of this book is that chemistry is a science based on observations and measurements, from which deductions and principles can be derived. Our view is that the study of chemistry should start with the visible, tangible world of macroscopic experience and move from there toward the simple models that help to organize and rationalize daily experience. As students become comfortable with

the chemist's way of viewing natural phenomena, the more fundamental (but also less evident) theoretical basis of the subject can be introduced gradually. It is our hope that students will come to look upon theory and experiment (or observation) as mutually reinforcing aspects of science. Each informs the other, and together they point the way to a deeper understanding of the workings of the real world. Our organization also allows students to begin meaningful laboratory work at an early stage.

Chapters 1 through 6 therefore begin by emphasizing macroscopic aspects of chemistry, moving from mass relationships in chemical reactions and the atomic theory of matter through the periodic table of the elements toward a description of the states of matter. The Lewis electron dot model is introduced early as a useful organizing principle directly connected to the structure of the periodic table. Chapters 7 through 14 then present a full treatment of chemical equilibrium and its thermodynamic and kinetic basis. In the next seven chapters, the emphasis shifts to a systematic development of the modern microscopic picture of the structure of matter: from electrons and nuclei through atoms and molecules to coordination complexes and extended solid structures. Chapters 22 through 25 then present an overview of chemical processes: industrial, atmospheric, geochemical, and biological. These chapters introduce no new principles; instead, they provide a useful review of principles from earlier chapters, such as the tradeoffs between thermodynamics and kinetics in process design. Our presentation of industrial chemistry has a strong historical component, illustrating the ways in which chemical processes are developed and improved. There is real value in concluding an introductory course with a broader look at chemistry in its "real-world" context, and that has been our goal in the final chapters.

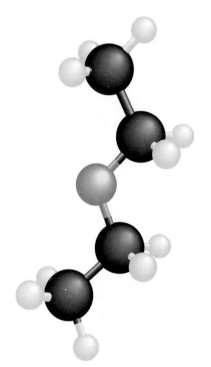

Alternative Teaching Options

We recognize that some instructors may prefer a different order of presentation from that just described. We have therefore made the book flexible enough to accommodate many alternative organizations. For example, the material on atomic structure and chemical bonding in Chapters 17 through 19 could be introduced after Chapter 3 or after Chapter 10. Chapters 15 and 19 through 21 could be postponed until later in the course or omitted if time constraints are severe. Some instructors choose to split the coverage of thermodynamics into two parts. Those wishing to do so could cover Chapter 10 anytime after Chapter 5, while postponing Chapter 11 until later in the course. Other instructors may choose to postpone the discussion of intermolecular forces in Section 6–1 until after the quantum theory has been introduced in Chapters 16 and 17.

A number of individual sections of chapters can be omitted without loss of continuity. For example, the last three sections of Chapter 9 (Dissolution and Precipitation Equilibria), while important for any course covering qualitative analysis of metal ions, are not essential to the material that follows them in the book. Chapter 15 (Nuclear Chemistry) could also be omitted or postponed until later in the course. Chapters 22 and 23 can be treated as "interchapters" and covered earlier. For example, Chapter 22 could be covered after Chapter 11 and Chapter 23 after Chapter 13. Chapters 24 and 25 provide a coherent introduction to organic chemistry (including polymers) at a level of detail appropriate to general chemistry and could be included after Chapter 18 or at the end of the course.

Appendices

The book contains six appendices. Appendix A treats scientific notation, experimental uncertainty (including the distinction between precision and accuracy), and significant figures. Appendix B gives an overview of the SI system of units and shows how units are interconverted. Appendix C treats some mathematical operations used in general chemistry, including the drawing and reading of graphs, the solution of quadratic equations, and the use of logarithms. These three appendices are placed at the back of the book for easy reference, but the material in them can be used by the instructor at the appropriate points in the course. All three contain problems for students to work to test their understanding of the material.

Appendices D and E consist of tables of thermodynamic and electrochemical data. Together with the tables inside the covers of the book, these provide quick access to data that are useful in solving problems in the text. Appendix F contains the answers to selected odd-numbered problems at the end of each chapter.

Acknowledgments

This third edition has benefited greatly from many incisive comments and suggestions from chemists around the country. Reviewers include

Lester Andrews, University of Virginia
E. David Cater, University of Iowa
Mary K. Campbell, Mt. Holyoke College
Timothy P. Hanusa, Vanderbilt University
David O. Harris, University of California, Santa Barbara
Stephen Quirk, Georgia Institute of Technology
Donald J. Wink, University of Illinois at Chicago

The staff at Saunders College Publishing has been extremely helpful in preparing this third edition. In particular, we would like to acknowledge the key role of our Developmental Editor, Kent Porter Hamann, who guided us toward the revisions in this third edition, and our Project Editor, Frank Messina, who kept a very tight schedule moving along smoothly. The continuing support and encouragement of Publisher John Vondeling is also gratefully acknowledged.

Rhonda Staudohar, Monica Rausa, and Kathy Wingate helped with printing, copying, and mailing. Our spouses and families once again supported our efforts in many ways, including putting up with our many hours away working on revisions of the manuscript.

DAVID W. OXTOBY
WADE A. FREEMAN
TOBY F. BLOCK
JUNE 1997

Supplements

Supporting Print Materials

Instructor's Manual by Wade A. Freeman. Included are lecture outlines, classroom demonstrations, and solutions to the even-numbered problems in the book. Also included are discussions of some conflicting usages in symbolism and terminology and the reasons for the choices that were made in the text.

Student Solutions Manual by Donald Wink, University of Illinois at Chicago. A manual of the odd-numbered problems in the book and their detailed solutions.

Student Study Guide by E. David Cater, University of Iowa. Includes additional worked-out examples, as well as questions and practice examinations, to help students gauge their mastery of each chapter and to encourage them to think about important chemical concepts.

Overhead Transparencies One hundred full-color figures from the text are available.

Problems Book by Ronald Ragsdale. Used as a supplement, this book helps students master problem-solving in general chemistry.

Printed Test Bank by Joel Tellinghuisen, Vanderbilt University. Contains over 1000 multiple-choice questions.

Laboratory Experiments for General Chemistry, Third Edition by Harold R. Hunt, Toby F. Block, and George M. McKelvy, Georgia Institute of Technology. A laboratory manual with 43 experiments, designed with careful regard for safety in the laboratory. An instructor's manual is also available.

Multimedia Materials

Saunders Interactive General Chemistry CD-ROM is a revolutionary interactive tool. This multimedia presentation serves as a companion to the text. Divided into chapters, the CD-ROM presents ideas and concepts with which the user can interact in several different ways, for example, by watching a reaction in progress, changing a variable in an experiment and observing the results, and listening to tips and suggestions for understanding concepts or solving problems. Students navigate through the CD-ROM using original animation and graphics, interactive tools, pop-up definitions, over 100 video clips of chemical experiments, which are enhanced by sound effects and narration, and over 100 molecular models and animations.

Cambridge Scientific ChemDraw and Chem3D are available shrink-wrapped with the text for a nominal fee. These software packages enable students to draw molecular structures using ChemDraw. Users draw with ChemDraw and then can transfer their structure into Chem3D, which allows them to create, manipulate, and view three-dimensional color models for a clearer image of a molecule's shape and reaction sites. Cambridge Scientific provides an accompanying User's Guide and Quick Reference Card, written exclusively for Saunders College Publishing.

f(g) Scholar–Spreadsheet/Graphing Calculator/Graphing Software f(g) Scholar is a powerful scientific/engineering spreadsheet software program with over 300 built-in math functions, developed by Future Graph, Inc. It uniquely integrates graphing calculator, spreadsheet, and graphing applications into one, and allows for quick and easy movement between the applications. Students will find many uses for f(g) Scholar across their science, math, and engineering courses, including working through their laboratories from start to finished reports. Other features include a programming language for defining math functions, curve fitting, three-dimensional graphing, and equation displaying. When bookstores order f(g) Scholar through Saunders College Publishing, they can pass on our exclusive low price to the student.

CalTech Chem istry Animation Project (CAP) is a set of six video units of unmatched quality and clarity that cover the chemical topics of Atomic Orbitals, Valence Shell Electron Pair Repulsion

Theory, Crystals and Unit Cells, Molecular Orbitals in Diatomic Molecules, Periodic Trends, and Hybridization and Resonance.

The Chemistry of Life Videodisc contains over 100 molecular model animations, chemical reaction videos, and approximately 2500 still images from a variety of Saunders College Publishing chemistry textbooks.

The Chemistry in Perspective Videodisc contains over 110 minutes of motion footage, including molecular model animations, chemical reaction videos, animated principles of chemistry, and videos demonstrating chemical principles at work in everyday life, as well as 2000 still images from Saunders College Publishing chemistry titles.

LectureActive™ Software for Macintosh and IBM formats accompanies The Chemistry of Life Videodisc and contains references to all video clips and still images. Instructors can create custom lectures quickly and easily. Lectures can be read directly from the computer screen or printed with barcodes that contain videodisc instructions.

Barcode Manual for The Chemistry of Life Videodisc contains complete descriptions, barcode labels, and reference numbers for every video clip and still image. The Manual also provides practical advice about The Chemistry of Life Videodisc and set-up instructions for first-time users.

ExaMaster+™ Computerized Test Bank is the software version of the printed Test Bank. Instructors can create thousands of questions in a multiple-choice format. A command reformats the multiple-choice question into a short-answer question. Adding or modifying existing problems, as well as incorporating graphics, can be done. ExaMaster has gradebook capabilities for recording and graphing students' grades.

KC?Discover Software (JCE: Software) developed by a team of chemists, is an extensive database allowing users to explore 48 different properties of the chemical elements, such as atomic radii, density, ionization energy, color, reactivity with air, water, and acids and bases. The HELP menu provides a reference for the source of data for each of the properties and the database. KC?Discover Software has the capability to correlate with the Periodic Table Videodisc. Chosen as the "official" database for this book, references appear in the Annotated Instructor's Edition of this text, and all the data tables in this book reflect information from KC?Discoverer. This program is available to qualified adopters or it may be purchased directly from JCE: Software, Department of Chemistry, University of Madison, Wisconsin, 1101 University Avenue, Madison, WI 53706.

Periodic Table Videodisc: Reactions of the Elements (JCE: Software) by Alton Banks, North Carolina State University, features still and live footage of the elements, their uses, and their reactions with air, water, acids, and bases. Users operate the videodisc from a videodisc player with a hand-controlled keypad, a barcode reader, or an interface to a computer running KC?Discoverer Software. It is particularly useful as a way to demonstrate chemical reactions in a large lecture room. Available to qualified adopters.

Shakhashiri Demonstration Videotapes feature well-known instructor Bassam Shakhashiri, University of Wisconsin, performing 50 three-to-five-minute chemical demonstrations. An accompanying Instructor's Manual describes each demonstration and includes discussion questions.

World of Chemistry Videotapes taken from the popular PBS television series and hosted by Nobel laureate Roald Hoffmann, are two 40-minute videos highlighting topics such as the mole, bonding, and acid-base chemistry and their applications. Order through the Annenberg Foundation at 1-800-LEARNER.

Saunders College Publishing may provide complimentary instructional aids and supplements or supplement packages to those adopters qualified under our adoption policy. Please contact your sales representative for more information. If as an adopter or potential user you receive supplements you do not need, please return them to your sales representative or send them to:

Attn: Returns Department
Troy Warehouse
465 South Lincoln Drive
Troy, MO 63379

To the Student

Chemistry is an immensely practical subject that, properly understood, furnishes answers to many important problems. Like all of the sciences, chemistry is an experimental discipline. Doing course work such as homework, quizzes, and examinations is very much like doing experiments. The difference is that someone else has asked the question (done the experiment) and collected the results in the form of data. It remains for you to devise a strategy for converting the data into an answer. How well you answer the question depends on how well you have "sized up" the problem.

Understanding exactly what is wanted is the first step. The second step is to call to mind the principle(s) that underlie the desired result. What factors are involved? Decide whether the information provided is relevant and discard anything that is not. Plan a method of attack and carry it out. Once you have an answer, consider whether it "makes sense." Is it about the right size? Are its units correct? Was there another approach you might have taken to work the problem? If so, does it give the same answer? (Often there are several ways to solve a given problem.)

As you gain experience in solving problems, you may find that your way of looking at new situations has changed. You may find yourself "looking for the facts," wondering what lies behind an occurrence, questioning the validity of a piece of information. Solving problems is a *transferable skill* that applies not only to working chemistry problems (and getting good grades on exams), but also to almost every activity in life.

As you are using this book, read thoughtfully for a few pages and then set the book aside. Reflect on what you've read. Is there a concept that might prove useful, a generalization that ties several ideas together, an analogy that is especially apt? Summarize it succinctly by making a note of it in your own words. The act of putting pen to paper is in itself a remarkable assist to the memory, and expressing the concept in your own words is a powerful study aid.

Another suggestion is to study with a classmate, taking turns instructing one another. There is no surer way to learn whether you understand a concept than to explain it to another person. Make up problems with which to test each other and, to overcome the "examination anxiety" that students sometimes experience, limit the time allotted to solve them. If you prefer, choose some of the odd-numbered problems at the end of the chapter you're studying. The answers are in Appendix F.

Above all, in your study of chemistry, develop a questioning, aggressive approach to what you read and hear. Ask for the evidence, and don't despair if you make mistakes. The wonderful edifice of science was built on trial and error, and persistence always wins!

About This Book

We have tried to make this a user-friendly book, and our principal concern has been to explain the subject in as clear and unambiguous a manner as possible. We have included a number of learning tools to help you increase your understanding of chemistry. We hope these features will stimulate your interest in the subject.

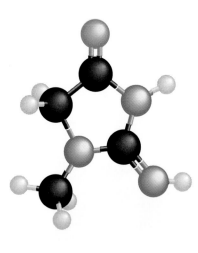

Figures

The book makes extensive use of **figures** to aid in explaining concepts and to illustrate chemical behavior. These include color photographs, diagrams, and drawings of molecular structures. Especially in the first 14 chapters, we use molecular structures with their figure captions to present brief "capsule summaries" of the properties and uses of important simple compounds. In these structural diagrams, carbon is black, hydrogen is light gray, oxygen is red, nitrogen is blue, sulfur is orange, phosphorus is purple, fluorine is yellow, chlorine is green, bromine is reddish brown, iodine is reddish purple, boron is brown, and xenon is blue-green.

Examples and Exercises

Each chapter contains numerous **worked examples** to illustrate the principles that have been discussed. Each example is followed by a related **exercise** together with its answer.

Highlighted Text

Throughout each chapter **key terms** are printed in boldface where they first appear. Important statements and equations are screened in blue, and answers to examples are screened in yellow. Acids are color-coded in red and bases in blue; formal charges are shown in blue circles, while oxidation numbers appear in red.

Marginal Notes

The occasional **notes in the margin** serve several purposes: to emphasize or elaborate on a point, to provide a different perspective on a statement made in the text, and to provide "signposts" directing you to other parts of the book where subjects have been or will be covered.

"Chemistry in Your Life" and "Chemistry in Progress"

Throughout the book we have included a number of **special boxed topics.** Those entitled "Chemistry in Your Life" focus on the chemistry of biological processes or else on the impact of chemistry in daily life. The boxes titled "Chemistry in Progress" show the ways in which chemistry contributes to and benefits from developments in science and technology that change the modern world.

Chapter Summaries

Each chapter ends with a **summary** of the material presented in that chapter. The purpose of this is to place the new material in context and to review it. These summaries should be especially useful when you are studying for an examination.

End-of-Chapter Problems

These **problems** are arranged in two groups. The first group consists of paired problems, placed in order of presentation of the topics in the chapter. The answer to the odd-numbered problem in the pair is given in Appendix F, allowing you to check the answer before undertaking the second problem, which parallels the first. These paired problems are followed by a group of **additional problems,** which may involve concepts from different parts of that chapter and which may draw on material from earlier chapters as well. More challenging problems are indicated with an asterisk.

Cumulative Problem

At the end of each of Chapters 1 through 20 there is a multipart **cumulative problem** that is built around a problem of chemical interest and draws on material from the entire chapter for its solution. Working through these exercises provides a useful review of material from the chapter, helps you to put principles into practice, and prepares you to solve the problems that follow. The answers to these problems appear in Appendix F.

Index/Glossary

The **index/glossary** at the back of the book gives a brief definition of each term and a reference to the pages on which that term appears. You should get in the habit of consulting it when you encounter a term that you do not fully understand or one you have encountered earlier in the book and need to review.

Contents Overview

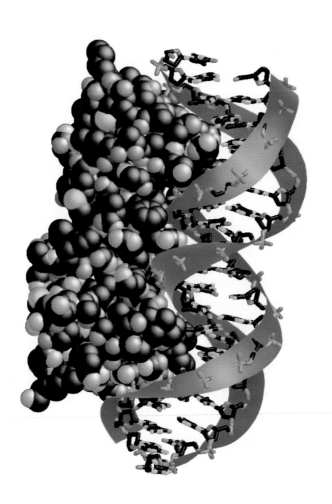

Contents

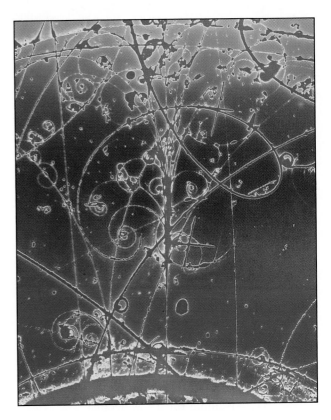

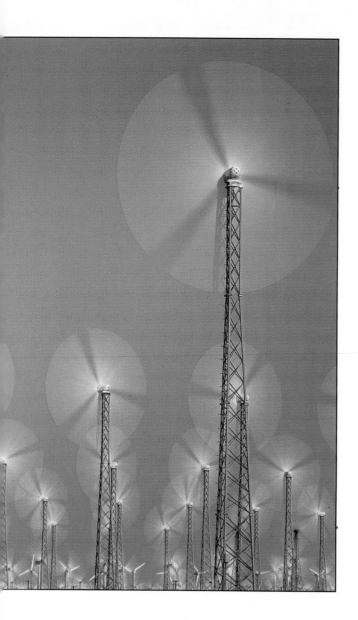

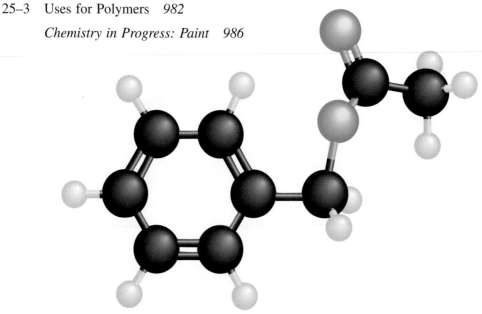

The Atomic Nature of Matter

The minerals orpiment (yellow, As_2S_3) and realgar (red, As_4S_4) in heterogenous mixture in a rock.

Chemical change occurs everywhere: in the processes of respiration that permit us to live, in the slow geological transformations that shape the earth's surface, in the formation of molecules deep in interstellar space. Some of these changes are slow, some fast; some are barely detectable, some strikingly dramatic. Chemistry is the study of these changes.

1-1 CHEMISTRY: SCIENCE OF CHANGE

A chemist carefully draws off a small amount of a yellow-green gas from a cylinder into a flask. The gas is dense, and it settles to the bottom of the flask, displacing the air up and out. The chemist then places a small piece of a soft metal into the flask, using a pair of tongs. Nothing happens. A tiny drop of water is then added, striking the metal. Immediately, flames and sparks shoot out as the metal reacts with the gas (Fig. 1–1); the flask becomes noticeably hotter. After the reaction dies down, a small amount of a white solid is found to have formed on the walls of the flask.

What has happened? The yellow-green gas is chlorine, a noxious irritant that can kill if inhaled. The metal is sodium, a substance so soft that it can be cut with a knife like butter. Like chlorine, it must be handled with care because it can burst into flame when placed in contact with water. The white solid that results from the combination of these two highly reactive materials is sodium chloride, the major constituent of ordinary table salt. Such transformations of substances into others with different properties are the subject of chemistry.

The first step in chemistry is observation—taking a close look at the properties of substances (color, shape, hardness, and so forth) and what happens to them when they are placed in contact with each other. Chemistry begins with the description of matter and change.

If the first question is "what?" then the second is "how much?" Chemists seek to make their knowledge quantitative—to understand how much sodium has reacted, how much sodium chloride was produced, and what volume of gaseous chlorine was consumed. The flask has become hotter. How can that heat be measured, and how does it compare with the heat evolved when, for example, iron instead of sodium reacts with chlorine?

The third question to be asked is "why?" Why does this particular mass of sodium react with this particular volume of chlorine? More fundamentally, why do sodium and chlorine react at all? Why is sodium so soft and sodium chloride so hard? Why does the drop of water start the reaction? Each question leads to further questions.

Although a single chemical reaction can raise many questions, each such reaction is also part of a far larger set of issues, such as why the sea contains so much salt and why a dietary excess of salt leads to hypertension and other serious health problems in humans. Here the overlap of chemistry with other fields such as geology and biology provides a fruitful source for new discoveries.

Chemistry is an extremely practical subject. The reaction we have just discussed is not a very useful one, because salt is readily available from salt deposits or from evaporated seawater. But if we were able to reverse the reaction and produce sodium and chlorine from sodium chloride, then we could use those products as starting materials for chemical reactions to make new substances. In fact, chemists have learned how to do this; sodium and chlorine can be obtained by passing an electric current through molten sodium chloride. Sodium has applications ranging from fill-

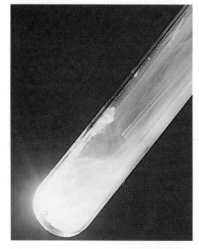

Figure 1–1 The reaction between sodium and chlorine proceeds very slowly until the addition of a small drop of water sets it off. Sparks then fly, the reaction vessel becomes hot, and a white powder of sodium chloride forms on the walls of the flask.

Figure 1-2 The bright yellow-orange glow from these street lights is given off by sodium atoms raised to high-energy states.

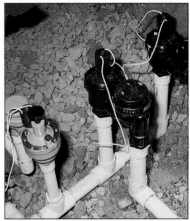

Figure 1-3 More than 56% by mass of this plastic pipe is chlorine.

ing sodium vapor lamps (Fig. 1–2) to producing titanium for aircraft hulls. Chlorine finds an equal range of uses, from treating wastewater to producing plastic piping (Fig. 1–3). As chemists seek to understand chemical reactions better, they also undertake to make new materials with useful properties.

1-2 THE COMPOSITION OF MATTER

Two traditions intertwine throughout the history of chemistry: **analysis** (taking things apart) and **synthesis** (putting things together). These two directions can be traced from the earliest periods of Greek science to the present time. The early Greek philosophers attempted to analyze the constituents of matter in terms of four elements: air, earth, fire, and water. Early artisans and metallurgists synthesized many new materials of practical importance, including bronze, glass, and cement, both by accident and by intent. The dual traditions of analysis and synthesis remain driving forces to the present day. When a useful product is discovered in a natural source (such as the antimalarial drug quinine in the bark of the cinchona tree), the first step is to analyze it to find its composition and structure. Then a strategy is designed to synthesize it from easily available starting materials. The procedure can be difficult. Penicillin was discovered in 1929 through the accidental contamination of a bacterial culture by mold; its laboratory synthesis was not achieved until 1957.

• Penicillin manufacture still uses a fermentation process, carried out in large vats. Direct chemical synthesis of penicillin is not economically competitive with this large-scale and efficient biological method.

CHEMISTRY IN YOUR LIFE

Semi-Synthesis, Total Synthesis, and Biosynthesis

The discovery in the early 1980s that paclitaxel, a compound extracted from the bark of the Pacific yew tree and trade-named Taxol, works astonishingly well in the treatment of even difficult cases of ovarian and breast cancer brought new hope to cancer sufferers. But there were problems. Only 0.1 to 0.2 g of Taxol can be extracted from a kilogram of bark, and the bark from three trees is required to treat a single patient for a year. The Pacific yew grows extremely slowly and is not abundant; stripping its bark kills it.

So, do human lives come first, even if it means depletion or extinction of an entire species? Should all the Pacific yew trees be sacrificed to help current cancer victims, with nothing being left for the future? Might the trees hold cures for other ailments that could be lost in a rush to extract Taxol? If authorities reserve some Pacific yews, is it right to buy black-market Taxol extracted from bark taken by "tree-rustlers"?

Fortunately, synthetic research by chemists soon turned these questions from personal to theoretical. A **semi-synthetic** preparation of Taxol starting from the needles of the English yew, a related species, was announced. In this procedure, a compound called 10-deacetyl-baccatin III is extracted from the needles and converted to Taxol by chemical reactions. This substance is much more concentrated in these needles than Taxol is in Pacific yew bark, and harvesting needles (within reason) does not harm the trees. This process is called semi-synthetic because the tree does most of the work, and the chemist just finishes up. Semi-synthesis has proved useful in preparing many drugs, including other anticancer agents.

In 1994, two groups of chemists managed to achieve a **total synthesis** of Taxol from stock starting materials—in effect, from the constituent chemical elements themselves. However, their multistep procedures, required by the extreme complexity of the structure of Taxol, give yields too poor for practical use in making Taxol.

One alternative to collecting yew needles as a cash crop is a **biosynthesis** of Taxol. Scientists are attempting

Figure 1–A   The bark of Pacific yew tree, *Taxus brevifolia*.

to isolate the genes that code for the synthesis of Taxol in the bark cells of the Pacific yew. They hope to insert them into the cells of plants that are easily cultured in the laboratory and activate them. The cells would then grow, divide repeatedly, and secrete Taxol for collection. Genetically engineered cells of baker's yeast already produce high-quality supplies of human insulin to control diabetes; biosynthesis will undoubtedly be adapted in the future to make more and more life-saving drugs.

Substances and Mixtures

Let us begin our study of chemistry with the process of analysis, which is the determination of what makes up matter (Fig. 1–4). Matter occupies space and has mass; it is all the stuff of the world around us. In daily life, we perceive matter either as objects or as materials, depending on our purposes. Materials endure, while objects come and go. Thus, chopping a wooden chair (an object) into splinters destroys the chair, but the wood (a material) remains and persists even if the splinters are sawn to dust. Chemical analysis focuses on materials, not objects.

Suppose then we have before us a sample of a material for study. We examine its various specific properties, or distinguishing characteristics. We start with **physical properties,** such as color, odor, taste, hardness, melting point, boiling point, electrical conductivity, magnetism, and density. Physical properties can be defined and discussed without reference to any other material and thus differ from **chemical properties,** which involve the behavior of the material toward other materials. If the sample is a piece of wood, we notice immediately that its properties are not uniform throughout the sample; some portions are darker than others, whereas others (the knots, for example) are harder and resist indentation better. If the properties of a material, like those of wood, change from one point to another within a sample, then the material is **heterogeneous.** Heterogeneous samples contain two or

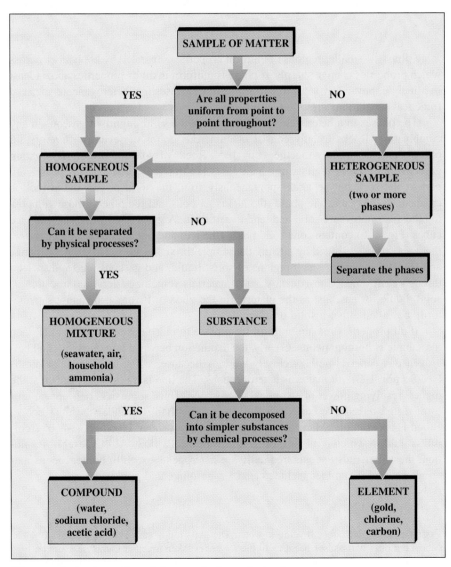

Figure 1–4 An outline of the steps in the analysis of matter.

Figure 1–5 Close examination of a piece of granite reveals that it is heterogeneous. At least four different kinds of grain can be distinguished in this close-up photo.

• The categories solid, liquid, and gas apply best to phases, not to materials in general. See Problem 73 at the end of this chapter.

more **phases**—smaller regions of uniform properties separated by boundaries across which properties change sharply. A phase is uniform in all its properties and, at least in principle, physically separable from the other phases in a heterogeneous material (practical separations are often not possible).

The phases in a heterogeneous material may be solid, liquid, or gaseous in any combination. Rocks usually consist of numerous grains of several different solid phases cemented tightly together (Fig. 1–5). Rocks of volcanic origin may contain gaseous inclusions—regions in which gases are trapped. Such gaseous phases are part of the rock but separate spontaneously from the solid phases if the rock is crushed. In dusty air, one phase (the air) is gaseous and the others (those composing the dust motes) are solid and tend to settle out. A glass filled with water and ice cubes (Fig. 1–6) contains one solid phase (the ice) and one liquid phase (the water) that are readily separated by passing the sample through a sieve. Pastes, gels, foams, and smokes consist of small regions of solid, liquid, and gaseous phases dispersed through each other. Consequently, such materials sometimes defy classification as solid, liquid, or gas, just as the material in the glass in Figure 1–6 cannot satisfactorily be called either a solid or a liquid.

If all properties remain uniform from place to place throughout a sample, then the sample is **homogeneous.** Clearly, the distinction between homogeneous and heterogeneous depends on how closely we look because, at the level of single atoms (see Section 1–3), *all* matter is heterogeneous. The line is traditionally drawn at the highest resolving power of an optical microscope. Thus, some materials appear uniform to the naked eye but betray heterogeneous character under the microscope. Blood is an example. Microscopic examination shows that blood consists of a colorless fluid in which a mixture of red and white cells float. At this level of resolution, most materials encountered daily are heterogeneous. Still, homogeneous materials abound. Examples include clean air, gasoline, and pure water.

Figure 1–6 Ice cubes floating in water. The glass holds many visible particles (the ice cubes) but only two phases because all the ice cubes taken together comprise a single phase. Within every ice cube the properties are uniform and equal to the properties within every other ice cube. The liquid water is the second phase. If the ice melts, the phase boundaries cease to exist, and the sample becomes a homogeneous liquid.

(a)   (b)   (c)

Figure 1–7 (a) A solid mixture of blue $Cu(NO_3)_2 \cdot 6\ H_2O$ and yellow CdS is added to water. (b) While the $Cu(NO_3)_2 \cdot 6\ H_2O$ dissolves readily and passes through the filter, the CdS remains largely undissolved and is held on the filter. (c) Evaporation of the solution leaves pure crystals of $Cu(NO_3)_2 \cdot 6\ H_2O$.

Even if a material consists of a single phase, it may be possible to separate it into two or more differing components by ordinary physical means. If so, it is a **homogeneous mixture,** or **solution.** Air, for example, is a homogeneous mixture of several gases (including oxygen, nitrogen, argon, and carbon dioxide). If we liquefy air and then warm it slowly, the gas having the lowest boiling point evaporates first from the mixture, leaving behind in the liquid the components with higher boiling points. Such a separation is not perfect, but we can repeat the process of liquefaction and evaporation on separated portions and improve the resolution of air into its component gases to any degree of purity we may require. We take advantage of the different physical properties (in this case, the boiling points) of the components of the mixture to separate them from one another.

Other physical operations can be used to separate both homogeneous mixtures and heterogeneous mixtures into components. Treating a mixture of solids with water or another solvent frequently allows the components to be separated on the basis of their different solubilities (Fig. 1–7). The salt and water in seawater can be separated to a considerable extent by slowly freezing the mixture; the first crystals to form are nearly pure water. Centrifuges are often used to separate a suspended solid from a liquid. "Molecular sieves" contain porous channels that permit small molecules to pass while larger molecules are blocked. Other materials trap and hold (adsorb) certain kinds of molecules on their surfaces. They are used to remove pollutants from gases or water supplies.

If all these physical procedures (and many more) fail to separate a material into portions that have different properties, we call the material a **chemical substance** (or simply a **substance**). Different samples of a chemical substance, when examined under the same conditions, have the same properties if they are pure.

What about the common material sodium chloride, which we call "table salt"? Is it a substance? The answer is yes if we use the term "sodium chloride," but no if we use the term "table salt." Table salt is a mixture that consists principally of sodium chloride, to which a small amount of sodium iodide has usually been added to provide for the requirements of our thyroid glands, along with a small quantity of magnesium carbonate that coats each grain of sodium chloride and prevents the salt from caking. Even if these two components were not deliberately added, table salt would contain other impurities that were not removed in its preparation; to that extent, table

• To a chemist, "pure" means "containing a single substance." Pure orange juice is unadulterated, but, chemically speaking, it is a mixture. It contains, among other substances, water, citric acid, ascorbic acid, and fructose.

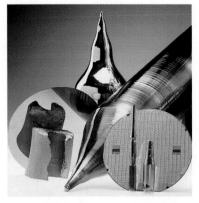

Figure 1-8 Nearly pure elemental silicon is produced by pulling out solid cylinders called "boules" from molten silicon. Most of the impurities are left behind. The boules are then sliced into thin wafers for use in integrated circuits.

• The word "copper" comes from the Greek name for the island of Cyprus in the Eastern Mediterranean, where the metal was found in large lumps.

salt is a mixture. When we refer to sodium chloride, however, we imply that all other materials are absent and that it is pure.

In practice, nothing is absolutely pure; the word "substance" is therefore an idealization. Among the most nearly pure materials ever prepared are silicon (Fig. 1–8) and germanium. For applications in electronic devices and solar cells, the electronic properties of these materials require very high purity (or precisely controlled concentrations of deliberately added impurities). By meticulous chemical and physical methods, germanium and silicon have been prepared with concentrations of impurities on the order of only one part in ten million.

Elements and Compounds

At present, literally millions of chemical substances have been discovered (or synthesized) and characterized (described in terms of their properties). Are these the fundamental building blocks of matter? If they were, classifying them, much less understanding them, would prove an insurmountable task. In fact, however, all these substances are found to be composed of a much smaller number of building blocks, called elements. **Elements** are substances that cannot be decomposed into two or more simpler substances by ordinary physical or chemical means. Here, the word "ordinary" is important, because we exclude the processes of radioactive decay (either natural or artificial) and of high-energy nuclear reactions that do transform one elementary substance into another. **Compounds** are chemical substances that contain two or more chemical elements combined in fixed proportions. Hydrogen and oxygen are elements because no further chemical separation has been found possible, whereas water is a compound because it can be separated into hydrogen and oxygen by passing an electric current through it (Fig. 1–9). Binary compounds are substances (such as water) that contain two elements. Ternary compounds contain three elements, quaternary compounds contain four, and so on.

At present, 112 chemical elements are known. A few have been known since before recorded history because they occur in nature in elemental form rather than in combination with one another in compounds. Gold, silver, lead, copper, and sulfur are chief among this group. Gold was found in streams in the form of little granules (placer gold) or nuggets in loosely consolidated rock. Sulfur was associated with volcanoes, and copper and, more rarely, lead could be found in their native states in shallow mines.

Each chemical element is designated both by a name and by a one- or two-letter symbol. The origin of these names and symbols is a fascinating subject. Many have a Latin root, such as gold (aurum, symbol Au), copper (cuprum, Cu), iron (ferrum, Fe), and mercury (hydrargyrum, Hg). Hydrogen (H) means "water-former." Potassium (kalium, K) takes its common name from potash (potassium carbonate), a useful chemical obtained in early times in impure form by leaching the ashes of wood fires with water. Many elements take their names from Greek and Roman mythology: cerium (Ce) from Ceres, goddess of plenty; tantalum (Ta) from Tantalus; and niobium (Nb) from Niobe, daughter of Tantalus. Some elements, such as eu-

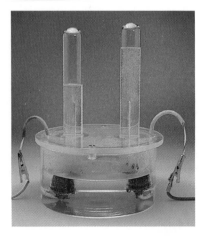

Figure 1-9 As electric current passes through water containing dissolved sulfuric acid, gaseous hydrogen and oxygen form as bubbles at the electrodes. This method of electrolysis can be used to prepare very pure gases. Note that the volume of hydrogen produced (on the left) is twice that of the oxygen (on the right). This is a consequence of the law of combining volumes, which is discussed in Section 1–4.

ropium (Eu) and americium (Am), are named for continents. Others are named after countries: germanium (Germany, Ge), francium (France, Fr), polonium (Poland, Po), and, less obviously, ruthenium (Russia, Ru). Cities provide the names of other elements: holmium (Stockholm, Ho) and berkelium (Berkeley, Bk). Some are named for the planets: uranium (U), neptunium (Np), and plutonium (Pu). Others take their names from colors: praseodymium (green, Pr), rubidium (red, Rb), and cesium (sky blue, Cs). The names of other elements honor great scientists: curium (Marie Curie, Cm), mendelevium (Dmitri Mendeleev, Md), fermium (Enrico Fermi, Fm), and einsteinium (Albert Einstein, Es).

• Ytterby, a village in Sweden, gave its name to four elements: erbium, terbium, ytterbium, and yttrium.

1-3 THE ATOMIC THEORY OF MATTER

The existence of elements does not resolve the question of the nature of matter on a microscopic scale. Two descriptions are possible, both of which can be traced back to ancient philosophy. One is a view of matter in which elements are continuous "qualities" that are mixed to form compounds. In this view, matter is divisible without limit, with the elements remaining mixed at every scale. The other view postulates the existence of **atoms** and holds that there is a lower limit to which substances can be divided before they are separated into the atoms of which they are composed (Fig. 1–10). Democritus (ca. 460–ca. 370 B.C.) postulated the existence of such unchangeable atoms of the elements, and he conceived them to undergo continuous random motion in the vacuum. The Greek philosophers were content to leave their views in the form of assertions, however, and they lacked the essentially modern scientific view that theories must be tested and refined by experiment.

In modern times, we are so used to speaking of atoms that we rarely stop to consider the experimental evidence of their existence. Twentieth-century science has developed a number of sophisticated techniques for measuring the properties of single atoms, and powerful electron microscopes even allow observation of their motion. In recent years, a new development, the invention of the scanning tunneling microscope (STM), has allowed direct verification of the existence of atoms. This device uses an incredibly fine-pointed electrically conducting probe that is passed over the surface of the sample being examined (Fig. 1–11). When it comes nearly

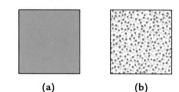

(a) (b)

Figure 1–10 (a) The printed color green appears uniform to the naked eye. (b) Magnification shows it to be made up of tiny yellow and blue dots that are perceived as their mixture, green, when viewed from a distance. The analogy to the behavior of a chemical substance like sodium chloride, which appears uniform until the atomic level is reached, must not be pushed too far. Yellow and blue in different proportions give different shades of green, but the proportions of the elements in a compound are fixed.

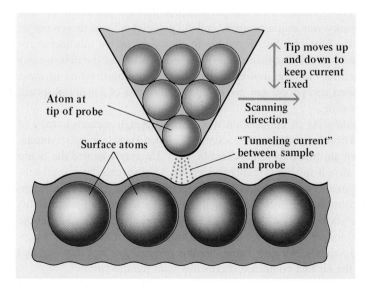

Figure 1–11 In a scanning tunneling microscope, an electric current passes through a single atom or small group of atoms in the probe tip and then into the surface of the sample being examined. As the probe moves over the surface, its distance is adjusted to keep the current constant, allowing a tracing out of the shapes of the atoms or molecules on the surface.

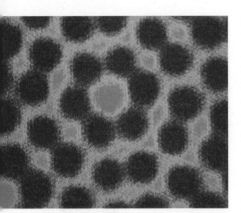

Figure 1–12 Atoms of silicon imaged with a scanning tunneling microscope.

in contact with the atoms of the sample, a small electric current (called the "tunneling current") can pass from the sample to the probe. The strength of this current is extraordinarily sensitive to the distance of the probe from the surface, typically falling off by a factor of 1000 as the probe moves out from the surface by 10^{-8} cm. Through a feedback circuit, the current is held constant while the probe is swept over the surface, moving in and out. The position of the probe is monitored, and the information is stored in a computer. Sweeping the probe tip along each of a series of closely spaced parallel tracks permits the construction of a three-dimensional image of the surface.

The images provided by the STM (Fig. 1–12) confirm visually many features, such as the sizes of atoms and the distances between them, known previously by other techniques. Much new information comes out as well. The STM images have revealed the positions and shapes of molecules undergoing chemical reactions on surfaces, helping to guide the search for new ways of carrying out such reactions. They have also been used to deduce the shape of the surface of molecules of the nucleic acid DNA, which plays a central role in genetics.

Well before direct observations of single atoms, accumulated indirect evidence allowed chemists to speak with confidence about atoms and the ways in which they combine to form molecules. Moreover, although the absolute mass of a single atom of oxygen or hydrogen was not measured until the early 20th century, some 50 years earlier chemists could assert (correctly) that the ratio of the two masses was close to 16:1. The chemical evidence for the existence of atoms and for the scale of relative atomic masses provides a fascinating and important story, both in its own right and as an illustration of the way in which science progresses. In the remainder of the chapter, we present this story.

The Law of Conservation of Mass

Significant progress on the microscopic nature of matter began in the 18th century. It arose out of an interest in the nature of heat and the way things burn, or undergo **combustion.** Fire of course had attracted speculation (and sometimes veneration) since its discovery, and the combustion of fuels was then, as now, essential for heat and light. It had been observed that wood, when burned, left behind a residue of ash, whereas metal, when heated in air, was transformed into a "calx," which we now call an oxide. The explanation for this phenomenon that was popular in the first part of the 18th century was that a property called "phlogiston" was driven out of wood or metal by the heat of a fire. From the modern perspective this seems absurd, because the ash weighs less than the original wood, whereas the calx weighs more than the metal. At the time, however, the principle of conservation of mass was not yet established, and people saw no reason why the mass of a material should not change upon heating.

A better understanding of combustion was gained through a careful study of the gases consumed or produced in such reactions. The great French chemist Antoine Lavoisier carried out an important experiment in 1775. He gently heated the liquid metal mercury in a closed apparatus that contained air. After several days, a red substance (mercury(II) oxide) was produced that was identical to the calx formed when mercury was burned in the open. He found that the gas remaining in the apparatus was reduced to five sixths of its original volume. Moreover, he discovered that the residual gas was no longer able to support combustion or life; it extinguished candles and suffocated animals. We now know that this residual gas was nitrogen and that the oxygen in the air had combined chemically with the mercury.

• The system of naming compounds that leads to the designation mercury(II) oxide is discussed in Section 3–8.

Lavoisier next took a carefully weighed amount of the red oxide of mercury and heated it strongly. This treatment transformed the red oxide back into mercury (Fig. 1–13), giving a mass[1] distinctly less than that of the red oxide. Lavoisier observed that the strong heating also generated a new gas. This gas exceeded air in its capacity to support respiration and caused a taper to burn very brightly. Also, when this new gas was added to the residual gas from the first part of the experiment, the mixture was indistinguishable from ordinary air. Lavoisier concluded that matter was not created or destroyed during the cycle of gentle and strong heating, but was only changed in form by combination with or loss of a definite substance, a gas we now know as oxygen. In further experiments, Lavoisier managed to weigh the gas produced by the decomposition of the red mercury(II) oxide and to show that its mass exactly compensated for the apparent loss of mass in the conversion of the red oxide to the metal. He was then able to state the **law of conservation of mass**:

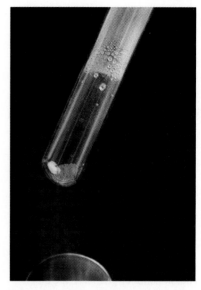

Figure 1–13   When the red solid mercury(II) oxide is heated, it decomposes to mercury and oxygen. Note the drops of liquid mercury condensing on the side of the test tube.

> In every chemical operation an equal quantity of matter exists before and after the operation.

Under this law, the small mass of ash left after the open burning of wood is explained by the loss of quantities of gases (principally carbon dioxide and water) that mix into the atmosphere. When these gases are captured and weighed, mass balance in the combustion reaction is restored.

Lavoisier was the first to observe that a chemical reaction is analogous to an algebraic equation. An algebraic equation involves equal quantities that are expressed in different mathematical terms; a chemical reaction involves equal masses that are present in different chemical forms. **Chemical equations** represent the conversion of one set of substances into a second set as it occurs in a chemical reaction. We would write Lavoisier's two reactions, which were each other's reverses, as

$$2\,Hg + O_2 \longrightarrow 2\,HgO \qquad \text{and} \qquad 2\,HgO \longrightarrow 2\,Hg + O_2$$

although at that time, the identity of the gas was not known. Lavoisier showed not only that mass was conserved but that combustion was the chemical combination of substances with oxygen from the air. Phlogiston did not exist.

The Law of Definite Proportions

Rapid progress ensued as chemists began to make accurate determinations of the masses of reactants and products. A controversy arose between two schools of thought, led by the French chemists Claude Berthollet and Joseph Proust. Berthollet believed that the proportions (by mass) of the elements in a particular compound were not fixed, but could vary over a certain range; thus water, for example, rather than containing 11.1% by mass of hydrogen, might have somewhat less, or somewhat more, than this mass percentage. Proust disagreed, and he showed that the apparent variation was due to the presence of impurities and experimental errors. He

---

[1]Chemists sometimes use the term "weight" in place of "mass." Strictly speaking, weight and mass are not the same. The mass of a body is an invariant quantity, but its weight is the force exerted upon it by gravitational attraction (usually by the earth). Newton's second law relates the two ($w = m \times g$, where g is the acceleration due to gravity). As g varies from place to place on the earth's surface, so also does the weight of a body. In chemistry, we deal mostly with ratios, which are the same for masses and weights. We shall use the term "mass" exclusively, but "weight" is still in common chemical use.

also stressed the difference between homogeneous mixtures and chemical compounds. Through his careful work, Proust was able to demonstrate the fundamental **law of definite proportions:**

> In a given chemical compound, the proportions by mass of the elements that compose it are fixed, independent of the origin of the compound or its mode of preparation.

Pure sodium chloride contains 39.34% sodium and 60.66% chlorine by mass, whether we obtain it from salt mines, crystallize it from waters of the oceans, or synthesize it from its elements, sodium and chlorine.[2] Of course, the key word in this sentence is "pure." We must be certain that no elements other than sodium and chlorine are present.

The discovery of the law of definite proportions was a crucial step in the development of modern chemistry; by 1808, Proust's conclusions had become widely accepted. In fact, we now recognize that this law is not strictly true in all cases. Certain solids exist over a range of compositions and are called **nonstoichiometric.** An example is wüstite, which has the nominal chemical formula FeO (with 77.73% iron by mass) but which in fact ranges continuously in composition from $Fe_{0.95}O$ (with 76.8% iron) down to $Fe_{0.85}O$ (74.8% iron), depending on the method of preparation. This illustrates a common pattern of scientific progress: experimental observation of parallel behavior leads to the establishment of a law or principle. Further, more accurate studies may then reveal exceptions to the general principle, the explanation of which leads to deeper understanding.

- Such compounds, also called berthollides in honor of Berthollet, are discussed further in Section 20–4.

Dalton's Atomic Theory

The English scientist John Dalton was by no means the first person to propose the existence of atoms. As we have seen, such ideas date back to classical times (the word "atom" comes from Greek *a-* (not) + *tomos* (cut), meaning "not divisible"). Dalton's major contribution was to marshal the evidence for their existence. He showed that the mass relationships found by Lavoisier and Proust could be interpreted most simply by postulating the existence of the atoms of the various elements.

In 1808, Dalton published *A New System of Chemical Philosophy*, in which the following five assumptions about the nature of matter were made. These are the postulates of Dalton's **atomic theory of matter:**

> 1. Matter consists of indivisible atoms.
> 2. All of the atoms of a given chemical element are identical in mass and in all other properties.
> 3. Different chemical elements have different kinds of atoms, and, in particular, such atoms have different masses.
> 4. Atoms are indestructible and retain their identity in chemical reactions.
> 5. The formation of a compound from its elements occurs through the combination of atoms of unlike elements in small whole-number ratios.

---

[2]This statement needs some qualification. As we see later in this chapter, many elements have several isotopes, which are species having atoms of essentially identical chemical properties but different masses. Natural variation in the abundances of isotopes leads to small variations in the mass proportions of elements in a compound. Larger variations can be induced by artificial isotopic enrichment.

The fourth postulate is clearly related to the law of conservation of mass, and the fifth is an attempt to explain the law of definite proportions.

Suppose that one rejected the atomic theory and adopted a view in which compounds are subdivisible without limit. What then ensures the constancy of composition of a substance like sodium chloride? On the other hand, if each sodium atom is matched by one chlorine atom in sodium chloride, then the constancy of composition can be understood. Note that it is not necessary to know how small the atoms of sodium and chloride are in order to understand why there should be a law of definite proportions. It is important merely that there be some lower bound to the subdivisibility of matter, because the moment such a lower bound arises, arithmetic steps in. Matter then becomes countable, and the units of counting are simply atoms. Believing in the law of definite proportions as an established experimental fact, Dalton postulated the atom.

New facts since 1808 have caused the modification of nearly all of the postulates of Dalton's original theory (see the following sections of this chapter and Problem 85 at the end of the chapter). The core of the theory, however, which is the essential graininess of matter, is stronger than ever. There is no doubt today of the real existence of atoms of the type postulated by Dalton.

1–4   CHEMICAL FORMULAS

The composition of a compound is shown by its **chemical formula.** A chemical formula gives the symbols of the elements in a compound together with numerical subscripts that show the relative number of atoms of that element present. When we write "H_2O" for water, we indicate that the substance water contains two atoms of hydrogen for each atom of oxygen. When such formulas were first written in the early 19th century, nothing was known about the organization of the atoms within substances. It is now known that in water each group of three atoms (two H and one O) is linked together by attractive forces strong enough to keep it together for a reasonable period of time. Such a grouping is called a **molecule,** a set of two or more atoms joined in a persistent combination. The numerical subscripts in a chemical formula may be used to convey such knowledge (see Chapter 2); however, the subscripts always give the relative numbers of atoms of the elements in the compound.

The determination of valid chemical formulas (and the accompanying determination of relative atomic masses) requires painstaking experiments to purify the compound under study and to analyze it by mass, that is, to arrive at its "elemental analysis." Finding the theoretical basis to move from these analytical data to accurate formulas was one of the major accomplishments of 19th-century chemistry, building on the atomic theory of Dalton.

The Law of Multiple Proportions

The simplest imaginable compound is one in which two elements combine, each element contributing equal numbers of atoms to the union to form **diatomic molecules,** which are molecules made up of two atoms. It was well known to 18th- and 19th-century chemists, however, that often two elements can combine with one another in different proportions to form more than one compound.

For example, the elements carbon and hydrogen form many different binary compounds (called hydrocarbons). Let us call two of the most common A and B. If we analyze them, we find that A contains about 6 (exactly 5.958) grams (g) of car-

• Dalton worked with both these compounds. He knew A as "olefiant gas" and B as "carburetted hydrogen."

Figure 1–14 The yellow mineral orpiment, As_2S_3, and the red mineral realgar, As_4S_4, are two relatively abundant sources of arsenic, often occurring in the same sample.

• Grams are units of mass in the "Système International d'Unités" (International System of Units, abbreviated SI). For discussion of these units and unit conversions, see Appendix B.

bon per 1.000 g of hydrogen and that B contains almost 3 (2.979) g of carbon per 1.000 g of hydrogen. Although we know nothing about the chemical formulas of these two hydrocarbons, we can immediately say that compound A contains twice as many atoms of carbon per hydrogen atom as does compound B. The evidence is that the ratio of the masses of carbon in A and B, for a fixed mass of hydrogen in each, is 5.958:2.979, or 2:1. If the formula of compound B were CH, then the formula of compound A would have to be C_2H, or C_4H_2, or C_6H_3, or some other multiple of C_2H. If compound B were formulated CH_2, then compound A would be C_2H_2 (or CH, C_3H_3, C_4H_4, and so on). At this point, we cannot say which of these (or an infinite number of other possibilities) are true formulas of compounds A and B, but we do know the *relationship* between the true formulas.

Consider another example. Arsenic (As) and sulfur (S) combine to form two sulfides in which the masses of sulfur per 1.000 g of arsenic are, respectively, 0.642 g and 0.428 g. The ratio of these sulfur masses is 0.642:0.428 = 1.50 = 3:2. We conclude that if the formula of the second compound is a multiple of AsS, then the formula of the first compound must be a multiple of As_2S_3.

These two examples illustrate the **law of multiple proportions:**

• The formula "$AsS_{3/2}$" which is As_2S_3 with both subscripts divided by 2, is avoided because it indicates a fractional atom.

> When two elements form a series of compounds, the masses of one element that combine with a fixed mass of the other element stand in a ratio of small whole numbers.

In the first example, the ratio of the masses of carbon in the two compounds for a given mass of hydrogen is 2:1. In the second example, the ratio of the masses of sulfur in the two compounds for a given mass of arsenic is 3:2. Today we know that compound A between carbon and hydrogen is C_2H_4 (called ethylene) and compound B is CH_4 (called methane), and that the arsenic sulfides are As_2S_3 and As_4S_4 (Fig. 1–14). Dalton could not have known this, however, because he had no information to decide how many atoms of carbon and hydrogen are in single molecules of the two hydrocarbons or how many atoms of arsenic and sulfur are in single molecules of the two arsenic–sulfur compounds.

EXAMPLE 1-1

Chlorine and oxygen form four different binary compounds. Analysis gives these results:

Compound	Mass of O Combined with 1.000 g of Cl
A	0.22564 g
B	0.90255 g
C	1.3539 g
D	1.5795 g

(a) Show that the law of multiple proportions holds for these compounds.
(b) If the formula of compound A is a multiple of Cl_2O (Cl_2O, or Cl_4O_2, or Cl_6O_3, and so forth), then determine the formulas of compounds B, C, and D.
(c) If the formula of compound A is a multiple of Cl_3O, then determine the formulas of compounds B, C, and D.

Solution

(a) The law of multiple proportions provides that in this series of binary compounds the masses of oxygen that combine with a fixed mass of chlorine stand in the ratio of small whole numbers. The simplest way to check is to form ratios by dividing each mass of oxygen by the smallest, which is 0.22564 g:

$$0.22564 \text{ g}:0.22564 \text{ g} = 1.0000 \text{ for compound A}$$
$$0.90255 \text{ g}:0.22564 \text{ g} = 4.0000 \text{ for compound B}$$
$$1.3539 \text{ g}:0.22564 \text{ g} = 6.0003 \text{ for compound C}$$
$$1.5795 \text{ g}:0.22564 \text{ g} = 7.0001 \text{ for compound D}$$

The ratios are whole numbers to a high degree of precision, and the law of multiple proportions is satisfied.
(b) If compound A has a formula that is some multiple of Cl_2O, then compound B is Cl_2O_4 (or ClO_2, or Cl_3O_6, and so forth) because it is four times richer in oxygen than compound A. Similarly, compound C, which is six times richer in oxygen than compound A, is Cl_2O_6 (or ClO_3, or Cl_3O_9, and so forth), and compound D, which is seven times richer in oxygen than compound A, is Cl_2O_7 (or a multiple thereof).
(c) The only difference here is the different starting assumption. If compound A is Cl_3O, or some multiple, then compound B is Cl_3O_4 (or Cl_6O_8, and so forth) because it is four times richer in oxygen than A. Compound C is then Cl_3O_6 (or ClO_2, Cl_2O_4, and so forth), and compound D is Cl_3O_7 or a multiple.

Exercise

Three different binary compounds of barium and nitrogen are known. The first contains 4.9021 g of barium, the second 9.8050 g of barium, and the third 14.7060 g of barium per 1.000 g of nitrogen. (a) If the formula of the third compound is Ba_3N_2, give possible formulas of the first and second compounds. (b) If the formula of the third compound is BaN, give possible formulas of the first and second compounds.

Answer: (a) BaN_2 and Ba_2N_2 (or BaN). (b) BaN_3 and Ba_2N_3.

Figure 1–15 Dalton's chemical symbols and some chemical equations he would have written. Note that some of his molecular formulas were incorrect, such as HO instead of H_2O. Modern designations for these substances are given in parentheses.

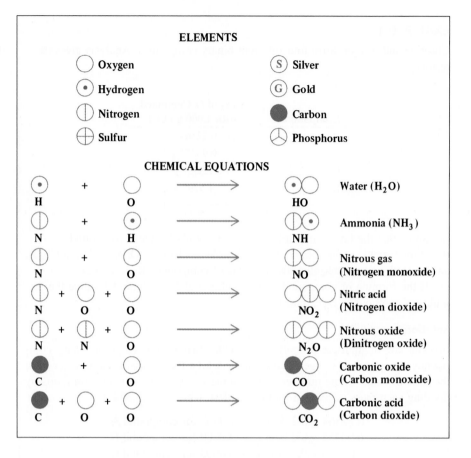

Dalton made a sixth assumption in order to resolve this dilemma of the absolute number of atoms present in a molecule; he called it the "rule of greatest simplicity." Dalton maintained that if two elements A and B form only a single compound, its molecules are diatomic, with the simplest possible formula: AB. Thus, he assumed that when hydrogen and oxygen combine to form water, the reaction is (Fig. 1–15).

$$H + O \longrightarrow HO$$

Unfortunately, Dalton was wrong, as we now know, and the correct reaction is

$$2\,H_2 + O_2 \longrightarrow 2\,H_2O$$

Law of Combining Volumes

In 1808, a French chemist, Joseph Gay-Lussac, drew attention to some important experiments on the volumes of gases that react with one another to form new gases. He announced the **law of combining volumes:**

When two gases react, the volumes that combine (at the same temperature and pressure) stand in the ratio of small integers. Moreover, the ratio of the volume of each product gas to the volume of either reacting gas is a ratio of small integers.

We consider three examples. It can be observed experimentally that:

2 volumes of hydrogen + 1 volume of oxygen $\longrightarrow$ 2 volumes of water vapor
1 volume of nitrogen + 1 volume of oxygen $\longrightarrow$ 2 volumes of nitrogen monoxide
1 volume of nitrogen + 3 volumes of hydrogen $\longrightarrow$ 2 volumes of ammonia

These results find ready interpretation in terms of combinations of atoms in ratios of small whole numbers and thus support Dalton's atomic theory. However, Dalton questioned their validity. For his own part, Gay-Lussac did not theorize on his findings. Shortly after their publication, an Italian chemist, Amedeo Avogadro, used them to formulate an important hypothesis.

Avogadro's Hypothesis

In 1811, Avogadro stated an idea that has since been known as **Avogadro's hypothesis:**

> Equal volumes of different gases (at the same temperature and pressure) contain equal numbers of particles.

The obvious question immediately arose: Are "particles" of the gaseous elements the same as Dalton's atoms? Avogadro took the point of view that they were not, but rather that gaseous elements (such as hydrogen, oxygen, or nitrogen) exist as diatomic molecules. This auxiliary assumption in combination with his main hypothesis allowed Avogadro to explain Gay-Lussac's law of combining volumes (Fig. 1–16). Thus, the reactions we wrote out in words become

$$2\ H_2 + O_2 \longrightarrow 2\ H_2O$$
$$N_2 + O_2 \longrightarrow 2\ NO$$
$$N_2 + 3\ H_2 \longrightarrow 2\ NH_3$$

in which the coefficients are in complete agreement with Gay-Lussac's experiments and in which the chemical formulas agree with modern results. Dalton, on the other hand, would have written

$$H + O \longrightarrow HO$$
$$N + O \longrightarrow NO$$
$$N + H \longrightarrow NH$$

in which the coefficients disagree with Gay-Lussac's observations.

One might think that Dalton would have welcomed Avogadro's brilliant hypothesis, but he did not. Dalton and others remained firm in their conviction that the gaseous elements contained only single, free atoms. One reason for this was the belief that molecules were held together by a force called "affinity," which expressed the attraction of opposites, just as we think of the attraction between positive and negative electric charges. If this were true, why should two like atoms be held together in a molecule? Moreover, if like atoms somehow did hold together in pairs, why should they not aggregate further? Hydrogen might then form H_3, H_4, H_5 molecules, and so forth. As a result, Avogadro's reasoning did not attract the attention it deserved. It was not until 50 years later that Avogadro was vindicated through the efforts of Stanislao Cannizzaro. Cannizzaro studied the elemental analyses of a large number of gaseous compounds and showed that their chemical formulas could be established in a consistent manner that used Avogadro's hypothesis but avoided any extra assumptions about molecular formulas. Gaseous hydrogen, oxygen, and nitrogen (and fluorine, chlorine, bromine, and iodine as well) indeed turn out to consist of di-

Figure 1–16 Each circle stands for a container of equal volume under the same conditions. These volumes hold different gases but contain equal numbers of molecules (as provided by Avogadro's hypothesis). If, as shown, hydrogen, oxygen, and nitrogen exist as diatomic molecules, then the combining volumes observed by Gay-Lussac in the three reactions are fully explained. The number of atoms never changes during a chemical reaction, but the number of molecules does.

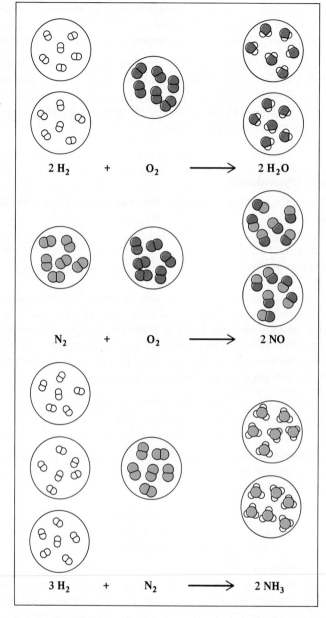

atomic molecules at ordinary conditions. Moreover, several other elements exist as polyatomic molecules: phosphorus (P_4), arsenic (As_4), carbon (C_{60} and others), sulfur (S_8 and others). Like atoms do indeed form molecules. Widespread acceptance of Avogadro's hypothesis came in the 1860s, after Cannizzaro published his treatise.

• Polyatomic molecules contain three or more atoms per molecule.

1–5 THE BUILDING BLOCKS OF THE ATOM

Persuasive evidence accumulated during the 19th century that atoms are not the indivisible tiny balls envisioned by Dalton in the first version of modern atomic theory. This evidence eventually led to the modern theory of chemical bonding discussed in Chapters 3 and 17. Here, we discuss the observations and experiments that revealed the interior structure of atoms.

In 1860, Robert Bunsen and Gustav Kirchhoff discovered the element cesium in certain mineral waters, and they named it for the sky-blue color of the light it emitted in a flame (Fig. 1–17). Other elements were also found to emit light of different characteristic color when heated to high temperatures. These observations strongly implied, although they did not prove, that atoms possess an internal structure of some kind. In other laboratories, scientists found that gases confined in a glass tube at low pressure were able to conduct electricity when a high voltage was applied across electrodes that had been sealed into the tube. It was evident that the electrical discharge was transported by charged particles of some kind. G. B. Stoney believed that the charge carriers in the gas-discharge tubes were derived from the molecules of the gas, and in 1874 he gave the name **electron** to the species that carried a negative charge. Because the originating atoms or molecules were electrically neutral, the very existence of negatively charged species, wrenched from atoms or molecules by the action of an electric field, meant that positively charged species must also exist in atoms.

Electrons

The passage of an electron current through a gas-discharge tube was observed to induce luminous regions in the confined gas, alternating with dark bands that lengthened toward the anode (positive electrode) as the pressure of the gas was reduced. At a sufficiently low pressure the luminous discharge disappeared altogether, but an electric current continued to flow (Fig. 1–18)!

The vehicle that carried the electric current through a vacuum was something of a mystery. The charge carriers could hardly be provided by gas molecules when essentially no gas was present. Instead, they appeared to come directly from the negative electrode and came to be known as **cathode rays.** Where the cathode rays impinged upon the glass tube a luminous spot appeared, and the rays possessed sufficient energy to heat a piece of metal foil in the tube to incandescence. Whatever these carriers were, they transferred a negative charge and could be deflected from their straight-line trajectories by both electric and magnetic fields. Some scientists believed that cathode rays were some kind of electromagnetic wave, a form of invisible light. Others believed that they were streams of charged particles—electrons, in fact. In 1897, J. J. Thomson, a British physicist at the Cavendish Laboratory of Cambridge University, settled the issue by proving that cathode rays are negatively charged particles. Thomson built an apparatus (Fig. 1–19) in which the deflection of a beam of cathode rays caused by a magnetic field was countered by an opposing deflection

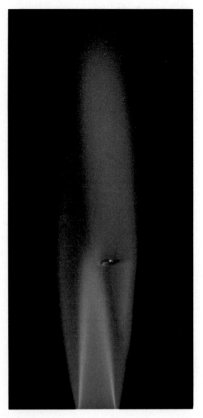

Figure 1–17 Cesium gives a purplish blue color to a flame.

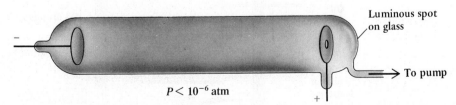

Luminous spot on glass

To pump

$P < 10^{-6}$ atm

Figure 1–18 An electric current passes through this high-voltage discharge tube despite a nearly total absence of matter between the two electrodes. Charge carriers apparently emanate from the negative electrode (the cathode). The hole in the positive electrode (the anode) allows some of them to pass through to excite a visible glow where they strike the wall of the tube.

Figure 1–19 Thomson's apparatus to measure the electron charge-to-mass ratio, e/m_e. Electrons (cathode rays) leave the cathode and accelerate toward the anode. Many pass through the hole in the anode and stream across the tube from left to right. The electric field alone deflects this beam down, and the magnetic field alone deflects it up. By adjusting the two field strengths, Thomson achieved zero net deflection. From the strengths of the two fields at this point, he could calculate the velocity of the beam. From that velocity and the deflection caused by the electric field alone, he arrived at the charge-to-mass ratio.

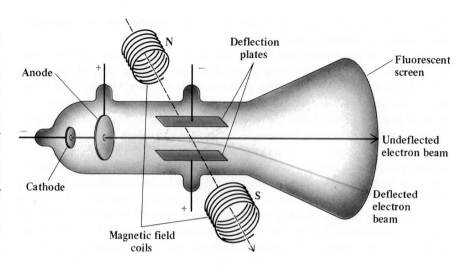

• The coulomb (C) is a unit for the measurement of quantity of electric charge. See Appendix B.

caused by an electric field. In careful experiments, he varied the fields so as to balance exactly the opposing deflections. From the conditions of balance, he demonstrated that cathode "rays" have mass and are truly particles. Indeed, he went further and measured the ratio of the charge of these particles (electrons) to their mass. The currently accepted value of the ratio, e/m_e, is 1.7588196×10^{11} C kg^{-1}.

Thomson's experiments were a milestone in the search to understand the composition of the atom, but they raised two obvious questions: What is the value of the electric charge carried by the electron, and what is the electron mass? The answers were not long in coming. In 1909, the American physicist Robert Millikan and his student, H. A. Fletcher, developed an elegant experiment that yielded e, the charge of the electron. Using a microscope, they observed the fall of tiny oil droplets that had acquired an electric charge (Fig. 1–20). If left to themselves, the droplets fell at a uniform rate, pulled down by gravity but slowed by the viscous drag of the air. When Millikan applied a vertical electric field in the region in which the oil droplets were observed, the field acted on their charge. Droplets could be slowed, held suspended, or even accelerated upward, just as a lock of dry hair can be suspended or moved upward by the "static electricity" (the electric field) surrounding a comb. Millikan could single out specific oil droplets and record their rates of de-

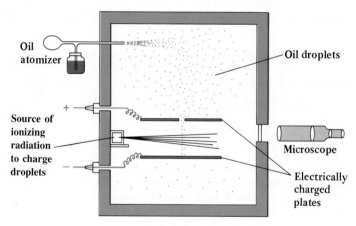

Figure 1–20 Millikan's apparatus to measure the charge (e) on an electron. By adjusting the strength of the electric field between the charged plates, Millikan could slow down, halt, or even reverse the fall of negatively charged oil drops. From the way the rate of fall depended on the strength of the field, he determined the net charges on the drops, which were always whole-number multiples of the charge on the electron.

scent in both the absence and the presence of the electric field. From these two rates and other pertinent facts (namely, the acceleration of gravity, the viscosity of the air, and the density of the oil), Millikan calculated the amount of charge that a single droplet carried. Repeated observations revealed that different oil drops bore different charges but that these charges were always integral multiples of some basic charge, which Millikan and Fletcher took to be the charge of a single electron. Their experimental value for the magnitude of this charge was 1.59×10^{-19} C, quite close to the currently accepted value of

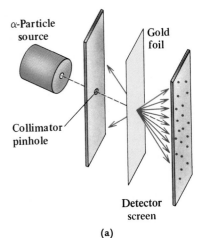

$$e = 1.6021773 \times 10^{-19} \text{ C}$$

Knowing e and the ratio e/m_e allows computation of the mass of the electron, which has the currently accepted value of

$$m_e = 9.109390 \times 10^{-31} \text{ kg}$$

The Rutherford Model of the Atom

In 1911, Ernest Rutherford reported experiments that revealed the arrangement of the charged components of the atom. Marie and Pierre Curie had recently isolated compounds of the radioactive elements radium and polonium and discovered that they emit positively charged particles, called alpha particles. Rutherford and his students characterized alpha particles as to charge and mass and then used them to probe atomic structure. They caused a beam of alpha particles from a radium source to bombard a thin gold foil and observed the deflections of the alpha particles. They detected the alpha particles by the scintillations they caused as they struck a fluorescent screen (Fig. 1–21a). Rutherford expected the alpha particles to experience only small deflections, and small deflections were in fact observed most of the time. Occasionally, however, a particle was deflected through a large angle. Even less frequently, an alpha particle was observed to rebound nearly directly backward! Rutherford was astounded by these rare events because the alpha particles were both relatively massive and fast moving. He remarked, "It was almost as incredible as if you fired a 15-inch shell at a piece of tissue paper and it came back and hit you." Some compact, heavy kernel must lie at the heart of the atom. A mathematical analysis of the frequency and angular distribution of the deflections confirmed that most of the mass of the gold atoms in the foil is concentrated in very dense, extremely small, positively charged particles, which were called **nuclei.** Rutherford's group estimated the radius of the nucleus in gold to be less than 10^{-14} m and the positive charge on the gold nucleus to be 100 times larger than the negative charge possessed by the electron (actual value, $+79e$). Bombardment studies using other elements as targets gave similar results but with different values for the positive charge on the nucleus.

Rutherford proposed the **nuclear model** of the atom to explain the results. He theorized that the massive nucleus of an atom possesses a net charge of $+Ze$, with Z negatively charged electrons (held by electrical attractions) surrounding the nucleus out to a distance of about 10^{-10} m. Thus, in the nuclear model, a gold atom has 79 extranuclear electrons (each with a charge of $-1e$) arranged around a nucleus of charge $+79e$. Nearly all of the volume of an atom is occupied by its electrons; nearly all of its mass is concentrated in the nucleus.

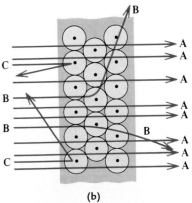

Figure 1–21   (a) Flashes of light mark the impact of alpha particles on a fluorescent screen after their encounters with the target foil. In the Rutherford experiment, the rate of hits on the screen varied from around 20 per minute at high angles to nearly 132,000 per minute at low angles. (b) The interpretation of Rutherford's experiment. Most of the alpha particles pass through the space between nuclei and undergo only small deflections (A). A few pass close to a nucleus and are more strongly deflected (B). Some are even scattered backward (C).

Protons and Neutrons

The Rutherford nuclear model quickly gained acceptance as a valid picture of atomic structure, and much subsequent research has focused on gaining an understanding of the structure of the central nucleus itself. Hydrogen has the smallest and simplest nucleus. Rutherford himself showed that the hydrogen nucleus is the fundamental unit of positive charge within the nuclei of heavier atoms. He proposed the name **proton** for the hydrogen nucleus in 1920. The different chemical elements contain different numbers of protons in the nuclei of their atoms. The value of Z associated with a nucleus in Rutherford's original work equals this number of protons and is now called the **atomic number** of the element. The chemical identity of an atom is determined by the number of protons in its nucleus.

In the same year that Rutherford proposed the term "proton," he suggested that nuclei also contain neutral particles of approximately the same mass as the proton. That particle, the **neutron,** was discovered in 1932 by James Chadwick, after a long series of unsuccessful experiments. Neutrons are essential to the stability of nuclei containing more than one proton. They prevent the electrostatic repulsions between protons from causing the nuclei to dissociate. Neutrons contribute to the mass of atoms but have zero electric charge and, for that reason, do not interact as strongly with matter as do protons, alpha particles, and electrons. They pass through great thicknesses of matter unabsorbed, which explains the delay in their discovery.

The number of neutrons in a nucleus is called the neutron number (N) of an atom, and the combined number of neutrons and protons is called the **mass number** (A) because the protons and neutrons packed into the tiny nucleus account for essentially all of the mass of atoms. Clearly, $A = Z + N$.

• Known nuclei have atomic numbers ranging from 1 to 112, corresponding to the 112 chemical elements currently known. Any new elements that may be discovered (or synthesized) in the future will have atomic numbers exceeding 112.

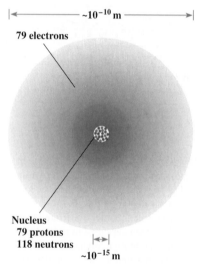

|←————— ~10^{-10} m —————→|

79 electrons

Nucleus
79 protons
118 neutrons |←→|
~10^{-15} m

This atom has $Z = 79$ and is therefore an atom of gold. Its mass number A equals 197, the sum of 79 and 118. In the real atom, the relative size of the nucleus is far smaller than shown.

1–6   ATOMIC MASSES

One of the most fundamental properties of an element is the characteristic mass of its individual atoms. Although for many years no method existed for measuring the actual masses of single atoms, chemistry flourished. Chemists realized that for most of their purposes it was sufficient to know the *relative* masses of atoms on a scale in which the mass of an atom of a particular element was assigned some arbitrary value. They established tables of **relative atomic masses** strictly by chemical synthesis and analysis, without knowing anything about atomic structure. More recently, knowledge of atomic structure has enabled incredibly precise determinations of relative atomic masses. In this section, we briefly illustrate the traditional chemical method and then take up the more precise physical method.

Relative Atomic Masses by Chemical Means

If the chemical formula of a compound is known (for example, through Avogadro's method), then an accurate analysis of the masses of the elements that make it up allows a determination of relative atomic masses.

As an example, consider water, which can be decomposed into hydrogen and oxygen by passing an electric current through it (see Fig. 1–9). The decomposition of 100.0 g of water yields 11.2 g of hydrogen and 88.8 g of oxygen. We can conclude that the water with which we started contained these same masses of hydrogen and oxygen bound together in chemical combination. How can we calculate the relative masses of the atoms of hydrogen and oxygen from this information, together with the chemical formula of water, which is H_2O?

We set up a simple ratio of masses:

$$\frac{\text{mass of oxygen in sample}}{\text{mass of hydrogen in sample}} = \frac{88.8\ \cancel{g}}{11.2\ \cancel{g}} = 7.93$$

This mass ratio is the same for any amount of water with which we start, and it should apply all the way down to a single molecule of water, for which we can write

$$\frac{\text{mass of one atom of oxygen}}{2 \times \text{mass of one atom of hydrogen}} = 7.93$$

where the factor of 2 comes from Avogadro's conclusion that for every atom of oxygen in water, there are two atoms of hydrogen. From this, by multiplying both sides of the preceding equation by 2, we find that

$$\frac{\text{mass of one atom of oxygen}}{\text{mass of one atom of hydrogen}} = 2 \times 7.93 = 15.9 \approx 16$$

For years, chemists lavished effort on determining the relative atomic masses of the elements by this method, choosing for convenience a scale that set the relative mass of oxygen at exactly 16. This gave hydrogen, the lightest element, a relative mass close to 1. Although very good results were obtained, the chemical method for relative masses is fundamentally limited by the accuracy with which substances can be weighed and the degree of completeness to which compounds can be resolved into their elements.

Mass Spectrometry and Isotopes

In the early 20th century the precise determination of relative atomic and molecular masses was significantly advanced by the work of J. J. Thomson, F. W. Aston, and others who developed the technique of **mass spectrometry.** Mass spectrometry relies on the manipulation of beams of **ions.** An ion results when an atom either loses or gains one or more electrons. Ionization occurs readily. The transferred electrons are furnished or taken away by the surroundings. Because electrons carry a negative charge, their loss converts atoms into positively charged ions, and their gain converts atoms into negatively charged ions. Ions are represented by the chemical symbol for the atom together with the net charge written as a right superscript. If chlorine (Cl) loses one electron, Cl^+ results; if Cl gains one electron, Cl^- results. Whole molecules can, without breaking apart, also lose or gain electrons to create **molecular ions,** the charges of which are indicated similarly.

All moving ions interact strongly with electric and magnetic fields. In a mass spectrometer (Fig. 1–22), the same number of electrons is removed from each atom or molecule in a sample. The resulting ions are accelerated by an electric field and then shot into a magnetic field, which causes their paths to curve. The curvature of the trajectories of the ions depends only on the ratios of their charges to their masses (once the strengths of the magnetic and electric fields are set). When the ions all have the same charge, the curvature depends only the mass of the ions: the paths of more massive ions curve less than the paths of less massive ions. This difference allows species of different mass to be separated and detected. Early experiments in mass spectrometry demonstrated, for example, a mass ratio of 16:1 for oxygen relative to hydrogen, confirming by physical techniques a relationship deduced originally on chemical grounds.

Further investigations began to produce some real surprises. Because the relative atomic mass of chlorine (atomic number $Z = 17$) determined by chemical means

Figure 1–22 A simplified representation of a modern mass spectrometer. Atoms or molecules of a substance are ionized and then accelerated by the electric field between the plates. The ion beam passes into a magnetic field, where it is separated into components, each containing particles having a characteristic charge-to-mass ratio. Here, the element chlorine, which has two naturally occurring isotopes, is under study. The spectrometer is adjusted to detect the less strongly deflected $^{37}Cl^+$ ions. By changing the magnitude of the electric or magnetic field, one can move the beam of $^{35}Cl^+$ ions to the exit slit so that it can be detected.

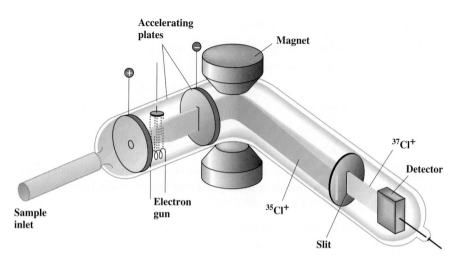

equaled 35.45, atoms of chlorine were expected to give a signal at the relative mass of 35.45. Instead, pure Cl gave *two* signals (peaks): the first at a relative mass near 35 and the second, about one third as intense, near 37 (Fig. 1–23). This meant that approximately three quarters of all chlorine atoms had a relative atomic mass of 35, and one quarter of all chlorine atoms a relative atomic mass of 37. Ordinary chlorine was revealed as a mixture of two types of atoms having nearly identical chemical properties but different masses! Such species are called **isotopes.** Further mass spectrometric experiments showed that most elements occur in nature in two or more isotopes. Dalton's assumption that all atoms of a given element are identical in mass was wrong.

Isotopes exist because atoms having the same number (Z) of protons may have different numbers (N) of neutrons in their nuclei. Neutrons change the mass of atoms but do not (except very slightly) affect their chemical properties. Because the chemical properties of different isotopes are extremely similar, their existence was not discovered until the development of the mass spectrometer. An isotope of an element is represented by its chemical symbol prefixed with a superscript to indicate its mass number. Thus, the two isotopes of chlorine are designated ^{35}Cl and ^{37}Cl. The mass number of an isotope is always close to its actual relative mass, simply because the relative masses of both the proton and neutron on our current scale equal approximately 1. For example, the relative atomic mass of ^{35}Cl equals 34.96885272 on a scale in which the relative mass of ^{12}C is defined as exactly 12 (see later in this chapter). The ten significant figures in this result illustrate the extremely high precision attainable in mass spectrometry, which is the most accurate method known for determining relative atomic masses.

• In spoken references to isotopes, the mass number is given after the name of the element: "chlorine-35" and "chlorine-37." Isotopes may also be designated this way in writing.

• Twenty elements have only one naturally occurring isotope. Some examples are fluorine (F), sodium (Na), aluminum (Al), phosphorus (P), and iodine (I). Tin (Sn) has ten naturally occurring isotopes, the most of any element.

EXAMPLE 1–2

An isotope detected in a mass spectrometer has atomic number 47 and relative mass 106.9051. Write the symbol for this isotope, and list the subatomic particles composing it.

Solution

The table on the inside back cover of this book reveals that element 47 is silver (Ag). The integer closest to the relative atomic mass of this isotope is 107, so the mass number is 107. The isotope therefore has the symbol ^{107}Ag. One atom of this type contains 47 electrons, 47 protons, and $107 - 47 = 60$ neutrons.

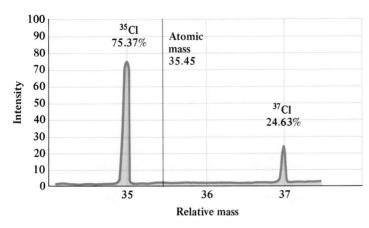

Figure 1–23 Naturally occurring chlorine has a relative atomic mass of 35.45. If all chlorine atoms had the same mass, a single peak would appear in the mass spectrum of chlorine. Instead, two peaks appear, one at a relative mass near 35 and the other near 37, showing that there are two naturally occurring isotopes of chlorine. The peak near 35 is about three times higher than the one near 37, meaning that the number of atoms of the lighter isotope is about three times greater than the number of atoms of the heavier isotope in chlorine.

Exercise

Write the symbol for the isotope of element 42 that has a relative mass of 97.9055. List the subatomic particles that compose this isotope.

Answer: ^{98}Mo has 42 electrons, 42 protons, and 56 neutrons.

The detector in a mass spectrometer in effect counts the arrivals of the individual ions making up the beam and so furnishes the **fractional abundance** as well as the relative masses of isotopes. The fractional abundance of an isotope in a sample of an element equals the number of atoms of that isotope divided by the total number of atoms. Ions of ^{35}Cl are detected at about triple the rate of ions of ^{37}Cl (Fig. 1–23), so the fractional abundances of these isotopes equal approximately 0.75 for ^{35}Cl and 0.25 for ^{37}Cl. These fractional abundances hold for chlorine isolated from a wide variety of terrestrial sources and are assumed to hold in all naturally occurring chlorine.

For some time, chemists used a scale of atomic masses in which the relative atomic mass of naturally occurring oxygen (a mixture of ^{16}O, ^{17}O, and ^{18}O) was set at 16. In 1961, by international agreement, the atomic mass scale was revised, with the adoption of exactly 12 as the relative atomic mass of ^{12}C. This scale continues to be used today. There are two stable isotopes of carbon: ^{12}C and ^{13}C (^{14}C and other isotopes of carbon are radioactive and have very low terrestrial abundance). In natural carbon, 98.892% of the atoms are ^{12}C and 1.108% are ^{13}C; this corresponds to fractional abundances of 98.892/100 = 0.98892 for ^{12}C and 1.108/100 = 0.01108 for ^{13}C.

The relative atomic masses of the elements as found in nature can be obtained as averages over the relative masses of the isotopes of each element, weighted by their observed fractional abundances. If an element consists of n isotopes, of relative masses $A_1, A_2, \ldots, A_n$ and fractional abundances $p_1, p_2, \ldots, p_n$, then the average relative atomic mass of the element is

$$A = A_1 p_1 + A_2 p_2 + \cdots + A_n p_n$$

Applying this formula to carbon gives

Isotope	Isotopic Mass × Fractional Abundance
^{12}C	$12.000000 \times 0.98892 = 11.867$
^{13}C	$13.003354 \times 0.01108 = \underline{\ 0.144}$
	Average relative atomic mass $= 12.011$

Chemists nearly always use elements with the isotopic compositions that occur naturally. For this reason, relative atomic masses computed by this formula are called "chemical relative atomic masses" and are tabulated as essential data on the chemical elements.

EXAMPLE 1–3

During 1982, the United States mint produced two kinds of one-cent pieces, one type with a mass of 2.52 g (light pennies) and one type with a mass of 3.09 g (heavy pennies). These coins, which have the same appearance and value, are monetary isotopes. In a group of 1000 1982 pennies, 220 pennies were light. Compute the average mass of a penny in this group.

Solution

If we were interested in the average mass of a 2.52-g coin and a 3.09-g coin, then adding up the masses and dividing by the number of coins (2) would quickly give the answer

$$\text{average mass} = \frac{2.52 + 3.09}{2} \text{ g} = \frac{1}{2}(2.52 \text{ g}) + \frac{1}{2}(3.09 \text{ g}) = 2.80 \text{ g}$$

The problem, however, concerns the average mass of unequal numbers of 2.52-g coins and 3.09-g coins. There are 220 light coins and, obviously, $1000 - 220 = 780$ heavy coins. We therefore weight the average to give extra emphasis to the more numerous coin:

$$\text{weighted average mass} = \frac{220}{1000}(2.52 \text{ g}) + \frac{780}{1000}(3.09 \text{ g}) = \boxed{2.96 \text{ g}}$$

Compare the forms of the two computations. In the first, the equal fractions (1/2 and 1/2) give equal importance to the two masses. In the second, a larger fraction (780/1000) gives more emphasis to the more numerous heavy penny, and a smaller fraction (220/1000) gives less emphasis to the less numerous light penny.

Exercise

The element europium has two naturally occurring isotopes. The first has a relative mass of 150.9196 and a fractional abundance of 0.47820, and the second has a relative mass of 152.9209 and a fractional abundance of 0.52180. Compute the chemical relative atomic mass of europium.

Answer: 151.96.

The number of significant figures in a chemist's table of relative atomic masses (see the inside back cover of this book) is limited not only by the accuracy of the mass spectrometric data but also by the slight variability in the natural abundances of the isotopes. If lead from one mine has a relative atomic mass of 207.18 and lead from another has a relative atomic mass of 207.23, then a value more precise than 207.2 cannot meaningfully be tabulated. In fact, geochemists now use small variations in the $^{16}O\!:\!^{18}O$ isotopic abundance ratio as a "thermometer" to deduce the temperatures at which different rocks were formed in the earth's crust over geological time scales. They also find anomalies in the oxygen isotopic composition of certain meteorites, implying that their origin lies outside our solar system.

• For a discussion of significant figures, accuracy, and precision in chemistry, see Appendix A.

1–7   COUNTING AND WEIGHING MOLECULES

The atomic masses discussed so far in this chapter have been relative atomic masses, in that they are all measured relative to the atomic mass of ^{12}C, set at exactly 12. *Relative atomic masses have no units* because they are ratios of two masses measured in whatever units we choose (grams, kilograms, pounds, and so forth). When we say that the relative atomic mass of ^{19}F is 18.9984, we mean that the ratio of the mass of one atom of ^{19}F to that of one atom of ^{12}C is $18.9884:12.0000 = 1.58320$. The **relative molecular mass** of a compound is the sum of the relative atomic masses of the elements that make it up, each one multiplied by the number of atoms of that element in a molecule. For example, the formula of water is H_2O, so its relative molecular mass is

$$\text{relative molecular mass of water} = 2(\text{atomic mass of H}) + \text{atomic mass of O}$$
$$= 2(1.00794) + 15.9994 = 18.0153$$

When the atoms of a substance are not organized as molecules, as is the case for salts (ionic compounds), it is common to speak of a relative **formula mass** to avoid implying that molecules are present. Computing a formula mass is identical to computing a molecular mass.

What are the actual masses of individual atoms? In other words, what is the connection between the macroscopic scale of masses used in the laboratory and the microscopic scale of individual atoms and molecules? The link between the two is provided by **Avogadro's number** (N_0), which is defined as the number of atoms in exactly 0.012 kg (12 g) of ^{12}C. Increasingly accurate techniques for determining N_0 have been developed, and the currently accepted value is

$$N_0 = 6.022137 \times 10^{23}$$

The mass of a single ^{12}C atom is then found by dividing exactly 12 g by N_0:

$$\text{mass of a } ^{12}C \text{ atom} = \frac{12.000000 \text{ g}}{6.022137 \times 10^{23}} = 1.992648 \times 10^{-23} \text{ g}$$

This is truly a very small mass, reflecting the very large number of atoms in a 12-g sample of carbon.

Avogadro's number is defined using ^{12}C because that atom is the reference for the modern scale of relative atomic masses, but it can be used with any substance. For example, sodium has a relative atomic mass of 22.98977, so a sodium atom is 22.98977/12 times as heavy as a ^{12}C atom. If N_0 atoms of ^{12}C have a mass of 12 g, then the mass of N_0 atoms of sodium must be

$$\frac{22.98977}{12}(12 \text{ g}) = 22.98977 \text{ g}$$

More generally, we can state:

The mass, in grams, of N_0 atoms of any element is numerically equal to the relative atomic mass of that element.

The same conclusion applies to molecules. From the relative molecular mass of water calculated above, the mass of N_0 molecules of water is 18.0153 g.

EXAMPLE 1–4

One of the heaviest atoms found in nature is ^{238}U. It has a relative atomic mass of 238.0508 on a scale on which 12 is the atomic mass of ^{12}C. Calculate the mass (in grams) of one atom of ^{238}U.

Solution

Because N_0 atoms of ^{238}U must have a mass of 238.0508 g and N_0 is 6.022137×10^{23}, one atom must have a mass equal to

$$\frac{238.0508 \text{ g of } ^{238}U}{N_0 \text{ atoms of } ^{238}U} = \frac{238.0508 \text{ g of } ^{238}U}{6.022137 \times 10^{23} \text{ atoms of } ^{238}U}$$

$$= 3.952929 \times 10^{-22} \text{ g}$$

Exercise

A single atom of a certain element has a mass of 2.10730×10^{-22} g. Compute the relative atomic mass of this atom on a scale on which 12 is the relative atomic mass of ^{12}C, and identify the element, assuming that it has only one isotope.

Answer: 126.904, iodine.

The Mole

Atoms and molecules have very small masses, and chemical experiments with measurable amounts of substances involve many of these atoms and molecules. It is convenient to group atoms or molecules in counting units that contain $N_0 = 6.022137 \times 10^{23}$ of them. We give this counting unit, which measures the **chemical amount** of a substance, the name **mole.**

One mole of a substance is the amount that contains Avogadro's number of atoms, molecules, or other entities.

In other words, one mole (mol) of ^{12}C contains N_0 ^{12}C atoms, one mole of water contains N_0 water molecules, and so forth. We must be careful in some cases because a term like "one mole of oxygen" is ambiguous. We should instead refer to "one mole of O_2" if we have N_0 oxygen molecules and "one mole of O" if we have N_0 oxygen atoms.

The mass of one mole of atoms of an element is called the **molar mass** (symbolized either $\mathcal{M}$ or M). The molar mass of an element is numerically equal to the dimensionless relative atomic mass of the element, but it has units of grams per mole. For example

$$\mathcal{M}_C = 12.00 \text{ g mol}^{-1} \text{ and } \mathcal{M}_{Au} = 196.97 \text{ g mol}^{-1}$$

The same relationship holds between the molar mass of a compound and its relative molecular mass or formula mass. Water has a relative molecular mass of 18.0153 and sodium chloride a relative formula mass of 58.4425, so the molar masses of the two are

$$\mathcal{M}_{H_2O} = 18.0153 \text{ g mol}^{-1} \text{ and } \mathcal{M}_{NaCl} = 58.4425 \text{ g mol}^{-1}$$

How do we determine the chemical amount, in moles, of a given substance in a sample? If we could count molecules, we would find the number of molecules in the sample N and then use

$$\text{chemical amount} = n = \frac{\text{number of molecules}}{\text{number of molecules per mole}} = \frac{N}{N_0}$$

But counting molecules in a 1-g sample would take millions of years, even with the help of the entire population of the world, each person counting at the rate of one molecule per second! Instead we use the chemist's most powerful tool, the laboratory balance. Suppose a sample of iron weighs 8.232 g; this mass can be converted to a chemical amount by multiplying by a chemical conversion factor, using the unit-factor method (see Appendix B). One mole of iron is equivalent to 55.85 g of iron, so that when we multiply

$$8.232 \text{ g Fe} \times \left(\frac{1 \text{ mol Fe}}{55.85 \text{ g Fe}} \right) = 0.1474 \text{ mol Fe}$$

we are, in effect, multiplying the 8.232 g Fe by one. The calculation can be turned around as well. Suppose we need a certain chemical amount, 0.2000 mol, of water to use in a chemical reaction. We have

$$(\text{chemical amount of water}) \times (\text{molar mass of water}) = \text{mass of water}$$

which becomes

$$0.2000 \text{ mol } H_2O \times \left(\frac{18.015 \text{ g } H_2O}{1 \text{ mol } H_2O} \right) = 3.603 \text{ g } H_2O$$

Thus, we simply measure out 3.603 g of water. In both cases, the molar mass is the conversion factor between mass of substance and chemical amount of substance. When mass in grams is changed to chemical amount in moles, the molar mass (in units of g mol^{-1}) appears in the denominator of the conversion factor, but when a chemical amount is converted to a mass, the molar mass appears in the numerator.

Although chemical amounts are frequently measured by weighing samples, it is preferable to think of a mole as a fixed number of particles (Avogadro's number) rather than as a fixed mass. The term "mole" is thus analogous to a term like "dozen": one dozen United States pennies weighs 26 g, substantially less than 60 g, which is the mass of one dozen nickels, but both contain 12 coins. Figure 1–24 shows mole quantities of several substances.

• Such "grouping terms" are common: a score is 20, a gross is 144, a great gross is 1728, and so forth.

Figure 1–24 One-mole quantities of several substances. Clockwise, from the lower right: oxalic acid ($H_2C_2O_4$), oxalic acid dihydrate ($H_2C_2O_4 \cdot 2H_2O$), and copper sulfate pentahydrate ($CuSO_4 \cdot 5H_2O$), and mercury(II) oxide (HgO). At the center is water (H_2O).

EXAMPLE 1–5

Nitrogen dioxide (NO_2) is a major component of urban air pollution. For a sample containing 1.000 g of NO_2, calculate (a) the chemical amount (the number of moles) of NO_2 and (b) the number of molecules of NO_2.

Solution

(a) We first calculate the molar mass of NO_2, using the tabulated molar masses of nitrogen (14.01 g mol^{-1}) and oxygen (16.00 g mol^{-1}):

$$\mathcal{M}_{NO_2} = \left(14.01 \ \frac{g \ N}{mol \ N}\right) + 2 \times \left(16.00 \ \frac{g \ O}{mol \ O}\right) = 46.01 \ \frac{g \ NO_2}{mol \ NO_2}$$

Because 46.01 g NO_2 is equivalent to one mole of NO_2, we use it as a chemical conversion factor from grams to moles:

$$n_{NO_2} = 1.000 \ g \ NO_2 \times \left(\frac{1 \ mol \ NO_2}{46.01 \ g \ NO_2}\right) = 0.02173 \ mol \ NO_2$$

(b) To convert from chemical amount to number of molecules, we multiply by Avogadro's number:

$$N_{NO_2} = 0.02173 \ mol \ NO_2 \times \left(\frac{6.0221 \times 10^{23} \ molecules \ NO_2}{1 \ mol \ NO_2}\right)$$
$$= 1.309 \times 10^{22} \ molecules \ NO_2$$

Exercise

Isoamyl acetate, a compound that contributes to the flavor of apples, has the molecular formula $C_7H_{14}O_2$. Calculate (a) how many moles and (b) how many molecules are present in 0.250 g of isoamyl acetate.

Answer: (a) 0.00192 mol. (b) 1.16×10^{21} molecules.

The Atomic Mass Unit

The absolute masses of single atoms are far too small to be expressed in kilograms or grams without the constant use of 10 raised to some large negative exponent. We need a unit of mass that is better sized for measuring on the atomic scale. The **unified atomic mass unit** (u) is defined as exactly one twelfth of the mass of a single atom of ^{12}C. We have seen that the mass of Avogadro's number (one mole) of ^{12}C atoms is exactly 12 g. Now we find from the definition that a *single* atom of ^{12}C has a mass of exactly 12 u. The definition of the atomic mass unit is contrived purposely to give this numerical equivalence, which extends to all elements and compounds: the molar mass of a substance, in grams per mole, is numerically equal to its molecular mass (or formula mass), in atomic mass units per molecule (or atomic mass units per formula unit). Thus, a mole of F weighs 18.9984 g, and a single atom of F weighs 18.9984 u; a mole of PF_3 weighs 87.9728 g, and a single molecule of PF_3 weighs 87.9728 u, and so forth. The essence of the relationship is that there are Avogadro's number of atomic mass units in a gram.

$$1 \ g = 6.022137 \times 10^{23} \ u$$

• The atomic mass unit is sometimes, particularly in biochemistry, called the dalton (abbreviated Da).

1–8 VOLUME AND DENSITY

The density of a sample of a material is the ratio of its mass to its volume:

$$\text{density} = \frac{\text{mass}}{\text{volume}}$$

In the International System of Units (SI, discussed in Appendix B), the base unit of mass is the kilogram (kg), but this is often inconveniently large for practical purposes in chemistry. We generally use grams instead. Several units for volume are in frequent use. The natural SI unit, the cubic meter (m^3), is quite unwieldy for laboratory purposes (1 m^3 of water weighs 1000 kg, or 1 metric ton). Therefore, chemists prefer as their units of volume the liter (L), which is the same as a cubic decimeter ($1 \text{ L} = 1 \text{ dm}^3 = 10^{-3} \text{ m}^3$), and the cubic centimeter, which is identical to the milliliter ($1 \text{ cm}^3 = 1 \text{ mL} = 10^{-3} \text{ L} = 10^{-6} \text{ m}^3$). Figure 1–25 shows some laboratory equipment used to measure volumes of liquids, and Table 1–1 lists the densities of a number of substances in units of g cm^{-3}.

It is important to recognize that the density of a substance depends on the conditions of pressure and temperature that exist at the time of the measurement. For some substances, especially gases and liquids, the volume may be more convenient to measure than the mass, and when the density is known, it provides the conversion factor to pass from volume to mass. This is illustrated by the following example.

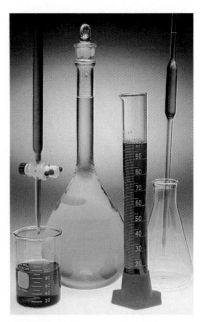

Figure 1–25 Some volumetric equipment: 150-mL beaker (green liquid); 25-mL buret (red); 1000-mL volumetric flask (yellow); 100-mL graduated cylinder (blue); and 10-mL volumetric pipet (green).

EXAMPLE 1–6

Near room temperature, liquid benzene (C_6H_6) has a density of 0.8765 g cm^{-3}. Suppose that 0.2124 L of benzene is measured into a container. Calculate (a) the mass and (b) the chemical amount of benzene in the container.

Solution

(a) Because $1 \text{ L} = 10^3 \text{ cm}^3$, the volume of the benzene sample is

$$0.2124 \text{ L benzene} \times \left(\frac{10^3 \text{ cm}^3}{1 \text{ L}}\right) = 212.4 \text{ cm}^3 \text{ benzene}$$

The density provides a bridge for converting between volume and mass. Because 0.8765 g of benzene is equivalent to 1 cm^3 of benzene, we can multiply by their ratio to find

$$212.4 \text{ cm}^3 \text{ benzene} \times \left(\frac{0.8765 \text{ g benzene}}{1 \text{ cm}^3 \text{ benzene}}\right) = 186.2 \text{ g benzene}$$

(b) The relative molecular mass of benzene is found from the relative atomic masses of carbon (12.011) and hydrogen (1.00794):

$$\text{molecular mass of benzene} = (6 \times 12.011) + (6 \times 1.00794) = 78.114$$

so that the molar mass is 78.114 g mol^{-1}. The chemical amount is then found by the unit-factor method:

$$n_{\text{benzene}} = 186.2 \text{ g benzene} \times \left(\frac{1 \text{ mol benzene}}{78.114 \text{ g benzene}}\right)$$
$$= 2.384 \text{ mol benzene}$$

Table 1–1 Densities of Some Substances[a]

Substance	Density (g cm^{-3})
Hydrogen	0.000082
Oxygen	0.00130
Water	1.00
Magnesium	1.74
Sodium chloride	2.16
Quartz	2.65
Aluminum	2.70
Iron	7.86
Copper	8.96
Silver	10.5
Lead	11.4
Mercury	13.5
Gold	19.3
Platinum	21.4

[a]Measured under room-temperature conditions and at average atmospheric pressure near sea level.

Exercise
The density of liquid mercury at 20°C is 13.594 g cm^{-3}. A chemical reaction requires 0.560 mol of mercury. What volume (in cubic centimeters) of mercury should be measured out?

Answer: 8.26 cm^3.

The density of a substance, together with Avogadro's number, gives the volume occupied by a molecule. In Example 1–6, 2.384 mol of benzene occupied a volume of 212.4 cm^3. The number of benzene molecules in this volume is

$$N_{\text{benzene}} = 2.384 \ \text{mol benzene} \times \left(\frac{6.022 \times 10^{23} \ \text{molecules}}{1 \ \text{mol benzene}} \right)$$

$$= 1.436 \times 10^{24} \ \text{molecules}$$

$$\text{volume per molecule} = \frac{212.4 \ \text{cm}^3}{1.436 \times 10^{24} \ \text{molecules}} = 1.479 \times 10^{-22} \ \frac{\text{cm}^3}{\text{molecule}}$$

This volume is not necessarily the same as the volume of the benzene molecule itself. It is the room available per molecule. The volume available per molecule increases greatly when a substance becomes a gas. At room conditions, a benzene molecule in gaseous benzene has about 4.1×10^{-20} cm^3 available to it, about 300 times more than the volume that we just computed. How can we interpret this fact on a microscopic scale? Note first that the volumes of liquids and solids do not change very much with changes in temperature or pressure, whereas gas volumes are quite sensitive to these changes. A hypothesis to explain this observation is that molecules in liquids and solids are close enough to touch one another, whereas in a gas they are separated by large expanses of empty space (Fig. 1–26). If this hypothesis is correct (and it has been borne out by further study), then the sizes of the molecules themselves can be estimated from the volume per molecule in the liquid or solid state. For the example of benzene presented above, this leads to a volume of about 1.5×10^{-22} cm^3, corresponding to a cube with edges about 5×10^{-8} cm long. This and other density measurements show that the characteristic size of atoms and small molecules is on the order of 10^{-8} cm. Avogadro's number provides the link between laboratory-scale measurements and the masses and volumes of single atoms and molecules.

• This is because the cube root of 1.5×10^{-22} cm^3 is about 5×10^{-8} cm.

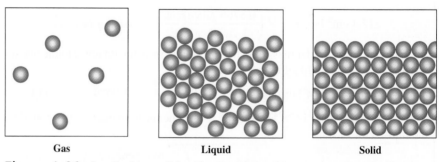

| Gas | Liquid | Solid |

Figure 1–26 In a liquid or solid, molecules (shown here as spheres) are in close contact with one another, so the volume per molecule fairly closely approximates the volume of the molecule itself. In a gas, the distances between molecules are greater and considerable empty space is present.

SUMMARY

1–1 Chemistry begins with the description of matter and change. Chemists ask, What reacts? How much of one substance is needed to react with another? How much product is produced? and, Why do substances have the characteristics that they possess? Chemists seek to understand reactions and to synthesize useful new products.

1–2 Chemists classify materials as elements, compounds, or mixtures. Substances are defined as pure **elements** or pure **compounds. Mixtures** contain more than one substance. Samples may be **homogeneous** or **heterogeneous.** Even a pure substance can exist in more than one **phase.**

1–3 Studies of combustion led to the formulation of the **law of conservation of mass,** and accurate analysis of compounds led to the **law of definite proportions.** Dalton's **atomic theory of matter** sought to explain these observed laws. Modern experimentation has confirmed the existence of atoms.

1–4 Dalton used the **law of multiple proportions** to find relative atomic masses and formulas of compounds. In order to do this, he had to guess or postulate the formulas of some compounds. Some of his guesses were wrong. Gay-Lussac's **law of combining volumes** and **Avogadro's hypothesis** provided the tools to correct Dalton's errors.

1–5 Experiments by Thomson, Millikan, and Rutherford in the late 19th and early 20th centuries led to the discovery of subatomic particles (**electrons, protons,** and **neutrons**) and a model of atomic structure: the negatively charged electrons surround a small, massive, positively charged **nucleus** composed of protons and neutrons.

1–6 Careful chemical analysis of a compound of known formula provides the **relative atomic masses** of the elements that make it up. **Mass spectrometry,** a physical method in which beams of ions are manipulated by electric and magnetic fields, provides more precise relative atomic masses. It also reveals the existence of **isotopes** (atoms that contain the same number of protons but different numbers of neutrons) in the majority of naturally occurring elements.

1–7 The **mole concept** and knowledge of **Avogadro's number** (6.022137×10^{23} mol^{-1}) permit one to "count" the number of atoms in a sample of a substance of known formula by weighing the sample. The molar mass $\mathcal{M}$ of a substance equals its relative atomic or molecular mass expressed in grams and contains Avogadro's number of formula units.

1–8 Consideration of the **density** of substances in the liquid or solid phase leads to an estimate of the sizes of atoms.

PROBLEMS

Note: Answers to blue-numbered problems are given in Appendix F. Problems that are more challenging are indicated with asterisks.

The Composition of Matter

1. Classify the following materials as homogeneous or heterogeneous: table salt, wood, mercury, air, water, seawater, sodium chloride, mayonnaise. Are they substances or mixtures? If they are substances, are they compounds or elements?

2. Classify the following materials as homogeneous or heterogeneous: absolute (pure) alcohol, milk (as purchased in a store), copper wire, rust, barium bromide, concrete, baking soda, baking powder. Are they substances or mixtures? If they are substances, are they compounds or elements?

3. A student asserts that salt from a salt mine is pure because it comes straight from the ground and that the salt sold in a carton in a store is impure because it has been processed.

Explain the difference between the definition of purity in this assertion and the chemical definition of purity.

4. Distillation is often used to separate liquids from solids and gases that might be dissolved in them. Discuss which of the following is purer: spring water direct from an artesian well, or spring water that has been triply distilled.

5. Give an example of a heterogeneous sample that would gradually become homogeneous if left to itself.

6. Give an example of a homogeneous sample that would gradually become heterogeneous if left to itself.

7. Apart from very minor impurities, vodka, a homogeneous material, contains three chemical elements: hydrogen, carbon, and oxygen. The text states that the term "compound" refers to substances that contain two or more chemical elements; however, vodka is not a compound. Explain.

8. Hydrogen, nitrogen, oxygen, fluorine, chlorine, bromine, and iodine all exist at room conditions as diatomic molecules (their formulas are H_2, N_2, O_2, F_2, Cl_2, Br_2, and I_2, respectively). Are these substances compounds or elements?

9. A 10.0-g sample of impure silicon is 99.44% silicon by mass, and the rest is carbon. Compute the mass of carbon present in the sample.

10. A sample of 94.4% pure calcium phosphate by mass contains 125.0 g of the compound. Determine the mass of the total sample.

The Atomic Theory of Matter

11. A 10.0-lb bundle of newspapers is burned in an incinerator. The ashes weigh 0.14 lb. Explain what has become of the rest of the mass.

12. A copper penny weighs 3.1041 g. It is heated in the air, cooled, and then reweighed. Its mass is 3.1063 g. Explain the gain in the penny's mass.

13. A 1.50-g sample of solid potassium sulfite, K_2SO_3, is placed in a flask containing 20.00 g of hydrochloric acid. The two react, generating gaseous sulfur dioxide, which escapes from the flask, and no other gases. After the reaction, the mass of the contents of the flask is 21.09 g. Determine the mass of the SO_2 produced.

14. Potassium chlorate ($KClO_3$) decomposes to potassium chloride and oxygen when heated. In one experiment, 100.0 g $KClO_3$ generates 36.9 g of O_2 and 57.3 g of KCl. What mass of $KClO_3$ remains unreacted?

15. A sample of ascorbic acid (vitamin C) is synthesized in the laboratory. It contains 30.0 g of carbon and 40.0 g of oxygen. Another sample of ascorbic acid, isolated from lemons (an excellent source of the vitamin), contains 12.7 g of carbon. Compute the mass of oxygen in the second sample.

16. A sample of a compound synthesized and purified in the laboratory contains 25.0 g of hafnium and 31.5 of tellurium. The identical compound is discovered in a rock formation. A sample from the rock formation contains 0.125 g of hafnium. Determine how much tellurium is in the sample from the rock formation.

Chemical Formulas

17. Determine the ratio of the number of oxygen atoms to the number of calcium atoms in each of the following compounds:
    (a) CaO
    (b) $Ca_3(PO_4)_2$
    (c) $Ca(H_2PO_2)_2$
    (d) $Ca(MnO_4)_2$
    (e) $Ca_5(PO_4)_3OH$

18. Write formulas for the following:
    (a) A ternary compound containing four atoms of sodium and two atoms of phosphorus for every seven atoms of oxygen.
    (b) A binary compound of nitrogen and oxygen in which the oxygen atoms are $2\frac{1}{2}$ times more numerous than the nitrogen atoms.
    (c) A ternary compound in which the numbers of carbon and hydrogen atoms equal each other and also equal ten times the number of iron atoms.

19. (See Example 1–1.) A binary compound between silicon and nitrogen contains 2.005 g of silicon per 1.000 g of N. A second compound between the same two elements contains 1.504 g of silicon per 1.000 g of N. Show that these data are consistent with the law of multiple proportions. If the second compound has the formula Si_3N_4, then what is the formula of the first compound?

20. (See Example 1–1.) A compound contains 31.813 g of sulfur (S) per 1.0000 g of hydrogen (H) and 23.810 g of oxygen (O) per 1.0000 g of hydrogen. A second compound contains 15.907 g of S per 1.0000 g of H, and 31.747 g of O per 1.0000 g of H.
    (a) Show that these data are consistent with the law of multiple proportions.
    (b) For each compound, compute the masses of hydrogen and sulfur that are present per 1.0000 g of oxygen.
    (c) If the formula of the first compound is $H_2S_2O_3$, what is the formula of the second compound?

21. MoS_3, MoS_4, and Mo_2S_3 are three binary compounds between molybdenum and sulfur, and molybdenum atoms have 3.0 times the mass of sulfur atoms.
    (a) Compute the mass of sulfur that combines with 1.0 g of molybdenum in each compound.
    (b) Compute the mass of molybdenum that combines with 1.0 g of sulfur in each compound.

22. Iodine (I) and fluorine (F) form a series of binary compounds with the following compositions:

Compound	Mass % I	Mass % F
1	86.979	13.021
2	69.007	30.993
3	57.191	42.089
4	48.829	51.171

(a) Compute in each case the mass of fluorine that combines with 1.0000 g of iodine.

(b) Show that these compounds satisfy the law of multiple proportions by figuring out small whole-number ratios among the four answers in part (a).

23. A liquid compound containing only hydrogen and oxygen is placed in a flask. Two electrodes are dipped into the liquid, and an electric current is passed between them. Gaseous hydrogen forms at one electrode and gaseous oxygen at the other. After a time, 14.4 mL of hydrogen (H_2) has evolved at the negative terminal and 14.4 mL of oxygen (O_2) has evolved at the positive terminal.
    (a) Assign a chemical formula to the liquid compound.
    (b) More than one formula is possible as the answer to part (a). Explain why.

24. A sample of liquid N_2H_4 is decomposed to give gaseous N_2 and gaseous H_2. The two gases are separated and the nitrogen occupies 13.7 mL at room conditions of pressure and temperature. Determine the volume of the hydrogen under the same conditions.

25. Pure nitrogen dioxide (NO_2) forms when dinitrogen oxide (N_2O) and oxygen (O_2) are mixed in the presence of a certain catalyst. What volumes of N_2O and oxygen are needed to produce 4.0 L of NO_2 if all gases are held at the same conditions of temperature and pressure?

26. Gaseous methanol (CH_3OH) reacts with oxygen (O_2) to produce water vapor and carbon dioxide. What volume of water vapor and carbon dioxide is produced from 2.0 L of methanol if all gases are held at the same conditions of temperature and pressure?

27. A chemist mixes 2.00 L of gaseous H_2S with 3.00 L of gaseous O_2. The two react completely (that is, no amount of either is left over) to give 2.00 L of gaseous SO_2 and 2.00 L of gaseous H_2O. All the volumes are measured at the same temperature and pressure; thus, 5.00 L of gas contracts to 4.00 L during the reaction. Explain how this happens and why it does not violate the law of conservation of mass.

28. In an experiment performed at room conditions of temperature and pressure, 11.71 mL of N_2 combines with 17.56 mL of O_2 to give 11.71 mL of a single gaseous product. No trace of either starting gas remains. What is the ratio of the number of nitrogen atoms to the number of oxygen atoms in this product?

The Building Blocks of the Atom

29. Complete the following table:

Symbol	Z	N	A	Number of Electrons
$^{12}C^+$	___	___	___	___
___	16	___	32	18
___	___	14	27	13
Pb	82	126	___	___

30. Complete the following table:

Symbol	Z	N	A	Number of Electrons
$^{228}Ra^{2+}$	___	___	___	___
___	53	___	127	54
___	___	20	40	20
N^{3-}	___	8	___	___

31. Suppose that an atomic nucleus were scaled up to the diameter of a baseball (about 10 cm). The size of the entire atom would then roughly equal the size of
    (a) a basketball
    (b) a city bus
    (c) a football stadium
    (d) Manhattan Island
    (e) the planet Earth
    Support your choice with a short calculation.

32. The mass of the proton is 1.672623×10^{-27} kg. How many electrons does it take to equal the mass of a single proton?

33. (See Example 1–3.) Small amounts of ^{241}Am are used in smoke detectors. Describe the composition of a neutral atom of ^{241}Am in terms of protons, neutrons, and electrons.

34. (See Example 1–3.) The heaviest element yet known was synthesized in February 1996, when researchers in Germany produced atoms having mass number 277 and atomic number 112. Describe the composition of these atoms in terms of protons, neutrons, and electrons.

35. Plutonium-239 is used for nuclear fission. Determine (a) the ratio of the number of neutrons in a ^{239}Pu nucleus to the number of protons and (b) the number of electrons in a single Pu^- ion.

36. The last "missing" element from the first six periods was promethium, which was finally discovered in 1947. Determine (a) the ratio of the number of neutrons in a ^{145}Pm nucleus to the number of protons and (b) the number of electrons in a single Pm^{3+} ion.

Atomic Masses

37. Before 1961, an atomic mass scale was used that was slightly different from the present one. It assigned exactly 16 as the atomic mass of the naturally occurring mixture of three oxygen isotopes. The relative atomic mass of this mixture on the present-day scale is 15.9994. Compute the atomic mass of ^{12}C on the pre-1961 scale.

38. Suppose that the atomic mass of naturally occurring phosphorus is assigned a value of exactly 10 "paulings." Determine the atomic mass of uranium in paulings.

39. (See Example 1–3.) In a certain collection of marbles, each red marble weighs 11.50 g, and each green marble weighs 12.70 g. A big sack contains 10,051 red marbles and 24,351 green marbles. Determine the mass of the average marble in this sack. Explain why the answer is not equal to (11.50 + 12.70)/2 = 12.10 g.

40. (See Example 1–3.) Naturally occurring chromium consists of four isotopes of the following relative masses and fractional abundances:

Isotope	Fractional Abundance	Atomic Mass
^{50}Cr	0.0431	49.9461
^{52}Cr	0.8376	51.9405
^{53}Cr	0.0955	52.9407
^{54}Cr	0.0238	53.9389

Calculate the relative atomic mass of naturally occurring chromium.

41. Only two isotopes of boron occur in nature, with the atomic masses and abundances given in the table below. Complete the table by computing the relative atomic mass of ^{11}B to four significant figures, taking the tabulated relative atomic mass of natural boron as 10.811.

Isotope	Fractional Abundance	Atomic Mass
^{10}B	0.1961	10.013
^{11}B	0.8039	——

42. Iridium has only two naturally occurring isotopes: ^{191}Ir, with a relative mass of 190.96058, and ^{193}Ir, with a relative mass of 192.96292. Compute the fractional abundance of ^{191}Ir, given that the chemical relative atomic mass of iridium equals 192.22.

43. Tin has isotopes with mass numbers ranging from 108 to 132, but only 10 of them occur naturally.
  (a) How many peaks should be observed in the mass spectrum of naturally occurring tin?
  (b) The fractional abundance of the most abundant isotope of tin is 0.3285; the fractional abundance of the least abundant is 0.0035. Compute the ratio of the sizes of the largest and smallest peaks in the mass spectrum of tin.

44. Suppose that a mass spectrum of gaseous H_2 is taken under conditions that prevent any of the H_2 molecules from breaking apart to give hydrogen atoms. The two naturally occurring isotopes of hydrogen are ^1H, with a relative atomic mass of 1.00783 and a fractional abundance of 0.99985, and ^2H, with a relative atomic mass of 2.01411 and a fractional abundance of 0.00015.
  (a) How many peaks does this mass spectrum have?
  (b) Give the relative atomic masses at each of these peaks.
  (c) Which peak is the largest and which is the smallest?

45. The chocolate-covered cherries from a certain candy kitchen all look alike, but 5.0% of them are missing the cherry and have a mass of only 4.0 g. Regular chocolate-covered cherries weigh 21.4 g. Compute the average mass of a box of 100 chocolate-covered cherries.

46. Atoms of indium all behave very similarly chemically, but 4.28% of them have a relative atomic mass of only 112.904. The rest have a relative atomic mass of 114.904. Compute the chemical relative atomic mass of an indium atom.

Counting and Weighing Molecules

47. Compute the relative molecular masses or formula masses of the following compounds on the ^{12}C scale:
  (a) P_4O_{10}
  (b) $BrCl$
  (c) $Ca(NO_3)_2$
  (d) K_2MnO_4
  (e) $(NH_4)_2SO_4$

48. Compute the relative molecular masses or formula masses of the following compounds on the ^{12}C scale:
  (a) $[Ag(NH_3)_2]Cl$
  (b) $Ca_3[Co(CO_3)_3]_2$
  (c) OsO_4
  (d) H_2SO_4
  (e) $Ca_3Al_2(SiO_4)_3$

49. Suppose that a person counts out gold atoms at the rate of one each second for the entire span of an 80-year life. Has the person counted enough atoms to detect with an ordinary balance? Explain.

50. A gold atom has a diameter of 2.88×10^{-10} m. Suppose that the atoms in 1.00 mol of gold atoms are arranged just touching their neighbors in a single straight line. Determine the length of the line.

51. (a) Determine the number of SF_2 molecules in 12.000 g of SF_2.
  (b) Compute the mass of sulfur present per 1.000 g of fluorine in SF_2.
  (c) Repeat parts (a) and (b) for the related compound S_2F_4.
  (d) Explain the similarities and differences between the answers for SF_2 and the answers for S_2F_4.

52. Arrange the following in order of increasing mass: 1.06 mol of SF_4; 117 g of CH_4; 8.7×10^{23} molecules of Cl_2O_7; 417×10^{23} atoms of argon (Ar).

53. The molecule of vitamin B_{12} has the formula $C_{63}H_{88}CoN_{14}O_{14}P$. Determine how many atoms (of all kinds) are present in 1.00 mol of vitamin B_{12}.

54. Vitamin A has the formula $C_{20}H_{30}O$ and that of vitamin A_2 has the formula $C_{20}H_{28}O$. Determine how many moles of vitamin A_2 contain the same number of atoms as 1.000 mol of vitamin A.

55. (See Example 1–4.) Compute the mass, in grams, of a single iodine atom if the relative atomic mass of iodine is 126.90447 on the accepted scale of atomic masses (based on 12 as the relative atomic mass of ^{12}C). Determine the mass, in atomic mass units, of a single iodine atom.

56. (See Example 1–4.) Determine the mass, in grams, of exactly 100 million atoms of fluorine if the relative atomic mass of fluorine is 18.998403 on a scale on which exactly 12 is the relative atomic mass of ^{12}C. Express the mass of the same 100 million atoms of fluorine in atomic mass units.

57. (See Example 1–5.) Ferrocene has the formula $Fe(C_5H_5)_2$.
    (a) Calculate the chemical amount of ferrocene in 100.0 g of ferrocene.
    (b) Calculate the number of molecules of ferrocene in 100.0 g of ferrocene.

58. (See Example 1–5.) The flavoring agent vanillin, $C_8H_8O_3$, occurs naturally in vanilla extract.
    (a) Calculate the chemical amount of vanillin in 10.00 g of vanillin.
    (b) Calculate the number of molecules of vanillin in 10.00 g of vanillin.
    (c) Compute the mass of one molecule of vanillin in atomic mass units.
    (d) Compute the mass of one molecule of vanillin in grams.

59. (a) Compute the mass (in grams) of lithium in 65.4 g of LiF.
    (b) Compute the mass (in pounds) of lithium in 65.4 lb of LiF.

60. (a) Compute the mass (in grams) of titanium in 91.2 g of TiO_2.
    (b) Compute the mass (in tons) of titanium in 91.2 tons of TiO_2.

61. You need 450,000 finishing nails for a big construction project you are managing. A supplier furnishes these nails, which weigh an average of 10 g each, in 100-lb kegs. How many kegs of nails should you order?

62. You get a job at a screw factory. As a joke, or test, one of the older workers tells you to get a count of the number of screws in a 250-lb lot of identical small machine screws. Explain how to get the answer without sitting down and counting all the screws.

63. A patient is found to have 5.1 nanograms (ng) of mercury per deciliter of blood (Note: 1 ng = 10^{-9} g; see Appendix B). Determine the mass of mercury in 500 mL (roughly a pint) of this patient's blood.

64. The United States Environmental Protection Agency (EPA) sets the maximum safe level for lead in the blood at 24 $\mu g \ dL^{-1}$. A 1.00-mL sample of a patient's blood contains 1.50×10^{-8} mol of lead. Express the patient's lead level in the unit used by the EPA to see if the patient is in danger of lead poisoning.

Volume and Density

65. (See Example 1–6.) Arsenic trifluoride (AsF_3) is an oily liquid with a density at room temperature of 2.163 g cm^{-3}. Compute the mass of a 5.12-cm^3 sample of arsenic trifluoride. Compute the chemical amount of AsF_3 in this sample.

66. (See Example 1–6.) At 20°C, the density of liquid methanol (CH_3OH) is 0.7914 g cm^{-3}. Compute the mass of a 9.33-cm^3 sample of methanol. Compute the chemical amount of CH_3OH in this sample.

67. A Nerf ball has a mass of 50.0 g and a volume of 549 cm^3. Compute its density.

68. Rhenium is the fourth densest element at room conditions; its density is 21.02 g cm^{-3}. Compute the volume of rhenium that has a mass of 1.000 g.

69. Mercury is traded by the "flask," a unit that has a mass of 34.5 kg. Determine the volume of a flask of mercury if the density of mercury is 13.6 g cm^{-3}.

70. Gold costs \$400 per troy ounce, and one troy ounce equals 31.1035 g. Determine the cost of 10.0 cm^3 of gold if the density of gold is 19.32 g cm^{-3} at room conditions.

71. Lutetium has a density of 9.84 g cm^{-3} under room conditions. Calculate the volume per atom in lutetium.

72. Lithium has a density of 0.534 g cm^{-3} under room conditions. Calculate the volume per atom in lithium.

Additional Problems

73. A schoolbook asks pupils to classify steel wool and sawdust (among other items) as solid, liquid, or gas. The official answer is that both are solids. Several pupils reject the official answer. They say, "Solids have no holes."; "Solids are not hollow."; "You can squeeze down a wad of steel wool, just like air in a balloon."; "You can pour sawdust, just like water." Identify the source of the misunderstanding and justify the official answer in words a schoolchild would understand.

74. In the 1640s, the natural philosopher van Helmont filled a large pot with 200 pounds of soil that had been dried in a furnace. He moistened the soil with rainwater and planted a small willow tree weighing 5 lb. He tended the tree for five years, adding only rainwater or distilled water and taking care to keep out dust and other solids. He then uprooted the tree, separated it from the soil, and found that it weighed 169 lb and about 3 oz. He recovered and dried all the soil used in the experiment and found it to weigh the same 200 lb he had started with, less about 2 oz. He then reported that "164 lb of wood, bark, and roots arose out of water only."
    (a) Is this conclusion valid? Why or why not?
    (b) Was van Helmont using the law of conservation of matter? Explain.

75. A naturopath swears by the merit of rose-hip tea (which contains much vitamin C) in warding off colds. This person contends that laboratory-synthesized vitamin C is ineffective in this role. Explain how the naturopath's contention could be correct.

76. Raisin Bran® cereal contains bran flakes and raisins. Crispix® cereal contains double-layered biscuits, each of which is rice on one side and corn on the other. Which is more analogous to a mixture and which to a compound? Explain.

77. A 17th-century chemist wrote of the "simple bodies which enter originally into the composition of mixtures and into which these mixtures resolve themselves or may be finally resolved." What is being discussed?

*78. The following statement appeared in a widely read scientific journal: "Chemists have managed so far to synthesize and measure at least some of the properties of more than seven million different molecules." Explain why this statement, taken literally, is nonsense.

79. Soft-wood chips weighing 17.2 kg are placed in an iron vessel and mixed with 150.1 kg of water and 22.43 kg of sodium hydroxide. A steel lid seals the vessel, which is then placed in an oven at 250°C for 6 hours. Much of the wood fiber decomposes under these conditions; the vessel and lid do not react.
   (a) Classify each of the materials mentioned as a substance or mixture. Subclassify all substances as elements or compounds.
   (b) Determine the mass of the contents of the iron vessel after the reaction.

80. A binary compound of nickel and oxygen contains 78.06% nickel by mass. Is this a stoichiometric or a nonstoichiometric compound? Explain.

81. Krulls weigh 4.60 times more than grebes, and grebes possess 0.780 the mass of stoats. The lightest of these objects weighs 1.000 g. How much does a collection of 17 krulls, 19 grebes, and 11 stoats weigh?

82. Suppose that the relative atomic mass of oxygen were 8 times larger than the relative atomic mass of hydrogen, instead of 16 times larger, as it actually is.
   (a) Write the empirical formula of water in such a case.
   (b) The true molecular formula of hydrogen peroxide is H_2O_2. What would its formula be in the case just described?

83. A chemistry student suggests that the formula of water is H_4O_2, that the formula of hydrogen is H_4, and that the formula of oxygen is O_4. Is this suggestion consistent with Gay-Lussac's observations on the combining volumes of hydrogen and oxygen? Explain.

84. Sir Humphrey Davy discovered several chemical elements but always insisted that any identification of a material as an element was tentative.
   (a) Explain how, in logic, the negative nature of the definition of a chemical element requires this approach.
   (b) In what sense did the discovery of isotopic forms of the chemical elements justify Davy's caution?

85. Dalton's 1808 version of the atomic theory of matter was based on five postulates (Section 1–3). According to modern understanding, four of these require amendment or extension. List the modifications that have been made to four of the five original postulates.

*86. Antimony-121 has an atomic mass of 120.9038, and antimony-123 has an atomic mass of 122.9041. You have 100.00-g specimens of each pure isotope, and you must prepare the largest possible sample of antimony with an average atomic mass of 121.5000. Compute how much (in grams) of each isotope you should put in this mixture.

87. Synthetic zircon, a gemstone, consists of a single compound.

One such gem contains 5.474 g of zirconium (Zr), 1.685 g of silicon (Si), and 3.840 g of oxygen (O). Determine the mass of zirconium in a second zircon that has a mass of 44.83 g.

88. You take a breath and inhale 1.16×10^{22} molecules of air, a mixture of several gases. Assume that the effective molar mass of air is 29.0 g mol^{-1}. Compute the mass of the air that you inhaled.

89. (a) Compute to five significant figures the average mass in grams of a molecule of UF_6.
   (b) Paradoxically, actual molecules of UF_6 never have exactly the mass that is the correct answer to part (a). Explain why.

90. Identify what is wrong or ambiguous about each of the following statements.
   (a) The molar mass of H_2O is 18.01.
   (b) 1.0 mol of bromine reacted.
   (c) The multiple proportions of the elements in H_2O are 2:1.
   (d) The mass of one atom of ^{12}C is 12.0000 g exactly.
   (e) A mole of water contains 6.022×10^{23} atoms.

91. Oxygen has three naturally occurring isotopes. In a sample of oxygen taken from the sea, the ratios of the numbers of atoms are

$$^{16}O{:}^{18}O = 498.7{:}1$$
$$^{17}O{:}^{18}O = 0.1832{:}1$$

On the basis of these data, how many out of every 10 million oxygen atoms are ^{16}O, ^{17}O, and ^{18}O?

92. Suppose 1.00 mol of "doped" silicon is prepared for use in a specialized electronic device by adding germanium to hyperpure silicon until the final product contains $0.62 \times 10^{-7}\%$ germanium on the basis of mass. Estimate (to two significant digits) the number of atoms of germanium present in this sample.

*93. The inhabitants of a planet in a distant galaxy measure mass in units of "margs," where 1 marg = 4.8648 g. Their scale of atomic masses is based on the isotope ^{32}S (atomic mass on earth = 31.972), so they define one "elom" of ^{32}S as the amount of sulfur in exactly 32 margs of ^{32}S. Furthermore, they define N_{or}, or "Ordagova's number" (after the well-known scientist Oedema Ordagova), as the number of atoms of sulfur in exactly 32 margs of ^{32}S.
   (a) Calculate the numerical value of N_{or} to four significant figures.
   (b) Calculate the atomic mass of phosphorus in margs per elom to five significant figures, assuming that the abundances of phosphorus isotopes are the same as those found on earth.

94. (a) Under room conditions, gaseous argon has a density of 1.64 g L^{-1}. Compute the volume (in liters) occupied by 1.00 mol of gaseous argon under these conditions.
   (b) The density of liquid argon at its normal boiling point is 1.40 g cm^{-3}. Compute the volume (in liters) occupied by 1.00 mol of liquid argon.
   (c) Assume that no vacant space exists between the atoms in liquid argon. Estimate the percentage of the volume of gaseous argon that is empty space at room conditions.

95. Iridium has a density of 22.56 g cm^{-3}, and hydrogen has a density of 8.24×10^{-5} g cm^{-3}, both at room conditions. In the following graph, identify which line corresponds to hydrogen and which to iridium. Explain your choice.

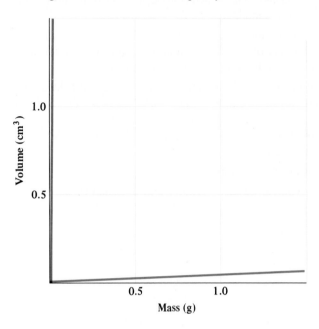

96. Sodium chloride (NaCl) has a density of 2.165 g cm^{-3} at 25°C. Calculate the number of atoms in 1.000 cm^3 of sodium chloride at 25°C.

97. A rough estimate of the radius of a nucleus is provided by the formula $r = kA^{1/3}$, where k ranges from 1.2×10^{-15} m to 1.3×10^{-15} m and A is the mass number of the nucleus. Estimate high and low values for the density of the nucleus of ^{127}I in grams per cubic centimeter.

98. In a neutron star, gravity causes the electrons to combine with protons to form neutrons. A certain neutron star has a mass half that of the sun compressed into a sphere of radius 20 km. If this neutron star contains 6.0×10^{56} neutrons, calculate its density in grams per cubic centimeter. Compare this with the density inside a ^{232}Th nucleus, in which 142 neutrons and 90 protons occupy a sphere of radius 9.1×10^{-13} cm.

99. Alpha particles turn out to be bare nuclei having a mass of 4.0015 u and a charge equal to $2e$, where e stands for the magnitude of the charge on the electron. Determine the mass number and atomic number of these nuclei. When alpha particles gain two electrons each from their surroundings, what chemical element is produced?

CUMULATIVE PROBLEM

Gallium Sulfides

Analysis of several samples of materials that are known to contain no elements other than gallium (Ga) and sulfur (S) gave these results:

Sample	Description	Mass of Ga	Mass of S	Total Mass	Volume of Sample
1	Yellow crystals	1.3509 g	0.9319 g	2.2828 g	0.625 cm^3
2	Yellow crystals	0.6289 g	0.2892 g	0.9181 g	0.238 cm^3
3	Gray powder	0.6441 g	0.1481 g	0.7922 g	0.189 cm^3
4	Yellow crystals	0.2228 g	0.1537 g	0.3765 g	0.103 cm^3
5	Yellow crystals	0.4949 g	0.2276 g	0.7225 g	0.187 cm^3

(a) Each sample is a chemical compound, not a mixture. Show that the mass data are consistent with this fact.

(b) Determine the minimum number of different compounds present.

(c) Assume that the formula of the compound in sample 1 is GaS. Give possible formulas of the compounds in samples 2 through 5.

(d) Assume that the formula of the compound in sample 1 is Ga$_2$S$_3$. Give possible formulas of the compounds in samples 2 through 5.

(e) If the relative atomic mass of sulfur is taken to be 32.07 and the formula of the compound in sample 1 is taken to be GaS, then what would be the relative atomic mass of gallium?

(f) The correct relative atomic masses of sulfur and gallium are 32.07 and 69.72, respectively. Determine the formulas of the compounds in samples 1 through 5.

(g) Sulfur occurs naturally in four isotopes: ^{32}S, ^{33}S, ^{34}S, and ^{36}S; gallium occurs naturally in two isotopes: ^{69}Ga and ^{71}Ga. Which isotope of sulfur is the most abundant; which isotope of gallium is more abundant?

(h) Use Avogadro's number and the correct relative atomic masses of sulfur and gallium to determine the number of atoms of gallium in sample 1, the number of atoms of sulfur in sample 1, and the ratio of these two numbers.

(i) Sample 1 liberates all of its sulfur in the form of gaseous H_2S if heated with an acid; sample 2 liberates all of its sulfur in the form of gaseous SO_2 if heated with oxygen. If sample 1 liberates 711 cm^3 of H_2S, what volume of SO_2 is liberated by sample 2? Assume that all volumes are measured under the same conditions of temperature and pressure.

(j) Calculate the densities of the five samples.

(k) Calculate the average volume (in cubic centimeters) occupied by an atom in each of the samples.

Chemical Equations and Reaction Yields

The smelting of copper involves a series of chemical reactions to convert the ore into elemental copper. Here, molten copper streams from an oven.

In Chapter 1, we showed how chemical and physical methods are used to establish chemical formulas and relative atomic and molecular masses. In this chapter, we take a further step and consider chemical reactions. We examine the balanced chemical equations that summarize these reactions and show how the masses of substances consumed and produced are related. This is an immensely practical and important subject. The questions of how much of a substance will react with a given amount of another substance and how much of a product will be produced are central to all chemical processes, whether industrial, geological, or biological.

2–1 EMPIRICAL AND MOLECULAR FORMULAS

Chemical formulas provide a succinct way to indicate the relative numbers of atoms of the different elements in any substance. Now we distinguish two types of chemical formulas: the molecular formula and the empirical formula. In a **molecular formula,** the subscripts specify the exact number of atoms of each element in one molecule of the substance; thus, the molecular formula of water is H_2O—each molecule of water contains two atoms of hydrogen linked to one atom of oxygen. Similarly, the molecular formula of the oxygen in the air is O_2—each molecule contains two atoms of oxygen bonded in a persistent combination; and the molecular formula of the sugar glucose is $C_6H_{12}O_6$—each glucose molecule contains 6 atoms of carbon, 6 atoms of oxygen, and 12 atoms of hydrogen. Meaningful molecular formulas can be written only for **molecular substances** such as H_2O, O_2, and $C_6H_{12}O_6$ that consist of collections of identical, well-defined, and persisting molecules. Strong attractions called **chemical bonds** hold the atoms in such molecules together. Other forces operate between neighboring molecules, but these **intermolecular forces** (discussed in Section 6–1) are weaker than full-fledged chemical bonds. The strength of the attractions *within* molecules always exceeds the strength of the attractions *between* molecules. Without this difference, molecular boundaries cannot be identified, which means that distinct molecules do not exist.

Indeed, many solid and liquid substances do not contain distinguishable molecules and therefore cannot be assigned molecular formulas. For example, sodium chloride, the main ingredient in table salt, consists as a solid of a collection of positively charged sodium ions (Na^+ ions) and negatively charged chloride ions (Cl^- ions) locked into a lattice constructed so that interactions between positive and negative ions are maximized while interactions between ions of like charge are minimized. Each ion is surrounded by six neighbors of opposite sign that are equivalent in every way. No special bond exists between any single sodium ion and any chloride ion. Thus, solid sodium chloride contains no definable molecules, and no molecular formula can be written for it.

• Empirical formulas are also called "simplest formulas."

By contrast, an empirical formula can be written for any substance. The **empirical formula** is the formula with the smallest whole-number subscripts that give the correct relative numbers of atoms of every kind in the substance. The formula "NaCl" is the empirical formula of sodium chloride. It indicates that equal numbers of sodium and chlorine atoms went into the compound. Similarly, the empirical formulas Li_2O and $CoCl_2$ state that twice as many lithium atoms as oxygen atoms went into making lithium oxide and that twice as many chlorine atoms as cobalt atoms went into cobalt(II) chloride.

Chemists use the term **formula unit** to refer to the grouping of atoms indicated by the empirical formula of a non-molecular substance or a substance that has an unknown microscopic organization. Saying "one formula unit of NaCl" avoids the incorrect implication of "one molecule of NaCl."

Figure 2-1 When cobalt(II) chloride crystallizes from solution, it brings with it six water molecules per formula unit, giving a red solid with the empirical formula $CoCl_2 \cdot 6H_2O$. This solid melts at 86°C; above approximately 110°C, it loses some water and forms a light purple solid.

The empirical formulas of molecular substances often differ from their molecular formulas. The empirical formula of glucose, for example, is CH_2O, indicating that the numbers of atoms of carbon, hydrogen, and oxygen stand in a ratio of 1:2:1, and the empirical formula of O_2 is simply O, indicating that molecular oxygen contains only the element oxygen.

Sometimes molecules are incorporated intact into solids, and the chemical formula is written to show this fact explicitly; thus, cobalt and chlorine form not only $CoCl_2$, but also, when enough water is available under the right conditions, $CoCl_2 \cdot 6H_2O$, in which six molecules of water are incorporated per $CoCl_2$ formula unit. The resulting compound is a distinct substance, differing in all its properties from $CoCl_2$ (Fig. 2–1). The dot in the formula $CoCl_2 \cdot 6H_2O$ sets off the water as a well-defined molecular component of the solid.

EXAMPLE 2-1

Ethylene is a gas under room conditions and is the chemical starting material for many plastics. Its molecular formula is C_2H_4.
(a) What is its empirical formula?
(b) Write other molecular formulas that correspond to the same empirical formula.

Solution

(a) The ratio of the number of carbon atoms to the number of hydrogen atoms in ethylene is 2:4. The smallest integers that have this ratio are 1 and 2, so the empirical formula is CH_2.
(b) Every integral multiple of CH_2 is a different molecular formula corresponding to this same empirical formula. Other molecular formulas include C_3H_6, C_4H_8, and so forth. Note that CH_2 itself is also a possible molecular formula.

Exercise

In the following list of six compounds of carbon, hydrogen, and oxygen, identify the pairs that have the same empirical formula and give that formula: acetic acid ($C_2H_4O_2$), sorbic acid ($C_6H_8O_2$), caproic acid ($C_6H_{12}O_2$), acrolein (C_3H_4O), propionaldehyde (C_3H_6O), and glucose ($C_6H_{12}O_6$).

Answer: Acetic acid and glucose (CH_2O), sorbic acid and acrolein (C_3H_4O), and caproic acid and propionaldehyde (C_3H_6O).

2–2   CHEMICAL FORMULA AND PERCENTAGE COMPOSITION

The empirical formula $Fe_2(SO_4)_3$ indicates that every 2 atoms of iron in this compound is accompanied by 3 atoms of sulfur and 12 of oxygen. Equivalently, one mole of $Fe_2(SO_4)_3$ contains 2 moles of iron atoms, 3 moles of sulfur atoms, and 12 moles of oxygen atoms; the mole ratios, like the atom ratios, are 2:3:12. Clearly, the empirical formula and the percentage composition by mass are related to one another, a connection we shall explore in this section.

• Why not write $Fe_2S_3O_{12}$ instead? Grouping symbols in parentheses implies that the atoms are grouped together in the compound. In this case, each sulfur atom is surrounded by four oxygen atoms to make a group called the sulfate ion (see Section 3–3).

Percentage Composition from Empirical Formula

If we know the empirical (or molecular) formula of a substance, we can deduce its percentage composition by mass. To do this, we calculate the mass of each element in one mole of the compound and then add those masses to find the total mass of one mole of the compound. The percentages by mass are then found by dividing the mass of each element by the total mass and multiplying by 100%. This is illustrated in the following example.

EXAMPLE 2–2

Iron forms a yellow crystalline compound with the empirical formula $Fe_2(SO_4)_3$, which is used in water and sewage treatment to aid in the removal of suspended impurities. Calculate the mass percentages of iron, sulfur, and oxygen in this compound.

Solution

We use the information in a table of atomic masses to calculate the mass of each of the elements in one mole of this compound.

$$m_{Fe} = 2 \text{ mol Fe} \times \left(\frac{55.847 \text{ g Fe}}{1 \text{ mol Fe}} \right) = 111.69 \text{ g Fe}$$

$$m_{S} = 3 \text{ mol S} \times \left(\frac{32.066 \text{ g S}}{1 \text{ mol S}} \right) = 96.198 \text{ g S}$$

$$m_{O} = 12 \text{ mol O} \times \left(\frac{15.999 \text{ g O}}{1 \text{ mol O}} \right) = 191.99 \text{ g O}$$

The total of these masses is 399.88 g, so the percentage composition is

$$\% \text{ Fe} = \left(\frac{111.69 \text{ g Fe}}{399.88 \text{ g compound}} \right) \times 100\% = 27.931\%$$

$$\% \text{ S} = \left(\frac{96.198 \text{ g S}}{399.88 \text{ g compound}} \right) \times 100\% = 24.057\%$$

$$\% \text{ O} = \left(\frac{191.99 \text{ g O}}{399.88 \text{ g compound}} \right) \times 100\% = 48.012\%$$

These percentages add up to 100.000%, as they should.

Exercise

Tetrodotoxin, a potent poison found in the ovaries and liver of the globe fish, has the empirical formula $C_{11}H_{17}N_3O_8$. Calculate the mass percentages of the four elements in this compound.

Answer: C, 41.38%; H, 5.37%; N, 13.16%; O, 40.09%.

Empirical Formula from Experiment

In principle, any compound can be broken down into its component elements and the elements then separated and weighed. Such an experiment, called an "elemental analysis," furnishes the composition by mass of the elements in the compound. Starting from these data, the procedure of Example 2–2 can be reversed to allow the empirical formula of the compound to be determined. This is illustrated by the following example.

EXAMPLE 2–3

A 60.00-g sample of a molecular compound used as a dry-cleaning fluid is found to contain 10.80 g of carbon, 1.36 g of hydrogen, and 47.84 g of chlorine. Using a table of atomic masses, determine the empirical formula of the compound.

Solution

The first step is to calculate the chemical amount (in moles) of each of the elements in the sample. Such quantities are represented with an n followed by the subscripted symbol of the element

$$\text{for carbon: } n_C = 10.80 \text{ g C} \times \left(\frac{1 \text{ mol C}}{12.011 \text{ g C}} \right) = 0.8992 \text{ mol C}$$

$$\text{for hydrogen: } n_H = 1.36 \text{ g H} \times \left(\frac{1 \text{ mol H}}{1.008 \text{ g H}} \right) = 1.35 \text{ mol H}$$

$$\text{for chlorine: } n_{Cl} = 47.84 \text{ g Cl} \times \left(\frac{1 \text{ mol Cl}}{35.453 \text{ g Cl}} \right) = 1.349 \text{ mol Cl}$$

The next step is to find the smallest set of integers that correctly represents the ratios of the chemical amounts of the elements in this compound. The chemical amounts (1.35 mol and 1.349 mol) of hydrogen and chlorine are essentially equal. The ratio of the chemical amount of chlorine (or hydrogen) to that of carbon is

$$\frac{1.349 \text{ mol Cl}}{0.8992 \text{ mol C}} = \frac{1.500 \text{ mol Cl}}{1.000 \text{ mol C}} = \frac{3 \text{ mol Cl}}{2 \text{ mol C}}$$

Here 3 and 2 appear as the smallest integers whose ratio is 1.500. We conclude that the numbers of moles stand in a whole-number ratio of 2:3:3, so the empirical formula is $C_2H_3Cl_3$. Further measurements would be necessary to find the actual molecular mass and the correct *molecular* formula from among $C_2H_3Cl_3$, $C_4H_6Cl_6$, and higher multiples $(C_2H_3Cl_3)_n$.

Exercise

Heating a 150.0-mg dose of a compound used to treat rheumatism decomposed it to its constituent elements, which were separated and then weighed. There were 60.29 mg of gold, 21.10 mg of sodium, 29.37 mg of oxygen, and 39.24 mg of sulfur. Determine the empirical formula of this compound.

Answer: $AuNa_3O_6S_4$.

Even if we know only *percentages* by mass of the elements in a compound (rather than the actual masses found in the analyzed sample), we can still obtain the empirical formula. In this case, which is very common, we begin by assuming some total mass (exactly 100 g is a convenient choice) and then compute the mass of each

Figure 2–2 Chromium(III) oxide (Cr_2O_3) is a green solid that can be reduced to shiny metallic, elemental chromium. The major use of chromium is in the plating of steel and other metals to give a brighter and more resistant finish, as in the chrome trim on the stapler.

element that this total would supply. From there on, the procedure follows the pattern just shown. If a compound has mass percentages of 40.04% sulfur and 59.96% oxygen, for example, we can immediately conclude that 100.00 g of the compound contains 40.04 g of sulfur and 59.96 g of oxygen. From these masses, we calculate the number of moles of the elements as in Example 2–3 (1.249 mol S and 3.748 mol O). From the ratio 3.748:1.249 = 3.0013, we conclude that the empirical formula of the compound is SO_3. Note that the percentage composition is independent of the actual number of grams in the sample, so we would get the same answer with any choice of total mass.

Sometimes, separating a particular element from a compound and weighing it is difficult to do accurately. By measuring the masses of all the other elements and using the fact that the total mass is conserved, we can still obtain the empirical formula of the compound. This tactic is particularly useful in cases in which a binary compound can react to give one of the elements in the pure state, as in the following example.

EXAMPLE 2–4

An oxide of chromium is reduced to elemental chromium (Fig. 2–2) by the action of hydrogen. Treatment of 6.245 g of the oxide in this way is found to yield 4.273 g of pure chromium. Determine the empirical formula of the oxide.

Solution

Because mass is conserved, the mass of oxygen in the original sample must have been

$$m_O = m_{compound} - m_{Cr} = 6.245 \text{ g} - 4.273 \text{ g} = 1.972 \text{ g}$$

Thus, the chemical amounts of chromium and oxygen atoms were

$$n_{Cr} = 4.273 \text{ g Cr} \times \left(\frac{1 \text{ mol Cr}}{51.996 \text{ g Cr}} \right) = 0.08218 \text{ mol Cr}$$

$$n_O = 1.972 \text{ g O} \times \left(\frac{1 \text{ mol O}}{15.999 \text{ g O}} \right) = 0.1233 \text{ mol O}$$

The ratio of these chemical amounts is

$$\frac{0.1233 \text{ mol O}}{0.08218 \text{ mol Cr}} = \frac{1.500 \text{ mol O}}{1.000 \text{ mol Cr}} = \frac{3 \text{ mol O}}{2 \text{ mol Cr}}$$

The empirical formula is Cr_2O_3 because the compound contains three moles of oxygen atoms for every two moles of chromium atoms.

Exercise

Moderate heating of 97.44 mg of a compound containing nickel, carbon, and oxygen and no other elements drives off all of the carbon and oxygen in the form of carbon monoxide (CO) and leaves 33.50 mg of metallic nickel behind. Determine the empirical formula of the compound.

Answer: NiC_4O_4.

Figure 2-3 A flame from the combustion of natural gas.

Hydrocarbons are compounds containing only carbon and hydrogen. The simplest hydrocarbon is methane (molecular formula CH_4), the major component of natural gas (Fig. 2–3). Two other hydrocarbons are the important industrial solvent benzene (molecular formula C_6H_6) and the compound styrene (molecular formula C_8H_8), which is used to produce polystyrene for packaging and insulation. The empirical formulas of hydrocarbons can be determined by analysis in the combustion train shown in Figure 2–4. In this apparatus, a weighed amount of the hydrocarbon is burned completely in oxygen, yielding carbon dioxide and water. The masses of water and carbon dioxide are then determined, and from these data the empirical formula is found, as in the following example.

• The primary source of these and many other hydrocarbons is petroleum.

EXAMPLE 2–5

A certain compound, used as a welding fuel, contains carbon and hydrogen only. Burning a small sample of it completely in oxygen gives 3.38 g of carbon dioxide, 0.690 g of water, and no other products. What is the empirical formula of the compound?

Solution

We first compute the chemical amounts of the carbon dioxide and water. Because all of the carbon has been converted to carbon dioxide and all of the hydrogen to water, we can immediately determine how much carbon and hydrogen were in the unburned gas.

$$n_C = 3.38 \text{ g CO}_2 \times \left(\frac{1 \text{ mol CO}_2}{44.01 \text{ g CO}_2}\right) \times \left(\frac{1 \text{ mol C}}{1 \text{ mol CO}_2}\right) = 0.0768 \text{ mol C}$$

$$n_H = 0.690 \text{ g H}_2\text{O} \times \left(\frac{1 \text{ mol H}_2\text{O}}{18.02 \text{ g H}_2\text{O}}\right) \times \left(\frac{2 \text{ mol H}}{1 \text{ mol H}_2\text{O}}\right) = 0.0766 \text{ mol H}$$

Note in the second equation that the number of moles of water is multiplied by two to determine the number of moles of hydrogen atoms because each water molecule contains two hydrogen atoms. Because 0.0766 is equal to 0.0768 within the precision of the data, we conclude that the compound contains equal chemical amounts (numbers of moles) of carbon and hydrogen. Its empirical formula is CH. Its molecular formula may be CH, or C_2H_2, or C_3H_3, and so on.

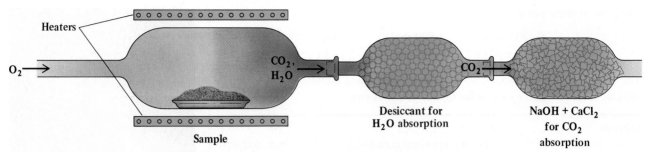

Figure 2–4   A combustion train for measuring the amounts of carbon and hydrogen in hydrocarbons. A sample is burned in a flow of oxygen to form water and carbon dioxide. These products pass into a chamber containing magnesium perchlorate ($Mg(ClO_4)_2$), where the water is absorbed, while the carbon dioxide passes on to a second chamber where it is absorbed on finely divided particles of sodium hydroxide (NaOH) mixed with calcium chloride ($CaCl_2$). By measuring the changes in mass of the two absorbers, the amounts of water and carbon dioxide produced can be found. The amounts of hydrogen and water are then computed as in Example 2–5.

Exercise

A sample of a liquid hydrocarbon weighing 142.70 mg is burned in a combustion train to give 477.00 mg of carbon dioxide, 111.60 mg of water, and no other products. What is the empirical formula of this hydrocarbon?

Answer: C_7H_8.

In this method of analysis, it is important that the hydrocarbon burn completely to carbon dioxide and water. For example, if some of the carbon remained incompletely burned as carbon monoxide, the analysis would be in error. It is to check for errors of this type that the mass of the original sample of hydrocarbon is invariably measured, even though the mass is not required in the calculations.

Molecular Formula from Empirical Formula

The empirical formula of a molecular substance does not furnish enough information to determine its molecular formula. The required missing fact is the approximate molar mass of the substance. Fortunately, molar masses are available from a variety of experiments. For example, many molecular substances can be vaporized and the intact molecules ionized in mass spectrometers. The molar masses are then determined by accelerating the ions and observing the degree of curvature of their paths through a magnetic field, as described in Section 1–6. Other experiments (described in Sections 5–4 and 6–6) give the approximate molar masses of other molecular substances. Once the molar mass and the empirical formula are known, the molecular formula follows quickly, as shown in the following example.

EXAMPLE 2–6

A molecular compound isolated from lemons is thought to confer important health benefits and is therefore studied. Its empirical formula is $C_5H_4O_2$. Mass spectrometry establishes that its molar mass is approximately 288 g mol^{-1}. Determine the molecular formula of the compound.

Solution

If the molecular formula were $C_5H_4O_2$, then the molar mass would be 96.08 g mol^{-1}. A molar mass of 288 g mol^{-1} is 288/96.08 = 3.00 times larger than this. Hence the molecular formula of the compound must be $C_{15}H_{12}O_6$, in which each subscript is triple the subscript in the empirical formula.

Exercise

A molecular compound called "anemone camphor" is isolated from the anemone plant. Its empirical formula is $C_5H_4O_2$, and its molar mass is approximately 192 g mol^{-1} according to mass spectrometry measurements. What is its molecular formula?

Answer: $C_{10}H_8O_4$.

Figure 2–5   When hot steel wool (which consists mostly of iron) is thrust into a flask filled with chlorine gas, it glows brightly as it reacts to release a cloud of brown iron(III) chloride.

2-3   WRITING BALANCED CHEMICAL EQUATIONS

Chemical reactions are events in which elements combine into compounds, compounds decompose back into elements, and existing compounds transform to new compounds. Because atoms are indestructible and retain their identity in chemical reactions, the same number of atoms (or moles of atoms) of each element must be present before and after any chemical reaction. The conservation of mass in a chemical change is represented in a balanced **chemical equation** for that process. Suppose we want to describe the reaction of iron with chlorine to give iron(III) chloride (Fig. 2–5). The balanced equation for this process is

$$2 \text{ Fe} + 3 \text{ Cl}_2 \longrightarrow 2 \text{ FeCl}_3$$

The formulas on the left-hand side of the arrow denote **reactants,** and those on the right-hand side denote **products.** This equation tells us that two moles of iron atoms react with three moles of chlorine molecules (the formula Cl_2 recognizes that the element chlorine exists in molecules of two atoms each) to give two moles of iron(III) chloride. The **coefficients** 2, 3, and 2 are essential in balancing the equation. They ensure that the two moles of Fe atoms and six moles of Cl atoms on the left side (present initially as reactants) appear on the right side as well, with no atom of any kind created or destroyed.

The coefficients required to balance an equation can often be found by simple stepwise reasoning once the formulas of all the reactants and products are established. Balancing an equation consists of adjusting coefficients only; the subscripts in the formulas are never changed. Changing a formula tampers with the experimental facts of what reacted and what formed, and is an error. As an example of balancing, consider the decomposition of ammonium nitrate (NH_4NO_3) upon gentle heating. Experimentation establishes that this chemical change generates dinitrogen oxide (N_2O) and water, and no other products. We can write the *unbalanced* equation for the reaction as

$$NH_4NO_3 \longrightarrow N_2O + H_2O$$

This equation is unbalanced because there are three moles of oxygen atoms and four moles of hydrogen atoms on the left but only two moles of oxygen atoms and two moles of hydrogen atoms on the right. How do we find the coefficients that balance

• It is admittedly a contradiction in terms to refer to an "unbalanced equation," but the practice is well entrenched in chemistry and causes no difficulties.

the equation? The first step is to assign 1 as the coefficient of one species, usually the one containing the most elements. Here, this is NH_4NO_3. Next, we seek out the elements that appear in only one other place in the equation and assign coefficients to balance the numbers of their atoms. Here, nitrogen appears in only one other substance (N_2O) and a coefficient of 1 for the N_2O ensures that there are two moles of nitrogen atoms on each side of the equation. Hydrogen appears in H_2O, and we must choose the coefficient to be 2 to balance the four moles of hydrogen atoms on the left side. This gives

$$1 \; NH_4NO_3 \longrightarrow 1 \; N_2O + 2 \; H_2O$$

We can verify that the last element, oxygen, is also balanced because there are three moles of oxygen atoms on each side. Coefficients of 1 (given explicitly for NH_4NO_3 and N_2O in the preceding) are generally omitted in chemical equations. Following this practice gives

$$NH_4NO_3 \longrightarrow N_2O + 2 \; H_2O$$

As another example, consider the reaction in which gaseous butane is burned in oxygen to form carbon dioxide and water. This is expressed in words as

"butane plus oxygen yields carbon dioxide plus water"

but this sentence conveys no idea of the quantitative relationships involved. If we replace the names of the compounds with chemical formulas, we have

$$\_\_ \; C_4H_{10} + \_\_ \; O_2 \longrightarrow \_\_ \; CO_2 + \_\_ \; H_2O$$

where spaces have been left in front of the formulas for the insertion of coefficients for the number of moles of each reactant and product. Suppose we start with one mole of butane (C_4H_{10}). It contains four moles of carbon atoms and must produce four moles of carbon dioxide molecules because the carbon can go nowhere else. We thus write 4 for the coefficient of CO_2. In the same way, the ten moles of hydrogen *atoms* must form five moles of water *molecules* because each water molecule contains two hydrogen atoms; thus, the coefficient of the H_2O is 5. Writing what we have so far gives

$$C_4H_{10} + \_\_ \; O_2 \longrightarrow 4 \; CO_2 + 5 \; H_2O$$

To determine the coefficient of the O_2, we observe that four moles of CO_2 contains eight moles of oxygen atoms and that five moles of H_2O contains five moles of oxygen atoms, giving a total of 13 moles of oxygen atoms. Thirteen moles of oxygen atoms (13 mol of O) is equivalent to $\frac{13}{2}$ moles of oxygen molecules (6.5 mol of O_2). Hence, the coefficient of O_2 is $\frac{13}{2}$. The balanced equation is then

$$C_4H_{10} + \frac{13}{2} \; O_2 \longrightarrow 4 \; CO_2 + 5 \; H_2O$$

There is nothing wrong with fractions like $\frac{13}{2}$ in a balanced equation because fractions of moles are perfectly meaningful. However, fractions must be eliminated if the equation is to be interpreted on the atomic level, as we do in Section 2–4. Here, multiplying all coefficients by 2 gives

$$2 \; C_4H_{10} + 13 \; O_2 \longrightarrow 8 \; CO_2 + 10 \; H_2O$$

Let us summarize the steps in balancing a chemical equation:

1. Assign 1 as the coefficient of one reactant or product. The best choice is the most complicated species, with the largest number of different elements.
2. Identify, in sequence, elements that appear in only one chemical species for which the coefficient is not yet determined. Choose that coefficient to balance the number of moles of atoms of that element. Continue until all coefficients are identified.
3. It is often desirable to eliminate fractional coefficients. To do so, multiply the entire equation by the smallest integer that eliminates the fractions.

This method of balancing chemical equations "by inspection" becomes unworkable when the equation involves several complex species. In Section 12–2, we shall present techniques for balancing more complex chemical equations.

EXAMPLE 2–7

The Hargreaves process is an industrial procedure for producing sodium sulfate (Na_2SO_4) for use in papermaking. The starting materials are sodium chloride (NaCl), sulfur dioxide (SO_2), water (H_2O), and oxygen (O_2); hydrogen chloride (HCl) is generated as a by-product. Write a balanced chemical equation for this process.

Solution

The unbalanced equation is

$$\_\_ NaCl + \_\_ SO_2 + \_\_ H_2O + \_\_ O_2 \longrightarrow \_\_ Na_2SO_4 + \_\_ HCl$$

We begin by assigning a coefficient of 1 to one reactant or product: Na_2SO_4 is a reasonable choice because it is the most complex species, composed of three different elements. Assigning this 1 means that there are two moles of sodium atoms on the right. To balance this we need the same number on the left, giving 2 as the coefficient of the NaCl. By a similar argument, the coefficient of the SO_2 must be 1 to balance the one mole of sulfur on the right. This gives

$$2 NaCl + SO_2 + \_\_ H_2O + \_\_ O_2 \longrightarrow Na_2SO_4 + \_\_ HCl$$

Next we note that there are two moles of chlorine atoms on the left-hand (reactant) side, so the coefficient of the HCl must be 2. Hydrogen is the next element to balance. With two moles of hydrogen atoms on the right-hand side, the coefficient for the H_2O must be 1:

$$2 NaCl + SO_2 + H_2O + \_\_ O_2 \longrightarrow Na_2SO_4 + 2 HCl$$

Finally, the oxygen atoms must be balanced. Four moles of oxygen atoms appear on the right; the left shows two moles from the SO_2 and one mole from the H_2O. Therefore, one mole of oxygen *atoms* must come from the O_2. The coefficient of the O_2 is $\frac{1}{2}$, because $\frac{1}{2}$ mole of O_2 molecules contains one mole of oxygen atoms:

$$2 NaCl + SO_2 + H_2O + \tfrac{1}{2} O_2 \longrightarrow Na_2SO_4 + 2 HCl$$

We multiply all coefficients in the equation by 2 to eliminate fractions, giving

$$4 NaCl + 2 SO_2 + 2 H_2O + O_2 \longrightarrow 2 Na_2SO_4 + 4 HCl$$

Note that in balancing the equation we considered oxygen last because it appears in several places on the left-hand side of the equation.

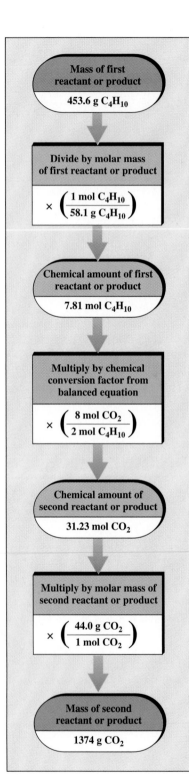

Isocyanic acid (HNCO) reacts with gaseous nitrogen monoxide (NO), an air pollutant, to form nitrogen, carbon dioxide, and water. The reaction may offer a feasible means to eliminate nitrogen monoxide from automobile exhaust. Write a balanced chemical equation to represent the reaction.

Answer: $4\ HNCO + 6\ NO \longrightarrow 5\ N_2 + 2\ H_2O + 4\ CO_2.$

The balancing of chemical equations is a routine, mechanical process of accounting once the reactants and products are known. The difficult part (and the part where *chemistry* comes in) is to know what substances react with each other and to determine what products are formed. This is a question to which we return many times throughout this book.

2-4   THE STOICHIOMETRY OF CHEMICAL REACTIONS

A balanced chemical equation provides a platform for a variety of quantitative statements about the relative amounts of the substances reacting and forming. The balanced chemical equation we wrote for the combustion of butane

$$2\ C_4H_{10} + 13\ O_2 \longrightarrow 8\ CO_2 + 10\ H_2O$$

can be interpreted as stating either that

2 molecules of C_4H_{10} + 13 molecules of $O_2 \longrightarrow$
$$8\ \text{molecules of}\ CO_2 + 10\ \text{molecules of}\ H_2O$$

or that

$$2\ \text{mol of}\ C_4H_{10} + 13\ \text{mol of}\ O_2 \longrightarrow 8\ \text{mol of}\ CO_2 + 10\ \text{mol of}\ H_2O$$

If we compute the molar masses of the different substances in the reaction and multiply each by the number of moles represented in the balanced equation, we obtain the mass of each substance involved in the reaction

$$m_{C_4H_{10}} = 2\ \text{mol}\ C_4H_{10} \times \left(\frac{58.1\ g\ C_4H_{10}}{1\ \text{mol}\ C_4H_{10}}\right) = 116.2\ g\ C_4H_{10}$$

$$m_{O_2} = 13\ \text{mol}\ O_2 \times \left(\frac{32.0\ g\ O_2}{1\ \text{mol}\ O_2}\right) = 416.0\ g\ O_2$$

$$m_{CO_2} = 8\ \text{mol}\ CO_2 \times \left(\frac{44.0\ g\ CO_2}{1\ \text{mol}\ CO_2}\right) = 352.0\ g\ CO_2$$

$$m_{H_2O} = 10\ \text{mol}\ H_2O \times \left(\frac{18.02\ g\ H_2O}{1\ \text{mol}\ H_2O}\right) = 180.2\ g\ H_2O$$

Figure 2-6   The steps in a stoichiometric calculation. In a typical calculation, the mass of one reactant or product is known, and the masses of one or more other reactants or products are to be calculated, using the balanced chemical equation and a table of relative atomic masses.

The equation then becomes, in terms of masses,

$$116.2 \text{ g of } C_4H_{10} + 416.0 \text{ g of } O_2 \longrightarrow 352.0 \text{ g of } CO_2 + 180.2 \text{ g of } H_2O$$

Note that the total mass of the products, 532.2 g, equals the total mass of the reactants, as it must under the law of conservation of mass.

A balanced chemical equation thus gives quantitative relationships among the masses of reactants and products. The study of these relationships is the branch of chemistry called **stoichiometry.** Stoichiometry is fundamental to all of chemistry. The preceding analysis explicitly established the masses of CO_2 and C_4H_{10} that are equivalent in the reaction: 352.0 g of CO_2 is produced for every 116.2 g of C_4H_{10} consumed. We can express this equivalence as a unit-factor

- The word "stoichiometry" comes from the Greek *stoicheion,* meaning "element," + *metron,* meaning "measure."

$$\left(\frac{352.0 \text{ g } CO_2}{116.2 \text{ g } C_4H_{10}} \right)$$

just as we express the equivalence of 1 km and 1000 m by the unit-factor (1 km/1000 m). Similar unit-factors can be constructed for every pair of substances taking part in the reaction. Suppose now that we need to know the mass of CO_2 that results when, say, 453.6 g (a pound) of C_4H_{10} burns according to this equation. (An environmental engineer might need to know this to compare emissions of CO_2 from burning different fuels.) We multiply the 453.6 g of C_4H_{10} by the unit-factor, observing the way that the units cancel out (shown in red)

- Recall that a unit-factor is a fraction in which the numerator (top) is equivalent to the denominator (bottom). See Appendix B–2 for a discussion of the unit-factor method.

$$m_{CO_2} = 453.6 \text{ g } C_4H_{10} \times \left(\frac{352.0 \text{ g } CO_2}{116.2 \text{ g } C_4H_{10}} \right) = 1374 \text{ g } CO_2$$

Usually it is not necessary to work out an explicit mass-to-mass unit-factor like the one just used. Instead, computations can be set up by stringing two or three other unit-factors together. For example, the preceding computation could be set up as a string of three unit-factors.

$$m_{CO_2} = 453.6 \text{ g } C_4H_{10} \times \left(\frac{1 \text{ mol } C_4H_{10}}{58.1 \text{ g } C_4H_{10}} \right) \times \left(\frac{8 \text{ mol } CO_2}{2 \text{ mol } C_4H_{10}} \right) \times \left(\frac{44.0 \text{ g } CO_2}{1 \text{ mol } CO_2} \right)$$

$$= 1374 \text{ g } CO_2$$

The first unit-factor converts from grams of C_4H_{10} to moles of C_4H_{10}. The second converts from moles of C_4H_{10} to moles of CO_2, using the coefficients assigned to the two substances in the balanced equation. The third converts from moles of CO_2 to grams of CO_2. The cancellation of units, indicated in red, confirms the correctness of every step. Figure 2–6 summarizes this pattern, which applies in all mass/mass stoichiometric calculations.

EXAMPLE 2–8

Boron carbide (B_4C) is a very hard, low-density solid with a high melting point. These properties make it useful in armor plates and other surfaces subject to wear and abrasion. Boron carbide is made by the reaction of boron oxide with carbon:

$$2 B_2O_3 + 4 C \longrightarrow B_4C + 3 CO_2$$

Calculate the chemical amount of carbon that reacts with 3.32 mol of B_2O_3, according to this equation, and the chemical amount of B_4C that is produced. Also, calculate the mass (in grams) of carbon needed to react with 232.0 g of B_2O_3.

Solution

According to the balanced equation, every two moles of B_2O_3 reacts with four moles of carbon. Therefore, 3.32 mol of B_2O_3 reacts with $2 \times 3.32 = 6.64$ mol of carbon. We can see this by setting up the ratio

$$\frac{4 \text{ mol C}}{2 \text{ mol B}_2O_3}$$

as a chemical unit-factor to convert between moles of B_2O_3 and moles of C.

$$n_C = 3.32 \text{ mol B}_2O_3 \times \left(\frac{4 \text{ mol C}}{2 \text{ mol B}_2O_3} \right) = \boxed{6.64 \text{ mol C}}$$

In the same way, we calculate the chemical amount of B_4C produced as

$$n_{B_4C} = 3.32 \text{ mol B}_2O_3 \times \left(\frac{1 \text{ mol B}_4C}{2 \text{ mol B}_2O_3} \right) = \boxed{1.66 \text{ mol B}_4C}$$

For the last calculation, we use a string of three unit-factors, as shown in Figure 2–6

$$m_C = 232.0 \text{ g B}_2O_3 \times \left(\frac{1 \text{ mol B}_2O_3}{69.62 \text{ g B}_2O_3} \right) \times \left(\frac{4 \text{ mol C}}{2 \text{ mol B}_2O_3} \right) \times \left(\frac{12.01 \text{ g C}}{\text{mol C}} \right)$$
$$= \boxed{80.04 \text{ g C}}$$

Exercise

At one point in the purification of silicon, gaseous $SiHCl_3$ reacts with gaseous H_2 to give gaseous HCl and solid Si.
(a) Determine the chemical amount (in moles) of H_2 required to react with 160.4 mol of $SiHCl_3$. (b) Determine the chemical amount of HCl that is produced. (c) Determine the mass (in grams) of Si that is produced.

Answer: (a) 160.4 mol of H_2. (b) 481.2 mol of HCl. (c) 4505 g of Si.

• Remember: the chemical unit-factors linking different species in a reaction depend on the coefficients in the *balanced* equation.

The solution to the last part of Example 2–8 uses a string of unit-factors to set up a single long multiplication, but each step could have been completed separately. We might have written

$$232.0 \text{ g B}_2O_3 \times \left(\frac{1 \text{ mol B}_2O_3}{69.62 \text{ g B}_2O_3} \right) = 3.3324 \text{ mol B}_2O_3$$

$$3.3324 \text{ mol B}_2O_3 \times \left(\frac{4 \text{ mol C}}{2 \text{ mol B}_2O_3} \right) = 6.665 \text{ mol C}$$

$$6.665 \text{ mol C} \times \left(\frac{12.01 \text{ g C}}{\text{mol C}} \right) = 80.04 \text{ g C}$$

The intermediate answers are interesting, but the labor saved and errors avoided in not transcribing intermediate results usually make it preferable to set up a string of unit-factors and do the arithmetic all at once.

EXAMPLE 2–9

Calcium hypochlorite ($Ca(OCl)_2$) is used as a bleaching agent (Fig. 2–7). It is produced by treating a mixed solution of sodium hydroxide and calcium hydroxide with gaseous chlorine. The subsequent reaction is described by the balanced equation

$$2 \text{ NaOH} + \text{Ca(OH)}_2 + 2 \text{ Cl}_2 \longrightarrow \text{Ca(OCl)}_2 + 2 \text{ NaCl} + 2 \text{ H}_2\text{O}$$

How many grams of chlorine and of sodium hydroxide react with 1067 g of $Ca(OH)_2$, and how many grams of $Ca(OCl)_2$ is produced according to this equation?

Solution

The chemical amount of $Ca(OH)_2$ consumed is

$$n_{Ca(OH)_2} = 1067 \text{ g Ca(OH)}_2 \times \left(\frac{1 \text{ mol Ca(OH)}_2}{74.09 \text{ g Ca(OH)}_2} \right) = 14.40 \text{ mol Ca(OH)}_2$$

where the molar mass of $Ca(OH)_2$ has been obtained from the molar masses of calcium, oxygen, and hydrogen as

$$\mathcal{M}_{Ca(OH)_2} = 40.08 + 2(15.999) + 2(1.0079) = 74.09 \text{ g mol}^{-1}$$

According to the balanced equation, one mole of $Ca(OH)_2$ reacts with two moles of NaOH and two moles of Cl_2 to produce one mole of $Ca(OCl)_2$. If 14.40 mol of $Ca(OH)_2$ reacts, then

$$n_{NaOH} = 14.40 \text{ mol Ca(OH)}_2 \times \left(\frac{2 \text{ mol NaOH}}{1 \text{ mol Ca(OH)}_2} \right) = 28.80 \text{ mol NaOH}$$

$$n_{Cl_2} = 14.40 \text{ mol Ca(OH)}_2 \times \left(\frac{2 \text{ mol Cl}_2}{1 \text{ mol Ca(OH)}_2} \right) = 28.80 \text{ mol Cl}_2$$

$$n_{Ca(OCl)_2} = 14.40 \text{ mol Ca(OH)}_2 \times \left(\frac{1 \text{ mol Ca(OCl)}_2}{1 \text{ mol Ca(OH)}_2} \right) = 14.40 \text{ mol Ca(OCl)}_2$$

From the chemical amounts and molar masses of reactants and products, the required masses can be found:

$$m_{NaOH} = 28.80 \text{ mol NaOH} \times \left(\frac{40.00 \text{ g NaOH}}{1 \text{ mol NaOH}} \right) = 1152 \text{ g NaOH reacting}$$

$$m_{Cl_2} = 28.80 \text{ mol Cl}_2 \times \left(\frac{70.91 \text{ g Cl}_2}{1 \text{ mol Cl}_2} \right) = 2042 \text{ g Cl}_2 \text{ reacting}$$

$$m_{Ca(OCl)_2} = 14.40 \text{ mol Ca(OCl)}_2 \times \left(\frac{142.98 \text{ g Ca(OCl)}_2}{1 \text{ mol Ca(OCl)}_2} \right)$$
$$= 2059 \text{ g Ca(OCl)}_2 \text{ produced}$$

Figure 2–7 A solution of calcium hypochlorite ($Ca(OCl)_2$) is a powerful bleaching agent. A few drops of a solution quickly destroy the color of the blue dye on the cloth (right).

Exercise

When heated in dry air, sodium reacts with oxygen to form sodium peroxide:

$$2 \text{ Na} + O_2 \longrightarrow Na_2O_2$$

What masses (in kilograms) of Na and of O_2 are required to prepare 457 kg of Na_2O_2 by this reaction?

Answer: 269 kg of Na; 188 kg of O_2.

Volume Relationships of Gases in Chemical Reactions

The law of combining volumes (see Section 1–4) states that the volumes of gaseous reactants and products stand in ratios of simple integers, as long as those volumes are measured at the same temperature and pressure. These simple integers turn out to be none other than the coefficients in the balanced chemical equation representing the reaction. A balanced chemical equation therefore provides relationships among *volumes* of gases reacting, as illustrated in the following.

EXAMPLE 2–10

Calcium cyanamide, $CaCN_2$, is a solid. It reacts with gaseous O_2 under strong heating to give CaO, which is a solid, and two gases:

$$2\ CaCN_2 + 7\ O_2 \longrightarrow 2\ CaO + 2\ CO_2 + 4\ NO_2$$

A run of this reaction consumes 1.40 L of gaseous O_2 (measured at room conditions of temperature and pressure). Predict the volume of CO_2 and the volume of NO_2 present when the products are examined at room conditions.

Solution

The ratios of the coefficients in the balanced equation equal the ratios of volumes of gases produced or reacting. As long as the temperature and pressure are the same after the reaction as before, then for every 7 volumes of O_2 consumed, 2 volumes of CO_2 and 4 volumes of NO_2 form. Therefore

$$V_{CO_2} = 1.40\ \text{L}\ O_2 \times \left(\frac{2\ \text{L}\ CO_2}{7\ \text{L}\ O_2}\right) = 0.40\ \text{L}\ CO_2$$

$$V_{NO_2} = 1.40\ \text{L}\ O_2 \times \left(\frac{4\ \text{L}\ NO_2}{7\ \text{L}\ O_2}\right) = 0.80\ \text{L}\ NO_2$$

Setting up ratios of volumes in this way works *only* for gases and *only* when the volumes are measured at the same temperature and pressure.

Exercise

Some gaseous H_2 and N_2 react to form 5.00 L of NH_3, also a gas, according to the equation $3\ H_2 + 2\ N_2 \longrightarrow 2\ NH_3$. What volume of H_2 reacted, assuming that the pressure and temperature are the same after the reaction as before?

Answer: 7.50 L of H_2.

(a)   (b)

Figure 2–8 (a) Aluminum, a shiny metal, reacts vigorously with bromine, a reddish brown liquid, to give aluminum bromide according to the equation

$$2\ Al + 3\ Br_2 \longrightarrow Al_2Br_6$$

(b) Some of the product is ejected, along with excess bromine, by the vigor of the reaction.

REACTANTS

PRODUCTS

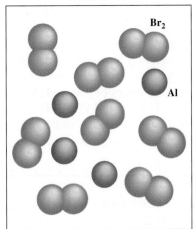

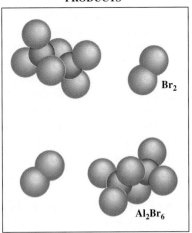

Figure 2–9 Every two atoms of aluminum react with three molecules of bromine (Br_2) to form one molecule of Al_2Br_6. In the case diagrammed here, bromine is present in excess, and some remains after the reaction ends. Aluminum is the limiting reactant; the addition of more aluminum would enable the further production of Al_2Br_6.

2–5 LIMITING REACTANT AND PERCENTAGE YIELD

In the cases considered in the previous section, the reactants were present in the exact ratios necessary for them to be completely consumed in forming products. It is rare in practice for things to work out so neatly. Usually one or more reactants are left over. Suppose we mix together several reactants in arbitrary amounts and allow them to react. The chemical reaction must stop when the supply of any one of the reactants gives out. The reactant that is used up first is the **limiting reactant.** The other reactants remain after the reaction stops; these reactants are present **in excess** (Figs. 2–8 and 2–9). An additional supply of the limiting reactant restarts the reaction and forms more products. This is not true of the other reactants.

In an industrial process, the limiting reactant is frequently the most expensive reactant, to ensure that none of it is wasted. In the preparation of silver chloride for photographic film by the reaction

$$AgNO_3 + NaCl \longrightarrow AgCl + NaNO_3$$

the silver nitrate is far more expensive than the sodium chloride; thus, it makes sense to carry out the reaction with an excess of sodium chloride, to ensure that as much as possible of the silver nitrate reacts to form products.

EXAMPLE 2–11

Cold drinks cost 60 cents from a (rather peculiar) vending machine that works only when one quarter, two dimes, and three nickels are deposited. Which coin runs out first if as many cold drinks as possible are purchased starting with 22 quarters, 36 nickels, and 30 dimes?

Solution

The "chemical equation" for the transaction is

$$1 \text{ quarter} + 3 \text{ nickels} + 2 \text{ dimes} \longrightarrow 1 \text{ cold drink}$$

If there were unlimited nickels and dimes, then, according to the 1:1 ratio in the equation, the 22 quarters would buy 22 cold drinks. If there were unlimited quarters and nickels, the 30 dimes would buy 30/2 = 15 cold drinks. If there were unlimited quarters and dimes, then the 36 nickels would buy 36/3 = 12 cold drinks. The nickels run out first because the first 12 purchases use up the entire supply of 36. At that point, ten quarters and six dimes remain in excess, but they cannot be used.

The problem can also be solved by direct comparison among the "reactants." The 22 quarters would require 66 nickels and 44 dimes for complete "conversion" (purchase of 22 cold drinks). The demand exceeds the supply of both nickels and dimes, so quarters are in excess. The 30 dimes would require 15 quarters and 45 nickels. The supply of nickels is only 36. This puts dimes in excess as well. Finally, the 36 nickels would require 12 quarters and 24 dimes. There are more than enough of both. Nickels are the limiting "reactant."

Exercise

Suppose that the vending machine in the example works only with one quarter, one nickel, and three dimes. Starting with the same 22 quarters, 36 nickels, and 30 dimes, find how many cold drinks can be purchased and which type of coin is used up first.

Answer: After ten purchases, the dimes run out.

The preceding example illustrates two ways to identify the limiting reactant when specific amounts of substances react in a known way:

1. Select any product. Use the balanced equation to calculate the amount of this product that would form if the entire supply of the first reactant were used up. Repeat for the same product with respect to every other reactant. The reactant giving the *smallest* amount of the selected product is the limiting reactant.

2. Select a reactant. Calculate the amounts of the other reactants required to consume all of the selected reactant, according to the balanced equation. Compare each answer to the actual supply. If any one calculated demand *exceeds* the actual supply, then the selected reactant is in excess. Repeat for each reactant; the one reactant not in excess is the limiting reactant.

EXAMPLE 2–12

Sulfuric acid (H_2SO_4) forms in the chemical reaction

$$2\ SO_2 + O_2 + 2\ H_2O \longrightarrow 2\ H_2SO_4$$

Suppose 400 g of SO_2, 175 g of O_2, and 125 g of H_2O are mixed and the reaction proceeds until one of the reactants is used up. Which is the limiting reactant? How many grams of H_2SO_4 are produced, and how many grams of the other reactants remain?

Solution

We first calculate the chemical amount of each reactant originally present by multiplying the masses by the appropriate unit-factors:

$$n_{SO_2} = 400\ \text{g SO}_2 \times \left(\frac{1\ \text{mol SO}_2}{64.06\ \text{g SO}_2} \right) = 6.24\ \text{mol SO}_2$$

$$n_{O_2} = 175 \text{ g } O_2 \times \left(\frac{1 \text{ mol } O_2}{32.00 \text{ g } O_2} \right) = 5.47 \text{ mol } O_2$$

$$n_{H_2O} = 125 \text{ g } H_2O \times \left(\frac{1 \text{ mol } H_2O}{18.02 \text{ g } H_2O} \right) = 6.94 \text{ mol } H_2O$$

To be used up completely, 6.24 mol SO_2 requires 3.12 mol O_2 and 6.24 mol H_2O. The oxygen and water are present in more than sufficient quantity. Therefore, the SO_2 is the limiting reactant.

Alternatively, one might proceed as follows. If all the SO_2 reacted, it would give

$$6.24 \text{ mol } SO_2 \times \left(\frac{2 \text{ mol } H_2SO_4}{2 \text{ mol } SO_2} \right) = 6.24 \text{ mol } H_2SO_4$$

and if all the O_2 reacted, it would give

$$5.47 \text{ mol } O_2 \times \left(\frac{2 \text{ mol } H_2SO_4}{1 \text{ mol } O_2} \right) = 10.9 \text{ mol } H_2SO_4$$

and if all the H_2O reacted, it would give

$$6.94 \text{ mol } H_2O \times \left(\frac{2 \text{ mol } H_2SO_4}{2 \text{ mol } H_2O} \right) = 6.94 \text{ mol } H_2SO_4$$

Therefore, SO_2 is the limiting reactant because the computation based on its amount gives the smallest amount of product (6.24 mol H_2SO_4). Oxygen and water are present in excess. The amount of each that remains after the reaction is the original amount minus the amount that reacted:

• This is true even though the mass of the SO_2 *exceeds* the masses of the other two reactants.

n_{O_2} (in excess) = 5.47 mol O_2 originally present − 3.12 mol O_2 used

= 2.35 mol O_2

n_{H_2O} (in excess) = 6.94 mol H_2O originally present − 6.24 mol H_2O used

= 0.70 mol H_2O

The masses at the conclusion of the reaction are

$$m_{H_2SO_4} = 6.24 \text{ mol } H_2SO_4 \times \left(\frac{98.07 \text{ g } H_2SO_4}{1 \text{ mol } H_2SO_4} \right) = 612 \text{ g } H_2SO_4 \text{ produced}$$

$$m_{H_2O} = 0.70 \text{ mol } H_2O \times \left(\frac{18.02 \text{ g } H_2O}{1 \text{ mol } H_2O} \right) = 13 \text{ g } H_2O \text{ remaining}$$

$$m_{O_2} = 2.35 \text{ mol } O_2 \times \left(\frac{32.00 \text{ g } O_2}{1 \text{ mol } O_2} \right) = 75 \text{ g } O_2 \text{ remaining}$$

The total mass at the end is 612 + 13 + 75 = 700 g, which is of course equal to the total mass originally present, 400 + 175 + 125 = 700 g, as required by the law of conservation of mass.

Exercise

Suppose that 1.00 g of sodium and 1.00 g of chlorine react to form sodium chloride (NaCl). Which of the reactants is in excess, and what mass (in grams) of it remains when all of the limiting reactant is consumed?

Answer: Sodium is in excess, and 0.352 g of it remains.

The amounts of products calculated so far have been **theoretical yields,** which were determined on the assumption that the reaction goes cleanly and completely. The **actual yield** of a product (the amount present after separating it from other products and reactants and purifying it) is usually less than the theoretical yield. There are several reasons for this. The reaction may stop short of completion, so that reactants remain unreacted. There may be competing reactions that give other products and so reduce the yield of the desired one. Finally, in the process of separation and purification, some of the product is invariably lost, although losses can be reduced by careful experimental techniques. The ratio of the actual yield to the theoretical yield (multiplied by 100%) gives the **percentage yield** for that product in the reaction.

Ingots of newly refined elemental zinc.

EXAMPLE 2–13

The sulfide ore of zinc (ZnS) is reduced to elemental zinc by "roasting" it (heating it in air) to give ZnO and then heating the ZnO with carbon monoxide. The equations for the two-step process are

$$ZnS + \tfrac{3}{2} O_2 \longrightarrow ZnO + SO_2$$

$$ZnO + CO \longrightarrow Zn + CO_2$$

Suppose 5.32 kg of ZnS is treated in this way and 3.30 kg of pure zinc is obtained. Calculate the theoretical yield of zinc and its percentage yield.

Solution

From the molar mass of ZnS (97.46 g mol^{-1}), we can calculate the chemical amount of ZnS initially present:

$$n_{ZnS} = 5320 \text{ g ZnS} \times \left(\frac{1 \text{ mol ZnS}}{97.46 \text{ g ZnS}} \right) = 54.6 \text{ mol ZnS}$$

Because each mole of ZnS gives one mole of ZnO in the first chemical equation, and each mole of ZnO then gives one mole of zinc, the theoretical yield of zinc is 54.6 mol. We can write this out in full equation form as

$$n_{Zn} = 54.6 \text{ mol ZnS} \times \left(\frac{1 \text{ mol ZnO}}{1 \text{ mol ZnS}} \right) \times \left(\frac{1 \text{ mol Zn}}{1 \text{ mol ZnO}} \right) = 54.6 \text{ mol Zn}$$

The theoretical yield of zinc in grams is

$$m_{Zn} = 54.6 \text{ mol Zn} \times \left(\frac{65.39 \text{ g Zn}}{1 \text{ mol Zn}} \right) = \boxed{3570 \text{ g Zn}}$$

where the molar mass of zinc, 65.39 g mol^{-1}, comes from a table of atomic masses. This equals 3.57 kg of zinc. The ratio of the actual yield to the theoretical yield, multiplied by 100%, gives the percentage yield of zinc:

$$\% \text{ yield} = \left(\frac{3.30 \text{ kg}}{3.57 \text{ kg}} \right) \times 100\% = \boxed{92.4\%}$$

Exercise

One step in the production of vanadium uses the reaction of V_2O_5 with an excess of aluminum at high temperature to give vanadium and Al_2O_3. In a test run, 38.4

kg of vanadium is isolated when 72.0 kg of V_2O_5 reacts. Calculate the percentage yield of vanadium.

Answer: 95.2%.

It is clearly desirable in chemical reactions to achieve as high a percentage yield of product as possible in order to reduce consumption of raw materials and prevent the generation of possibly noxious by-products. In some synthetic processes (especially in organic chemistry), the final product is the result of many successive reactions. In such processes, the yields in the individual steps must be quite high if the synthetic method is to be a practical success. Suppose, for example, ten consecutive reactions must be carried out to reach the product, and each of these has a percentage yield of 50% (a fractional yield of 0.5). Then the overall yield will be the product of the fractional yields of the steps:

$$\underbrace{(0.5) \times (0.5) \times \cdots \times (0.5)}_{10 \text{ terms}} = (0.5)^{10} = 0.001$$

This overall percentage yield of 0.1% makes the process useless for synthetic purposes. If all the individual percentage yields could be increased to 90%, however, the overall fractional yield would be $(0.9)^{10} = 0.35$, or a 35% overall yield. This is much more reasonable and might make the process feasible.

• Organic chemistry, the study of the chemistry of carbon compounds, is examined in Chapter 24.

2-6   THE SCALE OF CHEMICAL PROCESSES

Most experiments and lecture demonstrations in general chemistry are carried out on a scale of grams. Masses of substance ranging from 0.1 g to 100 g are easy to measure out and are appropriate for chemical operations performed in beakers and test tubes, on a laboratory bench, or in a fume hood. With proper care, the heat that is liberated is controllable, and the waste products can be disposed of in a straightforward and safe manner. Not all chemistry is carried out at this scale, however. In this section, we take a brief look at chemical processes on two different scales: microscale and industrial scale.

Microscale processes arise in a number of areas of chemistry, although they are particularly prevalent in biochemistry, the study of the chemistry of life processes. The reason for this is that certain natural products are effective in very small quantities. Hormones, for example, are chemical messengers that help to initiate, control, and coordinate chemical reactions in different organs in the body. Even at the level of parts per trillion, they can have significant effects on growth and development. Isolating a hormone and determining its structure require the separation of a very small amount of the substance from a large amount of starting material, followed by accurate analytical measurements and chemical transformations. One of the triumphs of such microscale chemistry was the isolation and characterization of the hormone insulin (used in the treatment of diabetes) by Frederick Sanger, for which he received the Nobel Prize in chemistry in 1957. Another hormone, cortisol, is synthesized in the adrenal glands. A synthetic derivative, cortisone, and other related chemicals are used in the treatment of arthritis and allergies and have annual world sales on the order of 1 billion dollars.

• If a hormone is present in a sample to the extent of one part per trillion by mass, then every 10^{12} g (or kg or lb) of the sample contains 1 g (or kg or lb) of that hormone. See Appendix C-4.

CHEMISTRY IN YOUR LIFE

Green Chemistry

Complete conversion of reactants to products is difficult to achieve. Reactants mix incompletely and never come in contact with each other; products spill or decompose while being purified. Unwanted side-reactions compete with the planned "official" reaction, reducing yields and creating impurities, which may be toxic or require the use of toxic solvents for removal.

In industry, unwanted by-products and "spent" solvents present huge problems. Although society benefits immensely from the desired products of chemistry (such as antibiotics, cheap durable cloth, and strong, corrosion-resistant materials), some have questioned whether "miracle" products are worth the price of polluted air and water or of houses made uninhabitable by seepage from toxic waste dumps. Because a successful transition from pollution disposal and remediation to pollution avoidance will require new fundamental discoveries, the United States National Science Foundation has funded research initiatives such as "Environmentally Conscious Manufacturing" and "Environmentally Benign Chemical Synthesis and Processing." Similarly, the Chemical Manufacturers Association has adopted a program called "Responsible Care," aimed at protecting the environment while the mod-

ern chemical goods we have come to depend upon continue to be produced.

The production of ethylene glycol, widely used as an antifreeze, at a new plant in Tennessee, exemplifies the type of "green" (environmentally friendly) chemical production that is being encouraged. The process starts with coal and water. It generates the desired product along with carbon dioxide and a mixture of waste solids. Carefully designed treatments make the plant's effluent cleaner than the river water it takes in. The waste solids are not carted off to dump sites for burial but are incorporated into construction materials. The carbon dioxide is not released into the atmosphere where it might contribute to global warming; it is cleaned and used for producing carbonated beverages.

Scientists working in academia are also turning their attention to this new green chemistry. Professor Barry Trost of Stanford University has introduced the notion of "atom economy" in the synthesis of compounds. By this he means that one should attempt to design synthetic processes in which all of the atoms in the reactants end up in the desired products. This concept favors addition reactions over displacement or decomposition reactions. (These and other types of chemical reactions are discussed in Chapter 4.)

• This quantity of 55.5 mol comes from the mass of 1 L of water (1.00×10^3 g) divided by the molar mass of water (18.02 g mol^{-1}).

EXAMPLE 2–14

Flavor chemists think that the characteristic "green herbaceous vegetative" aroma of Cabernet Sauvignon wine arises from the chemical 2-methoxy-3-(2-methylpropyl)-pyrazine, with chemical formula $C_9H_{14}ON_2$ and molar mass 166.2 g mol^{-1}. This substance is present in the wine to the extent of 2 molecules for every ten trillion (10^{13}) molecules of water. Taking the predominant component of wine to be water, with 55.5 mol of water per liter of wine, estimate (a) the mass of this flavor component in 1.00 L of wine and (b) the number of molecules of $C_9H_{14}ON_2$ in 1.00 L of wine.

Solution

(a) We can construct the unit-factor

$$\frac{2 \text{ mol } C_9H_{14}ON_2}{10^{13} \text{ mol } H_2O}$$

because the mole ratio of the two substances in the wine is the same as the ratio of the number of molecules: two molecules of $C_9H_{14}ON_2$ per ten trillion molecules of water. Therefore,

$$n_{C_9H_{14}ON_2} = 1.00 \text{ L wine} \times \left(\frac{55.5 \text{ mol } H_2O}{1 \text{ L wine}} \right) \times \left(\frac{2 \text{ mol } C_9H_{14}ON_2}{10^{13} \text{ mol } H_2O} \right)$$

$$= 1.1 \times 10^{-11} \text{ mol } C_9H_{14}ON_2$$

The mass of the $C_9H_{14}ON_2$ is

$$m_{C_9H_{14}ON_2} = 1.1 \times 10^{-11} \text{ mol } C_9H_{14}ON_2 \times \left(\frac{166.2 \text{ g } C_9H_{14}ON_2}{1 \text{ mol } C_9H_{14}ON_2} \right)$$

$$= 1.8 \times 10^{-9} \text{ g } C_9H_{14}ON_2$$

(b) The number of molecules is found by multiplying the number of moles by Avogadro's number:

$$N_{C_9H_{14}ON_2} = 1.1 \times 10^{-11} \text{ mol } C_9H_{14}ON_2 \times \left(\frac{6.022 \times 10^{23} \text{ molecules } C_9H_{14}ON_2}{1 \text{ mol } C_9H_{14}ON_2} \right)$$

$$= 6.6 \times 10^{12} \text{ molecules } C_9H_{14}ON_2$$

Even this tiny mass of compound contains several trillion molecules.

Exercise

The powerful antitumor medicine cisplatin, $PtCl_2(NH_3)_2$, is prepared for injection by suspending small particles of solid in iodinated poppy-seed oil as a vehicle. The spherical particles average 6.6×10^{-4} cm in diameter. Compute how many atoms of platinum are in the average particle if the density of cisplatin equals 3.86 g cm^{-3}.

Answer: 1.2×10^{12} atoms of platinum.

Dozens of distinct chemical substances contribute to the natural flavor and bouquet of wines.

Let us move to the opposite extreme and consider the industrial-scale processing of chemicals (Fig. 2–10). Chemical processes are developed in the laboratory and then must be scaled up to make them into economical sources of large quantities of bulk chemicals. The scaling-up process is not a simple one because there is no guarantee that expanding the size of the equipment will give a reaction that proceeds in the same way. One aspect of this is the different way in which disposal of heat must be carried out. In most reactions in beakers, any heat produced is dissipated rapidly to the surrounding air. In an industrial process using hundreds of kilograms of reactants, the heat can build up and lead to an explosion unless properly designed cooling equipment is used. A more extensive discussion of the nature of industrial-scale processes is presented in Chapter 22.

EXAMPLE 2–15

In the contact process for making sulfuric acid (described in Chapter 22), the overall reaction is

$$2 \text{ S} + 3 \text{ O}_2 + 2 \text{ H}_2\text{O} \longrightarrow 2 \text{ H}_2\text{SO}_4$$

A typical sulfuric-acid plant makes about 1000 (1.0×10^3) metric tons of sulfuric acid per day (1 metric ton is 10^3 kg). Calculate the volumes of sulfur and oxygen consumed by a plant per *year*, assuming the density of sulfur, a solid, to be 2.1 g cm^{-3} and the density of the gaseous oxygen to be 1.4 g L^{-1}.

Solution

A metric ton is 1000 kg, so writing

$$1.0 \times 10^3 \text{ metric ton} \times \left(\frac{1000 \text{ kg}}{1 \text{ metric ton}} \right) \times \left(\frac{1000 \text{ g}}{1 \text{ kg}} \right) = 1.0 \times 10^9 \text{ g}$$

Figure 2–10 A single chemical plant can produce more than 1 billion pounds of a product per year. This Texas plant produces ethylene (C_2H_4), a starting material for many plastics.

shows that 1000 metric tons of H_2SO_4 corresponds to 1.0×10^9 g of H_2SO_4. We convert this to moles:

$$n_{H_2SO_4} = 1.0 \times 10^9 \text{ g } H_2SO_4 \times \left(\frac{1 \text{ mol } H_2SO_4}{98.1 \text{ g } H_2SO_4} \right) = 1.0 \times 10^7 \text{ mol } H_2SO_4$$

Each *year*, 365 times this amount, of 3.7×10^9 mol H_2SO_4, is produced. The same chemical amount of sulfur is consumed because there is one S per H_2SO_4. The mass of sulfur consumed is

$$m_S = 3.7 \times 10^9 \text{ mol S} \times \left(\frac{32.07 \text{ g S}}{1 \text{ mol S}} \right) = 1.2 \times 10^{11} \text{ g S}$$

and its volume is

$$V_S = 1.2 \times 10^{11} \text{ g S} \times \left(\frac{1 \text{ cm}^3 \text{ S}}{2.1 \text{ g S}} \right) = 5.6 \times 10^{10} \text{ cm}^3 \text{ S} = 5.6 \times 10^4 \text{ m}^3 \text{ S}$$

This corresponds to a cube 38 m on a side (more than ten stories tall).
  The chemical amount of oxygen used annually by the plant is

$$n_{O_2} = 3.7 \times 10^9 \text{ mol } H_2SO_4 \times \left(\frac{3 \text{ mol } O_2}{2 \text{ mol } H_2SO_4} \right) = 5.55 \times 10^9 \text{ mol } O_2$$

so the volume of oxygen is

$$V_{O_2} = 5.55 \times 10^9 \text{ mol } O_2 \times \left(\frac{32.0 \text{ g } O_2}{1 \text{ mol } O_2} \right) \times \left(\frac{1 \text{ L } O_2}{1.4 \text{ g } O_2} \right) = 1.3 \times 10^{11} \text{ L } O_2$$

$$= 1.3 \times 10^8 \text{ m}^3 \text{ O}_2$$

which fills a cube greater than 500 m on a side.

Exercise

The production of ammonium perchlorate, which is used in solid fuels for rockets, can be represented by the equation

$$NH_3 + HCl + 4 H_2O \longrightarrow NH_4ClO_4 + 4 H_2$$

Calculate the volume of gaseous ammonia and the volume of gaseous hydrogen chloride (in liters) required to produce 8000 lb of NH_4ClO_4. Take the densities of the two gases as 0.759 and 1.63 g L^{-1}, respectively.

Answer: 6.9×10^5 L of each.

  Grams and moles are used so much in Sections 2–4 and 2–5 that one might think them essential to stoichiometry calculations. This is not so. Industrial chemists might choose to use megagrams (Mg) and megamoles (Mmol), or kilograms (kg) and kilomoles (kmol). If 12 g of carbon equals 1 mol of carbon, then clearly 12 kg of carbon equals 1 kmol of carbon. And if 44 g of CO_2 equals 1 mol of CO_2, then 44 kg of CO_2 equals 1 kmol of CO_2, and so on for every other substance. Inserting the prefix "kilo" scales up all quantities by a factor of 1000, just as inserting "mega" scales up by a factor of 1 million. Neither affects the real essential, which is the *ratio* of the masses. A biochemist might prefer to work with milligrams and millimoles or with micrograms and micromoles, and a radiochemist with nanograms and nanomoles. These prefixes scale quantities down by a factor of 1 thousand, 1 million, and 1 billion, respectively.

$$12 \text{ milligram (mg) C} = 1 \text{ millimole (mmol) C}$$
$$12 \text{ microgram } (\mu\text{g}) \text{ C} = 1 \text{ micromole } (\mu\text{mol}) \text{ C}$$
$$12 \text{ nanogram (ng) C} = 1 \text{ nanomol (nmol) C}$$

In every case, the essential ratios are preserved.

• Megagrams and megamoles could have been used advantageously in Example 2–15, which deals with metric tons of sulfuric acid.

SUMMARY

2–1 **Molecular formulas** give the actual number of atoms of each element in a molecule of compound, and **empirical formulas** give only the ratios. In many solids and liquids without distinct molecular subunits, only the empirical formula is meaningful.

2–2 If the empirical formula is known, the percentage composition of the elements in a compound can be determined (given a table of atomic masses) and vice versa. Finding the molecular formula from the empirical formula requires additional information: the approximate molar mass of the molecular compound.

2–3 If the reactants and products in a chemical reaction are known, a balanced **chemical equation** can be written for that reaction. Chemical equations are balanced by a logical stepwise process in which coefficients are chosen to ensure that the same number of atoms of each element (or moles of atoms) is represented on each side of the equation. The coefficients applying to gaseous reactants and products in balanced equations are the same simple integers mentioned in the law of combining volumes.

2–4 **Stoichiometry** is the study of mass relationships in chemical reactions. The coefficients in the balanced equation are used to construct chemical **unit-factors** (conversion factors) to relate the amount of any substance shown in the equation to the amounts of all the others.

2–5 The reactant that is used up first in a chemical reaction is called the **limiting reactant.** Reactants other than the limiting reactant are said to be **in excess.** The actual **yield** of a product of a reaction is determined by separating the product and weighing it. It usually is less than the **theoretical yield,** which is calculated on the assumption that the reaction goes cleanly and completely. The ratio of the actual yield of a product to its theoretical yield is multiplied by 100% to give the **percentage yield** for that product in a reaction.

2–6 In general chemistry laboratories, work is usually done with gram quantities, but the mole concept applies irrespective of scale. Thus, if 1 mol ^{12}C weighs exactly 12 g, 1 kmol ^{12}C weighs 12 kg, and 1 mmol ^{12}C weighs 12 mg.

PROBLEMS

Note: Answers to blue-numbered problems are given in Appendix F. Problems that are more challenging are indicated with asterisks.

Empirical and Molecular Formulas

1. (See Example 2–1.) Cyclohexane, a liquid at room conditions, has the molecular formula C_6H_{12}. Give the empirical formula of cyclohexane. Does cyclopropane, C_3H_6, a gas at room conditions, have the same empirical formula as cyclohexane? Explain why or why not.

2. (See Example 2–1). A compound called 4-phenylcyclohexene is used in carpet manufacture in the United States and is responsible for the characteristic odor of new carpets. The molecular formula of this compound is $C_{12}H_{14}$. Determine its empirical formula.

3. Cucurbituril is a compound with cage-like molecules big enough to surround and loosely trap smaller molecules. It has the molecular formula $C_{36}H_{36}N_{24}O_{12}$. Determine the empirical formula and the molar mass (in g mol^{-1}) of cucurbituril.

4. Quinine tartrate, a derivative of the drug quinine, has the molecular formula $C_{44}H_{54}N_4O_{10}$. Determine its molar mass (in g mol^{-1}) and empirical formula.

5. How many molecules of water are present in 55 g of $MgSO_4 \cdot 7H_2O$?

6. What chemical amount (in moles) of ammonia, NH_3, is present in 10.9 mg of $ZnCl_2 \cdot 4NH_3$?

7. The gold-containing drug disodium gold(I) thiomalate relieves suffering from rheumatoid arthritis. The molecular formula of the drug is $AuC_4H_4O_4Na_2S$. An injection of 50.0 mg of the drug is given each week for 20 weeks.
   (a) Compute the total mass of gold administered.
   (b) Assume that the molecular formula of the substance is not known. Show how to solve the problem knowing only the *empirical* formula of the drug.

8. The compound 3'-azido-3'-thymidine (AZT) helps in the treatment of AIDS. The molecular formula of AZT is $C_{10}H_{13}N_5O_5$. A patient weighs 60.5 kg. The daily experimental dose in a test is 3.00 mg per kilogram of body mass.
   (a) Compute the chemical amount (in moles) of AZT that this patient receives each day.
   (b) Explain why it is not possible to solve part (a) given only the empirical formula of AZT.

Chemical Formula and Percentage Composition

9. Benzo[a]pyrene is an environmental carcinogen found in auto exhaust and tobacco smoke. It has the empirical formula C_5H_3. Determine (to four significant figures) the mass percentage of hydrogen in this substance.

10. Octachlorooctaborane has the molecular formula B_8Cl_8 and tetrachlorotetraborane has the molecular formula B_4Cl_4. Calculate the mass percentage of boron in each of these compounds.

11. Arrange the following compounds from left to right in order of increasing percentage by mass of hydrogen: H_2O, $C_{12}H_{26}$, N_4H_6, LiH.

12. Arrange the following compounds from left to right in order of increasing percentage by mass of fluorine: HF, C_6HF_5, BrF, UF_6.

13. The male sex hormone testosterone has the molecular formula $C_{19}H_{28}O_2$. Compute, to four significant figures, the mass percentage of carbon in this hormone.

14. The female sex hormone estradiol has the molecular formula $C_{18}H_{24}O_2$.
   (a) Compute, to four significant figures, the mass percentage of carbon in this hormone.
   (b) The male sex hormone (problem 13) contains 19 atoms of carbon per molecule, and the female sex hormone contains only 18 atoms of carbon per molecule. Despite this, the female sex hormone contains the larger mass percentage of carbon. Explain how this is possible.

15. (See Example 2–2.) A newly synthesized compound has the molecular formula $ClF_2O_2PtF_6$. Compute, to four significant figures, the mass percentage of each of the four elements in this compound.

16. (See Example 2–2.) Acetaminophen, the generic name of the pain reliever in Tylenol and some other headache remedies, has the molecular formula $C_8H_9NO_2$. Compute, to four significant figures, the mass percentage of each of the four elements in acetaminophen.

17. Chrome yellow ($PbCrO_4$) is a yellow pigment used in oil paints. Chrome red is the related compound $PbCrO_4 \cdot PbO$. Compute the percentage by mass of lead in each of these pigments.

18. Compute the percentage by mass of magnesium in a mineral if the formula is given as $MgF_2 \cdot (MgSiO_3)_2$.

19. "Q-gas" is a mixture of 98.70% helium and 1.30% butane (C_4H_{10}) by mass. It is used as a filling for gas-flow Geiger counters. Compute the mass percentage of hydrogen in Q-gas.

20. A pharmacist prepares an antiulcer medicine by mixing 286 g of Na_2CO_3 with water, adding 150 g of glycine ($C_2H_5NO_2$), and stirring continuously at 40°C until a firm mass results. She heats the mass gently until all of the water has been driven away. No other chemical changes occur in this step. Compute the mass percentage of carbon in the resulting white crystalline medicine.

21. (See Example 2–3.) Zinc phosphate is used as a dental cement. A 50.00-mg sample is broken down into its constituent elements and gives 16.58 mg of oxygen, 8.02 mg of phosphorus, and 25.40 mg of zinc. Determine the empirical formula of zinc phosphate.

22. (See Example 2–3.) A 40.0-g sample of the compound bromoform was analyzed and determined to contain 37.94 g of bromine, 1.90 g of carbon, and 0.16 g of hydrogen. Determine the empirical formula of bromoform.

23. A binary compound of silicon and chlorine is 28.37% silicon

by mass. Determine the empirical formula of the compound.

24. A binary compound of silver and chlorine is 75.26% silver by mass. Determine the empirical formula of the compound.

25. Fulgurites are the products of the melting that occurs when lightning strikes the earth. Microscopic examination of a sand fulgurite reveals that it is a globule with variable composition containing some grains of the definite chemical composition Fe 46.01%, Si 53.99%. Determine the empirical formula of these grains.

26. A sample of a "suboxide" of cesium gives up 1.6907% of its mass as gaseous oxygen when gently heated, leaving pure cesium behind. Determine the empirical formula of this binary compound.

27. Barium and nitrogen form two binary compounds containing 90.745% and 93.634% barium, respectively. Determine the empirical formulas of these two compounds.

28. Carbon and oxygen form no fewer than five different binary compounds. The mass percentages of carbon in the five compounds are as follows: A: 27.29; B: 42.88; C: 50.02; D: 52.97; and E: 65.24. Determine the empirical formulas of the five compounds.

29. A mixture of precious metals is 48.0% gold, 45.0% platinum, and 7.0% silver by mass. Out of 1000 atoms of this alloy, how many, on the average, are atoms of gold?

30. Analysis of a pewter bowl reveals that it is made up of 92.1% tin, 6.1% antimony, and 1.8% copper by mass. Out of 1000 atoms in this bowl, how many, on the average, are atoms of copper?

31. (See Example 2–4.) A pure compound that weighs 65.32 g and contains only sodium, sulfur, and oxygen is found on analysis to contain 21.14 g of sodium and 14.75 g of sulfur. Calculate the empirical formula of the compound.

32. (See Example 2–4.) A white crystalline compound that contains only iodine, potassium, and oxygen is purified and weighed. A mass of 87.41 g of pure compound is found on analysis to contain 15.97 g of potassium and 51.83 g of iodine. What is the empirical formula of the substance?

33. (See Example 2–5.) Burning a sample of a hydrocarbon (a compound containing hydrogen and carbon only) in oxygen produced 3.701 g of CO_2 and 1.082 g of H_2O. Determine the empirical formula of the hydrocarbon.

34. (See Example 2–5.) Burning a compound of calcium, carbon, and nitrogen in oxygen in a combustion train generates calcium oxide (CaO), carbon dioxide (CO_2), nitrogen dioxide (NO_2), and no other substances. A small sample gives 2.389 g of CaO, 1.876 g of CO_2, and 3.921 g of NO_2. Determine the empirical formula of the compound.

35. A solid compound contains only chlorine, oxygen, and carbon. A sample of the compound breaks down when heated to give 32.21 g of carbon monoxide, 81.55 g of chlorine, and no other products. Determine the empirical formula of the compound.

36. A 317.4-mg sample of a compound containing only carbon, hydrogen, and chlorine was burned completely in a combustion train. The carbon dioxide weighed 369.78 mg, and the

water weighed 32.45 mg. Determine the empirical formula of the compound.

37. (See Example 2–6.) The empirical formula of squalene, which is isolated from shark liver oil, is C_3H_5. Its molar mass is 410.7 g mol^{-1}. Determine the molecular formula of squalene.

38. (See Example 2–6.) A newly synthesized compound has the empirical formula $C_{10}H_{12}N_2Cu_3I_5$. Write the molecular formula of this compound if its molar mass is 11,824 g mol^{-1}.

Writing Balanced Chemical Equations

39. (See Example 2–7.) Balance the following chemical equations:
(a) $N_2 + O_2 \longrightarrow NO$
(b) $N_2 + O_2 \longrightarrow N_2O$
(c) $K_2SO_3 + HCl \longrightarrow KCl + H_2O + SO_2$
(d) $NH_3 + O_2 + CH_4 \longrightarrow HCN + H_2O$
(e) $CaC_2 + CO \longrightarrow C + CaCO_3$

40. (See Example 2–7.) Balance the following chemical equations:
(a) $Al + FeO \longrightarrow Al_2O_3 + Fe$
(b) $Br_2 + I_2 \longrightarrow IBr_3$
(c) $PbO + NH_3 \longrightarrow Pb + N_2 + H_2O$
(d) $Cl_2O_7 + H_2O \longrightarrow HClO_4$
(e) $CaC_2 + CO_2 \longrightarrow C + CaCO_3$

41. (See Example 2–7.) Balance the following chemical equations:
(a) $H_2 + N_2 \longrightarrow NH_3$
(b) $K + O_2 \longrightarrow K_2O_2$
(c) $PbO_2 + Pb + H_2SO_4 \longrightarrow PbSO_4 + H_2O$
(d) $BF_3 + H_2O \longrightarrow B_2O_3 + HF$
(e) $KClO_3 \longrightarrow KCl + O_2$
(f) $CH_3COOH + O_2 \longrightarrow CO_2 + H_2O$
(g) $K_2O_2 + H_2O \longrightarrow KOH + O_2$
(h) $PCl_5 + AsF_3 \longrightarrow PF_5 + AsCl_3$

42. (See Example 2–7.) Balance the following chemical equations:
(a) $Al + HCl \longrightarrow AlCl_3 + H_2$
(b) $NH_3 + O_2 \longrightarrow NO + H_2O$
(c) $Fe + O_2 + H_2O \longrightarrow Fe(OH)_2$
(d) $HSbCl_4 + H_2S \longrightarrow Sb_2S_3 + HCl$
(e) $Al + Cr_2O_3 \longrightarrow Al_2O_3 + Cr$
(f) $XeF_4 + H_2O \longrightarrow Xe + O_2 + HF$
(g) $(NH_4)_2Cr_2O_7 \longrightarrow N_2 + Cr_2O_3 + H_2O$
(h) $NaBH_4 + H_2O \longrightarrow NaBO_2 + H_2$

43. (See Example 2–7.) Write balanced chemical equations to represent the following reactions. Avoid fractional coefficients in your answers.
(a) Benzene (C_6H_6) burns in oxygen to give carbon dioxide and water.
(b) Fluorine (F_2) reacts with water to give oxygen difluoride (OF_2) and hydrogen fluoride (HF).
(c) Calcium carbide (CaC_2) reacts with water to give calcium hydroxide ($Ca(OH)_2$) and acetylene (C_2H_2).
(d) Oxygen oxidizes ammonia (NH_3) to give nitrogen dioxide (NO_2) and water.
(e) Aluminum dissolves in aqueous sodium hydroxide (NaOH) with the release of hydrogen and the formation of sodium aluminate (Na_3AlO_3).

44. (See Example 2–7.) Write balanced chemical equations to represent the following reactions. Avoid fractional coefficients in your answers.

(a) The combustion of the hydrocarbon styrene (C_8H_8) in oxygen to give carbon dioxide and water.

(b) The reaction of dimanganese heptaoxide (Mn_2O_7) with calcium oxide (CaO) to give calcium permanganate ($Ca(MnO_4)_2$).

(c) The burning of magnesium in nitrogen to give magnesium nitride (Mg_3N_2).

(d) The oxidation of magnetite (Fe_3O_4) by oxygen to give iron(III) oxide (Fe_2O_3).

(e) The reduction of magnesium oxide (MgO) with carbon to give magnesium and carbon dioxide.

45. If Fe_2O_3 is heated in a stream of gaseous hydrogen, the products are iron and water.

(a) Express this fact in the form of a balanced chemical equation.

(b) The reaction is performed, but the temperature control is imperfect, and some Fe_3O_4 forms in addition to the iron and water. Write a balanced chemical equation for a side-reaction that gives Fe_3O_4.

46. If zinc sulfide (ZnS) is heated in the air, oxygen replaces sulfur in combination with the zinc to form zinc oxide (ZnO), and all the sulfur goes off in the form of gaseous sulfur dioxide (SO_2).

(a) Express these facts in the form of a balanced chemical equation.

(b) A student speculates that SO_3 forms instead of SO_2 under certain conditions. Write a balanced chemical equation to represent this reaction.

The Stoichiometry of Chemical Reactions

47. Take the coefficients in the following equations to give the number of moles of each substance. Determine the masses in grams of all products and reactants. Confirm that the total mass of reactants equals the total mass of products in each reaction.

(a) $CO_2 + C \longrightarrow 2\,CO$

(b) $2\,C_6H_{14} + 19\,O_2 \longrightarrow 12\,CO_2 + 14\,H_2O$

(c) $Mn_2O_7 + H_2O \longrightarrow 2\,HMnO_4$

48. Take the coefficients in the following equations to give the number of formula units of each substance. Determine the mass in atomic mass units of all products and reactants. Confirm that the total mass of reactants equals the total mass of products in each reaction. (*Note:* Compare to the previous problem.)

(a) $CO_2 + C \longrightarrow 2\,CO$

(b) $2\,C_6H_{14} + 19\,O_2 \longrightarrow 12\,CO_2 + 14\,H_2O$

(c) $Mn_2O_7 + H_2O \longrightarrow 2\,HMnO_4$

49. (See Examples 2–8.) The reaction of lead sulfide with hydrogen peroxide, H_2O_2, to give lead sulfate is described by the equation

$$PbS + 4\,H_2O_2 \longrightarrow PbSO_4 + 4\,H_2O$$

How many moles of hydrogen peroxide are required to give 2.47 mol $PbSO_4$?

50. (See Example 2–8.) The preparation of hydrogen cyanide from ammonia, oxygen, and methane is described by the equation

$$2\,NH_3 + 3\,O_2 + 2\,CH_4 \longrightarrow 2\,HCN + 6\,H_2O$$

How many moles of HCN can be prepared from 6.48 mol NH_3? What is the minimum number of moles of O_2 that would be consumed?

51. (See Example 2–9.) What mass in grams of the first reactant in each of the following reactions would be required to react completely with 1.000 g of the second reactant?

(a) $CH_4 + 2\,O_2 \longrightarrow CO_2 + 2\,H_2O$

(b) $TiCl_2 + TiCl_4 \longrightarrow 2\,TiCl_3$

(c) $2\,Na_3VO_4 + H_2O \longrightarrow Na_4V_2O_7 + 2\,NaOH$

(d) $2\,K_2O_2 + 2\,H_2O \longrightarrow 4\,KOH + O_2$

52. (See Example 2–9.) What mass in grams of the first reactant in each of the following reactions would be required to react completely with 1.000 g of the second reactant?

(a) $2\,C_5H_{12} + 16\,O_2 \longrightarrow 10\,CO_2 + 12\,H_2O$

(b) $XeF_4 + 2\,H_2O \longrightarrow Xe + 4\,HF + O_2$

(c) $K_2Cr_2O_7 + 2\,KOH \longrightarrow 2\,K_2CrO_4 + H_2O$

(d) $N_2H_4 + 2\,H_2O_2 \longrightarrow N_2 + 4\,H_2O$

53. (See Example 2–9.) Cryolite (Na_3AlF_6) is used in the production of aluminum from its ores. It is made by the reaction

$$6\,NaOH + Al_2O_3 + 12\,HF \longrightarrow 2\,Na_3AlF_6 + 9\,H_2O$$

Calculate the mass of cryolite that can be prepared by the complete reaction of 287 g of Al_2O_3.

54. (See Example 2–9.) Carbon disulfide (CS_2) is a liquid that is used in the production of rayon and cellophane. It is manufactured from methane and elemental sulfur by the reaction

$$CH_4 + 4\,S \longrightarrow CS_2 + 2\,H_2S$$

Calculate the mass of CS_2 that can be prepared by the complete reaction of 67.2 g of sulfur.

55. (See Example 2–9.) Potassium nitrate (KNO_3) is used as a fertilizer for certain crops. It is produced through the reaction

$$4\,KCl + 4\,HNO_3 + O_2 \longrightarrow 4\,KNO_3 + 2\,Cl_2 + 2\,H_2O$$

Calculate the minimum mass of KCl required to produce 567 g of KNO_3. What mass of Cl_2 is generated as well?

56. (See Example 2–9.) Elemental phosphorus can be prepared from calcium phosphate via the overall reaction

$$2\,Ca_3(PO_4)_2 + 6\,SiO_2 + 10\,C \longrightarrow$$
$$6\,CaSiO_3 + P_4 + 10\,CO$$

Calculate the minimum mass of $Ca_3(PO_4)_2$ required to produce 69.8 g of P_4. What mass of $CaSiO_3$ is generated as a by-product?

57. (See Example 2–11.) Gaseous acetylene (C_2H_2) reacts with hydrogen to produce gaseous ethane (C_2H_6):

$$C_2H_2 + 2\,H_2 \longrightarrow C_2H_6$$

Assuming that the volumes of reactants and products are measured at the same temperature and pressure, calculate the volume of ethane generated by the reaction of 14.2 L of hydrogen.

58. (See Example 2–11.) One step in the production of nitric acid is the gas-phase reaction

$$4\ NH_3 + 5\ O_2 \longrightarrow 4\ NO + 6\ H_2O$$

Assuming that the volumes of reactants and products are measured at the same temperature and pressure, calculate the volume of O_2 required to generate 16.4 L of NO.

59. An 18.6-g sample of K_2CO_3 was treated in such a way that all of its carbon was captured in the compound $K_2Zn_3[Fe(CN)_6]_2$. Compute the mass (in grams) of this product.

60. A chemist dissolves 1.406 g of pure platinum in an excess of a mixture of hydrochloric and nitric acids and then, after a series of subsequent steps involving several other chemicals, isolates a compound of molecular formula $Pt_2C_{10}H_{18}N_2S_2O_6$. Determine the maximum possible yield of this compound.

61. A hydrated compound of rhodium has the formula $Na_3RhCl_6 \cdot x\ H_2O$, where x is an integer. When heated to 120°C, 2.1670 g of the compound loses 0.7798 g of water, leaving 1.3872 g of Na_3RhCl_6. Determine the value of x.

62. A crystalline compound of formula $Na_2SiO_3 \cdot 9\ H_2O$ loses a portion of its water when it is heated to 100°C but undergoes no other chemical changes. When a 10.00-g sample is heated in this way, the residue weighs 6.198 g. What is the formula of the residue?

63. Disilane (Si_2H_6) is a gas that reacts with oxygen to give silica (SiO_2) and water. Calculate the maximum mass of silica that could be obtained if 25.0 cm^3 of disilane (with a density of 2.78×10^{-3} g cm^{-3}) reacted with excess oxygen.

64. Tetrasilane (Si_4H_{10}) is a liquid with a density of 0.825 g cm^{-3}. It reacts with oxygen to give silica (SiO_2) and water. Calculate the maximum mass of silica that could be obtained if 25.0 cm^3 of tetrasilane reacted completely with excess oxygen.

Limiting Reactant and Percentage Yield

65. (See Example 2–11). One jumbo cheeseburger contains two hamburger patties, one slice of cheese, four tomato slices, and one hamburger bun. A restaurant has 600 hamburger patties, 400 slices of cheese, 1000 tomato slices, and 350 hamburger buns. How many jumbo cheeseburgers can the restaurant make? How many hamburger patties will be left over?

66. (See Example 2–11.) Every airplane model kit must contain two wings, one fuselage, four engines, and six wheels. The manufacturer has 17,254 wings, 8,713 fuselages, 31,008 engines and 48,143 wheels. Determine how many complete kits can be assembled. Determine the number of leftover parts of each kind.

67. One kind of dining table consists of four legs, two leaves, one frame, and one top. At assembly there are 25.25 dozen legs, 20 dozen leaves, 11 dozen frames, and 7 dozen tops. How many tables can be assembled?

68. A ream is a bundle of paper consisting of 500 sheets. A booklet is to consist of 16 sheets of contents printed on "copy stock" weighing 1.628 kg per ream and a single folded binding sheet printed on "cover stock" weighing 3.071 kg per ream. There are 5.00 kg of the cover stock and 50.00 kg of the copy stock on hand. Determine how many booklets can be printed.

69. (See Example 2–12.) A mixture of 34.0 g of ammonia and 50.0 g of oxygen reacts according to the equation

$$4\ NH_3 + 3\ O_2 \longrightarrow 2\ N_2 + 6\ H_2O$$

Which is the limiting reactant? How many grams of water can form?

70. (See Example 2–12.) Given the equation

$$Si + C \longrightarrow SiC$$

determine the limiting reactant and the theoretical yield (in grams) of silicon carbide when 2.00 g of silicon reacts with 2.00 g of carbon.

71. Barium chloride and sodium sulfate react in aqueous solution to form insoluble barium sulfate, according to the equation

$$BaCl_2 + Na_2SO_4 \longrightarrow BaSO_4 + 2\ NaCl$$

If a solution that contains 6.500 g of barium chloride is mixed with a solution that contains 9.420 g of sodium sulfate, then
(a) identify which substance is the limiting reactant.
(b) calculate the maximum possible mass (in grams) of barium sulfate formed.

72. When hydrogen chloride is mixed with silver nitrate in aqueous solution, the following reaction occurs exclusively:

$$AgNO_3 + HCl \longrightarrow AgCl + HNO_3$$

A solution that contains 2.135 g of hydrogen chloride is combined with a solution that contains 11.563 g of silver nitrate.
(a) Identify the reactant that is present in excess.
(b) After the reaction goes as far as possible, what mass (in grams) of the excess reagent remains unreacted?

73. Exactly 2.000 g of hydrogen and 1.000 g of oxygen are mixed, and the two gases react to form water. Compute the theoretical yield (in grams) of water.

74. Compute the maximum mass of AgCl that can form when 35.453 g of NaCl and 107.87 g of $AgNO_3$ are mixed in water solution.

75. Gaseous C_2H_4 reacts with O_2 according to the equation

$$2\ C_2H_4 + 6\ O_2 \longrightarrow 4\ CO_2 + 4\ H_2O$$

Calculate the theoretical yield of CO_2, in liters, from a mixture of 40.0 L of C_2H_4 and 20.0 L of O_2. Assume that the final temperature and pressure equal the initial temperature and pressure.

76. A new process to produce N_2O is being tested. In it, 2.33 L of N_2 reacts with 2.70 L of O_2. Determine the theoretical yield of N_2O, in liters. Assume that the final temperature and pressure equal the initial temperature and pressure.

77. (See Example 2–13.) The iron oxide Fe_2O_3 reacts with carbon monoxide, CO, to give iron and carbon dioxide:

$$Fe_2O_3 + 3\ CO \longrightarrow 2\ Fe + 3\ CO_2$$

The reaction of 433.2 g of Fe_2O_3 with excess CO yields 254.3 g of iron. Calculate the theoretical yield of iron (assuming complete reaction) and its percentage yield.

78. (See Example 2–13.) Titanium dioxide (TiO_2) reacts with carbon and chlorine to give gaseous $TiCl_4$:

$$TiO_2 + 2\ C + 2\ Cl_2 \longrightarrow TiCl_4 + 2\ CO$$

The reaction of 7.39 kg of titanium dioxide with excess C and Cl_2 gives 14.24 kg of titanium tetrachloride. Calculate the theoretical yield of $TiCl_4$ (assuming complete reaction) and its percentage yield.

79. You promise to bake 200 dozen cookies and deliver them to a bake sale. Experience shows that you break (and then eat) 8.0% of your cookies during the process of making them.
(a) How many cookies should you buy ingredients for?
(b) How many cookies will you be eating?

80. Silicon nitride (Si_3N_4), a valuable ceramic, is made by the direct combination of silicon and nitrogen at high temperature. How much silicon must react with excess nitrogen to prepare 125 g of silicon nitride if the yield of the reaction is 95.0%?

The Scale of Chemical Processes

81. The chemical known as 2,3,7,8-tetrachlorodibenzo-p-dioxin (TCDD, commonly referred to as dioxin, with chemical formula $C_{12}H_4O_2Cl_4$), is a contaminant of some agricultural herbicides. It is extremely lethal to certain laboratory animals: an oral dose of just 0.6 μg per kilogram of body mass kills half the test population of guinea pigs. Assuming that the body of a guinea pig consists entirely of water molecules, what percentage of the molecules in the guinea pig are TCDD molecules after such a dose is administered?

82. Dioxin, or TCDD (see previous problem), appears to be much less lethal to human beings than to guinea pigs. An experiment with human volunteers involved treatment with 107,000 ng per kilogram of body mass, externally applied to the skin. No long-term symptoms were observed. If 20% of the TCDD entered a volunteer's body in such an experiment, what fraction of molecules in the body were TCDD molecules? Assume that essentially all of the molecules in the human body are water molecules.

83. A new chlor-alkali plant produces 660 metric tons of chlorine daily. The process uses the decomposition of an aqueous solution of sodium chloride (NaCl) to give the chlorine (Cl_2) and sodium hydroxide (NaOH) as a by-product. Estimate (in metric tons) how much sodium hydroxide is produced daily.

84. Portland cement is about 46% calcium and 9% silicon by mass, with most of the rest being oxygen. It is made from lime (CaO) derived from limestone. Estimate the mass of lime used in one year's world production of portland cement (800,000,000 metric tons).

85. How many kilomoles (kmol) of CO_2 molecules are contained in 88 kg of carbon dioxide? How many kilomoles of oxygen atoms are contained in the sample? How many atoms of oxygen are contained in the sample?

86. How many pounds of hydrogen (H_2) are needed to react completely with 8.0 lbs of oxygen (O_2) to produce water? How many pounds of water are produced as a result of this reaction?

Additional Problems

87. Fill in the blank spaces in the following table.

Name	Molecular Formula	Empirical Formula	Molar Mass (g mol^{-1})
Tetralin	$C_{10}H_{12}$	——	——
Quinine	——	$C_{10}H_{12}NO$	324.41
D-Ribose	——	CH_2O	150.13
Cubane	C_8H_8	——	——
Benzene	——	CH	78.11

88. Human parathormone has the impressive molecular formula $C_{691}H_{898}N_{125}O_{164}S_{11}$. Compute the mass percentages of all of the elements in this compound.

89. Osmium(VIII) oxide (OsO_4) is a volatile crystalline solid. It is hazardous to the eyes because it vaporizes readily (which means it can get into the eye) and reacts with organic matter (the fabric of the eye itself) to leave an opaque, black, permanent stain of osmium. Compute the mass of osmium in 17.6 mg of OsO_4.

90. A binary compound between erbium and boron is difficult to prepare and analyze. It is reported, however, that it has the chemical formula $ErB_{65 \pm 4}$. Compute the *range* of the mass percentage of erbium in this compound.

91. At one time, it was thought that indium formed a chloride of formula $InCl_2$. More recent work shows that the compound in question is actually $In_3[In_2Cl_9]$. Determine the percentage by mass of indium according to each formula and explain how the error could be made.

92. Plants provide many valuable compounds, including a group called essential oils. The desire for essential oils (herbs and spices) provided a major stimulus to early European exploration. Consider the molecular formulas of the following essential oils:

Name	Molecular Formula	Source
limonene	$C_{10}H_{16}$	citrus fruit
cedrene	$C_{15}H_{24}$	cedar wood
kaurene	$C_{20}H_{32}$	pine trees

What do these compounds have in common (besides their origin in plants)?

93. Natural rubber comes from latex, which is the sap of the tropical rubber tree (see Chapter 25). A reference book states that natural rubber has the formula $(C_5H_8)_n$, where n varies and has an average value of about 5000.
(a) Is natural rubber a chemical substance? Explain.
(b) Is it possible to compute the mass of one mole of natural rubber? Explain.
(c) What do essential oils (see the previous problem) and natural rubber have in common (besides coming from plants)?

94. Djenkolic acid is a sulfur-containing amino acid that is isolated from the djenkol and other beans and is possibly implicated in flatulence. It is 33.06% carbon, 5.55% hydrogen, 11.01% nitrogen, 25.16% oxygen, and 25.22% sulfur by mass. Each molecule contains two atoms of sulfur. Determine the molar mass of djenkolic acid and its molecular formula.

95. A binary compound between potassium and cesium is stable at temperatures below $-90°$ C but otherwise decomposes to the elements. A 0.03151-g sample of the compound yields 0.02346 g of cesium. Determine the empirical formula of this compound. (*Hint:* Take particular care with significant figures.)

*96. A white oxide of tungsten is 79.2976% tungsten by mass. A *blue* tungsten oxide also contains exclusively tungsten and oxygen, but it is 80.8473% tungsten by mass. Determine the empirical formulas of white tungsten oxide and blue tungsten oxide.

*97. A sample of an iron sulfide contains 62.613% iron by mass. Determine the empirical formula of this iron sulfide.

*98. A dark-brown binary compound contains oxygen and a metal. It is 13.38% oxygen by mass. Heating it moderately drives off some of the oxygen and gives a red binary compound that is 9.334% oxygen by mass. Strong heating drives off more oxygen and gives still another binary compound that is only 7.168% oxygen by mass.
(a) Compute the mass of oxygen that is combined with 1.000 g of the metal in each of these three oxides.
(b) Assume that the empirical formula of the first compound is MO_2 (where M stands for the metal). Give the empirical formulas of the second and third compounds.
(c) Name the metal.

99. A chemist finds that 0.7200 g of barium reacts when heated with a large excess of oxygen to form 0.8878 g of a solid compound.
(a) What is the empirical formula of the compound?
(b) Write a balanced equation for the reaction that forms the compound.

100. A new compound with excellent potential for making temperature-resistant plastics has the empirical formula $C_6N_2O_3$. When treated with water, it reacts to form a compound of empirical formula C_3HNO_2. It requires three moles of water to perform this conversion on one mole of the original compound, and no other products are formed. Determine the molecular formula and molar mass of the original compound.

101. Aspartame (molecular formula $C_{14}H_{18}N_2O_5$) is a sugar substitute in soft drinks. Under certain conditions, one mole of aspartame reacts with two moles of water to give one mole of aspartic acid (molecular formula $C_4H_7NO_4$), one mole of methanol (molecular formula CH_3OH), and one mole of phenylalanine. Determine the molecular formula of phenylalanine.

102. Verify that the following two chemical equations are balanced by counting the number and kind of atoms represented on the two sides of the arrow:

$$10 \text{ C} + 7 \text{ O}_2 \longrightarrow 6 \text{ CO} + 4 \text{ CO}_2$$
$$9 \text{ C} + 7 \text{ O}_2 \longrightarrow 4 \text{ CO} + 5 \text{ CO}_2$$

Explain why the *same* set of reactants and products can sometimes, as in this case, be balanced with *different* sets of coefficients.

103. Potassium oxide (K_2O) is prepared by carefully melting potassium nitrate (KNO_3) with elemental potassium:

$$2 \text{ KNO}_3 + 10 \text{ K} \longrightarrow 6 \text{ K}_2\text{O} + \text{N}_2$$

Determine the masses of KNO_3 and K necessary to produce 8.34 g of K_2O according to this equation.

104. Alkenes are hydrocarbons that have the general formula C_nH_{2n}, where n is a whole number. Suppose that 1.000 g of any alkene is burned in excess oxygen to give carbon dioxide and water. Compute the mass of carbon dioxide and the mass of water formed.

*105. 3'-Methylphthalanilic acid is used commercially as a "fruit set" to prevent premature drop of apples, pears, cherries, and peaches from the tree. It is 70.58% carbon, 5.13% hydrogen, 5.49% nitrogen, and 18.80% oxygen. If eaten, the fruit set reacts with water in the body to split into an innocuous product that contains carbon, hydrogen, and oxygen only and *m*-toluidine ($NH_2C_6H_4CH_3$), which causes anemia and kidney damage. Compute the mass of the fruit set that would produce 5.23 g of *m*-toluidine.

*106. If 1.000 g of a hydrated lanthanum phosphate is heated in a vacuum, it gives 0.2922 g of P_4O_{10}, 0.6707 g of La_2O_3, and 0.0371 g of H_2O.
(a) Determine the empirical formula of the hydrated lanthanum phosphate.
(b) Write a balanced equation to represent the decomposition described in part (a).

107. When 48.044 g of carbon reacts with 12.09 g of hydrogen to form a hydrocarbon, no part of either reactant is left in excess. Determine the empirical formula of the hydrocarbon.

108. A yield of 3.00 g of $KClO_4$ is obtained from the (unbalanced) reaction

$$\text{KClO}_3 \longrightarrow \text{KClO}_4 + \text{KCl}$$

when 4.00 g of the reactant is used. What is the percentage yield of the reaction?

109. A student tells you this way to identify the limiting reactant in a chemical reaction: "Figure out the number of moles of every reactant. Then divide each answer by the coefficient that reactant has in the balanced equation. The reactant that gives the *smallest* answer is the limiting reactant." Does this method work? Give a proof of your answer.

110. Potassium superoxide (KO_2) absorbs gaseous carbon dioxide from the air according to the (unbalanced) equation

$$\text{KO}_2 + \text{CO}_2 \longrightarrow \text{K}_2\text{CO}_3 + \text{O}_2$$

In one experiment, run at constant temperature and pressure, there was an excess of KO_2, the yield of O_2 was 98.5%

of the theoretical yield, and 117 L of gaseous O_2 was produced. Determine the volume (in liters) of CO_2 that was used.

111. If metallic niobium is covered with liquid niobium pentachloride in a sealed tube and heated to 300°C for a week, a chemical reaction occurs in which $NbCl_4$ is the only product.

(a) Write a balanced chemical equation for this reaction, taking $NbCl_5$ as the chemical formula of niobium pentachloride.

(b) The reaction is performed using 5.0 g of liquid niobium pentachloride and 0.50 g of niobium. Which reactant is the limiting reactant in this experiment?

(c) During the week of heating, it is learned that the molecular formula of niobium pentachloride is really Nb_2Cl_{10}, not $NbCl_5$ as suggested by the compound's name. How does this affect the conclusion in part (b)?

*112. The senior partner in your law firm calls you in on a case in which a drug dealer was arrested in possession of 5.0 U.S. gallons of piperidine (molecular formula $C_5H_{11}N$) and some other chemicals. Piperidine is a starting material in the synthesis of phencyclidine (molecular formula $C_{17}H_{25}N$), the notorious "angel dust." Your partner asks roughly how many kilograms of angel dust could be synthesized from 5.0 gal of piperidine.

(a) List the facts you would need to look up to find out first the mass (in kilograms) of piperidine and then the chemical amount (in moles) of piperidine that was seized.

(b) There is one N in the molecular formula of each substance, so it seems reasonable to assume that one mole of phencyclidine is produced for every mole of piperidine used. However, you are aware that this assumption could be wrong. Explain exactly *how* this assumption could be wrong.

(c) Use the assumption in part (b) and the facts in part (a) to answer your senior partner's question.

*113. A newspaper article about the danger of global warming from the accumulation of greenhouse gases such as carbon dioxide states that "reducing driving your car by 20 miles a week would prevent release of over 1000 lb of CO_2 per year into the atmosphere." Check whether this is a reasonable statement. Assume that gasoline is octane (molecular formula C_8H_{18}) and that it is burned completely to CO_2 and H_2O in the engine of your car. Facts (or reasonable guesses) about your car's gas mileage, the density of octane, and other factors are also needed.

114. The addition of oxygen-containing compounds to gasoline reduces the amount of pollutants (hydrocarbons and carbon monoxide) emitted in automobile exhaust. Compute the mass of methyl *t*-butyl ether (MTBE) that must be added to 1.0 kg of gasoline to raise the mass percentage of oxygen in the gasoline from 0.0% to 2.7%. The molecular formula of MTBE is $C_5H_{12}O$.

115. Existing stockpiles of the refrigerant Freon-12 (molecular formula CF_2Cl_2) have to be destroyed because it (and other Freons) leads to the depletion of the ozone layer when released into the atmosphere. One method is to pass gaseous Freon-12 through a bed of powdered sodium oxalate at 270°C. The following reaction occurs

$$CF_2Cl_2 + Na_2C_2O_4 \longrightarrow NaF + NaCl + C + CO_2$$

(a) Balance the equation.

(b) Determine the mass in kilograms of sodium oxalate needed to destroy 147×10^3 kg of Freon-12.

CUMULATIVE PROBLEM

Titanium

Metallic titanium and its alloys (especially those with aluminum and vanadium) combine the advantages of high strength and light weight and so are widely used in the aerospace industry. The major natural source for titanium is the ore rutile, which contains titanium dioxide (TiO_2).

(a) An intermediate in the preparation of elemental titanium from titanium dioxide is a volatile chloride of titanium (boiling point 136°C) that contains 25.24% titanium by mass. Determine the empirical formula of this compound.

(b) The titanium chloride in the previous part is produced by the reaction of chlorine with a hot mixture of titanium dioxide and coke (carbon), with carbon dioxide generated as a by-product. Write a balanced chemical equation for this reaction.

(c) What mass of chlorine is needed to produce 79.2 g of the titanium chloride?

(d) The titanium chloride then reacts with liquid magnesium at 900°C to give metallic titanium and magnesium chloride ($MgCl_2$). Write a balanced chemical equation for this step in the winning of titanium.

(e) Suppose the reaction chamber for part (d) contains 351 g of the titanium chloride and 63.2 g of liquid magnesium. Which is the limiting reactant? What maximum mass of titanium could result?

(f) World production of titanium by this process is approximately 1.0×10^5 metric tons per year. What mass of titanium dioxide is needed to produce this much titanium?

A titanium heat shield for a large aircraft engine.

3

Chemical Periodicity and the Formation of Simple Compounds

Samples of bromine, a non-metal (left); silicon, a semi-metal (top right); and copper, a transition metal (bottom) arrayed on a sheet of aluminum, a main-group metal.

In Chapter 2, we showed how the law of conservation of mass, through the mole concept, allows us to establish and use quantitative mass relationships in chemical reactions. In that discussion, we assumed as prior knowledge the chemical formulas of the reactants and products in each equation. We also assumed that the indicated reaction would actually occur. We turn now to the far more open-ended question of what determines chemical reactivity. Why are some elements and compounds fiercely reactive and others inert? Why are there compounds with the chemical formulas H_2O and $NaCl$, but never an H_3O or an $NaCl_2$? Why do hydrogen atoms cling together in closely bound H_2 molecules under most conditions, whereas helium atoms remain resolutely unassociated? These questions can be answered on various levels, and they are addressed throughout this book.

As discussed in Chapter 1, every atom contains one or more electrons, which are fundamental particles possessing both electrical charge and mass. The details of the behavior of electrons in atoms are complex, and we defer that discussion to Chapters 16 and 17. Our first concern in this chapter is to count electrons and to give a broad description of their arrangement in atoms. We do this because certain recurring similarities in the physical and chemical properties of the elements have their origin in similarities in the number of outer electrons, those farthest from the nucleus, in their atoms.

Much chemical reactivity involves the transfer or sharing of electrons among atoms. The number of an atom's outer electrons determines most of its chemistry because these electrons are the most strongly affected when other atoms approach. The formation of molecules can be described through Lewis structures, which illustrate the transfer and sharing of outer electrons in a particularly simple way. Our second concern in this chapter is to show how to set down, interpret, and evaluate these useful diagrams. We then extend the Lewis approach to the prediction of the shapes of molecules. Finally, we present a system of inorganic nomenclature that uses some of the concepts just developed and provides a language for describing the chemical reactions that we take up in Chapter 4.

3–1 GROUPS OF ELEMENTS

By the late 1860s, more than 60 chemical elements had been identified, and much descriptive information on the physical and chemical properties of these elements had been accumulated. Similarities among their properties suggested various natural groupings. Elements were classified as **metals** or **non-metals,** for example, depending on the presence (or absence) of a characteristic metallic luster, their good (or poor) ability to conduct electricity and heat, and their malleability (or brittleness). The classification of an element as a metal or non-metal was generally straightforward, but certain elements (antimony, arsenic, boron, silicon, and tellurium) resembled metals in some respects and non-metals in others and were called **semi-metals** or **metalloids.**

Close scrutiny of other properties of the elements, in particular the empirical formulas of their binary compounds with chlorine (their *chlorides*), with oxygen (their *oxides*), and with hydrogen (their *hydrides*), allowed smaller groups of three to six chemically similar elements to be identified. With some changes, these groups have significance to this day. We briefly describe the characteristics of seven of the groups and then show in Section 3–2 how they were brought together in a single pattern with the discovery of the **periodic law.**

• With the exception of mercury, the metallic elements are all solids at room temperature; the non-metallic elements are solids, liquids, and gases.

Seven Groups of Elements

The **alkali metals** (lithium, sodium, potassium, rubidium, and cesium) are relatively soft metals (Fig. 3–1) with low melting points; they form 1:1 binary compounds with chlorine with chemical formulas such as NaCl, RbCl, and so forth. The alkali metals react with water to liberate hydrogen. Potassium, rubidium, and cesium do this particularly vigorously, but all alkali metals must be handled with care, as the heat of their reaction with water can cause the hydrogen generated by them to burst into flame (Fig. 3–2). All of these metals form oxides in 2:1 proportions (with formulas such as Na_2O and K_2O), but all also combine with oxygen in other ratios to form reactive compounds such as KO_2 and Na_2O_2. The elements sodium and potassium are quite common in the form of dissolved NaCl and KCl in the oceans. Both are important in biological systems; a proper balance of dissolved sodium and potassium is essential in regulating the transport of molecules across cell walls. Dissolved lithium also can play a significant role in the cell; it is used in the treatment of manic-depressive behavior.

A second group of elements, the **alkaline earth metals** (beryllium, magnesium, calcium, strontium, barium, and radium), combine in a 1:2 atom ratio with chlorine to give compounds such as $MgCl_2$ and $CaCl_2$ and with oxygen in 1:1 proportions to give compounds such as MgO and CaO. Calcium is a rather abundant element, especially in the form of its carbonate, $CaCO_3$. This single compound takes on many different forms in nature, appearing as chalk, limestone, marble, or crystalline calcite and aragonite (Fig. 3–3) under different geological circumstances. When heated, it gives up carbon dioxide and leaves lime (CaO), a compound used for making glass, mortar, and cement. The carbonate of magnesium ($MgCO_3$) forms a plentiful mineral (called dolomite) with calcium carbonate. Beryllium is a much less abundant alkaline earth metal, although masses of the mineral beryl ($Be_3Al_2Si_6O_{18}$) that weigh as much as a ton have been found. Pure beryl is colorless, but impurities give

• The name "alkali" comes from the fact that some of these metals are found in compounds leached from wood ashes (*alqali* is Arabic for "ashes").

Figure 3–1 Elemental sodium is a soft metal that can easily be cut with a knife. It melts at a temperature of only 97.8°C (just below the normal boiling point of water). Sodium is stored under oil because it corrodes rapidly by reacting with water vapor in the air.

(a)

(b)

Figure 3–2   (a) A fragment of sodium added to water starts to react instantly. The reaction produces hydrogen and sodium hydroxide and enough heat to melt the unreacted sodium. The molten globule skitters over the surface of the water, propelled by evolving bubbles of gaseous hydrogen. Sometimes the hydrogen catches fire. Here, the red indicator shows where sodium hydroxide has formed. (b) The reaction of potassium with water is similar but more vigorous. Its heat generally ignites both the hydrogen that forms and the unreacted potassium, causing a burst of flames as they burn in air.

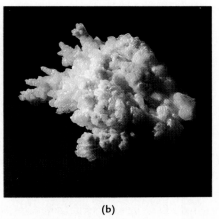

(a)   (b)

Figure 3-3 Calcium carbonate ($CaCO_3$) occurs in two crystalline forms: calcite (a) and aragonite (b). Limestone is an aggregate of small crystals of the two together with a minor percentage of impurities such as $MgCO_3$ and SiO_2.

it color. Chromium as an impurity makes beryl into emerald (Fig. 3–4), and iron makes it aquamarine (Fig. 3–5).

The **chalcogens** oxygen, sulfur, selenium, and tellurium form 1:1 compounds with the alkaline earth metals, but 2:1 compounds with the alkali metals; that is, the alkali metal chalcogenides have formulas such as Li_2O and Li_2S, whereas the alkaline earth chalcogenides have formulas such as CaO and CaS. The lightest chalcogen, oxygen, is the most abundant element at the earth's surface. It is found free (as O_2) in the atmosphere, in combination with hydrogen (in H_2O) in the oceans, and combined with other elements in the earth's crust. (The primary constituents of most rocks are oxygen, silicon, and aluminum.) Sulfur is also abundant. It is found in elemental form as a powdery or crystalline yellow solid, and in combination with metals in minerals such as iron pyrites (FeS_2, commonly known as "fool's gold," Fig. 3–6) and gypsum ($CaSO_4\cdot2H_2O$). Selenium and tellurium are quite rare, although it has been established that traces of selenium are essential in the human diet.

Fluorine, chlorine, bromine, and iodine are members of a family of elements called **halogens.** These four elements differ significantly in their physical proper-

Figure 3-4 An emerald is an impure crystal of beryl. Its deep green color depends on the presence of chromium atoms at about 20% of the beryllium atom sites in beryl. Crystals of beryl with no impurities are colorless.

Figure 3-5 Aquamarine gemstones can be green, yellow, or blue. Aquamarines are all crystals of beryl with atoms of iron lodged in different types of sites. Upon heating, a green aquamarine turns blue as chemical changes occur at some of the iron sites.

Figure 3-6 Iron pyrites, or fool's gold, has the chemical formula FeS_2. The mineral is an abundant and important source of elemental sulfur. Its luster makes iron pyrites appear metallic, but whereas malleable gold (a true metal) can be cut with a knife point, the brittle fool's gold shatters.

(a)

(b)

(c)

Figure 3–7   Chlorine (a), bromine (b), and iodine (c) differ in their colors and physical states, being, respectively, gas, liquid, and solid at room temperature. As members of the halogen family, these elements have closely related chemical properties.

• Goiter is an enlargement (visible at the throat) of the thyroid gland. It is associated with low production of the iodine-containing hormone thyroxine.

ties (the first two are gases at room temperature, whereas bromine is a liquid and iodine a solid, Fig. 3–7), but on the whole their chemical behaviors are similar. Any alkali metal element combines with any halogen in 1:1 proportions to form a compound such as LiF or RbI. As a group, these compounds are the "alkali-metal halides." They are relatively hard solids at room temperature and have high melting points; when dissolved in water, they give solutions that conduct electric current. Chlorine is quite abundant in seawater (as dissolved sodium chloride and potassium chloride); bromine is less abundant and iodine still less so. Iodine, however, is concentrated in seaweed and other forms of aquatic life and can be recovered economically from such sources. Iodine deficiency in humans causes goiter and is avoided through the use of iodized salt. The chemistry of fluorine often differs from that of the other halogens and is discussed in Chapter 23. Fluorine is found primarily in minerals such as fluorite, CaF_2 (Fig. 3–8).

Other elements fall into three additional groups that are less clearly defined in terms of their chemical and physical properties than those mentioned so far.

Figure 3–8   The mineral fluorite (calcium fluoride, CaF_2) occurs in various shades of purple. The color arises from a small number of defects (see Section 20–4) in which fluoride ions are missing from their sites in the crystal and their places are taken by electrons.

Figure 3-9 A small mountain of borax awaits shipment from a mine in the Mojave desert. Borax ($Na_2B_4O_5(OH)_4 \cdot 8H_2O$) is mined on a scale of millions of tons per year and is the major source of boron. Although boron is not abundant in the earth's crust, it occurs in high concentration in deposits associated with former volcanic activity and can be mined economically.

The first of these contains a semi-metal (boron) and four metals (aluminum, gallium, indium, and thallium). All form 1:3 chlorides (such as $GaCl_3$) and 2:3 oxides (such as Al_2O_3). This chemical similarity is the reason for the inclusion of boron in this group. Boron is found in large deposits of borax ($Na_2B_4O_5(OH)_4 \cdot 8H_2O$), especially in the California desert (Fig. 3–9). Of the metals of this group, aluminum is the most abundant, being found in many minerals in association with silicon, oxygen, and other elements.

Another group includes a non-metal as well as semi-metals and metals. It consists of the elements carbon, silicon, germanium, tin, and lead. All of these elements form 1:4 chlorides (such as $SiCl_4$), 1:4 hydrides (such as GeH_4), and 1:2 oxides (such as SnO_2). In this group, tin and lead are metals with low melting points, and silicon and germanium are semi-metals. Carbon, the non-metal, exists in several forms with dramatically different properties (Fig. 3–10): *graphite* is a soft, grayish

• The hydride of lead (PbH_4) is unstable and has not been characterized definitively.

(a)

(b)

Figure 3-10 Graphite and diamond (a), as well as buckminsterfullerene (b), are allotropic forms of carbon. The largest use of graphite is in electrodes for electrochemical cells, where advantage is taken of its ability to conduct electricity. Diamond does not conduct electricity. Smaller, non–gem-quality diamonds are used industrially in drill bits and in other instances in which the exceptional hardness of diamond is needed. Buckminsterfullerene is a recently discovered molecular form of carbon with the formula C_{60}. It and related fullerenes have many potential uses.

Figure 3-11 Fluorapatite ($Ca_5(PO_4)_3F$) is one of several apatite minerals that provide the commercial sources for phosphorus. Collectively, the apatites are mined on a scale of over a million tons per year and are massively used in the production of phosphate fertilizers. Major apatite deposits are found in Florida and in Morocco. Pure $Ca_5(PO_4)_3F$ is white. The purple in this sample results from manganese-containing impurities.

• The "lead" in a pencil is graphite.

black, flaky solid that conducts electricity; *diamond* consists of very hard, transparent crystals that are nonconductors of electricity; and the *fullerenes*, first synthesized in quantity in 1990, are dark-colored soft crystalline solids. These forms are **allotropes**—modifications of an element that differ because the atoms are organized in different ways. Silicon, the most abundant element of this group, is found in many minerals such as quartz (SiO_2). Carbon is less abundant overall, but it plays a crucial role in the biosphere. Life as we know it on Earth would not be possible without carbon.

A seventh and final group includes nitrogen, phosphorus, arsenic, antimony, and bismuth. These elements form binary compounds with hydrogen and oxygen with empirical formulas such as PH_3 and N_2O_5. The hydrides become increasingly unstable as their molar masses increase, and BiH_3 can be kept only below $-45°C$. A similar trend exists for the oxides, and Bi_2O_5 has never been obtained in pure form. Nitrogen and phosphorus are quite abundant, the former as the major component of the atmosphere and the latter in minerals such as the apatites (Fig. 3–11), with formula $Ca_5(PO_4)_3X$, where X stands for F, Cl, or OH. Arsenic and antimony, which are poisonous, are often troublesome impurities in the smelting of ores, but bismuth is less bothersome because it is less abundant. All three are used to make **alloys,** which are mixtures created by melting metals together. They tend to confer hardness, the property of expanding upon solidification (useful in making castings), and low melting points upon their alloys. The lighter members of this group are nonmetals (nitrogen and phosphorus); bismuth is a metal; and arsenic and antimony are classed as semi-metals.

• Several alloys of bismuth melt at temperatures below the boiling point of water. The melting of plugs of such alloys releases the water in fire-extinguishing sprinkler systems.

3-2   THE PERIODIC TABLE

The seven groups we have just outlined reflect modern understanding of the relationships among the elements. To a 19th-century chemist, some of the associations would have been evident (such as those in the alkali metal group), but others would have been much less obvious (such as placing nitrogen and bismuth in the same group). Moreover, many elements do not fall into these seven groups; several were well known in the 19th century. Silver, for example, resembles the alkali metals in forming a 1:1 chloride (AgCl) and a 2:1 oxide (Ag_2O), but its melting point is much higher (962°C, compared with 39°C for rubidium, for example), and it does not react with water. It therefore required considerable chemical judgment to determine which elements really belonged together in groups.

The Discovery of the Periodic Law

The next step came with the observation that there was a connection between the group properties and atomic masses. When the lightest 17 elements known at the time were arranged in order of increasing atomic mass,

Element:	(H)	Li	Be	B	C	N	O	F	Na	Mg	Al	Si	P	S	Cl	K	Ca
Atomic mass:	(1)	7	9	11	12	14	16	19	23	24	27	28	31	32	35.5	39	40

repetitive sequences of the groups appeared. An element of the halogen group, for example, was repeatedly followed by one of the alkali metal group and preceded by one of the chalcogens. The similarities among these elements occurred with a periodicity of seven; that is, there was a resemblance between the second and ninth elements (alkali metals), another between the third and tenth (alkaline earth metals), and so forth. When the preceding series was rewritten to begin a new row with every seventh element, elements of the same groups were automatically assembled in columns in a **periodic table:**

(H)						
(1)						
Li	Be	B	C	N	O	F
7	9	11	12	14	16	19
Na	Mg	Al	Si	P	S	Cl
23	24	27	28	31	32	35.5
K	Ca	etc.				
39	40					

• Obviously, the lightest element, hydrogen (H), does not fit the pattern.

The theoretical basis of this table was summarized in the original **periodic law:**

> The chemical and physical properties of the elements are periodic functions of their atomic masses.

More complete periodic tables were constructed and introduced (independently) by the German chemist Lothar Meyer and the Russian Dmitri Mendeleev in 1869 and 1870. Figure 3–12 gives Mendeleev's 1872 version of the table. At that time, one third of the naturally occurring chemical elements had not yet been discovered, and both chemists were farsighted enough to leave gaps where their analysis of periodic physical and chemical properties indicated that unknown elements should be located. Mendeleev was bolder than Meyer because he assumed that if a measured atomic mass put an element in the wrong place in the table, then the atomic mass was wrong. In some cases, this was true. Indium, for example, had previously been assigned a relative atomic mass near 76; however, there was no place for it in the periodic table between arsenic (with an atomic mass of 75) and selenium (with an

TABELLE II

REIHEN	GRUPPE I. — R^2O	GRUPPE II. — RO	GRUPPE III. — R^2O^3	GRUPPE IV. RH^4 RO^2	GRUPPE V. RH^3 R^2O^5	GRUPPE VI. RH^2 RO^3	GRUPPE VII. RH R^2O^7	GRUPPE VIII. — RO^4
1	H=1							
2	Li = 7	Be = 9,4	B = 11	C = 12	N = 14	O = 16	F = 19	
3	Na = 23	Mg = 24	Al = 27,3	Si = 28	P = 31	S = 32	Cl = 35,5	
4	K = 39	Ca = 40	— = 44	Ti = 48	V = 51	Cr = 52	Mn = 55	Fe = 56, Co = 59, Ni = 59, Cu = 63.
5	(Cu = 63)	Zn = 65	— = 68	— = 72	As = 75	Se = 78	Br = 80	
6	Rb = 85	Sr = 87	?Yt = 88	Zr = 90	Nb = 94	Mo = 96	— = 100	Ru = 104, Rh = 104, Pd = 106, Ag = 108.
7	(Ag = 108)	Cd = 112	In = 113	Sn = 118	Sb = 122	Te = 125	J = 127	
8	Cs = 133	Ba = 137	?Di = 138	?Ce = 140	—	—	—	— — —
9	(—)	—	—	—	—	—	—	
10	—	—	?Er = 178	?La = 180	Ta = 182	W = 184	—	Os = 195, Ir = 197, Pt = 198, Au = 199.
11	(Au = 199)	Hg = 200	Tl = 204	Pb = 207	Bi = 208	—	—	
12	—	—	—	Th = 231	—	U = 240	—	— — —

Figure 3–12 Dmitri Mendeleev published this early version of the periodic table in 1872. The German headings "GRUPPE" and "REIHEN" mean "group" and "rows," and the letter "R" is a generic symbol for all the elements of a group. Note how Mendeleev coped with the problem of the elements such as manganese (Mn) that do not fit well in the pattern of the seven main groups. He displayed them together with the main-group elements but pushed their symbols to the opposite side of the available space, seeking to establish a left/right alternation going down each group. The gap below boron was filled by the discovery of scandium, the gap below aluminum by gallium, and the gap below silicon by germanium. Mendeleev's superscripts mean the same as modern subscripts.

atomic mass of 79), and Mendeleev suggested that its relative atomic mass should be changed to $\frac{3}{2} \times 76 = 114$, near the currently accepted value of 114.82. In this position, its chemical properties fit the pattern defined by the known properties of aluminum and thallium. Subsequent work has shown that the elements are *not* strictly ordered in the periodic system by atomic mass and that there are four reversals. The chemical properties of tellurium, for example, demand that it come before iodine in the periodic table, even though its atomic mass is slightly greater. The other reversals are of argon and potassium, nickel and cobalt, and protactinium and thorium.

Mendeleev went further than Meyer in another respect: he predicted the properties of six elements yet to be discovered. A gap in his table just below aluminum, for example, implied the existence of an undiscovered element. Because he expected it to resemble aluminum in its properties, Mendeleev designated this element "eka-aluminum" (Sanskrit: *eka*, meaning "first") and predicted for it the set of properties given in Table 3–1. Just five years later, an element with the proper atomic mass was isolated and named "gallium" by its discoverer. The close correspondence (see Table 3–1) between the observed properties of gallium and Mendeleev's predictions for eka-aluminum lent strong support to the periodic law. Additional support came in 1885 when eka-silicon, which had also been described in advance by Mendeleev, was discovered and named "germanium."

The Noble Gases: An Eighth Group of Elements

The structure of the periodic table appeared to limit the number of possible elements. It was therefore quite surprising when John William Strutt, Lord Rayleigh,

• These four anomalies arise from differing distributions in the natural abundances of the isotopes of the two neighboring elements. For example, the most abundant isotope of potassium is ^{39}K, which weighs less than ^{40}Ar, the most abundant isotope of argon.

Table 3–1
Comparison of Eka-Aluminum and Gallium

Prediction by Mendeleev for Eka-Aluminum	Observed Properties of Gallium
Relative atomic mass 68	Relative atomic mass 69.72
Low melting metal	Metal, melting point = 29.8°C
Density 6.0 g cm^{-3}	Density 5.9 g cm^{-3}
Attacked slowly by acids and bases, does not oxidize readily in air	Observed as predicted
Formula of oxide E_2O_3, density of oxide 5.5 g gm^{-3}	Formula of oxide Ga_2O_3, density of oxide 5.88 g cm^{-3}
Oxide insoluble in water, soluble in bases or strong acids	Observed as predicted
Formula of chloride ECl_3, chloride volatile	Formula of chloride $GaCl_3$ ($GaCl_2$ is also found), melting point 78°C, boiling point 201°C

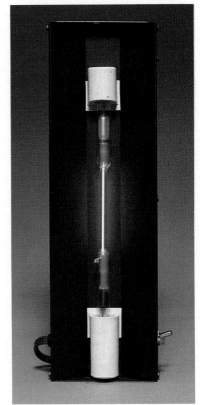

Figure 3–13 When an electrical discharge is passed through a tube filled with helium gas at low pressure, the helium atoms emit a characteristic violet light. The same characteristic color is part of the light from the sun, a fact that indicates the presence of helium in the sun.

discovered a gaseous element in 1894 that did not fit into the previous classification scheme. A century earlier, Cavendish had noted the existence of a residual gas when oxygen and nitrogen are removed from air, but its importance had not been realized. Together with William Ramsay, Rayleigh isolated the gas and named it "argon." Ramsay then studied a gas isolated from a uranium-containing mineral and discovered that it was helium. Helium, which also occurs in certain natural-gas deposits, had been detected (and named) previously, but only at long distance—by an additional line in the spectrum of sunlight (Fig. 3–13). Helium was not previously known on earth. Rayleigh and Ramsay observed that it resembled argon in its properties and postulated the existence of a new group in the periodic table to accommodate these new elements. In 1898 other members of the series (neon, krypton, and xenon) were isolated. These elements are referred to as **noble gases,** or sometimes as inert gases, because of their relative inertness toward chemical combination.

The Modern Periodic Table

To understand the ordering of elements in the periodic table, we recall that, contrary to Dalton's original idea, atoms are *not* hard, indestructible spheres but have internal structure (see Section 1–5). Every atom consists of a positively charged nucleus that is surrounded by a swarm of negatively charged electrons. Opposite charges attract, and like charges repel; the electrons in an atom repel each other and would fly apart except for the attraction of the nucleus. An atom as a whole is neutral because the total negative charge of its electrons exactly balances the positive charge of its nucleus. The charge of the nucleus comes from the protons it contains. Each proton carries a positive charge exactly equal in magnitude to the negative charge of the electron. This magnitude of charge, symbolized e, provides a convenient natural unit for measurements. An electron has charge equal to -1 in such units and a proton $+1$. The number of protons in the nucleus of an atom equals its atomic

• Recall that e equals $1.6021773 \times 10^{-19}$ coulombs (C). See Section 1–5 and Appendix B–1.

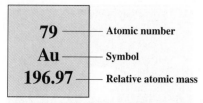

In the periodic table in the front of this book, the atomic number is placed directly above the element symbol and the relative atomic mass directly below.

• The lanthanide and actinide elements are usually placed below the rest of the table to conserve space.

• The International Union of Pure and Applied Chemistry (IUPAC) and the American Chemical Society have recommended a new system of group designation. In this system, the main-group elements comprise Groups 1, 2, and 13 to 18, and the transition elements make up Groups 3 to 12.

number Z. Hence, the charge on the nucleus of an atom equals $+Z$, and the atom must contain Z electrons if it is to be electrically neutral.

All atoms of a given element have the same atomic number and the same number of electrons, even though their masses vary from one isotope to another. The chemistry of an atom depends negligibly on its mass and almost entirely on its number of electrons, which is, as just explained, dictated by its atomic number. Thus, the atomic number is the fundamental determinant of chemical behavior for an atom. For this reason, the modern periodic table is arranged strictly according to atomic number. Hydrogen, the first element in the table, has atomic number 1 and one electron per atom. The electrically neutral atoms of the following elements have atomic numbers that are successively larger by 1 and have successively more electrons, up to element 112 with a nuclear charge of $+112$ and 112 electrons.

The periodic table, as we represent it today, encloses each element's symbol in a box and displays the boxes in **groups** (arranged vertically) and **periods** (arranged horizontally). The complete table is shown in Figure 3–14 and on the inside front cover of the book. There are eight groups of **representative elements** (or **main-group elements**), whose properties we have already sketched. There are ten groups of **transition elements** (or "transition metals"). We shall discuss certain unique aspects of these important elements in Chapter 19. Among the transition elements is a set of 15 consecutive elements (with atomic numbers 57 through 71) called the **lanthanide elements,** after the first of the series, lanthanum, or the "rare earth metals." Although it is now known that the rare-earth metals are not particularly rare (lanthanum and cerium, for example, are more abundant than tin), their remarkably similar chemical and physical properties still set them apart. They are usually found in association with one another and are difficult to separate. In the next period of the table, another set of 15 chemically similar elements (with atomic numbers 89 through 103) occurs. These **actinides** are all radioactive; that is, they have unstable nuclei and decay at varying rates to form less massive atoms. Only three of the actinides are found in nature; the rest must be produced artificially. The study of their chemistry poses special difficulties because they constantly contaminate themselves with decay products. Finally, elements beyond number 103 form a fourth period of transition elements. These radioactive elements are produced artificially and are so short-lived that few details of their chemistry are known.

The groups of representative elements are numbered (with Roman numerals) from I to VIII, with the letter "A" sometimes added to differentiate them from the transition-metal groups, which are labeled from IB to VIIIB. In this book, we use group numbers for the representative elements exclusively (and drop the "A"), and we refer to transition elements by the first element in the corresponding group. For example, we designate the elements in the carbon group (C, Si, Ge, Sn, Pb) as Group IV and the elements Cr, Mo, and W as the chromium group.

Physical as well as chemical properties of the elements vary systematically across the periodic table. Important physical properties include melting points and boiling points, thermal and electrical conductivity, density, and atomic size. In general, the elements on the left side of the table (especially in the later periods) are lustrous metallic solids and good conductors of both heat and electricity. On the right side (especially in earlier periods), the elements are non-metallic. In the solid and liquid states they lack the characteristic luster of metals, and many are gaseous at room temperature. They are poor conductors of heat and electricity. In between, the semi-metals form a zig-zag line of division between metal and non-metal (see the inside front cover of the book).

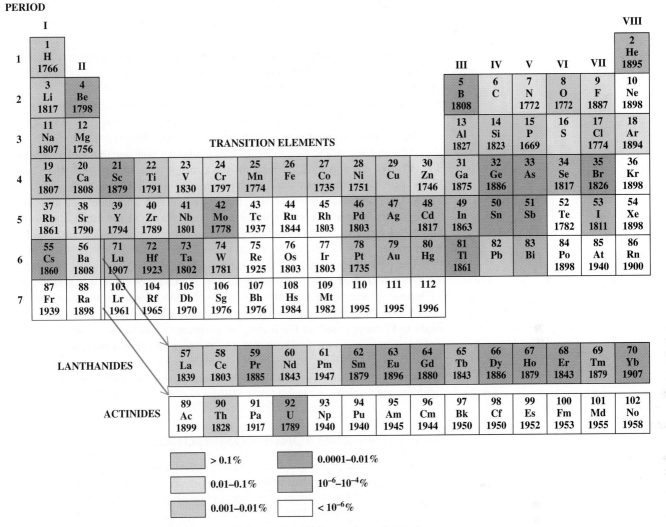

Figure 3–14 The modern periodic table of the elements. Below each symbol is the element's year of discovery; elements with no dates have been known since ancient times. Above each symbol is the atomic number. The color coding indicates the relative abundance by mass of the elements in the world around us (the atmosphere, oceans, freshwater lakes and rivers, and the earth's crust to a depth of 40 km). Oxygen alone makes up almost 50% of this mass, and silicon more than 25%.

Electronegativity, an Important Periodic Property

Two physical properties of atoms exert crucial influence on their chemistry: the energy change upon addition of an electron to a neutral atom and the energy change upon removal of an electron from a neutral atom. Metallic elements lose electrons (to form positive ions) more readily than non-metallic elements and gain electrons less readily; they are more **electropositive** than non-metallic elements. Non-metallic elements gain electrons (to form negative ions) more readily than metallic elements and lose electrons less readily; they are more **electronegative** than metallic elements. These differences in affinity carry into the chemical behavior of atoms and have led to this definition:

> The **electronegativity** of an atom is a measure of its power when in chemical combination to attract electrons to itself.

Large differences in electronegativity between two bonded atoms favor the transfer of electrons from the less electronegative to the more electronegative atom. Small differences are associated with the sharing of electrons between the atoms.

Although numerical electronegativities have been assigned to the elements (see Fig. 17–14), good use can be made of the concept on a wholly qualitative basis. With some exceptions, electronegativity increases from left to right across the periodic table and decreases from the top down the table. Elements of high electronegativity cluster toward the upper right-hand corner of the table; elements of low electronegativity (the most electropositive elements) occur toward the lower left-hand corner.

The periodic properties of the elements ultimately arise from the distribution of electrons in their atoms. For example, the chemistry of chlorine resembles that of bromine because the two elements have the same number of outer electrons arranged in similar spatial patterns. Our modern theoretical understanding of electrons in atoms is based on the principles of quantum mechanics, which are developed in Chapters 16 to 18. There, some of the regular patterns evident in the periodic table (such as the recurrence of eight main-group elements and ten transition elements per period) are related to the mathematical solutions of the equations of quantum mechanics.

3–3 IONS AND IONIC COMPOUNDS

The periodic table provides a useful way to describe systematically the properties of the elements, and we shall return to it frequently. In Chapter 17, we show how the structure of the table arises in the application of quantum mechanics to the electronic structures of atoms. Here we confine our rationalization of the chemical properties of the elements to a simple model, the **Lewis electron-dot model.** This model was proposed by the American chemist G. N. Lewis in 1916, well before modern quantum mechanics was fully developed.

Lewis Dot Symbols

The Lewis model begins by recognizing that not all the electrons in an atom participate directly in chemical bonding. Electrons can be divided into two groups: **core electrons,** which are held close to the nucleus and are not significantly involved in the formation of bonds, and **valence electrons,** which occupy the outer regions of the atom, the so-called **valence shell,** and take part in chemical bonds. Moreover, with the exception of helium, the number of valence electrons in a neutral atom of a representative element (those in Groups I through VIII) is equal to the element's group number in the periodic table. Atoms of bromine, for example, have seven valence electrons (like chlorine), and atoms of strontium have two (like beryllium). Similar chemical behavior within a group arises from the equal numbers of valence electrons in atoms of elements in that group. The maximum number of valence electrons is eight, and atoms with this full set of valence electrons (the noble gases from neon through xenon) are particularly stable and unreactive chemically. Helium is an

* *Valence* means "having to do with the capacity for chemical combination."

unreactive element with only two valence electrons. The Lewis model represents valence electrons by dots; core electrons are not shown. The first four dots are displayed singly around the four sides of the symbol of the element. If there are more than four valence electrons, then dots are paired with those already present. The result is a **Lewis dot symbol** for that atom. The Lewis notation for the elements of the first three periods is

• As we shall see, the number of un-paired dots tells the element's typical combining capacity in compounds.

$$\cdot H \qquad\qquad\qquad\qquad\qquad\qquad\qquad\qquad\qquad\qquad :He$$

$$\cdot Li \quad \cdot Be\cdot \quad \cdot \dot{B}\cdot \quad \cdot \dot{\underset{\cdot}{C}}\cdot \quad :\dot{N}\cdot \quad :\dot{\underset{\cdot}{O}}: \quad :\dot{\underset{\cdot}{F}}: \quad :\dot{\underset{\cdot\cdot}{Ne}}:$$

$$\cdot Na \quad \cdot Mg\cdot \quad \cdot \dot{Al}\cdot \quad \cdot \dot{\underset{\cdot}{Si}}\cdot \quad :\dot{P}\cdot \quad :\dot{\underset{\cdot}{S}}: \quad :\dot{\underset{\cdot\cdot}{Cl}}: \quad :\dot{\underset{\cdot\cdot}{Ar}}:$$

EXAMPLE 3-1

How many electrons does an atom of tellurium have? Of these, how many are valence electrons and how many core electrons? Draw the Lewis dot symbol for tellurium.

Solution

The atomic number of tellurium is given in the periodic table as 52, so each tellurium atom has 52 electrons. Because tellurium is in Group VI in the table, it has six valence electrons. This leaves $52 - 6 = 46$ core electrons. The Lewis dot symbol for tellurium resembles that of oxygen or sulfur:

$$:\dot{\underset{\cdot}{Te}}:$$

Exercise

How many electrons does an atom of germanium have? Of these, how many are valence electrons and how many core electrons? Draw the Lewis dot symbol for germanium.

Answer: 32, 4, 28; $\cdot \dot{\underset{\cdot}{Ge}}\cdot$

The Formation of Binary Ionic Compounds

A positively charged ion (called a **cation**) forms when an atom loses one or more electrons. A negatively charged ion (called an **anion**) forms when an atom adds electrons. Atoms lose and gain electrons fairly easily both in physical processes (such as mass spectrometry, see Section 1–6) and in chemical processes. The creation of ions is indicated by adding or removing the proper number of dots from the Lewis dot symbol and also by writing the net electric charge of the ion as a right superscript. This charge is understood to be expressed in units of the charge on the electron; thus, "2−" means that the ion has a net charge equal to the charge of two electrons. For example,

• Shuffling your feet across a carpet under the right conditions gives you a static electric charge: some of the atoms in your body are then ionized.

$Na\cdot$	Na^+	$\cdot Ca\cdot$	Ca^{2+}
sodium atom	sodium ion	calcium atom	calcium ion
$:\dot{\underset{\cdot\cdot}{F}}:$	$:\ddot{\underset{\cdot\cdot}{F}}:^-$	$:\dot{\underset{\cdot}{S}}:$	$:\ddot{\underset{\cdot\cdot}{S}}:^{2-}$
fluorine atom	fluoride ion	sulfur atom	sulfide ion

Special stability results when an atom, by either losing or gaining electrons, forms an ion with the same number of valence electrons as a noble-gas atom. Except for hydrogen and helium, which can have at most two valence electrons, atoms of the main-group elements of the periodic table have a maximum of eight valence electrons. Thus, we speak of a chloride ion ($: \overset{..}{\underset{..}{Cl}} : ^-$) or an argon atom ($: \overset{..}{\underset{..}{Ar}} :$) as having a completed **octet** of valence electrons. An ion such as Na^+ has the same number of electrons as the lower noble-gas atom Ne; however, dots to represent the octet of an Na^+ ion are not used. These are core electrons and were exposed only by the loss of valence electrons.

The tendency of atoms to attain valence octets explains much chemical reactivity. Atoms of the electropositive elements in Groups I and II achieve an octet by losing electrons to form cations, and atoms of the electronegative elements in Groups VI and VII achieve an octet by gaining electrons to form anions. The cations and anions then attract each other in **ionic bonds.** Ionic bonding is favored by large differences in electronegativity between the bonded atoms. If ionic bonds are the only or predominant type of bonds in a compound, it is an ionic compound. Reactions of the metallic elements on the left side of the periodic table with the non-metallic elements on the right side always transfer just enough electrons to form ions with completed octets. The following equations, in which e^- stands for an electron, use Lewis symbols to show the formation first of a cation and anion and then of a compound.

$$Na \cdot \longrightarrow Na^+ + e^- \qquad \text{loss of a valence electron}$$

$$e^- + : \overset{.}{\underset{..}{Cl}} : \longrightarrow : \overset{..}{\underset{..}{Cl}} : ^- \qquad \text{gain of a valence electron}$$

$$Na^+ + : \overset{..}{\underset{..}{Cl}} : ^- \longrightarrow Na^+ : \overset{..}{\underset{..}{Cl}} : ^- \qquad \text{combination to form the compound NaCl}$$

Another example is the formation of $CaBr_2$:

$$\cdot Ca \cdot + : \overset{.}{\underset{..}{Br}} : + : \overset{.}{\underset{..}{Br}} : \longrightarrow Ca^{2+} + : \overset{..}{\underset{..}{Br}} : ^- + : \overset{..}{\underset{..}{Br}} : ^- \longrightarrow Ca^{2+}(Br^-)_2$$

neutral atoms not having octets   positive ion with octet   negative ions with octets   compound

The model predicts a $1:1$ compound between Na and Cl and a $1:2$ compound between Ca and Br. Both predictions are correct.

EXAMPLE 3–2

Predict the formula of the compound between rubidium and sulfur. Give Lewis symbols for the elements both before and after chemical combination.

Solution

Rubidium, from Group I of the periodic table, has one valence electron. Sulfur, from Group VI, has six valence electrons. Their Lewis symbols are $Rb \cdot$ and $: \overset{.}{S} :$. The transfer of one electron from each of two rubidium atoms to a sulfur atom gives two Rb^+ ions and one $: \overset{..}{S} : ^{2-}$ ion, all having octets. The compound is Rb_2S or, in Lewis symbols, $[Rb^+]_2[: \overset{..}{\underset{..}{S}} : ^{2-}]$.

Exercise

Predict the formula of a compound between magnesium and nitrogen, which attains an octet by gaining three electrons. Use Lewis symbols to keep track of the valence electrons.

Answer: Mg_3N_2.

Ionic compounds are often called **salts,** by analogy with NaCl, the main component of common salt, which is everyone's typical ionic compound. Ionic compounds are solids at room conditions and generally have high melting and boiling points (for example, NaCl melts at 801°C and boils at 1413°C). Solid ionic compounds usually conduct electricity poorly, but their melts (the molten liquids) conduct well. Whether in the solid or liquid state, binary ionic compounds contain no molecules. Their chemical compositions are given by empirical, not molecular, formulas.

Names and Formulas of Ionic Compounds

Ionic compounds result from the combination of cations with anions. Their names consist of the name of the cation followed by a space and the name of the anion.

Ions can be polyatomic as well as monatomic. In polyatomic ions, or **molecular ions,** covalent bonds (see Section 3–4) maintain atoms in a tight group that carries a net charge. Because of the charge, the group can participate as a single entity in ionic bonding. Monatomic cations are named after the element from which they are derived. We have already encountered such examples as the sodium ion Na^+ and the calcium ion Ca^{2+}; ions of the other elements in Groups I and II are named in the same way. Polyatomic cations are most frequently named as derivatives of molecular compounds. For example, among the relatively few polyatomic cations of importance in inorganic chemistry are the ammonium ion NH_4^+ (obtained by adding H^+ to ammonia) and the hydronium ion H_3O^+ (obtained by adding H^+ to water). The names of polyatomic cations often end in *-onium* or *-ium*.

A monatomic anion is named by adding the suffix *-ide* to the first portion of the name of the element; thus, chlor*ine* gives the chlor*ide* ion, and ox*ygen* gives the ox*ide* ion. The other monatomic anions of Groups V, VI, and VII are named similarly. A large number of polyatomic anions exists, and the naming of these species is more complex. The names of the **oxoanions,** which contain oxygen in combination with a second element and are very prevalent, are derived by adding the ending *-ate* to the stem of the name of that second element. Several elements form two oxoanions. The *-ate* ending is then used for the oxoanion with the *larger* number of oxygen atoms (for example, NO_3^-, nitr*ate*), and the ending *-ite* is added for the name of the anion with the smaller number (for example, NO_2^-, nitr*ite*). For elements such as chlorine, which form more than two oxoanions, the additional prefixes *per-* (largest number of oxygen atoms) and *hypo-* (smallest number of oxygen atoms) are used. Oxoanions containing hydrogen as a third element include that word in their name. The HCO_3^- oxoanion, for example, is preferably called the "hydrogen carbonate ion" rather than its common (nonsystematic) name of "bicarbonate ion," and HSO_4^-, which is often called "bisulfate ion," is better designated as the hydrogen sulfate ion. Some of the most important anions are listed in Table 3–2. It is important to be able to recognize and name the ions and molecular ions from Table 3–2, not forgetting that the charge indicated by the right superscript is an essential part of the formula.

The composition of an ionic compound is determined by overall **charge neutrality:** the total positive charge on the cations must be exactly balanced by the total negative charge on the anions. The following names and formulas of ionic compounds illustrate this point:

sodium bromide	one +1 cation, one −1 anion	NaBr
potassium permanganate	one +1 cation, one −1 anion	$KMnO_4$
ammonium sulfate	two +1 cations, one −2 anion	$(NH_4)_2SO_4$
calcium dihydrogen phosphate	one +2 cation, two −1 anions	$Ca(H_2PO_4)_2$

Table 3–2
Formulas and Names of Some Common Anions

F^-	fluoride	SO_3^{2-}	sulfite
Cl^-	chloride	CO_3^{2-}	carbonate
Br^-	bromide	PO_4^{3-}	phosphate
I^-	iodide	HPO_4^{2-}	hydrogen phosphate
H^-	hydride	$H_2PO_4^-$	dihydrogen phosphate
O^{2-}	oxide	SiO_4^{4-}	silicate
S^{2-}	sulfide	CNO^-	cyanate
O_2^{2-}	peroxide	SCN^-	thiocyanate
O_2^-	superoxide	ClO_4^-	perchlorate
HCO_3^-	hydrogen carbonate	ClO_3^-	chlorate
HSO_4^-	hydrogen sulfate	ClO_2^-	chlorite
OH^-	hydroxide	ClO^-	hypochlorite
CN^-	cyanide	MnO_4^-	permanganate
NO_3^-	nitrate	CrO_4^{2-}	chromate
NO_2^-	nitrite	$Cr_2O_7^{2-}$	dichromate
SO_4^{2-}	sulfate		

This simple principle of charge neutrality, together with a knowledge of the places of elements in the periodic table, serves to predict the formulas of a large number of ionic compounds.

EXAMPLE 3–3

Give the chemical formulas of (a) calcium cyanide and (b) barium phosphate.

Solution

(a) Calcium cyanide is composed of Ca^{2+} and CN^- ions. For the overall charge to be zero, there must be two CN^- ions for each Ca^{2+} ion. Thus, the chemical formula of calcium cyanide is $Ca(CN)_2$.

(b) The ions present in this compound are Ba^{2+} and PO_4^{3-}. In order to ensure charge neutrality, there must be three Ba^{2+} ions (total charge $+6$) and two PO_4^{3-} ions (total charge -6) per formula unit. Thus, the chemical formula of barium phosphate is $Ba_3(PO_4)_2$.

Exercise

Give the chemical formulas of aluminum oxide, potassium sulfite, strontium chlorate, and lithium nitrite.

Answer: Al_2O_3, K_2SO_3, $Sr(ClO_3)_2$, and $LiNO_2$.

3–4 COVALENT BONDING AND LEWIS STRUCTURES

Elements of intermediate electronegativity form ionic compounds far less readily than highly electronegative or highly electropositive elements. Consider methane (CH_4), the simplest compound between carbon and hydrogen, which have comparable, intermediate electronegativities. Unlike any ionic compound, methane is a gas, not a solid, at room temperature. Cooling methane to low temperatures condenses it first to a liquid and then to a solid in which distinct molecules retain their identities. Unlike melted ionic compounds, liquid methane does not conduct electricity. It is thus not useful to think of methane as an ionic substance made up of C^{4-} and H^+ ions (or C^{4+} and H^- ions). It is a molecular substance. The bonding within the methane molecules arises from (approximately) equal sharing of electrons between the carbon and the hydrogen atoms, not from the transfer of electrons. Such bonds are **covalent bonds.**

Lewis Structures

The Lewis electron-dot model is quite effective in describing the covalent bonding in molecules formed by non-metallic main-group elements. Compounds that contain only covalent bonds are called **covalent compounds.** Hydrogen and chlorine combine, for example, to form the covalent compound hydrogen chloride. This can be indicated through a **Lewis structure** for the molecule of the product, in which the valence electrons from each atom are redistributed so that one electron from the hydrogen and one from the chlorine are shown as shared by the two atoms. The two dots representing this electron pair are placed between the symbols for the two elements:

$$\text{H} \cdot \ + \ \cdot \overset{\displaystyle ..}{\underset{\displaystyle ..}{\text{Cl}}} : \ \longrightarrow \ \text{H} : \overset{\displaystyle ..}{\underset{\displaystyle ..}{\text{Cl}}} :$$

The basic rule that governs the writing of Lewis structures is the **octet rule:**

> Whenever possible, the valence electrons in a covalent compound are distributed in such a way that each main-group element in a molecule (except hydrogen) is surrounded by eight electrons (an octet of electrons). Hydrogen should have two electrons in such a structure.

When the octet rule is satisfied, the atom attains the special stability of a noble-gas atom. In the structure for HCl shown above, the H nucleus, through sharing, is close to two valence electrons (like the noble gas helium), and the Cl has eight valence electrons near it (like the noble gas argon). Electrons that are shared between two atoms are counted as contributing to each atom.

Shared electrons hold the atoms together in covalent molecules. A covalent bond accordingly can be represented by a pair of dots positioned between the symbols of the atoms. A shared pair of electrons is also (and more frequently) represented by a short line (—):

$$\text{H} \!-\! \overset{\displaystyle ..}{\underset{\displaystyle ..}{\text{Cl}}} :$$

The unshared electron pairs around the chlorine atom in the Lewis structure are called **lone pairs,** and they make no contribution to the bond between the atoms. Lewis structures of some simple, but important, covalent compounds are

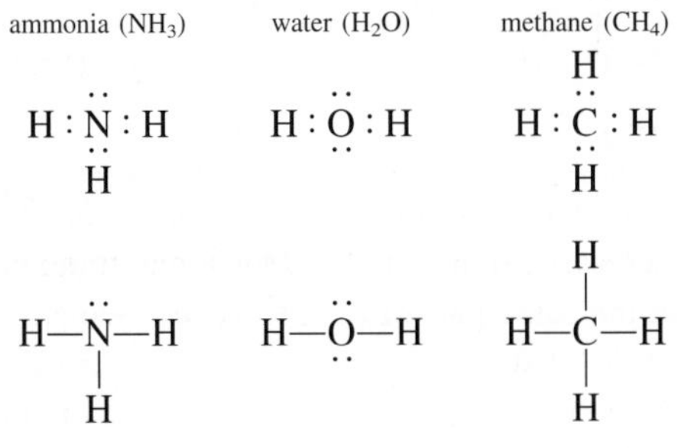

ammonia (NH$_3$)	water (H$_2$O)	methane (CH$_4$)
		H
H : N : H	H : O : H	H : C : H
H		H

		H
H—N—H	H—O—H	H—C—H
|		|
H		H

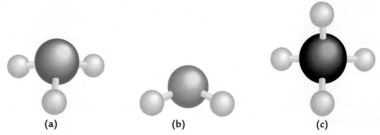

(a)   **(b)**   **(c)**

Figure 3–15   The molecules of three familiar substances are shown in ball-and-stick drawings. (a) Ammonia (NH_3) is produced in amounts approaching 100 million tons per year to meet the demand for nitrogen-based fertilizers. (b) Water (H_2O) fills the oceans, lakes, and rivers and accounts for most of the mass of the human body. (c) Methane (CH_4) is the major constituent of natural gas and is an important starting material for the synthesis of substances ranging from ammonia to plastics.

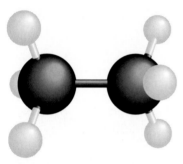

Ethane (C_2H_6) occurs with methane in natural gas. It can be burned in oxygen, and, if strongly heated, it reacts to form hydrogen and ethylene.

Lewis structures indicate the way bonds connect the atoms in a molecule, but they do not show the three-dimensional molecular geometry. The ammonia molecule, for example, is not planar but pyramidal, with the nitrogen atom at the apex. The water molecule is bent rather than straight. Three-dimensional geometry can be indicated by ball-and-stick models such as those in Figure 3–15.

Multiple Bonding

In some molecules, more than one pair of electrons is shared by two atoms. The oxygen atoms in O_2, for example, each contribute six valence electrons to the molecule. Drawing a single bond between them would not lead to valence octets. This can be accomplished only by drawing the Lewis structure as

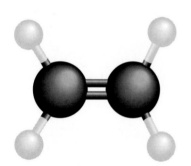

Ethylene (C_2H_4) is the largest-volume organic (carbon-containing) chemical produced. Its end uses are predominantly in polymers such as polyethylene plastic containers and polyvinyl chloride pipes and roofing materials.

$$\ddot{O} : : \ddot{O} \qquad or \qquad \ddot{O} = \ddot{O}$$

in which *two* pairs of electrons are shared between the oxygen atoms. That bond is again a covalent bond but is called a **double bond.** Similarly, the N_2 molecule has a **triple bond,** involving three shared electron pairs:

$$: N : : : N : \qquad or \qquad : N \equiv N :$$

As the examples show, double bonds are written as double lines and triple bonds as triple lines. Carbon-carbon bonds can involve the sharing of one, two, or three electron pairs. A progression from single to triple bonding is found in the three hydrocarbons ethane (C_2H_6), ethylene (C_2H_4), and acetylene (C_2H_2):

Acetylene (C_2H_2) has a triple bond that makes it highly reactive. The large amount of heat given off as it burns in oxygen in an oxy-acetylene torch makes the torch ideal for cutting and welding metals.

$$\begin{array}{cc} H & H \\ | & | \\ H-C-C-H \\ | & | \\ H & H \end{array} \qquad \begin{array}{c} H \qquad\qquad H \\ \diagdown \quad / \\ C = C \\ / \qquad \diagdown \\ H \qquad\qquad H \end{array} \qquad H-C\equiv C-H$$

Multiple bonding to attain an octet most often involves the elements carbon, nitrogen, oxygen, and, to a lesser extent, sulfur. Double and triple bonds are shorter than single bonds between the same pair of atoms (Table 3–3).

Table 3–3
Average Bond Lengths

C—C	1.54	N—N	1.45	C—H	1.10	O—O	1.48
C=C	1.34	N=N	1.25	N—H	1.01	O=O	1.21
C≡C	1.20	N≡N	1.10	O—H	0.96		
C—O	1.43	N—O	1.43	C—N	1.47		
C=O	1.20	N=O	1.18	C≡N	1.16		

All values are in units of 10^{-10} m.

Formal Charges

Let us consider a molecule of carbon monoxide (CO). This molecule has ten valence electrons (four from the C and six from the O). The only Lewis structure that gives octets to both atoms uses a triple bond:

$$:C::O:$$

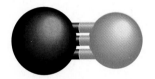

Carbon monoxide (CO) is a colorless, odorless, and toxic gas. It is produced by the incomplete burning of carbon-containing fuels in air. It is used in producing elemental metals from their oxide ores.

As far as the positioning of electrons is concerned, this Lewis structure is just like the one given for N_2. If we assume, however, that the six bonding electrons are shared equally between the carbon and oxygen atoms, then the carbon atom owns five valence electrons (one *more* than its group number), and the oxygen atom owns five valence electrons (one *less* than its group number). If the sharing of the electrons in the bond is equal, then formally the carbon atom has gained an electron, attaining a **formal charge** of -1, and the oxygen atom has lost an electron, attaining a formal charge of $+1$.

$$\overset{\textcircled{\scriptsize -1}}{}\;\overset{\textcircled{\scriptsize +1}}{}$$
$$:C::O:$$

• The formal charge of an atom is indicated by a circled number in blue next to that atom.

Carbon monoxide is a covalent compound, and the assignment of these formal charges does not make it ionic. These charges would exist only if the six bonding electrons were shared equally by the carbon and oxygen atoms. In fact, the bonding electrons in carbon monoxide are *not* equally shared because the more electronegative oxygen atom attracts the shared electrons to itself more strongly than the carbon atom does. This makes the true charges on the oxygen and carbon atoms nowhere near as large as the formal values of $+1$ and -1. Although they lack physical meaning, formal charges are useful in evaluating Lewis structures. A Lewis structure that gives large positive formal charges to some atoms and large negative formal charges to others is a poor model of covalent bonding. Often, two or more Lewis structures are possible for the same molecule. If so, the structure that assigns the lowest formal charges to the atoms generally provides the best description of the bonding. When non-zero formal charges appear, Lewis structures that assign negative formal charges to the more electronegative atoms and positive formal charges to the more electropositive atoms are preferable to the ones that do the reverse.

• Remember: a formal charge is determined only *after* a Lewis structure has been drawn.

The formal charge on an atom in a Lewis structure is simple to calculate. If the valence electrons are removed from an atom, it has a positive charge equal to its group number in the periodic table (for example, elements in Group VI, the chalcogens, have six valence electrons, and therefore a charge of $+6$ when these electrons are removed). To calculate the formal charge, first subtract from this positive charge the number of lone-pair valence electrons that an atom possesses in the Lewis structure, and then subtract one half of the number of bonding electrons associated with it:

$$\text{formal charge} = \text{group number} - \text{number of electrons in lone pairs} - \frac{1}{2}(\text{number of electrons in bonding pairs})$$

EXAMPLE 3–4

Compute the formal charges on the atoms in the following Lewis structure for the azide ion N_3^-.

$$\left[: \overset{..}{N} = N = \overset{..}{N} :\right]^-$$

Solution

Nitrogen is in Group V; hence, each nitrogen atom contributes five valence electrons to the bonding, and the negative charge on the ion contributes one more electron. The Lewis structure correctly represents 16 electrons. We now use the formula

$$\text{formal charge} = \text{group number} - \text{number of electrons in lone pairs} - \frac{1}{2}(\text{number of electrons in bonding pairs})$$

The group number of nitrogen is 5. The nitrogen on the left end of the Lewis structure has four electrons in lone pairs and four bonding electrons (which compose a double bond). Therefore,

$$\text{formal charge}_{(\text{left N})} = 5 - 4 - \frac{1}{2}(4) = \boxed{-1}$$

The nitrogen on the right end of the Lewis structure also has four electrons in lone pairs and four bonding electrons:

$$\text{formal charge}_{(\text{right N})} = 5 - 4 - \frac{1}{2}(4) = \boxed{-1}$$

The nitrogen in the center of the structure has no electrons in lone pairs. Its entire octet lies in the eight bonding electrons:

$$\text{formal charge}_{(\text{center N})} = 5 - 0 - \frac{1}{2}(8) = \boxed{+1}$$

The sum of the three formal charges is -1, which is the true overall charge on this molecular ion. If this check fails, there is an error in either the Lewis structure or the arithmetic.

Exercise

Determine the formal charges on the atoms in an alternative Lewis structure for the same ion:

$$\left[: \overset{..}{N} - N \equiv \overset{..}{N} :\right]^-$$

Answer: -2, $+1$, and 0, reading from left to right.

3–5 DRAWING LEWIS STRUCTURES

The drawing of Lewis structures strikes some as a game with no relation to chemical reality. In fact, however, Lewis structures work: they produce helpful predictions on matters of chemical importance. For example, double bonds are shorter and stronger than single bonds, so their placement in molecules affects geometry and energy. Also, chemical reactivity differs strongly from single to double and triple bonds, so Lewis structures help to explain or predict trends in reactivity. Throughout this section, think of Lewis structures as a *useful* even if imperfect description of reality.

Some molecules can exist as two or more **isomers**—structures containing the same set of atoms, connected to each other in different orders. Formal charges can sometimes be used to explore possible isomers and determine which are likely to be less stable because they involve considerable separation of formal charge. For example, of the two isomers of the molecule HCN

$$H:C:::N: \quad \text{and} \quad H:\overset{+1}{N}:::\overset{-1}{C}:$$

the one on the left (with zero formal charges on all atoms) is favored.

In evaluating isomers, it also helps to know that hydrogen and fluorine are always terminal atoms in Lewis structures, bonded to only one other atom. Several equally good isomeric arrangements of atoms may exist. The correct **molecular skeleton** (the plan of which atoms are connected to which other atoms) for a given substance must then be determined by other means. Given a skeleton, the following systematic procedure furnishes a valid Lewis structure, based on the octet rule.

1. Count up the total number of valence electrons *available* (symbolized by A) by first adding the group numbers of all the atoms present. If the species is a negative ion, *add* the absolute value of the total charge; if it is a positive ion, *subtract* it.
2. Calculate the total number of electrons *needed* (N) for each atom to have its *own* noble-gas set of electrons around it (two for hydrogen, eight for the elements from carbon on in the periodic table).
3. Subtract the number in step 1 from the number in step 2. This is the number of *shared* (or bonding) electrons present (S).

$$S = N - A$$

4. Assign two bonding electrons (as one shared pair) to each connection between two atoms in the molecule or ion.
5. If any of the electrons earmarked for sharing remain, assign them in pairs by making some of the bonds double or triple bonds. In some cases, there may be more than one way to do this. In general, double bonds form only between atoms of these elements: carbon, nitrogen, oxygen, and sulfur. Triple bonds are usually restricted to either carbon or nitrogen.
6. Assign the remaining electrons as lone pairs to the atoms, giving octets to all atoms except hydrogen.
7. Determine the formal charge on each atom, and write it next to that atom. Check that the formal charges add to give the correct total charge on the molecule or molecular ion.

The last step identifies structures that are undesirable because they imply large separations of negative from positive charge; it also catches mechanical errors (using the wrong number of dots and so forth).

The use of the rules is illustrated by the following examples.

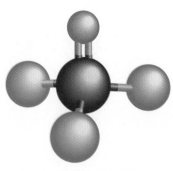

Phosphoryl chloride ($POCl_3$) is a reactive compound used to introduce phosphorus into organic molecules. The products of such reactions range from insecticides to flame retardants.

EXAMPLE 3-5

Write a Lewis electron-dot structure for phosphoryl chloride ($POCl_3$), which consists of a central phosphorus atom bonded to three chlorine atoms and one oxygen atom. Assign formal charges to all of the atoms.

Solution

The first step is to calculate the total number of valence electrons available in the molecule. For $POCl_3$, it is

$$A = 5 \text{ (from P)} + 6 \text{ (from O)} + 3 \times 7 \text{ (from Cl)} = 32$$

Recall that the number of valence electrons on an atom is equal to its group number. Next, we calculate how many electrons would be necessary if each atom were to have its *own* noble-gas shell of electrons around it. Because there are five atoms (none of them hydrogen), this would require

$$N = 40$$

in the present case. The difference of these numbers is

$$S = N - A = 40 - 32 = 8$$

so we conclude that each atom can achieve a valence octet only if eight electrons are shared between pairs of atoms. Eight electrons correspond to four electron pairs, so we foresee that each of the four linkages in $POCl_3$ must be a single bond. (If the number of shared electron pairs were *larger* than the number of linkages, double or triple bonds would be present.)

We are now ready to draw the Lewis structure, assigning the other 24 valence electrons as lone pairs to the atoms in such a way that each atom obtains an octet. The result is

$$
\begin{array}{c}
\overset{\textstyle -1}{\underset{\textstyle\cdot\cdot}{\cdot\cdot\,}}\overset{\displaystyle:O:}{}\,\\
\overset{0}{:}\underset{\cdot\cdot}{Cl}\!-\!\underset{\cdot\cdot}{\overset{+1}{P}}\!=\!\overset{0}{\underset{\cdot\cdot}{Cl}}: \\
\underset{\cdot\cdot}{\overset{0}{|}}\\
\underset{\cdot\cdot}{:Cl:}
\end{array}
$$

Formal charges are already indicated in this diagram. The formal charge of $+1$ on the central phosphorus atom was computed by

$$
\underset{\text{group number of P}}{5} \quad - \quad \underset{\text{lone-pair electrons}}{0} \quad - \quad \underset{\text{bonding electrons}}{\frac{1}{2}(8)} \quad = +1
$$

All three chlorine atoms have formal charges of 0, computed by

$$
\underset{\text{group number of Cl}}{7} \quad - \quad \underset{\text{lone-pair electrons}}{6} \quad - \quad \underset{\text{bonding electrons}}{\frac{1}{2}(2)} \quad = 0
$$

The formal charge on the oxygen atom is -1, because $6 - 6 - \frac{1}{2}(2) = -1$. We verify that the formal charges add to zero, as they must for this electrically neutral species.

Exercise

Draw a Lewis dot structure for thionyl chloride ($SOCl_2$). Sulfur is the central atom. Indicate formal charges on all atoms.

Answer:

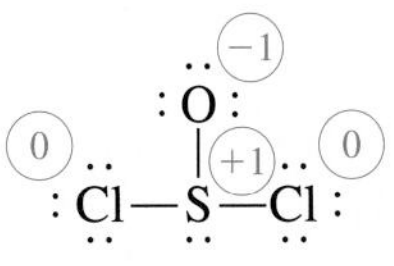

$$
\begin{array}{c}
\overset{\overset{\textstyle\cdot\cdot}{\textstyle}}{\;:\!\overset{\cdot\cdot}{O}\!:}\;\overset{(-1)}{} \\
\underset{\textstyle}{\underset{\cdot\cdot}{:}\overset{(0)}{Cl} - \underset{\textstyle}{\overset{(+1)}{S}} \overset{(0)}{-} \overset{\cdot\cdot}{Cl}:}
\end{array}
$$

EXAMPLE 3–6

Draw a Lewis structure for the molecular ion NO_2^+, which has a central nitrogen atom. Estimate the bond lengths in this ion, using Table 3–3.

Solution

Counting the valence electrons in one N and two O's gives 5 (from N, Group V) $+2 \times 6$ (from O, Group VI) $= 17$. From this, we *subtract* 1 because of the net charge of $+1$ on the ion (one electron has been *removed* from NO_2). This leaves $A = 16$ valence electrons available.

Giving each atom its *own* noble-gas shell would require $N = 3 \times 8 = 24$ valence electrons. Subtracting 16 from 24 gives $S = 8$ shared electrons. After two are assigned to each of the two N-to-O links, four bonding electrons remain. These are used to make both bonds into double bonds:

$$O::N::O$$

We have used 8 of the 16 available valence electrons. The other 8 are placed on the two oxygen atoms to give them octet configurations (the nitrogen atom already has an octet). The result is

$$[:\overset{\cdot\cdot}{O}::N::\overset{\cdot\cdot}{O}:]^+ \quad \text{or} \quad [:\overset{\cdot\cdot}{O}=N=\overset{\cdot\cdot}{O}:]^+$$

Finally, we assign the formal charges:

$$\text{formal charge on O atoms} = 6 - 4 - \frac{1}{2}(4) = 0$$

$$\text{formal charge on N atom} = 5 - 0 - \frac{1}{2}(8) = +1$$

These add up to the total charge on the ion, $+1$. The bond lengths for the two N=O bonds should be approximately 1.18×10^{-10} m.

Note that we did *not* choose to make one single and one triple bond

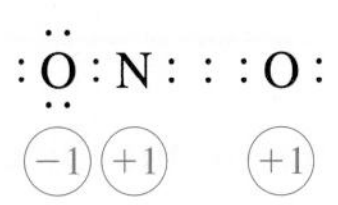

$$
:\overset{\cdot\cdot}{\underset{\cdot\cdot}{O}}:N:::O:
$$
$$
\;\;(-1)(+1)\quad\;\;(+1)
$$

This Lewis structure is undesirable because it displays a larger separation of formal charge than the other and forces a positive formal charge onto an oxygen, which is more electronegative than nitrogen.

Exercise

Draw Lewis structures for both ions in ammonium sulfate. Show all formal charges.

Answer:

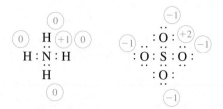

EXAMPLE 3–7

Draw a Lewis structure for the cyanide ion CN⁻. Estimate the bond length.

Solution

The carbon and nitrogen atoms contribute $4 + 5 = 9$ valence electrons. To this, we *add* 1 because of the net negative charge of -1 on the ion, giving $A = 10$ valence electrons. The number of electrons to make two separate octets would be $N = 16$, so there are $S = 16 - 10 = 6$ shared electrons. Thus, there must be a triple bond between the two atoms:

$$C : : : N$$

The remaining four electrons form lone pairs on the two atoms:

$$[: C : : : N :]^- \quad \text{or} \quad [: C \equiv N :]^-$$

The formal charges are:

$$\text{formal charge on C} = 4 - 2 - \frac{1}{2}(6) = -1$$

$$\text{formal charge on N} = 5 - 2 - \frac{1}{2}(6) = 0$$

The net negative charge of -1 is formally on carbon, giving

$$\overset{\boxed{-1}}{:C} : : : \overset{\boxed{0}}{N} :.$$

This structure puts the required negative formal charge on carbon rather than on the more electronegative nitrogen. Although this is undesirable, it is not possible to do otherwise without breaking the octet rule. The bond length (Table 3–3) should be approximately 1.16×10^{-10} m.

Exercise

Draw two possible Lewis structures for the thiocyanate ion SCN⁻.

Answer:

$$\overset{\boxed{-1}}{:S} : \overset{\boxed{0}}{C} : : : \overset{\boxed{0}}{N} : \quad \text{or} \quad \overset{\boxed{0}}{:S} : : \overset{\boxed{0}}{C} : : \overset{\boxed{-1}}{N} :$$

Resonance Structures

For certain molecules or molecular ions, two or more equivalent Lewis structures can be drawn on the same skeleton. An example is sulfur dioxide (SO_2), for which there are two Lewis structures that satisfy the octet rule:

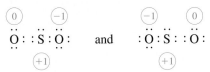

Each of these structures suggests that one S—O bond is a single bond and the other a double bond, so that the molecule would be asymmetric. In fact, the lengths of the S—O bonds are found experimentally to be equal. The Lewis approach has apparently failed. The difficulty is remedied by saying that the true structure is a **resonance hybrid** of the two Lewis structures, in which each of the bonds is intermediate between a single and a double bond. This is diagrammed by drawing both structures and connecting them with a double-headed arrow.

Sulfur dioxide (SO_2) is produced by burning sulfur as the first step in the production of sulfuric acid, the most widely used industrial acid. Sulfur dioxide is also emitted in the burning of coal, oil, and natural gas, all of which contain small amounts of sulfur. In the atmosphere it contributes significantly to the formation of acid rain.

The term "resonance" does *not* mean that the molecule resonates, or jumps, back and forth from one structure to the other. Rather, the true structure is a hybrid that *simultaneously* includes all the bonding features of the contributing Lewis structures. For some molecules, numerous resonance contributors are possible and must be diagrammed. This awkwardness can be avoided by treating bonds using molecular orbitals, as we show in Chapter 18.

EXAMPLE 3–8

Draw three resonance forms for sulfur trioxide (SO_3), in which the central sulfur atom is bonded to the three oxygen atoms and the octet rule is satisfied for all atoms.

Solution

In SO_3, there are $A = 24$ valence electrons. For each atom to have its own octet would require $N = 4 \times 8 = 32$ electrons. Therefore, $S = 32 - 24 = 8$ electrons must be shared between atoms, implying a total of four bonding pairs. These can be distributed in one double and two single bonds, leading to the equivalent resonance structures

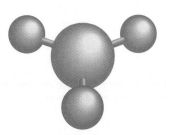

The two singly bonded oxygen atoms carry formal charges of -1, the doubly bonded oxygen atom carries a formal charge of 0, and the formal charge of the sulfur atom is $+2$. Experimentally, SO_3 is found to have three equivalent S—O bonds.

Sulfur trioxide (SO_3) is a reactive and corrosive gas that is produced by the oxidation of sulfur dioxide. It reacts with water to give sulfuric acid.

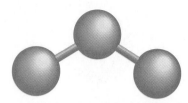

Ozone (O_3) is a pale blue gas with a pungent odor that condenses to a deep blue liquid below $-112°C$. In the upper atmosphere, it plays a crucial role in shielding the earth's surface from the full effect of the sun's radiation.

Exercise

The ozone molecule contains three oxygen atoms. Draw two resonance structures that jointly represent the bonding in this molecule.

Answer:

$$\left\{ \overset{(0)}{\underset{(+1)}{\ddot{O}}} : : \overset{(-1)}{\ddot{O}} : \ddot{O} : \quad \longleftrightarrow \quad : \ddot{O} : \overset{(-1)}{\underset{(+1)}{\ddot{O}}} : : \overset{(0)}{\ddot{O}} \right\}$$

Breakdown of the Octet Rule

Lewis structures constructed using the octet rule are useful in predicting whether a proposed molecule will be stable under ordinary conditions of temperature and pressure; for example, we can write a simple Lewis structure for water (H_2O):

$$H : \ddot{O} : H$$

in which each atom has a noble-gas configuration. This is impossible to do for OH or for H_3O, suggesting that these species are either unstable or highly reactive. There are several situations, however, in which the octet rule is *not* satisfied, but the molecule or molecular ion is still stable.

Case 1: Odd-Electron Molecules. The electrons in a Lewis structure that satisfies the octet rule must occur in pairs: bonding pairs or lone pairs. No molecule with an odd number of electrons can satisfy the octet rule on all of its atoms. Most stable molecules have an even number of electrons, but a few have an odd number. In this case, the best one can do is to give up the octet rule for one of the atoms by leaving an unpaired lone electron on it and to reduce the separation of formal charge as much as possible. All bonding electrons should be kept paired in such Lewis structures.

• Just leaving off the troublesome odd electrons from Lewis structures is *not* legal.

An example is nitrogen monoxide (NO), a stable (although reactive) molecule that is an important factor in air pollution. It has 11 valence electrons. In the two Lewis structures that can be drawn for it

$$\overset{(0)}{:}\overset{.}{N}=\overset{..}{\overset{(0)}{O}}: \quad \text{and} \quad \overset{(-1)}{:}\overset{..}{N}=\overset{.}{\overset{(+1)}{O}}:$$

only one of the atoms has an octet of electrons. Of these two structures, the second, in which the odd electron resides on the oxygen atom, should be less favored because it leads to a separation of formal charge with positive formal charge assigned to the more electronegative atom. Some other stable odd-electron molecules (and molecular ions) are chlorine monoxide (ClO), nitrogen dioxide (NO_2), and the superoxide ion (O_2^-).

Case 2: Octet-Deficient Molecules. Covalent compounds of beryllium and boron are frequently octet deficient: that is, sufficient electrons are not available for each atom to achieve an octet, consistent with other constraints on the bonding. For example, applying the standard rules to BF_3 leads to the Lewis structure

• This "deficiency" is from the point of view of simple bonding theory.

$$\begin{array}{c} : \ddot{F} : \\ | \\ : \ddot{F}-\underset{(-1)}{B}=\overset{(+1)}{\ddot{F}} \end{array}$$

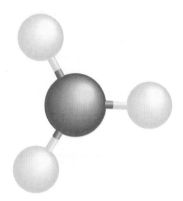

Boron trifluoride (BF$_3$) is a highly reactive gas that condenses to a liquid at $-100°C$. Its major use is in speeding up a large category of reactions involving carbon compounds.

but experimental evidence strongly indicates that there are no double bonds in BF$_3$ (fluorine never forms double bonds). Moreover, the placement of a positive formal charge on the very electronegative fluorine atom is quite undesirable. Placing only six electrons around the central (boron) atom is more nearly correct:

$$: \overset{\displaystyle ..}{F} :$$
$$|$$
$$: \overset{..}{\underset{..}{F}} — B — \overset{..}{\underset{..}{F}} :$$

Although this Lewis structure denies an octet to the boron atom, it does at least assign zero formal charges to all atoms.

● The B in BF$_3$ still "wants" an octet. It reacts vigorously with NH$_3$ to form F$_3$B—NH$_3$, in which the new bond, which uses electrons from a lone pair on the N in NH$_3$, makes its octet complete.

Case 3: Valence-Shell Expansion.

Lewis structures become more complex in the compounds of elements from the third and subsequent periods of the periodic table. Sulfur, for example, forms some compounds that are easily described by Lewis structures that satisfy the octet rule. An example is hydrogen sulfide (H$_2$S), which is analogous to water in its Lewis representation. Other sulfur compounds cannot be described in this way. In sulfur hexafluoride (SF$_6$), the central sulfur atom is bonded to six fluorine atoms. A Lewis structure can be drawn only if more than eight electrons are allowed around the sulfur atom, a process called **valence-shell expansion.** The resulting Lewis structure is written as:

$$\begin{array}{ccc} & : \overset{..}{F} : \\ : \overset{..}{\underset{..}{F}} & | & \overset{..}{\underset{..}{F}} : \\ & S & \\ : \overset{..}{\underset{..}{F}} & | & \overset{..}{\underset{..}{F}} : \\ & : \overset{..}{\underset{..}{F}} : \end{array}$$

in which the outer atoms have octets, and the central sulfur atom shares a total of 12 electrons.

The need for valence-shell expansion is signaled in the standard procedure for writing Lewis structures when the value of S calculated for the number of shared electrons is not large enough to place a bonding pair between each pair of atoms that are supposed to be bonded. In sulfur hexafluoride, for example, we have $A = 48$ electrons available and $N = 56$ needed. This means that $S = N - A = 8$ electrons. Four electron pairs are not sufficient to make even single bonds between the central sulfur atom and the six outer fluorine atoms. In this case, rule 4 (assign one bonding pair to each bond in the molecule or ion) is still followed, even though doing so uses more than S electrons. Rule 5 becomes irrelevant because there are no extra shared electrons, and rule 6 is replaced with a new rule:

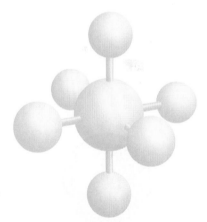

Sulfur hexafluoride (SF$_6$) is an extremely stable, dense, and unreactive gas. It is used as an insulator in high-voltage generators and switches.

> **6'.** Assign lone pairs to the *outer* atoms to give them octets. If any electrons still remain, assign them to the central atoms as lone pairs.

The effect of rule 6' is to abandon the octet rule for the central atoms but preserve it for the outer atoms.

EXAMPLE 3–9

Write a Lewis structure for the linear I_3^- (triiodide) ion.

Solution

We calculate $A = 22$ (there are seven valence electrons from each iodine atom, plus one from the overall charge of the ion). $N = 3 \times 8 = 24$, so 24 electrons are needed to give each atom its *own* octet. Hence $S = 24 - 22 = 2$. Because two electrons are not sufficient to make two different bonds, valence expansion is necessary.

We therefore place a pair of electrons as each of the two bonds. We then proceed to rule 6' and complete the octets of the two outer iodine atoms. This gives

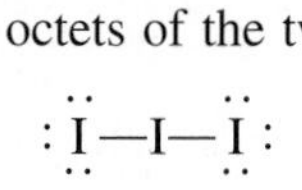

$$: \ddot{I} — I — \ddot{I} :$$

At this stage, we have used a total of 16 valence electrons. There are six left, which are then placed as lone pairs on the central iodine atom. A formal charge of -1 then resides on this atom:

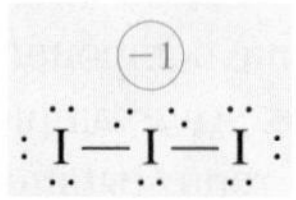

$$: \ddot{I} \overset{(-1)}{—} \ddot{I} — \ddot{I} :$$

Note the valence expansion on the central atom: it shares or owns a total of ten electrons, rather than the eight required by obedience to the octet rule.

Exercise

Write a Lewis structure for the XeF_4 molecule, with a central xenon atom.

Answer:

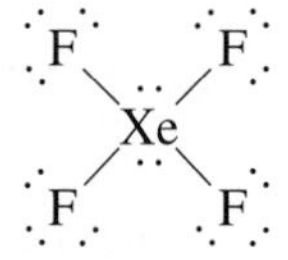

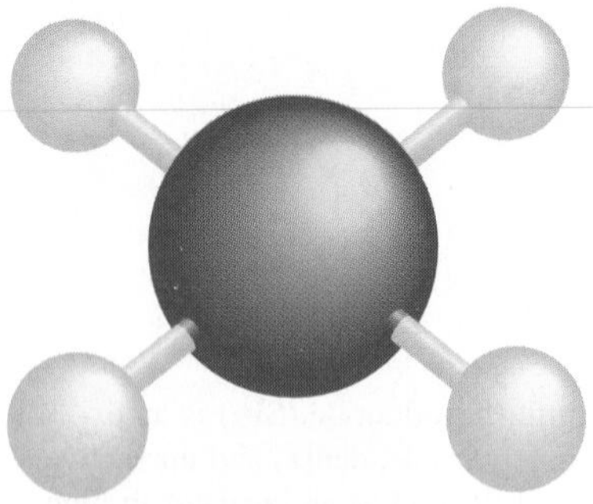

$$
\begin{array}{ccc}
: \ddot{F} & & \ddot{F} : \\
 & \diagdown \; \diagup & \\
 & Xe & \\
 & \diagup \; \diagdown & \\
: \ddot{F} & & \ddot{F} : \\
\end{array}
$$

Xenon tetrafluoride (XeF_4) is a white crystalline solid that melts at 117°C. It is one of several compounds of xenon with fluorine and other elements, whose preparation since 1962 has demonstrated that the noble gases are *not* chemically inert, despite their having complete octets of electrons.

Valence-shell expansion can also be used to avoid non-zero formal charges in certain Lewis structures. The bonding in $POCl_3$ can be described as in Example 3–5, or it can be written as:

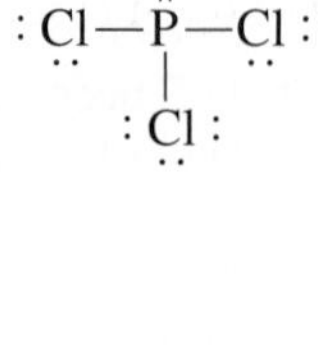

$$
\begin{array}{c}
: \ddot{O} : \\
\| \\
: \ddot{Cl} — P — \ddot{Cl} : \\
| \\
: \ddot{Cl} : \\
\end{array}
$$

in which all formal charges are zero, but the central phosphorus atom is surrounded by ten, rather than eight, valence electrons. In a similar fashion, we can write the Lewis structure of the sulfate ion, SO_4^{2-}, as

These two structures offer a choice between building up significant formal charges on some atoms and valence-shell expansion on the central sulfur atom. Which is preferable: formal charges or valence-shell expansion? It is hard to say. Valence expansion should certainly not be used for second-period elements, but it may play a role in the bonding of elements in later periods. It is probably best to regard the true structure of the sulfate ion as a resonance hybrid to which *both* of the structures shown contribute. In fact, this type of question cannot be answered very well within the simple Lewis model. Its resolution requires the more powerful methods of quantum mechanics that we introduce in Chapters 16 through 18.

• Bond lengths do not help much in determining whether the S-to-O bond is single or double because comparison requires clear-cut examples of S—O and S=O that do not exist.

3–6 THE SHAPES OF MOLECULES

Molecules are three-dimensional objects. Although Lewis structures show which atoms are connected to which and even provide a basis to estimate the distances between bonded atoms (as in the use of Table 3–3), they reveal nothing about the way the atoms are situated in space. An extension of Lewis theory is required to predict the angles defined at an atom by two or more covalently bonded neighbors.

The VSEPR Theory

The VSEPR theory (**valence shell electron-pair repulsion** theory) starts with the fundamental idea that electron pairs in the valence shell of an atom repel each other on a spherical surface formed by the underlying core of the atom. The repulsions involve both lone pairs, which are localized on the atom and are not involved in bonding, and bonding pairs, which are covalently shared with other atoms. The electron pairs occupy sites on the valence shell of each atom in such a way as to minimize their repulsions. They position themselves as far apart from each other as possible. The molecular geometry, which is defined by the relative positions of the *atoms*, is then traced from the relative locations of the electron pairs.

• Picture electron pairs as bugs who hate each other but are forced to reside on the surface of the same billiard ball.

The arrangements that best minimize repulsions naturally depend on the number of electron pairs. Figure 3–16 shows the configuration of minimum energy for two to six electron pairs around a central atom. Two electron pairs locate themselves on opposite sides of the atom in a linear arrangement, three pairs form a trigonal planar structure, four arrange themselves at the corners of a tetrahedron, five define a trigonal bipyramid, and six form an octahedron. To find which geometry applies, one determines the **steric number** (*SN*) of the central atom. Steric means "having to do with space." The steric number of an atom in a molecule can be determined by drawing the Lewis structure of the molecule and adding the number of atoms that are bonded to it and the number of lone pairs that it has. The Lewis structure

• A central atom is one that has two or more other atoms bonded to it. The steric numbers of non-central atoms are not worth calculating because they do not influence the actual positions of atoms in molecules but only the (unobservable) positions of lone pairs.

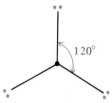

2: **Linear**

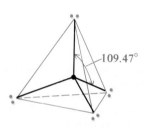

3: **Trigonal planar**

4: **Tetrahedral**

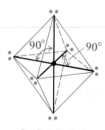

5: **Trigonal bipyramidal**

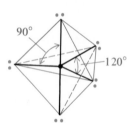

6: **Octahedral**

Figure 3–16 Diagrams and names of the geometries favored for various numbers of mutually repelling electron pairs around a central atom. The angles between the electron pairs are indicated.

need not be complete because it is not necessary to consider resonance structures and formal charges. Indeed, the Lewis structure often need not even be drawn out. All that is really required is to find out how many atoms are bonded to the central atom (the atom in question) and how many of the valence electrons on the central atom take part in covalent bonds. The remaining electrons are then parceled off into lone pairs, and the steric number is calculated

> SN = (number of atoms bonded to central atom) +
> (number of lone pairs on central atom)

EXAMPLE 3–10

Calculate steric numbers for iodine in IF_4^- and for bromine in BrO_4^-. These molecular ions have central I or Br atoms bonded to the other four atoms.

Solution

The central I^- has eight valence electrons. Each fluorine atom has seven valence electrons of its own and needs to share one of the electrons from the I^- to achieve a noble-gas configuration. Four of the I^- valence electrons thus take part in covalent bonds, leaving the remaining four to form two lone pairs. The steric number is given by

$$SN = 4 \text{ (bonded atoms)} + 2 \text{ (lone pairs)} = 6$$

In BrO_4^-, each oxygen atom needs to share two of the electrons from the Br^- in order to achieve a noble-gas configuration. Because this accounts for all eight of the Br^- valence electrons, there are no lone pairs on the central atom and

$$SN = 4 \text{ (bonded atoms)} + 0 \text{ (lone pairs)} = 4$$

Exercise

Calculate the steric numbers for the central Se in $SeOF_4$ and the central N in NO_2^-.

Answer: 5 and 3.

In determining the steric number, double-bonded or triple-bonded atoms count the same as single-bonded atoms. In the case of CO_2, for example, two double-bonded oxygen atoms are attached to the central carbon and no lone pairs are on that atom, so $SN = 2$.

The steric number is used to predict molecular geometries. For the simplest case, molecules XY_n in which there are no lone pairs on the central atom X,

$$SN = \text{number of bonded atoms} = n \quad \text{(no lone pairs)}$$

The n bonding electron pairs (and, therefore, the outer atoms) locate themselves as shown in Figure 3–16 to minimize electron-pair repulsion. Thus, CO_2 is predicted (and found experimentally) to be linear, BF_3 trigonal planar, CH_4 tetrahederal, PCl_5 trigonal bipyramidal, and SF_6 octahedral (Fig. 3–17). In all cases, lone pairs are absent from the central atom.

When lone pairs are present, the situation is more complicated. Three different types of repulsions are now possible: bonding pair against bonding pair, bonding

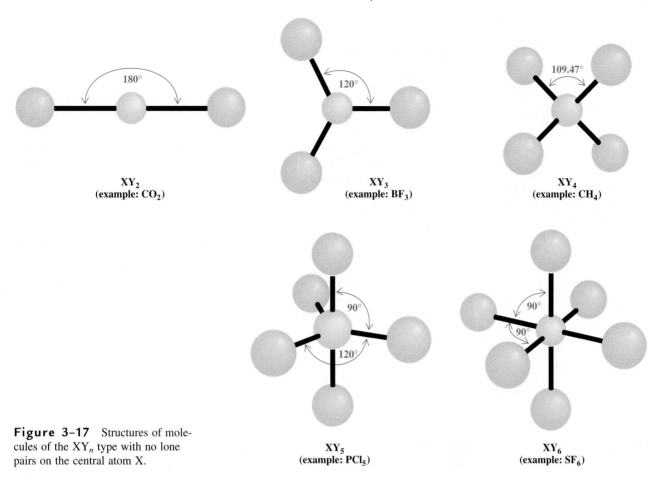

Figure 3–17 Structures of molecules of the XY_n type with no lone pairs on the central atom X.

XY_2
(example: CO_2)

XY_3
(example: BF_3)

XY_4
(example: CH_4)

XY_5
(example: PCl_5)

XY_6
(example: SF_6)

pair against lone pair, and lone pair against lone pair. Consider the molecule of ammonia (NH_3), for example. It has three bonding electron pairs and one lone pair (Fig. 3–18). The steric number is 4, and the four electron pairs arrange themselves into an approximately tetrahedral structure. It is found that lone pairs tend to occupy more space than bonding pairs (because they are held closer to the central atom). As a consequence, the angles between bonds opposite to lone pairs are reduced. The geometry of the *molecule*, as distinct from that of the electron pairs, is named for the sites occupied by actual *atoms*. In describing the molecular geometry, no consideration is given to lone pairs that may be present on the central atom, even though their presence affects the geometry. The structure of the ammonia molecule is thus predicted to be a trigonal pyramid in which the H—N—H bond angle is smaller than the tetrahedral angle of 109.5°. The observed structure is a trigonal pyramid with an H—N—H bond angle of 107.3°.

The greater steric demands of lone pairs imply that repulsive forces among valence pairs diminish in the following order:

lone pair vs. lone pair > lone pair vs. bonding pair >
bonding pair vs. bonding pair

Figure 3–18 Ammonia (NH_3) has a pyramidal structure in which the bond angles are smaller than 109.5°.

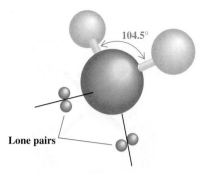

Figure 3–19 Water (H_2O) has a bent structure with a bond angle smaller than 109.5°.

EXAMPLE 3–11

Predict the structure of water using VSEPR theory.

Solution

Here the steric number is 4, with two bonds and two lone pairs on the central oxygen atom. The electron pairs around the oxygen atom define a distorted tetrahedron (Fig. 3–19).

The water molecule itself is bent, with its bond angle somewhat less than the value of 109.5° found in an undistorted tetrahedron because the two lone pairs require more space. The measured value is 104.5°.

Exercise

Determine the steric number of the sulfur atom in SO_2, and predict the molecular structure of SO_2.

Answer: The steric number of the sulfur is 3 (two bonded atoms and one lone pair). The SO_2 molecule is bent, with a bond angle slightly less than 120°.

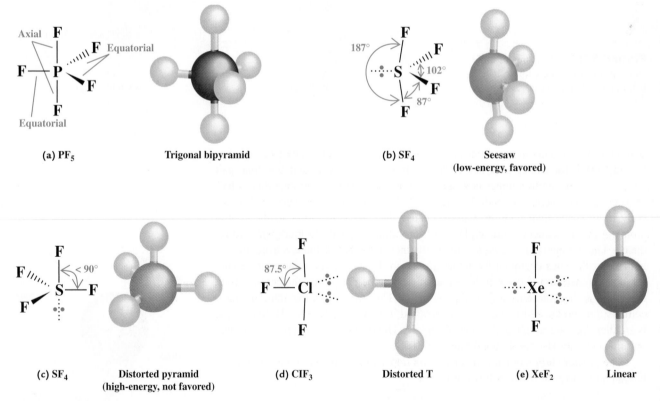

Figure 3–20 Molecules with steric number 5. All five of the molecular geometries shown here derive from the trigonal bipyramid, but only one of them includes those words in its name. Molecular geometries are named for the sites occupied by the atoms, not for the underlying distribution of the electron pairs.

Molecules of the fluorides PF_5, SF_4, ClF_3, and XeF_2 all have steric number 5 but have different numbers of lone pairs (0, 1, 2, and 3, respectively). What shapes do these molecules have? We have already mentioned that PF_5 is trigonal bipyramidal. Two of the fluorine atoms in PF_5 occupy **axial** sites (Fig. 3–20a), and the other three occupy **equatorial** sites. Because the two kinds of sites are not equivalent, there is no reason for all of the P—F bond lengths to be equal. Experiment shows that the equatorial P—F bond length is 1.534×10^{-10} m, shorter than the axial P—F lengths, which are 1.577×10^{-10} m.

Sulfur tetrafluoride has four bonded atoms and one lone pair. Does the lone pair occupy an axial or an equatorial site? In VSEPR theory, electron pairs repel each other more strongly when they form a 90° angle with respect to the central atom than when the angle is larger. A single lone pair therefore finds a position that minimizes the number of 90° repulsions it has with bonding electron pairs. It occupies an equatorial position with two 90° repulsions (Fig. 3–20b) rather than an axial position with three 90° repulsions (Fig. 3–20c). The direction of axial S—F bonds is skewed slightly away from the lone pair, so the molecular structure of SF_4 is a distorted seesaw. A second lone pair (for example, in ClF_3) also takes an equatorial position, leading to a distorted T-shaped molecular structure (Fig. 3–20d). A third lone pair (for example, in XeF_2 or I_3^-) occupies the third equatorial position, and the molecular geometry is linear (Fig. 3–20e). To summarize this discussion:

> Lone pairs occupy equatorial positions in preference to axial positions in structures of steric number 5.

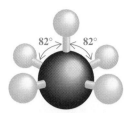

Figure 3–21 The molecular structure of iodine pentafluoride (IF_5). The deviations of the F—I—F bond angles from the nominal value of 90° are caused by the spatial requirements of the lone pair of electrons (not shown) at the bottom.

EXAMPLE 3–12

Predict the geometry of the following molecules and ions:

(a) ClO_3^+ (c) SiH_4
(b) ClO_2^+ (d) IF_5

Solution

(a) All of the valence electrons on the central chlorine atom (six, because of the net positive charge on the ion) take part in bonds to the surrounding three oxygen atoms; there are no lone pairs. The steric number of the central Cl atom is 3. The central Cl atom should be surrounded by the three oxygen atoms in a trigonal planar structure.

(b) The central Cl atom in this ion has a steric number of 3; it has two bonded atoms and a single lone pair. The prediction is a bent species with a bond angle somewhat less than 120°.

(c) The central Si atom has a steric number of 4 and no lone pairs. The molecular geometry should consist of the silicon atom surrounded by a regular tetrahedron of hydrogen atoms.

(d) Iodine has seven valence electrons, five of which are shared in bonding pairs with fluorine atoms. This leaves two electrons to form a lone pair, so the steric number of the I is 5 (bonded atoms) + 1 (lone pair) = 6. The structure is based on the octahedron of electron pairs from Figure 3–16, with five fluorine atoms and one lone pair. The lone pair can be placed on any one of the six equivalent sites and causes the four more closely neighboring fluorine atoms to bend away from it toward the more distant fifth fluorine atom, giving the distorted structure shown in Figure 3–21.

> **Exercise**
> The ionic compound $(S_2N^+)(AsF_6^-)$ was recently isolated and studied. The geometry of the molecular anion and of the molecular cation both conform to the predictions of the VSEPR theory. What are those geometries?
>
> **Answer:**   The cation is linear and the anion is octahedral.

The VSEPR theory is a remarkably powerful model for predicting the geometries and approximate bond angles of molecules that have a central atom. In fact, in many cases it is more successful than theories that are much more elaborate or that require extensive calculation.

Dipole Moments

Two bonded atoms share electrons unequally whenever they differ in electronegativity. Thus, in hydrogen chloride (HCl), the highly electronegative Cl atom attracts the bonding electrons more strongly than does the H atom. The Cl end of the HCl molecule consequently carries a slight negative electric charge and the H end car-

---

CHEMISTRY IN YOUR LIFE

Dry Ice Versus Wet Ice

Perhaps you have amused yourself at picnics by throwing the blocks of solid carbon dioxide (Dry Ice) that kept the ice cream cold into a pond or lake and watching the water "boil" up and spew clouds of white vapor. Dipole moments help explain what was going on.

As VSEPR theory predicts, molecules of carbon dioxide are linear. This means (Fig. 3–22a) that the bond dipoles in O=C=O cancel out. The molecule is nonpolar. Nonpolar molecules attract each other weakly. Consequently, substances composed of them are hard to condense to liquids or solids. Carbon dioxide is gaseous under room conditions and requires considerable cooling to solidify. Only at $-78°C$ do the weak attractions between its nonpolar molecules take hold sufficiently for Dry Ice to form. In contrast, the bend in water molecules means that the two O—H bond dipoles partially reinforce each other (Fig. 3–22c). The dipolar molecules in water attract their neighbors rather strongly. Water freezes at $0°C$, a much higher temperature than $-78°C$.

Dry Ice is far colder than the water in a pond. When it is thrown in, the surrounding water heats it rapidly. The solid CO_2 converts to gaseous CO_2, which churns to the surface, where it is still cold enough to cause the moisture in the air to condense as trails of white fog. Soon the activity abates, not because the Dry Ice is gone, but be-

Figure 3–A   Bubbles of gaseous carbon dioxide (CO_2) erupt when Dry Ice is put into water.

cause it has frozen an insulating layer of wet ice (water ice) all around it.

If H—O with H attached were somehow linear, it would be nonpolar. Its melting point would be far lower than $0°C$ and probably lower than $-78°C$. Linear water could not support life as we known it, but some other kind of life might then enjoy watching hot Dry Ice melt cold H—O—H away.

ries a positive charge of equal magnitude. The HCl molecule is an electric **dipole.** Dipolar (or **polar**) molecules are said to possess a **dipole moment** (symbolized μ) because they experience a torque in electric fields. They tend to rotate to align their negative ends as close as possible to the positive side of the field and their positive ends as close as possible to the negative side of the field. Hydrogen chloride is diatomic, but polyatomic molecules also often possess dipole moments. This molecular property greatly influences the bulk properties of substances because dipolar molecules interact strongly and directionally with each other.

* Torques are called moments or turning-moments in physics.

We can predict the presence or absence of a dipole moment in a molecule. First we draw the Lewis structure and figure out the molecular geometry using VSEPR theory. Next, we assign a *bond dipole* (shown as an arrow) to each bond in the molecule, with the arrow pointing from the more electronegative to the more electropositive atom and the length corresponding to the magnitude of the bond dipole. Only identical bonds (those having the same atoms at each end) have bond dipole arrows of identical length. Finally, the total dipole moment is obtained by adding the arrows of the bond dipoles vectorially, that is, by placing the arrows head to tail (without rotating them) and examining the net arrow that results.

In CO_2, for example, each C—O has a bond dipole pointing from O to C (Fig. 3–22a). Because the bonds are identical, the bond dipoles are equal in magnitude. They point in opposite directions, so the total molecular dipole moment vanishes. In OCS, which is also linear with a central carbon atom, the C—O and C—S bond dipole moments also point in opposite directions but have different magnitudes, leaving a net non-zero dipole moment for the OCS molecule (Fig. 3–22b). The water molecule has an overall dipole moment (Fig. 3–22c) because the two bond dipole moments, although equal in magnitude, do not directly oppose each other. They add vectorially to give a non-zero molecular dipole moment. The more symmetric molecule CCl_4, on the other hand, has no net dipole moment (Fig. 3–22d). Even though each of the four C—Cl bonds does have a dipole moment (pointing from the four corners of a tetrahedron), when the four arrows are placed head to tail, their sum is

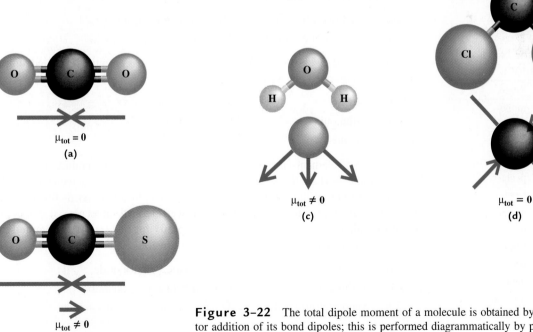

Figure 3–22   The total dipole moment of a molecule is obtained by vector addition of its bond dipoles; this is performed diagrammatically by placing them head to tail. (a) CO_2, (b) OCS, (c) H_2O, (d) CCl_4.

zero. Molecules such as H_2O and OCS with non-zero dipole moments are polar, whereas those such as CO_2 and CCl_4 with no net dipole moment are nonpolar, even though they contain polar bonds.

EXAMPLE 3–13

Predict whether the molecules of NH_3 and SF_6 have dipole moments.

Solution

The NH_3 molecule has a dipole moment because the three N—H bond dipoles add vectorially to give a net dipole pointing downward from the nitrogen atom to the base of the pyramid defined by the three H atoms. The octahedral SF_6 molecule does not have a dipole moment because each S—F bond dipole is balanced by an equal bond dipole pointing in the opposite direction on the other side of the molecule.

Exercise

Predict whether the molecules of BF_3 and SO_2 have dipole moments.

Answer: BF_3 does not have a dipole moment; SO_2 does have one.

3–7   NAMING BINARY MOLECULAR COMPOUNDS

How do we name molecular compounds? If a pair of elements forms only one compound, we use the name of the more electropositive element, which generally appears first in the chemical formula, followed by the second element with the suffix *-ide* added to the first portion of its name. This is analogous to the naming of ionic compounds. Just as we call NaBr "sodium bromide," so we give the following names to typical molecular compounds:

HBr	hydrogen bromide	H_2S	hydrogen sulfide
$BeCl_2$	beryllium chloride	BN	boron nitride

A number of well-established nonsystematic names that continue to be used simply need to be memorized. These include:

H_2O	water	N_2H_4	hydrazine
NH_3	ammonia	PH_3	phosphine
AsH_3	arsine		

• Even the strictest chemist does not call water "dihydrogen oxide"!

If a pair of elements forms more than one molecular compound, then Greek prefixes (Table 3–4) are used to specify the number of atoms of each element in the molecular formula of the compound (*di-* for two, *tri-* for three, *tetra-* for four, and so forth). If the compound is a solid without well-defined molecules, it is the empirical formula that is named in this way. The prefix for one (*mono-*) is omitted except when its presence is necessary to avoid confusion.

SO_2	sulfur dioxide	IF_5	iodine pentafluoride
SO_3	sulfur trioxide	As_4O_6	tetraarsenic hexaoxide
XeF_4	xenon tetrafluoride	P_4O_{10}	tetraphosphorus decaoxide

Table 3–4
Prefixes Used for Naming Binary Compounds

Number	Prefix
1	mono-
2	di-
3	tri-
4	tetra-
5	penta-
6	hexa-
7	hepta-
8	octa-
9	nona-
10	deca-
11	undeca-
12	dodeca-

EXAMPLE 3-14

Name the following binary molecular compounds between non-metals:

(a) Cl_2O_7 (b) SiO_2 (c) NF_3

Solution

Using the prefixes from Table 3–4 gives the following:

(a)   Dichlorine heptaoxide
(b)   Silicon dioxide
(c)   Nitrogen trifluoride

Exercise

Write chemical formulas for the following binary compounds: (a) dichlorine hexaoxide. (b) oxygen difluoride. (c) tetraiodine dodecaoxide.

Answer: (a) Cl_2O_6. (b) OF_2. (c) I_4O_{12}.

The need for such a systematic way of naming compounds is illustrated by the many oxides of nitrogen, which are shown in Figure 3–23.

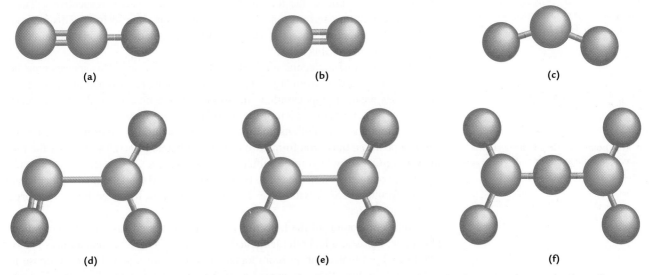

(a)   (b)   (c)

(d)   (e)   (f)

Figure 3–23   (a) Dinitrogen monoxide (N_2O, also called "nitrous oxide") is a colorless and rather unreactive gas that has a linear structure. It is used as an anesthetic and is referred to as "laughing gas."
(b) Nitrogen monoxide (NO, also called "nitric oxide") is another colorless gas. It has an odd number of electrons, and when warmed at high pressure it reacts to form N_2O and NO_2.
(c) Nitrogen dioxide (NO_2) is a brown gas formed when NO reacts with oxygen. Like NO, it is an odd-electron compound. Together with NO, NO_2 is a significant factor in urban air pollution.
(d) Dinitrogen trioxide (N_2O_3) exists in pure form only in the solid state (in which it is a very pale blue) at temperatures below about 100°C. At higher temperatures it breaks down extensively to NO and NO_2.
(e) Dinitrogen tetraoxide (N_2O_4) is a colorless compound formed when pairs of NO_2 molecules link up (a process called dimerization) at low temperature.
(f) Dinitrogen pentaoxide (N_2O_5) is a white solid made up of nitronium (NO_2^+) and nitrate (NO_3^-) ions. The symmetric N_2O_5 molecule shown here exists in the vapor, although it dissociates fairly rapidly to form nitrogen dioxide and oxygen.

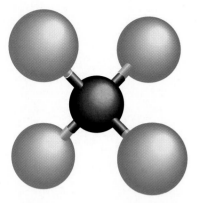

Figure 3-24 The tetrachlorides of titanium ($TiCl_4$), lead ($PbCl_4$), and carbon (CCl_4) have tetrahedral structures, although the central atom comes from three quite different places in the periodic table. All are liquids at room temperature, with very similar freezing points: $TiCl_4$ ($-24°C$), $PbCl_4$ ($-15°C$), CCl_4 ($-23°C$). Lead tetrachloride is the least stable of the three, decomposing to $PbCl_2$ and Cl_2 above 50°C. Carbon tetrachloride boils at 76.5°C, and $TiCl_4$ boils at 136.5°C.

• This point has already been emphasized for main-group metals in Section 3–3.

3-8   ELEMENTS FORMING MORE THAN ONE ION

The transition elements and the metals found late in Groups III, IV, and V differ from the Groups I and II metals discussed in Section 3–3 in that they often form several stable ions in compounds and in solution. Although calcium compounds never contain Ca^{3+} ions (always Ca^{2+}), the element iron forms both Fe^{2+} and Fe^{3+} ions, and thallium forms both Tl^+ and Tl^{3+} ions. Furthermore, the transition elements differ from the main-group elements (see Section 3–4) in that no simple octet rule allows one to write Lewis structures for their compounds. The additional ions of these metals offer rich possibilities for chemical combination, but their existence is not predicted by simple ideas concerning the periodic table.

Naming Compounds of Metals Having More than One Stable Ion

When a metal forms ions of more than one charge, we distinguish them by placing a Roman numeral in parentheses immediately after the name of the metal (with no space) to indicate the ionic charge:

Cu^+	copper(I)	Fe^{2+}	iron(II)	Sn^{2+}	tin(II)
Cu^{2+}	copper(II)	Fe^{3+}	iron(III)	Sn^{4+}	tin(IV)

An earlier method for distinguishing between pairs of ions such as Fe^{2+} and Fe^{3+} used the suffixes *-ous* and *-ic*, added to the root of the name (often the Latin name) of the metal, to indicate the ions of lower and higher charge, respectively. Thus, Fe^{2+} was called the "ferrous ion," Fe^{3+} the "ferric ion," and so forth. This method, although still sometimes used, is too limited for systematic nomenclature and does not appear further in this book.

One interesting molecular ion is Hg_2^{2+}, the mercury(I) ion. This species must be carefully distinguished from Hg^{2+}, the mercury(II) ion. The Roman numeral "I" in parentheses means in this case that the average charge on each of the two mercury atoms is +1. A monatomic ion of the form Hg^+ is not found.

In compounds of the transition metals, charge neutrality determines the chemical composition; thus, iron forms two different sulfates. Iron(III) sulfate has the formula $Fe_2(SO_4)_3$ in which three sulfate ions (total charge $3 \times (-2) = -6$) balance the charge on the two iron ions ($2 \times (+3) = +6$). Iron(II) sulfate, on the other hand, has the formula $FeSO_4$, with the +2 charge on the iron balanced by a -2 charge on the sulfate ion.

Some compounds of the transition elements have bonds that are more covalent than ionic. For example, both titanium(IV) chloride ($TiCl_4$) and vanadium(IV) chloride (VCl_4) are molecular; at room temperature they are liquids that decompose in water. Despite this, they are named as if they contained stable Ti^{4+} or V^{4+} ions, that is, by the same system that gives the names iron(II) chloride, iron(III) chloride, and copper(II) chloride, which are solids with more or less typical ionic properties. Covalent character occurs in compounds of the heavier elements of Groups III, IV, and V as well as the transition metals. Lead(II) chloride, for example, is a white ionic solid that dissolves (sparingly) in water to give Pb^{2+} and Cl^- ions. Lead(IV) chloride, on the other hand, is an oily yellow liquid that decomposes in water. Its bonds are largely covalent, and it somewhat resembles the molecular compound carbon tetrachloride (CCl_4) in its physical properties (Fig. 3–24). Roman numerals used in naming covalent compounds never state the *true* charge on the atom in question but rather refer to what that charge would be *if* the compound were ionic. They report what is called the **oxidation state** or **oxidation number** of the element in the compound. The oxidation state coincides with the actual ionic charge on a metal atom only when the charge is low; thus, the compound Mn_2O_7 has the name manganese(VII) oxide, but the $+7$ oxidation state of manganese in this compound does not indicate that this is an ionic compound of Mn^{7+} and O^{2-} ions. Rather, there is substantial covalent character to the bonds in this compound. Much important chemistry involves changes in the oxidation states of elements. We discuss this more extensively in Chapters 4 and 13.

EXAMPLE 3–15

Write chemical formulas for each of the following compounds:
(a) Osmium(VI) fluoride
(b) Vanadium(V) oxide
(c) Copper(I) sulfide

Solution

(a) Each fluoride ion carries a -1 charge, so there must be six such ions to balance the hypothetical $+6$ charge on the osmium. The formula is thus OsF_6.
(b) Each oxide ion has a -2 charge, and the vanadium ion has a hypothetical $+5$ charge. If each vanadium atom were actually to give up five electrons, they could be transferred to $\frac{5}{2} = 2\frac{1}{2}$ oxygen atoms to form oxide ions. To avoid fractions, we can multiply by 2 and say that *two* vanadium ions give up *ten* electrons, transferring them to *five* oxygen atoms. The formula is therefore V_2O_5.
(c) The sulfide ion has a -2 charge. This must be balanced by two copper(I) ions (Cu^+), so the chemical formula is Cu_2S.

Exercise

Write the systematic names corresponding to the following chemical formulas:
(a) CrO_3. (b) $TlCl_3$. (c) Mn_3N_2.

Answer: (a) Chromium(VI) oxide. (b) Thallium(III) chloride. (c) Manganese(II) nitride.

Coordination Complexes

Many transition-metal ions form **coordination complexes** in solution or in the solid state. These species consist of a metal ion surrounded by a group of anions or neutral molecules, called **ligands.** The interaction involves the sharing by the metal ion of a lone pair on each ligand molecule; the sharing creates a partially covalent bond with that ligand. Such complexes frequently have intense colors. When exposed to gaseous ammonia, greenish white crystals of copper sulfate $(CuSO_4)$ give a deep blue crystalline solid with the chemical formula $Cu(NH_3)_4SO_4$. The anions in the solid are still sulfate ions (SO_4^{2-}), but the cations are now coordination complexes of the central Cu^{2+} ion with four ammonia molecules, indicated as $[Cu(NH_3)_4]^{2+}$. The ammonia molecules coordinate to the copper ion through their lone-pair electrons, as is shown in the Lewis structure

• Coordination complexes are discussed more extensively in Chapter 19. Systematic names for these complexes are presented there.

$$H_3N : \overset{\overset{\displaystyle NH_3}{\cdot\cdot}}{\underset{\underset{\displaystyle NH_3}{\cdot\cdot}}{\left(Cu^{2+}\right)}} : NH_3$$

Figure 3–25 The very different colors of solutions of $FeCl_3$ (pale yellow) and $K_3[Fe(CN)_6]$ (reddish brown) show the dramatic effect that complex ion formation can have on a physical property.

The octet rule applies to the ligands but not to the central metal ion in coordination complexes. Also, VSEPR theory encounters difficulty in predicting the geometry around the central metal ions in coordination complexes, although it works for the ligands. Additional ideas are needed to explain the bonding and molecular geometry in coordination complexes (see Chapter 19).

When solid $Cu(NH_3)_4SO_4$ is dissolved in water, the deep blue color remains. This is evidence that the complex persists in water, because when ordinary $CuSO_4$ (without ammonia ligands) is dissolved in water, a much paler blue color results. Similar color changes are seen in forming other complex ions. The soluble ionic compound $Fe(NO_3)_3$ has a pale yellow color in solution. Upon addition of colorless potassium cyanide, the complex ion $[Fe(CN)_6]^{3-}$ forms and makes the solution reddish brown (Fig. 3–25). The complex responsible for this color consists of an Fe^{3+} ion surrounded by six CN^- ligands, coordinated through the lone-pair electrons on the carbon atoms.

Charge, Formal Charge, and Oxidation State

Before closing this chapter, let us look back at the meanings and roles of the different measures of charge on ions and on atoms within molecules. An isolated ion in free space has a well-defined and measurable electric charge; this charge can be positive or negative, but it is always a whole-number multiple of the charge carried by an electron. This simple behavior carries over to ionic compounds formed between elements in the early groups of the periodic table and those in late groups: NaCl is very well described as an ionic compound composed of Na^+ and Cl^- ions.

Other compounds are not purely ionic but involve electron sharing between atoms. In compounds with covalent bonds, it is not possible to associate a measurable charge with a particular atom. In a purely artificial way, we define the formal charge to be the charge an atom would have if electrons were divided exactly equally between the two atoms that share them. In other words, when the bonding is covalent, we simply assume that it is *perfectly* covalent (equal sharing). If the result is a Lewis structure with large positive or negative formal charges on atoms, then something is wrong. The artificial assignment of formal charge is a check for consistency because, in a true covalent compound, the atoms should not be highly charged. If they turn out to be that way in our representation, we either look for a new Lewis structure or predict that the compound will not be stable. Formal charge also allows conclusions about the nature of bonding. A C—N single bond, for example, has a different length, bond strength, and reactivity than a C=N double bond, so the choice of the best Lewis structure (based on formal charge) is often a very useful guide to molecular properties.

Oxidation state is just as arbitrary as formal charge, but it is defined in a completely different way and has a different purpose. To determine oxidation state, we suppose that the compound is purely *ionic* and ask what charge results on the different atoms. If the compound *is* ionic, then oxidation state coincides with the true charge, but if there is covalent character, this no longer is true. The concept of oxidation state is useful in understanding the chemistry of transition-metal elements that form a variety of compounds of mixed ionic and covalent character. It leads to predictions of chemical and physical properties; for example, compounds with elements in high oxidation states tend to have lower boiling points than compounds of the same elements in low oxidation states. The concept of oxidation state is also useful in naming compounds.

If every compound could be classed as purely ionic or purely covalent, then neither formal charge nor oxidation state would be necessary. In purely covalent molecules, formal charges would always be zero. In purely ionic compounds, oxidation state would always correspond to ionic charge and there would be no point in defining it separately. The beauty of chemistry, however, is that borderline situations abound. Most bonds have partial ionic and partial covalent character. Formal charge and oxidation state, although they are arbitrary concepts, help to describe and predict real chemical behavior.

SUMMARY

3–1 Elements are classified as metals, non-metals, or semi-metals and also fall naturally into **groups** based on similarities in chemical and physical properties. The groups include (among others) **the alkali metals, the alkaline earth metals,** the **chalcogens,** and the **halogens.**

3–2 The modern periodic law states that the chemical and physical properties of the elements are periodic functions of their atomic numbers. The periodic table displays elements in groups (columns) and **periods** (rows). The eight **main groups** (designated I through VIII) are the **representative elements.** An additional 10 groups of **transition** elements (all metals) include the **actinides** and **lanthanides.** The **electronegativity** (the power of an atom when in chemical combination to attract electrons to itself) is a periodic property; the relative electronegativity of two bonded atoms determines the type of bonding (ionic versus covalent) between them.

3–3 In ionic compounds **(salts),** metals (electropositive elements) lose electrons to form **cations,** and non-metals (electronegative elements) gain electrons to form **anions;** the two kinds of ions then interact. Lewis dot symbols show the outer (or **valence**) electrons of a species as dots. Atoms tend to lose or gain electrons to achieve an **octet** of valence electrons (doublet for hydrogen). Because transfers of electrons preserve **charge neutrality,** the formulas of ionic compounds can be predicted. The naming of simple ionic compounds consists of pairing the name of the cation and anion in that order; the names of the ions derive from the names of the elements they contain.

3–4 Non-metals combine by sharing valence electrons in **covalent** bonds. Compounds that contain only covalent bonds are **covalent compounds.** A single bond is a pair of shared electrons; bonds may be single or multiple. Valence electrons not involved in bonding appear in **lone pairs. Formal charges** associated with atoms in **Lewis structures** can help one to decide whether a proposed structure is meaningful.

3-5 Drawing a Lewis structure requires first a plan of the connections of the atoms. Then count the valence electrons; join all covalently bonded atoms by single bonds; subtract the number of electrons used in this way from the total count; and distribute the remaining electrons in lone pairs and multiple bonds to give each atom a valence octet (a doublet for hydrogen). For some compounds and molecular ions, covalent bonding can be rationalized by assuming that two or more Lewis structures contribute as **resonance hybrids.** Exceptions to the octet rule include odd-electron molecules (an odd number of valence electrons), octet-deficient molecules (fewer than eight electrons around a non-H atom), and molecules showing valence-shell expansion (more than eight electrons around an atom).

3-6 Molecular geometries can be predicted by the **valence shell electron-pair repulsion (VSEPR)** theory: bonded atoms and lone pairs around a central atom position themselves as far from one another as possible; the steric number of an atom equals the total number of such atoms and pairs; a different three-dimensional geometry is associated with each steric number. The geometries for different steric numbers are 2, linear; 3, trigonal planar; 4, tetrahedral; 5, trigonal bipyramidal; 6, octahedral, each giving different sets of angles between the electron pairs.

3-7 The names of binary molecular compounds derive from the names of the elements, with prefixes to indicate the number of atoms of each element and the suffix *-ide* added to the name of the more electronegative element.

3-8 The transition elements tend to exhibit mixed ionic and covalent behavior. Most of them occur in more than one **oxidation state,** and compounds of the higher oxidation states have more covalent character than those of the lower. Many transition-metal ions form **coordination complexes** with molecules or ions that serve as **ligands** by sharing lone pairs with the metal ion. The resulting ligand-metal bonds are partially covalent.

PROBLEMS

Note: Answers to blue-numbered problems are given in Appendix F. Problems that are more challenging are indicated with asterisks.

The Periodic Table

1. Hydrogen forms compounds with C, N, O, and F that have the formulas CH_4, NH_3, H_2O, and HF. Predict the formulas of the compounds hydrogen forms with P, Cl, Si, and S.

2. Carbon monoxide reacts with many metals to give metal carbonyls. Examples are chromium carbonyl ($Cr(CO)_6$), manganese carbonyl ($Mn_2(CO)_{10}$), iron carbonyl ($Fe(CO)_5$), and cobalt carbonyl ($Co_2(CO)_8$). Use the periodic table to predict the molecular formulas of tungsten carbonyl, molybdenum carbonyl, ruthenium carbonyl, and rhenium carbonyl.

3. Based on the periodic table, write the formula expected for binary compounds between the following elements:
   (a) Potassium and polonium
   (b) Strontium and chlorine
   (c) Barium and sulfur

4. Based on the periodic table, write the formula expected for binary compounds between the following elements:
   (a) Magnesium and oxygen
   (b) Cesium and chlorine
   (c) Aluminum and bromine

5. Plot the melting points (on the vertical axis) of the elements Li through Ca against their atomic numbers (on the horizontal axis). The melting points in °C: Li(181), Be(1283), B(2300), C(3550), N(−210), O(−218), F(−220), Ne(−249), Na(98), Mg(649), Al(660), Si(1410), P(44), S(113), Cl(−101), Ar(−189), K(64), Ca(839). Summarize the periodic pattern in a brief statement.

6. Plot the melting points (on the vertical axis) of the elements Li through Ca against their atomic *masses* (on the horizon-

tal axis). Refer to problem 5 for data. Discuss any violations of the periodic pattern in this plot.

7. A certain element M is a main-group metal that reacts with chlorine to give a compound with chemical formula MCl_2 and with oxygen to give the compound MO.
   (a) To which group in the periodic table does element M belong?
   (b) The chloride contains 44.7% chlorine by mass. Name the element M.

8. A certain nonmetallic element Q reacts with potassium to give a compound with chemical formula K_2Q and with calcium to give the compound CaQ.
   (a) To which group in the periodic table does element Q belong?
   (b) The compound K_2Q contains 70.9% potassium by mass. Name the element Q.

9. (a) Name six elements that are gases at room temperature.
   (b) Do these elements cluster in some part of the periodic table? Explain.

10. Metals are found on the left side or in the lower part of the periodic table. Most are solids at room temperature. Name one metallic element that is a liquid at room temperature.

11. Before the element scandium was discovered in 1879, it was known as "eka-boron." Predict the properties of scandium from averages of the corresponding properties of its neighboring elements in the periodic table, as listed below.

Element	Symbol	Melting Point (°C)	Boiling Point (°C)	Density (g cm^{-3})
Calcium	Ca	839	1484	1.55
Titanium	Ti	1660	3287	4.50
Scandium	Sc	?	?	?

12. The element technetium (Tc) is not found in nature but has been produced artificially through nuclear reactions. Use the data for several neighboring elements (given in the table below) to estimate the melting point, boiling point, and density of technetium.

Element	Symbol	Melting Point (°C)	Boiling Point (°C)	Density (g cm^{-3})
Manganese	Mn	1244	1962	7.2
Molybdenum	Mo	2610	5560	10.2
Rhenium	Re	3180	5627	20.5
Ruthenium	Ru	2310	3900	12.3

13. Use the group structure of the periodic table to predict the empirical formulas for the binary compounds that hydrogen forms with each of the following elements: antimony, bromine, tin, and selenium.

14. Use the group structure of the periodic table to predict the empirical formulas for the binary compounds that hydrogen forms with each of the following elements: germanium, fluorine, tellurium, and bismuth.

Ions and Ionic Compounds

15. (See Example 3–1.) State the number of core electrons and valence electrons in an atom of each of the following elements. Draw a Lewis dot symbol for each of these atoms:
    (a) P (b) I (c) Al (d) Sr

16. (See Example 3–1.) State the number of core electrons and valence electrons in an atom of each of the following elements. Draw a Lewis dot symbol for each of these atoms.
    (a) Xe (b) Sn (c) B (d) Bi

17. Give the Lewis dot symbols for Kr, Te^{2-}, S^+, and Mg^+.

18. Give the Lewis dot symbols for O^-, Ca^+, K, and Cs^-.

19. For each of the atoms or ions in problem 17, give the total number of electrons, the number of valence electrons, and the number of core electrons.

20. For each of the atoms or ions in problem 18, give the total number of electrons, the number of valence electrons, and the number of core electrons.

21. (See Example 3–2.) Give the name and formula of a compound involving only the elements in each pair listed below. Write Lewis symbols for the elements both before and after chemical combination.
    (a) Strontium and selenium   (d) Chlorine and gallium
    (b) Iodine and rubidium   (e) Francium and oxygen
    (c) Sodium and sulfur

22. (See Example 3–2.) Give the name and formula of a compound involving only the elements in each pair listed below. Write Lewis symbols for the elements both before and after chemical combination.
    (a) Lithium and tellurium
    (b) Bromine and barium
    (c) Magnesium and astatine
    (d) Aluminum and chlorine
    (e) Cesium and sulfur

23. Give systematic names to the following compounds:
    (a) Al_2O_3 (d) $Ca(NO_3)_2$
    (b) Rb_2Se (e) Cs_2SO_4
    (c) $(NH_4)_2S$ (f) $KHCO_3$

24. Give systematic names to the following compounds:
    (a) KNO_2 (d) NaH_2PO_4
    (b) $Sr(MnO_4)_2$ (e) $BaCl_2$
    (c) $MgCr_2O_7$ (f) $NaClO_3$

25. (See Example 3–3.) Write the chemical formulas for the following compounds:
    (a) Silver cyanide
    (b) Calcium hypochlorite
    (c) Potassium chromate
    (d) Gallium oxide
    (e) Potassium superoxide
    (f) Barium hydrogen carbonate

26. (See Example 3–3.) Write the chemical formulas for the following compounds:
    (a) Cesium sulfite
    (b) Strontium thiocyanate
    (c) Lithium hydride

(d) Sodium peroxide

(e) Ammonium dichromate

(f) Rubidium hydrogen sulfate

27. Trisodium phosphate (TSP) is a heavy-duty cleaning agent. Write its chemical formula. What is the systematic name for this ionic compound?

28. Monoammonium phosphate is a compound made up of NH_4^+ and $H_2PO_4^-$ ions that is employed as a flame retardant (its use for this purpose was first suggested by Gay-Lussac in 1821). Write its chemical formula. What is the systematic chemical name of this compound?

29. The phosphate ion has the formula PO_4^{3-}, and K_3PO_4 is potassium phosphate. Name the compound KH_2PO_4.

30. The tartrate anion has the formula $C_4H_4O_6^{2-}$, and $Na_2C_4H_4O_6$ is sodium tartrate. Name the compound $NaHC_4H_4O_6$.

Covalent Bonding and Lewis Structures

31. By counting electrons, determine what is wrong with each of the following Lewis structures:

    (a) IF_2 (b) OCN (c) CH_2O

32. By counting electrons, determine what is wrong with each of the following Lewis structures:

    (a) CO_2^- (b) $HBCl_3$ (c) C_2H_2

33. (See Example 3–4.) Assign formal charges to all atoms in the following Lewis structures:

    (a) SO_4^{2-} (b) $S_2O_3^{2-}$

    (c) SbF_3 (d) SCN^-

34. (See Example 3–4.) Assign formal charges to all atoms in the following Lewis structures

    (a) ClO_4^- (b) SO_2

    (c) BrO_2^- (d) NO_3^-

35. (See Example 3–4.) Determine the formal charge on all the atoms in the following Lewis structures:

    $$H-N=O \quad \text{and} \quad H-O=N$$

    Which one would probably be favored as the structure of the molecule HNO?

36. (See Example 3–4.) Determine the formal charge on all the atoms in the following Lewis structures:

    $$:Cl-Cl-O: \quad \text{and} \quad :Cl-O-Cl:$$

    Which one would probably be favored as the structure of the molecule Cl_2O?

37. In each of the following Lewis structures, Z represents a main-group element. Name the group to which Z belongs in each case, and give an example of such a compound or ion that actually exists.

    (a)   (b)

    (c)   (d)

38. In each of the following Lewis structures, Z represents a main-group element. Name the group to which Z belongs in each case, and give an example of such a compound or ion that actually exists.

(a)   (b)

$$[:C\equiv Z:]^-$$

$$\left[\begin{array}{c}:\ddot{O}:\\|\\:\ddot{O}-Z-\ddot{O}:\\|\\:\ddot{O}:\end{array}\right]^-$$

(c)   (d)

$$\left[\begin{array}{c}:\ddot{O}:\\|\\:\ddot{O}-Z-\ddot{O}:\end{array}\right]^{2-}$$ $$\begin{array}{ccc}&\ddot{} & \ddot{}\\ H-Z-Z-H\\ &|&|\\ &H&H\end{array}$$

Drawing Lewis Structures

39. Draw Lewis electron-dot structures for the following species:
    (a) H_2S
    (b) AsH_3
    (c) HOCl
    (d) $CNCN^{2-}$ (atoms arranged in a row, as written)

40. Draw Lewis electron-dot structures for the following species:
    (a) Methane   (c) Phosphorus trichloride
    (b) Carbon dioxide   (d) Perchlorate ion

41. Draw a Lewis structure for OF_2, with O as the central atom.

42. Draw a Lewis structure for $ClBr_2^-$, with Cl as the central atom.

43. Urea is an important chemical fertilizer, and it is also excreted in urine. Its chemical formula is ($H_2NCO(NH_2)$), and the carbon atom is bonded to both nitrogen atoms and to the oxygen atom. Draw its Lewis structure, and use Table 3–3 to estimate its bond lengths.

44. Acetic acid is the active ingredient of vinegar. Its chemical formula is CH_3COOH, and the second carbon atom is bonded to the first carbon atom and to both oxygen atoms. Draw its Lewis structure, and use Table 3–3 to estimate its bond lengths.

45. (See Examples 3–5 through 3–7.) Draw the Lewis structure for each of the following ions, and assign formal charges to all atoms:
    (a) ClO_3^- (b) $AlCl_4^-$ (c) XeF^+

46. (See Examples 3–5 through 3–7.) Draw the Lewis structure for each of the following ions, and assign formal charges to all atoms:
    (a) OCN^- (b) PH_4^+ (c) $PO_2Cl_2^-$ (central P atom)

47. (See Example 3–8.) Draw Lewis structures for the two resonance forms of the nitrite ion, NO_2^-. In what range do you expect the nitrogen-oxygen bond length to fall? (*Hint*: Use Table 3–3.)

48. (See Example 3–8.) Draw Lewis structures for the three resonance forms of the carbonate ion, CO_3^{2-}. In what range do you expect the carbon-oxygen bond length to fall? (*Hint*: Use Table 3–3.)

49. The compound with molecular formula ONF may exist in two isomers: one with N as the central atom, and the other

with O as the central atom in a three-atom chain. Draw Lewis structures to represent the bonding in both isomers, and indicate formal charges.

50. Some evidence suggests that the molecule S_3 exists, possibly in a bent geometry analogous to that of ozone (see Exercise 3–8) and also possibly in a *cyclic* form (an equilateral triangle of S atoms). Draw Lewis structures (with formal charge where necessary) to represent the bonding in both these isomers of S_3.

51. The stable form of sulfur under certain conditions consists of rings of eight sulfur atoms. Draw the Lewis structure for such a ring.

52. White phosphorus (P_4) consists of four phosphorus atoms arranged at the corners of a tetrahedron. Draw in the valence electrons on this structure to give a Lewis structure that satisfies the octet rule.

53. (See Example 3–9.) Draw Lewis structures for the following compounds. In the formulas, the symbol of the central atom is given first. (*Hint*: The valence octet may be expanded for the central atom.)
    (a) PF_5 (b) SF_4 (c) XeO_2F_2

54. (See Example 3–9.) Draw Lewis structures for the following ions. In the formulas, the symbol of the central atom is given first. (*Hint*: The valence octet may be expanded for the central atom.)
    (a) BrO_4^- (b) PCl_6^- (c) XeF_3^+

The Shapes of Molecules

55. (See Example 3–10.) Give the steric number for the central atom in each of the following molecules. In each case, the central atom is listed first, and the other atoms are all bonded directly to it.
    (a) CBr_4 (d) $SOCl_2$
    (b) SO_3 (e) ICl_3
    (c) SeF_6

56. (See Example 3–10.) For each of the following molecules or molecular ions, give the steric number of the central atom. In each case, the central atom is listed first and the other atoms are all bonded directly to it.
    (a) PF_3 (d) ClO_2^-
    (b) SO_2Cl_2 (e) GeH_4
    (c) PF_6^-

57. (See Example 3–12.) For each of the molecules in problem 55, sketch and name the approximate molecular geometry.

58. (See Example 3–12.) For each of the molecules or molecular ions in problem 56, sketch and name the approximate molecular geometry.

59. For each of the following molecules or molecular ions, give the steric number of the central atom, sketch and name the approximate molecular geometry, and describe the direction of any *distortions* from the approximate geometry due to lone pairs. In each case, the central atom is listed first, and the other atoms are all bonded directly to it.
    (a) ICl_4^- (c) BrO_3^-
    (b) OF_2 (d) CS_2

60. For each of the following molecules or molecular ions, give the steric number of the central atom, sketch and name the approximate molecular geometry, and describe the direction of any *distortions* from the approximate geometry due to lone pairs. In each case, the central atom is listed first, and the other atoms are all bonded directly to it.
    (a) TeH_2 (c) PCl_4^+
    (b) AsF_3 (d) XeF_5^+

61. Give an example of a molecule or ion having the following formula type and structure:
    (a) AB_3 (planar)   (c) AB_2^- (bent)
    (b) AB_3 (pyramidal)   (d) AB_3^{2-} (planar)

62. Give an example of a molecule or ion having the following formula type and structure:
    (a) AB_4^- (tetrahedral)   (c) AB_6^- (octahedral)
    (b) AB_2 (linear)   (d) AB_3^- (pyramidal)

63. (See Example 3–13.) For each of the answers in problem 55, state whether the species is *polar* or *nonpolar*.

64. (See Example 3–13.) For each of the answers in problem 56, state whether the species is *polar* or *nonpolar*.

65. The molecules of a certain compound contain one atom each of nitrogen, fluorine, and oxygen. Two structures, NOF (O as central atom) and ONF (N as central atom), are under consideration as the correct structure. Does the information that the molecule is bent allow a decision between these two? Explain.

66. Mixing $SbCl_3$ and $GaCl_3$ in a one-to-one molar ratio (using liquid sulfur dioxide as a solvent) gives a solid ionic compound of empirical formula $GaSbCl_6$. A controversy erupts over whether this compound is $(SbCl_2^+)(GaCl_4^-)$ or $(GaCl_2^+)(SbCl_4^-)$.
    (a) Predict the molecular structures of the two anions.
    (b) It is learned that the cation in the compound has a bent structure. Based on this fact, which formulation is more likely to be correct?

67. (a) Use the VSEPR theory to predict the structure of the NNO molecule.
    (b) The substance NNO has a small dipole moment. Which end of the molecule is more likely to be the positive end, given that O is more electronegative than N.

68. Ozone (O_3) has a non-zero dipole moment. In molecules of O_3, one of the oxygen atoms is directly bonded to the other two, which are not bonded to each other.
    (a) Based on this information, which of the following structures are possible for the ozone molecule: symmetric linear, unsymmetric linear (for example, different O—O bond lengths), and bent? (*Note:* Even an O—O bond can have a bond dipole if the two oxygen atoms are bonded to different atoms, or if only one of the oxygen atoms is bonded to a third atom.)
    (b) Use VSEPR theory to predict which of the structures of part (a) should be observed.

Naming Binary Molecular Compounds

69. Give chemical formulas for the following molecular compounds:
    (a) Dinitrogen pentaoxide   (c) Boron nitride
    (b) Hydrogen iodide   (d) Phosphine

70. Give chemical formulas for the following molecular compounds:
    (a) Hydrogen selenide   (c) Arsine
    (b) Dinitrogen tetrafluoride (d) Phosphorus pentafluoride

71. (See Example 3–14.) Give the names of the following molecular compounds:
    (a) H_2Te (c) NH_3
    (b) PCl_3 (d) S_4N_4

72. (See Example 3–14.) Give the names of the following molecular compounds:
    (a) As_4O_6 (c) N_2H_4
    (b) HF (d) Cl_2O

73. The last step in the industrial preparation of phosphoryl chloride (see Example 3–5) is the reaction of tetraphosphorus decaoxide with phosphorus pentachloride to give $POCl_3$ as the sole product. Write a balanced chemical equation for this reaction.

74. Dichlorine heptaoxide is a colorless, oily liquid that boils at 81°C. Upon further heating, it can decompose explosively to chlorine trioxide and chlorine tetraoxide. Write a balanced chemical equation for this reaction.

Elements Forming More than One Ion

75. Give names for the following compounds of transition-metal or late main-group elements:
    (a) Fe_2O_3 (d) $PbCl_4$
    (b) $TiBr_4$ (e) MnF_3
    (c) WO_3

76. Give names for the following compounds of transition-metal or late main-group elements:
    (a) Mo_2S_5 (c) Hg_2Cl_2
    (b) NiF_3 (d) $NbCl_4$

77. (See Example 3–15.) Write chemical formulas for the following compounds of transition-metal or late main-group elements:
    (a) Tin(II) fluoride   (d) Tungsten(V) chloride
    (b) Rhenium(VII) oxide   (e) Copper(II) nitrate
    (c) Cobalt(III) fluoride

78. (See Example 3–15.) Write chemical formulas for the following compounds of transition-metal or late main-group elements:
    (a) Molybdenum(III) iodide
    (b) Thallium(I) chloride
    (c) Manganese(IV) oxide
    (d) Niobium(IV) oxide
    (e) Copper(I) oxide

79. Rhenium(VII) sulfide reacts with hydrogen to give rhenium(VI) sulfide and hydrogen sulfide. Write a balanced chemical equation for this process.

80. Vanadium(V) oxide reacts with carbon monoxide to give vanadium(III) oxide and carbon dioxide. Write a balanced chemical equation for this process.

Additional Problems

81. Trends in physical and chemical properties are not always smooth across the periodic table.
    (a) Estimate the boiling point of zinc, based on the boiling points of its neighbors copper (2567°C) and gallium (2403°C). Compare your estimate with the measured value of 907°C.
    (b) Gold and thallium are solids at room temperature, whereas mercury is a liquid. Can you draw a general conclusion about the boiling temperatures of elements in the zinc group?
    (c) Where would you predict the boiling point of cadmium to lie relative to those of silver and indium?

82. Indium forms a pale yellow oxide that is 82.712% indium and 17.288% oxygen by mass.
    (a) At the time of Mendeleev, it was believed that the relative atomic mass of oxygen was 16 and the relative atomic mass of indium was 76. At that time, what formula was given to the pale yellow indium oxide?
    (b) Mendeleev suggested that the correct atomic mass of indium was 114 ($\frac{3}{2}$ of 76). What formula was he simultaneously suggesting for the pale yellow oxide?

83. In a recent science-fiction novel, the author has one character say, "You don't realize it, but oxygen is a corrosive gas. It's in the same chemical family as chlorine and fluorine, and hydrofluoric acid is the most corrosive acid known." Criticize this statement. Give the name and the formula of the compound of oxygen that is most analogous to hydrofluoric acid (HF).

84. Only some of the elements in the seventh period have been discovered.
    (a) Suppose scientists are able to prepare element 113 and study its properties. What elements would you predict it to resemble in chemical properties? Do you expect it to be a metal?
    (b) What other elements should resemble element 117? Predict the chemical formula of the compound of element 117 with potassium.

85. Which of the following do you expect to exist as stable ionic compounds? For each unstable compound, write the empirical formula of a stable compound containing the same ions.

Name the five compounds.
    (a) Na_3CO_3
    (b) $MgCl$
    (c) $(NH_4)_2SO_4$
    (d) $BaCN$
    (e) $Sr(OH)_2$

86. Predict the chemical formula of sodium selenate, by comparison with the chemical formula of sodium sulfate.

87. Many important fertilizers are ionic compounds containing the elements nitrogen, phosphorus, and potassium, because these are frequently the limiting nutrients in soil for plant growth.
    (a) Write the chemical formulas for the following chemical fertilizers: ammonium phosphate, potassium nitrate, ammonium sulfate.
    (b) Calculate the mass percentage of nitrogen, phosphorus, and potassium for each of the compounds in part (a).

88. Two possible Lewis structures for sulfine (H_2CSO) are

    (a) Compute the formal charges on all atoms.
    (b) Draw a Lewis structure for which all the atoms in sulfine have formal charges of zero.

89. A hydrogen ion, H^+, consists of a bare nucleus with no electrons. To satisfy the Lewis criterion, it needs an electron pair. Predict what happens when an H^+ ion encounters an H_2O molecule or an NH_3 molecule. Write the Lewis structures of the products. Does your analysis explain the common existence of NH_4^+, the ammonium ion?

90. Consider the three molecules N_2, N_2H_2, and N_2H_4. Each contains a central nitrogen-nitrogen bond. Write Lewis structures for all three molecules, and use Table 3–3 to estimate bond lengths for all the bonds.

91. There is persuasive evidence for the brief existence of the unstable molecule O—P—Cl.
    (a) Draw a Lewis structure for this molecule in which the octet rule is satisfied on all atoms and the formal charges on all atoms are zero.
    (b) The compound OPCl reacts with oxygen to give O_2PCl. Draw a Lewis structure of O_2PCl for which all formal charges are equal to zero. Draw a Lewis structure in which the octet rule is satisfied on all atoms.

92. A common error is to mix up the subscripts in a binary compound (for example, to write HO_2 when H_2O is intended). Sometimes both the compounds represented in such mix-ups actually exist.
    (a) Draw Lewis structures for ammonia (NH_3) and hydrazoic acid (HN_3). In HN_3, the three nitrogen atoms are attached in a row, with the hydrogen atom on one end.
    (b) Draw Lewis structures for sulfur dioxide (SO_2) and disulfur oxide (S_2O). Both have central sulfur atoms. Do not overlook the possibility of resonance.

* **93.** The compound SF_3N has been synthesized.
  - (a) Draw the Lewis structure of this molecule, supposing that the three F's and the N surround the S. Indicate the formal charges. Repeat, but assume that the three F's and the S surround the N.
  - (b) On the basis of the results in part (a), speculate about which arrangement is more likely to correspond to the actual molecular structure.

94. In nitryl chloride (NO_2Cl), the chlorine atom and the two oxygen atoms are bonded to a central nitrogen atom, and all the atoms lie in a plane. Draw the two resonance structures that satisfy the octet rule and that together are consistent with the fact that the two nitrogen-oxygen bonds are equivalent.

95. The peroxonitrite ion is an isomer of the nitrate ion. It has the same charge and the same number and kind of atoms but uses a different molecular skeleton. In the peroxonitrite ion, the atoms are connected as follows.

$$O \cdots N \cdots O \cdots O$$

Draw a Lewis structure for this ion, and assign formal charges to the atoms.

96. The element xenon (Xe) is by no means chemically inert; it forms a number of chemical compounds with highly electronegative elements such as fluorine and oxygen. The reaction of xenon with varying amounts of fluorine produces XeF_2 and XeF_4. Subsequent reaction of one or the other of these compounds with water produces (depending on conditions) XeO_3, XeO_4, and H_4XeO_6, as well as mixed compounds such as $XeOF_4$. Predict the structures of these six xenon compounds, using VSEPR theory.

97. The ion $SbCl_6^{3-}$ is prepared from $SbCl_5^{2-}$ by treating it with the Cl^- ion.
  - (a) Predict the geometry of the $SbCl_5^{2-}$ ion, using the VSEPR method.
  - (b) Determine the steric number of the central antimony atom in $SbCl_6^{3-}$, and discuss the extension of the VSEPR theory to make a prediction of its molecular geometry.

98. Predict the arrangement of the atoms around the sulfur atom in $F_4S{=}O$, assuming that double-bonded atoms require more space than single-bonded atoms.

99. The "dicyanomethanide ion" has the formula $(NC-C-CN)^{2-}$; that is, it has a central carbon atom with two $-CN$ groups attached. A recent publication reports that this ion is bent at the central C atom. Determine the most likely steric number for the central C atom, and predict the bond angle at that atom.

100. Draw Lewis structures and predict the geometries of the following molecules. State which are polar and which nonpolar:
  - (a) ONCl   (d) SCl_4
  - (b) O_2NCl (e) CHF_3
  - (c) XeF_2

* **101.** The reaction between H_2S and SO_2 under certain conditions gives a compound with the molecular formula $H_2S_2O_2$. One researcher suggests that the skeleton of the compound is

$$H-O-S-S-O-H,$$

but another thinks that it is

Draw Lewis structures, including resonance forms if they exist, for both of these isomers of $H_2S_2O_2$. What guidance do the Lewis structures offer in deciding which isomer is more likely to have been isolated?

* **102.** The dimerization of nitrogen dioxide to give dinitrogen tetraoxide is mentioned in Figure 3–23.
  - (a) NO_2 is an odd-electron compound. Draw the best Lewis structures you can for it, recognizing that one atom cannot achieve an octet configuration. Use formal charges to decide whether that should be the nitrogen atom or one of the oxygen atoms.
  - (b) From the structure given in Figure 3–23 for N_2O_4, draw resonance structures for this molecule that obey the octet rule.

103. Although magnesium and the alkaline earth metals located below it in the periodic table form ionic chlorides, beryllium chloride ($BeCl_2$) is a covalent compound.
  - (a) Follow the usual rules to write a Lewis structure for $BeCl_2$, in which each atom attains an octet configuration. Indicate formal charges.
  - (b) The Lewis structure that results from part (a) is an extremely unlikely structure because of the double bonds and formal charges it shows. By relaxing the requirement of placing an octet on the beryllium atom, show how a Lewis structure with all formal charges equal to zero can be written. (*Note:* The compound $BeCl_2$, like BF_3, in fact lacks an octet on the central atom.)

104. (a) The first compound of a noble gas, prepared by Neil Bartlett in 1962, was an orange-yellow ionic solid consisting of XeF^+ and $Pt_2F_{11}^-$ ions. Draw a Lewis structure for the XeF^+ ion.
  - (b) Shortly after the preparation of the ionic compound discussed in part (a), it was found that the irradiation of mixtures of xenon and fluorine with sunlight produced white crystalline XeF_2. Draw a Lewis structure for this molecule, allowing valence expansion on the central xenon atom.

* **105.** Molecules of "iodine trioxide" actually have the molecular formula I_4O_{12}. The following structural diagram appears in a research report.
  - (a) Compute the formal charge on each atom in this structure, assuming that each line that connects two atoms represents a single shared pair of electrons.

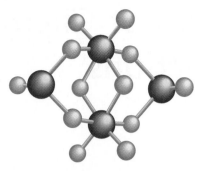

(b) Draw the Lewis structure in which the electrons are redistributed so that all of the oxygen atoms have exactly two bonds. Calculate the formal charges on all the atoms in this structure.

106. Chlorine dioxide can be produced on either a laboratory or an industrial scale through the reaction of sodium chlorite with chlorine. Sodium chloride is a by-product. Write a balanced chemical equation for this reaction.

107. When indium and sulfur are dissolved in molten tin at a temperature between 650°C and 1100°C, they react to form In_5S_4.

(a) Write a balanced chemical equation for this reaction.

(b) Name the product of the reaction, as if both elements were non-metals.

108. At elevated temperatures, nitrogen trifluoride reacts with copper to give the colorless, reactive gas dinitrogen tetrafluoride and copper(I) fluoride. Write a balanced chemical equation for this reaction.

109. In the following coordination complexes, identify the transition metals and the ligands: $Na_2[CuCl_4]$, $[Co(H_2O)_6]Br_2$, $K_2[Pt(CN)_4]$, $[Cr(CO)_6]$.

110. At one time, the bonding in coordination complexes was not understood, and they were often formulated as "double salts"; for example, $K_2[Pt(CN)_4]$ was written as $Pt(CN)_2 \cdot 2$ KCN. Translate the formulas of the following complexes from the outmoded double-salt notation to current notation: $CuCl_2 \cdot 2KCl$, $CoF_3 \cdot 3NaF$.

*111. The nitrate ion and the peroxonitrite ion (see problem 95) are isomers. They are the only two species of molecular formula NO_3^- so far known to exist. Amazingly, no fewer than eight different skeletons are conceivable by which a single N atom and three O atoms might bond together. Draw Lewis structures for all eight NO_3^- isomers. (*Hint*: The eight include four skeletons in which atoms are linked in a cycle.)

CUMULATIVE PROBLEM

Compounds of Oxygen

Consider the four compounds KO_2, BaO_2, TiO_2, and H_2O_3. All contain oxygen but in quite different bonding situations.

(a) The oxygen in TiO_2 occurs as O^{2-} ions. Give the Lewis dot symbol for this ion. How many total electrons does this ion have? How many of these are valence and how many core electrons? What is the chemical name for TiO_2?

(b) Recall that Group II elements form stable +2 ions. By referring to Table 3–2, identify the oxygen-containing ion in BaO_2, and give the name of the compound. Draw a Lewis structure for the oxygen-containing ion, and show formal charges for all atoms. Is the bond in this ion a single or a double bond?

(c) Recall that Group I elements form stable +1 ions. By referring to Table 3–2, identify the oxygen-containing ion in KO_2 and give the name of the compound. Show that the oxygen-containing ion is an odd-electron species. Draw the best Lewis structures you can for it.

(d) H_2O_3 is a molecular compound. Name this compound. Draw a Lewis structure for H_2O_3 in which every atom has a formal charge of zero. *Hint*: Use a skeleton that links the three oxygen atoms in a chain with hydrogen atoms at each end.

(e) By referring to Table 3–3, predict the bond distances in H_2O_3.

(f) Use VSEPR theory to predict the three bond angles in H_2O_3.

On the right is a sample of rutile, TiO_2, which is the primary ore of titanium. Behind it is a large quartz crystal through which slender rutile hairs penetrate. Crystals of this type are called Venus hairstone.

4

Types of Chemical Reactions

Hot steel wool is oxidized by pure oxygen to form iron(III) oxide.

$\mathbf{A}$t the heart of chemistry are chemical reactions, the dynamic processes that lead to the formation of new substances. No one can predict the course of chemical events when substances interact in all possible combinations under all conditions. However, patterns of reactivity do exist. In this chapter, we classify chemical reactions, particularly those taking place in solution, with an eye to predicting their outcomes. We also extend the methods of stoichiometry to deal with reactions in solution.

4–1   DISSOLUTION REACTIONS

In **dissolution,** two (or more) substances spread out, or disperse, into each other at the level of individual atoms, molecules, or ions. As this happens, formerly distinct phases merge into a new single phase. The result is a homogeneous mixture called a **solution.** Thus, solid sucrose (sugar) disperses into liquid water to give a new liquid (sugar-water); liquid octane and liquid methanol dissolve each other to give a new liquid fuel; and gaseous hydrogen infiltrates into solid palladium to give a new solid.

Usually, one component in a solution is present in greater amount than any other. The major component is the **solvent,** and the minor components are **solutes.** Dilute solutions have smaller proportions of solute; concentrated solutions have larger proportions of solute. The physical and chemical properties of a solution differ from those of the pure solvent or pure solute and change if the relative amounts of the components are changed.

The phenomenon of dissolution is intermediate between a physical change and a chemical change. During dissolution, the attractions among the particles in the original phases (solvent-to-solvent and solute-to-solute attractions) are broken up and replaced, at least in part, by new solvent-to-solute attractions. In this respect, dissolution is a distinctly chemical event. On the other hand, a solution, unlike a compound, has its components in *indefinite* proportions and cannot be represented by the usual kind of chemical formula. On this basis, dissolution is a physical change, like the mixing of salt and sand. Equations for dissolution reactions simply omit the solvent as a reactant. They show the change by indicating the state of the solute in parentheses on the reactant side of the arrow and the name of the solvent in parentheses on the product side. Thus, solid (*s*) sucrose dissolves in water to give an **aqueous** (*aq*) solution of sucrose

$$C_{12}H_{22}O_{11}(s) \longrightarrow C_{12}H_{22}O_{11}(aq)$$

In principle, the solute and solvent can be any combination of solid (*s*), liquid (ℓ), and gaseous (*g*) phases. In practice, solid–solid dissolutions are mostly too slow to worry about, mixtures of gases are treated in a different way (see Section 5–6), and the best solvents are liquids. Liquid water is indisputably the world's best-known, most important, most nearly universal solvent; it readily dissolves many disparate solids, liquids, and gases. We therefore emphasize aqueous solutions with the understanding that dissolution also occurs in a host of other solvents, both inorganic ones such as liquid ammonia (NH_3) and organic ones such as acetone (CH_3COCH_3), ethanol (C_2H_5OH), diethyl ether ($C_2H_5OC_2H_5$), benzene (C_6H_6), and chloroform ($CHCl_3$).

• In a true solution the particles are generally smaller than about 10^{-9} m in diameter. Dispersions with bigger particles are colloids (see Section 6–8). The dividing line between solution and colloid is necessarily vague.

• Dissolution also takes place in supercritical fluids, which are phases that are neither gaseous nor liquid. See Section 6–5.

Ionic Compounds in Water

All ionic compounds are solids in the range of temperature in which water is a liquid. They have rigid lattices in which strong forces (ionic bonds) pin the constituent

Figure 4–1 When an ionic solid (in this case, K_2SO_4) dissolves in water, the ions move away from their sites in the solid, where they are attracted strongly by ions of opposite electric charge. New strong attractions replace those lost as each ion becomes surrounded by a group of water molecules. In a precipitation reaction, the process is reversed.

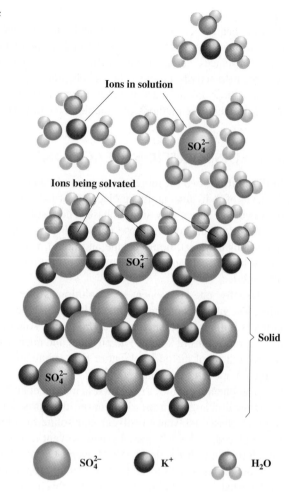

Ions in solution

SO_4^{2-}

Ions being solvated

SO_4^{2-}

Solid

SO_4^{2-}

SO_4^{2-}

SO_4^{2-} K^+ H_2O

• If the solvent is not water, the more general term "solvation" is used.

ions in place. The high melting points of ionic compounds indicate that a good deal of energy must be supplied to destroy the lattice and produce a liquid (molten) form in which the ions move more freely. How is it then that many ionic compounds dissolve freely in water? How does water overcome the forces of attraction (the solute–solute forces) among the ions in soluble ionic solids such as NaCl or K_2SO_4? The answer is that it replaces those forces with strong and numerous new solvent–solute attractions, a process called **aquation.** Molecules of water are electrically dipolar (having sides bearing opposite but equal electric charges, see Fig. 3–22). When water touches an ionic compound such as K_2SO_4, the negative poles of some of its molecules attract the positive ions while the positive poles of others attract the negative ions. The combined strength of these electrostatic attractions pulls the ions out of the ionic lattice. Layers of water molecules cling to the freed ions, which then wander off, more or less independently of each other, into the aqueous phase (Fig. 4–1). The solute is said to **dissociate** into ions or to **ionize** upon dissolution. Equations for the dissolution of ionic substances recognize dissociation by showing aquated ions as products

$$K_2SO_4(s) \longrightarrow 2\ K^+(aq) + SO_4^{2-}(aq)$$

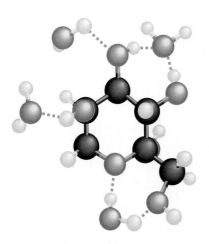

Figure 4-2 A molecule of fructose in aqueous solution. Note the attractions between the hydroxyl (O—H) groups of the fructose and molecules of water. The fructose molecule is aquated; the exact number and arrangement of the aquating molecules fluctuate.

Molecular Substances in Water

Molecular substances contain no ions to which water molecules can adhere, but their molecules are often polar—the differing electronegativities of the atoms lead to build-ups of negative and positive charge in different molecular regions. Sugars are notable examples of molecular substances having polar molecules. Members of a broader class of compounds, known as the carbohydrates ("hydrates of carbon"), sugars have the general formula, $C_m(H_2O)_n$. Typical sugars include sucrose, $C_{12}H_{22}O_{11}$ (table sugar); fructose, $C_6H_{12}O_6$ (fruit sugar); and ribose, $C_5H_{10}O_5$ (a subunit in the biomolecules known as ribonucleic acids).

Despite the appearance of the general formula (and the name, carbohydrate), sugars do not contain water molecules. They do contain O—H (hydroxyl) groups bonded to carbon atoms. Hydroxyl groups are bent at the oxygen atom and are therefore polar. They provide sites for interactions with the polar water molecule that favor dissolution (Fig. 4–2). The molecules remain intact, so the dissolution of fructose in water is represented

• The O—H group (uncharged) is not the same as the hydroxide ion (OH^-). The latter has one more electron.

$$C_6H_{12}O_6(s) \longrightarrow C_6H_{12}O_6(aq)$$

The dissolution of many other molecular substances in water follows the same pattern. No ions form, but the solute molecules are sufficiently polar to support strong attractions to the water molecules. These attractions replace existing solute–solute attractions, and aquated molecules move off into solution.

Electrolytes and Non-Electrolytes

Major evidence for our ideas about dissolution in water comes from studies on the electrical conductance of solutions. Pure water barely conducts electricity. Numerous solutes, called **electrolytes,** increase the conductance of water when they dissolve. Electrolytes are mainly ionic compounds such as potassium sulfate (K_2SO_4) and sodium chloride (NaCl), but some important molecular compounds are also electrolytes. These include acids and bases (see Section 4–3) such as hydrochloric acid (HCl), nitric acid (HNO_3), sulfuric acid (H_2SO_4), and ammonia (NH_3). Other solutes, called **non-electrolytes,** do not increase the conductance of water when they dissolve. Non-electrolytes are generally molecular substances. Examples are acetone (CH_3COCH_3), the alcohols ethanol (C_2H_5OH) and methanol (CH_3OH), and sugars such as fructose.

• Water from the tap contains sufficient dissolved electrolytes to make it a rather good conductor of electricity.

Figure 4–3 An aqueous solution of potassium sulfate conducts electricity better than does pure water. When metallic plates (electrodes) charged by a battery are inserted into the solution, positive ions (K^+) migrate toward the negative plate and negative ions (SO_4^{2-}) migrate toward the positive plate. Thus, the solution conducts electricity. Potassium sulfate is a strong electrolyte.

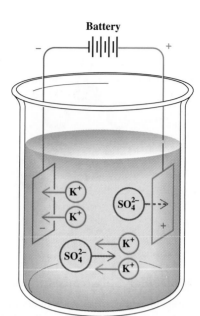

Battery

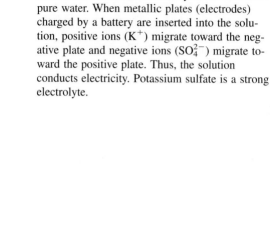

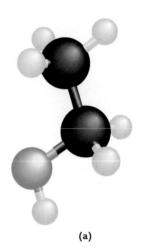

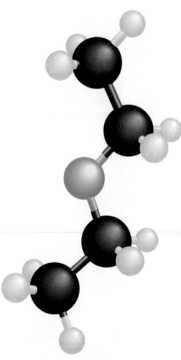

(a)

(b)

(a) In ethanol (C_2H_5OH), one of the H atoms in water is replaced by a C_2H_5 group (the ethyl group). (b) In diethyl ether ($H_5C_2OC_2H_5$), both of the H atoms in water are replaced by C_2H_5 groups.

We can easily explain the observation of high electrical conductance in solutions of ionic compounds by reasoning that aquated cations and anions are freely mobile and migrate to transport a current (Fig. 4–3). If electrical conductance in solutions relies on the presence of mobile ions as charge-carriers, molecular substances that are electrolytes must generate ions by reacting with water in some fashion as they dissolve. Such reactions have been identified and in fact are characteristic of compounds that are acids and bases (see Section 4–3).

Electrolytes increase the conductance of water to different degrees. **Strong electrolytes** increase conductance greatly, whereas **weak electrolytes** increase conductance only slightly. The electrical conductance of a solution of calcium chloride ($CaCl_2$), a strong electrolyte, is well over 100 times greater than that of a similar solution of mercury(II) chloride ($HgCl_2$), a weak electrolyte, under the same conditions. Other weak electrolytes are lead acetate ($Pb(CH_3COO)_2$), cadmium iodide (CdI_2), mercury(II) cyanide ($Hg(CN)_2$), acetic acid (CH_3COOH), and ammonia (NH_3). Solutions of weak electrolytes conduct electricity poorly because they furnish far fewer ions when dissolved than the same amount of a strong electrolyte furnishes under the same conditions. Their dissociation is incomplete.

Experimental studies on solutions of both strong and weak electrolytes show that the conductance from one mole of an electrolyte in a dilute solution always *exceeds* the conductance from the same mole of the electrolyte in a concentrated solution. Among strong electrolytes, the conductance per mole starts high and increases moderately but quite unmistakably with dilution. The early interpretation of this fact was that dissociation into ions remains incomplete in strong electrolytes as well as in weak electrolytes, unless the solutions are quite dilute. Current theory however takes the view that dissociation into ions *is* complete, but the ions do not act in complete independence in solution. Oppositely charged aquated ions attract each other and even form loosely bound, temporary **ion-pairs.** The attractions exert a "drag" on the mobility of the ions when they are called upon to conduct electricity. Dilution increases the conductance per mole of a strong electrolyte by increasing the average distance between ions, not by releasing more ions.

Solubilities

Some substances dissolve in each other in all proportions. For example, ethanol (C_2H_5OH) and water form solutions of all possible compositions, ranging from slightly wet alcohol through an equal mixture (at which point the two trade places as solute and solvent) on to slightly alcoholic water. Such pairs are **miscible** in all proportions. More frequently, however, substances are only partially miscible. Mixing diethyl ether (($C_2H_5)_2O$) with water gives two phases, one clear liquid floating on top of another. The bottom layer consists mainly of water with some dissolved diethyl ether, and the top layer consists mainly of diethyl ether with some dissolved water. To be more exact, at 25°C the water phase holds 6.41 g of ether per 100 g of water, and the ether phase holds 1.36 g of water per 100 g of ether. The water is **saturated** with respect to ether, and the ether is saturated with respect to water. Neither liquid can hold any more of the other.

• It is common to report solubilities in units of grams of dissolved substance (solute) per 100 g of solvent or moles per liter of solution; other units occur as well.

The **solubility** of a substance is the largest amount that can dissolve in a given amount of a solvent at a particular temperature. By this definition, the solubility of ethanol in water is infinite (saturation is impossible for fully miscible substances), but the solubility of diethyl ether in water is 6.41 g per 100 g, and the solubility of water in diethyl ether is 1.36 g per 100 g at 25°C. If additional solute is added to a saturated solution, it does not dissolve but forms a new phase that can be separated by decantation (pouring off), filtration, or other physical means.

Predicting Dissolution Reactions

Dissolution is a complex phenomenon, and absolute rules to predict solubilities do not exist. Some imperfect generalizations about solubility can be made based on experimental solubility data.

The most general statement is *"like dissolves like."* Substances having molecules with similar polarity tend to be mutually soluble. Water, which has markedly

(a)

(b)

"Like dissolves like." (a) Ethylene glycol, the main ingredient in antifreeze, has polar molecules and dissolves readily in water. (b) Oil and water are nearly completely immiscible. The molecules in water are quite polar; those in oil are not. Here, the less-dense oil rises to float on top of the water.

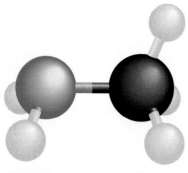

Methyl amine (CH_3NH_2) is ammonia (NH_3) with one of its H atoms replaced by a methyl (CH_3) group. It is water soluble.

polar molecules, dissolves ionic and polar-molecular substances such as salt, ethanol, and the simple sugars but does not dissolve benzene (C_6H_6) and common oils, which have nonpolar molecules. Carbon tetrachloride (CCl_4), which has nonpolar molecules, dissolves benzene, oils, and other nonpolar or slightly polar substances, but not water and ionic compounds. Some additional generalities follow:

- No gaseous or solid substances are infinitely soluble in (miscible in all proportions with) water; many liquids are.
- Common inorganic acids are soluble in water; organic acids (see Section 4–3) of low molar mass are soluble in water. Some are completely miscible.
- Organic compounds of low molar mass containing the OH group (these include sugars and alcohols such as methanol and ethanol) or the NH_2 group (the amine group) are usually soluble in water; several are completely miscible.
- The solubilities of gases in water and other solvents *decrease* as the temperature increases; the solubilities of solids and liquids usually *increase* with temperature.
- The solubilities of ionic compounds in water vary. Nearly all ionic compounds containing lithium, sodium, potassium, rubidium, cesium, or ammonium ions dissolve well in water and less well in other solvents. Table 4–1 lists aqueous solubilities of other ionic compounds.

Table 4–1
Solubilities of Ionic Compounds in Water

Anion	Soluble[a]	Slightly Soluble	Insoluble
NO_3^- (nitrate)	All	—	—
CH_3COO^- (acetate)	Most	—	$Be(CH_3COO)_2$
ClO_3^- (chlorate)	All	—	—
ClO_4^- (perchlorate)	Most	$KClO_4$	—
F^- (fluoride)	Group I,[b] AgF, BeF_2	SrF_2, BaF_2, PbF_2	MgF_2, CaF_2
Cl^- (chloride)	Most	$PbCl_2$	$AgCl$, Hg_2Cl_2
Br^- (bromide)	Most	$PbBr_2$, $HgBr_2$	$AgBr$, Hg_2Br_2
I^- (iodide)	Most	—	AgI, Hg_2I_2, PbI_2, HgI_2
SO_4^{2-} (sulfate)	Most	$CaSO_4$, Ag_2SO_4, Hg_2SO_4	$SrSO_4$, $BaSO_4$, $PbSO_4$
S^{2-} (sulfide)	Groups I and II,[c] $(NH_4)_2S$	—	Most
CO_3^{2-} (carbonate)	Group I, $(NH_4)_2CO_3$	—	Most
SO_3^{2-} (sulfite)	Group I, $(NH_4)_2SO_3$	—	Most
PO_4^{3-} (phosphate)	Group I, $(NH_4)_3PO_4$	Li_3PO_4	Most
OH^- (hydroxide)	Group I, $Ba(OH)_2$	$Sr(OH)_2$, $Ca(OH)_2$	Most

[a]Soluble compounds are defined as those that dissolve to the extent of 1 g or more per 100 g water, slightly soluble as 0.01 to 1 g per 100 g water, and insoluble as less than 0.01 g per 100 g water at room temperature.

[b]Compounds of elements from the first column in the periodic table: Li, Na, K, Rb, Cs.

[c]Compounds of elements from the second column in the periodic table: Be, Mg, Ca, Sr, Ba.

EXAMPLE 4–1

Predict whether the following substances are soluble in water:
(a) Cobalt(II) chloride. (b) Lanthanum(III) perchlorate. (c) Maltose ($C_{12}H_{22}O_{11}$).

Solution

(a) According to Table 4–1, "most chlorides" are soluble in water, and $CoCl_2$ is not among the listed exceptions. The prediction is that $CoCl_2$ is soluble.
(b) "Most perchlorates" are soluble, and $La(ClO_4)_3$ is not a listed exception. The prediction is that $La(ClO_4)_3$ dissolves.
(c) From its name and formula we expect maltose to be a sugar (like glucose, fructose, and sucrose). Maltose should be soluble, like other low-molar-mass sugars.

Exercise

Predict whether the following substances are soluble in water: (a) Calcium carbonate. (b) Mercury(II) sulfide. (c) Isopropanol (C_3H_7OH).

Answer: (a) Insoluble. (b) Insoluble. (c) Soluble.

4–2   PRECIPITATION REACTIONS

Whenever the concentration of a substance in solution exceeds its solubility, a new phase starts to separate. A new solid phase is usually dense enough to sink to the bottom of a liquid solution. That is, solids usually **precipitate** ("fall down") from solutions when their solubility is exceeded. Precipitation of solids can be effected in several ways:

1. By removing some solvent. Removal is easy when the solvent evaporates readily but the solute does not, a very common case. Evaporation of seawater causes sodium chloride (and other salts) to deposit on the sides and bottom of the container. A saturated solution that deposits a precipitate in this way is called the "mother-liquor" of the precipitate. The equation for the precipitation of $NaCl(s)$ from water is

$$Na^+(aq) + Cl^-(aq) \longrightarrow NaCl(s)$$

Note that this equation for precipitation is exactly the reverse of the equation for dissolution of NaCl.

2. By changing the solvent. At 25°C, the solubility of sodium chloride is 36 g/100 g of water, but only 0.12 g/100 g of ethanol. Ethanol and water are miscible in all proportions. Stirring ethanol into an aqueous solution of sodium chloride creates a "mixed solvent" with less power to dissolve NaCl than water. Unless the original aqueous solution of NaCl was very dilute, the added ethanol causes NaCl to precipitate

$$Na^+(ethanol,aq) + Cl^-(ethanol,aq) \longrightarrow NaCl(s)$$

3. By changing the temperature. The solubility of silver nitrate ($AgNO_3$) in water is 654 g/100 g of water at 90°C, but only 235 g/100 g of water at 25°C. Cooling a solution that is saturated in $AgNO_3$ at 90°C to 25°C causes $AgNO_3$ to precipitate.

All three of these techniques are much used in chemistry.

Other precipitation reactions approach the limit of solubility in a different way. As Table 4–1 shows, most ionic compounds dissolve in water to some extent.

Figure 4–4 The amounts of different substances that dissolve in 1 L of water at 20°C. Clockwise from the front are borax ($Na_2B_4O_5(OH)_4 \cdot 8H_2O$), potassium permanganate ($KMnO_4$), lead(II) chloride ($PbCl_2$), sodium phosphate hydrate ($Na_3PO_4 \cdot 12H_2O$), calcium oxide (CaO), and potassium dichromate ($K_2Cr_2O_7$).

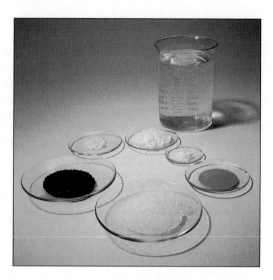

• Barium sulfate is routinely swallowed in suspended form in a "barium cocktail" to improve contrast in x-ray photographs of the alimentary tract. Soluble barium compounds are quite toxic, but $BaSO_4$ is harmless because nearly all of the Ba^{2+} ion stays tightly bound up with sulfate ion and is not absorbed by the body.

However, actual solubilities differ (Fig. 4–4). For example, the solubilities at 25°C of $K_2SO_4(s)$ and $BaCl_2(s)$ differ by a factor of three: 12 g/100 g and 38 g/100 g of water, respectively. These compounds are both water-soluble. Meanwhile, the solubility of $BaSO_4$ is a mere 0.12 mg/100 g of water at the same temperature, about 100,000 times less. $BaSO_4$ is "insoluble" in water. Suppose a solution of K_2SO_4 is mixed with a solution of $BaCl_2$. Both solutes are ionized in solution. The solubility of barium sulfate is quickly exceeded as barium ions ($Ba^{2+}(aq)$) encounter sulfate ions ($SO_4^{2-}(aq)$). Solid barium sulfate precipitates (Fig. 4–5):

$$BaCl_2(aq) + K_2SO_4(aq) \longrightarrow BaSO_4(s) + 2\ KCl(aq)$$

This reaction can be viewed as a switching of ionic partners driven by the precipitation of the insoluble barium sulfate: the (s) on the $BaSO_4$ tells why the reaction goes. It is a **metathesis** reaction (a reaction in which atoms or groups of atoms are interchanged).

• Metathesis reactions can be driven by processes other than precipitation, as discussed subsequently in this chapter.

Ionic Equations and Net Ionic Equations

Often, chemical equations are written to recognize the presence of aquated ions among the reactants or products. Rewriting the previous equation as an **ionic equation**

$$Ba^{2+}(aq) + 2\ Cl^-(aq) + 2\ K^+(aq) + SO_4^{2-}(aq) \longrightarrow$$
$$BaSO_4(s) + 2\ K^+(aq) + 2\ Cl^-(aq) \quad \text{(ionic equation)}$$

reveals which ions take part in the reaction (the aquated barium and sulfate ions start out moving at large in solution but end up tied together in the precipitate) and which are **spectator ions** (the aquated potassium ions and chloride ions are unaffected by the reaction).

Cancellation of species that appear on both the left and right sides of an ionic equation results in a **net ionic equation.** For example

$$Ba^{2+}(aq) + SO_4^{2-}(aq) \longrightarrow BaSO_4(s) \quad \text{(net ionic equation)}$$

Figure 4–5 A solution of potassium sulfate is added to one of barium chloride. A cloud of white particles of solid barium sulfate forms and settles out (precipitates); the potassium chloride remains in solution.

Net ionic equations include only the ions (and compounds) that actually react. They provide excellent models for the prediction of reactions because removal of the spectator ions focuses attention on what really drives the chemical change. In this case, the driving force is the insolubility of $BaSO_4$ in water. The net ionic equation emphasizes that a solution of *any* soluble barium salt mixed with a solution of *any* soluble sulfate salt gives the same result: a precipitate of barium sulfate.

EXAMPLE 4-2

An aqueous solution of sodium carbonate is mixed with an aqueous solution of calcium chloride, and a white precipitate immediately forms. Write a net ionic equation to account for this precipitate.

Solution

The aqueous sodium carbonate contains $Na^+(aq)$ and $CO_3^{2-}(aq)$ ions, and the aqueous calcium chloride contains $Ca^{2+}(aq)$ and $Cl^-(aq)$ ions because the dissolution of both of these compounds involves their breaking up into ions. Mixing the two solutions places $Na^+(aq)$ and $Cl^-(aq)$ ions and also $Ca^{2+}(aq)$ and $CO_3^{2-}(aq)$ ions in contact for the first time. The precipitate forms by the reaction

$$Ca^{2+}(aq) + CO_3^{2-}(aq) \longrightarrow CaCO_3(s)$$

The $Na^+(aq)$ and $Cl^-(aq)$ ions are omitted because they are simply nonreacting spectators; sodium chloride is known to be soluble in water.

Exercise

Write a net ionic equation to represent the formation of the precipitate observed when aqueous solutions of $CaCl_2$ and NaF are mixed. Identify the spectator ions in this process.

Answer: $Ca^{2+}(aq) + 2\ F^-(aq) \longrightarrow CaF_2(s)$. The spectator ions are $Na^+(aq)$ and $Cl^-(aq)$.

Weak electrolytes also take part in precipitation reactions. When a solution of the weak electrolyte mercury(II) chloride and a solution of potassium iodide are mixed, insoluble mercury(II) iodide forms immediately:

$$HgCl_2(aq) + 2\ KI(aq) \longrightarrow 2\ KCl(aq) + HgI_2(s) \quad \text{(overall equation)}$$

$$HgCl_2(aq) + 2\ K^+(aq) + 2\ I^-(aq) \longrightarrow$$
$$2\ K^+(aq) + 2\ Cl^-(aq) + HgI_2(s) \quad \text{(ionic equation)}$$

$$HgCl_2(aq) + 2\ I^-(aq) \longrightarrow HgI_2(s) + 2\ Cl^-(aq) \quad \text{(net ionic equation)}$$

The precipitation of insoluble $HgI_2(s)$ pulls the Hg^{2+} ion out of its association with Cl^- ions.

Predicting Precipitation Reactions

Precipitation is the reverse of dissolution. The various solubility rules (see Table 4–1) therefore serve to predict precipitations. To know the result when two ionic compounds are mixed in aqueous solution, simply check the solubilities of the compounds formed by switching partners:

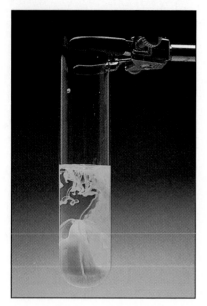

Figure 4-6 A yellow precipitate of lead iodide, PbI_2, is formed when KI solution is added to a solution of $Pb(NO_3)_2$.

- If either of the new compounds is insoluble (or slightly soluble), then precipitation occurs.
- If *both* new compounds are insoluble, then *two* precipitation reactions occur concurrently.
- If both new compounds are soluble, no precipitation occurs.

As an example of the last, consider mixing aqueous sodium chloride with aqueous calcium nitrate. "Exchanging partners" suggests the formation of sodium nitrate and calcium chloride

$$2\,NaCl(aq) + Ca(NO_3)_2(aq) \longrightarrow 2\,NaNO_3(aq) + CaCl_2(aq) \quad \text{(overall equation)}$$

The solubility rules, however, reveal that sodium nitrate and calcium chloride are both soluble. Thus,

$$2\,Na^+(aq) + 2\,Cl^-(aq) + Ca^{2+}(aq) + 2\,NO_3^-(aq) \longrightarrow$$
$$2\,Na^+(aq) + 2\,NO_3^-(aq) + Ca^{2+}(aq) + 2\,Cl^-(aq) \quad \text{(ionic equation)}$$

in which the left and right sides of the equation are identical. Nothing visible happens. This is often indicated by writing "NR" (for no reaction) after the arrow.

$$NaCl(aq) + Ca(NO_3)_2(aq) \longrightarrow NR$$

Writing a net ionic equation is not possible because no net change occurs.

EXAMPLE 4-3

Predict the result when a solution of KI is added to one of $Pb(NO_3)_2$.

Solution

The first step is to find the chemical formulas of the possible precipitates by requiring charge neutrality, following the approach of Example 3–3. Combining lead (Pb^{2+}) and iodide (I^-) ions results in a formula for lead iodide of PbI_2 in which two iodide ions with charge -1 balance the $+2$ charge on the lead ion. According to Table 4–1, PbI_2 is insoluble, and the other possible precipitate, KNO_3, is soluble. Thus, a precipitate of PbI_2 should appear (Fig. 4–6):

$$Pb^{2+}(aq) + 2\,I^-(aq) \longrightarrow PbI_2(s)$$

Exercise

Suppose that a solution of sodium sulfate is added to one of potassium nitrate. Predict the result.

Answer: The solutions mix completely, but no precipitate forms.

In dissolution and precipitation, molecules or ions retain their identities and are simply reshuffled in their associations. They leave one phase to take up a position in another, or the reverse. These reactions tend to be less vigorous (generating less heat, for example) than the acid–base and redox reactions that we examine in Sections 4–3 through 4–5.

4–3 ACIDS AND BASES AND THEIR REACTIONS

Acids and bases have been known and characterized since early times. Many fruits contain acids that contribute to their characteristic flavors, such as tartaric acid ($C_4H_6O_6$) in grapes, citric acid ($C_6H_8O_7$) in lemons, and malic acid ($C_4H_6O_5$) in apples. These acids are responsible for the sharp, sour taste of the fruit, just as acetic acid (CH_3COOH) is responsible for the tang of vinegar. The ashes of plants yield substances, called **bases,** that counteract, or neutralize, the properties of acids. Potassium hydroxide (KOH), called caustic potash, is an important base. It was first prepared from potassium carbonate (K_2CO_3), obtained by leaching plant ashes with water and evaporating the solution in large pots (whence the name). Both potassium hydroxide and the closely related base sodium hydroxide (NaOH, caustic soda) are quite corrosive and destructive of organic tissue; they convert animal fats to soaps, a property exploited in commercial soap-making (see Section 25–1). Although most bases are less caustic than these two, bases generally have a soapy feel, as they act on oils in the skin, and a bitter taste. Bases are also called **alkalis.**

• The word "acid" comes from the Latin *acidus,* meaning "sharp."

Many natural pigments signal the presence of an acid or base by changing color. They are acid–base **indicators.** Usually, vegetable blues turn red in acid and vegetable reds turn blue. This happens with grape juice (Fig. 4–7). Similarly, blueberry juice is dark red in its normal (acidic) state but turns blue in a neutral or basic environment. This explains why blueberry pie stains the tongue blue (saliva is nearly neutral) but stains a napkin red. Strips of paper impregnated with litmus, a blue pigment obtained from lichens, have long been used to test solutions: litmus turns red in acid and blue in base.

• The term "litmus test" now means a decisive test on any issue.

The ease of recognizing acids and bases by taste, feel, or the color of a natural indicator and their simple reactivity (acid + base ⟶ neutral) led to an early acceptance of acid and base as fundamental chemical types. Once the atomic theory was accepted, chemists sought a structural explanation for acid and base behavior. A great many acids, including all of the organic acids just listed and several important inorganic acids, such as nitric acid (HNO_3) and sulfuric acid (H_2SO_4), in-

Figure 4–7 Grape juice is a natural acid–base indicator, changing from red in acidic solution to greenish blue in basic solution.

The lye in this can is sodium hydroxide.

• Oxygen got its name, which comes from the Greek for "acid-giver," during this period.

clude oxygen in their chemical formulas. For a time in the late 18th century it was thought that oxygen was an essential constituent of any acid. Then, in 1810 the British chemist Sir Humphrey Davy analyzed hydrochloric acid (HCl dissolved in water) and showed that this notion was wrong. From then on, the role of *hydrogen* in the properties of acids was increasingly stressed, and fresh understanding of acid–base reactivity developed. Note that hydrogen is the only element common to HNO_3, H_2SO_4, HCl, and acetic acid, CH_3COOH.

Arrhenius Acids and Bases

Over the years, chemists have used a variety of definitions of the terms "acid" and "base." The most recent (see Sections 8–1 and 8–8) aim for generality and succeed in encompassing earlier ideas as special cases. The earlier concepts, however, provide major insights. One still useful pair of definitions derives from work performed by the Swedish chemist Svante Arrhenius in the late 19th century. These definitions apply only for "wet chemistry"—that is, work carried out in aqueous solutions. Although acids and bases react in solvents other than water, and, indeed, in the absence of solvent, aqueous solutions are so important in everyday life that we restrict our attention to them at this point.

In pure liquid water, small but equal amounts of aquated hydrogen ions $H^+(aq)$ and aquated hydroxide ions ($OH^-(aq)$) are present. The source of these ions is the dissociation of a very small proportion of the H_2O molecules into ions. These ions become aquated, just as do ions from a soluble ionic compound. The reaction is

$$H_2O(\ell) \longrightarrow H^+(aq) + OH^-(aq)$$

One might view the process as water dissolving in itself. In any case, this **autoionization** is an intrinsic property of water and cannot be prevented. Water always contains $H^+(aq)$ and $OH^-(aq)$ ions. Arrhenius took these two ions as essential for acid and base behavior in aqueous solution:

• The term "proton-donor" is also often used for acids.

> An **Arrhenius acid** is a substance that, when dissolved in water, dissociates to produce hydrogen ions; it is a **hydrogen-ion donor.**
>
> An **Arrhenius base** is a substance that, when dissolved in water, dissociates to produce hydroxide ions; it is a **hydroxide-ion donor.**

Clearly, insoluble substances and those containing no hydrogen atoms can satisfy neither definition. Also, the definitions elevate water, which donates hydrogen ion and hydroxide ion equally, to a unique status: water is simultaneously and equally both an acid and a base. A substance having both acidic and basic properties is said to be **amphoteric.**

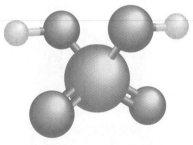

Sulfuric acid (H_2SO_4) is the industrial chemical produced on the largest scale in the world, in amounts exceeding 100 million tons per year. It is used for fertilizer production, as an intermediate in the manufacture of other chemicals, and in the metals and petroleum-refining industries.

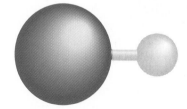

Hydrogen chloride (HCl) is produced directly from the elements or as a by-product in making chlorinated polymers from petroleum and chlorine. The largest use of the aqueous acid is in the pickling of steel and other metals to remove oxide layers on the surface.

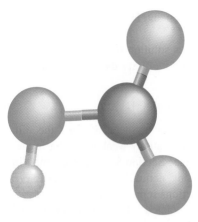

Nitric acid (HNO_3) is made largely from ammonia; once produced, most of it is reacted further with ammonia to make ammonium nitrate for use as a fertilizer.

Arrhenius acids and bases both release ions into solution. They thereby increase the electrical conductance of water and must be electrolytes. **Strong acids,** such as hydrochloric acid (HCl), nitric acid (HNO_3), and sulfuric acid (H_2SO_4), and **strong bases,** such as sodium hydroxide (NaOH) and potassium hydroxide (KOH), are strong electrolytes. Strong acids and strong bases dissociate essentially completely in water

$$HCl(g) \longrightarrow H^+(aq) + Cl^-(aq)$$
$$NaOH(s) \longrightarrow Na^+(aq) + OH^-(aq)$$

The Arrhenius model views the reaction between an acid and a base (called a **neutralization reaction**) as a metathesis. The H^+ from the acid and the OH^- from the base leave their previous partners to combine into neutral H_2O. The previous partners form a salt, which stays in solution or precipitates, depending on its solubility, or gets involved in other reactions. The following three chemical equations all represent the neutralization reaction between hydrochloric acid and sodium hydroxide.

$$HCl(aq) + NaOH(aq) \longrightarrow H_2O(\ell) + NaCl(aq) \quad \text{(overall equation)}$$
$$H^+(aq) + Cl^-(aq) + Na^+(aq) + OH^-(aq) \longrightarrow$$
$$H_2O(\ell) + Na^+(aq) + Cl^-(aq) \quad \text{(ionic equation)}$$
$$H^+(aq) + OH^-(aq) \longrightarrow H_2O(\ell) \quad \text{(net ionic equation)}$$

The driving force in Arrhenius neutralization is the formation of the molecular substance water by the combination of H^+ and OH^- ions.

Weak Acids

Only a handful of acids are strong electrolytes. They include the seven common strong acids: nitric acid (HNO_3), sulfuric acid (H_2SO_4), chloric acid ($HClO_3$), perchloric acid ($HClO_4$), hydrochloric acid (HCl), hydrobromic acid (HBr), and hydroiodic acid (HI). Other common acids are all weak. Important weak acids include hydrofluoric acid (HF), acetic acid (CH_3COOH), and formic acid (HCOOH), all of which can donate one hydrogen ion per molecule; oxalic acid ($H_2C_2O_4$), which can donate two hydrogen ions per molecule; and phosphoric acid (H_3PO_4), which can donate three hydrogen ions per molecule. Acetic acid has four hydrogens but only one "acidic hydrogen." Acidic hydrogens, the ones that can be donated, are gener-

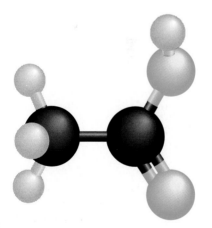

Acetic acid (CH_3COOH) is a weak organic acid. It is the active ingredient in vinegar, which is a 3% to 5% solution of acetic acid in water. The single acidic hydrogen atom in acetic acid is the one bonded to the oxygen. Pure acetic acid is called "glacial acetic acid" because it freezes into layered crystals that resemble glaciers.

ally bonded to oxygen or another electronegative element. Hydrogens bonded to carbon are rarely acidic.

Weak acids produce hydrogen ions in aqueous solution but to a limited extent that depends on their concentration (greater dilution *increases* dissociation). For example, the dissociation of acetic acid

$$CH_3COOH(aq) \longrightarrow CH_3COO^-(aq) + H^+(aq)$$

• Many organic acids contain the carboxylic acid group

$$-C\underset{\cdot\cdot O-H}{\overset{\cdot\cdot O}{\diagdown}}$$

in which the hydrogen atom is always acidic, and their formulas are written to show it.

stops with less than 10% of the acetic acid molecules dissociated unless the solution is quite dilute. Naturally, weak acids are not all equally weak but differ in their degree of dissociation at the same concentration (see Section 8–3). Despite their incomplete dissociation, weak acids, like strong acids, neutralize strong bases to produce water and salts. The equations for such reactions represent the actual situation better if they show the acidic hydrogen as undissociated. The neutralization of the acetic acid by sodium hydroxide is thus represented

$$CH_3COOH(aq) + NaOH(aq) \longrightarrow H_2O(\ell) + NaCH_3COO(aq) \quad \text{(overall equation)}$$

$$CH_3COOH(aq) + Na^+(aq) + OH^-(aq) \longrightarrow$$
$$H_2O(\ell) + Na^+(aq) + CH_3COO^-(aq) \quad \text{(ionic equation)}$$

$$CH_3COOH(aq) + OH^-(aq) \longrightarrow H_2O(\ell) + CH_3COO^-(aq) \quad \text{(net ionic equation)}$$

The coefficients in these balanced equations show that the neutralization of a mole of base does *not* require more weak acid than strong acid. Weakness and strength in Arrhenius acids relate to degree of dissociation, not to the ultimate amount of $H^+(aq)$ ion that can be furnished.

Naming Acids

Table 4–2 lists the names and formulas of a number of important acids. They are grouped in three categories: **binary acids,** which are formed by hydrogen with a second element; **oxoacids,** which are formed from hydrogen, oxygen, and a third element; and **organic acids,** which are based on compounds of carbon. The name of an

Table 4–2
Names of Common Acids

Binary Acids	Oxoacids	Organic Acids
HF, hydrofluoric acid	$HClO_4$, perchloric acid	HCOOH, formic acid
HCl, hydrochloric acid	$HClO_3$, chloric acid	CH_3COOH, acetic acid
HBr, hydrobromic acid	$HClO_2$, chlorous acid	C_6H_5COOH, benzoic acid
HI, hydroiodic acid	HClO, hypochlorous acid	HOOC—COOH, oxalic acid
HCN,[a] hydrocyanic acid	H_2SO_4, sulfuric acid	
H_2S, hydrosulfuric acid	H_2SO_3, sulfurous acid	
	HNO_3, nitric acid	
	HNO_2, nitrous acid	
	H_3PO_4, phosphoric acid	
	H_2CO_3, carbonic acid	

[a]Contains three elements but is named as a binary acid.

anion (see Table 3–2) is closely connected to that of the corresponding acid obtained by combining the anion with enough H^+ to give an electrically neutral compound:

1. *For binary acids.* Add *-ic* to the stem of the element (in place of *-ide* for the ion) and precede the stem by the prefix *hydro-*. For example, the stem of the element chlorine is *chlor-*:

• Hydrocyanic acid (HCN) is a ternary acid in this group.

chloride ion (Cl^-)   hydrochloric acid (HCl)

2. *For oxoacids.* Add *-ic* to the stem of the element to give the oxoacid corresponding to an oxoanion whose name ends in *-ate*. Add *-ous* to the stem for the oxoacid corresponding to an oxoanion whose name ends in *-ite*. For example, the stem of the element nitrogen is *nitr-*:

nitrate ion (NO_3^-)   nitric acid (HNO_3)
nitrite ion (NO_2^-)   nitrous acid (HNO_2)

The prefixes *per-* (largest number of oxygen atoms in the oxoacid) and *hypo-* (smallest number) are used, just as for oxoanions, in cases in which an element forms more than two oxoacids.

3. *For organic acids.* The stems for the common organic acids are nonsystematic and must be memorized. The same relation as for oxoacids holds between the names of the ion and the acid. For example:

acetate ion (CH_3COO^-)   acetic acid (CH_3COOH)

In naming an acid based on the element sulfur, the full stem *sulfur-* is used (for example, H_2S is hydrosulfuric acid, rather than "hydrosulfic acid," as the ion name "sulfide" might suggest). Acids based on phosphorus are named using the stem *phosphor-*. The binary compound H_2S is referred to as hydrogen sulfide in pure form (as a gas, liquid, or solid) but as hydrosulfuric acid when its acidic properties in aqueous solution are being discussed. This pattern holds for the other binary acids as well.

EXAMPLE 4–4

Give systematic names for the following acids: H_2Te, H_2CrO_4. Write the overall, ionic, and net ionic equations for the neutralization of H_2Te by potassium hydroxide (KOH), a strong base.

Solution

H_2Te is a binary acid based on the element tellurium (stem *tellur-*). Its name is thus hydrotelluric acid.

Note from Table 3–2 that the oxoanion corresponding to H_2CrO_4 is called the "chromate ion" (CrO_4^{2-}). The oxoacid is therefore called chromic acid. Like other inorganic acids (see Section 4–1), H_2Te dissolves in water. It is a weak acid (not being one of the seven strong acids). We therefore represent it among the reactants as undissociated and aquated. Neutralization gives water and a soluble salt (nearly all potassium salts are soluble (see Table 4–1). The equations are

$$H_2Te(aq) + 2\ KOH(aq) \longrightarrow 2\ H_2O(\ell) + K_2Te(aq) \quad \text{(overall equation)}$$

$$H_2Te(aq) + 2\ K^+(aq) + 2\ OH^-(aq) \longrightarrow$$
$$2\ H_2O(\ell) + 2\ K^+(aq) + Te^{2-}(aq) \quad \text{(ionic equation)}$$

$$H_2Te(aq) + 2\ OH^-(aq) \longrightarrow 2\ H_2O(\ell) + Te^{2-}(aq) \quad \text{(net ionic equation)}$$

Exercise

Give the systematic names for the acids H_3PO_3 and H_2SeO_3. Write the overall, ionic, and net ionic equations for the complete neutralization of H_2SeO_3 by sodium hydroxide.

Answer: The acids are phosphorous acid and selenous acid. The equations are:

$$H_2SeO_3(aq) + 2\ NaOH(aq) \longrightarrow 2\ H_2O(\ell) + Na_2SeO_3(aq) \quad \text{(overall equation)}$$

$$H_2SeO_3(aq) + 2\ Na^+(aq) + 2\ OH^-(aq) \longrightarrow$$
$$2\ H_2O(\ell) + 2\ Na^+(aq) + SeO_3^{2-}(aq) \quad \text{(ionic equation)}$$

$$H_2SeO_3(aq) + 2\ OH^-(aq) \longrightarrow 2\ H_2O(\ell) + SeO_3^{2-}(aq) \quad \text{(net ionic equation)}$$

Weak Bases

Any ionic compound containing hydroxide ions serves as an Arrhenius base simply by dissolving in water. However, the only soluble ionic hydroxides are $Ba(OH)_2$ and the hydroxides of the alkali metals (most notably NaOH and KOH). All of these compounds dissociate in water when they dissolve. They are strong bases. Other ionic hydroxides are either slightly soluble or insoluble. They donate OH^- ion poorly because they keep it mostly tied up in the solid. They are weak bases. Experimentally, poorly soluble weak bases are recognized by their ability to neutralize acids. Magnesium hydroxide ($Mg(OH)_2$), for example, dissolves only very slightly in water (about 0.001 g/100 g of water at 25°C). A fine powder of magnesium hydroxide suspended in water ("milk of magnesia") is an effective antacid, however, and is drunk to counteract excess stomach acidity. The small proportion of $Mg(OH)_2$ that dissolves indeed dissociates to Mg^{2+} and OH^- ions, but these ions are too few to raise the conductance of the water significantly. Hydroxides that are poorly soluble in water dissolve in acidic solutions by undergoing neutralization reactions. Equations for the reaction of magnesium hydroxide with hydrochloric acid are

$$2\ HCl(aq) + Mg(OH)_2(s) \longrightarrow 2\ H_2O(\ell) + MgCl_2(aq) \quad \text{(overall equation)}$$

$$2\ H^+(aq) + 2\ Cl^-(aq) + Mg(OH)_2(s) \longrightarrow$$
$$2\ H_2O(\ell) + Mg^{2+}(aq) + 2\ Cl^-(aq) \quad \text{(ionic equation)}$$

$$2\ H^+(aq) + Mg(OH)_2(s) \longrightarrow 2\ H_2O(\ell) + Mg^{2+}(aq) \quad \text{(net ionic equation)}$$

White solid $Mg(OH)_2$ disappears into the clear aqueous solution of HCl as this reaction proceeds.

Modifying the Arrhenius Model

Many compounds, such as ammonia (NH_3), methylamine (CH_3NH_2), and pyridine (C_5H_5N), have a bitter taste and a soapy feel and neutralize acids, but *do not contain hydroxide ions*. To include these substances as bases, the Arrhenius definition is expanded as follows:

An Arrhenius base is a substance that, when dissolved in water, increases the amount of hydroxide ion over what is present in the pure solvent.

Ammonia, like the other compounds just named, increases the number of hydroxide ions by reacting with water

$$NH_3(aq) + H_2O(\ell) \longrightarrow NH_4^+(aq) + OH^-(aq)$$

The reaction occurs to only a slight extent. Although ammonia is quite soluble, it is a weak base; conductance measurements confirm that it is a weak electrolyte.

The definition of an Arrhenius acid is modified similarly:

An Arrhenius acid is a substance that, when dissolved in water, increases the amount of hydrogen ion over what is present in the pure solvent.

The neutralization of aqueous ammonia by hydrochloric acid

$$NH_3(aq) + HCl(aq) \longrightarrow NH_4Cl(aq) \quad \text{(overall equation)}$$
$$NH_3(aq) + H^+(aq) + Cl^-(aq) \longrightarrow NH_4^+(aq) + Cl^-(aq) \quad \text{(ionic equation)}$$
$$NH_3(aq) + H^+(aq) \longrightarrow NH_4^+(aq) \quad \text{(net ionic equation)}$$

produces a salt, but no water. This fact does not fit neatly into the Arrhenius model. An early attempt to repair the difficulty was the suggestion that dissolved NH_3 reacts completely with water to form soluble "ammonium hydroxide" (NH_4OH) that is then neutralized according to

$$NH_4OH(aq) + HCl(aq) \longrightarrow H_2O(\ell) + NH_4Cl(aq) \quad \text{(overall equation)}$$

Although concentrated ammonia solutions are still marketed under the name "ammonium hydroxide," few NH_4^+ and OH^- ions are present in them. Moreover, the reaction between NH_3 and HCl occurs even in the complete absence of water (between the two gases) to produce the same salt (ammonium chloride) as a solid and no water. No amount of stretching can make this gas-phase reaction into an Arrhenius neutralization. Other acid–base concepts, some of which are explored in Chapter 8, accommodate it easily.

Acid and Base Anhydrides

The modification of the Arrhenius definitions qualifies numerous additional substances as acids and bases. Of particular interest and importance are the binary oxides. For example, the freely soluble gas sulfur trioxide (SO_3) gives solutions in which the amount of hydrogen ion far exceeds the amount present in pure water; sulfur trioxide fits the broadened definition of an acid well. In fact, SO_3 reacts with water as it dissolves, in what is called a **hydration** reaction, to give sulfuric acid

$$SO_3(g) + H_2O(\ell) \longrightarrow H_2SO_4(aq)$$

The sulfuric acid, which is a strong oxoacid, then donates hydrogen ions. As for bases, sodium oxide (Na_2O) becomes an Arrhenius base under the modified definition because it too undergoes a hydration reaction

$$Na_2O(s) + H_2O(\ell) \longrightarrow 2\,NaOH(aq)$$

and the product, sodium hydroxide, is an excellent hydroxide-ion donor.

Many other oxides behave in a similar way. They are termed **acid anhydrides** and **base anhydrides** ("anhydrous" means without water), depending on whether they give acidic or basic solutions. In general, ionic oxides are base anhydrides, and covalent or partially covalent oxides are acid anhydrides. Therefore, with few ex-

Figure 4–8 Among the oxides of the main-group elements, behavior as an acid anhydride tends to strengthen from left to right and from bottom to top in the periodic table; behavior as a base anhydride does the reverse. Acidity also increases with increasing oxidation number of the element. Oxygen difluoride, however, has only weakly acidic properties. The oxides listed here have the element in the maximum oxidation state.

Increasing acidity ⟶

I	II	III	IV	V	VI	VII
Li_2O	BeO	B_2O_3	CO_2	N_2O_5	(O_2)	OF_2
Na_2O	MgO	Al_2O_3	SiO_2	P_4O_{10}	SO_3	Cl_2O_7
K_2O	CaO	Ga_2O_3	GeO_2	As_2O_5	SeO_3	Br_2O_7
Rb_2O	SrO	In_2O_3	SnO_2	Sb_2O_5	TeO_3	I_2O_7
Cs_2O	BaO	Tl_2O_3	PbO_2	Bi_2O_5	PoO_3	At_2O_7

Increasing basicity (↓, left)   Increasing acidity (↑, right)

⟵ Increasing basicity

• Non-metals have electronegativities close to that of the non-metal oxygen, and so their oxides have predominantly covalent bonding character.

ceptions, the oxides of the non-metals are acid anhydrides, and the oxides of alkali metals and alkaline-earth metals are base anhydrides (Fig. 4–8). Oxides of semi-metals, such as boron, and oxides in which a non-metal is in a lower oxidation state are acid anhydrides, but the hydration reaction generates weakly acidic solutions.

$$Cl_2O(g) + H_2O(\ell) \longrightarrow 2\ HOCl(aq)$$

oxide of a non-metal in a low oxidation state   weak acid

Oxides that have bonding on the borderline between ionic and covalent are frequently amphoteric (intermediate between acid and base) in their behavior as anhydrides and indeed dissolve to only a limited extent in water. Transition-metal oxides show a range of acidic and basic properties, with acidity usually increasing with the oxidation state of the element.

The formulas of acid and base anhydrides can be generated in a purely formal operation from the chemical formulas of oxoacids and hydroxides by removing molecules of water until only an oxide remains. This corresponds to the **dehydration** of the compound. Thus, *subtracting* a molecule of water from the formula H_2CO_3 (carbonic acid) identifies CO_2 (carbon dioxide) as an acid anhydride

$$H_2CO_3 - H_2O \longrightarrow CO_2$$

When an oxoacid or hydroxide contains an odd number of hydrogen atoms in its chemical formula, the formula is doubled and *then* water molecules are removed until no hydrogen atoms remain.

EXAMPLE 4–5

What is the acid anhydride of nitrous acid (HNO_2)?

Solution

The dehydration of nitrous acid can be represented by

$$2\ HNO_2 \longrightarrow N_2O_3 + H_2O$$

Consequently, dinitrogen trioxide (N_2O_3) is the acid anhydride of nitrous acid.

The subtraction of water molecules from an oxoacid or hydroxide can sometimes be carried out experimentally, but not always. Although carbonic acid dehydrates readily to its acid anhydride,

$$H_2CO_3(aq) \longrightarrow CO_2(g) + H_2O(\ell)$$

any attempt to dehydrate liquid perchloric acid by heating it to induce the reaction

$$2\,HClO_4(\ell) \longrightarrow Cl_2O_7(\ell) + H_2O(\ell)$$

would not only fail but most likely cause an explosion. Other, less direct ways of dehydrating perchloric acid must be used.

Figure 4–9 When acetic acid is added to sodium carbonate (washing soda), bubbles of carbon dioxide form:

$2\,CH_3COOH(aq) + Na_2CO_3(s) \longrightarrow$
$2\,Na^+(aq) + 2\,CH_3COO^-(aq) +$
$CO_2(g) + H_2O(\ell)$

Further Reactions of Acids and Bases

So far we know that solutions of Arrhenius acids and bases neutralize each other to give water plus a salt and cause color changes in indicators. What else do they do? The following typical reactions of acids can be used to predict the course and products of large numbers of reactions:

1. Acids act on carbonates and hydrogen carbonates (such as Na_2CO_3 or $NaHCO_3$) to liberate gaseous CO_2 and produce a salt and water (Fig. 4–9). A typical reaction is

   $NaHCO_3(s) + HCl(aq) \longrightarrow NaCl(aq) + H_2O(\ell) + CO_2(g)$ (overall equation)

   $NaHCO_3(s) + H^+(aq) \longrightarrow Na^+(aq) + H_2O(\ell) + CO_2(g)$ (net ionic equation)

2. Acids react with the oxides of metals to form salts and water. A typical reaction is

   $CuO(s) + H_2SO_4(aq) \longrightarrow CuSO_4(aq) + H_2O(\ell)$ (overall equation)

   $CuO(s) + 2\,H^+(aq) \longrightarrow Cu^{2+}(aq) + H_2O(\ell)$ (net ionic equation)

3. Acids react with zinc, iron, and many other metallic elements to liberate gaseous H_2 and form a salt (Fig. 4–10).

   $Zn(s) + 2\,HCl(aq) \longrightarrow ZnCl_2(aq) + H_2(g)$ (overall equation)

   $Zn(s) + 2\,H^+(aq) \longrightarrow Zn^{2+}(aq) + H_2(g)$ (net ionic equation)

The first reaction is easy to summarize: acids make carbonates fizz. Visualize a changing of ionic partners. Hydrogen ions from the acid join to the carbonate (or hydrogen carbonate) to produce carbonic acid (H_2CO_3). This unstable compound quickly decomposes to give water and carbon dioxide (CO_2), which is not very soluble in water. The fizzing occurs as gaseous CO_2 bubbles out of solution.

In the second reaction, hydrogen ions from the acid are attracted by the negative charge on the oxide ions. They link to the oxides, forming the stable molecular compound H_2O. This frees the metal ion to go into solution or to take part in further reactions on its own.

Figure 4–10 As zinc dissolves in dilute hydrochloric acid, bubbles of hydrogen appear, and zinc chloride forms in solution.

The third reaction is the displacement of hydrogen ions from solution by "active metals." This reaction is discussed along with other oxidation–reduction reactions in Section 4–4.

EXAMPLE 4–6

Pickling is the removal of a surface oxide layer on a metal by its reaction with an acid or base to give a soluble salt. Write a balanced chemical equation representing the reaction between hydrochloric acid and nickel(II) oxide, $NiO(s)$.

Solution

Hydrochloric acid is an aqueous solution of hydrogen chloride. We predict the reaction of nickel oxide with $HCl(aq)$ according to the pattern set by other metal oxides and acids. The reaction is

$$NiO(s) + 2\ HCl(aq) \longrightarrow NiCl_2(aq) + H_2O(\ell) \quad \text{(overall equation)}$$
$$NiO(s) + 2\ H^+(aq) \longrightarrow Ni^{2+}(aq) + H_2O(\ell) \quad \text{(net ionic equation)}$$

The salt formed in such a reaction might precipitate. That precipitation would be represented by a separate equation. The nickel(II) chloride ($NiCl_2$) in this example is water-soluble. What is observed is the dissolution of the green-black $NiO(s)$ to give a green solution.

Exercise

One of the constituents of acid rain is nitric acid. Marble (used in monuments and statues) has the chemical composition of calcium carbonate. Write balanced chemical equations representing how rain containing nitric acid dissolves marble statues.

Answer: $CaCO_3(s) + 2\ HNO_3(aq) \longrightarrow$
$$Ca(NO_3)_2(aq) + H_2O(\ell) + CO_2(g) \quad \text{(overall equation)}$$
$$CaCO_3(s) + 2\ H^+(aq) \longrightarrow Ca^{2+}(aq) + H_2O(\ell) + CO_2(g) \quad \text{(net ionic equation)}$$

Bases also have typical reactions:

---

1. With the exception of ammonia itself, bases act on ammonium salts, such as NH_4Cl and $(NH_4)_2SO_4$, to generate gaseous ammonia, a salt, and water. For instance:

$$NH_4Cl(s) + NaOH(aq) \longrightarrow NaCl(aq) + NH_3(g) + H_2O(\ell) \quad \text{(overall equation)}$$
$$NH_4Cl(s) + OH^-(aq) \longrightarrow Cl^-(aq) + NH_3(g) + H_2O(\ell) \quad \text{(net ionic equation)}$$

2. Bases react with the oxides of non-metals to produce salts and water. For example:

$$SO_3(g) + 2\ NaOH(aq) \longrightarrow Na_2SO_4(aq) + H_2O(\ell) \quad \text{(overall equation)}$$
$$SO_3(g) + 2\ OH^-(aq) \longrightarrow SO_4^{2-}(aq) + H_2O(\ell) \quad \text{(net ionic equation)}$$

The first reaction starts with a metathesis to give NH_4OH, which, if it ever actually exists, instantly breaks down to $NH_3(aq)$ and H_2O. Although ammonia is actually quite soluble in water, some (but not enough to fizz) still wafts out of the solution. Even small amounts of ammonia are easily detected because of ammonia's penetrating odor.

Acids and Bases in and Around You

Acids and bases are everywhere. Insoluble carbonates (in limestone) form the foundation stones of houses; dilute nitric acid in the rain (from the reaction of atmospheric nitrogen with oxygen during lightning flashes) weathers the stones away. Amino acids contain a basic group (the amino group NH_2) and an acidic group (the carboxylic acid group COOH) in every molecule; they are the building blocks of the proteins that regulate the operation of living cells. Nucleic acids make up the genetic material in cells (DNA stands for "deoxyribonucleic acid," and RNA stands for "ribonucleic acid"; both are derivatives of phosphoric acid). Most food is acidic; soaps, detergents, and cleansers for washing dishes and for cleaning the stove are basic. Citrus fruits, which all contain citric acid, furnish vitamin C, which is also called "ascorbic acid." Cola drinks contain phosphoric acid or lactic acid (read the labels), which give them their tartness; salad dressings contain vinegar (dilute acetic acid). Window-washing solutions contain the base ammonia, and the active ingredient in spray-on oven cleaners is caustic sodium hydroxide. Both promote the conversion of sticky, greasy materials into soapy, soluble residues that can be easily wiped away. Cooking tomato sauce in an aluminum sauce pan can pit the metal—tomatoes are loaded with acids, and aluminum is an active metal.

The body depends on a complex web of acid–base chemistry. Normal blood remains nearly neutral thanks to several interacting acid–base reactions. Some conditions, such as severe diarrhea, diabetes, and prolonged starva-

Figure 4–A Many common foods are acidic; most cleaning agents are basic.

tion, can lead to *acidosis*, a state in which the blood is too acidic. One treatment for acidosis is intravenous administration of solutions containing hydrogen carbonate (HCO_3^-) ions, which react with the acid. Prolonged vomiting or the ingestion of excessive amounts of hydrogen carbonate ion can lead to *alkalosis*, a state in which the blood is too basic. This sometimes occurs in self-treatment of stomach pain with sodium hydrogen carbonate ($NaHCO_3$, "bicarbonate of soda"). Carefully monitored intravenous administration of aqueous ammonium chloride ($NH_4Cl(aq)$) is sometimes used to treat alkalosis. Both acidosis and alkalosis are life-threatening medical emergencies. The treatments for both are straightforward uses of the typical reactions of acids and bases.

The second reaction completes a symmetry: bases react with the oxides of *non-metals* to give a salt and water just as acids react with the oxides of *metals* to give a salt and water.

EXAMPLE 4–7

Predict the products of a reaction between solid ammonium sulfate and aqueous potassium hydroxide.

Solution

The reaction resembles that between NH_4Cl and $NaOH$ on page 144.

$$(NH_4)_2SO_4(s) + 2\ KOH(aq) \longrightarrow$$
$$K_2SO_4(aq) + 2\ NH_3(g) + 2\ H_2O(\ell) \quad \text{(overall equation)}$$

$$(NH_4)_2SO_4(s) + 2\ OH^-(aq) \longrightarrow$$
$$SO_4^{2-}(aq) + 2\ NH_3(g) + 2\ H_2O(\ell) \quad \text{(net ionic equation)}$$

These products are aqueous potassium sulfate, ammonia, and water.

> **Exercise**
>
> In a certain combustion train used for the determination of carbon and hydrogen (see Fig. 2–4), the gaseous carbon dioxide produced by combustion is absorbed in a reaction with barium hydroxide. Predict the products of this reaction.
>
> **Answer:** $BaCO_3$ and H_2O.

4–4   OXIDATION–REDUCTION REACTIONS

In Chapter 3, we learned to predict the formulas of ionic compounds that result from the direct combination of two elements. The reactions that give these compounds are all **oxidation–reduction reactions,** an extensive and important class that is characterized by the transfer of electrons.

When sodium burns in chlorine to give sodium chloride (see Fig. 1–1),

$$2\,Na(s) + Cl_2(g) \longrightarrow 2\,NaCl(s)$$

the sodium is **oxidized:** it *gives up* electrons as the charge on its atoms *increases* from zero (in elemental Na) to $+1$ (in NaCl). Some species must always be present to *accept* these electrons. This electron-accepting species is then said to be **reduced.** In this case, chlorine is reduced, as the charge on its atoms *decreases* from zero to -1. We can indicate the transfer of electrons by means of vertical arrows:

$$\overset{0}{2\,Na} + \overset{0}{Cl_2} \longrightarrow \overset{+1\ -1}{2\,NaCl}$$
$$2 \times 1\,e^- = 1 \times 2 \times 1\,e^-$$

Note the following conventions: the arrows point *away* from the species being oxidized (giving up electrons) and *toward* the species being reduced (accepting electrons). The charges are indicated above the species. The electron bookkeeping beneath the equation ensures that the same number of electrons is taken up by chlorine as is given up by sodium:

number of electrons lost by sodium = number of electrons gained by chlorine

$$2\,Na\ atoms \times \frac{1\,e^-\ lost}{1\,Na\ atom} = 1\,Cl_2\ molecule \times \frac{2\,Cl\ atoms}{1\,Cl_2\ molecule} \times \frac{1\,e^-\ gained}{1\,Cl\ atom}$$

$$2\ electrons\ lost = 2\ electrons\ gained$$

Originally, the term "oxidation" referred only to reactions of a substance with oxygen, such as the burning of magnesium in air (Fig. 4–11).

$$\overset{0}{2\,Mg(s)} + \overset{0}{O_2(g)} \longrightarrow \overset{+2\ -2}{2\,MgO(s)}$$
$$2 \times 2\,e^- = 1 \times 2 \times 2\,e^-$$

Figure 4–11   The burning of magnesium in air gives off an extremely bright light. This led to the incorporation of magnesium into the flash powder used in early photography. Magnesium powder is still used in fireworks and flares.

At that time, the term "reduction" had a narrower meaning as well—the winning of a metallic element from chemical combination with non-metals such as oxygen or sulfur in an ore. Iron(III) oxide, for example, is reduced by carbon monoxide to elemental iron in the reaction

$$\overset{+3}{Fe_2O_3(s)} + 3\,\overset{+2}{CO(g)} \longrightarrow 2\,\overset{0}{Fe} + 3\,\overset{+4}{CO_2}$$

The terms "oxidation" and "reduction" are now used to describe reactions of all species, as long as electron transfer takes place.

Oxidation Number

Sodium chloride and other compounds that are combinations between Group I or II elements and Group VI or VII elements are ionic. They are legitimately described as mixtures of ions with integral charges, and, as just shown, there is no difficulty in tracing the transfer of electrons in the formation of these compounds. In contrast, compounds formed from elements in the middle of the periodic table have mixed covalent and ionic character and are not composed of well-defined ionic units. Thus, tin(IV) chloride ($SnCl_4$) is a molecular compound that is very poorly described as a grouping of Sn^{4+} and Cl^- ions. When tin and chlorine combine to form $SnCl_4$, many valence electrons end up being shared in covalent bonds. What proportion (if any) of these electrons should be counted as transferred? This question is answered by the concept of the **oxidation number.** Oxidation numbers (also called **oxidation states**) are determined for the atoms in covalently bonded compounds by applying the following simple set of rules:

• Oxidation states were briefly introduced in Section 3–8.

> **1.** The oxidation numbers of the atoms in a neutral molecule must add up to zero, and those in an ion must add up to the charge on the ion.
> **2.** Alkali metal (Group I) atoms have oxidation number $+1$, and alkaline earth (Group II) atoms have oxidation number $+2$ in their compounds; atoms of Group III elements usually have oxidation number $+3$ in their compounds.
> **3.** Fluorine always has an oxidation number of -1 in its compounds. The other halogens have oxidation number -1 in their compounds, *except* in compounds with oxygen and with other halogens, in which they can have positive oxidation numbers.
> **4.** Hydrogen is assigned an oxidation number of $+1$ in its compounds, *except* in metal hydrides such as LiH, in which rule 2 takes precedence and hydrogen has an oxidation number of -1.
> **5.** Oxygen is assigned an oxidation number of -2 in compounds. There are two exceptions: in compounds with fluorine, rule 3 takes precedence, and in compounds that contain O—O bonds, rules 2 and 4 take precedence. Thus, the oxidation number of oxygen in OF_2 is $+2$; in peroxides (such as H_2O_2 and Na_2O_2), its oxidation number is -1, and in superoxides (such as KO_2), its oxidation number is $-\frac{1}{2}$.

• Fractional oxidation numbers, although not common, are allowed and in fact are necessary to be consistent with this set of rules.

Rule 1 is essential if the total number of electrons is to remain constant in chemical reactions. This rule also requires that the oxidation numbers of atoms in the uncombined (free) forms of the elements be zero. Rules 2 through 5 are conventions based on the principle that in ionic compounds the sum of the oxidation numbers must equal the charge on the ion.

Oxidation numbers must not be confused with the formal charges of Chapter 3. Their resemblance to formal charges arises because both are assigned by arbitrary rules to symbols in formulas for specific purposes. The purposes are different, however. Formal charges are used solely in connection with Lewis structures. Oxidation numbers are used in nomenclature, as in the names iron(II) chloride and iron(III) chloride (see Section 3–8) and in identifying electron transfer in oxidation-reduction reactions. In Chapter 17 and later chapters, we see how oxidation numbers help in exploring trends in chemical reactivity across the periodic table.

With the preceding rules in hand, chemists can assign oxidation numbers to the atoms in the vast majority of compounds. Apply rules 2 through 5 as listed above, noting the exceptions given. Then assign oxidation numbers to the other elements in such a way that rule 1 is always obeyed. Note that rule 1 applies not only to free ions but also to the ions making up an ionic solid. Chlorine has oxidation number -1 not only as a free Cl^- ion but also in the ionic solid $AgCl$. It is important to recognize common ionic species (especially molecular ions) and to know the total charge they carry. Table 3–2 lists the names and formulas of many common anions. Inspection of the table reveals that several elements exhibit different oxidation numbers in different anions. In Ag_2S, for example, sulfur appears as the sulfide ion and has oxidation number -2, but in Ag_2SO_4 it appears as part of a sulfate (SO_4^{2-}) group and must have oxidation number $+6$ in order to satisfy rule 1:

$$\text{(oxidation number of S)} + 4 \times \text{(oxidation number of O)} = \text{total charge on ion}$$
$$+6 + 4(-2) = -2$$

A convenient way to indicate the oxidation number of an atom is to write it directly above the corresponding symbol in the formula of the compound:

$$\overset{+1\,-2}{N_2O} \qquad \overset{+1\,-1}{LiH} \qquad \overset{0}{O_2} \qquad \overset{+6\,-2}{SO_4^{2-}}$$

EXAMPLE 4–8

Assign oxidation numbers to all the elements in the following chemical compounds and ions:

(a) $NaCl$
(b) ClO^-
(c) $Fe_2(SO_4)_3$
(d) SO_2
(e) I_2
(f) $KMnO_4$
(g) CaH_2

Solution

(a) $\overset{+1\,-1}{NaCl}$, from rules 2 and 3.

(b) $\overset{+1\,-2}{ClO^-}$, from rules 1 and 5.

(c) $\overset{+3\,+6\,-2}{Fe_2(SO_4)_3}$, from rules 1 and 5. Note that this is solved by recognizing the presence of sulfate (SO_4^{2-}) groups.

(d) $\overset{+4\,-2}{SO_2}$, from rules 1 and 5.

(e) $\overset{0}{I_2}$, from rule 1. I_2 is an element.

(f) $\overset{+1\,+7\,-2}{KMnO_4}$, from rules 1, 2, and 5. Note that this can also be solved by recognizing the presence of permanganate (MnO_4^-) groups.

(g) $\overset{+2\,-1}{CaH_2}$, from rules 1 and 4 (metal hydride case).

Exercise

Determine the oxidation number of carbon in each of the following compounds: CH_4, CH_2O, CO_2, and C_6H_6.

Answer: -4, 0, $+4$, and -1.

We can now make the following definitions:

> An atom is **oxidized** (loses electrons) if its oxidation number increases in a chemical reaction; an atom is **reduced** (gains electrons) if its oxidation number decreases.

The amount of the change tells how many electrons are lost or gained. If any atom in a chemical reaction changes its oxidation state, then the reaction is a redox reaction.

Predicting Redox Reactions

In discussing redox reactions, chemists speak of the relative strengths of oxidizing and reducing agents. An **oxidizing agent** (or oxidant) causes the oxidation of another species by accepting electrons from it. It is itself reduced in the process. A **reducing agent** (or reductant) gives electrons to another species and is itself oxidized. A strong oxidizing agent can take electrons away from a poor reducing agent; a strong reducing agent can force electrons onto a poor oxidizing agent. Predicting the occurrence and outcome of redox reactions requires facts about the relative strengths of different atoms, molecules, and ions as oxidizing or reducing agents. Experiments have provided much information of this kind along with two realizations:

- Both the strength of an oxidizing or reducing agent and the identity of its products can change drastically with conditions. In particular, the course of a redox reaction in aqueous solution depends on whether the solution is acidic or basic.
- Redox strength is not the same as *rapidity* of action. An oxidizing agent (or reducing agent) that is weaker may sometimes react more rapidly than one that is stronger.

The first point is illustrated by some reactions of a well-known, potent oxidizing agent: the aqueous permanganate ion ($MnO_4^-(aq)$). If iron(II) sulfate and an acidic solution of potassium permanganate are mixed, they react according to the net ionic equation

$$5\ Fe^{2+}(aq) + MnO_4^-(aq) + 8\ H^+(aq) \longrightarrow$$
$$5\ Fe^{3+}(aq) + Mn^{2+}(aq) + 4\ H_2O(\ell) \quad \text{(acidic solution)}$$

The oxidation number of manganese starts at $+7$ (in the permanganate ion) and drops to $+2$ (in the manganese(II) ion), showing that the permanganate ion is reduced, while the oxidation number of iron rises from $+2$ to $+3$, showing that it is oxidized. The same two reactants give different products under neutral and basic conditions:

$$3\ Fe^{2+}(aq) + MnO_4^-(aq) + 4\ H^+(aq) \longrightarrow$$
$$3\ Fe^{3+}(aq) + MnO_2(s) + 2\ H_2O(\ell) \quad \text{(neutral solution)}$$

$$Fe^{2+}(aq) + MnO_4^-(aq) + 3\ OH^-(aq) \longrightarrow$$
$$Fe(OH)_3(s) + MnO_4^{2-}(aq) \quad \text{(basic solution)}$$

Figure 4–12 Redox strength is *not* the same as rapidity of action. Copper (left) reduces nitric acid more rapidly than aluminum, although aluminum is a stronger reducing agent.

A similar diversity of products is found with other oxidizing and agents with reducing agents as well. Nitrogen in ammonia (NH_3) has an oxidation number of -3. When ammonia serves as a reductant, N ends up in oxidation states ranging from -2 to $+5$, depending on the conditions. Such factors clearly complicate the prediction of the products even when a redox reaction is certain to take place.

The quite different rates at which nitric acid attacks aluminum and copper illustrate the second point. Aluminum ($Al(s)$) is fundamentally a far stronger reducing agent than copper ($Cu(s)$). Yet copper dipped into nitric acid immediately begins a vigorous redox reaction, whereas aluminum remains unaffected (Fig. 4–12).

We examine the relative strengths of oxidizing and reducing agents in more detail in Chapters 12 and 13. In the meantime, we describe some important types of redox reactions and offer guidelines to decide whether they occur.

Types of Redox Reactions

Redox Combination and Decomposition Reactions

* Many of these reactions are vigorous. Figure 2–8 shows the metal aluminum and the non-metal bromine as they combine.

Most metallic elements react with most non-metallic elements to give ionic compounds. The metals are oxidized; the non-metals are reduced. Simple Lewis theory (see Section 3–3) predicts the products of some of these **combination** reactions. If the metal forms two or more ions (see Section 3–8), multiple products are usually possible. For example, the reaction of iron and oxygen gives iron(II) oxide (FeO) or iron(III) oxide (Fe_2O_3), depending on conditions. Usually, a higher oxidation state in the metal is favored by an excess of the non-metal.

Non-metallic elements also can combine directly with other non-metallic elements. Covalent compounds result. In these reactions the oxidation number of the less electronegative element in the final compound depends on the conditions. For example, phosphorus reacts vigorously (Fig. 4–13) with chlorine. If the relative amount of chlorine is limited, the product is phosphorus trichloride (PCl_3)

$$\overset{0}{P_4}(s) + 6\ \overset{0}{Cl_2}(g) \longrightarrow 4\ \overset{+3\ -1}{PCl_3}(\ell)$$

but with an excess of chlorine, phosphorus is oxidized all the way to the +5 state, and the product is phosphorus pentachloride

$$\overset{0}{P_4}(s) + 10\ \overset{0}{Cl_2}(g) \longrightarrow 4\ \overset{+5\ -1}{PCl_5}(\ell)$$

Similarly, carbon burns in oxygen to give both carbon dioxide (CO_2) and carbon monoxide (CO). Although an excess of oxygen favors CO_2, some CO almost always forms.

Combination reactions can be reversed by a proper choice of conditions. Such reversed reactions are **decompositions** because a more complex substance breaks down to two or more simpler substances. In Lavoisier's preparation of oxygen (see Section 1–3), strong heating decomposed mercury(II) oxide to its elements

$$2\ \overset{+2\ -2}{HgO}(s) \longrightarrow 2\ \overset{0}{Hg}(\ell) + \overset{0}{O_2}(g)$$

although mild heating of the two elements generated the compound in the first place. Similar decomposition reactions occur when mercury(I) oxide and silver oxide are heated moderately

$$2\ \overset{+1\ -2}{Hg_2O}(s) \longrightarrow 4\ \overset{0}{Hg}(\ell) + \overset{0}{O_2}(g)$$
$$2\ \overset{+1\ -2}{Ag_2O}(s) \longrightarrow 4\ \overset{0}{Ag}(s) + \overset{0}{O_2}(g)$$

A sufficiently high temperature causes the redox breakdown of any compound to its constituent elements, although the temperatures required to decompose some strongly bonded compounds are truly extreme.

Often, decomposition to an element and a compound occurs upon heating. Hydrogen peroxide decomposes to give oxygen and water

$$2\ \overset{+1\ -1}{H_2O_2}(\ell) \longrightarrow 2\ \overset{+1\ -2}{H_2O}(\ell) + \overset{0}{O_2}(g)$$

at low temperature, but the H_2O resists further decomposition. Moderate heating of sulfuryl chloride ($SO_2Cl_2(g)$) decomposes it to sulfur dioxide ($SO_2(g)$) and $Cl_2(g)$; no O_2 or S is produced unless the temperature is much higher. Decomposition products can sometimes be predicted by looking for the formulas of simple stable compounds (such as H_2O, HCl, CO_2, SO_2, or NaCl) embedded in the formula of the substance being heated. On this basis, one predicts that heating $KClO_3$ would break it down to KCl (a simple stable compound) and O_2

$$2\ \overset{+1\ +5\ -2}{KClO_3}(\ell) \longrightarrow 2\ \overset{+1\ -1}{KCl}(s) + 3\ \overset{0}{O_2}(g)$$

which is exactly what happens. Similarly, heating the salt ammonium nitrate melts it and then drives out the stable compound water

$$\overset{-3\ +1\ +5\ -2}{NH_4NO_3}(\ell) \longrightarrow \overset{+1\ -2}{N_2O}(g) + 2\ \overset{+1\ -2}{H_2O}(g)$$

• Burning charcoal (mostly carbon) in a brazier for heat in a closed room has led to many deaths by inhalation of the poisonous minor product carbon monoxide.

Figure 4–13 White phosphorus (P_4) reacts vigorously with chlorine to give phosphorus pentachloride (PCl_5). If the relative amount of chlorine is restricted, the reaction gives the trichloride PCl_3.

Some decompositions in which one compound breaks down to two or more new compounds are not redox reactions. This includes, for example, all dehydration reactions (see Section 4–3). The point is easily checked by determining all oxidation numbers on both sides of the equation.

Oxygenation

Oxygen occurs in two forms: ordinary atmospheric oxygen ("dioxygen" consisting of O_2 molecules) and the less stable ozone ("trioxygen," consisting of O_3 molecules). Both are powerful oxidants. In both forms, oxygen combines directly with most other elements and attacks many compounds to give binary oxides—compounds in which oxygen is in a negative oxidation state and everything else is in some positive oxidation state and combined with oxygen. These are **oxygenation** reactions (and also, of course, combination reactions). Examples are

$$\overset{0}{4\,Li(s)} + \overset{0}{O_2(g)} \longrightarrow \overset{+1\ -2}{2\,Li_2O(s)}$$

and

• This reaction appears in Example 2-13.

$$\overset{+2\,-2}{2\,ZnS(s)} + \overset{0}{3\,O_2(g)} \longrightarrow \overset{+2\,-2}{2\,ZnO(s)} + \overset{+4\,-2}{2\,SO_2(g)}$$

The electropositive elements sodium and barium can be more thoroughly oxygenated if oxygen is plentiful to form the **peroxides** Na_2O_2 and BaO_2 in reactions such as

$$\overset{0}{Ba(s)} + \overset{0}{O_2(g, \text{ in excess})} \longrightarrow \overset{+2\ -1}{BaO_2(s)}$$

Potassium, rubidium, and cesium are even more avid for oxygen and form the **superoxides** KO_2, RbO_2, and CsO_2 through reactions such as

$$\overset{0}{K(s)} + \overset{0}{O_2(g, \text{ in excess})} \longrightarrow \overset{+1\,-\frac{1}{2}}{KO_2(s)}$$

Oxides are more stable than peroxides or superoxides and usually are obtained either by allowing the peroxides and superoxides to react with additional metal or oxygenating the metal with a restricted supply of oxygen.

Practically every binary compound involving hydrogen can be oxidized with oxygen to water and an oxide. For example

$$\overset{-3\ +1}{4\,PH_3(g)} + \overset{0}{8\,O_2(g)} \longrightarrow \overset{+5\ -2}{P_4O_{10}(s)} + \overset{+1\,-2}{6\,H_2O(g)}$$

Most reactions of O_2 are slow at room temperature but become rapid at high temperature. Once O_2 and another substance are combining rapidly enough to generate light and heat, the process becomes a combustion. Heating oxygen-containing mixtures often ignites combustion. Complete combustion of hydrocarbons (organic compounds containing C and H only) yields carbon dioxide and water. For example, octane burns according to the equation

Molten sulfur burns in oxygen to give sulfur dioxide: $S(\ell) + O_2(g) \longrightarrow$ $SO_2(g)$. The higher oxide, sulfur trioxide, forms under similar conditions in the presence of platinum.

• The combustion of hydrocarbon fuels refined from petroleum is essential to modern civilization.

$$\overset{-\frac{9}{4}\ +1}{C_8H_{18}(\ell)} + \overset{0}{\tfrac{25}{2}\,O_2(g)} \longrightarrow \overset{+4\ -2}{8\,CO_2(g)} + \overset{+1\,-2}{9\,H_2O(g)}$$

Similar combustions take place with carbohydrates, with other compounds that contain C, H, and O only, and with a large number of compounds containing predominantly C and H. In all these cases, complete combustion produces CO_2 and H_2O. Incomplete combustion occurs when the supply of oxygen is limited. It gives rise to new compounds containing either more oxygen (if the oxygen joins into the compound) or less hydrogen (if the oxygen takes out some hydrogen as water) or both.

Hydrogenation

Elemental hydrogen (H_2) is a good reducing agent. It **hydrogenates** other substances. Complete hydrogenation of a compound or element usually leads to new binary compounds having hydrogen in the $+1$ oxidation state and the other elements in negative oxidation states. For example, the hydrogenation of phosphorus gives PH_3, the hydrogenation of sulfur gives H_2S, and the hydrogenation of bromine gives HBr. These hydrogenations are also combination reactions, and the formulas of the products are readily predicted using Lewis theory.

In some hydrogenation reactions, H_2 is reduced, not oxidized. These involve very strong reducing agents, such as the alkali and heavier alkaline earth metals, and they give compounds in which hydrogen is in the -1 oxidation state. Hydrogen reacts at high pressures and elevated temperatures with liquid sodium, for example, to give sodium hydride

$$\overset{0}{2\ Na(\ell)} + \overset{0}{H_2(g)} \longrightarrow \overset{+1\ -1}{2\ NaH(s)}$$

The product is a *saline* (salt-like) hydride, an ionic solid composed of Na^+ cations and *negatively* charged hydride ions (H^-). Alkaline earth hydrides (such as CaH_2) are 1:2 ionic compounds in which two H^- ions balance the charge of each $+2$ ion. Saline hydrides are oxidized by oxygen in air and by water. The reaction of NaH with water is quite violent, but that of CaH_2 is more controllable and can be used as a convenient laboratory source for hydrogen (Fig. 4–14):

$$\overset{+2\ -1}{CaH_2(s)} + \overset{+1\ -2}{2\ H_2O(\ell)} \longrightarrow \overset{0}{2\ H_2(g)} + \overset{+2}{Ca^{2+}(aq)} + \overset{-2\ +1}{2\ OH^-(aq)}$$

Note that CaH_2 is a base because it increases the amount of hydroxide ion when put in water.

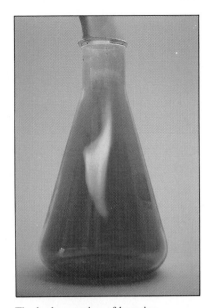

The hydrogenation of bromine. Bromine, the brown gas in the flask, reacts vigorously as gaseous hydrogen is passed in through a tube.

Figure 4–14 Calcium hydride reacts vigorously with water to give hydrogen.

The hydrogenation of metal oxides usually leads to the removal of the oxygen in the form of water and the appearance of the uncombined metal

$$\overset{+3\ -2}{Fe_2O_3(g)} + 3\ \overset{0}{H_2(g)} \longrightarrow 2\ \overset{0}{Fe(s)} + 3\ \overset{+1-2}{H_2O(g)}$$

or

$$\overset{+1\ -2}{Ag_2O(s)} + \overset{0}{H_2(g)} \longrightarrow \overset{0}{Ag(s)} + \overset{+1\ -2}{H_2O(g)}$$

while the hydrogenation of non-metal oxides can give water and the non-metal combined with hydrogen

$$\overset{+4-2}{SO_2(g)} + 3\ \overset{0}{H_2(g)} \longrightarrow \overset{+1-2}{H_2S(g)} + 2\ \overset{+1-2}{H_2O(g)}$$

$$\overset{+2-2}{CO(g)} + 3\ \overset{0}{H_2(g)} \longrightarrow \overset{-4+1}{CH_4(g)} + \overset{+1\ -2}{H_2O(g)}$$

or else water and the free element

$$\overset{+4-2}{SO_2(g)} + 2\ \overset{0}{H_2(g)} \longrightarrow \overset{0}{S(s)} + 2\ \overset{+1-2}{H_2O(g)}$$

depending on conditions.

Many organic compounds containing double and triple bonds take up hydrogen to give new compounds containing single bonds only. Hydrogen adds "across a multiple bond." This means that H atoms from H_2 appear on either side of the multiple bond as it becomes a single bond

$$\overset{+1-2\ \ -2+1}{H_2C{=}CH_2(g)} + \overset{0}{H_2(g)} \longrightarrow \overset{+1-3\ \ -3+1}{H_3C{-}CH_3(g)}$$

$$\overset{+1-1\ \ -1+1}{HC{\equiv}CH(g)} + 2\ \overset{0}{H_2(g)} \longrightarrow \overset{+1-3\ \ -3+1}{H_3C{-}CH_3(g)}$$

$$\overset{+1\ 0\ \ \ -2}{H_2C{=}O(g)} + \overset{0}{H_2(g)} \longrightarrow \overset{+1-2\ \ -2+1}{H_3C{-}OH(g)}$$

Although hydrogen is a strong reducing agent, hydrogenation reactions are usually slow at room conditions. Rapid and complete hydrogenation is favored by the addition of a catalyst, by the use of high-pressure hydrogen, by raising the temperature, or by a combination of all three.

The production of water from its elements

$$2\ \overset{0}{H_2(g)} + \overset{0}{O_2(g)} \longrightarrow 2\ \overset{+1-1}{H_2O(g)}$$

which is a violent event once ignited, can be described as both the hydrogenation of O_2 and the oxygenation of H_2. Hydrides (from hydrogenation) and oxides (from oxygenation) are particularly common compounds. Water ("dihydrogen oxide") is the most common compound of all.

Displacement Reactions

Often one element displaces another from a compound. Thus, copper forces silver out of chemical combination and assumes its place, as shown in Figure 4–15, according to the equation

• Catalysts accelerate chemical reactions without themselves being consumed. See Section 14-7.

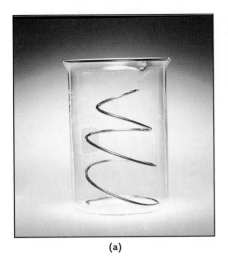

(a)

(b)

(c)

Figure 4–15 Copper displaces silver ion from solution. (a) A length of clean copper wire. (b) After addition of a solution of silver nitrate, the $Cu(s)$ reduces $Ag^+(aq)$ ions to metallic silver, which appears as granular crystals clinging to the wire. (c) The other product, $Cu(NO_3)_2(aq)$, makes the solution blue.

$$\overset{+1\,+5\,-2}{2\ AgNO_3(aq)} + \overset{0}{Cu(s)} \longrightarrow \overset{+2\ +5\,-2}{Cu(NO_3)_2(aq)} + \overset{0}{2\ Ag(s)} \quad \text{(overall equation)}$$

The nitrate ion (NO_3^-) plays no part in this reaction. Omitting it gives

$$\overset{+1}{2\ Ag^+(aq)} + \overset{0}{Cu(s)} \longrightarrow \overset{+2}{Cu^{2+}(aq)} + \overset{0}{2\ Ag(s)} \quad \text{(net ionic equation)}$$

Displacement reactions always involve a change in oxidation numbers. Electropositive elements displace each other from compounds in aqueous solution according to the **activity series** on the top line in Table 4–3. Activity (ability to displace) increases from left to right, and the strongest *reducing* agent is at the far right. The second line gives the oxidation state to which the displacing element advances. The position of hydrogen is important. Elements more active than hydrogen (to the right of H_2 in the series) are "active metals." They can displace hydrogen from acidic solutions. The most active metals (Na, Ca, K, Li) can displace hydrogen from water

$$\overset{0}{2\ Na(s)} + \overset{+1\,-2}{2\ H_2O(\ell)} \longrightarrow \overset{+1\,-2\,+1}{2\ NaOH(aq)} + \overset{0}{H_2(g)}$$

The activity series in Table 4–3 was constructed for aqueous solutions at 25°C with all solutes having a concentration of 1 mol L^{-1}. However, it provides a good guide for predicting displacement reactions under any conditions. The activities of electronegative elements can be compared in a similar way. Among the halogens,

Table 4–3
Activity Series for Some Electropositive Elements

					Increasing Ability to Displace ⟶														
Au	Pt	Hg	Ag	Cu	H_2	Pb	Sn	Ni	Co	Cd	Fe	Cr	Zn	Mn	Al	Mg	Na	Ca	K Li
Au^{3+}	Pt^{2+}	Hg^{2+}	Ag^+	Cu^{2+}	H^+	Pb^{2+}	Sn^{2+}	Ni^{2+}	Co^{2+}	Cd^{2+}	Fe^{2+}	Cr^{3+}	Zn^{2+}	Mn^{2+}	Al^{3+}	Mg^{2+}	Na^+	Ca^{2+}	K^+ Li^+

for example, the elements higher up in the periodic table displace elements lower down from compounds

	Electronegative Element	**Product**
	F_2	F^-
	Cl_2	Cl^-
	Br_2	Br^-
	I_2	I^-

Increasing Ability to Displace ↑

In this table, the strongest *oxidizing* agent is at the top of the left column. The table indicates that chlorine displaces iodide ion from, for example, potassium iodide by oxidizing it to iodine.

$$\overset{0}{Cl_2}(g) + 2 \overset{+1-1}{KI}(aq) \longrightarrow \overset{0}{I_2}(s) + 2 \overset{+1-1}{KCl}(aq) \quad \text{(overall equation)}$$

$$\overset{0}{Cl_2}(g) + 2 \overset{-1}{I^-}(aq) \longrightarrow \overset{0}{I_2}(s) + 2 \overset{-1}{Cl^-}(aq) \quad \text{(net ionic equation)}$$

The net ionic equation helps focus attention on the species that actually gain and lose electrons.

EXAMPLE 4-9

Predict the products of the following redox reactions. If no reaction occurs, write NR in place of products. Attach a descriptive name to each reaction.
(a) $Rb(s) + Cl_2(g) \longrightarrow$
(b) $Br_2(\ell) + NaCl(s) \longrightarrow$
(c) $Br_2(\ell) + NaI(s) \longrightarrow$
(d) $Fe_2O_3(s) + H_2(g) \longrightarrow$ (with heating)

Solution

(a) $2 \overset{0}{Rb}(s) + \overset{0}{Cl_2}(g) \longrightarrow 2 \overset{+1 \; -1}{RbCl}(s)$

We expect rubidium to be very active electropositive element by analogy to other alkali metals such as sodium. Sodium is known to be capable of reducing chlorine to chloride. This is a combination reaction.
(b) NR; bromine cannot displace chlorine from compounds.

(c) $\overset{0}{Br_2}(\ell) + 2 \overset{+1-1}{NaI}(s) \longrightarrow \overset{0}{I_2}(s) + 2 \overset{+1-1}{NaBr}(s)$

Bromine displaces iodine from the compound.

(d) $\overset{+3-2}{Fe_2O_3}(s) + \overset{0}{H_2}(g) \longrightarrow \overset{0}{Fe}(s) + \overset{+1-2}{H_2O}(g);$

A metal oxide is reduced to give the metal plus water. This is both a displacement (hydrogen displaces iron from a compound) and a hydrogenation reaction.

Exercise

Predict the products of the following redox reactions, and attach a descriptive name to each.

(a) $P_4(s) + O_2(g) \longrightarrow$

(b) $Te(s) + H_2(g) \longrightarrow$

(c) $Fe_2O_3(s) + Al(s) \longrightarrow$

Answer: (a) P_2O_5 in a combination reaction that is also a combustion. (b) H_2Te in a combination reaction that is a hydrogenation. (c) Al_2O_3 and Fe in a displacement reaction.

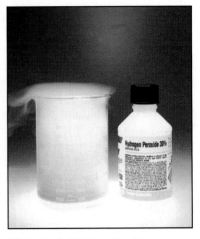

Figure 4–16 Pure hydrogen peroxide, a colorless liquid at room temperature, is hard to handle because minor impurities trigger its violent disproportionation. Here, a 30% mixture of hydrogen peroxide with water disproportionates vigorously enough to heat the water to boiling.

Disproportionation

A single molecule or ion can, under the proper conditions, undergo a reaction in which a portion of one of the elements that it contains is oxidized and the rest of that same element is reduced. The species takes part in a redox reaction with itself. This phenomenon is called **disproportionation.** The decomposition of hydrogen peroxide (Fig. 4–16) to water and oxygen

$$\overset{+1\ -1}{2\ H_2O_2(\ell)} \longrightarrow \overset{+1\ -2}{2\ H_2O(\ell)} + \overset{0}{O_2(g)}$$

is a disproportionation. Oxygen starts out on the left in the -1 oxidation state (see rule 5 concerning oxidation states on page 147). The reaction reduces half of it to the -2 state (in H_2O) and oxidizes the other half to the zero state (in O_2). Disproportionation is a possibility whenever a substance can act as both an oxidizing and a reducing agent; that is, whenever both lower and higher oxidation states are available to any element in the substance. Disproportionation is fairly common. Metal ions in intermediate oxidation states often disproportionate. For example

$$\overset{+1}{2\ Cu^+(aq)} \longrightarrow \overset{0}{Cu(s)} + \overset{+2}{Cu^{2+}(aq)} \quad \text{(net ionic equation)}$$

$$\overset{+3}{Mn^{3+}(aq)} + \overset{+1\ -2}{2\ H_2O(\ell)} \longrightarrow \overset{+2}{Mn^{2+}(aq)} + \overset{+4\ -2}{MnO_2(s)} + \overset{+1}{4\ H^+(aq)} \quad \text{(net ionic equation)}$$

Some non-metallic elements disproportionate in basic water

$$\overset{0}{P_4(s)} + \overset{-2\ +1}{3\ OH^-(aq)} + \overset{+1\ -2}{3\ H_2O(\ell)} \longrightarrow \overset{-3\ +1}{PH_3(g)} + \overset{+1\ +1\ -2}{3\ H_2PO_2^-(aq)} \quad \text{(net ionic equation)}$$

$$\overset{0}{I_2(s)} + \overset{-2\ +1}{2\ OH^-(aq)} \longrightarrow \overset{-1}{I^-(aq)} + \overset{+1\ -2}{IO^-(aq)} + \overset{+1\ -2}{H_2O(\ell)} \quad \text{(net ionic equation)}$$

The first reaction is slow until the water is boiled. A single substance may disproportionate in more than one way, depending on conditions. Thus, iodine may disproportionate in basic water according to

$$\overset{0}{3\ I_2(s)} + \overset{-2\ +1}{6\ OH^-(aq)} \longrightarrow \overset{-1}{5\ I^-(aq)} + \overset{+5\ -2}{IO_3^-(aq)} + \overset{+1\ -2}{3\ H_2O(\ell)} \quad \text{(net ionic equation)}$$

as well as the way shown in the preceding equation.

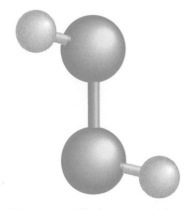

The structure of hydrogen peroxide (H_2O_2).

CHEMISTRY IN PROGRESS

Separation and Storage of Hydrogen

The hydrogenation of transition metals often gives non-stoichiometric compounds (see Section 1–3), with formulas such as $TiH_{1.7}$ or $ZrH_{1.9}$, in which the bonding is complex and not fully understood. Lanthanum reacts directly with hydrogen under room conditions to form LaH_2, and additional hydrogen can be put in until a composition of LaH_3 is reached. These hydrides retain the metallic character of the pure elements, although they are more brittle. Their electrical conductivity decreases with increasing hydrogen content.

The entry of hydrogen into transition metals and their alloys is often reversible. Gaseous hydrogen is taken up under pressure and flows out again if the pressure is reduced. This fact has led to proposals to store "dissolved" hydrogen in these metals. Hydrogen is attractive as a fuel because it burns cleanly in air to give an environmentally harmless product, water. Storage as a gas is impractical, however, because very large tanks would be required. Storage as a liquid is expensive because low temperatures

are needed (the normal boiling point of hydrogen is $-253°C$). Therefore, the high hydrogen densities that are possible under room conditions in transition metals appear attractive. Researchers have focused particular attention on the alloys FeTi and $LaNi_5$.

Although the use of transition metals for hydrogen storage is largely speculative right now, one transition metal (palladium) is used for the *separation* of hydrogen from other gases in multiton quantities today. This method dates from the observation by Thomas Graham in 1866 that palladium can absorb up to 935 times its own volume of hydrogen (giving a hydrogen density approaching that of pure liquid hydrogen). Moreover, the hydrogen diffuses rapidly through the metal and the metal remains ductile. No other gases (not even helium) enter palladium in this way; thus, hydrogen can be separated from a gas mixture by passing it through palladium. Industrial plants produce volumes in excess of 250 million L (20,000 kg) of pure hydrogen per day.

EXAMPLE 4–10

Complete and balance the following chemical equations, and state whether each represents a redox, acid–base, dissolution, or precipitation reaction.
(a) $H_2(g) + Cl_2(g) \longrightarrow$
(b) $HI(aq) + NH_3(aq) \longrightarrow$
(c) $RbH(s) + H_2O(\ell) \longrightarrow$

Solution

(a) This is analogous to the reaction of H_2 with I_2, and the balanced equation is

$$\overset{0}{H_2}(g) + \overset{0}{Cl_2}(g) \longrightarrow \overset{+1 \ -1}{2\ HCl}(g)$$

It is a redox reaction because the oxidation states of H and Cl change from 0 to $+1$ and -1, respectively.
(b) This is an acid–base reaction between the strong Arrhenius acid HI (hydroiodic acid) and the Arrhenius base NH_3 (ammonia).

$$\overset{+1 \ -1}{HI}(aq) + \overset{-3 +1}{NH_3}(aq) \longrightarrow \overset{-3 +1 \ -1}{NH_4I}(aq) \quad \text{(overall equation)}$$
$$H^+(aq) + NH_3(aq) \longrightarrow NH_4^+(aq) \quad \text{(net ionic equation)}$$

Note that no oxidation numbers change.
(c) This is analogous to the reaction of sodium hydride with water and is written

$$\overset{+1 \ -1}{RbH}(s) + \overset{+1 \ -2}{H_2O}(\ell) \longrightarrow \overset{+1}{Rb^+}(aq) + \overset{-2 +1}{OH^-}(aq) + \overset{0}{H_2}(g) \quad \text{(net ionic equation)}$$

It is a redox reaction because one hydrogen atom changes oxidation state from -1 to 0 and one changes oxidation state from $+1$ to 0. The reaction also increases the concentration of hydroxide ions in water.

Exercise

Complete and balance the following chemical equations, and state whether each represents a redox, acid–base, dissolution, or precipitation reaction.

(a) $NH_3(g) + H_2O(\ell) \longrightarrow$

(b) $Li(\ell) + H_2(g) \longrightarrow$

(c) $C_4H_{10}(g) + O_2(g) \longrightarrow$

Answer: (a) $NH_3(g) \longrightarrow NH_3(aq)$ is a dissolution reaction.

(b) $2\,Li(\ell) + H_2(g) \longrightarrow 2\,LiH(s)$ is a redox reaction.

(c) $2\,C_4H_{10}(g) + 13O_2\,(g) \longrightarrow 8CO_2(g) + 10\,H_2O\,(g)$ is a redox reaction.

4–5   THE STOICHIOMETRY OF REACTIONS IN SOLUTION

Liquid solutions find constant use in practical laboratory work in chemistry for several reasons. Dissolved substances mix rapidly and uniformly on the molecular or ionic level when stirred together; intimate contact among reactive particles is assured. Handling and storage of chemicals are often easier in solution than in a pure state. Dilution in a solvent can moderate or nullify unpleasant or dangerous properties of a pure substance. Finally, substances in solution are easily dispensed in known quantities because measuring liquid volumes is easy.

• The other way to dispense known quantities of substances is to weigh them out. Weighing is easy too but usually less rapid and convenient.

The actual amount of a solute in any given volume of solution depends on how concentrated or dilute the solution happens to be. This is expressed by a ratio: the **concentration** c of a solute in a solution equals the chemical amount of the solute (n) divided by the volume (V) of the entire solution

$$c_{\text{solute}} = \frac{n_{\text{solute}}}{V}$$

• The term "concentration" by international agreement refers exclusively to ratios having the volume of the solution in the denominator; the word is often used more loosely, however, to refer to any measure of the composition of a solution.

When the chemical amount of the solute is measured in moles and the volume of solution is measured in liters, the resulting unit of concentration is called the **molarity:**

$$\text{molarity} = \frac{\text{moles solute}}{\text{liters solution}} = \text{mol L}^{-1}$$

This is the most useful unit of concentration for routine laboratory work. The symbol M stands for "moles per liter of solution." A 1.0 M solution of fructose has 1.0 mol of fructose per liter of solution and is said to be "1.0 molar in fructose." This fact is also expressed by writing

$$[\text{fructose}] = 1.0 \text{ mol L}^{-1}$$

The brackets are understood to mean the molarity of the species whose formula or name they enclose. A solution might contain several solutes. If so, it might be 1.0 M in solute number 1, 3.0 M in solute number 2, and so forth. Although the molarity is the most used unit of solution composition in chemistry, it has the disadvantage for accurate measurements that it depends slightly on temperature. If a solution is heated or cooled, its volume changes, so the number of moles of solute per liter of solution also changes.

EXAMPLE 4–11

Solutions of magnesium chloride are sometimes administered as purgatives. A solution is prepared by mixing 10.0 g of magnesium chloride with 90.0 mL of water to produce 92.3 mL of solution. Find the molarity of this solution.

Solution

Begin by converting the mass of magnesium chloride to its chemical amount by using the molar mass of $MgCl_2$ as a conversion factor.

$$10.0 \text{ g MgCl}_2 \times \left(\frac{1 \text{ mol MgCl}_2}{95.21 \text{ g MgCl}_2} \right) = 0.105 \text{ mol MgCl}_2$$

Now, obtain the ratio of the chemical amount to the volume of the solution in liters.

$$\left(\frac{0.105 \text{ mol MgCl}_2}{92.3 \text{ mL}} \right) \times \left(\frac{1000 \text{ mL}}{1 \text{ L}} \right) = 1.14 \text{ mol L}^{-1}$$

Exercise

Calculate the molarity of a solution prepared by dissolving 10.0 g of $Al(NO_3)_3$ in enough water to make 250.0 mL of solution.

Answer: 0.188 mol L^{-1}.

In Example 4–11, the bottle holding the solution would be labeled 1.14 M $MgCl_2$ but would *not* contain 1.14 mol L^{-1} of molecules of $MgCl_2$. The strong electrolyte $MgCl_2$ is dissociated in aqueous solution. Complete dissociation gives $[Mg^{2+}] = 1.14$ mol L^{-1}, $[Cl^-] = 2.28$ mol L^{-1}, and $[MgCl_2] = 0$ mol L^{-1} because two chloride ions form for every magnesium ion that appears.

The Preparation of Solutions of Accurately Known Molarity

If asked to prepare a 1.000 M solution of ethanol (C_2H_5OH) in water, a beginner might dissolve 1.000 mol (46.07 g) of pure ethanol in 1.000 L (1000 mL) of water. This naive approach gives a solution with the wrong molarity. It fails because it ignores the volume that the solute contributes; the total volume of the solution after mixing exceeds 1000 mL, and the molarity is accordingly less than 1.000 M. The correct method of mixing is not even to try to pre-measure the volume of solvent, but instead to dissolve the required chemical amount of solute in somewhat less solvent than required for the target molarity, and then to add just enough further solvent with continuous stirring to bring the volume "up to the mark," that is, to reach the required total volume. Stirring (or swirling) in preparing solutions ensures thorough mixing and speeds dissolution. The latter is important because many dissolution reactions are discouragingly slow otherwise. Experienced laboratory workers also usually grind up solids before attempting to dissolve them; fine powders dissolve faster than large chunks. Gentle heating sometimes works to hasten dissolution but can cause problems. Undesired reactions may occur, and both solution and container definitely change volume. Because an accurate final volume is crucial in

Figure 4–17 To prepare a solution of ammonium chromate ($NH_4Cr_2O_7$) of accurately known concentration, weigh out an amount of the solid, transfer it to a volumetric flask, dissolve it in somewhat less water than required, and dilute to the total volume marked on the neck of the volumetric flask. The molarity equals the number of moles of solute dissolved divided by the total volume in liters.

knowing the molarity, hot solutions must be fully cooled before adding the final portion of solvent that brings the volume to the required total. For accurate work in preparing solutions, a **volumetric flask** is used (Fig. 4–17). These flasks are calibrated to contain precisely known volumes of solution at a specified temperature when filled to the mark etched on their necks.

Dilutions

On occasion, it is necessary to prepare a dilute solution of specified concentration from a more concentrated solution by adding pure solvent to it. Suppose that the initial concentration is c_i and the initial solution volume is V_i. The chemical amount of solute is

$$c_i \times V_i = n_{solute}$$

$$\left(\frac{\text{moles solute}}{\cancel{\text{L solution}}} \right) \times (\cancel{\text{L solution}}) = (\text{moles solute})$$

The amount of solute does not change upon dilution to a final solution volume (V_f) because only solvent, and not solute, is being added. Thus, $c_iV_i = c_fV_f$, and the final concentration c_f

$$c_f = \frac{\text{moles solute}}{\text{final solution volume}} = \frac{c_iV_i}{V_f}$$

This equation can be used to calculate the final concentration after dilution to a known final volume. Alternatively, we can solve for V_f to determine what final volume must be aimed for to achieve a desired final concentration.

EXAMPLE 4–12

(a) Describe how to prepare 0.500 L of a 0.0250 M aqueous solution of potassium dichromate ($K_2Cr_2O_7$).
(b) Describe how to dilute this solution to obtain a solution with a final concentration of 0.0140 M $K_2Cr_2O_7$.

Solution

(a) Start with the molarity and volume of the desired solution and calculate the mass of $K_2Cr_2O_7$ that must be present in the final solution.

$$0.500 \text{ L solution} \times \left(\frac{0.0250 \text{ mol } K_2Cr_2O_7}{1 \text{ L solution}}\right) \times$$

$$\left(\frac{294.2 \text{ g } K_2Cr_2O_7}{1 \text{ mol } K_2Cr_2O_7}\right) = 3.68 \text{ g } K_2Cr_2O_7$$

To prepare the solution, weigh out 3.68 g of $K_2Cr_2O_7$, dissolve it in a small amount of water, and add enough water to make the final volume of the solution equal to 0.500 L. This is *not* the same as weighing out 3.68 g of $K_2Cr_2O_7$ and mixing it with 0.500 L of water.

(b) Rearranging the dilution equation gives

$$V_f = \frac{c_i}{c_f} V_i = \frac{0.0250 \text{ mol L}^{-1}}{0.0140 \text{ mol L}^{-1}} \times 0.500 \text{ L} = 0.893 \text{ L}$$

Dilute the 0.500 L of solution from part (a) to a total volume of 0.893 L by adding water. Once again, this is not the same as adding 0.393 L of water!

Exercise

A flask contains 625 mL of 3.05 M nitric acid (HNO_3) solution. What volume of 15.8 M HNO_3 contains the same amount of HNO_3 as this solution?

Answer: 0.121 L (121 mL) of 15.8 M HNO_3.

Figure 4–18 Glass piping transfers reddish brown elemental bromine in a chemical plant in Arkansas where brominated flame retardants are synthesized.

• This is a displacement reaction. The bromide ion is oxidized and the chlorine in aqueous solution (the "chlorine water") is reduced.

Using the Molarity in Calculations

The procedure for stoichiometric calculations given in Figure 2–6 is readily modified to deal with concentrations of solutions. Instead of converting between masses and chemical amounts (using the molar mass as a conversion factor), the approach is to convert between *solution volumes* and chemical amounts, using the molarity as the conversion factor.

Consider as an example a reaction that is used commercially to prepare elemental bromine (Fig. 4–18) from its salts:

$$2 \text{ Br}^-(aq) + Cl_2(aq) \longrightarrow 2 \text{ Cl}^-(aq) + Br_2(aq)$$

Suppose we have 50.0 mL of a 0.0600 M solution of NaBr. What volume of a 0.0500 M solution of Cl_2 is needed to react completely with the Br^-? First find the chemical amount of Br^- available. Because every mole of NaBr furnishes one mole of Br^- in solution upon dissociation, we have:

$$0.0500 \text{ L solution} \times \left(\frac{0.0600 \text{ mol NaBr}}{1 \text{ L solution}}\right) \times \left(\frac{1 \text{ mol Br}^-}{1 \text{ mol NaBr}}\right) = 3.00 \times 10^{-3} \text{ mol Br}^-$$

Next use the chemical conversion factor of one mole of Cl_2 per two moles of Br^- to find the chemical amount of Cl_2 reacting:

$$n_{Cl_2} = 3.00 \times 10^{-3} \text{ mol Br}^- \left(\frac{1 \text{ mol } Cl_2}{2 \text{ mol Br}^-}\right) = 1.50 \times 10^{-3} \text{ mol } Cl_2$$

Finally, find the volume of aqueous chlorine needed:

$$1.50 \times 10^{-3} \text{ mol } Cl_2 \times \left(\frac{1 \text{ L } Cl_2 \text{ solution}}{0.0500 \text{ mol } Cl_2}\right) = 3.00 \times 10^{-2} \text{ L } Cl_2 \text{ solution}$$

In this last step, the molarity of the Cl_2 solution appears "upside-down" in the conversion factor because the unit "mol Cl_2" must cancel out. The reaction requires 3.00×10^{-2} L, or 30.0 mL, of the Cl_2 solution. In practice, an excess of Cl_2 solution would be used to ensure more nearly complete conversion of the Br^- to Br_2.

We might also want to know the Cl^- concentration after the reaction has been completed. Because each mole of Br^- that reacts gives one mole of Cl^- in the products, the chemical amount of Cl^- that is produced is 3.00×10^{-3} mol. In calculating the concentration of Cl^-, however, we must take care. The final volume of the solution is the *total* volume, approximately equal to the sum of the 50.0 mL originally present and the 30.0 mL added, giving 80.0 mL = 0.800 L. The final concentration of Cl^- is then

$$[Cl^-] = \frac{3.00 \times 10^{-3} \text{ mol}}{0.0800 \text{ L}} = 0.0375 \text{ M}$$

• Turning the ratio upside-down is legitimate: a solution with 0.0500 mol Cl_2 per liter also has 1 l of solution per 0.0500 mol Cl_2.

where we have used the notation of square brackets around a chemical symbol to refer to molarity.

In the preceding discussion, volumes were given in milliliters, not liters. Sometimes it is easier when working with milliliters to change concentration units from mol L^{-1} to mmol mL^{-1}. Because a millimole equals 0.001 mol and a milliliter equals 0.001 L, the numerical value stays the same. The first step in the preceding process then becomes

$$50.0 \text{ mL solution} \times \left(\frac{0.0600 \text{ mmol NaBr}}{1 \text{ mL solution}}\right) \times \frac{(1 \text{ mmol Br}^-)}{(1 \text{ mmol NaBr})} = 3.00 \text{ mmol Br}^-$$

At the heart of any stoichiometric calculation is a relation between the chemical amounts (the numbers of moles) of two substances taking part in the reaction. This central relation comes from the coefficients assigned to the two substances in the balanced chemical equation. The first and last steps then involve conversions between chemical amount and other measures of amounts of substance, such as mass (see Section 2–4), gas volume (see Section 5–5), or solution volume (this section).

EXAMPLE 4–13

When the orange salt potassium dichromate is added to a solution of concentrated hydrochloric acid, it reacts according to the net ionic equation

$$K_2Cr_2O_7(s) + 14 \text{ HCl}(aq) \longrightarrow$$
$$2 K^+(aq) + 2 Cr^{3+}(aq) + 8 Cl^-(aq) + 7 H_2O(\ell) + 3 Cl_2(g)$$

producing a mixed solution of chromium(III) chloride and potassium chloride, and evolving gaseous chlorine. Suppose that 6.20 g of $K_2Cr_2O_7$ reacts completely in this way in a solution with a total volume of 100.0 mL. Calculate the final concentration of $Cr^{3+}(aq)$ ion that results and the chemical amount (in moles) of chlorine produced.

Solution

The first step is to convert the amount of $K_2Cr_2O_7$ that reacts from grams to moles using the molar mass of potassium dichromate, 294.19 g mol^{-1}.

$$n_{K_2Cr_2O_7} = 6.20 \text{ g } K_2Cr_2O_7 \times \left(\frac{1 \text{ mol } K_2Cr_2O_7}{294.19 \text{ g } K_2Cr_2O_7}\right) = 0.0211 \text{ mol } K_2Cr_2O_7$$

The balanced equation shows that 1 mol of $K_2Cr_2O_7$ reacts to give 2 mol of Cr^{3+} and 3 mol of Cl_2. Employing these facts in unit-factors gives

$$n_{Cr^{3+}} = 0.0211 \text{ mol } K_2Cr_2O_7 \times \left(\frac{2 \text{ mol } Cr^{3+}}{1 \text{ mol } K_2Cr_2O_7} \right)$$

$$= 0.0422 \text{ mol } Cr^{3+}$$

$$n_{Cl_2} = 0.0211 \text{ mol } K_2Cr_2O_7 \times \left(\frac{3 \text{ mol } Cl_2}{1 \text{ mol } K_2Cr_2O_7} \right)$$

$$= 0.0633 \text{ mol } Cl_2$$

The reaction generates 0.0422 mol of $Cr^{3+}(aq)$ in a solution with a volume of 0.100 L. The molarity of the $Cr^{3+}(aq)$ is its chemical amount in moles divided by this volume:

$$[Cr^{3+}] = \frac{0.0422 \text{ mol } Cr^{3+}(aq)}{0.100 \text{ L solution}} = 0.422 \text{ M}$$

Exercise

Suppose the chlorine remained dissolved in solution. Calculate its concentration. (The resulting concentration exceeds the solubility of chlorine in water; most of the chlorine must leave the solution.)

Answer: 0.633 M $Cl_2(aq)$.

Titrations

Analytical chemists determine the amounts of substances in solution by means of titrations. A titration consists of a controlled addition of measured volumes of a solution of known concentration to a second solution of unknown concentration under conditions in which the solutes react cleanly (without side reactions), completely, and rapidly. A titration is complete when the second solute is used up. Completion is signaled by a change in some physical property, such as the color of the reacting mixture or the color of an indicator that has been added to it.

As an illustration, consider the **standardization** of a solution of the strong base sodium hydroxide. Standardization is the accurate determination of the concentration of a "stock solution" (a solution for use in later experiments). It consumes a portion of the stock but gives essential information about the remainder. This standardization employs a neutralization reaction between the aqueous NaOH and an accurately known amount of an acid. Many acids would work, but we choose the easily handled and purified weak acid potassium hydrogen phthalate, $KHC_8H_4O_4$, a solid that dissolves to give $K^+(aq)$ and $HC_8H_4O_4^-(aq)$ ions. The latter donates hydrogen ions to neutralize aqueous NaOH according to the equation

$$OH^-(aq) + HC_8H_4O_4^-(aq) \longrightarrow H_2O(\ell) + C_8H_4O_4^{2-}(aq)$$

* Recall that indicators undergo color changes that depend on the acidity or basicity of their surroundings. Phenolphthalein is colorless in acid solution and pink in basic solution.

A known mass of potassium hydrogen phthalate is dissolved in a flask, and a few drops of a solution of the indicator phenolphthalein is added. The solution is colorless. Some of the NaOH solution is poured into a **buret,** a tube with a delivery valve (a stopcock) and markings for measuring solution volumes (Fig. 4–19). The initial level of the solution in the buret is noted, and the titration is begun (Fig. 4–20). Small quantities of NaOH solution are dispensed from the buret into the flask containing the acid. The flask is swirled after each addition of base, and its interior walls are washed occasionally with water to flush any splashed droplets of solution into the main body of solution. As the NaOH joins the solution in the titration flask, regions of pink appear and then fade. This is the indicator responding to locally high concentrations of NaOH. As long as the original chemical amount of $KHC_8H_4O_4$ exceeds the total chemical amount of NaOH added, the color fades with swirling. As more and more NaOH solution is added, the pink color fades more and more slowly. Finally, when all the acid has just been used up, the whole of the reaction mixture stays pink. This is the **end-point.** The final level of the solution in the buret is noted, and the volume of NaOH solution added during the titration is calculated by subtraction.

In a good titration, the observed end-point approximates the theoretical **equivalence point** very closely. The equivalence point in this titration is the point at which the chemical amount of NaOH dispensed exactly equals the chemical amount of $KHC_8H_4O_4$ originally put into the reaction flask. The concentration of the NaOH solution equals this number of moles divided by the volume of solution delivered from the buret to reach the equivalence point.

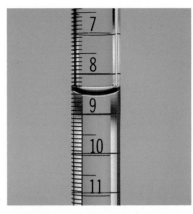

Figure 4–19 The tube of this buret is marked in milliliters with 10 subdivisions between each main marking. Readings are taken at the bottom of the lens-like meniscus. Careful workers attempt to estimate where the level lies between the finest markings. The reading here is 8.53 or 8.54 mL.

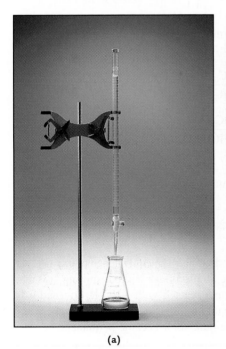

(a)

(b)

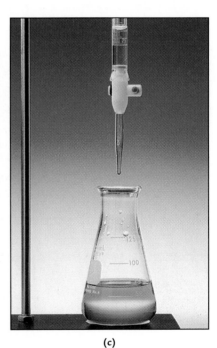

(c)

Figure 4–20 The standardization of a solution of NaOH. (a) A buret is filled with some of the NaOH solution. (b) The solution is dispersed slowly into a flask containing a known chemical amount of an acid; the two neutralize each other. (c) A change in the color of the indicator signals the end-point. In a carefully performed titration, the color change becomes complete with the addition of a single drop (or less) from the buret.

EXAMPLE 4–14

It requires 37.65 mL of a solution of NaOH to titrate 0.6135 g of dissolved potassium hydrogen phthalate to the phenolphthalein end-point. Compute the molarity of the NaOH solution, and list the assumptions behind the calculation.

Solution

The chemical amount of $KHC_8H_4O_4$ used in the titration equals the mass used divided by the molar mass of $KHC_8H_4O_4$

$$n_{KHC_8H_4O_4} = 0.6135 \text{ g } \cancel{KHC_8H_4O_4} \times \left(\frac{1 \text{ mol } KHC_8H_4O_4}{204.2 \text{ g } \cancel{KHC_8H_4O_4}} \right) =$$

$$3.004 \times 10^{-3} \text{ mol } KHC_8H_4O_4$$

At the equivalence point, the chemical amounts of NaOH and $KHC_8H_4O_4$ are equal

$$n_{NaOH} = 3.004 \times 10^{-3} \text{ mol } \cancel{KHC_8H_4O_4} \times \left(\frac{1 \text{ mol NaOH}}{1 \text{ mol } \cancel{KHC_8H_4O_4}} \right) =$$

$$3.004 \times 10^{-3} \text{ mol NaOH}$$

This chemical amount of sodium hydroxide was contained in 37.65 mL of solution. Therefore

$$M_{NaOH} = \left(\frac{3.004 \times 10^{-3} \text{ mol NaOH}}{37.65 \text{ mL}} \right) \times \left(\frac{1000 \text{ mL}}{1 \text{ L}} \right) = 0.07980 \text{ mol L}^{-1}$$

The calculation can also be performed in terms of milligrams, millimoles, and milliliters:

$$0.6135 \text{ g } \cancel{KHC_8H_4O_4} \times \left(\frac{1000 \text{ mg } KHC_8H_4O_4}{1 \text{ g } \cancel{KHC_8H_4O_4}} \right) = 613.5 \text{ mg } KHC_8H_4O_4$$

$$613.5 \text{ mg } \cancel{KHC_8H_4O_4} \times \left(\frac{1 \text{ mmol } KHC_8H_4O_4}{204.2 \text{ mg } \cancel{KHC_8H_4O_4}} \right) = 3.004 \text{ mmol } KHC_8H_4O_4$$

$$n_{NaOH} = 3.004 \text{ mmol } \cancel{KHC_8H_4O_4} \times \left(\frac{1 \text{ mmol NaOH}}{1 \text{ mmol } \cancel{KHC_8H_4O_4}} \right) =$$

$$3.004 \text{ mmol NaOH}$$

$$M_{NaOH} = \frac{3.004 \text{ mmol NaOH}}{37.65 \text{ mL}} = 0.07980 \text{ mmol mL}^{-1} = 0.07980 \text{ mol L}^{-1}$$

The crucial assumptions are (1) that the end-point is the equivalence point; (2) that all the $KHC_8H_4O_4$ was transferred into the titration flask without spilling; (3) that every drop of NaOH solution from the buret reacted with $KHC_8H_4O_4$ (and was not splashed out of the flask or reacted with an impurity); and (4) that the addition of base was stopped at the instant the pink color became permanent.

Exercise

Compute the molarity of a solution of sodium hydroxide if 25.64 mL of solution must be added to a solution containing 0.5333 g of $KHC_8H_4O_4$ to reach the phenolphthalein end-point.

Answer: 0.1019 mol L^{-1}.

Once a sodium hydroxide solution has been standardized, it can be used in the titration of acid solutions. For instance, to determine the concentration of a solution of acetic acid, we add a few drops of phenolphthalein solution to the acid solution, fill a clean buret with standardized sodium hydroxide solution, and titrate to the phenolphthalein end-point. The base added from the buret reacts with the acetic acid according to the net ionic equation

$$CH_3COOH(aq) + OH^-(aq) \longrightarrow CH_3COO^-(aq) + H_2O(\ell)$$

EXAMPLE 4–15

A production lot of vinegar is being tested for its acetic acid content. A 50.0-mL sample is measured out and titrated with aqueous NaOH. It takes 31.66 mL of 1.3057 M NaOH to reach the phenolphthalein end-point. Calculate the molarity of acetic acid in the vinegar.

Solution

The chemical amount of NaOH that reacts is found by multiplying the volume of solution (31.66 mL = 0.03166 L) by its concentration (1.3057 M).

$$n_{NaOH} = 0.03166 \ \cancel{L} \times \left(\frac{1.3057 \ \text{mol NaOH}}{1 \ \cancel{L}} \right) = 4.134 \times 10^{-2} \ \text{mol NaOH}$$

Because one mole of acid reacts with one mole of $OH^-(aq)$, the chemical amount of acetic acid originally present must also have been 4.134×10^{-2} mol. Its concentration was then

$$[CH_3COOH] = \frac{4.134 \times 10^{-2} \ \text{mol}}{0.0500 \ \text{L}} = 0.827 \ \frac{\text{mol}}{\text{L}} = \boxed{0.827 \ \text{M}}$$

Exercise
The indicator methyl red turns from yellow to red when the medium in which it is dissolved changes from basic to acidic. A 25.00-mL volume of a sodium hydroxide solution is titrated with 0.8367 M HCl. It takes 22.48 mL of this acid to reach a methyl-red end-point. Find the molarity of the sodium hydroxide solution.

Answer: 0.7524 M.

Many (but by no means all) acid–base reactions have 1:1 stoichiometry; that is, the coefficients in the balanced neutralization equation are all 1's. At the equivalence point in a 1:1 acid–base titration, the number of moles of base reacted equals the number of moles of acid reacted

$$n_{acid} = n_{base} \quad \text{(at equivalence)}$$

These chemical amounts equal the concentrations of the solutions multiplied by the volumes used. Therefore

$$c_{acid}V_{acid} = c_{base}V_{base} \quad \text{(at equivalence)}$$

This equation is very useful. It applies in Example 4–15 as follows

$$c_{CH_3COOH}V_{CH_3COOH} = c_{NaOH}V_{NaOH} \quad \text{(at equivalence)}$$
$$(c_{CH_3COOH}) \times (50.0 \ \text{mL}) = (1.3057 \ \text{mol L}^{-1}) \times (31.66 \ \text{mL})$$
$$c_{CH_3COOH} = 0.827 \ \text{mol L}^{-1} = 0.827 \ \text{M}$$

• This relation resembles the equation for dilutions ($c_i V_i = c_f V_f$). However, it applies only at the equivalence point of a titration in which substances react in a 1:1 ratio.

• The solution must be kept acidic to prevent the precipitation of insoluble $Fe(OH)_2(s)$.

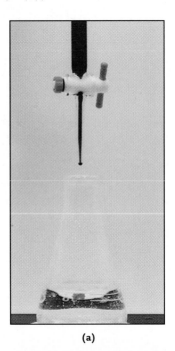

(a)

(b)

Figure 4–21 (a) Addition of a small amount of potassium permanganate from the buret gives a dash of purple color to the solution of Fe^{2+}. The color soon disappears as the permanganate ion is used up in oxidizing Fe^{2+} to Fe^{3+}. (b) At the end-point, all the Fe^{2+} ions have just been consumed. Additional drops of permanganate solution give a lasting pale purple color to the solution.

Redox reactions are often the basis of titrations. Consider the determination of the amount (or concentration) of Fe^{2+} ions in a solution. Under acidic conditions, the Fe^{2+} ions react quantitatively with the permanganate ion in the redox reaction

$$MnO_4^-(aq) + 5\ Fe^{2+}(aq) + 8\ H^+(aq) \longrightarrow Mn^{2+}(aq) + 5\ Fe^{3+}(aq) + 4\ H_2O(\ell)$$

A key aspect of this reaction is that a solution of the MnO_4^- ion has a deep purple color, whereas a solution containing Mn^{2+} and Fe^{3+} has a very pale yellow color. We choose the soluble salt, potassium permanganate, as our source of permanganate ion. As potassium permanganate is slowly dispensed from a buret into the iron solution, a dash of purple color from the permanganate solution appears in the flask (Fig. 4–21a), but it disappears as the MnO_4^- and Fe^{2+} ions react to give the nearly colorless Mn^{2+} and Fe^{3+} products. Eventually, with continued addition of $MnO_4^-(aq)$ ion, the last of the Fe^{2+} ions is consumed. The addition of just one or two drops of permanganate solution *beyond* this point imparts a pale purple color to the solution from the excess permanganate, and this color does *not* disappear (Fig. 4–21b). The titration is at its end-point.

Suppose that the solution of potassium permanganate had a concentration of 0.09625 M and that 26.34 mL (0.02634 L) of solution brought the titration to the end-point. The methods of solution stoichiometry then provide a way to find out how much iron(II) was present in the original solution. First,

$$n_{MnO_4^-} = 0.02634\ \text{L} \times \left(\frac{0.09625\ \text{mol}\ MnO_4^-}{1\ \text{L}} \right)$$
$$= 2.535 \times 10^{-3}\ \text{mol}\ MnO_4^-$$

Then, from the balanced chemical equation, each mole of MnO_4^- reduced causes the oxidation of five moles of Fe^{2+}:

$$n_{Fe^{2+}} = 2.535 \times 10^{-3}\ \text{mol}\ MnO_4^- \times \left(\frac{5\ \text{mol}\ Fe^{2+}}{1\ \text{mol}\ MnO_4^-} \right)$$
$$= 1.268 \times 10^{-2}\ \text{mol}\ Fe^{2+}$$

What might be the purpose of this titration? Suppose we have a sample of rock for which we wish to determine the total iron content. The iron may be in various forms, such as hematite (Fe_2O_3), limonite ($Fe_2O_3 \cdot H_2O$), or magnetite (Fe_3O_4). We would crush a weighed sample of the rock and add concentrated hydrochloric acid to bring the iron into solution as a mixture of Fe^{2+} and Fe^{3+}. Reaction with tin(II) chloride reduces all the Fe^{3+} to Fe^{2+}:

$$Sn^{2+}(aq) + 2\ Fe^{3+}(aq) \longrightarrow Sn^{4+}(aq) + 2\ Fe^{2+}(aq)$$

Excess tin(II) is then removed from the solution by the addition of mercury(II) chloride

$$Sn^{2+}(aq) + 2\ HgCl_2(aq) \longrightarrow Sn^{4+}(aq) + Hg_2Cl_2(s) + 2\ Cl^-(aq)$$

Finally, we would titrate as just described to find the amount of Fe^{2+} present. This would tell us the amount of iron in the original sample, provided that the preparative reactions proceeded in close to 100% yield.

Many analytical determinations are indirect and rely on preparative reactions of the sample before the titration can be started; for example, a solution containing an unknown amount of a soluble calcium salt does not react with potassium permanganate. Instead, we add a solution of ammonium oxalate ($(NH_4)_2C_2O_4$), which causes the quantitative precipitation of the calcium as insoluble calcium oxalate:

$$Ca^{2+}(aq) + C_2O_4^{2-}(aq) \longrightarrow CaC_2O_4(s)$$

After filtering and washing the solid, we dissolve it in sulfuric acid to form oxalic acid:

$$CaC_2O_4(s) + 2\ H^+(aq) \longrightarrow Ca^{2+}(aq) + H_2C_2O_4(aq)$$

Finally, we titrate the oxalic acid with permanganate solution, taking advantage of the redox reaction

$$2\ MnO_4^-(aq) + 5\ H_2C_2O_4(aq) + 6\ H^+(aq) \longrightarrow$$
$$2\ Mn^{2+}(aq) + 10\ CO_2(g) + 8\ H_2O(\ell)$$

In other words, the calcium reacts in the preparation stage, and the titration involves a second species (here, oxalic acid). The development and use of analytical techniques require a broad knowledge of reaction chemistry—the three reactions in this example are a precipitation reaction, an acid–base reaction, and a redox reaction, respectively.

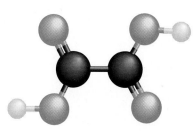

Oxalic acid ($H_2C_2O_4$) is an acid that can give up two hydrogen ions in aqueous solution. Spinach and rhubarb leaves contain large amounts of potassium oxalate ($K_2C_2O_4$) and calcium oxalate (CaC_2O_4).

SUMMARY

4–1 In **dissolution,** ions or molecules disperse from a pure phase into a homogeneous mixture. The polarity of the water molecule makes water an excellent solvent for ionic substances and **polar** molecular substances. Some substances, strong **electrolytes** or weak electrolytes, dissolve in water by complete or partial **ionization** (**dissociation** into ions); other substances (non-electrolytes) dissolve without forming ions. Electrolyte solutions conduct electricity better than pure water.

4–2 Precipitation is the reverse of dissolution. It occurs when the solubility of a solid solute is exceeded. A precipitation resulting from the mixing of solutions of ionic compounds is a **metathesis.** It can be represented by an ordinary balanced equation, by an **ionic equation,** or by a **net ionic equation.**

4–3 Acids are characterized by sour taste, ability to change the colors of indicators, and **neutralization** by bases. **Bases** are characterized by bitter taste, soapy feel, ability to change the colors of indicators, and neutralization by acids. Neutralization of an acid by a base gives a salt plus water. **Arrhenius acids** donate hydrogen ions in aqueous solution; **Arrhenius bases** donate hydroxide ions. Strong Arrhenius acids and bases are strong electrolytes; weak acids and bases are weak electrolytes. Metal oxides are **base anhydrides,** whereas non-metal oxides are **acid anhydrides. Hydration** of an anhydride gives an acid or base; **dehydration** of an oxoacid or a hydroxide gives an anhydride.

4–4 Oxidation–reduction (redox) reactions involve the transfer of electrons from one species to another. An atom is oxidized if its **oxidation number** increases; it is reduced if its oxidation number decreases. In a redox reaction, an **oxidizing agent** takes electrons from another species and is itself reduced; a **reducing agent** furnishes electrons to another species and is itself oxidized. Oxidizing agents and reducing agents have relative strengths, just as acids and bases do. An **activity series** gives the relative strengths of series of oxidizing agents or reducing agents. Redox reactions include the **combination** of elements into compounds; the **decomposition**

of compounds to elements; the **oxygenation** (gaining of oxygen) and the **hydrogenation** (gaining of hydrogen) by both compounds and elements; **displacement** reactions, in which more-active elements displace less-active elements from compounds; and **disproportionation** reactions, in which a species undergoes a redox reaction with itself.

4–5 The **molarity** of a solution equals the chemical amount in moles of solute dissolved per liter of solution. Solutions of known molarity are prepared by weighing out the desired chemical amount of solute and adding enough solvent to bring the solution to desired final volume. **Titration** is an analytical technique to determine the amount of a substance in solution by measuring the volume of a second solution (of known concentration) required to consume it in a well-defined reaction.

PROBLEMS

Note: Answers to blue-numbered problems are given in Appendix F. Problems that are more challenging are indicated with asterisks.

Dissolution Reactions

1. Define and distinguish strong electrolyte, weak electrolyte, and non-electrolyte.

2. Ethanol (C_2H_5OH) is completely miscible with water, but the solutions do not conduct electricity. Write an equation for the dissolution of liquid ethanol in water.

3. Magnesium perchlorate has a strong affinity for water and is used as a drying agent to remove water vapor from gases. Write a chemical equation for the dissolution of $Mg(ClO_4)_2$ in water.

4. Some toothpastes contain soluble stannous fluoride (systematic name: tin(II) fluoride) as an aid in combating tooth decay. Write two chemical equations for the dissolution of this compound, one assuming that it is a strong electrolyte and the other assuming that it is a weak electrolyte.

5. (See Example 4–1.) Predict whether the following compounds dissolve in water:
   (a) deoxyribose ($C_5H_{10}O_4$) (This compound is related to ribose by the removal of one oxygen atom.)
   (b) $KClO_4$
   (c) ethylene glycol ($HOCH_2CH_2OH$)

6. (See Example 4–1.) Predict whether the following compounds dissolve in water:
   (a) formic acid ($HCOOH$)
   (b) heptane (C_7H_{16})
   (c) potassium rubidium sulfate ($KRbSO_4$)

Precipitation Reactions

7. A solution is prepared by dissolving 1 mol of $NaCl(s)$ and 1 mol of $KBr(s)$ in 10 L of water. The solution is then left in an open container until all of the water evaporates. A solid residue remains. Name the compounds present in this mixture.

8. A solution is prepared by dissolving 0.1 mol of $KClO_4(s)$ and 0.1 mol of $NaCl(s)$ in sufficient water to get both the solids in solution. The solution is then allowed to evaporate until a solid starts to precipitate. Name the solid.

9. Lithium chloride is soluble in ethanol, but rubidium chloride is not. Suggest a method based on this fact to separate LiCl from RbCl if the two are mixed together in aqueous solution.

10. An experiment requires water that contains little or no dissolved oxygen. Suggest a way to remove dissolved oxygen (taken up from the air) from a supply of water.

11. Rewrite the following balanced equations as net ionic equations:
    (a) $NaCl(aq) + AgNO_3(aq) \longrightarrow AgCl(s) + NaNO_3(aq)$
    (b) $K_2CO_3(s) + 2\,HCl(aq) \longrightarrow$
    $$2\,KCl(aq) + CO_2(g) + H_2O(\ell)$$
    (c) $2\,Cs(s) + 2\,H_2O(\ell) \longrightarrow 2\,CsOH(aq) + H_2(g)$
    (d) $2\,KMnO_4(aq) + 16\,HCl(aq) \longrightarrow$
    $$5\,Cl_2(g) + 2\,MnCl_2(aq) + 2\,KCl(aq) + 8\,H_2O(\ell)$$

12. Rewrite the following balanced equations as net ionic equations:
    (a) $Na_2SO_4(aq) + BaCl_2(aq) \longrightarrow$
    $$BaSO_4(s) + 2\,NaCl(aq)$$
    (b) $6\,NaOH(aq) + 3\,Cl_2(g) \longrightarrow$
    $$NaClO_3(aq) + 5\,NaCl(aq) + 3\,H_2O(\ell)$$
    (c) $Hg_2(NO_3)_2(aq) + 2\,KI(aq) \longrightarrow$
    $$Hg_2I_2(s) + 2\,KNO_3(aq)$$
    (d) $3\,NaOCl(aq) + KI(aq) \longrightarrow$
    $$NaIO_3(aq) + 2\,NaCl(aq) + KCl(aq)$$

13. (See Example 4–2.) The silver bromide that is used in photographic film is made by mixing sodium bromide and silver nitrate in aqueous solution. Write a balanced overall equation for this reaction, and write a net ionic equation.

14. (See Example 4–2.) Nickel carbonate ($NiCO_3$) is a light green solid used to color glass. It is made by mixing aqueous sodium carbonate with nickel(II) nitrate (under controlled acidic conditions). Write a balanced equation for this reaction, and write a net ionic equation.

15. Explain how a saturated solution prepared by dissolving solid barium sulfate in water differs from a saturated solution of barium sulfate prepared by mixing aqueous barium chloride with aqueous sodium sulfate.

16. Solid sodium chloride precipitates from water when $HCl(g)$, which is quite soluble in water, is bubbled into an aqueous solution of sodium chloride. Explain why.

17. (See Example 4–3.) Predict the products of the following reactions. Complete and balance the equations, and write a net ionic equation for each reaction. If no reaction occurs, write NR on the right side of the equation.
(a) $Zn(NO_3)_2(aq) + K_2S(aq) \longrightarrow$
(b) $AgClO_4(aq) + CaCl_2(aq) \longrightarrow$
(c) $NaOH(aq) + Fe(NO_3)_3(aq) \longrightarrow$
(d) $Ba(CH_3COO)_2(aq) + Na_3PO_4(aq) \longrightarrow$

18. (See Example 4–3.) Predict the products of the following reactions. Complete and balance the equations, and write a net ionic equation for each reaction. If no reaction occurs, write NR on the right side of the equation.
(a) $CaI_2(aq) + NH_4F(aq) \longrightarrow$
(b) $Hg_2(NO_3)_2(aq) + ZnSO_4(aq) \longrightarrow$
(c) $NaBr(aq) + Mg(CH_3COO)_2(aq) \longrightarrow$
(d) $NH_4I(aq) + Pb(ClO_3)_2(s) \longrightarrow$

19. (See Example 4–3.) Use the information in Table 4–1 to write net ionic equations showing the result of mixing aqueous solutions of the following. If no reaction occurs, write NR.
(a) beryllium nitrate and sodium acetate
(b) barium nitrate and silver sulfate
(c) sodium hydroxide and calcium chloride
(d) sodium chloride and potassium phosphate

20. (See Example 4–3.) Use the information in Table 4–1 to write net ionic equations showing the result of mixing aqueous solutions of the following. If no reaction occurs, write NR.
(a) ammonium phosphate and calcium nitrate
(b) copper(II) chloride and potassium sulfate
(c) cesium hydroxide and barium bromide
(d) lead chloride and potassium iodide

21. An aqueous solution of $NaNO_3$ is mixed with a solution of KCl. The reaction

$$NaNO_3 + KCl \longrightarrow NaCl + KNO_3$$

is anticipated, but nothing happens. Explain why. Use net ionic equations and the solubility data in Table 4–1.

22. Silver fluoride (AgF) and calcium chloride ($CaCl_2$) are both soluble in water. An aqueous solution of the first is mixed with an aqueous solution of the second. Write two net ionic equations to represent the *two* precipitation reactions that occur. Use the solubility data in Table 4–1.

23. A mixture of magnesium nitrate and barium nitrate is dissolved in water (both are quite soluble). A chemist wants to add a second solution in order to obtain a solid barium salt without simultaneous precipitation of any magnesium. What should the second solution contain? Refer to Table 4–1.

24. A student accidentally contaminates a solution of sodium acetate by adding silver acetate. What can be done to remove the silver ions from the solution? Refer to Table 4–1.

Acids and Bases and Their Reactions

25. (See Example 4–4.) Name the following acids:
(a) H_2S (c) H_2CO_3
(b) HIO_4 (d) HBr

26. (See Example 4–4.) Name the following acids:
(a) HIO (c) HNO_2
(b) CH_3COOH (d) H_2Se

27. (See Example 4–5.) Write the chemical formula and give the name of the anhydride corresponding to each of the following acids or bases, and identify it as an acid or base anhydride.
(a) HIO_3 (c) H_2CrO_4
(b) $Ba(OH)_2$ (d) $H_2N_2O_2$

28. (See Example 4–5.) Write the chemical formula and give the name of the anhydride corresponding to each of the following acids or bases, and identify it as an acid or base anhydride.
(a) H_3AsO_4 (c) $RbOH$
(b) H_2MoO_4 (d) H_2SO_3

29. Identify each of the following oxides as an acid or base anhydride. Write the chemical formula, and give the name of the acid or base formed upon reaction with water.
(a) MgO (c) SO_3
(b) Cl_2O (d) Cs_2O

30. Identify each of the following oxides as an acid or base anhydride. Write the chemical formula, and give the name of the acid or base formed upon reaction with water.
(a) N_2O_3 (c) P_4O_{10}
(b) Li_2O (d) BaO

31. (See Example 4–7.) Predict the products of the following reactions. If no reaction occurs, write NR.
(a) $HNO_3(aq) + K_2CO_3(s) \longrightarrow$
(b) $HBr(aq) + Zn(s) \longrightarrow$
(c) $H_2SO_4(aq) + Zn(OH)_2(s) \longrightarrow$

32. (See Example 4–7.) Predict the products of the following reactions. If no reaction occurs, write NR.
(a) $HNO_3(aq) + LiOH(aq) \longrightarrow$
(b) $Ni(s) + HCl(aq) \longrightarrow$
(c) $(NH_4)_2CO_3(aq) + HI(aq) \longrightarrow$

33. (See Example 4–7.) Predict the products of the following reactions. If no reaction occurs, write NR.
(a) $HClO_3(aq) + KOH(aq) \longrightarrow$
(b) $N_2O_5(g) + NaOH(aq) \longrightarrow$
(c) $(NH_4)_2SO_4(aq) + Ba(OH)_2(aq) \longrightarrow$

34. (See Example 4–7.) Predict the products of the following reactions. If no reaction occurs, write NR.
(a) $NH_4Cl(aq) + RbOH(aq) \longrightarrow$
(b) $SeO_3(s) + KOH(aq) \longrightarrow$
(c) $HF(aq) + CsOH(aq) \longrightarrow$

35. Write balanced chemical equations to represent the acid–base reactions between aqueous solutions of the following:
(a) Hydrobromic acid and calcium hydroxide
(b) Ammonia and sulfuric acid
(c) Lithium hydroxide and nitric acid

36. Write balanced chemical equations to represent the acid–base reactions between aqueous solutions of the following:
(a) Benzoic acid and ammonia
(b) Perchloric acid and magnesium hydroxide
(c) Nitric acid and cesium hydroxide

37. Write balanced equations for acid–base reactions that lead to the production of each of the following salts. Name the acid, base, and salt.
(a) CaF_2
(b) Rb_2SO_4
(c) $Zn(NO_3)_2$
(d) KCH_3COO

38. Write balanced equations for acid–base reactions that lead to the production of each of the following salts. Name the acid, base, and salt.
(a) Na_2SO_3 (c) $PbSO_4$
(b) $Ca(C_6H_5COO)_2$ (d) $CuCl_2$

39. The gaseous compound PF_3 reacts slowly with water to give H_3PO_3 and HF.
(a) Name the three compounds given as formulas.
(b) Write and balance an equation representing this reaction.

40. The acid $HClO_4$ reacts with P_4O_{10} to give Cl_2O_7 and H_3PO_4.
(a) Name all four of these compounds.
(b) Write and balance an equation representing this reaction.

41. Hydrogen sulfide can be removed from natural gas by reaction with excess sodium hydroxide. Name the salt that is produced in this reaction. (*Note:* Hydrogen sulfide donates two hydrogen ions in the course of this reaction.)

42. During the preparation of viscose rayon, cellulose is dissolved in a bath containing sodium hydroxide and later reprecipitated as rayon by the addition of a solution of sulfuric acid. Name the salt that is a by-product of this process. Rayon production is in fact a significant commercial source for this salt.

43. Your tomato sauce eats a hole in an aluminum saucepan. Explain why, using a net ionic equation.

44. Manganese(II) oxide (MnO) is treated with concentrated $HCl(aq)$ and dissolves. Name the products, and write a balanced chemical equation for this reaction.

Oxidation–Reduction Reactions

45. (See Example 4–8.) Assign oxidation numbers to each element in the following compounds:
(a) Formaldehyde (CH_2O)
(b) Carbonic acid (H_2CO_3)
(c) Rubidium hydride (RbH)
(d) Dinitrogen pentaoxide (N_2O_5)

46. (See Example 4–8.) Assign oxidation numbers to each element in the following compounds:
(a) Ethane (C_2H_6)
(b) Acetic acid ($C_2H_4O_2$)
(c) Ammonium nitrate (NH_4NO_3)
(d) Potassium superoxide (KO_2)

47. A useful analytical technique for determining the amount of arsenic in a solution relies on its oxidation by iodine through reactions such as

$$As_2O_3(aq) + 2\ I_2(s) + 2\ H_2O(\ell) \longrightarrow$$
$$As_2O_5(aq) + 4\ HI(aq)$$

Write the oxidation number above the symbol for each atom that changes oxidation state in the course of this reaction, identify which is reduced and which is oxidized, and verify that the total decrease in oxidation number is equal to the total increase in oxidation number.

48. When calcium sulfate is heated with coke (carbon) above 1200°C, it reacts according to

$$2\ CaSO_4(s) + C(s) \longrightarrow 2\ CaO(s) + 2\ SO_2(g) + CO_2(g)$$

This reaction is used industrially on a small scale as a source for SO_2, which can be converted into sulfuric acid. Write the oxidation number above the symbol for each atom that changes oxidation state in the course of this reaction, identify which is reduced and which is oxidized, and verify that the total decrease in oxidation number is equal to the total increase in oxidation number.

49. For each of the following balanced equations, write the oxidation number above the symbol of each atom that changes oxidation state in the course of the reaction:
(a) $2\ PF_2I(\ell) + 2\ Hg(\ell) \longrightarrow P_2F_4(g) + Hg_2I_2(s)$
(b) $2\ KClO_3(s) \longrightarrow 2\ KCl(s) + 3\ O_2(g)$
(c) $4\ NH_3(g) + 5\ O_2(g) \longrightarrow 4\ NO(g) + 6\ H_2O(g)$
(d) $2\ As(s) + 6\ NaOH(\ell) \longrightarrow 2\ Na_3AsO_3(s) + 3\ H_2(g)$

50. For each of the following balanced equations, write the oxidation number above the symbol of each atom that changes oxidation state in the course of the reaction:
(a) $N_2O_4(g) + KCl(s) \longrightarrow NOCl(g) + KNO_3(s)$
(b) $H_2S(g) + 4\ O_2F_2(s) \longrightarrow SF_6(g) + 2\ HF(g) + 4\ O_2(g)$
(c) $2\ POBr_3(s) + 3\ Mg(s) \longrightarrow 2\ PO(s) + 3\ MgBr_2(s)$
(d) $4\ BCl_3(g) + 3\ SF_4(g) \longrightarrow$
$$4\ BF_3(g) + 3\ SCl_2(\ell) + 3\ Cl_2(g)$$

51. For each of the reactions in problem 49, identify the species that is oxidized and the one that is reduced. Verify that the total decrease in oxidation number is equal to the total increase in oxidation number.

52. For each of the reactions in problem 50, identify the species that is oxidized and the one that is reduced. Verify that the total decrease in oxidation number is equal to the total increase in oxidation number.

53. Selenic acid (H_2SeO_4) is a powerful oxidizing acid that dissolves not only silver (as does the related acid H_2SO_4) but also gold, through the reaction

$$2\ Au(s) + 6\ H_2SeO_4(aq) \longrightarrow$$
$$Au_2(SeO_4)_3(aq) + 3\ H_2SeO_3(aq) + 3\ H_2O(\ell)$$

Determine the oxidation numbers of the atoms in this equation. Which species is oxidized and which is reduced?

54. Diiodine pentaoxide reacts with carbon monoxide under room conditions

$$I_2O_5(s) + 5\ CO(g) \longrightarrow I_2(s) + 5\ CO_2(g)$$

This can be used in an analytical method to measure the amount of carbon monoxide in a sample of air. Determine the oxidation numbers of the atoms in this equation. Which species is oxidized and which is reduced?

55. Aluminum(III) oxide is caused to react with carbon to give elemental aluminum and carbon dioxide. Write a balanced equation for this process, and indicate which species is oxidized and which is reduced.

56. During the recharging of a lead-acid storage battery (used in automobiles), lead(II) sulfate reacts with water to give metallic lead, lead(IV) oxide, and aqueous sulfuric acid. Write a balanced equation for the reaction that occurs, and identify which species is oxidized and which is reduced.

57. (See Example 4–10.) Give likely products of the following reactions. Balance the chemical equations in each case, and give a descriptive name for the reaction.
    (a) $HCl(g) + O_2(g) \longrightarrow$
    (b) $H_2C{=}O(g) + H_2(g) \longrightarrow$
    (c) $Mg(s) + HCl(aq) \longrightarrow$
    (d) $N_2(g) + H_2(g) \longrightarrow$

58. (See Example 4–10.) Give likely products of the following reactions. Balance the chemical equations in each case, and give a descriptive name for the reaction.
    (a) $Br_2(g) + H_2(g) \longrightarrow$
    (b) $NH_2NH_2(g) + H_2(g) \longrightarrow$
    (c) $CH_4(g) + O_2(g) \longrightarrow$
    (d) $Ca(s) + Br_2(g) \longrightarrow$

59. Chloric acid ($HClO_3$) cannot be prepared outside of solution. Even in solution, it is not stable and can decompose in several ways, depending on conditions. The modes of decomposition include

$$HClO_3(aq) \longrightarrow ClO_2(g) + O_2(g) + H_2O(\ell)$$
$$HClO_3(aq) \longrightarrow HCl + O_2(g)$$
$$HClO_3(aq) \longrightarrow HClO_4(aq) + HCl(aq)$$

   (a) Balance all three of the equations, and assign oxidation numbers to the elements in every substance.
   (b) Which of these reactions is a disproportionation?

60. Aqueous nitrous acid reacts to give nitric acid, gaseous nitrogen monoxide, and water. Write a balanced chemical equation for this reaction, and show that it is a disproportionation.

61. Heating ammonium nitrite (NH_4NO_2) gives water and nitrogen (N_2).
   (a) Write and balance a chemical equation to represent this change.
   (b) Determine the oxidation number of nitrogen in the ammonium ion (NH_4^+) and in the nitrite ion (NO_2^-); determine the "overall" oxidation number of nitrogen in this compound. Discuss whether this is a redox reaction.

62. Heating lead(II) nitrate gives lead(II) oxide, oxygen, and one other product, the gas X, which contains no lead. The volume of X is four times the volume of the oxygen when the two are held at the same temperature and pressure. Name X. Is this a redox reaction?

63. Write balanced equations for the formation of barium hydride from the elements and the reactions of barium hydride with water, with hydrochloric acid, and with aqueous zinc sulfate.

64. Write balanced equations for the formation of hydrogen bromide from the elements and the reactions of hydrogen bromide with water, with potassium carbonate, and with magnesium.

65. (See Example 4–10.) Complete and balance each of the following chemical equations, and state whether each represents a redox, acid–base, dissolution, or precipitation reaction or does not fit in any of these categories:
    (a) $H_2Te(g) + H_2O(\ell) \longrightarrow$
    (b) $SrO(s) + CO_2(g) \longrightarrow$
    (c) $HI(aq) + CaCO_3(s) \longrightarrow$
    (d) $Na_2O(s) + NH_4Br(aq) \longrightarrow$

*66. (See Example 4–10.) Complete and balance each of the following chemical equations, and state whether each represents a redox, acid–base, dissolution, or precipitation reaction or does not fit in any of these categories:
    (a) $HBr(aq) + ZnO(s) \longrightarrow$
    (b) $Cs(s) + O_2(g) \longrightarrow$
    (c) $MgO(s) + SO_3(g) \longrightarrow$
    (d) $NaH(s) + Cu(NO_3)_2(aq) + H_2O(\ell) \longrightarrow$

The Stoichiometry of Reactions in Solution

67. Determine the chemical amount, in moles or millimoles, of sucrose contained in 225 mL of 0.25 M sucrose.

68. What mass, in grams or milligrams, of silver chloride should precipitate when 25.0 mL of 0.150 M silver nitrate is mixed with 15.0 mL of 0.100 M calcium chloride?

69. (See Example 4–12.) Suppliers sell concentrated hydrochloric acid that is nominally 12 M. How much of this concentrated acid should be added to water to produce 10.0 L of a solution that is approximately 0.1 M HCl?

70. What is the final concentration in a solution prepared by mixing 500.0 mL of 0.600 M fructose with 300.0 mL of 0.400 M fructose? State any assumptions you must make in obtaining your answer.

71. A 100.0-mL sample of 0.2516 M aqueous barium chloride was treated with an excess of sulfuric acid. The resulting barium sulfate precipitate was filtered, dried, and found to have a mass of 4.9852 g. Compute the percentage yield of barium sulfate in this process.

72. Potassium iodate reacts with potassium iodide and hydrochloric acid to produce iodine, water, and potassium chloride.
   (a) Write a balanced chemical equation for this reaction.
   (b) Determine the limiting reagent when 50.00 mL of 0.01000 M potassium iodate is treated with 1.00 g of potassium iodide and 10.00 mL of 3 M hydrochloric acid.

73. (See Example 4–13.) Hydrazine is made by the reaction of ammonia with sodium hypochlorite in aqueous solution:

$$2 \, NH_3(aq) + OCl^-(aq) \longrightarrow$$
$$N_2H_4(aq) + Cl^-(aq) + H_2O(\ell)$$

What mass of hydrazine is produced from 51.6 L of a 0.650 M solution of sodium hypochlorite that reacts completely according to this equation with an excess of ammonia?

74. (See Example 4–13.) When the blue liquid dinitrogen trioxide is added to a solution of sodium hydroxide, it reacts to give sodium nitrite:

$$N_2O_3(\ell) + 2 \, OH^-(aq) \longrightarrow 2 \, NO_2^-(aq) + H_2O(\ell)$$

What is the concentration of sodium nitrite if 2.13 g of N_2O_3 is added to an excess of aqueous sodium hydroxide and the resulting solution is diluted to a total volume of 856 mL?

75. When treated with acid, lead(IV) oxide is reduced to a lead(II) salt, liberating oxygen:

$$2 \, PbO_2(s) + 4 \, HNO_3(aq) \longrightarrow$$
$$2 \, Pb(NO_3)_2(aq) + 2 \, H_2O(\ell) + O_2(g)$$

What volume of a 7.91 M solution of nitric acid is just sufficient to react with 15.9 g of lead(IV) oxide according to this equation?

76. Phosphoric acid is made industrially by the reaction of fluorapatite ($Ca_5(PO_4)_3F$) in phosphate rock with sulfuric acid:

$$Ca_5(PO_4)_3F(s) + 5 \, H_2SO_4(aq) + 10 \, H_2O(\ell) \longrightarrow$$
$$3 \, H_3PO_4(aq) + 5(CaSO_4 \cdot 2H_2O)(s) + HF(aq)$$

What volume of 6.3 M phosphoric acid is generated by the reaction of 2.2 metric tons (220 kg) of fluorapatite?

77. The carbon dioxide produced (together with hydrogen) from the industrial-scale oxygenation of methane in the presence of nickel is removed from the gas mixture in a scrubber containing an aqueous solution of potassium carbonate:

$$CO_2(g) + H_2O(\ell) + K_2CO_3(aq) \longrightarrow 2 \, KHCO_3(aq)$$

Calculate the mass (in kilograms) of carbon dioxide that reacts with 187 L of a 1.36 M potassium carbonate solution.

78. Nitrogen monoxide can be generated on a laboratory scale by the reaction of dilute sulfuric acid with aqueous sodium nitrite:

$$6 \, NaNO_2(aq) + 3 \, H_2SO_4(aq) \longrightarrow$$
$$4 \, NO(g) + 2 \, HNO_3(aq) + 2 \, H_2O(\ell) + 3 \, Na_2SO_4(aq)$$

What volume of 0.646 M aqueous $NaNO_2$ should be used in this reaction to generate 6.07 g of nitrogen monoxide?

79. Phosphorus trifluoride is a highly toxic gas that reacts slowly with water to give a mixture of phosphorous acid and hydrofluoric acid. Determine the concentrations (in moles per liter) of each of the acids that result from the reaction of 0.077 g of phosphorus trifluoride with water to give a solution volume of 872 mL.

80. Phosphorus pentachloride reacts violently with water to give a mixture of phosphoric acid and hydrochloric acid.
    (a) Write a balanced chemical equation for this reaction.
    (b) Determine the concentration (in moles per liter) of each of the acids that result from mixing 0.0293 g of phosphorus pentachloride with enough water to give a solution with a final volume of 697 mL.

81. (See Examples 4–14 and 4–15.) In order to determine the concentration of a certain solution of nitric acid, a 100.0-mL sample is placed in a flask and titrated with a 0.1279 M solution of potassium hydroxide. A volume of 37.85 mL is required to reach the phenolphthalein end-point. Calculate the molarity of nitric acid in the original sample.

82. (See Examples 4–14 and 4–15.) It requires 34.98 mL of a certain sodium hydroxide solution to titrate 0.6753 g of dissolved potassium hydrogen phthalate, $KHC_8H_4O_4(aq)$, to the phenolphthalein end-point but only 21.49 mL to titrate 20.00 mL of a solution of sulfuric acid to the phenolphthalein end-point. Determine the molarity of the sodium hydroxide solution and of the sulfuric acid solution. The products of the reaction between sodium hydroxide and sulfuric acid are water and sodium sulfate.

83. Potassium dichromate in acidic solution is used to titrate solutions containing Fe^{2+} ion. The reaction is

$$Cr_2O_7^{2-}(aq) + 6 \, Fe^{2+}(aq) + 14 \, H^+(aq) \longrightarrow$$
$$2 \, Cr^{3+}(aq) + 6 \, Fe^{3+}(aq) + 7 \, H_2O(\ell)$$

A potassium dichromate solution is prepared by dissolving 5.134 g of $K_2Cr_2O_7$ in water and diluting to a total volume of 1.00 L. A total of 34.26 mL of this solution is required to reach the end-point in a titration of a 500.0-mL sample containing $Fe^{2+}(aq)$. Determine the concentration of Fe^{2+} in the original solution.

84. Cerium(IV) ions are strong oxidizing agents in acidic solution, oxidizing arsenous acid to arsenic acid according to the equation

$$2 \, Ce^{4+}(aq) + H_3AsO_3(aq) + H_2O(\ell) \longrightarrow$$
$$2 \, Ce^{3+}(aq) + H_3AsO_4(aq) + 2 \, H^+(aq)$$

A sample of As_2O_3 weighing 0.217 g is dissolved in basic solution and then acidified to make H_3AsO_3. Titration with a solution of acidic cerium(IV) sulfate requires 21.47 mL. Determine the original concentration of $Ce^{4+}(aq)$ in the titrating solution.

Additional Problems

85. All of the following substances are soluble in water. State whether the solutions conduct electricity well, poorly, or not at all.
    (a) acetic acid   (c) $HgCl_2$
    (b) acetone   (d) $NaClO_3$

86. The solubility of lithium chlorate ($LiClO_3$) is 315 g/100 g of water at room temperature. It is said to be one of the most soluble of all salts.

(a) Compute the ratio of the number of water molecules to the number of ions in this solution, assuming complete ionization.

(b) Is $LiClO_3$ completely dissociated in a saturated solution? Why or why not?

87. When ammonia and carbon dioxide are added to water, an aqueous solution of ammonium carbonate forms. This solution reacts with calcium sulfate to give a solid and an electrolyte solution. Identify the solid and the solution, and write a balanced chemical equation for the reaction. This reaction is carried out industrially on a small scale; the salt in the electrolyte solution is a useful fertilizer.

88. About 0.6 g of mercury(I) sulfate dissolves in 1.0 L of water at room temperature. Suppose that a saturated solution of mercury(I) sulfate is mixed with aqueous barium chloride. Tell what will be seen, and write net ionic equations to represent the reaction or reactions taking place.

89. The autoionization of water

$$H_2O(\ell) \longrightarrow H^+(aq) + OH^-(aq)$$

cannot be prevented. This means that even the purest water contains quantities of $OH^-(aq)$ and $H^+(aq)$ ions. Does water then fulfill the definition of a substance given in Chapter 1? Explain.

90. Aluminum sulfate hydrate $(Al_2(SO_4)_3 \cdot 18H_2O)$ is used in two major ways: by the paper industry to coat paper pulp and give it a hard surface (its aqueous solutions are referred to as "papermaker's alum") and in water treatment to remove bacteria and impurities.

(a) Propose a method for making this hydrate from the aluminum ore bauxite $(Al(OH)_3)$. Write a balanced chemical equation for the reaction.

(b) In aqueous solution, aluminum sulfate reacts with calcium hydroxide to form a loose solid hydroxide precipitate (called a "floc") that removes impurities from the drinking water as it settles. Write a balanced equation for this reaction.

91. A researcher adds a solution of potassium hydroxide to one of perchloric acid in order to neutralize the strong acid. The reaction generates a large amount of heat, which was expected, and a white precipitate appears, which surprises the researcher. Write chemical equations that explain both observations.

92. Workers in paint factories making the pigment white lead $(Pb(OH)_2 \cdot 2PbCO_3)$ at one time customarily added dilute sulfuric acid to their drinking water for protection against lead poisoning. Explain the chemistry behind this practice.

93. A compound is said to be *hygroscopic* if it has a strong affinity for water. A very hygroscopic compound such as P_4O_{10} dehydrates many oxoacids to give the corresponding acid anhydrides (while it in turn forms a phosphorus oxoacid). Write a balanced chemical equation for the reaction between P_4O_{10} and nitric acid. In this case, the oxoacid of phosphorus produced is HPO_3 rather than the fully hydrated H_3PO_4.

94. Oxygen difluoride does not react with water to give an oxoacid. Rather, its reaction is described by the equation

$$OF_2(g) + H_2O(\ell) \longrightarrow O_2(g) + 2\ HF(aq)$$

(a) Is this a redox reaction? If so, name the species being oxidized and the species being reduced.

(b) The oxoacid that might be predicted to form from the addition of water to OF_2 was first prepared in 1968 through the reaction of F_2 with water at very low temperatures and under special conditions. It is a white solid that melts to a pale yellow liquid at $-117°C$. Give its chemical formula and its name, based on the formula and name of the parallel oxoacid of chlorine.

95. A total of 222 mL of 0.0450 M hydrobromic acid (HBr) is mixed with 125 mL of a 0.0540 M solution of the base barium hydroxide $(Ba(OH)_2)$. They react.

(a) Determine which reactant is the limiting reactant.

(b) Determine the additional volume of acid or base that must be added to exactly neutralize the excess base or acid.

96. Bismuth forms an ion with the formula Bi_5^{3+}. Arsenic and fluorine form a complex ion $[AsF_6]^-$, with fluorine atoms arranged around an arsenic atom. Assign oxidation numbers to each of the atoms in the bright yellow crystalline solid with the formula $Bi_5(AsF_6)_3 \cdot 2SO_2$.

*97. The best method to prepare very pure oxygen on a small scale is the decomposition of $KMnO_4$ in a vacuum above $215°C$:

$$2\ KMnO_4(s) \longrightarrow K_2MnO_4(s) + MnO_2(s) + O_2(g)$$

Assign oxidation numbers to each atom, and verify that the total decrease in oxidation number equals the total increase in oxidation number.

*98. (a) Determine the oxidation number of lead in each of the following oxides: PbO, PbO_2, Pb_2O_3, Pb_3O_4.

(b) The only known lead ions are Pb^{2+} and Pb^{4+}. How can you reconcile this statement with your answer to part (a)?

99. The hydrogen xenate ion reacts with base according to the equation

$$2\ HXeO_4^-(aq) + 2\ OH^-(aq) \longrightarrow$$
$$XeO_6^{4-}(aq) + Xe(g) + O_2(g) + 2\ H_2O(\ell)$$

Give the oxidation numbers for all the atoms in this equation. Which species are being reduced? Which oxidized?

100. In some forms of the periodic table, hydrogen is placed in Group I; in others, it is placed in Group VII. Give arguments in favor of each location.

101. The following descriptive names of reactions have all been used in the chemical literature. Use a balanced equation to give a likely example of each type of reaction. Compare and contrast each with those discussed in the chapter.

(a) chlorination

(b) dehydrogenation

(c) hydrochlorination

*102. The following descriptive names of processes have all been used in the chemical literature. Give an example of each. Compare and contrast each with those discussed in the chapter.
(a) deoxygenation
(b) reproportionation
(c) fluoridation

103. Barium perxenate (Ba_2XeO_6) is a colorless salt of the perxenate ion.
(a) Americium perxenate has also been prepared. Write its chemical formula, given that the stable ion formed by americium is Am^{3+}.
(b) The corresponding oxoacid, perxenic acid, has not been prepared because it is spontaneously reduced in aqueous solution. Give its chemical formula.
(c) Give the chemical formula and name for the anhydride of perxenic acid.

104. Arsenic forms two important oxoacids, with chemical formulas H_3AsO_3 and H_3AsO_4.
(a) Give the chemical formulas and names of the acid anhydrides of these two acids.
(b) Follow the usual naming procedure for an element that forms two oxoacids to determine the names of the oxoacids of arsenic.
(c) Arsenic is in the same group as nitrogen. How are the chemical formulas of the oxoacids of arsenic and nitrogen different?
(d) Can you give an explanation for the result in part (c) based on Lewis structures of the arsenic oxoacids? (*Hint*: Recall that second-period elements readily form multiple bonds, but multiple bonding is unusual in the fourth period.)

105. When coal that contains sulfur is burned to generate electricity, sulfur dioxide and sulfur trioxide are produced. Because these are air pollutants that contribute significantly to acid rain, their removal in the stacks of power plants would be very desirable. One possible way to do this is to line the stacks with lime (CaO). What would be the products of its reaction with the sulfur oxides in the stacks?

106. Balance the following equations, and classify them as representing dissolution, precipitation, acid–base, or redox reactions or as reactions that do not fit any of the categories listed:
(a) $H_3PO_4(aq) + NaOH(aq) \longrightarrow$
$$NaH_2PO_4(aq) + H_2O(\ell)$$
(b) $Fe_3O_4(s) + C(s) \longrightarrow Fe(s) + CO(g)$
(c) $Ca(NO_3)_2(aq) + K_3PO_4(aq) \longrightarrow$
$$Ca_3(PO_4)_2(s) + KNO_3(aq)$$
(d) $K_2Cr_2O_7(aq) + HI(aq) \longrightarrow$
$$CrI_3(s) + I_2(s) + KI(aq) + H_2O(\ell)$$

107. Balance the following equations, and classify them as representing dissolution, precipitation, acid–base, or redox reactions or as reactions that do not fit any of the categories listed:
(a) $KClO_3(s) \longrightarrow KCl(s) + KClO_4(s)$
(b) $H_2S(g) \longrightarrow H^+(aq) + HS^-(aq)$
(c) $H_2O_2(g) + HI(g) \longrightarrow I_2(g) + H_2O(g)$
(d) $P_4O_{10}(s) + C(s) \longrightarrow P_4(g) + CO(g)$

108. A chemical company hands out cards printed with the following table of concentrated commercial reagents:

Reagent	Concentration (M)	Molar Volume (mL)	Density (g mL^{-1})
HCl	12.1	82.6	1.187
HNO$_3$	15.8	63.3	1.415
H$_2$SO$_4$	18.0	55.6	1.835
CH$_3$COOH	17.4	57.5	1.044
NH$_3$	14.8	67.6	0.900

(a) Explain how to compute the molar volumes (third column) from the concentrations (second column).
(b) You need 2.50 L of 6.5 M hydrochloric acid. You have a bottle of concentrated hydrochloric acid (first entry in the table) and various calibrated containers. Tell what to do to prepare the solution you need.
(c) Four of these five commercial reagents consist of the chemical mixed with water; one of the five contains no water. Determine which one of the five contains no water.

CUMULATIVE PROBLEM

The Solvay Process

The Solvay process is used industrially to make sodium carbonate from the inexpensive starting materials sodium chloride and calcium carbonate. It can be represented in the following six steps, which are discussed in greater detail in Chapter 23:

1. $2 NH_3(aq) + 2 CO_2(g) + 2 H_2O(\ell) \longrightarrow 2 NH_4HCO_3(aq)$
2. $2 NH_4HCO_3(aq) + 2 NaCl(aq) \longrightarrow 2 NaHCO_3(s) + 2 NH_4Cl(aq)$
3. $2 NaHCO_3(s) \longrightarrow Na_2CO_3(s) + H_2O(g) + CO_2(g)$
4. $CaCO_3(s) \longrightarrow CaO(s) + CO_2(g)$
5. $CaO(s) + H_2O(\ell) \longrightarrow Ca(OH)_2(aq)$
6. $2 NH_4Cl(aq) + Ca(OH)_2(aq) \longrightarrow 2 NH_3(aq) + CaCl_2(aq) + 2 H_2O(\ell)$

Answer the following questions about these reaction steps:

(a) Which oxoacid has CO_2 as its anhydride?

(b) What type of reaction is the second step? Write a net ionic equation for this step.

(c) Is step 3 an acid–base reaction? If so, what is the acid reacting and what is the base?

(d) What is the base of which CaO is the anhydride? In which step is this base prepared?

(e) Write a net ionic equation for step 6.

(f) Are any of the steps redox reactions? If so, which ones?

(g) Add the six reactions, canceling chemical species that appear on both sides of the equation. The result is the overall reaction. Suppose the two products were mixed in aqueous solution. Predict the results.

Most of the chemicals involved in the Solvay process are found in the home, including Na_2CO_3 (sodium carbonate, washing soda) and $NaHCO_3$ (sodium hydrogen carbonate, baking soda).

5 The Gaseous State

Hot air is less dense than cool air. The difference gives these hot air balloons their buoyancy.

$\mathbf{T}$here are two major ways of classifying matter. The first is to study the *chemical constitution* of a material and categorize it as an element, compound, or mixture of substances. Chapter 1 uses this approach. The second scheme of classification considers the *physical state* of a material, calling it a gas, a liquid, or a solid (see Fig. 3–7). We recognize gases, liquids, and solids quite easily because an important example of each is all around us every day: air, water, and the Earth. Gases and liquids are fluid, but solids are rigid. A gas expands to fill any container it occupies; a liquid has a fixed volume but flows to conform to the shape of its container; a solid has a fixed volume *and* a fixed shape that resists deformation. These simple properties allow us to classify nearly all substances, at least under ordinary terrestrial conditions. Changes of state are also matters of everyday experience—that is, ice melts and water freezes; water boils and steam condenses. Similar transitions occur with thousands of less familiar substances and are crucial in chemistry.

In this chapter and the one that follows, our goal is to develop a deeper understanding of the states of matter and the transitions between them. This allows us later to explore in a quantitative fashion the extents and rates of chemical reactions. We begin with gases, both because they have played a central role throughout the history of chemistry and because their physical properties are simpler than those of liquids and solids.

5–1 THE CHEMISTRY OF GASES

The Greeks considered air to be one of the four fundamental elements in nature. As early as the 17th century, its physical properties (such as resistance to compression) were being studied. The chemical composition of air (Table 5–1) was not appreciated until late in the 18th century, when Priestley, Lavoisier, and others showed that it consists primarily of two different substances: oxygen and nitrogen. Oxygen is characterized by its ability to support life. Once the oxygen in air is used up (by the burning of a candle in a closed container, for example), the nitrogen that remains no longer keeps animals alive. More than 100 years elapsed before a careful reanalysis of air showed that oxygen and nitrogen account for only about 99% of the total volume, with most of the remaining 1% consisting of a new gas, which was given the name "argon." The other noble gases (helium, neon, krypton, and xenon) are present in air to a lesser extent.

Gases are found at and under the earth's surface as well. Methane (CH_4) is produced by bacterial processes, especially in swampy areas. It is a major constituent of natural gas deposits formed over many millennia by the decay of plant matter under the earth's surface. The recovery of methane from municipal landfills for use as a fuel gas is now a commercially feasible process. Helium is found in natural gas sources as well, and the radioactive gas radon, which arises from the decay of trace amounts of actinide elements in certain rocks, can cause hazardous concentrations of radioactivity in the air in unventilated basements. The evaporation of liquids also produces gases. All liquids are **volatile** to some extent and therefore evaporate; the most familiar example is water. Water vapor rising from lakes and oceans causes the humidity of the air.

Gases are often formed by chemical reactions. It is quite common, for example, for solids to decompose to give one or more gases when heated. We have already mentioned (see Fig. 1–13) the decomposition of mercury(II) oxide to mercury and oxygen:

- Methane has long been known as "marsh gas" owing to its presence in swampy areas.

- "Volatile" means liable to evaporate at normal temperatures. It does *not* mean explosive, toxic, noxious, or dangerous.

$$2 \text{ HgO}(s) \xrightarrow{\text{heat}} 2 \text{ Hg}(\ell) + \text{O}_2(g)$$

Table 5–1
The Composition of Air[a]

Constituent	Formula	Volume Percent
Nitrogen	N_2	78.110
Oxygen	O_2	20.953
Argon	Ar	0.934
Neon	Ne	0.001818
Helium	He	0.000524
Krypton	Kr	0.000114
Xenon	Xe	0.0000087
Hydrogen	H_2	0.00005
Methane	CH_4	0.0002
Dinitrogen monoxide	N_2O	0.00005

[a]Air also contains other constituents, the abundance of which is quite variable in the atmosphere. Examples are water (H_2O), 0–7%; carbon dioxide (CO_2), 0.01–0.1%; ozone (O_3), 0–0.000007%; carbon monoxide (CO), 0–0.000002%; nitrogen dioxide (NO_2), 0–0.000002%; sulfur dioxide (SO_2), 0–0.0001%.

Carbon dioxide (CO_2) dissolved in aqueous solution gives soft drinks their fizz. It also serves as an inexpensive acid through its reaction with water to form $H_2CO_3(aq)$. Solid carbon dioxide (Dry Ice) is used for refrigeration.

• The role of these compounds in air pollution is considered further in Section 18–4.

This reaction was used by Antoine Lavoisier in establishing the law of conservation of mass. Even earlier (1756), Joseph Black had shown that marble, which is composed primarily of calcium carbonate ($CaCO_3$), would decompose on heating to give quicklime (CaO) and carbon dioxide:

$$CaCO_3(s) \xrightarrow{\text{heat}} CaO(s) + CO_2(g)$$

The salt ammonium chloride (NH_4Cl) can be decomposed by heat to produce both gaseous ammonia and gaseous hydrogen chloride:

$$NH_4Cl(s) \xrightarrow{\text{heat}} NH_3(g) + HCl(g)$$

Many ammonium salts give similar reactions. Some gas-forming reactions proceed explosively; the decomposition of nitroglycerin is a detonation in which all the products are gaseous:

$$4\ C_3H_5(NO_3)_3(\ell) \longrightarrow 6\ N_2(g) + 12\ CO_2(g) + O_2(g) + 10\ H_2O(g)$$

Recall from Section 4–4 that several of the elements react with oxygen to form gaseous oxides. Carbon dioxide is produced during animal respiration but also, to a much greater extent, by burning as fuels materials such as coal and oil that contain compounds of carbon. Oxides of sulfur are produced by burning elemental sulfur, and oxides of nitrogen are produced in the oxidation of elemental nitrogen that occurs inside the engines of motor vehicles:

$$S(s) + O_2(g) \longrightarrow SO_2(g)$$
$$2\ SO_2(g) + O_2(g) \longrightarrow 2\ SO_3(g)$$
$$N_2(g) + O_2(g) \longrightarrow 2\ NO(g)$$
$$2\ NO(g) + O_2(g) \longrightarrow 2\ NO_2(g)$$

CHEMISTRY IN YOUR LIFE

How an Air-Bag Works

This is a true story. One rainy day, a motorist hit a bump at an intersection hard enough to cause the air-bag in her car to deploy. No one was hurt, and damage to the car was minor, but a police officer noticed some strange white powder on the air-bag. He immediately cleared the area and called in the bomb squad! Traffic was snarled until the bomb squad determined that no danger existed.

Why did the police officer fear an explosion? He had learned at a training session that the "air" that inflates air-bags is actually nitrogen from the decomposition of sodium azide

$$2 \text{ NaN}_3(s) \longrightarrow 2\text{Na}(s) + 3 \text{ N}_2(g)$$

He had also learned that elemental sodium reacts violently with water

$$2 \text{ Na}(s) + 2 \text{ H}_2\text{O}(\ell) \longrightarrow 2\text{NaOH}(s) + \text{H}_2(g)$$

If the white powder were elemental sodium from the first reaction that had escaped the air-bag, then the rain might set off the second reaction, with disastrous results. Therefore, he took drastic action. Good chemistry as far as it went . . . but incomplete.

The material that generates the gas in an air-bag is not pure sodium azide, but a mixture of sodium azide with (usually) potassium nitrate (KNO_3) and silicon dioxide (SiO_2). When sensors detect a collision and send an electric current through this mixture, the NaN_3 experiences a "deflagration," or slow detonation, to give gaseous nitrogen and elemental sodium, just as the police officer was told. The sodium, however, reacts nearly instantly with the potassium nitrate to generate more nitrogen:

$$10 \text{ Na}(s) + 2 \text{ KNO}_3(s) \longrightarrow$$
$$\text{K}_2\text{O}(s) + 5 \text{ Na}_2\text{O}(s) + \text{N}_2(g)$$

Thus, no elemental sodium accumulates when an air-bag deploys. The potassium oxide and sodium oxide from the second reaction are anhydrides of strong bases and might

Figure 5-A Deployment of an air-bag. In this crash test, a real man, not a dummy, is at the wheel. He was unhurt.

prove injurious if released. The acid anhydride SiO_2 neutralizes them to form silicates. Consequently, the only chemical products in an air-bag deployment are the nitrogen that inflates the bag and a harmless glassy mixture of the salts Na_2SiO_3 and K_2SiO_3.

Although the products of an air-bag deployment are innocuous, the gas-generating mixture is not; sodium azide is quite toxic. Because drivers try hard *never* to deploy an air-bag, the release of sodium azide during the eventual junking of cars with unused air-bags may pose a real problem. One possible solution is to develop inflator modules that are recyclable from old cars into new ones.

What about the suspicious white powder that fired the police officer's concern? It was cornstarch that had been sprinkled between the folds of the air-bag to prevent sticking and ensure smooth inflation.

Another important class of reactions that form gases is the action of acids on certain ionic solids. In Section 4–3, we mentioned the production of carbon dioxide from carbonates (Fig. 5–1)

$$\text{CaCO}_3(s) + 2 \text{ HCl}(aq) \longrightarrow \text{CaCl}_2(aq) + \text{CO}_2(g) + \text{H}_2\text{O}(\ell)$$

as a characteristic reaction for acids. Other examples of this type of reaction are:

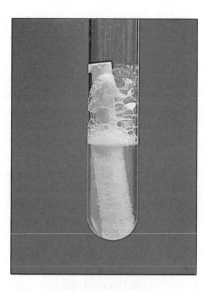

Figure 5-1 Chalk is one especially familiar source of carbon dioxide. Here the calcium carbonate in a stick of chalk reacts with an aqueous solution of hydrochloric acid, producing bubbles of carbon dioxide.

$$Na_2S(s) + 2\,HCl(aq) \longrightarrow 2\,NaCl(aq) + H_2S(g)$$

$$K_2SO_3(s) + 2\,HCl(aq) \longrightarrow 2\,KCl(aq) + SO_2(g) + H_2O(\ell)$$

$$NaCl(s) + H_2SO_4(aq) \longrightarrow NaHSO_4(aq) + HCl(g)$$

In these reactions, the identity of the metal ion (sodium, calcium, and potassium) is immaterial; similar reactions of the carbonates, sulfides, sulfites, and chlorides of other metals also occur.

The gases we have considered show extremely varied chemical behavior. Some, like HCl and SO_3, are very reactive, corrosive, and acidic, whereas others, like N_2O and N_2 are much less reactive. Although the chemical properties of gases vary significantly, their physical properties are much simpler to understand. At sufficiently low densities, all gases behave in the same way. Their properties approach the physical properties of an "ideal" gas, which we consider in the next sections.

• An interesting generalization: all colored gases are poisonous. It is *not* true that all colorless gases are non-poisonous!

5-2 PRESSURE AND BOYLE'S LAW

The force exerted by a gas on a unit area of the walls of its container is called the **pressure** of the gas. It is hard to realize that the air around us exerts a pressure on every square centimeter of our skin and every other surface that it touches, even when no wind blows. We grow up under this pressure and are accustomed to it. Atmospheric pressure was first measured in an ingenious experiment devised by Evangelista Torricelli (1608–1647), an Italian scientist who had been an assistant to Galileo. He sealed a long glass tube at one end, filled it with mercury, closed the open end with his thumb, turned it upside-down, and immersed it in a dish of mercury (Fig. 5–2a), taking care that no air leaked in. When he removed his thumb, the level of mercury in the tube fell, leaving a nearly perfect vacuum at the closed end. All of the mercury did not flow out of the tube. It stopped when its top was at a level about 76 cm above the level of the mercury in the dish. Torricelli showed that the exact height varied somewhat from day to day and from place to place.

This simple device, called a **barometer,** works like a balance, one arm of which is loaded with the mass of mercury in the tube, the other with a column of air of

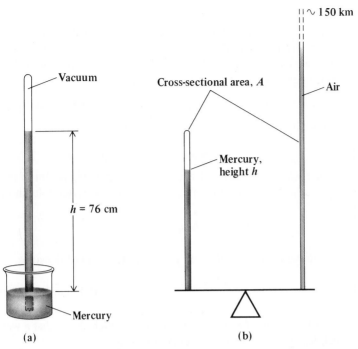

Figure 5-2 (a) In Torricelli's barometer, the height of the top of the mercury in the tube is approximately 76 cm above that in the open dish. (b) The mass of mercury in height h exactly balances that of a column of air of the same diameter extending to the top of the atmosphere.

the same cross-sectional area that extends to the top of the earth's atmosphere, approximately 150 km up (Fig. 5–2b). The height of the mercury column adjusts itself so that the two masses (and thus the two forces on the surface of the mercury in the dish) become equal. Day-to-day changes in the height of the column occur as the force exerted by the atmosphere varies with the weather. Barometers at high elevations have a portion of the atmosphere below them and always give lower readings than those at sea level, if the effect of the weather is excluded.

The relationship between the height, h, of the column of liquid in a barometer and the pressure (P) exerted by the atmosphere is

$$P = gdh$$

In this equation, g stands for the acceleration of gravity at the surface of the earth, 9.80665 m s^{-2}, and d stands for the density of the liquid. This equation implies that a pressure that raises mercury in a barometer to a height of, say, 50 cm would support a shorter column if the strength of the earth's gravitational field were somehow increased, and a longer one if the effect of gravity were somehow decreased. Furthermore, this same pressure could raise a column of a less dense liquid higher than 50 cm. In principle, a barometer can be constructed using almost any liquid if a long enough tube is available. Mercury is the best liquid for barometers because it does not evaporate very rapidly, it is fairly unreactive, and its high density keeps the instrument small. The last point is illustrated by the following example.

EXAMPLE 5–1

A barometer is constructed using water instead of mercury. Estimate the height of the column of water in this barometer on a day when the atmosphere supports a 76.0-cm column of mercury. The density of mercury is approximately 13.6 times greater than the density of water.

Solution

The equation $P = gdh$ applies equally to a water and to a mercury barometer, and the pressure on each is the same. The acceleration of gravity, g, is also the same in the two cases, so that only the density d and height h differ. Because the P's in the two equations

$$P = gd_{water}h_{water} \quad \text{and} \quad P = gd_{mercury}h_{mercury}$$

equal each other, it follows that

$$d_{water}h_{water} = d_{mercury}h_{mercury}$$

Because water has a lower density, the corresponding height of the column of liquid must be greater. Solving for the height of the water column gives

$$h_{water} = \frac{d_{mercury}}{d_{water}}h_{mercury}$$

Substituting $h_{mercury} = 76.0$ cm and using the ratio of densities, 13.6, gives

$$h_{water} = \frac{13.6d_{water}}{d_{water}} \times 76.0 \text{ cm} = 1030 \text{ cm} = \boxed{10.3 \text{ m}}$$

This is the height of a three-story building, which shows that a water barometer is much less practical than a mercury barometer.

Exercise

The pressurized habitat of a moon colony contains air at a pressure sufficient to raise a 76.0-cm column of mercury in a barometer on earth. Estimate the height of the column of mercury in a barometer that operates inside the habitat. The acceleration of the moon's gravity is only 0.165 times that of the earth's.

Answer: 460 cm.

With the equation $P = gdh$, we can calculate the pressure exerted by the atmosphere in SI units. The density of mercury at 0°C is 13.596 g cm^{-3}. Let us first convert this to a combination of base SI units:

$$d \text{ (of mercury)} = \frac{13.596 \text{ g}}{\text{cm}^3} \times \left(\frac{10^6 \text{ cm}^3}{1 \text{ m}^3}\right) \times \left(\frac{1 \text{ kg}}{10^3 \text{ g}}\right) = 1.3596 \times 10^4 \frac{\text{kg}}{\text{m}^3}$$

As Torricelli observed, the height of the mercury column under ordinary atmospheric conditions near sea level is close to 76 cm. Let us use exactly this value (which is 760.0 mm, or 0.7600 m) for the height of the column in our computation:

$$P = gdh = 9.80665 \frac{\text{m}}{\text{s}^2} \times 1.3596 \times 10^4 \frac{\text{kg}}{\text{m}^3} \times 0.7600 \text{ m} = 1.0133 \times 10^5 \frac{\text{kg}}{\text{m s}^2}$$

Pressure can be expressed in various units. Because pressure is a force divided by an area, its natural SI unit is the SI unit of force, the *newton* (N), divided by the SI unit of area, the square meter. This unit, the N m^{-2}, is exactly equivalent to the unit that arose in the above computation, the kg m^{-1} s^{-2}. Both are called the *pascal* (Pa) (see Appendix B). One **standard atmosphere** (1 atm) is defined as exactly 1.01325×10^5 Pa. The standard atmosphere is a useful unit because of the inconveniently small size of the pascal and also because of the importance of atmospheric

• One newton is about the gravitational force exerted by the earth on an apple; a penny resting flat on a table exerts a pressure (averaged over its surface) of about 100 Pa.

Table 5–2
Units for Pressure

Unit	Definition or Relationship
Pascal (Pa)	$1 \text{ kg m}^{-1} \text{ s}^{-2} = 1 \text{ N m}^{-2}$ (the SI unit)
Bar	1×10^5 Pa
Standard atmosphere (atm)	101,325 Pa
Torr	1/760 atm
760 mm Hg (at 0°C)	1 atm
14.6960 pounds of force per square inch (lb in^{-2}, or psi)	1 atm

pressure as a standard of reference. Pressures must be converted to pascals if calculations are to be carried out in SI units.

Unfortunately, units in common use for pressure have proliferated. Although we shall work primarily with the standard atmosphere, it is important to be able to recognize other units and convert among them. Weather reports express the atmospheric pressure in terms of the height (in millimeters or inches) of the column of mercury it can support. A pressure of 1 atm supports a column of mercury 760 mm or 29.92 inches high at 0°C, so we often speak of 1 atm pressure as 760 mm (of mercury); 0.20 atm corresponds to 152 mm, and so on. Because the density of mercury depends slightly on temperature, we must specify that temperature (0°C) and make the proper corrections for accurate work. A more precise term is the *torr*, which is defined as 1 torr = 1/760 atm (or 760 torr = 1 atm) at *any* temperature. Only at 0°C do the torr and the mm Hg equal each other exactly. In the American system of units (used in measuring tire pressure, for example), one atmosphere is equal to approximately 14.7 pounds of force per square inch (psi). Finally, the *bar* is defined as 10^5 Pa, so that it is close in value to the standard atmosphere. Pressures in the kilobar (1 kbar = 10^3 bar) and even megabar (1 Mbar = 10^6 bar) range can be achieved experimentally. Table 5–2 summarizes the units of pressure.

• An important caution: What is measured by a tire pressure gauge is the *gauge pressure*, the pressure in excess of atmospheric pressure. Thus, a gauge pressure of 35 lb in^{-2} corresponds to an absolute pressure of 35 + 14.7 = 50 lb in^{-2}. The abbreviation for the pressure measured by a tire pressure gauge is psig, and it corresponds to the absolute pressure minus the atmospheric pressure.

EXAMPLE 5–2

Water starts to boil at 21°C (room temperature) if the pressure is reduced to 18.65 torr. Express this pressure in atmospheres, pascals, and pounds per square inch.

Solution

The question requires conversion among the various units for pressure. The necessary conversion factors are in Table 5–2. To convert to atmospheres:

$$18.65 \text{ torr} \times \left(\frac{1 \text{ atm}}{760 \text{ torr}} \right) = \boxed{0.02454 \text{ atm}}$$

To convert to pascals, link two conversion factors from the table:

$$18.65 \text{ torr} \times \left(\frac{1 \text{ atm}}{760 \text{ torr}} \right) \times \left(\frac{101,325 \text{ Pa}}{1 \text{ atm}} \right) = \boxed{2.486 \times 10^3 \text{ Pa}}$$

A similar computation converts the pressure to pounds per square inch:

$$18.65 \text{ torr} \times \left(\frac{1 \text{ atm}}{760 \text{ torr}} \right) \times \left(\frac{14.6960 \text{ psi}}{1 \text{ atm}} \right) = \boxed{0.3606 \text{ psi}}$$

Exercise

Barium is a solid at 510°C. If the pressure on a sample of barium at this temperature is increased to 61 kbar, the barium melts. Further compression (still at 510°C) to 68 kbar causes the barium to solidify once again. Express these two transition pressures in megapascals (MPa) and in torr.

Answer: 6.1×10^3 MPa and 6.8×10^3 MPa; 4.6×10^7 torr and 5.1×10^7 torr.

Boyle's Law: The Effect of Pressure on Gas Volume

Robert Boyle, an English natural philosopher and theologian, studied the properties of confined gases. He noted the familiar fact that when a gas is compressed or expanded, it tends to spring back to its original volume. These properties are much like those of metal springs, which were being investigated by his collaborator Robert Hooke. Boyle's experiments on the compression and expansion of air were reported in 1662 in a monograph entitled "The Spring of the Air and Its Effects."

Boyle worked with a simple piece of apparatus—a J-tube, in which air was trapped at the closed end by a column of mercury (Fig. 5–3). If the difference in height (h) between the two mercury levels in such a tube is zero, then the pressure of the air in the closed part exactly balances that of the atmosphere. The pressure

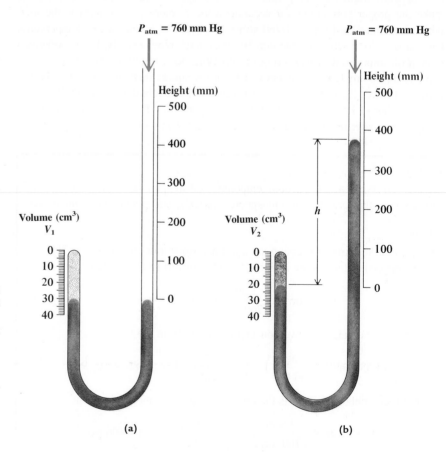

Figure 5–3 Boyle's J-tube. When the height of mercury on the two sides of the tube is the same (*left*), the pressure of the confined gas must be equal to that of the atmosphere, 1 atm or 760 mm Hg. After mercury has been added (*right*), the pressure of the gas is increased by the number of millimeters of mercury in the height difference h. The compression of the gas causes it to occupy a smaller volume.

(a)   (b)

(P) on the trapped air is therefore 1 atm, or 760 mm Hg. If mercury is poured into the open end of the tube, the pressure of the confined air is increased by 1 mm Hg for every millimeter by which the level of mercury on the open side of the tube exceeds the level on the closed side. That is,

$$P = (760 + h) \text{ mm Hg}$$
$$P = (760 + h) \text{ mm Hg} \times \left(\frac{1 \text{ atm}}{760 \text{ mm Hg}} \right) = \frac{(760 + h)}{760} \text{ atm}$$

The volume of the confined air can be read off simply from the previously calibrated tube. We certainly expect squeezing a gas (increasing the pressure) to force it into a smaller volume, and Boyle confirmed just such an *inverse* relationship. His data allowed a generalization:

> The product of the pressure and volume, $P \times V$, of a sample of gas is a constant at a constant temperature. $PV = C$ (fixed temperature and fixed amount of gas).

This result is known as **Boyle's law** (Fig. 5–4).

The conditions on the constancy of C in Boyle's law are as important as the equation itself. If either the temperature or the amount of gas changes during the experiment, then Boyle's law does not apply. The value of C does *not* depend, however, on the identity of the gas trapped in the J-tube. At 0°C and for one mole of gas (for example, 32.0 g of O_2, 28.0 g of N_2, or 2.02 g of H_2), C is 22.4 L atm, so that under these circumstances we have

$$PV = 22.4 \text{ L atm} \quad \text{(at 0°C for 1 mol)}$$

for all of these gases. If the pressure is exactly 1 atm, the volume of each gas should be 22.4 L at 0°C; if P is 4.00 atm, V should be 22.4/4.00 = 5.60 L, and so forth. Conditions of atmospheric pressure ($P = 1.00$ atm) and 0°C are referred to as **standard temperature and pressure** (STP). The volume occupied by one mole of any ideal gas at STP is 22.4 L.

In many practical cases, the actual value of C is not needed to compute useful results, as in the manometer of Figure 5–5. Suppose that a gas has a pressure P_1 and a volume V_1. We adjust the pressure to a new value, P_2, keeping the temperature and the amount of gas unchanged. Boyle's law applies both before and after the adjustment; that is, $P_1V_1 = C$ and $P_2V_2 = C$. The value of C is the same in the two cases, so it follows that

$$P_1V_1 = P_2V_2 \quad \text{at a fixed temperature and for a fixed amount of gas}$$

This useful form of Boyle's law also applies if the volume of the gas is adjusted from V_1 to V_2 at constant temperature for a fixed amount of gas.

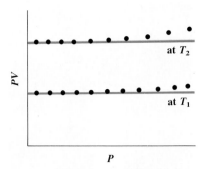

Figure 5–4 If the product of the pressure (P) and volume (V) of a sample of gas is plotted against P at a fixed temperature T_1, a horizontal straight line results; that is, $PV = C$. At a higher temperature T_2, the value of PV is larger, but the line is again straight. At both temperatures, deviations from the straight line set in at higher pressures.

• Visualize the volume 22.4 L as the contents of a cube that is about 28 cm (11 in) on each edge.

EXAMPLE 5–3

Soft drinks and soda water contain carbon dioxide dissolved in water. They are bottled under pressure, so the gas above the solution expands when the cap is removed. (Gas bubbles out of the solution as well, but we do not consider this here.) Suppose the volume of gas above the soft drink in a bottle is 1.6 cm^3 and it increases to a volume of 7.2 cm^3 when it expands to a pressure of 1.0 atm without changing its temperature. Calculate its original pressure, using Boyle's law.

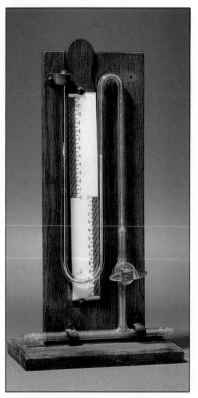

Figure 5–5 A manometer is used to measure the pressure of a gas. The glass tube at the bottom is connected to the gas sample, and the difference in height between the levels of the mercury in the two arms of the tube is read off on a calibrated scale. This simple modern device has essentially the same design as Boyle's J-tube.

• Recall that to convert a temperature from degrees Celsius to degrees Fahrenheit, one first multiplies by 9/5 and then adds 32. To convert a temperature from degrees Fahrenheit to degrees Celsius, one first subtracts 32 and then multiplies by 5/9. That is:
$t_F = 9/5\ t_C + 32$ and
$t_C = 5/9\ (t_F - 32)$. See Appendix B.

Solution

Boyle's law requires that

$$P_1 V_1 = P_2 V_2$$

where P_1 and P_2 are the initial and final pressures, and V_1 and V_2 are the initial and final volumes of a sample of gas at constant temperature. Substituting for the final pressure (1.0 atm) and the initial and final volumes of gas gives

$$P_1\ (1.6\ cm^3) = (1.0\ atm)\ (7.2\ cm^3)$$
$$P_1 = 1.0\ atm\ \left(\frac{7.2\ cm^3}{1.6\ cm^3}\right) = \boxed{4.5\ atm}$$

Exercise

The long cylinder of a bicycle pump has a volume of 1131 cm^3 and is filled with air at a pressure of 1.02 atm. The outlet valve is sealed shut and the pump handle is pushed down until the volume of the air is 517 cm^3. Compute the pressure inside the pump.

Answer: 2.23 atm.

Careful measurements for real gases near 1 atm pressure reveal small deviations from the predictions of Boyle's law. For gases at high pressure (beyond 50 to 100 atm), substantial deviations are found (see Fig. 5–4). Boyle's law is an idealization that real gases follow closely at low pressure but increasingly poorly as the pressure is raised.

5–3 TEMPERATURE AND CHARLES'S LAW

Temperature is one of those elusive properties that is readily understood in a general way but is very difficult to pin down in a quantitative fashion. We have an instinctive feeling (through the sense of touch) of *hot* and *cold*; for example, water at its freezing point is colder than at its boiling point and thus should have a lower temperature. When the Celsius scale of temperature was devised, the freezing point of water was assigned a temperature of 0°C and the boiling point (at 1 atm pressure) a temperature of 100°C. On the Fahrenheit scale, the same two temperatures are 32°F and 212°F.

Assigning two fixed points in this way does not tell how to assign other temperatures; for example, simple observation of a sample of ether at 1 atm pressure quickly establishes that it is warmer than 0°C but cooler than 100°C when it boils. But what temperature should be assigned to this boiling point? Further arbitrary choices are certainly not the answer. The problem is that temperature is not a mechanical quantity like pressure; thus, it is more difficult to define.

Fortunately, some mechanical properties *depend* on temperature; for example, liquid mercury expands by an easily measured amount as it is heated from 0°C to 100°C. This change in volume could be used as the basis for the definition of a temperature scale. Temperature could be *assumed* to be linearly related to the volume of mercury, so that temperatures could be read simply from the volume of mercury in a tube (a mercury thermometer). The problem with this definition, of course, is that it is tied to the properties of a single substance, mercury. If we measure the volume of some other substance, such as butyl alcohol, on the mercury scale, it is *not* linearly related to the temperature. Why should mercury be chosen rather than another substance? How do we define temperature below the freezing point or above the boiling point of mercury? We shall soon see how these questions can be resolved, but for the time being, let us simply adopt a *provisional* temperature scale based on a mercury thermometer. We can then study how the properties of gases change with temperature.

Charles's Law: The Effect of Temperature on Gas Volume

Boyle observed that the product of the pressure and volume of a confined gas changes upon heating and thus depends on temperature (Fig. 5–6), but the first quantitative experiments were performed by the French scientist Jacques Charles more than a century later. Charles confined gases in such a way that they could be heated or cooled at constant pressure (Fig. 5–7). Under these conditions, the gases expanded in volume when heated and contracted when cooled. A plot of V, the volume of the gas, versus t, its temperature, gave a straight line (Fig. 5–8).

Figure 5-6 When a balloon is cooled with liquid nitrogen, its volume shrinks drastically as the air inside cools.

> At constant pressure, the volume of a sample of gas is a linear function of its temperature (**Charles's law**).

Heating a sample of N_2 from the freezing point of water to the boiling point of water causes it to expand to 1.366 times its original volume; the same 36.6% increase in volume is found for O_2, CO_2, and other gases. In fact, at sufficiently low pressures, *all* gases expand by the same relative amount if they have equal initial temperatures and are heated to the same final temperature. To investigate this striking result further, we write the following mathematical statement of Charles's law

$$V = V_0 + \alpha V_0 t \quad \text{at a fixed pressure and for a fixed amount of gas}$$

In this equation, V is the observed volume of a gas, V_0 is its volume at 0°C (the "ice point"), t is the temperature measured in degrees Celsius, and α (the Greek letter "alpha") represents a new quantity, the "coefficient of thermal expansion." All straight-line relationships have the mathematical form "$y = mx + b$," where the con-

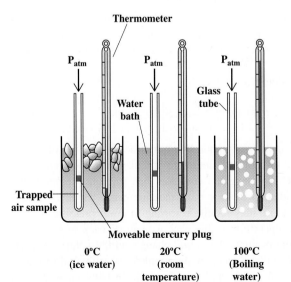

Figure 5-7 To investigate Charles's law, a glass tube is sealed at one end and mounted vertically with the open end up. A small amount of mercury is placed down the opening to trap a bubble of gas. From then on, the pressure on the gas bubble remains equal to 760 mm Hg (atmospheric pressure) plus the height of the mercury "plug" above it. The temperature of the gas bubble is varied by immersing the tube in baths of various temperatures, and its volume is read from previously made calibration marks on the tube.

Figure 5–8 The results of a Charles's law experiment for two samples of gas. The first (red line) has a volume of 1.0 L at a temperature of 0°C. It shrinks as it is cooled at constant pressure. The second (blue line) held at the same pressure takes up more volume at 0°C but shrinks faster as it is cooled. Its *percentage* change in volume is the same as that of the first sample for every degree of temperature change. Extrapolating the trends, as shown by the dashed lines, predicts that the volumes of both samples go to zero at a temperature of −273.15°C.

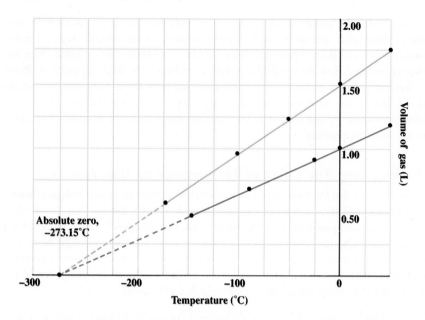

stant m is the slope of the line and the constant b is its y-intercept (see Appendix C). Our mathematical statement of Charles's law fits this pattern, having V_0 as its intercept and αV_0 as its slope. The observation that all gases expand by the same relative amount when heated between a set pair of temperatures at sufficiently low pressures now translates to saying that the constant α has the same value for all gases. In 1802, Gay-Lussac reported a value of $1/267$ $(°C)^{-1}$ for α. Subsequent experiments have refined this result to give $\alpha = 1/273.15$ $(°C)^{-1}$. Among liquids and solids, by contrast, the coefficient of thermal expansion varies widely from substance to substance.

The simple and universal behavior shown by gases at low pressures suggests an improved way of measuring temperatures. Instead of relying on the thermal expansion of the single substance mercury, we can turn Charles's law around and use the thermal expansion of an entire class of substances, the gases, to define temperature (in degrees Celsius). We solve Charles's law for t

$$t = \frac{V - V_0}{\alpha V_0} = \frac{\dfrac{V}{V_0} - 1}{\alpha}$$

and substitute the observed value of α to give

$$t = 273.15°C \left[\frac{V}{V_0} - 1 \right]$$

• Gas thermometers are more than mere theoretical toys. They can give excellent practical results.

In a gas thermometer, a measurement is made by noting the volume (V) of a low-pressure sample of gas at the unknown temperature and also at the ice point (V_0) under the same low pressure. Applying the equation then gives the temperature. Atmospheric pressure is sufficiently low for most measurements, but highly accurate determinations of temperature may require resorting to lower pressures.

Once we have defined temperature in this way, we can go back to our provisional scale of temperature and evaluate the *actual* changes in the volume of mercury with temperature. Suppose that a mercury thermometer is calibrated to match the gas thermometer at 0°C and 100°C and that the scale in between is divided evenly into 100 parts to mark off degrees. A temperature of exactly 40°C on the gas thermometer would read as 40.11°C on the mercury thermometer, and discrepancies would be found at other temperatures as well: the relation between the volume of mercury (and other liquids) and the temperature is *not* exactly linear.

Absolute Temperature

Charles's law suggests a very interesting lower limit to the temperatures that can be obtained physically. We cast the law in yet another form by solving the previous equation for V:

$$V = V_0 \left(1 + \frac{t}{273.15°C}\right)$$

Negative temperatures on the Celsius scale correspond to temperatures below the freezing point of water and, of course, are physically meaningful. But what happens as t in this equation approaches $-273.15°C$? The term in parentheses on the right-hand side of the equation approaches zero, and the volume (V) approaches zero as well (Fig. 5–8). If t could go below $-273.15°C$, the volume would become negative, clearly an impossible result. We therefore surmise that $t = -273.15°C$ is a fundamental limit, a floor below which temperatures cannot be lowered. All real gases condense to a liquid or solid before they reach this **absolute zero** of temperature, but more rigorous arguments show that *no* substance (gas, liquid, *or* solid) can be cooled below $-273.15°C$. In fact, it becomes increasingly difficult to cool a substance as absolute zero is approached. The coldest temperature reached to date is several microdegrees (1 microdegree = 10^{-6} °C) above absolute zero.

• A 130-g sample of copper was recently chilled to a record low temperature and maintained there for several days using a special magnetic cooling process.

The absolute zero of temperature is a compellingly logical choice as the zero point of a temperature scale. The easiest way to create such a new temperature scale is to add 273.15°C to the Celsius temperature. This gives the **Kelvin** temperature scale:

$$T \text{ (Kelvin)} = 273.15 + t \text{(Celsius)}$$

The capital T signifies that this is an absolute scale. The unit on this scale is named the *kelvin* (K). Clearly, a kelvin is equal in size to a Celsius degree; thus, a temperature of 25.00°C corresponds to $(273.15°C + 25.00°C) \times (1 \text{ K}/1°C) = 298.15$ K (Fig. 5–9).

• The unit is not the "degree Kelvin," but simply the kelvin. It is named after William Thomson, Lord Kelvin, who first pointed out the existence of absolute zero.

On an absolute temperature scale, Charles's law can be expressed as

$$V \propto T \qquad \text{or} \qquad V = aT$$

where the proportionality constant, a, is determined by the pressure and the amount of gas present. Ratios of the volumes occupied at two different temperatures by a fixed amount of gas at constant pressure have the simple and useful form

$$\frac{V_1}{V_2} = \frac{T_1}{T_2} \qquad \text{at a fixed pressure and for a fixed amount of gas}$$

EXAMPLE 5–4

A scientist sets about to study the properties of gaseous hydrogen at low temperatures. She starts with a volume of 2.50 L of hydrogen at atmospheric pressure and a temperature of 25.00°C and cools the gas at constant pressure to $-200.00°C$. Predict the volume of the hydrogen at $-200.00°C$.

Solution

The first step is always to convert temperatures into kelvins:

$$t_1 = 25.00°C \Rightarrow T_1 = 273.15 + 25.00 = 298.15 \text{ K}$$
$$t_2 = -200.00°C \Rightarrow T_2 = 273.15 - 200.00 = 73.15 \text{ K}$$

We can now set up a ratio in Charles's law:

$$\frac{V_1}{T_1} = \frac{V_2}{T_2}$$

$$\frac{2.50 \text{ L}}{298.15 \text{ K}} = \frac{V_2}{73.15 \text{ K}}$$

$$V_2 = \frac{(73.15 \text{ K})(2.50 \text{ L})}{298.15 \text{ K}} = \boxed{0.613 \text{ L}}$$

Exercise

The gas in a gas thermometer that has been placed in an oven has a volume that is 2.56 times larger than the volume that it occupies at 100°C. Determine the temperature in the oven (in degrees Celsius).

Answer: 682°C.

Although we now have a practical way of defining and measuring temperature, we have come no closer to understanding its *meaning*. What happens on a microscopic scale when a substance is heated? The answer to this question lies in the **kinetic theory of matter.** According to this theory, the molecules in all forms of matter (gases, liquids, and solids) are in a constant state of motion; the higher the temperature, the faster the motion. This kinetic theory makes many predictions concerning the behavior of matter in general and of gases in particular. We consider some of them in Section 5–7.

5–4   THE IDEAL GAS LAW

We have presented two empirical properties of gases at low pressures. The first, Boyle's law, tells how the volume of a gas depends on the pressure if the temperature and amount of the gas do not change:

$$V \propto \frac{1}{P} \quad \text{at constant temperature, fixed amount of gas}$$

The second, Charles's law, tells how the volume of a gas depends on the temperature if the pressure and amount of the gas do not change,

$$V \propto T \quad \text{at constant pressure, fixed amount of gas}$$

where T must be the absolute temperature. A third equation should relate the volume of a gas to its amount, at constant temperature and pressure. Under these conditions, the volume of a sample of *any* substance (whether solid, liquid, or gas) de-

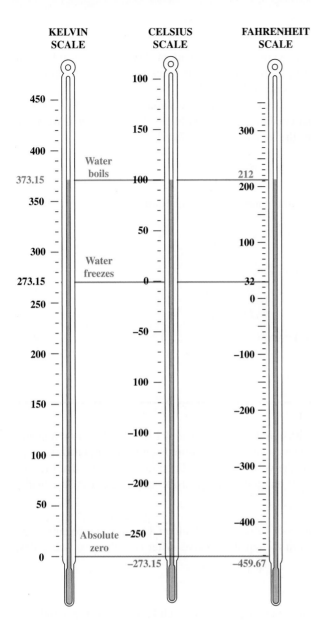

Figure 5–9 The Kelvin, the Celsius, and the Fahrenheit scales all find common use. The size of the degree is the same on the Kelvin and Celsius scales, but the zero points are offset by 273.15°. Fahrenheit degrees are smaller by a factor of 5/9 than kelvins or Celsius degrees, so a 9°F change is counted for every 5°C or 5 K change in temperature. Also, the Fahrenheit zero is offset from the Celsius zero by 32°F.

pends only on the amount of it present. If this were not so, then substances would not have the constant densities that are observed. Therefore, we can certainly conclude that for a gas

$$V \propto n \qquad \text{at constant temperature and pressure}$$

where n is the chemical amount of the gas.

These three statements may be combined in the form

$$V \propto \frac{nT}{P}$$

We now introduce a constant of proportionality, which we call R:

$$V = R\frac{nT}{P} \qquad \text{or} \qquad PV = nRT$$

The highlighted equations are two forms of the **ideal gas law,** an empirical law that holds approximately for all gases near atmospheric pressure and that becomes increasingly accurate as the pressure is decreased.

Suppose a gas undergoes a change from some initial condition (described by P_1, V_1, T_1, and n_1) to a final condition (described by P_2, V_2, T_2, and n_2). We solve the ideal gas equation under the initial conditions for R

$$R = \frac{P_1 V_1}{n_1 T_1}$$

and do the same for the equation under the final conditions

$$R = \frac{P_2 V_2}{n_2 T_2}$$

Because R is a constant, we conclude

$$\frac{P_1 V_1}{n_1 T_1} = \frac{P_2 V_2}{n_2 T_2}$$

a useful alternative form of the ideal gas law.

EXAMPLE 5–5

A weather balloon filled with helium has a volume of 1.0×10^4 L at 1.00 atm and 30°C. It ascends to an altitude at which the pressure is only 0.60 atm and the temperature is −20°C. What is the volume of the balloon then? Assume that the balloon stretches in such a way that the pressure inside stays close to that outside.

Solution

Because the number of moles of helium in the balloon does not change during the ascent, we can set $n_1 = n_2$ in the equation above. After canceling on both sides,

$$\frac{P_1 V_1}{T_1} = \frac{P_2 V_2}{T_2}$$

The initial and final pressures and temperatures are all known, as is the initial volume V_1. The only unknown quantity is the final volume V_2. It is then convenient to rearrange the equation so that only V_2 appears on the left-hand side:

$$V_2 = V_1 \left(\frac{P_1}{P_2}\right)\left(\frac{T_2}{T_1}\right)$$

Before the ideal gas law is used, temperatures must be converted into kelvins. Here, $T_1 = 273.15 + 30 = 303$ K and $T_2 = 273.15 + (-20) = 253$ K. Inserting these absolute temperatures, the pressures, and the initial volume into the equation for V_2 gives

$$V_2 = 1.0 \times 10^4 \text{ L} \left(\frac{1.00 \text{ atm}}{0.60 \text{ atm}}\right)\left(\frac{253 \text{ K}}{303 \text{ K}}\right) = \boxed{1.4 \times 10^4 \text{ L}}$$

This result implies an increase by about 12% in the diameter of the balloon. The balloon expands because the pressure surrounding it drops with increasing altitude. This is partially offset by a contraction attributable to the cooling.

Exercise

At one point during its ascent, the same weather balloon has the same volume as when it was launched, although its temperature is $-10°C$. Compare the pressure at that point.

Answer: 0.868 atm.

The approach outlined in Example 5–5 can be extended to cases in which other combinations of the four variables (P, V, T, and n) remain constant (Fig. 5–10). For example, when n and T remain constant, they can be canceled from the equation, and Boyle's law, which was used in solving Example 5–3, results. When n and P remain constant, this relationship reduces to Charles's law, used in Example 5–4.

We are not quite finished. We have shown that for a given gas, PV/nT is a constant, called R, but we have not yet established whether R is a *universal* constant or whether it changes from one gas to another (for example, from oxygen to nitrogen to hydrogen). To deal with this issue, we invoke Avogadro's hypothesis, now perhaps better called Avogadro's law because its validity was established by the success of the laws of chemical combination considered in Chapter 1. According to this law, fixed volumes of different gases at the same temperature and pressure contain the same number of molecules (and therefore of moles). In other words, for fixed V, P, and T, the value of n is fixed as well and is independent of the particular gas studied. Therefore, R is a **universal gas constant** that is the same for all gaseous substances.

The numerical value of R depends on the units chosen for P, V, n, and T. We use the Kelvin scale exclusively for T, and the unit for n is the mole. At the ice point ($T = 273.15$ K), the product PV for 1 mol of gas approaches the value 22.4141 L

Figure 5–10 In a hot-air balloon, the volume remains nearly constant (the balloon is rigid) and the pressure is nearly constant as well (unless the balloon rises very high). Thus, n is inversely proportional to T: as the air inside the balloon is heated, its amount decreases and the density of the air in the balloon falls. This lower density gives the balloon its lift.

atm at low pressures; hence, R has the value

$$R = \frac{PV}{nT} = \frac{(22.4141 \text{ L atm})}{(1.0000 \text{ mol})(273.15 \text{ K})} = 0.082058 \text{ L atm mol}^{-1} \text{ K}^{-1}$$

If P is measured in SI units of pascals (newton m^{-2}) and V in m^3, then R has the value

$$R = \frac{(1.01325 \times 10^5 \text{ newton m}^{-2})(22.4141 \times 10^{-3} \text{ m}^3)}{(1.0000 \text{ mol})(273.15 \text{ K})}$$
$$= 8.3145 \text{ newton m mol}^{-1} \text{ K}^{-1}$$

One newton-meter is the work done by a force of one newton acting through a distance of one meter and is defined to be one *joule* (the SI unit of energy, or work, abbreviated J) (see Appendix B). The gas constant is then

$$R = 8.3145 \text{ J mol}^{-1} \text{ K}^{-1}$$

EXAMPLE 5–6

What mass of helium is needed to fill the weather balloon from Example 5–5?

Solution

First, we solve the ideal gas law for n:

$$PV = nRT$$
$$n = \frac{PV}{RT}$$

Because P is given in atmospheres and V in liters, we choose to use $R = 0.08206$ L atm mol^{-1} K^{-1} to cause the units to cancel.

$$n_{\text{He}} = \frac{(1.00 \text{ atm})(1.00 \times 10^4 \text{ L})}{(0.08206 \text{ L atm mol}^{-1} \text{ K}^{-1})(303.15 \text{ K})} = 402 \text{ mol He}$$

The molar mass of helium is 4.00 g mol^{-1}; hence, the mass of helium required is

$$m_{\text{He}} = 402 \text{ mol He} \left(\frac{4.00 \text{ g He}}{1 \text{ mol He}} \right) = 1610 \text{ g} = 1.61 \text{ kg He}$$

Exercise

The weather balloon in Example 5–5 is filled with H_2 instead of He. What mass of H_2 is needed to fill it to the same volume under the same conditions?

Answer: 0.810 kg.

Gas Density and Molar Mass

Example 5–6 showed how the mass m of a sample of a gas can be computed using the ideal gas law and given the temperature, pressure, volume, and the molar mass $\mathcal{M}$ of the gas. Using the fact that $n = m/\mathcal{M}$, the ideal gas law can be rewritten as

$$PV = \frac{m}{\mathcal{M}} RT$$

• Memorizing the many useful forms of the ideal gas law is unwise. It is far better to understand how to derive what is needed for a given problem from the basic equation $PV = nRT$.

Recall that the density d of a substance is its mass divided by its volume. Rearranging this equation gives

$$d = \frac{m}{V} = \frac{P}{RT} \mathcal{M}$$

The following example illustrates the calculation of gas densities.

EXAMPLE 5–7

Calculate the density of gaseous carbon tetrafluoride, CF_4, at room pressure (1.00 atm) and 50°C.

Solution

Combining the molar masses of C and F from a table of atomic masses gives a molar mass for CF_4 of 88.00 g mol^{-1}, and the absolute temperature is $T = 273.15 + 50 = 323.15$ K. Because the gas pressure is given in atmospheres, the appropriate value of R to use has units of L atm mol^{-1} K^{-1}. All these values are inserted into the equation just given for the density:

$$d = \frac{P}{RT} \mathcal{M} = \left(\frac{1.00 \text{ atm}}{0.08206 \text{ L atm mol}^{-1} \text{ K}^{-1} \times 323.15 \text{ K}} \right) \left(\frac{88.00 \text{ g } CF_4}{1 \text{ mol } CF_4} \right)$$

$$= 3.32 \text{ g L}^{-1}$$

Exercise

Calculate the density of gaseous hydrogen at a pressure of 1.32 atm and a temperature of −45°C.

Answer: 0.142 g L^{-1}.

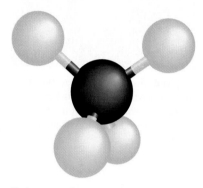

Carbon tetrafluoride, CF_4, is an exceptionally stable gas that condenses to a liquid at −128.5°C.

The above procedure can be reversed to calculate the molar mass of a gas from an experimental measurement of its density. Recall from Section 2–1 that the molecular formula of a substance differs from its empirical formula; it contains more information. The molar mass is exactly the extra fact needed to get a molecular formula from an empirical formula. This is illustrated in the following example.

EXAMPLE 5–8

The welding fuel in Example 2–5 was found to have empirical formula CH. A volume of 10.0 L of this gas at a pressure of 0.986 atm and a temperature of 65.0°C is found to weigh 9.25 g. Determine the approximate molar mass of the gas, assuming that it behaves as an ideal gas, and assign its molecular formula.

Solution

The chemical amount of gas in the sample can be calculated by inserting the pressure, the volume, and the absolute temperature (in this case, $273.15 + 65.0 = 338.15$ K) into the ideal gas law:

$$n = \frac{PV}{RT} = \frac{0.986 \text{ atm} \times 10.0 \text{ L}}{0.08206 \text{ L atm mol}^{-1} \text{ K}^{-1} \times 338.15 \text{ K}} = 0.3553 \text{ mol}$$

Figure 5–11 An oxy-acetylene torch is used to weld metals because of the high temperature of the flame produced by burning acetylene in oxygen.

Dividing the mass m of a substance by its chemical amount n gives its molar mass. In this case, it is

$$\mathcal{M} = \frac{m}{n} = \frac{9.25 \text{ g}}{0.3553 \text{ mol}} = \boxed{26.0 \text{ g mol}^{-1}}$$

The formula mass of CH is $12.01 + 1.01 = 13.02$ g mol^{-1}. Because the approximate molar mass of the compound (26.0 g mol^{-1}) is twice this, we conclude that there must be two CH units in each molecule of compound. By this reasoning, the molecular formula is $\boxed{C_2H_2}$. The compound is acetylene (Fig. 5–11).

Exercise

Fluorocarbons are compounds of fluorine and carbon. A 45.60-g sample of a gaseous fluorocarbon contains 7.94 g of carbon and 37.66 g of fluorine and occupies 7.40 L at STP ($P = 1.00$ atm and $T = 273.15$ K). Determine the approximate molar mass of the fluorocarbon and give its molecular formula.

Answer: The approximate molar mass is 138 g mol^{-1}, and the molecular formula is C_2F_6.

5–5 CHEMICAL CALCULATIONS FOR GASES

One of the most important applications of the gas laws in chemistry is in calculating the volumes of gases consumed or produced in chemical reactions. If the conditions of pressure and temperature are known, then the ideal gas law can be used to convert between chemical amount and gas volume. Instead of working with the mass of a gas taking part in a reaction, we can use its volume, which is usually much easier to measure. This is illustrated in the following example.

EXAMPLE 5–9

A key step in the production of sulfuric acid is the oxidation of sulfur dioxide:

$$2 \text{ SO}_2(g) + \text{O}_2(g) \longrightarrow 2 \text{ SO}_3(g)$$

Suppose that 6.37 L of O_2 (measured at 420°C and a pressure of 1.22 atm) reacts completely with an excess of SO_2. The SO_3 produced is cooled to 80.0°C at $P = 0.970$ atm. Calculate the volume of the SO_3.

Solution

The chemical amount of oxygen that reacts is found from the ideal gas law (recall that the temperature must first be expressed in kelvins by adding 273.15 to 420°C):

$$n_{O_2} = \frac{PV}{RT} = \frac{1.22 \text{ atm} \times 6.37 \text{ L}}{0.08206 \text{ L atm mol}^{-1} \text{ K}^{-1} \times (420 + 273.15 \text{ K})} = 0.1366 \text{ mol}$$

The balanced equation gives the chemical conversion factor between the amount of O_2 consumed and the amount of SO_3 produced:

$$0.1366 \text{ mol } O_2 \times \left(\frac{2 \text{ mol } SO_3}{1 \text{ mol } O_2} \right) = 0.2732 \text{ mol } SO_3$$

Finally, the volume of this chemical amount of SO_3 is calculated by using its temperature $(80.0 + 273.15 = 353.15 \text{ K})$ and pressure:

$$V = \frac{nRT}{P} = \frac{0.2732 \text{ mol} \times 0.08206 \text{ L atm mol}^{-1} \text{ K}^{-1} \times 353.15 \text{ K}}{0.970 \text{ atm}} = 8.16 \text{ L}$$

Exercise

Ethylene, C_2H_4, burns completely in air to produce carbon dioxide and water vapor:

$$C_2H_4(g) + 3 \, O_2(g) \longrightarrow 2 \, CO_2(g) + 2 \, H_2O(g)$$

Suppose that the ethylene reacting has a volume of 3.51 L at a temperature of 25°C and a pressure of 4.63 atm and that the water vapor is collected at a temperature of 130°C and a pressure of 0.955 atm. Calculate the volume of water vapor generated by the reaction.

Answer: 46.0 L

The convenience of measuring volumes of gases instead of masses is available even when some of the reactants and products in a reaction are solids or liquids. This is shown in the next example.

EXAMPLE 5-10

Concentrated nitric acid acts on copper to give nitrogen dioxide and dissolved copper ions (Fig. 5–12) according to the balanced net ionic equation

$$Cu(s) + 4 \, H^+(aq) + 2 \, NO_3^-(aq) \longrightarrow 2 \, NO_2(g) + Cu^{2+}(aq) + 2 \, H_2O(\ell)$$

Suppose that 6.80 g of copper is consumed in this reaction and that the NO_2 is collected at a pressure of 0.970 atm and a temperature of 45°C. Calculate the volume of NO_2 produced.

Solution

The first step (as in Fig. 2–6) is to convert from the known mass of a reactant or product (in this case 6.80 g of Cu) to its chemical amount (in moles) by using the appropriate molar mass (in this case 63.55 g mol^{-1}):

$$n_{Cu} = 6.80 \text{ g Cu} \times \left(\frac{1 \text{ mol Cu}}{63.55 \text{ g Cu}} \right) = 0.107 \text{ mol Cu}$$

Next we find the chemical amount of NO_2 generated in the reaction by forming the unit factor "2 mol of NO_2 per 1 mol of Cu" from the balanced equation and multiplying the chemical amount of Cu by it:

$$n_{NO_2} = 0.107 \text{ mol Cu} \times \left(\frac{2 \text{ mol } NO_2}{1 \text{ mol Cu}} \right) = 0.214 \text{ mol } NO_2$$

Figure 5-12   When concentrated nitric acid is added to a copper penny, the copper is oxidized and an aqueous solution of copper(II) nitrate is formed. In addition, some of the nitrate ion is reduced to brown gaseous nitrogen dioxide, which bubbles off.

Finally, we use the ideal gas law to find the volume of NO_2 from its chemical amount:

$$V_{NO_2} = \frac{nRT}{P} = \frac{(0.214 \text{ mol})(0.08206 \text{ L atm mol}^{-1} \text{ K}^{-1})(273.15 + 45 \text{ K})}{0.970 \text{ atm}}$$

$$= 5.76 \text{ L}$$

Therefore, 5.76 L of $NO_2(g)$ is produced under these conditions.

Exercise

The compound hydrazine (N_2H_4) is used as a rocket fuel. It is prepared by the reaction of ammonia with a solution of sodium hypochlorite:

$$2 \text{ NH}_3(g) + \text{NaOCl}(aq) \longrightarrow \text{N}_2\text{H}_4(aq) + \text{NaCl}(aq) + \text{H}_2\text{O}(\ell)$$

What volume of gaseous ammonia at a temperature of 10°C and a pressure of 3.63 atm is required to produce 15.0 kg of hydrazine?

Answer: 5.99×10^3 L.

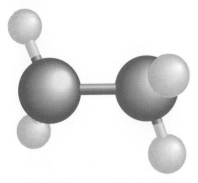

Hydrazine (N_2H_4) is a colorless liquid that boils at 113.5°C and freezes at 2°C. In addition to its use as a rocket fuel, it is a versatile reducing agent. It is used extensively to remove oxygen from boiler water, giving water and nitrogen as products: $N_2H_4(aq) + O_2(aq)$ $\longrightarrow 2 \text{ H}_2\text{O}(\ell) + \text{N}_2(g)$.

5-6   MIXTURES OF GASES

Suppose several gases are mixed in a single container and do not react with each other. We define the **partial pressure** of each gas as the pressure it would exert if it were the *only* gas present in the container. **Dalton's law** then states that the total pressure measured is the sum of the partial pressures of the individual gases. This law, discovered by John Dalton, holds under the same conditions as the ideal gas law itself: it is approximate at moderate pressures and becomes increasingly accurate as the pressure is lowered.

- This is the same Dalton who proposed the atomic theory.

At low pressures, each gas independently obeys the ideal gas law, so that for a mixture of two gases A and B in a container of volume V,

$$P_A V = n_A RT$$
$$P_B V = n_B RT$$

The total pressure is

$$P_{\text{tot}} = P_A + P_B = \frac{n_A RT}{V} + \frac{n_B RT}{V} = (n_A + n_B)\frac{RT}{V}$$

Because the total number of moles, n_{tot}, is $n_A + n_B$, we can rewrite this equation as

$$P_{\text{tot}} = n_{\text{tot}}\frac{RT}{V}$$

- The best example of a mixture of gases is the air.

so that the ideal gas law may be applied to the total pressure of a gas mixture, provided that the total chemical amount, the sum of the chemical amounts of the components, is used.

Mole Fraction

We have seen that for a gas mixture in which A is one component,

$$P_A = n_A \frac{RT}{V} \quad \text{and} \quad P_{tot} = n_{tot} \frac{RT}{V}$$

Dividing the first equation by the second gives

$$\frac{P_A}{P_{tot}} = \frac{n_A}{n_{tot}} \quad \text{or} \quad P_A = \frac{n_A}{n_{tot}} P_{tot}$$

We define

$$X_A = \frac{n_A}{n_{tot}}$$

as the **mole fraction** of A in the mixture: the number of moles of A divided by the total number of moles present. Then

$$P_A = X_A P_{tot}$$

The partial pressure of any component in a mixture of gases is the total pressure multiplied by the mole fraction of that component.

EXAMPLE 5–11

The manufacture of sulfuric acid starts with the burning of sulfur in oxygen to produce sulfur dioxide. Suppose that the resulting gas mixture contains 23.2 g of oxygen and 53.1 g of sulfur dioxide and has a total pressure of 2.13 atm. Calculate the mole fraction of the sulfur dioxide in this mixture, and its partial pressure.

Solution

The first step is to calculate the chemical amounts of oxygen and sulfur dioxide in the mixture, from their molar masses.

$$n_{O_2} = 23.2 \ g \ O_2 \times \left(\frac{1 \ mol \ O_2}{32.00 \ g \ O_2} \right) = 0.725 \ mol \ O_2$$

$$n_{SO_2} = 53.1 \ g \ SO_2 \times \left(\frac{1 \ mol \ SO_2}{64.06 \ g \ SO_2} \right) = 0.829 \ mol \ SO_2$$

The total number of moles of the two gases is

$$n_{tot} = n_{O_2} + n_{SO_2} = 0.725 + 0.829 = 1.554 \ mol$$

The mole fraction of SO_2 is

$$X_{SO_2} = \frac{n_{SO_2}}{n_{tot}} = \frac{0.829 \ mol}{1.554 \ mol} = 0.533$$

and its partial pressure is

$$P_{SO_2} = X_{SO_2} P_{tot} = 0.533 \times 2.13 \ atm = \boxed{1.14 \ atm}$$

Exercise

A solid hydrocarbon is burned in air in a closed container, and the resulting gas mixture has a total pressure of 3.34 atm. Analysis of the mixture shows it to contain 0.340 g of water vapor, 0.792 g of carbon dioxide, 0.288 g of oxygen, and 3.790 g of nitrogen. Calculate the mole fraction and partial pressure of carbon dioxide in this mixture.

Answer: $X_{CO_2} = 0.0993$; $P_{CO_2} = 0.332$ atm.

5–7 THE KINETIC THEORY OF GASES

The ideal gas law summarizes certain physical properties of gases at low pressures. It is an empirical law, the consequence of experimental observations, but its simplicity and generality make us ask whether it has some underlying microscopic explanation that involves the properties of the atoms and molecules making up a gas. This explanation must allow other properties of gases at low pressures to be predicted and must point the way toward understanding why real gases *deviate* from the ideal gas law to a small but measurable extent. Such a theory was developed in the 19th century, notably by the physicists Rudolf Clausius, James Clerk Maxwell, and Ludwig Boltzmann, who worked in Germany, England, and Austria, respectively. The **kinetic theory of gases** is one of the great milestones of science, and its success provides strong evidence for the atomic theory of matter discussed in Chapter 1.

The underlying assumptions of the kinetic theory of gases are very simple:

---

1. A pure gas consists of a large number of identical molecules separated by distances that are large compared with their size.
2. The molecules of a gas are constantly moving in random directions (Fig. 5–13) with a distribution of speeds.
3. The molecules of a gas exert no forces on one another except during collisions, so that between collisions they move in straight lines with constant velocities.
4. The collisions of the molecules with each other and with the walls of the container are elastic; no energy is lost during a collision.

Temperature and Molecular Motion

We now use the kinetic theory of gases to establish a relationship between the pressure and volume of an ideal gas and the mass and speed of its molecules as they fly about in their postulated random motion. Comparing this relationship with the ideal gas law ($PV = nRT$) then provides insight into the meaning of temperature. Because the goal is not mathematical rigor but physical understanding, we use arguments based on proportionality, without worrying about the specific numerical constants that appear in the equations.

We begin by recalling that the pressure of a gas is the total force exerted by its molecules on its containing walls divided by the area of the walls. In the kinetic theory, this force results from a ceaseless series of abrupt *impulses* as molecules bombard the container walls. The push (force) that a molecule exerts on a wall during its collision starts from zero, rises to a peak, and then falls back to zero a short

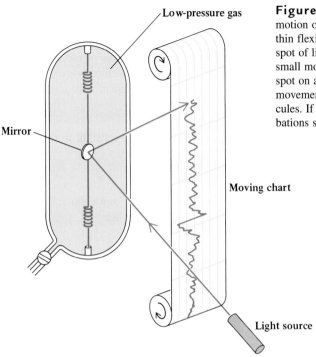

Low-pressure gas

Mirror

Moving chart

Light source

Figure 5-13 A visible demonstration of the existence and random motion of molecules in a gas. A small mirror is suspended from a very thin flexible fiber inside a glass container of a gas at low pressure. A spot of light shines on the mirror and is reflected at an angle. Even very small motions of the mirror are detectable as movements of the reflected spot on a distant screen. The graph shows how the regular twisting movement of the mirror is perturbed by the random impact of gas molecules. If the gas is completely removed from the container, these perturbations stop.

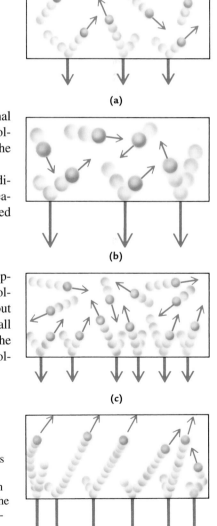

(a)

(b)

(c)

(d)

time later as the molecule bounces away. The impulse per collision is proportional to the *momentum* ($m \times u$) carried by the molecule, where m is the mass of the molecule and u is its speed. The heavier the molecule and the faster it is moving, the greater its impulse when it strikes the wall.

The collision rate (the number of collisions per unit time) under most conditions is so large that the effect becomes that of a smooth push on the walls. We reason, then, that the pressure is proportional to the impulse per collision multiplied by the rate of collisions of molecules with the walls:

pressure ∝ (impulse per collision) × (rate of collisions with the walls)

We must now relate the rate of collision of molecules with the walls to properties of the molecules. First, this rate must be proportional to the number of molecules per unit volume (N/V) because increasing the number of molecules without changing the volume of the container would certainly increase the number of wall collisions in direct proportion. In addition, the rate of collision is proportional to the molecule speed (u) because the faster the molecules move, the more often they collide with the wall. We can put all this together (Fig. 5-14) and conclude that

$$P \propto (m \times u) \times \left[\left(\frac{N}{V}\right) \times u\right] = \frac{Nmu^2}{V} \quad \text{or} \quad PV \propto Nmu^2$$

Figure 5-14 (a) Collisions of the molecules of a gas with its walls lead to impulses (shown by arrows where each molecule hits the bottom wall). The effect of the impulses over a time interval is the exertion of pressure by the gas. (b) If each molecule doubles in mass, the impulses are twice as large: pressure is proportional to molecular mass. (c) If the gas density is doubled, then twice as many collisions occur with the wall: pressure is proportional to gas density. (d) If the average molecular speed is doubled, there are twice as many collisions over the time interval *and* each impulse is twice as large: pressure is proportional to the square of the molecular speed.

We must clarify one point before proceeding. We have referred to the speed of the molecules as u, but actually there is no reason that all molecules should have the *same* speed. We should interpret our equation as applying not to a single molecule, but to the whole collection of molecules in an *average* sense. We therefore replace u^2 by the average of the squares of the speeds, the **mean-square speed** of all the molecules, denoted with an overbar: $\overline{u^2}$. The last equation becomes

$$PV \propto Nm\overline{u^2}$$

A more detailed argument shows that the proportionality constant in this relationship is $\frac{1}{3}$, so that

$$PV = \tfrac{1}{3}Nm\overline{u^2}$$

This equation is the desired relationship between pressure and volume of a gas and the motion of its molecules. Comparison to the ideal gas law

$$PV = nRT$$

allows us to write

$$\tfrac{1}{3}Nm\overline{u^2} = nRT$$

which has considerable interest: it is a relationship between the speeds of the molecules of a gas and the temperature of the gas. We can tidy things a little, however. The equation has the number of molecules (N) on the left and the number of moles (n) on the right. Because N is just n multiplied by Avogadro's number (N_0), we can divide both sides by n and write

$$\tfrac{1}{3}N_0m\overline{u^2} = RT$$

Let us look at this result in two different ways. First, note that the **kinetic energy** of a molecule of mass m moving at velocity u is given by the laws of physics as $\frac{1}{2}mu^2$, so the *average* kinetic energy of N_0 molecules (one mole) is $\frac{1}{2}N_0m\overline{u^2}$. This is exactly the left side of our equation, with the factor $\frac{1}{2}$ instead of $\frac{1}{3}$:

$$\text{average kinetic energy per mole} = \tfrac{1}{2}N_0m\overline{u^2} = \tfrac{3}{2} \times (\tfrac{1}{3}N_0m\overline{u^2}) = \tfrac{3}{2}RT$$

This means that the average kinetic energy of the molecules of a gas depends only on the temperature and is independent of the mass of the molecules or the gas density.

A second way to look at the equation is to recall that if m is the mass of a single molecule, then N_0m is the mass of one mole of molecules: the molar mass ($\mathcal{M}$). We can then solve the equation for the mean-square speed:

$$\overline{u^2} = \frac{3RT}{\mathcal{M}}$$

The higher the temperature and the less massive the molecules, the greater is the mean-square speed.

Distribution of Molecular Speeds

We define the **root-mean-square speed** (u_{rms}) as the square root of the mean-square speed $3RT/\mathcal{M}$:

$$u_{\text{rms}} = \sqrt{\overline{u^2}} = \sqrt{\frac{3RT}{\mathcal{M}}}$$

This equation can be used successfully only if all the terms in it are expressed in a coherent set of units such as the SI units. The value for R in SI units is

$$R = 8.3145 \text{ J mol}^{-1} \text{ K}^{-1}$$

Recall that in terms of the base units of the SI (see Appendix B),

$$1 \text{ J} = 1 \text{ kg (m/s)}^2 = 1 \text{ kg m}^2 \text{ s}^{-2}$$

where s is the abbreviation for second. R then has units of $\text{kg m}^2 \text{ s}^{-2} \text{ mol}^{-1} \text{ K}^{-1}$ when expressed in SI base units. We usually think of molar masses in *grams* per mole; here we must convert to *kilograms* per mole before using the equation:

$$\frac{\text{kg}}{\text{mol}} = \frac{\cancel{\text{g}}}{\text{mol}} \times \left(\frac{1 \text{ kg}}{1000 \cancel{\text{g}}} \right)$$

By putting all of the units into the equation for u_{rms} we can confirm that the final result comes out in the expected SI unit for a speed, which is meters per second. This is illustrated in the following example.

EXAMPLE 5–12

Calculate u_{rms} for (a) a helium atom and (b) an oxygen molecule at 298 K.

Solution

Because the factor $3RT$ appears in both expressions for u_{rms}, let us calculate it first.

$$3RT = (3)\left(8.3145 \, \frac{\text{kg m}^2}{\text{s}^2 \text{ mol } \cancel{\text{K}}} \right)(298 \, \cancel{\text{K}}) = 7.43 \times 10^3 \, \frac{\text{kg m}^2}{\text{s}^2 \text{ mol}}$$

The molar masses of helium and oxygen are 4.00 g mol^{-1} and 32.00 g mol^{-1}, respectively. These molar masses are converted to 4.00×10^{-3} kg mol^{-1} and 32.00×10^{-3} kg mol^{-1} and inserted together with the value for $3RT$ into the formula for u_{rms}.

$$u_{\text{rms}}(\text{He}) = \sqrt{ \frac{7.43 \times 10^3 \, \frac{\cancel{\text{kg}} \, \text{m}^2}{\text{s}^2 \, \cancel{\text{mol}}}}{4.00 \times 10^{-3} \, \frac{\cancel{\text{kg}}}{\cancel{\text{mol}}}} } = 1360 \text{ m s}^{-1}$$

$$u_{\text{rms}}(\text{O}_2) = \sqrt{ \frac{7.43 \times 10^3 \, \frac{\cancel{\text{kg}} \, \text{m}^2}{\text{s}^2 \, \cancel{\text{mol}}}}{32.00 \times 10^{-3} \, \frac{\cancel{\text{kg}}}{\cancel{\text{mol}}}} } = 482 \text{ m s}^{-1}$$

At the same temperature, the helium and oxygen molecules have the same average kinetic energy. The calculation shows that lighter molecules move faster, which compensates for their lesser mass. The average molecule in a gas moves along quite rapidly at room temperature (298 K): these rms speeds convert to 3050 and 1080 miles per hour, respectively.

Exercise

At a certain temperature, the root-mean-square speed of the molecules of H_2 in a sample of gas is 1055 m s^{-1}. Compute the rms speed of molecules of O_2 at the same temperature.

Answer: 265.8 m s^{-1}.

Figure 5–15 The Maxwell–Boltzmann distribution of molecular speeds in nitrogen at three temperatures. The peak in each curve gives the most probable speed (u_{mp}), which is slightly smaller than the root-mean-square speed (u_{rms}). The average speed (u_{av}) (obtained simply by adding the speeds and dividing by the total number of molecules in the sample) lies in between. All three measures give comparable estimates of typical molecular speeds. All three increase with increasing temperature.

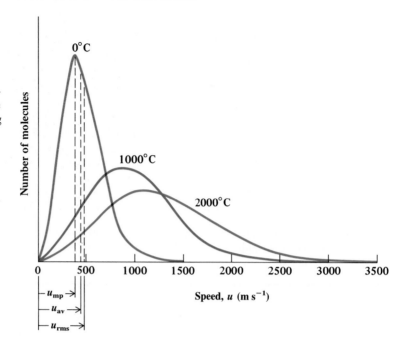

• An average income of $50,000 a year among ten families does not necessarily mean that all ten are financially comfortable; one family could make $491,000 and the other nine make only $1,000 each.

Figure 5–16 This device can be used to measure the distribution of molecular speeds experimentally. Only those molecules with the correct speed to pass through *both* rotating sectors reach the detector, where their numbers can be counted. By changing the rate of rotation of the sectors, the speed distribution can be determined.

The root-mean-square speed gives some idea of the typical speed of molecules in a gas, but averages of any type can be deceptive. We must learn the *distribution* of molecular speeds in a gas before we can draw many conclusions of chemical interest. The necessary distribution was independently deduced from kinetic theory by Maxwell (in 1860) and Boltzmann (in 1872) and confirmed in a direct experiment in 1955. It is called the **Maxwell–Boltzmann speed distribution.** Figure 5–15, which presents the distribution for several temperatures, shows that as the temperature is raised, the whole distribution of molecular speeds shifts toward higher values. As is evident from the figure, the number of very slow and very fast molecules is small, and the most probable speed is an intermediate value. The speed distribution of the molecules in a gas can be measured experimentally by a device such as that shown in Figure 5–16.

For each temperature, there is a unique Maxwell–Boltzmann distribution curve that defines the concept "temperature" in the kinetic theory of gases.

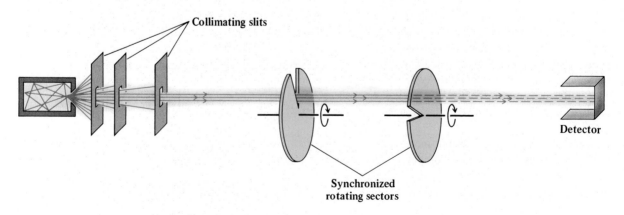

Temperature has meaning in a system of gaseous molecules only when their distribution of speeds is the Maxwell–Boltzmann distribution.

Consider a closed container filled with molecules having a distribution of speeds that is not "Maxwellian." Such a situation is possible (for example, just after an explosion), but it cannot persist for very long. Any other distribution of speeds quickly becomes a Maxwell–Boltzmann distribution through collisions of molecules, which exchange energy. Once attained, the Maxwell–Boltzmann distribution persists indefinitely. The gas molecules have come to **thermal equilibrium** with one another, and a temperature can be defined only if the condition of thermal equilibrium exists. Finally, we cannot associate a temperature with a single molecule, or even a group of 10 or 20 molecules. Temperature is a property of collections of molecules having enough members to give meaning to a distribution of speeds.

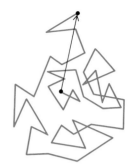

Figure 5–17 A molecule in a gas at ordinary conditions follows a straight-line path only for a very short time before undergoing a collision. Its overall path is a zig-zag one.

Mean Free Path and Diffusion

The root-mean-square speeds of molecules in gases are high at ordinary temperatures, on the order of hundreds or thousands of meters per second (see Example 5–12). Suppose an odorous gas is released in one corner of a room. If the molecules followed straight-line trajectories, the odor would reach other points in the room almost instantly. Instead, we observe a delay before the smell arrives. Why does it take so long for the molecules of odorant to spread out through the large volume?

The explanation is that molecules undergo many collisions with one another, even under conditions at which the gas follows the ideal gas law quite closely. A molecule in a typical gas at atmospheric pressure experiences about 10^{10} collisions per second; in other words, molecules move only for a time of about 10^{-10} second between collisions. We define the **mean free path** as the average distance traveled between collisions. It is the average speed multiplied by the average time between collisions, and it is on the order of 10^{-7} m (10^{-5} cm). Although this is large compared with the size of the molecules (several times 10^{-8} cm, as shown in Section 1–8), it is small compared with the size of a typical container. Molecules follow a zig-zag course (Fig. 5–17) as their frequent collisions incessantly redirect them, and they take far longer to travel a given distance from their starting point than if there were no collisions. This type of irregular motion of molecules is called **diffusion.**

• The average time between repeating events (such as collisions) is the reciprocal of their average frequency. If a bell sounds 20 times per hour, the average duration between rings is 1/20 hour (3 minutes).

Gaseous Effusion

In 1846, Thomas Graham showed experimentally that the rates of effusion of different gases through a small hole into a vacuum (Fig. 5–18) are inversely proportional to the square roots of the molar masses if the gases are all studied at the same temperature and pressure. The kinetic theory of gases can be used to explain this result, now called **Graham's law of effusion.** By Avogadro's law, at a given temperature and pressure, a volume contains the same number of molecules of the different gases; thus, the only factor affecting their rates of effusion is their different speeds. We expect the rate of effusion to be proportional to the average speed (and to the root-mean-square speed), so we can write (for two different gases A and B)

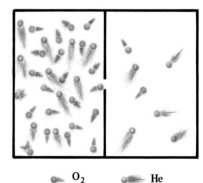

O₂   **He**

Figure 5–18 A small hole in the box permits molecules to effuse out into a vacuum. The lighter particles (here, helium atoms) effuse at a greater rate than the heavier oxygen molecules because their speeds are greater on the average.

rate of effusion of A $\propto u_{\text{rms}}(A)$ and   rate of effusion of B $\propto u_{\text{rms}}(B)$

Dividing the first proportion by the second gives

$$\frac{\text{rate of effusion of A}}{\text{rate of effusion of B}} = \frac{u_{\text{rms}}(A)}{u_{\text{rms}}(B)}$$

When we insert the formulas for u_{rms}, the factors of $3RT$ cancel out, leaving

$$\frac{\text{rate of effusion of A}}{\text{rate of effusion of B}} = \frac{\sqrt{\dfrac{3RT}{\mathcal{M}_A}}}{\sqrt{\dfrac{3RT}{\mathcal{M}_B}}} = \sqrt{\frac{\mathcal{M}_B}{\mathcal{M}_A}}$$

This predicts that the rate of effusion varies as the inverse square root of the molar mass—exactly what was observed by Graham. The agreement provides additional support for the kinetic theory of gases.

Graham's law applies to the effusion of a mixture of two gases through a small hole. The gas that emerges is enriched in the lighter component because lighter molecules effuse more rapidly than heavier ones. If B is heavier than A, the **enrichment factor** is $\sqrt{\mathcal{M}_B/\mathcal{M}_A}$.

EXAMPLE 5-13

Calculate the enrichment factor from effusion for a mixture of $^{235}UF_6$ and $^{238}UF_6$, uranium hexafluoride with two different uranium isotopes (^{235}U has a relative atomic mass of 235.04, and ^{238}U has an atomic mass of 238.05). The relative atomic mass of fluorine is 19.00.

Solution

The two molar masses are

$$\mathcal{M}(^{238}UF_6) = 238.05 + 6(19.00) \text{ g mol}^{-1} = 352.05 \text{ g mol}^{-1}$$
$$\mathcal{M}(^{235}UF_6) = 235.04 + 6(19.00) \text{ g mol}^{-1} = 349.04 \text{ g mol}^{-1}$$
$$\text{enrichment factor} = \sqrt{\frac{\mathcal{M}(^{238}UF_6)}{\mathcal{M}(^{235}UF_6)}} = \sqrt{\frac{352.05 \text{ g mol}^{-1}}{349.04 \text{ g mol}^{-1}}} = 1.0043$$

Exercise

A gas mixture contains equal numbers of molecules of N_2 and SF_6. Some of it is passed through a gaseous effusion apparatus. Calculate how many molecules of N_2 are present in the product gas for every 100 molecules of SF_6.

Answer: 228.

Graham's law of effusion holds true only if the opening in the vessel is small enough and the pressure low enough that most molecules follow straight-line trajectories through the opening without undergoing collisions with one another. A related but more complex phenomenon is **gaseous diffusion** through a porous barrier. This differs from effusion in that molecules undergo many collisions with one another and with the barrier during their passage through it. Interestingly, the rate of gaseous diffusion is *also* inversely proportional to the square root of the molar mass of the gas. If a mixture of a heavier gas B and a lighter gas A is placed in contact with a porous barrier, the gas passing through is enriched in the lighter component by a factor of $\sqrt{\mathcal{M}_B/\mathcal{M}_A}$, and the remaining gas is enriched in the heavier component.

During World War II, it was necessary to separate the isotope ^{235}U from ^{238}U in order to develop the atomic bomb. Because ^{235}U has a natural abundance of only 0.7%, its isolation in nearly pure form posed a daunting task. The procedure that

Figure 5–19   A gaseous diffusion plant for separation of uranium isotopes in uranium hexafluoride at Oak Ridge National Laboratory. If the uranium is to be used as fuel in nuclear power plants, a total enrichment over many stages, from 0.7% to 3% ^{235}U, is necessary.

succeeded was to allow uranium to react with fluorine to form the relatively volatile compound UF_6 (boiling point 56°C), which was enriched in ^{235}U by passing it through a porous barrier. As established in Example 5–13, each passage gives an enrichment factor of only 1.0043, so thousands of such barriers in succession were necessary to provide sufficient enrichment of the ^{235}U component. Such a gaseous diffusion plant was constructed in a very short time at the Oak Ridge National Laboratories, and the enriched ^{235}U was used in the Hiroshima atomic bomb. Similar methods are still used today to separate the isotopes of uranium (Fig. 5–19).

5–8   REAL GASES

The ideal gas law

$$PV = nRT$$

is a particularly simple example of an **equation of state:** an equation relating the pressure, temperature, chemical amount, and volume to each other. Such equations of state can be obtained either from theory or from experiments. They are useful not only for ideal gases but for real gases, liquids, and solids.

We have stressed that real gases obey the ideal gas law only at sufficiently low pressures. Deviations appear in a variety of forms. Boyle's law (PV = constant) is no longer satisfied at high pressures, and Charles's law ($V \propto T$) begins to break down when the temperature becomes low. Avogadro's law also does not strictly hold for real gases at moderate pressures. Consider some concrete data. The volume of one mole of ideal gas at exactly one atmosphere of pressure and 273.15 K equals 22.414 L. The *observed* volumes of one-mole samples of several real gases under these conditions are:

H_2	22.433 L	O_2	22.397L
He	22.434 L	CO_2	22.260 L
N_2	22.404 L	NH_3	22.079 L

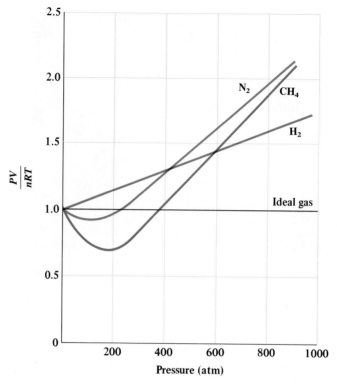

Figure 5–20 A plot of *PV/nRT* against pressure shows deviations from the ideal gas law quite clearly. For an ideal gas, the plot is the straight horizontal line. The behavior of several real gases at 25°C deviates as shown.

None equals 22.414 L: the volumes of hydrogen and helium deviate to the high side, and those of carbon dioxide and ammonia are significantly low (by 0.7% for CO_2 and 1.5% for NH_3).

Figure 5–20 displays deviations from ideality in a different way. It shows a plot of *PV/nRT*, which equals 1 for an ideal gas, for some real gases as the pressure increases at 25°C. The curve for hydrogen (red) never goes below the horizontal black line that represents ideal behavior. Hydrogen experiences a **positive deviation** from ideality at this temperature. Methane and nitrogen experience **negative deviations** from ideality at moderate pressure, but positive deviations grow in and predominate at higher pressure. These observations show that the ideal gas law is inadequate at moderate and high pressures. Indeed, it is exactly correct for real gases only in the limit of zero pressure. Any improved equation of state should account for the two competing kinds of deviations.

Van der Waals Equation of State

One of the earliest and most important improvements on the ideal gas equation of state was proposed by the Dutch physicist Johannes van der Waals in 1873. The **van der Waals equation of state** has the form

$$\left(P + a\frac{n^2}{V_2}\right)(V - nb) = nRT$$

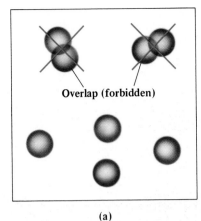
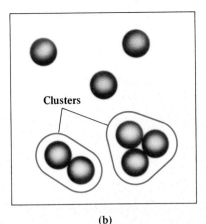

(a)   (b)

Figure 5–21 (a) In a real gas at moderate or high densities, a test particle is excluded from some of the volume in the container because repulsive forces between molecules prevent it from sitting on top of a second molecule. (b) A second effect arises from attractive forces. When temporary associations of molecules form through these attractions, there are in effect fewer particles present in the gas and the pressure is lower.

Comparison with the ideal gas law shows two modifications: an addition to P and a subtraction from V. These arise from the existence of forces between molecules (Fig. 5–21), which are repulsive at short distances and attractive at large distances. Repulsive forces keep molecules from overlapping. Each molecule denies other molecules entry into the volume that it occupies, so the effective volume available to any test molecule is not V, the true volume of the container, but $V - nb$, where b is a constant describing the volume excluded per mole of molecules. Repulsive forces cause positive deviations from ideality.

Attractive forces confer a degree of "stickiness" on molecules. Sticky molecules linger together a while after collisions. Any tendency to group together reduces the effective number of independent molecules in the gas and therefore reduces the number of collisions made with the walls of the container. Fewer wall collisions means a lower pressure than the gas would exert if there were no attractions. Van der Waals argued that this effect, because it depends on attractions between *pairs* of molecules, should be proportional to the *square* of the number of molecules per unit volume $(N/V)^2$ or, equivalently, to the square of the number of moles per unit volume $(n/V)^2$. The effect therefore reduces the pressure by an amount $a \times (n^2/V^2)$, where a is a positive constant that depends on the strength of the attractive forces. By lowering P from its hypothetical ideal value, attractive forces cause negative deviations from ideality. Putting these two corrections into the ideal gas law gives the van der Waals equation.

The only way to get values for the constants a and b is to fit experimental PVT data for each and every real gas—the two constants differ from gas to gas (Table 5–3). They have the units

$$a: \quad \text{atm L}^2 \text{ mol}^{-2}$$

$$b: \quad \text{L mol}^{-1}$$

when R has units of L atm mol^{-1} K^{-1}. For nitrogen it is found that $b = 0.03913$ L mol^{-1} and $a = 1.390$ atm L^2 mol^{-2}. The constant b is the volume excluded by one mole of molecules; it should therefore be close to the volume per mole in the liquid state, where molecules are essentially in contact with each other. The density of

Table 5-3
Van der Waals Constants of Several Gases

Name	Formula	a (atm L^2mol^{-2})	b (L mol^{-1})
Ammonia	NH_3	4.170	0.03707
Argon	Ar	1.345	0.03219
Carbon dioxide	CO_2	3.592	0.04267
Hydrogen	H_2	0.2444	0.02661
Hydrogen chloride	HCl	3.667	0.04081
Methane	CH_4	2.253	0.04278
Nitrogen	N_2	1.390	0.03913
Nitrogen dioxide	NO_2	5.284	0.04424
Oxygen	O_2	1.360	0.03183
Sulfur dioxide	SO_2	6.714	0.05636
Water	H_2O	5.464	0.03049

liquid nitrogen is 0.808 g cm^{-3}. One mole of N_2 weighs 28.0 g, so

$$\text{volume per mole of } N_2(\ell) = \frac{1 \text{ cm}^3 \text{ } N_2(\ell)}{0.808 \text{ g } N_2(\ell)} \times \left(\frac{28.0 \text{ g } N_2}{1 \text{ mol } N_2} \right) \times \left(\frac{1 \text{ L}}{1000 \text{ cm}^3} \right)$$

$$= 0.0347 \text{ L mol}^{-1}$$

This is reasonably close to the van der Waals b parameter of 0.03913 L mol^{-1}.

EXAMPLE 5-14

A sample consists of 8.00 kg of gaseous nitrogen and fills a 100-L flask at 300°C. What is the pressure of the gas, using the van der Waals equation of state? What pressure would be predicted by the ideal gas equation?

Solution

We rewrite the van der Waals equation as

$$P + a \frac{n^2}{V^2} = \frac{nRT}{V - nb}$$

$$P = \frac{nRT}{V - nb} - a \frac{n^2}{V^2}$$

Because the molar mass of nitrogen is 28.0 g mol^{-1},

$$n = 8.00 \times 10^3 \text{ g } N_2 \times \left(\frac{1 \text{ mol } N_2}{28.0 \text{ g } N_2} \right) = 286 \text{ mol } N_2$$

The absolute temperature is $T = 300 + 273.15 = 573$ K. Using $R = 0.08206$ L atm mol^{-1} K^{-1} and the van der Waals constants for nitrogen given in Table 5-3, we calculate $P = 151 - 11 = $ 140 atm. If instead we use the ideal gas law $P = $

nRT/V, we compute a pressure of 134 atm. The deviations of real gases from the predictions of the ideal gas law become sizeable at higher pressures.

Exercise

Compare the pressure exerted by 1.50 mol of an ideal gas in a 2.00-L container at a temperature of 600 K and the pressure exerted by 1.50 mol of sulfur dioxide in a 2.00-L container at the same temperature according to the van der Waals equation, by computing their difference.

Answer: The pressure exerted by the ideal gas is larger by 2.15 atm.

SUMMARY

5–1 The sources and chemical reactions of gases are quite varied. Air consists of a mixture of several gaseous substances, principally oxygen and nitrogen.

5–2 The physical behavior of gases at low density follows simple laws. The **pressure** of any gas is the force that it exerts on a unit area of the walls of its container; the air around us exerts a pressure that can be measured with a **barometer. Boyle's law** states that the pressure exerted by a sample of a gas is inversely proportional to its volume at fixed temperature.

5–3 **Charles's law** states that when a sample of gas is heated or cooled at constant pressure, its volume depends linearly on its temperature. Charles's law allows us to define a universal temperature scale based on ratios of gas volumes. An **absolute zero** of temperature exists below which it is impossible to cool any substance, and it is logical to measure temperatures upward from that point. The **kelvin,** which is equal in size to a Celsius degree, is the unit of temperature for measurements referred to absolute zero.

5–4 Boyle's law, Charles's law, and Avogadro's law (which states that a given volume contains the same number of molecules of any gas at fixed pressure and temperature) can all be combined into a single mathematical statement, the **ideal gas law** ($PV = nRT$). The proportionality constant in this law is the **universal gas constant,** R.

5–5 The ideal gas law is the basis for many calculations of practical importance in chemistry because it relates the volumes of gases in reactions or other processes to chemical amounts or masses.

5–6 **Dalton's law** extends the ideal gas law to mixtures. Each component in an ideal gas mixture has a **partial pressure** equal to the product of its **mole fraction** and the total pressure. The partial pressures add up to the total pressure because the mole fractions must add up to unity.

5–7 The **kinetic theory of gases** provides a microscopic picture that accounts for properties of gases at low densities. It assumes that the molecules in a gas move in random directions with a distribution of speeds and travel in straight lines between collisions with other molecules or with the walls of their container. From this kinetic theory, in combination with the ideal gas law, it can be shown that the average kinetic energy of the molecules of a gas is proportional to the absolute temperature; therefore, the **mean-square speed** is as well. The full distribution of molecular speeds in a gas at **thermal equilibrium** is given by the **Maxwell-Boltzmann equation.** The kinetic theory can be used to estimate the diffusive motion of molecules in a gas, including their **mean free path,** and to account for the relative rates of **effusion** of gases through small holes. It is found under the same conditions of temperature and pressure that lighter molecules effuse more rapidly than heavier ones by a factor equal to the inverse of the ratio of the square root of their molar masses **(Graham's law).**

5–8 Real gases obey the ideal gas law only at sufficiently low pressures. At higher pressures, the **van der Waals equation of state** accounts for deviations from ideality in an approximate fashion by considering the repulsions and attractions among molecules. Calculations using the van der Waals equation give better agreement with experiment for gases at high pressures.

PROBLEMS

Note: Answers to blue-numbered problems are given in Appendix F. Problems that are more challenging are indicated with asterisks.

The Chemistry of Gases

1. Predict the gaseous product of the reaction of ammonium hydrosulfide ($NH_4HS(s)$) with the acid $HCl(aq)$.

2. Predict the gaseous product of the reaction of $NH_4HS(s)$ with an aqueous solution of the base NaOH.

3. Solid ammonium hydrosulfide (NH_4HS) decomposes entirely to gases when it is heated. Write a chemical equation representing this change.

4. Solid ammonium carbamate ($NH_4CO_2NH_2$) decomposes entirely to gases when it is heated. Write a chemical equation representing this change.

5. Ammonia (NH_3) is an important and useful gas. Suggest a way to generate it from ammonium bromide (NH_4Br). Include a balanced chemical equation.

6. Hydrogen cyanide (HCN) is a poisonous gas. Explain why solutions of potassium cyanide (KCN) should never be acidified. Include a balanced chemical equation.

7. Barium peroxide (BaO_2) gives BaO(s) and a gas when it is heated. Identify the gas and write an equation to represent the reaction.

8. Ammonium dichromate (($NH_4)_2Cr_2O_7$) reacts when heated to give $Cr_2O_3(s)$, water vapor, and a second gas. Write a balanced chemical equation to represent this change.

Pressure and Boyle's Law

9. (See Example 5–1.) Estimate (in standard atmospheres) the pressure of a gas in a container if the pressure supports a column of water 44.5 cm high.

10. (See Example 5–1.) The metallic element gallium is a liquid with a density of 6.114 g cm^{-3} at 30°C. A barometer is constructed using liquid gallium in a vertical tube. On a particular day when the temperature is 30°C, the height of the column of gallium is 168 cm. Compute that day's atmospheric pressure in standard atmospheres.

11. Determine the height of a column of heavy oil (density = 1.60 g cm^{-3}) in a tube if its mass is the same as that of a 305-mm-long column of mercury (density = 13.6 g cm^{-3}) in an identical tube.

12. Calcium dissolved in the ocean is used by marine organisms to form $CaCO_3(s)$ in skeletons and shells. When the organisms die, their remains fall to the bottom. The amount of calcium carbonate that can be dissolved in seawater depends on the pressure. At great depths, where the pressure exceeds about 414 atm, the shells slowly redissolve. This prevents all the world's calcium from being tied up as insoluble $CaCO_3(s)$ at the bottom of the sea. Estimate the depth (in meters) of water that exerts a pressure great enough to dissolve seashells.

13. Suppose that all the air of the earth were replaced by an ocean of mercury. How deep would this ocean have to be to exert the same pressure as the air?

14. Suppose that the atmosphere were perfectly uniform, with a density throughout equal to that of air at STP, 1.3 g L^{-1}. Calculate the depth of such an atmosphere that would cause a pressure of exactly one standard atmosphere at the earth's surface.

15. (See Example 5–2.) The "critical pressure" of mercury is 172.00 MPa. Above this pressure, mercury cannot be liquefied, no matter what the temperature (see Section 6–5). Express this pressure in atmospheres and in bars.

16. (See Example 5–2.) Experimental studies of solid surfaces and the chemical reactions that occur on them require very low gas pressures to avoid surface contamination. A high-vacuum apparatus for such experiments can routinely reach pressures of 5×10^{-10} torr. Express this pressure in atmospheres and in pascals.

17. A sample of gaseous N_2 is trapped in a J-tube (as in Fig. 5–3). The level of the mercury on the side open to the atmosphere is 67.5 mm higher than the level on the other side of the tube. The atmospheric pressure in the room is 1.000 atm and the temperature is 0.00°C. Determine the pressure of the trapped N_2 (in units of atmospheres).

18. A sample of 12.07 cm³ of gaseous CO_2 is trapped in a J-tube like the one in the previous problem. The pressure on the CO_2 is 1.106 atm. More mercury is poured into the open side of the tube, and the mercury level rises on both sides. The volume of the gas is reduced to 11.51 cm³. Compute the pressure (in atm) of the CO_2 after the new mercury has been added.

19. (See Example 5–3.) Some nitrogen is held in a 2.00-L tank at a pressure of 3.00 atm. The tank is connected to a 5.00-L tank that is completely empty (evacuated), and a valve is opened to connect the two. No temperature change occurs in the process. Determine the total pressure in this two-tank system after the nitrogen stops flowing.

20. (See Example 5–3.) The Stirling engine, a heat engine invented by a Scottish minister, has been considered for use in automobile engines because of its efficiency. In such an engine, an ideal gas goes through a four-step cycle of (1) expansion at constant T, (2) cooling at constant V, (3) compression at constant T back to its original volume, and (4) heating at constant V back to its original temperature. Suppose an ideal gas starts at a pressure of 1.23 atm, and its volume changes from 0.350 L to 1.31 L during the expansion at constant T. Calculate the pressure of the gas at the end of this step in the cycle.

21. Each of the following sets of pressure/volume data applies to a sample of an ideal gas held at constant temperature. In each case, calculate the missing quantity.
    (a) $V = 0.235$ L at 1.00 atm; $V =$ _____ at 2.37 atm
    (b) $V = 785$ mL at 1.50 atm; $V =$ _____ at 110.0 kPa
    (c) $V = 604$ mL at 995 torr; $V = 406$ mL at _____ torr
    (d) $V = x$ mL at 2.00 atm; $V = x/2$ mL at _____ atm

22. Each of the following sets of pressure/volume data applies to a sample of an ideal gas held at constant temperature. In each case, calculate the missing quantity.
    (a) $V = 485$ L at 512 torr; $V =$ _____ at 50.0 kPa
    (b) $V = 781$ mL at 1.50 atm; $V =$ _____ at 110.0 kPa
    (c) $V = 60.0$ cm³ at 995 torr; $V = 90.0$ mL at _____ atm
    (d) $V = y$ mL at 2.00 atm; $V = 3.25y$ mL at _____ kPa

Temperature and Charles's Law

23. Ordinary internal body temperature in a healthy human being is 98.6°F. Express this temperature on the Celsius and Kelvin temperature scales.

24. Temperature readings on the Celsius and Fahrenheit scales are numerically equal at only one temperature. Determine that temperature.

25. The absolute temperature of a 4.00-L sample of gas is doubled at constant pressure. Determine the volume of the gas after this change.

26. The Celsius temperature of a 4.00-L sample of gas is doubled from 20.0°C to 40.0°C at constant pressure. Determine the volume of the gas after this change.

27. (See Example 5–4.) The gill is an obscure unit of volume. If some $H_2(g)$ has a volume of 17.4 gills at 100°F, what volume would it have if the temperature were reduced to 0°F, assuming that its pressure stays constant?

28. (See Example 5–4.) A gas originally at a temperature of 26.5°C is cooled at constant pressure. Its volume decreases from 5.40 L to 5.26 L. Determine its new temperature in degrees Celsius.

29. (See Example 5–4.) Calcium carbide (CaC_2) reacts with water to produce acetylene (C_2H_2) according to the equation

$$CaC_2(s) + 2\ H_2O(\ell) \longrightarrow Ca(OH)_2(s) + C_2H_2(g)$$

A certain mass of CaC_2 reacts completely with water to give 64.5 L of C_2H_2 at 50°C and $P = 1$ atm. If the same mass of CaC_2 reacts completely at 400°C and $P = 1$ atm, what volume of C_2H_2 can be collected at the higher temperature?

30. (See Example 5–4.) A convenient laboratory source for oxygen of very high purity is the decomposition of potassium permanganate at 230°C.

$$2\ KMnO_4(s) \longrightarrow K_2MnO_4(s) + MnO_2(s) + O_2(g)$$

Suppose 3.41 L of oxygen is needed at atmospheric pressure and a temperature of 20°C. What volume of oxygen should be collected at 230°C and the same pressure in order to give this volume when cooled?

31. Each of the following sets of volume/temperature data applies to a sample of an ideal gas held at constant pressure. In each case, calculate the missing quantity.
    (a) $V = 185$ L at 0.0°C; $V =$ _____ at 100.0°C
    (b) $V = 18$ quarts at 400 K; $V =$ _____ at 200 K
    (c) $V = 1.000 \times 10^3$ mL at 1000 K; $V =$ _____ mL at 100°C
    (d) $V = y$ mL at 200°C; $V = 3.25y$ mL at _____ °C

32. Each of the following sets of volume/temperature data applies to a sample of an ideal gas held at constant pressure. In each case, calculate the missing quantity.
    (a) $V = 24.2$ L at 25.0°C; $V =$ _____ at 0.0°C
    (b) $V = 781$ mL at 150 K; $V =$ _____ at 450 K
    (c) $V = 781$ cm³ at 150 K; $V =$ _____ mL at 150°C
    (d) $V = y$ mL at 200 K; $V = 3.25y$ mL at _____ K

The Ideal Gas Law

33. (See Example 5–5.) A gas is in a container with movable walls under a pressure of 5.50 atm. The volume of the container is increased by a factor of 3.25 by moving its walls. The ab-

solute temperature is simultaneously doubled. Calculate the pressure of the gas after these changes.

34. (See Example 5–5.) Some gaseous helium is kept under a pressure of 20.0 atm in a steel tank at a temperature of 20.0°C. The tank is moved outside on a cold day when the temperature is 0.0°C. Determine the pressure of helium inside the tank after it cools down to 0.0°C.

35. (See Example 5–5.) A 20.6-L sample of "pure" air is collected in Greenland at a temperature of −20.0°C and a pressure of 1.01 atm and is forced into a 1.05-L bottle for shipment to Europe for analysis.
 (a) Compute the pressure inside the bottle just after it is filled.
 (b) Compute the pressure inside the bottle as it is opened in the 21.0°C comfort of the European laboratory.

36. (See Example 5–5.) Iodine heptafluoride (IF_7) can be made at elevated temperatures by the reaction

$$I_2(g) + 7 F_2(g) \longrightarrow 2 IF_7(g)$$

Suppose 63.6 L of gaseous IF_7 is made by this reaction at 300°C and a pressure of 0.459 atm. Calculate the volume this gas occupies when heated to 400°C at a pressure of 0.980 atm, assuming that no decomposition or other reactions take place.

37. Certain thin-walled glass bottles generally withstand a pressure difference across their walls of 1.25 atm before they burst. A typical bottle of this type is sealed on a day when the temperature is 68°F and the pressure is 0.98 atm, and it is immediately heated in an oven. Estimate the temperature in degrees Fahrenheit at which it explodes.

38. The pressure of a poisonous gas inside a sealed container is 1.47 atm at 20°C. If the barometric pressure is 0.96 atm, to what temperature (in degrees Celsius) must the container and its contents be cooled so that the container can be opened without any risk of gas spurting out?

39. (See Example 5–6.) At what temperature (in kelvins) does 29.8 g of O_2 gas have a pressure of 2.00 atm in a 10.0-liter tank?

40. (See Example 5–6.) What mass (in grams) of neon exerts a pressure of 2.00 atm in a 35.0-liter tank at 25.0°C?

41. According to a reference handbook, "The weight of one liter [of $H_2Te(g)$ is] 6.234 g." Tell why this information is nearly valueless. Assume that $H_2Te(g)$ is an ideal gas, and calculate the temperature (in degrees Celsius) at which this statement is true if the pressure is 1.00 atm.

42. A scuba diver's tank contains 0.30 kg of oxygen (O_2) compressed into a volume of 2.32 L.
 (a) Use the ideal gas law to estimate the gas pressure inside the tank at 5°C, and express it in atmospheres and in pounds per square inch.
 (b) What volume would this oxygen occupy at 30°C and a pressure of 0.98 atm?

43. In clean, dry air, 1140 out of every 1 billion molecules (or approximately one in a million) is a krypton (Kr) atom. Compute the number of atoms of krypton present in 15 cm³ (a gasp) of clean, dry air at 21.0°C and a pressure of 0.95 atm. The molar mass of air is 29 g mol⁻¹.

44. Like carbon dioxide, methane exerts a "greenhouse effect" in the atmosphere. Although the increase in atmospheric carbon dioxide concentration from the burning of fossil fuels attracts more attention, methane levels are also on the rise. In a recent survey, the average worldwide concentration of methane in dry air in the lower 7 to 10 miles of the atmosphere was 1.657 parts per million by volume. Estimate (to within a power of ten) the average number of molecules of methane in one liter of this portion of the atmosphere.

45. (See Example 5–7.) Calculate the density of gaseous SF_6 at a temperature of 15°C and a pressure of 0.942 atm.

46. (See Example 5–7.) Calculate the density of gaseous N_2H_4 at a temperature of 350°C and a pressure of 1.47 atm.

47. (See Example 5–8.) The empirical formula of a gaseous fluorocarbon is CF_2. If 1.55 g of this compound occupies 0.174 L at STP, then determine the molecular formula of this compound.

48. (See Example 5–8.) A sample of a gaseous binary compound of boron and chlorine weighing 2.842 g occupies 0.153 L at STP. This sample is decomposed to give solid boron and gaseous chlorine (Cl_2). At STP, the chlorine occupies 0.688 L. Determine the molecular formula of the compound.

Chemical Calculations for Gases

49. (See Example 5–10.) The classic method for manufacturing hydrogen chloride—still in use today to a small extent—involves the reaction of sodium chloride with excess sulfuric acid at elevated temperatures. The overall equation for this process is

$$NaCl(s) + H_2SO_4(aq) \longrightarrow NaHSO_4(s) + HCl(g)$$

What volume of hydrogen chloride is produced at 550°C and a pressure of 0.97 atm from the reaction of 2500 kg of sodium chloride?

50. (See Example 5–10.) The laboratory-scale preparation of pure oxygen can be carried out by the Brin process, which consists of decomposing barium peroxide (BaO_2) by heating it:

$$2 BaO_2 \longrightarrow 2 BaO + O_2$$

Suppose that 5.00 g of BaO_2 is decomposed according to this equation:
 (a) What volume of oxygen is produced (at 25°C and 1.00 atm)?
 (b) What mass of BaO is formed?

Note: The reaction can be reversed by heating BaO in pressurized air, and the cyclic process may be repeated indefinitely. In this way the element oxygen can be separated from the mixture (air).

51. (See Example 5–10.) Hydrogen is produced by the complete reaction of 6.24 g of sodium with an excess of gaseous hydrogen chloride.
 (a) Write a balanced chemical equation for the reaction that occurs.
 (b) How many liters of hydrogen are produced at a temperature of 50.0°C and a pressure of 0.850 atm?

52. (See Example 5–10.) In 1783, the French physicist Jacques Charles supervised and took part in the first human flight in a hydrogen balloon. Such balloons rely on the lower density of hydrogen relative to air to give them their buoyancy. In Charles's balloon ascent, the hydrogen was produced (together with iron(II) sulfate) from the action of aqueous sulfuric acid on iron filings.
    (a) Write a balanced chemical equation for this reaction.
    (b) What volume of hydrogen is produced at 300 K and a pressure of 1.0 atmosphere when 300 kg of sulfuric acid is consumed in this reaction?

53. Elemental chlorine was first produced by Scheele in 1774 using the reaction of pyrolusite (MnO_2) with sulfuric acid and sodium chloride:

$$4 \ NaCl(s) + 2 \ H_2SO_4(aq) + MnO_2(s) \longrightarrow$$
$$2 \ Na_2SO_4(s) + MnCl_2(s) + 2 \ H_2O(\ell) + Cl_2(g)$$

Calculate the minimum mass of MnO_2 required to generate 5.32 L of gaseous chlorine, measured at a pressure of 0.953 atm and a temperature of 33°C.

54. Aluminum reacts with excess aqueous hydrochloric acid to produce hydrogen.
    (a) Write a balanced chemical equation for the reaction. (*Hint:* Water-soluble $AlCl_3$ is the stable chloride of aluminum.)
    (b) Calculate the mass of pure aluminum that furnishes 10.0 L of hydrogen at a pressure of 0.750 atm and a temperature of 30.0°C.

55. Elemental sulfur can be recovered from gaseous hydrogen sulfide (H_2S) through the reaction

$$2 \ H_2S(g) + SO_2(g) \longrightarrow 3 \ S(s) + 2 \ H_2O(\ell)$$

    (a) What volume of H_2S (in liters at STP) is required to produce 2.00 kg (2000 g) of sulfur according to this equation?
    (b) What minimum mass and volume (at STP) of SO_2 are required to produce 2.00 kg of sulfur according to this equation?

56. When ozone (O_3) is placed in contact with dry, powdered KOH at −15°C, the red-brown solid potassium ozonide (KO_3) forms, according to the balanced equation

$$5 \ O_3(g) + 2 \ KOH(s) \longrightarrow 2 \ KO_3(s) + 5 \ O_2(g) + H_2O(s)$$

Calculate the minimum volume of ozone needed (at a pressure of 0.134 atm and −15°C) to produce 4.69 g of KO_3.

57. A gaseous binary compound of carbon and hydrogen, in a volume of 300.0 cm^3 at a temperature of 300.0 K and a pressure of 1.200 atm, furnishes 1.054 g of H_2O and 1.930 g of CO_2 when burned completely in an excess of oxygen. Determine the molecular formula of this hydrocarbon.

58. A gaseous hydrocarbon, in a volume of 25.4 L at 400 K and a pressure of 3.40 atm, reacts in an excess of oxygen to give 47.4 g of H_2O and 231.6 g of CO_2. Determine the molecular formula of the hydrocarbon.

Mixtures of Gases

59. The partial pressure of $O_2(g)$ in a mixture of $O_2(g)$ and $N_2(g)$ is 1.02 atm. The total pressure of the mixture is 2.43 atm. Determine the partial pressure of $N_2(g)$, and determine the mole fraction of each gas in the mixture.

60. In the gas mixture described in the previous problem, exactly half of the $O_2(g)$ is removed. No change in temperature or volume occurs. Determine the partial pressure of the remaining $O_2(g)$ and its mole fraction.

61. (See Example 5–11.) Sulfur dioxide reacts with oxygen in the presence of platinum to give sulfur trioxide:

$$2 \ SO_2(g) + O_2(g) \longrightarrow 2 \ SO_3(g)$$

Suppose at one stage in the reaction 263 g SO_2, 837 g O_2, and 179 g SO_3 are present in the reaction vessel at a total pressure of 0.950 atm. Calculate the mole fraction of SO_3 and its partial pressure.

62. (See Example 5–11.) The synthesis of ammonia from the elements is carried out at high pressures and temperatures.

$$N_2(g) + 3 \ H_2(g) \longrightarrow 2 \ NH_3(g)$$

Suppose at one stage in the reaction 149 g NH_3, 765 g N_2, and 164 g H_2 are present in the reaction vessel at a total pressure of 210 atm. Calculate the mole fraction of NH_3 and its partial pressure.

63. The atmospheric pressure at the surface of Mars is 5.92×10^{-3} atm. The Martian atmosphere is 95.3% CO_2 and 2.7% N_2 by volume, with small amounts of other gases also present. Compute the mole fraction and partial pressure of N_2 in the atmosphere of Mars.

64. The atmospheric pressure at the surface of Venus is 90.8 atm. The Venusian atmosphere is 96.5% CO_2 and 3.5% N_2 by volume, with small amounts of other gases also present. Compute the mole fraction and partial pressure of N_2 in the atmosphere of Venus.

65. A gas mixture at room temperature contains 10.0 mol of CO and 12.5 mol of O_2.
    (a) Compute the mole fraction of CO in the mixture.
    (b) The mixture is then heated, and the CO starts to react with the O_2 to give CO_2:

$$CO(g) + \tfrac{1}{2} O_2(g) \longrightarrow CO_2(g)$$

The reaction is stopped when 3.0 mol of CO_2 is present. Determine the mole fraction of CO in the new mixture.

66. A gas mixture contains 4.5 mol of Br_2 and 33.1 mol F_2.
    (a) Compute the mole fraction of Br_2 in the mixture.
    (b) The mixture is heated above 150°C and starts to react to give BrF_5:

$$Br_2(g) + 5 \ F_2(g) \longrightarrow 2 \ BrF_5(g)$$

The reaction is stopped when 2.2 mol of BrF_5 is present. Determine the mole fraction of Br_2 in the mixture at that point.

The Kinetic Theory of Gases

67. (See Example 5–12.)
    (a) Compute the root-mean-square speed of H_2 molecules in hydrogen at a temperature of 300 K.
    (b) Repeat for SF_6 molecules in gaseous sulfur hexafluoride at 300 K.

68. (See Example 5–12.) Researchers report that the temperature of a gas consisting of 500 sodium atoms confined in a tiny atomic trap is 0.00024 K. Compute the root-mean-square speed of the atoms in this confinement.

69. Compare the root-mean-square speed of helium atoms near the surface of the sun, where the temperature is approximately 6000 K, with that of helium atoms in an interstellar cloud, where the temperature is 100 K.

70. The "escape velocity" for objects to leave the gravitational field of the earth is 11.2 km s^{-1}. Calculate the ratio of the escape velocity to the root-mean-square speed of helium, argon, and xenon atoms at 2000 K. Does the result help to explain the low abundance of the light gas helium in the atmosphere? Explain.

71. Chlorine dioxide (ClO_2) is used for bleaching wood pulp. In a gaseous sample held at thermal equilibrium at a particular temperature, 35.0% of the molecules have speeds exceeding 400 m s^{-1}. The sample is heated slightly. Is the percentage of molecules with speeds in excess of 400 m s^{-1} then greater than or less than 35%? Explain.

72. The ClO_2 from Problem 71 is heated further until it explodes, yielding Cl_2, O_2, and other gaseous products. The mixture is then cooled until the original temperature is reached. Is the percentage of *chlorine* molecules with speeds in excess of 400 m s^{-1} greater than or less than 35%? Explain.

73. Methane (CH_4) effuses through a small opening in the side of a container at the rate of 1.30×10^{-8} mol s^{-1}. An unknown gas effuses through the same opening at the rate of 5.42×10^{-9} mol s^{-1} when maintained at the same temperature and pressure as the methane. Determine the molar mass of the unknown gas.

74. (See Example 5–13.) Equal chemical amounts of two gases, fluorine and bromine pentafluoride, are mixed. Determine the ratio of the rates of effusion of the two gases through a very small opening in their container.

Real Gases

75. (See Example 5–14.) Oxygen is supplied to hospitals and chemical laboratories under pressure in large steel cylinders. Typically, such a cylinder has an internal volume of 28.0 L and contains 6.80 kg of oxygen. Use the van der Waals equation to estimate the pressure inside such a cylinder at 20°C. Express it in atmospheres and in pounds per square inch.

76. (See Example 5–14.) Steam at high pressures and temperatures is used to generate electrical power in utility plants. A large utility boiler has a volume of 2500 m^3 and contains 140 metric tons (1 metric ton = 10^3 kg) of steam at a tempera-

ture of 540°C. Use the van der Waals equation to estimate the pressure of the steam under these conditions, in atmospheres and in pounds per square inch.

77. Careful measurements show that 0.7500 mol of CO_2 gas confined in a volume of 1.000 L at a temperature of 240.0 K exerts a pressure of 12.5655 atm.
    (a) Compute the ideal-gas pressure (the pressure that an ideal gas would exert under these conditions).
    (b) Compute the van der Waals pressure (see Table 5–3 for the van der Waals constants for CO_2).
    (c) Explain why neither of these pressures equals the observed pressure.

78. A careful experiment establishes that a 0.5000-mol sample of CO_2 occupies a volume of 0.5866 L at a temperature of 250.0 K and a pressure of 14.804 atm. These values do not exactly fit in the van der Waals equation. Suppose the van der Waals constant b remains unchanged at 0.04267 L mol^{-1}. Compute the value of a that makes the van der Waals equation exactly correct for these particular conditions.

Additional Problems

79. After a flood fills a basement to a depth of 9.0 ft and completely saturates the surrounding earth, the owner buys an electric pump and quickly pumps the water out of the basement. Suddenly, a basement wall collapses, the structure is severely damaged, and mud oozes in. Explain this event by estimating the difference between the outside pressure at the base of the basement walls and the pressure inside the drained basement. Assume that the mud has a density of 4.9 g cm^{-3}. Report the answer both in atmospheres and in psi (pounds per square inch).

80. The density of mercury is 13.5955 g cm^{-3} at 0.0°C but only 13.5094 g cm^{-3} at 35°C. Suppose that a mercury barometer is read on a hot summer day when the temperature is 35°C. The column of mercury is 760.0 mm long. Correct for the expansion of the mercury, and compute the true pressure in standard atmospheres.

81. The barometric pressure in the center of one of the worst hurricanes of the 20th century was reported to be 886 millibars. Express this in standard atmospheres.

82. Aqueous acetic acid and sodium hydrogen carbonate react to give carbon dioxide and sodium acetate:

$$CH_3COOH(aq) + NaHCO_3(s) \longrightarrow$$
$$NaCH_3COO(aq) + H_2O(\ell) + CO_2(g)$$

Some solid $NaHCO_3$ is placed in a large Ziploc plastic bag along with a small bottle of a solution of acetic acid ($CH_3COOH(aq)$). The plastic bag is sealed shut. Working through the walls of the plastic bag, an experimenter opens up the bottle of acid so that it can mix with the $NaHCO_3(s)$. The whole assembly is then immediately placed on a sensitive balance with a digital readout. As the reaction proceeds, the plastic bag puffs out (as $CO_2(g)$ builds up), and the digital readout indicates a steadily diminishing mass! Is the law of conservation of mass being violated? Explain what is happening. (*Hint*: Think about what happens to the air above the balance as the bag puffs up, and its effect on the measured mass.)

83. The following arrangement of flasks is set up. Assuming no temperature change, determine the final pressure inside the system after all stopcocks are opened. Ignore the volume of the connecting tube.

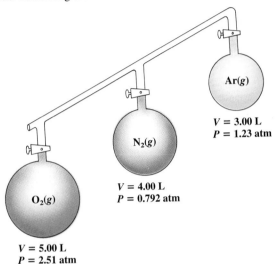

Ar(g)

$V = 3.00$ L
$P = 1.23$ atm

N₂(g)

$V = 4.00$ L
$P = 0.792$ atm

O₂(g)

$V = 5.00$ L
$P = 2.51$ atm

84. When a certain sample of a gas is cooled at constant pressure, it is found that the volume decreases linearly:

$$V = 209.4 \text{ L} + \left(0.456 \frac{\text{L}}{°\text{F}}\right) \times t_F$$

where t_F is the temperature in degrees Fahrenheit. Estimate the absolute zero of temperature in degrees Fahrenheit from this relationship.

85. The Tire Industry Safety Council annually reminds motorists that tires can become dangerously underinflated when temperatures begin to dip below freezing. They say that every time the outside temperature drops by 10°F, the air pressure in a car's tires goes down about 1 pound per square inch (psi). Recall that 14.7 psi equals 1.00 atm.
    (a) Explain why the Council's rule about tire pressure is at best only a rule of thumb.
    (b) Assume that a typical tire with a constant volume is filled with an ideal gas at 70°F to a gauge pressure of 28 psi. The temperature then falls to 20°F. What is the new pressure in the tire?

86. Amontons's law relates pressure to absolute temperature. Consider the ideal gas law and then write a statement of Amontons's law in a form that is analogous to the statements of Charles's law and Boyle's law in the text.

87. When 1 volume of a gaseous compound that contains only carbon, hydrogen, and nitrogen is burned in oxygen, 2 volumes of CO₂(g), 4 volumes of water vapor, and 2 volumes of

NO₂(g) are produced.
    (a) What volume of O₂ is required for the combustion?
    (b) What is the molecular formula of the compound?

88. Complete combustion (burning) of 2.00 volumes of a gaseous hydrocarbon to CO₂(g) and H₂O(g) required 9.00 volumes of pure O₂(g). Both volumes were measured at the same temperature and pressure. The reaction yielded 0.135 g of H₂O and 0.330 g of CO₂. Determine the molecular formula of the hydrocarbon.

89. Compute the minimum masses (in grams) of potassium nitrate and silicon dioxide that must be mixed with 125.0 g of NaN₃ to ensure that no Na, Na₂O, or K₂O remains unreacted when the mixture is detonated in an automobile air-bag.

90. Determine the mass (in grams) of NaN₃ that would have to be decomposed to Na and N₂ to generate 70.0 L of nitrogen in an automobile air-bag at 25°C and 1 atm pressure.

91. A sample of limestone (calcium carbonate, CaCo₃) is heated at 950 K until it is completely converted into calcium oxide (CaO) and CO₂. The CaO is then all converted into calcium hydroxide by addition of water, yielding 8.47 kg of solid Ca(OH)₂. Calculate the volume of CO₂ produced in the first step, assuming it to be an ideal gas at 950 K and a pressure of 0.976 atm.

92. A gas exerts a pressure of 0.740 atm in a certain container. Suddenly a chemical change occurs that consumes half of the molecules originally present and forms two new molecules for every three consumed. Determine the new pressure in the container if the volume of the container and the temperature are unchanged.

93. Exactly 1.0 lb of Hydrone, an alloy of sodium with lead, yields 2.6 ft³ (at 0.0°C and 1.00 atm) of hydrogen when it is treated with water. All of the sodium reacts according to the equation

$$2 \text{ Na}_{\text{in alloy}} + 2 \text{ H}_2\text{O}(\ell) \longrightarrow 2 \text{ NaOH}(aq) + \text{H}_2(g)$$

and the lead does not react with water. Compute the percentage by mass of sodium in the alloy.

94. Baseball reporters say that long fly balls that would have carried for home runs in July "die" in the cool air of October and are caught. The idea behind this reasoning is that a baseball carries better when the air is less dense. Dry air is a mixture of gases with an effective molar mass of 29.0 g mol⁻¹.
    (a) Compute the density of dry air on a July day when the temperature is 95.0°F and the pressure is 1.00 atm.
    (b) Compute the density of dry air on an October evening when the temperature is 50.0°F and the pressure is 1.00 atm.
    (c) Suppose that the humidity on the July day is 100%, so that the air is saturated with water vapor. Would the density of the hot, moist air be less than, equal to, or more than the density of the hot, dry air computed in part (a)? In other terms, does high humidity favor the home run?

95. A mixture of CS₂(g) and excess O₂(g) in a 10.0-L reaction vessel at 100.0°C is under a pressure of 3.00 atm. When the

mixture is ignited by a spark, it explodes. The vessel successfully contains the explosion, in which all of the $CS_2(g)$ reacts to give $CO_2(g)$ and $SO_2(g)$. The vessel is cooled back to its original temperature of $100.0°C$, and the pressure of the mixture of the two product gases and the unreacted $O_2(g)$ is found to be 2.40 atm. Calculate the mass (in grams) of $CS_2(g)$ originally present.

96. Researchers recently reported the first optical atomic trap. In this device, beams of laser light replace the physical walls of conventional containers. The laser beams are tightly focused. They briefly (for 0.5 s) exert enough pressure to confine 500 sodium atoms in a volume of 1.0×10^{-15} m^3. The temperature of this gas is 0.00024 K.
(a) Assume ideal gas behavior to compute the pressure exerted on the "walls" of the optical bottle in this experiment.
(b) In this gas, the mean free path (the average distance that the sodium atoms travel between collisions) is 3.9 m. Compare this with the mean free path of the atoms in gaseous sodium at room conditions.

97. Molecules of UF_6 are approximately 175 times more massive than H_2 molecules; however, Avogadro's number of H_2 molecules confined at a set temperature exert the same pressure on the walls of the container as the same number of UF_6 molecules. Explain how this is possible.

98. The number density of atoms (chiefly hydrogen) in interstellar space is about 10 per cubic centimeter, and the temperature is about 100 K.
(a) Calculate the pressure of the gas in interstellar space, and express it in atmospheres.
(b) Under these conditions, an atom of hydrogen collides with another atom once every 1×10^9 seconds (that is, once every 30 years). By using the root-mean-square speed, estimate the distance traveled by a hydrogen atom between collisions. Compare this distance with the distance from the earth to the sun (150 million km).

99. Naturally occurring fluoride is entirely ^{19}F, but suppose that it were 50% ^{19}F and 50% ^{20}F. Discuss whether gaseous diffusion of UF_6 would then work to separate ^{235}U from ^{238}U.

A space shuttle taking off.

CUMULATIVE PROBLEM

Ammonium Perchlorate

Ammonium perchlorate (NH_4ClO_4) is a solid rocket fuel used in the space shuttle. When heated above 200°C, it decomposes to several gaseous products, of which the most important are nitrogen, chlorine, oxygen, and water vapor.

(a) Write a balanced chemical equation for the decomposition of NH_4ClO_4, assuming that the products listed are the only ones generated.

(b) The sudden appearance of hot gaseous products in a small initial volume leads to a rapid increase in pressure and temperature that gives the rocket its thrust. What total pressure of gas would be produced at 800°C by igniting 7.00×10^5 kg of NH_4ClO_4 (a typical charge of the booster rockets in the space shuttle) and allowing it to expand to fill a volume of 6400 m^3 (6.40×10^6 L)? Use the ideal gas law.

(c) Calculate the mole fraction of chlorine and its partial pressure in the mixture of gases produced.

(d) The van der Waals equation applies only to pure real gases, not to mixtures. For a mixture like that resulting from the reaction of part (a), it may still be possible to define effective a- and b-parameters to relate total pressure, volume, temperature, and total chemical amount. Suppose the gas mixture has $a = 4.00$ atm L^2 mol^{-2} and $b = 0.0330$ L mol^{-1}. Recalculate the pressure of the gas mixture in part (b) using the van der Waals equation.

(e) Calculate and compare the root-mean-square speeds of water and chlorine molecules under the conditions of part (b).

(f) The gas mixture from part (b) cools and expands until it reaches a temperature of 200°C and a pressure of 3.20 atm. Calculate the volume occupied by the gas mixture after this expansion has occurred. Assume ideal gas behavior.

Condensed Phases and Phase Transitions

Solid iodine is converted directly to a vapor (sublimes) when warmed. Here, the purple iodine vapor is redeposited as solid on the cooled surface above.

In the condensed phases of matter, attractive forces hold the constituent particles much closer together than in a gas. In this chapter we consider some of the properties of the condensed phases in the light of the kinetic–molecular theory.

Of particular interest are transitions between phases. These meltings, boilings, freezings, and so forth occur constantly in daily life, in industrial processes, and in the research laboratory. We examine them because of their practical importance and because of what they reveal about fundamental structure. Under the right conditions, two phases of a substance, or even three phases, can coexist. We discuss the dynamic processes that prevail when this happens and introduce a standard graphical display of how temperature and pressure affect phase transitions.

Dissolved species also affect phase transitions because they alter the play of the forces among the constituent particles. We discuss some important practical applications of these effects, such as the measurement of the molar masses of unknown substances and purification through distillation.

6-1 INTERMOLECULAR FORCES: WHY CONDENSED PHASES EXIST

Matter consists of vast assemblies of molecules, atoms, and other particles that exert all kinds of forces on each other. In our development of the kinetic theory of gases (see Section 5–7) we ignored these **intermolecular forces,** stating ". . . molecules of a gas exert no forces on one another except during collisions." This approach succeeds for gases at low pressure and low density. It starts to fail for gases at higher pressures, but fairly simple modifications such as those used in the van der Waals equation (see Section 5–8), can compensate. It collapses entirely for liquids and solids. Comparisons of **molar volume,** the space occupied per mole of substance, illustrate the problem. A gas at room conditions occupies about 24,000 cm^3 mol^{-1}, but a solid or a liquid under the same conditions occupies only 10 to 100 cm^3 mol^{-1}. The same number of particles occupies much less room because their mutual attractions hold them close together. This is why liquids and solids are called **condensed phases.**

We distinguish *inter*molecular forces from *intra*molecular forces; the latter are the chemical bonds discussed in Chapter 3. Covalent bonds among atoms establish and maintain the structure of discrete molecules and are *strong, directional,* and comparatively *short range* in their effect. Ionic bonds are as strong as (or stronger than) covalent bonds but nondirectional and take hold at longer range than covalent bonds.

Intermolecular forces:

- are always weaker than chemical bonds, usually much weaker.
- are less directional than covalent chemical bonds, but often more directional than ionic bonds.
- operate at longer range than covalent chemical bonds, but at shorter range than ionic bonds.

Potential Energy Curves

Graphs of the potential energy of pairs of particles illustrate the distinction between chemical bonds and intermolecular interactions. The existence of attractions among the particles in an assembly means that they have **potential energy** (energy of po-

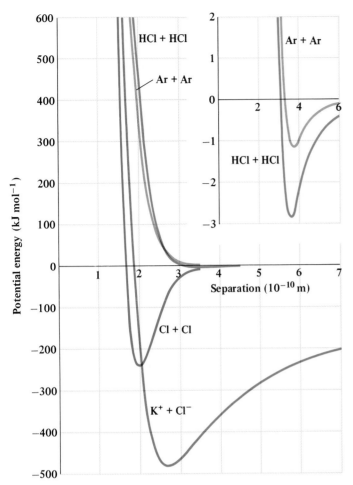

Figure 6-1 The potential energy of a pair of atoms, ions, or molecules depends on the distance between the members of a pair. Here, the potential energy at large separations (to the right side of the graph) is set to zero. As two particles approach each other, the potential energy becomes negative because attractive forces come into effect. The lowest point in each curve occurs at the distance at which attractive and repulsive forces exactly balance. The potential energy at this point is a measure of the strength of the attractive forces. Note the very shallow potential energy minimum for HCl and Ar. The inset shows these same two curves with the vertical scale expanded by a factor of 100. (The HCl–HCl curve was computed for a pair of molecules lying side by side with the H end of one opposite the Cl end of the other.)

sition) in addition to their kinetic energy (energy of motion). Consider the moon and the earth, which attract each other. To move the moon farther from the earth would require energy to overcome the gravitational attraction between the two. After the move, the pair of bodies would have a higher potential energy. Moving the moon closer to the earth would lower the earth–moon potential energy. The lowering of potential energy with closer approach would ultimately be interrupted, however. If the surface of the moon were to contact the surface of the earth, a very strong repulsion would set in to oppose the closer approach of the bodies.

Figure 6–1 shows how the potential energy depends on the center-to-center distance for several pairs of ions, atoms, and molecules. The curves all have a central dip, but the dips differ in width and depth. At infinite separation, the potential energy of any pair of attracting particles equals zero. The dips show that the potential energy becomes *negative* as the two particles approach each other: it requires energy from outside to separate the two against their mutual attraction, and so their potential energy as a pair is less. Eventually, the particles "touch," and the potential energy shoots up as repulsive forces become dominant. The repulsive forces, which are very strong and short range, come from the extreme reluctance of core (non-valence) electrons of different particles to occupy the same region of space.

The bottom of the dip in each curve represents a state in which attractive and repulsive forces balance one another. If left alone, a pair of particles settles down "in contact" at this separation. The deeper the *potential energy well* (as the dip is known), the stronger the interaction between the members of the pair. The ion–ion interaction of K^+ with Cl^- (the red line in Fig. 6–1) starts at longest range and is the strongest on the graph, stronger even than the covalent interaction in Cl_2 (the green line). Both rate as true chemical bonds. The HCl–HCl and Ar–Ar interactions are hundreds of times weaker. They are **nonbonded attractions.** Most nonbonded attractions have potential wells no deeper than about 5 kJ mol^{-1}. Routine collisions with other particles at room temperature easily supply enough energy to lift interacting pairs out of such shallow wells (and so end the interaction). Chemical bonds have potential wells ranging from 100 to more than 1000 kJ mol^{-1} deep. Collisions at room temperature do not break bonds this strong.

Types of Nonbonded Attractions

We distinguish types of nonbonded interactions just as we distinguish types of chemical bonds.

Dipole–Dipole Interactions

Polar molecules interact through dipole–dipole forces. As shown in Figure 6–2, two polar molecules may repel each other as well as attract, depending on their orientation. Dipole–dipole forces are moderately strong but weaker than the very strong

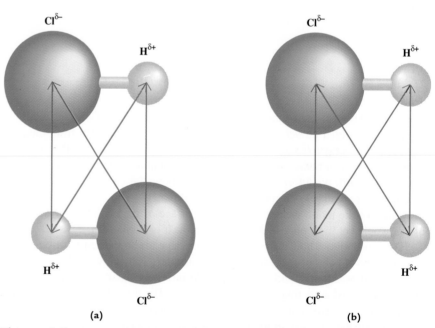

Figure 6–2 A molecule of HCl is a dipole; it has a small net negative charge on the Cl end (symbolized δ^-), and a small net positive charge on the H (symbolized δ^+). The interaction between two HCl molecules depends on their orientations. In (a), the oppositely charged ends (blue arrows) are closer than the ends with the same charge (red arrows). This gives a net attraction. In (b), the opposite is true, and the molecules repel each other.

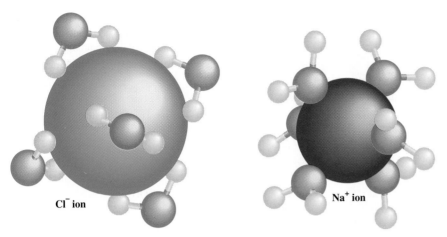

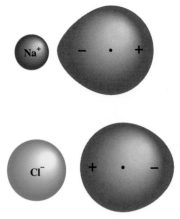

Figure 6–3 Both positive and negative ions are attracted to neighboring water molecules in aqueous solution by ion–dipole forces. The orientations of the water molecules are reversed in the two cases, however. Oxygen atoms in molecules of water bear small negative charges; hydrogen atoms bear small positive charges.

Figure 6–4 As an ion approaches an atom or molecule, its electrostatic field distorts the distribution of the outer electrons. Here the average electron densities are shown by shading. The effect of this distortion is to create a dipole moment that exerts an attractive force back on the ion.

attractions between oppositely charged ions that lead to ionic bonding. Dipole–dipole forces fall off more rapidly with distance than do ion–ion forces.

Ion–Dipole Interactions

A polar molecule interacts with both positive and negative ions: positive ions are attracted by the negative end of the dipole and repelled by the positive end, and the reverse is true for negative ions. These are ion–dipole interactions. The attraction between a polar solvent molecule, such as that of water, and a dissolved ion is the most common case of ion–dipole interaction. In an aqueous solution of NaCl, for example, the Cl^- is surrounded by a group of H_2O dipoles whose positive ends point toward the Cl^- ion, and the Na^+ ion is similarly solvated by a set of H_2O dipoles that are turned the other way around (Fig. 6–3).

• Ion-dipole forces require a mixture of two (or more) substances: one to provide the ions, another to provide the polar molecules.

Induced Dipole Attractions

The electrons in a nonpolar molecule or atom are distributed symmetrically, but an approaching electric charge can distort the symmetry and so induce a dipole. An argon atom, for example, has no dipole moment, but an approaching Na^+ ion (with its positive charge) attracts the electrons on its near side more strongly than the ones on its far side. By tugging the nearer electrons harder, the Na^+ ion induces polarity in the argon atom where none existed before (Fig. 6–4). Once the induced dipole is present, the situation resembles the ion–dipole case just described. Negative ions and other dipoles also induce dipoles in nonpolar molecules. Induced dipole attractions are weak and are effective only at short range.

Dispersion Forces

Helium atoms are uncharged and nonpolar, so nothing mentioned so far explains the observation that helium atoms attract each other weakly but measurably. Such attractions derive from dispersion forces, which exist between all atoms and molecules. They arise from fluctuations over time in the distribution of the electrons on two neighboring atoms or molecules. A fluctuation in one molecule (making a tem-

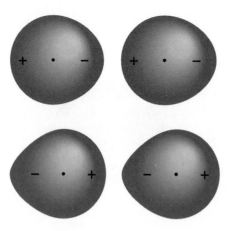

Figure 6–5 A temporary fluctuation of the electron distribution on one atom induces a temporary dipole moment on a neighboring atom. The two dipole moments interact to give a net attractive force, called a "dispersion force."

porary dipole) induces a second temporary dipole in the other. The interaction of these two dipoles then causes an attractive force between the two molecules. Figure 6–5 provides an impression of the source of this interaction. The strength of dispersion forces increases with the number of electrons in the atom or molecule, so larger atoms or molecules interact more strongly than smaller ones. Among dispersion-"bonded" diatomic systems such as He_2, Ne_2, and Ar_2, the interaction grows stronger going down a group in the periodic table. This contrasts with true covalently bonded diatomic molecules, in which the bonding generally becomes weaker going down a group.

The whole set of nonbonded interactions is often referred to as **van der Waals forces** in recognition of the early contributions of Johannes van der Waals to understanding how intermolecular forces influence the physical behavior of solids, liquids, and gases.

Hydrogen Bonding

One particularly strong type of dipole–dipole attraction is the so-called **hydrogen bond.** Hydrogen bonds form when a hydrogen atom that is covalently bonded to a nitrogen, oxygen, or fluorine atom interacts with the lone electron pair of a second such atom nearby. Hydrogen bonds are the strongest kind of nonbonded attractions but are still much weaker than regular covalent and ionic bonds.

• A few cases of hydrogen bonding do involve hydrogen bonded to atoms other than N, O, or F.

For example, water molecules in the gas phase form a weakly associated $(H_2O)_2$ dimer (double molecule) by means of hydrogen bonding. Experimental studies have established the structure of this entity. As Figure 6–6 shows, a hydrogen atom from one water molecule slightly penetrates the "hard edge" of the oxygen atom of the other. This hydrogen in the middle keeps a regular covalent bond with its own oxygen atom but forms a longer, weaker bond (the hydrogen bond) with the second oxygen atom. It lies almost exactly on a straight line joining the two oxygen atoms and significantly closer to the second oxygen than would be expected from the usual steepness of repulsive forces. In the H_2O–H_2O hydrogen bond, the hydrogen atom has a fractional positive charge and is attracted to the fractional negative charge on the neighboring oxygen atom, as would be expected in any dipole–dipole attraction. The interaction is stronger than other dipole–dipole attractions for two reasons. First, the nucleus of a bonded H atom in water is only poorly shielded by electrons because, unlike all other elements, hydrogen has no core electrons. As a result, the bonded H atom can approach very close to the lone-pair electrons on the oxygen atom of a second molecule, leading to a large electrostatic interaction between the two. Second, a small amount of covalent bonding arises from the sharing of electrons between the two oxygen atoms and the intervening hydrogen atom. Covalent bonding explains why the hydrogen atom penetrates the oxygen atom. These details are typical of most hydrogen bonds.

Hydrogen bonds may be intermolecular (between molecules) or intramolecular (within a molecule). The latter shape the structures of many important and complex biological substances, as discussed in Chapter 25. The former are important in the liquid-state and solid-state structures of many simple substances.

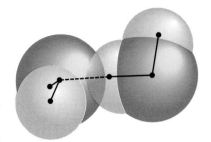

Figure 6–6   A single hydrogen bond linking two otherwise isolated molecules of water. Hydrogen bonds are weaker than ordinary covalent bonds but stronger than other non-bonded attractions: it takes about 23 kJ to break a mole of the bonds shown here. Upwards of 150 kJ is required to break a mole of covalent bonds, but only 2 to 3 kJ is required to overcome a mole of typical dipole–dipole non-bonded attractions.

EXAMPLE 6–1

State which attractive intermolecular forces are likely to predominate in the associations among molecules in the following substances:

(a) $F_2(s)$ (c) $NH_4Cl(s)$

(b) $HBr(\ell)$ (d) $HF(\ell)$

Solution

(a) Molecules of F_2 are nonpolar, so the predominant attractive forces between molecules in $F_2(s)$ are dispersion forces.

(b) The HBr molecule has a permanent dipole moment. The predominant forces between molecules are dipole–dipole. Dispersion forces also contribute to associations, especially because Br is a rather heavy atom.

(c) The ammonium ions are attracted to the chloride ions primarily by ion–ion forces.

(d) Liquid hydrogen fluoride has hydrogen bonds (dipole–dipole attractions) between HF molecules. Dispersion forces contribute as well.

Exercise

List the types of attractive intermolecular forces that exist among the particles in each of these substances: (a) NaH (b) ClBr (c) Rn (d) NH_3.

Answer: (a) Ion–ion forces and dispersion forces. (b) Dipole–dipole forces and dispersion forces. (c) Dispersion forces. (d) Hydrogen bonding (dipole–dipole) and dispersion forces.

6–2   THE KINETIC THEORY OF LIQUIDS AND SOLIDS

The term "intermolecular attractions" is used (somewhat loosely) to refer to all attractions among the constituent particles of a material, whether they are molecules, atoms, or ions. Such attractions are responsible for the existence of solids and liquids. At very high temperatures, the high kinetic energy of the molecules overpowers all attractions: all materials are gaseous at a high enough temperature. Lowering the temperature lowers the average kinetic energy, and intermolecular attractions, which do not change with temperature, gain significance. They cause particles to congregate in small, loose clusters that grow slowly in size until, as the temperature drops further, droplets of liquid suddenly condense. At low enough temperatures, liquids congeal into solids. The temperatures at which liquefaction and freezing occur depend on the strengths of the intermolecular attractions, which naturally vary from substance to substance.

Because each molecule in a liquid or solid interacts strongly with many neighboring molecules, simple equations of state such as the ideal gas equation or the van der Waals equation are not available. A vivid picture of motions at the molecular level in solids and gases has, however, emerged from experiment and calculation.

The Structure of Solids and Liquids

The similar molar volumes of solid and liquid forms of the same substance suggest that the distances between neighboring molecules in the two states must be approximately the same. Density measurements (see Section 1–8) show that the intermolecular contacts, the distances between the nuclei of atoms at the far edge of one molecule and the near edge of a neighbor, usually range between 3×10^{-10} m and 5×10^{-10} m in solids and liquids. At these distances, longer-range attractive forces and shorter-range repulsive forces just balance one another, giving a minimum in the potential energy. Although these intermolecular separations are significantly longer than most chemical bond distances (which range from 0.5 to 2.5 $\times 10^{-10}$ m), they are much shorter than the intermolecular separations in gases, which average around 30×10^{-10} m under room conditions. The distinction is shown schematically in Figure 6–7.

Gases change their volumes dramatically with changes in temperature, but liquids and solids do not. This observation is consistent with the presence of strong intermolecular attractions in the condensed states and their absence in the gaseous state. An increase in volume in a solid or liquid requires that attractive forces between molecules and their neighbors be partially overcome. A solid or liquid has stronger intermolecular attractions and does not expand as much for a given temperature increase as a gas, in which attractions play a much smaller role.

The long stretches of empty space between gas molecules explain why gases are so easily compressed compared with liquids and solids. It costs relatively little in energy to bring widely separated molecules closer together. Attempts to compress

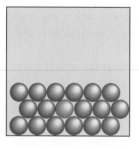 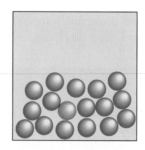 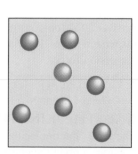

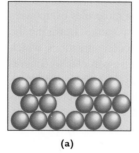 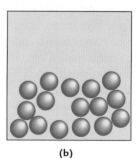 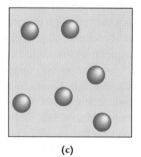

Figure 6–7 Intermolecular forces create structure in liquids and solids. If a single atom is removed from a snapshot of the atomic arrangement in a solid (a), it is easy to figure out exactly where to put it back in. For a liquid (b), the choices are limited. If a single atom is removed from a gas (c), no clue remains to tell where it came from.

(a)   (b)   (c)

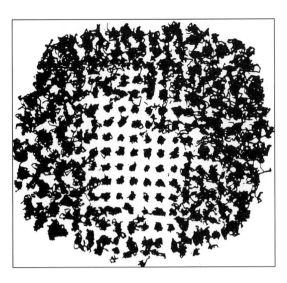

Figure 6–8 In this computer-simulated picture of the motions of atoms in a tiny melting crystal, the atoms at the center (in the solid) move erratically around particular sites. The atoms at the surface (in the liquid) move over much greater distances.

liquids and solids, however, must operate against the strong repulsive forces that come into effect once the molecules are in contact, which accounts for the nearly total incompressibility of liquids and solids.

Motion in Solids and Liquids

Figure 6–7 shows a "snapshot" of a liquid and a solid, fixing the positions of the particles that make it up at a particular instant in time. The paths followed by the particles in these two states can also be examined over a short time interval (Fig. 6–8). In liquids, molecules have the freedom to travel through the sample, changing neighbors constantly in the course of their wanderings. In a solid, the molecules vibrate around their home positions, but mostly they stay close to that position. This explains why liquids flow more or less readily in response to an external stress: their molecules can quickly change neighbors and tumble past each other. The rigidity of solids against external stress suggests, in contrast, a durable arrangement of neighbors around any given particle. This is the crucial difference between the solid and liquid states.

In a liquid, individual molecules experience interactions with neighbors that lead, at any instant, to a local environment that may closely resemble that in a solid, but then they quickly move on. Their trajectories consist of "rattling" motions in a temporary cage formed by neighbors and superimposed upon erratic displacements over larger distances. In this respect, a liquid is intermediate between a gas and a solid. A gas (see Fig. 5–17) provides no temporary cages, so each molecule of a gas travels a longer distance before colliding with a second molecule. Consequently, diffusion is faster in a gas than in a liquid. In a solid, the cages are nearly permanent, so diffusion is very slow. When a solid is heated sufficiently, the thermal energy increases the amplitude of vibration of the molecules in the solid to the point that they are set free to make major excursions; the solid becomes a liquid. This process is called **melting** or **fusion.**

These ideas on motion in liquids and solids fit nicely with the observed rapidity with which liquids can dissolve substances compared with solids. Individual molecules of a liquid quickly wander into contact with those of an added, second substance, and new attractions between the unlike particles have an early chance to replace those existing originally in the pure liquid.

6-3 PHASE EQUILIBRIUM

Recall that a **phase** is a sample of matter that is uniform throughout, both in its chemical constitution and in its physical state (see Section 1–2). Our classification and discussion of the states of matter of a substance have so far focused on one phase at a time (solid, liquid, or gas), but two or more phases can coexist as well (for example, an ice cube floating in liquid water, or mercury vapor confined above a column of liquid mercury in a thermometer). In this section, we examine the properties of coexisting phases of substances and the dynamic processes by which phase equilibrium is reached.

Vapor Pressure

Suppose that we put a small quantity of liquid water in an evacuated flask. We hold its temperature at 25.0°C by placing the flask in a constant-temperature bath, and we use a pressure gauge to monitor how the pressure inside the flask changes with time. Immediately after the water enters the flask, the pressure in the space above it begins to rise from zero. It increases rapidly at first but then levels off at a value of 0.03126 atm, which is the **vapor pressure** of water at 25°C. Liquid water remains visible inside the flask, which now contains water vapor in addition. Nothing new happens as long as the flask is left to itself at 25°C. The **system** (the contents of the flask) has reached **equilibrium,** a condition in which no further changes in macroscopic properties tend to occur. The passage toward equilibrium that was just described is a spontaneous process, occurring in the closed system without any external prompting. If we remove some of the water vapor that has formed, the vapor pressure in the flask is temporarily less than its equilibrium value, but it increases with time to re-establish itself at $P_{vap}(H_2O) = 0.03126$ atm. If we admit air to the flask and then reseal it, $P_{vap}(H_2O)$ stays very close to 0.03126 atm.

What happens on a microscopic scale to cause the spontaneous movement of the system toward equilibrium? According to kinetic theory, the molecules of water in the liquid are in a constant state of thermal motion. Some of those near the surface move fast enough to escape from the liquid into the space above it; this process of **evaporation** leads to the observed increase in pressure. Evaporation is the familiar process by which water (and other liquids) left out in open containers eventually disappear completely. Of course, the water evaporated from a puddle in the sunshine does not cease to exist; rather, it becomes widely dispersed in the air. In our experiment, the closed container keeps the evaporated water in the space above the liquid. As the number of water molecules in the vapor increases, some inevitably are recaptured by the liquid when they chance to strike its surface; this is **condensation.** As the pressure of the vapor increases, the rate of condensation increases until it balances the rate of evaporation from the surface (Fig. 6–9). Once this occurs, there is no further net flow of matter from one phase to the other. The system has reached an equilibrium that is characterized by a particular value of the vapor pressure. Water molecules continue to evaporate from the surface of the liquid, but other water molecules return to the liquid from the vapor at an equal rate.

The vapor pressure of the water is independent of the size and shape of the container. If the experiment is duplicated in a larger (or smaller) flask, then a greater (or lesser) *amount* of water evaporates on the way to equilibrium, but the final pressure in the flask at 25°C is still 0.03126 atm as long as some liquid water is pre-

• It is essential to distinguish between two similar terms: the "pressure of a vapor: and the "vapor pressure of a liquid." At a given temperature, the vapor pressure of a liquid has a unique value (see Chapter 7), but the pressure of a gas may have any one of an infinite number of values.

Figure 6-9 The approach to equilibrium in evaporation and condensation. Initially, the pressure above the liquid is very low, and many more molecules leave the surface of the liquid than return to it. As time passes, more molecules fill the gas phase until the equilibrium vapor pressure (P_{vap}) is approached; the rates of evaporation and condensation then become equal.

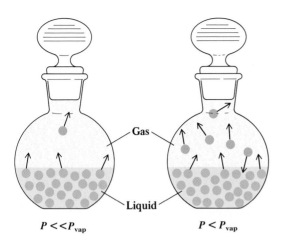

Gas
Liquid

$P \ll P_{vap}$ $P < P_{vap}$ $P = P_{vap}$

• Both of these solids have aromatic vapors and are used in mothballs.

sent. If the experiment is repeated at a temperature of 30.0°C, everything happens as just described, except that the pressure in the space above the water settles at a final value of 0.04187 atm. A higher temperature corresponds in the kinetic theory to a larger average kinetic energy among the water molecules. A new balance between the rates of evaporation and condensation is struck, but at a higher vapor pressure. The vapor pressure of water, and of all other substances, increases with increasing temperature (Fig. 6–10).

The random jostlings of thermal motion in time give a few molecules in a liquid *or* solid whatever speed and direction they need to evaporate. Molecules in all liquids and solids thus have escaping tendencies, and, in principle, all liquids and solids possess vapor pressures. *Volatile* liquids and solids have vapor pressures large enough that they evaporate from open containers reasonably soon at the temperature in question. At room temperature, the liquids water and gasoline are volatile, as are the solids *para*-dichlorobenzene and naphthalene and many other liquids and

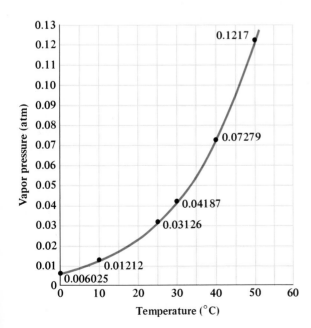

Figure 6-10 The vapor pressure of water, like that of all other substances, increases exponentially with increasing temperature.

solids. Odors come from evaporated molecules that reach the nose, so liquids or solids with a distinguishable smell are volatile or contain a volatile component. Liquid mercury has a vapor pressure at room temperature of only 1.6×10^{-6} atm. Because this is enough to create hazardous local concentrations of mercury vapors, mercury is treated as volatile (kept in tightly capped bottles) despite the fact that open containers of it last for years at room temperature.

Nonvolatile liquids and solids have low vapor pressures. Inorganic salts (NaCl, Na_2SO_4, $CaBr_2$, and so forth) are generally nonvolatile at room conditions, as are most metals and minerals. Some vapor pressures are extremely low. One of the least volatile of all substances is tungsten, which has an immeasurably low vapor pressure at room temperature and a vapor pressure of only 2×10^{-25} atm at 1000°C. This property means that tungsten hardly evaporates even when used as the glowing filament in a light bulb. Very high temperatures, of course, volatilize even tungsten.

• This vapor pressure corresponds to 1 atom of tungsten in a volume of 868 L.

The Dynamic Nature of Phase Equilibrium

The equilibrium coexistence between two phases, such as liquid and vapor, is the time-independent state achieved after the initial effects of the method of preparation of the system have died away. Once a system is at equilibrium, no further macroscopic changes can be seen, but activity continues at the molecular level. Phase equilibrium is a *dynamic* process that is quite different from the static equilibrium reached as a marble rolls to a stop after being spun into a bowl. In the equilibrium between liquid water and water vapor, the partial pressure levels off, not because evaporation and condensation stop, but rather because their rates become equal. In the equilibrium between ice and water, molecules likewise continue to pass between the two phases, although no change is observable macroscopically.

An elegant experiment demonstrates the dynamic nature of phase equilibrium. We inject some liquid water at 25°C into an evacuated flask; at the same time, we inject enough water vapor to give a pressure of 0.03126 atm. (This particular value is chosen because it is the equilibrium vapor pressure of water at this temperature.) We use ordinary liquid water, but instead of ordinary water vapor, which has a natural abundance of only 0.2% of the isotope ^{18}O, we use water vapor that is artificially enriched to 100% in ^{18}O. This "labelled" water is so similar to ordinary water in its physical properties that the equilibrium water-vapor pressure is still 0.03126 atm. By labelling the molecules, however, we are able to determine where they go as time progresses (Fig. 6–11).

After letting the system equilibrate (start toward equilibrium) for a few minutes, we withdraw a sample of water vapor and one of liquid water and use a mass spectrometer (see Section 1–6) to determine their isotopic compositions. We find that the percentage of water labelled with ^{18}O has decreased from its original value in the vapor phase, but the percentage in the liquid phase has increased. If we wait long enough, the two percentages become almost equal. The exchange of labelled molecules between the phases proves that equilibrium is not a static condition in which evaporation and condensation cease; the two processes continue unabated but balance each other at equilibrium.

• Not *exactly* equal because there is a very slight difference between the vapor pressures of $H_2^{16}O$ and $H_2^{18}O$. A larger effect on vapor pressure is seen if hydrogen is replaced by its heavy isotope deuterium.

Another important feature of equilibrium that can be observed from the liquid–vapor system is that its properties are independent of the direction from which it is approached. If we inject enough water vapor into an empty flask to make the initial pressure *higher* than the vapor pressure of liquid water $P_{vap}(H_2O)$, then liquid water condenses until $P_{vap}(H_2O)$ is achieved (0.03126 atm at 25°C). The equi-

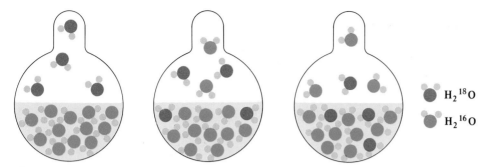

Figure 6–11 In this experiment, the liquid water has the ordinary isotopic composition with only 0.2% ^{18}O atoms, whereas the vapor consists entirely of water with ^{18}O (*left*). As time goes on, the fraction of ^{18}O in the liquid phase increases, whereas that in the gas phase decreases, until eventually the two become almost equal (*right*).

librium is then indistinguishable from one set up starting with liquid water. Of course, if we do not use enough water vapor to exceed a pressure of $P_{vap}(H_2O)$, all the water remains in the vapor phase, and the two-phase equilibrium does not occur.

Correcting for the Vapor Pressure of Water in "Wet" Gases

The vapor pressure of the volatile liquid water has one very practical effect on the quantitative analysis of reaction products. Suppose a chemical reaction is carried out in aqueous solution and a gas is produced. It appears not as "dry" gas but as humid or "wet" gas, containing a partial pressure of gaseous water that equals the equilibrium vapor pressure of liquid water at the temperature of the experiment. The amount of gas produced by the reaction is then properly determined not by the total gas pressure, but by the partial pressure of the product gas, which can be computed by using Dalton's law (see Section 5–6). It equals the total gas pressure minus the vapor pressure of water. This correction for the vapor pressure of water is significant in quantitative work involving the collection of gases over water.

• The equilibrium vapor pressure of water depends only on the temperature. Accurate values are widely available (often quoted in torr or mm Hg). An abridged table follows.

EXAMPLE 6–2

Zinc reacts in dilute aqueous sulfuric acid to produce hydrogen:

$$Zn(s) + 2\ H^+(aq) \longrightarrow Zn^{2+}(aq) + H_2(g)$$

A volume of gas equal to 17.2 mL is collected from the reaction products at a total pressure of 0.9962 atm at 30°C. At this temperature, the vapor pressure of water (assumed to be equal to that of the aqueous solution under study) is 0.0419 atm. Calculate the partial pressure of the hydrogen and the mass of zinc consumed in producing this hydrogen.

Solution

The gas contains hydrogen and gaseous water, so the partial pressure of H_2 is the difference between the total pressure and the vapor pressure of water:

$$P_{H_2} = P_{total} - P_{H_2O} = 0.9962\ atm - 0.0419\ atm = \boxed{0.9543\ atm}$$

Vapor Pressure of Water at Various Temperatures

Temperature (°C)	Vapor Pressure (atm)
15.0	0.01683
17.0	0.01912
19.0	0.02168
21.0	0.02454
23.0	0.02772
25.0	0.03126
30.0	0.04187
50.0	0.1217

From the ideal gas law, the chemical amount of hydrogen in a volume of 17.2 mL ($= 0.0172$ L) at this temperature ($30°C = 303$ K) is

$$n_{H_2} = \frac{PV}{RT} = \frac{(0.9543 \text{ atm})(0.0172 \text{ L})}{0.08206 \dfrac{\text{L atm}}{\text{mol K}}(303 \text{ K})} = 6.60 \times 10^{-4} \text{ mol H}_2$$

The mass of zinc consumed is then

$$m_{Zn} = 6.60 \times 10^{-4} \text{ mol H}_2 \times \left(\frac{1 \text{ mol Zn}}{1 \text{ mol H}_2} \right) \times \left(\frac{65.39 \text{ g Zn}}{1 \text{ mol Zn}} \right)$$

$$= 4.32 \times 10^{-2} \text{ g Zn} = 43.2 \text{ mg Zn}$$

If the vapor pressure of water is ignored, the (incorrect) answer is 45.1 mg, which is about 4% too high.

Exercise

When an electric current is passed through a dilute solution of sodium sulfate in water, hydrogen and oxygen are produced and can be collected:

$$2 \text{ H}_2\text{O}(\ell) \longrightarrow 2 \text{ H}_2(g) + \text{O}_2(g)$$

If this reaction is carried out at 25°C under a total pressure of 0.9969 atm, what mass of oxygen is present in 1.00 L of the wet product gases? Assume that the hydrogen and oxygen are collected in a single container. At 25°C, the vapor pressure of water is 0.0313 atm.

Answer: 0.421 g.

6–4   PHASE TRANSITIONS

Suppose one mole of gaseous sulfur dioxide is compressed, with the temperature fixed at 30.00°C (quite near room temperature). The volume of the gas is measured at each pressure and plotted (Fig. 6–12). At low pressures, the graph shows the inverse dependence that is predicted by the ideal gas law:

$$V = \frac{nRT}{P} = \frac{(1 \text{ mol})\left(0.08206 \dfrac{\text{L atm}}{\text{mol K}} \right)(303.15 \text{ K})}{P} = 24.88 \text{ L atm} \left(\frac{1}{P} \right)$$

As the pressure increases, deviations appear because the gas is nonideal. At a temperature of 30°C, attractive forces are more important in SO_2 than repulsions; therefore, the volume falls below the ideal gas value. It approaches 4.74 L (rather than 5.50 L) as the pressure approaches 4.52 atm.

    For pressures up to 4.52 atm, this behavior is quite regular and can be described by the van der Waals equation. At that pressure, the volume of the sample drops abruptly by a factor of 100 and remains low as the pressure is increased further. What has happened? The gas has been liquefied solely by the application of pressure. Further compression causes another abrupt (but smaller) change in volume as the liquid freezes to a solid (still at a temperature of 30°C). Both events are called **phase transitions.** Phase transitions also occur with changes in temperature at constant pressure. If steam (water vapor) is cooled at 1 atm pressure, it condenses to

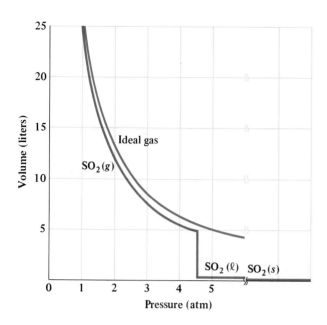

Figure 6–12 As one mole of gaseous SO_2 is compressed at a constant temperature of 30°C, the volume at first falls somewhat below its ideal gas value. Then at 4.52 atm, the volume falls abruptly as the gas condenses to a liquid. At a much higher pressure, a further transition to the solid takes place.

liquid water at 100°C and freezes to solid ice at 0°C. Six phase transitions are conceivable among the three usual states of matter, and all occur (Fig. 6–13).

Boiling and Melting Points

Typically, solids melt to give liquids when they are heated, and liquids boil to give gases. In **boiling,** bubbles of gaseous substance form actively throughout the body of a liquid and rise up to escape at the surface. Boiling is an extension of evaporation; in evaporation, the vapor escapes from the surface only. Only when the vapor pressure of a liquid exceeds the external pressure under which it is being held, can the liquid start to boil. The **boiling point** is the temperature at which the vapor pressure of a liquid equals the external pressure. The external pressure influences boiling points quite strongly; for example, water boils merrily at 25°C if the external pressure is reduced below 0.03126 atm (recall that the vapor pressure of water at 25°C is just this number), but it requires a temperature of 121°C to boil under an

• Cold tap water contains dissolved air. Heating it expels the air as very small bubbles. This is not boiling, although it is sometimes mistaken for boiling.

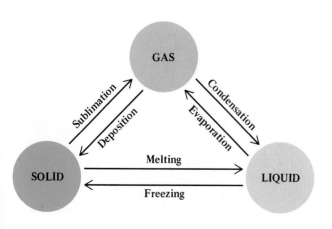

Figure 6–13 Direct transitions between all of the three common states of matter are possible and are observed in everyday life.

external pressure of 2.0 atm. The **normal boiling point** is defined as the temperature at which the vapor pressure of the liquid equals 1 atm.

Melting is the conversion of a solid to a liquid. The **normal melting point** of a solid is the temperature at which solid and liquid are in equilibrium under a pressure of 1 atm of air. The normal melting point of ice is 0.00°C. Liquid water and ice coexist indefinitely (are in equilibrium) at this temperature at a pressure of 1 atm. If the temperature is lowered by even a small amount, then all of the water eventually freezes; if the temperature is raised infinitesimally, then all of the ice eventually melts.

The normal melting and boiling points are members of a standard group of properties that are routinely determined to characterize substances. Determining the melting or boiling point of an unknown can aid in its identification through reference to tables. Caution is required because small amounts of impurities in substances can markedly depress their melting points and elevate their boiling points (see Section 6–6). Another effect of impurities is to blur the sharp melting or boiling points exhibited by pure substances into melting or boiling *ranges*. For this reason, a simple examination of melting or boiling behavior is often a helpful check of the purity of a substance.

Phase transitions are not always prompt but can require some time to take place. This makes it possible to overshoot a phase boundary when changing the pressure or temperature of a sample. An example is the **superheating** of a liquid. Liquid water, for instance, can reach a temperature somewhat above 100°C if heated rapidly. When vaporization of a superheated liquid *does* occur, it can be quite violent, with liquid thrown out of the container. To avoid such unpleasant events in the laboratory, small porous particles (called "boiling chips") may be added to the liquid before it is heated. They help to initiate boiling as soon as the normal boiling point is reached by providing sites where gas bubbles can form. **Undercooling** of liquids below their freezing points is also possible. In very careful experiments, undercooled liquid water has been studied at temperatures below −30°C (at room pressure).

Many materials react chemically when heated, before they have a chance to melt or boil. Such materials include household goods such as wood, cloth, paper, and some plastics (Fig. 6–14). Substances that change their chemical identity before changing state do not have normal melting points or boiling points (or both). Cooking oil, for example, smokes and darkens (signs of chemical decomposition) when heated strongly at room pressure, but it does not boil. Sucrose (table sugar) melts, but it quickly begins to darken and eventually chars at its melting point (Fig. 6–15). A temperature hot enough to overcome the intermolecular attractions in sugar is also sufficient to break apart the molecules of sugar themselves.

• The qualifying term "normal" is often omitted in talking about melting points because they depend only weakly on the pressure.

Figure 6–14 These materials have no melting or boiling points. They undergo chemical changes (decomposition) when heated, before undergoing physical changes (melting or boiling).

• Undercooling is also called "supercooling."

Figure 6–15 When sugar is heated at room pressure, it melts and simultaneously partly decomposes to give a dark-colored "caramelized" mixture.

Intermolecular Forces and Phase Transitions

The recorded normal melting and boiling points of substances have a prodigious range. Among the elements, the lowest melting point is near absolute zero (for helium) and the highest boiling point is over 5600°C (for tungsten).

> The stronger the intermolecular attractions in a liquid, the lower the vapor pressure at any temperature and the higher the boiling point of that liquid.

Substances with strong intermolecular attractions require high temperatures to make their vapor pressures equal 1 atm; therefore, they have high boiling points. For this reason, ionic liquids generally boil at higher temperatures than polar liquids, which in turn generally boil at higher temperatures than nonpolar liquids. In a series of related substances in which the molar mass increases, such as the noble gases, the normal boiling point increases with the increasing strength of the dispersion forces. Thus, radon has the highest boiling point of the noble gases and helium the lowest.

    Melting points depend on molecular shapes and on details of the molecular interactions more than do boiling points, so they vary less systematically with the strength of the attractive forces.

• All of these categories are very large, so there are some exceptions: a high-end nonpolar liquid might have a higher boiling point than a low-end polar liquid.

The Special Case of Water

The physical and chemical properties of water shape every detail of the terrestrial environment. Many of these properties, especially those involving phase transitions, differ sharply from what would be expected by comparison with the properties of other substances of low molar mass. In particular:

1. Water expands when it freezes. Consequently, ice has a lower density than liquid water and floats in it. Most substances contract when they freeze.

2. Water has a boiling point much higher than would be expected from the trend set by the boiling points of H_2S, H_2Se, and H_2Te. It also has an abnormally high melting point.

3. The boiling of water (converting liquid water to steam) requires input of an exceptionally large amount of heat per mole (about 45 kJ mol^{-1}). The reverse transition, the condensation of steam (or water vapor) to liquid water, releases the same large amount of heat.

These properties can be traced to the network of hydrogen bonds in ice and liquid water. A water molecule can accept two hydrogen bonds at its oxygen end and also donate two hydrogen bonds. If all possible hydrogen bonds form in a mole (N_0 molecules) of pure water, then every molecule is surrounded tetrahedrally by four hydrogen atoms—its own two and two from neighboring molecules. The hydrogen bonds number 2 N_0 and form a three-dimensional network. The small number of water molecules surrounding a given water molecule in such a network gives ice an unusually open structure (Fig. 6–16) and a lower density than would be expected in the absence of hydrogen bonds. When ice melts, some hydrogen bonds break, and the open structure of Figure 6–16 partially collapses. The result is a liquid with a smaller volume (higher density), as diagrammed in Figure 6–17. The reverse transformation, the freezing of water, is accompanied by an increase in volume. The 8.3% expansion of water (Fig. 6–18) upon freezing can cause bursting of water pipes and freeze–thaw cracking of rocks and concrete. It has beneficial aspects as well, however. If ice were more dense than water, the winter ice that forms at the surface of

• Not $4N_0$. Consider that every hydrogen bond has a water molecule at each of its ends.

Figure 6-16 The structure of ice. One H atom lies along each of the four lines connecting a given oxygen atom to its neighbors; the oxygen "owns" two of these H's and is hydrogen-bonded to the other two. The structure is quite open because of the hydrogen bonding.

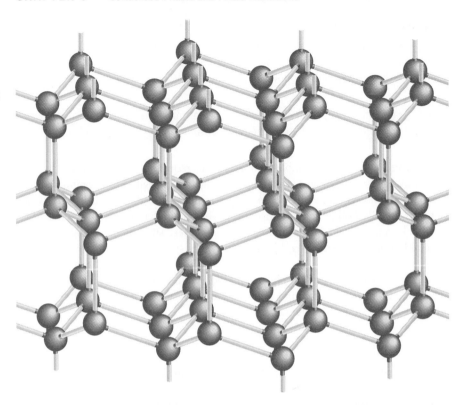

a lake would sink to the bottom, and the lake would freeze solid from the bottom up. Instead, the ice remains at the surface and the liquid water near the bottom achieves a stable wintertime temperature near 4°C, which permits the survival of aquatic life.

Figure 6-17 The density of water rises to a maximum as it is cooled to 3.98°C, and then slowly starts to decrease. Undercooled water (water chilled below its freezing point but not yet converted to ice) continues the smooth decrease in density. When liquid water freezes, the density drops suddenly (and the volume increases). For most substances the density of the liquid increases steadily as temperature is lowered, and the density of the solid is *greater* than that of the liquid.

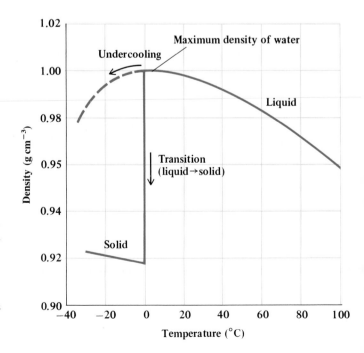

(a)

(b)

Figure 6–18   Unlike most substances, liquid water expands as it freezes. (a) A tightly sealed, brimful jar of water. (b) The expansion caused by freezing the water has shattered the jar. Although this volume change has profound significance in daily life, it still amounts to only about 8.3% of the total volume.

Hydrogen bonding also accounts for the anomalously high boiling point of water: a high temperature is required to overcome the strong attractions among the molecules in liquid water and launch them independently of each other into the gaseous state. An extrapolation of the trend from hydrides that lack hydrogen bonds gives a boiling point for "water without hydrogen bonds" near 150 K ($-123°C$). Life as we know it would not exist if water boiled at this low temperature. The boiling points of the compounds NH_3 and HF are also high compared with those of the hydrides of other elements of Groups V, VI, and VII (Fig. 6–19), and for the same reason.

Finally, hydrogen bonding is responsible for the large amount of heat required to boil or evaporate a quantity of water compared with the same amount of other substances. The extra energy goes to overcome the hydrogen bonds in liquid water,

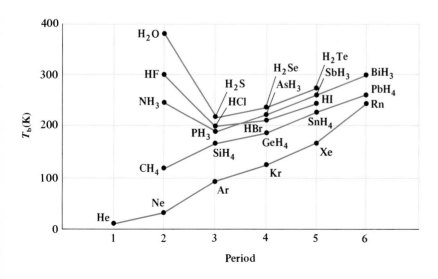

Figure 6–19   Trends in the boiling points of hydrides of some elements of the main group and of the noble gases. The boiling points of HF, H_2O, and NH_3 are all higher than would be expected from the trend in boiling point among related compounds.

many of which persist right up to the boiling point. Evaporation of water from oceans and lakes consumes a large fraction of the sun's energy and protects the environment against major swings in temperature. When large amounts of water condense all at once (as in a rainstorm), the energy previously absorbed in evaporation is released, often into a restricted region of the atmosphere. This energy fuels the violent winds of tornadoes and hurricanes.

EXAMPLE 6–3

Predict the order of increase in the normal boiling points of the following substances from Example 6–1: F_2, HBr, NH_4Cl, and HF.

Solution

As an ionic substance, NH_4Cl should have the highest boiling point of the four. HF should have a higher boiling point than HBr because its molecules form hydrogen bonds (Fig. 6–19) that are stronger than the dipole–dipole interactions in HBr. Fluorine, F_2, is nonpolar and contains light atoms and so should have the lowest boiling point of the four substances.

Exercise

Rank the following from lowest to highest in terms of expected normal boiling point: KCl, Ar, HCl.

Answer: Ar < HCl < KCl.

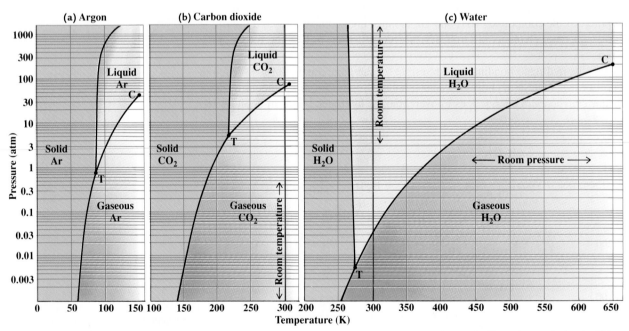

Figure 6–20 In these phase diagrams, the pressure increases by a factor of 10 at regular intervals along the vertical axis. This method of graphing allows large ranges of pressure to be plotted. The marked horizontal and vertical lines are at a pressure of 1 atm and a temperature of 298.15 K, or 25°C (room conditions). Argon and carbon dioxide are gases at room conditions, but water is a liquid. The letter "T" marks the triple points of the substances, and the letter "C" marks their critical points. The region of stability of the liquid is larger for water than for either carbon dioxide or argon.

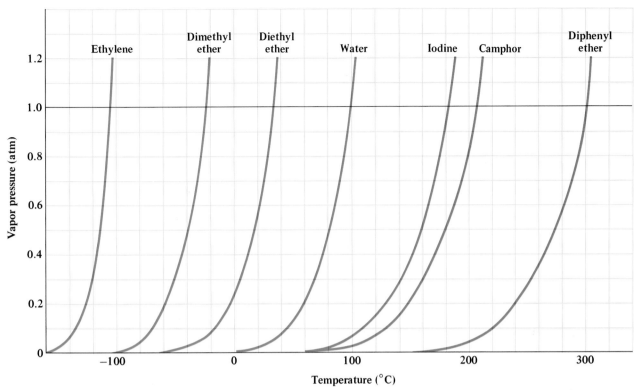

Figure 6–21 The vapor pressure of a solid or liquid depends strongly on temperature. The temperature at which the vapor pressure becomes equal to 1 atm defines the normal boiling point of the liquid and the normal sublimation point of a solid.

6–5 PHASE DIAGRAMS

By keeping the temperature of a substance constant, changing its pressure, and watching for phase transitions, we can determine the range of pressures over which each state of matter (gas, liquid, or solid) of the substance is stable. Repeating the process at many different temperatures gives the data necessary to create the **phase diagram** of that substance: a graph of pressure against temperature that shows the state of matter that is stable for every pressure–temperature combination. The phase diagrams for argon, carbon dioxide, and water are shown in Figure 6–20. A great deal of information can be read from such diagrams. Each substance has a unique combination of pressure and temperature at which the gas, liquid, and solid phases coexist in equilibrium. This combination is called the **triple point** (marked T in Fig. 6–20). Extending from the triple point are three lines, each of which denotes the conditions for the equilibrium coexistence of two phases: solid and gas, solid and liquid, and liquid and gas. The areas between lines denote regions of pressure and temperature in which only one phase exists at equilibrium.

The gas–liquid coexistence curve extends upward in temperature and pressure from the triple point. This line, stretching from T to C in the phase diagrams, is the vapor-pressure curve of the liquid substance. It displays the increasing pressure observed in an enclosed space above the liquid as it is warmed up very slowly. (The slow warming ensures that the liquid and its vapor stay in equilibrium.) Eventually, the vapor pressure reaches 1 atm. This occurs at a temperature equal to the normal boiling point of the liquid. Confined liquids, however, do not boil at their normal boiling points. If they did, the pressure above them would instantly exceed 1 atm and stop the boiling; instead, the vapor pressure just continues to rise. Vapor-pressure curves showing this are collected in Figure 6–21, in which normal boiling points are also indicated.

• It is quite possible for the chemical decomposition of the liquid to intervene and prevent this.

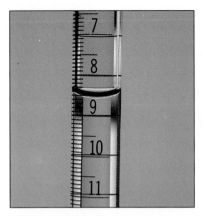

Figure 6–22 When a gas and liquid coexist, the interface between them is clearly visible as a lens-like meniscus. The meniscus between coexisting liquid and gaseous forms of a pure substance disappears at the critical point.

A liquid boils when its vapor pressure *exceeds* the surrounding pressure, which may be artificially maintained at any value (for example, below 1 atm by pumping away the vapor, or above 1 atm by applying an excess pressure). At high elevations, in the mountains, the pressure of the atmosphere is lower than 1 atm, so water boils at a temperature that is less than 100°C. As a result, food cooks more slowly in boiling water at high altitudes. In contrast, the use of a pressure cooker increases the boiling temperature and speeds the cooking of food.

EXAMPLE 6–4

Atop a tall mountain, the atmospheric pressure is 0.50 atm. Use the vapor-pressure curve of water to estimate the boiling point of water under these conditions.

Solution

Find the position on the pressure axis corresponding to 0.5 atm in Figure 6–21 and move horizontally across to the curve for water. Reading directly down from this point locates a temperature of 82°C.

Exercise

Estimate the boiling temperature of diethyl ether ($C_2H_5OC_2H_5$) atop the same mountain (under a pressure of 0.50 atm), using the vapor pressure curve of diethyl ether in Figure 6–21.

Answer: 18°C.

The gas–liquid coexistence curve does not continue forever but instead terminates at a point called the **critical point** (point C in Fig. 6–20). The pressure at the critical point of a substance is its **critical pressure,** and the temperature is its **critical temperature.** All along the liquid–gas coexistence curve, there is an abrupt, discontinuous change in the density and other properties between the two phases. However, the differences between the properties of the liquid and the gas become smaller as the critical point is approached, and they disappear altogether at that point. If a liquid is put in a closed container and gradually heated, a **meniscus** initially appears at the boundary between liquid and gas (Fig. 6–22); at the critical point, this meniscus disappears! In other terms, at temperatures above the critical temperature (373.9°C for water), the distinction between gas and liquid ceases to exist because no amount of compression can cause liquid to appear. When both T and P exceed their critical values, a substance becomes a **supercritical fluid,** so called because "fluid" includes both gases and liquids and "supercritical" means beyond the critical point.

The liquid–solid coexistence curve does not terminate, as the gas–liquid curve does at a critical point, but continues to indefinitely high pressures. In practice, such curves shoot up almost vertically because very large changes in pressure are necessary to change the freezing temperature of a liquid. For most substances, this curve inclines slightly to the right (see Fig. 6–20a, b); an increase in pressure increases the freezing point of the liquid. Another way of saying this is that at constant temperature, an increase in pressure leads to the formation of a phase with higher density (smaller volume for a given mass), and for most substances the solid is denser than the liquid. Water and a few other substances are anomalous (see Fig. 6–20c); for them, the liquid–solid coexistence curve slopes up slightly to the *left*, meaning that an increase in pressure causes the solid to melt. This follows from the anomalous densities of the liquid and solid phases; ice is less dense than water (see Fig. 6–17), so when ice is compressed at 0°C, it melts.

For most substances, including water, atmospheric pressure occurs between the triple-point pressure and the critical pressure. Consequently, in our ordinary experience, we see all three phases—gas, liquid, and solid—if we raise and lower the temperature enough. For a few substances, however, the triple-point pressure lies *above* $P = 1$ atm. In this case, the phase transition called **sublimation** (see Fig. 6–13) occurs as the substance is warmed at atmospheric pressure to take it directly from solid to gas without passing through the liquid state. Carbon dioxide is such a substance (see Fig. 6–20b): its triple-point pressure is 5.117 atm (at a triple-point temperature of $-56.57°C$). Solid carbon dioxide (Dry Ice) sublimes directly to gaseous carbon dioxide at atmospheric pressure. In our daily experience, ordinary ice melts before it evaporates. Ice does sublime at pressures below its triple-point pressure of 0.0060 atm, however. This fact is used in freeze drying, in which foods are frozen and then put in a vacuum chamber at a pressure of less than 0.0060 atm. The ice crystals that formed on freezing then sublime, leaving a dried food (with a much lower mass) that can be reconstituted by adding water.

Many substances exhibit more than one solid phase as the temperature and pressure are varied (Fig. 6–23). At ordinary pressures, the stable state of carbon is graphite, a rather soft dark solid, but at high enough pressures, the hard, transparent form of carbon called diamond is stable. Below 13.2°C (and at room pressure), elemental tin undergoes a slow transformation from the metallic white form to a powdery gray form, a process referred to as "tin disease." No fewer than nine solid forms of ice are known, some of which exist only over a very limited range of temperatures and pressures.

• Some substances other than water that expand on freezing are the elements silicon, gallium, bismuth, and plutonium, and the compound gallium arsenide (GaAs).

EXAMPLE 6–5

Consider a sample of carbon dioxide at 250 K and 500 atm (see Fig. 6–20b).
(a) What phase(s) is (are) present at equilibrium?
(b) Suppose this carbon dioxide is cooled at constant pressure. What happens?
(c) Describe a procedure to convert carbon dioxide originally at 250 K and 500 atm into gas without changing the temperature.

Solution

(a) This combination of T and P is well within the region of the phase diagram labelled liquid, so the sample is a liquid at equilibrium.
(b) A decrease in temperature at constant pressure corresponds to moving to the left in the figure. The liquid carbon dioxide freezes to a solid at approximately 225 K.
(c) If the pressure is lowered sufficiently at constant temperature (below the triple-point pressure of 5.1 atm, for example) the carbon dioxide is converted completely to gas.

Exercise

A sample of solid argon is heated at a constant pressure of 20 atm from 50 K to 150 K. Describe any phase transitions, and give the approximate temperatures at which they occur.

Answer: The solid argon melts completely to a liquid near 85 K, and the liquid vaporizes completely to a gas near 125 K.

Figure 6–23 Red phosphorus and white phosphorus (stored under water in the tube at the top) are two different solid phases of the single substance phosphorus. White and red exist at room conditions because their phase transitions to black phosphorus are very slow. The most stable form of phosphorus under room conditions is yet a third form, black phosphorus (not shown).

An Exotic (But Useful) State of Matter

No known substance has a critical point at or even near atmospheric pressure; all occur at elevated pressures (Fig. 6–A). Consequently, supercritical fluids strike us as peculiar at first and exotic. However, the required temperature and pressure are not hard to attain for many substances; supercritical fluids are routinely prepared and have important uses. Like liquids, supercritical fluids can act as solvents to draw out and dissolve particular substances from mixtures. Moreover, their properties as solvents can be manipulated over a wide range by changing the temperature and pressure. This offers a considerable advantage over ordinary solvents: supercritical fluids easily "de-solvate" from a product when the temperature and

Figure 6–B A rack of synthetic quartz crystals grown from silicon dioxide dissolved in supercritical water.

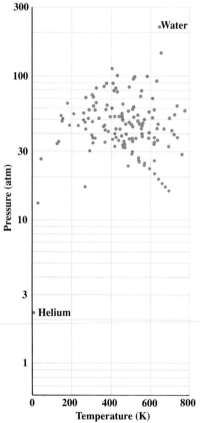

Figure 6–A Each point on this pressure-versus-temperature graph is the critical point of a different substance. Note the critical point of water, which has strong intermolecular attractions, and the critical point of helium, which has the weakest intermolecular attractions of any substance known. All the critical points are at pressures exceeding 1 atm, ordinary room pressure. This fact makes supercritical fluids alien to our experience.

pressure are returned to ordinary values, but traditional solvents are frequently difficult to separate.

In recent years, supercritical fluid extraction has been widely developed in the food, drug, and chemical industries. Caffeine is extracted from green coffee beans by immersing them in supercritical fluid carbon dioxide. (Carbon dioxide has a critical temperature of 31°C and a critical pressure of 73 atm.) Careful control of the temperature and pressure allow nearly complete removal of caffeine into the carbon dioxide while the flavorsome components remain in the coffee. A variation of this technique can be used to extract cholesterol from eggs and fats from potato chips and chicken.

Oxygen dissolves readily in supercritical water. Water, which becomes a *non*polar solvent when it is supercritical, also dissolves many organic pollutants. This makes supercritical water an excellent medium for the destruction of such pollutants by oxygenation. Silicon dioxide, the major component of ordinary sand, dissolves to an extent in supercritical water. Large specimens of crystalline silicon dioxide (quartz) are prepared from these solutions for use in the electronics industry (Fig. 6–B).

6–6 COLLIGATIVE PROPERTIES OF SOLUTIONS

The addition of a solute alters the play of forces among the molecules in a pure solvent. This results in changes in the solvent's vapor pressure, melting point, and boiling point. Remarkably, for dilute solutions, the sizes of these changes depend entirely on the number of solute particles and not at all on the chemical identity of the solute. To a reasonable approximation, every kind of dissolved particle changes the binding together of a solvent's molecules by the same amount! We discuss four **colligative properties** of solutions: the three just mentioned and a fourth, the osmotic pressure. First, however, we introduce some additional measures of solution composition to aid the discussion. These units differ from the molarity (see Section 4–5) but are related to it.

• "Colligative" comes from the Latin word meaning to bind together, fasten, or attach.

Mass Percentage, Mole Fraction, and Molality

Mass percentage (also called weight percentage) is perhaps the simplest way to state the composition of a mixture. For each component

$$\text{mass percentage of the component} = \frac{\text{mass of the component}}{\text{total mass of mixture}} \times 100\%$$

For example, a solution of KCl in water that is 4.3% KCl by mass contains 4.3 g KCl for every 100.0 g of solution.

The **mole fraction** of a given substance in a mixture is the chemical amount (the number of moles) of that substance divided by the total number of moles of all substances present in the mixture. This term is used in the discussion of gas mixtures and Dalton's law in Section 5–6. In a binary mixture containing n_1 mol of species 1 and n_2 mol of species 2, the mole fractions X_1 and X_2 are

$$X_1 = \frac{n_1}{n_1 + n_2}$$

$$X_2 = \frac{n_2}{n_1 + n_2} = 1 - X_1$$

The mole fractions of all the species present in a solution (or any other mixture) must add up to 1. When it is possible to make a clear distinction between solvent and solutes, the custom is to apply the subscript 1 to the solvent and to use higher subscripts to refer to the solutes. A clear distinction is often not possible. If comparable amounts of two liquids like water and ethanol are mixed, the assignment of the labels 1 and 2 is arbitrary.

EXAMPLE 6–6

An antifreeze solution is prepared by mixing 457 g of ethylene glycol ($C_2H_6O_2$) with 521 g of water. Calculate the mass percentage and the mole fraction of ethylene glycol in the mixture. The molar mass of ethylene glycol is 62.07 g mol^{-1}, and the molar mass of water is 18.02 g mol^{-1}.

Solution

The total mass of the mixture is $457 + 521 = 978$ g. The desired mass percentage is

Antifreeze being added to a car's radiator.

$$\text{mass percentage } C_2H_6O_2 = \frac{457 \text{ g}}{978 \text{ g}} \times 100\% = \boxed{46.7\%}$$

The chemical amounts of the ethylene glycol and water are

$$n_{C2H6O2} = 457 \text{ g } C_2H_6O_2 \times \left(\frac{1 \text{ mol } C_2H_6O_2}{62.07 \text{ g } C_2H_6O_2}\right) = 7.363 \text{ mol } C_2H_6O_2$$

$$n_{H2O} = 521 \text{ g } H_2O \times \left(\frac{1 \text{ mol } H_2O}{18.02 \text{ g } H_2O}\right) = 28.91 \text{ mol } H_2O$$

The mole fraction of ethylene glycol is then

$$X_{C2H6O2} = \frac{7.363 \text{ mol}}{7.363 + 28.91 \text{ mol}} = 0.203$$

Exercise

A solution is prepared by mixing 37.9 g of methanol (CH_3OH) with 103.4 g of water. Calculate the mass percentage and mole fraction of methanol.

Answer: 26.8%, 0.171.

• Using *m* for "molality" unfortunately invites confusion with the use of *m* for "mass." To overcome this, using *b* for "molality" has been officially suggested. It is rarely used.

Although the mole fraction can be used to discuss colligative properties, another unit, the **molality,** is more common. The molality is defined as the number of moles of solute per kilogram of *solvent.*

$$\text{molality}_{\text{solute}} = m_{\text{solute}} = \frac{\text{moles solute}}{\text{kilograms solvent}} = \text{mol kg}^{-1}$$

• The similarity between the terms molality and molarity (and molar and molal) often causes confusion.

A 1.0 molal solution of HCl has 1.0 mol of HCl per kilogram of *solvent.* It is *not* the same as a 1.0 *molar* (1.0 M) solution, which contains 1.0 mol of HCl per *liter of solution.* Molarities work admirably for solution stoichiometry (see Section 4–5) but change with temperature (because solutions change their volume when heated or cooled). Molalities, which involve a ratio of masses, do not change with temperature.

Because water has a density of 1.00 g cm^{-3} at 20°C, one liter of water weighs 1.00×10^3 g, or 1.00 kg. It follows that, in dilute aqueous solutions, the number of moles of solute per liter of solution is *approximately* the same as the number of moles per kilogram of water. In such solutions, the molarity and molality have nearly equal values. For nonaqueous solutions, or for concentrated aqueous solutions, this approximate equality fails.

EXAMPLE 6–7

A solution for washing glass is prepared by dissolving 84.6 g of ammonia (NH_3) in 822 g of water, giving a solution with a total volume of 0.945 L. Calculate the molality and the molarity of ammonia (molar mass 17.03 g mol^{-1}) in this solution.

Solution

The chemical amount of ammonia in the solution is

$$n_{NH_3} = 84.6 \text{ g NH}_3 \times \left(\frac{1 \text{ mol NH}_3}{17.03 \text{ g NH}_3} \right)$$

$$= 4.968 \text{ mol NH}_3$$

The molality of the NH_3 is

$$\text{molality}_{NH_3} = m_{NH_3} = \frac{4.968 \text{ mol NH}_3}{822 \text{ g solvent}} \times \left(\frac{1000 \text{ g solvent}}{1 \text{ kg solvent}} \right)$$

$$= \frac{6.04 \text{ mol NH}_3}{\text{kg solvent}}$$

and the molarity of the NH_3 is

$$\text{molarity}_{NH_3} = c_{NH_3} = \frac{4.968 \text{ mol NH}_3}{0.945 \text{ L solution}} = \frac{5.26 \text{ mol NH}_3}{\text{L solution}} = 5.26 \text{ M}$$

As this example shows, the molarity of a concentrated aqueous solution differs markedly from its molality.

Exercise

Suppose that 32.6 g of acetic acid, CH_3COOH, is dissolved in 83.8 g of water, giving a total solution volume of 112.1 mL. Calculate the molality and molarity of acetic acid (molar mass 60.05 g mol^{-1}) in this solution.

Answer: 6.48 molal and 4.84 molar.

Converting Among Mass Percentage, Mole Fraction, Molality, and Molarity

We now have four measures of solution composition: mass percentage, mole fraction, molarity (see Section 4–5), and molality. Figure 6–24 shows how to convert from one measure to any of the others. For example, the masses that define the mass percentage can be converted into chemical amounts by using the molar masses (see center of figure) and the chemical amounts then used to compute the mole fractions as in Example 6–6. The molality, on the other hand, is obtained by setting up a ratio of a chemical amount (the number of moles of solute) to a mass (the mass of the solvent in kilograms).

EXAMPLE 6–8

A concentrated solution of dextrose ($C_6H_{12}O_6$) in water is 75.3% glucose by mass. Calculate the mole fraction and the molality of dextrose in this solution.

Solution

The mole fraction and molality are the same regardless of the size of the sample. For convenience, then, consider 100.0 g of the solution. It contains 75.3 g of dextrose and 24.7 g of water. The chemical amounts of each are

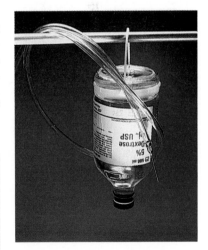

An intravenous feeding liquid is a saline (salt) solution containing dextrose.

$$n_{C_6H_{12}O_6} = 75.3 \text{ g } C_6H_{12}O_6 \times \left(\frac{1 \text{ mol } C_6H_{12}O_6}{180.16 \text{ g } C_6H_{12}O_6} \right) = 0.418 \text{ mol } C_6H_{12}O_6$$

$$n_{H_2O} = 24.7 \text{ g } H_2O \times \left(\frac{1 \text{ mol } H_2O}{18.02 \text{ g } H_2O} \right) = 1.371 \text{ mol } H_2O$$

The total number of moles is $1.371 + 0.418 = 1.789$ mol, and the mole fraction of dextrose is

$$X_{C_6H_{12}O_6} = \frac{0.418 \text{ mol}}{1.789 \text{ mol}} = 0.234$$

The mole fraction of water is $1.000 - 0.234 = 0.766$. Although dextrose is the main component of the mixture on a mass basis, water is the main component on the basis of chemical amount.

The molality of the dextrose is

$$m_{C_6H_{12}O_6} = \frac{0.418 \text{ mol } C_6H_{12}O_6}{0.0247 \text{ kg solvent}} = \frac{16.9 \text{ mol } C_6H_{12}O_6}{\text{kg solvent}}$$

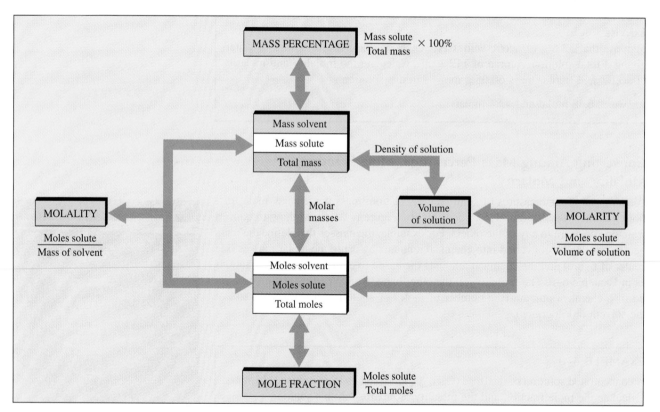

Figure 6–24 Conversions between different measures of solution composition. Three of these (mass percentage, mole fraction, and molality) are completely interconvertible: once one is known, the other two can be calculated by following the arrows. Calculations relating molarity to the other concentration units require a knowledge of the density of the solution (not the solvent) as well. In a dilute aqueous solution, this density can be taken to be 1.00 g cm^{-3}.

Exercise

An aqueous solution of sulfuric acid is 96.00% H_2SO_4 by mass. Calculate the mole fraction of H_2O in the solution and the molality of the H_2O (taking sulfuric acid as the solvent).

Answer: 0.185, 2.31 mol kg^{-1}.

Conversions between molarity and the other three measures of composition are slightly more difficult because, as Figure 6–24 shows, the volume of the solution is needed, not the mass. Knowing the density of the solution allows one to calculate its volume from its mass, but solution densities must be measured; they cannot be deduced from how the solution was put together. Conversions to and from molarity are illustrated by the following examples.

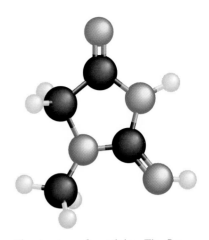

The structure of creatinine. The five-membered ring is nearly flat. The molecule is polar enough that creatinine is substantially soluble in water.

EXAMPLE 6–9

Creatinine ($C_4H_7N_3O$) is a component of muscle tissue and blood. An 8.00% (by mass) solution of creatinine in water is prepared, and measurements establish its density to be 1.0194 g cm^{-3}. Calculate the molarity of creatinine (which has a molar mass of 113.12 g mol^{-1}) in this solution.

Solution

Imagine that 100.00 g of solution is prepared. This requires 8.00 g of creatinine and $100.00 - 8.00 = 92.00$ g of H_2O. Molarity is defined as the ratio of the chemical amount of solute in moles to the volume of the solution in liters. Obtain the first of these two quantities by converting the mass of the creatinine to its chemical amount:

$$8.00 \text{ g creatinine} \times \left(\frac{1 \text{ mol creatinine}}{113.12 \text{ g creatinine}} \right) = 0.0707 \text{ mol creatinine}$$

Next, calculate the volume of the solution in liters by using its mass and density:

$$100.0 \text{ g solution} \times \left(\frac{1 \text{ cm}^3 \text{ solution}}{1.0194 \text{ g solution}} \right) \times \left(\frac{1 \text{ mL solution}}{1 \text{ cm}^3 \text{ solution}} \right) \times$$

$$\left(\frac{1 \text{ L solution}}{1000 \text{ mL solution}} \right) = 0.09810 \text{ L}$$

The molarity of creatinine is then the first of these two quantities divided by the second:

$$c_{\text{creatinine}} = \frac{0.0707 \text{ mol creatinine}}{0.09810 \text{ L solution}} = 0.721 \text{ mol L}^{-1}$$

Exercise

A solution prepared by mixing 20.00 g of $CdCl_2$ with 80.00 g of water has a density at 20°C of 1.1988 g cm^{-3}. Compute the molarity and molality of $CdCl_2$ in this solution.

Answer: 1.308 mol L^{-1} and 1.364 mol kg^{-1}.

EXAMPLE 6–10

A 4.98 M aqueous solution of sulfuric acid has a density of 1.2855 g cm^{-3}. Calculate the molality of sulfuric acid in this solution.

Solution

Suppose that we have exactly one liter of solution. The mass of this solution is

$$1.0000 \text{ L solution} \times \left(\frac{1000 \text{ cm}^3 \text{ solution}}{1 \text{ L solution}} \right) \times \left(\frac{1.2855 \text{ g solution}}{1 \text{ cm}^3 \text{ solution}} \right) = 1285.5 \text{ g}$$

The chemical amount of H_2SO_4 in the one liter of solution is 4.98 mol, so its mass is

$$\text{mass}_{H_2SO_4} = 4.98 \text{ mol } H_2SO_4 \times \left(\frac{98.08 \text{ g } H_2SO_4}{1 \text{ mol } H_2SO_4} \right) = 488.4 \text{ g } H_2SO_4$$

The solution consists entirely of H_2O and H_2SO_4. The mass of H_2O is then the total mass minus the mass of the H_2SO_4:

$$\text{mass}_{H_2O} = 1285.5 \text{ g} - 488.4 \text{ g} = 797.1 \text{ g}$$

We know the chemical amount of the solute (H_2SO_4) and the mass of the solvent. The final step is to apply the definition of molality:

$$\text{molality}_{H_2SO_4} = \frac{4.98 \text{ mol } H_2SO_4}{0.7971 \text{ kg solvent}} = \frac{6.25 \text{ mol } H_2SO_4}{\text{kg solvent}}$$

Exercise

An aqueous solution is 3.34 M in HNO_3 and has a density of 1.1087 g cm^{-3}. Compute the molality of the HNO_3 in this solution.

Answer: 3.72 mol kg^{-1}.

Ideal Solutions and Raoult's Law

Consider a solution made by dissolving a single nonvolatile solute in a solvent. By "nonvolatile," we mean that the vapor pressure of the solute above the solution is negligible; an example is a solution of sucrose (cane sugar) in water, in which the vapor pressure of sucrose above the solution is almost zero.

The *solvent* vapor pressure is not zero and can be studied as the composition of the solution is changed at a fixed temperature. If the mole fraction of solvent (X_1) is 1, then the vapor pressure is P_1°, the vapor pressure of pure solvent at the temperature of the experiment. When X_1 approaches zero (which corresponds to pure solute), the vapor pressure P_1 of the solvent must go to zero also, because solvent is no longer present. As the mole fraction X_1 changes from 1 to zero, P_1 drops from P_1° to zero. What is the shape of the curve of P_1 versus X_1?

The French chemist François Marie Raoult found that for some solutions a plot of the vapor pressure of the solvent against its mole fraction comes very close to being a straight line (Fig. 6–25). Solutions that conform to this straight-line relationship obey the simple equation.

• We consider the case of a volatile solute in Section 6-7.

$$P_1 = X_1 P_1^\circ$$

which is known as **Raoult's law.** Such solutions are called **ideal solutions.** Only ideal solutions obey Raoult's law over the whole range of X_1, but *all* solutions with nondissociating solutes approach the behavior predicted by Raoult's law as X_1 approaches 1 (that is, as X_2 approaches zero), just as all real gases obey the ideal gas law at sufficiently low densities. Solutions that deviate from straight-line behavior are called **nonideal solutions.** They may show negative deviations (with vapor pressures lower than those predicted by Raoult's law) or positive deviations (with higher vapor pressures).

On a molecular level, an ideal solution is one in which the solute–solvent attractions are equal in strength to the solvent–solvent attractions that existed before any solute was added. Negative deviations from ideality arise when the solute attracts solvent molecules especially strongly, reducing their tendency to escape into the vapor phase. Positive deviations result from the opposite case, in which solvent and solute molecules are not strongly attracted to each other, and the attractive forces between solvent molecules are disrupted by the presence of solute.

Raoult's law helps in explaining the four colligative properties.

Lowering of Vapor Pressure

The *change* in the vapor pressure of the solvent when a nonvolatile solute is added is

$$\Delta P_1 = P_1 - P_1^\circ$$

Raoult's law consists of an expression for P_1. Substituting it into the expression for ΔP_1 gives

$$\Delta P_1 = X_1 P_1^\circ - P_1^\circ$$

Because $X_1 = 1 - X_2$, we can rewrite this as

$$\Delta P_1 = X_1 P_1^\circ - P_1^\circ = (X_1 - 1)P_1^\circ = -X_2 P_1^\circ$$

The negative sign appears because the vapor pressure of solvent above a dilute solution is always *lower* than that above the pure solvent; that is, ΔP_1 is always negative.

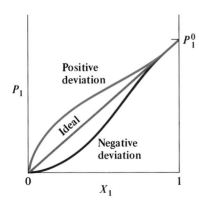

Figure 6–25 A graph of the vapor pressure of the solvent (P_1) versus the mole fraction of solvent (X_1) is a straight line for an ideal solution. Nonideal solutions behave differently; examples of positive and negative deviations from the behavior of an ideal solution are shown. The vapor pressure of pure solvent is P_1°.

EXAMPLE 6–11

At 25°C, the vapor pressure of pure benzene is $P_1^\circ = 0.1252$ atm. Suppose 6.40 g of the solid hydrocarbon naphthalene, $C_{10}H_8$ (molar mass 128.17 g mol^{-1}) is dissolved in 78.0 g of pure benzene (molar mass 78.0 g mol^{-1}). Calculate the vapor pressure of benzene over the solution.

Solution

The chemical amount of solvent n_1 is 1.00 mol in this case (because we have used 78.0 g = 1.00 mol benzene), and the chemical amount of solute is

$$n_2 = 6.40 \text{ g } C_{10}H_8 \times \left(\frac{1.00 \text{ mol } C_{10}H_8}{128.17 \text{ g } C_{10}H_8} \right) = 0.0499 \text{ mol } C_{10}H_8$$

so the mole fraction X_1 of solvent is

$$X_1 = \frac{n_1}{n_1 + n_2} = \frac{1.00 \text{ mol}}{(1.00 + 0.0499) \text{ mol}} = 0.9525$$

According to Raoult's law, then, the vapor pressure of benzene above the solution is

$$P_1 = X_1 P_1^\circ = 0.9525 \times 0.1252 \text{ atm} = \boxed{0.1193 \text{ atm}}$$

Exercise

The vapor pressure of water at 80°C is 0.4672 atm. Determine how much sucrose ($C_{12}H_{22}O_{11}$) must be dissolved in 180.2 g of water at 80°C to reduce its vapor pressure to 0.4640 atm.

Answer: 24 g of sucrose.

Elevation of the Boiling Point

The normal boiling point of a pure liquid or a solution is defined as the temperature at which its vapor pressure reaches 1 atm. Because a dissolved solute reduces the vapor pressure, the temperature of the solution must be increased to bring it to a boil (Fig. 6–26); that is, the boiling point of a solution of a nonvolatile solute in a volatile solvent always exceeds that of the pure solvent.

For sufficiently dilute solutions, the elevation of the boiling point ΔT_b is proportional to the concentration of solute in the solution. Because temperature changes are involved, molality m (moles of solute per kilogram of solvent), which is independent of temperature, is used to express the composition of the solution. We write

$$\Delta T_b = K_b m$$

in which K_b is a constant of proportionality. This equation is quite useful because K_b is a property of the solvent only; it does not depend on the identity of the solute. As a result, K_b can be obtained by measuring the elevation of the boiling point for dilute solutions of known molality and tabulated for later use. The following example shows how a K_b is determined.

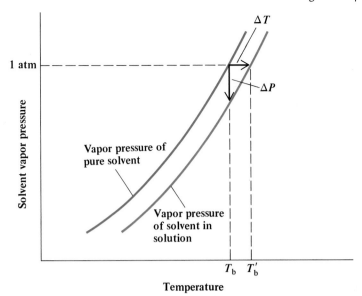

Figure 6-26 The vapor pressure of the solvent above a dilute solution is lower than that of the pure solvent at all temperatures. As a result, in order for the solution to boil (that is, for the vapor pressure to reach 1 atm), a higher temperature is required than for the pure solvent. This amounts to elevation of the boiling point.

EXAMPLE 6-12

When 5.50 g of biphenyl ($C_{12}H_{10}$), a nonvolatile compound, is dissolved in 100.0 g of benzene, the boiling point of the solution exceeds the boiling point of pure benzene by 0.903°C. What is K_b for benzene?

Solution

Because the molar mass of biphenyl is 154 g mol^{-1}, 5.50 g of biphenyl contains

$$5.50 \text{ g biphenyl} \times \left(\frac{1 \text{ mol biphenyl}}{154 \text{ g biphenyl}} \right) = 0.0357 \text{ mol biphenyl}$$

The molality of the biphenyl

$$m_{C_{12}H_{10}} = \frac{\text{mol solute}}{\text{kg solvent}} = \frac{0.0357 \text{ mol}}{0.1000 \text{ kg}} = 0.357 \text{ mol kg}^{-1}$$

$$K_b = \frac{\Delta T_b}{m} = \frac{0.903 \text{ K}}{0.357 \text{ mol kg}^{-1}} = \boxed{2.53 \text{ K kg mol}^{-1}} \text{ for benzene}$$

Exercise

When 4.58 g of nonvolatile picric acid ($C_6H_3N_3O_7$) dissolves in 240.0 g of chloroform, the normal boiling point is elevated by 0.302°C. Determine K_b for chloroform.

Answer: 3.63 K kg mol^{-1}.

Table 6–1
Boiling-Point Elevation and Freezing-Point Depression Constants

Solvent		$T_b(°C)$	K_b (K kg mol^{-1})	$T_f(°C)$	K_f (K kg mol^{-1})
Acetic acid	CH_3COOH	118.1	3.07	17	3.9
Benzene	C_6H_6	80.1	2.53	5.5	4.9
Carbon tetrachloride	CCl_4	76.7	5.03	−22.9	32
Diethyl ether	$C_4H_{10}O$	34.7	2.02	−116.2	1.8
Ethanol	C_2H_5OH	78.4	1.22	−114.7	—
Naphthalene	$C_{10}H_8$	—	—	80.5	6.8
Water	H_2O	100.0	0.512	0.0	1.86

Table 6–1 lists values of K_b for a number of solvents. Once K_b has been found, it can be used to predict the elevations of the boiling point caused by solutes of known molar mass. A far more important application is the determination of the molar masses of unknown substances from measurements of how much they elevate the boiling point when they go into solution. This is illustrated in the following example.

EXAMPLE 6–13

When 6.30 g of a nonvolatile hydrocarbon of unknown molar mass is dissolved in 150.0 g of benzene, the boiling point of the solution is 80.696°C. The boiling point of pure benzene is 80.099°C. What is the molar mass of the hydrocarbon?

Solution

Calculate the change in the boiling point and use the fact that any *change* in temperature is numerically the same in °C and K.

$$\Delta T_b = 80.696°C - 80.099°C = 0.597°C = 0.597 \text{ K}$$

Solving $\Delta T_b = K_b m$ for m gives

$$m = \frac{\Delta T_b}{K_b} = \frac{0.597 \text{ K}}{2.53 \text{ K kg mol}^{-1}} = 0.236 \text{ mol kg}^{-1}$$

The product of the molality of the solution and the mass of solvent is the chemical amount of solute:

$$\left(\frac{0.236 \text{ mol solute}}{\text{kg solvent}} \right) \times 0.150 \text{ kg solvent} = 0.0354 \text{ mol solute}$$

Finally, the molar mass of the solute is its mass divided by its chemical amount in the solution:

$$\mathcal{M}_{solute} = \frac{6.30 \text{ g solute}}{0.0354 \text{ mol solute}} = \frac{178 \text{ g solute}}{\text{mol solute}}$$

The unknown hydrocarbon might be anthracene ($C_{14}H_{10}$), which has a molar mass of 178.24 g mol^{-1}.

The Effect of Dissociation

So far we have considered only nondissociating solutes. If a solute dissociates (as does NaCl, when it dissolves in water to furnish Na^+ and Cl^- ions), then the effective number of solute particles increases. Because colligative properties depend on the number of dissolved particles, the equations describing them must be adjusted to take dissociation into account. The required change is the insertion of an additional factor, the **van't Hoff i.** The equation for boiling point elevation becomes

$$\Delta T_b = i K_b\, m$$

The i equals the number of particles released into solution per formula unit of solute. Its minimum value equals 1 (no dissociation); its maximum (symbolized by ν, the Greek letter nu) equals the number of particles generated by complete dissociation of a formula unit of solute. Values of ν for common electrolytes are usually obvious from the formula and name; thus ν clearly equals 2 for NaCl and 3 for K_2SO_4. Because aquated ions of opposite charge associate into ion-pairs, often to a significant extent, the van't Hoff i for an electrolyte in water is usually less than ν. The amount of association varies with the concentration of the solution, so i must be determined experimentally.

• The van't Hoff i is named after Jacobus van't Hoff, the Dutch chemist who introduced it.

EXAMPLE 6–14

Lanthanum(III) chloride ($LaCl_3$), dissociates into ions as it dissolves in water:

$$LaCl_3(s) \longrightarrow La^{3+}(aq) + 3\,Cl^-(aq)$$

Suppose 0.2453 g of $LaCl_3$ is dissolved in 100.0 g of H_2O. What is the boiling point of the solution at atmospheric pressure, assuming no association among ions and that the solution behaves ideally?

Solution

Use the molar mass of $LaCl_3$ to convert from the mass of $LaCl_3$ to its chemical amount:

$$n_{LaCl_3} = 0.2453 \text{ g } LaCl_3 \times \frac{1 \text{ mol } LaCl_3}{245.3 \text{ g } LaCl_3} = 0.001000 \text{ mol } LaCl_3$$

This amount of $LaCl_3$ is dissolved in 0.1000 kg of water. The molality of $LaCl_3$ is therefore

$$\text{molality}_{LaCl_3} = m_{LaCl_3} = \frac{0.001000 \text{ mol } LaCl_3}{0.1000 \text{ kg solvent}} = \frac{0.01000 \text{ mol } LaCl_3}{\text{kg solvent}}$$

The equation for the dissolution shows that ν is 4 (each mole of $LaCl_3$ gives a maximum of four moles of particles, three of Cl^- ions and one of La^{3+} ions), and we assume that dissociation is complete. Hence $i = \nu = 4$. Substitution in the equation for boiling point elevation gives

$$\Delta T_b = i\,K_b\,m = 4(0.512 \text{ K kg mol}^{-1})(0.01000 \text{ mol kg}^{-1}) = 0.0205 \text{ K}$$

The predicted boiling point of the solution is ΔT_b (converted from K to °C) added to exactly 100°C, or 100.0205°C. The actual boiling point is slightly lower than this because the solution is nonideal, and ion-pairs in fact do form.

Exercise

Lead(II) nitrate dissociates to Pb^{2+} and NO_3^- ions in aqueous solution. Determine ν, and, assuming no association among ion and that the solution is ideal, calculate the boiling point of a solution prepared by mixing 0.194 mol of $Pb(NO_3)_2$ and 1.171 kg of water.

Answer: $\nu = 3$; boiling point = 100.254°C.

Depression of the Freezing Point

The addition of a solute causes the boiling point of a solution to increase, but it causes the freezing point to *decrease*. We consider here only the simplest case of freezing-point depression, that in which the solid freezing out of solution is the pure solvent. Cases in which solute crystallizes out in combination with solvent are more complicated.

For sufficiently small concentrations of a nondissociating solute, the freezing-point depression ΔT_f is related to the molality by the equation

$$\Delta T_f = -K_f m$$

where K_f is another positive constant that depends only on the properties of the solvent (Table 6–1). This formula is quite similar to the one describing boiling-point elevation; the minus sign appears here because the change in freezing point ΔT_f (defined as the freezing point of the solution minus that of the pure solvent) is always negative; a solution always has a *lower* freezing temperature than the pure solvent. For dissociating solutes, the preceding equation becomes

$$\Delta T_f = -iK_f m \quad \text{(for dissociating solutes)}$$

where the i again accounts for the extra particles created in solution by the dissociation.

The phenomenon of freezing-point depression explains the fact that seawater, which contains dissolved salts, has a lower freezing point than fresh water. Concentrated salt solutions have still lower freezing points. Spreading salt on an icy road creates a solution with a lower freezing point than pure water and causes the ice to melt. Antifreeze added to a car radiator lowers the freezing point of the coolant and keeps the cooling system from freezing (and perhaps cracking the engine block) in winter.

Measurements of the depression of the freezing point, like those of the elevation of the boiling point, can be used to determine molar masses of unknown substances.

EXAMPLE 6-15

The chemical amounts of the major dissolved species in a 1.000-L sample of seawater are given below. Estimate the freezing point of the seawater, given that $K_f = 1.86$ K kg mol^{-1} for water and assuming that no association occurs among the dissolved ions.

$$Na^+, 0.458 \text{ mol} \quad Cl^-, 0.533 \text{ mol}$$
$$Mg^{2+}, 0.052 \text{ mol} \quad SO_4^{2-}, 0.028 \text{ mol}$$
$$Ca^{2+}, 0.010 \text{ mol} \quad HCO_3^-, 0.002 \text{ mol}$$
$$K^+, 0.010 \text{ mol} \quad Br^-, 0.001 \text{ mol}$$
$$\text{Neutral species, } 0.001 \text{ mol}$$

Solution

Because water has a density of 1.00 g cm^{-3}, one liter of water weighs 1.00 kg. For dilute aqueous solutions, the number of moles per kilogram of solvent (the molality) is therefore approximately equal to the number of moles per liter. The total molality, obtained by adding up the molalities of the individual species given above, is $m = 1.095$ mol kg^{-1}. Then

$$\Delta T_f = -K_f m = -(1.86 \text{ K kg mol}^{-1})(1.095 \text{ mol kg}^{-1}) = -2.04 \text{ K}$$

The freezing point of pure water is 0.00°C, so the seawater should freeze at approximately −2°C. The actual freezing point is slightly *higher* than this because the solution is nonideal and association among the ions reduces the effective number of particles somewhat.

Exercise

An aqueous solution is 0.040 molal in $CaCl_2$, 0.030 molal in NaCl, and 0.100 molal in sucrose. Assuming no association among the ions, predict the freezing point of this solution.

Answer: −0.52°C.

Both freezing-point depression and boiling-point elevation can be used to determine whether a species of known molar mass dissociates in solution (see Fig. 6–27), as the following example shows.

EXAMPLE 6-16

When 0.494 g of $K_3Fe(CN)_6$ is dissolved in 100.0 g of water (Fig. 6–28), the freezing point is found to be −0.093°C. How many ions are present for each formula unit of $K_3Fe(CN)_6$ dissolved?

Solution

The problem in effect asks for computation of i for this solution. Because the molar mass of $K_3Fe(CN)_6$ is 329.25 g mol^{-1}, the chemical amount in solution is

$$0.494 \text{ g } K_3Fe(CN)_6 \times \left(\frac{1 \text{ mol } K_3Fe(CN)_6}{329.25 \text{ g } K_3Fe(CN)_6} \right) = 0.00150 \text{ mol}$$

and

$$m_{K_3Fe(CN)_6} = \frac{0.00150 \text{ mol}}{0.1000 \text{ kg solvent}} = 0.0150 \text{ mol kg}^{-1}$$

We solve the freezing-point depression equation for i and substitute the values of K_f, the change in temperature, and the molality

$$i = \frac{-\Delta T_f}{K_f m} = -\frac{(-0.093 \text{ K})}{1.86 \text{ K kg mol}^{-1}(0.0150 \text{ mol kg}^{-1})} = 3.3$$

Thus, 1 formula unit of $K_3Fe(CN)_6$ generates an apparent 3.3 mol of ions in this solution. In fact, $K_3Fe(CN)_6$ dissolves as follows

$$K_3[Fe(CN)_6](s) \longrightarrow 3 \text{ K}^+(aq) + [Fe(CN)_6]^{3-}(aq)$$

According to this equation, the coordination complex of iron with six cyanide ions does not dissociate, and ν equals 4. The fact that i is less than ν means that the aquated ions associate.

Exercise

Mannitol has a molar mass of 182.17 g mol^{-1}. When 1.00 g of mannitol is dissolved in 99.00 g of water, the freezing point of the resulting solution is $-0.103°C$. Determine how many moles of particles are formed in solution by mannitol for each mole dissolved.

Answer: One.

Figure 6–27 The heavy, blue lines give the *observed* depression of the freezing point of water by acetic acid, NaCl, and FeCl$_3$ as the molalities of the solutions increase. Straight black lines sketch the predicted ideal behavior for $\nu = 1, 2, 3,$ and 4. The experimental curve for NaCl (which gives *two* particles per mole dissolved) stays close to the ideal straight line for $\nu = 2$; the experimental curve for FeCl$_3$ (which gives four particles per mole dissolved) stays fairly close to the ideal straight line for $\nu = 4$. The pattern suggests that acetic acid dissolves to give one mole of particles per mole of solute.

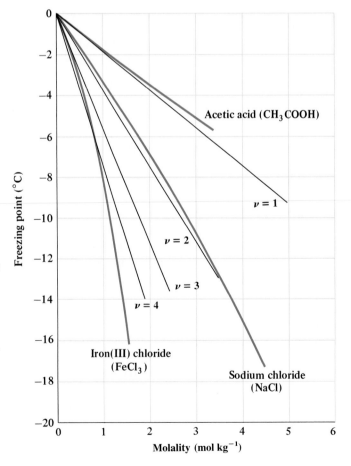

The fractional **degree of dissociation** α of a dissociating solute is given by the equation

$$\alpha = \frac{(i-1)}{(\nu-1)}$$

The denominator $(\nu-1)$ of the fraction equals the number of *extra* particles released by complete dissociation of one formula unit of the solute. "Extra" does not count the one particle released per formula unit in the case of no dissociation. The numerator $(i-1)$ equals the effective number of extra particles actually present. The closer $i-1$ comes to $\nu-1$, the more complete the dissociation, that is, the less ion pairing is taking place. In Example 6–16,

$$\alpha = \frac{(i-1)}{(\nu-1)} = \frac{(3.3-1)}{(4-1)} = 0.76$$

Values of the van't Hoff i and the degree of dissociation are given in Table 6–2 for several electrolytes at different concentrations. The data confirm that dilution works against the association of ions in solution.

Figure 6–28 A solution of potassium ferricyanide ($K_3Fe(CN)_6$) has a reddish color. By contrast, a solution of potassium ferrocyanide ($K_4Fe(CN)_6$) has a lemon yellow color. Iron is in oxidation states $+3$ and $+2$, respectively, in these compounds.

Osmotic Pressure

The fourth colligative property, osmotic pressure, plays a vital role in the transport of molecules across cell membranes in living organisms. Such membranes are **semipermeable.** They allow small molecules, such as those of water, to pass through, but they block the passage of large molecules, such as those of proteins or carbohydrates. Semipermeable membranes are fairly common natural and synthetic materials; a sheet of cellophane is an example. An ideal semipermeable membrane passes solvent molecules only and prevents the passage of every kind of solute molecules. Real membranes are less than ideally efficient but still can be used to separate solutes from solvents.

Suppose that one end of an open-ended glass tube is covered with a semipermeable membrane, clamped vertically with the membrane end down, and filled with a solution. When the membrane end of the tube is inserted in a beaker of pure sol-

	Maximum Number of Particles	Molality (mol kg^{-1})					
		0.1		*0.01*		*0.001*	
Solute	ν	i	α	i	α	i	α
NaCl	2	1.87	0.87	1.94	0.94	1.97	0.97
MgSO$_4$	2	1.21	0.21	1.53	0.53	1.82	0.82
K$_2$SO$_4$	3	2.32	0.66	2.70	0.85	2.84	0.92
K$_3$[Fe(CN)$_6$]	4	2.85	0.62	3.36	0.79	3.82	0.94

Table 6–2
Effect of Increasing Dilution on the van't Hoff i and the Degree of Dissociation α in Some Electrolyte Solutions^a

^{a}at 25°C

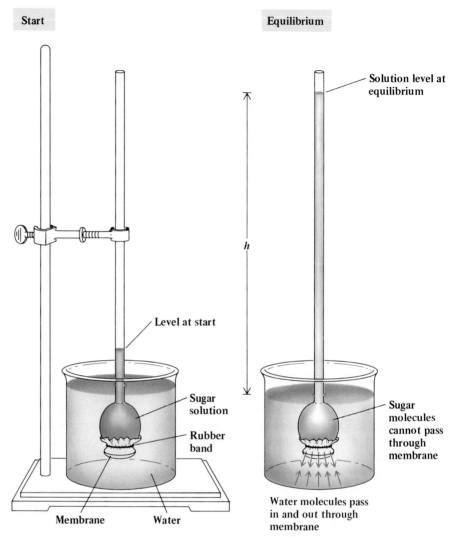

Figure 6–29 In this device to measure osmotic pressure, the semipermeable membrane allows solvent, but not solute, molecules to pass through. This results in a net flow of solvent into the tube until an equilibrium is achieved, with the level of the solution at a height h above that of the solvent in the beaker. Once this happens, the solvent molecules pass through the membrane at the same rate in both directions.

vent, solvent flows from the beaker into the tube (Fig. 6–29). The driving force for the flow is the tendency of the pure solvent to mix with the solution and dilute it, just as any solutions of differing composition tend to mix to a uniform composition when put in contact with each other. Because molecules of solute cannot pass down through the semipermeable membrane, the tendency can be expressed only by solvent molecules passing up. The solution rises within the tube to some final height (h) above the level of the pure solvent in the beaker. The pressure on the solution side of the membrane is greater than the atmospheric pressure on the surface of the pure solvent by an amount that is the **osmotic pressure** (Π) of the solution contained in the vertical tube. This osmotic pressure is computed from h by the equation

$$\Pi = gdh$$

where d is the density of the solution in the tube and g is the acceleration due to the gravitational field of the earth. To use this equation in consistent SI units, d must be in kg m^{-3}, h in meters, and g in m s^{-2} (the measured value of g is 9.81 m s^{-2}).

Typical dilute aqueous solutions are able to raise columns 10 to 30 cm high in this experiment. The densities of such solutions are close to the density of pure water, which is 1.00 g cm^{-3}, or 1000 kg m^{-3}. Suppose, then, that the final height of the column of solution in the tube is 0.22 m. The osmotic pressure of this solution is

$$\Pi = \left(\frac{9.81 \text{ m}}{\text{s}^2}\right) \times \left(\frac{1000 \text{ kg}}{\text{m}^3}\right) \times (0.22 \text{ m}) = 2.16 \times 10^3 \frac{\text{kg}}{\text{m s}^2} = 2.2 \times 10^3 \text{ Pa}$$

Converting from pascals to atmospheres gives

$$\Pi = 2.16 \times 10^3 \text{ Pa} \times \left(\frac{1 \text{ atm}}{1.01325 \times 10^5 \text{ Pa}}\right) = 0.021 \text{ atm}$$

This computation shows that easily measured heights of aqueous solutions correspond to quite small osmotic pressures.

In 1887 van't Hoff discovered an important approximate relationship between a solution's osmotic pressure (Π), its concentration (c), and its absolute temperature (T):

$$\Pi = cRT$$

where R is the gas constant. Using c in mol L^{-1}, R in L atm mol^{-1} K^{-1}, and T in K in this equation gives Π in atmospheres. In the case of a dissociating solute, the van't Hoff i is inserted to compensate, just as with the other colligative properties.

$$\Pi = icRT \qquad \text{(for dissociating solutes)}$$

Because $ic = n/V$, where n is the chemical amount of particles arising from the solute and V is the volume of the solution, van't Hoff's equation can be rewritten as

$$\Pi V = nRT$$

which has exactly the form of the ideal gas law.

Armed with these relationships, we can determine the molar mass of a dissolved substance from a measurement of its osmotic pressure, as the next example shows.

• The osmotic pressure of a solution measures its ability to pull additional solvent molecules into itself by osmosis.

EXAMPLE 6–17

A chemist dissolves 2.00 g of a (nondissociating) protein in 0.100 L of water and finds the osmotic pressure to be 0.021 atm at 25°C. What is the approximate molar mass of the protein?

Solution

The concentration in moles per liter is

$$c = \frac{\Pi}{RT} = \frac{0.021 \text{ atm}}{(0.08206 \text{ L atm mol}^{-1} \text{ K}^{-1})(298 \text{ K})} = 8.6 \times 10^{-4} \frac{\text{mol}}{\text{L}}$$

Now, 2.00 g dissolved in 0.100 L gives the same concentration as 20.0 g in 1.00 L. Therefore, 8.6×10^{-4} mol of the protein must weigh 20.0 g, and the molar mass is

$$\mathcal{M} = \frac{20.0 \text{ g}}{8.6 \times 10^{-4} \text{ mol}} = \boxed{23{,}000 \text{ g mol}^{-1}}$$

Exercise

A dilute aqueous solution of a nondissociating compound contains 1.19 g of the compound per liter of solution and has an osmotic pressure of 0.0288 atm at a temperature of 37.0°C. Compute the molar mass of the compound.

Answer: 1.05×10^3 g mol^{-1}.

Measurements of osmotic pressure are a good way to determine the molar masses of substances, such as proteins, that have large molecules and low solubilities. For the protein of Example 6–17, the height (h) is 22 cm, an easily measured distance. By contrast, the other three colligative properties show very small effects in this case. For the same solution,

$$\text{lowering of vapor pressure} = 4.8 \times 10^{-7} \text{ atm}$$
$$\text{elevation of boiling point} = 0.00044°C$$
$$\text{depression of freezing point} = 0.0016°C$$

All of these changes are too small for accurate measurement.

Osmosis has other important uses. In some parts of the world, potable water is a precious commodity. It can be obtained much more economically by desalinizing brackish waters than by distillation (see Section 6–7), through a process called **reverse osmosis.** When a pressure in excess of its osmotic pressure is applied to a solution that is in contact with a semipermeable membrane, water of quite high purity passes through. Reverse osmosis is being used increasingly in both domestic and industrial applications, especially for control of water pollution.

6–7 MIXTURES AND DISTILLATION

The concept of an ideal solution developed in Section 6–6 extends to solutions of two or more components, each of which is volatile. An ideal solution is then one in which the vapor pressure of *each* component is proportional to its mole fraction over the whole range of mole fraction.

For an ideal solution of two volatile substances (1 and 2), the vapor pressure of component 1 is

$$P_1 = X_1 P_1^\circ$$

from Raoult's law, and that of component 2 is

$$P_2 = X_2 P_2^\circ = (1 - X_1)P_2^\circ$$

The total vapor pressure exerted by the solution is, from Dalton's law, the sum of these two partial pressures.

$$P_{\text{tot}} = P_1 + P_2 = X_1 P_1^\circ + (1 - X_1)P_2^\circ$$

CHEMISTRY IN YOUR LIFE

Osmosis, Dialysis, and Your Kidneys

The walls of living cells are semipermeable membranes through which the solvent (water) can pass. As the concentration of dissolved species inside and outside the cell changes, the osmotic pressure difference across the wall of the cell changes. This causes water to flow into or out of the cell, and the cell swells or shrinks (Fig. 6–C).

Red blood cells can be regarded as little bags that hold a solution of approximately 0.15 molal NaCl. They circulate in a blood fluid having the same osmotic pressure. Consequently, the proper medium for intravenous injections is 0.15 molal aqueous NaCl, *not* water. This ensures that the osmotic pressure of the blood fluid stays equal to the osmotic pressure inside the red cells. Injecting pure water into the bloodstream causes the red cells to swell and perhaps burst as water osmoses into them to dilute their contents to the same concentration as their surroundings. The consequences are catastrophic.

It is sometimes said that the elimination of wastes from the bloodstream proceeds by osmosis. In fact, kidney function is more akin to **dialysis,** in which some solute molecules as well as solvent molecules pass through a membrane. A dialyzing membrane has pores large enough to allow water, small ions, and small molecules to pass through, but too small for large molecules. The kidneys contain millions of structures called nephrons, each consisting of a cluster of capillaries along a long narrow tube. The walls of the capillaries serve as dialyzing membranes. Blood pressure, which is twice as high in the capillaries of the nephrons as in the rest of the body, forces ions and small molecules (including the waste products of metabolism) through the capillary walls and into the tubes of the nephrons. Large molecules, such as proteins, stay in the blood. Molecules of waste compounds are then selectively

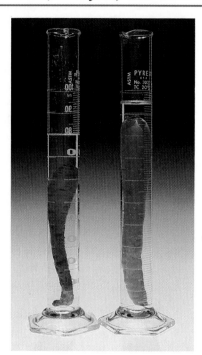

Figure 6–C   When a carrot is immersed in saltwater (*left*), the osmotic pressure outside the cells of the vegetable is higher than that inside. Water flows out into the solution, causing the carrot to shrink. A carrot left to stand in pure water (*right*) does not shrivel.

removed by dialytic transport into the cells of the tubular wall and ultimately secreted in the urine. Most of the water and solutes that have use in the body, such as glucose, amino acids, and ions of various salts, are reabsorbed from the nephron through the capillaries and returned to the bloodstream.

The vapor pressures for such an ideal solution are graphed as a function of concentration in Figure 6–30. In real solutions, negative or positive deviations from this straight-line behavior are seen, depending on the relative strengths of the solute–solvent and solvent–solvent forces.

At sufficiently low mole fraction X_2, the vapor pressure of component 2 (even in a nonideal solution) is proportional to X_2:

$$P_2 = k_H X_2$$

This equation states **Henry's law:** the vapor pressure of a volatile solute is proportional to the mole fraction of that substance in solution (Fig. 6–31). The constant of

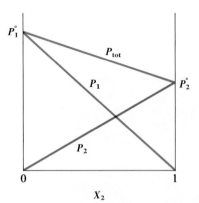

Figure 6–30 In an ideal solution of two volatile liquids, the vapor pressures P_1 and P_2 of each component, as well as their sum, vary linearly with mole fraction. P_1° and P_2° are the vapor pressures of the pure component liquids.

proportionality k_H is called the **Henry's law constant.** This constant differs for every solute–solvent pair and depends on the temperature as well. Henry's law is most often applied to solutions of gases in water. An example is the partial pressure of carbon dioxide above a carbonated soft drink, which is proportional to the mole fraction of dissolved carbon dioxide:

$$P_{CO_2} = k_H X_{CO_2}$$

The k_H for CO_2 in water equals 1650 atm at 25°C.

Raoult's law and Henry's law are identical for ideal solutions; in those cases, the Henry's law constant k_H is equal to P_2°. For nonideal solutions, Henry's law, like Raoult's law, is a *limiting law* that is valid only for dilute solutions. Whenever Raoult's law is valid for the solvent, Henry's law is valid for the solute.

EXAMPLE 6–18

The Henry's law constant for oxygen dissolved in water is 4.34×10^4 atm at 25°C. If the partial pressure of oxygen in air is 0.20 atm under ordinary atmospheric conditions, calculate the concentration (in moles per liter) of dissolved oxygen in water in equilibrium with air at 25°C.

Solution

We first use Henry's law to calculate the mole fraction of oxygen in the solution:

$$X_{O2} = \frac{P_{O2}}{k_H} = \frac{0.20 \text{ atm}}{4.34 \times 10^4 \text{ atm}} = 4.6 \times 10^{-6}$$

Next, we need to change from mole fraction to molarity. One liter of water has a mass of 1000 g, so it contains

$$1000 \text{ g water} \times \left(\frac{1 \text{ mol water}}{18.02 \text{ g water}} \right) = 55.5 \text{ mol water}$$

Because X_{O_2} is so small, $n_{H_2O} + n_{O_2}$ is very close to n_{H_2O}, and we can write

$$X_{O_2} = \frac{n_{O_2}}{n_{H_2O} + n_{O_2}} \approx \frac{n_{O_2}}{n_{H_2O}}$$

$$4.6 \times 10^{-6} = \frac{n_{O_2}}{55.5 \text{ mol}}$$

Thus, the chemical amount of oxygen in one liter of water is

$$n_{O_2} = (4.6 \times 10^{-6})(55.5 \text{ mol}) = 2.6 \times 10^{-4} \text{ mol}$$

and the concentration of dissolved O_2 is 2.6×10^{-4} M.

Exercise

When the partial pressure of nitrogen over a sample of water at 19.4°C is 9.20 atm, then the concentration of nitrogen in the water is 5.76×10^{-3} M. Compute the Henry's law constant for nitrogen in water at this temperature.

Answer: 8.86×10^4 atm.

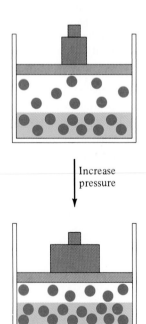

Increase pressure

Figure 6–31 According to Henry's law, as the pressure of a gas above a solution is increased, the mole fraction of the gas in solution increases proportionately.

Distillation

The vapor pressures of the pure components of a solution usually differ from each other, and for this reason the composition of the vapor above a solution usually differs from the composition of the solution itself. We can best see this in an example.

Hexane (C_6H_{14}) and heptane (C_7H_{16}) form a nearly ideal solution over the whole range of mole fractions. At 25°C, the vapor pressure of pure hexane is $P_1^\circ = 0.198$ atm and that of pure heptane is $P_2^\circ = 0.0600$ atm. Suppose that a solution contains 4.00 mol of hexane and 6.00 mol of heptane, so that its mole fractions are $X_1 = 0.400$ and $X_2 = 0.600$. Raoult's law gives the two partial pressures:

$$P_{hexane} = P_1 = X_1 P_1^\circ = (0.400)(0.198 \text{ atm}) = 0.0792 \text{ atm}$$

$$P_{heptane} = P_2 = X_2 P_2^\circ = (0.600)(0.0600 \text{ atm}) = 0.0360 \text{ atm}$$

We now use Dalton's law, adding the two pressures to find the total pressure:

$$P_{tot} = P_1 + P_2 = 0.1152 \text{ atm}$$

Let X_1' and X_2' represent the mole fractions of hexane and heptane in the vapor. Then

$$X_1' = \frac{0.0792 \text{ atm}}{0.1152 \text{ atm}} = 0.688$$

$$X_2' = 1 - X_1' = \frac{0.0360 \text{ atm}}{0.1152 \text{ atm}} = 0.312$$

The liquid and the vapor with which it is in equilibrium have different compositions (Fig. 6–32), and the vapor is enriched in the more volatile component.

Such behavior is general. It underlies the technique of separating a mixture into its pure components by **fractional distillation.** In this process a solution is boiled to raise vapors, the vapors are collected and condensed, and the new solution composed of the condensed vapors is taken into a fresh cycle of boiling and condensation. The details of fractional distillation are best illustrated by transforming the vapor pressure–mole fraction plot into a plot of boiling temperature versus mole fraction, which is shown in Figure 6–33. Note that component 2, which has the lower vapor pressure, has the higher boiling point, $T_{b,2}$. If a solution is heated until it reaches the temperature given by the lower curved line in the plot, it starts to boil. The vapor in equilibrium with the boiling solution is richer in the more volatile

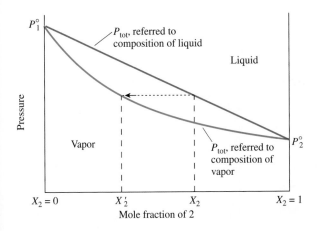

Figure 6–32 The composition of the vapor above a solution differs from the composition of the liquid with which it is in equilibrium. Here, the straight blue line is the total pressure of the vapor in equilibrium with an ideal solution having mole fraction X_2 of component 2. By moving horizontally from that line to a point of intersection with the red curve, we can locate the mole fraction X_2' of component 2 in the vapor.

Figure 6–33 The boiling temperature of an ideal solution varies with the composition of the solution. The blue curve is the boiling temperature referred to the liquid composition, and the red curve is the boiling temperature referred to the vapor composition. The vapors boiling off a solution are enriched in the more volatile component to an extent given by moving horizontally from the blue line to the red line.

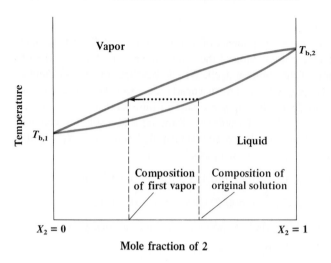

component, 1. The composition of the vapor is determined from the plot by extending a horizontal constant-temperature line across the page until it intersects the upper curve, the equilibrium vapor curve, and then reading down from the point of intersection to the composition axis.

We can vaporize the liquid in different ways. We could simply boil it until it all boils away. In this case, the final composition of the collected vapor is obviously the same as that of the original liquid, although the mixture boils over a range of temperatures rather than at a single temperature like a pure liquid. If we stop the boiling before all the liquid has evaporated, collect the vapor that has boiled off,

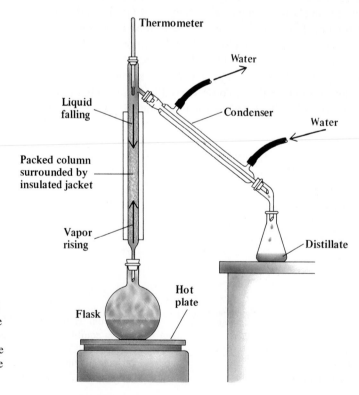

Figure 6–34 In a distillation column, the temperature decreases moving up the column. The less volatile components condense and fall back to the flask, but the more volatile ones continue up the column to pass into the water-cooled condenser, where they condense and are recovered in the receiver.

and recondense it, the result is not the same. The condensed liquid (*distillate*) is now richer in component 1 than the original liquid. By starting over with this distillate and repeating the process again and again, mixtures successively richer in component 1 can be obtained. This is the principle behind the **distillation column** (Fig. 6–34). Throughout the length of the column, such vaporizations and recondensations take place and cause mixtures to be separated into their constituent substances. Fractional distillation is used industrially to separate nitrogen and oxygen in air: the air is liquefied and then distilled, with the lower boiling nitrogen ($T_b = -196°C$) vaporizing before the oxygen ($T_b = -183°C$).

6-8 COLLOIDAL DISPERSIONS

Colloidal dispersions resemble solutions in some ways, but the dispersed particles in a colloid are larger than the solvated molecules or ions found in true solutions. They range upward from 10^{-9} m to about 1000×10^{-9} m in diameter and consist of aggregates of many smaller molecules and ions or sometimes of single very large molecules (for example, some protein molecules have diameters on the order of 500×10^{-9} m).

The dispersed substance and the background medium in a colloid may be any combination of gas, liquid, or solid. Example of colloids are extremely varied: aerosol sprays (liquid dispersed in gas), smoke (solid particles in air), milk (fat droplets and solids in water), mayonnaise (water droplets in oil), paint (solid pigment particles in oil for oil-based paints, or pigment and oil dispersed in water for latex paints), and biological fluids (proteins and fats in water).

Colloids with larger particles look translucent, cloudy, or milky, but those with smaller particles look clear, even under a microscope. The best evidence for the presence of colloidal particles is the **Tyndall effect:** rays of visible light passing into a colloid are scattered, making the path of a beam of light through the dispersion visible from the side. A familiar example is the visible cone marking the passage of light from a movie projector though a dispersion of small dust particles in air. The gemstone opal has remarkable optical properties that arise from colloidal water suspended in the solid silicon dioxide that makes up most of the mass of the stone.

The dispersed particles in colloids usually acquire electric charges—some positive, some negative. Because all the particles of a certain kind of colloid acquire the same kind of charge, they tend to repel each other. This helps to keep them dispersed in the suspension and prevents them from settling. Some colloids settle out into two separate phases if left standing long enough; others can persist indefinitely. A suspension of gold particles in water prepared by Michael Faraday in 1857 shows no settling to date. The settling out of colloids is speeded by dissolving salts in the background medium. The added ions reduce the repulsive electrostatic forces between the dispersed particles, leading to their **coagulation** (aggregation into larger particles) and sedimentation (Fig. 6–35). An interesting example occurs in river deltas. Where river water that contains suspended clay particles meets the salty water of the ocean, the colloidal clay coagulates and settles out as flocculent (open, low-density) sediments. Flocculating agents are deliberately added to paints so that the pigment settles in a loosely packed sediment. When the paint is stirred, the pigment is then easily redispersed through the medium. In the absence of such agents, the suspended particles tend to form compact sediments that are difficult to resus-

• A colloidal dispersion is also called a colloid. The word is derived from a Greek word meaning "glue-like."

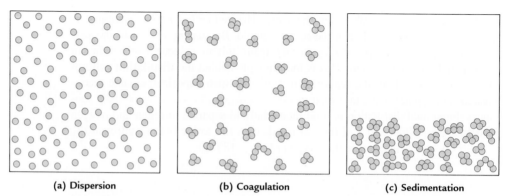

(a) Dispersion   (b) Coagulation   (c) Sedimentation

Figure 6–35   When a salt is added to a colloidal dispersion (a), the repulsive forces between the colloidal particles are reduced and coagulation (sticking together) occurs (b). Eventually, the aggregated particles fall to the bottom of the container as a sediment or precipitate (c).

pend. Reversal of coagulation, a process called **peptization,** occurs frequently upon stirring or other treatment of coagulated colloids.

In some cases, such as in the precipitation of a solid from solution (see Section 4–2) the formation of a colloid is not desirable. If the solid particles of a newly formed precipitate are quite large, they fall quickly to the bottom of the vessel, where they can easily be separated from the liquid by filtration. Sometimes (especially with metal sulfides) the solid precipitate may appear as a colloidal dispersion, with particles small enough to pass through ordinary filter paper (Fig. 6–36). If this happens, separation can be achieved only by coagulation, by centrifugation, or by forcing the dispersion through a membrane such as cellophane that permits passage of only the small solvent molecules.

Many biological fluids are colloids. Blood plasma consists of quantities of colloidal lipids (fatty, water-insoluble compounds) and colloidal albumin, a protein, as well as conventionally dissolved salts, glucose, amino acids, vitamins, and hormones. The cytosol, which is the viscous, semifluid, semitransparent content of a cell, is a conventional solution of sugars, soluble proteins, and potassium, sodium, and other salts of small-molecule organic acids. This medium then contains water-insoluble proteins and other large biomolecules in colloidal dispersion.

Figure 6–36   The colloidal dispersion of $PbCrO_4$ in (a) appears cloudy. After coagulation, the particles of $PbCrO_4$ soon settle to the bottom (b).

(a)   (b)

Colloidal particles are in a constant state of motion called **Brownian motion** (after Robert Brown, a Scottish botanist who used a microscope to observe the motion of pollen particles in water). This motion results from the constant random buffeting of the particles by molecules of solvent. In 1905, Albert Einstein showed how the motion of Brownian particles could be described on a microscopic level; his work provided one of the most convincing verifications of the molecular hypothesis and of the kinetic theory of matter and led to a fairly accurate determination of Avogadro's number.

SUMMARY

6-1 Intermolecular forces give the molecules in an assembly **potential energy** in addition to their kinetic energy; such forces are responsible for the existence of condensed phases. **Ion–ion forces** dominate the interactions of ionic solids and liquids, whereas weaker **dipole–dipole forces** are important for **polar** molecules, in which the two ends of a molecule exhibit opposite charges. **Dispersion forces,** which arise from fluctuating dipole moments on neighboring atoms or molecules, cause attractions among atoms and nonpolar molecules. **Hydrogen bonds** are particularly strong, dipole–dipole interactions between certain hydrogen-containing molecules such as those of water. Hydrogen bonding causes many of the unexpected properties of water. Any two ions, atoms, or molecules exhibit **repulsive forces** at sufficiently short distances.

6-2 A kinetic model for liquids and solids that incorporates intermolecular forces can describe the qualitatively different structures and dynamics of the three states of matter.

6-3 Two phases of a pure substance can coexist in **phase equilibrium.** At a given temperature, a liquid or solid coexists with a vapor inside a closed container, and the **vapor pressure** is the final pressure that persists indefinitely above the solid or liquid when the rates of **evaporation** from and **condensation** onto a liquid or a solid become equal. The vapor pressure of the water above an aqueous solution must be taken into account in accurate measurements of the amounts of gas produced in a chemical reaction when the gas is collected over water. The **normal boiling point** is the temperature at which the vapor pressure of a liquid equals 1 atm. The **normal melting point** is the temperature at which solid and liquid coexist indefinitely at a pressure of 1 atm of air.

6-4 Melting and **freezing** are the solid-to-liquid and liquid-to-solid phase transitions, respectively. **Evaporation** and **condensation** describe the liquid-to-gas transition and its reverse, and **sublimation** and **deposition** describe the direct solid-to-gas transition and *its* reverse. The melting and boiling points of water are abnormally high compared with those of other substances of low molar mass. Also, the density of ice is less than that of liquid water, and the heat required to vaporize water is abnormally large. These exceptional characteristics derive from the strong intermolecular forces (hydrogen bonds) acting between molecules of water.

6-5 A **phase diagram** of a substance shows which phase is stable at any given pressure and temperature. The lines in such a diagram define conditions of coexistence of two phases in equilibrium, and a **triple point** represents conditions that allow the coexistence of three phases in equilibrium. The fluid–solid coexistence curve continues to arbitrarily high pressures, but the gas–liquid curve terminates at the **critical point;** at temperatures or pressures above its **critical temperature** and **crit-**

ical pressure, a substance is **supercritical;** that is, the distinction between gas and liquid has ceased to exist. For many substances, more than one solid phase appears in the phase diagram, especially at elevated pressures.

6-6 The presence of a solute changes the properties of a solvent. It **lowers the vapor pressure** of the solvent below that of the pure solvent in proportion to the mole fraction of solute **(Raoult's law).** Also, a solute **elevates the boiling point** and **depresses the freezing point** by amounts proportional to its molality, a fact that permits the measurement of solute molar masses. The **osmotic pressure** of a solution is the pressure difference that exists across a semipermeable membrane between the solution and a supply of pure solvent. It, too, depends on the concentration of solute. Its measurement is useful for determining the molar masses of solutes. These **colligative properties** depend only on the ratio of the number of solute particles to the number of solvent particles, and not on the identity of the solute. When solutes dissociate, additional particles form; the effect of solute dissociation on the colligative properties is taken into account by insertion of the **van't Hoff i** into the appropriate equations.

6-7 The laws governing the properties of ideal solutions can be extended to mixtures of two volatile substances. **Henry's law** states that the partial pressure of a volatile solute is proportional to its mole fraction in solution at sufficiently low mole fractions. When two components in a liquid mixture have different vapor pressures, they can be separated by a process of repeated vaporization and condensation called **fractional distillation.**

6-8 Colloids are mixtures of substances in which one phase is dispersed as small particles (between about 10^{-6} and 10^{-9} m in diameter) in a second phase. Colloids exhibit the **Tyndall effect** and **Brownian motion.** Colloidal particles tend not to settle out, but colloids can be **coagulated** (often by the addition of solutions of salts). The reverse of coagulation is **peptization.** Colloidal particles can also be separated from the dispersing medium by centrifugation.

PROBLEMS

Note: Answers to blue-numbered problems are given in Appendix F. Problems that are more challenging are indicated with asterisks.

Intermolecular Forces: Why Condensed Phases Exist

1. (a) Use Figure 6–1 to estimate the length of the covalent bond in Cl_2 and the length of the ionic bond in K^+Cl^-. *Note:* The latter corresponds to the distance between the atoms in an isolated single molecule of K^+Cl^-, not in $KCl(s)$ (solid potassium chloride).
   (b) A book states: "the shorter the bond, the stronger the bond." What features of Figure 6–1 show that this is not always true?

2. True or false: Any two atoms held together by nonbonded attractions must be farther apart than any two atoms held together by a chemical bond. Explain.

3. (See Example 6–1.) Name the types of attractive forces that contribute to the interactions between atoms, molecules, or ions in the following substances. Indicate the one(s) you expect to predominate.
   (a) KF   (d) Rn
   (b) HI   (e) N_2
   (c) CH_3OH

4. (See Example 6–1.) Name the types of attractive forces that contribute to the interactions between atoms, molecules, or ions in the following substances. Indicate the one(s) you expect to predominate.

(a) Ne   (d) $BaCl_2$
(b) ClF   (e) NH_2OH
(c) F_2

5. Predict whether a sodium ion is most strongly attracted to a bromide ion, a molecule of hydrogen bromide, or an atom of krypton.

6. Predict whether an atom of argon is most strongly attracted to another atom of argon, to an atom of neon, or to an atom of krypton.

7. As a vapor, methanol exists to an extent as a tetramer,

$(CH_3OH)_4$. In the tetramer, four $H—\overset{\displaystyle H}{\underset{\displaystyle H}{C}}—O—H$ molecules

are held together by hydrogen bonds. Propose a reasonable structure for this tetramer.

8. Hypofluorous acid (HOF) is the simplest possible compound that allows comparison between fluorine and oxygen in their ability to form hydrogen bonds. Solid HOF unexpectedly contains no H···F hydrogen bonds, but many H···O hydrogen bonds. Propose and draw a structure for chains of HOF molecules linked by O···H hydrogen bonds. The bond angle in HOF is 101°.

The Kinetic Theory of Liquids and Solids

9. The "diffusion constant" is a measure of the rapidity of mixing between substances caused by the chaotic motions of their molecules. Do you expect the diffusion constant to increase or decrease as the density of a liquid is increased (by compressing it) at constant temperature? Explain. What happens to the diffusion constant of a gas and a solid as the density increases?

10. Do you anticipate that the diffusion constant (see previous problem) would increase as the temperature of a liquid increases at constant pressure? Why or why not? Will the diffusion constant increase with temperature for a gas and a solid? Explain.

11. The mean free path of the molecules in a gas equals the average distance they travel between collisions (see Chapter 5). Compare the mean free path of molecules in a liquid to the mean free path of molecules in a gas at the same temperature.

12. Surface tension, which makes the surfaces of liquids behave like weak elastic skins, is caused by intermolecular interactions. Is the surface tension of molten sodium chloride higher than or lower than that of liquid carbon tetrachloride? Explain.

Phase Equilibrium

13. Hydrogen at a pressure of 1 atm condenses to a liquid at 20.3 K and solidifies at 14.0 K. The vapor pressure of liquid hydrogen is 0.213 atm at 16.0 K. Calculate the volume of 1.00 mol of H_2 vapor under these conditions, and compare it with the volume of 1.00 mol of H_2 at STP.

14. Helium condenses to a liquid at 4.224 K under atmospheric pressure and remains a liquid down to the absolute zero of temperature. The vapor pressure of liquid helium at 2.20 K is 0.05256 atm. Calculate the volume occupied by 1.000 mol of helium vapor in equilibrium with liquid helium under these conditions, and compare it with the volume of the same amount of helium at STP.

15. (See Example 6–2.) Calcium carbide reacts with water to produce acetylene (C_2H_2) and calcium hydroxide. The acetylene is collected over water at 40.0°C under a total pressure of 0.9950 atm. The vapor pressure of water at this temperature is 0.0728 atm. Calculate the mass of acetylene per liter of "wet" acetylene collected in this way, assuming ideal gas behavior.

16. (See Example 6–2.) When an excess of sodium hydroxide is added to an aqueous solution of ammonium chloride, gaseous ammonia is produced:

$$NaOH(aq) + NH_4Cl(aq) \longrightarrow$$
$$NaCl(aq) + NH_3(g) + H_2O(\ell)$$

Suppose 3.68 g of ammonium chloride reacts in this way at 30°C and a total pressure of 0.9884 atm. At this temperature, the vapor pressure of water is 0.0419 atm. Calculate the volume of ammonia saturated with water vapor that is produced under these conditions, assuming no leaks or other losses of gas.

Phase Transitions

17. (See Example 6–3.) In Chapter 3, it was stated that, other things being equal, ionic character in compounds of metals decreases with increasing oxidation number. Rank the following compounds from lowest to highest in normal boiling point: $SnBr_2$, $SnBr_4$, and Br_2.

18. (See Example 6–3.) Rank the following from lowest to highest in normal boiling point: AsF_5, SbF_3, SbF_5, F_2, and Cl_2.

19. High in the Andes, an explorer notes that the water for tea is boiling vigorously at a temperature of 90°C. Use Figure 6–21 to estimate the atmospheric pressure at the altitude of the camp. What fraction of the earth's atmosphere lies below the level of the explorer's camp?

20. The total pressure in a pressure cooker filled with water rises to 4.0 atm when it is heated, and this pressure is maintained by the periodic operation of a relief valve. Use Figure 6–20c to estimate the temperature of the water in the pressure cooker.

21. It is essential to boil a sample of a certain organic liquid. Unfortunately, the liquid starts to smoke and char before it boils. Suggest a way to cause the liquid to boil at a lower temperature and avoid this decomposition.

22. Aluminum melts at a temperature of 660°C and boils at 2470°C, whereas thallium melts at a temperature of 304°C and boils at 1460°C. Which metal is more volatile at room temperature?

Phase Diagrams

23. At its melting point (624°C), solid plutonium has a density of 16.24 g cm^{-3}. The density of liquid plutonium is 16.66 g cm^{-3}. A small sample of liquid plutonium at 625°C is strongly compressed. Predict what phase changes, if any, occur.

24. Phase changes occur between different solid forms, as well as from solid to liquid, liquid to gas, and solid to gas. When white tin at 1.00 atm is cooled below 13.2°C, it spontaneously changes (over a period of weeks) to gray tin. The density of gray tin is *less* than the density of white tin (5.75 g cm^{-3} versus 7.31 g cm^{-3}). Some white tin is compressed to a pressure of 2.00 atm. At this pressure, should the temperature be higher or lower than 13.2°C for the conversion to gray tin to take place? Explain your reasoning.

25. (See Example 6–5.) Determine whether argon is a solid, a liquid, or a gas at each of the following combinations of temperature and pressure (use Fig. 6–20).
    (a) 50 atm and 100 K
    (b) 8 atm and 150 K
    (c) 1.5 atm and 25 K
    (d) 0.25 atm and 120 K

26. (See Example 6–5.) Some water starts out at a temperature of 298 K and a pressure of 1 atm. It is compressed to 500 atm at constant temperature and then heated to 750 K at constant pressure. Next it is decompressed at 750 K back to 1 atm and finally cooled to 400 K at constant pressure.
    (a) What is the state (solid, liquid, or gas) of the water at the start of the experiment?
    (b) What is the state (solid, liquid, or gas) of the water at the end of the experiment?
    (c) Do any phase transitions occur during the four steps described? If so, at what temperature and pressure do they occur? (*Hint:* Trace out the various changes on the phase diagram of water [see Fig. 6–20]).

27. The vapor pressure of solid acetylene at −84.0°C is 760 torr.
    (a) Does the triple-point temperature lie above or below −84.0°C? Explain.
    (b) Suppose a sample of solid acetylene is held under an external pressure of 0.80 atm and heated from 10 K to 300 K. What phase change(s), if any, occurs?

28. The triple point of hydrogen occurs at a temperature of 13.8 K and a pressure of 0.069 atm.
    (a) What is the vapor pressure of solid hydrogen at 13.8 K?
    (b) Suppose a sample of solid hydrogen is held under an external pressure of 0.030 atm and heated from 5 K to 300 K. What phase change(s), if any, occurs?

29. The density of nitrogen at its critical point is 0.3131 g cm^{-3}. At a very low temperature, 0.3131 g of solid nitrogen is sealed into a thick-walled glass tube with a volume of 1.000 cm^3. Describe what happens inside the tube as the tube is warmed past the critical temperature, 126.19 K.

30. At its critical point, ammonia has a density of 0.235 g cm^{-3}. You have a special thick-walled glass tube that has a 10.0-mm outside diameter, a wall thickness of 4.20 mm, and a length of 155 mm. How much ammonia must you seal into the tube if you wish to observe the disappearance of the meniscus as you heat the tube and its contents past 132.23°C, the critical temperature?

Colligative Properties of Solutions

31. (See Examples 6–6 and 6–8.) A disinfectant solution is made by dissolving 4.50 g of hydrogen peroxide (H_2O_2) in 143.2 g of water. Calculate the mass percentage, mole fraction, and molality of hydrogen peroxide in the solution.

32. (See Examples 6–6 and 6–8.) A solution of purified vinegar is determined to contain 29.3 g of acetic acid (CH_3COOH), 575 g of water, and no other components. Calculate the mass percentage, mole fraction, and molality of acetic acid in the solution.

33. (See Examples 6–8 and 6–9.) An aqueous solution of hydrochloric acid is 38.00% HCl by mass. The solution has a density of 1.1886 g cm^{-3} at 20°C. Compute the molality and the molarity of the HCl.

34. (See Examples 6–8 and 6–9.) An aqueous solution of acetic acid is 20.00% CH_3COOH by mass. The solution has a density of 1.0250 g cm^{-3} at a certain temperature. Compute the molality and molarity of the acetic acid.

35. (See Example 6–10.) Concentrated aqueous phosphoric acid sold for use in the laboratory contains 12.2 mol of H_3PO_4 per liter of solution at 25°C and has a density of 1.33 g cm^{-3}. Calculate the molality of the phosphoric acid.

36. (See Example 6–10.) A 9.20 M perchloric acid ($HClO_4$) solution has a density of 1.54 g cm^{-3} at 25°C. Calculate the molality of the perchloric acid.

37. (See Example 6–11.) The vapor pressure of pure acetone (CH_3COCH_3) at 30°C is 0.3720 atm. Suppose 15.0 g of benzophenone ($C_{13}H_{10}O$), which does not dissociate in acetone, is dissolved in 50.0 g of acetone. Calculate the vapor pressure of acetone above the resulting solution.

38. (See Example 6–11.) The vapor pressure of diethyl ether (molar mass 74.12 g mol^{-1}) at 30°C is 0.8517 atm. Suppose 1.800 g of maleic acid ($C_4H_4O_4$), which does not dissociate in ether, is dissolved in 100.0 g of diethyl ether at 30°C. Calculate the vapor pressure of diethyl ether above the resulting solution.

39. The vapor pressure of carbon tetrachloride (CCl_4) at 50°C is 0.437 atm. When 7.42 g of a pure, nonvolatile, and nondissociating substance is dissolved in 100.0 g of carbon tetrachloride, the vapor pressure of the solution at 50°C is 0.411 atm. Calculate the mole fraction of solute in the solution.

40. At 25°C, the vapor pressure of benzene (C_6H_6) is 0.1252 atm. When 10.00 g of an unknown nonvolatile substance is dissolved in 100.0 g of benzene, the vapor pressure of the solution at 25°C is 0.1199 atm. Calculate the mole fraction of solute in the solution, assuming no dissociation by the solute.

41. (See Example 6–13.) When 39.8 g of a nondissociating, non-volatile sugar is dissolved in 200.0 g of water, the boiling point of the water is raised by 0.30°C. Estimate the molar mass of the sugar.

42. (See Example 6–13.) When 2.60 g of a nondissociating, non-volatile compound of indium and chlorine is dissolved in 50.0 g of tin(IV) chloride, the normal boiling point of the tin(IV) chloride is raised from 114.1°C to 116.3°C. If $K_b = 9.43$ K kg mol^{-1} for $SnCl_4$, what are the approximate molar mass and the molecular formula of the solute?

43. (See Example 6–15.) The following substances are all soluble in benzene and do not dissociate in solution. In separate experiments 1.00 g of each is added to 100.0 g of benzene. Which substance gives a solution with the lowest freezing point?
(a) C_6H_5Cl　　(b) N_2H_4　　(c) $(C_2H_5)_2O$　　(d) CCl_4

44. (See Example 6–15.) The substances NaCl, $CaCl_2$, and urea (NH_2CONH_2) have been used for melting ice on streets. Estimate the lowering of the freezing point caused by dissolving 50.0 g of each in 1.00 kg of solvent (water). Assume that urea does not dissociate significantly in aqueous solution and that the two ionic substances dissociate completely into ions.

45. Ice cream is made by freezing a liquid mixture that, to a first approximation, can be considered a solution of sucrose ($C_{12}H_{22}O_{11}$) in water. Estimate the temperature at which the first ice crystals begin to appear in a mix that consists of 34% (by mass) sucrose in water. As ice crystallizes out, the remaining solution becomes more concentrated. What happens to its freezing point?

46. The solution to problem 45 shows that to make homemade ice cream, temperatures ranging downward from −3°C are needed. Ice cubes from a freezer have a temperature of about −12°C (+10°F), which is cold enough, but contact with the warmer ice-cream mixture causes them to melt to liquid at 0°C, which is too warm. To obtain a liquid that is cold enough, salt (NaCl) is dissolved in water and ice is added to the salt water. The salt lowers the freezing point of the water enough so that it can freeze liquid ice-cream mixture. The instructions for an ice-cream maker say to add one part of salt to eight parts water (by mass). What is the freezing point of this solution (in degrees Celsius and degrees Fahrenheit)? Assume that the NaCl dissociates fully into ions and that the solution is ideal.

47. (See Example 6–16.) A solution is prepared by dissolving 1.55 g of acetone (CH_3COCH_3) in 50.00 g of water. Its freezing point is measured to be −0.970°C. Does acetone dissociate into ions when dissolved in water? Explain your answer.

48. (See Example 6–16.) When 2.02 g of baking soda (sodium hydrogen carbonate, $NaHCO_3$) is dissolved in 200 g of water, the freezing point is −0.396°C. Determine the van't Hoff i for this solution. Write a single chemical equation to represent the dissolution of sodium hydrogen carbonate in water.

49. (See Example 6–16.) An aqueous solution is 0.8402 molal in Na_2SO_4. It has a freezing point of −4.218°C.
(a) Determine the effective number of particles arising from each Na_2SO_4 formula unit in this solution.
(b) Determine the fractional degree of dissociation of the Na_2SO_4 in this solution.

50. (See Example 6–16.) The freezing-point depression constant of pure H_2SO_4 is 6.12°C kg mol^{-1}. When 2.3 g of ethanol (C_2H_5OH) is dissolved in 1.00 kg of pure sulfuric acid, the freezing point of the solution is 0.92°C lower than the freezing point of pure sulfuric acid. Assuming that dissociation is essentially complete, determine how many particles are formed as one molecule of ethanol goes into solution in sulfuric acid.

51. (See Example 6–17.) A polymer of large molar mass is dissolved in water at 15°C, and the resulting solution rises to a final height of 15.2 cm above the level of the pure water, as water molecules pass through a semipermeable membrane into the solution. If the solution contains 4.64 g polymer per liter, calculate the molar mass of the polymer.

52. (See Example 6–17.) Suppose 0.125 g of a protein is dissolved in 10.0 cm^3 of ethyl alcohol (C_2H_5OH), whose density at 20°C is 0.789 g cm^{-3}. The solution rises to a height of 26.3 cm in an osmometer (an apparatus for measuring osmotic pressure). What is the approximate molar mass of the protein?

53. Use the information in Tables 6–1 and 6–2 to predict the normal freezing point, normal boiling point, and osmotic pressure (at 25°C) of a 0.010 mol kg^{-1} solution of $MgSO_4$ in water. State any assumptions you must make in performing your calculations.

54. The fractional degree of dissociation of a 0.15 mol kg^{-1} solution of an ionic solute in water equals 0.76. If $\nu = 3$ for this solute, compute the freezing point of the solution.

Mixtures and Distillation

55. (See Example 6–18.) The Henry's law constant at 25°C for carbon dioxide dissolved in water is 1.65×10^3 atm. If a carbonated beverage is bottled under a carbon dioxide pressure of 5.0 atm:
(a) Calculate the chemical amount of carbon dioxide dissolved per liter of water under these conditions, using 1.00 g cm^{-3} as the density of water.
(b) Explain what happens on a microscopic level after the bottle cap is removed.

56. (See Example 6–18.) The Henry's law constant at 25°C for nitrogen dissolved in water is 8.57×10^4 atm; the constant for oxygen is 4.34×10^4 atm; and the constant for helium is 1.7×10^5 atm.

(a) Calculate the number of moles of nitrogen and oxygen dissolved per liter of water at 25°C in equilibrium with air at 25°C. Use Table 5–1.

(b) Air is dissolved in blood and other bodily fluids. As a deep-sea diver descends, the pressure increases and the concentration of dissolved air in the diver's blood increases. If the diver returns to the surface too quickly, gas bubbles out of solution within the body so rapidly that it can cause a dangerous condition called "the bends." Show, using Henry's law, why divers sometimes use helium in combination with oxygen in place of compressed air.

57. At 25°C, some water is added to a sample of gaseous methane (CH_4) at 1.00 atm pressure in a closed vessel and shaken until as much methane as possible dissolves. Then 1.00 kg of the solution is removed and boiled to expel the methane, yielding a volume of 3.01 L of $CH_4(g)$ at STP. Determine the Henry's law constant for methane in water. (*Note:* The solution is dilute enough that 1.00 kg of solution contains 1.00 kg of water.)

58. When exactly the same procedure as in problem 57 is carried out using benzene (C_6H_6) in place of water, the volume of methane that results is 0.510 L at STP. Determine the Henry's law constant for methane in benzene.

59. At 20°C, the vapor pressure of toluene is 0.0289 atm and the vapor pressure of benzene is 0.0987 atm. Equal chemical amounts (equal numbers of moles) of toluene and benzene are mixed and form an ideal solution. Compute the mole fraction of benzene in the vapor in equilibrium with this solution.

60. At 90°C, the vapor pressure of toluene is 0.534 atm and the vapor pressure of benzene is 1.34 atm. Benzene (0.400 mol) is mixed with toluene (0.900 mol). The two form an ideal solution. Compute the mole fraction of toluene in the vapor in equilibrium with this solution.

61. At 40°C, the vapor pressure of pure carbon tetrachloride (CCl_4) is 0.293 atm, and the vapor pressure of pure dichloroethane ($C_2H_4Cl_2$) is 0.209 atm. A nearly ideal solution is prepared by mixing 30.0 g of carbon tetrachloride with 20.0 g of dichloroethane.

(a) Calculate the mole fraction of CCl_4 in the solution.

(b) Calculate the total vapor pressure of the solution at 40°C.

(c) Calculate the mole fraction of CCl_4 in the vapor in equilibrium with the solution.

62. At 300 K, the vapor pressure of pure benzene (C_6H_6) is 0.1355 atm, and the vapor pressure of pure hexane (C_6H_{14}) is 0.2128 atm. Mixing 50.0 g of benzene with 50.0 g of hexane gives a solution that is nearly ideal.

(a) Calculate the mole fraction of benzene in the solution.

(b) Calculate the total vapor pressure of the solution at 300 K.

(c) Calculate the mole fraction of benzene in the vapor in equilibrium with the solution.

Colloidal Dispersions

63. What is the difference between a solution and a colloidal suspension? Give examples of each, and show how, in some cases, it may be difficult to classify a mixture as one or the other.

64. When hot water is mixed with a concentrated aqueous solution of iron(III) chloride, a colloid of hydrated iron(III) oxide forms. Attached to the surface of the colloidal particles are iron(III) ions. As the number of iron(III) ions on the surface of the colloidal particles increases, does the rate of coagulation increase or decrease?

65. The water in your beaker is boiling furiously. As you examine it, you exclaim, "Wow, these white clouds of steam are really hot." Your chemistry teacher hears you and comments that the white clouds, although hot, are not steam. You think it over and say, "All right, the white clouds of water vapor are really hot." "Wrong again," says your teacher. Explain what the white clouds above the kettle really are.

66. In a modern sewage treatment plant, waste water is allowed to stand in settling pools so that sand and other fine particles can settle out. "Floc" is then added to the pool. What purpose does the "floc" serve?

Additional Problems

67. Liquid hydrogen chloride dissolves many ionic compounds. Diagram how molecules of hydrogen chloride tend to distribute themselves around a negative ion and around a positive ion in such solutions.

*68. We saw in Section 5–8 that the van der Waals constant b (with units of L mol^{-1}) is related to the volume per molecule in the liquid and thus to the sizes of the molecules. The combination of van der Waals constants a/b has units of L atm mol^{-1}. Because the liter atmosphere is a unit of energy (1 L atm = 101.325 J), a/b is proportional to the energy per mole for interacting molecules and thus to the strength of the attractive forces between molecules, as shown in Figure 6–1. By using the van der Waals constants in Table 5–3, rank the attractive forces in the following from strongest to weakest: N_2, H_2, SO_2, and HCl.

69. At 25°C, the equilibrium vapor pressure of water is 0.03126 atm. A humidifier is placed in a room of volume 110 m^3 and is operated until the air becomes saturated with water vapor. Assuming that no water vapor leaks out of the room and that initially there was no water vapor in the air, calculate the number of grams of water that have passed into the air.

70. The air over an unknown liquid is saturated with the vapor of that liquid at 25°C and a total pressure of 0.980 atm. Suppose that a sample of 6.00 L of the saturated air is collected, and the vapor of the unknown liquid is removed from that sample by cooling and condensation. The pure air remaining occupies a volume of 3.75 L at −50°C and 1.000 atm. Calculate the vapor pressure of the unknown liquid at 25°C.

71. If it is true that all solids and liquids have vapor pressures, then at low enough external pressures, every substance should start to boil. In space, there is effectively zero external pressure. Explain why spacecraft do not just boil away as vapors when placed in orbit.

72. Oxygen melts at 54.8 K and boils at 90.2 K at atmospheric pressure. At the normal boiling point, the density of the liquid is 1.14 g cm^{-3}, and the vapor can be approximated as an ideal gas. The critical point is defined by $T_c = 154.6$ K, $P_c = 49.8$ atm, and (density) $d_c = 0.436$ g cm^{-3}. The triple point is defined by $T_t = 54.4$ K, $P_t = 0.0015$ atm, a liquid density equal to 1.31 g cm^{-3}, and a solid density of 1.36 g cm^{-3}. At 130 K, the vapor pressure of the liquid is 17.25 atm. Use this information to construct a phase diagram showing P versus T for oxygen. You need not draw the diagram to scale, but you should give numerical labels to as many points as possible on both axes.

73. It can be shown that if a gas obeys the van der Waals equation, its critical temperature, its critical pressure, and its molar volume at the critical point are given by the equations

$$T_c = \frac{8a}{27\,Rb}, \quad P_c = \frac{a}{27b^2}, \quad \text{and} \quad \left(\frac{V}{n}\right)_c = 3b$$

where a and b are the van der Waals constants of the gas. Use the van der Waals constants for oxygen, carbon dioxide, and water (from Table 5–3) to estimate the critical point properties of these substances. Compare with the observed values given in Figure 6–20 and in problem 72.

74. Each increase in pressure of 100 atm decreases the melting point of water by about 1.0°C.
    (a) Estimate the temperature at which liquid water freezes under a pressure of 400 atm.
    (b) One possible explanation of why a skate moves smoothly over ice is that the pressure exerted by the skater on the ice lowers its freezing point and causes it to melt. The pressure exerted by an object is the force (its mass times the acceleration of gravity, 9.8 m s^{-2}) divided by the area of contact. Calculate the change in the freezing point of ice when a skater with a mass of 75 kg stands on a blade of area 8.0×10^{-5} m^2 in contact with the ice. Is this sufficient to explain the ease of skating at a temperature of, for example, −5°C (23°F)?

75. Sulfur is soluble in carbon disulfide, but iron is not. A mixture of powdered sulfur and powdered iron weighs 195.4 g. The components are separated by treatment with 1171 g of carbon disulfide. This gives a solution that is 0.0304 molal in sulfur.
    (a) Compute the mass of the iron in the original mixture.
    (b) Carbon disulfide is smelly and hard to handle. Suggest an alternate means of separating these components.

76. A recipe for a sodium carbonate solution calls for the use of either 1 part by mass of Na_2CO_3 or 2.7 parts by mass of $Na_2CO_3 \cdot 10H_2O$ dissolved in 5 parts by mass of water. Compute the molality of the solution in each case.

77. Silver nitrate is prepared by the oxidation of silver with hot nitric acid:

$$Ag(s) + 2\,HNO_3(aq) \longrightarrow$$
$$AgNO_3(aq) + NO_2(g) + H_2O(\ell)$$

A drop of a 1% (by mass) $AgNO_3$ solution is placed in the eyes of newborn babies to prevent infection. Calculate the mass of silver required to prepare 1.00 L of a 1.00% $AgNO_3$ solution. The density of this solution is 1.007 g cm^{-3}.

*78. Veterinarians use Donovan's solution to treat skin diseases in animals. The solution is prepared by mixing 1.00 g of $AsI_3(s)$, 1.00 g of $HgI_2(s)$, and 0.900 g of $NaHCO_3(s)$ in enough water to make a total volume of 100.0 mL.
    (a) Compute the total mass of iodine per liter of Donovan's solution in units of g L^{-1}.
    (b) You need a lot of Donovan's solution to treat an outbreak of rash in an elephant herd. You have plenty of mercury(II) iodide and sodium hydrogen carbonate, but the only arsenic(III) iodide you can find is 1.50 L of a 0.100 M aqueous solution. Explain how to prepare 3.50 L of Donovan's solution starting with these materials.

79. A saturated aqueous solution of NaCl in contact with excess NaCl is placed in an enclosed space at a temperature of 25°C. The presence of this solution maintains the relative humidity inside the space at 75.3%. If pure water replaces the NaCl solution, the relative humidity inside the enclosed space is maintained at 100%. Explain why there is a difference.

80. The vapor pressure of pure liquid CS_2 is 0.3914 atm at 20°C. When 40.0 g of rhombic sulfur is dissolved in 1.00 kg of CS_2, the vapor pressure of CS_2 falls to 0.3868 atm. Determine the molecular formula for the sulfur molecules dissolved in CS_2.

*81. A certain solvent has a freezing point of −22.465°C. Dilute (0.050 molal) solutions of four common acids are prepared in this solvent, and their freezing points are measured, with these results:

Acid	Freezing Point
HCl	−22.795°C
H_2SO_4	−22.788°C
$HClO_4$	−22.791°C
HNO_3	−22.632°C

Determine K_f for this solvent and advance a reason why one of the acids differs so much from the others in its power to depress the freezing point.

82. The expressions for boiling-point elevation and freezing-point depression apply accurately to *dilute* solutions only. A saturated aqueous solution of NaI (sodium iodide) in water has a boiling point of 144°C. The mole fraction of NaI in the solution is 0.390. Compute the molality of this solution. Compare the boiling point elevation predicted by the expressions in this chapter with the elevation actually observed.

83. You take a bottle of a carbonated soft drink out of your refrigerator. The contents are liquid and stay liquid, even when you shake them. Presently, you remove the cap, and the liquid freezes solid. Offer a possible explanation for this observation.

*84. Mercury(II) chloride ($HgCl_2$) freezes at $276.1°C$ and has a freezing-point depression constant K_f of 34.3 K kg mol^{-1}. When 1.36 g of solid mercury(I) chloride (empirical formula $HgCl$) is dissolved in 100 g of $HgCl_2$, the freezing point is lowered by $0.99°C$. Calculate the molar mass of the dissolved solute species, and give its molecular formula.

85. A new compound has the empirical formula $GaCl_2$. This surprises some chemists, who expect a chloride of gallium, based on the position of gallium in the periodic table, to have the formula $GaCl_3$ or possibly $GaCl$. They suggest that the "$GaCl_2$" is really $Ga[GaCl_4]$, in which the bracketed group behaves as a unit with a -1 charge. Suggest experiments to test this hypothesis.

86. Ethylene glycol (CH_2OHCH_2OH) is used in antifreeze because, when mixed with water, it lowers the freezing point below $0°C$. What mass percentage of ethylene glycol in water must be used to reduce the freezing point of the mixture to $-5.0°C$, assuming ideal solution behavior?

87. Table 6–2 shows less association among ions in a 0.01 mol kg^{-1} solution of NaCl than in a 0.01 mol kg^{-1} solution of $MgSO_4$. Suggest a reason why NaCl (for which $\nu = 2$) associates less than $MgSO_4$ (for which $\nu = 2$) in solutions of equal concentration.

88. A thermometer put into a bucket of water mixed with ice gives a temperature of $0°C$. When some salt is added to the mixture, the temperature in the bucket drops several degrees. Your brother in junior high school is astonished to see this. He says, "They put salt on icy streets in the winter to melt ice! Why does adding salt lower the temperature in the bucket? It should raise it." Explain.

89. The formula for osmotic pressure, $\Pi = cRT$ is only approximate, so the formula $\Pi = m'RT$ is derived and offered as an improvement (m' is the "volume molality," the number of moles of solute per liter of solvent). The measured osmotic pressure of a 0.50 molal solution of sucrose in water is 12.75 atm at $20°C$. Compute the osmotic pressure of this solution according to both formulas to see which is better. The density of the sucrose solution is 1.061 g cm^{-3}.

*90. Two solutions are prepared. One contains 5 g of NaCl and 95 g of water, and the other contains 10 g of NaCl and 90 g of water. The two are placed in open beakers next to each other and are enclosed under a bell jar that is just big enough to fit over the beakers. After two weeks, which beaker has more liquid in it? Explain. (*Hint:* Think about the analogy to osmotic pressure.)

*91. The walls of erythrocytes (red blood cells) are permeable to water but not to sodium or chloride ions. In an experiment at $25°C$, an aqueous solution of NaCl that has a freezing point of $-0.406°C$ causes erythrocytes neither to swell nor to shrink, indicating that the osmotic pressure of their contents is equal to that of the NaCl solution. Calculate the osmotic pressure of the solution inside the erythrocytes under these conditions, assuming that its molarity and molality are equal.

92. In the days before refrigeration, people relied on pickling in brine (concentrated salt solution) to preserve meat and vegetables; they also preserved fruit in syrup that contained lots of sugar.
 (a) Is a pickled cucumber larger or smaller than a fresh cucumber. Why?
 (b) Why does pickling work to preserve food?

93. The Henry's law constant for the dissolution of $N_2(g)$ in $H_2O(\ell)$ is 8.57×10^4 atm, and for the dissolution of $N_2(g)$ in $CCl_4(\ell)$, it is 3.52×10^9 atm. Determine the mole fraction of saturated $N_2(aq)$ and saturated $N_2(CCl_4)$ if the pressure of $N_2(g)$ is 5.0 atm.

94. Henry's law is important in environmental chemistry, where it predicts the distribution of pollutants between water and the atmosphere. Benzene (C_6H_6) emitted in wastewater streams, for example, can pass into the air, where it is degraded by processes induced by light from the sun. The Henry's law constant for benzene in water at $25°C$ is 301 atm. Calculate the partial pressure of benzene vapor in equilibrium with a solution of 2.0 g of benzene per 1000 L of water. How many benzene molecules are present in each cubic centimeter of air?

*95. Methyl butyrate and ethyl acetate are soluble in each other in all proportions and form ideal solutions. At $50°C$, methyl butyrate (molar mass 102.13 g mol^{-1}) has a vapor pressure of 0.1443 atm. At the same temperature, ethyl acetate (molar mass 88.10 g mol^{-1}) has a vapor pressure of 0.3713 atm. The total pressure at $50°C$ of a certain solution of the two substances is 0.2400 atm.
 (a) What is the mole fraction of ethyl acetate in the solution?
 (b) What is the mole fraction of ethyl acetate in the vapor phase?

*96. Refer to the data of problem 60. Calculate the mole fraction of toluene in a mixture of benzene and toluene that boils at $90°C$ under atmospheric pressure.

*97. In a highly nonideal solution, the dependence of boiling temperature on composition can differ strongly from that shown in Figure 6–33. An example of the type of behavior that results (for HCl in water) is shown on the next page. For this mixture, there is a maximum in the boiling temperature at an intermediate mole fraction called the *azeotropic* composition. Suppose an equimolar mixture of water and hydrochloric acid is fractionally distilled. What are the compositions of the two liquid fractions obtained?

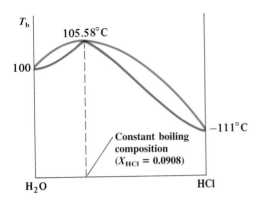

98. Gold–ruby glass has a deep red color produced by colloidal gold particles suspended in the glass. To make it, a small amount of gold is dissolved in molten glass, and the glass is then cooled to a temperature of 600 to 700°C, at which point the gold is not soluble and forms a suspension of small spherical gold crystals. Suppose a sample of gold–ruby glass contains 0.0080% by mass of gold. The glass has a density of 2.8 g cm^{-3}, and the density of the gold itself is 19.3 g cm^{-3}. If the gold spheres have a radius of 3.0 nm (3.0 × 10^{-7} cm), calculate the number of colloidal particles per cubic centimeter of glass.

CUMULATIVE PROBLEM

Chlorine and Chlorides

At room conditions, chlorine (Cl$_2$) is a yellow-green, poisonous gas with a suffocating odor. Chlorides, obtained from the reduction of chlorine, are more innocuous. Indeed, the aqueous chloride ion is essential in living tissues. Chlorine melts at −101°C and boils at −34°C at a pressure of 1 atm. Its critical point is at 143.8°C and 78.9 atm, and its triple point is at −101.5°C and 0.012 atm.

(a) What type of forces operate among neighboring Cl$_2$ molecules in liquid chlorine?

(b) Use Figure 6–1 to estimate the depth of the potential energy well (in kJ mol^{-1}) as two Cl$_2$ molecules approach each other. Estimate the approximate distance of closest approach.

(c) Chlorine is transported under high pressure in steel cylinders. You buy a cylinder and attach a take-off valve. Surprisingly, the pressure of chlorine at the valve does not drop as you remove gas! Only when nearly all the chlorine has been removed does the pressure finally drop. Explain.

(d) A sample of chlorine at 25°C is heated to 150°C at a constant pressure of 1 atm, compressed to 100 atm and then cooled back to 25°C at a constant pressure of 100 atm. What phase changes, if any, are observed?

(e) At 25°C and under a pressure of 1 atm, a maximum of 0.641 g of Cl$_2$ dissolves in 100.00 g of water. The density of this saturated "chlorine water" is 1.022 g cm^{-3}. Compute the mass percentage, mole fraction, molality, and molarity of the Cl$_2(aq)$, assuming no dissociation or other reaction in water.

(f) Use the data in part (e) to estimate the Henry's law constant for Cl$_2$ in water at 25°C, again assuming no dissociation or other reaction in water.

(g) Aqueous chlorine in fact does react with the solvent:

$$Cl_2(aq) + H_2O(\ell) \longrightarrow HOCl(aq) + H^+(aq) + Cl^-(aq)$$

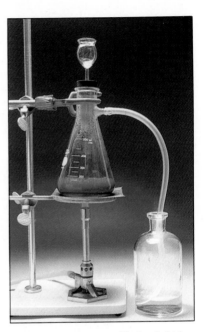

Yellow-green gaseous chlorine bubbles into water as it is produced from the oxidation of NaCl with K$_2$Cr$_2$O$_7$ in acidic solution.

The reaction stops after generating 0.326 mol of each product per 1.000 mol of $Cl_2(aq)$ originally dissolved. Determine ν and i for this solution of Cl_2. How does this situation differ from the case of the dissociation of a salt or a strong acid or base in water?

(h) Estimate the osmotic pressure of saturated $Cl_2(aq)$ at 25°C.

(i) Adding NaOH(s) to the chlorine water drives the reaction given in part (g) to completion and neutralizes the acidic HOCl(aq) and HCl(aq). The result is a solution of sodium chloride and sodium hypochlorite with neither unreacted Cl_2 nor NaOH left over. Determine the freezing point of this solution, assuming that the NaCl and NaOCl are completely dissociated in Na^+, Cl^-, and OCl^- ions.

Chemical Equilibrium

"Soda–straw" stalactites. Stalactites such as these form in caves as the loss of dissolved carbon dioxide shifts a heterogeneous equilibrium to favor precipitation of calcium carbonate from groundwater.

Balanced chemical equations provide a compact and efficient means of representing chemical reactions (see Section 2–3). Balanced equations allow us to calculate the quantities of products to be expected when known quantities of reactants undergo chemical change. The assumption behind such stoichiometric calculations is that the chemical reaction goes "to completion." Although such calculations are important and useful, we must now acknowledge that many chemical reactions do not actually reach completion. Instead, they stop short at an intermediate state of **chemical equilibrium,** in which unconsumed reactants are mixed with the products but no further net change occurs (Fig. 7–1).

Important questions immediately come to mind. What is the nature of this equilibrium? What governs the extent of a reaction? Can it be altered? How can we tell a reaction that has truly come to equilibrium from one that is still proceeding, but only very slowly? We consider some of these questions in this chapter; others must wait for the development of thermodynamic and kinetic principles in Chapters 10 through 14. In this chapter, we examine the way in which chemical equilibrium is established and show how the use of equilibrium constants allows us to calculate the amounts of reactants and products present at equilibrium. In Chapters 8 and 9, we focus on two particularly important types of chemical equilibria in aqueous solution—those involving acids and bases, and those between ionic solids and their dissolved ions.

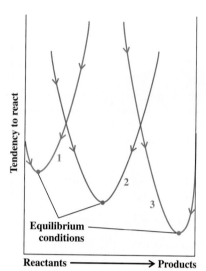

Figure 7–1 Chemical reactions slow down and stop at intermediate points in their progress, as their products build up. The low points on the lines represent conditions in which no tendency remains to react in either direction. This is the equilibrium state. In case 1, the reaction barely begins before reaching equilibrium and stopping. In case 2, the reaction gets about halfway to completion. In case 3, the reaction very nearly reaches completion.

• This equilibrium is a phase equilibrium. See Section 6-3.

7–1 CHEMICAL REACTIONS AND EQUILIBRIUM

Suppose that some liquid water is sealed in a container and held at constant temperature. The water immediately starts to evaporate as fast-moving molecules break free of the attractions of their neighbors in the bulk of the liquid and enter the space above. The pressure of this vapor grows rapidly at first, but then more slowly as molecules start to make the reverse transfer, from vapor into liquid. When the rate of condensation equals the rate of evaporation, the vapor pressure settles to an easily measured final value that depends solely on the temperature. This pressure holds unchanged, as long as the container and its contents are left alone. The liquid/vapor mixture has come to an **equilibrium,** a condition in which all tendency for change has been exhausted.

We represent this condition by the equilibrium chemical equation:

$$H_2O(\ell) \rightleftharpoons H_2O(g)$$

The double arrows emphasize the dynamic nature of the equilibrium: liquid water continues to evaporate to form water vapor while the vapor simultaneously condenses to give liquid, even though the pressure of the water vapor does not change.

Subsequently, comparison between the equilibrium pressure recorded in this experiment and the vapor pressure above *any* enclosed sample of water held at the same temperature tells everything that there is to know about the progress of the sample toward equilibrium. The shape and size of the container[1] do not matter nor do the amounts of liquid and vapor, as long as some of each is present. Rather, liquid vaporizes as long as the pressure is less than the experimental equilibrium vapor pressure; vapor condenses whenever its pressure exceeds the equilibrium value.

[1] If the container is very large, then all of the water might evaporate without raising the pressure of the vapor up to the equilibrium value. The result is not an equilibrium between the liquid and vapor because one of the essential components (the liquid) is absent.

Remarkably, a similar point applies to *all* chemical reactions:

The equilibrium condition for *every* reaction can be summed up in a single equation in which a number, the **equilibrium constant (K)** of the reaction, equals an **equilibrium expression,** a function of properties of the reactants and products.

The equilibrium condition for the evaporation of water may be written as

$$\text{vapor pressure of } H_2O = P_{H_2O} = K$$

The value of K is numerically equal to the vapor pressure (in atmospheres) of pure water at the temperature in question. Equilibrium constants depend on temperature. At 25°C, K for this reaction is 0.03126, whereas at 30°C, it is 0.04187:

$$H_2O(\ell) \rightleftharpoons H_2O(g) \text{ at } 25°C \qquad K = 0.03126$$
$$H_2O(\ell) \rightleftharpoons H_2O(g) \text{ at } 30°C \qquad K = 0.04187$$

In a full, correct treatment of equilibrium, equilibrium constants must be dimensionless (unitless) numbers. How can a dimensionless quantity, K, equal a pressure, which has units of atmospheres? Strictly speaking, it cannot. To make the equation $P_{H_2O} = K$ entirely correct, the vapor pressure of the water must be replaced by a *ratio* of its value to a reference pressure $P_{ref} = 1$ atm, so that the equation reads

$$\frac{P_{H_2O}}{P_{ref}} = K$$

in which both sides are dimensionless. Because P_{ref} is numerically equal to 1, provided that we express all pressures in atmospheres, its presence serves only to make the equation dimensionally correct.

The convention in this book is to express all pressures in atmospheres and to omit factors of P_{ref} because their value is unity. K is a pure number.

The equilibrium between a liquid and its vapor is so simple that introducing the equilibrium constant seems unnecessary. Chemical equilibria are usually more complicated than simple evaporation, however. Equilibrium equations provide a unified understanding of how even complex **reaction systems** (mixtures of many species taking part in concurrent chemical reactions) move toward equilibrium. Even in the most complex reaction systems, the physical principles parallel those at work in the evaporation and condensation of water.

Figure 7-2 Cooling a mixture of NO_2 and N_2O_4 (right) causes its color to fade as red-brown NO_2 is converted to colorless N_2O_4. Heating such a mixture (left) reverses the conversion and intensifies the color. The two tubes contain the same mass of substance, which is distributed in different ways between NO_2 and N_2O_4.

• This includes chemical equilibria and phase equilibria of all types.

A Simple Example of Chemical Equilibrium

When gaseous nitrogen dioxide (NO_2) is cooled, its red-brown color fades (Fig. 7-2). The reason for this fading of color is the chemical conversion of NO_2 to dinitrogen tetraoxide (N_2O_4), a change represented by the equation

$$2\,NO_2(g) \longrightarrow N_2O_4(g)$$

Because $N_2O_4(g)$ is colorless and $NO_2(g)$ is reddish brown, the color of their mixture tells the extent of the chemical change between the two. The reverse reaction

$$N_2O_4(g) \longrightarrow 2\,NO_2(g)$$

• The partial pressure of N_2O_4 present at a given partial pressure of NO_2 falls with increasing temperature. Evidently, a smaller proportion of the NO_2 molecules *dimerize* (form a molecule with a doubled formula) to N_2O_4 at higher temperature.

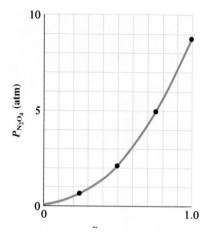

Figure 7-3 A graph of the partial pressure of N_2O_4 at equilibrium versus that of NO_2 traces an arc of a parabola if the temperature is constant. The points on this curve are found by placing different amounts of pure NO_2 in a flask, cooling to 25°C, and measuring the partial pressure of each substance.

restores the color when the temperature is raised. At high temperatures, the intense red-brown color means that NO_2 predominates.

If a flask is filled with a known amount of NO_2 at a high temperature and then cooled, the total pressure drops below that given by the ideal gas law for NO_2 alone. The reason is that for each molecule of N_2O_4 formed, *two* molecules of NO_2 are consumed, decreasing the total number of molecules and thereby the number of collisions with the walls. In a series of such experiments, in which the high temperature and low temperature are kept the same but the amount of NO_2 placed into the flask is varied, an interesting relationship emerges between the observed partial pressures of the NO_2 and the N_2O_4. A plot of $P_{N_2O_4}$ against P_{NO_2} (Fig. 7-3) has the simple form

$$P_{N_2O_4} = K \times (P_{NO_2})^2$$

where K is a constant whose value depends only on the particular temperature at which the final measurements are made. This relationship, which is the equation for a parabola, can be rearranged to the form

$$\frac{P_{N_2O_4}}{(P_{NO_2})^2} = K$$

The experimental value of K is 8.8 at 25°C, provided that the partial pressures of N_2O_4 and NO_2 are expressed in atmospheres. The NO_2-N_2O_4 reaction mixture is in chemical equilibrium

$$2\ NO_2(g) \rightleftharpoons N_2O_4(g)$$

and $K = 8.8$ is its equilibrium constant at 25°C.

Chemical equilibrium, like the equilibrium between phases, is a dynamic process. A spontaneous approach to equilibrium is observed when the initial partial pressures of the reactants are high, and collisions between their molecules cause product molecules to form. Once the partial pressures of the products have increased sufficiently, the reverse reaction, which regenerates the reactants from the products, begins to occur.

> As the equilibrium state is approached, the forward and backward rates of reaction approach equality. At equilibrium the rates are equal, and no further net change occurs in the partial pressures of reactants or products.

* Similar experiments show the dynamic nature of equilibrium in the coexistence of the liquid and vapor phases, as discussed in Section 6-3.

Just as the equilibrium between liquid water and water vapor is an active, dynamic process on the molecular scale, with evaporation and condensation taking place simultaneously, so the chemical equilibrium between NO_2 and N_2O_4 occurs through the continuous formation of molecules of N_2O_4 from NO_2 molecules and their dissociation back into NO_2 molecules, with equal rates. This fact has been verified for all types of chemical reactions by means of isotopic labelling experiments: if isotopically labelled nitrogen dioxide ($^{15}NO_2$) is mixed with unlabelled N_2O_4 in their exact equilibrium proportions, the labelled nitrogen atoms become distributed between the two compounds, even though the amounts of the two gases present in the equilibrium mixture do not change. This shows that the forward and reverse reactions must continue after equilibrium is reached. Chemical equilibrium is not a static condition, although macroscopic properties such as partial pressures do stop changing when equilibrium is attained. Rather, equilibrium is the consequence of a dynamic balance between forward and backward reactions.

The same equilibrium state will be reached whether starting with the reactants or with the products. Suppose that the study of NO_2 and N_2O_4 had begun not with pure NO_2 at high temperature but with an equal mass of pure N_2O_4 at low temperature (Fig. 7–4). Upon heating the N_2O_4 to 25°C, dissociation to NO_2 would begin. The ultimate equilibrium state would be absolutely indistinguishable from the one established starting from NO_2 at high temperature and cooling it to 25°C. This characteristic of the equilibrium state is used to test whether a chemically reacting mixture is truly in equilibrium. Such a test is important because many reactions proceed exceedingly slowly. A very slow reaction may require months or even years before displaying any visible signs of change. In the meantime, it is far from equilibrium, and its temporary changelessness is deceptive. If the same state is reached beginning with either reactants or products, it may safely be concluded that it is a true equilibrium state.

We summarize the four fundamental characteristics of equilibrium states in systems that are isolated from outside interference:

1. They display no macroscopic evidence of change.
2. They are reached through spontaneous processes.
3. They show a dynamic balance of forward and reverse processes.
4. They are the same regardless of the direction from which they were approached.

The Form of Equilibrium Expressions

The relationship between the partial pressures of NO_2 and N_2O_4 at equilibrium

$$\frac{P_{N_2O_4}}{(P_{NO_2})^2} = K$$

is a particular example of a general law that was formulated in 1864 by two Norwegian scientists, C. M. Guldberg and P. Waage. This law states that the partial pressures of reacting gases present at equilibrium are not independent of each other but are related through the temperature-dependent equilibrium constant. Similar relationships have been determined experimentally for other gas-phase chemical reactions. For example, the reaction between hydrogen and nitrogen to produce ammonia

$$N_2(g) + 3\,H_2(g) \rightleftharpoons 2\,NH_3(g)$$

is characterized by the equilibrium law

$$\frac{(P_{NH_3})^2}{(P_{N_2})(P_{H_2})^3} = K$$

where K is equal to 6.8×10^5 at 25°C and 1.9×10^{-4} at 400°C, with pressures expressed in atmospheres. Notice that the form of this expression differs significantly from the one discovered for the equilibrium between NO_2 and N_2O_4. Is there any way other than experimentation to know the mathematical form of the relationship between the partial pressures and the equilibrium constant? The answer fortunately is "yes." Its details are given in this generalization:

High temperature

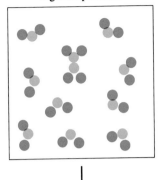

Cool

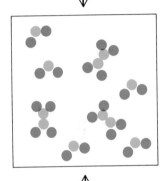

Heat

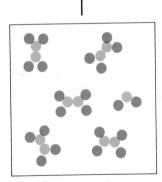

Low temperature

Figure 7–4 The same equilibrium state is reached at a given temperature, regardless of whether the initial state consists of NO_2 molecules, which dimerize as the gas is cooled from a high temperature, or of the same mass of N_2O_4 molecules, which dissociate as the gas is heated from a low temperature.

In a chemical reaction in which a moles of species A and b moles of species B react to form c moles of species C and d moles of species D,

$$a\text{A} + b\text{B} \rightleftharpoons c\text{C} + d\text{D}$$

the partial pressures at equilibrium are related through

$$\frac{P_C^c P_D^d}{P_A^a P_B^b} = K$$

provided that all species are present as low-pressure gases.

The highlighted statement of the form of equilibrium expressions applies whenever two reactants give two products. If more products and reactants than this appear in the chemical equation, then additional partial pressures are inserted into the equilibrium expression so that there is one for every product and reactant. The partial pressures of *products* appear in the numerator, and those of *reactants* appear in the denominator. Each partial pressure is raised to a power that is equal to its coefficient in the balanced chemical equation. In Section 7–6, we extend this law to include solutions and heterogeneous equilibria; this more general form is called the **law of mass action.**

• *Heterogeneous* equilibria involve chemical reactants and products in two or more phases.

Equilibrium equations hold for low-density gas mixtures, no matter how complex they are. For example, the relationship between the partial pressures of NO_2 and N_2O_4 at equilibrium

$$\frac{P_{N_2O_4}}{(P_{NO_2})^2} = K$$

holds not only for binary mixtures of $NO_2(g)$ and $N_2O_4(g)$ but also for mixtures containing many other species, even if those species react with the $NO_2(g)$ or $N_2O_4(g)$. Thus, although $N_2(g)$, $O_2(g)$, and $NO(g)$ may all be present, with additional equilibrium equations relating their partial pressures to those of $NO_2(g)$ and $N_2O_4(g)$, the relationship between $P_{N_2O_4}$ and P_{NO_2} and the numerical value of K remains unaltered. Equilibrium equations continue to hold despite the presence of other species or the operation of simultaneous equilibria, provided that the temperature is unchanged. At high enough densities, when the gas mixture is not ideal and Dalton's law begins to fail, deviations do appear.

EXAMPLE 7–1

Write equilibrium expressions for each of the following gas-phase chemical equilibria:
(a) $PCl_5(g) \rightleftharpoons PCl_3(g) + Cl_2(g)$
(b) $2\,NOCl(g) \rightleftharpoons 2\,NO(g) + Cl_2(g)$
(c) $CO(g) + 1/2\,O_2(g) \rightleftharpoons CO_2(g)$

Solution

(a) $\dfrac{(P_{PCl_3})(P_{Cl_2})}{P_{PCl_5}} = K$

Recall that products appear in the numerator and reactants in the denominator.

(b) $\dfrac{(P_{NO})^2(P_{Cl_2})}{(P_{NOCl})^2} = K$

The two exponents come from the two coefficients of 2 in the balanced equation.

(c) $\dfrac{P_{CO_2}}{(P_{CO})(P_{O_2})^{1/2}} = K$

Fractional exponents appear in the equilibrium expression whenever fractional co-efficients are present in the balanced equation.

Exercise

Write equilibrium expressions for the reactions defined by the following equations:

(a) $3\,H_2(g) + SO_2(g) \rightleftharpoons H_2S(g) + 2\,H_2O(g)$

(b) $2\,C_2F_5Cl(g) + 4\,O_2(g) \rightleftharpoons Cl_2(g) + 4\,CO_2(g) + 5\,F_2(g)$

Answer:

(a) $\dfrac{P_{H_2S}P_{H_2O}^2}{P_{H_2}^3 P_{SO_2}} = K$ (b) $\dfrac{P_{Cl_2}P_{CO_2}^4 P_{F_2}^5}{P_{C_2F_5Cl}^2 P_{O_2}^4} = K$

7-2 CALCULATING EQUILIBRIUM CONSTANTS

If the equilibrium partial pressures of all of the reactants and products can be mea-sured, then the equilibrium constant can be calculated by writing the equilibrium expression and simply substituting the experimental values (in atmospheres) into it. Suppose that at a certain temperature a gas mixture has reached equilibrium with the following partial pressures:

$$P_{H_2} = 0.12 \text{ atm}, \; P_{N_2} = 0.70 \text{ atm, and } P_{NH_3} = 0.84 \text{ atm}$$

Then, at that temperature, the equilibrium constant for the reaction

$$N_2(g) + 3\,H_2(g) \rightleftharpoons 2\,NH_3(g)$$

has the value

$$K = \dfrac{(P_{NH_3})^2}{(P_{N_2})(P_{H_2})^3} = \dfrac{(0.84)^2}{(0.70)(0.12)^3} = 5.8 \times 10^2$$

In most cases, it is not practical to measure the equilibrium partial pressure of each separate reactant and product directly. Fortunately, the equilibrium constant can usually be derived from other available data. Consider, for example, the for-mation of phosgene, $COCl_2$, from CO and Cl_2:

$$CO(g) + Cl_2(g) \rightleftharpoons COCl_2(g)$$

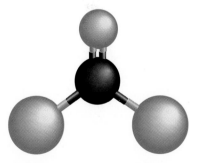

Phosgene ($COCl_2$) is an important chemical intermediate. Although it is highly toxic, observing standard safety precautions reduces the danger in han-dling it to a minimum. Phosgene is used in making polyurethane, a kind of polymer employed both in foams (for furniture cushions and bedding) and as a coating for wood.

This reaction reaches equilibrium rapidly at 600°C. Suppose that a mixture of CO and Cl_2 is prepared at 600°C that has initial partial pressures (before reaction) of 0.60 atm for CO and 1.10 atm for Cl_2. After the reaction mixture has reached equilibrium, the partial pressure of $COCl_2(g)$ is measured to be 0.10 atm. We wish to calculate the equilibrium constant.

In this case, we have the equilibrium partial pressure of $COCl_2(g)$, but only the *initial* partial pressures of the other two gases. To find the equilibrium constant, we need to determine the equilibrium partial pressures of CO and Cl_2. To do this, we set up a table detailing the changes occurring in coming to equilibrium:

	$CO(g)$ +	$Cl_2(g)$ $\rightleftharpoons$	$COCl_2(g)$
Initial partial pressure (atm)	0.60	1.10	0
Change in partial pressure (atm)	?	?	+0.10
Equilibrium partial pressure (atm)	?	?	0.10

Note that the first two lines must add to give the third in all three columns. To fill in the blanks in the table, we use the relationships built into the balanced chemical equation. Every mole of $COCl_2$ that is produced consumes exactly one mole of CO and exactly one mole of Cl_2. (This simple case arises because all coefficients in the balanced equation are 1.) According to the ideal gas law, the partial pressures of gases are proportional to their chemical amounts as long as the volume and temperature are held fixed; therefore, the change in partial pressure of each gas must be proportional to the change in its chemical amount as the mixture goes toward equilibrium. If the partial pressure of $COCl_2$ *increases* by 0.10 atm through reaction, the partial pressures of CO and Cl_2 must each *decrease* by 0.10 atm. Inserting these values into the table gives

	$CO(g)$ +	$Cl_2(g)$ $\rightleftharpoons$	$COCl_2(g)$
Initial partial pressure (atm)	0.60	1.10	0
Change in partial pressure (atm)	−0.10	−0.10	+0.10
Equilibrium partial pressure (atm)	0.50	1.00	0.10

where we have obtained entries in the last line by adding those in the first two.

We can now calculate the equilibrium constant by inserting the equilibrium partial pressures into the equilibrium expression:

$$K = \frac{P_{COCl_2}}{(P_{CO})(P_{Cl_2})} = \frac{(0.10)}{(0.50)(1.00)} = 0.20$$

The calculation of K for the production of phosgene is particularly simple because the coefficients in the balanced equation are all 1. When this is not true, care is required to derive correct expressions for the changes in partial pressure of the products and reactants, as in the following example.

EXAMPLE 7-2

Consider the equilibrium

$$4 NO_2(g) \rightleftharpoons 2 N_2O(g) + 3 O_2(g)$$

The three gases are introduced into a container at partial pressures of 3.6 atm (for NO_2), 5.1 atm (for N_2O), and 8.0 atm (for O_2) and react to reach equilibrium at a fixed temperature. The equilibrium partial pressure of the NO_2 is measured to be 2.4 atm. Calculate the equilibrium constant of the reaction at this temperature, assuming that no competing reactions occur.

Solution

Set up the recommended table of changes:

	$4 NO_2(g)$	$\rightleftharpoons 2 N_2O(g)$	$+ 3 O_2(g)$
Initial partial pressure (atm)	3.6	5.1	8.0
Change in partial pressure (atm)	$-4x$	$+2x$	$+3x$
Equilibrium partial pressure (atm)	2.4	$5.1 + 2x$	$8.0 + 3x$

If the partial pressure of NO_2 decreases by $4x$ atm, that of $N_2O(g)$ must increase by $2x$ atm because of the coefficients of 4 and 2 in the balanced equation. That of O_2 must increase by $3x$ atm. The changes in partial pressures of products have the opposite sign (positive instead of negative) because as reactants disappear, products appear. Their relative magnitudes depend on the coefficients in the balanced equation.

From the first column of the table, $3.6 - 4x = 2.4$ atm, so $x = 0.3$ atm. Inserting this value of x into the second and third columns shows that the equilibrium partial pressure of N_2O must be $5.1 + 2(0.3) = 5.7$ atm and that of O_2 must be $8.0 + 3(0.3) = 8.9$ atm. The equilibrium constant is then

$$K = \frac{(P_{N_2O})^2(P_{O_2})^3}{(P_{NO_2})^4} = \frac{(5.7)^2(8.9)^3}{(2.4)^4} = 6.9 \times 10^2$$

Exercise

The little-known compound $GeWO_4(g)$ forms at high temperature in the reaction

$$2 GeO(g) + W_2O_6(g) \rightleftharpoons 2 GeWO_4(g)$$

Some $GeO(g)$ and $W_2O_6(g)$ are mixed. Before they start to react, the partial pressure of each is 1.000 atm. After their reaction at constant temperature and volume, the equilibrium partial pressure of $GeWO_4(g)$ is 0.980 atm. Assuming that this is the only reaction that takes place, (a) determine the equilibrium partial pressures of $GeO(g)$ and of $W_2O_6(g)$, and (b) determine the equilibrium constant for the reaction.

Answer: (a) $P_{GeO} = 0.020$ atm; $P_{W_2O_6} = 0.510$ atm. (b) $K = 4.7 \times 10^3$.

Relationships Among Equilibrium Expressions

Consider these two chemical equations, which are accompanied by equilibrium-constant expressions:

$$2\,H_2(g) + O_2(g) \rightleftharpoons 2\,H_2O(g) \qquad \frac{P_{H_2O}^2}{P_{H_2}^2 P_{O_2}} = K_1$$

$$2\,H_2O(g) \rightleftharpoons 2\,H_2(g) + O_2(g) \qquad \frac{P_{H_2}^2 P_{O_2}}{P_{H_2O}^2} = K_2$$

According to the first equation, hydrogen and oxygen combine to form water vapor. According to the second equation, water vapor decomposes into hydrogen and oxygen: the second is clearly the reverse of the first. The expressions for the equilibrium constants K_1 and K_2 are just as clearly inverses of each other—the numerator of the first is the denominator of the second and vice versa—so their product $K_1 K_2$ equals 1.

> The equilibrium constant for a reverse reaction is always the reciprocal of the equilibrium constant for the corresponding forward reaction.

Suppose that we multiply all of the coefficients in the first of these equations by a constant. The result of multiplying by $\frac{1}{2}$ is

$$H_2(g) + \tfrac{1}{2} O_2(g) \rightleftharpoons H_2O(g)$$

This is a perfectly satisfactory way of writing the equation for the chemical reaction; it simply states that one mole of hydrogen reacts with one-half mole of oxygen to yield one mole of water vapor. The corresponding equilibrium expression is

$$\frac{P_{H_2O}}{P_{H_2} P_{O_2}^{1/2}} = K_3$$

Comparison with the expression for K_1 shows that

$$K_3 = K_1^{1/2} = \sqrt{K_1}$$

> When the coefficients in a balanced chemical equation are all multiplied by a constant factor, the corresponding equilibrium constant is raised to a power equal to that factor.

Further operations can be carried out on chemical equations: we can add two equations together to make a third or subtract one equation from another to make a third. In these two cases, the equilibrium constants for the resultant equation are, respectively, the product and the quotient of the equilibrium constants of the first two equations. As examples, take two equations that represent reactions among the halogens:

$$2\,BrCl(g) \rightleftharpoons Cl_2(g) + Br_2(g) \qquad \frac{P_{Cl_2} P_{Br_2}}{P_{BrCl}^2} = K_1 = 0.45 \text{ at } 25°C$$

$$Br_2(g) + I_2(g) \rightleftharpoons 2\,IBr(g) \qquad \frac{P_{IBr}^2}{P_{Br_2}P_{I_2}} = K_2 = 0.051 \text{ at } 25°C$$

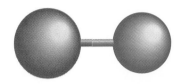

Adding the two chemical equations gives

$$2\,BrCl(g) + Br_2(g) + I_2(g) \rightleftharpoons 2\,IBr(g) + Cl_2(g) + Br_2(g)$$

The term $Br_2(g)$ appears on both sides of this equation. Removing it from both sides gives

$$2\,BrCl(g) + I_2(g) \rightleftharpoons 2\,IBr(g) + Cl_2(g) \qquad \frac{P_{IBr}^2 P_{Cl_2}}{P_{BrCl}^2 P_{I_2}} = K_3$$

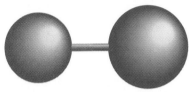

and inspection of the expressions for K_1, K_2, and K_3 confirms

$$K_3 = K_1 K_2 = (0.45)(0.051) = 0.023 \text{ at } 25°C$$

If a second equation is subtracted from the first, the resulting equilibrium constant is that of the first divided by that of the second. In summary,

Bromine chloride (BrCl) and iodine bromide (IBr) are two of a rather large group of compounds called the *interhalogens*, which are combinations of two or more halogen elements. Bromine chloride dissociates so readily into Br_2 and Cl_2 that it cannot be prepared in pure form. Iodine bromide is more stable, forming a black crystalline solid that melts at 41°C.

> The operations of addition and subtraction, when applied to chemical equations, lead to the operations of multiplication and division of their corresponding equilibrium expressions and equilibrium constants.

EXAMPLE 7-3

The atmospheric concentrations of the oxides of nitrogen are monitored in air-pollution studies. At 25°C, the equilibrium constant for the reaction

$$NO(g) + \tfrac{1}{2}O_2(g) \rightleftharpoons NO_2(g)$$

is

$$\frac{P_{NO_2}}{P_{NO}P_{O_2}^{1/2}} = K_1 = 1.3 \times 10^6$$

and that for

$$\tfrac{1}{2}N_2(g) + \tfrac{1}{2}O_2(g) \rightleftharpoons NO(g)$$

is

$$\frac{P_{NO}}{P_{N_2}^{1/2}P_{O_2}^{1/2}} = K_2 = 6.5 \times 10^{-16}$$

Find the equilibrium constant K_3 for the reaction

$$N_2(g) + 2\,O_2(g) \rightleftharpoons 2\,NO_2(g)$$

Solution

Adding the chemical equations for the first two reactions gives

$$\tfrac{1}{2}N_2(g) + O_2(g) \rightleftharpoons NO_2(g)$$

The equilibrium constant for this reaction, K'_3, is just the product of K_1 and K_2, or $K_1 K_2$. The constant we seek, K'_3, is defined by a chemical equation that is twice this, so K'_3 must be raised to the power 2 (that is, squared) to give K_3.

$$K_3 = (K'_3)^2 = (K_1 K_2)^2 = \boxed{7.1 \times 10^{-19}}$$

Exercise

At 100°C, the equilibrium constant for the reaction

$$CF_4(g) + 2\,H_2O\,(g) \rightleftharpoons CO_2(g) + 4\,HF(g)$$

is 5.9×10^{23}, and the equilibrium constant for the reaction

$$CO(g) + \tfrac{1}{2}\,O_2(g) \rightleftharpoons CO_2(g)$$

is 1.3×10^{35}. Compute the equilibrium constant at 100°C for the reaction

$$CF_4(g) + 4\,H_2O(g) \rightleftharpoons 2\,CO_2(g) + 8\,HF(g) + O_2(g)$$

Answer: 2.1×10^{-23}.

7–3   THE REACTION QUOTIENT

The law of mass action gives information not only about the nature of the equilibrium state, but also about the direction in which a reaction tends to proceed when its reactants and products are mixed together in arbitrary amounts. For the general gas-phase reaction

$$a\mathrm{A} + b\mathrm{B} \rightleftharpoons c\mathrm{C} + d\mathrm{D}$$

we define the **reaction quotient Q** through the expression

$$Q = \frac{P_C^c P_D^d}{P_A^a P_B^b}$$

in which the partial pressures do *not* necessarily have equilibrium values. The distinction between Q and K is crucial. The equilibrium constant K is determined by the partial pressures of reactants and products *at equilibrium* and is a constant, depending only on the temperature. The reaction quotient Q is determined by the instantaneous partial pressures, whatever they may be; thus, Q changes with time. As the reaction approaches equilibrium, Q approaches K. The initial partial pressures $P_{A(init)}$, $P_{B(init)}$, $P_{C(init)}$, and $P_{D(init)}$ give an initial reaction quotient $Q_{(init)}$, the magnitude of which, relative to K, determines the direction in which the reaction proceeds to reach equilibrium. If $Q_{(init)}$ is *less* than K, then Q must *increase* as time goes on. This requires an increase in the product partial pressures and a decrease in reactant partial pressures; in other words, the reaction proceeds from left to right. If $Q_{(init)}$ is *greater* than K, similar reasoning shows that the reaction must proceed from right to left, with Q *decreasing* with time until it becomes equal to K (Fig. 7–5).

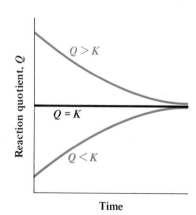

Figure 7–5   If nitrogen and hydrogen are mixed in 1:3 proportions, together with an arbitrary amount of ammonia, then the following occurs, either as written or in reverse:

$$N_2(g) + 3\,H_2(g) \rightleftharpoons 2\,NH_3(g)$$

If the initial value of Q, $Q_{(init)}$, is less than K, Q increases with time to reach K (red line); if $Q_{(init)}$ is greater than K, Q decreases with time to reach K (blue line). Only in the special case that $Q_{(init)}$ equals K is there no change (black line).

EXAMPLE 7–4

The reaction between nitrogen and hydrogen to produce ammonia

$$N_2(g) + 3\,H_2(g) \rightleftharpoons 2\,NH_3(g)$$

is essential in making nitrogen-containing fertilizers. This reaction has an equilibrium constant equal to 1.9×10^{-4} at 400°C. Suppose that 1.0 mol of N_2, 0.20 mol of H_2, and 0.40 mol of NH_3 are sealed in a 1.00-L vessel at 400°C. In which direction does the reaction proceed?

Solution

With the aid of the ideal-gas law, we calculate the initial partial pressures $P_{(\text{init})} = n_{(\text{init})}\,RT/V$ to be

$$P_{N_2(\text{init})} = 55 \text{ atm},\quad P_{H_2(\text{init})} = 11 \text{ atm, and } P_{NH_3(\text{init})} = 22 \text{ atm}$$

The initial value of Q is therefore

$$Q_{(\text{init})} = \frac{P_{NH_3}^2}{P_{N_2}P_{H_2}^3} = \frac{(22)^2}{(55)(11)^3} = 6.6 \times 10^{-3}$$

Because $Q_{(\text{init})} > K$, the reaction proceeds from right to left. As it does, Q diminishes toward K. Ammonia decomposes until equilibrium is reached, at which point Q equals K.

Exercise

The equilibrium constant for the reaction

$$P_4(g) \rightleftharpoons 2\,P_2(g)$$

is 1.39 at 400°C. Suppose that 1.40 mol of $P_4(g)$ and 1.25 mol of $P_2(g)$ are mixed in a closed 25.0-L container at 400°C. Compute Q at the moment of mixing, $Q_{(\text{init})}$, and state the direction in which the reaction proceeds.

Answer: $Q_{(\text{init})} = 2.47$. The reaction proceeds from right to left, generating P_4.

The magnitude of the equilibrium constant gives direct qualitative and quantitative information about the nature of the equilibrium. If K is very large, then at equilibrium the partial pressures and concentrations of the product species are large compared with those of the reactant species. In this case only, stoichiometry alone may be used to calculate the chemical amounts or the masses of the products formed. If K is very small, the reactant partial pressures are large compared with those of the products, and the extent of reaction is very small. For intermediate values of K, both reactants and products are present in significant proportions at equilibrium.

7–4   CALCULATION OF GAS-PHASE EQUILIBRIA

The law of mass action provides an equilibrium expression for any gas-phase reaction. Such expressions can then be used to predict quantitatively the results of chemical reactions. The details of arriving at such predictions depend on the experimental data available. Suppose, for example, that we know all the equilibrium partial pressures but one. We can use the equilibrium constant to calculate the last partial pressure. This is illustrated in the following example.

EXAMPLE 7-5

Hydrogen is made from natural gas (methane) for immediate consumption in industrial processes, such as the production of ammonia for fertilizers. The first step in this process is the "steam reforming of methane."

$$CH_4(g) + H_2O(g) \rightleftharpoons CO(g) + 3\,H_2(g)$$

Assume that the equilibrium constant for this reaction is 0.172 at 900 K. The partial pressures of $CH_4(g)$ and $H_2O(g)$ in an equilibrium mixture of gases at 900 K are both 0.400 atm, and the partial pressure of $CO(g)$ is 0.872 atm. Determine the partial pressure of $H_2(g)$.

Solution

The equilibrium expression is

$$\frac{(P_{CO})(P_{H_2})^3}{(P_{CH_4})(P_{H_2O})} = K$$

Inserting the three known partial pressures gives

$$\frac{(0.872)(P_{H_2})^3}{(0.400)(0.400)} = 0.172$$

Solving for $(P_{H_2})^3$ gives

$$(P_{H_2})^3 = \frac{(0.400)(0.400)}{0.872} \times 0.172 = 0.0316$$

Taking the cube root of both sides of this equation gives

$$P_{H_2} = 0.316 \text{ atm}$$

Exercise

The carbon monoxide produced in the reforming reaction of Example 7–5 can react further according to the so-called shift reaction to generate additional hydrogen. This reaction is represented by the chemical equation

$$CO(g) + H_2O(g) \rightleftharpoons CO_2(g) + H_2(g)$$

At 900 K, the equilibrium constant for this reaction is 0.64. Suppose the partial pressures of three of the gases at equilibrium are measured to be

$$P_{CO} = 2.00 \text{ atm}, \quad P_{CO_2} = 0.80 \text{ atm}, \quad \text{and} \quad P_{H_2} = 0.48 \text{ atm}$$

Calculate the partial pressure of the last reactant, H_2O, under these conditions.

Answer: $P_{H_2O} = 0.30$ atm.

Another frequently encountered type of calculation involves determining the equilibrium partial pressures of reactants or products from measurements of the partial pressures *before* any reaction occurs, as in the following example.

EXAMPLE 7–6

Suppose $I_2(g)$ and $H_2(g)$ are sealed in a flask at $T = 400$ K with partial pressures $P_{I_2} = 1.320$ atm and $P_{H_2} = 1.140$ atm. The sealed flask is then heated to 600 K, a temperature at which the gases speedily reach equilibrium.

$$I_2(g) + H_2(g) \rightleftharpoons 2\,HI(g)$$

Assume that the equilibrium constant for the reaction is 92.6 at 600 K:

$$\frac{P_{HI}^2}{P_{I_2}P_{H_2}} = 92.6$$

(a) What are the equilibrium values of P_{I_2}, P_{H_2}, and P_{HI} at 600 K?
(b) What percentage of the I_2 originally present has reacted when equilibrium is reached?

Solution

(a) Suppose I_2 and H_2 did *not* react at 600 K. From the ideal gas law at constant volume, their partial pressures would be

$$P_{I_2(init)} = 1.320 \text{ atm} \times \left(\frac{600 \text{ K}}{400 \text{ K}}\right) = 1.980 \text{ atm}$$

$$P_{H_2(init)} = 1.140 \text{ atm} \times \left(\frac{600 \text{ K}}{400 \text{ K}}\right) = 1.710 \text{ atm}$$

Of course, the two gases *do* react, and the extent of the reaction can be calculated. To do this, we set up a table of changes as in Example 7–2.

	$I_2(g)$ +	$H_2(g)$ $\rightleftharpoons$	$2\,HI(g)$
Initial partial pressure (atm)	1.980	1.710	0
Change in partial pressure (atm)	$-x$	$-x$	$+2x$
Equilibrium partial pressure (atm)	$1.980 - x$	$1.710 - x$	$2x$

If the partial pressure of I_2 decreases by x atm as the reaction proceeds, then the partial pressure of H_2 must also decrease by x atm because each mole of I_2 reacts with one mole of H_2. By similar reasoning, the partial pressure of HI increases by $2x$ atm: two moles of HI forms from each mole of I_2. Inserting the equilibrium partial pressures into the equilibrium expression results in the equation

- Because $P_{gas} = n_{gas}\,(RT/V)$, the partial pressures of the gases are directly proportional to their chemical amounts in the mixture as long as T and V do not change.

$$\frac{(2x)^2}{(1.980 - x)(1.710 - x)} = 92.6$$

Multiplying and collecting terms gives

$$88.6x^2 - 341.694x + 313.525 = 0$$

Solving for x using the quadratic formula (see Appendix C) gives

$$x = \frac{-(-341.694) \pm \sqrt{(341.694)^2 - 4(88.6)(313.525)}}{2(88.6)}$$

$$x = 1.5044 \text{ atm} \quad \text{or} \quad 2.35322 \text{ atm}$$

The second root is not physical because it leads to negative answers for the equilibrium partial pressures of the $H_2(g)$ and $I_2(g)$. Discarding it leaves

$$P_{HI} = 2 \times 1.5044 \text{ atm} = 3.0088 \text{ atm}$$
$$P_{I_2} = 1.980 \text{ atm} - 1.5044 \text{ atm} = 0.4756 \text{ atm}$$
$$P_{H_2} = 1.710 \text{ atm} - 1.5044 \text{ atm} = 0.2056 \text{ atm}$$

It is a good idea to check such results by inserting the calculated equilibrium partial pressures back into the equilibrium expression and making sure that the known value of K comes out

$$\text{Check:} \quad \frac{(3.0088)^2}{(0.4756)(0.2056)} = 92.6$$

As our final step, we round off each answer to the correct number of significant digits: $P_{HI} = 3.01$ atm, $P_{I_2} = 0.48$ atm, $P_{H_2} = 0.21$ atm. Rounding off sooner makes the K calculated in the check differ excessively from 92.6.

(b) The fraction of I_2 that has *not* reacted equals the chemical amount of I_2 present at equilibrium divided by the amount that was present at the start. Because neither volume nor temperature has changed, this ratio equals the ratio of the final partial pressure of I_2 (0.48 atm) to its initial partial pressure (1.980 atm).

$$\text{fraction of } I_2 \text{ unreacted} = \frac{0.48 \text{ atm}}{1.980 \text{ atm}} = 0.24$$

The percentage of unreacted I_2 is then $0.24 \times 100\%$, or 24%, and the percentage of I_2 that *has* reacted is 76%.

Exercise

At 100°C, the equilibrium constant for the reaction

$$H_2(g) + Si_2H_6(g) \rightleftharpoons 2 \, SiH_4(g)$$

equals 89.6. The two reactant gases are mixed in a sealed flask at 100°C and, before any reaction takes place, the partial pressure of each is 0.560 atm. (a) Compute the partial pressure of $SiH_4(g)$ when the reaction comes to equilibrium. (b) What percentage of the Si_2H_6 remains unreacted at equilibrium?

Answer: (a) 0.925 atm. (b) 17.4%.

The calculation of gas-phase equilibria occasionally leads to equations that are difficult to solve. Chemical insight, coupled with numerical approximation, often provides quicker answers in such cases than standard algebra, as illustrated in the following example.

EXAMPLE 7–7

At 330°C, the reaction

$$2 \, NOCl(g) \rightleftharpoons 2 \, NO(g) + Cl_2(g)$$

has an equilibrium constant of 4.0×10^{-4}. Find the equilibrium partial pressures of all three gases if a reaction vessel is initially charged with 2.00 atm of $NO(g)$ and 1.50 atm of $Cl_2(g)$ at 330°C and if this is the only reaction that takes place.

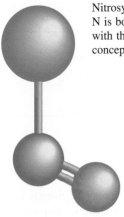

Nitrosyl chloride (NOCl) is a corrosive reddish yellow gas. The N is bonded to the O and Cl and the molecule is bent, in accord with the valence-shell electron-pair repulsion concepts developed in Chapter 3.

Solution

The reaction runs in reverse to reach equilibrium because only products were put into the container. We recognize this in a table of changes (constructed on the pattern defined in Example 7–2) by putting negative signs on the changes in the partial pressures of the products and a positive sign on the change in the partial pressure of the reactant:

	2 NOCl(g) $\rightleftharpoons$	2 NO(g) +	Cl$_2$(g)
Initial partial pressure (atm)	0	2.00	1.50
Change in partial pressure (atm)	$+2x$	$-2x$	$-x$
Equilibrium partial pressure (atm)	$2x$	$2.00 - 2x$	$1.50 - x$

Writing the equilibrium law and inserting K and the three equilibrium partial pressures

$$K = \frac{(P_{NO})^2 \, (P_{Cl_2})}{(P_{NOCl})^2}$$

$$4.0 \times 10^{-4} = \frac{(2.00 - 2x)^2 \, (1.50 - x)}{(2x)^2}$$

Rearrangement gives a cubic equation in x that is difficult to solve. We can avoid tedious algebra by thinking about chemistry. The K is small, indicating that the equilibrium mixture contains mostly NOCl; hence $2x$ must be relatively large and $(2.00 - 2x)$ or $(1.50 - x)$ relatively small. Also, x cannot exceed 1.00, because an x greater than 1.00 would give a negative partial pressure of NO. With these facts in mind and a calculator in hand, *guessing* plausible values of x and checking whether the expression equals K (see Appendix C–2) soon provides the answer. If $x = 0.90$, then

$$\frac{(2.00 - 2x)^2 \, (1.50 - x)}{(2x)^2} = \frac{(2.00 - 1.80)^2(1.50 - 0.90)}{(1.80)^2} = 0.0074 = Q$$

where 0.0074 is labelled Q because it is in fact a reaction quotient. Because Q exceeds K ($0.0074 > 0.00040$), the equilibrium lies farther to the left than guessed. A choice of $x = 0.99$ gives $Q = 5.2 \times 10^{-5}$, which is smaller than K. The two Q's have bracketted K. Further intelligent guesses soon lead to $x = 0.9732$, corresponding to $Q = K = 4.0 \times 10^{-4}$. Thus, the equilibrium values of the partial pressures are

$$P_{NO} = 2.00 - 2x = 2.00 - 2(0.9732) = \boxed{0.05 \text{ atm}}$$
$$P_{Cl_2} = 1.50 - x = 1.50 - 0.9732 = \boxed{0.53 \text{ atm}}$$
$$P_{NOCl} = 2x = \boxed{1.95 \text{ atm}}$$

Exercise

Find the equilibrium partial pressures of NO, NOCl, and Cl_2 if the flask in the preceding example is initially charged with 1.000 atm of NOCl(g) at 330°C.

Answer: $P_{NO} = 0.088$ atm; $P_{Cl_2} = 0.044$ atm; $P_{NOCl} = 0.912$ atm.

Concentrations of Gases in Equilibrium Calculations

Sometimes the data for a calculation of gas-phase equilibrium may be given in terms of concentrations rather than partial pressures. In such a case, there are two options: either to convert all concentrations to partial pressures before carrying out the calculations or to rewrite the equilibrium expression in terms of concentration variables. Both options require the use of the ideal gas equation, which relates the concentration [A] of an ideal gas A to its partial pressure P_A as follows:

• Recall from Section 4-5 that square brackets denote concentration in moles per liter.

$$[A] = \frac{n_A}{V} = \frac{P_A}{RT}$$

This allows us to write

$$P_A = RT[A]$$

Let us substitute one such relation for every species appearing in the equilibrium expression for a specific reaction. We choose the familiar $2\,NO_2 \rightleftharpoons N_2O_4$ equilibrium. It is best to put the factors of $P_{ref} = 1$ atm back into the equilibrium expression in order to examine the units of the resulting equations. With these factors, the equilibrium expression is

$$\frac{(P_{N_2O_4}/P_{ref})}{(P_{NO_2}/P_{ref})^2} = K$$

Substituting $P_{NO_2} = RT[NO_2]$ and $P_{N_2O_4} = RT[N_2O_4]$ gives

$$\frac{[N_2O_4](RT/P_{ref})}{[NO_2]^2(RT/P_{ref})^2} = \frac{[N_2O_4]}{[NO_2]^2} \times \left(\frac{RT}{P_{ref}}\right)^{-1} = K$$

Multiplying both sides of this equation by RT/P_{ref} gives

$$\frac{[N_2O_4]}{[NO_2]^2} = K\left(\frac{RT}{P_{ref}}\right)$$

For the general gas-phase reaction

$$aA\,(g) + bB(g) \rightleftharpoons cC(g) + dD(g)$$

• In some books, the expression on the right side of this equation is called K_c (c for "concentration"), and what we call K is called K_p (p for "pressure"). This is misleading, however, because there is only one equilibrium constant, K.

a similar algebraic manipulation gives

$$\frac{[C]^c[D]^d}{[A]^a[B]^b} = K\left(\frac{RT}{P_{ref}}\right)^{a+b-c-d}$$

In the general case, RT/P_{ref} is raised to a power that is equal to the sum of the number of moles of gas-phase reactant molecules minus the sum of the number of moles of gas-phase product molecules in the balanced chemical equation. This relationship allows us to work with concentrations of gas-phase species instead of partial pressures. Because partial pressures are given in atmospheres and concentrations in moles per liter, the appropriate value for R is

$$R = 0.08206 \text{ L atm mol}^{-1} \text{ K}^{-1}$$

Note that if $a + b - c - d = 0$ (that is, if there is no change in the total number of moles of gases in the reaction mixture as reactants convert to products), then the right-hand side of the equilibrium expression reduces to K, because $(RT/P_{ref})^0 = 1$.

EXAMPLE 7–8

At elevated temperatures, PCl_5 dissociates extensively according to the equation

$$PCl_5(g) \rightleftharpoons PCl_3(g) + Cl_2(g)$$

At 300°C, the equilibrium constant for this reaction is $K = 11.5$. The concentrations of PCl_3 and Cl_2 at equilibrium in a container at 300°C are both 0.0100 mol L^{-1}. Calculate the concentration of PCl_5 at equilibrium.

Solution

In the chemical equation as it is written, two moles of gases appear for each mole of gas that is consumed, so RT/P_{ref} must be raised to the power $1 - 2 = -1$. Hence,

$$\frac{[PCl_3][Cl_2]}{[PCl_5]} = K\left(\frac{RT}{P_{ref}}\right)^{-1} = K\left(\frac{P_{ref}}{RT}\right)$$

$$= 11.5 \times \frac{1 \text{ atm}}{(0.08206 \text{ L atm mol}^{-1} \text{ K}^{-1})(573 \text{ K})} = \frac{0.245 \text{ mol}}{L}$$

Solving this equation for $[PCl_5]$ gives

$$[PCl_5] = \frac{[PCl_3][Cl_2]}{0.245 \dfrac{\text{mol}}{L}} = \frac{\left(0.0100 \dfrac{\text{mol}}{L}\right)\left(0.0100 \dfrac{\text{mol}}{L}\right)}{0.245 \dfrac{\text{mol}}{L}} = 4.08 \times 10^{-4} \frac{\text{mol}}{L}$$

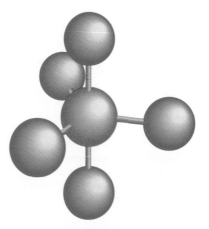

Phosphorus pentachloride, PCl_5, has a trigonal bipyramidal geometry. It is made on an industrial scale by the reaction of Cl_2 with PCl_3. It reacts violently with an excess of water to form HCl and H_3PO_4; when PCl_5 and water are mixed in a 1:1 molar ratio, $POCl_3$ is formed.

Exercise

The concentrations of $H_2(g)$, $CO(g)$, and $H_2O(g)$ in an equilibrium mixture of gases at 900 K are all 0.00642 mol L^{-1}. Calculate the concentration of $CH_4(g)$ in the mixture. The equilibrium constant for the reaction

$$CH_4(g) + H_2O(g) \rightleftharpoons CO(g) + 3 H_2(g)$$

equals 0.172 at this temperature.

Answer: 0.00839 mol L^{-1}.

7–5 EFFECTS OF EXTERNAL STRESSES ON EQUILIBRIA: LE CHATELIER'S PRINCIPLE

If a system that is at equilibrium is left to itself, it simply stays there, unchanging; this is what being at equilibrium means. Now, suppose that such a system is subjected to some stress that comes from its surroundings, such as a change in volume or temperature or a change in the concentration or partial pressure of one of the reactants or products. How does the system respond? The qualitative answer is embodied in a principle stated by Henri Le Chatelier in 1884:

> A system in equilibrium that is subjected to a stress reacts in a way that counteracts the stress.

Le Chatelier's principle provides a way to predict the response of an equilibrium system to an external perturbation.

Effects of Adding or Removing Reactants or Products

Consider what happens when a small quantity of a reactant is added to an equilibrium mixture. The presence of additional reactant lowers the value of the reaction quotient Q below K, and reaction takes place in the forward direction, converting reactants to products, until Q again equals K. In this way, the system counteracts the stress (the increase in the quantity of one of the reactants) and eventually attains a new equilibrium state. If, instead of a reactant, one of the *products* is added to an equilibrium mixture, Q temporarily becomes *greater* than K, and a net *back* reaction occurs, counteracting the imposed stress by reducing the partial pressures of the products and increasing those of the reactants (Fig. 7–6).

• The system can never return to the way it was before the addition of the extra reactant. The perturbing stress is counteracted, not canceled.

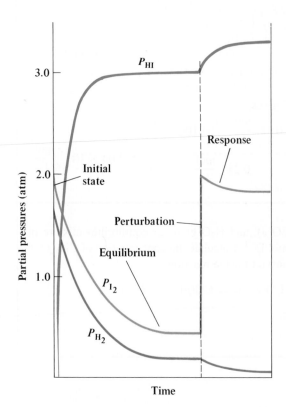

Figure 7–6 Partial pressures versus time for the equilibrium

$$I_2(g) + H_2(g) \rightleftharpoons 2\,HI(g)$$

The left part of the figure shows the attainment of equilibrium starting from the initial conditions of Example 7–6. Then the equilibrium state is abruptly perturbed by an increase in the partial pressure of I_2 to 2.000 atm. In accordance with Le Chatelier's principle, the system responds (Example 7–9) in such a way as to decrease the partial pressure of I_2—that is, to counteract the perturbation that moved it away from equilibrium in the first place.

EXAMPLE 7–9

An equilibrium mixture of $I_2(g)$, $H_2(g)$, and $HI(g)$ at 600 K has

$$P_{I_2} = 0.4756 \text{ atm}, \ P_{H_2} = 0.2056 \text{ atm}, \text{ and } P_{HI} = 3.009 \text{ atm}$$

This is essentially the final equilibrium state of Example 7–6. Enough I_2 is added and vaporized to increase the partial pressure of I_2 to 2.000 atm at 600 K before any reaction takes place. The mixture then once again reaches equilibrium at 600 K. What are the final partial pressures of the three gases?

Solution

We begin by setting up our usual table, where "initial" refers to the moment after the addition of the new I_2, but before it reacts further:

	$I_2(g)$	+	$H_2(g)$	$\rightleftharpoons$	2 HI(g)
Initial partial pressure (atm)	2.000		0.2056		3.009
Change in partial pressure (atm)	$-x$		$-x$		$+2x$
Equilibrium partial pressure (atm)	$2.000 - x$		$0.2056 - x$		$3.009 + 2x$

We know from Le Chatelier's principle that, after the addition of the I_2, a net reaction consumes I_2. We have used this fact in assigning a negative sign to the change in the partial pressure of the I_2 in the table.

Substituting the equilibrium partial pressures into the equilibrium law gives

$$\frac{(3.009) + 2x)^2}{(2.000 - x)(0.2056 - x)} = 92.6$$

Expanding this expression results in the quadratic equation

$$88.60x^2 - 216.275x + 29.023 = 0$$

which can be solved to give

$$x = 0.142 \quad \text{or} \quad 2.30$$

The second root would lead to negative partial pressures of H_2 and I_2 and is therefore not physically possible. Substituting the first root into the expressions from the set-up table gives

$$P_{I_2} = 2.000 - 0.142 = 1.86 \text{ atm}$$
$$P_{H_2} = 0.2056 - 0.142 = 0.063 \text{ atm}$$
$$P_{HI} = 3.009 + 2(0.142) = 3.29 \text{ atm}$$

Check: $\dfrac{(3.29)^2}{(1.86)(0.063)} = 92.4$

Exercise

A sealed flask contains this equilibrium mixture of gases:

$$P_{H_2} = 0.0975 \text{ atm}, \ P_{Si_2H_6} = 0.0975 \text{ atm, and } P_{SiH_4} = 0.925 \text{ atm}$$

at 100°C. Enough hydrogen is suddenly added to raise the partial pressure of $H_2(g)$ momentarily to 1.000 atm. Compute the partial pressure of $SiH_4(g)$ after the equilibrium

$$H_2(g) + Si_2H_6(g) \rightleftharpoons 2\ SiH_4(g)$$

is re-established. For this equilibrium, K is 90.0.

Answer: 1.091 atm.

If one of the products is *removed* from an equilibrium mixture, the reaction occurs in the *forward* direction to compensate partially by increasing the partial pressures of products. Removing the products as they form causes a reaction to be driven to completion. Most industrial operations are designed in such a way that products can be removed continuously in order to achieve high overall yields even for reactions with small equilibrium constants.

Effects of Changing the Volume of the System

Le Chatelier's principle also predicts the effect of a change in volume on a gas-phase equilibrium. An externally imposed decrease in the volume of a gaseous system causes an increase in the total pressure of the reaction mixture. The position of the equilibrium then shifts in the direction that reduces the total pressure exerted by the gases taking part in the equilibrium, if that is possible. Such a shift counteracts (at least in part) the stress imposed by the decrease in volume. In the equilibrium

$$2\ P_2(g) \rightleftharpoons P_4(g)$$

for example, the reaction shifts in the forward direction when the volume is decreased. This occurs because for every two P_2 molecules that react, only one P_4 molecule forms, and fewer molecules exert a lower pressure. The stress caused by the decrease in volume is thereby diminished. An *increase* in the volume of this system, on the other hand, favors reactants over products, and some P_4 dissociates to form P_2 (Fig. 7–7). If the total number of gas-phase molecules on the two sides of the equation is the same, then a change in volume has no effect upon the equilibrium.

When the volume of a system is decreased, its total pressure increases. Another way to increase the total pressure is to leave the size of the container alone but add an inert gas such as argon to the reaction mixture. The effect on the equilibrium is now entirely different. Because the partial pressures of the reactant and product gases are unchanged by the addition of an inert gas and because the equilibrium law is independent of total pressure, addition of the inert gas at constant volume has no effect on the position of the equilibrium.

Effects of Changing the Temperature

As a rule, chemical reactions are accompanied by either an absorption of heat or a liberation of heat. The former are called **endothermic** reactions, and the latter are called **exothermic** reactions. When a reaction is endothermic, an increase in the

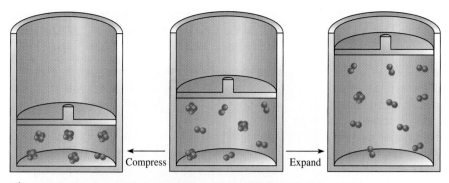

Figure 7–7   An equilibrium mixture of P_2 and P_4 (*middle figure*) is forced into a smaller volume (*left*). Some P_2 molecules combine to give P_4 molecules, in order to reduce the total number of molecules and thus the total pressure. If the volume is increased (*right*), some P_4 molecules dissociate to pairs of P_2 molecules to increase the total number of molecules and thereby the pressure.

temperature (as heat is put into the system) causes its equilibrium to shift toward the formation of more products. By Le Chatelier's principle, the external stress from the addition of heat is partially relieved by a net reaction to take up some of that heat. On the other hand, raising the temperature of an exothermic reaction that is at equilibrium shifts it in the direction of the reactants. Reducing the temperature has the reverse effects, favoring the products of exothermic reactions and the reactants of endothermic reactions. Heat acts upon the equilibrium state of a reaction as though it were a material species appearing on either the left side of the equation (for endothermic reactions) or the right side of the equation (for exothermic reactions).

Another way to think about the effect of temperature changes is to focus on the change they cause in the equilibrium constant K for the reaction. In an exothermic reaction, K always decreases as the temperature increases, so the equilibrium yield of products decreases. In an endothermic reaction, K and the equilibrium yield of products always increase as the temperature increases.

Driving Reactions to Completion

In the chemical industry, high yields translate into less waste and pollution. Le Chatelier's principle guides efforts to bring about shifts in the positions of equilibria to favor the desired product.

As an example, let us consider the exothermic reaction

$$N_2(g) + 3 H_2(g) \rightleftharpoons 2 NH_3(g)$$

which is the basis of the industrial synthesis of ammonia. Because the reaction is exothermic, the yield of ammonia is increased by working at as low a temperature as possible (Fig. 7–8a). At too low a temperature, however, the reaction is too slow, so a compromise temperature near 500°C is usually employed. Because the number of moles of gas decreases from the left side of the reaction to the right side, the yield of product is enhanced by decreasing the volume of the reaction vessel. Typically, total pressures of 150 to 300 atm are used (Fig. 7–8b), although some plants work at up to 900 atm of pressure. Even at high pressures, however, the yield of ammonia is usually only 15 to 20% because of the unfavorably small equilib-

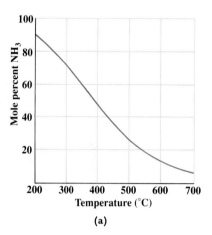

(a)

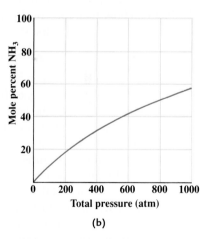

(b)

Figure 7–8   (a) The equilibrium mole percentage of ammonia in a 1:3 mixture of N_2 and H_2 varies with temperature; low temperatures favor a high yield of NH_3. The data shown correspond to a fixed total pressure of 300 atm. (b) At a fixed temperature (here, 500°C), the equilibrium amount of NH_3 increases with increasing total pressure.

CHEMISTRY IN PROGRESS

Gas Synthesis

The production of ammonia described in Section 7–5 requires large supplies of hydrogen and nitrogen. Nitrogen is readily obtained from the distillation of air, as explained in Section 6–7. Hydrogen is another story. Currently its principal source is the methane in natural gas, and the method of production involves two major steps. The first is the **reforming reaction:**

$$CH_4(g) + H_2O(g) \rightleftharpoons CO(g) + 3 H_2(g)$$

This reaction is endothermic and proceeds with an increase in the chemical amount of gas from two moles to four. It is accordingly carried out at high temperature (750 to 1000°C) and low total pressure to increase yield. An excess of steam is also added to drive the reaction toward completion. The gas mixture that it produces is called **synthesis gas** and is used directly for the production of methanol (CH_3OH, Problem 7–44) and other chemicals. If a feedstock other than methane is used, synthesis gas is obtained with different ratios of carbon monoxide to hydrogen. With propane, for example, the reforming reaction is

$$C_3H_8(g) + 3 H_2O(g) \rightleftharpoons 3 CO(g) + 7 H_2(g)$$

The second step in the production of hydrogen is the **shift reaction.** In it, additional steam reacts with the carbon monoxide from the first step:

$$CO(g) + H_2O(g) \rightleftharpoons CO_2(g) + H_2(g)$$

This reaction is exothermic so that low temperature favors the products. As is frequently the case with exothermic equilibria, however, the temperature cannot be lowered too far because the reaction then becomes too slow. The best compromise comes at a temperature of about 350°C. The shift reaction furnishes hydrogen beyond that from the reforming reaction, and also carbon dioxide as a by-product. It is in fact the major commercial source of carbon dioxide, which is frozen for refrigeration (Dry Ice), liquefied to fill fire extinguishers, or dissolved in water to make sparkling beverages. Gaseous CO_2 is also used in chemical synthesis, as in the production of salicylic acid for aspirin (see Fig. 22–2).

At present, natural gas is the primary raw material for making hydrogen and carbon dioxide. As reserves of nat-

Figure 7–A Coal reserves are much more extensive than those of petroleum or natural gas. The production of gas from coal for use as a fuel or in chemical synthesis is being studied in this demonstration plant in Texas.

ural gas are consumed, other sources must be developed. One older process that is again attracting attention is the reaction of coal with steam. Coal is a complex mixture that contains a large proportion of carbon by mass; for our purposes it is sufficient to represent it as entirely carbon. The reaction with steam is:

$$C(s) + H_2O(g) \rightleftharpoons CO(g) + H_2(g)$$

This mixture of product gases is called **water gas** and burns with a blue flame because of its carbon monoxide content. It has less hydrogen proportionately than the synthesis gas produced from methane or higher hydrocarbons. Water gas can react further, in the shift reaction, to give additional hydrogen and carbon dioxide. Once a mixture of CO and H_2 is prepared with the appropriate proportions, the reforming reaction given at the beginning of this section can be reversed to make methane for use as a fuel; the overall process is referred to as the gasification of coal (Fig. 7–A).

rium constant. To overcome this, ammonia plants use a cyclic process in which the gas mixture is cooled so that the ammonia liquefies (its boiling point is much higher than those of nitrogen and hydrogen) and is removed. Continuous removal of products helps to drive the reaction to completion. We discuss the industrial production of ammonia further in Chapter 22.

A high-pressure converter used in the synthesis of ammonia from nitrogen and hydrogen.

EXAMPLE 7–10

For each of the following reactions, state whether an increased equilibrium yield of products is favored by high or low total volume, and by high or low temperature.
(a) $PCl_3(g) + Cl_2(g) \rightleftharpoons PCl_5(g)$ (exothermic)
(b) $C_2H_5OH(g) \rightleftharpoons CO(g) + 3 H_2(g)$ (endothermic)
(c) $N_2(g) + O_2(g) \rightleftharpoons 2 NO(g)$ (endothermic)

Solution

Higher yields of product in reaction (a) are favored by a low temperature because that reaction is exothermic. High product yields in reactions (b) and (c) are favored by a high temperature. The product in reaction (a) is favored by low total volume because the reaction causes a decrease in the number of gas-phase molecules. The products of reaction (b) are favored by high total volume because the reverse is true. The equilibrium in reaction (c) is unaffected by changes in the total volume because the reaction causes no change in the total number of gas-phase molecules.

Exercise

State the effect of an increase in temperature and also of a decrease in volume on the equilibrium yield of the products of each of the following reactions:

(a) $CH_3OCH_3(g) + H_2O(g) \rightleftharpoons 2 CH_4(g) + O_2(g)$ (endothermic)
(b) $H_2O(g) + CO(g) \rightleftharpoons HCOOH(g)$ (exothermic)

Answer: (a) Higher temperature increases yield, and lower volume decreases yield.
(b) Higher temperature decreases yield, and lower volume increases yield.

Le Chatelier's principle is a qualitative way of summarizing certain properties of equilibrium states; in this regard, it is a precursor to the more complete thermodynamic analysis that we carry out in Chapters 10 and 11. At that point, we shall also quantify the effect of temperature on equilibrium constants.

7–6 HETEROGENEOUS EQUILIBRIUM

We turn now to the study of equilibria that involve solids, liquids, and dissolved species as well as gases. Let us begin by recalling the phase equilibrium between liquid water and water vapor that was considered at the beginning of this chapter:

$$H_2O(\ell) \rightleftharpoons H_2O(g) \qquad P_{H_2O} = K$$

Figure 7–9 Burning lime. The conversion of limestone to lime is important in the chemical industry because calcium oxide (lime) is the cheapest available base. Lime is used in making chemicals as well as building materials such as cement and glass. The reaction is carried out in giant rotary lime kilns, ranging up to 500 feet in length and 17 feet in diameter.

As long as *some* liquid water is in the container, the pressure of water vapor at 25°C adjusts to an equilibrium value of 0.03126 atm. The position of this equilibrium is not affected by the amount of liquid water present, and no mention of the liquid appears in the equilibrium expression. The reason for this is that the activity of pure liquid water as a reactant is constant at a given temperature and pressure. An analogous situation is observed in the equilibrium between solid iodine and iodine dissolved in aqueous solution:

$$I_2(s) \rightleftharpoons I_2(aq) \qquad [I_2(aq)] = K$$

The equilibrium concentration of $I_2(aq)$ at 25°C is always 1.34×10^{-3} M, regardless of the amount of solid iodine present, as long as there is *some*; the pure solid does not appear in the experimental equilibrium expression. Dissolved species such as $I_2(aq)$ enter equilibrium expressions through their *concentrations*, which are expressed in moles per liter. These "concentrations" are really dimensionless ratios equal to the molarities divided by a reference concentration of 1.00 M. They are similar to the dimensionless ratios obtained when the pressures of gases in atmospheres are divided by a reference pressure of 1 atm.

Consider now a third heterogeneous equilibrium, one involving a chemical reaction. Solid calcium carbonate decomposes to another solid and a gas:

$$CaCO_3(s) \rightleftharpoons CaO(s) + CO_2(g) \qquad P_{CO_2} = K$$

If calcium carbonate is heated to a sufficiently high temperature, its decomposition to calcium oxide (lime) and carbon dioxide (Fig. 7–9) is extensive; the reverse reaction is favored by low temperatures. The equilibrium has been studied experimentally, and it is found that at a given temperature the equilibrium pressure of $CO_2(g)$ is a constant, independent of the amounts of $CaCO_3(s)$ and $CaO(s)$, as long as some of each is present (Fig. 7–10). The two pure solids are represented by unity in the equilibrium expression, and the equilibrium constant is numerically equal to the partial pressure of carbon dioxide (expressed in units of atmospheres).

We are now in a position to state the **law of mass action** in a general form:

---

1. Gases enter equilibrium expressions as partial pressures, in atmospheres.
2. Dissolved species enter as concentrations, in moles per liter.
3. Pure solids and pure liquids are represented in equilibrium expressions by the number 1 (unity); a solvent taking part in a chemical reaction is represented by unity, provided that the solution is dilute.
4. Partial pressures and concentrations of products appear in the numerator and those of reactants in the denominator. Each is raised to a power equal to its coefficient in the balanced chemical equation.

Figure 7–10 As long as *some* of both CaO(s) and $CaCO_3(s)$ are present at equilibrium in a closed container at a given temperature, the partial pressure of $CO_2(g)$ does not depend on the relative amounts of the two solids present.

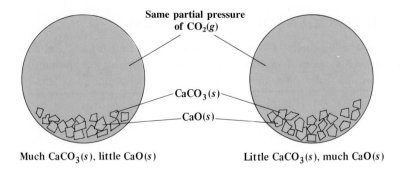

Same partial pressure of $CO_2(g)$

$CaCO_3(s)$
$CaO(s)$

Much $CaCO_3(s)$, little CaO(s)   Little $CaCO_3(s)$, much CaO(s)

EXAMPLE 7–11

Hypochlorous acid (HOCl) can be produced by bubbling chlorine through an agitated suspension of mercury(II) oxide (HgO) in water. The chemical equation for this process is

$$2 Cl_2(g) + 2 HgO(s) + H_2O(\ell) \rightleftharpoons HgO{\cdot}HgCl_2(s) + 2 HOCl(aq)$$

Write the equilibrium expression for this reaction.

Solution

$$\frac{(1)[HOCl]^2}{P_{Cl_2}^2(1)^2(1)} = \frac{[HOCl]^2}{P_{Cl_2}^2} = K$$

The reactant HgO and the product HgO·HgCl$_2$ are represented in the expression by unity because they are pure solids; water is represented by unity because it is the solvent, present in high and essentially constant concentration. The partial pressure in atmospheres of chlorine (divided by 1 atm) enters the expression because chlorine is a gas. The concentration in moles per liter of HOCl (divided by 1 mole per liter) appears because HOCl is a dissolved species. The concentration of HOCl and the partial pressure of Cl$_2$ are raised to the second power because their coefficients are 2 in the balanced chemical equation.

Exercise

Write equilibrium expressions for the following reactions:

(a) $Si_3N_4(s) + 4 O_2(g) \rightleftharpoons 3 SiO_2(s) + 2 N_2O(g)$
(b) $O_2(g) + 2 H_2O(\ell) \rightleftharpoons 2 H_2O_2(aq)$
(c) $CaH_2(s) + 2 C_2H_5OH(\ell) \rightleftharpoons Ca(OC_2H_5)_2(s) + 2 H_2(g)$

Answer: (a) $\dfrac{P_{N_2O}^2}{P_{O_2}^4} = K$ (b) $\dfrac{[H_2O_2]^2}{P_{O_2}} = K$ (c) $P_{H_2}^2 = K$

These rules give equilibrium equations that are valid to reasonable accuracy when the pressures of gases do not exceed several atmospheres and the concentrations of solutes in solution do not exceed 0.1 M. At higher pressures or concentrations, where gases and solutions depart seriously from ideality, more advanced methods of treatment are required.

(continued on page 308)

CHEMISTRY IN YOUR LIFE

The cells of the human body need oxygen to live, and your bloodstream satisfies this need with oxygen transported from the lungs. Whole blood can carry as much as 0.01 mol of O_2 per liter because the compound hemoglobin (Hb) in the red blood cells binds O_2 chemically. By contrast, blood plasma, which contains no hemoglobin, dissolves only about 0.0001 mol of O_2 per liter, a value that is close to the solubility of O_2 in ordinary water at room conditions (see Example 6–18). As highly oxygenated blood from the lungs passes through the capillaries, the binding of oxygen to hemoglobin loosens, and the freed O_2 is taken up by a different oxygen-binding compound, called myoglobin (Mb), that is found in near-by cells. Myoglobin then carries the oxygen on toward its ultimate fate, which is reduction in the cell to carbon dioxide.

How can hemoglobin bind oxygen so aggressively in the lungs yet release it so meekly to myoglobin in the capillaries? The answer starts with a comparison of the heterogeneous equilibria by which hemoglobin and myoglobin bind oxygen. Hemoglobin consists of a large protein (the globin part) in which four iron-containing **heme** groups are embedded (see Fig. 25–11). Oxygen can bind at each heme group, so one molecule of hemoglobin binds up to four molecules of O_2 in a series of *consecutive* equilibria

$$Hb(aq) + O_2(g) \rightleftharpoons Hb(O_2)(aq)$$

$$K_1 = \frac{[Hb(O_2)]}{[Hb]P_{O_2}} = 31$$

$$Hb(O_2)(aq) + O_2(g) \rightleftharpoons Hb(O_2)_2(aq)$$

$$K_2 = \frac{[Hb(O_2)_2]}{[Hb(O_2)] P_{O_2}} = 26$$

$$Hb(O_2)_2(aq) + O_2(g) \rightleftharpoons Hb(O_2)_3(aq)$$

$$K_3 = \frac{[Hb(O_2)_3]}{[Hb(O_2)_2] P_{O_2}} = 4.3$$

$$Hb(O_2)_3(aq) + O_2(g) \rightleftharpoons Hb(O_2)_4(aq)$$

$$K_4 = \frac{[Hb(O_2)_4]}{[Hb(O_2)_3] P_{O_2}} = 2100$$

The four K's are *not* equal, and K_4 is far larger than the other three.

Myoglobin is different. The globin portion of myoglobin embeds just one heme group so that each myoglobin molecule can bind just one O_2

$$Mb(aq) + O_2(g) \rightleftharpoons Mb(O_2)(aq)$$

$$K = \frac{[Mb(O_2)]}{[Mb]P_{O_2}} = 29$$

According to Le Chatelier's principle, changes in the partial pressure of O_2 (P_{O_2}) should affect the myoglobin equilibrium and all four hemoglobin equilibria similarly: increasing P_{O_2} should shift all of the equilibria to the right; decreasing P_{O_2} should shift them to the left. One way to check this is to plot the fraction of the binding sites occupied by O_2 as a function of P_{O_2}. Figure 7–B shows such **fractional saturation** plots for hemoglobin and myoglobin. Clearly, a reduction in P_{O_2} in fact causes less O_2 to be bound. But hemoglobin and myoglobin respond to the same reduction to quite different degrees. At $P_{O_2} = 0.13$ atm (100 torr), both hemoglobin and myoglobin are more than 95% saturated with oxygen at equilibrium. In contrast, at $P_{O_2} = 0.040$ atm (30 torr), the saturation of hemoglobin is sharply less (55%) whereas that of myoglobin stays high (above 90%). The first of these partial pressures prevails in arterial blood from the lungs and the second in venous blood. Thus, as P_{O_2} diminishes in the capillaries, hemoglobin releases a substantial amount of oxygen, but myoglobin in the nearby cells remains able to take it up. The outcome is a transfer of O_2 from blood (hemoglobin) to cells (myoglobin).

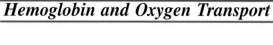

Hemoglobin and Oxygen Transport

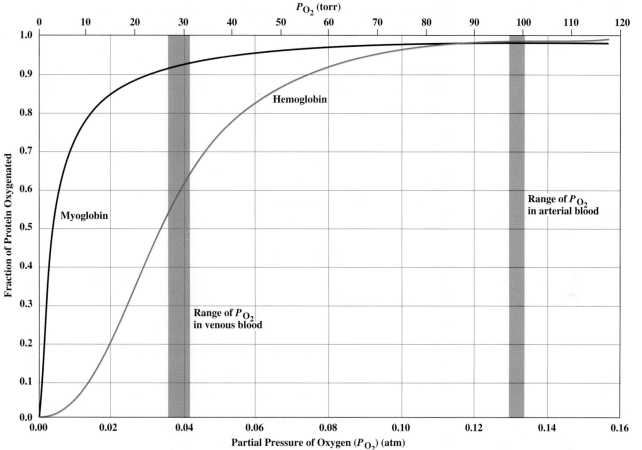

Figure 7–B A plot of the fraction of the binding sites of hemoglobin (red) and myoglobin (black) that are occupied by O_2 as a function of the partial pressure of O_2.

The S-shaped curve for hemoglobin in Figure 7–B can be derived mathematically from the different K's for the four equilibria by which hemoglobin binds O_2. The large K_4 means that the affinity of hemoglobin for oxygen *rises* with the uptake of oxygen. Reversing the perspective, we see that the *loss* of the first O_2 from $Hb(O_2)_4$ sets up the easier loss of more O_2, and the fractional saturation curve for Hb dips below the curve for Mb. This phenomenon, called **positive cooperativity,** derives from structural changes within the hemoglobin molecule as it binds successive molecules of oxygen.

EXAMPLE 7–12

Graphite (a form of solid carbon) is added to a vessel that contains $CO_2(g)$ at a pressure of 0.824 atm at a certain high temperature. The pressure rises because a reaction occurs that produces $CO(g)$. The total pressure reaches an equilibrium value of 1.366 atm. Write a balanced chemical equation for the process, and give the value of its equilibrium constant.

Solution

The reaction can only be the oxidation of C by CO_2 during which the CO_2 is itself reduced (to CO). It can be written as

	$C(s)$ + $CO_2(g)$ $\rightleftharpoons$ 2 $CO(g)$	
Initial partial pressure (atm)	0.824	0
Change in partial pressure (atm)	$-x$	$+2x$
Equilibrium partial pressure (atm)	$0.824 - x$	$2x$

The total pressure at equilibrium is

$$P_{tot} = 0.824 \text{ atm} - x + 2x = 0.824 + x = 1.366 \text{ atm}$$

Solving for x gives

$$x = 1.366 - 0.824 = 0.542 \text{ atm}$$

The equilibrium partial pressures of the two gases are

$$P_{CO} = 2x = 1.084 \text{ atm}$$
$$P_{CO_2} = 0.824 - 0.542 = 0.282 \text{ atm}$$

and the equilibrium constant of the reaction is therefore

$$K = \frac{(1.084)^2}{0.282} = 4.17$$

Exercise

A vessel holds pure $CO(g)$ at a pressure of 1.282 atm and a temperature of 354 K. A quantity of nickel is added, and the partial pressure of $CO(g)$ drops to an equilibrium value of 0.709 atm because of the reaction

$$Ni(s) + 4 CO(g) \rightleftharpoons Ni(CO)_4(g)$$

Compute the equilibrium constant for this reaction at 354 K.

Answer: $K = 0.567$.

7–7 EXTRACTION AND SEPARATION PROCESSES

A very important type of heterogeneous equilibrium involves the partitioning of a solute between two mutually insoluble solvent phases. Such equilibria are used in many separation processes in chemical research and in industry.

Suppose that the two immiscible liquids water and carbon tetrachloride are placed together in a container. **Immiscible** means "mutually insoluble." Immiscible liquids, if shaken together, mix temporarily but eventually separate into distinct phases, with the denser on the bottom and the less dense on the top. In this case, water floats as a layer on top of the denser carbon tetrachloride. A visible boundary, the meniscus, separates the two phases. If a small quantity of iodine is added and the vessel is shaken in order to distribute the iodine thoroughly, some of the iodine dissolves in the carbon tetrachloride and the rest in the water (Fig. 7–11). Moreover, an equilibrium law governs the distribution of the iodine between the two immiscible solvents. Thus, if a little more iodine is added to the container (which is again shaken thoroughly), the iodine concentration in each solvent increases, but the *ratio* of the two concentrations remains the same as long as the temperature remains constant. The ratio of the concentrations of iodine in the two phases is the **partition coefficient,** the equilibrium constant K for the process

- A familiar example is the separation of the oil from the water phase in salad dressing. In salad dressing, the water phase is more dense, and the oil floats on top.

$$I_2(aq) \rightleftharpoons I_2(CCl_4)$$

It can be written as

$$\frac{[I_2]_{CCl_4}}{[I_2]_{aq}} = K$$

in which $[I_2]_{CCl_4}$ and $[I_2]_{aq}$ are the concentrations (in moles per liter) of I_2 in the CCl_4 and aqueous phases, respectively. At 25°C, K has the value 85 for this equilibrium. The fact that K is larger than unity shows that iodine is more soluble in CCl_4 than it is in water.

Extraction takes advantage of the partitioning of a solute between two immiscible solvents to remove that solute from one solvent into another. Suppose that iodine is present as a contaminant in water that also contains other solutes insoluble in carbon tetrachloride. In such a case, most of the iodine could be removed by shaking the aqueous solution with CCl_4, allowing the two phases to separate, and then pouring off (decanting) the water layer from the heavier layer of carbon tetrachloride. The larger the equilibrium constant for the partition of a solute from the original solvent into the extracting (that is, added) solvent, the more complete such a separation is.

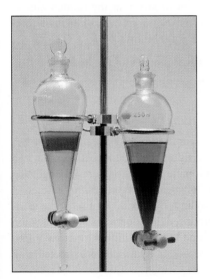

Figure 7–11 Iodine is dissolved in water and poured on top of carbon tetrachloride in a separatory funnel (*left*). After the funnel is shaken (*right*), the iodine reaches a partition equilibrium between the upper (aqueous) phase and the lower (CCl_4) phase. The deeper color in the lower phase indicates that iodine dissolves preferentially in the denser CCl_4 phase.

EXAMPLE 7–13

An aqueous solution has an iodine concentration of 2.00×10^{-3} M. Calculate the percentage of iodine removed by extraction of 0.100 L of this aqueous solution with 0.050 L of CCl_4 at 25°C.

Solution

The chemical amount of I_2 originally present is

$$n_{I_2} = \left(\frac{2.00 \times 10^{-3} \text{ mol } I_2}{\text{L water}} \right) \times 0.100 \text{ L water} = 2.00 \times 10^{-4} \text{ mol } I_2$$

• It is important to note that the changes in *concentrations* of I_2 in the two phases do *not* have the same magnitude because the volumes are different. It is only the changes in *numbers of moles* that can be written as $-y$ and $+y$.

Adding the 0.050 L of CCl_4 allows some of this I_2 to redistribute from the aqueous phase to the CCl_4 phase. If y mol of I_2 transfers to the CCl_4, then, by subtraction, $2.00 \times 10^{-4} - y$ mol of I_2 remains in the aqueous phase. Then

$$K = 85 = \frac{[I_2]_{CCl_4}}{[I_2]_{aq}} = \frac{\dfrac{y}{0.050}}{\dfrac{2.00 \times 10^{-4} - y}{0.100}}$$

Note that the volumes of the two solvents in the calculation are unchanged because of their immiscibility. Rearranging gives

$$\frac{2.0y}{2.00 \times 10^{-4} - y} = 85$$

from which $y = 1.954 \times 10^{-4}$ mol I_2. The fraction removed from the aqueous phase is $1.954 \times 10^{-4}/2.0 \times 10^{-4} = 0.98$, which is 98%. A second extraction with a fresh 0.050 L of CCl_4 would remove 98% of the 2% left after the first extraction.

Exercise

Advantage is taken of the immiscibility of molten zinc and molten lead in the Parkes process, in which silver is extracted from lead into zinc. A 105-L sample of molten lead that contains 0.0306 mol L^{-1} of dissolved silver is shaken with 21.0 L of molten zinc. Although the actual mode of extraction is quite complicated, assume that the equilibrium concentration of silver in the molten zinc is 3.0×10^2 times that in the lead. What mass of silver (in grams) is originally present and how much of it remains in the lead at equilibrium?

Answer: 347 g of silver is present; 5.7 g of silver remains in the lead.

The most important use of extraction is separating a desired solute from one or more unwanted solutes that have different partition coefficients between the two solvent phases. In favorable cases, almost complete separations are attained: the desired solute is extracted and the others are left behind. A process of extraction is used industrially on a large scale to purify sodium hydroxide for use in the manufacture of rayon. The sodium hydroxide produced by electrolysis (see Section 23–2) typically contains 1% sodium chloride and 0.1% sodium chlorate as impurities. If a concentrated aqueous solution of this sodium hydroxide is extracted with

Figure 7-12 In this demonstration of chromatography, a solvent travels upward along a strip of paper, separating the colored components of a sample of ink.

liquid ammonia, the NaCl and $NaClO_3$ are partitioned into the ammonia phase in preference to the aqueous phase. The heavier aqueous phase is added to the top of an extraction vessel that is filled with liquid ammonia, and equilibrium is reached as droplets of it settle through the ammonia to the bottom. The concentrations of the impurities in the sodium hydroxide solution are reduced to about 0.08% NaCl and 0.0002% $NaClO_3$ by this procedure.

Chromatography

Partition equilibria form the basis for an important class of separation techniques called **chromatography.** This word comes from the Greek root *chroma*, meaning "color," and was chosen because the first chromatographic separations involved colored substances (Fig. 7–12). The principle behind chromatography is one of a continuous extraction process in which solute species are exchanged between two phases. One phase, the *mobile phase*, moves with respect to a second *stationary phase*. There are several types of chromatography that differ in the mobile and stationary phases used; a few of the more important ones are listed in Table 7–1. Some of the techniques listed are used principally to analyze complex mixtures in order to learn the identities and amounts of their components; others are well suited for the isolation of a substance from a mixture.

Column chromatography (Fig. 7–13) uses a tube packed with a porous material, frequently a silica gel on which water has been adsorbed. Water is the stationary phase in this case. The mixture to be separated is introduced at the top of the column. Another solvent is used as the mobile phase and is then added from the

TABLE 7-1
Chromatographic Separation Techniques[a]

Name	Mobile Phase	Stationary Phase
Gas–liquid	Gas	Liquid adsorbed on a porous solid in a tube
Gas–solid	Gas	Porous solid in a tube
Column	Liquid	Liquid adsorbed on a porous solid in a tubular column
Paper	Liquid	Liquid held in the pores of a thick paper
Thin layer	Liquid	Liquid or solid; solid is held on a glass plate and liquid may be adsorbed on it
Ion exchange	Liquid	Solid (finely divided ion-exchange resin) in a tubular column

[a]Adapted from D. A. Skoog and D. M. West, *Analytical Chemistry*, Saunders College Publishing, 1980, Table 18–1.

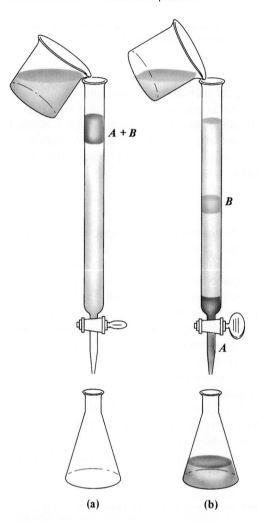

Figure 7–13 (a) In a column chromatograph, the top of the column is loaded with a mixture of solutes to be separated (green). (b) With addition of solvent, the different solutes travel at different rates, giving rise to bands. In this case, both solutes are colored. The separate fractions can be collected in different flasks for use or analysis.

top. As the mobile phase passes down the column, the components of the solute mixture are removed from the stationary phase and move with the mobile phase, but they are retarded to differing extents by their interactions with the water adsorbed on the packing. The components separate as bands along the column and reach the bottom of the column at different times, where they are collected separately for analysis or use. In some procedures, it is most efficient to use different solvents in succession as mobile phases. The partition coefficient K of a solute such as acetic acid (CH_3COOH) is defined as the ratio of its concentration in the stationary phase (for example, water) to that in the mobile phase (for example, benzene):

$$\frac{[CH_3COOH]_{aq}}{[CH_3COOH]_{benzene}} = K$$

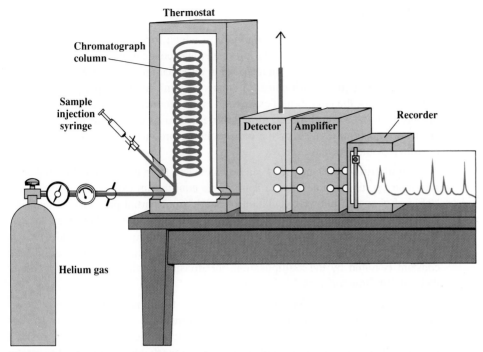

Thermostat

Chromatograph column

Sample injection syringe

Detector

Amplifier

Recorder

Helium gas

Figure 7-14   In a gas–liquid chromatograph, the sample (which may be a complex mixture) is vaporized and passes through a column, carried in a stream of an inert gas such as helium or nitrogen. The residence time of any substance on the column depends on its partition coefficient from the vapor to the liquid in the column. A species leaving the column at a given time can be detected by a variety of techniques. The result is a gas chromatogram, with a peak corresponding to each substance in the mixture.

As the mobile benzene passes over the stationary water, the acetic acid molecules move between the two phases. True equilibrium is never fully established in column chromatography because the motion of the mobile phase continually brings fresh solvent into contact with the stationary phase. Nevertheless, the partition coefficient K provides a guide to the behavior of a particular solute on the column. The larger K is, the greater is the affinity of the solute for the stationary phase and therefore the slower is its progress down the chromatographic column.

Gas–liquid chromatography (Fig. 7–14) is one of the most important separation techniques in modern chemical research. The stationary phase is once again a liquid adsorbed on a porous solid or held as a film on the inner wall of a quartz or metal capillary tube. The solutes are substances that have appreciable vapor pressures, and the mobile phase is a gas that flows through the capillary tube. Gas–liquid chromatography is widely used for separating the products of organic reactions. It also finds much use in determining the purity of substances, because even very small amounts of impurities appear clearly as separate peaks in chromatograms. The technique is important in separating and identifying possibly toxic substances in trace amounts in a variety of environmental and biological samples. Amounts on the order of parts per trillion by mass (10^{-12} g in a 1-g sample) can be detected and identified.

SUMMARY

7–1 Equilibrium is a dynamic condition in which forward and reverse reactions occur at equal rates so that partial pressures and concentrations remain constant. The equilibrium condition is summed up in an equation in which K, the temperature-dependent **equilibrium constant** equals an **equilibrium expression.** For gas-phase reactions, the equilibrium expression is the product of the partial pressures of the gases produced, each raised to a power equal to its coefficient in the balanced chemical equation, divided by the corresponding product of powers of the partial pressures of the reactants.

7–2 Equilibrium constants (K's) can be calculated from experimental partial pressure data by substitution into the appropriate equilibrium expression. Setting up a "table of changes" allows calculation of equilibrium partial pressures of reactants and products in a reaction system when initial partial pressures and the K are known. If a chemical equation is obtained from other chemical equations by adding them, subtracting them, or multiplying them by a constant, the corresponding equilibrium constant is found by the multiplication, the division, or the raising to a power of their equilibrium constants.

7–3 A value of the **reaction quotient (Q)** results from inserting any arbitrary partial pressures into the form of the equilibrium expression. As a reaction mixture approaches equilibrium, its reaction quotient approaches the equilibrium constant. If Q is less than K, the reaction shifts to the right to reach equilibrium; if Q is greater than K, the reaction shifts to the left.

7–4 Equilibrium laws can be used to compute the partial pressure of one reactant or product from those of the others, or to calculate the final partial pressures reached from a given initial state.

7–5 Le Chatelier's principle describes in qualitative terms how a system at equilibrium responds to external perturbation. If a reactant or product gas is added to an equilibrium mixture, reaction tends to take place to reduce the partial pressure of that gas. If the total pressure is increased by a decrease in the volume of the container, and if the numbers of moles of reactants and products in the balanced chemical equation are different, the reaction reduces the total number of gas molecules and thus reduces the total pressure. When the temperature of an equilibrium mixture is raised, the equilibrium constant increases if the reaction absorbs heat as it proceeds (if it is **endothermic**) and decreases if it liberates heat as it proceeds (if it is **exothermic**).

7–6 Heterogeneous equilibria involve reactants and products in two or more phases. The **law of mass action** governs the form of equilibrium expressions for all equilibria, including heterogeneous equilibria: gases enter equilibrium expressions as partial pressures; dissolved species enter as concentrations; pure solids, pure liquids, and solvents taking part in a reaction enter as the number 1. Partial pressures and concentrations of products appear in the numerator, and those of reactants in the denominator, with each raised to a power equal to its coefficient in the balanced chemical equation.

7–7 One particularly important type of heterogeneous equilibrium involves the **partition** of a solute species between two immiscible solvents. The equilibrium constant for such a process is called a **partition coefficient.** Procedures for the **extraction** of a dissolved species from one solvent into another depend on such equilibria, which also form the basis of **chromatography.** Numerous chromatographic techniques (including **column chromatography** and **gas–liquid chromatography**) allow chemical compounds to be separated and identified.

PROBLEMS

Note: Answers to blue-numbered problems are given in Appendix F. Problems that are more challenging are indicated with asterisks.

Chemical Reactions and Equilibrium

1. (See Example 7–1.) Write the equilibrium expression for each of the following gas-phase reactions:
   (a) $2 H_2(g) + O_2(g) \rightleftharpoons 2 H_2O(g)$
   (b) $Xe(g) + 3 F_2(g) \rightleftharpoons XeF_6(g)$
   (c) $2 C_6H_6(g) + 15 O_2(g) \rightleftharpoons 12 CO_2(g) + 6 H_2O(g)$

2. (See Example 7–1.) Write the equilibrium expression for each of the following gas-phase reactions:
   (a) $2 Cl_2(g) + O_2(g) \rightleftharpoons 2 Cl_2O(g)$
   (b) $N_2(g) + O_2(g) + Br_2(g) \rightleftharpoons 2 NOBr(g)$
   (c) $C_3H_8(g) + 5 O_2(g) \rightleftharpoons 3 CO_2(g) + 4 H_2O(g)$

3. At a moderately elevated temperature, phosphoryl chloride ($POCl_3$) can be produced in the vapor phase from the gaseous elements. Write a balanced chemical equation and an equilibrium expression for this system. Note that gaseous phosphorus consists of P_4 molecules, at least at moderate temperatures.

4. If confined at high temperature, ammonia and oxygen quickly react and come to equilibrium with their products, water vapor and nitrogen monoxide. Write a balanced chemical equation and an equilibrium expression for this system.

Calculating Equilibrium Constants

5. At 454 K, $Al_2Cl_6(g)$ reacts to form $Al_3Cl_9(g)$ according to the equation
$$3 Al_2Cl_6(g) \rightleftharpoons 2 Al_3Cl_9(g)$$
In an experiment at this temperature, the equilibrium partial pressure of $Al_2Cl_6(g)$ is 1.00 atm, and the equilibrium partial pressure of $Al_3Cl_9(g)$ is 1.02×10^{-2} atm. Compute the equilibrium constant of the above reaction at 454 K.

6. At 298 K, $F_3SSF(g)$ decomposes partially to $SF_2(g)$. At equilibrium, the partial pressures of $SF_2(g)$ and $F_3SSF(g)$ are 1.1×10^{-4} atm and 0.0484 atm, respectively.
   (a) Write a balanced equilibrium equation to represent this reaction.
   (b) Compute the equilibrium constant corresponding to this equation.

7. The compound 1,3-di-*t*-butylcyclohexane exists in two forms that are known as the "chair" and "boat" conformations because the molecular structures resemble these objects. Equilibrium exists between these forms, represented by the equation
$$\text{chair} \rightleftharpoons \text{boat}$$
At 580 K, 6.42% of the molecules are in the chair form. Calculate the equilibrium constant for the reaction as written above.

8. At 248°C and a total pressure of 1.000 atm, the fractional dissociation of $SbCl_5$ is 0.718 for the reaction

$$SbCl_5(g) \rightleftharpoons SbCl_3(g) + Cl_2(g)$$
This means that 718 of every 1000 molecules of $SbCl_5$ originally present have dissociated. Calculate the equilibrium constant.

9. (See Example 7–2.) An experiment is run at 425°C to determine the equilibrium constant for the reaction
$$H_2(g) + I_2(g) \rightleftharpoons 2 HI(g)$$
at that temperature. Some $H_2(g)$ at a partial pressure of 5.73 atm is mixed with $I_2(g)$ at the same initial partial pressure. When the reaction comes to equilibrium, the partial pressure of HI is measured to be 9.00 atm. Compute the equilibrium constant, assuming that no side reactions occur.

10. (See Example 7–2.) A second experiment is run at 425°C to check the equilibrium constant for the reaction
$$H_2(g) + I_2(g) \rightleftharpoons 2 HI(g)$$
obtained in the previous problem. Some $H_2(g)$ at a partial pressure of 2.75 atm is mixed with $I_2(g)$ at a partial pressure of 1.50 atm. The reaction comes to equilibrium, and the partial pressure of HI is measured to be 2.79 atm. Compute the equilibrium constant, assuming that no side reactions occur.

11. Sulfuryl chloride (SO_2Cl_2) is a colorless liquid that boils at 69°C. Above this temperature, the vapors dissociate into sulfur dioxide and chlorine:
$$SO_2Cl_2(g) \rightleftharpoons SO_2(g) + Cl_2(g)$$
This reaction is slow at 100°C, but it is accelerated by the presence of some $FeCl_3$ (which does not affect the final position of the equilibrium). In an experiment, 3.174 g of $SO_2Cl_2(g)$ and a small amount of solid $FeCl_3$ are put into an evacuated 1.000-L flask, which is then sealed and heated to 100°C. The total pressure in the flask at that temperature is found to be 1.30 atm.
   (a) Calculate the partial pressure of each of the three gases present.
   (b) Calculate the equilibrium constant at this temperature.

12. A certain amount of $NOBr(g)$ is sealed in a flask, and the temperature is raised to 350 K. The following equilibrium is established:
$$NOBr(g) \rightleftharpoons NO(g) + 1/2 Br_2(g)$$
The total pressure in the flask when equilibrium is reached at this temperature is 0.675 atm, and the vapor density is 2.219 g L^{-1}.
   (a) Calculate the partial pressure of each species. (*Hint:* Note that vapor density cannot change in a gas-phase reaction if the volume is fixed.)
   (b) Calculate the equilibrium constant at this temperature.

13. At a certain temperature, the value of the equilibrium constant for the reaction
$$CS_2(g) + 3 O_2(g) \rightleftharpoons CO_2(g) + 2 SO_2(g)$$

is K_1. How is K_1 related to the equilibrium constant K_2 for the related equilibrium

$$1/3\ CS_2(g) + O_2(g) \rightleftharpoons 1/3\ CO_2(g) + 2/3\ SO_2(g)$$

at the same temperature?

14. At 25°C, the equilibrium constant for the reaction

$$6\ ClO_3F(g) \rightleftharpoons 2\ ClF(g) + 4\ ClO(g) + 7\ O_2(g) + 2\ F_2(g)$$

is 32.6. Calculate the equilibrium constant at 25° C for the reaction

$$1/3\ ClF(g) + 2/3\ ClO(g) + 7/6\ O_2(g) + 1/3\ F_2(g) \rightleftharpoons$$
$$ClO_3F(g)$$

15. (See Example 7–3.) Suppose that K_1 and K_2 are the respective equilibrium constants for the two reactions

$$XeF_6(g) + H_2O(g) \rightleftharpoons XeOF_4(g) + 2\ HF(g)$$
$$XeO_4(g) + XeF_6(g) \rightleftharpoons XeOF_4(g) + XeO_3F_2(g)$$

Give the equilibrium constant for the reaction

$$XeO_4(g) + 2\ HF(g) \rightleftharpoons XeO_3F_2(g) + H_2O(g)$$

in terms of K_1 and K_2.

16. (See Example 7–3). At 1330 K, germanium (II) oxide (GeO) and tungsten (VI) oxide (W_2O_6) are both gases. The following *two* equilibria are established simultaneously:

$$2\ GeO(g) + W_2O_6(g) \rightleftharpoons 2\ GeWO_4(g)$$
$$GeO(g) + W_2O_6(g) \rightleftharpoons GeW_2O_7(g)$$

The equilibrium constants for the two are, respectively, 7.0×10^3 and 38×10^3.

Compute K for the reaction

$$GeO(g) + GeW_2O_7(g) \rightleftharpoons 2\ GeWO_4(g)$$

The Reaction Quotient

17. (See Example 7–4.) The reaction

$$2\ NO(g) + Br_2(g) \rightleftharpoons 2\ NOBr(g)$$

has an equilibrium constant of 116.6 at 25°C. In each of the following mixtures, compute the reaction quotient Q and use it to decide the direction the reaction takes (right or left) to come to equilibrium. The initial partial pressures (all in atm) in the gas mixture are

(a) $P_{NO} = 0.105$; $P_{Br_2} = 0.400$; $P_{NOBr} = 0.969$

(b) $P_{NO} = 0.400$; $P_{Br_2} = 0.105$; $P_{NOBr} = 0.969$

(c) $P_{NO} = 0.210$; $P_{Br_2} = 0.800$; $P_{NOBr} = 1.94$

18. (See Example 7–4.) Suppose the reaction

$$SbF_5(g) + 4\ Cl_2(g) \rightleftharpoons SbCl_3(g) + 5\ ClF(g)$$

has an equilibrium constant of 0.0200 at 300°C. Portions of $SbF_5(g)$, $Cl_2(g)$, $SbCl_3(g)$, and $ClF(g)$ are mixed at 300°C and have the following partial pressures (all in atm) immediately after mixing but before reaction. In each case, compute the reaction quotient Q, and use it to decide the direction the reaction proceeds (right or left) to come to equilibrium. Assume that this is the only reaction that takes place.

(a) $P_{SbCl_3} = 0.655$; $P_{ClF} = 0.410$; $P_{SbF_5} = 1.20$;
    $P_{Cl_2} = 0.200$

(b) $P_{SbF_5} = 0.400$; $P_{Cl_2} = 10.0$; $P_{SbCl_3} = 0.125$;
    $P_{ClF} = 1.20$

(c) $P_{SbF_5} = 0.800$; $P_{Cl_2} = 0.800$; $P_{SbCl_3} = 0.800$;
    $P_{ClF} = 0.800$

19. (See Example 7–4.) Some Al_2Cl_6 (at a partial pressure of 0.473 atm) is placed in a closed container at 454 K with some Al_3Cl_9 (at a partial pressure of 1.02×10^{-2} atm).

(a) Calculate the initial reaction quotient for the reaction

$$3\ Al_2Cl_6(g) \rightleftharpoons 2\ Al_3Cl_9(g)$$

(b) As the gas mixture reaches equilibrium, is there net production or consumption of Al_3Cl_9? (Use the data given in problem 5.)

20. Some SF_2 (at a partial pressure of 2.3×10^{-4} atm) is placed in a closed container at 298 K with some F_3SSF (at a partial pressure of 0.0484 atm).

(a) Calculate the initial reaction quotient for the decomposition of F_3SSF to SF_2.

(b) As the gas mixture reaches equilibrium, is there net formation or dissociation of F_3SSF? (Use the data given in problem 6.)

21. The progress of the reaction

$$H_2(g) + Br_2(g) \rightleftharpoons 2\ HBr(g)$$

can be monitored visually by following changes in the color of the reaction mixture (Br_2 is reddish brown, and H_2 and HBr are colorless). A gas mixture is prepared at 700 K, in which 0.40 atm is the initial partial pressure of both H_2 and Br_2 and 0.90 atm is the initial partial pressure of HBr. The color of this mixture then fades as the reaction progresses toward equilibrium. Give a condition that must be satisfied by the equilibrium constant K (for example, it must be greater than or smaller than a given number).

22. Gaseous NO_2 is brown. At elevated temperatures, it reacts with CO according to

$$NO_2(g) + CO(g) \rightleftharpoons NO(g) + CO_2(g)$$

The other three gases taking part in this reaction are colorless. When a gas mixture is prepared at 500 K, in which 3.4 atm is the initial partial pressure of both NO_2 and CO and 1.4 atm is the partial pressure of both NO and CO_2, the brown color of the mixture is observed to fade as the reaction progresses toward equilibrium. Give a condition that must be satisfied by the equilibrium constant K (for example, it must be greater than or smaller than a given number).

Calculation of Gas-Phase Equilibria

23. (See Example 7–5.) Phosgene ($COCl_2(g)$) is produced by the reaction

$$CO(g) + Cl_2(g) \rightleftharpoons COCl_2(g)$$

for which the equilibrium constant equals 0.20 at 600°C. Calculate the partial pressure of phosgene that is in equilibrium with a mixture of 0.025 atm of carbon monoxide and 0.0035 atm of chlorine at 600°C.

24. (See Example 7–6). At 1107 K, the equilibrium constant of the reaction

$$H_2(g) + I_2(g) \rightleftharpoons 2\,HI(g)$$

equals 38.6. Calculate the partial pressure of hydrogen that must be present at equilibrium in this system if the partial pressure of iodine is 0.0252 atm and the partial pressure of hydrogen iodide is 0.427 atm.

25. (See Example 7–6). The reaction

$$SO_2Cl_2(g) \rightleftharpoons SO_2(g) + Cl_2(g)$$

has an equilibrium constant at 100°C of 2.40.

(a) Suppose that the initial partial pressure of SO_2Cl_2 in a reaction tank with rigid walls is 1.500 atm and that no other species are present. Calculate the reaction quotient Q, and state whether the total pressure increases or decreases as the reaction tends toward equilibrium.

(b) Calculate the partial pressures of SO_2Cl_2, SO_2, and Cl_2 when equilibrium is reached.

26. (See Example 7–6.) The equilibrium constant for the gas-phase reaction of bromine with chlorine to give bromine monochloride at 25°C is 58.0.

$$Br_2(g) + Cl_2(g) \rightleftharpoons 2\,BrCl(g)$$

(a) Suppose bromine and chlorine are mixed to give partial pressures of 0.100 atm for each at 25°C. Calculate the reaction quotient Q just after mixing but before any reaction occurs.

(b) Calculate the partial pressure of BrCl after the reaction mixture from part (a) reaches equilibrium at 25°C.

27. The dehydrogenation of benzyl alcohol to make the flavoring agent benzaldehyde is an equilibrium process described by the equation

$$C_6H_5CH_2OH(g) \rightleftharpoons C_6H_5CHO(g) + H_2(g)$$

At 523 K, K for the reaction equals 0.558.

(a) Suppose 1.20 g of benzyl alcohol is placed in a 2.00-L vessel and heated to 523 K. What is the partial pressure of benzaldehyde when equilibrium is attained?

(b) What fraction of benzyl alcohol is dissociated into products at equilibrium at 523 K?

28. Heating isopropyl alcohol causes it to break down into acetone and hydrogen:

$$(CH_3)_2CHOH(g) \rightleftharpoons (CH_3)_2CO(g) + H_2(g)$$

At 179°C, the equilibrium constant for this dehydrogenation reaction is 0.444.

(a) If 10.00 g of isopropyl alcohol is placed in a 10.00-L vessel and heated to 179°C, what is the partial pressure of acetone when equilibrium is attained?

(b) What fraction of isopropyl alcohol is dissociated at equilibrium?

29. A weighed quantity of $PCl_5(s)$ is sealed in a 100.0-cm³ glass bulb to which a pressure gauge is attached. The bulb is heated to 250°C, and the gauge shows that the pressure in the bulb rises from 0 atm to 0.895 atm. At this temperature, the solid PCl_5 is all vaporized and also partially dissociated into $Cl_2(g)$ and $PCl_3(g)$ according to the equation

$$PCl_5(g) \rightleftharpoons PCl_3(g) + Cl_2(g)$$

At 250°C, $K = 2.15$ for this reaction. Assume that this is the only reaction that takes place and that the contents of the bulb are at equilibrium, and calculate the partial pressures of the three different chemical species in the vessel.

30. Suppose 93.0 g of $HI(g)$ is placed in a glass vessel and heated to 1107 K. At this temperature, equilibrium is quickly established between $HI(g)$ and its decomposition products, $H_2(g)$ and $I_2(g)$:

$$2\,HI(g) \rightleftharpoons H_2(g) + I_2(g)$$

and no other reactions take place. The equilibrium constant at 1107 K is 0.0259, and the total pressure at equilibrium is observed to equal 6.45 atm. Calculate the equilibrium partial pressures of $HI(g)$, $H_2(g)$, and $I_2(g)$.

31. The equilibrium constant at 350 K for the reaction

$$Br_2(g) + I_2(g) \rightleftharpoons 2\,IBr(g)$$

has a value of 322. Bromine at an initial partial pressure of 0.0500 atm is mixed with iodine at an initial partial pressure of 0.0400 atm and held at 350 K until equilibrium is reached. Calculate the equilibrium partial pressure of each of the gases.

32. The equilibrium constant for the reaction of fluorine and oxygen to form oxygen difluoride (OF_2) is 40.1 at 298 K:

$$F_2(g) + 1/2\,O_2(g) \rightleftharpoons OF_2(g)$$

Suppose that some OF_2 is introduced into an evacuated container at 298 K and dissociates according to the above equation until its partial pressure reaches an equilibrium value of 1.00 atm. Calculate the equilibrium partial pressures of F_2 and O_2 in the container.

33. (See Example 7–7.) Find the equilibrium partial pressures of $Al_2Cl_6(g)$ and $Al_3Cl_9(g)$ in the reaction

$$3\,Al_2Cl_6(g) \rightleftharpoons 2\,Al_3Cl_9(g)$$

if K = 50 and a flask is charged with enough Al_2Cl_6 to exert an initial partial pressure of 0.050 atm. Assume that no other reactions take place.

34. (See Example 7–7.) The reaction

$$3\,S_8(g) \rightleftharpoons 4\,S_6(g)$$

has $K = 1.6 \times 10^{-4}$ at 500 K. Suppose that a container initially holds 0.150 atm of $S_8(g)$ at 500 K and that this is the only reaction that takes place. Compute the partial pressure of $S_8(g)$ when equilibrium is reached.

35. For the mixtures of gases in problem 17, compute the initial concentration of each gas in moles per liter.

36. For the mixtures of gases in problem 18, compute the initial concentration of each gas in moles per liter.

37. (See Example 7–8.) Calculate the concentration (in mol L^{-1}) of phosgene ($COCl_2$) that is present at 600°C in equilibrium with carbon monoxide (at a concentration of 2.3×10^{-4} mol L^{-1}) and chlorine (at a concentration of 1.7×10^{-2} mol L^{-1}). (Use the data of problem 23.)

38. (See Example 7–8.) Calculate the concentration (in mol L^{-1}) of SO_2 that is present at 100°C in equilibrium with SO_2Cl_2 (at a concentration of 3.6×10^{-4} mol L^{-1}) and chlorine (at a concentration of 6.9×10^{-3} mol L^{-1}. (Use the data of problem 25.)

Effects of External Stresses on Equilibria: Le Chatelier's Principle

39. (See Example 7–9.) Some $H_2(g)$ at a partial pressure of 5.73 atm is mixed with $I_2(g)$ at that same initial partial pressure and the two react:

$$H_2(g) + I_2(g) \rightleftharpoons 2\ HI(g)$$

The equilibrium constant of this reaction equals 53.7 at the temperature of the experiment. After the reaction comes to equilibrium, the partial pressure of $HI(g)$ is reduced by 2.00 atm. Subsequently the equilibrium re-establishes itself. Compute the partial pressure of the $HI(g)$ when the equilibrium is re-established.

40. (See Example 7–9.) Some $H_2(g)$ at a partial pressure of 5.73 atm is mixed with $I_2(g)$ at that same initial partial pressure and the two react:

$$H_2(g) + I_2(g) \rightleftharpoons 2\ HI(g)$$

The equilibrium constant of this reaction equals 53.7 at the temperature of the experiment. After the reaction comes to equilibrium, the partial pressure of $H_2(g)$ is reduced by 1.00 atm. Subsequently the equilibrium re-establishes itself. Compute the partial pressure of the $HI(g)$ when the equilibrium is re-established.

41. (See Example 7–10.) The following reaction is exothermic:

$$3\ NO(g) \rightleftharpoons N_2O(g) + NO_2(g)$$

Explain the effect of each of the following stresses on the position of the equilibrium.

(a) $N_2O(g)$ is added to the equilibrium mixture without change of volume or temperature.

(b) The volume of the equilibrium mixture is reduced at constant temperature.

(c) The temperature of the equilibrium mixture is decreased.

(d) Gaseous argon (which does not react) is added to the equilibrium mixture while keeping both the total gas pressure and the temperature constant.

(e) Gaseous argon is added to the equilibrium mixture without changing the volume.

42. (See Example 7–10.) The following reaction is endothermic:

$$SO_3(g) \rightleftharpoons SO_2(g) + 1/2\ O_2(g)$$

Explain the effect of each of the following stresses on the position of equilibrium.

(a) $O_2(g)$ is added to the equilibrium mixture without change of volume or temperature.

(b) The mixture is compressed at constant temperature.

(c) The equilibrium mixture is cooled.

(d) An inert gas is pumped into the equilibrium mixture while keeping the total gas pressure and the temperature constant.

(e) An inert gas is added to the equilibrium mixture without changing the volume.

43. (See Example 7–10.) The most extensively used organic compound in the chemical industry is ethylene (C_2H_4). The two equations

$$C_2H_4(g) + Cl_2(g) \rightleftharpoons C_2H_4Cl_2(g)$$
$$C_2H_4Cl_2(g) \rightleftharpoons C_2H_3Cl(g) + HCl(g)$$

represent the way in which vinyl chloride (C_2H_3Cl) is synthesized for eventual use in polymeric plastics (polyvinyl chloride, PVC). The by-product of the reaction, HCl, is now most cheaply made by this and similar reactions, rather than by the direct combination of H_2 and Cl_2. Heat is given off in the first reaction and taken up in the second. Describe how you would design an industrial process to maximize the yield of vinyl chloride.

44. (See Example 7–10.) Methanol is made from synthesis gas via the exothermic reaction

$$CO(g) + 2\ H_2(g) \longrightarrow CH_3OH(g)$$

Describe how you would control the temperature and pressure to maximize the yield of methanol.

45. In a gas-phase reaction, it is observed that the equilibrium yield of products is increased by lowering the temperature and by reducing the volume.

(a) Is the reaction exothermic or endothermic?

(b) Is there a net increase or a net decrease in the number of gas molecules in the reaction?

46. The equilibrium constant of a gas-phase reaction is observed to increase as the temperature is increased. When the nonreacting gas neon is admitted to the reaction mixture (holding the temperature and the total pressure fixed and increasing the volume of the reaction vessel), the product yield is observed to decrease

(a) Is the reaction exothermic or endothermic?

(b) Is there a net increase or a net decrease in the number of gas molecules in the reaction?

Heterogeneous Equilibrium

47. (See Example 7–11.) Using the law of mass action, write the equilibrium expression for each of the following reactions:

(a) $8\ H_2(g) + S_8(s) \rightleftharpoons 8\ H_2S(g)$

(b) $C(s) + H_2O(\ell) + Cl_2(g) \rightleftharpoons COCl_2(g) + H_2(g)$

(c) $CaCO_3(s) \rightleftharpoons CaO(s) + CO_2(g)$

(d) $3\ C_2H_2(g) \rightleftharpoons C_6H_6(\ell)$

48. (See Example 7–11.) Using the law of mass action, write the equilibrium expression for each of the following reactions:

(a) $3\ C_2H_2(g) + 3\ H_2(g) \rightleftharpoons C_6H_{12}(\ell)$

(b) $CO_2(g) + C(s) \rightleftharpoons 2\ CO(g)$

(c) $CF_4(g) + 2\ H_2O(\ell) \rightleftharpoons CO_2(g) + 4\ HF(g)$

(d) $K_2NiF_6(s) + TiF_4(s) \rightleftharpoons K_2TiF_6(s) + NiF_2(s) + F_2(g)$

49. (See Example 7–11.) Using the law of mass action, write the equilibrium expression for each of the following reactions:

(a) $Zn(s) + 2\ Ag^+(aq) \rightleftharpoons Zn^{2+}(aq) + 2\ Ag(s)$

(b) $VO_4^{3-}(aq) + H_2O(\ell) \rightleftharpoons VO_3(OH)^{2-}(aq) + OH^-(aq)$

(c) $2\ As(OH)_6^{3-}(aq) + 6\ CO_2(g) \rightleftharpoons$
$$As_2O_3(s) + 6\ HCO_3^-(aq) + 3\ H_2O(\ell)$$

50. (See Example 7–11.) Using the law of mass action, write the equilibrium expression for each of the following reactions:

(a) $6\ I^-(aq) + 2\ MnO_4^-(aq) + 4\ H_2O(\ell) \rightleftharpoons$
$$3\ I_2(aq) + 2\ MnO_2(s) + 8\ OH^-(aq)$$

(b) $2 Cu^{2+}(aq) + 4 I^-(aq) \rightleftharpoons 2 CuI(s) + I_2(aq)$

(c) $1/2 O_2(g) + Sn^{2+}(aq) + 3 H_2O(\ell) \rightleftharpoons$
$\qquad SnO_2(s) + 2 H_3O^+ \quad (aq)$

51. N_2O_4 is soluble in the solvent cyclohexane; however, dissolution does not prevent N_2O_4 from breaking down to give NO_2 according to the equation

$$N_2O_4 \; (cyclohexane) \rightleftharpoons 2 NO_2 \; (cyclohexane)$$

An effort to compare this solution equilibrium to the similar equilibrium in the gas phase gave the following experimental data at 20°C:

$[N_2O_4]$ (mol L^{-1})	$[NO_2]$ (mol L^{-1})
0.190×10^{-3}	2.80×10^{-3}
0.686×10^{-3}	5.20×10^{-3}
1.54×10^{-3}	7.26×10^{-3}
2.55×10^{-3}	10.4×10^{-3}
3.75×10^{-3}	11.7×10^{-3}
7.86×10^{-3}	17.3×10^{-3}
11.9×10^{-3}	21.0×10^{-3}

(a) Graph the *square* of the concentration of NO_2 versus the concentration of N_2O_4.

(b) Compute the average equilibrium constant of this reaction.

52. NO_2 is soluble in carbon tetrachloride (CCl_4). As it dissolves, it dimerizes to give N_2O_4 according to the equation

$$2 NO_2(CCl_4) \rightleftharpoons N_2O_4(CCl_4)$$

A study of this equilibrium gave the following experimental data at 20°C:

$[N_2O_4]$ (mol L^{-1})	$[NO_2]$ (mol L^{-1})
0.192×10^{-3}	2.68×10^{-3}
0.721×10^{-3}	4.96×10^{-3}
1.61×10^{-3}	7.39×10^{-3}
2.67×10^{-3}	10.2×10^{-3}
3.95×10^{-3}	11.0×10^{-3}
7.90×10^{-3}	16.6×10^{-3}
11.9×10^{-3}	21.4×10^{-3}

(a) Graph the concentration of N_2O_4 versus the *square* of the concentration of NO_2.

(b) Compute the average equilibrium constant of this reaction.

53. The equilibrium constant for the "water gas" reaction

$$C(s) + H_2O \; (g) \rightleftharpoons CO(g) + H_2(g)$$

is $K = 2.6$ at a temperature of 1000 K. Calculate the reaction quotient Q for each of the following conditions (all partial pressures are given in atm), and state which direction the reaction shifts in coming to equilibrium:

(a) $P_{H_2O} = 0.600$; $P_{CO} = 1.525$; $P_{H_2} = 0.805$

(b) $P_{H_2O} = 0.724$; $P_{CO} = 1.714$; $P_{H_2} = 1.383$

54. The equilibrium constant for the reaction

$$H_2S(g) + I_2(g) \rightleftharpoons 2 HI(g) + S(s)$$

at 110°C is equal to 0.0023. Calculate the reaction quotient Q for each of the following conditions, and determine whether solid sulfur is consumed or produced as the reaction comes to equilibrium:

(a) $P_{I_2} = 0.461$ atm; $P_{H_2S} = 0.050$ atm; $P_{HI} = 0.0$ atm

(b) $P_{I_2} = 0.461$ atm; $P_{H_2S} = 0.050$ atm; $P_{HI} = 9.0$ atm

55. (See Example 7–12.) Pure solid NH_4HSe is placed in an evacuated container at 24.8°C. Eventually, the pressure above the solid reaches 0.0184 atm, thanks to the reaction

$$NH_4HSe(s) \rightleftharpoons NH_3(g) + H_2Se(g)$$

This pressure is the equilibrium pressure in the container.

(a) Calculate the equilibrium constant of this reaction at 24.8°C.

(b) In a different container, the partial pressure of $NH_3(g)$ in equilibrium with $NH_4HSe(s)$ at 24.8°C is 0.0252 atm. What is the partial pressure of $H_2Se(g)$?

56. (See Example 7–12.) The total pressure of the gases in equilibrium with solid sodium hydrogen carbonate at 110°C is 1.648 atm, corresponding to the reaction

$$2 NaHCO_3(s) \rightleftharpoons Na_2CO_3(s) + H_2O(g) + CO_2(g)$$

($NaHCO_3$ is used in dry chemical fire extinguishers because the products of this decomposition reaction smother the fire.)

(a) Calculate the equilibrium constant at 110°C.

(b) What is the partial pressure of water vapor in equilibrium with $NaHCO_3(s)$ at 110°C if the partial pressure of $CO_2(g)$ is 0.800 atm?

57. The equilibrium constant at 25°C for the reaction

$$CO_2(g) + H_2(g) \rightleftharpoons H_2O(\ell) + CO(g)$$

is 3.22×10^{-4}. What initial pressures of CO_2 and H_2 must be mixed to produce CO at a pressure of 0.100 atm if the pressures of CO_2 and H_2 are required to be equal? Assume that all the gases obey the ideal gas law.

58. At 25°C, the equilibrium constant for the formation of urea from ammonia and carbon dioxide according to the equation

$$2 NH_3(g) + CO_2(g) \rightleftharpoons CO(NH_2)_2(s) + H_2O(g)$$

equals 0.615. What are the equilibrium partial pressures of ammonia and carbon dioxide at 25°C if crystalline urea is placed in a closed container in which water vapor is maintained at a constant pressure of 0.03126 atm?

Extraction and Separation Processes

59. An aqueous solution, initially 1.00×10^{-2} M in iodine (I_2), is shaken with an equal volume of an immiscible organic solvent, CCl_4. The iodine distributes itself between the aqueous and CCl_4 layers, and when equilibrium is reached at 27°C, the concentration of I_2 in the aqueous layer is 1.30×10^{-4} M. Calculate the partition coefficient K at 27°C for the reaction

$$I_2(aq) \rightleftharpoons I_2(CCl_4)$$

60. An aqueous solution, initially 2.50×10^{-2} M in iodine (I_2), is shaken with an equal volume of an immiscible organic solvent, CS_2. The iodine distributes itself between the aqueous and CS_2 layers, and when equilibrium is reached at 25°C, the concentration of I_2 in the aqueous layer is 4.16×10^{-5} M. Calculate the partition coefficient K at 25°C for the reaction

$$I_2(aq) \rightleftharpoons I_2(CS_2)$$

61. Benzoic acid (C_6H_5COOH) dissolves in water to the extent of 2.00 g L^{-1} at 15°C and in diethyl ether to the extent of 6.6×10^2 g L^{-1} at the same temperature.
    (a) Calculate the equilibrium constants at 15°C for the two reactions

    $$C_6H_5COOH(s) \rightleftharpoons C_6H_5COOH(aq)$$

    and

    $$C_6H_5COOH(s) \rightleftharpoons C_6H_5COOH(ether)$$

    (b) From your answers to part (a), calculate the partition coefficient K for the reaction

    $$C_6H_5COOH(aq) \rightleftharpoons C_6H_5COOH(ether)$$

62. Citric acid ($C_6H_8O_7$) dissolves in water to the extent of 720 g L^{-1} at 15°C and in diethyl ether to the extent of 22 g L^{-1} at the same temperature.
    (a) Calculate the equilibrium constants at 15°C for the two reactions

    $$C_6H_8O_7(s) \rightleftharpoons C_6H_8O_7(aq)$$

    and

    $$C_6H_8O_7(s) \rightleftharpoons C_6H_8O_7(ether)$$

    (b) From your answers to part (a), calculate the partition coefficient K for the reaction

    $$C_6H_8O_7(aq) \rightleftharpoons C_6H_8O_7(ether)$$

Additional Problems

63. At 298 K, unequal amounts of $BCl_3(g)$ and $BF_3(g)$ were mixed in a container. The gases reacted to form $BFCl_2(g)$ and $BClF_2(g)$. When equilibrium was finally reached, the four gases were present in these relative chemical amounts:

    $BCl_3(90)$, $BF_3(470)$, $BClF_2(200)$, $BFCl_2(45)$.

    (a) Determine the equilibrium constants at 298 K of the two reactions

    $$2\ BCl_3(g) + BF_3(g) \rightleftharpoons 3\ BFCl_2(g)$$
    $$BCl_3(g) + 2\ BF_3(g) \rightleftharpoons 3\ BClF_2(g)$$

    (b) Determine the equilibrium constant of the reaction

    $$BCl_3(g) + BF_3(g) \rightleftharpoons BFCl_2(g) + BClF_2(g)$$

    and explain why knowing this equilibrium constant really adds nothing to what you knew in part (a).

64. Methanol can be synthesized by means of the equilibrium reaction

    $$CO(g) + 2\ H_2(g) \rightleftharpoons CH_3OH(g)$$

    for which the equilibrium constant at 225°C is 6.08×10^{-3}. Assume that the ratio of the pressures of $CO(g)$ and $H_2(g)$ is 1:2. What values should they have if the partial pressure of methanol is to be 0.500 atm?

65. At equilibrium at 425.6°C, a sample of *cis*-1-methyl-2-ethyl-cyclopropane is 73.6% converted into the *trans* form:

    $$cis \rightleftharpoons trans$$

    (a) Compute the equilibrium constant K for this reaction.
    (b) Suppose that 0.525 mol of the *cis* compound is placed in a 15.00-L vessel and heated to 425.6°C. Compute the equilibrium partial pressure of the *trans* compound.

66. The equilibrium constant for the reaction

    $$(CH_3)_3COH(g) \rightleftharpoons (CH_3)_2CCH_2(g) + H_2O(g)$$

    is equal to 2.42 at 450 K.

    (a) A pure sample of the reactant, which is named "*tert*-butanol," is confined in a container of fixed volume at a temperature of 450 K and at an original pressure of 0.100 atm. Calculate the fraction of this starting material that is converted to products at equilibrium.
    (b) A second sample of the reactant is confined, this time at an original pressure of 5.00 atm. Again, calculate the fraction of the starting material that is converted to products at equilibrium.

67. Consider the equilibrium

    $$2\ SO_2(g) + O_2(g) \rightleftharpoons 2\ SO_3(g)$$

    The equilibrium constant of this reaction rises from 0.25 at 1100 K to 0.70 at 523 K. A container holds a mixture of SO_2 and O_2, initially at partial pressures of 2.0 and 3.0 atm, respectively.

    (a) Find the equilibrium partial pressure of SO_3 in the flask if it is heated to 1100 K and and its contents come to equilibrium at that temperature.
    (b) Repeat the calculation for a temperature of 523 K.
    (c) Show that the percent of SO_2 that has reacted is greater at 523 K than it is at 1100 K.

68. At elevated temperatures, phosphorus pentachloride decomposes to give phosphorus trichloride and chlorine

    $$PCl_5(g) \rightleftharpoons PCl_3(g) + Cl_2(g)$$

    An equilibrium mixture of the three gases, obtained from the decomposition of initially pure phosphorus pentachloride in a vessel of fixed volume, has a density of 1.46 g L^{-1} at 523 K and 0.569 atm. Find the value of K for the reaction at 523 K.

69. When $I_2(g)$, $Br_2(g)$, and $IBr(g)$ are mixed at 400 K such that their initial partial pressures are 2.00 atm, 4.00 atm, and 3.00 atm, respectively, they take part in the reaction

    $$I_2(g) + Br_2(g) \rightleftharpoons 2\ IBr(g)$$

    The equilibrium constant is 131. Calculate the partial pressures of the three gases when equilibrium is reached.

70. Acetic acid in the vapor phase consists of both monomeric and dimeric forms in equilibrium:

    $$2\ CH_3COOH(g) \rightleftharpoons (CH_3COOH)_2(g)$$

    At 110°C the equalibrium constant for this reaction is 3.72.

    (a) Calculate the partial pressure of the dimer when the total pressure is 0.725 atm at equilibrium.
    (b) What percentage of the acetic acid is dimerized under these conditions?

71. Much of the content of Le Chatelier's principle can be understood easily by using the reaction quotient Q. Consider, for example, the reaction used to make cyclohexane (C_6H_{12}), a starting material in the production of nylon. In the presence of a small amount of nickel or platinum, cyclohexane is produced from benzene and hydrogen:

$$C_6H_6(g) + 3\,H_2(g) \rightleftharpoons C_6H_{12}(g)$$

Write the expression for the reaction quotient Q for this reaction, and then rewrite it (using the ideal gas law) in terms of numbers of moles, volume, and temperature. At equilibrium, $Q = K$ for the reaction.

   (a) Now suppose that, starting from equilibrium, a small chemical amount of H_2 is added while volume and temperature are held fixed. Does Q become larger or smaller than K? In which direction does the reaction shift to restore equilibrium?

   (b) Suppose instead that the temperature is held fixed and the volume is decreased. Does Q become larger or smaller than K? In which direction does the reaction shift to restore equilibrium?

72. Write the equilibrium expression for the oxidation of gold in a basic solution containing cyanide ions, a reaction used in the recovery of gold from mine tailings:

$$4\,Au(s) + 8\,CN^-(aq) + O_2(g) + 2\,H_2O(\ell) \rightleftharpoons$$
$$4\,Au(CN)_2^-(aq) + 4\,OH^-(aq)$$

*73. At 298 K, chlorine is only slightly soluble in water. Thus, under a pressure of 1.00 atm of $Cl_2(g)$, 1.00 L of water at equilibrium dissolves just 0.091 mol of Cl_2.

$$Cl_2(g) \rightleftharpoons Cl_2(aq)$$

In such solutions, the $Cl_2(aq)$ concentration is 0.061 M and the concentrations of $Cl^-(aq)$ and $HOCl(aq)$ are both 0.030 M. These two additional species are formed by the reaction:

$$Cl_2(aq) + H_2O(\ell) \rightleftharpoons$$
$$H^+(aq) + Cl^-(aq) + HOCl(aq)$$

There are no other Cl-containing species. Compute the equilibrium constants for the two reactions.

74. At 400°C, the reaction

$$BaO_2(s) + 4\,HCl(g) \rightleftharpoons BaCl_2(s) + 2\,H_2O(g) + Cl_2(g)$$

has an equilibrium constant equal to K_1. How is K_1 related to the equilibrium constant K_2 of the reaction

$$2\,Cl_2(g) + 4\,H_2O(g) + 2\,BaCl_2(s) \rightleftharpoons$$
$$8\,HCl(g) + 2\,BaO_2(s)$$

at 400°C ?

75. Ammonium hydrogen sulfide, a solid, decomposes to give $NH_3(g)$ and $H_2S(g)$. At 25°C, some $NH_4HS(s)$ is placed in an evacuated container. A portion of it decomposes, and the total pressure at equilibrium is 0.659 atm. Extra $NH_3(g)$ is then injected into the container and, when equilibrium is re-established, the partial pressure of $NH_3(g)$ is 0.750 atm.

   (a) Compute the equilibrium constant for the decomposition of ammonium hydrogen sulfide.

   (b) Determine the final partial pressure of $H_2S(g)$ in the container.

76. The equilibrium constant for the reaction

$$KOH(s) + CO_2(g) \rightleftharpoons KHCO_3(s)$$

is 6×10^{15} at 25°C. Suppose that 7.32 g of KOH and 9.41 g of $KHCO_3$ are placed in a closed evacuated container and allowed to reach equilibrium. Calculate the pressure of $CO_2(g)$ at equilibrium.

77. The equilibrium constant for the reduction of nickel(II) oxide to nickel at 754°C is 255.4, corresponding to the reaction

$$NiO(s) + CO(g) \rightleftharpoons Ni(s) + CO_2(g)$$

If the total pressure of the system at 754°C is 2.50 atm, calculate the partial pressures of $CO(g)$ and $CO_2(g)$.

78. Both glucose (corn sugar) and fructose (fruit sugar) taste sweet, but fructose tastes sweeter. Each year in the United States, tons of corn syrup destined to sweeten food are treated to convert glucose as fully as possible to the sweeter fructose. The reaction is an equilibrium:

$$glucose \rightleftharpoons fructose$$

   (a) A 0.2564 M solution of pure glucose is treated at 25°C with an enzyme (catalyst) that causes the above equilibrium to be reached quickly. The final concentration of fructose is 0.1175 M. In another experiment at the same temperature, a 0.2666 M solution of pure fructose is treated with the same enzyme, and the final concentration of glucose is 0.1415 M. Compute an average equilibrium constant for the above reaction.

   (b) At equilibrium under these conditions, what percentage of glucose is converted to fructose?

79. Stearic acid tends to dimerize when dissolved in hexane:

$$2\,C_{17}H_{35}COOH(hex) \rightleftharpoons (C_{17}H_{35}COOH)_2(hex)$$

The equilibrium constant for this reaction is 40 at 48°C. Stearic acid (15.0 g) is dissolved in 1.250 L of hexane, and the dimerization comes to equilibrium at 48°C. Calculate the concentrations of both the monomer and the dimer.

80. Superheated steam ($H_2O(g)$) at a pressure of 22.2 atm is pumped into an iron vessel at 800 K (527°C). At this temperature, the equilibrium

$$3\,Fe(s) + 4\,H_2O(g) \rightleftharpoons Fe_3O_4(s) + 4\,H_2(g)$$

has an equilibrium constant of 12.3. If the steam is given enough time to come to equilibrium with the walls of the vessel, what partial pressure of hydrogen is generated?

81. Polychlorinated biphenyls (PCBs) are a major environmental problem. These oily substances have many uses, but they resist breakdown by bacterial action when spilled in the environment and, being fat soluble, can accumulate to dangerous concentrations in the fatty tissues of fish and other animals. One little-appreciated complication in controlling the problem is that there are 209 different PCBs, all now in the environment. The various PCBs are generally similar, but their solubilities in fats differ considerably. The best measure of this is K_{OW}, the equilibrium constant for the partition of a PCB between the fat-like solvent octanol and water.

$$PCB(aq) \rightleftharpoons PCB(octanol)$$

An equimolar mixture of PCB-2 and PCB-11 in water

is treated with an equal volume of octanol. Determine the ratio between the amounts of PCB-2 and PCB-11 in the water at equilibrium. At room temperature, K_{OW} is 3.98×10^4 for PCB-2 and 1.26×10^5 for PCB-11.

82. Refer to the data in problems 61 and 62. Suppose that 2.00 g of a solid consisting of 50.0% benzoic acid and 50.0% citric acid by mass is added to 100.0 mL of water and 100.0 mL of diethyl ether, and the whole assembly is shaken. When the immiscible layers are separated and the solvents are removed by evaporation, two solids result. Calculate the percentages (by mass) of the major component in each solid.

83. At 25°C, the partition coefficient for the equilibrium

$$I_2(aq) \rightleftharpoons I_2(CCl_4)$$

has the value $K = 85$. To 0.100 L of an aqueous solution,

which is initially 2×10^{-3} M in I_2, we add 0.025 L of CCl_4. The mixture is shaken in a separatory funnel and allowed to separate into two phases, and the CCl_4 phase is withdrawn.
(a) Calculate the fraction of the I_2 remaining in the aqueous phase.
(b) Suppose that the remaining aqueous phase is shaken with another 0.025 L of CCl_4 and again separated. What fraction of the I_2 from the original aqueous solution is *now* in the aqueous phase?
(c) Compare your answer with that of Example 7–13, in which the same total amount of CCl_4 (0.050 L) was used in a *single* extraction. For a given total amount of extracting solvent, which is the more efficient way to remove iodine from water?

Sulfur mined for conversion to sulfuric acid.

CUMULATIVE PROBLEM

Sulfuric Acid

Sulfuric acid is produced in larger volume than any other chemical. It has applications ranging from fertilizer manufacture to metal treatment and the synthesis of organic and medicinal chemicals. Its production, properties, and uses are discussed in detail in Chapter 22.

The modern industrial production of sulfuric acid takes place in three steps, for which the balanced chemical equations are

1. $S(s) + O_2(g) \rightleftharpoons SO_2(g)$
2. $SO_2(g) + 1/2\ O_2(g) \rightleftharpoons SO_3(g)$
3. $SO_3(g) + H_2O(\ell) \rightleftharpoons H_2SO_4(\ell)$

(a) Write equilibrium expressions for each of the three steps above, with equilibrium constants K_1, K_2, and K_3.

(b) If these reactions could be carried out at 25°C, the equilibrium constants would be 3.9×10^{52}, 2.6×10^{12}, and 2.2×10^{14}. Write a balanced equation for the overall reaction, and calculate its equilibrium constant at 25°C.

(c) Although the products of all three equilibria are strongly favored at 25°C (part (b)), reactions 1 and 2 occur too slowly to be practical; they must be carried out at elevated temperatures. At 700°C, the partial pressures of SO_2, O_2, and SO_3 in an equilibrium mixture are measured to be 2.23 atm, 1.14 atm, and 6.26 atm, respectively. Calculate K_2 at 700°C.

(d) At 600°C, K_2 has the value 9.5. Is reaction 2 exothermic or endothermic?

(e) Some SO_2 is placed in a flask and heated with oxygen to 600°C. At equilibrium, 62% of it has reacted to give SO_3. Calculate the partial pressure of oxygen at equilibrium in this reaction mixture.

(f) Equal chemical amounts of SO_2, O_2, and SO_3 are mixed and heated quickly to 600°C. Their total pressure before reaction is 0.090 atm. Does reaction 2 occur from right to left or from left to right? Does the total pressure increase or decrease during the course of the reaction?

(g) Reactions 1 and 3 are both exothermic. State the effects on equilibria 1 and 3 of an increase in temperature and of a decrease in volume. (*Note:* A change in volume has little effect on liquids and solids taking part in a reaction.)

Acid–Base Equilibria

Fabrication of a specialized electrode for use in pH measurements.
(Courtesy of Orion Research, Inc.)

Acid–base reactions comprise one of the major categories of chemical action established in Chapter 4. In this chapter, we apply our ideas about chemical equilibrium and the equilibrium constant (developed in Chapter 7) to an especially important aspect of acid–base chemistry—calculating how far such reactions proceed before coming to equilibrium.

Although many acid–base reactions take place in the gaseous and solid states and in nonaqueous solutions, the focus in this chapter is on acid–base reactions in aqueous solutions. There are two reasons. First, aqueous acids and bases are overwhelmingly important in the daily events of life, from the (perhaps mundane) properties of vinegar or ammonia-containing cleansers to the biochemical roles of the amino acids in proteins and the nucleic acids in genes. Second, acid–base reactions attain equilibrium rapidly in aqueous solutions. One never need worry about waiting for a real-life system to settle down to equilibrium because aqueous acid–base reactions reach equilibrium moments after the reactants are mixed.

8-1 BRØNSTED–LOWRY ACIDS AND BASES

Pure water contains equal, small concentrations of hydrogen ions ($H^+(aq)$) and hydroxide ions ($OH^-(aq)$) generated by autoionization

$$H_2O(\ell) \rightleftharpoons H^+(aq) + OH^-(aq)$$

Acidic solutions have higher concentrations of hydrogen ions than does pure water; basic solutions have higher concentrations of hydroxide ions. Svante Arrhenius accordingly defined acids as $H^+(aq)$ donors and bases as $OH^-(aq)$ donors (see Section 4–3). However, many substances that contain neither hydrogen nor oxygen (and which cannot donate H^+ or OH^-) are found to give acidic or basic solutions. For example, aqueous $CuCl_2$ is acidic, and aqueous $NaCN$ is basic. Such observations prompted a broadening of the original definition. Acids became defined as substances that increased the concentration of hydrogen ions in water and bases as substances that increased the concentration of hydroxide ions. The increases might come directly, by donation of H^+ or OH^-, or indirectly, through reaction with water, as in

$$Cu^{2+}(aq) + H_2O(\ell) \rightleftharpoons CuOH^+(aq) + H^+(aq)$$

and

$$CN^-(aq) + H_2O(\ell) \rightleftharpoons HCN(aq) + OH^-(aq)$$

The enlarged definition was still not entirely satisfactory. It did not apply in nonaqueous solvents; it did not apply in the absence of a solvent; it suggested nothing about *why*, in terms of bonding and structure, some substances are strongly acidic whereas others are strongly basic. Two new definitions helped deal with these issues.

The first, the Brønsted–Lowry definition (named for its originators, Johannes Brønsted and Thomas Lowry, who worked independently), identifies the transfer of H^+ ions from acid to base as the essence of the acid–base reaction. The second, the Lewis definition (named for *its* originator, Gilbert Lewis), identifies bases as electron-pair donors and acids as electron-pair acceptors and focuses on the making and breaking of bonds involving lone-pair electrons. We take up the Brønsted–Lowry concept now and the Lewis concept in Section 8–8.

The Brønsted–Lowry definition has proved extraordinarily helpful in organizing information about the relative strengths of acids and bases and finds wide use to this day:

A **Brønsted–Lowry acid** is a substance that can *donate* a hydrogen ion; a **Brønsted–Lowry base** is a substance that can *accept* a hydrogen ion.

• Brønsted-Lowry acids and bases are very frequently called **proton donors** and **proton acceptors.** This is somewhat inexact because a proton is an $^1H^+$ ion, but ordinary hydrogen contains both ^2H and ^1H.

Acids and bases occur in the Brønsted–Lowry picture as **conjugate acid–base pairs.** The formula of a conjugate base is obtained by subtracting a hydrogen ion (H^+) from the formula of the acid, and the formula of a conjugate acid is obtained by adding H^+ to the formula of the base. For example

$$OH^- \qquad\qquad H_2O \qquad\qquad H_3O^+$$

conjugate base of H_2O conjugate acid of OH^- conjugate acid of H_2O
and
conjugate base of H_3O^+

The conjugate acid of water, H_3O^+, is called the **oxonium** or **hydronium ion.** It consists of three H atoms bonded to a central O atom with a net charge of $+1$.

Acid–base reactions in the Brønsted–Lowry scheme occur as the transfer of an H^+ ion from donor to acceptor. The products are the conjugate base of the donor (the acid) and the conjugate acid of the acceptor (the base). For example, as acetic acid (CH_3COOH) dissolves in water, H_2O acts as a Brønsted–Lowry base to accept a hydrogen ion donated by the acetic acid:

$$\underset{\text{acid}_1}{CH_3COOH(aq)} + \underset{\text{base}_2}{H_2O(\ell)} \rightleftharpoons \underset{\text{acid}_2}{H_3O^+(aq)} + \underset{\text{base}_1}{CH_3COO^-(aq)}$$

The subscripts 1 and 2 designate the two conjugate acid–base pairs. The reaction creates a new acid, hydronium ion, and a new base, acetate ion (CH_3COO^-). Depending on the desired emphasis, one can say that CH_3COO^- is the conjugate base of CH_3COOH or that CH_3COOH is the conjugate acid of CH_3COO^-.

The autoionization of water fits the Brønsted–Lowry definition. To see this, consider the reaction

$$\underset{\text{acid}_1}{H_2O(\ell)} + \underset{\text{base}_2}{H_2O(\ell)} \rightleftharpoons \underset{\text{acid}_2}{H_3O^+(aq)} + \underset{\text{base}_1}{OH^-(aq)}$$

in which one H_2O functions as an acid and the other as a base. Now compare the formulas $H_3O^+(aq)$ and $H^+(aq)$. The first singles out one molecule of water among the several (the exact number varies) that aquate the H^+ and shows it explicitly. The second lumps all the aquating molecules into the "(aq)." Both represent the same thing, a dissolved hydrogen ion. Hence, the preceding equation is equivalent to

$$H_2O(\ell) \rightleftharpoons H^+(aq) + OH^-(aq)$$

The H_3O^+ notation is usually used when discussing the Brønsted–Lowry theory because it highlights transfers of H^+ to H_2O, a common event. At other times, the simpler notation $H^+(aq)$ is preferred.

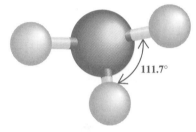

The structure of the hydronium ion (H_3O^+).

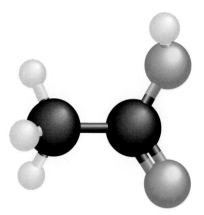

Acetic acid (CH_3COOH) is a weak organic acid. It is the active ingredient in vinegar, which is a 3 to 5% solution of acetic acid in water. Pure acetic acid is called "glacial acetic acid" because it freezes into layered crystals that resemble glaciers.

EXAMPLE 8–1

When carbon dioxide is dissolved in water to make a carbonated beverage, it reacts with the water to form carbonic acid (H_2CO_3). Give the name and formula of the conjugate base of H_2CO_3.

Solution

Conjugate bases are formed by the removal of a hydrogen ion. We simply subtract H^+ from H_2CO_3 to leave HCO_3^-. This is the hydrogen carbonate ion.

Exercise

Trimethylamine (C_3H_9N) is a soluble weak base with a foul odor (it contributes to the smell of rotten fish). Write the formula of its conjugate acid.

Answer: The conjugate acid is $C_3H_9NH^+$.

In a Brønsted–Lowry acid–base reaction, a hydrogen ion is transferred from an acid to a base. A fruitful alternative view of this interaction is that it is a *competition* between two bases for a hydrogen ion. In the reaction of ammonia with water,

$$\underset{\text{acid}_1}{H_2O(\ell)} + \underset{\text{base}_2}{NH_3(aq)} \rightleftharpoons \underset{\text{acid}_2}{NH_4^+(aq)} + \underset{\text{base}_1}{OH^-(aq)}$$

the two bases $NH_3(aq)$ and $OH^-(aq)$ compete for hydrogen ions. The outcome of the competition depends on the relative strengths of the two. The weakness of ammonia as a base means that the forward reaction occurs only to a limited extent, so that dissolving one mole of NH_3 leads to the production of less than one mole of OH^- ion in solution.

EXAMPLE 8–2

Sodium cyanide is a highly toxic compound that dissolves in water to give a solution that turns litmus blue. Other indicators confirm that aqueous NaCN is basic. Write equations explaining these observations.

Solution

As the sodium cyanide dissolves, it dissociates into $Na^+(aq)$ and $CN^-(aq)$ ions in the dissolution reaction

$$NaCN(s) \longrightarrow Na^+(aq) + CN^-(aq)$$

By taking a hydrogen ion away from the solvent

$$CN^-(aq) + H_2O(\ell) \rightleftharpoons OH^-(aq) + HCN(aq)$$

$CN^-(aq)$ would generate its own conjugate acid (HCN) and force OH^- out into the solution. Evidently, the CN^- ion has at least some success in winning hydrogen ions away from water molecules because the solution is basic. Cyanide ion acts as a Brønsted–Lowry base in this reaction.

Exercise

A solution of NH_4Cl in water gives colors with indicators that show that it is acidic. Write a net ionic equation that explains these observations.

Answer: $NH_4^+(aq) + H_2O(\ell) \rightleftharpoons NH_3(aq) + H_3O^+(aq)$

The Brønsted–Lowry approach is not limited to aqueous solutions. With liquid ammonia as the solvent, for example, we can write

$$\underset{\text{acid}_1}{HCl(NH_3)} + \underset{\text{base}_2}{NH_3(\ell)} \rightleftharpoons \underset{\text{acid}_2}{NH_4^+ (NH_3)} + \underset{\text{base}_1}{Cl^-(NH_3)}$$

NH_3 acts here as a base, even though the hydroxide ion OH^- is not present.

Another advantage of the Brønsted–Lowry approach is the insight it gives into the phenomenon of **amphoterism.** An amphoteric molecule or ion can function as either an acid or a base, depending on reaction conditions (see Section 4–3). A good example is water itself. In the presence of ammonia, water serves as an acid, donating a hydrogen ion to NH_3; in the presence of acetic acid it serves as a base, accepting a hydrogen ion from CH_3COOH. In the same way, the hydrogen carbonate ion can act as an acid

$$HCO_3^-(aq) + H_2O(\ell) \rightleftharpoons H_3O^+(aq) + CO_3^{2-}(aq)$$

or as a base:

$$HCO_3^-(aq) + H_2O(\ell) \rightleftharpoons H_2CO_3(aq) + OH^-(aq)$$

The Brønsted–Lowry definition of acid implies that any molecule or ion that contains hydrogen can in principle serve as an acid if a strong enough base is available to accept (or extract) a hydrogen ion from it. Further, any molecule or ion can act as a base if a strong enough acid is in the vicinity to force a hydrogen ion upon it. Thus, acetic acid acts as a base when dissolved in anhydrous sulfuric acid

$$\underset{\text{acid}_1}{H_2SO_4(\ell)} + \underset{\text{base}_2}{CH_3COOH(H_2SO_4)} \rightleftharpoons \underset{\text{acid}_2}{CH_3COOH_2^+(H_2SO_4)} + \underset{\text{base}_1}{HSO_4^-(H_2SO_4)}$$

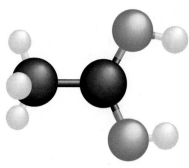

The structure of the $CH_3COOH_2^+$ ion, the product when acetic acid is forced to act as a base and accept a hydrogen ion. The H^+ ion is accepted at an oxygen atom.

8–2 WATER AND THE pH SCALE

The autoionization of water proceeds to only a slight extent under ordinary conditions, but it always happens and is responsible for a small but measurable presence of $H_3O^+(aq)$ and $OH^-(aq)$ ions in even the purest water. The equilibrium expression for this reaction is

$$\frac{[H_3O^+][OH^-]}{(1)^2} = [H_3O^+][OH^-] = K_w$$

in accordance with the rules given in Section 7–6. The equilibrium constant K_w is a small number because the autoionization reaction does not proceed far to the right before reaching equilibrium. As with all equilibrium constants, K_w depends on the temperature (Table 8–1). Its value at 25°C is worth memorizing, because room-temperature water is so common.

$$K_w = 1.0 \times 10^{-14} \quad \text{at 25°C}$$

• From this point on in this chapter, 1's representing pure solids, pure liquids, and dilute solvents in equilibrium expressions will simply be omitted.

• In all problems in this chapter, a temperature of 25°C may be assumed unless otherwise stated.

Table 8–1
Temperature Dependence of K_w

Temperature (°C)	K_w
0	0.114×10^{-14}
10	0.292×10^{-14}
20	0.681×10^{-14}
25	1.01×10^{-14}
30	1.47×10^{-14}
40	2.92×10^{-14}
50	5.47×10^{-14}
60	9.61×10^{-14}

Figure 8-1 The H_3O^+–OH^- equilibrium in water operates like a seesaw: if $[OH^-]$ is up, then $[H_3O^+]$ is down, and vice versa.

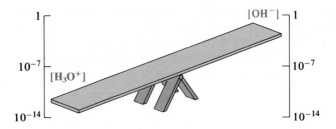

Pure water contains no ions other than $H_3O^+(aq)$ and $OH^-(aq)$. Because bulk samples of water are electrically neutral, the number of positive charges they contain must equal the number of negative charges. Consequently,

$$[H_3O^+] = [OH^-] = y$$

Substituting into the equilibrium expression gives

$$y^2 = K_w = 1.0 \times 10^{-14}$$
$$y = 1.0 \times 10^{-7}$$

so that in pure water at 25°C, the concentrations of $H_3O^+(aq)$ and $OH^-(aq)$ ions are 1.0×10^{-7} M. If something is dissolved in water that raises the concentration of $H_3O^+(aq)$ above 1.0×10^{-7}, then the concentration of $OH^-(aq)$ immediately drops below 1.0×10^{-7} so that the product $[H_3O^+] \times [OH^-]$ stays equal to 1.0×10^{-14} (Fig. 8–1).

Strong Acids and Bases

An aqueous solution is acidic if it contains an excess of H_3O^+ over OH^- ions. Acidic species generate excesses of $H_3O^+(aq)$ by reactions with water. A **strong acid** is one that reacts essentially completely with water to produce $H_3O^+(aq)$. Hydrochloric acid (HCl) is a strong acid. When it is put in water, essentially every molecule transfers a hydrogen ion to a water molecule to form $H_3O^+(aq)$:

$$HCl(aq) + H_2O(\ell) \longrightarrow H_3O^+(aq) + Cl^-(aq) \quad \text{(reaction essentially complete)}$$

Another strong acid is perchloric acid ($HClO_4$). Dissolving 0.10 mol of either of these acids in enough water to make 1.0 L of solution gives a final concentration of 0.10 M for $H_3O^+(aq)$. The acid–base properties of aqueous solutions are dictated by their concentration of $H_3O^+(aq)$, so these two strong acids have the same apparent strength in water despite the difference in their intrinsic abilities to donate hydrogen ions. The same point applies to other strong acids as well. Water has a **leveling effect** on strong acids; the reactions of strong acids with water all lie so far to the right at equilibrium that the differences between the acids are undetectably small. The concentration of H_3O^+ in a 0.10 M solution of *any* strong acid that donates one hydrogen ion per molecule is simply 0.10 M; the OH^- concentration is

• We can safely ignore the contribution of H_3O^+ from the autoionization of water in these cases because it is far smaller than 0.10 mol L^{-1}.

$$[OH^-] = \frac{K_w}{[H_3O^+]} = \frac{1.0 \times 10^{-14}}{0.10} = 1.0 \times 10^{-13} \text{ M}$$

• Because H^- and NH_2^- ions are electrically charged, they are available only in combination with positive ions, as in the compounds NaH and KNH_2.

A **strong base** is defined, in an analogous fashion, as a base that reacts essentially completely to give $OH^-(aq)$ ion when put in water. The amide ion, NH_2^-, and the hydride ion, H^-, are both strong bases. For every mole of either of these species that is added to one liter of solution, one mole per liter of $OH^-(aq)$ is formed. The other products are NH_3 and H_2, respectively:

$$H_2O(\ell) + NH_2^-(aq) \longrightarrow NH_3(aq) + OH^-(aq) \quad \text{(reaction essentially complete)}$$
$$H_2O(\ell) + H^-(aq) \longrightarrow H_2(aq) + OH^-(aq) \quad \text{(reaction essentially complete)}$$

The important base NaOH, a solid, increases the OH^- concentration in water by dissociating completely, as discussed in Section 4–3:

$$NaOH(s) \longrightarrow Na^+(aq) + OH^-(aq) \quad \text{(reaction essentially complete)}$$

For every mole of NaOH that dissolves in water, essentially one mole of $OH^-(aq)$ forms; this makes NaOH a strong base, too. Strong bases are leveled in aqueous solution in the same way that strong acids are leveled, and for the same reason—the interaction between a strong base and water is so extensive that the intrinsic differences between strong bases are drowned out.

Calculations to determine the concentrations of $OH^-(aq)$ and $H_3O^+(aq)$ in solutions of strong bases and strong acids are straightforward because the chemical amount of OH^- or H_3O^+ produced depends only on the chemical amount of strong base or strong acid put into the water. If 0.10 mol of NaOH or NH_2^- ion or H^- ion is put into enough water to make 1.0 L of solution, then in every case

$$[OH^-] = 0.10 \text{ M}$$
$$[H_3O^+] = \frac{1.0 \times 10^{-14}}{0.10} = 1.0 \times 10^{-13} \text{ M}$$

The OH^- contribution from the autoionization of water is negligible in these cases. Of course, if only a very small amount of strong base or acid were added to pure water (for example, 10^{-7} mol of base or acid per liter), then the autoionization of water would have to be taken into account.

The pH Function

In aqueous solution, the concentration of hydronium ion can range from 10 M to 10^{-15} M. It is convenient to compress this enormous range by introducing a logarithmic scale for the intensity of acidity in aqueous solutions: the **pH** scale. The pH of an aqueous solution is given by the equation

$$pH = -\log_{10}[H_3O^+]$$

Pure water at 25°C has $[H_3O^+] = 1.0 \times 10^{-7}$ M, so that

$$pH = -\log_{10}(1.0 \times 10^{-7}) = -(-7.00) = 7.00$$

A 0.10 M solution of HCl has $[H_3O^+] = 0.10$ M, so

$$pH = -\log_{10}(0.10) = -\log_{10}(1.0 \times 10^{-1}) = -(-1.00) = 1.00$$

whereas a 0.10 M solution of NaOH has

$$pH = -\log_{10}\frac{1.0 \times 10^{-14}}{0.10} = -\log_{10}(1.0 \times 10^{-13}) = -(-13.00) = 13.00$$

As the examples show, calculating the pH is especially easy when the concentration of H_3O^+ is a power of 10, because the logarithm is then just the power to which 10 is raised. In other cases, a calculator is needed.

• The "concentration" of the H_3O^+ in this calculation is really a dimensionless ratio equal to 1.0×10^{-7} M divided by a reference concentration of 1 M. See Section 7-6.

• See Appendix C-3 for a discussion of significant figures when logarithms are involved.

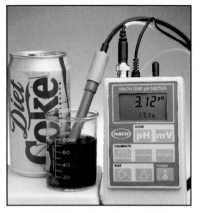

Figure 8-2 The pH of a soft drink is being checked with a pH meter equipped with a digital read-out.

EXAMPLE 8-3

Calculate the pH (at 25°C) of an aqueous solution that has an $OH^-(aq)$ concentration of 1.2×10^{-6} M.

Solution

The concentration of $H_3O^+(aq)$ in this sample is

$$[H_3O^+] = \frac{K_w}{[OH^-]} = \frac{1.0 \times 10^{-14}}{1.2 \times 10^{-6}} = 8.3 \times 10^{-9} \text{ M}$$

Using a calculator, we find that

$$\log_{10}(8.3 \times 10^{-9}) = -8.08$$

The pH of the solution is defined as the negative of this, or 8.08.

Exercise

Compute the pH of an aqueous solution in which the hydrogen-ion concentration is twice the hydroxide-ion concentration: $[H_3O^+] = 2.0 \times [OH^-]$

Answer: pH = 6.85.

Solutions with $H_3O^+(aq)$ concentrations greater than 1 M have negative pH's; for example, if $[H_3O^+] = 2.0$ M, then pH = -0.30. However, $H_3O^+(aq)$ concentrations are most frequently lower than 1 M, so the negative sign in the definition of pH gives a number that is positive for most solutions. A *high* pH signifies a *low* concentration of H_3O^+, and vice versa. A change in pH by one unit implies a change in the concentrations of H_3O^+ and OH^- by a factor of 10 (that is, by one order of magnitude). In terms of pH,

pH < 7	acidic solution	$[H_3O^+] > [OH^-]$
pH = 7	neutral solution	$[H_3O^+] = [OH^-]$
pH > 7	basic solution	$[H_3O^+] < [OH^-]$

• These ranges apply strictly only for aqueous solutions at 25°C. Variations of a few degrees in the temperature are usually not important, however.

The pH is most directly measured through the use of a **pH meter** (Fig. 8–2). In this instrument the solution under test is made a part of an electrochemical cell having an output voltage that is sensitive to the hydronium-ion concentration, and the voltage is measured. The principle behind electrochemical measurements is considered in Chapter 13. The values of the pH for several common fluids are shown in Figure 8–3.

If the pH is given, the concentration of $H_3O^+(aq)$ can be calculated by raising 10 to the power (−pH). This is done on a calculator by using the 10^x key or the INV LOG key.

EXAMPLE 8-4

The pH of some grape juice at 25°C is 2.85. Calculate $[H_3O^+]$ and $[OH^-]$.

Solution

$$pH = 2.85 = -\log_{10}[H_3O^+]$$
$$[H_3O^+] = 10^{-pH} = 10^{-2.85}$$

This can be evaluated by using a calculator to give

$$[H_3O^+] = 1.4 \times 10^{-3} \text{ M}$$

$$[OH^-] = \frac{1.0 \times 10^{-14}}{1.4 \times 10^{-3}} = 7.1 \times 10^{-12} \text{ M}$$

Exercise

Calculate $[H_3O^+]$ and $[OH^-]$ in saliva that has a pH of 6.60.

Answer: $[H_3O^+] = 2.5 \times 10^{-7}$ M; $[OH^-] = 4.0 \times 10^{-8}$ M.

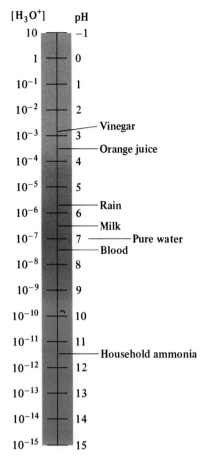

Figure 8–3 Many everyday materials are acidic or basic aqueous solutions with a wide range of pH values.

The prefix "p" in pH signals us to take the negative logarithm of the hydronium-ion concentration. This usage has been generalized to give definitions like

$$pOH = -\log_{10}[OH^-] \quad \text{and} \quad pK_w = -\log_{10}K_w$$

The notation has advantages. For example, the product of $[H_3O^+]$ and $[OH^-]$ in water always equals K_w; that is,

$$[H_3O^+][OH^-] = K_w$$

Taking the logarithm of both sides and multiplying through by -1 gives

$$-\log_{10}[H_3O^+] - \log_{10}[OH^-] = -\log_{10}K_w$$

These three negative logarithms are "p" quantities, so the equation becomes

$$pH + pOH = pK_w$$

At 25°C $pK_w = -\log_{10}(1.0 \times 10^{-14}) = 14.00$. Hence, at room temperature,

$$pH + pOH = 14.00 \text{ at } 25°C$$

8-3   THE STRENGTHS OF ACIDS AND BASES

Acids are classified as strong or weak depending on whether their reactions with water to give $H_3O^+(aq)$ go to completion or reach an equilibrium somewhere short of completion. These reactions are often called **acid ionization** reactions, because many acids create new ions in solution as they transfer H^+ ions to water, or **acid dissociation** reactions, because the acid breaks apart as it donates an H^+ ion. The proportion of an acid that has reacted with water in a particular solution is called the **fraction ionized** (or fraction dissociated) of that acid. The fraction ionized of a strong acid is essentially equal to 1; the fraction ionized of a weak acid is substantially less than 1. There are just seven common strong acids, and they are listed in Section 4–3 and at the top of Table 8–2.

Nearly all other acids are weak acids. A compound that is a weak acid is also a weak electrolyte (see Section 4–1). An aqueous solution of a weak electrolyte conducts electricity less well than a strong electrolyte of the same concentration does because fewer ions are present. Among solutions of the same concentration, the weaker the acid, the lower the concentration of hydronium ions and the higher the pH.

The Brønsted–Lowry model provides a basis for a quantitative scale of acid strength. The ionization of a generic acid (which we symbolize as "HA") in aqueous solution can be written as

$$HA(aq) + H_2O(\ell) \rightleftharpoons H_3O^+(aq) + A^-(aq)$$

• Weak electrolytes also depress the freezing point of solutions less than do strong electrolytes at the same concentration, as illustrated in Figure 6–27.

Table 8–2
Acidity Constants in Water at 25°C

Acid	Formula	Conjugate Base	K_a	pK_a
Hydriodic	HI	I^-	$\sim 10^{11}$	~ -11
Hydrobromic	HBr	Br^-	$\sim 10^9$	~ -9
Perchloric	$HClO_4$	ClO_4^-	$\sim 10^7$	~ -7
Hydrochloric	HCl	Cl^-	$\sim 10^7$	~ -7
Chloric	$HClO_3$	ClO_3^-	$\sim 10^3$	~ -3
Sulfuric (1)	H_2SO_4	HSO_4^-	$\sim 10^2$	~ -2
Nitric	HNO_3	NO_3^-	~ 20	~ -1.3
Hydronium ion	H_3O^+	H_2O	1	0.0
Urea acidium ion	$NH_2CONH_3^+$	$(NH_2)_2CO$ (urea)	6.6×10^{-1}	0.18
Iodic	HIO_3	IO_3^-	1.6×10^{-1}	0.80
Oxalic (1)	$H_2C_2O_4$	$HC_2O_4^-$	5.9×10^{-2}	1.23
Sulfurous (1)	H_2SO_3	HSO_3^-	1.5×10^{-2}	1.82
Sulfuric (2)	HSO_4^-	SO_4^{2-}	1.2×10^{-2}	1.92
Chlorous	$HClO_2$	ClO_2^-	1.1×10^{-2}	1.96
Phosphoric (1)	H_3PO_4	$H_2PO_4^-$	7.5×10^{-3}	2.12
Arsenic (1)	H_3AsO_4	$H_2AsO_4^-$	5.0×10^{-3}	2.30
Chloroacetic	$CH_2ClCOOH$	CH_2ClCOO^-	1.4×10^{-3}	2.85
Hydrofluoric	HF	F^-	6.6×10^{-4}	3.18
Nitrous	HNO_2	NO_2^-	4.6×10^{-4}	3.34
Formic	HCOOH	$HCOO^-$	1.8×10^{-4}	3.74
Benzoic	C_6H_5COOH	$C_6H_5COO^-$	6.5×10^{-5}	4.19
Oxalic (2)	$HC_2O_4^-$	$C_2O_4^{2-}$	6.4×10^{-5}	4.19
Hydrazoic	HN_3	N_3^-	1.9×10^{-5}	4.72
Acetic	CH_3COOH	CH_3COO^-	1.8×10^{-5}	4.74
Propionic	CH_3CH_2COOH	$CH_3CH_2COO^-$	1.3×10^{-5}	4.89
Pyridinium ion	$HC_5H_5N^+$	C_5H_5N	5.6×10^{-6}	5.25
Carbonic (1)	H_2CO_3	HCO_3^-	4.3×10^{-7}	6.37
Sulfurous (2)	HSO_3^-	SO_3^{2-}	1.0×10^{-7}	7.00
Arsenic (2)	$H_2AsO_4^-$	$HAsO_4^{2-}$	9.3×10^{-8}	7.03
Hydrosulfuric	H_2S	HS^-	9.1×10^{-8}	7.04
Phosphoric (2)	$H_2PO_4^-$	HPO_4^{2-}	6.2×10^{-8}	7.21
Hypochlorous	HClO	ClO^-	3.0×10^{-8}	7.52
Ammonium ion	NH_4^+	NH_3	5.6×10^{-10}	9.25
Hydrocyanic	HCN	CN^-	6.2×10^{-10}	9.21
Carbonic (2)	HCO_3^-	CO_3^{2-}	4.8×10^{-11}	10.32
Methylammonium ion	$CH_3NH_3^+$	CH_3NH_2	2.3×10^{-11}	10.64
Arsenic (3)	$HAsO_4^{2-}$	AsO_4^{3-}	3.0×10^{-12}	11.52
Hydrogen peroxide	H_2O_2	HO_2^-	2.4×10^{-12}	11.62
Phosphoric (3)	HPO_4^{2-}	PO_4^{3-}	2.2×10^{-13}	12.66
Water	H_2O	OH^-	1.0×10^{-14}	14.00

where A^- is the conjugate base of HA. The equilibrium expression for this chemical reaction is

$$\frac{[H_3O^+][A^-]}{[HA]} = K_a$$

where the subscript "a" in K_a stands for "acidity." For example, if the symbol "HA" refers to hydrogen cyanide (HCN), we write

$$\frac{[H_3O^+][CN^-]}{[HCN]} = K_a$$

where K_a is the **acidity constant** (also called the *acid ionization constant* or *acid dissociation constant*) of hydrogen cyanide in water and has a numerical value of 6.2×10^{-10} at 25°C. Table 8–2 gives values of K_a for a number of important acids. The useful quantity $pK_a = -\log_{10}K_a$ is also given.

An acidity constant is a quantitative measure of the strength of the acid in a given solvent (in this case, water). In a very weak acid, the products of the reaction with the solvent have low concentrations and K_a is small. A strong acid gives the products essentially completely, so [HA] in the denominator is small and K_a is large. It is difficult to determine the values of K_a for strong acids in water because the amount of unreacted acid at equilibrium is too small to be measured accurately. Relative acid strengths can be estimated in these difficult cases by replacing water with a solvent that is a weaker base. If the chosen solvent is a poor enough base (that is, has a much weaker tendency than H_2O to accept H^+ ions), then even a relatively strong acid may donate hydrogen ions incompletely. For example, HCl donates H^+ ions less extensively than $HClO_4$ in the solvent diethyl ether ($C_2H_5OC_2H_5$); therefore, HCl is a weaker acid than $HClO_4$.

Hydrogen cyanide (HCN) is a highly toxic gas that dissolves in water to form equally toxic solutions of hydrocyanic acid. Its conjugate base, CN^-, forms complexes with gold and silver and is used in the treatment of mine tailings to extract these precious metals by bringing them into solution.

Base Strength

The strength of a base is inversely related to the strength of its conjugate acid; the weaker the acid, the stronger its conjugate base, and vice versa. The equation representing the transfer of H^+ ion from water to ammonia in aqueous solution is

$$H_2O(\ell) + NH_3(aq) \rightleftharpoons NH_4^+(aq) + OH^-(aq)$$

This reaction has the equilibrium expression

$$\frac{[NH_4^+][OH^-]}{[NH_3]} = K_b$$

where the equilibrium constant K_b is a **basicity constant.** Recognizing that $[OH^-]$ and $[H_3O^+]$ are related through the water autoionization equilibrium expression,

$$[H_3O^+][OH^-] = K_w$$

allows us to rewrite the K_b expression as

$$K_b = \frac{[NH_4^+]}{[NH_3]} \times [OH^-] = \frac{[NH_4^+]}{[NH_3]} \times \frac{K_w}{[H_3O^+]} = \frac{K_w}{K_a}$$

where K_a is the acidity constant for NH_4^+, the conjugate acid of the base NH_3. This

Hydrogen fluoride (HF) is a colorless liquid that boils at 19.5°C. It reacts with water as a weak acid. This contrasts with the other hydrogen halides, which are strong acids in water. Hydrogen fluoride and its aqueous solution, hydrofluoric acid, are highly corrosive materials that can etch glass.

kind of relationship between the K_b of a base and the K_a of its conjugate acid is completely general and can also be written

$$K_a K_b = K_w$$

Taking the logarithm of both sides of this equation and multiplying through by -1 gives

$$-\log K_a - \log K_b = -\log K_w$$

Using the p-notation (a prefix "p" on any quantity stands for the negative logarithm of that quantity), this equation becomes

$$pK_a + pK_b = pK_w$$

These relationships apply for any conjugate acid–base pair dissolved in water. Both show that the weaker the acid (K_a small and pK_a large), the stronger the conjugate base (K_b large and pK_b small). A helpful consequence is that we do not need to tabulate K_b or pK_b values separately from K_a or pK_a values. When required, they are easily calculated through one of the preceding expressions. Figure 8–4 shows the way in which the strengths of members of conjugate acid–base pairs are related.

Stronger bases generate more $OH^-(aq)$ ion in solution than equal amounts of weaker bases. The base PO_4^{3-} is stronger than CN^-, which is stronger than Cl^-: 1.0 mol of PO_4^{3-} per liter of solution produces 0.19 mol of $OH^-(aq)$; 1.0 mol of CN^- produces 0.0040 mol of $OH^-(aq)$; and 1.0 mol of Cl^- ions produces 0.00 mol of $OH^-(aq)$. Although PO_4^{3-} and CN^- outperform Cl^- as bases, both are outdone by NaOH, which produces the maximum possible $OH^-(aq)$ ion: 1.0 mole of OH^- per mole of NaOH dissolved.

If two bases compete for hydrogen ions, the stronger base "wins"; that is, it holds the larger proportion of hydrogen ions and *less* of that base is present when equilibrium is reached. To see this, consider the equilibrium

$$HF(aq) + CN^-(aq) \rightleftharpoons HCN(aq) + F^-(aq)$$

with equilibrium expression

$$\frac{[HCN][F^-]}{[HF][CN^-]} = K$$

In this case, the two bases $F^-(aq)$ and $CN^-(aq)$ compete for hydrogen ions. It is only an experimental detail whether we start the competition by mixing HF(aq) and $CN^-(aq)$ or $F^-(aq)$ and HCN(aq). In the first case action starts from the left side of the equation, and in the second case it starts from the right, but the equilibrium outcome is the same. To find out which base competes more effectively, we need the equilibrium constant for the competition reaction. To obtain it, we write the chemical equations and equilibrium expressions for the acid ionization of HF(aq) and HCN(aq), taking values for K_a from Table 8–2:

• Recall that this is a fundamental characteristic of the equilibrium state (see Section 7-1).

$$HF(aq) + H_2O(\ell) \rightleftharpoons H_3O^+(aq) + F^-(aq) \qquad \frac{[H_3O^+][F^-]}{[HF]} = K_a = 6.6 \times 10^{-4}$$

$$HCN(aq) + H_2O(\ell) \rightleftharpoons H_3O^+(aq) + CN^-(aq)$$

$$\frac{[H_3O^+][CN^-]}{[HCN]} = K_a' = 6.2 \times 10^{-10}$$

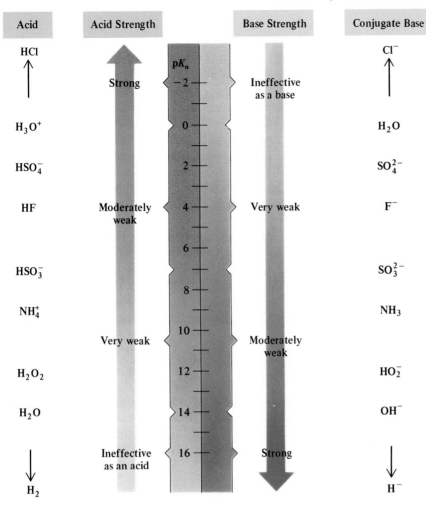

Acid	Acid Strength		Base Strength	Conjugate Base

Figure 8-4 The relative strengths of some acids and their conjugate bases. Strong acids have weak conjugate bases; strong bases have weak conjugate acids. The "one side high/other side low" pattern recalls the seesaw of Figure 8–1.

If the second equation is subtracted from the first, the result equals the chemical equation for the reaction of interest. Accordingly, the first equilibrium expression is *divided* by the second to give the equilibrium expression for the competition reaction. The numerical value of K is

• This is the same as *reversing* the second equation and *adding* it to the first.

$$K = \frac{K_a}{K_a'} = \frac{6.6 \times 10^{-4}}{6.2 \times 10^{-10}} = 1.1 \times 10^6$$

Because HCN(aq) is a weaker acid than HF(aq), K_a' is smaller than K_a. The K just calculated is plainly larger than 1, and the competition reaction lies far to the right at equilibrium. Most of the hydrogen ions end up associated with the CN$^-$(aq) in preference to the F$^-$(aq). The CN$^-$(aq) wins the competition with F$^-$(aq) because it is the stronger base. The dominant dissolved species at equilibrium are HCN(aq), which is a weaker acid than HF(aq), and F$^-$(aq), which is a weaker base than CN$^-$(aq).

Similar manipulation of other chemical equations and their associated equilibrium expressions allows the direction of net transfer of hydrogen ions to be predicted for any number of reactions between acids and bases.

Indicators

An indicator is a soluble compound, generally an organic dye, that changes its color noticeably over a fairly short range of pH. The typical indicator is a weak organic acid that has a different color than its conjugate base (Fig. 8–5). Methyl red, for example, is red when a single H^+ has been accepted (to form the acid) but yellow when that H^+ has been donated (leaving the base). Litmus changes color from red to blue as its acidic form is converted to its basic form. Good indicators have such intense colors that just two or three drops of a dilute solution of indicator suffice to give a strong color to the solution under investigation. This makes changes in color easy to see, and because the concentration of indicator molecules is then very small, the pH of the solution under study is practically unaffected by the presence of the indicator. The color changes of the indicator reflect the effects of the *other* acids and bases present in the solution.

Let us represent the acid form of a given indicator as HIn, and the conjugate base form as In^-. The acid–base equilibrium between the two is

$$HIn(aq) + H_2O(\ell) \rightleftharpoons H_3O^+(aq) + In^-(aq) \qquad \frac{[H_3O^+][In^-]}{[HIn]} = K_a$$

where K_a is the acidity constant for the indicator. This expression can be rearranged to give

$$\frac{[H_3O^+]}{K_a} = \frac{[HIn]}{[In^-]}$$

If the concentration of hydronium ion $[H_3O^+]$ is large relative to K_a, this ratio is large and [HIn] is large compared with $[In^-]$. The solution is the color of the acid form of the indicator because most molecules of the indicator are in the acid form. The indicator litmus, for example, has a K_a near 10^{-7}. If the pH is 5, then

$$\frac{[H_3O^+]}{K_a} = \frac{10^{-5}}{10^{-7}} = 100$$

so 100 times more indicator molecules are in the acid form than in the base form, and the solution is red, the color of the acid form.

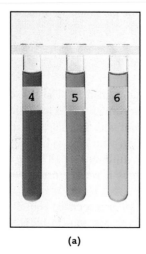
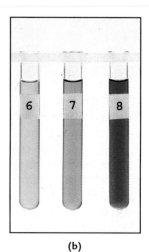
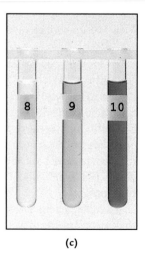

Figure 8–5 Color changes in three indicators. Methyl red (a) goes from red at low pH to orange at about pH 5 to yellow at high pH. Bromothymol blue (b) is yellow at low pH, blue at high pH, and green at about pH 7. Phenolphthalein (c) goes from colorless to pink at about pH 9.

(a)   (b)   (c)

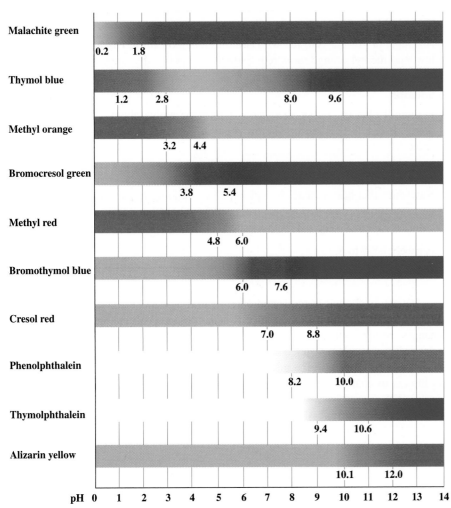

Figure 8-6 Indicators change their colors at very different values of the pH, so the best choice of indicator in a titration depends on what is being titrated, on what the titrant is, and on the particular experimental conditions.

As the concentration of hydronium ion is reduced, more molecules of the acid form HIn lose an H^+ ion and go over to the base form In^-. When $[H_3O^+]$ is near K_a, almost equal amounts of the two forms are present and the color is an almost equal mixture of the colors of HIn and In^-. A further decrease in $[H_3O^+]$ to a value much smaller than K_a then leads to a predominance of the base form, with the corresponding color being observed.

Different indicators have different values for K_a and thus change color at different pH values (Fig. 8–6). The weaker an indicator is as an acid, the higher is the pH at which the color change takes place. The color changes of indicators occur over a range of 1 to 2 pH units. Methyl red, for example, is red when the pH is below 4.8 and yellow when the pH is above 6.0; shades of orange are seen for intermediate pH values. This limits the accuracy to which the pH can be determined through the use of indicators; for accurate work, a pH meter must be used.

• For example, the intermediate color of litmus is purple, the sum of red (the color of the acid form) and blue (the color of the base).

EXAMPLE 8–5

A sample of vinegar is tested with two different indicators. Thymol blue turns yellow, and methyl orange turns red. Estimate the pH of the vinegar.

Solution

If thymol blue is yellow, then the pH must be above about 2.8, but if methyl orange is red, the pH must be below about 3.2. The pH is therefore 3.0, to within about 0.2 pH units.

Exercise

Rainwater that has leached through a pile of building materials and puddled underneath turns bromothymol blue from yellow to blue, but it leaves phenolphthalein colorless. Estimate the pH of this water.

Answer: 7.6 to 8.2.

Indicators abound in foods and other natural products (see Figs. 4–7 and 8–7). When a small amount of lemon juice is added to a cup of tea, the color of the tea changes from brown to light yellow. This change has nothing to do with the rather faint color of the lemon juice itself but arises from the decrease in pH of the tea caused by the citric acid the juice contains. The acid acts on the natural indicators in the tea to change their color. The single naturally occurring indicator cyanidin causes both the red color of the poppy and the blue color of the cornflower; the sap of the poppy is sufficiently acidic to turn it red, but the sap of the cornflower is

(a)   (b)   (c)

Figure 8–7   (a) Red cabbage extract is a natural pH indicator. In acidic solution it is red. (b) Upon addition of base, several colors, including green and blue, appear as incomplete mixing causes different regions of the sample to have different pH. (c) Once a uniformly basic pH is established, the color of the cabbage extract is blue.

Figure 8-8 An anthocyanin dye gives these strawberries their characteristic red color.

basic and makes it blue. Cyanidin in its acidic form is also the coloring matter in rhubarb. Related natural dyes called anthocyanins (Fig. 8–8) contribute to the colors of raspberries, blueberries, cherries, oranges, and red cabbage.

8–4 EQUILIBRIA INVOLVING WEAK ACIDS AND BASES

Weak acids and bases react only partially with water before coming to equilibrium. Consequently, calculations of the pH of their solutions involve the use of K_a or K_b and the laws of chemical equilibrium. Calculating $[H_3O^+]$ or $[OH^-]$ follows the pattern set in solving Example 7–6 for gas equilibria. In that case, we knew the initial partial pressures of the reactants, and we calculated the equilibrium pressures of the products of the incomplete reaction. Now, we know the initial concentration of an acid or base, and we calculate the equilibrium concentrations of the products of its incomplete reaction with water.

Weak Acids

A weak acid in water has a K_a that is smaller than 1, which is the K_a of the hydronium ion (see Table 8–2). The weaker the acid, the smaller its K_a, and the *larger* its pK_a. Values of pK_a start at zero for the strongest weak acid and range upward. If the pK_a of a compound or ion is greater than 14, then that species is ineffective as an acid in aqueous solution.

When a weak acid HA is mixed in water, the original concentration of acid is almost always known, but reaction with water consumes some HA molecules and generates A^- and H_3O^+ ions.

• Again, "HA" stands for a general weak acid.

$$HA(aq) + H_2O(\ell) \rightleftharpoons H_3O^+(aq) + A^-(aq)$$

We wish to calculate the concentrations of H_3O^+, A^-, and HA at equilibrium. The only complication relative to examples from Chapter 7 is that one of the products (H_3O^+) has a second source, the autoionization of the solvent, water. In cases of practical interest, this second effect is almost always small and can be neglected; however, it is wise to confirm this at the end of each calculation by verifying that the $[H_3O^+]$ from the acid ionization alone exceeds 10^{-7} M by at least one order of magnitude.

EXAMPLE 8-6

Acetic acid (CH_3COOH) in water donates a single hydrogen ion with a K_a of 1.8×10^{-5} at 25°C. Suppose that 1.00 mol of it is dissolved in enough water to give 1.00 L of solution. Calculate the pH and the fraction of acetic acid ionized at equilibrium.

Solution

The initial concentration of acetic acid is 1.00 M. If y mol per liter ionizes, then

$CH_3COOH(aq) + H_2O(\ell) \rightleftharpoons$	$H_3O^+(aq) +$	$CH_3COO^-(aq)$
Initial concentration (M)	1.00	≈ 0	0
Change in concentration (M)	$-y$	$+y$	$+y$
Equilibrium concentration (M)	$1.00 - y$	y	y

This set-up ignores the autoionization of water as a source of H_3O^+ because the concentration from that source is small compared to y for all but the weakest of acids or the most dilute of solutions. Next, from the equilibrium expression comes the equation

$$\frac{[H_3O^+][CH_3COO^-]}{[CH_3COOH]} = \frac{y^2}{1.00 - y} = K_a = 1.8 \times 10^{-5}$$

We could solve this equation using the quadratic formula, as in Chapter 7. A quicker way uses chemical insight: acetic acid is weak, so its fraction ionized is likely to be small at equilibrium. We therefore try the approximation that y is small relative to 1.00 (that is, that the final, equilibrium concentration of CH_3COOH is close to its initial concentration). This gives

$$\frac{y^2}{1.00 - y} \approx \frac{y^2}{1.00} = 1.8 \times 10^{-5}$$
$$y = 4.24 \times 10^{-3}$$

so that

$$[H_3O^+] = y = 4.24 \times 10^{-3} \text{ M}$$

where the final "4," a non-significant digit, is retained because rounding off (to 4.0×10^{-3}) at this intermediate stage might change the final answer.

We now check the approximations we have made. First, y is indeed much smaller than the original concentration of 1.00 M (by a factor of 200), so neglecting it in the denominator is justified. What is the tolerance for "much smaller"? A good rule of thumb is that the quantity that is neglected must come out to be less than 4 to 5% of the quantity from which it is subtracted (or to which it is added) in the set-up of the equilibrium calculation. Second, the concentration of H_3O^+ from acetic acid (4.2×10^{-3} M) is quite large compared with that coming from the autoionization of water, so the neglect of the autoionization is also justified. Therefore, we conclude that

$$pH = -\log_{10}(4.24 \times 10^{-3}) = 2.37$$

The fraction ionized of the acetic acid is the ratio of the concentration of CH_3COO^- present at equilibrium to the concentration of CH_3COOH that was present in the first place:

$$\frac{4.2 \times 10^{-3} \text{ M}}{1.00 \text{ M}} = 4.2 \times 10^{-3}$$

The percentage of the acetic acid that is ionized is 100 times this, or 0.42%. Fewer than one in 100 of the molecules of this typical weak acid dissociate into ions in this solution.

Exercise

Propionic acid (CH_3CH_2COOH) has $K_a = 1.3 \times 10^{-5}$ at 25°C. (a) Compute the pH at equilibrium in a 0.65 M solution of propionic acid. (b) Compute the fraction of the propionic acid ionized at equilibrium.

Answer: (a) pH = 2.54. (b) Fraction ionized = 0.0045.

If additional water is poured into a solution of a weak acid, the equilibrium concentration of H_3O^+ decreases, moving toward 10^{-7} M as the solution more nearly approximates pure water. Simultaneously, the fraction of the weak acid that is ionized *increases*. The following example shows how this comes about.

• This seeming contradiction is an example of the response of an equilibrium system to a perturbation, in accord with Le Chatelier's principle (see Section 7-5).

EXAMPLE 8–7

Suppose that 0.00010 mol of acetic acid is used instead of the 1.00 mol used in the previous example. Calculate the pH and the fraction of acetic acid ionized.

Solution

Once again, assume (subject to a later check) that the contribution of the autoionization of water to $[H_3O^+]$ is negligible. Therefore,

	$CH_3COOH(aq)$ + $H_2O(\ell)$ $\rightleftharpoons$	$H_3O^+(aq)$ +	$CH_3COO^-(aq)$
Initial concentration (M)	0.00010	≈0	0
Change in concentration (M)	$-y$	$+y$	$+y$
Equilibrium concentration (M)	$0.00010 - y$	y	y

$$\frac{y^2}{0.00010 - y} = 1.8 \times 10^{-5}$$

Solving quickly by assuming that y is much smaller than 0.00010 and neglecting it in the denominator gives $y = 4.2 \times 10^{-5}$, corresponding to ionization of 42% of the CH_3COOH molecules. Thus, y turns out *not* to be small relative to 0.00010 M because a substantial fraction of the acetic acid ionizes.

One way to deal with such a case is to multiply out the equilibrium equation and solve the quadratic equation that results. Multiplication and rearrangement give

$$y^2 + (1.8 \times 10^{-5})y - (1.8 \times 10^{-9}) = 0$$

Then substitution in the quadratic formula gives

$$y = \frac{-1.8 \times 10^{-5} \pm \sqrt{(1.8 \times 10^{-5})^2 - 4(1)(-1.8 \times 10^{-9})}}{2(1)}$$

As in other such computations, only one of the two roots of this equation has physical meaning. It is

$$y = 3.44 \times 10^{-5}$$

Referring back to the definition of y allows us to write

$$[H_3O^+] = [CH_3COO^-] = 3.44 \times 10^{-5} \text{ M}$$

Because $[H_3O^+] \gg 10^{-7}$ M, neglect of water autoionization is justified.

$$pH = \log_{10}(3.44 \times 10^{-5}) = \boxed{4.46}$$

$$\text{fraction ionized} = \frac{3.44 \times 10^{-5} \text{ M}}{0.00010 \text{ M}} = \boxed{0.34}$$

A larger fraction of acetic acid ionizes than in Example 8–6, but because the original concentration of the acid in the solution is much less, a smaller concentration of H_3O^+ results at equilibrium.

Another method of solution is numerical analysis, as used in Example 7–7. We guess, starting with likely values, a y to fit the equation. Then we check and refine the guess. Let us start with $y = 4.2 \times 10^{-5}$, the wrong answer that we just rejected. Substitution gives

$$\frac{(4.2 \times 10^{-5}) \times (4.2 \times 10^{-5})}{(0.00010 - 4.2 \times 10^{-5})} = 3.0 \times 10^{-5} = Q$$

This choice of y leads to a Q higher than K_a. Therefore, we try a smaller y. If $y = 2.1 \times 10^{-5}$ (one half the first guess)

$$\frac{(2.1 \times 10^{-5}) \times (2.1 \times 10^{-5})}{(0.00010 - 2.1 \times 10^{-5})} = 0.56 \times 10^{-5} = Q$$

Now Q is *less* than K_a. The correct y has been bracketted between 4.2×10^{-5} and 2.1×10^{-5}. Guessing $y = 3.2 \times 10^{-5}$ gives a Q that is slightly less than K_a, guessing $y = 3.6 \times 10^{-5}$ gives a Q that is slightly larger, and so on until $Q = K_a$ to the desired number of significant figures, and the last guess becomes the answer.

Exercise

For propionic acid (CH_3CH_2COOH), K_a is 1.3×10^{-5} at 25°C. (a) Calculate the pH of a 0.00010 M solution of propionic acid in water. (b) Calculate the fraction of the propionic acid ionized.

Answer: (a) pH = 4.52. (b) Fraction ionized = 0.30.

Numerical methods are often faster to apply than the quadratic formula. It is important to remember that the accuracy of a result is limited both by the accuracy of the input data (the values of K_a and the initial concentrations) and by the non-ideality of solutions, which causes deviations from the mass action law itself. It is pointless, and in fact wrong to calculate equilibrium concentrations to any higher degree of accuracy than 1 to 3% without evaluating and correcting for the effects of nonideality.

• Such corrections fall outside the scope of this text.

Weak Bases

The definition and description of weak acids, their K_a's, and the production of $H_3O^+(aq)$ ion when they are dissolved in water can be translated to apply to weak bases, their K_b's, and the production of $OH^-(aq)$ ion. A weak base such as ammonia reacts incompletely with water to produce $OH^-(aq)$:

• Recall that a strong base reacts essentially completely to give $OH^-(aq)$.

$$H_2O(\ell) + NH_3(aq) \rightleftharpoons NH_4^+(aq) + OH^-(aq) \qquad \frac{[NH_4^+][OH^-]}{[NH_3]} = K_b$$

The K_b of a weak base is smaller than 1, and the weaker the base, the smaller the K_b. If the K_b of a compound or ion is smaller than 1×10^{-14}, then that species is ineffective as a base in aqueous solution.

The analogy to weak acids holds as well in the calculation of the aqueous equilibria of weak bases, as the following example shows.

EXAMPLE 8-8

Calculate the pH of a solution made by dissolving 0.010 mol of NH_3 in enough water to give 1.00 L of solution at 25°C. The K_b for aqueous ammonia is 1.8×10^{-5}.

Solution

First, write the equilibrium expression for this reaction, equating it to K_b:

$$\frac{[NH_4^+][OH^-]}{[NH_3]} = K_b = 1.8 \times 10^{-5}$$

Second, set up a table of the changes in concentrations that occur as the reaction goes to equilibrium. The small contribution to $[OH^-]$ from the autoionization of water is neglected:

	$H_2O(\ell) +$ $NH_3(aq)$	$\rightleftharpoons$ $NH_4^+(aq) +$	$OH^-(aq)$
Initial concentration (M)	0.010	0	≈ 0
Change in concentration (M)	$-y$	$+y$	$+y$
Equilibrium concentration (M)	$0.010 - y$	y	y

Finally, substitution gives

$$\frac{y^2}{0.010 - y} = K_b = 1.8 \times 10^{-5}$$

which can be solved for y by either the quadratic formula or numerically:

$$y = 4.15 \times 10^{-4}$$
$$[OH^-] = [NH_4^+] = 4.15 \times 10^{-4} \text{ M}$$

The product of $[H_3O^+]$ and $[OH^-]$ is always K_w. Hence,

$$[H_3O^+] = \frac{K_w}{[OH^-]} = \frac{1.0 \times 10^{-14}}{4.15 \times 10^{-4}} = 2.4 \times 10^{-11} \text{ M}$$
$$pH = -\log(2.4 \times 10^{-11}) = \boxed{10.62}$$

The pH is greater than 7, as expected for a solution of a base.

Exercise

Cocaine is a weak base with a K_b of 2.6×10^{-7}. It is not very soluble in water, and a nearly saturated solution contains $0.0055 \text{ mol L}^{-1}$. Compute the pH of such a solution.

Answer: 10.07.

Hydrolysis

Most of the acids considered up to this point have been uncharged species, with the general formula HA. In the Brønsted–Lowry picture, however, there is no reason why acids must consist of electrically neutral particles. If NH_4Cl, which is a salt, dissolves in water, then $NH_4^+(aq)$ ions are present. Some of the ammonium ions transfer hydrogen ions to molecules of water, a straightforward Brønsted–Lowry acid–base reaction:

$$\underset{\text{acid}_1}{NH_4^+(aq)} + \underset{\text{base}_2}{H_2O(\ell)} \rightleftharpoons \underset{\text{acid}_2}{H_3O^+(aq)} + \underset{\text{base}_1}{NH_3(aq)}$$

$$\frac{[H_3O^+][NH_3]}{[NH_4^+]} = K_a = 5.6 \times 10^{-10}$$

The $NH_4^+(aq)$ acts as an acid here, just as acetic acid acted as an acid in Examples 8–6 and 8–7. Because K_a is considerably smaller than 1, the reaction is far from complete. The $NH_4^+(aq)$ ion is a weak acid, but it is nonetheless an acid, and a solution of ammonium chloride has a pH below 7 (Fig. 8–9).

The term **hydrolysis** is applied to reactions of aquated ions that change the pH from 7. Despite the special term, there is no reason to treat hydrolysis in any special way. As we have just seen, the hydrolysis of the ammonium ion is adequately described as a Brønsted–Lowry reaction in which water acts as a base and $NH_4^+(aq)$ acts as an acid.

Similarly, the hydrolysis of anions such as the F^- ion is simply another Brønsted–Lowry acid–base reaction, with water now acting as an acid (hydrogen-ion donor)

$$H_2O(\ell) + F^-(aq) \rightleftharpoons HF(aq) + OH^-(aq)$$

This reaction increases the OH^- concentration and so raises the pH.

(a)

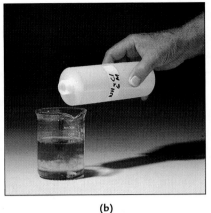

(b)

(c)

Figure 8-9 Proof that ammonium chloride is an acid. (a) A solution of sodium hydroxide, with the pink color of the indicator phenolphthalein indicating its basic character. (b, c) As aqueous ammonium chloride is added, a neutralization reaction takes place and the indicator turns colorless. The change proceeds from the bottom up because the mixture is not well stirred and the solution of the ammonium chloride is denser than that of the sodium hydroxide.

EXAMPLE 8-9

Suppose that 0.100 mol of $NaCH_3COO$ is dissolved in enough water to make 1.00 L of solution. What is the pH of the solution?

Solution

The K_a for CH_3COOH, the conjugate acid of CH_3COO^-, is 1.8×10^{-5}. Therefore, K_b for the acetate ion is

$$\frac{[CH_3COOH][OH^-]}{[CH_3COO^-]} = K_b = \frac{K_w}{K_a} = 5.6 \times 10^{-10}$$

The equilibrium here is

	$H_2O(\ell) + CH_3COO^-(aq) \rightleftharpoons$	$CH_3COOH(aq) +$	$OH^-(aq)$
Initial concentration (M)	0.100	0	≈ 0
Change in concentration (M)	$-y$	$+y$	$+y$
Equilibrium concentration (M)	$0.100 - y$	y	y

Substituting into the equilibrium expression gives

$$\frac{y^2}{0.100 - y} = K_b = 5.6 \times 10^{-10}$$
$$y = 7.5 \times 10^{-6}$$

We verify that y is small relative to 0.100 (it is much less than 1% of it) and large relative to 10^{-7}. Then $[OH^-] = 7.5 \times 10^{-6}$ M, and the computation of $[H_3O^+]$ and the pH follows the established pattern:

$$[H_3O^+] = \frac{K_w}{[OH^-]} = \frac{1.0 \times 10^{-14}}{7.5 \times 10^{-6}} = 1.3 \times 10^{-9} \text{ M}$$
$$pH = -\log_{10}(1.3 \times 10^{-9}) = 8.89$$

Exercise

Compute the pH of a 0.095 M aqueous solution of NaF. The K_a of HF is 6.6×10^{-4}.

Answer: pH = 8.08.

Most ionic species hydrolyze to a detectable extent. The hydrolysis of anions typically raises the pH; the hydrolysis of cations typically lowers the pH. When the cation and anion of a dissolved salt simultaneously hydrolyze, the outcome for the pH depends on the relative strengths of the cation as an acid (given by its K_a) and the anion as a base (given by its K_b).

The list of anions that have K_b's effectively equal to 0 and so do *not* raise the pH by hydrolysis is short: chloride (Cl^-), bromide (Br^-), iodide (I^-), hydrogen sulfate (HSO_4^-), nitrate (NO_3^-), chlorate (ClO_3^-), and perchlorate (ClO_4^-). These non-hydrolyzing anions are the conjugate bases of the seven common strong acids identified in Section 4–3 and Table 8–2.

The list of common nonhydrolyzing cations is also short, consisting of only ten: $Li^+, Na^+, K^+, Rb^+, Cs^+, Mg^{2+}, Ca^{2+}, Sr^{2+}, Ba^{2+}$, and Ag^+ ion. Most other cations, such as $Ni^{2+}(aq)$ and $Fe^{2+}(aq)$ ion, hydrolyze detectably.

To understand how a species that possesses no hydrogen can raise the concentration of H_3O^+ in solution, recall that all ions are aquated in water solution. A positive ion attracts electrons to itself from the neighboring water molecules, as shown in Figure 6–3. If it attracts them sufficiently, it weakens the O—H bonds in the water molecules. This causes those hydrogen atoms to be more readily donated than the ones in ordinary molecules of water; that is, it makes those hydrogen atoms acidic. Consider the dissolved Ni^{2+} ion. The central Ni^{2+} is closely associated with six molecules of water in a species formulated $Ni(H_2O)_6^{2+}$. This ion acts as a Brønsted–Lowry acid according to the equation

$$Ni(H_2O)_6^{2+}(aq) + H_2O(\ell) \rightleftharpoons Ni(H_2O)_5(OH)^+(aq) + H_3O^+(aq)$$

The equilibrium can be treated in the same way as the acid ionization of $HCN(aq)$ or $NH_4^+(aq)$ ion.

• Additional aspects of the interaction of metal ions with water are discussed in Section 9–5.

Knowing which ions do and which do not hydrolyze allows one to predict whether a salt gives an acidic, basic, or neutral solution. Aqueous NaCl gives neutral solutions because neither Na^+ nor Cl^- ion hydrolyzes. Aqueous Na_3PO_4 is basic (Na^+ does not hydrolyze, but PO_4^{3-} ion does), and aqueous $NiCl_2$ is acidic (Ni^{2+} hydrolyzes but Cl^- does not). When both the cation and anion in a salt hydrolyze (as in solutions of NH_4CN), the pH is raised if K_b for the anion exceeds K_a for the cation and lowered if K_b is less than K_a.

8–5 BUFFER SOLUTIONS

A **buffer solution** (or **buffer**) is any solution that maintains an approximately constant pH despite small additions of acid or base. Typically, a buffer solution contains a weak acid and a weak base that are conjugate to one another. Buffer solutions play important roles in controlling the solubility of ions in solution and in maintaining constant pH in biochemical and physiological processes. Many life processes are sensitive to pH and require regulation within a small range of H_3O^+ and OH^- concentrations. Organisms have built-in buffers to protect them against large changes in pH. Human blood, for example, has a pH near 7.4 that is maintained by a combination of carbonate, phosphate, and protein buffers. A blood pH below 7.0 or above 7.8 leads quickly to death.

The pH of Buffer Solutions

Consider a typical weak acid, formic acid (HCOOH), and its conjugate base, formate ion ($HCOO^-$). The latter can be obtained by dissolving a salt such as sodium formate (NaHCOO) in water. The acid–base equilibrium between these species is represented by

$$HCOOH(aq) + H_2O(\ell) \rightleftharpoons H_3O^+(aq) + HCOO^-(aq)$$

with an acidity constant:

$$\frac{[H_3O^+][HCOO^-]}{[HCOOH]} = K_a = 1.8 \times 10^{-4}$$

In Section 8–4, we show how to compute the pH of a solution of a weak acid such as HCOOH or of a weak base such as $HCOO^-$ ion. Suppose now that a weak acid and its conjugate base are *both* present initially. The required calculations differ only slightly. They resemble closely the calculation in Example 7–9, in which equilibrium is established starting from a mixture in which both reactants and products are present.

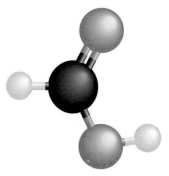

Formic acid (HCOOH) is the simplest carboxylic acid, with only a hydrogen atom attached to the —COOH group. It is used in curing leather, in dyeing, and in finishing paper and textiles. It is also one of the irritants in bee stings.

EXAMPLE 8–10

Suppose that 1.00 mol of HCOOH and 0.500 mol of NaHCOO are added to water and diluted to 1.00 L. Calculate the pH of the solution.

Solution

The NaHCOO dissolves by dissociation, giving 0.500 M $Na^+(aq)$, which (because it does not hydrolyze) plays no role in determining the pH, and 0.500 M $HCOO^-(aq)$. The HCOOH dissolves to give 1.00 M HCOOH, which donates some H^+ ions to H_2O in a reaction that comes to equilibrium far short of completion. The $HCOO^-(aq)$ ions from the NaHCOO affect the position of this equilibrium:

	$HCOOH(aq)$ + $H_2O(\ell)$ $\rightleftharpoons$	$H_3O^+(aq)$ +	$HCOO^-(aq)$
Initial concentration (M)	1.00	≈ 0	0.500
Change in concentration (M)	$-y$	$+y$	$+y$
Equilibrium concentration (M)	$1.00 - y$	y	$0.500 + y$

The equilibrium expression is

$$\frac{y(0.500 + y)}{1.00 - y} = K_a = 1.8 \times 10^{-4}$$

Because y is likely to be small relative to 1.00 (and to 0.500), this equation can be approximated as

$$\frac{y(0.500)}{1.00} \approx 1.8 \times 10^{-4}$$
$$y = 3.6 \times 10^{-4} \text{ M} = [H_3O^+]$$

A glance verifies that y is indeed small relative to 1.00 and 0.500. Then

$$pH = -\log_{10}(3.6 \times 10^{-4}) = \boxed{3.44}$$

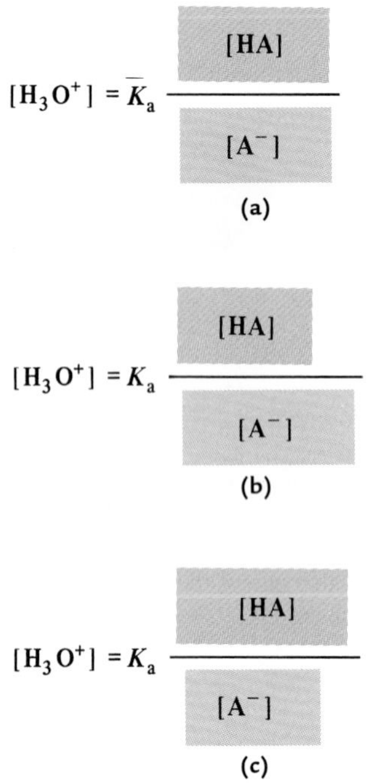

$$[H_3O^+] = \overline{K}_a \frac{[HA]}{[A^-]}$$

(a)

$$[H_3O^+] = K_a \frac{[HA]}{[A^-]}$$

(b)

$$[H_3O^+] = K_a \frac{[HA]}{[A^-]}$$

(c)

Figure 8-10 The $[H_3O^+]$ in a buffer depends on the ratio $[HA]:[A^-]$. If both the numerator and denominator of this fraction are large, then small additions of base or acid change the ratio only very slightly. (a) Initial ratio. (b) The ratio after addition of a small amount of base. (c) The ratio after the addition of a small amount of acid.

Exercise

Household bleach contains hypochlorous acid mixed with sodium hypochlorite. Suppose that 0.88 mol of hypochlorous acid (HClO) and 1.2 mol of sodium hypochlorite (NaClO) are mixed in enough water to give 1.50 L of solution. Compute the pH of the solution. The K_a of hypochlorous acid is 3.0×10^{-8}.

Answer: 7.66.

How Buffer Solutions Work

Writing the equilibrium expression for the ionization of a weak acid HA in the form

$$[H_3O^+] = K_a \frac{[HA]}{[A^-]}$$

shows that the concentration of hydronium ion depends on the *ratio* of the concentration of the weak acid to the concentration of its conjugate base. A solution acts as a buffer if both of these concentrations are fairly large. Adding a small amount of base to an effective buffer takes away only a few percent of the HA molecules by converting them into A^- ions and adds only a few percent to the amount of A^- that was originally present. The ratio $[HA]/[A^-]$ decreases, but only slightly. Added acid consumes a small fraction of the base A^- to generate a bit more HA. The ratio $[HA]/[A^-]$ now increases but, again, the change is only slight. Because the concentration of H_3O^+ is tied directly to this ratio, it also changes only slightly. This effect is diagrammed in Figure 8–10. The following example shows the quantitative workings of buffer action.

EXAMPLE 8–11

Suppose 0.10 mol of a strong acid such as HCl is included as the solution in Example 8–10 is mixed. Calculate the pH of the resulting solution.

Solution

The 0.10 mol of the strong acid HCl is clearly not enough to neutralize 0.500 mol of the base, formate ion. Let us suppose then that *all* of the hydrogen ions from HCl are taken up by formate ions, giving formic acid, after which some formic acid ionizes back to formate ion and H_3O^+. This route to equilibrium is simply one of many that might be imagined. It is not a statement of the sequence of reactions that actually occurs. The position of the final equilibrium does not depend on the route by which it is attained. Because 0.10 mol of H^+ ions from HCl would consume 0.10 mol of $HCOO^-$, the concentrations of $HCOO^-$ and HCOOH *after* complete reaction with HCl but *before* ionization are

$$[HCOO^-]_0 = 0.50 - 0.10 = 0.40 \text{ M}$$
$$[HCOOH]_0 = 1.00 + 0.10 = 1.10 \text{ M}$$

We set up a table of changes to calculate the concentrations at equilibrium:

	$HCOOH(aq)$ + $H_2O(\ell)$ $\rightleftharpoons$	$H_3O^+(aq)$	+ $HCOO^-(aq)$
Initial concentration (M)	1.10	≈ 0	0.40
Change in concentration (M)	$-y$	$+y$	$+y$
Equilibrium concentration (M)	$1.10 - y$	y	$0.40 + y$

The equilibrium expression then becomes

$$\frac{y(0.40 + y)}{1.10 - y} = 1.8 \times 10^{-4}$$

Because y is likely to be small relative to both 0.40 and 1.10,

$$y \approx \left(\frac{1.10}{0.40}\right)(1.8 \times 10^{-4}) = 4.9 \times 10^{-4}$$

$$pH = -\log_{10}(4.9 \times 10^{-4}) = 3.31$$

Even though 0.10 mol of a strong acid was added, the pH changed only slightly, from 3.44 to 3.31. By contrast, the same amount of hydrochloric acid in a liter of pure water would drop the pH from 7 to 1.

• The small change is in the direction of lower pH because acid was added.

Exercise

A total of 0.20 mol of HCl is added to the bleach solution from Exercise 8–10. Compute the final pH.

Answer: 7.49.

A similar calculation can be performed if a strong base such as $OH^-(aq)$ is added instead of a strong acid. The base reacts with formic acid to produce formate ions. Adding 0.10 mol $OH^-(aq)$ to the $HCOOH/HCOO^-$ buffer of Example 8–10 increases the pH only to 3.58. In the absence of the buffer system, the same amount of $OH^-(aq)$ would raise the pH of the liter of water to 13.00.

Designing Buffers

Controlling the pH is vital in synthetic and analytical chemistry, just as it is in living organisms. Fortunately, by the proper choice of a weak acid and the ratio in which it is mixed with its conjugate base, it is possible to prepare buffer solutions that maintain the pH close to any desired value. We now consider how to choose the best conjugate acid–base system and how to calculate the required acid–base ratio.

In a buffer consisting of a weak acid and its conjugate base, a competition exists between the tendency of the acid to donate hydrogen ions to water (decreasing the pH) and the tendency of the conjugate base to accept hydrogen ions from water (increasing the pH). The outcome depends on the magnitude of K_a. If K_a is large compared with 10^{-7}, the acid ionization wins out, and the natural pH range of the buffer is on the acid side (pH $<$ 7), as the case of the $HCOOH/HCOO^-$ buffer illustrates. Clearly then, a basic buffer (pH $>$ 7) results when the K_a of the acid in the conjugate–acid–base pair is smaller than 10^{-7}. These considerations provide a good general guideline for choosing a buffer system. Now, what about maintaining a specific value of the pH?

In Examples 8–10 and 8–11, the equilibrium concentrations of the acid and conjugate base in the buffer systems were close to their initial concentrations. When this is the case, the calculation of the pH of the buffer is simplified greatly. We can then write

$$K_a = \frac{[H_3O^+][A^-]}{[HA]} \approx \frac{[H_3O^+][A^-]_0}{[HA]_0}$$

Solving this equation for $[H_3O^+]$ gives

$$[H_3O^+] \approx \left(\frac{[HA]_0}{[A^-]_0}\right)K_a$$

Taking the negative logarithm of both sides of the equation and then using the definitions of pH and pK_a gives

$$pH \approx pK_a - \log_{10}\frac{[HA]_0}{[A^-]_0}$$

• Use care when employing this equation because it is approximate. It works only if *both* $[H_3O^+]$ and $[OH^-]$ (and, therefore, the extent of ionization) are small relative to the original concentrations $[HA]_0$ and $[A^-]_0$.

In biochemistry, this equation is known as the Henderson–Hasselbalch equation. It relates pH to pK_a and the concentrations of the acid and conjugate base and can be used to design buffers to maintain any desired pH. It can be verified that the equation gives the correct result for the pH in Examples 8–10 and 8–11.

An optimal buffer is one in which the acid and its conjugate base are as nearly equal in concentration as possible; if the difference between their concentrations is too great, the buffer becomes less resistant to the effects of additional acid or base. To select a buffer system, we should first look for an acid with a pK_a that is as close as possible to the desired pH. The concentrations of acid and conjugate base can then be adjusted slightly from the most desirable 1:1 ratio to give exactly the desired pH.

EXAMPLE 8–12

Design a buffer system with pH 4.60.

Solution

From a list of pK_a values (Table 8–2), we find that the pK_a for acetic acid is 4.74, so the CH_3COOH/CH_3COO^- buffer is a suitable one. The concentrations required to give the desired pH are related by

$$pH = 4.60 = pK_a - \log_{10}\frac{[CH_3COOH]_0}{[CH_3COO^-]_0}$$

$$\log_{10}\frac{[CH_3COOH]_0}{[CH_3COO^-]_0} = pK_a - pH = 4.74 - 4.60 = 0.14$$

$$\frac{[CH_3COOH]_0}{[CH_3COO^-]_0} = 10^{0.14} = 1.4$$

Such a ratio could be established by dissolving 0.100 mol of sodium acetate and 0.140 mol of acetic acid in one liter of water, or 0.200 mol $NaCH_3COO$ and 0.280 mol CH_3COOH in the same volume, and so on. As long as the ratio of the concentrations is 1.4 (and the concentrations are not too small), the solution is buffered at close to pH 4.60.

Exercise

Design a buffer to maintain a pH of 9.20, and suggest how to prepare it.

Answer: A NH_3/NH_4^+ buffer with a $[NH_4^+]$ to $[NH_3]$ concentration ratio of 1.12 would work. The buffer solution could be prepared by dissolving 0.100 mol of NH_3 and 0.112 mol of NH_4Cl in a reasonable quantity of water.

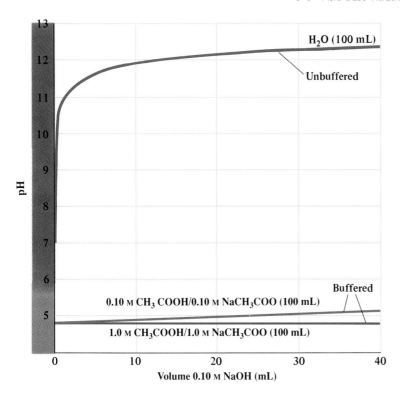

Figure 8–11   Addition of a given volume of base to buffered and unbuffered solutions causes a much larger change in the pH of the un-buffered solution. The two buffered solutions both resist pH changes, but the one with the higher concentration of buffer does so more effectively than the one with the lower concentration.

As the previous example shows, the *absolute* concentrations of acid and con-jugate base in a buffer are much less important than their ratio in determining the pH. However, the absolute concentrations *do* affect the capacity of the solution to resist changes in pH induced by added acid or base. The higher the absolute con-centrations of the buffering species, the smaller the change in pH when a fixed amount of a strong acid or base is added. In Example 8–10, for example, changing the concentrations of the buffering species from 1.00 M and 0.500 M to 0.500 M and 0.250 M, respectively, does not alter the original pH of 3.44 because the ratio of acid to base concentrations is unchanged. The pH after 0.100 mol of HCl is added (see Example 8–11) does change: it is 3.14 rather than 3.31. The buffer at lower con-centration resists pH change less well (Fig. 8–11). The buffering capacity of *any* buffer is exhausted if sufficient strong acid (or strong base) is added to consume all of the weak base (or the weak acid) originally present.

8–6   ACID-BASE TITRATION CURVES

Recall from Section 4–5 that an acid–base titration consists of adding carefully me-tered volumes of a basic solution of known concentration to an acidic solution of un-known concentration (or adding acid to base) to reach an end-point. The end-point is signaled by the color change of an indicator or a sudden jump in pH. Naturally, the pH of the reaction mixture changes constantly over the course of an acid–base titration. A graph of the pH versus the volume of titrating solution is a **titration curve.** The shape of an acid–base titration curve depends on the ionization constants of the acid and base and on their concentrations. The concepts of acid–base equilib-ria provide tools to predict the exact shapes of titration curves when K_a and K_b and

Figure 8–12 A titration curve for the titration of a strong acid by a strong base. The curve tracks the rise in pH as 0.1000 M NaOH is added to 100.0 mL of 0.1000 M HCl.

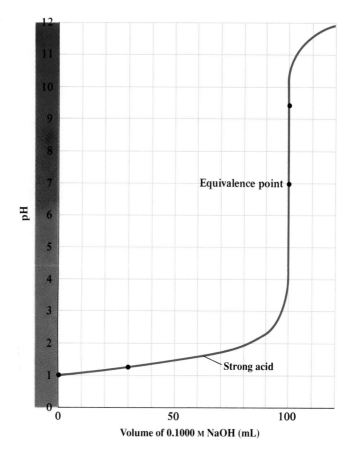

the concentrations of the acid and base are all known. The same concepts allow the calculation of K_a, K_b, and the unknown concentration from an experimental titration curve. We consider three categories of titrations: strong acid reacting with strong base; weak acid reacting with strong base; and strong acid reacting with weak base. We do not discuss titrations involving a weak acid and a weak base because they are not useful for analytical purposes.

Titration of a Strong Acid with a Strong Base

The addition of a strong base to a strong acid (or the reverse) is the simplest type of titration. The chemical reaction is the neutralization.

$$H_3O^+(aq) + OH^-(aq) \rightleftharpoons 2\,H_2O(\ell)$$

This reaction is the reverse of the autoionization of water. Its equilibrium constant therefore equals $1/K_w = 1/10^{-14} = 10^{+14}$. Clearly, neutralization lies far to the right at equilibrium.

Suppose we start with a solution of 100.0 mL (0.1000 L) of 0.1000 M HCl and titrate it with 0.1000 M NaOH. What does the titration curve look like? Perhaps the easiest way to find out is to perform the titration using a pH meter to generate the curve experimentally. The result is given in Figure 8–12. We also can draw a theoretical curve by figuring out the pH of the reaction mixture at many different points during the addition of the NaOH solution and plotting the results. Let us examine several of these calculations.

1. $V = 0$ mL NaOH added

Initially, $[H_3O^+] = 0.1000$ M, so the pH is 1.000. The chemical amount of H_3O^+ present initially is

$$n_{H_3O^+} = [H_3O^+] \times (\text{volume of acid solution})$$
$$= \left(0.1000 \frac{\text{mol}}{L}\right) \times (0.1000\,L)$$
$$= 1.000 \times 10^{-2} \text{ mol } H_3O^+$$

2. $V = 30.00$ mL NaOH added

30.00 mL of 0.1000 M NaOH solution contains

$$\frac{0.1000 \text{ mol NaOH}}{1\text{ L solution}} \times 0.03000 \text{ L solution} = 3.000 \times 10^{-3} \text{ mol NaOH}$$

and, therefore, 3.000×10^{-3} mol of OH^- ion. This reacts with (and neutralizes) an equal chemical amount of the H_3O^+ ion present initially. The chemical amount of hydronium ion that remains is

$$n_{H_3O^+} = 1.000 \times 10^{-2} \text{ mol} - 3.000 \times 10^{-3} \text{ mol}$$
$$= 7.00 \times 10^{-3} \text{ mol } H_3O^+$$

In addition (and this is very important to remember), the volume of the titration mixture has increased from 100.0 to 130.0 mL (that is, from 0.1000 to 0.1300 L). The *concentration* of H_3O^+ at this point in the titration is

$$[H_3O^+] = \frac{n_{H_3O^+}}{V_{\text{total}}} = \frac{7.00 \times 10^{-3} \text{ mol}}{0.1300 \text{ L}} = 0.0538 \text{ M}$$
$$pH = 1.269$$

• Over the course of a titration, the titrant reduces the concentration of the species being titrated in two ways: by removing some of it by reaction, and by diluting what remains.

It is often convenient to perform such calculations in terms of millimoles and milliliters:

$$n_{H_3O^+} = \left(\frac{0.1000 \text{ mmol}}{\text{mL}}\right) \times 100.0 \text{ mL} - \left(\frac{0.1000 \text{ mmol}}{\text{mL}}\right) \times 30.00 \text{ mL}$$
$$= 7.00 \text{ mmol } H_3O^+$$
$$[H_3O^+] = \frac{7.00 \text{ mmol } H_3O^+}{130.00 \text{ mL}} = 0.0538 \text{ M}$$

3. $V = 100.00$ mL NaOH added

This is called the **equivalence point**, the point in this titration at which the chemical amount of base added equals the chemical amount of acid originally present. It corresponds to the titration end-point of Section 4–5. The volume of base required to reach the equivalence point is the **equivalent volume** (V_e). At the equivalence point in the titration of a strong acid with a strong base, the concentrations of OH^- and H_3O^+ must be equal, and the pH is 7, due to the autoionization of water. At this point, the titration mixture is a solution of a salt (in this case, NaCl). The pH is 7 at the equivalence point *only* in the titration of a strong acid with a strong base (or vice versa). It differs from 7 if a weak acid or weak base takes part in the titration, as we shall see.

4. $V = 100.05$ mL

Beyond the equivalence point, we are adding OH^- to an unbuffered solution. The OH^- concentration can be found from the chemical amount of OH^- added after the equivalence point has been reached.

The volume beyond the equivalence point in this case is 0.05 mL, or 5×10^{-5} L. This is the volume of approximately one drop of solution let in from the buret. The chemical amount of OH^- in this volume is

$$n_{OH^-} = \left(\frac{0.1000 \text{ mol NaOH}}{1 \text{ L NaOH solution}} \right) \times 5 \times 10^{-5} \text{ L NaOH solution} \times$$
$$\left(\frac{1 \text{ mol } OH^-}{1 \text{ mol NaOH}} \right) = 5 \times 10^{-6} \text{ mol } OH^-$$

Meanwhile, the total volume of the titration mixture is

0.1000 L HCl solution + 0.10005 L NaOH solution = 0.20005 L solution

so the concentration of OH^- is

$$[OH^-] = \frac{\text{moles } OH^-}{\text{total volume}} = \frac{5 \times 10^{-6} \text{ mol}}{0.20005 \text{ L}} = 2.5 \times 10^{-5} \text{ M}$$

$$[H_3O^+] = \frac{K_w}{[OH^-]} = \frac{1.0 \times 10^{-14}}{2.5 \times 10^{-5}} = 4 \times 10^{-10} \text{ M; pH} = 9.4$$

As the titration curve shows (see Fig. 8–12), the pH increases dramatically in the immediate vicinity of the equivalence point: the concentration of H_3O^+ changes by four orders of magnitude between 99.98 mL and 100.02 mL NaOH! Any indicator whose color changes between pH = 5.0 and pH = 9.0 signals the end-point of the titration to an accuracy of ± 0.02 mL in 100.0 mL, or $\pm 0.02\%$.

The titration of a strong base by a strong acid is entirely parallel. In such a titration, the pH starts at a higher value and *drops* through a pH of 7 at the equivalence point. The roles of acid and base, and of H_3O^+ and OH^-, are reversed in the equations and calculations given above.

Titration of a Weak Acid with a Strong Base

We turn now to the titration of a *weak* acid with a strong base (the titration of a weak base with a strong acid is analogous). Consider the addition of 0.1000 M NaOH to 100.0 mL of 0.1000 M CH_3COOH. The concept of the equivalence point applies in this titration just as previously: at the equivalence point, the chemical amount of NaOH added equals the chemical amount of acetic acid originally present

$$n_{NaOH} = n_{CH_3COOH} \qquad \text{(at the equivalence point)}$$

Recall that the chemical amount of a solute in a solution equals its concentration multiplied by the volume of the solution, $n = cV$. Applying this fact to the two solutions gives

$$c_{NaOH}V_{NaOH} = c_{CH_3COOH}V_{CH_3COOH} \qquad \text{(at the equivalence point)}$$

The c_{CH_3COOH} is the *original* concentration, and V_{CH_3COOH} is the *original* volume of the acid (before any base was added); c_{NaOH} is the concentration of the titrating base, and V_{NaOH} is the volume of titrant required to reach the equivalence point. This volume, as explained in the previous section, is the equivalent volume V_e. In this titration three of these four quantities are known. Hence

$$V_e = \frac{c_{CH_3COOH}}{c_{NaOH}}V_{CH_3COOH} = \left(\frac{0.1000 \text{ M}}{0.1000 \text{ M}} \right)(100.0 \text{ mL}) = 100.0 \text{ mL}$$

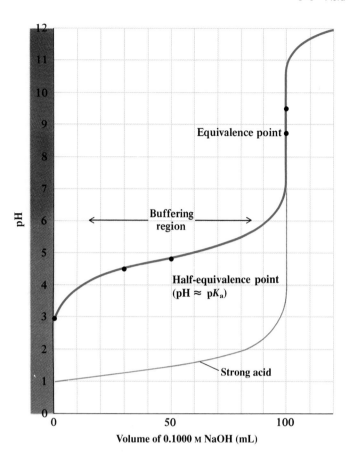

Figure 8–13 A titration curve for the titration of a weak acid by a strong base (*upper curve*). The curve shown is for 100.0 mL of 0.1000 M CH_3COOH titrated with NaOH. The five marked points are points at which the pH is calculated in this section. For comparison, the thin red line shows the titration curve for a strong acid of the same amount and concentration.

The titration has four distinct ranges, each of which corresponds to a type of calculation that we have considered already. Figure 8–13 shows the pH-versus-volume curve for this titration. In the following sections, we detail the calculation of four typical points on this curve.

1. Before any NaOH is added

At the beginning of this titration we simply have the ionization of a weak acid, which we considered in Section 8–4. Proceeding as in Example 8–6 gives the pH at the starting point as 2.88.

2. Less than the equivalent volume of NaOH has been added

In this range, the acid has been partially neutralized by added NaOH solution. Because OH^- is a stronger base than acetate ion, it competes against the acetate ion with great success to gain hydrogen ions; that is, it reacts essentially completely with the acetic acid originally present:

$$CH_3COOH(aq) + OH^-(aq) \rightleftharpoons H_2O(\ell) + CH_3COO^-(aq)$$

$$K = \frac{1}{K_b} = \frac{K_a}{K_w} = 2 \times 10^9 \gg 1$$

To be specific, suppose that 30.00 mL of 0.1000 M NaOH has been added. The 30.00 mL of NaOH contains

$$(0.1000 \text{ mol } L^{-1}) \times (0.03000 \, L) = 3.000 \times 10^{-3} \text{ mol } OH^-$$

and the original solution contained

$$(0.1000 \text{ mol L}^{-1}) \times (0.1000 \text{ L}) = 1.000 \times 10^{-2} \text{ mol CH}_3\text{COOH}$$

The neutralization reaction generates one $CH_3COO^-(aq)$ ion for every OH^- ion added. Hence, 3.000×10^{-3} mol of $CH_3COO^-(aq)$ ion is generated. The amount of acetic acid that remains unreacted is

$$1.000 \times 10^{-2} - 3.000 \times 10^{-3} = 7.00 \times 10^{-3} \text{ mol}$$

Because the total volume is now 130.0 mL (= 0.1300 L), the nominal concentrations after reaction are

* Again, it is perfectly acceptable (and often convenient) to obtain the molarities by dividing millimoles by milliliters.

$$[CH_3COOH] \approx \frac{7.00 \times 10^{-3} \text{ mol}}{0.1300 \text{ L}} = 5.38 \times 10^{-2} \text{ M}$$

$$[CH_3COO^-] \approx \frac{3.00 \times 10^{-3} \text{ mol}}{0.1300 \text{ L}} = 2.31 \times 10^{-2} \text{ M}$$

This is nothing other than a buffer solution containing acetic acid at a concentration of 5.38×10^{-2} M and sodium acetate at a concentration of 2.31×10^{-2} M. Its pH can be found by the method used in Example 8–10:

$$\frac{[H_3O^+][CH_3COO^-]}{[CH_3COOH]} = K_a$$

$$\frac{[H_3O^+](2.31 \times 10^{-2} + [H_3O^+])}{(5.38 \times 10^{-2} - [H_3O^+])} = 1.8 \times 10^{-5}$$

$$\frac{[H_3O^+](2.31 \times 10^{-2})}{5.38 \times 10^{-2}} \approx 1.8 \times 10^{-5}$$

$$[H_3O^+] \approx 1.8 \times 10^{-5} \times \frac{5.38 \times 10^{-2}}{2.31 \times 10^{-2}} = 4.2 \times 10^{-5} \text{ M}$$

$$pH = 4.38$$

This region of the titration shows clearly the buffering action of a mixture of a weak acid with its conjugate base. At the **half-equivalence point** $V = V_e/2$, we have $[CH_3COOH] = [CH_3COO^-]$, which corresponds to an equimolar buffer; at this point, pH $\approx pK_a = 4.74$. In the vicinity of this point, the pH rises relatively slowly as the NaOH solution is added.

3. The equivalent volume of NaOH has been added

At the equivalence point, $c_{NaOH}V_e$ moles of OH^- has been added to an equal chemical amount of the acid CH_3COOH. An identical solution could have been prepared simply by mixing 1.000×10^{-2} mol of the base CH_3COO^- (in the form of $NaCH_3COO$) in enough water to give 0.2000 L of solution. The calculation of the pH at the equivalence point, therefore, is identical to the calculation of the pH in the reaction of CH_3COO^- with water shown in Example 8–9. We obtain pH = 8.72 at the equivalence point.

When a weak acid is titrated with a strong base, the equivalence point does *not* come at pH 7 but at a higher (more basic) value. Similarly, the equivalence point in the titration of a weak base by a strong acid occurs at a pH lower than 7.

4. More than the equivalent volume of NaOH has been added

Beyond the equivalence point, we are adding $OH^-(aq)$ to a solution of the weak base CH_3COO^-. The $[OH^-]$ comes almost entirely from the hydroxide ion added beyond the equivalence point; very little comes from the equilibrium

$$CH_3COO^-(aq) + H_2O(\ell) \rightleftharpoons CH_3COOH(aq) + OH^-(aq)$$

because the added $OH^-(aq)$ ion shifts this equilibrium to the left. The hydrolysis of the CH_3COO^- ion is "suppressed" by the excess OH^- ion.

Beyond V_e, the pH for the titration of a weak acid by a strong base is very close to that for a strong acid by a strong base.

If the equivalent volume V_e were not known, it could be readily determined in the laboratory by using an indicator that changes color near pH = 8.72, the pH at the equivalence point of the acetic acid titration. A suitable choice would be phenolphthalein, which changes from colorless to red over a pH range from 8.2 to 10.0. In titrations of weak acids, the pH changes a bit less sharply near the equivalence point than it does for strong acids. This makes determinations of the equivalent volume (and of the original concentration of the weak acid) somewhat less precise.

The use of titrations to determine the concentrations of solutions and the K_a or K_b of weak acids or bases is illustrated in the following example.

EXAMPLE 8-13

Exactly 50.00 mL of a solution of the weak acid propionic acid of unknown concentration is titrated with a 0.1000 M solution of NaOH. The equivalence point is reached after 39.30 mL of NaOH solution has been added. At the half-equivalence point (19.65 mL), the pH is 4.85. Calculate the original concentration of the acid and its acidity constant, K_a.

Solution

The chemical amount of acid originally present is equal to the chemical amount of base added to reach the equivalence point, so that

$$c_{acid} = \left(\frac{V_e}{V_{acid}}\right)c_{NaOH} = \left(\frac{39.30 \text{ mL}}{50.00 \text{ mL}}\right) \times 0.1000 \text{ M} = 0.07860 \text{ M}$$

This is the original concentration of propionic acid.

Let the propionic acid be represented simply as HA. The acid ionization has the usual equilibrium expression

$$K_a = \frac{[H_3O^+][A^-]}{[HA]}$$

At the half-equivalence point, $[A^-] \approx [HA]$ because half of the original amount of HA has been converted to A^-. Substitution into the preceding equation gives

$$K_a \approx [H_3O^+]$$
$$K_a = 10^{-pH} = 10^{-4.85} = 1.4 \times 10^{-5}$$

This is quite close to the value tabulated for propionic acid (CH_3CH_2COOH) in Table 8–2.

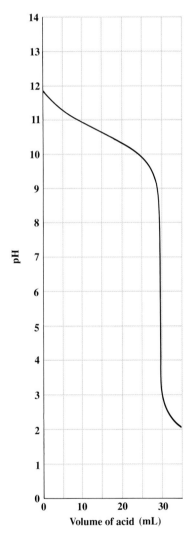

Curve for the titration of 25.00 mL of methylamine solution with 0.1000 M HCl (see exercise on the next page).

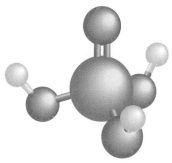

Phosphoric acid (H_3PO_4) is a triprotic acid that is used in the manufacture of phosphate fertilizers. Because it is odorless and nontoxic, phosphoric acid has many uses in the food industry. Deoxyribonucleic acid (DNA) is a derivative of phosphoric acid (see Fig. 25–20).

• The successive acidity constants for polyprotic acids are designated K_{a1}, K_{a2}, and so forth.

Exercise

Exactly 25.00 mL of a solution of the weak base methylamine (CH_3NH_2) is titrated with 0.1000 M hydrochloric acid. The equivalence point is reached after 29.64 mL of HCl solution has been added. At the half-equivalence point (14.82 mL), the pH is 10.64. Calculate the original concentration of the methylamine and its base ionization constant, K_b.

Answer: $c_0 = 0.1186$ M; $K_b = 4.4 \times 10^{-4}$.

8–7 POLYPROTIC ACIDS

So far we have considered **monoprotic acids**, acids capable of donating only a single hydrogen ion to acceptors. **Polyprotic acids** also exist; these acids can donate two or more hydrogen ions. Sulfuric acid is a familiar and important example. It reacts with water in two stages, first

$$H_2SO_4(aq) + H_2O(\ell) \rightleftharpoons H_3O^+(aq) + HSO_4^-(aq) \qquad K_a = K_{a1} \approx 100$$

to give the hydrogen sulfate ion, and then

$$HSO_4^-(aq) + H_2O(\ell) \rightleftharpoons H_3O^+(aq) + SO_4^{2-}(aq) \qquad K_a = K_{a2} = 1.2 \times 10^{-2}$$

The hydrogen sulfate ion, which appears on the right side of the first equation and the left side of the second, is amphoteric: it is a base in the first reaction (its conjugate acid is H_2SO_4) and an acid in the second (its conjugate base is SO_4^{2-}). In its first acid ionization, H_2SO_4 is a strong acid, but the product, HSO_4^-, is itself only a weak acid.

The H_3O^+ present in a solution of H_2SO_4 therefore comes primarily from the first ionization. When a solution of H_2SO_4 reacts with a strong base, its neutralizing power is twice as great as that of a monoprotic acid of the same concentration because each mole of sulfuric acid can react with and neutralize two moles of hydroxide ion. Thus the OH^- ion removes two H^+ ions essentially completely from sulfuric acid, although water does not.

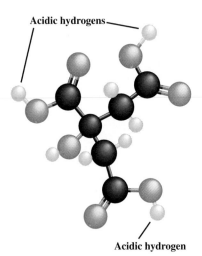

Acidic hydrogens

Acidic hydrogen

Citric acid ($C_6H_8O_7$) gives citrus fruits their tart taste. Only three of the eight hydrogens are acidic. Acidic hydrogens, the ones that can be donated, are generally bonded to oxygen or another electronegative element. Hydrogens bonded to carbon are rarely acidic.

Weak Polyprotic Acids

Polyprotic acids donate two or more hydrogen ions in stages. Examples are carbonic acid ($H_2CO_3(aq)$) formed from aquated CO_2, or carbonated water (Fig. 8–14), and phosphoric acid (H_3PO_4). Carbonic acid can give up one hydrogen ion to form HCO_3^- (hydrogen carbonate ion) or two hydrogen ions to form CO_3^{2-} (carbonate ion). Phosphoric acid gives up H^+ in three stages, forming successively $H_2PO_4^-$, HPO_4^{2-}, and PO_4^{3-}. A weak acid with several hydrogen atoms in its formula does not automatically donate them all, just because carbonic acid and phosphoric acid do. Citric acid ($C_6H_8O_7$) is polyprotic, but it donates a maximum of three hydrogen ions (not eight). The formula of citric acid is often written $H_3C_6H_5O_7$, to segregate the acidic and non-acidic hydrogen atoms. Similarly, acetic acid ($C_2H_4O_2$) is a monoprotic (not a tetraprotic) weak acid. Writing the formula $HC_2H_3O_2$ or CH_3COOH is meant to convey this.

Two simultaneous equilibria are involved in the interaction of a diprotic acid like H_2CO_3 with water:

$$H_2CO_3(aq) + H_2O(\ell) \rightleftharpoons H_3O^+(aq) + HCO_3^-(aq)$$

$$\frac{[H_3O^+][HCO_3^-]}{[H_2CO_3]} = K_{a1} = 4.3 \times 10^{-7}$$

and

$$HCO_3^-(aq) + H_2O(\ell) \rightleftharpoons H_3O^+(aq) + HCO_3^{2-}(aq)$$

$$\frac{[H_3O^+][CO_3^{2-}]}{[HCO_3^-]} = K_{a2} = 4.8 \times 10^{-11}$$

We establish two important points at the outset:

1. The H_3O^+ produced in the first equilibrium is absolutely indistinguishable from that produced in the second; the $[H_3O^+]$ in the two equilibrium expressions is thus one and the same.
2. K_{a2} is invariably smaller than K_{a1} because the negative charge left behind by the loss of a hydrogen ion in the first stage causes the second hydrogen ion to be more tightly bound.

Exact calculations of simultaneous equilibria can be quite complex. Considerable simplification results when the original acid concentration is not too small and the acidity constants K_{a1} and K_{a2} differ substantially in magnitude (by a factor of 100 or more). The latter condition is almost always satisfied. Under such conditions, the two equilibria can be treated one at a time, as in the following example.

• In fact, most dissolved CO_2 actually remains as $CO_2(aq)$, and only a small fraction reacts with water to give $H_2CO_3(aq)$. Therefore, $[H_2CO_3]$ indicates the total concentration of both of these species. Approximately 0.034 mol of CO_2 dissolves per liter of water at 25°C under a $CO_2(g)$ pressure of 1 atm.

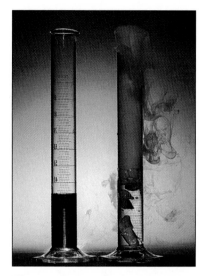

Figure 8-14 The indicator bromothymol blue is blue in basic solution (*left*). When solid carbon dioxide is placed in the bottom of the cylinder (*right*) it dissolves to form carbonic acid. This causes the indicator to change to its colorless acidic form.

EXAMPLE 8-14

Calculate the concentrations at equilibrium of H_2CO_3, HCO_3^-, CO_3^{2-}, and H_3O^+ in a saturated aqueous solution of CO_2 in which the original concentration of H_2CO_3 is 0.034 M.

Solution

The H_3O^+ ion arises from the ionization of H_2CO_3, from the subsequent acid dissociation of HCO_3^-, and, as always, from the autoionization of water. Because $K_{a2} \ll K_{a1}$, and also $K_w \ll K_{a1}$, a reasonable start is to ignore the H_3O^+ contributed by the acid dissociation of HCO_3^- and the autoionization of water. We also ignore the lowering of the concentration of HCO_3^- caused by its acid dissociation. These approximations are checked later in the calculation.

If y moles per liter of H_2CO_3 ionizes, then we have

	$H_2CO_3(aq)$ + $H_2O(\ell)$ $\rightleftharpoons$	$H_3O^+(aq)$ +	$HCO_3^-(aq)$
Initial concentration (M)	0.034	≈ 0	0
Change in concentration (M)	$-y$	$+y$	$+y$
Equilibrium concentration (M)	$0.034 - y$	y	y

where equating both $[HCO_3^-]$ and $[H_3O^+]$ to y invokes the assumption that the K_{a2} reaction has only a small effect on the concentration of the HCO_3^-. Then we may write

$$\frac{[H_3O^+][HCO_3^-]}{[H_2CO_3]} = K_{a1}$$

$$\frac{y^2}{0.034 - y} = 4.3 \times 10^{-7}$$

We solve this equation for y to find

$$y = 1.2 \times 10^{-4} \text{ M} = [H_3O^+] = [HCO_3^-]$$
$$[H_2CO_3] = 0.034 - y = 0.034 \text{ M}$$

We now write the second equilibrium as

$$\frac{[H_3O^+][CO_3^{2-}]}{[HCO_3^-]} = K_{a2} = \frac{(1.2 \times 10^{-4})[CO_3^{2-}]}{1.2 \times 10^{-4}} = 4.8 \times 10^{-11}$$
$$[CO_3^{2-}] = 4.8 \times 10^{-11} \text{ M}$$

The concentration of the base produced by the second equilibrium, $[CO_3^{2-}]$, is numerically equal to K_{a2}.

We now check our assumptions. Because

$$[CO_3^{2-}] = 4.8 \times 10^{-11} \text{ M} \ll 1.2 \times 10^{-4} \text{ M} = [HCO_3^-]$$

we were justified in neglecting the effect of the second equilibrium on the concentrations of HCO_3^- and of H_3O^+. The additional concentration of H_3O^+ furnished by HCO_3^- is only 4.8×10^{-11} M. Finally, $[H_3O^+]$ is much larger than 10^{-7} M, so neglecting the water autoionization also was justified.

Exercise

Salicylic acid ($H_2C_7H_4O_3$) is a diprotic acid with $K_{a1} = 1.1 \times 10^{-3}$ and $K_{a2} = 3.6 \times 10^{-14}$ (at 25°C). It is sometimes taken as an analgesic drug instead of aspirin (acetylsalicylic acid), but its greater acidity can cause bleeding in the stomach. Calculate the concentrations at equilibrium of $H_2C_7H_4O_3(aq)$, $HC_7H_4O_3^-(aq)$, $C_7H_4O_3^{2-}(aq)$, $H_3O^+(aq)$, and $OH^-(aq)$ in 0.065 M salicylic acid.

Answer: $[H_2C_7H_4O_3] = 0.057$ M; $[HC_7H_4O_3^-] = [H_3O^+] = 7.9 \times 10^{-3}$ M; $[C_7H_4O_3^{2-}] = 3.6 \times 10^{-14}$ M; $[OH^-] = 1.3 \times 10^{-12}$ M.

If we had been working with a triprotic acid such as H_3PO_4, we could have gone on to calculate the concentration of the base (PO_4^{3-}) resulting from the third acid dissociation step.

An analogous procedure works in studying the reactions of a base that is capable of accepting two or more hydrogen ions. In a solution of sodium carbonate (Na_2CO_3), for example, the carbonate ion reacts as a base with water to form first HCO_3^- and then, in a second stage, H_2CO_3:

$$H_2O(\ell) + CO_3^{2-}(aq) \rightleftharpoons HCO_3^-(aq) + OH^-(aq)$$

$$\frac{[OH^-][HCO_3^-]}{[CO_3^{2-}]} = K_{b1} = \frac{K_w}{K_{a2}} = 2.1 \times 10^{-4}$$

• Note that the *first* base constant, K_{b1}, equals K_w divided by the *second* acid constant, K_{a2}, and vice versa.

$$H_2O(\ell) + HCO_3^-(aq) \rightleftharpoons H_2CO_3(aq) + OH^-(aq)$$

$$\frac{[OH^-][H_2CO_3]}{[HCO_3^-]} = K_{b2} = \frac{K_w}{K_{a1}} = 2.3 \times 10^{-8}$$

In this case $K_{b1} \gg K_{b2}$, so we assume that essentially all the OH^- arises from the first reaction. Of course, there is only one $OH^-(aq)$ concentration in the solution, and it is common to both reactions. The ensuing calculation for the concentrations of species present at equilibrium is just like that in Example 8–14.

Effect of pH on Solution Composition

Changing the pH shifts the positions of all acid–base equilibria in a solution, including those involving polyprotic acids. Le Chatelier's principle allows only qualitative predictions of the effects of a change in pH. To calculate the actual amount of the change, we use acid–base equilibrium expressions and equilibrium constants. For example, we can rewrite the two equilibrium equations that apply for H_2CO_3-HCO_3^--CO_3^{2-} solutions as

$$\frac{[HCO_3^-]}{[H_2CO_3]} = \frac{K_{a1}}{[H_3O^+]} \qquad \frac{[CO_3^{2-}]}{[HCO_3^-]} = \frac{K_{a2}}{[H_3O^+]}$$

At a given pH, the right-hand sides are known, and the relative amounts of the three carbonate species can be calculated. This is illustrated in the following example.

EXAMPLE 8–15

Calculate the fractions of the total carbonate present as H_2CO_3, HCO_3^-, and CO_3^{2-} at pH 10.00

Solution

At this pH, $[H_3O^+] = 1.0 \times 10^{-10}$ M. Inserting that value and the values of K_{a1} and K_{a2} in the equations above gives

$$\frac{[HCO_3^-]}{[H_2CO_3]} = \frac{4.3 \times 10^{-7}}{1.0 \times 10^{-10}} = 4.3 \times 10^3$$

$$\frac{[CO_3^{2-}]}{[HCO_3^-]} = \frac{4.8 \times 10^{-11}}{1.0 \times 10^{-10}} = 0.48$$

It is most convenient to rewrite these ratios with the same species (say, HCO_3^-) in the denominator. The first equation becomes

$$\frac{[H_2CO_3]}{[HCO_3^-]} = \frac{1}{4.3 \times 10^3} = 2.3 \times 10^{-4}$$

The fraction of the total carbonate present in each of the three forms equals the concentration of that form divided by the sum of the three concentrations. For H_2CO_3, for example,

$$\text{fraction present as } H_2CO_3 = \frac{[H_2CO_3]}{[H_2CO_3] + [HCO_3^-] + [CO_3^{2-}]}$$

The right side of the equation is evaluated by dividing numerator and denominator by $[HCO_3^-]$ and inserting the ratios just calculated.

$$\text{fraction present as } H_2CO_3 = \frac{\dfrac{[H_2CO_3]}{[HCO_3^-]}}{\dfrac{[H_2CO_3]}{[HCO_3^-]} + 1 + \dfrac{[CO_3^{2-}]}{[HCO_3^-]}}$$

$$= \frac{2.3 \times 10^{-4}}{2.3 \times 10^{-4} + 1 + 0.48} = 1.6 \times 10^{-4}$$

In the same way, we calculate the fractions of HCO_3^- and CO_3^{2-} present as 0.68 and 0.32, respectively.

Exercise

Calculate the fractions of salicylic acid ($H_2C_7H_4O_3(aq)$), hydrogen salicylate ion ($HC_7H_4O_3^-(aq)$), and salicylate ion ($C_7H_4O_3^{2-}(aq)$) present when the drug is at equilibrium in a stomach that has a pH of 1.50. Use the K_{a1} and K_{a2} values from Exercise 8–14.

Answer: Fraction of $H_2C_7H_4O_3(aq) = 0.966$; fraction of $HC_7H_4O_3^-(aq) = 0.034$; fraction of $C_7H_4O_3^{2-}(aq) = 3.8 \times 10^{-14}$.

If we repeat the calculation of Example 8–15 at a series of different pH values, we obtain the graph shown in Figure 8–15. At high pH, CO_3^{2-} predominates; at low pH, H_2CO_3 predominates. At intermediate pH (near 8, the approximate pH of seawater), the hydrogen carbonate ion HCO_3^- is most prevalent.

The variation in composition of sedimentary rocks from different locations can be traced back to the effect of pH on the equilibrium composition of carbonate-containing waters. Sediments that contain carbonates were formed from alkaline (high pH) lakes and oceans in which CO_2 was present mainly as CO_3^{2-} ion. Sediments deposited from waters with intermediate pH are hydrogen carbonates or mixtures of carbonates and hydrogen carbonates. An example of the latter is trona ($2\ Na_2CO_3 \cdot NaHCO_3 \cdot 2H_2O$), an ore found in the western United States that is an important source of both carbonates of sodium. Acidic waters do not deposit carbonates but instead release $CO_2(g)$ to the atmosphere.

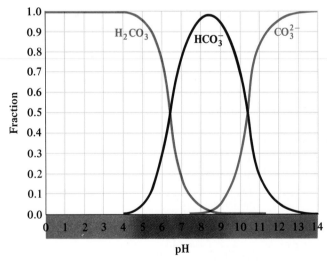

Figure 8–15 The equilibrium fractions of H_2CO_3, HCO_3^-, and CO_3^{2-} present in aqueous solution at different values of the pH. Low pH favors H_2CO_3, and high pH favors CO_3^{2-}.

8-8 LEWIS ACIDS AND BASES

Suppose that we use Lewis structures to describe a Brønsted–Lowry acid–base equilibrium. We can write the reaction of hydrogen chloride with water as

$$: \ddot{C}l\text{—}H \; + \; : \ddot{O} : \; \rightleftharpoons \; : \ddot{C}l : ^- \; + \; \left[H : \ddot{O} : \right]^+$$

The hydrogen ion, initially bound to the Cl^- ion, is transferred to a nearby water molecule. Two bases (Cl^- and water, in this case) compete with one another for hydrogen ions. The two bases both contain lone pairs of electrons and act as electron donors toward the electron acceptor, H^+. This suggested to Gilbert Lewis a more general theory of acids and bases. We now say:

> A **Lewis base** is *any* species that donates electrons through coordination to its lone pairs, and a **Lewis acid** is *any* species that accepts such electron pairs.

All the acids and bases we have considered so far fit into this description (with the Lewis acid, H^+, coordinating with various Lewis bases, the electron-pair donors). Other reactions that do *not* involve hydrogen-ion transfer can still be considered Lewis acid–base reactions. An example is the reaction between electron-deficient BF_3 and electron-rich NH_3:

$$: \ddot{F}\text{—}B \; + \; : N\text{—}H \; \longrightarrow \; : \ddot{F}\text{—}B\text{—}N\text{—}H$$

Ammonia, the Lewis base, donates lone-pair electrons in order to coordinate with BF_3, the Lewis acid or electron acceptor. The bond that forms is a **coordinate covalent bond**, in which both electrons are supplied by a lone pair on the Lewis base. "Neutralization" of a Lewis acid by a Lewis base proceeds with the formation of a coordinate covalent bond.

Octet-deficient compounds of elements from Group III such as boron and aluminum are often strong Lewis acids, because Group III atoms can achieve octet configurations by accepting an electron pair and forming coordinate covalent bonds. Atoms and ions from Groups V through VII have the necessary lone pairs to act as Lewis bases. Compounds of main-group elements from the later periods can also act as Lewis acids through valence expansion. In such a reaction, the central atom accepts a share in additional lone pairs beyond the eight electrons needed to satisfy the octet rule. For example, $SnCl_4$ acts as a Lewis acid to accept electrons from chloride-ion lone pairs in the reaction

$$SnCl_4(\ell) + 2\,Cl^-(aq) \longrightarrow SnCl_6^{2-}(aq)$$

$$: \ddot{C}l : Sn : \ddot{C}l : \; + \; 2 : \ddot{C}l : ^- \; \longrightarrow \; \left[: \ddot{C}l\text{—}Sn\text{—}\ddot{C}l : \right]^{2-}$$

After the reaction, each atom of tin is surrounded by 12 rather than 8 valence electrons.

The formation of a coordination complex by a transition-metal ion is a final example of a Lewis acid–base reaction. In the reaction

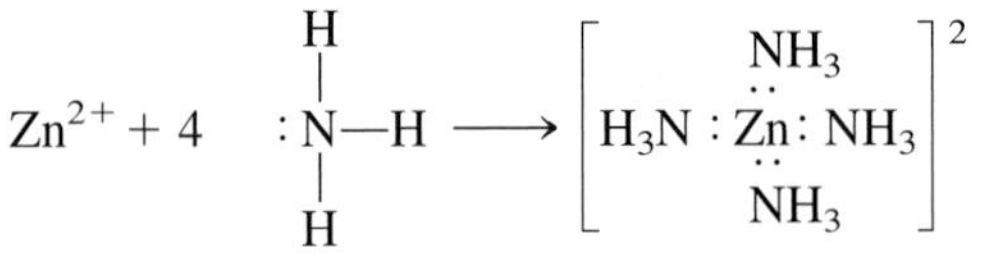

$$Zn^{2+} + 4 \quad :\!N\!-\!H \longrightarrow \left[H_3N : Zn : NH_3 \atop {NH_3 \atop NH_3}\right]^2$$

the zinc ion acts as a Lewis acid, forming coordinate covalent bonds with the nitrogen atoms by accepting a share in the lone pairs on four molecules of the Lewis base, ammonia.

EXAMPLE 8–16

In the following reactions, identify the Lewis acid and the Lewis base.
(a) $AlCl_3 + Cl^- \longrightarrow AlCl_4^-$
(b) $CH_3COOH(aq) + NH_3(aq) \longrightarrow CH_3COO^-(aq) + NH_4^+(aq)$
(c) $Co^{3+}(aq) + 6F^-(aq) \longrightarrow [CoF_6]^{3-}(aq)$

Solution

(a) In $AlCl_3$ the Al atom has only six valence electrons around it. The $AlCl_3$ acts as a Lewis acid, accepting electrons at the Al atom from lone pairs on the Lewis base, Cl^-.
(b) The Lewis base is NH_3, whose lone-pair electrons coordinate with the Lewis acid, H^+. (Note that it is not correct to call acetic acid (CH_3COOH) a Lewis acid because the full molecule does not accept a lone pair. Only the H^+ portion of it is a Lewis acid.)
(c) The transition-metal ion Co^{3+} is a Lewis acid, forming a coordinate covalent bond by sharing electron pairs donated by the Lewis base, F^-.

Exercise
Gaseous PCl_5 molecules condense to form an ionic solid containing PCl_4^+ and PCl_6^- ions. Identify the Lewis acid and base in this process.

Answer: Lewis acid: one of the PCl_5 molecules. Lewis base: Cl^- ion.

SUMMARY

8–1 In the Brønsted–Lowry definition, an acid is a donor of a hydrogen ion and a base is an acceptor of a hydrogen ion. When a species acts as an acid, it gives rise to its **conjugate base;** when a species acts as a base, it gives rise to its **conjugate acid.** An acid–base reaction can be viewed as a competition between two bases for a hydrogen ion.

8–2 In its **autoionization,** water acts simultaneously as both acid and base, which leads to small concentrations of **hydronium** (H_3O^+) and hydroxide (OH^-) ions in water at equilibrium. The product of these two concentrations is K_w, the autoionization constant for water. At 25°C, K_w equals 1.0×10^{-14}. When a **strong acid**

dissolves in water, it reacts essentially completely, increasing the concentration of hydronium ion over that in pure water. A **strong base** reacts essentially completely to give the hydroxide ion. The hydronium-ion concentration is conveniently represented on the logarithmic **pH scale** (pH = $-\log_{10}[H_3O^+]$). Pure water has a pH of 7 at 25°C. Lower values of pH indicate acidic solutions, and higher values indicate basic solutions.

8–3 The strength of an acid in water is shown by its **acidity constant,** K_a, the equilibrium constant for the reaction in which that acid donates a hydrogen ion to water. The larger the value of K_a, the stronger the acid and the weaker its conjugate base. The strength of a base is shown by its **basicity constant,** K_b. The product of K_a for an acid and K_b for its conjugate base equals K_w. In acid–base equilibria, the stronger acid donates hydrogen ions to the stronger base, giving more of the weaker acid and the weaker base as the reaction goes toward equilibrium. An **indicator** is a weak acid (or weak base) that has a noticeably different color from its conjugate weak base (or weak acid). Indicators can be used to estimate the pH of solutions.

8–4 The pH and concentration of dissolved species can be calculated at equilibrium from the initial concentration of a weak acid and its acidity constant K_a. It is usually a good approximation to neglect the contribution to $[H_3O^+]$ from the autoionization of water. Calculations of pH for solutions of weak bases proceed analogously to those of weak acids. **Hydrolysis** refers to the acid or base reaction of a dissolved cation or anion with water.

8–5 A **buffer solution** results from mixing a weak acid with its conjugate weak base. Such a solution resists changes in pH; adding a small amount of either strong acid or base has only a small effect on the buffer's pH. In the design of a buffer solution, the best choice of acid is one having a pK_a that is close to the desired pH of the buffer.

8–6 An **acid–base titration curve** is a plot of pH versus the volume of titrant added. An **equivalence point** on a titration curve is a point at which the chemical amount of base is equal to the chemical amount of acid. The volume of titrant to reach an equivalence point is the **equivalent volume.** If the acid and base are both strong, the pH at the equivalence point is 7 at 25°C. When a weak acid is titrated with a strong base, a buffering region exists in the vicinity of the half-equivalence point over which the pH changes slowly with added base, and the pH at the equivalence point is greater than 7. The computation of pH in the four distinct ranges of a titration curve corresponds to standard pH calculations. Acid–base titration is useful for determining concentrations and values of K_a for solutions of acids (or bases) of unknown concentration or unknown identity.

8–7 Polyprotic acids donate H^+ ion in several steps. Here it is usually a good approximation to assume that the pH is determined entirely by the first step; the extent of subsequent steps can then be calculated sequentially. From the successive acidity constants, the equilibrium concentrations of all the species present can be calculated at any value of the pH.

8–8 A **Lewis base** is *any* species that donates electrons through coordination to its lone pairs, and a **Lewis acid** is *any* species that accepts such electron pairs. The bond that forms between a Lewis acid and base is a **coordinate covalent bond,** in which both electrons are supplied by a lone pair on the Lewis base. Neutralization between a Lewis acid and base proceeds with the formation of such a bond.

PROBLEMS

Note: Answers to blue-numbered problems are given in Appendix F. Problems that are more challenging are indicated with asterisks.

Brønsted–Lowry Acids and Bases

1. (See Example 8–1.) Which of the following can act as Brønsted–Lowry acids? Give the formula of the conjugate Brønsted–Lowry base for each of them.
   (a) Cl^- (d) NH_3
   (b) HSO_4^- (e) H_2O
   (c) NH_4^+

2. (See Example 8–1.) Which of the following can act as Brønsted–Lowry bases? Give the formula of the conjugate Brønsted–Lowry acid for each of them.
   (a) F^- (d) OH^-
   (b) SO_4^{2-} (e) H_2O
   (c) O^{2-}

3. (See Example 8–2.) When tested with indicators, the following solutions give colors that show them to be acidic. Write net ionic equations to explain these observations.
   (a) $NH_4Br(aq)$ (c) $(NH_4)_2SO_4(aq)$
   (b) $H_2S(aq)$

4. (See Example 8–2.) When tested with indicators, the following solutions give colors that show them to be basic. Write net ionic equations to explain these observations.
   (a) $Na_2CO_3(aq)$ (c) $Na_2HPO_4(aq)$
   (b) $RbOH(aq)$

Water and the pH Scale

5. (See Example 8–3.) The concentration of H_3O^+ in a sample of wine is 2.0×10^{-4} M. Calculate the pH and pOH of the wine.

6. (See Example 8–3.) The concentration of OH^- in a solution of household bleach is 3.6×10^{-2} M. Calculate the pOH and pH of the bleach.

7. (See Example 8–4.) The pH of normal human urine is in the range of 5.5 to 6.5. Compute the range of the concentration of H_3O^+ and of OH^-.

8. (See Example 8–4.) The pH of normal human blood is in the range of 7.35 to 7.45. Compute the range of the concentration of H_3O^+ and of OH^- in normal blood.

9. A solution of nitric acid with pH 2.32 at 25°C is diluted with water to eight times its original volume. What is the pH of the resulting solution? What is the concentration of OH^-?

10. A solution of potassium hydroxide with pH 11.65 at 25°C is diluted with water to six times its original volume. What is the pH of the resulting solution? What is the concentration of OH^- after the dilution?

11. Suppose that 11.74 mL of 0.071 M NaOH is added to 15.78 mL of 0.094 M HCl. Determine the pH of the resulting solution, assuming that the volumes are additive.

12. Suppose that 264.9 mL of 0.065 M KOH is added to 127.3 mL of 0.073 M HCl. Determine the pOH and pH of the resulting solution, assuming that the volumes are additive.

The Strengths of Acids and Bases

13. Which of the following is the weakest acid?
    (a) Formic acid, $K_a = 1.8 \times 10^{-4}$
    (b) Benzoic acid, $K_a = 6.5 \times 10^{-5}$
    (c) Ammonium ion, $K_a = 5.6 \times 10^{-10}$

14. Which of the following is the weakest acid?
    (a) Pyruvic acid, $pK_a = 2.49$
    (b) Lactic acid, $pK_a = 3.85$
    (c) Hypochlorous acid, $pK_a = 7.53$

15. Ephedrine ($C_{10}H_{15}ON$) is a base that is used in nasal sprays as a decongestant.
    (a) Write an equation for its equilibrium reaction with water.
    (b) The K_b for ephedrine is 1.4×10^{-4}. Calculate the K_a for its conjugate acid.
    (c) Is ephedrine a weaker or a stronger base than ammonia?

16. Niacin (C_5H_4NCOOH), one of the B vitamins, is an acid.
    (a) Write an equation for its equilibrium reaction with water.
    (b) The K_a for niacin is 1.5×10^{-5}. Calculate the K_b for its conjugate base.
    (c) Is the conjugate base of niacin a stronger or a weaker base than pyridine (the conjugate base of the pyridinium ion in Table 8–2)?

17. (a) Write a balanced equation for the acid–base equilibrium that results when sodium hypochlorite (NaClO) is dissolved in water.
    (b) Calculate the equilibrium constant for the reaction in part (a) using Table 8–2.

18. (a) Write a balanced equation for the acid–base equilibrium that results when potassium nitrite (KNO_2) is dissolved in water.
    (b) Calculate the equilibrium constant for the reaction in part (a) using Table 8–2.

19. Phenol (C_6H_5OH) has a K_a of 1.1×10^{-10}
    (a) Write the formula of the conjugate base of phenol, the phenolate ion.
    (b) Is the phenolate ion a weak or a strong base?
    (c) Write balanced equations to represent the events when sodium phenolate is mixed with water.

20. The pK_a of ethane (C_2H_6), a gas at room conditions, was recently reported to be 50.6.
    (a) Is ethane a weak or a strong acid? Is that what you would expect, based on the formula of ethane?
    (b) Write the formula of the conjugate base of ethane.
    (c) Write balanced equations to represent the events when NaC_2H_5 is mixed with water.

21. Use the data in Table 8–2 to determine the equilibrium constant for the reaction.

$$HClO_2(aq) + NO_2^-(aq) \rightleftharpoons HNO_2(aq) + ClO_2^-(aq)$$

22. Use the data in Table 8–2 to determine the equilibrium constant for the reaction

$$HPO_4^{2-}(aq) + HCO_3^-(aq) \rightleftharpoons PO_4^{3-}(aq) + H_2CO_3(aq)$$

23. (See Example 8–5.) (a) Which is the stronger acid: the acidic form of the indicator bromocresol green or the acidic form of methyl orange?
    (b) A solution is prepared in which bromocresol green is green and methyl orange is orange. Estimate the pH of this solution.

24. (See Example 8–5.) (a) Which is the stronger base: the basic form of the indicator cresol red or the basic form of thymolphthalein?
    (b) A solution is prepared in which cresol red is red and thymolphthalein is colorless. Estimate the pH of this solution.

Equilibria Involving Weak Acids and Bases

25. (See Example 8–6.) Find the pH and fraction dissociated of the acid in 0.20 M solutions of
    (a) Formic acid
    (b) Benzoic acid
    (c) Ammonium ion
    How does the pH vary in going from the strongest acid to the weakest? How does the fraction of acid dissociated in these solutions vary in going from the strongest acid to the weakest?

26. (See Example 8–6.) Find the pH and fraction dissociated of the acid in 0.20 M solutions of
    (a) Pyruvic acid, $pK_a = 2.49$
    (b) Lactic acid, $pK_a = 3.85$
    (c) Hypochlorous acid, $pK_a = 7.53$
    How does the pH vary in going from the strongest acid to the weakest? How does fractional dissociation vary in going from the strongest to the weakest?

27. (See Example 8–7.) The active component of aspirin is acetylsalicylic acid, $HC_9H_7O_4$, which has a K_a of 3.0×10^{-4}.
    (a) Calculate the pH of a solution made by dissolving 0.65 g of acetylsalicylic acid in water and diluting to 50.0 mL.
    (b) Repeat the calculation, assuming that the same mass of acetylsalicylic acid is dissolved and diluted to 100.0 mL of solution.
    (c) Is the pH lower or higher in the more dilute solution? Which solution has the higher fraction of its aspirin ionized?

28. (See Example 8–7.) Vitamin C is ascorbic acid ($HC_6H_7O_6$), for which K_a is 8.0×10^{-5}. Calculate the pH of a solution made by dissolving a 500-mg tablet of pure vitamin C in water and diluting to 100.0 mL.

29. (a) Calculate the pH of a 0.35 M solution of propionic acid at 25°C.
    (b) How many moles of formic acid must be dissolved per liter of solution to obtain the same pH as the solution from part (a)?

30. (a) Calculate the pH of a 0.027 M solution of benzoic acid at 25°C.
    (b) How many moles of acetic acid must be dissolved per liter of solution to obtain the same pH as the solution from part (a)?

31. (See Example 8–7.) Iodic acid (HIO_3) is fairly strong for a weak acid, having a K_a equal to 0.16 at 25°C. Compute the pH of a 0.100 M solution of HIO_3.

32. (See Example 8–7.) At 25°C, the K_a of pentafluorobenzoic acid (C_6F_5COOH) is 0.033. Suppose that 0.100 mol of pentafluorobenzoic acid is dissolved in 1.00 L of water. What is the pH of this solution?

33. Papaverine hydrochloride (papHCl) is a drug used as a muscle relaxant. It is a weak acid. At 25°C, a 0.205 M solution of papHCl has a pH of 3.31. Compute the K_a of papHCl.

34. The compound 2-germaacetic acid (GeH_3COOH) is an unstable weak acid derived structurally from acetic acid (CH_3COOH) by having a germanium atom replace one of the carbon atoms. At 25°C, a 0.050 M solution of 2-germaacetic acid has a pH of 2.42. Compute the K_a of 2-germaacetic acid and compare it with that of acetic acid.

35. (See Example 8–8.) Morphine is a weak base for which K_b is 8×10^{-7}. Calculate the pH of a solution made by dissolving 0.0400 mol of morphine in water and diluting to 600 mL.

36. (See Example 8–8.) Methylamine is a weak base for which K_b is 4.4×10^{-4}. Calculate the pH of a solution made by dissolving 0.070 mol of methylamine in water and diluting to 800 mL.

37. Determine the concentration of $H_3O^+(aq)$ and $NO_2^-(aq)$ in a solution in which $HNO_2(aq)$ is 4.50% ionized. Calculate the concentration of $HNO_2(aq)$ at equilibrium in this solution.

38. When 0.0145 mol of a certain weak base is placed in 1.000 L of water, it is found that 0.012% of the molecules have reacted, at equilibrium, to give $OH^-(aq)$ ion. Compute K_b of this weak base.

39. Predict whether an aqueous solution of the following is acidic, basic, or neutral.
    (a) KCN   (c) $RbClO_4$
    (b) $CuBr_2$ (d) $Mg(NO_3)_2$

40. Predict whether an aqueous solution of the following is acidic, basic, or neutral.
    (a) $AlCl_3$ (c) AgF
    (b) NH_4ClO_4 (d) $SrCl_2$

41. In aqueous solutions of NH_4CN, the $NH_4^+(aq)$ ion is weakly acidic and the $CN^-(aq)$ ion is weakly basic. Use data from

Table 8–2 to determine the equilibrium constant for the reaction between the two according to the equation

$$NH_4^+(aq) + CN^-(aq) \rightleftharpoons NH_3(aq) + HCN(aq)$$

42. In aqueous solutions of NH_4F, the $NH_4^+(aq)$ ion is weakly acidic and the $F^-(aq)$ ion is weakly basic. Use data from Table 8–2 to determine the equilibrium constant for the reaction between the two according to the equation

$$NH_4^+(aq) + F^-(aq) \rightleftharpoons NH_3(aq) + HF(aq)$$

43. You have 50.00 mL of a solution that is 0.100 M in acetic acid, and you neutralize it by adding 50.00 mL of a solution that is 0.100 M in sodium hydroxide. The pH of the resulting solution is not 7.00. Explain why. Determine whether the pH of the solution is greater than or less than 7.

44. A 75.00-mL portion of a solution that is 0.0460 M in $HClO_4$ is treated with 150.00 mL of 0.0230 M $KOH(aq)$. State whether the pH of the resulting mixture is greater than, less than, or equal to 7.0. Explain.

Buffer Solutions

45. (See Examples 8–10 and 8–11.)   (a) Calculate the equilibrium concentration of H_3O^+ and the pH in a solution prepared by dissolving 0.060 mol of formic acid and 0.045 mol of sodium formate in water and adjusting the volume to 500 mL.
(b) Suppose 0.010 mol of NaOH is added to the buffer from part (a). Calculate the pH of the solution that results.

46. (See Examples 8–10 and 8–11.)   (a) Calculate the equilibrium concentration of H_3O^+ in a solution prepared by dissolving 3.62 g of NH_4Cl in 400 mL of a solution that is 0.100 M in ammonia (NH_3) and diluting to a total volume of 600 mL.
(b) Suppose 0.040 mol of HCl is added to the buffer from part (a). Calculate the pH of the solution that results.

47. "Tris" is short for tris(hydroxymethyl)aminomethane. This weak base is widely used in biochemical research for the preparation of buffers. It offers low toxicity and a pK_b (equal to 5.92 at 25°C) that is convenient for the control of pH in clinical applications. A buffer is prepared by mixing 0.050 mol of tris with 0.025 mol of HCl in a volume of 2.00 L. Compute the pH of the solution.

48. "Bis" is short for bis(hydroxymethyl)aminomethane. It is a weak base that is closely related to tris (see problem 47) and has similar properties and uses. Its pK_b is 8.8 at 25°C. A buffer is prepared by mixing 0.050 mol of bis with 0.025 mol of HCl in a volume of 2.00 L (the same proportions as in the previous problem). Compute the pH of the solution.

49. (See Example 8–12.) A physician wishes to prepare a buffer solution at pH = 3.82 that efficiently resists changes in pH yet contains only small concentrations of the buffering agents. Which one of the following weak acids, together with its

sodium salt, would probably be best to use: *m*-chlorobenzoic acid, $K_a = 1.0 \times 10^{-4}$; *p*-chlorocinnamic acid, $K_a = 3.9 \times 10^{-5}$; 2,5-dihydroxybenzoic acid, $K_a = 1.1 \times 10^{-3}$; acetoacetic acid, $K_a = 2.6 \times 10^{-4}$? Explain.

50. (See Example 8–12.) Suppose you were designing a buffer system for imitation blood, and you wanted the buffer to maintain the blood at the realistic pH of 7.40. All other things being equal, which buffer system would be preferable: H_2CO_3/HCO_3^- or $H_2PO_4^-/HPO_4^{2-}$? Explain.

Acid–Base Titration Curves

51. Suppose 100.0 mL of a 0.3750 M solution of the strong base $Ba(OH)_2$ is titrated with a 0.4540 M solution of the strong acid $HClO_4$. The neutralization reaction is

$$Ba(OH)_2(aq) + 2\ HClO_4(aq) \longrightarrow$$
$$Ba(ClO_4)_2(aq) + 2\ H_2O(\ell)$$

Compute the pH of the titration solution before any acid is added, when the tritration is 1.00 mL short of the equivalence point, when the titration is at the equivalence point, and when the titration is 1.00 mL past the equivalence point. (*Caution*: Remember that each mole of $Ba(OH)_2$ gives *two* moles of OH^- in solution.)

52. A sample consisting of 26.38 mL of 0.1439 M HBr is titrated with a solution of NaOH having a molarity of 0.1219 M. Compute the pH of the titration solution before any base is added, when the titration is 1.00 mL short of the equivalence point, when the titration is at the equivalence point, and when the titration is 1.00 mL past the equivalence point.

53. A sample consisting of 50.00 mL of 0.1000 M hydrazoic acid (HN_3) is being titrated with 0.1000 M sodium hydroxide. Compute the pH before any NaOH is added, after the addition of 25.00 mL of NaOH, after the addition of 50.00 mL of NaOH, and after the addition of 51.00 mL of NaOH.

54. A 140.0-mL sample of a 0.175 M solution of aqueous ammonia is titrated with a 0.106 M solution of the strong acid HCl. The reaction is

$$NH_3(aq) + HCl(aq) \longrightarrow NH_4^+(aq) + Cl^-(aq)$$

Compute the pH of the titration solution before any HCl is added, when the titration is at the half-equivalence point, when the titration is at the equivalence point, and when the titration is 1.00 mL past the equivalence point.

55. What is the mass of codeine ($C_{18}H_{21}O_3N$) in 100.0 mL of an aqueous solution if it requires 15.90 mL of 0.0750 M HCl to titrate it to the equivalence point? What is the pH at the equivalence point if $K_b = 9.0 \times 10^{-7}$? What would be a suitable indicator for the titration?

56. A chemist who works in the process laboratory of the Athabasca Alkali Company makes frequent analyses for ammonia recovered from the Solvay process for making sodium

carbonate. What is the pH at the equivalence point if she titrates the aqueous ammonia solution (approximately 0.10 M) with a strong acid of comparable concentration? Select an indicator that would be suitable for the titration.

57. (See Example 8–13.) 40.00 mL of a solution of cacodylic acid of unknown concentration is titrated with a 0.2000 M solution of KOH. The equivalence point is reached after 23.28 mL of the KOH solution has been added. At the half-equivalence point (11.64 mL) the pH was 6.19. Calculate the original concentration of the acid and its acidity constant K_a.

58. (See Example 8–13.) 50.00 mL of a solution of aniline (a base) of unknown concentration is titrated with a 0.1800 M solution of HCl. The equivalence point is reached after 37.32 mL of the HCl solution has been added. At the half-equivalence point (18.66 mL) the pH was 10.67. Calculate the original concentration of the aniline and its base ionization constant K_b.

59. Examine the titration curve below.

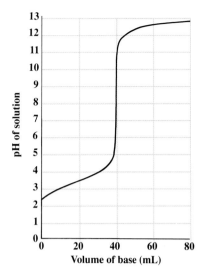

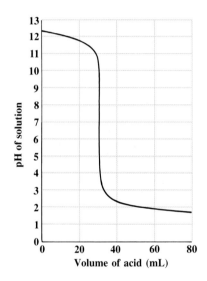

(a) Which of the following titrations could it represent: HCl by KOH, RbOH by HBr, NH_3 by HNO_3?

(b) Choose a suitable indicator for signaling the end-point of the titration. Justify your answer.

(c) Suppose that the figure represents the titration of 100.0 mL of NaOH(aq) by a solution of 0.065 M HNO_3. Calculate the concentration of the NaOH in the original solution.

60. Examine the titration curve at the beginning of the next column.

(a) Which of the following titrations could it represent: HCl by NaOH, NH_3 by HBr, HNO_2 by KOH?

(b) Choose a suitable indicator for signaling the end-point of the titration. Justify your answer.

(c) Estimate the K_a of the species in the original solution.

Polyprotic Acids

61. (See Example 8–14.) Arsenic acid (H_3AsO_4) is a weak triprotic acid. Given the three acidity constants from Table 8–2, calculate the equilibrium concentrations of H_3AsO_4, $H_2AsO_4^-$, $HAsO_4^{2-}$, AsO_4^{3-}, and H_3O^+ in a solution prepared by dissolving 0.100 mol of H_3AsO_4 in 1.000 L of solution.

62. (See Example 8–14.) Phthalic acid ($H_2C_8H_4O_4$, abbreviated H_2Ph) is a diprotic acid. Its ionization in water at 25°C takes place in two steps:

$$H_2Ph(aq) + H_2O(\ell) \rightleftharpoons H_3O^+(aq) + HPh^-(aq)$$
$$K_{a1} = 1.26 \times 10^{-3}$$
$$HPh^-(aq) + H_2O(\ell) \rightleftharpoons H_3O^+(aq) + Ph^{2-}(aq)$$
$$K_{a2} = 3.10 \times 10^{-6}$$

If 0.0100 mol of phthalic acid is dissolved per liter of solution, calculate the equilibrium concentrations of H_2Ph, HPh^-, Ph^{2-}, and H_3O^+.

63. A solution as initially prepared contains 0.050 mol per liter of phosphate ion (PO_4^{3-}) at 25°C. Given the three acidity constants from Table 8–2, calculate the equilibrium concentrations of PO_4^{3-}, HPO_4^{2-}, $H_2PO_4^-$, H_3PO_4, and OH^-.

64. Oxalic acid ($H_2C_2O_4$) ionizes in two stages in aqueous solution (Table 8–2). Calculate the equilibrium concentrations of $C_2O_4^{2-}$, $HC_2O_4^-$, $H_2C_2O_4$, and OH^- in a 0.10 M solution of sodium oxalate ($Na_2C_2O_4$).

65. (See Example 8–15.) The pH of a normal raindrop is 5.6. Compute the concentrations of $H_2CO_3(aq)$, $HCO_3^-(aq)$, and $CO_3^{2-}(aq)$ in this raindrop if the total concentration of dissolved carbonates is 1.0×10^{-5} M.

66. (See Example 8–15.) The pH of a drop of acid rain is 4.00. Compute the concentration of $H_2CO_3(aq)$, $HCO_3^-(aq)$, and $CO_3^{2-}(aq)$ in the acid raindrop, if the total concentration of dissolved carbonates is 3.6×10^{-5} M.

Lewis Acids and Bases

67. (See Example 8–16.) Indicate which species is the Lewis acid and which the Lewis base in each of the following reactions:
(a) $Ag^+(aq) + 2\,NH_3(aq) \longrightarrow Ag(NH_3)_2^+(aq)$
(b) $N(CH_3)_3(g) + BF_3(g) \longrightarrow F_3BN(CH_3)_3(g)$
(c) $B(OH)_3(aq) + OH^-(aq) \longrightarrow B(OH)_4^-(aq)$

68. (See Example 8–16.) Indicate which species is the Lewis acid and which the Lewis base in each of the following reactions:
(a) $HgI_2(s) + 2\,I^-(aq) \longrightarrow HgI_4^{2-}(aq)$
(b) $Co^{2+} + 6\,H_2O \longrightarrow [Co(H_2O)_6]^{2+}$
(c) $Fe^{3+}(aq) + 6\,CN^-(aq) \longrightarrow [Fe(CN)_6]^{3-}(aq)$

69. An important step in many industrial processes (including the Solvay process discussed in the Cumulative Problem in Chapter 4) is the slaking of lime, in which water is added to calcium oxide to make calcium hydroxide.
(a) Write the balanced equation for this process.
(b) Can this be considered a Lewis acid–base reaction? If so, what is the Lewis acid and what is the Lewis base?

70. Silica (SiO_2) is an impurity that must be removed from a metal oxide or sulfide ore when the ore is being reduced to elemental metal. To do this, lime (CaO) is added. It reacts with the silica to form a slag of calcium silicate ($CaSiO_3$), which can be separated and removed from the ore.
(a) Write the balanced equation for this process.
(b) Can this be considered a Lewis acid–base reaction? If so, what is the Lewis acid and what is the Lewis base?

71. Chemists working with fluorine and its compounds sometimes find it helpful to think in terms of acid–base reactions in which the fluoride ion (F^-) is donated and accepted.
(a) Is the acid in this system the fluoride donor or fluoride acceptor?
(b) Identify the acid and base in these reactions:

$$ClF_3O_2 + BF_3 \longrightarrow ClF_2O_2 \cdot BF_4$$
$$TiF_4 + 2\,KF \longrightarrow K_2[TiF_6]$$

72. Researchers working with ceramics often think of acid–base reactions in terms of oxide-ion donors and oxide-ion acceptors. The oxide ion is O^{2-}.
(a) In this system, is the base the oxide donor or the oxide acceptor?
(b) Identify the acid and base in each of these reactions:

$$2\,CaO + SiO_2 \longrightarrow Ca_2SiO_4$$
$$Ca_2SiO_4 + SiO_2 \longrightarrow 2\,CaSiO_3$$
$$Ca_2SiO_4 + CaO \longrightarrow Ca_3SiO_5$$

Additional Problems

73. Calculate the pH of a solution obtained by mixing 1.00 L of 1.00×10^{-5} M NaOH, which has a pH of 9.00, and 1.00 L of 1.00×10^{-6} M NaOH, which has a pH of 8.00. The answer pH = 8.50 is incorrect.

74. When placed in water, potassium starts to react instantly and continues to react with great vigor. On the basis of this in-

formation, select the better of the following two equations to represent the reaction:

$$2\,K(s) + 2\,H_2O(\ell) \longrightarrow 2\,KOH(aq) + H_2(g)$$
$$2\,K(s) + 2\,H_3O^+(aq) \longrightarrow 2\,K^+(aq) + H_2(g) + 2\,H_2O(\ell)$$

State the reason for your preference.

75. $Cl_2(aq)$ reacts with $H_2O(\ell)$ as follows:

$$Cl_2(aq) + 2\,H_2O(\ell) \rightleftharpoons$$
$$H_3O^+(aq) + Cl^-(aq) + HOCl(aq)$$

For an experiment to succeed, $Cl_2(aq)$ must be present, but the amount of $Cl^-(aq)$ in the solution must be minimized. For this purpose, should the pH of the solution be high, low, or neutral? Explain.

76. In pure nitric acid at 25°C, this autoionization reaction occurs:

$$2\,HNO_3(\ell) \rightleftharpoons NO_2^+(HNO_3) +$$
$$NO_3^-(HNO_3) + H_2O\ (HNO_3)$$

At equilibrium at 25°C, the molality of the NO_3^- ion is 0.25 mol kg^{-1}.
(a) Calculate the molarity of the NO_3^- ion. The density of pure nitric acid is 1.503 g cm^{-3}.
(b) Compute the autoionization constant of nitric acid. (This is "K_{HNO_3}" and is the constant for nitric acid that is analogous to K_w for water.)

77. Compute the pH of pure water at 60°C. See the K_w values in Table 8–1.

78. Calculate the pH of neutral water at three different temperatures (taking data from Table 8–1) and use the results, in conjunction with Le Chatelier's principle, to decide whether the autoionization of water is exothermic or endothermic.

79. Use the values of K_a in Table 8–2 to calculate equilibrium constants for the following equilibria at 25°C. In each case, identify the stronger Brønsted–Lowry acid and the stronger Brønsted–Lowry base.
(a) $H_2SO_3(aq) + CH_3COO^-(aq) \rightleftharpoons$
$$CH_3COOH(aq) + HSO_3^-(aq)$$
(b) $HF(aq) + CH_2ClCOO^-(aq) \rightleftharpoons$
$$CH_2ClCOOH(aq) + F^-(aq)$$
(c) $HCN(aq) + NO_2^-(aq) \rightleftharpoons HNO_2(aq) + CN^-(aq)$

80. Exactly 1.0 L of a solution of acetic acid gives the same color with methyl red as 1.0 L of a solution of hydrochloric acid. Which solution neutralizes the greater amount of 0.10 M NaOH(aq)? Explain.

81. Urea (NH_2CONH_2) is a component of urine. Compute the equilibrium concentration of urea in a solution that starts out containing no urea but instead 0.15 mol L^{-1} of the urea acidium ion, the conjugate acid of urea.

82. For each of the following compounds, indicate whether a 0.100 M aqueous solution is acidic (pH < 7), basic (pH > 7), or neutral (pH = 7): HCl, NH_4Cl, KNO_3, Na_3PO_4, $NaCH_3COO$.

83. Which of these procedures would *not* make a pH = 4.74 buffer?

(a) Mix 50.0 mL of 0.10 M acetic acid and 50.0 mL of 0.10 M sodium acetate.

(b) Mix 50.0 mL of 0.20 M acetic acid and 50.0 mL of 0.10 M NaOH.

(c) Start with 50.0 mL of 0.20 M acetic acid and add a solution of strong base until the pH equals 4.74.

(d) Start with 50.0 mL of 0.20 M HCl and add a solution of strong base until the pH equals 4.74.

(e) Start with 100.0 mL of 0.20 M sodium acetate and add 50.0 mL of 0.20 M HCl.

84. A buffer solution is 0.683 M in HN_3 (hydrazoic acid) and 0.593 M in NaN_3 (sodium azide).

(a) Calculate the pH of the solution.

(b) Write the chemical equation that describes how the buffer responds when a small amount of NaOH(aq) is added.

(c) What happens to the pH of the solution when a small amount of NaOH(aq) is added?

* 85. A buffer solution is prepared by mixing 1.00 L of 0.050 M pentafluorobenzoic acid (C_6F_5COOH) and 1.00 L of 0.060 M sodium pentafluorobenzoate (NaC_6F_5COO). The K_a of this weak acid is 0.033. Determine the pH of the buffer solution.

86. It takes 4.71 mL of 0.0410 M NaOH to titrate a 50.00-mL sample of flat (no CO_2) GG's Cola to a pH of 4.9. At this point, the addition of one more drop (0.02 mL) of NaOH raises the pH to 6.1. The only acid in GG's Cola is phosphoric acid. Compute the concentration of phosphoric acid in this cola. Assume that the 4.71 mL of base removes only the first hydrogen from the H_3PO_4; that is, assume that the reaction is:

$$H_3PO_4(aq) + OH^-(aq) \longrightarrow H_2O(\ell) + H_2PO_4^-(aq)$$

87. To 202.3 mL of 0.451 M NaOH is added 159.4 mL of 0.251 M H_2SO_4. Determine the pH of the resulting solution, assuming that the volumes are additive.

88. The chief chemist of Victory Vinegar Works, Ltd., interviews two chemists for employment. He states, "Quality control requires that our high-grade vinegar contain $5.00 \pm 0.01\%$ acetic acid by mass. How would you analyze our product to ensure that it meets this specification?"

Anne Dalton says, "I would titrate a 50.00-mL sample of the vinegar with 1.000 M NaOH, using phenolphthalein to detect the equivalence point to within ± 0.02 mL of base."

Charlie Cannizzarro says, "I would use a pH meter to determine the pH to ± 0.01 pH units and interface it with a computer to print out the mass percentage of acetic acid digitally." Which candidate did the chief chemist hire? Why?

89. Sodium benzoate, the sodium salt of benzoic acid, is used as a food preservative. A sample containing solid sodium benzoate mixed with sodium chloride is dissolved in 50.0 mL of 0.500 M HCl, giving an acidic mixture (benzoic acid mixed with HCl). This mixture is then titrated with 0.393 M NaOH. After the addition of 46.50 mL of the NaOH solution, an endpoint (sudden rise in pH) is observed. Further addition of the NaOH solution to a total of 63.61 mL causes a *second* end-

point. Calculate the mass of sodium benzoate (NaC_6H_5COO) in the original sample. (*Hint*: Decide which acid is neutralized first when NaOH is added.)

* 90. An antacid tablet (like Tums or Rolaids) weighs 1.3259 g. The only acid-neutralizing ingredient in this brand of antacid is $CaCO_3$. When placed in 12.07 mL of 1.070 M HCl, the tablet fizzes merrily as $CO_2(g)$ is given off. After all of the CO_2 has left the solution, an indicator is added, followed by 11.74 mL of 0.5310 M NaOH. The indicator shows that, at this point, the solution is definitely basic. Addition of 5.12 mL of 1.070 M HCl makes the solution acidic again. Then, 3.17 mL of the 0.5310 M NaOH brings the titration exactly to an end-point, as signaled by the indicator. Compute the percentage by mass of $CaCO_3$ in the tablet.

91. Baking soda (sodium hydrogen carbonate, $NaHCO_3$) reacts with acids in foods to form carbonic acid (H_2CO_3), which in turn decomposes to water and carbon dioxide. In a batter, the carbon dioxide appears as gas bubbles that cause the bread or cake to rise.

(a) A rule of thumb in cooking is that $\frac{1}{2}$ teaspoon of baking soda is neutralized by 1 cup of sour milk. The acid component of sour milk is lactic acid ($HC_3H_5O_3$). Write an equation for the neutralization reaction.

(b) If the density of baking soda is 2.16 g cm^{-3}, calculate the concentration of lactic acid in the sour milk, in moles per liter. Take 1 cup = 236.6 mL = 48 teaspoons.

(c) Calculate the volume of carbon dioxide that is produced at 1 atm pressure and 350°F (177°C) from the reaction of $\frac{1}{2}$ teaspoon of baking soda.

92. Egg whites contain dissolved carbon dioxide and water, which react to give carbonic acid (H_2CO_3). In the first days after an egg is laid, it loses carbon dioxide through its shell. Does the pH of the egg white increase or decrease during this period?

93. Phosphonocarboxylic acid

effectively inhibits the replication of the herpes virus. Structurally, it is a combination of phosphoric acid and acetic acid. Each molecule can donate three hydrogen ions. The stepwise acidity constants are $K_{a1} = 1.0 \times 10^{-2}$, $K_{a2} = 7.8 \times 10^{-6}$, and $K_{a3} = 2.0 \times 10^{-9}$. Enough phosphonocarboxylic acid is added to blood (pH 7.40) to make its total concentration 1.0×10^{-5} M. The pH of the blood does not change. Determine the concentrations of all four forms of the acid in this mixture.

94. A reference book states that a saturated aqueous solution of potassium hydrogen tartrate is a buffer with a pH of 3.56. Write two chemical equations that show the buffer action of this solution. (Tartaric acid is a diprotic acid with the formula $H_2C_4H_4O_6$. Potassium hydrogen tartrate is $KHC_4H_4O_6$.)

*95. A solution of 1.0×10^{-8} M HNO_3 does *not* have a pH of 8.00 at 25°C.

(a) Explain why 1.0×10^{-8} M HNO_3 must have a pH below 7.

(b) Find the pH of the solution using these facts: the nitric acid is essentially completely dissociated; the total concentration of positive ions equals the total concentration of negative ions; that is, $[H_3O^+] = [OH^-] + [NO_3^-]$; the K_w equilibrium law holds.

96. Aluminum hydroxide is said to be amphoteric because it dissolves in both acidic and basic solutions. Write balanced equations for these reactions and show that they are acid–base reactions according to the Lewis model. The aluminum species formed when aluminum hydroxide dissolves in basic solution is $Al(OH)_4^-$.

Power plant emissions cause acid rain.

CUMULATIVE PROBLEM

Acid Rain

Acid rain is a major environmental problem throughout the industrialized world. It results largely from the burning of fossil fuels (coal, oil, and natural gas) that contain sulfur. The combustion of sulfur gives sulfur dioxide, the acid anhydride of the weak acid sulfurous acid, H_2SO_3. What is more serious, sulfur dioxide may be oxidized to sulfur trioxide, the acid anhydride of the strong acid sulfuric acid, H_2SO_4. Both anhydrides react with water droplets in the air to increase the acidity of the rain, which damages trees, kills fish in lakes, and eats away stone and metal surfaces.

(a) A sample of rainwater is tested for pH by the use of two indicators. Addition of methyl orange to half of the sample gives a yellow color, and addition of methyl red to the other half gives a red color. Estimate the pH of the sample.

(b) The pH in acid rain can drop to 3 or even a bit lower in heavily polluted areas. Calculate the concentrations of H_3O^+ and OH^- in a raindrop at pH 3.30 at 25°C.

(c) When sulfur dioxide (SO_2) dissolves in water to form sulfurous acid ($H_2SO_3(aq)$), that acid can donate a hydrogen ion to water. Write a balanced chemical equation for this reaction, and identify the stronger Brønsted–Lowry acid and base in the equation.

(d) Ignore the further ionization of HSO_3^-, and calculate the pH of a solution in which the initial concentration of H_2SO_3 is 4.0×10^{-4} M. (*Hint*: Use the quadratic equation or numerical analysis in this case.)

(e) Now suppose that all the dissolved SO_2 from part (d) is oxidized to SO_3, so that 4.0×10^{-4} mol of H_2SO_4 is dissolved per liter. Calculate the pH in this case. (*Hint*: Because the first ionization of H_2SO_4 is that of a strong acid, the concentration of H_3O^+ can be written as 4.0×10^{-4} plus the amount of hydrogen ion from the ionization of $HSO_4^-(aq)$.)

(f) Lakes have a natural buffering capacity, especially in regions where limestone gives rise to dissolved calcium carbonate. Write an equation for the reaction of a small amount of acid rain containing sulfuric acid falling into a lake containing carbonate (CO_3^{2-}) ions. Discuss how the lake resists further pH changes. What happens if a large excess of acid rain is deposited in the lake?

(g) A sample of 1.00 L of rainwater known to contain only sulfurous (and not sulfuric) acid is titrated with 0.0100 M NaOH. The equivalence point of the H_2SO_3/HSO_3^- titration is reached after 31.6 mL has been added. Calculate the original concentration of sulfurous acid in the sample, again ignoring any effect of SO_3^{2-} on the equilibria.

(h) Calculate the pH at the half-equivalence point. (*Note*: Sulfurous acid is a strong enough acid that the Henderson–Hasselbalch equation cannot be used. Use the quadratic equation or numerical analysis.)

9

Dissolution and Precipitation Equilibria

Dissolution of sugar in water.

Dissolution and precipitation reactions, by which solids pass into and out of solution, are the reverse of each other. Together they comprise a major category of chemical reactions (as discussed in Chapter 4). Their practical importance matches that of acid–base reactions. The dissolving and reprecipitation of solids permit chemists to isolate single products from reaction mixtures and to purify impure solid samples. Understanding the mechanisms of these reactions helps engineers to prevent scale from forming in boilers and doctors to reduce the incidence of painful kidney stones. Dissolution and precipitation control the formation of mineral deposits and profoundly affect the ecologies of rivers, lakes, and oceans.

In this chapter, we consider some quantitative aspects of the equilibria that govern dissolution and precipitation. Our theme throughout the chapter is the manipulation of equilibria to control the solubilities of ionic solids.

9–1 THE NATURE OF SOLUBILITY EQUILIBRIA

Often, the major task in synthetic chemistry is not making a compound but purifying it. Side-reactions can generate serious amounts of impurities when a compound is prepared. Other impurities enter with the starting materials, fall in accidentally during the reaction, or are put in deliberately to make the reaction go faster. Running a reaction may take only hours, but the *work-up* (separating crude product) and subsequent purification may require weeks. **Recrystallization** is one of the most powerful methods for purifying solids. It relies on differences between the solubilities of the desired substance and its contaminants. An impure product is dissolved and reprecipitated, repeatedly if necessary. A successful purification by recrystallization depends on closely controlling the factors that influence solubility. Manipulating solubility in turn requires an understanding of the equilibria that exist between an undissolved substance and its solution.

In recrystallization, a solution begins to deposit a compound when it is brought to the point of saturation with respect to that compound. A **saturated solution** is one in which a dissolution–precipitation (solubility) equilibrium exists between the solid substance and its dissolved form. For example, the equilibrium between solid iodine and iodine dissolved in CCl_4 can be written as

$$I_2(s) \rightleftharpoons I_2(CCl_4)$$

In dissolution, the solvent attacks the solid and **solvates** it at the level of individual particles. To continue the example, this means that I_2–CCl_4 attractions replace the I_2–I_2 interactions formerly present in the solid, and the iodine molecules go off into the bulk of the CCl_4. In precipitation, the reverse occurs; solute-to-solute attractions are re-established as the solute leaves the solution. Often, solute-to-solvent attractions persist right through the process of precipitation, and solvent incorporates itself into the solid. When lithium sulfate (Li_2SO_4) precipitates from water, for example, it brings with it into the solid one molecule of water per formula unit:

$$2\,Li^+(aq) + SO_4^{2-}(aq) + H_2O(\ell) \longrightarrow Li_2SO_4 \cdot H_2O(s)$$

Such loosely bound solvent is **solvent of crystallization**: lithium sulfate precipitates from water with one water of crystallization. Dissolving and then reprecipitating a compound may thus furnish material with a changed chemical formula.

Solubility equilibria resemble the equilibria that volatile liquids (or solids) establish with their own vapors in a closed container. In both cases, particles from a condensed phase tend to escape and to spread throughout a larger but limited volume. In

• The first separation of the element ytterbium (Yb) from the chemically very similar element lutetium (Lu) was achieved by a laborious series of 15,000 cycles of dissolution and reprecipitation of a mixture of $Yb(NO_3)_3$ and $Lu(NO_3)_3$, using aqueous nitric acid as the solvent.

• A saturated solution was defined previously (see Section 4–1) in slightly different, but equivalent, terms.

both cases, equilibrium is a dynamic compromise in which the rate of escape of particles from the condensed phase is equal to their rate of return. In a vaporization–condensation equilibrium, the vapor above the condensed phase is assumed to be an ideal gas. The analogous starting assumption for a dissolution–precipitation reaction is that the solution above the undissolved solid is an ideal solution.

• An ideal solution is a solution that obeys Raoult's law (see Section 6–6).

Figure 9–1 The evaporating dish originally contained an aqueous solution of $(NH_4)_2Cr_2O_7$. It was loosely covered to keep out dust and allowed to stand. As the water evaporated, the solution became saturated and deposited solid $(NH_4)_2Cr_2O_7$ in the form of long needle-like crystals.

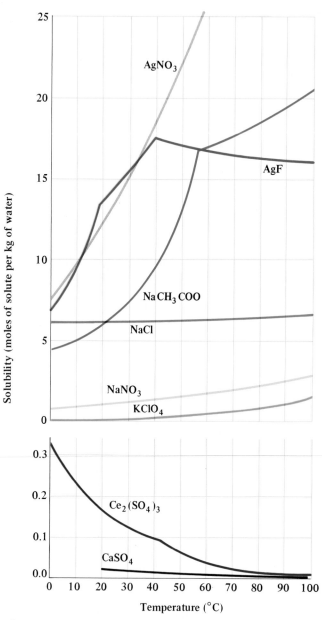

Figure 9–2 Most solubilities of solids increase with increasing temperature, but some decrease. Note that solubilities do not necessarily increase or decrease smoothly as the temperature changes.

The simplest possible equilibrium expression for a dissolution–precipitation reaction has a form such as

$$[I_2]_{CCl_4} = K$$

This expression states that the molarity of a saturated solution is a constant. It closely resembles an equilibrium law for vaporization–condensation, which states that the pressure of the vapor above a volatile substance is a constant. The undissolved solute does not appear in the expression because the amount of undissolved solid present at the bottom of a flask has no effect on the position of the equilibrium, as long as some of it is present. The equilibrium expressions for many dissolution–precipitation reactions are more complex than this, as we shall see.

Le Chatelier's principle applies to these equilibria, as it does to all equilibria. One way to exert a stress on a solubility equilibrium is to change the amount of solvent. Increases are particularly easy; just pour in some more. Adding solvent reduces the concentration of dissolved substance. More solid then tends to dissolve (according to Le Chatelier's principle) to restore the concentration of the dissolved substance to its equilibrium value. If so much solvent is added that all of the solid dissolves, then obviously solid cannot simultaneously be dissolving and precipitating. The solubility equilibrium ceases to exist, and the solution is **unsaturated.** This corresponds in a vaporization–condensation equilibrium to the complete evaporation of the condensed phase. Removing solvent from an already saturated solution forces additional solid to precipitate in order to maintain a constant concentration. A volatile solvent is often removed by simply letting a solution stand uncovered; the solvent evaporates. When conditions are right, the solid forms as crystals on the bottom and sides of the container (Fig. 9–1).

The equilibrium constants for dissolution–precipitation reactions depend on the temperature. If the dissolution reaction is exothermic, then raising the temperature of a saturated solution shifts the equilibrium to favor the solid, reducing the concentration of solute in solution. If the dissolution reaction is endothermic, then raising the temperature causes more solid to dissolve at saturation—that is, it shifts the equilibrium toward the side of the products. In a survey of more than 500 solid inorganic compounds, 86% were found to increase in solubility in water with increasing temperature, and only 7% were found to decrease in solubility. (The remaining 7% had solubilities that increased and then decreased again, or vice versa.) In practice, when the behavior of a substance is not known, the odds favor heating the solution to dissolve more solid and cooling it to increase the amount of precipitate. Figure 9–2 displays the dependence of solubility on temperature for several substances.

• Thus, changes in temperature act on these equilibria in the same way that they act on other kinds of equilibria (see Section 7-5).

Dissolution–precipitation reactions frequently come to equilibrium slowly because it takes time to transfer material across the phase boundary between solid and solution. It can require days or even weeks of shaking a solid in contact with a solvent before the solution becomes saturated. Moreover, solutions sometimes become **supersaturated,** a condition in which the concentration of dissolved solid *exceeds* its equilibrium value. The delay then is in forming, rather than dissolving, the solid. Supersaturated solutions may persist for months or years and require extraordinary measures to bring them to equilibrium, although the tendency toward equilibrium is always there (Fig. 9–3). The generally sluggish approach to equilibrium in dissolution–precipitation reactions contrasts sharply with the rapid rates at which acid–base reactions come to equilibrium.

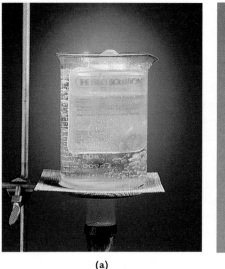

(a)   (b)   (c)

Figure 9-3   (a) A concentrated solution of aqueous sodium acetate (NaCH₃COO) is contained in a plastic pouch. At the temperature of boiling water, all of the sodium acetate dissolves. (b) After cooling to room temperature, the sodium acetate *tends* to precipitate, but the process is quite slow. The solution becomes supersaturated, and the pouch can be stored almost indefinitely, with no solid forming. (c) Flexing a metal disk that is in contact with the solution initiates precipitation. Much solid then forms, until the remaining solution is saturated. The change generates a comfortable heat for soothing minor aches.

9-2 THE SOLUBILITY OF SALTS

When an ionic solid dissolves in water, it dissociates more or less completely into aquated ions (see Section 4–1). We recognize dissociation in equilibrium equations by showing ions as products. For example, the equilibrium between solid and dissolved cesium chloride (CsCl) is written

$$CsCl(s) \rightleftharpoons Cs^+(aq) + Cl^-(aq)$$

If dissolution formed aquated CsCl molecules, then the equation would be

$$CsCl(s) \rightleftharpoons CsCl(aq)$$

Salts have a wide range of solubilities (see Table 4–1). For a highly soluble salt (such as CsCl), the concentrations of the ions in saturated aqueous solution are so large that the ions inevitably associate. The solution is nonideal. Ion pairs (temporary pairs of oppositely charged ions, such as $Cs^+ \ldots Cl^-$) and larger ion clusters as well exist in such solutions. A full description of dissolution would require several equilibrium equations, one for each dissolved species. We therefore restrict our attention to sparingly soluble and "insoluble" salts, for which the concentrations in a saturated solution are 0.1 mol per liter or less. Although these solutions are saturated, they remain dilute enough that associations among the aquated ions are negligible. A single equilibrium equation, one that assumes complete dissociation into ions, describes them adequately.

The Solubility Product

The sparingly soluble salt silver chloride establishes the following equilibrium when placed in water:

$$AgCl(s) \rightleftharpoons Ag^+(aq) + Cl^-(aq)$$

The equilibrium law for this reaction is written by following the general rules for heterogeneous equilibria from Section 7–6:

$$\frac{[Ag^+][Cl^-]}{1} = [Ag^+][Cl^-] = K_{sp}$$

where the subscript "sp," which stands for **solubility product,** distinguishes the K as referring to the dissolution of a slightly soluble ionic solid in water. At 25°C, K_{sp} has the numerical value 1.6×10^{-10} for silver chloride. Like all equilibrium constants, K_{sp} is a dimensionless number, and the concentrations are each divided by a reference concentration, defined to be 1.0 M. As in other equilibrium expressions, we omit the normalizing reference concentrations.

In the AgCl solubility-product expression, the concentrations of the two ions produced are raised to the first power because their coefficients are 1 in the chemical equation. AgCl, a pure solid, is accounted for by the 1 in the denominator. Obviously, dividing by 1 does not change an algebraic expression. In chemical terms, this means that the AgCl(s) does not affect the equilibrium as long as *some* is present. If *no* solid is present, then the product of the two ion concentrations is no longer constrained by the solubility-product expression.

• From this point on in this chapter, the 1's representing pure solids, pure liquids, and dilute solvents in equilibrium expressions will simply be omitted.

EXAMPLE 9–1

Write the expression for the solubility product for the dissolution of lead iodide ($PbI_2(s)$) in water.

Solution

One mole of $PbI_2(s)$ produces one mole of Pb^{2+} ions and *two* moles of I^- ions in solution

$$PbI_2(s) \rightleftharpoons Pb^{2+}(aq) + 2\,I^-(aq)$$

so that the concentration of I^- is raised to the second power in its solubility-product expression

$$[Pb^{2+}][I^-]^2 = K_{sp}$$

The pure solid $PbI_2(s)$ does not appear explicitly in the expression.

Exercise

Write the K_{sp} equation for the dissolution of aluminum hydroxide ($Al(OH)_3$) in water.

Answer: $[Al^{3+}][OH^-]^3 = K_{sp}$

Solubility and K_{sp}

The solubility of a salt in water is not the same as its solubility-product constant, but there is often a simple relation between them. For example, if y mol of AgCl(s) dissolves in one liter of solution at 25°C, then $y = [Ag^+] = [Cl^-]$ is the molarity of each ion at equilibrium. We then write

$$[Ag^+][Cl^-] = y^2 = K_{sp} = 1.6 \times 10^{-10}$$

Taking the square root of both sides of the equation gives

$$y = 1.26 \times 10^{-5} \text{ mol L}^{-1}$$

which rounds off to 1.3×10^{-5} mol L^{-1}. This is the molar solubility of AgCl in water. We convert the molar solubility to a gram solubility as follows:

$$\frac{1.25 \times 10^{-5} \text{ mol AgCl}}{\text{L solution}} \times \left(\frac{143.3 \text{ g AgCl}}{1 \text{ mol AgCl}} \right) = \frac{1.8 \times 10^{-3} \text{ g AgCl}}{\text{L solution}}$$

Therefore, 1.8×10^{-3} g of AgCl dissolves per liter of saturated solution at 25°C.

Solubility-product constants (and solubilities) can depend strongly on temperature. At 100°C, the K_{sp} for silver chloride is 2.2×10^{-8}; hot water dissolves about 12 times more silver chloride than water at 25°C. Temperatures are always quoted in listings of K_{sp} values. Refer to Table 9–1 for the solubility-product constants at 25°C of a number of important sparingly soluble salts.

Table 9–1
Solubility-Product Constants (K_{sp}'s) at 25°C

Fluorides		Iodates	
BaF$_2$	1.7×10^{-6}	AgIO$_3$	3.1×10^{-8}
CaF$_2$	3.9×10^{-11}	Cu(IO$_3$)$_2$	1.4×10^{-7}
MgF$_2$	6.6×10^{-9}	Pb(IO$_3$)$_2$	2.6×10^{-13}
PbF$_2$	3.6×10^{-8}		
SrF$_2$	2.8×10^{-9}	**Chromates**	
		Ag$_2$CrO$_4$	1.9×10^{-12}
Chlorides		BaCrO$_4$	2.1×10^{-10}
AgCl	1.6×10^{-10}	PbCrO$_4$	1.8×10^{-14}
CuCl	1.0×10^{-6}		
PbCl$_2$	1.6×10^{-5}	**Carbonates**	
Hg$_2$Cl$_2$	2×10^{-18}	Ag$_2$CO$_3$	6.2×10^{-12}
		BaCO$_3$	8.1×10^{-9}
Bromides		CaCO$_3$	8.7×10^{-9}
AgBr	7.7×10^{-13}	PbCO$_3$	3.3×10^{-14}
CuBr	4.2×10^{-8}	MgCO$_3$	4.0×10^{-5}
Hg$_2$Br$_2$	1.3×10^{-21}	SrCO$_3$	1.6×10^{-9}
Iodides		**Oxalates**	
AgI	1.5×10^{-16}	CuC$_2$O$_4$	2.9×10^{-8}
CuI	5.1×10^{-12}	FeC$_2$O$_4$	2.1×10^{-7}
PbI$_2$	1.4×10^{-8}	MgC$_2$O$_4$	8.6×10^{-5}
Hg$_2$I$_2$	1.2×10^{-28}	PbC$_2$O$_4$	2.7×10^{-11}
		SrC$_2$O$_4$	5.6×10^{-8}
Hydroxides			
AgOH	1.5×10^{-8}	**Sulfates**	
Al(OH)$_3$	3.7×10^{-15}	BaSO$_4$	1.1×10^{-10}
Fe(OH)$_3$	1.1×10^{-36}	CaSO$_4$	2.4×10^{-5}
Fe(OH)$_2$	1.6×10^{-14}	PbSO$_4$	1.1×10^{-8}
Mg(OH)$_2$	1.2×10^{-11}		
Mn(OH)$_2$	2×10^{-13}		
Zn(OH)$_2$	4.5×10^{-17}		

EXAMPLE 9-2

The K_{sp} of calcium fluoride is 3.9×10^{-11}. Calculate the concentrations of calcium and fluoride ions in a saturated solution of CaF_2 at 25°C, and determine the solubility of CaF_2 in grams per liter of solution.

Solution

The solubility equilibrium is

$$CaF_2(s) \rightleftharpoons Ca^{2+}(aq) + 2\,F^-(aq)$$

for which the equilibrium expression is

$$[Ca^{2+}][F^-]^2 = K_{sp}$$

Note that $[F^-]$ is squared because $F^-(aq)$ has a coefficient of 2 in the equation.
   If y moles of CaF_2 dissolves in one liter, the equilibrium concentration of Ca^{2+} is $[Ca^{2+}] = y$. The concentration of F^- is $[F^-] = 2y$ because each mole of CaF_2 that dissolves produces *two* moles of fluoride ions. Therefore,

$$[Ca^{2+}][F^-]^2 = y \times (2y)^2 = 4y^3 = K_{sp} = 3.9 \times 10^{-11}$$

Solving for y^3 gives

$$y^3 = \frac{1}{4}(3.9 \times 10^{-11})$$

By taking the cube root of both sides of this equation we arrive at

$$y = 2.14 \times 10^{-4}$$

The equilibrium concentrations are therefore

$$[Ca^{2+}] = y = 2.1 \times 10^{-4} \text{ mol L}^{-1}$$
$$[F^-] = 2y = 4.3 \times 10^{-4} \text{ mol L}^{-1}$$

Because the molar mass of CaF_2 is 78.1 g mol^{-1}, the gram solubility is

$$\text{gram solubility} = \left(\frac{2.14 \times 10^{-4} \text{ mol}}{L}\right) \times \left(\frac{78.1 \text{ g}}{1 \text{ mol}}\right) = \frac{0.017 \text{ g}}{L}$$

The experimental value is 0.018 g L^{-1}.

Exercise
Determine the mass of lead(II) iodate dissolved in 2.50 L of a saturated aqueous solution of $Pb(IO_3)_2$ at 25°C. The K_{sp} of $Pb(IO_3)_2$ is 2.6×10^{-13}.

Answer: 0.056 g.

It is possible, of course, to reverse the procedure outlined above to determine the value of K_{sp} from measured solubilities, as the following example illustrates.

• Values of K_{sp} can also be obtained from thermodynamic or electrochemical studies, as shown in Chapters 11 and 13.

EXAMPLE 9-3

Silver chromate (Ag_2CrO_4) is a red solid that dissolves to the extent of 0.029 g L^{-1} at 25°C. $PbCl_2$ is a white solid that dissolves to the extent of 8.67 g L^{-1} at 25°C. Estimate the K_{sp} of each of these salts, and compare the estimates with the values in Table 9–1.

Solution

We convert the solubility of silver chromate to moles per liter as follows

$$\left(\frac{0.029 \text{ g Ag}_2\text{CrO}_4}{\text{L solution}}\right) \times \left(\frac{1 \text{ mol Ag}_2\text{CrO}_4}{331.73 \text{ g Ag}_2\text{CrO}_4}\right) = \frac{8.7 \times 10^{-5} \text{ mol Ag}_2\text{CrO}_4}{\text{L solution}}$$

Each mole of dissolved Ag_2CrO_4 gives two moles of Ag^+ ion and one mole of CrO_4^{2-} ion

$$[Ag^+] = 2 \times (8.7 \times 10^{-5}) \text{ mol L}^{-1} \qquad [CrO_4^{2-}] = 8.7 \times 10^{-5} \text{ mol L}^{-1}$$

Substitution of these numbers in the equilibrium expression gives K_{sp}

$$[Ag^+]^2[CrO_4^{2-}] = (2 \times 8.7 \times 10^{-5})^2 (8.7 \times 10^{-5}) = K_{sp} = \boxed{2.6 \times 10^{-12}}$$

This is fairly close to the tabulated value of 1.9×10^{-12}.

Estimation of the K_{sp} of $PbCl_2$ proceeds similarly

$$\left(\frac{8.67 \text{ g PbCl}_2}{\text{L solution}}\right) \times \left(\frac{1 \text{ mol PbCl}_2}{278.1 \text{ g PbCl}_2}\right) = \frac{0.0312 \text{ mol PbCl}_2}{\text{L solution}}$$

$$[Pb^{2+}] = 0.0312 \text{ mol L}^{-1} \qquad [Cl^-] = 2 \times (0.0312) \text{ mol L}^{-1}$$

$$[Pb^{2+}][Cl^-]^2 = (0.0312)(2 \times 0.0312)^2 = K_{sp} = \boxed{1.2 \times 10^{-4}}$$

This exceeds substantially 1.6×10^{-5}, the tabulated K_{sp}.

Exercise

Compute the K_{sp} of calcium sulfate if its solubility is 0.67 g L^{-1}.

Answer: 2.4×10^{-5}.

Computing a K_{sp} from a solubility, or a solubility from a K_{sp}, is valid if the solution is ideal and no side-reactions reduce the concentrations of the ions after they enter solution. If such reactions occur, they cause higher solubilities than the K_{sp} predicts (or, conversely, apparent K_{sp}'s exceeding the true values). Thus, the formation of the species $PbOH^+(aq)$ and $PbCl^+(aq)$ causes the K_{sp} estimated in Example 9–3 to exceed the true K_{sp} of $PbCl_2$ (as compiled in Table 9–1) by a factor of about 7.5. We defer further discussion of these side-reactions to Section 9–5.

9–3   PRECIPITATION AND THE SOLUBILITY PRODUCT

Up to now, we have considered only cases in which a single slightly soluble salt attains an equilibrium with its component ions in water. In such solutions, the relative concentrations of the cations and anions echo the relative number of moles of each in the original salt. Thus, when AgCl is dissolved, equal numbers of $Ag^+(aq)$ and $Cl^-(aq)$ ions result, and when Ag_2SO_4 is dissolved, twice as many $Ag^+(aq)$ ions as $SO_4^{2-}(aq)$ ions are produced. A solubility-product relationship such as

$$[Ag^+][Cl^-] = K_{sp}$$

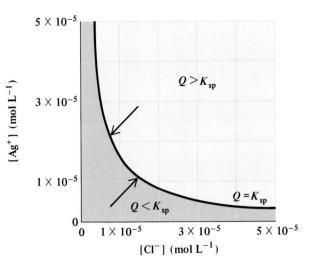

Figure 9-4 Some solid silver chloride is in contact with a solution containing $Ag^+(aq)$ and $Cl^-(aq)$ ions. If a solubility equilibrium exists, then the product Q of the concentrations of the ions $[Ag^+] \times [Cl^-]$ is a constant, K_{sp} (*curved line*). If Q exceeds K_{sp}, solid silver chloride tends to precipitate until equilibrium is attained. If Q is less than K_{sp}, then additional solid tends to dissolve. If no solid is present, Q remains less than K_{sp}.

is more general than this, however, and continues in force even if the relative chemical amounts of the two ions in solution differ from those in the pure solid compound. Such situations often result when two solutions are mixed to give a precipitate, or when another salt is present that contains an ion common to the salt under consideration.

Precipitation from Solution

Suppose that a solution of one soluble salt, such as $AgNO_3$, is mixed with a solution of a second, such as NaCl. Does a precipitate of silver chloride form? To answer this, we use the concept of the reaction quotient Q that was first introduced in connection with gas-phase equilibria (see Section 7–3).

The initial reaction quotient $Q_{(init)}$, evaluated when the mixing of the solutions is complete but before any reaction occurs, is

$$Q_{(init)} = [Ag^+]_{(init)}[Cl^-]_{(init)}$$

If $Q_{(init)} < K_{sp}$, no solid silver chloride can appear. On the other hand, if $Q_{(init)} > K_{sp}$, solid silver chloride precipitates until the reaction quotient Q reaches K_{sp} (Fig. 9–4).

• The reaction quotient in this case has a denominator equal to 1, which is omitted.

EXAMPLE 9-4

Suppose 500 mL of a solution of NaCl with a chloride ion concentration of 0.0080 mol L^{-1} is added to 300 mL of a 0.040 mol L^{-1} solution of $AgNO_3$. Does solid AgCl form at equilibrium?

Solution

The "initial" concentrations to be used in calculating $Q_{(init)}$ are those *before* reaction but *after* dilution through mixing the two solutions. The initial concentration of $Ag^+(aq)$ after dilution from 300 to 800 mL of solution is

$$[Ag^+]_{(init)} = 0.040 \text{ mol } L^{-1} \times \left(\frac{300 \text{ mL}}{800 \text{ mL}}\right) = 0.015 \text{ mol } L^{-1}$$

and that of $Cl^-(aq)$ after dilution from 500 to 800 mL is

$$[Cl^-]_{(init)} = 0.0080 \text{ mol L}^{-1} \times \left(\frac{500 \text{ mL}}{800 \text{ mL}}\right) = 0.0050 \text{ mol L}^{-1}$$

The initial reaction quotient is

$$Q_{(init)} = [Ag^+]_{(init)}[Cl^-]_{(init)} = (0.015)(0.0050) = 7.5 \times 10^{-5}$$

Because $Q_{(init)} > K_{sp}$, solid silver chloride forms at equilibrium.

Exercise

The K_{sp} of thallium(I) iodate ($TlIO_3$) is 3.1×10^{-6} at 25°C. Suppose that 555 mL of a 0.0022 M solution of $TlNO_3$ is mixed with 445 mL of a 0.0022 M solution of $NaIO_3$. Does $TlIO_3$ precipitate at equilibrium?

Answer: $TlIO_3$ does not precipitate.

The equilibrium concentrations of ions after mixing two solutions to give a precipitate are most easily calculated by supposing that the precipitation reaction first goes to completion (using up one of the ions) and that subsequent dissolution of the solid then restores some of that ion to solution. This is the same approach used in Example 8–11 to treat the case of the addition of a strong acid to a buffer solution.

EXAMPLE 9–5

Calculate the equilibrium concentrations of silver and chloride ions resulting from the precipitation reaction of Example 9–4.

Solution

In this case, the silver ion is clearly in excess, so the chloride ion is the limiting reactant. If all of the chloride ion were used up to make solid AgCl, the concentration of the remaining silver ion would be

$$[Ag^+] = 0.015 - 0.0050 = 0.010 \text{ mol L}^{-1}$$

We set up a table of changes starting at this point

	AgCl(s) $\rightleftharpoons$ Ag$^+$(aq)	+	Cl$^-$(aq)
Initial concentration (mol L^{-1})	0.010		0
Change in concentration (mol L^{-1})	+ y		+ y
Equilibrium concentration (mol L^{-1})	0.010 + y		y

Inserting the values in the last line into the solubility-product relationship gives the equation

$$(0.010 + y)y = K_{sp} = 1.6 \times 10^{-10}$$

Although this quadratic equation can be solved by use of the quadratic formula, it is far easier to make the approximation that y is much smaller than 0.010 and can be neglected when added to 0.010. Then the equation simplifies to

$$(0.010)y \approx 1.6 \times 10^{-10}$$
$$y \approx 1.6 \times 10^{-8} \text{ mol L}^{-1} = [Cl^-]$$

The assumption about the size of y is justified. The equilibrium concentration of silver ion is

$$[Ag^+] = 0.010 \text{ mol L}^{-1}$$

Exercise

The K_{sp} of thallium(I) iodate ($TlIO_3$) is 3.1×10^{-6} at 25°C. Suppose that 125 mL of a 0.0100 M solution of $TlNO_3$ is mixed with 375 mL of a 0.0250 M solution of $NaIO_3$. Determine the concentration of $Tl^+(aq)$ in the solution at equilibrium.

Answer: 1.9×10^{-4} M.

The Common-Ion Effect

Suppose that a small amount of NaCl(s) is added to a saturated solution of AgCl. What happens? Sodium chloride is quite soluble in water. It dissolves to give $Na^+(aq)$ and $Cl^-(aq)$ ions, raising the concentration of chloride ion. The reaction quotient $Q = [Ag^+][Cl^-]$ then exceeds the K_{sp} of silver chloride, and silver chloride precipitates until the concentrations of $Ag^+(aq)$ and $Cl^-(aq)$ are sufficiently reduced that the solubility-product expression once again is satisfied (Fig. 9–5). Silver chloride also precipitates if a small amount of the soluble salt $AgNO_3$ is added to the saturated solution of AgCl.

The same equilibrium may be approached from the other direction. The amount of AgCl(s) that can be dissolved in a solution that contains either $Ag^+(aq)$ or $Cl^-(aq)$ is less than the amount that can be dissolved in the same volume of pure water. Figure 9–5 shows why. Extra $Cl^-(aq)$ reduces the maximum concentration of $Ag^+(aq)$ permitted and lowers the solubility of the AgCl(s); extra $Ag^+(aq)$ reduces

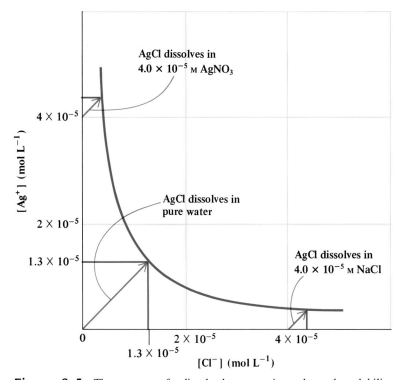

Figure 9–5 The presence of a dissolved common ion reduces the solubility of a salt in solution. In all three cases shown, as the AgCl dissolves, the concentrations of the ions follow the paths shown by the red arrows until they reach the blue equilibrium line. The molar solubilities are proportional to the lengths of the green lines shown: 1.3×10^{-5} mol L^{-1} for AgCl in pure water, but only 0.37×10^{-5} mol L^{-1} in either 4.0×10^{-5} M $AgNO_3$ or 4.0×10^{-5} M NaCl.

the concentration of $Cl^-(aq)$ permitted at equilibrium and also reduces the solubility of AgCl. This effect, which occurs with all salts, is referred to as the **common-ion effect:**

> If a solution and a solid salt to be dissolved in it have an ion in *common*, then the solubility of the salt is depressed.

Let us examine the quantitative consequences of the common-ion effect. Suppose $AgCl(s)$ is added to a 0.100 M NaCl solution. What is its solubility? If y moles of AgCl dissolve per liter, the concentration of $Ag^+(aq)$ is y moles per liter, and that of $Cl^-(aq)$ is

$$[Cl^-] = (0.100 + y) \text{ mol L}^{-1}$$

because the chloride ion has two sources: the 0.100 M NaCl and the dissolution of AgCl. The expression for the solubility product is

$$[Ag^+][Cl^-] = y(0.100 + y) = K_{sp} = 1.6 \times 10^{-10}$$

The solubility of $AgCl(s)$ in this solution must be smaller than it is in pure water, which is much smaller than 0.100 moles per liter. That is,

$$y < 1.3 \times 10^{-5} \text{ mol L}^{-1} \ll 0.100 \text{ mol L}^{-1}$$

Thus, we can approximate $(0.100 + y)$ by 0.100 (as in Example 9–5), giving

$$(0.100)y \approx 1.6 \times 10^{-10}$$
$$y \approx 1.6 \times 10^{-9} \text{ mol L}^{-1}$$

We see immediately that $y = 1.6 \times 10^{-9}$ is indeed much smaller than 0.100 mol L^{-1}, so our approximation was a very good one. Therefore, at equilibrium

$$[Ag^+] = y = 1.6 \times 10^{-9} \text{ mol L}^{-1}$$
$$[Cl^-] = 0.100 \text{ mol L}^{-1}$$

The gram solubility of AgCl here is

$$\frac{1.6 \times 10^{-9} \text{ mol AgCl}}{1 \text{ L solution}} \times \left(\frac{143.3 \text{ g AgCl}}{1 \text{ mol AgCl}}\right) = 2.3 \times 10^{-7} \frac{\text{g AgCl}}{\text{L solution}}$$

The solubility of AgCl in 0.100 M NaCl is lower than that in pure water by a factor of about 8000.

EXAMPLE 9–6

What is the solubility of $CaF_2(s)$ in a 0.100 M solution of NaF?

Solution

Again, we represent the solubility as y. The only source of the $Ca^{2+}(aq)$ in the solution at equilibrium is the dissolution of CaF_2, but the $F^-(aq)$ has two sources, the CaF_2 and the NaF. Hence,

	$CaF_2(s) \rightleftharpoons$	$Ca^{2+}(aq)$	$+$	$2 F^-(aq)$
Initial concentration (mol L^{-1})		0		0.100
Change in concentration (mol L^{-1})		$+ y$		$+ 2y$
Equilibrium concentration (mol L^{-1})		y		$0.100 + 2y$

Let us try approximating $(0.100 + 2y)$ as 0.100. Then,

$$[Ca^{2+}][F^-]^2 = K_{sp}$$
$$y(0.100)^2 = 3.9 \times 10^{-11}$$
$$y = 3.9 \times 10^{-9} \text{ mol L}^{-1}$$

We now verify that

$$2\,y = 7.8 \times 10^{-9} \text{ mol L}^{-1} \ll 0.100 \text{ mol L}^{-1}$$

so that our assumption was justified. The gram solubility of CaF_2 in the 0.100 M NaF solution is

$$3.9 \times 10^{-9} \frac{\text{mol } CaF_2}{L} \times \left(\frac{78.1 \text{ g } CaF_2}{1 \text{ mol } CaF_2}\right) = 3.0 \times 10^{-7} \frac{\text{g } CaF_2}{L}.$$

Thus, the solubility in this case is reduced by a factor of 50,000 from that in pure water (see Example 9–2).

Exercise

The K_{sp} of thallium(I) iodate ($TlIO_3$) is 3.1×10^{-6} at 25°C. Determine the molar solubility of $TlIO_3$ in 0.050 M KIO_3.

Answer: 6.2×10^{-5} mol L^{-1}.

9-4 THE EFFECTS OF pH ON SOLUBILITY

Some solids are only weakly soluble in water but dissolve readily in more acidic solutions. The sulfides of copper and nickel in ores, for example, can be brought into solution with strong acids, a fact that aids greatly in the separation and recovery of these valuable metals in elemental form. A less desirable effect of pH on solubility can be seen dramatically in the damage caused by acid rain to buildings and monuments (Fig. 9–6). Both marble and limestone are made up of small crystals of

Figure 9-6 The calcium carbonate in marble and limestone is very slightly soluble in neutral water. Its solubility is much larger in acidic water. Objects carved of these materials dissolve relatively rapidly in areas where rain, snow, and fog are acidified from air pollution. Shown here is the damage to a gargoyle at the Lincoln Cathedral in England between 1910 (left) and 1984 (right).

calcite ($CaCO_3$), which has a slow rate of dissolution in "natural" rainwater (with a pH of about 5.6) but which dissolves much more rapidly as the rainwater becomes more acidic. The reaction

$$CaCO_3(s) + H_3O^+(aq) \longrightarrow Ca^{2+}(aq) + HCO_3^-(aq) + H_2O(\ell)$$

causes this increase. In this section, we examine the role of pH in solubility.

Solubility of Hydroxides

A direct effect of pH on solubility occurs with the numerous metal hydroxides. The concentration of OH^- appears explicitly in the expression for the solubility product of such compounds. Thus, for the dissolution of $Zn(OH)_2(s)$,

$$Zn(OH)_2(s) \rightleftharpoons Zn^{2+}(aq) + 2\,OH^-(aq)$$

we have

$$[Zn^{2+}][OH^-]^2 = K_{sp} = 4.5 \times 10^{-17}$$

As the solution is made more acidic, the concentration of hydroxide ion decreases, causing the equilibrium concentration of $Zn^{2+}(aq)$ ion to increase. Zinc hydroxide is thus more soluble in acidic solution than in pure water.

EXAMPLE 9–7

Compute the solubility of $Zn(OH)_2$ in pure water at 25°C (in g L^{-1}) and compare it with that in a solution buffered at pH 6.0.

Solution

In pure water, we have the usual solubility-product calculation

$$[Zn^{2+}] = y$$
$$[OH^-] = 2y$$
$$y(2y)^2 = 4y^3 = K_{sp} = 4.5 \times 10^{-17}$$
$$y = 2.2 \times 10^{-6} \text{ mol } L^{-1} = [Zn^{2+}]$$

so the solubility is 2.2×10^{-6} mol per liter, or 2.2×10^{-4} g per liter. The saturated solution of $Zn(OH)_2$ is found from

$$[OH^-] = 2y = 4.5 \times 10^{-6} \text{ M}$$

to have a pH of 8.65.

In the second case, we must assume that the solution is buffered sufficiently that the pH remains 6.0 after dissolution of the zinc hydroxide. Then pOH = 8.0 and

$$[OH^-] = 1.0 \times 10^{-8} \text{ M}$$
$$[Zn^{2+}] = \frac{K_{sp}}{[OH^-]^2} = \frac{4.5 \times 10^{-17}}{(1.0 \times 10^{-8})^2} = 0.45 \text{ M}$$

so that 0.45 mol per liter or 45 g per liter, should dissolve. When ionic concentrations are this high, the simple form of the solubility expression is likely to break down, but the qualitative conclusion remains valid: $Zn(OH)_2$ is far more soluble at pH 6.0 than in pure water.

> **Exercise**
> Estimate the molar solubility of $Fe(OH)_3$ in a solution that is buffered to a pH of 2.9.
>
> **Answer:** 2.2×10^{-3} mol L^{-1}.

Solubility of Salts of Bases

Metal hydroxides can be described as salts of a strong base, the hydroxide ion. Salts in which the anion is a different weak or strong base are also affected by pH in their solubility. For example, in a saturated solution of calcium fluoride the solubility equilibrium is

$$CaF_2(s) \rightleftharpoons Ca^{2+}(aq) + 2\ F^-(aq) \qquad K_{sp} = 3.9 \times 10^{-11}$$

If the solution is made more acidic, then some of the fluoride ion reacts with hydronium ion through

$$H_3O^+(aq) + F^-(aq) \rightleftharpoons HF(aq) + H_2O(\ell)$$

Because this reaction is just the reverse of the acid ionization of HF, its equilibrium constant is the reciprocal of K_a for HF, or $1/(3.5 \times 10^{-4}) = 2.9 \times 10^3$. As acid is added, the concentration of fluoride ion is reduced, so the calcium ion concentration must increase in order to maintain the solubility-product equilibrium for CaF_2. As a result, the solubility of calcium fluoride increases in acidic solution. The same applies to other ionic substances in which the anion is either a weak or a strong base. By contrast, the solubility of a salt like AgCl is only very slightly affected by a decrease in pH. The reason is that HCl is a strong acid, so that Cl^-, its conjugate base, is ineffective as a base. That is, the reaction

$$H_3O^+(aq) + Cl^-(aq) \longrightarrow HCl(aq) + H_2O(\ell)$$

occurs to a negligible extent in acidic solution.

Selective Precipitation of Ions

The analysis of a mixture of elements usually requires separation of the mixture into its components. One way to do this is to exploit the differences in the solubilities of compounds of the elements. To separate barium ion from calcium ion, for example, a search is made for compounds of these different elements that (1) have a common anion and (2) have widely different solubilities. As Table 9–1 shows, the fluorides BaF_2 and CaF_2 are two such compounds because barium fluoride is much more soluble in water than calcium fluoride. The solubility equilibria for the two are:

$$BaF_2(s) \rightleftharpoons Ba^{2+}(aq) + 2\ F^-(aq) \qquad K_{sp} = 1.7 \times 10^{-6}$$
$$CaF_2(s) \rightleftharpoons Ca^{2+}(aq) + 2\ F^-(aq) \qquad K_{sp} = 3.9 \times 10^{-11}$$

Consider a solution that is 0.10 M in both Ba^{2+} and Ca^{2+}. Is it possible to add enough F^- to precipitate almost all the Ca^{2+} ions but to leave all the Ba^{2+} ions in solution? If so, an essentially quantitative separation of the two species can be achieved.

In order for Ba^{2+} to remain in solution, the reaction quotient for the first reaction must remain smaller than K_{sp}: $Q = [Ba^{2+}][F^-]^2 < K_{sp}$. Inserting K_{sp} and the concentration of Ba^{2+} into this expression and rearranging gives

$$[F^-]^2 < \frac{K_{sp}}{[Ba^{2+}]} = \frac{1.7 \times 10^{-6}}{0.10} = 1.7 \times 10^{-5}$$

Taking the square root on both sides of the inequality gives

$$[F^-] < 4.1 \times 10^{-3} \text{ mol L}^{-1}$$

Hence, as long as the fluoride ion concentration remains smaller than 0.0041 mol L^{-1}, no BaF$_2$ should precipitate. In order to reduce the *calcium* concentration in solution as far as possible (that is, to precipitate out as much calcium fluoride as possible), the fluoride ion concentration should be kept as high as possible while not exceeding 0.0041 mol L^{-1}. If exactly this concentration of F$^-(aq)$ is chosen, then at equilibrium,

$$[Ca^{2+}] = \frac{K_{sp}}{[F^-]^2} = \frac{3.9 \times 10^{-11}}{0.0041^2} = 2.3 \times 10^{-6} \text{ mol L}^{-1}$$

At an equilibrium concentration of F$^-$ equal to 0.0041 mol L^{-1}, the concentration of Ca^{2+} is reduced from 0.10 to 2.3×10^{-6} mol L^{-1}. All the Ba^{2+} ions remain in solution, but almost all the Ca^{2+} ions are at the bottom of the container in the calcium fluoride precipitate. A good separation has been achieved.

Metal Sulfides

Controlling the solubility of metal sulfides has important applications. According to Table 4–1, most metal sulfides are "insoluble" in water. This means only a very small amount of a compound such as ZnS(s) dissolves in water. Although it is tempting to write the resulting equilibrium as

$$ZnS(s) \rightleftharpoons Zn^{2+}(aq) + S^{2-}(aq) \qquad K_{sp} = ?$$

in analogy to the equilibria for other weakly soluble salts, this is not correct. The S^{2-} ion, like the O^{2-} ion, is a very strong base (stronger than OH$^-$) and reacts almost quantitatively with water:

$$S^{2-}(aq) + H_2O(\ell) \longrightarrow HS^-(aq) + OH^-(aq) \quad \text{(reaction essentially complete)}$$

Recent research has shown that K_b for this reaction is on the order of 10^5, so virtually no S^{2-} is present in aqueous solution. A better representation of the dissolution reaction for zinc sulfide is found by adding the two preceding equations:

$$ZnS(s) + H_2O(\ell) \rightleftharpoons Zn^{2+}(aq) + OH^-(aq) + HS^-(aq)$$

for which the equilibrium expression is

$$[Zn^{2+}][OH^-][HS^-] = K = 2 \times 10^{-25}$$

Table 9–2 gives the equilibrium constants for the comparable dissolution reactions of some other metal sulfides. In all of these reactions, as the pH decreases (as the solution becomes more acidic), the concentration of OH$^-$ decreases. Simultaneously, the concentration of HS$^-$ decreases as the equilibrium

$$HS^-(aq) + H_3O^+(aq) \rightleftharpoons H_2S(aq) + H_2O(\ell)$$

is shifted to the right by the additional H$_3$O$^+$. If both [OH$^-$] and [HS$^-$] decrease, then the concentration of the metal ion must increase in order to maintain a constant value for the product [M^{2+}][OH$^-$][HS$^-$]. As a result, the solubilities of metal sulfides increase as the pH of the solution decreases.

The quantitative calculation of how the solubility of metal sulfides depends on pH requires treating several simultaneous equilibria, as the following example and exercise illustrate. As shown, a single adjustment of the pH (maintained by the use

Table 9–2
Equilibrium Constants for the Dissolution of Some Metal Sulfides at 25°C

Metal Sulfide	K^a
CuS	5×10^{-37}
PbS	3×10^{-28}
CdS	7×10^{-28}
Sns	9×10^{-27}
ZnS	2×10^{-25}
FeS	5×10^{-19}
MnS	3×10^{-14}

$^a K$ is the equilibrium constant for the reaction MS(s) + H$_2$O(ℓ) $\rightleftharpoons$ M^{2+}(aq) + OH$^-$(aq) + HS$^-$(aq)

of a buffer) keeps the ions of one metal (Fe^{2+} in the example) entirely in solution and causes those of a second (Cd^{2+} in the exercise) to precipitate almost entirely as metal sulfide. Setting the right pH is crucial for separating metal ions in qualitative analysis, as is described in Section 9–6.

EXAMPLE 9–8

In saturated solutions of $H_2S(aq)$ at 25°C, the concentration of $H_2S(aq)$ equals 0.1 mol L^{-1}. Calculate the molar solubility of $FeS(s)$ in such a solution, if it is buffered at pH 2.0.

Solution

If the pH is 2.0, then the pOH is 12.0 and

$$[OH^-] = 1 \times 10^{-12} \text{ mol L}^{-1}$$

The H_2S reacts with water as a weak acid according to the equilibrium

$$H_2S(aq) + H_2O(\ell) \rightleftharpoons H_3O^+(aq) + HS^-(aq) \qquad K_a = 9.1 \times 10^{-8}$$

with the K_a taken from Table 8–2. Substituting $[H_2S] = 0.1$ mol L^{-1} for a saturated solution of H_2S and $[H_3O^+] = 1 \times 10^{-2}$ mol L^{-1} (at pH 2.0) into the equilibrium expression for this reaction gives

$$\frac{[H_3O^+][HS^-]}{[H_2S]} = \frac{(1 \times 10^{-2})[HS^-]}{0.1} = K_a = 9.1 \times 10^{-8}$$

Solving for $[HS^-]$ gives

$$[HS^-] = 9 \times 10^{-7} \text{ mol L}^{-1}$$

For the reaction

$$FeS(s) + H_2O(\ell) \rightleftharpoons Fe^{2+}(aq) + HS^-(aq) + OH^-(aq)$$

the equilibrium equation is

$$[Fe^{2+}][HS^-][OH^-] = K = 5 \times 10^{-19}$$

where K comes from Table 9–2. Substituting the values of $[HS^-]$ and $[OH^-]$ and solving for $[Fe^{2+}]$ gives

$$[Fe^{2+}](9 \times 10^{-7})(1 \times 10^{-12}) = 5 \times 10^{-19}$$
$$[Fe^{2+}] = \boxed{0.6 \text{ mol L}^{-1}}$$

Exercise

Calculate the solubility of $CdS(s)$ in a solution that is saturated with H_2S and buffered at a pH of 2.0.

Answer: 8×10^{-10} mol L^{-1}.

9–5   COMPLEX IONS AND SOLUBILITY

In a **complex ion,** a central metal ion is bound to one or more ligand molecules or ions that are usually capable of independent existence in solution. In $[Ag(NH_3)_2]^+$, for example, a central Ag^+ ion is **coordinated** to the lone-pair electrons of two different ammonia molecules. In $[Cu(NH_3)_4]^{2+}$, each Cu^{2+} ion is surrounded by *four* ammonia molecules (Fig. 9–7). We now examine the effects of the formation of

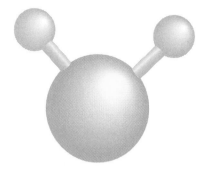

Hydrogen sulfide (H_2S) is a poisonous, foul-smelling gas. It arises in nature in the gases vented from volcanoes and also from the action of bacteria; it is most familiar as the odor of rotten eggs. When dissolved in water, it gives a weak acid—hydrosulfuric acid.

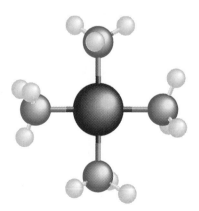

Figure 9–7 The structures of the complex ions $Ag(NH_3)_2^+$ (a) and $Cu(NH_3)_4^{2+}$(b).

• Coordination complexes are introduced in Section 3–8.

complex ions on equilibria in aqueous solutions. The structure and bonding of complex ions are considered in Chapter 19.

Complex-Ion Equilibria

Complex ions form by the stepwise addition of ligands to metal ions. When a soluble silver salt is dissolved in an aqueous solution containing ammonia, the following equilibria are established:

$$Ag^+(aq) + NH_3(aq) \rightleftharpoons Ag(NH_3)^+(aq) \qquad \frac{[Ag(NH_3)^+]}{[Ag^+][NH_3]} = K_1 = 2.1 \times 10^3$$

$$Ag(NH_3)^+(aq) + NH_3(aq) \rightleftharpoons Ag(NH_3)_2^+(aq) \qquad \frac{[Ag(NH_3)_2^+]}{[Ag(NH_3)^+][NH_3]} =$$
$$K_2 = 8.2 \times 10^3$$

Because K_1 and K_2 for the formation of silver–ammonia complexes are both large, we anticipate that if a silver salt is dissolved in water that contains an excess of ammonia, most of the silver ends up as the complex ion $[Ag(NH_3)_2]^+$ at equilibrium. Let us work out a quantitative example.

EXAMPLE 9–9

Suppose that 0.100 mol of $AgNO_3$ is mixed with 1.00 L of a 1.00 M solution of NH_3. Calculate the concentrations of the Ag^+, $Ag(NH_3)^+$, and $Ag(NH_3)_2^+$ ions present at equilibrium.

Solution

We assume that most of the Ag^+ is present as $Ag(NH_3)_2^+$ (this will be checked later). This starting assumption suggests that we conceive of the two-stage complexation reaction as going to completion (forming $Ag(NH_3)_2^+$ until *all* of the Ag^+ is tied up), and then reversing to a small extent to form first $Ag(NH_3)^+$ and then Ag^+. The imagined state after complete reaction but before dissociation of the complex is then the "initial" state, for the purposes of equilibrium calculations. The concentration of $Ag(NH_3)_2^+$ in this state is

$$[Ag(NH_3)_2^+]_{(init)} = 0.100 \text{ mol L}^{-1}$$

Because each silver ion reacts with two ammonia molecules in reaching this state, the "initial" ammonia concentration is

$$[NH_3]_{(init)} = 1.00 - (2 \times 0.100) = 0.80 \text{ mol L}^{-1}$$

The two stages of the dissociation of the $Ag(NH_3)_2^+$ ion are the reverses of the complexation reactions, and their equilibrium constants are the reciprocals of K_2 and K_1, respectively

$$Ag(NH_3)_2^+(aq) \rightleftharpoons Ag(NH_3)^+(aq) + NH_3(aq)$$
$$\frac{[Ag(NH_3)^+][NH_3]}{[Ag(NH_3)_2^+]} = \frac{1}{K_2} = \frac{1}{8.2 \times 10^3}$$

$$Ag(NH_3)^+(aq) \rightleftharpoons Ag^+(aq) + NH_3(aq)$$
$$\frac{[Ag^+][NH_3]}{[Ag(NH_3)^+]} = \frac{1}{K_1} = \frac{1}{2.1 \times 10^3}$$

Suppose that y moles per liter of $Ag(NH_3)_2^+$ dissociates at equilibrium according to the first equation. The table of changes to reach equilibrium is then

	$Ag(NH_3)_2^+(aq)$	$\rightleftharpoons$	$Ag(NH_3)^+(aq)$	$+$	$NH_3(aq)$
Initial concentration (mol L^{-1})	0.100		0		0.80
Change in concentration (mol L^{-1})	$-y$		$+y$		$+y$
Equilibrium concentration (mol L^{-1})	$0.100 - y$		y		$0.80 + y$

and the equilibrium expression becomes

$$\frac{y(0.80 + y)}{0.10 - y} = \frac{1}{K_2} = \frac{1}{8.2 \times 10^3}$$

$$y = 1.5 \times 10^{-5} \text{ mol L}^{-1} = [Ag(NH_3)^+]$$

We can then calculate the concentration of uncomplexed Ag^+ ions from the equilibrium equation for the second step of the dissociation of the complex ion:

$$\frac{[Ag^+][NH_3]}{[Ag(NH_3)^+]} = \frac{1}{K_1} = \frac{1}{2.1 \times 10^3}$$

$$\frac{[Ag^+](0.80)}{1.5 \times 10^{-5}} = \frac{1}{2.1 \times 10^3}$$

$$[Ag^+] = 9 \times 10^{-9} \text{ mol L}^{-1}$$

It is clear that most of the silver present is tied up in the $Ag(NH_3)_2^+$ complex, whose concentration remains 0.100 mol L^{-1}.

Exercise

Suppose that 0.25 mol of $CuSO_4$ is added to 1.0 L of 2.4 M NH_3 solution. The equilibrium constants for the addition of four successive NH_3 ligands to Cu^{2+} are 1×10^4, 2×10^3, 5×10^2, and 9×10^1. Determine the concentrations of $Cu(NH_3)_4^{2+}$, $Cu(NH_3)_3^{2+}$, $Cu(NH_3)_2^{2+}$, $Cu(NH_3)^{2+}$, and Cu^{2+} at equilibrium.

Answer: 0.25, 0.0020, 2.8×10^{-6}, 9.9×10^{-10}, and 7×10^{-14} M.

The calculation in Example 9–9 was simplified by two circumstances: an excess of ligand (ammonia) was present in solution, and the successive ligand association constants K_1 and K_2 were large compared to 1. The latter means that $1/K_2$ and $1/K_1$ are small compared to 1 and allows us to treat the equilibria one at a time, just as we treat the steps in the ionization of a polyprotic acid separately. When the association constants are not large (as often happens), a full calculation of the coupled simultaneous equilibria is necessary.

The formation of coordination complexes can have a large effect on the solubility of a compound in water. Silver bromide is very poorly soluble in water,

$$AgBr(s) \rightleftharpoons Ag^+(aq) + Br^-(aq) \qquad K_{sp} = 7.7 \times 10^{-13}$$

but addition of thiosulfate ion ($S_2O_3^{2-}$) to water in contact with solid AgBr allows the complex ion $Ag(S_2O_3)_2^{3-}$ to form.

$$AgBr(s) + 2\,S_2O_3^{2-}(aq) \rightleftharpoons Ag(S_2O_3)_2^{3-}(aq) + Br^-(aq) \qquad K = 22$$

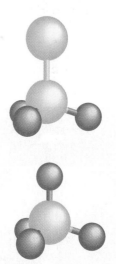

The thiosulfate ion ($S_2O_3^{2-}$) is related to the sulfate ion (SO_4^{2-}) by the replacement of one oxygen atom by one sulfur atom. It is prepared, however, by the reaction of elemental sulfur with sulfite ion (SO_3^{2-}).

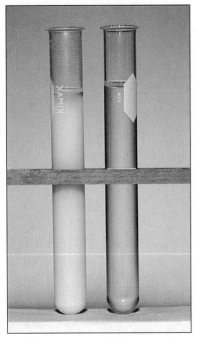

Figure 9–8 An illustration of the effect of complex-ion formation on solubility. Both test tubes contain 2.0 g of AgBr, but the one on the right also contains dissolved thiosulfate ($S_2O_3^{2-}$) ion, which forms a complex ion with Ag^+. Almost none of the white solid AgBr has dissolved in pure water, but all of it has dissolved in the solution containing thiosulfate.

This greatly increases the solubility of the silver bromide, which goes into solution (Fig. 9–8). The formation of this complex ion is an important step in the development of photographic images; thiosulfate ion is a component of the "fixer" that brings silver bromide into solution from the unexposed portion of the film.

Another interesting effect of complex ions on solubilities is illustrated by the addition of iodide ion to a solution containing mercury(II) ion. After a moderate amount of iodide ion has been added, an orange precipitate forms (Fig. 9–9), according to the equation.

$$Hg^{2+}(aq) + 2\,I^-(aq) \rightleftharpoons HgI_2(s)$$

With further addition of iodide ion, however, the orange solid redissolves because complex ions form:

$$HgI_2(s) + I^-(aq) \rightleftharpoons HgI_3^-(aq)$$
$$HgI_3^-(aq) + I^-(aq) \rightleftharpoons HgI_4^{2-}(aq)$$

In the same way, silver chloride dissolves in a concentrated solution of sodium chloride by forming $AgCl_2^-$ complex ions.

Hydrolysis and Amphoterism of Complex Ions

When dissolved in water, many metal ions act as weak acids, decreasing the pH of the solution (see Section 8–4). The iron(III) ion is an example. Each dissolved Fe^{3+} ion is aquated by six water molecules, leading to a complex ion $Fe(H_2O)_6^{3+}$. This complex ion can act as a Brønsted–Lowry acid, donating hydrogen ions to the solvent, water:

$$\underset{\text{acid}_1}{Fe(H_2O)_6^{3+}(aq)} + \underset{\text{base}_2}{H_2O(\ell)} \rightleftharpoons \underset{\text{acid}_2}{H_3O^+(aq)} + \underset{\text{base}_1}{Fe(H_2O)_5OH^{2+}(aq)}$$

$$\frac{[H_3O^+][Fe(H_2O)_5OH^{2+}]}{[Fe(H_2O)_6^{3+}]} = K_a = 7.7 \times 10^{-3}$$

Metal-ion hydrolysis therefore fits into the general scheme of the Brønsted–Lowry acid–base reaction. Table 9–3 gives values of the pH for 0.1 M solutions of several metal ions. Those that form strong complexes with hydroxide ion have low pH, whereas those that do not form such complexes give neutral solutions (pH 7).

EXAMPLE 9–10

Calculate the pH of a solution that is 0.100 M in $Fe(NO_3)_3$.

Solution

The iron(III) is present as $Fe(H_2O)_6^{3+}$, which reacts as a weak acid

$$Fe(H_2O)_6^{3+}(aq) + H_2O(\ell) \rightleftharpoons Fe(H_2O)_5OH^{2+}(aq) + H_3O^+(aq)$$

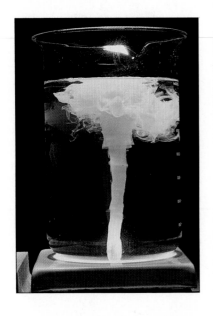

Figure 9–9 The "orange tornado" is a striking demonstration of the effects of complexation on solubility. A solution is prepared with an excess of $I^-(aq)$ over $Hg^{2+}(aq)$, so that the Hg^{2+} is complexed as $HgI_3^-(aq)$ and $HgI_4^{2-}(aq)$. A magnetic stirrer is used to create a vortex in the solution. Addition of a solution containing Hg^{2+} down the center of the vortex then causes the orange solid HgI_2 to form (by the reaction $HgI_4^{2-}(aq) + Hg^{2+}(aq) \longrightarrow 2\,HgI_2(s)$, for example) in the layer at the edges of the vortex, giving a tornado effect.

with K_a equal to 7.7×10^{-3}. If y moles per liter of $[Fe(H_2O)_6]^{3+}$ reacts, then (neglecting the autoionization of water)

$$[H_3O^+] = [Fe(H_2O)_5OH^{2+}] = y$$
$$[Fe(H_2O)_6^{3+}] = 0.100 - y$$

The equilibrium expression has the form

$$\frac{y^2}{0.100 - y} = 7.7 \times 10^{-3}$$

$$y = 2.4 \times 10^{-2} \text{ mol L}^{-1} = [H_3O^+]$$

so the pH is 1.62. Solutions of salts of iron(III) are strongly acidic.

Table 9-3 pH of 0.1 M Aqueous Metal Nitrate Solutions at 25°C	
Metal Nitrate	**pH**
$Fe(NO_3)_3$	1.6
$Pb(NO_3)_2$	3.6
$Cu(NO_3)_2$	4.0
$Zn(NO_3)_2$	5.3
$Ca(NO_3)_2$	6.7
$NaNO_3$	7.0

Exercise

The acidity constant for the $Cu(H_2O)_4^{2+}$ ion is $K_a = 1 \times 10^{-7}$. Calculate the pH of a 0.100 M solution of $Cu(NO_3)_2$.

Answer: 4.0.

Another acceptable way to write the reaction that makes iron(III) solutions acidic is

$$Fe^{3+}(aq) + 2 H_2O(\ell) \rightleftharpoons H_3O^+(aq) + FeOH^{2+}(aq)$$

in which the specific mention of the six waters of hydration is omitted. The $FeOH^{2+}(aq)$ complex ions are brown, but $Fe^{3+}(aq)$ ions are almost colorless. This reaction occurs to an extent sufficient to make a solution of $Fe(NO_3)_3$ in water pale brown. When strong acid is added, the equilibrium is driven back to the left, and the color fades.

Various cations behave differently as water ligands are replaced by hydroxide ions in an increasingly basic solution. A particularly interesting example is Zn^{2+} ion. It forms a series of hydroxo complex ions:

$$Zn^{2+}(aq) + OH^-(aq) \rightleftharpoons ZnOH^+(aq)$$
$$ZnOH^+(aq) + OH^-(aq) \rightleftharpoons Zn(OH)_2(s)$$
$$Zn(OH)_2(s) + OH^-(aq) \rightleftharpoons Zn(OH)_3^-(aq)$$
$$Zn(OH)_3^-(aq) + OH^-(aq) \rightleftharpoons Zn(OH)_4^{2-}(aq)$$

In Brønsted–Lowry language, the polyprotic Brønsted–Lowry acid, $Zn(H_2O)_4^{2+}(aq)$, donates hydrogen ions in succession to make all the product species. The second product, $Zn(OH)_2$, is amphoteric (capable of reacting as both an acid and a base) and is only slightly soluble in pure water (its K_{sp} is only 1.9×10^{-17}). If enough acid is added to solid $Zn(OH)_2$, OH^- ligands are removed, forming the soluble Zn^{2+} ion; if enough base is added, OH^- ligands are added to form the soluble $Zn(OH)_4^{2-}$ (zincate) ion. Thus, $Zn(OH)_2$ is soluble in strongly acidic *or* strongly basic solutions, but it is only slightly soluble at intermediate pH values (Fig. 9–10). This amphoterism can be used to separate Zn^{2+} from other cations that do not share the property. For example, Mg^{2+} ion adds a maximum of only two OH^- ions to form $Mg(OH)_2$, a sparingly soluble hydroxide. If a mixture of Mg^{2+} and Zn^{2+} ions is

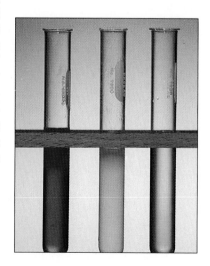

Figure 9–10 Zinc hydroxide is insoluble in water as is evident from its milky appearance (*center*), but it dissolves readily in acid (*left*) and base (*right*). The indicator used is bromocresol red, which is red in acid and yellow in base.

made sufficiently basic, the Mg^{2+} precipitates as $Mg(OH)_2$, but the zinc remains in solution as $Zn(OH)_4^{2-}$, allowing the two to be separated.

9-6 CONTROLLING SOLUBILITY IN QUALITATIVE ANALYSIS

The goal of the **qualitative analysis** of inorganic substances in aqueous solution is to determine the identities of ions that are present, without concern for the exact amounts of each. In this section, we show how the principles of solubility control underlie a scheme for the qualitative analysis of the common metal cations in aqueous solution. The same principles can be applied to identify anions. Qualitative analysis impresses some as a musty 19th-century technique that has no place in modern chemistry. No one today (or even 100 years ago) outside of introductory chemistry laboratories carries out a full qualitative analysis of the type we present. So why do it? The answer is simple: learning qualitative analysis is the best way to study the chemistry of ions in solution, which is an immensely practical subject in modern research and technology. In this section, we can only begin to relate the basic ideas behind qualitative analysis to the principles of solubility presented earlier in the chapter. An entire book would be required to do justice to the subject. It would be even more instructive to spend a few weeks in a laboratory exploring the qualitative behavior of inorganic compounds.

We begin by recalling the three basic tools that are used to separate cations from one another.

1. **Different solubilities of cations with the same anion.** For example, if HCl is added to a solution containing both $Ag^+(aq)$ and $Ni^{2+}(aq)$ ions, AgCl precipitates but $NiCl_2$ does not.

2. **pH control of relative solubilities.** Strong acids can bring insoluble hydroxides, carbonates, and many sulfides back into solution, and pH control by means of buffers at intermediate pH can separate metal ions from one another.

3. **Complex-ion formation that brings a metal ion back into solution.** The dissolution of AgCl in an aqueous solution of ammonia, for example, takes place through the formation of the $Ag(NH_3)_2^+$ complex ion.

Once a cation or a group of cations has precipitated from a solution, it must be physically separated from the residual solution and the cations that remain in it. This is accomplished either by **filtration,** in which a filter catches the solid but lets the solution pass through, or by **centrifugation,** in which rapid spinning of the solution causes the solid to collect in a compact form at the bottom of a tube, allowing the **supernatant solution** above it simply to be poured off or withdrawn with a pipet.

In any qualitative analysis scheme, the first step is to separate the ions into groups that contain a smaller number of ions (typically from three to seven). A different attack can then be applied to each group to separate and identify its ions, after which a confirmatory test, specific to the ion in question, can be carried out to verify the tentative conclusion from the separation process. One important point to remember is that group separations work only when done in sequence. If a group of ions is only partially removed by precipitation in an early step, it can wreak havoc in subsequent separations. Reactions must therefore be brought as close to completion as possible. A second point is that reactants added to a solution remain there unless they are chemically removed. The procedure must be viewed as a whole, therefore, and if sulfide, for example, will interfere with a subsequent chloride reaction, the sulfide must be quantitatively removed before proceeding. In spite of its

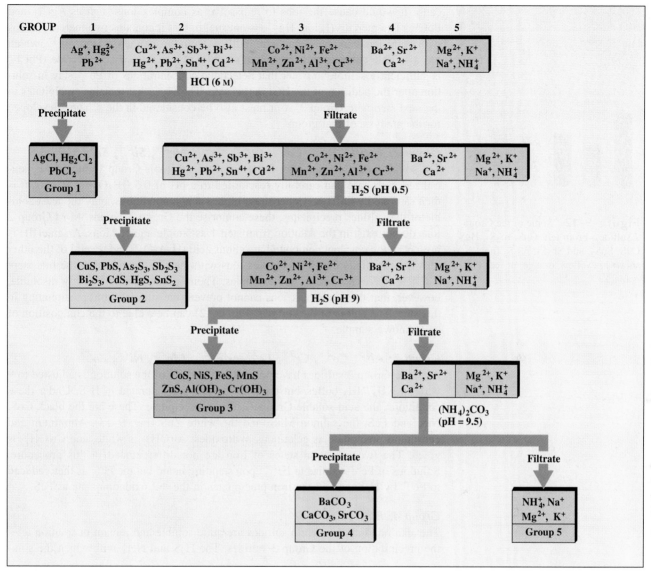

Figure 9–11 A general outline of a qualitative analysis scheme. It is important to avoid confusing the groups in the periodic table with the five groups in this scheme of qualitative analysis. The two are unrelated.

name, qualitative analysis relies strongly on quantitative separations and on the precise control of reaction conditions.

An Overview of Group Separations

The first step in the scheme of qualitative analysis that we present here is to separate the cations that may be present in solution into five groups, each of which contains a smaller number of ions (Fig. 9–11). Each group can then be subjected to further analysis.

Group 1: Ag^+, Hg_2^{2+}, Pb^{2+}

Addition of 6 M HCl to the mixture of cations listed in Figure 9–11 causes the precipitation of three insoluble white chlorides: AgCl, $PbCl_2$, and Hg_2Cl_2. All other cations remain in solution. A slight excess of chloride ion is used to force out as many of these three ions from solution as possible, but a large excess is avoided be-

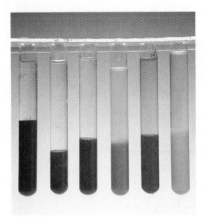

Figure 9–12 Some of the Group 2 sulfides. From left: CuS, Bi_2S_3, HgS, CdS, SnS_2, and Sb_2S_3. As_2S_3 (not shown) is yellow, orange, or red; PbS is black.

cause it would cause the ions to redissolve as complex ions such as $AgCl_2^-$ and $PbCl_3^-$. The mercury(I) ion, Hg_2^{2+}, is unusual in that it contains two mercury atoms with a covalent bond between them. It differs from the mercury(II) ion, Hg^{2+}, which has a soluble chloride and remains in solution at this stage. Lead chloride ($PbCl_2$) is sufficiently soluble in water that detectable concentrations of Pb^{2+} stay in solution after the addition of the HCl in this step. The Pb^{2+} that remains precipitates in the next step with Group 2, so it must also be considered in the scheme developed for the analysis of that group.

Group 2: As(III), Bi^{3+}, Cd^{2+}, Cu^{2+}, Pb^{2+}, Hg^{2+}, Sb^{3+}, Sn^{4+}

The solution that remains after filtration of the insoluble Group 1 chlorides is neutralized with base and carefully reacidified to a pH of 0.5 ($[H_3O^+] = 0.3$ M). It is then saturated with H_2S. Under these highly acidic conditions, only the least soluble of the sulfides precipitates; these comprise the Group 2 cations. Most Group 2 ions do not exist in the solution from step 1 as simple aquated ions. Arsenic(III) is present in the covalent compound arsenious acid (H_3AsO_3), and several of the other ions form partially covalent complex ions with the Cl^- ion added in the first step: $BiCl_4^-$, $SbCl_4^-$, and $SnCl_6^{2-}$ are examples. Their sulfides are sufficiently insoluble, however, that these complex ions cannot prevent the sulfides from precipitating in this step. The colors of the sulfides (Fig. 9–12) can be a clue to the composition of an unknown sample.

Group 3: Al^{3+}, Co^{2+}, Cr^{3+}, Fe^{3+} or Fe^{2+}, Mn^{2+}, Ni^{2+}, Zn^{2+}

After the Group 2 sulfides have precipitated, the pH of the solution is adjusted to 9 with an NH_4^+/NH_3 buffer, but the solution remains saturated in H_2S. Under these conditions, the acid-soluble Group 3 sulfides precipitate. These are the black CoS, NiS, and FeS, the salmon MnS, and the white ZnS (Fig. 9–13). Aluminum and chromium precipitate as gelatinous hydroxides: $Al(OH)_3$ is white, and $Cr(OH)_3$ is green. The two oxidation states of iron are not differentiated in this procedure. Solutions of Fe^{2+} oxidize to Fe^{3+} upon standing in air, but the Fe^{3+} is then reduced to Fe^{2+} by H_2S, and all the iron precipitates in the +2 oxidation state as FeS.

Group 4: Ba^{2+}, Ca^{2+}, Sr^{2+}

The alkali and alkaline earth sulfides are quite soluble and remain in solution after the precipitation of the Group 3 sulfides. The H_2S and NH_4^+ with which the solu-

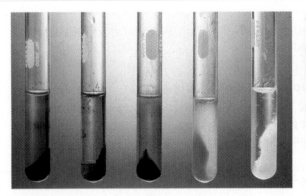

Figure 9–13 Five Group 3 cations precipitate as sulfides when a solution saturated in H_2S is made basic. From left: NiS, CoS, FeS, MnS, and ZnS. Two precipitate as hydroxides: $Al(OH)_3$ and $Cr(OH)_3$.

tion has been treated must now be removed to prevent them from interfering with subsequent steps. The H_2S is removed by adding a few drops of 12 M HCl and evaporating almost to dryness. Then some 16 M HNO_3 is added, and strong heating removes the ammonium ion by decomposing the NH_4Cl and NH_4NO_3 to gaseous products. The resulting solid is then redissolved; it contains Group 4 and Group 5 cations. When the pH is adjusted to 9.5 and $(NH_4)_2CO_3$ is added, the three Group 4 cations precipitate as carbonates. Because $MgCO_3$ is more soluble than the heavier alkaline-earth carbonates, it remains in solution under the usual conditions of qualitative analysis.

Group 5: K^+, Mg^{2+}, Na^+, NH_4^+

The alkali-metal cations remain in solution after this last step, together with Mg^{2+} and some ammonium ion, which is often also included in a qualitative-analysis unknown. This solution is not used, however, because it is almost impossible to avoid contamination with sodium ions in the reagents added in the earlier separation steps; of course, ammonium ion has been deliberately added as well. To identify Group 5 cations, several portions of the original unknown are subjected to tests that are specific to the particular ion in question, even in the presence of all the other cations in the qualitative-analysis procedure. For example, a positive identification of sodium can be made by adding a solution of magnesium uranyl acetate and observing the precipitation of $NaMg(UO_2)_3(CH_3COO)_9 \cdot 9H_2O(s)$, one of the few salts of sodium that is only slightly soluble.

EXAMPLE 9–11

An unknown solution may contain ions from Groups 1 through 5. Addition of 6 M HCl gives a white precipitate. Saturation of the solution separated from this precipitate with H_2S at pH 0.5 gives no further precipitation, but when the pH is increased to 9 a precipitate forms. After this precipitate is filtered off, the H_2S and NH_4^+ ion are removed from the solution and the pH is adjusted to 9.5. No precipitate then forms upon addition of $(NH_4)_2CO_3$. What ions may be present in the original solution from each qualitative-analysis group?

Solution

Group 1: Either Ag^+ or Hg_2^{2+} (or both) could be present, giving a white chloride precipitate, but Pb^{2+} cannot be present, because it would show up in Group 2 as well.

Group 2: None present (would have precipitated as sulfides in 0.3 M H_3O^+).

Group 3: Co^{2+}, Ni^{2+}, Fe^{2+}, Fe^{3+}, Mn^{2+}, Zn^{2+}, Al^{3+}, and Cr^{3+} may all be present because a precipitate forms at pH 9.

Group 4: None present (would have a carbonate precipitate with $(NH_4)_2CO_3$).

Group 5: Any Group 5 cation could be present.

Exercise

A solution may contain ions from Groups 1 through 5. Neither the addition of 6 M HCl nor the addition of H_2S at a pH of 0.5 causes a precipitate. What ions may be present from each group?

Answer: No Group 1 or Group 2 ions are present. Any ions from Groups 3 through 5 may be present.

Analysis for Group 1

The systematic analysis of each of the five groups is not described here. Instead, we discuss only Group 1, the members of which precipitate as chloride salts in acid solution (Fig. 9–14). The members of this group, Ag^+, Hg_2^{2+}, and Pb^{2+}, differ sufficiently in their chemistry that their separation is straightforward.

Lead(II) chloride ($PbCl_2$) is the most soluble of the three chlorides, and its solubility increases strongly with temperature. When hot water (near 100°C) is added to the mixed chloride precipitate, the lead chloride dissolves, accompanied by only small amounts of the other two chlorides. If the solution is centrifuged while still hot, the solid contains the silver and mercury(I) chlorides, and the lead chloride is in the supernatant liquid. A good test for lead ion is to add a soluble chromate salt to the supernatant liquid after pouring it off from the solids. If lead ion is present, it precipitates as bright yellow lead chromate (Fig. 9–15):

$$Pb^{2+}(aq) + CrO_4^{2-}(aq) \rightleftharpoons PbCrO_4(s)$$

To separate and identify the silver and mercury(I) that may be present, the precipitate that remains after extraction with hot water is treated with a concentrated

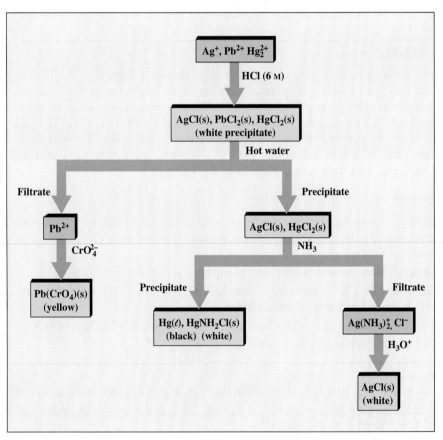

Figure 9–14 A scheme for the qualitative analysis of the Group 1 cations. Precipitates at each stage are shown in blue boxes.

solution of ammonia. Silver chloride dissolves in aqueous ammonia by complex ion formation:

$$AgCl(s) + 2\ NH_3(aq) \rightleftharpoons Ag(NH_3)_2^+(aq) + Cl^-(aq)$$

The equilibrium constant for this reaction is not very large ($K = 3 \times 10^{-3}$), so the dissolution is helped by using the highest possible concentration of ammonia (15 M) to drive the equilibrium to the right.

By contrast, any mercury(I) chloride that may be present in the chloride precipitate does not dissolve when treated with a solution of ammonia, but reacts according to the equation

$$Hg_2Cl_2(s) + 2\ NH_3(aq) \longrightarrow HgNH_2Cl(s) + Hg(\ell) + NH_4^+(aq) + Cl^-(aq)$$

to leave a dark, solid residue. A gray to black residue after the addition of the ammonia is a positive indication for Hg_2^{2+}.

The final step is to centrifuge the solution to remove the mercury residue, if it has formed, and to test the supernatant liquid for silver. Any silver that is present is tied up in the silver–ammonia complex. As the solution is made acidic through addition of 6 M HNO_3, the equilibrium

$$NH_3(aq) + H_3O^+(aq) \rightleftharpoons NH_4^+(aq) + H_2O(\ell)$$

is shifted to the right. Because this reduces the concentration of the ammonia, some dissociation of the metal-ion complex takes place to replenish it; that is, the equilibrium

$$Ag(NH_3)_2^+(aq) \rightleftharpoons Ag^+(aq) + 2\ NH_3(aq)$$

is also shifted to the right. The silver ion released reacts with the Cl^- ion still in the solution to precipitate silver chloride:

$$Ag^+(aq) + Cl^-(aq) \rightleftharpoons AgCl(s)$$

so that the appearance of a white solid at this stage confirms that Ag^+ is present. Several simultaneous equilibria are involved in this process; an overall equation is obtained by adding twice the first equation to the second and third equations, giving

$$Ag(NH_3)_2^+(aq) + Cl^-(aq) + 2\ H_3O^+(aq) \rightleftharpoons AgCl(s) + 2\ NH_4^+(aq) + 2\ H_2O(\ell)$$

Figure 9–15 Brilliant yellow lead(II) chromate $PbCrO_4$ (the artist's pigment chrome yellow) precipitates when a solution of a soluble chromate salt is added to one of a soluble lead(II) salt.

EXAMPLE 9–12

A solution containing one or more Group 1 cations is treated with hydrochloric acid and gives a white precipitate. Hot water is added, and some insoluble white solid is separated from the hot solution. Treatment of this hot solution with aqueous K_2CrO_4 yields no precipitate. The white solid is then treated with 15 M ammonia and dissolves completely. What metal ions are present in the unknown?

Solution

No Pb^{2+} is present because no yellow lead chromate precipitates. The white solid contains no Hg_2Cl_2 because it dissolves completely in 15 M ammonia (Hg_2Cl_2 would give an insoluble gray to black residue when treated with aqueous ammonia). Because *some* Group I ion must be present in the white solid in the first place, Ag^+ is present. This conclusion can be checked by treating the white solid with concentrated aqueous ammonia; it should dissolve.

Table 9–4
Colors in the Flame Test

Element	Color
Li	Red
Na	Yellow
K	Violet
Rb	Red
Cs	Blue
Ca	Brick red
Cu	Green
Sr	Crimson
Ba	Green

Exercise

A solution containing one or more Group 1 cations is treated with 6 M hydrochloric acid that is heated to 95°C. It gives a white precipitate that, after cooling to room temperature, dissolves completely in 15 M ammonia. What ions are definitely present and what ions are definitely absent from the solution?

Answer: Ag^+ is definitely present; Hg_2^{2+} and Pb^{2+} are definitely absent.

Other Methods of Analysis

The procedure we have outlined for qualitative analysis is a "classical" one, based on solubility differences under particular conditions of pH with particular reactants present. Although portions of this procedure are still in use in modern qualitative analysis, other methods are more widely used today.

One of the most powerful methods of qualitative analysis is **emission spectroscopy,** in which a sample is heated to a temperature high enough to vaporize it, and the wavelengths of the light emitted are recorded. As we shall see in Chapter 16, each element emits light of a characteristic set of wavelengths. The light emitted can be analyzed (automatically, using modern spectrometers) to find which elements are present. The method is useful for all the elements in the periodic table. Emission spectroscopy is a 20th-century elaboration of a much earlier technique for qualitative analysis—the flame test. When certain elements are placed in a flame, they give off characteristic colors of visible light (Table 9–4 and Fig. 9–16). Visual observation then provides a confirmatory test for certain elements: crimson for strontium, green for barium, and pale violet for potassium are examples. Sodium gives a very intense yellow flame in this test.

A second method for qualitative analysis uses **ion-exchange chromatography.** In this type of column chromatography (see Section 7–7), a column is packed with an ion-exchange resin, an insoluble substance of high molar mass that possesses negatively or positively charged sites that can interact with ions in solution. **Cation-exchange resins** have negatively charged sites that can attract cations. **Anion-ex-**

Lithium   Sodium   Potassium

Figure 9–16 Flame tests.

CHEMISTRY IN YOUR LIFE

Lead All Around You

Like several other heavy metals, lead is quite toxic. Part of the reason appears in Table 9–2. The small K_{sp} for the dissolution of PbS(s) means that lead(II) has a strong affinity for sulfide ions. This affinity extends to sulfur in the -2 oxidation state in other compounds as well, including many important proteins in the body that contain —SH groups. Once mobilized into a living system, lead bonds at —SH groups in proteins, disrupting their normal functions. Sufferers of lead poisoning experience neurological symptoms, anemia, liver and kidney damage, and hearing loss. Death can ensue. Lead poisoning in children slows mental and physical development and can cause learning and behavior problems.

Lead contamination is encountered throughout the home—in paint, house dust, and drinking water. Paints manufactured before 1960 often used white lead (Pb$_3$(OH)$_2$(CO$_3$)$_2$) as a pigment. Remnants of such paint, which contained as much as 50% lead by mass, are still flaking from the walls of older buildings. Lead continues to be used in paint, but in much reduced amounts, because it makes the paint last longer. The U. S. Environmental Protection Agency allows up to 600 ppm (parts per million) by mass of lead in household paint.

Window sills often gather dust that contains lead from crumbling paint and from the emissions of smelters, incinerators, and foundries. Such dust can get on children's hands and toys and then into their mouths through normal behavior, such as thumb-sucking. This makes children particularly susceptible to lead poisoning.

Lead leaches into the water supply from old lead pipes in city systems and home plumbing and from the lead-containing solder used to connect pipes. Because the solubility of many lead compounds increases with temperature (see Fig. 9–2), health agencies urge against drinking hot water straight from the tap. They advise that hot water for cooking, baking, or brewing be drawn cold from the tap and heated on a stove.

Metallic lead (sometimes called "black lead") is a malleable gray-black substance that is coated with an oxide that rubs off easily. People who work with metallic lead dirty their hands with the oxide and may absorb lead

The claim on this bean can lid appears because joints in cans are sometimes closed with lead-containing solder.

through their skin. When lead-containing objects (for example, fishing sinkers, sash weights, certain plumbing supplies) are handled, skin contact should be minimized and the hands washed thoroughly before food is prepared or eaten.

Recent findings suggest that lead poisoning is harmful at blood levels once thought safe. Lower IQ scores, slower development, and more attention deficits than normal have been observed in children with concentrations in their blood as low as 10^{-4} g L^{-1}. Lead concentrations in the blood exceeding about 7×10^{-4} g L^{-1} constitute a medical emergency. The treatment is "chelation therapy." A solution of a ligand having a very strong affinity for Pb^{2+} ion is administered. The lead is tied up in the form of a complex ion, and, by Le Chatelier's principle, the concentration of lead in the blood and tissues is reduced. Successful therapy naturally requires dealing with the ill effects of the complexing agent as well.

change resins have positively charged sites that attract negative ions such as Cl$^-$ or negatively charged complex ions such as [AsCl$_4$]$^-$. As a solution under analysis percolates down such a column, the ions in it displace whatever species occupy the sites on the resin with varying degrees of effectiveness. They spend different amounts of time on the column and can be separated and identified by confirmatory tests.

SUMMARY

9–1 When a solution is in equilibrium with undissolved solute, then it is **saturated** with respect to that solute. In an **unsaturated** solution, the concentration of the solute is less than its equilibrium value; in a **supersaturated** solution, the concentration of the solute exceeds its equilibrium value.

9–2 The **solubility-product** expression is used for equilibrium calculations on the saturated aqueous solutions of slightly soluble ionic solids. It relates the concentrations of the cations and anions that make up a dissolved salt to an equilibrium constant, K_{sp}, and can be used to calculate the amount of the salt that dissolves provided that no side-reactions are present.

9–3 A solubility-product expression can be used to predict whether precipitation occurs when two solutions are mixed: if the initial reaction quotient $Q_{(init)}$ exceeds K_{sp}, precipitation is favored, but if $Q_{(init)}$ is less than K_{sp}, none can occur. The K_{sp} expression can also be used in quantitative calculations of the **common-ion effect:** the reduction in solubility of a salt in a solution in which one of its component ions is already present.

9–4 The pH of the solution often affects the solubility of salts. Salts in which the anion is a base, such as metal hydroxides and sulfides, increase in solubility as acid is added. The anion is removed by reaction with hydronium ion, and more of the salt dissolves to maintain the solubility-product equilibrium. The pH dependence of the solubility of metal sulfides is particularly useful in permitting the quantitative separation of different metal ions in aqueous solution through pH control.

9–5 The formation of complex ions increases the solubility of salts by tying up a product of the solubility equilibrium with ligands. Certain metal ions that are strongly hydrated act as Brønsted–Lowry acids by donating hydrogen ions to surrounding water molecules and thereby reducing the pH of the solution.

9–6 The guiding principle behind the **qualitative analysis** for ions in aqueous solution is the separation of different ionic species from one another on the basis of their different solubilities, followed by specific tests to determine which are present in a sample. The separation relies on the different solubilities of different metal ions with the same anion, on pH control of solubility, and on the use of ligands to bring metal ions selectively into solution in the form of complex ions. Many different schemes are possible, depending on the ions that may be present in a sample. Modern qualitative analysis emphasizes two techniques in addition to the classical methods of solution chemistry: **emission spectroscopy** and **ion-exchange chromatography.**

PROBLEMS

Note: Answers to blue-numbered problems are given in Appendix F. Problems that are more challenging are indicated with asterisks.

The Nature of Solubility Equilibria

1. Copper selenate precipitates from solution as $CuSeO_4 \cdot 5H_2O$. When heated above 150°C, all the water of crystallization is lost. What mass of solid results from heating 64.8 g of copper selenate pentahydrate above 150°C?

2. Gypsum has the formula $CaSO_4 \cdot 2H_2O$. Plaster of paris has the chemical formula $CaSO_4 \cdot \frac{1}{2}H_2O$. In making wall plaster, water is added to plaster of paris, and the mixture then hardens into solid gypsum. How much water (in liters, at a density of 1.00 kg L^{-1}) should be added to a 25.0-kg sack of plaster of paris to turn it into gypsum, assuming no loss due to evaporation?

3. When heated above 420°C, the mineral bieberite loses its wa-

ter of crystallization to become cobalt(II) sulfate. When 38.4 g of bieberite is heated in this way, 21.2 g of $CoSO_4$ results. Give the chemical formula of bieberite.

4. A1.00-g sample of magnesium sulfate is dissolved in water, and the water is then evaporated away until the residue is bone dry. If the temperature of the water is kept between 48°C and 69°C, the solid that remains weighs 1.898 g. If the experiment is repeated with the temperature held between 69°C and 100°C, however, the solid has a mass of 1.150 g. Determine how many waters of crystallization per $MgSO_4$ are in each of these two solids.

5. The following graph shows the solubility of KBr in water in units of grams of KBr per 100 g of H_2O. If 80 g of KBr is added to 100 g of water at 10°C and the mixture is heated slowly, at what temperature does the last KBr dissolve?

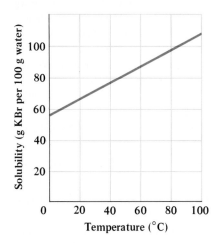

6. Figure 9–2 shows the solubility of $AgNO_3$ in water in units of moles of $AgNO_3$ per kg of H_2O. If 255 g of $AgNO_3$ is mixed with 100 g of water at 95°C and cooled slowly, at what temperature does the solution become saturated?

7. The dissolution of ammonium chloride in water is quite endothermic. Predict the effect of increased temperature on the solubility of ammonium chloride.

8. Codeine has the molecular formula $C_{18}H_{21}NO_3$. It is used in cough medicines but has sometimes been abused, leading to addiction. It is soluble in water to the following extents: 1.00 g per 120 mL of water at room temperature, 1.00 g per 60 mL of water at 80°C. Compute the molal solubility (the solubility in moles of codeine per kg of water) at both temperatures, taking the density of water to be fixed at 1.00 g cm^{-3}. Is the dissolution of codeine in water endothermic or exothermic?

The Solubility of Salts

9. A solution consists of 0.156 mol of $SrBr_2$ dissolved in 4.15 L of solution. Assuming that dissociation is complete:
   (a) Determine the molarity of $SrBr_2$ in this solution.
   (b) Determine the molarity of Br^- in this solution.

10. A total of 5.0 g of $La_2(SO_4)_3$ is dissolved in 40.0 L of water. Determine the total concentration (in mol L^{-1}) of charged particles (negative ions and positive ions) in the solution, assuming complete dissociation.

11. (See Example 9–1.) Write balanced chemical equations to describe the dissolution of the following substances in water. Also, write an expression describing the equilibrium established between undissolved solid and the resulting solution.
    (a) $SrI_2(s)$ (c) $Ca_3(PO_4)_2(s)$
    (b) $I_2(s)$ (d) $BaCrO_4(s)$

12. (See Example 9–1.) Write balanced chemical equations to describe the dissolution of the following substances in water. Also, write an expression describing the equilibrium established between undissolved solid and the resulting solution.
    (a) $Ag_2SO_4(s)$ (c) $Cu(OH)_2(s)$
    (b) $Br_2(\ell)$ (d) $CaF_2(s)$

13. (See Example 9–1.) Iron(III) sulfate ($Fe_2(SO_4)_3$) is a yellow compound that is used as a coagulant in water treatment. Write a balanced chemical equation and a solubility-product expression for its dissolution in water.

14. (See Example 9–1.) Lead antimonate ($Pb_3(SbO_4)_2$) is used as an orange pigment in oil-base paints and in glazes. Write a balanced chemical equation and a solubility-product expression for its dissolution in water.

15. The solubility-product constant of mercury(I) iodide is 1.2×10^{-28} at 25°C. Estimate the concentration of Hg_2^{2+} and I^- in equilibrium with solid Hg_2I_2.

16. The solubility-product constant of $BaSO_4$ is 1.1×10^{-10} at 25°C. Estimate the concentration of Ba^{2+} and SO_4^{2-} in equilibrium with solid $BaSO_4$ at 25°C.

17. (See Example 9–2.) Thallium(I) iodate ($TlIO_3$) is only slightly soluble in water. Its K_{sp} at 25°C is 3.07×10^{-6}. Estimate the solubility of thallium(I) iodate in water in units of grams per 100.0 mL of solution.

18. (See Example 9–2.) Thallium thiocyanate (TlSCN) is only slightly soluble in water. Its K_{sp} at 25°C is 1.82×10^{-4}. Estimate the solubility of thallium thiocyanate in units of grams per 100.0 mL of solution.

19. The solubility-product constant for $PbCl_2$ is given in Table 9–1.
    (a) Estimate the concentration of Pb^{2+} and Cl^- in equilibrium with solid $PbCl_2$ at 25°C.
    (b) Is the concentration of Pb^{2+} in saturated $PbCl_2$ (aq) low enough that the solution might legally be flushed down the drain? (The Environmental Protection Agency says water released into sewers must contain less than 50 ppm Pb by mass.) (Note: "ppm" stands for "parts per million"; the water must contain less than 50 g of Pb per 10^6 g H_2O, which is equivalent to 0.05 g of Pb per liter of water.

20. The solubility-product constant for Hg_2Cl_2 is 2×10^{-18}.
    (a) Estimate the concentration of Hg_2^{2+} and Cl^- in equilibrium with solid Hg_2Cl_2 at 25°C. List the assumptions made in this calculation.

(b) Express the concentration of Hg_2^{2+} in saturated $Hg_2Cl_2(aq)$ in units of ppb. (The unit "ppb" stands for "parts per billion." A concentration of 5 ppb Hg_2^{2+} indicates that the solution contains 5 g Hg_2^{2+} per 10^9 g H_2O.

21. (See Example 9–3.) The solubility of silver chromate (Ag_2CrO_4) in 500 mL of water at 25°C is 0.0129 g. Estimate its solubility-product constant.

22. (See Example 9–3.) At 25°C, 400 mL of water can dissolve 0.00896 g of lead iodate ($Pb(IO_3)_2$). Estimate K_{sp} for lead iodate.

23. (See Example 9–3.) At 100°C, water dissolves 1.8×10^{-2} g of AgCl per liter. Estimate the K_{sp} of AgCl at this temperature.

24. (See Example 9–3.) A mass of 0.017 g of silver dichromate ($Ag_2Cr_2O_7$) dissolves in 300 mL of water at 25°C. Estimate the solubility-product constant K_{sp} of silver dichromate.

Precipitation and the Solubility Product

25. A solution of barium chromate ($BaCrO_4$) is prepared by dissolving 6.3×10^{-3} g of this yellow solid in 1.00 L of hot water. Does solid barium chromate precipitate upon cooling to 25°C, according to the solubility-product expression? Explain.

26. A solution is prepared by dissolving 0.090 g of PbI_2 in 1.00 L of hot water and cooling the solution to 25°C. Does solid precipitate result from this process, according to the solubility-product expression? Explain.

27. (See Example 9–4.) A solution is prepared by mixing 250.0 mL of 2.0×10^{-3} M $Ce(NO_3)_3$ and 150.0 mL of 1.0×10^{-2} M KIO_3 at 25°C. Determine whether $Ce(IO_3)_3(s)$ ($K_{sp} = 1.9 \times 10^{-10}$) tends to precipitate from this mixture.

28. (See Example 9–4.) 100.0 mL of a 0.0010 M $CaCl_2$ solution is added to 50.0 mL of a 6.0×10^{-5} M NaF solution at 25°C. Determine whether $CaF_2(s)$ tends to precipitate from this mixture.

29. (See Example 9–4.) Suppose that 140 mL of 0.0010 M $Sr(NO_3)_2$ is mixed with enough 0.0050 M NaF to make 1000 mL of solution. Does $SrF_2(s)$ precipitate at equilibrium? Explain.

30. (See Example 9–4.) Suppose that 140 mL of 0.0010 M $Sr(NO_3)_2$ is mixed with enough 0.0050 M Na_2CrO_4 to make 1000 mL of solution. Does $SrCrO_4(s)$ ($K_{sp} = 3.6 \times 10^{-5}$) precipitate at equilibrium? Explain.

31. (See Example 9–5.) 50.0 mL of a 0.0500 M solution of $Pb(NO_3)_2$ is mixed with 40.0 mL of a 0.200 M solution of $NaIO_3$ at 25°C. Calculate the $[Pb^{2+}]$ and $[IO_3^-]$ when the mixture comes to equilibrium. At this temperature, K_{sp} for $Pb(IO_3)_2$ is 2.6×10^{-13}.

32. (See Example 9–5.) Silver iodide (AgI) is used in place of silver chloride for the fastest photographic film because it is more sensitive to light and so can form an image in a very short exposure time. A silver-iodide emulsion is prepared by adding 6.60 L of 0.10 M NaI solution to 1.50 L of 0.080 M $AgNO_3$ solution at 25°C. Calculate the concentration of silver ion remaining in solution when the mixture comes to equilibrium, and its chemical amount relative to the amount present initially.

33. (See Example 9–5.) When 50.0 mL of 0.100 M $AgNO_3$ and 30.0 mL of 0.0600 M Na_2CrO_4 are mixed, a precipitate of silver chromate (Ag_2CrO_4) is formed. Calculate the $[Ag^+]$ and $[CrO_4^{2-}]$ remaining in solution at equilibrium at 25° C.

34. (See Example 9–5.) When 40.0 mL of 0.0800 M $Sr(NO_3)_2$ and 80.0 mL of 0.0500 M KF are mixed, a precipitate of strontium fluoride (SrF_2) is formed. The solubility product K_{sp} of strontium fluoride in water at 25°C is 2.8×10^{-9}. Calculate the $[Sr^{2+}]$ and F^-] remaining in solution at equilibrium, neglecting all other interactions between ions.

35. (See Example 9–6.) Calculate the solubility (in mol L^{-1}) of $CaF_2(s)$ at 25°C in a 0.040 M aqueous solution of NaF.

36. (See Example 9–6.) Calculate the mass of AgCl that can dissolve in 100 mL of 0.150 M NaCl solution.

37. (See Example 9–6.) The solubility product of nickel(II) hydroxide ($Ni(OH)_2$) at 25°C is $K_{sp} = 1.6 \times 10^{-16}$.
    (a) Calculate the molar solubility of $Ni(OH)_2$ in pure water at 25°C.
    (b) Calculate the molar solubility of $Ni(OH)_2$ in 0.100 M NaOH.

38. (See Example 9–6.) Silver arsenate (Ag_3AsO_4) is a slightly soluble salt
$$Ag_3AsO_4(s) \rightleftharpoons 3\,Ag^+(aq) + AsO_4^{3-}(aq)$$
$$K_{sp} = 1.0 \times 10^{-22} \text{ at } 25°C.$$
    (a) Estimate the molar solubility of silver arsenate in pure water at 25°C.
    (b) Estimate the molar solubility of silver arsenate in 0.100 M $AgNO_3$.

The Effects of pH on Solubility

39. (See Example 9–7.) Compare the molar solubility of AgOH in pure water to that in a solution buffered at pH 7. Note the difference between the two: when AgOH is dissolved in pure water, the pH does not remain at 7.

40. (See Example 9–7.) Compare the molar solubility of $Mg(OH)_2$ in pure water to that in a solution buffered at pH 9.

41. To what pH must a solution be adjusted to maintain the solubility of $Fe(OH)_2$ at 0.001 mol L^{-1}?

42. To what pH must a solution be adjusted to maintain the solubility of AgOH at 0.001 mol L^{-1}?

43. An aqueous solution at 25°C is 0.10 M in both Mg^{2+} and Pb^{2+} ions. We wish to separate the two metal ions by taking advantage of the different solubilities of their oxalates, MgC_2O_4 and PbC_2O_4.
    (a) What is the highest possible oxalate-ion concentration such that only one solid oxalate salt is present at equilibrium? Which ion is present in the solid—Mg^{2+} or Pb^{2+}?
    (b) What fraction of the less-soluble ion still remains in solution under the conditions of part (a)?

44. An aqueous solution at 25°C is 0.10 M in Pb^{2+} and 0.50 M in Ag^+ ions. We wish to separate the two by taking advantage of the different solubilities of their chlorides, $PbCl_2$ and $AgCl$.

(a) What is the highest possible chloride-ion concentration such that only one solid chloride salt is present at equilibrium? Which ion is present in the solid—Pb^{2+} or Ag^+?

(b) What fraction of the less-soluble ion still remains in solution under the conditions of part (a)?

45. (See Example 9–8.) Calculate the $[Zn^{2+}]$ in a solution that is in equilibrium with $ZnS(s)$ and in which $[H_3O^+] = 1.0 \times 10^{-5}$ M and $[H_2S] = 0.10$ M.

46. (See Example 9–8.) Calculate the $[Cu^{2+}]$ in a solution that is in equilibrium with $CuS(s)$ and in which $[H_3O^+] = 1.0 \times 10^{-3}$ M and $[H_2S] = 0.10$ M.

47. What is the highest pH at which 0.10 M Fe^{2+} remains entirely in a solution that is saturated with H_2S, at a concentration of $[H_2S] = 0.10$ M? At this pH, what would be the concentration of $Pb^{2+}(aq)$ in a similar solution in equilibrium with solid PbS?

48. What is the highest pH at which 0.050 M Mn^{2+} remains entirely in a solution that is saturated with H_2S, at a concentration of $[H_2S] = 0.10$ M? At this pH, what would be the concentration of $Cd^{2+}(aq)$ in a similar solution in equilibrium with solid CdS?

Complex Ions and Solubility

49. (See Example 9–9.) 0.10 mol of $Cu(NO_3)_2$ and 1.5 mol of NH_3 are dissolved in water and diluted to a total volume of 1.00 L. Calculate the concentrations of $Cu(NH_3)_4^{2+}$ and of Cu^{2+} at equilibrium, using the equilibrium constants given in Exercise 9–9.

50. (See Example 9–9.) The equilibrium constant for the formation of the $TlCl_4^-$ complex ion from Tl^{3+} and Cl^- is 1×10^{18}. Suppose that 0.080 mol of $Tl(NO_3)_3$ is dissolved in 1.00 L of a 0.50 M solution of NaCl. Calculate the concentration at equilibrium of $TlCl_4^-$ and of Tl^{3+}. Assume that almost all the Tl^{3+} is bound up as $TlCl_4^-$ at equilibrium.

51. The pH of a 0.2 M solution of $CuSO_4$ is 4.0. Write chemical equations to explain why a solution of this salt is neither basic (from the reaction of $SO_4^{2-}(aq)$ with water) nor neutral, but acidic.

52. Is a 0.05 M solution of $FeCl_3$ acidic, basic, or neutral? Explain your answer by writing chemical equations to describe any reactions taking place.

53. (See Example 9–10.) The acidity constant for $Co(H_2O)_6^{2+}(aq)$ is 3×10^{-10}. Calculate the pH of a 0.10 M solution of $Co(NO_3)_2$.

54. (See Example 9–10.) The acidity constant for $Fe(H_2O)_6^{2+}(aq)$ is 3×10^{-6}. Calculate the pH of a 0.10 M solution of $Fe(NO_3)_2$, and compare it with the pH of the corresponding iron(III) nitrate solution from Table 9–3.

55. A 0.15 M aqueous solution of the chloride salt of the complex ion $Pt(NH_3)_4^{2+}$ is found to be weakly acidic with a pH

of 4.92. This is initially puzzling because the Cl^- ion in water is not acidic and NH_3 in water is *basic*, not acidic. Finally, it is suggested that the $Pt(NH_3)_4^{2+}$ ion as a group donates hydrogen ions. Compute the K_a of this acid, assuming that just one hydrogen ion is donated.

56. The pH of a 0.10 M solution of $Ni(NO_3)_2$ is 5.0. Calculate the acidity constant of $Ni(H_2O)_6^{2+}(aq)$.

Controlling Solubility in Qualitative Analysis

57. The cations in an aqueous solution that contains 0.100 M $Hg_2(NO_3)_2$ and 0.0500 M $Pb(NO_3)_2$ are to be separated by taking advantage of the difference in the solubilities of their iodides. What should be the concentration of iodide for the best separation? In the "best" separation, one of the cations remains entirely in solution, and the other precipitates as fully as possible.

58. The cations in an aqueous solution that contains 0.150 M $Ba(NO_3)_2$ and 0.0800 M $Ca(NO_3)_2$ are to be separated by taking advantage of the difference in the solubilities of their sulfates. What should be the concentration of sulfate ion for the best separation?

59. Cesium nitrate and barium nitrate both form colorless crystals with similar densities. Suggest two ways of deciding the correct chemical identity of a crystal known to be one or the other.

60. A colorless crystal is known to be either potassium nitrate or calcium nitrate. Suggest two ways of deciding which is the correct chemical composition.

61. An aqueous solution contains one or more of the following types of metal ions (and no others): Hg^{2+}, Ni^{2+}, Sr^{2+}. Outline a procedure to determine which are present and which are not.

62. An aqueous solution contains one or more of the following types of metal ions (and no others): Ag^+, Cr^{3+}, Hg_2^{2+}. Outline a procedure to determine which are present and which are not.

63. (See Example 9–11.) A solution containing ions from the scheme outlined in Section 9–6 is treated with hydrochloric acid, and no precipitate forms. The pH is adjusted to 9, and the solution is then saturated with H_2S, causing a precipitate to appear. State and justify any conclusions that can be drawn about which qualitative-analysis groups are present and which are not.

64. (See Example 9–11.) A solution containing ions from the scheme outlined in Section 9–6 is treated with hydrochloric acid, and a white precipitate forms. This precipitate is removed, and the remaining solution is saturated with H_2S at pH 9. No precipitate forms. After removing ammonium chloride and H_2S, the pH of the solution is adjusted to 9.5, and ammonium carbonate is added, at which point another white precipitate forms. State and justify any conclusions that can be drawn about which groups are present and which are not present.

Additional Problems

65. Write a chemical equation for the dissolution of mercury(I) chloride in water, and give its solubility-product expression.

66. Magnesium ammonium phosphate has the formula $MgNH_4PO_4 \cdot 6H_2O$. It is only slightly soluble in water (its K_{sp} is 2.3×10^{-13}). Write a chemical equation and the corresponding equilibrium expression for the dissolution of this compound in water.

67. Soluble barium compounds are poisonous, but barium sulfate is routinely ingested as a suspended solid in a "barium cocktail" to improve the contrast in x-ray images. Calculate the concentration of dissolved barium per liter of water in equilibrium with solid barium sulfate.

68. The K_{sp} of SrF_2 is 2.8×10^{-9} and that of $SrSO_4$ is 2.8×10^{-7}. Although the K_{sp} of SrF_2 is 100 times smaller than that of $SrSO_4$, its molar and gram solubility are both larger than those of $SrSO_4$! Compute the solubilities of both compounds, and write a brief explanation of the apparent paradox.

69. The solubilities of the carbonates of the Group II elements in water at 20°C are

$CaCO_3$	14×10^{-3} g L^{-1}
$SrCO_3$	10×10^{-3} g L^{-1}
$BaCO_3$	17×10^{-3} g L^{-1}

Note how the solubility goes down and then up again moving down the column of Group II elements in the periodic table; there is no clear periodic trend in these data.

(a) Compute the molar solubilities of the three compounds, and comment on the presence or absence of a periodic trend in the results.

(b) Compute K_{sp} for each of the compounds from these data. Why do the values for K_{sp} computed here differ somewhat from those in Table 9–1? Is there a periodic trend in the K_{sp}'s?

70. A saturated aqueous solution of silver perchlorate ($AgClO_4$) contains 84.8% by mass $AgClO_4$, but a saturated solution of $AgClO_4$ in 60% aqueous perchloric acid contains only 5.63% by mass $AgClO_4$. Explain this large difference, using chemical equations.

71. The K_{sp} of $KClO_4$ at 25°C equals 1.07×10^{-2}. Estimate the molar solubility of $KClO_4$ at that temperature in
(a) pure water.
(b) 0.075 M aqueous K_2SO_4.
(c) 0.075 M aqueous $AgClO_4$.
(d) 0.075 M aqueous NaCl.

72. The concentration of calcium ion in a town's supply of drinking water is 0.0020 M. This water would be referred to as hard water (because it contains such a large concentration of Ca^{2+}). Suppose that the water is to be fluoridated by the addition of NaF for the purpose of reducing tooth decay. What is the maximum concentration of fluoride ion that can be achieved in the water before precipitation of CaF_2 begins? Can the water supply attain the concentration of fluoride ion recommended by the U. S Public Health Service, of about 5×10^{-5} M (1 mg fluorine per liter)?

73. The two solids CuBr(s) and AgBr(s) are only very slightly soluble in water. $K_{sp}(CuBr) = 4.2 \times 10^{-8}$ and $K_{sp}(AgBr) = 7.7 \times 10^{-13}$. Some CuBr(s) and AgBr(s) are both mixed into a quantity of water that is then stirred until it is saturated with respect to both solutes. Next, a small amount of KBr is added and dissolves completely. Compute the ratio of $[Cu^+]$ to $[Ag^+]$ after the system re-establishes equilibrium.

74. The two salts $BaCl_2$ and Ag_2SO_4 are both far more soluble in water than either $BaSO_4$ ($K_{sp} = 1.1 \times 10^{-10}$) or AgCl ($K_{sp} = 1.6 \times 10^{-10}$) at 25°C. Suppose that 50.0 mL of 0.040 M $BaCl_2(aq)$ is added to 50.0 mL of 0.020 M $Ag_2SO_4(aq)$. Calculate the concentrations of $SO_4^{2-}(aq)$, $Cl^-(aq)$, $Ba^{2+}(aq)$, and $Ag^+(aq)$ that remain in solution at equilibrium.

*75. The Mohr method is a technique for determining the amount of chloride ion in an unknown sample. It is based on the difference in solubility between silver chloride and silver chromate. In using this method, one adds a small amount of chromate ion to a solution with unknown chloride concentration. By measuring the volume of $AgNO_3$ added before the appearance of the red silver chromate, one can determine the amount of Cl^- originally present. Suppose a solution is 0.100 M in Cl^- and 0.00250 M in CrO_4^{2-}. If 0.100 M $AgNO_3$ solution is added dropwise, does AgCl or Ag_2CrO_4 precipitate first? When $Ag_2CrO_4(s)$ first appears, what fraction of the Cl^- that was originally present remains in solution?

76. Oxide ion, like sulfide ion, is a strong base. Write an equation for the dissolution of CaO in water, and give its equilibrium constant expression. Write the corresponding equation for the dissolution of CaO in an aqueous solution of a strong acid, and relate its equilibrium constant to the previous one.

77. (a) Only about 0.16 mg of AgBr(s) dissolves in one liter of water (this volume of solid is smaller than the head of a pin). In a solution of ammonia that contains 0.10 mol ammonia per liter of water, there are about 555 water molecules for every molecule of ammonia. In this solution, however, more than 400 times as much AgBr dissolves (68 mg). Explain how such a tiny change in the composition of the solution can have such a large effect on the solubility of AgBr.

(b) Explain why AgCl is soluble in 6 M NH_3 while AgI is not.

78. Potassium ion reacts with "cryptands," which are large organic molecules that enfold potassium ions in a pocket. Let the abbreviation "crypt" represent a certain cryptand. The cryptand and the potassium ion (K^+) interact according to the equilibrium

$$crypt + K^+ \rightleftharpoons Kcrypt^+$$

At 25°C, the equilibrium constant for this reaction is 3.1×10^{11}. Compute the equilibrium concentration of K^+ if 0.10 mol of K^+ and 0.10 mol of crypt are mixed in 1.0 L of solvent at 25°C.

79. The organic compound "18-crown-6" binds with the alkali metals in aqueous solution by wrapping around and enfolding the ion. It presents a niche that nicely accommodates the K^+ ion but is too small for the Rb^+ ion and too large for the Na^+ ion. The values of the equilibrium constants tell the story

$$Na^+(aq) + 18\text{-crown-}6(aq) \rightleftharpoons Na\text{-crown}^+(aq) \quad K = 6.6$$
$$K^+(aq) + 18\text{-crown-}6(aq) \rightleftharpoons K\text{-crown}^+(aq) \quad K = 111.6$$
$$Rb^+(aq) + 18\text{-crown-}6(aq) \rightleftharpoons Rb\text{-crown}^+(aq) \quad K = 36$$

An aqueous solution is initially 0.0080 M in 18-crown-6(aq) and also 0.0080 M in $K^+(aq)$. Compute the equilibrium concentration of free K^+. ("Free" means not tied up with the 18-crown-6.) Compute the concentration of free Na^+ if the solution contains 0.0080 M $Na^+(aq)$ instead of $K^+(aq)$.

80. (a) Use the data in Table 9–3 to estimate the K_a of $Pb^{2+}(aq)$ ion at 25°C.

(b) Estimate the pH of a solution that is 0.20 M in $Pb(NO_3)_2$.

81. An aqueous solution of $K_2[Pt(OH)_6]$ has a pH greater than 7. Explain this fact by writing an equation showing the $Pt(OH)_6^{2-}$ ion acting as a Brønsted–Lowry base and accepting a hydrogen ion from water.

CUMULATIVE PROBLEM

Carbonate Minerals

The carbonates are among the most abundant and important minerals in the earth's crust. When these minerals come into contact with fresh water or seawater, solubility equilibria are established that greatly affect the chemistry of these natural waters. Calcium carbonate ($CaCO_3$), the most important natural carbonate, makes up limestone and other rocks such as marble. Other carbonate minerals include dolomite ($CaMg(CO_3)_2$) and magnesite ($MgCO_3$). These compounds are sufficiently soluble that their solutions are nonideal, so calculations based on solubility-product expressions have only approximate validity.

(a) The rare mineral nesquehonite contains $MgCO_3$ together with water of crystallization. A sample containing 21.7 g of nesquehonite is acidified and heated, and the volume of $CO_2(g)$ produced is measured to be 3.51 L at STP. Assuming that all the carbonate has reacted to form CO_2, give the chemical formula for nesquehonite.

(b) Write a chemical equation and a solubility-product expression for the dissolution of dolomite in water.

(c) In a sufficiently basic solution, the carbonate ion does not react significantly with water to form hydrogen carbonate ion. Calculate the solubility (in grams per liter) of limestone (calcium carbonate) in a 0.10 M solution of sodium hydroxide. Use the K_{sp} from Table 9–1.

(d) A solution of Na_2CO_3 is also strongly basic and has a CO_3^{2-} concentration of 0.10 M. What is the gram solubility of limestone in this solution? Compare your answer with that to part (c).

(e) In a mountain lake having a pH of 8.1, the total concentration of carbonate species, $[CO_3^{2-}] + [HCO_3^-]$, is measured to be 9.6×10^{-4} M, and the concentration of Ca^{2+} is 3.8×10^{-4} M. Calculate the concentration of CO_3^{2-} in this lake, using $K_a = 4.8 \times 10^{-11}$ for the acidity constant of HCO_3^-. Is the lake unsaturated, saturated, or supersaturated with respect to $CaCO_3$?

(f) Does acid rainfall into the lake increase or decrease the solubility of limestone rocks in the lake's bed?

(g) Seawater contains a high concentration of Cl^- ions, which can form weak complexes with calcium ions such as $CaCl^+$. Does the presence of such complexes increase or decrease the equilibrium solubility of $CaCO_3$ in seawater?

(h) Among the major cationic impurities found in limestone rock are Fe^{2+} and Mn^{2+}. Outline a scheme of analysis to determine whether either or both of these ions are present in a given rock sample.

Carbonate minerals. Calcite (*top*) and aragonite (*middle*) are both $CaCO_3$, and smithsonite (*bottom*) is $ZnCO_3$.

10 Thermochemistry

The production of steel is made possible by the coupling of exothermic reactions to drive endothermic chemical reactions.

Heat plays a central role in chemistry. Some chemical reactions, such as the combustion of oil or natural gas in a furnace or the dramatic reaction of iron(III) oxide with aluminum in the thermite process (Fig. 10–1), release large amounts of heat. Other reactions require a supply of heat to occur. The production of calcium carbide, an intermediate in making acetylene,

$$CaO(s) + 3 C(s) \longrightarrow CaC_2(s) + CO(g)$$

requires a large supply of heat at very high temperatures in a carbon arc furnace to make the lime and coke react. The measurement and prediction of such heat effects are the subject matter of **thermochemistry.**

Heat is one of the forms (the other is work) in which energy can be transferred between a **system** and its **surroundings.** Here we use the term "system" to refer to the substance, body, reacting mixture, or even region of space upon which we focus attention; the surroundings are everything that lies outside the system. In this chapter, we develop the relationship between heat and work as manifestations of energy and use it to formulate the first law of thermodynamics, which states that energy is conserved in all processes.

This slab of steel is being cast at temperatures at which it is red hot.

10–1 HEAT

Our routine use of fire and ice gives us an intuitive notion of temperature and the nature of heat. Everyone discovers in childhood that objects of different temperature reach a common temperature when placed in contact with each other. A piece of metal held over a flame becomes red hot; when it is plunged into chilled water, the metal cools down and the water warms up until both come to the same intermediate temperature.

(a)   (b)   (c)

Figure 10–1 The thermite reaction: $2 Al(s) + Fe_2O_3(s) \longrightarrow 2 Fe + Al_2O_3$ is among the most exothermic of all reactions, liberating approximately 16 kJ of heat for every gram of aluminum that reacts. The iron(III) oxide and aluminum are mixed in finely divided form. Like the combustion of paper, the reaction requires a source of ignition, but then continues on its own.

Early scientists tried to explain these commonplace observations by describing heat as a fluid. This fluid, referred to from the Middle Ages through the 19th century as *caloric*, was assumed to flow from a hot to a cold body until the amounts in the two bodies became equal. The problem with this simple picture was that attempts to isolate caloric or establish its exact nature failed. All that was known was that caloric did not have mass; objects had the same weight whether hot or cold. A significant experiment was carried out by Benjamin Thompson, Count Rumford, in 1798. In the course of his work as military advisor to the King of Bavaria, Thompson observed that boring out cannon barrels produced heat. Heat could be produced indefinitely, as long as the cannon was rotated against a cutting tool. He concluded (correctly) that heat was not a substance contained in the brass of the cannon but was a form of motion generated by friction. Nonetheless, throughout the first half of the 19th century, the prevailing view was that heat was a fluid.

In the second half of the 19th century, the kinetic theory of gases showed that the temperature of a gas is related to the kinetic energy of the atoms or molecules of which it is composed. When two bodies are placed in contact, the collisions of the molecules cause kinetic energy to be exchanged, and eventually the temperatures are equalized. In this theory, which is the one accepted today, heat is a form of motion at the molecular level by which energy is transferred between materials. "Heat flow" is merely a convenient way of summarizing the effects of the many collisions that take place when molecules interact.

• The joule is quite a small unit for practical chemistry: the combustion of only 2×10^{-5} g of natural gas, which occupies a volume of 0.03 cm^3 at STP, produces a joule of heat.

If heat is a form in which energy is transferred, the appropriate unit for it is the same as that for energy itself, the joule. We shall measure heat in joules in our development of thermochemistry. In many chemical processes, large amounts of heat are involved, and the kilojoule ($1 \text{ kJ} = 10^3 \text{ J}$) and the megajoule ($1 \text{ MJ} = 10^6 \text{ J}$) become convenient units.

Sources of Heat

The sun has always been the most important source of energy for the earth, maintaining the temperatures at which life can exist. As we explain in Chapter 15, the source of solar energy lies in nuclear reactions that take place inside the sun and produce radiation that strikes the earth and transfers heat to it. Although the amount of heat reaching any locality varies daily with the weather and the seasons, the average over the year is fairly predictable.

• Nuclear reactions are discussed in Chapter 15.

Fuels, materials with energy stored in the bonds of molecules, allow the local production of heat. In most cases, the ultimate source of this heat is also the sun (Fig. 10–2). The combustion of wood releases as heat the chemical potential energy that was accumulated from light by the tree as it grew. Burning coal extracts energy from deposits of carbonaceous material derived from plants buried for millennia and modified by geological processes. Heat can be produced by passing an electric current through a resistor, but the source of the electrical energy itself is most likely a coal-fired generating plant. Even hydroelectric power comes from the sun, which evaporates seawater that later condenses as rain. Rainfall in the mountains gathers in streams whose flow can be harnessed to operate electrical generators. The sun provides, directly or indirectly, almost all of the heat that we use today. One important exception is the harnessing of nuclear sources of energy in the last 50 years. Nuclear fission involves splitting the nuclei of certain heavy elements and releasing large amounts of energy as heat. This energy was deposited in the nuclei as they were synthesized in the explosions of stars that existed before the sun and the solar system formed. The heavy elements were subsequently incorporated into the rocks of the earth.

Figure 10-2 One of the oldest sources of heat energy (a wood fire) and one of the newest (wind turbines). The energy for both can be traced ultimately to the sun. Trees use light energy from the sun to grow, and air currents spring into motion because the solar heating varies at different localities on the surface of the earth.

As we shall see in the rest of this chapter and in the next, heat is only one form in which energy is used. Sources of heat are also sources of work: to run machines, to lift weights, and to build buildings. They are used to store chemical energy: to make compounds with high energy content that can be released upon demand through controlled chemical reactions. All of these forms of energy are interconnected, and when we speak of an "energy crisis," we are referring to the depletion of sources of energy that can easily be harnessed in these various ways.

10-2   CALORIMETRY

When a body absorbs heat, its temperature rises. We consider here only processes at constant (usually atmospheric) pressure. The **heat capacity** C_P of a body is defined as the amount of heat required to raise its temperature by one degree at constant pressure. The greater the heat capacity of a body, the more heat must be added to it to cause a given rise in temperature. Heat capacity equals the ratio of the heat absorbed, q, to the resulting temperature change, ΔT.

$$\text{heat capacity} = C_P = \frac{q}{\Delta T}$$

The heat capacity is always positive, because q and ΔT are either both positive or both negative. A negative sign for q means that heat is evolved from the object, lowering its temperature, whereas a positive value for q means that heat is absorbed by the object, raising its temperature. The unit of heat capacity is that of energy (heat is a form of energy) divided by temperature change: the joule per kelvin (J K^{-1}).

• The Greek letter Δ (delta) indicates the change in a quantity: the final value minus the initial value. Thus, ΔT equals $T_f - T_i$ and is the change in temperature; ΔT can be positive or negative, indicating an increase or a decrease in the temperature.

The **molar heat capacity** of a substance is the heat capacity of a sample divided by the chemical amount of substance comprising the sample:

$$c_P = \frac{C_P}{n}$$

• We follow the convention that *molar* heat capacities are indicated with the lowercase letter *c*, reserving C_P for the *total* heat capacity of a body at constant pressure.

Note the lowercase *c*. The SI unit of c_P is the joule per kelvin per mole (J K^{-1} mol^{-1}). Dividing by the chemical amount removes the effect of sample size; a molar heat capacity is the amount of heat causing a given temperature change in a *set number* (one mole) of atoms or molecules of substance, just as a molar mass is the mass of a set number of atoms or molecules. If we know the chemical amount *n* of substance and its molar heat capacity c_P, then the relation between the amount of heat added and the resulting temperature change is

$$q = nc_P \, \Delta T$$

Another useful quantity is the **specific heat capacity.** The word "specific" before the name of a physical quantity very often means "divided by the mass." Accordingly, the specific heat capacity of a body is its heat capacity at constant pressure divided by its mass

$$c_s = \frac{C_P}{m}$$

The division again takes sample size into account, although now in a different way. The specific heat capacity c_s has the natural SI unit J K^{-1} kg^{-1}. The J K^{-1} g^{-1} is more widely used, however, because the specific heat capacities of common materials are on the order of 1 J K^{-1} g^{-1} (see Table 10–1). Using this unit, the specific heat capacity is the amount of heat required to raise the temperature of one gram of material by one kelvin (at constant pressure). For mass *m* of material, the relation between the amount of heat added and the temperature change is

$$q = mc_s \, \Delta T$$

Solving for C_P in the equations defining c_s and c_P and setting them equal to each other provides a relationship between molar heat capacity and specific heat capacity. If we employ the fact that *n*, the chemical amount of a substance, equals its mass divided by its molar mass ($\mathcal{M}$), we can write

$$C_P = mc_s = nc_P = \left(\frac{m}{\mathcal{M}}\right)c_P$$

$$c_s = \frac{c_P}{\mathcal{M}}$$

The molar heat capacity of a substance is its specific heat capacity (c_s) multiplied by its molar mass. The molar heat capacity is not defined for materials that are not substances (for example, for mixtures like seawater or wood) because molar masses are not meaningful for mixtures. The specific heat capacity is still useful for such materials.

Table 10–1 shows that the specific heat capacity of water (4.18 J K^{-1} g^{-1}) is relatively large. In particular, it is several times larger than that of compounds such as CaCO$_3$ and SiO$_2$, which are the major constituents of rocks at the earth's sur-

Table 10–1
Specific Heat Capacities at 25°C

Substance	Specific Heat Capacity (J K^{-1} g^{-1})
Hg(ℓ)	0.140
Cu(s)	0.385
Fe(s)	0.449
SiO$_2$(s)	0.739
CaCO$_3$(s)	0.818
O$_2$(g)	0.917
H$_2$O(ℓ)	4.18

face. Water can absorb or give up a large amount of heat with a smaller change in its temperature than can an equal mass of rock. The temperatures at the shores of oceans and large lakes consequently tend to be warmer in winter and cooler in summer than inland. Water can also be used in conjunction with solar panels to store large amounts of thermal energy from the sun.

Campers fill this bag with water and leave it in the sun all day. It absorbs enough solar energy to provide a comfortably warm evening shower.

EXAMPLE 10–1

(a) Using Table 10–1, calculate the molar heat capacity of quartz, which is one form of SiO_2.
(b) Large beds of rock store thermal energy in some solar-heated homes. Calculate the amount of heat required to raise the temperature of 45.0 kg of rocks by 15.0°C. Assume the specific heat of the rocks to equal that of quartz.

Solution

(a) The molar mass of SiO_2 is 60.08 g mol^{-1}; therefore, the molar heat capacity is

$$c_P = \left(\frac{60.08 \text{ g}}{1 \text{ mol}}\right) \times \frac{0.739 \text{ J}}{\text{K g}} = \boxed{\frac{44.4 \text{ J}}{\text{K mol}}}$$

(b) The heat transferred is calculated using the relationship

$$q = mc_s \, \Delta T$$
$$= 45.0 \times 10^3 \text{ g} \times \frac{0.739 \text{ J}}{\text{K g}} \times 15.0 \text{ K}$$
$$= 5.00 \times 10^5 \text{ J} = \boxed{500 \text{ kJ}}$$

Note that a temperature *change* is the same in degrees Celsius and in Kelvins.

Exercise
Exactly 500.0 kJ of heat is absorbed at a constant pressure by a sample of gaseous helium. The temperature increases by 15.0 K. (a) Compute the heat capacity of the sample. (b) The mass of the helium sample is 6.42 kg. Compute the specific heat capacity and the molar heat capacity of helium.

Answer: (a) 3.3×10^4 J K^{-1} (b) 5.19 J K^{-1} g^{-1} and 20.8 J K^{-1} mol^{-1}

Another unit for heat is in common use: the **calorie.** During the period when the connection between heat and energy was not fully understood, the calorie was defined as the amount of heat needed to raise the temperature of one gram of water by one degree Celsius. In other words, the specific heat capacity of water was taken to be 1.0 cal (°C)$^{-1}$ g^{-1} (which equals 1.0 cal K^{-1} g^{-1} because a *change* of 1°C equals a *change* of 1 K). A separate unit for heat became unnecessary when the nature of heat as a form of energy was understood, and the thermochemical calorie was redefined in terms of the joule:

$$\boxed{1 \text{ calorie} = 4.184 \text{ J exactly}}$$

• The calorie used in measuring food energy is the "great calorie," or Calorie (with a capital "C"). It equals 1000 small calories, or 4184 J. (See Chemistry in Your Life: Food Calorimetry, later in this chapter.)

Equilibration Temperatures

When bodies at two different temperatures are placed in contact with one another, energy is exchanged between them in the form of heat until they reach a common final temperature. If we know the heat capacities of the bodies, we can compute that final temperature. The computation relies on the fact that all the heat lost by the hotter body must be gained by the cooler body as long as none is exchanged with (lost to or absorbed from) the surroundings; that is,

heat gained by the cooler body = heat lost by the warmer body

We have defined the symbol q as the heat *gained* by a body. Under this convention, a heat loss is written as a negative gain, just as a bookkeeper might treat a dollar loss as equivalent to the acquisition of a debt. The q's, the amounts of heat gained by body 1 and body 2, are thus equal in magnitude but opposite in sign:

$$q_1 = -q_2$$

Each of these q's can be related to the molar heat capacity of the substance composing the body and the temperature change of that body, giving

$$n_1\, c_{P1}\, \Delta T_1 = -n_2\, c_{P2}\, \Delta T_2$$
$$n_1\, c_{P1}\, (T_f - T_i)_1 = -n_2\, c_{P2}\, (T_f - T_i)_2$$

Equivalent equations can be written in terms of the masses and the specific heat capacities of the two bodies:

$$m_1\, c_{s1}\, \Delta T_1 = -m_2\, c_{s2}\, \Delta T_2$$
$$m_1\, c_{s1}(T_f - T_i)_1 = -m_2\, c_{s2}(T_f - T_i)_2$$

The final temperatures of body 1 and body 2 are equal at the temperature of thermal equilibrium T_f. This is the only unknown quantity if the amounts of substance, the molar or specific heat capacities, and the initial temperatures are known. The following example illustrates the calculation of T_f.

EXAMPLE 10–2

A piece of iron with a mass of 72.4 g is heated to 100.0°C and plunged into 100.0 g of water that is initially at 10.0°C. Calculate the final temperature that is reached, assuming no heat loss to the surroundings.

Solution

Because the data involve masses, it is easier to work with specific heat capacities (see Table 10–1) than with molar heat capacities. If t_f is the final temperature, then the equation for heat balance gives

$$m_1\, c_{s1}\, \Delta t_1 = -m_2\, c_{s2}\, \Delta t_2$$

$$100.0 \text{ g } H_2O \times \frac{4.18 \text{ J}}{°C \text{ g } H_2O} \times (t_f - 10.0)°C =$$

$$-72.4 \text{ g Fe} \times \frac{0.449 \text{ J}}{°C \text{ g Fe}} \times (t_f - 100.0)°C$$

This is a linear equation in the unknown temperature t_f. We can solve it in the following algebraic steps:

$$418\, t_f - 4180 = -32.51\, t_f + 3251$$
$$450.5\, t_f = 7431$$
$$t_f = \boxed{16.5°C}$$

Note that exchanging the labels (so that the water is body 2 and the iron is body 1) does not change the answer. In this calculation, we have used the fact that spe-

> • Lowercase t's appear in this equation because we reserve T to represent absolute temperatures.

cific heat capacities are numerically the same whether expressed in $J\ K^{-1}\ g^{-1}$ or in $J\ (°C)^{-1}\ g^{-1}$ (because the degree Celsius and the kelvin have the same size). Converting 10.0°C and 100.0°C to kelvins and using specific heats in units of $J\ K^{-1}\ g^{-1}$ gives $T_f = 289.7$ K, an equivalent answer.

Exercise

The specific heat capacities of hafnium and ethanol are 0.146 $J\ K^{-1}\ g^{-1}$ and 2.45 $J\ K^{-1}\ g^{-1}$, respectively. A piece of hot hafnium weighing 15.6 g and at a temperature of 160.0°C is dropped into 125 g of ethanol that has an initial temperature of 20.0°C. Calculate the final temperature that is reached, assuming no heat loss to the surroundings.

Answer: 21.0°C

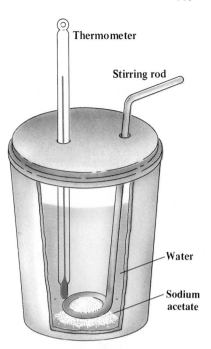

Figure 10–3 A weighed amount of lithium chloride (LiCl) has been put in the water in this Styrofoam-cup calorimeter. As the lithium chloride dissolves in the water, it releases heat to its surroundings, which are primarily the water in the cup. From the temperature change of the water, the amount of heat evolved can be determined.

Calorimeters

Calorimetry concerns itself with the accurate measurement of amounts of heat and the accompanying temperature changes. If we know the heat capacity of a body, we can calculate the amount of heat absorbed by it by measuring its temperature change *if* no heat is lost to the surroundings. The essence of an accurate calorimeter is effective thermal insulation of its interior, where heat is being generated and absorbed, from its surroundings. A perfect calorimeter would have walls that are completely impassable to heat. Real calorimeters are designed to minimize and make reproducible the inevitable leakage of heat to (or from) the surroundings.

A simple calorimeter is shown in Figure 10–3. It uses a Styrofoam coffee cup to minimize heat exchange because conduction through Styrofoam is slow. It is important to insulate the top of the cup as well as the sides. This type of calorimeter is useful for studying reactions in aqueous solution, such as the heat absorbed or evolved when a salt is dissolved in water. The water in the cup is the primary source or acceptor of heat, so the heat capacity of the calorimeter is approximately that of the water in it. Accurate measurements require accounting for the heat absorbed by the thermometer and the walls of the container as well. This is done empirically by recognizing that the relationship

$$q = C\ \Delta T$$

applies to the calorimeter as a whole, where C is the heat capacity of the entire calorimetry assembly, also called the **calorimeter constant.** If a known amount of heat is generated inside the calorimeter (for example, by passing a known electric current through a calibrated resistance coil for a measured period of time) and the temperature rise is measured, the constant C can be found. The calibrated calorimeter can then be used for further measurements.

EXAMPLE 10–3

(a) A Styrofoam-cup calorimeter contains 150 g of water. The generation of 1430 J of heat inside the calorimeter causes the temperature to increase by 1.93°C. Calculate the calorimeter constant.
(b) The dissolution of a small amount of lithium chloride in the water in this calorimeter causes a temperature rise of 3.46 K. Calculate the amount of heat evolved in the dissolution of the lithium chloride.

Solution

(a) The temperature change is equivalent to 1.93 K, so the calorimeter constant is

$$C = \frac{q}{\Delta T} = \frac{1430 \text{ J}}{1.93 \text{ K}} = 741 \text{ J K}^{-1}$$

- This calorimeter constant is a heat capacity measured at constant pressure, a C_P.

(b) All of the heat from the dissolution reaction goes to heat up the calorimeter by 3.46 K. Using the calorimeter constant from part (a) gives

$$q = C \, \Delta T = (741 \text{ J K}^{-1})(3.46 \text{ K}) = 2.56 \times 10^3 \text{ J}$$

Exercise

The dissolution of 1.00 g of $NH_4NO_3(s)$ in 150 g of water absorbs 321 J of heat. (a) Calculate the temperature change that dissolving 1.00 g of $NH_4NO_3(s)$ would cause in the Styrofoam-cup calorimeter described in the example. (b) Calculate the temperature change upon the dissolution of 1.00 g of $NH_4NO_3(s)$ in 150 g of water that is contained in a hypothetical cup having a cover and walls that absorb zero heat.

Answer: (a) $\Delta T = -0.433$ K (b) $\Delta T = -0.511$ K

Figure 10–4   The combustion calorimeter is also called a "bomb calorimeter"; it confines a combustion reaction to a fixed volume as it is carried out within.

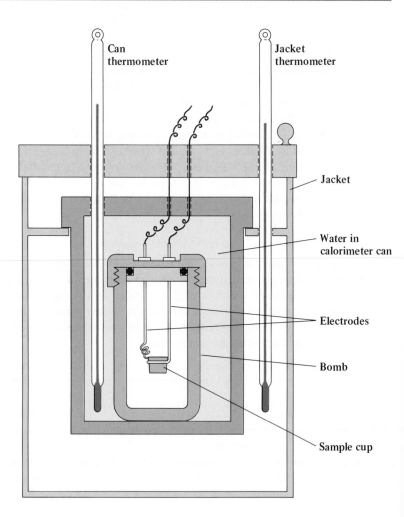

Calorimeters designed to study the heat evolved or absorbed in other types of chemical reactions, such as combustion reactions, are more elaborate but use the same fundamental principles. A combustion calorimeter is shown in Figure 10–4.

10–3 ENTHALPY

Heat is a form in which energy is transferred between systems, not a property of the systems themselves. To refer to the "heat content" of a system therefore has no meaning. There is one important class of processes, however, for which the heat transferred *is* a measure of a change in a fundamental property of the system itself. For processes carried out at constant pressure, the heat absorbed by a system equals the change in its **enthalpy** (H):

$$q_P = \Delta H \qquad \text{(constant pressure)}$$

where the subscript "P" on the q reminds us that the heat transfer occurs at constant pressure and ΔH symbolizes the change in enthalpy. Much chemistry is carried out at constant (atmospheric) pressure. In a chemical change under such conditions, the change in the enthalpy of a system is directly measurable through calorimetry.

The difference between heat and enthalpy must be stressed because the two terms are frequently confused. Enthalpy is called a **state property** of a system because it depends only on the state of the system (for example, its temperature, pressure, volume, and composition) and not on its history (the path followed in reaching the present state). When a system is brought from some initial state to some final state (by heating, cooling, expansion, or compression, for example), the enthalpy change depends only on the initial and final states, and not on the particular path followed between them (Fig. 10–5). The heat transfer, on the other hand, is *not* a state property. The amount of heat transferred depends on just how the change was accomplished; it equals the enthalpy change only in the special case of a constant-pressure path, which is the reason for the subscript "P" in the preceding equation.

• As before, the Greek letter Δ denotes the final value minus the initial value. In this case, $\Delta H = H_f - H_i$.

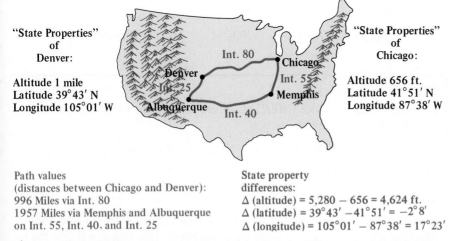

"State Properties" of Denver:

Altitude 1 mile
Latitude 39°43′ N
Longitude 105°01′ W

"State Properties" of Chicago:

Altitude 656 ft.
Latitude 41°51′ N
Longitude 87°38′ W

Path values
(distances between Chicago and Denver):
996 Miles via Int. 80
1957 Miles via Memphis and Albuquerque
on Int. 55, Int. 40, and Int. 25

State property differences:
Δ (altitude) = 5,280 − 656 = 4,624 ft.
Δ (latitude) = 39°43′ − 41°51′ = −2°8′
Δ (longitude) = 105°01′ − 87°38′ = 17°23′

Figure 10–5 Differences in state properties (such as the difference in altitude between two points) are independent of the path traced going from one to the other. Other properties (such as the total distance traveled) depend on the particular path that is followed.

(a)

(b)

Figure 10–6 (a) When mixed in a flask, the two solids $Ba(OH)_2 \cdot 8H_2O(s)$ and $NH_4NO_3(s)$ undergo an acid–base reaction:

$$Ba(OH)_2 \cdot 8H_2O(s) +$$

$$2\ NH_4NO_3(s) \longrightarrow Ba(NO_3)_2(s) +$$

$$2\ NH_3(aq) + 10\ H_2O(\ell)$$

(b) The water produced dissolves excess ammonium nitrate in an endothermic reaction. This dissolution absorbs so much heat that the water on the surface of the wet wooden block freezes to the bottom of the flask, and the block can be lifted up with the flask.

Enthalpy of Reaction

If a chemical reaction occurs at constant pressure, the heat absorbed equals the enthalpy change of the system. This is the **enthalpy of reaction** of that process. Enthalpies of reaction may be positive or negative. If a chemical reaction evolves heat, the system loses heat, and this heat loss at constant pressure is its decrease in enthalpy ($\Delta H < 0$). Such a reaction, with a negative ΔH, is **exothermic.** The combustion of methane (a component of natural gas) is a strongly exothermic reaction:

$$CH_4(g) + 2\ O_2(g) \longrightarrow CO_2(g) + 2\ H_2O(\ell) \qquad \Delta H < 0,\ \text{exothermic}$$

The products of this reaction have lower enthalpy than the reactants. In an **endothermic** reaction, heat is absorbed by the reaction from the surroundings, making q_P and ΔH both positive (Fig. 10–6). An example of an endothermic reaction is the formation of nitrogen dioxide from the elements, a reaction that occurs in automobile engines.

$$N_2(g) + 2\ O_2(g) \longrightarrow 2\ NO_2(g) \qquad \Delta H > 0,\ \text{endothermic}$$

EXAMPLE 10–4

When ammonium chloride (NH_4Cl) dissolves in water, the temperature of the water drops. Is the dissolution reaction

$$NH_4Cl(s) \longrightarrow NH_4^+(aq) + Cl^-(aq)$$

endothermic or exothermic?

Solution

Because the temperature of the water drops, heat is being *removed from* the surrounding water (q for the surroundings is negative). This heat is *added to* the reactants as they go to give the products, so that ΔH is positive. The reaction is endothermic.

Let us now consider the *magnitude* of ΔH and the amount of heat evolved or
absorbed in a chemical reaction. When carbon monoxide is burned in oxygen to carbon dioxide, heat is evolved. Both the q_P of the reaction and its enthalpy change
have negative signs: the reaction is exothermic. When the reaction is enclosed in a
calorimeter, this heat is captured within the calorimeter and causes its internal temperature to rise. Repeated calorimetric measurements show that if 1.000 mol of
$CO(g)$ burns in this way at 25°C, and if reactants and products are all held at constant pressures of 1 atm, then the enthalpy change is

$$\Delta H = q_P = -2.830 \times 10^5 \text{ J} = -283.0 \text{ kJ}$$

We can write the balanced chemical equation as

$$CO(g) + \tfrac{1}{2}O_2(g) \longrightarrow CO_2(g) \qquad \Delta H = -283.0 \text{ kJ}$$

An enthalpy change written next to a balanced chemical equation refers to a reaction in which the chemical amounts of reactants and products equal their coefficients in the equation. The enthalpy change is -283.0 kJ when one mole of $CO(g)$
and one-half mole of $O_2(g)$ are converted to one mole of $CO_2(g)$. If the equation is
multiplied through by a factor of 2, the enthalpy change must also be doubled because twice as much material is then involved:

$$2\,CO(g) + O_2(g) \longrightarrow 2\,CO_2(g) \qquad \Delta H = -566.0 \text{ kJ}$$

We are not limited in this kind of manipulation to whole numbers of moles for the
reacting species. Suppose that 0.832 mol of $CO(g)$ reacts. Then, because -283.0 kJ
is the enthalpy change associated with the conversion of *one* mole of CO, we write

$$\Delta H = 0.832 \text{ mol CO} \times \left(\frac{-283.0 \text{ kJ}}{1 \text{ mol CO}}\right) = -235 \text{ kJ}$$

The factor in parentheses in this equation is a unit-factor, just like the ones used in
stoichiometry problems. We emphasize the comparison by computing the chemical
amount of O_2 that is required for the conversion of 0.832 mol of CO to CO_2:

$$n_{O_2} = 0.832 \text{ mol CO} \times \left(\frac{\tfrac{1}{2} \text{ mol } O_2}{1 \text{ mol CO}}\right) = 0.416 \text{ mol } O_2$$

Under these conditions of temperature and pressure, an enthalpy change of -283.0
kJ is just as essential for the conversion of one mole of CO to CO_2 as providing the
one-half mole of oxygen.

Figure 10–7   Red phosphorus reacts vigorously and exothermically
with liquid bromine. The rising fumes
are a mixture of the product PBr_3 and
unreacted bromine that has boiled off.

• See Section 2-4 on the use of unit
factors in stoichiometric calculations.

EXAMPLE 10-5

Red phosphorus reacts with liquid bromine in an exothermic reaction (Fig. 10–7):

$$2\,P(s) + 3\,Br_2(aq) \longrightarrow 2\,PBr_3(g) \qquad \Delta H = -243 \text{ kJ}$$

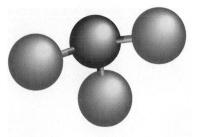

The structure of phosphorus tribromide (PBr_3).

Calculate the enthalpy change when 2.63 g of phosphorus reacts with an excess of bromine in this way.

Solution

The first step is to convert the amount of phosphorus from grams to moles, using the molar mass of phosphorus ($\mathcal{M}_P$) = 30.97 g mol^{-1}.

$$n_P = 2.63 \text{ g P} \times \left(\frac{1 \text{ mol P}}{30.97 \text{ g P}}\right) = 0.0849 \text{ mol P}$$

Knowing that an enthalpy change of -243 kJ is associated with two moles of phosphorus, we can construct a unit-factor and use it to calculate the enthalpy change associated with 0.0849 mol of phosphorus

$$\Delta H = 0.0849 \text{ mol P} \times \left(\frac{-243 \text{ kJ}}{2 \text{ mol P}}\right) = -10.3 \text{ kJ}$$

At constant pressure 10.3 kJ of heat is released for this amount of phosphorus. The "2 mol P" in the denominator of the conversion factor traces back to the coefficient of 2 in the balanced chemical equation.

Exercise

Hydrazine reacts with chlorine according to the equation

$$N_2H_4(\ell) + 2 \text{ Cl}_2(g) \longrightarrow 4 \text{ HCl}(g) + N_2(g) \qquad \Delta H = -420 \text{ kJ}$$

Calculate the enthalpy change (a) when 25.4 g of hydrazine reacts with excess chlorine; (b) when 1.45 mol of HCl(g) is generated by this reaction.

Answer: (a) $\Delta H = -333$ kJ. (b) $\Delta H = -152$ kJ.

If we reverse the direction of a chemical reaction, the enthalpy change (ΔH) reverses sign, because heat that was evolved must now be absorbed, or vice versa. Thus, for the reaction

$$CO_2(g) \longrightarrow CO(g) + \tfrac{1}{2}O_2(g) \qquad \Delta H = +283.0 \text{ kJ}$$

the sign of the enthalpy change is the reverse of that of the reaction of CO with O_2. If this were not so, the enthalpy could not be a state function. The decomposition of CO_2 into CO and O_2 is very difficult to carry out in the laboratory, but the reverse reaction offers no major practical obstacles. Our thermochemical concepts allow us to predict the ΔH of the decomposition reaction with complete confidence, even if a clean calorimetric experiment is never actually carried out.

Phase Changes

Although phase changes are not chemical reactions, the enthalpy changes that accompany them are analogous to enthalpies of reaction. The archetypal phase change is the melting of ice. Heat must be transferred to ice in order to melt it to liquid water, so the "reaction" is endothermic, with ΔH positive:

$$H_2O(s) \longrightarrow H_2O(\ell) \qquad \Delta H_{fus} = +6.007 \text{ kJ}$$

Table 10–2
Enthalpy Changes of Fusion and Vaporization[a]

Substance	ΔH_{fus} (kJ mol^{-1})	ΔH_{vap} (kJ mol^{-1})
NH_3	5.65	23.35
HCl	1.992	16.15
CO	0.836	6.04
CCl_4	2.5	30.0
H_2O	6.007	40.66
NaCl	28.8	170

[a]The enthalpy changes are measured at the normal melting point and the normal boiling point, respectively.

Here ΔH_{fus} is the **enthalpy of fusion,** the heat that is absorbed when the melting occurs at constant pressure. If the water freezes, the reaction is reversed, and an equal amount of heat is *given off* to the surroundings; that is, $\Delta H_{freez} = -\Delta H_{fus}$. Enthalpies of fusion have been measured for many thousands of substances and are quoted in reference works, in units of kJ mol^{-1} or cal mol^{-1}, as **molar enthalpies of fusion,** the enthalpy change that accompanies the melting of one mole of the substance.

• Enthalpies of fusion are also often recorded in units of calories per gram or kilojoules per gram.

The vaporization of a quantity of a liquid at constant pressure and temperature requires an amount of heat called the **enthalpy of vaporization** (ΔH_{vap}). For water, the reaction and the associated ΔH are

$$H_2O(\ell) \longrightarrow H_2O(g) \qquad \Delta H_{vap} = +40.7 \text{ kJ}$$

The condensation of a liquid from a vapor is an exothermic process, with $\Delta H_{cond} = -\Delta H_{vap}$. **Molar enthalpies of vaporization** are also characteristic properties of substances that have been measured and compiled.

Table 10–2 lists the molar enthalpies of fusion and vaporization of several compounds.

EXAMPLE 10–6

The vaporization of 100.0 g of carbon tetrachloride (CCl_4) requires 19.5 kJ of heat at its normal boiling point, 76.7°C, and a pressure of 1 atm. Calculate the molar enthalpy of vaporization of CCl_4 under these conditions.

Solution

The molar mass of CCl_4 is 153.8 g mol^{-1}, so the chemical amount in 100.0 g is

$$n_{CCl_4} = 100.0 \text{ g } CCl_4 \times \left(\frac{1 \text{ mol } CCl_4}{153.8 \text{ g } CCl_4} \right) = 0.6502 \text{ mol } CCl_4$$

The enthalpy change per mole of CCl_4, which is the molar enthalpy of vaporization of the compound, is

$$\Delta H_{vap} = \frac{19.5 \text{ kJ}}{0.6502 \text{ mol}} = \boxed{30.0 \text{ kJ mol}^{-1}}$$

This can also be expressed by writing the balanced equation (with coefficients of 1) and its associated ΔH:

$$CCl_4(\ell) \longrightarrow CCl_4(g) \qquad \Delta H = +30.0 \text{ kJ}$$

Exercise

The melting of 0.140 g of Br_2 (which occurs at $-7.2°C$ at 1 atm pressure) absorbs 9.43 J of heat. Compute the enthalpy change of 2.00 mol of Br_2 when it *freezes*.

Answer: $\Delta H = -2.15 \times 10^4 \text{ J}.$

Hess's Law

• Needing to know this ΔH is not a whim. The reaction occurs in many processes that use coal or coke (an example is steelmaking; see Section 12–1). Designing such processes requires detailed knowledge about the heat that this reaction evolves (or absorbs).

Suppose we need to know the enthalpy change for burning graphite, a form of carbon, to carbon monoxide in oxygen:

$$C(\textit{graphite}) + \tfrac{1}{2}O_2(g) \longrightarrow CO(g) \qquad \Delta H = ?$$

It is not possible to carry out this reaction in a laboratory calorimeter. Heating one mole of graphite with one-half mole of oxygen causes almost half of the carbon to burn to $CO_2(g)$ and leaves the remainder unreacted. It is possible, however, to perform other calorimetry experiments and from their results to *calculate* the heat that would be evolved if the experiment could be performed. Here is how. We can readily measure the heat evolved when one mole of graphite is burned with one mole of oxygen to form carbon dioxide at 25°C:

$$C(\textit{graphite}) + O_2(g) \longrightarrow CO_2(g) \qquad \Delta H = -393.5 \text{ kJ}$$

• The actual calorimetric experiment is performed by running the *reverse* of this reaction and measuring the amount of heat evolved. See the discussion following Example 10–5.

From other measurements we know the enthalpy change for the decomposition of CO_2 to CO and O_2:

$$CO_2(g) \longrightarrow CO(g) + \tfrac{1}{2}O_2(g) \qquad \Delta H = +283.0 \text{ kJ}$$

We now play our ace: the ΔH of any reaction is independent of the path followed from reactants to products (enthalpy is a state property, Fig. 10–8). We imagine a path in which one mole of C is burned with O_2 to CO_2 (with $\Delta H = -393.5$ kJ) and then the CO_2 is converted to CO and O_2 (with $\Delta H = +283.0$ kJ). The total ΔH is

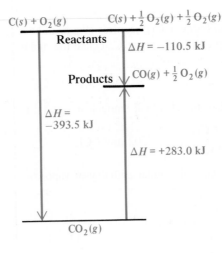

Figure 10–8 Because the enthalpy is a state property, the enthalpy change for the reaction of carbon with oxygen to give carbon monoxide (*red arrow*) can be determined through measurements along a more indirect (but easier to study) path (*blue arrows*). The enthalpy change being sought equals the enthalpy change in burning a mole of carbon to carbon dioxide plus the enthalpy change in converting a mole of carbon dioxide to a mole of carbon monoxide and a half mole of oxygen.

the sum of the two known enthalpy changes, $-393.5 + 283.0 = -110.5$ kJ. This is shown most clearly by writing out the reactions and then adding them:

$$C(graphite) + O_2(g) \longrightarrow CO_2(g) \qquad \Delta H_1 = -393.5 \text{ kJ}$$
$$CO_2(g) \qquad\qquad \longrightarrow CO(g) + \tfrac{1}{2}O_2(g) \quad \Delta H_2 \quad = +283.0 \text{ kJ}$$
$$\overline{C(graphite) + \tfrac{1}{2}O_2(g) \longrightarrow CO(g) \quad \Delta H = \Delta H_1 + \Delta H_2 = -110.5 \text{ kJ}}$$

A similar procedure always works, no matter how complicated the imagined path may be.

> If two or more chemical equations are combined by addition or subtraction to give a new chemical equation, then adding or subtracting their enthalpy changes in a parallel operation gives the enthalpy change of the reaction represented by that new equation.

This statement is known as **Hess's law.**

EXAMPLE 10–7

Hydrazine (N_2H_4) has been used as a rocket fuel because its reaction with oxygen is extremely exothermic:

$$N_2H_4(\ell) + O_2(g) \longrightarrow N_2(g) + 2\,H_2O(g) \qquad \Delta H = -534 \text{ kJ}$$

Calculate the enthalpy change that would result if the water were produced in liquid rather than gaseous form:

$$N_2H_4(\ell) + O_2(g) \longrightarrow N_2(g) + 2\,H_2O(\ell) \qquad \Delta H = ?$$

Use 40.7 kJ mol^{-1} for the enthalpy of vaporization of water.

Solution

Imagine that the reaction takes place in two steps. First, the hydrazine reacts to give nitrogen and water vapor:

$$N_2H_4(\ell) + O_2(g) \longrightarrow N_2(g) + 2\,H_2O(g) \qquad \Delta H_1 = -534 \text{ kJ}$$

Then the water vapor condenses to liquid water:

$$2\,H_2O(g) \longrightarrow 2\,H_2O(\ell) \qquad \Delta H_2 = -81.4 \text{ kJ}$$

We calculated ΔH_2 from

$$\Delta H_2 = (2 \text{ mol}) \times (-\Delta H_{vap}) = (2 \text{ mol}) \times (-40.7 \text{ kJ mol}^{-1}) = -81.4 \text{ kJ}$$

The factor of 2 was inserted in anticipation of canceling the two moles of water in the first equation. Also, the sign of ΔH_{vap} is changed because the process here is one of *condensation* to the liquid. Adding the two equations and their enthalpy changes gives

$$N_2H_4(aq) + O_2(g) \longrightarrow N_2(g) + 2\,H_2O(\ell) \qquad \Delta H = \Delta H_1 + \Delta H_2 = \boxed{-615 \text{ kJ}}$$

Exercise

Suppose hydrazine and oxygen react to give dinitrogen pentaoxide and water vapor:

$$2\,N_2H_4(\ell) + 7\,O_2(g) \longrightarrow 2\,N_2O_5(s) + 4\,H_2O(g)$$

Calculate the ΔH of this reaction, given that the reaction

$$N_2(g) + \tfrac{5}{2}O_2(g) \longrightarrow N_2O_5(s)$$

has a ΔH of -43 kJ.

Answer: -1154 kJ.

10-4   STANDARD-STATE ENTHALPIES

Calorimetric experiments at constant pressure measure only *changes* in enthalpy. Because it is not possible to measure the absolute enthalpy of a sample of a substance, we must select arbitrary reference states in order to compile enthalpy data, just as we select an arbitrary reference level for a list of altitude data (sea level is usually chosen). A hasty statement such as "water is assigned a enthalpy of zero" does not work because the enthalpies of substances vary with the temperature and pressure. To cope with this, chemists define **standard states** for chemical substances as follows:

- For solids and liquids, the standard state is the state of the liquid or solid under a pressure of 1 atm and at a specified temperature.
- For gases, the standard state is the gaseous phase under a pressure of 1 atm, at a specified temperature, and exhibiting ideal-gas behavior.
- For dissolved species, the standard state is a 1 M solution under a pressure of 1 atm, at a specified temperature, and exhibiting ideal-solution behavior.

Standard-state values of enthalpy and other quantities are designated by attaching a superscript ° (pronounced "naught") to the symbol for the quantity and writing the temperature as a subscript. For example, the density d of a substance in its standard state at 400 K is symbolized d°_{400}. Any temperature may be chosen as the "specified temperature." The most common choice is 298.15 K (25°C exactly); if the temperature of a standard state is not explicitly indicated, 298.15 K should be assumed.

• Our discussion of gases (Chapter 5) sets the standard state pressure (1 atm) and a temperature of 273.15 K (0°C) as "standard temperature and pressure" (STP). However, 298.15 K (25°C) is more comfortable for experimental work than 0°C and is preferred in thermochemistry.

Gases cannot actually be prepared in their standard states because real gases do not behave ideally. Non-ideality causes gases to condense to liquids or solids, if the temperature is low enough. For example, $H_2O(g)$ at a temperature of 25°C and a pressure of 1 atm immediately condenses to $H_2O(\ell)$ because of the attractions among its molecules. The standard state of $H_2O(g)$ at 25°C is the hypothetical state that the substance would be in if the intermolecular attractions were somehow magically switched off. The hypothetical nature of the standard states of gases presents no problems because it is always possible to relate the enthalpy of the real substance to the enthalpy of the substance in the idealized standard state.

We now set the enthalpies of selected reference substances in standard states arbitrarily equal to zero. This is completely analogous to assigning zero as the altitude at sea level. Chemists have agreed as follows: *the chemical elements in standard states at 298.15 K have zero enthalpy.* A complication immediately arises. Some elements have two or more allotropic forms. For example, oxygen can be prepared as $O_2(g)$ or $O_3(g)$ (ozone), and carbon exists in numerous allotropic modifications that include graphite, diamond, and the fullerenes. Allotropic forms of an element differ in structure and all physical properties, including their enthalpy. Which

• Figures 20-25 and 20-30 show the structures of diamond and graphite; Figure 18-A shows the structure of the fullerene C_{60}.

form is assigned an enthalpy of zero? The rule is to select the form that is the most stable at 298.15 K and 1 atm. Thus, $O_2(g)$, the most stable form of oxygen under these conditions, is assigned an enthalpy of zero in its standard state at 298.15 K, and $O_3(g)$ in its standard state at 298.15 K has a non-zero enthalpy. The most stable form of carbon at 298.15 K and 1 atm is graphite. Hence, graphite in its standard state at 298.15 K is assigned an enthalpy of zero, and diamond and the many fullerenes all have non-zero enthalpies. In summary:

$H° = 0$ for each element in its most stable form at 298.15 K

We now complete the groundwork for a table of relative enthalpies by defining the **standard molar enthalpy of formation** ($\Delta H_f°$) of a substance to equal the enthalpy change for the reaction that produces one mole of the substance in a standard state from the most stable forms of its constituent elements, also in standard states. The standard molar enthalpy of formation of carbon dioxide, for example, is the enthalpy change of the reaction

$$C(graphite) + O_2(g) \longrightarrow CO_2(g) \qquad \Delta H° = -393.5 \text{ kJ}$$

$$\Delta H_f° (CO_2(g)) = \frac{-393.5 \text{ kJ}}{1 \text{ mol } CO_2(g)} = -393.5 \text{ kJ mol}^{-1}$$

The superscript $°$ indicates that the reactants and products are in standard states and the subscript "f" stands for *formation*.

The $\Delta H_f°$ of the most stable form of an element in a standard state at 298.15 K is exactly zero, because forming an element from itself involves no change at all:

$$H_2(g) \longrightarrow H_2(g) \qquad \Delta H_f° = 0 \text{ kJ}$$

However, the standard enthalpy of formation of a mole of *atoms* of an element is frequently a large positive quantity. Breaking up $H_2(g)$ molecules into $H(g)$ atoms, for example, is a strongly endothermic process.

$$H_2(g) \longrightarrow 2 H(g) \qquad \Delta H° = +435.92 \text{ kJ}$$

This change produces two moles of $H(g)$. Because the standard enthalpy of formation of $H(g)$ is defined as the enthalpy change to produce one mole of $H(g)$, the $\Delta H_f°$ of $H(g)$ is just half of 435.92 kJ or

$$\Delta H_f° (H(g)) = \frac{435.92 \text{ kJ}}{2 \text{ mol } H(g)} = 217.96 \frac{\text{kJ}}{\text{mol}}$$

Introducing standard enthalpies of formation means that we need only tabulate $\Delta H_f°$ for compounds (a long list at that) rather than $\Delta H°$ for all the conceivable reactions those compounds may undergo with one another (an infinitely long list!). We assert this because we can imagine any reaction at 1 atm pressure to take place in two steps: the complete "un-formation" (decomposition) of the reactants in standard states to give the constituent elements in standard states followed by the formation of the products in standard states from those same elements. Hess's law governs the combining of the enthalpy changes, which are all $-\Delta H_f°$ or $\Delta H_f°$, associated with the two steps. The following example shows how this important application of Hess's law works in finding the standard enthalpy changes of reactions without having to run them in a calorimeter.

• A table of standard enthalpies of formation at 298.15 K is given in Appendix D.

EXAMPLE 10–8

Using the table of ΔH_f° values in Appendix D, calculate ΔH° for the following reactions at 25°C

$$2\ NO(g) + O_2(g) \longrightarrow 2\ NO_2(g)$$

Solution

Because enthalpy is a function of state, we can calculate ΔH° along any path we choose (Fig. 10–9). We select a path consisting of two steps for which ΔH°'s are readily available. Step 1 is the decomposition of the reactants into the most stable forms of the elements in standard states:

$$2\ NO(g) \longrightarrow N_2(g) + O_2(g) \qquad \Delta H_1 = -(2\ \text{mol NO}) \times \Delta H_f^\circ\ (NO)$$

The minus sign appears because the process selected is the *reverse* of the formation of NO; the factor of 2 appears because two moles of NO are involved.

Step 2 is the combination of the elements to form *products*.

$$N_2(g) + 2\ O_2(g) \longrightarrow 2\ NO_2(g) \qquad \Delta H_2 = (2\ \text{mol NO}_2) \times \Delta H_f^\circ\ (NO_2)$$

From Hess's law, the enthalpy change of the overall reaction is the sum of these two enthalpy changes:

$$\begin{aligned}
\Delta H^\circ &= \Delta H_1 + \Delta H_2 \\
&= -(2\ \text{mol NO}) \times \Delta H_f^\circ\ (NO) + (2\ \text{mol NO}_2) \times \Delta H_f^\circ\ (NO_2)
\end{aligned}$$

Using standard enthalpies of formation from Appendix D gives

$$\Delta H^\circ = -2\ \text{mol NO} \times \left(\frac{90.25\ kJ}{1\ \text{mol NO}} \right) + 2\ \text{mol NO}_2 \times \left(\frac{33.18\ kJ}{1\ \text{mol NO}_2} \right)$$

$$= -114.14\ kJ$$

Exercise

Using the table of ΔH_f° values in Appendix D, calculate ΔH° for the following reaction at 25°C:

$$2\ H_2S(g) + 3\ O_2(g) \longrightarrow 2\ SO_2(g) + 2\ H_2O(g)$$

Answer: $\Delta H^\circ = -1036\ kJ$.

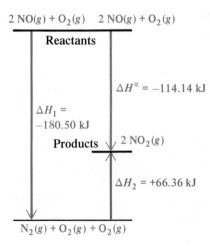

Figure 10–9 The oxidation of nitrogen monoxide to nitrogen dioxide can be carried out directly (*red arrow*). Alternatively, one can imagine a path in which the reactants are decomposed to the elements in their standard states, which then combine to give products (*blue arrows*). The enthalpy change is the same regardless of which path is used.

A general pattern emerges from the example. For a reaction, $\Delta H°$ is the sum of the $\Delta H_f°$'s of the *products* (multiplied by their coefficients in the balanced chemical equation) minus the sum of the $\Delta H_f°$'s of the *reactants* (also multiplied by their coefficients). For a reaction of the form

$$a\text{A} + b\text{B} \longrightarrow c\text{C} + d\text{D}$$

the standard enthalpy change is

$$\Delta H° = c\ \Delta H_f°(\text{C}) + d\ \Delta H_f°(\text{D}) - a\ \Delta H_f°(\text{A}) - b\ \Delta H_f°(\text{B})$$

The direct combination of the elements to give a specific compound is almost never possible in a practical calorimetry experiment. Lists of standard enthalpies of formation are instead worked out by Hess's law calculations from calorimetry data gathered on reactions that *are* practical. Such a calculation is illustrated by the following example.

EXAMPLE 10-9

Propane ($C_3H_8(g)$) burns in oxygen according to the equation

$$C_3H_8(g) + 5\ O_2(g) \longrightarrow 3\ CO_2(g) + 4\ H_2O(g)$$

When 14.64 g of propane is burned in an excess of oxygen in a calorimeter at 25°C and atmospheric pressure, 678.6 kJ of heat is evolved. Calculate the standard molar enthalpy of formation of propane, using -393.5 kJ mol^{-1} and -241.82 kJ mol^{-1} for the standard enthalpies of formation of $CO_2(g)$ and $H_2O(g)$, respectively.

Solution

The first step is to find the enthalpy change when *one mole* of propane is burned. The chemical amount of propane that reacts is

$$14.64 \text{ g propane} \times \left(\frac{1 \text{ mol propane}}{44.096 \text{ g propane}}\right) = 0.3320 \text{ mol propane}$$

The standard enthalpy change $\Delta H°$ for burning *one* mole of gaseous propane is

$$\Delta H° = 1 \text{ mol propane} \times \left(\frac{-678.6 \text{ kJ}}{0.3320 \text{ mol propane}}\right) = -2044 \text{ kJ}$$

The minus sign means heat is given off by the reaction. We can now use Hess's law to relate this to the standard enthalpies of formation, using the general equation just preceding this example:

$$\Delta H° = 3 \text{ mol } CO_2(g) \times \Delta H_f°\ (CO_2(g)) + 4 \text{ mol } H_2O(g) \times$$
$$\Delta H_f°\ (H_2O(g)) - 1 \text{ mol } C_3H_8(g) \times \Delta H_f°\ (C_3H_8(g))$$

Everything in this equation is known except $\Delta H_f°\ (C_3H_8(g))$:

$$-2044 \text{ kJ} = (3 \text{ mol}) \times \left(\frac{-393.51 \text{ kJ}}{\text{mol}}\right) + (4 \text{ mol}) \times \left(\frac{-241.82 \text{ kJ}}{\text{mol}}\right) -$$
$$(1 \text{ mol}) \times \Delta H_f°\ (C_3H_8(g))$$

Solving for $\Delta H_f°\ (C_3H_8(g))$ gives

$$\Delta H_f°\ (C_3H_8(g)) = \ -104 \text{ kJ mol}^{-1}$$

Exercise

Calcium sulfate reacts with carbon (in the form of graphite) as follows:

$$2\,CaSO_4(s) + C(graphite) \longrightarrow 2\,CaO(s) + 2\,SO_2(g) + CO_2(g)$$

The reaction of 204.2 g of $CaSO_4(s)$ with excess graphite at 25°C and a constant pressure of 1 atm absorbs 456 kJ of heat. Calculate the standard molar enthalpy of formation of $CaSO_4(s)$. For the other compounds, use the standard enthalpies of formation in Appendix D.

Answer: -1433 kJ mol^{-1}.

10–5   BOND ENTHALPIES

Chemical change occurs with the breaking and re-making of chemical bonds. In the previous two sections, we considered only the overall change from the starting materials to end products. Chemists have developed methods to study highly reactive *intermediates*, species in which bonds have been broken and not yet reformed. For example, a hydrogen atom can be removed from a methane molecule,

$$CH_4(g) \longrightarrow CH_3(g) + H(g)$$

creating two fragments, neither of which has a stable valence electron structure in the Lewis electron-dot picture. Both rapidly go on to react with other molecules or fragments and eventually form the stable products of the reaction. During the short time that they are present, however, many of their properties can be measured.

One important quantity that has been measured is the enthalpy change when a bond is broken in the gas phase, called the **bond enthalpy.** This is invariably positive, because heat must be added to a collection of stable molecules to break their bonds. The bond enthalpy of a C—H bond in methane, for example, is 438 kJ mol^{-1}. It is the measured standard enthalpy change for the reaction

$$CH_4(g) \longrightarrow CH_3(g) + H(g) \qquad \Delta H° = +438 \text{ kJ}$$

in which 6.022×10^{23} C—H bonds (Avogadro's number) are broken, one in each molecule in a mole of methane. The following gas-phase reactions each involve the breaking of one mole of C—H bonds:

$$
\begin{aligned}
C_2H_6(g) &\longrightarrow C_2H_5(g) + H(g) & \Delta H° &= +410 \text{ kJ} \\
CHF_3(g) &\longrightarrow CF_3(g) + H(g) & \Delta H° &= +429 \text{ kJ} \\
CHCl_3(g) &\longrightarrow CCl_3(g) + H(g) & \Delta H° &= +380 \text{ kJ} \\
CHBr_3(g) &\longrightarrow CBr_3(g) + H(g) & \Delta H° &= +377 \text{ kJ}
\end{aligned}
$$

The approximate constancy of the measured enthalpy changes (all lie within 8% of their average value) persists over hundreds of compounds containing C—H bonds. A similar near constancy develops in experiments on the disruption of O—H, N—H, and other types of bonds. It is therefore useful to tabulate *average* bond enthalpies from such measurements (Table 10–3). Any given bond enthalpy may differ somewhat from the average values, but deviations are mostly small.

Certain atoms (including C, N, O, and S) sometimes form double or triple bonds in order to achieve octets as their valence-electron configurations. In their Lewis structures, two such atoms may share two or three pairs of electrons rather than the single pair shared in a single bond. Experiment shows that the average bond enthalpy of a double bond significantly exceeds that of a single bond between atoms

Table 10-3
Average Bond Enthalpies

	Bond Enthalpy[a] ($kJ\ mol^{-1}$)								
	H—	C—	C=	C≡	N—	N=	N≡	O—	O=
H	436	413			391			463	
C	413	348	615	812	292	615	891	351	728
N	391	292	615	891	161	418	945		
O	463	351	728					139	498
S	339	259	477						
F	563	441			270			185	
Cl	432	328			200			203	
Br	366	276							
I	299	240							

[a]From L. Pauling, *The Nature of the Chemical Bond*, 3rd ed., Cornell University Press, Ithaca, New York, 1960.

of the same two elements, and that the average bond enthalpy of a triple bond is higher still. For example, the average bond enthalpy of a C—C single bond from Table 10–3 is 348 kJ mol^{-1}, but that of a C=C double bond is 615 kJ mol^{-1} and that of a C≡C triple bond is 812 kJ mol^{-1}. The larger number of shared electron pairs leads to stronger bonds.

Applications of Bond Enthalpies

The bond enthalpies of Table 10–3 can be used to estimate standard enthalpy changes ($\Delta H°$) for gas-phase reactions. We can visualize any such reaction as a two-step process. In the first step, the bonds in all the reactant molecules are broken to give free atoms in the gas phase. The enthalpy change for this step can be calculated by adding the bond enthalpies from Table 10–3. In the second step, the product molecules are formed from the atoms. The enthalpy change for this step can again be estimated from the bond enthalpies of Table 10–3, which must now be taken with minus signs because the bonds are being *formed* instead of *broken*. The standard enthalpy change in the reaction is the sum of the enthalpy changes of the two steps. This is most easily seen by working an example.

EXAMPLE 10–10

Use bond enthalpies to estimate the standard enthalpy change for the reaction

$$CCl_2F_2(g) + 2\ H_2(g) \longrightarrow CH_2Cl_2(g) + 2\ HF(g)$$

Dichlorodifluoromethane ($CCl_2F_2(g)$) has the Lewis structure:

$$: \ddot{F} :$$
$$|$$
$$: \ddot{F} - C - \ddot{Cl} :$$
$$|$$
$$: \ddot{Cl} :$$

and CH_2Cl_2 has the same structure with H-atoms replacing F-atoms.

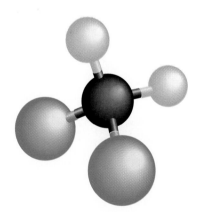

Dichlorodifluoromethane (CCl_2F_2) is also known as Freon-12. Its low reactivity and high volatility make it useful as a refrigerant. Production of Freon-12 and related chlorofluorocarbons (CFCs) has been largely phased out because they accelerate depletion of the ozone layer in the outer atmosphere when lost to the air.

Figure 10–10 The enthalpy ΔH for the overall reaction (*red arrow*) is the sum of the enthalpy changes for bond breaking (*black, green, and yellow arrows*) and the enthalpy changes for bond formation (*blue, orange, and purple arrows*). The final result is only approximate because average bond enthalpies have been used.

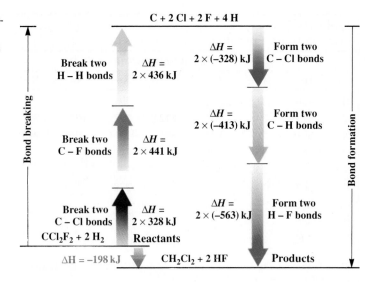

Solution

We can replace this reaction by a hypothetical two-step process (Fig. 10–10). The bonds in all the reactant molecules are first broken, and then the atoms are combined to make the products. In the first step,

$$CCl_2F_2(g) + 2\ H_2(g) \longrightarrow C(g) + 2\ Cl(g) + 2\ F(g) + 4\ H(g)$$

two moles of C—Cl bonds, two moles of C—F bonds, and two moles of H—H bonds are broken. The enthalpy change ΔH_1 for this step can be estimated from the bond enthalpies of Table 10–3:

$$\Delta H_1 \approx \left(2\ \text{mol} \times \frac{328\ \text{kJ}}{\text{mol}}\right) + \left(2\ \text{mol} \times \frac{441\ \text{kJ}}{\text{mol}}\right) + \left(2\ \text{mol} \times \frac{436\ \text{kJ}}{\text{mol}}\right) = 2410\ \text{kJ}$$

The second step is

$$C(g) + 2\ Cl(g) + 2\ F(g) + 4\ H(g) \longrightarrow CH_2Cl_2(g) + 2\ HF(g)$$

This step involves the formation (with release of heat and, therefore, negative enthalpy change) of two moles of C—Cl bonds, two moles of C—H bonds, and two moles of H—F bonds. The net ΔH for this step is

$$\Delta H_2 \approx -\left(2\ \text{mol} \times \frac{328\ \text{kJ}}{\text{mol}}\right) - \left(2\ \text{mol} \times \frac{413\ \text{kJ}}{\text{mol}}\right) - \left(2\ \text{mol} \times \frac{563\ \text{kJ}}{\text{mol}}\right) = -2608\ \text{kJ}$$

The total enthalpy change is then

$$\Delta H_f^\circ = \Delta H_1 + \Delta H_2 = 2410 - 2608 = \boxed{-198\ \text{kJ}}$$

This compares fairly well with the experimental value of -158 kJ. To expect better agreement is unrealistic because tabulated bond enthalpies are, after all, only average values; the true bond enthalpies do vary from molecule to molecule. The overall enthalpy change in the reaction is the difference of two large numbers; small errors in each can give a large error in the final result.

Exercise

Estimate the standard enthalpy change for the gas-phase reaction that forms methanol from methane and water

$$CH_4(g) + H_2O(g) \longrightarrow CH_3OH(g) + H_2(g)$$

and compare it with the $\Delta H°$ obtained from the data in Appendix D.

Answer: Estimated: $+89$ kJ; using Appendix D: $+116$ kJ.

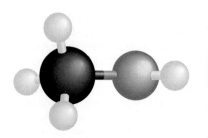

The structure of methanol.

Bond enthalpies are particularly useful in situations in which hard calorimetric data are unavailable. An example is determination of the enthalpy required to break a specific bond or group of bonds in a complex biological molecule, such as the hydrogen bonds that hold the two strands of a DNA molecule together (see Fig. 25–20).

10-6   THE FIRST LAW OF THERMODYNAMICS

When heat is gained by a body, its temperature rises as its molecules begin to move at greater speeds. Transfer of heat leads to an increase in the kinetic energy of the molecules in the body, but energy gains may take other forms (Fig. 10–11). When a pitcher hurls a baseball, the ball gains overall kinetic energy that is not related to a change in temperature or a transfer of heat. When a person or a machine lifts a heavy object, the *potential* energy of the object increases (it is called "potential" energy because it has the potential to be converted into other forms of energy; for example, if the raised object is dropped, it falls and gains kinetic energy). To account for these energy transfers that do not involve heat, we need to introduce the concept of *work*. Then we must establish the connection between heat transfers, work transfers, and energy changes. This connection is the first law of thermodynamics, an empirical law

• Potential energy is also discussed in Section 6-1.

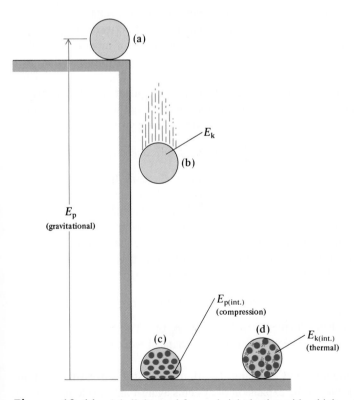

Figure 10–11   A ball dropped from a height begins with a high *potential energy* (a) that is associated with the gravitational attraction of the earth. As it falls (b), this potential energy is converted into *kinetic energy* of motion of the ball. After impact with the ground (c), the molecules near the surface of the ball are pushed against each other, increasing the *internal potential energy* of the ball. As the ball stops bouncing, the molecules readjust their positions, but they move a little faster (d). The ball has a higher *internal kinetic energy* and a higher temperature.

• The remarkable way in which the results of calorimetry experiments are
connected to the values of equilibrium
constants is discussed in Chapter 11.

that is essential to **chemical thermodynamics.** Thermodynamics is the study of heat,
work, and energy; chemical thermodynamics is that part of the field in which calorimetric data are used to predict the positions of equilibria in chemical reactions.

Work

Work is defined in mechanics as the product of the external force F on a body multiplied by the displacement d the force causes the body to undergo. A force is best
visualized as a push. Pushing on a massive object (like a building) that does not
move is work in a physiological sense because the effort tires the muscles.
Mechanical work does not occur, however, because no motion (displacement) occurs. If a body moves over a distance d impelled by a constant force F applied along
the direction of the path, then the work done on the body is

$$w = F \times d \quad \text{(force along direction of path)}$$

Pushing a body in this way causes it to go faster. If the body is already in motion
and if the push (force) opposes the direction of the motion, then the relationship includes a minus sign

$$w = -F \times d \quad \text{(force opposite to direction of path)}$$

and the push slows the body down.

What is the relationship between work and energy in mechanical systems?
Expending mechanical work on a body in the manner just described changes its kinetic energy, E_k. The kinetic energy of a body is its energy of motion, defined as
$E_k = \frac{1}{2}mv^2$, where m is the mass of the body and v is its velocity. A push that does
work on such a body either speeds it up or slows it down, which means that the
push changes the velocity of the body and therefore changes its kinetic energy. In
the absence of opposing forces (such as friction), mechanical work expended on a
body is entirely manifested in a change in the kinetic energy of the body:

$$w = \Delta E_k = \Delta(\tfrac{1}{2}mv^2) \quad \text{(no frictional forces)}$$

Energy, including kinetic energy, is measured in the Système International (SI) in
units of joules, which are units of work as well. Indeed, the newton (the SI unit of
force) multiplied by the meter (the SI unit of distance) equals the joule.

Next, consider the work done in lifting objects in a gravitational field. Here an
opposing force operates. To raise a body of mass m from height h_1 to height h_2, an
upward force sufficient to counteract the downward force of gravity (mg) must be
exerted. Thus, the displacement is $h_2 - h_1$, and the work performed on the object is

$$w = mg(h_2 - h_1) = mg\,\Delta h$$

• Potential energy is the energy a body
has by virtue of being attracted to or
repelled by other bodies (see Section
6-1).

This work leads to a change in the *potential* energy E_p of the object ($\Delta E_p = mg\,\Delta h$),
and we see once again a connection between mechanical work done and a change
in energy.

The most important kind of mechanical work in chemistry is **pressure–volume
work.** This work occurs in the expansion of systems against an outside pressure or
their compression under the influence of an outside pressure. Large changes in volume occur when gases are consumed or generated in chemical reactions. The expansion of newly generated hot gases drives the pistons when fuel is burned in the
cylinders of an automobile engine, for example. Let us analyze the pressure–vol-

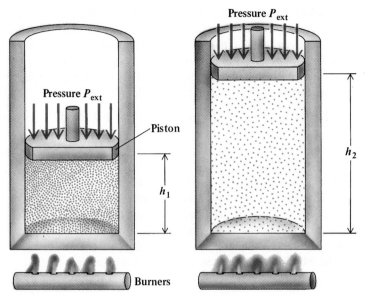

Figure 10–12 Heating the gas inside the cylinder causes it to expand, pushing the piston against the pressure exerted by the gas outside (P_{ext}). As the piston is displaced over a distance $h_2 - h_1 = \Delta h$, the volume of the cylinder increases by an amount $A\Delta h$, where A is the surface area of the piston.

ume work accompanying the volume changes of a gaseous system in a highly idealized frictionless-cylinder piston engine. We imagine that the gas has pressure P_1 and is confined in the cylinder by a piston of cross-sectional area A and of negligible mass (Fig. 10–12). The force exerted on the piston face by the gas is $F_1 = P_1A$ (obtained by rearranging $P_1 = F_1/A$, the defining equation for pressure). Suppose that there is a gas on the other side of the piston with pressure P_{ext} ("ext" for "external"). If $P_{ext} = P_1$, the piston experiences equal forces on its two sides, and it does not move; if P_{ext} exceeds P_1, the gas in the cylinder is compressed and the piston moves inward; if P_{ext} is less than P_1, the gas in the cylinder expands and the piston moves outward. Suppose that expansion of the confined gas lifts the piston from h_1 to h_2. The work in this case is

$$w = -F_{ext}(h_2 - h_1) = -F_{ext}\,\Delta h$$

The negative sign is inserted because the force from the gas outside *opposes* the expansion of the gas inside the cylinder. Substitution for F_{ext} gives

$$w = -P_{ext}A\,\Delta h$$

But the product $A \times \Delta h$ is equal to the volume change of the system, ΔV (Fig. 10–12). Hence, the work is

$$w = -P_{ext}\,\Delta V$$

For an expansion, $\Delta V > 0$, so that $w < 0$, and the system performs work: it pushes back the surroundings. If, instead of expanding, the gas had been compressed (because P_{ext} exceeded P_1), the surroundings would have done work *on* the system. Again, the applicable equation is $w = -P_{ext}\,\Delta V$, but now $\Delta V < 0$ so that $w > 0$. If no volume change occurs, then $\Delta V = 0$ and no pressure–volume work is done. Finally, if there is no mechanical link to the surroundings (that is, if $P_{ext} = 0$), then once again pressure–volume work cannot be performed.

We must also consider the *units* for pressure–volume work. If the pressure P_{ext} is expressed in newtons per square meter (pascals) and the volume change in cubic meters, then their product is in joules. These choices correspond to the natural SI units. For many purposes, however, it is more convenient to express pressures in atmospheres and volumes in liters; then the product of the two, the liter-atmosphere (L atm), appears as a unit of work. It is related to the joule as follows:

$$1 \text{ L atm} = (10^{-3} \text{ m}^3)(1.01325 \times 10^5 \text{ newton m}^{-2}) = 101.325 \text{ J}$$

EXAMPLE 10–11

Suppose that 2.00 L of gas is confined in a cylinder (with frictionless walls) at a pressure of 1.00 atm. The external pressure is also 1.00 atm. The gas is then heated slowly. Because the piston slides perfectly freely, the pressure of the gas remains close to 1.00 atm as the volume of the gas increases (according to the ideal gas law, $PV = nRT$). Suppose the heating continues until a final volume of 3.50 L is reached. Calculate the work done on the gas, and express it in joules.

Solution

This is an expansion of a system from 2.00 L to 3.50 L against a constant external pressure of 1.00 atm. The work done on the system is then

$$\begin{aligned} w &= -P_{ext}\Delta V \\ &= -(1.00 \text{ atm})(3.50 \text{ L} - 2.00 \text{ L}) \\ &= -1.50 \text{ L atm} \end{aligned}$$

Converting to joules gives

$$w = -1.50 \text{ L atm} \times \frac{101.325 \text{ J}}{\text{L atm}} = -152 \text{ J}$$

Because w is negative, we can say either that -152 J of work is done *on* the gas, or that $+152$ J of work is done *by* the gas.

Exercise

Suppose that the gas in the cylinder described in the example is cooled slowly at a constant pressure of 1.00 atm from an initial volume of 2.00 L to a final volume of 1.20 L. Calculate the work done on the gas in joules.

Answer: +81.0 J.

The First Law

As Figure 10–11 emphasizes, energy comes in several forms. In chemistry (unlike physics), we are usually not interested in the *overall* kinetic and potential energies of our system. If we are studying the chemical reactions inside an automobile engine, for example, we do not care whether the automobile is moving or is stopped (that is, has high or low total kinetic energy), nor do we care whether it is at the top or the bottom of a hill (that is, has high or low total potential energy). The forms of energy we care about involve the *internal* kinetic and potential energies of the molecules: how fast they are moving *relative* to the container that encloses them and how they interact with each other and the walls. We therefore focus our discussion exclusively on changes in the **internal energy** of systems, to which we give the symbol ΔE.

Both heat and work are forms in which energy is transferred into and out of a system. If a change in internal energy is caused by the establishment of *mechanical* contact between a system and its surroundings, then *work* is done. If the change in internal energy is caused only by a *thermal* contact (leading to an equalization of temperature differences between system and surroundings), then *heat* is transferred. In many processes, both heat and work cross the boundary of the system, and the internal energy change is the sum of the two contributions. This statement, called the **first law of thermodynamics,** takes the mathematical form[1]

$$\Delta E = q + w$$

We cannot speak of a system as "containing" work or "containing" heat because both work and heat refer not to *states* of the system but to *processes* that transform one state into another. In the 1840s, the English physicist James Joule showed that the temperature of water could be increased by doing work on it as well as by adding heat to it. Figure 10–13 shows an apparatus in which a paddle, driven by a falling weight, churns the water in an insulated tank. Work is performed on the water, and the temperature increases. The work done is $-mg\,\Delta h$, where Δh is the (negative) change in the height of the weight and m is its mass. If all of this work goes to increase the temperature of the water (that is, if none is lost to friction in the mechanism), then the heat capacity of the water can be found. Two easily visualized paths connect the same initial state (cold water) and final state (warm water). On one of these paths (simple heating), heat is transferred to the system without work being done. On the second path (Joule's experiment), work is done without heat being transferred. Additional imaginable paths involve various combinations of heat and work transfer. Because q and w depend on the particular process (or path) connecting the states, they are not state properties. Their sum ($\Delta E = q + w$), however, is independent of path, so that internal energy *is* a state property.

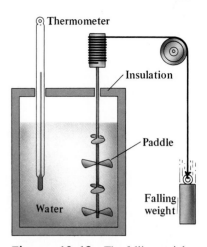

Figure 10–13   The falling weight turns a paddle that does work on the system (the water), causing an increase in its temperature.

> The fundamental physical content of the first law of thermodynamics is the observation that q and w depend individually on the path followed between a given pair of states, but their sum does not.

In any process, the heat *added to* the system is *removed from* the surroundings, so that

$$q_{sys} = -q_{surr}$$

In the same way, the work done *on* the system is done *by* the surroundings, so that

$$w_{sys} = -w_{surr}$$

---

[1]In some books, especially older ones, work is defined as positive when it is done *by* the system. The reason is that, in many engineering applications, one's main interest is the work done by a particular heat engine, so "work done by" is defined as positive. Following this convention, the work is given by

$$w = P_{ext}\Delta V \quad \text{(older convention)}$$

and the first law reads

$$\Delta E = q - w \quad \text{(older convention)}$$

We shall not use this convention in this text, but when using other books, be careful to check which convention is employed.

If we add these two, we conclude that

$$\Delta E_{sys} = -\Delta E_{surr}$$

so that the energy changes of system and surroundings have the same magnitude but opposite signs. The energy change of the universe (system plus surroundings) is then

$$\Delta E_{univ} = \Delta E_{sys} + \Delta E_{surr} = 0$$

> In any process, the total energy of the universe remains unchanged: energy is conserved.

This is another statement of the first law of thermodynamics.

Enthalpy and Energy

As long as the volume of a system is held constant, no pressure–volume work can be done either *on* that system or *by* that system, as we have seen. We can write the first law of thermodynamics for the constant-volume case as

$$\Delta E = q + w = q + 0 \quad \text{(at constant volume)}$$

$$\Delta E = q_V \quad \text{(constant volume)}$$

• The subscript "V" emphasizes that the volume of the system is constant during the change—the pressure may change. The subscript "P" means that the pressure of the system is constant, but its volume may change.

We have already stated that

$$\Delta H = q_P \quad \text{(constant pressure)}$$

The similarity of these two equations provokes immediate questions about the connection between enthalpy and energy. In fact, the enthalpy is *defined* as

$$H = E + PV$$

To see that this definition is consistent with the previous relations, consider an enthalpy change ΔH. According to the definition, that change is given by

$$\Delta H = \Delta E + \Delta(PV)$$

• Here we employ a useful mathematical fact—namely, that if a quantity that changes (here, the volume, V) is multiplied by a constant (here, the pressure, P_{ext}), that constant can be "pulled in front of" the Δ (delta) symbol: $\Delta(P_{ext}V) = P_{ext} \Delta V$.

If the pressure is a constant external pressure (such as that exerted by the atmosphere), this can be written as

$$\Delta H = \Delta E + P_{ext} \Delta V \quad \text{(constant pressure)}$$

If we substitute the first-law relation for ΔE, we find

$$\Delta H = (q_P + w) + P_{ext} \Delta V \quad \text{(constant pressure)}$$

Now, we put in the definition of pressure–volume work for this case, namely that $w = -P_{ext}\Delta V$, giving

$$\Delta H = q_P - P_{ext} \Delta V + P_{ext} \Delta V = q_P \quad \text{(constant pressure)}$$

The definition of enthalpy ensures that it is a state property and that heat transfers that are measured under conditions of constant pressure give enthalpy changes.

For experiments carried out at constant volume, the internal energy is the natural quantity to use because changes in internal energy equal the amount of heat gained at constant volume. Experiments at constant pressure are easier to describe using enthalpy, because enthalpy changes are equal to the amount of heat gained at constant pressure. Because constant-pressure conditions predominate in chemistry, the enthalpy is the more useful property.

We might wish to know the change in the internal energy (ΔE) during a reaction as well as the enthalpy change (ΔH). We have seen, for example, that the enthalpy change for the combustion of one mole of carbon in the form of graphite to carbon monoxide at a pressure of 1 atm is

$$C(graphite) + \tfrac{1}{2}O_2(g) \longrightarrow CO(g) \qquad \Delta H = -110.5 \text{ kJ}$$

How large is the change in internal energy (ΔE)? To answer, we use the relation

$$\Delta H = \Delta E + \Delta(PV)$$

in the rearranged form:

$$\Delta E = \Delta H - \Delta(PV)$$

Graphite is a solid, and its volume is negligible compared with the volumes of the gases, so its contribution to the change in PV can be ignored. We can reasonably expect the gases to obey the ideal gas law, so we write

$$\Delta(PV) = \Delta(nRT) = RT \, \Delta n_g$$

because the temperature is constant. Here, Δn_g is the change in the total chemical amount of gases in the reaction:

$$\Delta n_g = \text{total moles of product gases} - \text{total moles of reactant gases}$$
$$= 1 \text{ mol} - \tfrac{1}{2} \text{ mol} = \tfrac{1}{2} \text{ mol}$$

Hence, if the temperature is 25°C (298.15 K),

$$\Delta(PV) = RT \, \Delta n_g = \left(\frac{8.315 \text{ J}}{\text{K mol}} \right) \times 298 \text{ K} \times \tfrac{1}{2} \text{ mol} = 1.24 \times 10^3 \text{ J} = 1.24 \text{ kJ}$$

Therefore,

$$\Delta E = -110.5 - 1.24 \text{ kJ} = -111.7 \text{ kJ}$$

Note that R must be expressed in J K^{-1} mol^{-1} if a result in joules is desired. For reactions in which only liquids and solids take part, or those in which the chemical amount of gas does not change (those for which $\Delta n_g = 0$), the changes in enthalpy and internal energy are so nearly equal that their difference can be neglected. Even when the chemical amount of gas does change (as in this example), the effect is usually not very large, and changes in the internal energy and the enthalpy of reaction are nearly equal.

EXAMPLE 10–12

Calculate the internal energy of vaporization of one mole of CCl_4 at its normal boiling point, 349.9 K, and compare it with the enthalpy of vaporization at the same temperature (determined in Example 10–6).

CHEMISTRY IN YOUR LIFE

Your body is fueled by the metabolic oxidation of the carbohydrates, fats, and proteins that make up food. All "calorie count" data on different food packages derive directly or indirectly from calorimetric measurements. A portion of food (perhaps even an entire hamburger) or an edible chemical compound is sealed in a bomb calorimeter (see Fig. 10–4), and oxygen is pumped in to a pressure of perhaps 30 atm. Combustion is initiated with a spark, and the food burns. The quantity of heat that is generated is computed from the change in the temperature and the known heat capacities of the water and other materials surrounding the reaction. For example, the complete combustion of 1.00 mol of the carbohydrate sucrose (table sugar) liberates 5640 kJ in a bomb calorimeter

$$C_{12}H_{22}O_{11}(s) + 12\ O_2(g) \longrightarrow$$
$$12\ CO_2(g) + 11\ H_2O(\ell) \quad \Delta E° = -5640\ \text{kJ}$$

The $\Delta H°$ of this reaction also equals -5640 kJ because $\Delta n_g = 0$. The Calorie (1 Calorie = 1000 calories = 4.184 kilojoules) is the unit most used in measurements of food energy in the United States, and food is commonly sold by mass. Converting to these units gives

$$\left(\frac{-5640\ \text{kJ}}{\text{mol sucrose}}\right) \times \left(\frac{1\ \text{mol sucrose}}{342.3\ \text{g sucrose}}\right) \times \left(\frac{1\ \text{Calorie}}{4.184\ \text{kJ}}\right)$$

$$= \frac{-3.94\ \text{Calorie}}{\text{g sucrose}}$$

This is a specific enthalpy because it is on a per-unit-mass basis.

Calorimetry experiments have shown that specific enthalpies of combustion of carbohydrates stay constant, within 5 to 10%, from one compound to the next. The carbohydrate starch, a very common polymeric food (see Fig. 25–18a) that has properties quite different from those of sucrose, liberates 4.18 Calories per gram when completely oxygenated in a bomb calorimeter. This is only a 6% difference from sucrose. Food carbohydrates release an average of 4.15 Calories per gram when burned to carbon dioxide and water.

Calorimetry has established that the specific enthalpies of combustion of fats and proteins are approximately constant as well. Fats of all kinds liberate fairly close to 9.45 Calories per gram when burned completely in a calorimeter; proteins liberate fairly close to 5.65 Calories per gram.

Your body does not actually derive this much energy from the carbohydrates, fats, and proteins that you eat for two reasons. First, the gastrointestinal tract is slightly inefficient in absorbing food. Carbohydrate absorption in a healthy adult is estimated at 97%, fat absorption at 95%, and protein absorption at 92%, on the basis of calorimetric measurements on fecal matter. Second, although the human metabolism converts absorbed carbohydrates and fats essentially completely to carbon dioxide and water, it does *not* convert protein to the same products (namely, $CO_2(g)$, $H_2O(\ell)$, $N_2(g)$, and $H_2SO_4(aq)$) that form when protein is

Solution

This phase change can be described by the equation

$$CCl_4(\ell) \longrightarrow CCl_4(g)$$

Here the change in the number of moles of gas is $\Delta n_g = 1$, because an increase of one mole of gaseous products occurs for each mole of liquid that is vaporized. Therefore,

$$\Delta E = \Delta H_{\text{vap}} - \Delta(PV) = \Delta H_{\text{vap}} - RT\ \Delta n_g$$

Inserting $T = 349.9$ K, $\Delta H_{\text{vap}} = 30.0$ kJ, and $\Delta n_g = 1$ mol gives

$$\Delta E = 30.0\ \text{kJ} - \left(\frac{8.315\ \text{J}}{\text{K mol}}\right) \times \left(\frac{10^{-3}\ \text{kJ}}{1\ \text{J}}\right) \times 349.9\ \text{K} \times 1\ \text{mol}$$

Food Calorimetry

Nutrition Facts

Serving Size 1 Container (227g)

Amount Per Serving

Calories 240	Calories from Fat 25

	% **Daily Value***
Total Fat 3g	5%
Saturated Fat 1.5g	8%
Cholesterol 15mg	5%
Sodium 150mg	6%
Potassium 450mg	13%
Total Carbohydrate 45g	15%
Dietary fiber 1g	4%
Sugars 42g	
Protein 9g	

Vitamin A 2%	•	Vitamin C 6%
Calcium 35%	•	Iron 0%

*Percent Daily Values are based on a 2,000 calorie diet.

KEEP REFRIGERATED

Figure 10–A   Facts about yogurt.

combusted in a bomb calorimeter. The body discards compounds that have caloric value, such as urea and creatinine, in the urine. Calorimetric measurements on these wastes show that the enthalpy change in protein metabolism in human beings equals only 77% of the enthalpy change in protein combustion. The average effective calorie values of the three categories of food thus come out to be

Carbohydrates	$0.97 \times 4.15 = 4.0$ Calories g^{-1}
Fats	$0.95 \times 9.45 = 9.0$ Calories g^{-1}
Proteins	$0.92 \times 0.77 \times 5.65 = 4.0$ Calories g^{-1}

A "Nutritional Facts" panel from a container of yogurt is reproduced in Figure 10–A. The data were obtained by analyzing the yogurt for carbohydrates, fats, and proteins. The fats, for example, were extracted from the aqueous yogurt mixture into diethyl ether or a similar solvent, as discussed in Section 7–7, and then weighed separately. One 227-g serving was found to contain 45 g of carbohydrates, 3 g of fat, and 9 g of protein. The effective calorie content in human consumption of the serving of yogurt is then

$$45 \text{ g carbohydrate} \times \left(\frac{4.0 \text{ Calorie}}{\text{g carbohydrate}} \right) +$$

$$3 \text{ g fat} \times \left(\frac{9.0 \text{ Calorie}}{\text{g fat}} \right) + 9 \text{ g protein} \left(\frac{4.0 \text{ Calorie}}{\text{g protein}} \right)$$

$$= 243 \text{ Calories}$$

This rounds off to 240 Calories, the value stated on the information panel.

$$\Delta E = (30.0 - 2.9) \text{ kJ} = \boxed{27.1 \text{ kJ}}$$

Thus, of the 30.0 kJ of energy transferred from the surroundings in the form of heat, 27.1 kJ is used to increase the internal energy of the system (ΔE), and 2.9 kJ is used to expand the resulting vapor ($\Delta(PV)$).

Exercise

Compute the internal energy of vaporization per mole of water, and compare it with the enthalpy of vaporization from Table 10–2. Water boils at 373.15 K.

Answer: $\Delta E = 40.66 \text{ kJ} - 3.10 \text{ kJ} = 37.56 \text{ kJ}$; ΔE is about 8% less than ΔH.

SUMMARY

10-1 Thermochemistry is the study of the heat effects associated with chemical reactions. **Heat** is a form of motion at the molecular level by which energy is transferred between materials. The units for heat flow are the same as the units of energy; important units are the joule, kilojoule, and calorie.

10-2 Calorimetry is the quantitative study of heat: the measurement of how much heat is evolved or absorbed in chemical reactions or phase changes, or changes in temperature. The **heat capacity** of any body is the amount of heat needed to raise its temperature by one unit of temperature. The **molar heat capacity** equals the heat capacity per mole of substance; the **specific heat capacity** equals the heat capacity per unit mass of material. These quantities allow calculation of the temperature change caused by flow of a quantity of heat to or from a body and determination of the final temperature reached when two bodies at different temperatures are placed in contact with one another and insulated from their surroundings.

10-3 When heat is transferred to an object at constant pressure, the amount absorbed is equal to the change in the **enthalpy** ΔH of that object. Enthalpy is a **state property.** It depends only on the state of the system and not on the path followed to reach that state. Reactions and other processes in which heat is evolved at constant pressure are **exothermic,** and ΔH is negative. **Endothermic** reactions absorb heat from the surroundings (at constant pressure) and have a positive ΔH. Enthalpy changes can be treated using the principles of stoichiometry and are proportional to the amount of substance reacting. **Hess's law** states that when several chemical equations are added or subtracted to give an overall equation, addition or subtraction of the corresponding enthalpy changes gives the enthalpy change associated with the overall equation.

10-4 The **standard molar enthalpy of formation** of a compound is the enthalpy change for the formation of one mole of that compound from the elements, with both reactants and products in **standard states,** which are states at 1 atm pressure and a specified temperature, usually 25°C. Tables of standard molar enthalpies of formation allow the calculation of standard enthalpy changes of reactions with the use of Hess's law.

10-5 Average **bond enthalpies** can be used to estimate enthalpy changes in gas-phase reactions.

10-6 In classical mechanics, **work** done on a body increases its energy. This energy can take many forms, such as kinetic energy (as when an object is accelerated to a greater velocity by a force acting on it) or potential energy (as when an object is lifted in a gravitational field). One important type of work in chemistry is **pressure-volume work,** the expansion of a system against, or its compression by, an external pressure. The **first law of thermodynamics** states that the change in **internal energy** of a system is the sum of the work done on it and the heat transferred to it. Both heat and work depend on the way the change occurs and thus are not functions of state, but their sum depends only on the initial and final states of the system. The first law is equivalent to the statement that the total energy of the universe remains constant, because energy changes in a system are compensated for by equal and opposite changes in the surroundings: energy is conserved. Enthalpy is defined by the equation $H = E + PV$, but only *changes* in enthalpy can be measured or calculated. This makes $\Delta H = \Delta E + \Delta(PV)$ a more useful form of the definition. For many reactions, ΔH and ΔE are almost equal. If the total chemical amount of gases n_g changes in the course of a reaction, $\Delta(PV)$ may be large. At ordinary pressure $\Delta(PV) = \Delta(n_g RT)$, an equation that becomes $\Delta(PV) = RT \, \Delta n_g$ at constant temperature.

PROBLEMS

Note: Answers to blue-numbered problems are given in Appendix F. Problems that are more challenging are indicated by asterisks.

Heat

1. While using an electric drill to bore a hole in a board, you smell burning wood and see a plume of smoke coming from the hole. Where does the heat to start this fire come from?

2. Petroleum deposits in porous rocks under the earth's surface currently supply the fuel for the heating of many homes and buildings. The petroleum is thought to have come from decaying organic matter. Can the heat from oil be traced back to the sun? Explain

Calorimetry

3. Rank the following from lowest heat capacity to highest: 1 kg water, 2 kg water, a 1-kg iron ball, a 2-kg iron plate.

4. Before the days of electricity, flatirons were literally made of iron. They were heated on a stove and had to be quite heavy, despite the fact that anyone could show that it was the heat and not the weight of a flatiron that took the wrinkles out of clothes. Explain why irons had to be heavy.

5. (See Example 10–1). The specific heat capacities of $Li(s)$, $Na(s)$, $K(s)$, $Rb(s)$, and $Cs(s)$ at 25°C are 3.57, 1.23, 0.756, 0.363, and 0.242 $J\ K^{-1}\ g^{-1}$, respectively. Compute the molar heat capacities of these elements, and identify any periodic trend. If there is a trend, use it to predict the molar heat capacity of $Fr(s)$.

6. (See Example 10–1.) The specific heat capacities of $F_2(g)$, $Cl_2(g)$, $Br_2(g)$, and $I_2(g)$ are 0.824, 0.478, 0.225, and 0.145 $J\ K^{-1}\ g^{-1}$, respectively. Compute the molar heat capacities of these elements and identify any periodic trend. If there is a trend, use it to predict the molar heat capacity of $At_2(g)$.

7. The specific heat capacities of the metals nickel, zinc, rhodium, tungsten, gold, and uranium at 25°C are 0.444, 0.388, 0.243, 0.132, 0.129, and 0.116 $J\ K^{-1}\ g^{-1}$, respectively. Calculate the molar heat capacities of these six metals. Note how closely the molar heat capacities for these metals, which were selected at random, cluster around a value of 25 $J\ K^{-1}$ mol^{-1}. The rule of Dulong and Petit states that the molar heat capacities of the metallic elements are approximately 25 $J\ K^{-1}\ mol^{-1}$.

8. Use the empirical rule of Dulong and Petit stated in problem 7 to estimate the specific heat capacities of vanadium, gallium, and silver.

9. The specific heat capacity of white phosphorus, $P_4(s)$, at 25°C is 0.757 $J\ K^{-1}\ g^{-1}$. Calculate the molar heat capacity of white phosphorus at this temperature.

10. The molar heat capacity of uranium at 25°C is 27.49 $J\ K^{-1}$ mol^{-1}. Calculate the specific heat capacity of uranium at this temperature.

11. (See Example 10–2.) A piece of zinc at 20.0°C that weighs 60.0 g is dropped into 200.0 g of water at 100.0°C. The spe-cific heat capacity of zinc is 0.389 $J\ K^{-1}\ g^{-1}$, and that of water near 100°C is 4.22 $J\ K^{-1}\ g^{-1}$. Calculate the final temperature reached by the zinc and the water.

12. (See Example 10–2.) Iron pellets weighing 17.0 g at a temperature of 92.0°C are mixed in an insulated container with 17.0 g of water at a temperature of 20.0°C. The specific heat capacity of water is 9.31 times greater than the specific heat capacity of iron. What is the final temperature inside the container?

13. (See Example 10–3.) A bomb calorimeter used to study combustion is calibrated.
    (a) It is found that when an electric current is passed through a resistance coil to generate 5682 J of heat inside the calorimeter, the temperature increases by 2.31°C. Calculate the calorimeter constant.
    (b) When 0.258 g of benzene ($C_6H_6(\ell)$) is burned in oxygen in the calorimeter, the temperature rises by 4.40°C. Calculate the heat evolved in the combustion of this mass of benzene. The heat absorbed by the products of the reaction is negligible.

14. (See Example 10–3.) A Styrofoam-cup calorimeter is calibrated by adding heat to the water in it with an electrical heater.
    (a) The temperature increases by 1.67°C when 1770 J of heat is generated in this calorimeter. Calculate the calorimeter constant.
    (b) When 0.988 g of $NaOH(s)$ is dissolved in the water in the calorimeter, the temperature rises by 1.04°C. Calculate the amount of heat evolved in the dissolution of this mass of sodium hydroxide.

Enthalpy

15. (See Example 10–4.) When slaked lime is added to water in the production of cement, the reaction

$$Ca(OH)_2(s) \longrightarrow Ca^{2+}(aq) + 2\ OH^-(aq)$$

takes place, causing the temperature to increase. Determine the sign of the enthalpy change ΔH for this process.

16. (See Example 10–4.) A commercial hand-warmer (see Fig. 9–3) consists of a plastic pouch containing a chemical dissolved in water. A small piece of flexible metal is also in the pouch. When the metal piece is flexed, crystals start to form near it and soon spread throughout the solution. At the same time, the pouch and its contents become pleasantly warm. Determine the sign of the enthalpy change ΔH for the process taking place in the pouch at room temperature.

17. (See Example 10–5.) For each of the following reactions, the enthalpy change written is that measured when the numbers of moles of reactants and products taking part in the reaction are as given by the coefficients in the equation. Calculate the

enthalpy change when 1.00 *gram* of the underlined substance is consumed or produced.

(a) $4 Na(s) + O_2(g) \longrightarrow \underline{2 Na_2O(s)}$ $\Delta H = -828$ kJ

(b) $CaMg(CO_3)_2(s) \longrightarrow \underline{CaO(s)} + MgO(s) + 2 CO_2(g)$
$\Delta H = +302$ kJ

(c) $H_2(g) + \underline{2 CO(g)} \longrightarrow H_2O_2(\ell) + 2 C(s)$
$\Delta H = +33.3$ kJ

18. (See Example 10–5.) For each of the following reactions, the enthalpy change written is that measured when the numbers of moles of reactants and products taking part in the reaction are as given by the coefficients in the equation. Calculate the enthalpy change when 1.00 *gram* of the underlined substance is consumed or produced.

(a) $Ca(s) + \underline{Br_2(\ell)} \longrightarrow CaBr_2(s)$ $\Delta H = -683$ kJ

(b) $6 Fe_2O_3(s) \longrightarrow 4 \underline{Fe_3O_4(s)} + O_2(g)$ $\Delta H = +472$ kJ

(c) $2 \underline{NaHSO_4(s)} \longrightarrow 2 NaOH(s) + 2 SO_2(g) + O_2(g)$
$\Delta H = +806$ kJ

19. Liquid bromine dissolves readily in aqueous NaOH:

$Br_2(\ell) + 2 NaOH(aq) \longrightarrow$
$NaBr(aq) + NaOBr(aq) + H_2O(\ell)$

Suppose that 2.88×10^{-3} mol of $Br_2(\ell)$ is sealed in a glass capsule that is then immersed in a solution containing excess $NaOH(aq)$. The capsule is broken, the mixture is stirred, and a measured 121.3 J of heat evolves. In a separate experiment, simply breaking an empty capsule and stirring the solution in the same way evolves 2.34 J of heat. Compute the heat evolved as 1.00 mol of $Br_2(\ell)$ dissolves in excess $NaOH(aq)$.

20. A chemist mixes 1.00 g of $CuCl_2$ with an excess of $(NH_4)_2HPO_4$ in dilute aqueous solution. He measures the evolution of 670 J of heat as the two substances react to give $Cu_3(PO_4)_2(s)$. Compute the ΔH for the reaction of 1.00 mol of $CuCl_2$ with an excess of $(NH_4)_2HPO_4$.

21. Predict the sign of the enthalpy change for the following processes:

(a) $H_2O(\ell) \longrightarrow H_2O(g)$

(b) $Cl_2(g) \longrightarrow Cl_2(s)$

(c) $Hg(\ell) \longrightarrow Hg(s)$

22. Predict the sign of the enthalpy change for the following processes:

(a) $CO_2(s) \longrightarrow CO_2(g)$

(b) $OsO_4(\ell) \longrightarrow OsO_4(g)$

(c) $BaCl_2(s) \longrightarrow BaCl_2(\ell)$

23. (See Example 10–6.) The vaporization of 10.00 g of liquid carbon tetrabromide (CBr_4) requires 1.359 kJ at its normal boiling point and a pressure of 1 atm. Calculate ΔH for the vaporization of 1.000 mol of $CBr_4(\ell)$.

24. (See Example 10–6.) The condensation of 1.00 g of methane (CH_4) to liquid methane at its normal boiling point and a pressure of 1 atm generates 555 J.

(a) Calculate ΔH for the condensation of 2.00 mol of $CH_4(g)$ to $CH_4(\ell)$.

(b) Calculate ΔH for the vaporization of 2.00 mol of $CH_4(\ell)$.

25. Calculate the enthalpy change when 2.38 g of carbon monox-

ide (CO) vaporizes at its normal boiling point. Use data from Table 10–2.

26. Molten sodium chloride is used for making elemental sodium and chlorine. Suppose that the electrical power to a vat containing 56.2 kg of molten sodium chloride is cut off and the salt crystallizes (without changing its temperature). Calculate the enthalpy change, using data from Table 10–2.

27. (See Example 10–7.) When 1.00 mol of *solid* cesium ($Cs(s)$) reacts with $O_2(g)$ to give $CsO_2(s)$, the ΔH equals -266.1 kJ. When 1.00 mol of *liquid* cesium ($Cs(\ell)$) reacts with $O_2(g)$ to give the same product, ΔH equals -268.8 kJ. Determine ΔH for the process $Cs(\ell) \longrightarrow Cs(s)$.

28. (See Example 10–7.) The following data are found in a reference book:

$P_4(g) + 5 O_2(g) \longrightarrow P_4O_{10}(s)$ $\Delta H = -3035.7$ kJ
$4 P(\ell) + 5 O_2(g) \longrightarrow P_4O_{10}(s)$ $\Delta H = -2977.0$ kJ

Determine ΔH for the process of $4 P(\ell) \longrightarrow P_4(g)$.

29. The measured enthalpy change for the burning of ketene (CH_2CO)

$$CH_2CO(g) + 2 O_2(g) \longrightarrow 2 CO_2(g) + H_2O(g)$$

is $\Delta H_1 = -981.1$ kJ at 25°C. The enthalpy change for the burning of methane

$$CH_4(g) + 2 O_2(g) \longrightarrow CO_2(g) + 2 H_2O(g)$$

is $\Delta H_2 = -802.3$ kJ at 25°C. Estimate the enthalpy change at 25°C for the reaction

$$2 CH_4(g) + 2 O_2(g) \longrightarrow CH_2CO(g) + 3 H_2O(g)$$

30. Given the following two reactions and corresponding enthalpy changes,

$CO(g) + SiO_2(s) \longrightarrow SiO(g) + CO_2(g)$
$\Delta H = +520.9$ kJ
$8 CO_2(g) + Si_3N_4(s) \longrightarrow 3 SiO_2(s) + 2 N_2O(g) + 8 CO(g)$
$\Delta H = +461.05$ kJ

compute the ΔH of the reaction

$5 CO_2(g) + Si_3N_4(s) \longrightarrow 3 SiO(g) + 2 N_2O(g) + 5 CO(g)$

31. The enthalpy change to make diamond from graphite is 1.88 kJ mol^{-1}. Which gives off more heat when burned—a pound of diamonds or a pound of graphite? Explain.

32. The enthalpy change of combustion of monoclinic sulfur to $SO_2(g)$ is -9.376 kJ g^{-1}. Under the same conditions, the enthalpy change of combustion of the rhombic form of sulfur to $SO_2(g)$ is -9.293 kJ g^{-1}. Compute the ΔH of the reaction

$$S(monoclinic) \longrightarrow S(rhombic)$$

per gram of sulfur reacting.

Standard-State Enthalpies

33. (See Example 10–8.) Calculate the standard enthalpy change $\Delta H°$ at 25°C for the reaction

$$N_2H_4(\ell) + 3\ O_2(g) \longrightarrow 2\ NO_2(g) + 2\ H_2O(\ell)$$

using the standard enthalpies of formation (ΔH_f°) of reactants and products at 25°C from Appendix D.

34. (See Example 10–8.) Using the data in Appendix D, calculate ΔH° for each of the following reactions:
    (a) $2\ NO(g) + O_2(g) \longrightarrow 2\ NO_2(g)$
    (b) $C(s) + CO_2(g) \longrightarrow 2\ CO(g)$
    (c) $2\ NH_3(g) + \frac{7}{2}\ O_2(g) \longrightarrow 2\ NO_2(g) + 3\ H_2O(g)$
    (d) $C(s) + H_2O(g) \longrightarrow CO(g) + H_2(g)$

35. Zinc is commonly found in nature in the form of the mineral sphalerite (ZnS). A step in the smelting of zinc is the roasting of sphalerite with oxygen to produce zinc oxide:

    $$2\ ZnS(s) + 3\ O_2(g) \longrightarrow 2\ ZnO(s) + 2\ SO_2(g)$$

    (a) Calculate the standard enthalpy change ΔH° for this reaction, using data from Appendix D.
    (b) Calculate the heat absorbed when 3.00 metric tons (1 metric ton = 10^3 kg) of sphalerite is roasted under constant-pressure conditions.

36. The thermite process (see Fig. 10–1) is used for welding steel rails. In this reaction, aluminum reduces iron(III) oxide to metallic iron:

    $$2\ Al(s) + Fe_2O_3(s) \longrightarrow 2\ Fe(s) + Al_2O_3(s)$$

    Igniting a small charge of barium peroxide mixed with aluminum triggers the reaction of a mixture of aluminum powder and iron(III) oxide; molten iron is produced, flows into the space between the rail ends, and solidifies.
    (a) Calculate the standard enthalpy change ΔH° for this reaction, using data from Appendix D.
    (b) Calculate the heat given off when 3.21 g of iron(III) oxide is reduced by aluminum under conditions at a constant pressure of 1 atm.

37. The dissolution of calcium chloride in water

    $$CaCl_2(s) \longrightarrow Ca^{2+}(aq) + 2\ Cl^-(aq)$$

    is used in first-aid hot packs. In these packs, an inner pouch containing the salt is broken, allowing the salt to dissolve in the surrounding water.
    (a) Calculate the standard enthalpy change ΔH° for this reaction, using data from Appendix D.
    (b) Suppose that 20.0 g of $CaCl_2$ is dissolved in 0.100 L of water at 20.0°C. Calculate the temperature reached by the solution, assuming it to be an ideal solution with a heat capacity close to that of 100 g of pure water (418 J K^{-1}).

38. Ammonium nitrate dissolves in water according to the reaction

    $$NH_4NO_3(s) \longrightarrow NH_4^+(aq) + NO_3^-(aq)$$

    (a) Calculate the standard enthalpy change ΔH° for this reaction, using data from Appendix D.
    (b) Suppose 15.0 g of NH_4NO_3 is dissolved in 0.100 L of water at 20.0°C. Calculate the temperature reached by the solution, assuming it to be an ideal solution with a heat capacity close to that of 100 g of pure water (418 J K^{-1}).

(c) From a comparison with the results of problem 37, can you suggest a practical application of this dissolution reaction?

39. (See Example 10–9.) The standard enthalpy change of combustion (to $CO_2(g)$ and $H_2O(\ell)$) at 25°C of the organic liquid cyclohexane ($C_6H_{12}(\ell)$) is -3923.7 kJ mol^{-1}. Determine the ΔH_f° of $C_6H_{12}(\ell)$. Use data from Appendix D.

40. (See Example 10–9.) The standard enthalpy change of combustion (to $CO_2(g)$ and $H_2O(\ell)$) at 25°C of the organic liquid cyclohexene ($C_6H_{10}(\ell)$) is -3731.7 kJ mol^{-1}. Determine the ΔH_f° of $C_6H_{10}(\ell)$.

41. The standard molar enthalpy of formation of liquid chloroform at 25°C is -134.5 kJ mol^{-1}, and that of gaseous chloroform at the same temperature is -103.1 kJ mol^{-1}. Determine the standard enthalpy change in condensing 2.5 mol of chloroform at 25°C.

42. The standard molar enthalpy of formation of solid UF_6 at 25°C is -2197.0 kJ mol^{-1}, and that of gaseous UF_6 at the same temperature is -2063.7 kJ mol^{-1}. Determine the standard enthalpy change in the sublimation (solid-to-gas transition) of 0.5610 mol of UF_6 at 25°C.

Bond Enthalpies

43. A second chlorofluorocarbon used as a refrigerant and in aerosols (besides that discussed in Example 10–10) is CCl_3F. Use the average bond enthalpies from Table 10–3 to estimate the enthalpy change when one mole of this compound is broken into separate atoms in the gas phase.

44. Because it decomposes more quickly in the atmosphere and is much less liable to reduce the concentration of ozone in the stratosphere, the compound CF_3CHCl_2 (with a C—C bond) is being used as a substitute for CCl_3F and CCl_2F_2. Use average bond enthalpies from Table 10–3 to estimate the enthalpy change when one mole of CF_3CHCl_2 is broken into atoms in the gas phase.

45. (See Example 10–10.) Use average bond enthalpies (Table 10–3) to estimate the ΔH° of these reactions:
    (a) $H{-}C{\equiv}N(g) \longrightarrow H(g) + C(g) + N(g)$
    (b) $H{-}C{\equiv}N(g) \longrightarrow \frac{1}{2}\ H_2(g) + C(g) + \frac{1}{2}\ N_2(g)$
    (c) $H{-}C{\equiv}N(g) \longrightarrow H{-}N{\equiv}C(g)$

46. (See Example 10–10.) Use average bond enthalpies (Table 10–3) to estimate the ΔH° of these reactions:
    (a) $3\ H_2(g) + N_2(g) \longrightarrow 2\ NH_3(g)$
    (b) $2\ H_2(g) + N_2(g) \longrightarrow H_2N{-}NH_2(g)$
    (c) $H_2(g) + N_2(g) \longrightarrow HN{=}NH(g)$

47. (See Example 10–10.) Propane (C_3H_8) has the structure $H_3C{-}CH_2{-}CH_3$.
    (a) Write the Lewis structure for propane.
    (b) Use average bond enthalpies from Table 10–3 to estimate the change in enthalpy (ΔH°) for the following reaction:

    $$C_3H_8(g) + 5\ O_2(g) \longrightarrow 3\ CO_2(g) + 4\ H_2O(g)$$

48. (See Example 10–10.)
    (a) Draw Lewis structures for ethylene (C_2H_4) and ethane (C_2H_6) (refer to Section 3–4 if necessary). Draw a Lewis structure for chloroethane (C_2H_5Cl) in which one hydrogen atom from ethane is replaced by a chlorine atom.
    (b) Use average bond enthalpies from Table 10–3 to estimate the change in enthalpy ($\Delta H°$) for the following reactions:

    $$CH_4(g) + CH_3Cl(g) \longrightarrow C_2H_5Cl(g) + H_2(g)$$
    $$C_2H_4(g) + H_2(g) \longrightarrow C_2H_6(g)$$

49. The reaction

    $$BBr_3(g) + BCl_3(g) \longrightarrow BBr_2Cl(g) + BCl_2Br(g)$$

    has a $\Delta H°$ very close to zero. Sketch the Lewis structures of the four compounds, and explain why $\Delta H°$ is so small.

50. At 381 K, the following reaction takes place:

    $$Hg_2Cl_4(g) + Al_2Cl_6(g) \longrightarrow 2\ HgAlCl_5(g) \quad \Delta H° = +10\ kJ$$

    (a) Offer an explanation for the very small $\Delta H°$ for this reaction in terms of the known structures of the compounds

    (b) Explain why the small $\Delta H°$ in this reaction is evidence against

    as the structure of $Hg_2Cl_4(g)$.

The First Law of Thermodynamics

51. (See Example 10–11.) Some nitrogen used to synthesize ammonia is heated slowly, maintaining the external pressure close to the internal pressure of 50.0 atm, until its volume has increased from 542 L to 974 L. Calculate the work done on the nitrogen as it is heated, and express it in joules.

52. (See Example 10–11.) The gas mixture inside one of the cylinders of an automobile engine expands against a constant external pressure of 0.98 atm, from an initial volume of 150 mL (at the end of the compression stroke) to a final volume of 800 mL. Calculate the work done on the gas mixture during this process, and express it in joules.

53. (See Example 10–11.) A constant pressure of 1.14 atm is exerted on a gas. The gas is cooled and contracts from an initial volume of 4.00 L to a final volume of 0.40 L. Compute the work done on the gas during the contraction and express it in liter-atmospheres and in joules.

54. (See Example 10–11.) A 1.25-L sample of a gas is heated and expands against a constant pressure of 0.86 atm to a final volume of 3.75 L. Compute the work done on the gas during the expansion and express it in liter-atmospheres and in joules.

55. (See Example 10–11.) Suppose that 2.00 L of a gas is confined at a pressure of 1.00 atm. The external pressure is zero atmospheres. The gas is heated and expands to a final volume of 3.50 L. Determine the work done on the gas during the expansion.

56. (See Example 10–11.) A 7.00-L sample of a gas is confined in a strong steel vessel with inflexible walls. It is heated strongly so that its pressure increases from 1.50 atm to 407 atm. Determine the work done on the gas during this process.

*57. A chemical system is sealed in a strong, rigid container at room temperature and then heated vigorously.
    (a) State whether ΔE, q, and w of the system are positive, negative, or zero during the heating process.
    (b) Next, the container is cooled to its original temperature. Determine the signs of ΔE, q, and w for the cooling process.
    (c) Designate heating as step 1 and cooling as step 2. Determine the signs of $(\Delta E_1 + \Delta E_2)$, $(q_1 + q_2)$ and $(w_1 + w_2)$, if possible.

58. A battery harnesses a chemical reaction to extract energy in the form of useful electrical work.
    (a) A certain battery runs a toy truck and becomes partially discharged. In the process, it performs a total of 117.0 J of work on its immediate surroundings. It also gives off 3.0 J of heat, which the surroundings absorb. No other work or heat is exchanged with the surroundings. Compute q, w, and ΔE of the battery, making sure each quantity has the proper sign.
    (b) The same battery is now recharged exactly to its original condition. This requires 210.0 J of electrical work from an outside generator. Determine q for the battery in this process. Explain why q has the sign that it does.

59. Calculate the difference between the enthalpy and internal energy of 2.00 mol of $N_2(g)$ at 400 K. Assume that $N_2(g)$ is an ideal gas.

60. Trinitrotoluene (TNT) is a solid. When it explodes, it forms several gases:

    $$2\ C_7H_5N_3O_6(s) \longrightarrow$$
    $$3\ N_2(g) + 7\ CO(g) + 5\ H_2O(g) + 7\ C(s)$$

    Compute the *difference* between $\Delta H°$ and $\Delta E°$ for this reaction.

61. A sample of pure solid naphthalene ($C_{10}H_8$) weighing 0.6410 g is burned completely with oxygen to $CO_2(g)$ and $H_2O(\ell)$ in a constant-volume calorimeter at 298.15 K. The amount of heat evolved is observed to be 25.79 kJ.
    (a) Write and balance the chemical equation for the combustion reaction.
    (b) Calculate the standard change in internal energy ($\Delta E°$) for the combustion of 1.000 mol of naphthalene to $CO_2(g)$ and $H_2O(\ell)$.

(c) Calculate the standard enthalpy change ($\Delta H°$) for the same reaction as in part (b).

(d) Calculate the standard molar enthalpy of formation of naphthalene, using data for the standard enthalpies of formation of $CO_2(g)$ and $H_2O(\ell)$ from Appendix D.

62. A sample of solid benzoic acid (C_6H_5COOH) that weighs 0.800 g is burned in an excess of oxygen to $CO_2(g)$ and $H_2O(\ell)$ in a constant-volume calorimeter at 25°C. The temperature rise is observed to be 2.15°C. The heat capacity of the calorimeter and its contents is known to be 9382 J K^{-1}.

(a) Write and balance the equation for the combustion of benzoic acid.

(b) Calculate the standard change in internal energy ($\Delta E°$) for the combustion of 1.000 mol of benzoic acid to $CO_2(g)$ and $H_2O(\ell)$ at 25°C.

(c) Calculate the standard enthalpy change ($\Delta H°$) for the same reaction as in part (b).

(d) Calculate the standard molar enthalpy of formation of benzoic acid, using data for the standard enthalpies of formation of $CO_2(g)$ and $H_2O(\ell)$ from Appendix D.

Additional Problems

63. A reference book quotes the specific heat capacity (at constant pressure) of aluminum as 0.215 cal K^{-1} g^{-1}.

(a) Determine the molar heat capacity of aluminum in J K^{-1} mol^{-1}.

(b) Compute the heat capacity of 1000 kg (a metric ton) of aluminum in J K^{-1}.

64. At one time, it was thought that the molar mass of indium was near 76 g mol^{-1}. Use the rule of Dulong and Petit (problem 7) to show how the measured specific heat capacity of metallic indium, 0.233 J K^{-1} g^{-1}, makes this value unlikely.

65. Suppose 61.0 g of hot metal, which is initially at 120.0°C, is plunged into 100.0 g of water that is initially at 20.00°C. The metal cools down and the water heats up until they reach a common temperature of 26.39°C. Calculate the specific heat capacity of the metal, using 4.18 J K^{-1} g^{-1} as the specific capacity heat of the water.

66. In their *Memoir on Heat*, published in 1783, Lavoisier and Laplace report (in translation): "The heat necessary to melt ice is equal to three quarters of the heat that can raise the same mass of water from the temperature of the melting ice to that of boiling water." Use this 18th-century observation to compute the amount of heat (in joules) needed to melt 1.00 g of ice. Assume that heating 1.00 g of water requires 4.18 J of heat for each 1.00°C throughout the range from 0°C to 100°C.

67. When glucose, a sugar, reacts fully with oxygen, carbon dioxide and water are produced:

$$C_6H_{12}O_6(s) + 6\,O_2(g) \longrightarrow 6\,CO_2(g) + 6\,H_2O(\ell)$$
$$\Delta H° = -2820 \text{ kJ}$$

Suppose that a person weighing 50 kg (mostly water, with specific heat capacity 4.18 J K^{-1} g^{-1}) eats a candy bar containing 14.3 g of glucose. If all the glucose reacts with oxygen and the heat produced is used entirely to increase the person's body temperature, what temperature increase would be produced? (In fact, most of the heat produced is lost to the surroundings before such a temperature increase results.)

*68. In walking a kilometer, you expend 100 kJ of energy. Compare this to the energy change in driving a car that gets 8.0 km L^{-1} of gasoline (19 miles per gallon) the same distance. The density of gasoline is 0.68 g cm^{-3}, and its standard enthalpy of combustion is -48 kJ g^{-1}.

*69. An ice cube weighing 36.0 g and having a temperature of 0.0°C is dropped into 360 g of water that has a temperature of 20.0°C. Calculate the final temperature that is reached by the mixture, assuming no heat loss to the surroundings. The enthalpy of fusion of ice is $\Delta H_{fus} = 333$ J g^{-1}, and the specific heat capacity of water is 4.18 J K^{-1} g^{-1}. (*Hint*: Consider first the heat absorbed as the ice melts and its effect on the temperature of the surrounding water. Then calculate the final temperature attained from the mixture of liquid water at two different temperatures.)

*70. You have a supply of ice at 0.0°C and a glass containing 150 g of water at 25°C. The enthalpy of fusion for ice is $\Delta H_{fus} = 333$ J g^{-1}, and the specific heat capacity of water is 4.18 J K^{-1} g^{-1}. How many grams of ice must be added to the glass (and melted) to reduce the temperature of the water to 0°C?

71. A given substance can have more than one standard state at 1 atm and 298.15 K. True or false? Explain.

72. Which is larger: $H°_{300}$ of $H_2O(g)$ or $H°_{400}$ of $H_2O(g)$?

73. Liquid helium and liquid nitrogen are both used as coolants; He(ℓ) boils at 4.21 K, and $N_2(\ell)$ boils at 77.35 K. The specific heat capacity of liquid helium near its boiling point is 4.25 J K^{-1} g^{-1}, and the specific heat capacity of liquid nitrogen near *its* boiling point is 1.95 J K^{-1} g^{-1}. The enthalpy of vaporization of He(ℓ) is 25.1 J g^{-1}, and the enthalpy of vaporization of $N_2(\ell)$ is 200.3 J g^{-1}. Discuss which liquid is the better coolant (on a per-gram basis) near its boiling point, and which is better *at* its boiling point.

74. The enthalpy change to form one mole of $Hg_2Br_2(s)$ from the elements at 25°C is -206.77 kJ mol^{-1}, and that to form one mole of $HgBr(g)$ is 96.23 kJ mol^{-1}. Compute the enthalpy change for the decomposition of one mole of $Hg_2Br_2(s)$ to two moles of $HgBr(g)$:

$$Hg_2Br_2(s) \longrightarrow 2\,HgBr(g)$$

*75. The gas most commonly used in welding is acetylene ($C_2H_2(g)$). When acetylene is burned in oxygen, the reaction that takes place is

$$C_2H_2(g) + \tfrac{5}{2}O_2(g) \longrightarrow 2\,CO_2(g) + H_2O(g)$$

(a) Using data from Appendix D, calculate $\Delta H°$ for this reaction.

(b) Calculate the total heat capacity of 2.00 mol of $CO_2(g)$ and 1.00 mol of $H_2O(g)$, using $c_P(CO_2(g)) = 37$ J K^{-1} mol^{-1} and $c_P(H_2O(g)) = 36$ J K^{-1} mol^{-1}.

(c) When this reaction is carried out in an open flame, almost all the heat produced in part (a) goes to raise the temperature of the products. Calculate the maximum flame temperature attainable in an open flame burning acetylene in oxygen. Actual flame temperatures are lower than this because of heat losses to the surroundings.

76. The standard molar enthalpy of formation of $(NH_4)_2PtF_6(s)$ at 25°C is given in a table. Write the chemical equation for the reaction that would have to occur in the calorimeter in a direct measurement of this value.

77. Silicon nitride ($Si_3N_4(s)$) has physical, chemical, and mechanical properties that make it a useful industrial material. It is crucial to know its standard molar enthalpy of formation. A clever experiment allows the direct determination of the $\Delta H°$ of the reaction

$$3 CO_2(g) + Si_3N_4(s) \longrightarrow 3 SiO_2(s) + 2 N_2(g) + 3 C(s)$$

State what additional data must either be looked up or measured before the $\Delta H_f°$ of $Si_3N_4(s)$ can be computed.

78. (a) Draw Lewis structures for O_2, CO_2, H_2O, CH_4, C_8H_{18}, and C_2H_5OH. In C_8H_{18}, the carbon atoms form a straight chain with single bonds; in C_2H_5OH the two carbon atoms are bonded to one another.

Using average bond enthalpies from Table 10–3, estimate the enthalpy change in each of the following reactions.

(b) $CH_4(g) + 2 O_2(g) \longrightarrow CO_2(g) + 2 H_2O(g)$
(burning methane, or natural gas)

(c) $C_8H_{18}(g) + \frac{25}{2} O_2(g) \longrightarrow 8 CO_2(g) + 9 H_2O(g)$
(burning octane, in gasoline)

(d) $C_2H_5OH(g) + 3 O_2(g) \longrightarrow 2 CO_2(g) + 3 H_2O(g)$
(burning ethanol, in gasohol)

79. When a ball of mass m is dropped through a height difference Δh, its potential energy changes by an amount $mg\Delta h$, where g is the acceleration of gravity, equal to 9.81 m s^{-2}. Suppose that all that energy is converted into heat, increasing the temperature of the ball, when the ball hits the ground. If the specific heat capacity of the fabric of the ball is 0.850 J K^{-1} g^{-1}, calculate the height from which the ball must be dropped to increase the temperature of the ball by 1.00°C. (*Hint*: The mass of the ball does not matter, because it cancels out once the problem is set up. Make sure to convert properly between grams and kilograms.)

80. During his honeymoon in Switzerland, James Joule is said to have used a thermometer to measure the temperature difference between the water at the top and at the bottom of a waterfall. Take the height of the waterfall to be Δh and the acceleration of gravity g to be 9.81 m s^{-2}. Assuming that all

the potential energy change $mg\Delta h$ of a mass m of water is used to heat that water by the time it reaches the bottom, calculate the temperature difference between the top and the bottom of a waterfall 100 m high. Take the specific heat capacity of water to be 4.18 J K^{-1} g^{-1}.

81. The gas inside a cylinder expands against a constant external pressure of 1.00 atm from a volume of 5.00 L to a volume of 13.00 L. In doing so, it turns a paddle immersed in 1.00 L of water. Calculate the temperature rise of the water, assuming no loss of heat to the surroundings or frictional losses in the mechanism. Take the density of water to be 1.00 g cm^{-3} and its specific heat capacity to be 4.18 J K^{-1} g^{-1}.

* 82. One mole of argon (assumed to be an ideal gas) is confined in a strong, rigid container of volume 22.41 L at 273.15 K. The system is heated until 3.000 kJ (3000 J) of heat has been added. The molar heat capacity of the gas does not change during the heating and equals 12.47 J K^{-1} mol^{-1}.

(a) Calculate the original pressure inside the vessel (in atmospheres).

(b) Determine q for the system during the heating process.

(c) Determine w for the system during the heating process.

(d) Compute the temperature in degrees Celsius of the gas after the heating. Assume that the container has zero heat capacity.

(e) Compute the pressure (in atmospheres) inside the vessel after the heating.

(f) Compute ΔE of the gas during the heating process.

(g) Compute ΔH of the gas during the heating process.

(h) The correct answer to part (g) exceeds 3.000 kJ. The increase in enthalpy (which at one time was misleadingly called the "heat content") in this system exceeds the amount of heat actually added. Why is this not a violation of the law of conservation of energy?

83. A runner generates 100 Calories (418 kJ) of energy per mile from the oxidation of food. The runner's body must dispose of essentially all this energy to avoid overheating.

(a) Suppose that the evaporation of sweat is the only way that energy is lost. Estimate the volume of sweat (in liters) that must be evaporated by the runner in the course of a marathon (just over 26 miles). Take the enthalpy of vaporization of water at the runner's body temperature as 44 kJ mol^{-1}.

(b) Does the answer seem high or low (based on experience viewing or running footraces)? List three other mechanisms that the body uses to dispose of energy.

CUMULATIVE PROBLEM

Methanol as a Gasoline Substitute

Methanol (CH_3OH) is used as a substitute for gasoline in certain high-performance vehicles. Its use as a fuel in ordinary cars has been seriously considered. Obviously, the thermochemistry of methanol must be thoroughly understood in order to design engines to burn it efficiently.

A methanol-powered bus.

(a) Methanol in an automobile engine must be in the gas phase before it can react. Calculate the heat (in kilojoules) that must be added to 1.00 kg of liquid methanol to raise its temperature from 25.0°C to its normal boiling point, 65.0°C. The molar heat capacity of liquid methanol is 81.6 J K^{-1} mol^{-1}.

(b) Once methanol has reached its boiling point, it must be vaporized. The molar enthalpy of vaporization of methanol is 38 kJ mol^{-1}. How much heat must be added to vaporize 1.00 kg of methanol?

(c) Once it is in the vapor phase, the methanol can react with oxygen in the air according to

$$CH_3OH(g) + \tfrac{3}{2}O_2(g) \longrightarrow CO_2(g) + 2\,H_2O(g)$$

Use average bond enthalpies to estimate the enthalpy change in this reaction, for one mole of methanol reacting. (*Note*: The structure of methanol is given on page 432.)

(d) Use data from Appendix D to calculate the standard enthalpy change in this reaction, assuming it to be the same at 65°C as at 25°C.

(e) Calculate the amount of heat released when 1.00 kg of gaseous methanol is burned in air at constant pressure. Use the more accurate result of part (d) rather than that of part (c).

(f) Calculate the *difference* between the change in enthalpy and the change in internal energy when 1.00 kg of gaseous methanol is oxidized to gaseous CO_2 and H_2O at 65°C.

(g) Suppose now that the methanol is burned inside the cylinder of an automobile engine. Taking the radius of the cylinder to be 4.0 cm and the distance the piston moves during one stroke to be 12 cm, calculate the work done *on* the gas per stroke as it expands against an external pressure of 1.00 atm. Express your answer in liter-atmospheres and in joules.

11

Spontaneous Change and Equilibrium

CHAPTER OUTLINE

Highly organized symmetrical objects — snow crystals — form spontaneously from water vapor in clouds.

A **spontaneous** change is one that occurs by itself, given enough time, without outside intervention. One of the most striking features of spontaneous change in nature is that it has a direction: gases expand into empty containers but never spontaneously contract out of a container to leave it empty. Heat flows from a hot body to a cold one when the two come into thermal contact but never flows spontaneously from cold to hot. When a teakettle is started on a gas flame, the water always gets hotter; it never cools or freezes while the flame gets hotter. When a spark touches a mixture of hydrogen and oxygen, the two gases spontaneously (and explosively) react to produce water, but water never decomposes spontaneously to hydrogen and oxygen. One of the goals of thermodynamics must be to account for this *directionality* of spontaneous change (Fig. 11–1).

In this chapter, we introduce two state properties. The first, the entropy, is connected with the extent of disorder present in a system and its surroundings. An examination of the way in which this property changes during real and hypothetical processes leads to valid predictions about the direction of spontaneous change and the equilibrium state that is ultimately achieved. The second property, the Gibbs function, makes it much easier to apply the new concept to phase changes and chemical reactions.

11–1 ENTHALPY AND SPONTANEOUS CHANGE

Mechanical systems reach equilibrium in the state of lowest energy: a ball bearing thrown into a bowl rolls around until it dissipates its kinetic energy and comes to rest in a state of lowest potential energy at the bottom of the bowl. During the 19th century, chemists sought a similar principle that would predict the equilibrium state of a chemical reaction.

Two chemists, Marcellin Berthelot in France and Julius Thomsen in Denmark, suggested that the enthalpy could play this role in processes occurring at constant pressure. By 1878, they had become convinced that spontaneous chemical reactions

Figure 11–1 A bullet hitting a steel plate at a speed of 1600 ft/sec melts as its kinetic energy is converted to heat. Molten lead sprays in all directions. These three photographs make sense only in the order shown; the reverse process is unmistakably impossible.

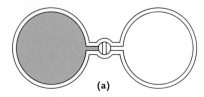

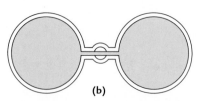

Figure 11–2 The free expansion of a gas into a vacuum. (a) Stopcock closed; all gas is in the left bulb. (b) Stopcock opened; half of the gas is in each bulb.

"tend toward the production of the body or the system of bodies that sets free the most heat." They were wrong, however. If their principle were valid, one could predict the spontaneity of a reaction at constant pressure from the sign of ΔH. If the reaction were exothermic (ΔH negative), then it would be spontaneous. Spontaneous endothermic reactions would not be permitted.

Most spontaneous chemical reactions *are* exothermic, and, to this extent, the principle of Berthelot and Thomsen is a valid summary of much experimental data. There are many exceptions, however. The dissolution of ammonium chloride is endothermic,

$$NH_4Cl(s) \longrightarrow NH_4^+(aq) + Cl^-(aq) \qquad \Delta H° = +14.8 \text{ kJ}$$

but the reaction is nevertheless spontaneous. The acid–base reaction (see Fig. 10–6)

$$Ba(OH)_2 \cdot 8H_2O(s) + 2 NH_4^+(aq) \longrightarrow Ba^{2+}(aq) + 2 NH_3(g) + 10 H_2O(\ell)$$

is also endothermic, with a $\Delta H°$ of $+119$ kJ, yet it is spontaneous. Some reactions that are non-spontaneous at ordinary temperature become spontaneous at elevated temperatures, even though their enthalpy change remains positive. An example is the observation by Joseph Black in 1755 that chalk (a form of limestone) decomposes to lime and carbon dioxide at high temperature:

$$CaCO_3(s) \longrightarrow CaO(s) + CO_2(g) \qquad \Delta H° = +178.3 \text{ kJ}$$

This reaction becomes spontaneous above about 800°C even though the enthalpy change remains positive.

Phase changes are also spontaneous processes that do not always obey the principle of Berthelot and Thomsen. It is true that, below 100°C, water vapor spontaneously condenses to liquid in an exothermic process:

$$H_2O(g) \longrightarrow H_2O(\ell) \qquad \Delta H = -40.66 \text{ kJ}$$

However, the reverse process, the boiling of liquid water

$$H_2O(\ell) \longrightarrow H_2O(g) \qquad \Delta H = +40.66 \text{ kJ}$$

becomes spontaneous *above* 100°C, even though it is endothermic. The melting of a solid is another endothermic process that can occur spontaneously.

It is clear, then, that the sign of the enthalpy change is not a satisfactory criterion for the spontaneity of a chemical or a physical change. The exothermicity of a process is one factor that favors spontaneous change, but there is another. As the temperature becomes higher, this second factor grows in importance, and the predictive power of the principle of Berthelot and Thomsen diminishes. A new state property, the entropy, is needed to establish a criterion for spontaneity that is valid at all temperatures.

11–2 DISORDER AND ENTROPY

To search for this new criterion, let us consider first a spontaneous process in which neither energy nor enthalpy plays a role: the free expansion of an ideal gas into a vacuum, which is shown in Figure 11–2. The gas is initially held in the left bulb in a volume $V/2$ while the right bulb is evacuated. After the stopcock is opened, the gas expands spontaneously to fill the entire volume V. Let us examine this free expansion from a microscopic point of view. If we could follow the path of some particular molecule, we would see that, after spending some time on the left side, it crosses to the right, then back to the left, and so forth. Over the long term, the molecule spends equal time on the two sides because there is no reason for it to prefer one side to the other. Put in other terms, the *probability* that the molecule is on the left side at any moment is 1/2, the same as the probability that it is on the right side.

This concept of probability provides the key to understanding spontaneous change. The gas expands to fill the whole volume because it is overwhelmingly probable that it do so. We never see a gas spontaneously compress itself into a smaller volume because such a process is overwhelmingly improbable. Let us calculate exactly how improbable a spontaneous compression of one mole of an ideal gas into a particular half of the available volume is. The probability that a specific molecule is on the left side in Figure 11–2 at a given instant is clearly one in two, or 1/2. A second specific molecule may be either on the left or the right, so the probability that both molecules are on the left is $1/2 \times 1/2 = 1/4$. Four specific molecules are on the left $1/2 \times 1/2 \times 1/2 \times 1/2 = 1/16$ of the time, on the average (Fig. 11–3). Continuing this argument for all $N_0 = 6.0 \times 10^{23}$ molecules, the probability that all N_0 molecules are on the left is

$$\text{probability} = \underbrace{\frac{1}{2} \times \frac{1}{2} \times \frac{1}{2} \times \cdots \times \frac{1}{2}}_{6.0 \times 10^{23} \text{ terms}} = \left(\frac{1}{2}\right)^{6.0 \times 10^{23}} = \frac{1}{2^{6.0 \times 10^{23}}}$$

In scientific notation, this becomes

$$\text{probability} = \frac{1}{10^{1.8 \times 10^{23}}}$$

This is a vanishingly small probability, because $10^{1.8 \times 10^{23}}$ is an unimaginably large number. It is vastly larger than the number 1.8×10^{23}, which is itself a large number. To realize this, think about writing out these numbers. The number 1.8×10^{23} is fairly easy. It contains 22 zeros:

$$180,000,000,000,000,000,000,000$$

But $10^{1.8 \times 10^{23}}$ is 1 followed by 1.8×10^{23} zeros. To write out these zeros would require the whole human race—man, woman, and child—to write a zero every second for over a million years!

Nothing in the laws of mechanics prevents a gas from compressing spontaneously, but we do not see such an occurrence because it is overwhelmingly improbable.

> The directionality of spontaneous change is a consequence of the random behavior of the large number of molecules in macroscopic systems.

These are the systems in which we see inexorable, directional change every day.

EXAMPLE 11–1

Suppose that the gas had only 6 molecules instead of 6×10^{23}. What is the probability that all six are in the volume on the left side of the stopcock? Would you be surprised if a "snapshot" showed them all to be on the left?

• A probability is a likelihood or chance. The probability of an event is a fraction ranging from 0 (utterly impossible) to 1 (absolutely certain).

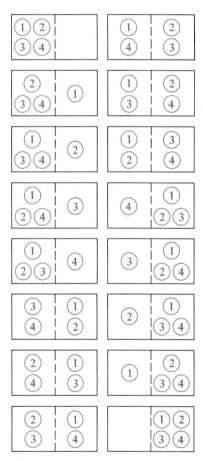

Figure 11–3 The 16 possible states for a system of four molecules that may occupy either side of a container. In only one of these are all four molecules on the left side.

Solution

The probability that any given molecule is on the left is 1/2. By the argument shown in Figure 11–3, the probability that all six are on the left is

$$\frac{1}{2} \times \frac{1}{2} \times \frac{1}{2} \times \frac{1}{2} \times \frac{1}{2} \times \frac{1}{2} = \left(\frac{1}{2}\right)^6 = \frac{1}{64}$$

so the probability is one in sixty-four. Although this is not a *high* probability, it is not small enough for us to be especially surprised if occasionally we were to see all six on the left. We *would* be surprised to see 6×10^{23} molecules all on one side!

Exercise

Calculate the probability that all the molecules of a 20-molecule gas are on the right side of the stopcock in Figure 11–2.

Answer: 1 in 2^{20} or 1 in 1,048,576.

Entropy

Spontaneous processes occur when constraints are removed from a system. In the free expansion of a gas, the molecules are initially constrained to be in one part of the container. After the constraint is removed (the stopcock is opened), they are free to stay on the left-hand side, but they also have other possibilities open to them; the number of states available to the system increases. We use the term "microscopic state" or **microstate** to specify each particular way of arranging molecules among the positions accessible to them while keeping the total energy fixed. In a spontaneous process, the number of available microstates increases. Figure 11–3 shows a schematic of the 16 microstates for a four-molecule system, an increase over the 1 microstate that the system has if all the molecules are constrained to be on the left side.

The **entropy** S is a quantitative measure of the number of microstates available to the molecules of a system. The entropy of an isolated system increases as the number of microscopic states increases. A common way to describe this phenomenon is to relate the entropy to disorder: a highly ordered system has low entropy because its molecules are constrained to occupy only certain positions in space. As the constraints are removed and the molecules are freed to occupy more locations, the disorder increases and the entropy rises. An isolated system is not observed to become ordered spontaneously because the number of all possible states (each equally probable) is overwhelmingly greater than the number of ordered ones (Fig. 11–4).

Melting a solid is an excellent example of this. In a solid, the atoms or molecules are constrained by strong attractions to their neighbors to stay near fixed positions. In a liquid, however, they can move far away from these fixed positions. The liquid is more disordered than the crystal because many more microscopic molecular arrangements correspond to the observable state labelled "liquid" than to the observable state labelled "solid." The entropy of the liquid is higher than the entropy of the solid, and the entropy change (ΔS) of a substance upon melting is positive. When a liquid evaporates, the disorder and the entropy increase again. The molecules in a liquid mostly remain at the bottom of their container, but those in a

• We must be careful about our use of the terms "order" and "disorder." What is an ordered arrangement to one person may appear very disordered to another. It is preferable to think of entropy in terms of the number of alternative arrangements available to a system.

• Section 6–2 identifies the *durability* of the arrangement of the neighbors around any given molecule as the crucial difference between the solid and liquid states of matter.

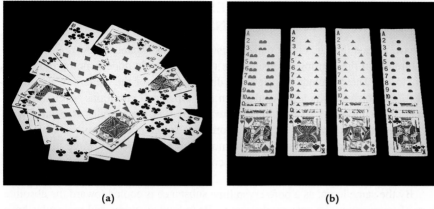

(a)   **(b)**

Figure 11–4   (a) When a deck of cards is thrown down, the cards almost always land in a disorderly heap. The number of ordered arrangements as in (b) of a deck of cards is far smaller than the number of disordered ones.

gas move freely throughout the whole enclosed volume. A gaseous macroscopic state has a much greater number of microscopic arrangements than a liquid one, so the entropy change (ΔS) on vaporization is also positive.

Although the concepts of order and disorder are helpful if used with understanding, the more fundamental factor determining the entropy is the number of available microstates. In some cases (such as in certain transformations between two solid phases), it is difficult to say which phase is more ordered. A calculation of the relative number of microstates available, however, predicts correctly the sign of the entropy change in a transformation from one such phase to the other.

EXAMPLE 11–2

Predict whether the entropy change of the system (ΔS) is positive or negative for each of the following processes:
(a) Steam condenses to liquid water.
(b) Oxygen and nitrogen are contained in separate volumes at the same pressure with a membrane between the two. The membrane is pierced, and the gases mix in the combined volume.

Solution

(a) The ΔS is negative because the number of microstates available in a liquid is less than in a gas.
(b) The ΔS is positive because each molecule can be found in *either* volume after the constraint is removed; the number of available microstates is greater.

Exercise
Predict whether the entropy change is positive or negative for the system in each of the following processes: (a) compression of $O_2(g)$ at 1 atm to $O_2(g)$ at 5 atm; (b) shuffling a deck of cards that was arranged by suit.

Answer: (a) negative; (b) positive.

Figure 11–5 The three types of molecular motion of a diatomic molecule.

Translation

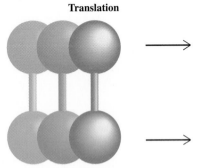

Rotation

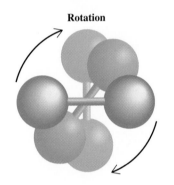

Vibration

11–3 ABSOLUTE ENTROPIES AND CHEMICAL REACTIONS

We have argued that the entropy increases when a gas expands. How does the temperature affect the entropy? Consider the types of motion undergone by molecules (Fig. 11–5). These include translations (displacements of the centers of the molecules), rotations (in which molecules spin like tops), and vibrations (stretching and compression of bonds). As the temperature is increased, the energies associated with all of these types of motion increase. Moreover, the number of ways in which this energy can be parcelled out among the molecules also increases; more microstates become available to the molecules. The conclusion is clear: the entropy of a body increases with increasing temperature.

By the same logic, as a pure crystalline substance is cooled toward the absolute zero of temperature, its entropy decreases. Its individual atoms and molecules have fewer and fewer alternatives in terms of their positions and energies, reducing the number of available microstates. In a perfectly ordered sample in which atoms can occupy only certain specific sites, there would be only one microstate, and the entropy would be zero. From such considerations, we can state the following:

> The entropy of a crystalline substance at equilibrium approaches zero as the absolute zero of temperature is approached.

This is one form of the **third law of thermodynamics,** which was discovered in the late 19th century by the German chemist Walther Nernst. It gives the baseline from which entropies at other temperatures can be measured.[1]

We have discussed in some detail the qualitative changes in entropy that result from heating, expanding, melting, and vaporizing a substance. How can we make this more quantitative and assign numerical values to the entropies of substances? A graphical method often used by chemists is illustrated in Figure 11–6. We measure how the molar heat capacity (c_P) of a substance depends on temperature, and

The fundamental relationship between entropy (S) and number of microstates (W) was derived by Ludwig Boltzmann in 1868. The equation he obtained "$S = k \log W$" is carved on his tombstone in Vienna. The constant k, now called Boltzmann's constant, is the ratio of the universal gas constant R to Avogadro's number, N_0, and the logarithm is the base e log, which is more usually written as ln.

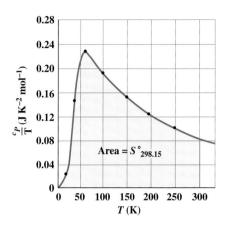

Figure 11–6   A graph of c_P/T for platinum. The black dots represent experimental measurements at $P = 1$ atm. The area under the curve up to any temperature is proportional to the molar entropy at that temperature. Here the area up to 298.15 K corresponding to $S°$, the standard molar entropy at 298.15 K, has been shaded.

determine the area under a graph of c_P/T against T from 0 K to any desired temperature. This area turns out to equal the molar entropy change going from 0 K to T. If a substance melts, boils, or undergoes other phase changes before reaching the temperature T, we must add the entropy change associated with those phase transitions to the result from the measurement of the area. The ΔS of a phase transition is given by the ΔH of the transition divided by the temperature at which it takes place. In the vaporization of a liquid at its normal boiling point T_b, for example,

$$\Delta S_{vap} = \frac{\Delta H_{vap}}{T_b}$$

The absolute entropy of one mole of substance in a standard state at 298.15 K is its **standard molar entropy** $(S°)$. Values of $S°$ are tabulated for a number of elements and compounds in Appendix D. The standard molar entropy has the unit $J\ K^{-1}\ mol^{-1}$. Figure 11–7 displays the $S°$ values of several groups of elements. Note that absolute entropies of all substances are positive (because entropy starts at zero at 0 K and increases with temperature).

• This unit is the same as the unit of the molar heat capacity ($J\ K^{-1}\ mol^{-1}$).

Entropies of Reaction

A perusal of the list of entropies in Appendix D and a glance at Figure 11–7 show that the entropies of gases tend to be higher than those of liquids and solids, as we anticipate from the greater number of microstates in the gaseous state. As a result, the entropy change in a reaction in which a gas is produced, such as

$$CaCO_3(s) \longrightarrow CaO(s) + CO_2(g)$$

is usually positive (Fig. 11–8). In reactions in which the total chemical amount of gas increases, the entropy almost always increases. Thus, the entropy change for the reaction

$$2\ H_2O(g) \longrightarrow 2\ H_2(g) + O_2(g)$$

is positive because there is a net increase from two moles of gas to three moles. Exceptions to this general statement exist, but they are rare.

Dissolution and precipitation reactions are major types of chemical reactions. It might seem that dissolution should always lead to an increase in the entropy and

[1]This form of the third law is actually due to the German physicist Max Planck. Nernst stated only that entropy *changes* become zero at the absolute zero of temperature. Although historically the third law was discovered after the second law, we introduce it earlier in our treatment. The second law follows soon.

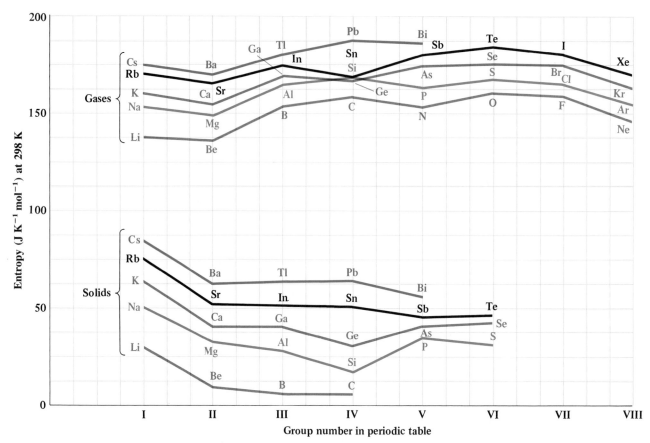

Figure 11-7 The standard entropies of the elements, like so many other physical properties, display smooth trends across and down the periodic table. Notice, however, the very large difference in entropy between the solid and gaseous states of the same element at 25°C.

precipitation to a decrease in the entropy of a system because molecules or ions certainly have fewer microstates organized in a crystal than dispersed through a solution. For many compounds, this is true: dissolved sodium chloride and other alkali halides have higher entropies than the solid plus pure water for exactly this reason. In other cases, however, dissolution leads to a *decrease* in entropy of the solution. An example is the dissolution of magnesium chloride:

$$MgCl_2(s) \longrightarrow Mg^{2+}(aq) + 2\ Cl^-(aq) \qquad \Delta S^\circ = -114.7\ \text{J K}^{-1}$$

The reason for the negative ΔS° is that the ions impose order on the water molecules around them in solution (recall Fig. 6–3). When this effect is large enough, it outweighs the entropy increase accompanying the dissolution of the highly ordered salt itself and leads to an overall decrease in the entropy of the system.

The tabulated entropies from Appendix D can be combined to calculate standard entropies of reaction. For the general reaction

$$aA + bB \longrightarrow cC + dD$$

the standard entropy change is

$$\Delta S^\circ = cS^\circ(C) + dS^\circ(D) - aS^\circ(A) - bS^\circ(B)$$

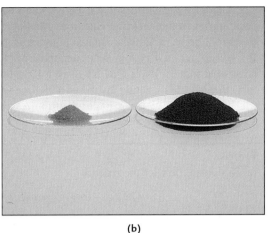

(a)   **(b)**

Figure 11–8   (a) The decomposition of the orange solid $(NH_4)_2Cr_2O_7$ provides a spectacular example of a spontaneous reaction in which gases are produced from a solid starting material

$$(NH_4)_2Cr_2O_7(s) \longrightarrow N_2(g) + 4\,H_2O(g) + Cr_2O_3(s)$$

The green dust that forms in the course of the reaction, together with the sparks and smoke, make this reacting system resemble a volcano. (b) The solid reactant and product.

That is, it equals the total standard entropy of the products minus the total standard entropy of the reactants. A calculation of $\Delta S°$ follows the pattern set by the Hess's law calculations of $\Delta H°$ from tables of $\Delta H_f°$ (Section 10–4) with one important difference: the standard molar entropies of the most stable forms of the elements (unlike their standard molar enthalpies of formation ($\Delta H_f°$) are *not* zero at 298.15 K. This point is illustrated in the following example.

• Room temperature lies far above absolute zero. A substance in a standard state at 298.15 K, whether a compound or an element, therefore has plenty of disorder. The entropy of a system can approach zero only at temperatures near absolute zero.

EXAMPLE 11–3

(a) Using the table of standard molar entropies in Appendix D, calculate the entropy change for the chemical reaction

$$N_2(g) + 2\,O_2(g) \longrightarrow 2\,NO_2(g)$$

with reactants and products in standard states at a temperature of 25°C.
(b) Calculate the change in entropy under the same circumstances for the conversion of 40.00 g of $N_2(g)$ and 91.38 g of $O_2(g)$ to 131.38 g of $NO_2(g)$.

Solution

(a) From the table,

$$S°(N_2) = 191.50 \text{ J K}^{-1} \text{ mol}^{-1}$$
$$S°(O_2) = 205.03 \text{ J K}^{-1} \text{ mol}^{-1}$$
$$S°(NO_2) = 239.95 \text{ J K}^{-1} \text{ mol}^{-1}$$

$$\Delta S° = 2 \text{ mol} \times \left(\frac{239.95 \text{ J}}{\text{K mol}}\right) - 1 \text{ mol} \times \left(\frac{191.50 \text{ J}}{\text{K mol}}\right) - 2 \text{ mol} \times \left(\frac{205.03 \text{ J}}{\text{K mol}}\right)$$

$$\underset{\text{for the NO}_2}{} \qquad \underset{\text{for the N}_2}{} \qquad \underset{\text{for the O}_2}{}$$

$$= -121.66 \text{ J K}^{-1}$$

The factor of 2 multiplies the $S°$ of NO_2 and the $S°$ of O_2 because two moles of each appear in the chemical equation. The decrease in entropy is consistent with the 3-to-2 decrease in the chemical amount of gases in the reaction.

(b) Now, the initial system is not one mole of $N_2(g)$ and two moles of $O_2(g)$; it is somewhat larger: 40.00 g of $N_2(g)$ is 1.428 mol, and 91.38 g of $O_2(g)$ is 2.856 mol. Because this reaction involves 1.428 times more material than the one in part (a), its $\Delta S°$ is 1.428 times larger than the $\Delta S°$ in part (a). Multiplying -121.66 J K^{-1} by 1.428 gives -173.7 J K^{-1}.

Exercise

(a) Calculate $\Delta S°$ for the reaction

$$CH_4(g) \longrightarrow C(graphite) + 2\ H_2(g)$$

(b) Calculate $\Delta S°$ of the system when 26.71 g of $CH_4(g)$ reacts to give 20.00 g of $C(graphite)$ and 6.71 g of $H_2(g)$ at 25°C. Use the data in Appendix D.

Answer: (a) $+80.73$ J K^{-1}. (b) $+134.4$ J K^{-1}.

• A negative absolute entropy implies more order in the system than perfect order—an obvious impossibility.

Some $S°$ values in Appendix D, such as that of 1 M $Mg^{2+}(aq)$ and other ions in solution, are negative. Because ions in solution are mixtures, they do not become pure crystals at absolute zero. A negative absolute entropy is impossible, and these values are not absolute entropies but are referenced to another standard. This has no effect on the method of calculation of $\Delta S°$, as is shown in the following example.

EXAMPLE 11–4

Compute $\Delta S°$ for the dissolution of $MgCl_2(s)$ in water. Use data from Appendix D.

Solution

The chemical reaction is

$$MgCl_2(s) \longrightarrow Mg^{2+}(aq) + 2\ Cl^-(aq)$$

From Appendix D,

$$S°(Mg^{2+}(aq)) = -138.1 \text{ J K}^{-1} \text{ mol}^{-1}$$
$$S°(Cl^-(aq)) = 56.5 \text{ J K}^{-1} \text{ mol}^{-1}$$
$$S°(MgCl_2(s)) = 89.62 \text{ J K}^{-1} \text{ mol}^{-1}$$

$$\Delta S° = 2 \text{ mol} \times \left(\frac{56.5 \text{ J}}{\text{K mol}}\right) + 1 \text{ mol} \times \left(\frac{-138.1 \text{ J}}{\text{K mol}}\right) - 1 \text{ mol} \times \left(\frac{89.62 \text{ J}}{\text{K mol}}\right)$$

$$\underbrace{}_{\text{for the } Cl^-(aq)} \qquad \underbrace{}_{\text{for the } Mg^{2+}(aq)} \qquad \underbrace{}_{\text{for the } MgCl_2(s)}$$

$$= -114.7 \text{ J K}^{-1}$$

Exercise

Compute $\Delta S°$ for the reaction $HCl(g) \longrightarrow H^+(aq) + Cl^-(aq)$.

Answer: $\Delta S° = -130.3$ J K^{-1}.

11-4 THE SECOND LAW OF THERMODYNAMICS

It was shown in Section 11–1 that the sign of the change in enthalpy during a process fails as a criterion for spontaneity, although some link between exothermicity and spontaneous change surely does exist. Does the entropy provide this criterion? At first, it appears that it does not. Although some spontaneous processes, such as the free expansion of a gas, are clear examples of increases in the randomness of the system, others are clear counterexamples. Below 0°C, water undergoes a spontaneous phase transition, freezing, and *decreases* its entropy. All freezing and condensation transitions have negative ΔS's. The entropy change from reactants to products is also negative for many spontaneous chemical reactions. When solutions of silver nitrate and sodium chloride are mixed, solid silver chloride precipitates spontaneously

$$\text{Ag}^+(aq) + \text{Cl}^-(aq) \longrightarrow \text{AgCl}(s)$$

even though the entropy of $\text{AgCl}(s)$ is lower than that of the original ions (the $\Delta S°$ of this reaction is -33 J K^{-1}).

To arrive at a general criterion for spontaneous change, we need to broaden our view to consider not only the system (the water that freezes in a beaker, or the solid silver chloride and its aqueous solution) but the surroundings as well. When water freezes or silver chloride precipitates, heat is liberated from the system into the surroundings (the beaker, the table, and the surrounding air, for example). The entropy of the surroundings thereby changes, as well as that of the system. The **second law of thermodynamics** states:

> In any spontaneous process, the entropy of the universe (that is, of system plus surroundings) increases.

Put in the form of an equation, the second law reads

$$\Delta S_{\text{univ}} = \Delta S_{\text{sys}} + \Delta S_{\text{surr}} > 0 \quad \text{(spontaneous process)}$$

Any process leading to a *decrease* in the entropy of the universe is forbidden. Its *reverse* occurs spontaneously, however, because if the forward process has a negative ΔS_{univ}, the reverse automatically has a positive ΔS_{univ}. If $\Delta S_{\text{univ}} = 0$, the forward and reverse processes occur equally well, and the process is at equilibrium.

With the aid of the second law of thermodynamics, we can begin to understand the different types of spontaneous processes discussed in this chapter. In the free expansion of an ideal gas, no heat is exchanged with the surroundings. Because the surroundings are unaffected, their entropy change is zero, and the direction of spontaneous change is determined entirely by the entropy of the system—that is, by the direction in which the randomness of the system increases. In an exothermic chemical reaction or phase change, heat flows to the surroundings, increasing their entropy. When a positive ΔS_{surr} is large enough, it compensates for a decrease in entropy of the system, as in the freezing of water. Using the second law, we understand why the principle of Berthelot and Thomsen linking spontaneous change to exothermicity sometimes worked and sometimes failed. Under the second law, spontaneous endothermic changes may occur, despite an entropy decrease in the surroundings, if there is sufficient compensation from an entropy increase in the system.

The second law of thermodynamics profoundly affects the way in which we look at nature and at physical processes. It provides an arrow to time: a direction for the evolution of physical systems. Energy is conserved in both forward and reverse processes, so a process and its reverse are equally well allowed as far as the first law of thermodynamics is concerned. The entropy of the universe, however, increases in only one of the two directions. Real processes that decrease ΔS_{univ} are impossible or, rather, improbable beyond all conception.

The second law, like the first, has enormous practical importance. If a process is impossible, either because energy is not conserved or because it would lead to a decrease in the entropy of the universe, there is no reason to waste time trying to make it happen. Attempts by inventors to build perpetual-motion machines always fail because the machines violate one or another of the laws of thermodynamics. This, sadly, has not deterred other inventors, ignorant of the laws of thermodynamics, from trying their hand.

11-5 THE GIBBS FUNCTION

In the previous section, the sign of the change in entropy of a system plus its surroundings (that is, the change of total entropy ΔS_{univ}) emerged as the ultimate criterion for deciding if a process can occur without outside intervention:

$$\Delta S_{univ} > 0 \quad \text{(spontaneous)}$$
$$\Delta S_{univ} = 0 \quad \text{(equilibrium)}$$
$$\Delta S_{univ} < 0 \quad \text{(non-spontaneous)}$$

Using this criterion requires calculating the entropy change of the surroundings as well as the entropy change of the system, something that is usually impracticable. It would certainly be desirable to have a state property that gives the feasibility of a process without reference to the surroundings. If we limit consideration to changes that go on at constant temperature and pressure (a most important type of change in chemistry), such a property exists. It is the **Gibbs function (G)** which is defined as

$$G = H - TS$$

• The Gibbs function is named after J. Willard Gibbs, a pioneering mathematical physicist at Yale University who developed the field of statistical thermodynamics in the late 1800s.

The Gibbs function is measured in joules, just like the enthalpy (H) and internal energy (E). It is also called the *Gibbs free energy* and the *free enthalpy*.

To see how the Gibbs function provides a criterion for spontaneity at constant pressure and temperature, we focus on the *change* in the Gibbs function of a system during a process. From the preceding definition

$$\Delta G_{sys} = \Delta H_{sys} - \Delta(T_{sys}S_{sys})$$

If the temperature of the system stays constant and equal to the temperature of the surroundings, then T_{sys} can be taken outside the parentheses and abbreviated to T:

$$\Delta G_{sys} = \Delta H_{sys} - T\,\Delta S_{sys}$$

The connection between ΔG_{sys}, and ΔS_{univ}, the total entropy change, is made by noting (although we shall not derive the equation here) that at constant pressure and temperature

$$\Delta S_{surr} = -\frac{\Delta H_{sys}}{T}$$

The minus sign is important in this equation. If the process is exothermic, ΔH_{sys} is negative, and the entropy change of the surroundings is positive. The total entropy change in a spontaneous process is then

$$\Delta S_{univ} = \Delta S_{sys} + \Delta S_{surr} = \Delta S_{sys} - \frac{\Delta H_{sys}}{T} > 0$$

If we multiply both sides of the inequality by the absolute temperature T, which is always positive, it becomes

$$T \, \Delta S_{sys} - \Delta H_{sys} > 0 \qquad \text{(spontaneous process at constant } T \text{ and } P\text{)}$$

or, equivalently,

$$\Delta H_{sys} - T \, \Delta S_{sys} = \Delta G_{sys} < 0 \qquad \text{(spontaneous process at constant } T \text{ and } P\text{)}$$

- The direction of the inequality changes when we multiply it through by -1 in this step.

This means that if the Gibbs function of a system is forecast (by a computation perhaps) to decrease during a process at constant temperature and pressure, then the process is spontaneous. If the Gibbs function is forecast to increase, the process cannot occur under the conditions (because then ΔS_{univ} would be negative), but the reverse of the process is spontaneous. If the Gibbs function is forecast to stay constant, then ΔS_{univ} equals zero for the process and no change in either direction can occur under the conditions. This amounts to a state of equilibrium. In summary:

For a change at constant temperature and pressure
$$\begin{cases} \Delta G_{sys} < 0 & \text{(spontaneous)} \\ \Delta G_{sys} = 0 & \text{(equilibrium)} \\ \Delta G_{sys} > 0 & \text{(non-spontaneous, but reverse spontaneous)} \end{cases}$$

The Gibbs Function and Phase Transitions

To see the use of the Gibbs function, consider the freezing of one mole of a liquid to form a solid. The most familiar example is the water-to-ice transition:

$$H_2O(\ell) \longrightarrow H_2O(s)$$

Examine first what happens when this process is carried out at the freezing point of water under atmospheric pressure, 273.15 K. The measured enthalpy change (the heat absorbed at constant pressure) associated with freezing one mole of water is

$$\Delta H_{273.15} = -6007 \text{ J}$$

and the entropy change is

- Why is $\Delta H_{273.15}$ negative? Heat leaves the water when it freezes; the heat that it *absorbs* is therefore negative.

$$\Delta S_{273.15} = \frac{\Delta H_{273.15}}{T_{freez}} = \frac{-6007 \text{ J}}{273.15 \text{ K}} = \frac{-21.99 \text{ J}}{\text{K}}$$

The change in the Gibbs function of the system is then

$$\Delta G_{273.15} = \Delta H_{273.15} - T \, \Delta S_{273.15} = -6007 \text{ J} - (273.15 \text{ K}) \times \left(\frac{-21.99 \text{ J}}{\text{K}} \right) = 0 \text{ J}$$

Finding that $\Delta G_{273.15}$ equals 0 is no great surprise. At the normal freezing point, the change in the Gibbs function equals 0 because liquid and solid are in equilibrium at this temperature.

Now consider what happens as water is cooled below 273.15 K to 263.15 K ($-10.00°C$). Liquid water under these conditions can exist temporarily; it is undercooled (see Section 6–4). To calculate the change in the Gibbs function of the water as it freezes at this lower temperature, we assume that neither ΔH nor ΔS for the freezing process changes significantly with the drop in temperature. Then

$$\Delta G_{263.15} = -6007 \text{ J} - (263.15 \text{ K})\left(\frac{-21.99 \text{ J}}{\text{K}}\right)$$
$$= -220 \text{ J}$$

Because $\Delta G < 0$, liquid water tends to freeze without outside intervention at 263.15 K. A similar calculation using any temperature *greater* than 273.15 gives a ΔG greater than zero, showing that freezing of the liquid is non-spontaneous in that range of temperature. This accords with daily experience. Water at atmospheric pressure never freezes if the temperature exceeds 273.15 K (equivalent to 0°C or 32°F); rather, the reverse process occurs, and ice melts.

Writing the change in the Gibbs function as

$$\Delta G = \Delta H - T \, \Delta S \qquad \text{(\textit{T} and \textit{P} constant)}$$

reveals that a negative value of ΔG is favored by a *negative* value of ΔH and a *positive* value of ΔS. In the freezing of a liquid, ΔH is negative; however, ΔS for freezing is *also* negative, rather than positive, because the solid has lower entropy than the liquid. Whether a liquid freezes at a given temperature and pressure depends on the outcome of a competition between two factors: an enthalpy effect that favors freezing and an entropy effect that opposes it (Fig. 11–9). At low temperatures (below T_{freez}) the enthalpy effect dominates, and the liquid freezes spontaneously. At higher temperatures (above T_{freez}) the entropy effect dominates, and freezing does not occur. At T_{freez} the Gibbs functions of the solid and liquid are equal (remember that the change in going from one to the other, ΔG, is zero) and the two phases coexist in a state of equilibrium. Similar analysis can be carried out for other phase transitions, such as boiling a liquid.

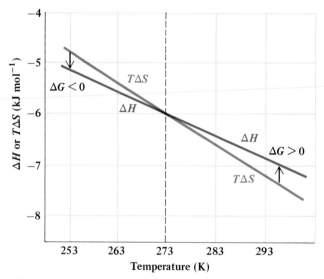

Figure 11–9 Plots of ΔH and $T \, \Delta S$ versus temperature for the freezing of water. At 273.15 K, the two curves cross, so at this temperature $\Delta G = 0$ and ice and water coexist. Below this temperature, the freezing of water to ice is spontaneous; above it, the reverse process, the melting of ice to water, is spontaneous.

EXAMPLE 11-5

The normal boiling point of benzene (C_6H_6) is 80.1°C, and the molar enthalpy of vaporization of benzene at that temperature is 30.8 kJ mol^{-1}.
(a) Calculate the entropy of vaporization of 1.00 mol of benzene.
(b) Calculate the change in the Gibbs function of the system if 1.00 mol of benzene vaporizes at 60°C and a pressure of 1 atm. Does benzene boil at this temperature and pressure?

Solution

(a) The change under consideration is

$$C_6H_6(\ell) \longrightarrow C_6H_6(g)$$

By analogy to the case of fusion, the entropy of vaporization is the enthalpy change divided by the boiling temperature in kelvins (80.1 + 273.15 = 353.25 K), so that

$$\Delta S_{vap} = \frac{\Delta H_{vap}}{T_b} = \frac{30.8 \times 10^3 \text{ J}}{353.25 \text{ K}} = +87.2 \text{ J K}^{-1}$$

(b) At 60°C (or 333.15 K), we have

$$\begin{aligned} \Delta G_{vap} &= \Delta H_{vap} - T\,\Delta S_{vap} \\ &= 30{,}800 \text{ J} - (333.15 \text{ K})(87.2 \text{ J K}^{-1}) \\ &= 1749 \text{ J} = +1.7 \text{ kJ} \end{aligned}$$

Because ΔG is positive, benzene does not boil at this temperature and a pressure of 1 atm.

Exercise
Given the data

	ΔH_f° (kJ mol^{-1})	S° (J K^{-1} mol^{-1})
$PCl_3(g)$	−287.0	311.7
$PCl_3(\ell)$	−319.7	217.1

(a) Calculate ΔG for the vaporization of 1.00 mol of $PCl_3(\ell)$ to give $PCl_3(g)$ at 25°C and 1 atm. Does PCl_3 boil under these conditions? (b) Estimate the normal boiling point of $PCl_3(\ell)$.

Answer: (a) $\Delta G = +4.5$ kJ; PCl_3 does not boil under these conditions. (b) 346 K (which equals 73°C).

Trouton's Rule

Remarkably, experiment shows that most liquids have about the same molar entropy of vaporization at their normal boiling points: the increase in disorder in changing a mole of a liquid to a gas is nearly the same for every substance. **Trouton's rule** states the magnitude of this entropy change:

$$\Delta S_{vap} = 88 \pm 5 \text{ J K}^{-1} \text{ mol}^{-1}$$

Note that the ΔS_{vap} of benzene, calculated in Example 11–5, is well within this range. Trouton's rule exists because vaporizations occur with an increase in molar volume that is huge and also nearly the same from substance to substance. The entropy increase arising from the increased volume is big enough to drown out other contributions to ΔS. Because of the relationship between the entropy and enthalpy of vaporization,

$$\Delta S_{vap} = \frac{\Delta H_{vap}}{T_b}$$

the constancy of ΔS_{vap} means that ΔH_{vap} and T_b, which vary widely from substance to substance, must do so in the same proportion. Trouton's rule finds use in making rough estimates of enthalpies of vaporization from measurements of boiling points. Exceptions exist among substances that have especially large amounts of order as liquids. The ΔS_{vap} of water, for instance, is 109 J K^{-1} mol^{-1} because extensive hydrogen bonding organizes the liquid (see Section 6–1).

11–6   THE GIBBS FUNCTION AND CHEMICAL REACTIONS

The sign of the change in the Gibbs function tells whether a proposed process can occur at constant temperature and pressure (Fig. 11–10). We now present methods for computing changes in the Gibbs function for chemical reactions.

The Standard Gibbs Function of Formation

The change in the Gibbs function for a chemical reaction taking place at constant temperature is

$$\Delta G = \Delta H - T\,\Delta S$$

where ΔH is the enthalpy change of the reaction, and ΔS is its entropy change. Because we cannot know the *absolute* value of the Gibbs function of a substance (just as we cannot know the absolute value of its enthalpy H), we must define a **standard molar Gibbs function of formation** (ΔG_f°).

• ΔG_f° is analogous to the standard molar enthalpy of formation (ΔH_f°), which was introduced in Section 10–4.

We use the same definitions of standard states as previously (see Section 10–4): for solids and liquids, the state of the liquid or solid under a pressure of 1 atm and at a specified temperature; for gases, the gaseous phase under a pressure of 1 atm, at a specified temperature, and exhibiting ideal-gas behavior; for dissolved species, a 1 M solution under a pressure of 1 atm, at a specified temperature, and exhibiting ideal-solution behavior. The ΔG_f° of a compound is the change in Gibbs function for the reaction in which one mole of pure compound in a standard state at 298.15 K forms from the most stable forms of its constituent elements, also in standard states at 298.15 K. For example, ΔG_f° for $CO_2(g)$ is the change in Gibbs function accompanying the reaction

$$C(graphite) + O_2(g) \longrightarrow CO_2(g)$$

We can calculate ΔG_f° at 25°C from ΔH° and ΔS° for this reaction at the same temperature. Here ΔH° is simply ΔH_f° for $CO_2(g)$

$$\Delta H^{\circ} = \Delta H_f^{\circ}(CO_2(g)) = -393.51 \text{ kJ}$$

and ΔS° can be obtained from the absolute entropies of the substances involved at

25°C and a pressure of 1 atm (both elements *and* compounds, because the absolute entropy $S°$ of an element in its standard state does *not* equal zero).

$$\Delta S° = S°(CO_2(g)) - S°(C(graphite)) - S°(O_2(g))$$

$$\Delta S° = 1 \text{ mol} \times \left(\frac{213.63 \text{ J}}{\text{K mol}}\right) - 1 \text{ mol} \times \left(\frac{5.74 \text{ J}}{\text{K mol}}\right) - 1 \text{ mol} \times \left(\frac{205.03 \text{ J}}{\text{K mol}}\right) = 2.86 \frac{\text{J}}{\text{K}}$$

for the $CO_2(g)$ for the C(s, gr)   for the $O_2(g)$

The $\Delta G_f°$ for $CO_2(g)$ at 25°C is then

$$\Delta G_f° = \Delta H_f° - T \Delta S°$$

$$= -393.51 \text{ kJ} - 298.15 \text{ K} \times \left(2.86 \frac{\text{J}}{\text{K}}\right) \times \left(\frac{1 \text{ kJ}}{10^3 \text{J}}\right) = -394.36 \text{ kJ}$$

From our definition, the $\Delta G_f°$ for an *element* in its most stable form at 298.15 K is zero: preparing an element from itself is no change at all.

Because G is a state property, we can add chemical equations together, combining their associated $\Delta G_f°$'s in the same way that we do for changes in enthalpy. Thus, for the reaction

$$aA + bB \longrightarrow cC + dD$$

we have

$$\Delta G° = c\Delta G_f°(C) + d\Delta G_f°(D) - a\Delta G_f°(A) - b\Delta G_f°(B)$$

This is the difference between the Gibbs function of the pure products in standard states and the Gibbs function of the pure reactants in standard states. A relatively short table of values of $\Delta G_f°$, such as that in Appendix D, can be used to calculate the $\Delta G°$ of a host of chemical reactions at 298.15 K.

EXAMPLE 11-6

Calculate $\Delta G°$ for the following reaction at 25°C, using tabulated values for $\Delta G_f°$ from Appendix D.

$$3 \text{ NO}(g) \longrightarrow \text{N}_2\text{O}(g) + \text{NO}_2(g)$$

Solution

$$\Delta G° = \Delta G_f°(\text{N}_2\text{O}) + \Delta G_f°(\text{NO}_2) - 3 \Delta G_f°(\text{NO})$$

$$\Delta G° = 1 \text{ mol} \times \left(\frac{104.18 \text{ kJ}}{\text{mol}}\right) + 1 \text{ mol} \times \left(\frac{51.29 \text{ kJ}}{\text{mol}}\right) - 3 \text{ mol} \times \left(\frac{86.55 \text{ kJ}}{\text{mol}}\right)$$

for the N_2O for the NO_2 for the NO

$$= -104.18 \text{ kJ}$$

Exercise

Calculate $\Delta G°$ for the following reaction at 25°C using $\Delta G_f°$ values from Appendix D:

$$\text{Fe}_2\text{O}_3 \text{ (hematite)} + 3 \text{ Co}(s) \longrightarrow 3 \text{ CoO}(s) + 2 \text{ Fe}(s)$$

Answer: +99.5 kJ.

• The standard Gibbs function of formation computed in this way must, and in fact does, equal the value given in Appendix D.

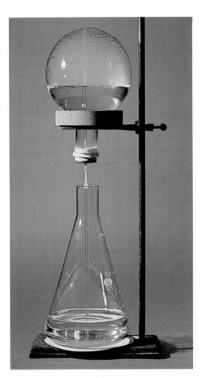

Figure 11-10 The dissolution of hydrogen chloride in water

$$\text{HCl}(g) \longrightarrow \text{H}^+(aq) + \text{Cl}^-(aq)$$

is a spontaneous process, with $\Delta G° = -35.9 \text{ kJ}$. In this demonstration, the upper flask is filled with gaseous hydrogen chloride and a small amount of water is injected into it. As the hydrogen chloride dissolves spontaneously, its pressure drops. The resulting partial vacuum in the upper flask draws water up the tube from the lower flask, allowing more hydrogen chloride to dissolve. The change is so fast that a vigorous fountain of water plays into the upper flask. The change in the Gibbs function of the process appears as work, raising the water.

Effects of Temperature on $\Delta G°$

Values of $\Delta G°$ calculated from Appendix D apply only at the temperature 298.15 K (25°C). For *other* temperatures, we can still use the relation

$$\Delta G° = \Delta H° - T\,\Delta S°$$

together with tabulated values of entropies and standard enthalpies of formation to estimate the change in Gibbs function for a chemical reaction carried out at a pressure of 1 atm. The estimate is close if $\Delta H°$ and $\Delta S°$ do not themselves depend too strongly on temperature, which is usually the case (Fig. 11–11).

Changes in Gibbs function depend strongly on the absolute temperature because T appears explicitly in the preceding equation. Four different types of behavior are possible (Fig. 11–12), depending on the relative signs of $\Delta H°$ and $\Delta S°$. If $\Delta H°$ is negative and $\Delta S°$ is positive, then $\Delta G°$ is always negative, and the standard reaction is spontaneous at all temperatures. If the reverse is true ($\Delta H°$ is positive and $\Delta S°$ is negative), then the standard reaction is never spontaneous. The other two cases are more interesting. In both, there is a temperature

$$T = \frac{\Delta H°}{\Delta S°}$$

at which $\Delta G°$ equals zero. At this temperature, the reactants and products, at partial pressures of 1 atm or concentrations of 1 mol L^{-1}, are in equilibrium. If $\Delta H°$ and $\Delta S°$ are both negative, the reaction is spontaneous at temperatures below this crossover point and not spontaneous above. If both are positive, the reaction is spontaneous above this temperature and not spontaneous below. An example of the latter is the decomposition of $CaCO_3$,

$$CaCO_3(s) \longrightarrow CaO(s) + CO_2(g)$$

in which both $\Delta H°$ and $\Delta S°$ are positive. The reaction occurs spontaneously only at elevated temperatures, for a CO_2 gas pressure of 1 atm.

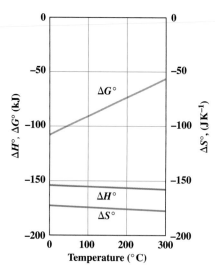

Figure 11–11 The standard entropy change of the reaction $3\,NO(g) \longrightarrow N_2O(g) + NO_2(g)$ varies less than 5% between 0°C and 300°C; the standard enthalpy of reaction is even closer to being constant. The change in the standard Gibbs function for the reaction, however, grows rapidly more positive with increasing temperature.

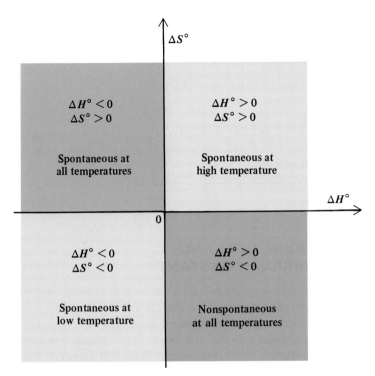

Figure 11–12 The spontaneity of a reaction at different temperatures depends on the signs of $\Delta H°$ and $\Delta S°$. When both signs are the same, the temperature determines the spontaneity of reaction.

EXAMPLE 11-7

A slow reaction in the earth's interior combines dolomite with quartz to produce diopside and carbon dioxide:

$$CaMg(CO_3)_2(s) + 2\ SiO_2(s) \longrightarrow CaMgSi_2O_6(s) + 2\ CO_2(g)$$

$$\text{dolomite} \qquad \text{quartz} \qquad\qquad \text{diopside} \qquad \text{carbon dioxide}$$

The standard enthalpy and entropy changes of this reaction at 298.15 K are

$$\Delta H° = +155 \text{ kJ}$$
$$\Delta S° = +331 \text{ J K}^{-1}$$

Assume that the CO_2 produced is at atmospheric pressure (which is not accurate for the interior of the earth), and estimate the temperature ranges in which this reaction is and is not spontaneous, if $\Delta H°$ and $\Delta S°$ do not change with temperature.

Solution

The temperature at which $\Delta G°$ is zero should be approximately

$$T = \frac{\Delta H°}{\Delta S°} = \frac{155,000 \text{ J}}{331\ \dfrac{\text{J}}{\text{K}}} = 468 \text{ K} = 195°C$$

Because $\Delta H°$ and $\Delta S°$ are both positive, the reaction becomes spontaneous *above* this temperature and remains non-spontaneous below this temperature.

Exercise

Compute $\Delta H°$ and $\Delta S°$ at 298.15 K (obtain the data from Appendix D), and use them to estimate the temperature ranges in which the reaction

$$C(graphite) + H_2O(g) \longrightarrow CO(g) + H_2(g)$$

is and is not spontaneous.

Answer: $\Delta H° = +131.30$ kJ, and $\Delta S° = +133.67$ J K^{-1}. The reaction is spontaneous above approximately 980 K and non-spontaneous below 980 K when the three gases have pressures of 1 atm.

11-7 THE GIBBS FUNCTION AND THE EQUILIBRIUM CONSTANT

The standard change in the Gibbs function $\Delta G°$ for a chemical reaction applies only to the case in which separated pure reactants are converted into separated pure products, and all reactants and products are in their standard states (gases at a pressure of 1 atm and dissolved species at a concentration of 1 mol L^{-1}) at a specified temperature. This is a very artificial situation. What is the change in the Gibbs function when the reactants and products have *nonstandard* partial pressures and concentrations? To answer this question, recall that the reaction quotient (Q), which gives the current position of a reaction, becomes equal to the equilibrium constant (K) at equilibrium. This establishes the direction that the reaction takes in coming to equilibrium: if $Q < K$, then the reaction shifts toward the product side to reach equilibrium; if $Q > K$, it shifts toward the reactant side. It is therefore not surprising to discover a connection between ΔG and Q (Table 11–1).

The exact relationship between the two is given by the equation

$$\Delta G = \Delta G° + RT \ln Q$$

- The natural (or base e) logarithm of a number is written ln. Scientific calculators have an "ln x" key for natural logarithms and an "e^x" key (or the equivalent) for the inverse operation, calculating the exponential of x (Appendix C).

Let us examine this important equation closely. If all the reactants and products are in standard states, the reaction quotient Q is equal to 1. Because the logarithm of 1 is zero, the last term vanishes, and we have $\Delta G = \Delta G°$, as we would expect from the definitions of standard states. When at least one partial pressure is not 1 atm or at least one concentration is not 1 mol L^{-1}, we simply insert the actual values into the expression for Q and proceed to evaluate the sign and magnitude of ΔG. At equilibrium, ΔG becomes zero, because there is no tendency for reaction to occur in either direction. At the same time, Q becomes equal to the equilibrium constant K. If we substitute these two values into the preceding equation, we obtain

- Note that $\Delta G°$ (unlike ΔG) depends only on the intrinsic properties of the products and reactants and does not change during the course of the reaction.

$$0 = \Delta G° + RT \ln K$$

or

$$-\Delta G° = RT \ln K$$

- This linkage of the results of thermochemical experiments to the nature of the equilibrium state is at the heart of chemical thermodynamics.

This result provides a way to calculate equilibrium constants for chemical reactions from tabulated standard Gibbs functions of formation. On a deeper level, it shows that the extent that any chemical reaction must proceed to reach equilibrium can be deduced from calorimetric data alone.

Table 11-1
Criteria for Spontaneity

Spontaneous Processes	Equilibrium Processes	Non-spontaneous Processes	Conditions
$\Delta S_{univ} > 0$	$\Delta S_{univ} = 0$	$\Delta S_{univ} < 0$	All conditions
$\Delta G_{sys} < 0$	$\Delta G_{sys} = 0$	$\Delta G_{sys} > 0$	Constant P and T
$Q < K$	$Q = K$	$Q > K$	Constant P and T

EXAMPLE 11-8

The $\Delta G°$ at 25°C for the chemical reaction

$$3 \text{ NO}(g) \longrightarrow \text{N}_2\text{O}(g) + \text{NO}_2(g)$$

was calculated in Example 11–6. Now, calculate the equilibrium constant of this reaction at 25°C.

Solution

The $\Delta G°$ during conversion of three moles of NO to one mole of N_2O and one mole of NO_2 was found to be -104.18 kJ. We use $-\Delta G° = RT \ln K$ to compute K. To keep the units of the calculation correct, we rewrite $\Delta G°$ as -104.18 kJ mol^{-1}, where the "per mole" signifies per mole of the reaction as it is written—that is, per three moles of NO, one mole of N_2O, and one mole of NO_2, the chemical amounts in the balanced equation. Substitution then gives

$$\ln K = \frac{-\Delta G°}{RT} = -\frac{\left(\dfrac{-104{,}180 \text{ J}}{\text{mol}}\right)}{\left(\dfrac{8.315 \text{ J}}{\text{K mol}}\right) \times 298.15 \text{ K}} = 42.02$$

$$K = \text{antiln } 42.02 = e^{42.02} = 1.8 \times 10^{18}$$

If we had written the reaction with doubled coefficients (6 NO $\longrightarrow$ 2 N_2O + 2 NO_2), then $\Delta G°$ would have been doubled, to $-208{,}360$ J per mole of reaction (that is, six moles of NO, two moles of N_2O, and two moles of NO_2). Consequently, $\ln K$ would have been doubled and K would have been squared, exactly suiting the equilibrium expression that is written for the equation with doubled coefficients.

Exercise

Calculate the equilibrium constants at 25°C for the reaction that is the reverse of the above reaction and also for the reaction that is the reverse reaction multiplied through by 1/2:

$$\text{N}_2\text{O}(g) + \text{NO}_2(g) \longrightarrow 3 \text{ NO}(g) \quad \text{and} \quad \tfrac{1}{2} \text{N}_2\text{O}(g) + \tfrac{1}{2} \text{NO}_2(g) \longrightarrow \tfrac{3}{2} \text{NO}(g)$$

State the relationship between these answers and the answer to Example 11–8.

Answer: 5.6×10^{-19} and 7.5×10^{-10}, respectively. The first is the reciprocal of the answer in Example 11–8; the second is the square root of the reciprocal of that answer.

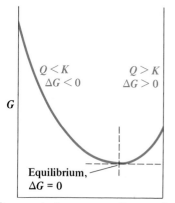

$Q < K$
$\Delta G < 0$

$Q > K$
$\Delta G > 0$

G

Equilibrium,
$\Delta G = 0$

Pure reactants   **Pure products**

Figure 11–13 The Gibbs function of a reaction system is plotted against its progress from pure reactants (*left*) to pure products (*right*). The quantity ΔG is best understood as the slope of this line. Equilibrium comes at the lowest point of the line, where $\Delta G = 0$. To the reactant side of equilibrium, the slope is negative, $\Delta G < 0$, and $Q < K$. To the product side, the slope is positive, $\Delta G > 0$, and $Q > K$.

The conversion of $NO(g)$ to $N_2O(g)$ plus $NO_2(g)$ is spontaneous when all three gases are present at 1 atm and at 25°C, and its reverse is therefore non-spontaneous. The forward reaction is scarely observed because it is so slow at room temperature, but its equilibrium constant can nonetheless be calculated! Such calculations often have enormous practical impact. All three of the nitrogen oxides in Example 11–8, for example, play roles in air pollution, but N_2O is perhaps the least noxious. Such a calculation would ensure that a plan to exploit the reaction to cut down the amount of NO in cooled automobile exhaust would not be thermodynamically doomed at the outset. The fundamental reaction tendency is there. Success of the plan would hinge on making the reaction go faster.

The relationship between ΔG and Q can be used to calculate the change in the Gibbs function for a reaction under nonstandard conditions. Substituting the expression connecting $\Delta G°$ to K into the equation for ΔG gives

$$\Delta G = \Delta G° + RT \ln Q = -RT \ln K + RT \ln Q = RT(\ln Q - \ln K)$$

If we now use the property of logarithms that

$$\ln a - \ln b = \ln \frac{a}{b}$$

we can rewrite this as

$$\Delta G = RT \ln \frac{Q}{K}$$

Here the connection between the two different criteria for spontaneous reaction is quite clear. It is summarized in Figure 11–13. If the reaction quotient, Q, is *less* than K, then the quantity Q/K is obviously less than 1. The logarithm of any number that is less than 1 is negative, so ΔG is less than 0. The reaction tends to proceed as written, from left to right. If Q exceeds K, then ΔG is greater than 0, and the *reverse* reaction (right to left) tends to occur until equilibrium is reached. The reaction-quotient criterion for the direction of reaction is a consequence of the second law of thermodynamics.

EXAMPLE 11–9

Calculate the change in Gibbs function (ΔG) in the reaction from Example 11–8 at 25°C

$$3\ NO(g) \longrightarrow N_2O(g) + NO_2(g)$$

but under conditions in which the partial pressures of the three gases are all 0.0010 atm rather than 1 atm.

Solution

Under conditions in which all partial pressures are 0.0010 atm, the reaction quotient is

$$Q = \frac{(P_{N_2O})(P_{NO_2})}{(P_{NO})^3} = \frac{(0.0010)(0.0010)}{(0.0010)^3} = 1.0 \times 10^3$$

The change in Gibbs function is then

$$\Delta G = RT \ln \frac{Q}{K}$$

$$= \frac{8.315 \text{ J}}{\text{K mol}} \times 298.15 \text{ K} \times \ln \left(\frac{1.0 \times 10^3}{1.8 \times 10^{18}} \right) = -8.71 \times 10^4 \frac{\text{J}}{\text{mol}}$$

$$= \frac{-87.1 \text{ kJ}}{\text{mol}}$$

This is less negative than under atmospheric-pressure conditions, so the reaction has a lesser tendency to occur at these lower pressures.

We may think of $-\Delta G$ as the driving force for the reaction; low-pressure conditions reduce the magnitude of the driving force of this reaction. This statement corresponds with the conclusion from Le Chatelier's principle (see Chapter 7) that in a reaction in which the chemical amount of gas decreases, an increase in volume tends to drive the equilibrium back toward the reactants.

Exercise

Calculate the change in Gibbs function (ΔG) at 25°C in the reaction

$$N_2O(g) + NO_2(g) \longrightarrow 3 \text{ NO}(g)$$

under conditions in which the partial pressures of the three gases are all 2.00 atm rather than 1 atm.

Answer: +105.9 kJ for the reaction as written.

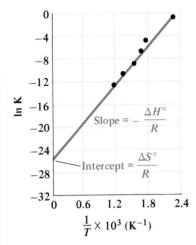

Figure 11–14 The temperature dependence of the equilibrium constant for the reaction

$$N_2(g) + 3 \text{ H}_2(g) \rightleftharpoons 2 \text{ NH}_3(g)$$

Experimental data are shown by points.

11-8 THE TEMPERATURE DEPENDENCE OF EQUILIBRIUM CONSTANTS

Engineers and chemists must often estimate in advance the value of an equilibrium constant for a reaction that is to take place at a constant temperature other than 25°C. Equilibrium constants are strongly affected by temperature, so using $K_{298.15}$ is unacceptable. Fortunately, $\Delta H°$ and $\Delta S°$ do *not* vary much, at least over a limited temperature range (recall Fig. 11–11). To the extent that the temperature dependence of $\Delta H°$ and $\Delta S°$ can be neglected, ln K in a linear function of $1/T$:

$$\ln K = -\frac{\Delta G}{RT} = -\frac{\Delta H°}{RT} + \frac{\Delta S°}{R}$$

• Appendix C reviews the slope-intercept form of the equation for a straight line.

A graph of ln K against $1/T$ is approximately a straight line with a slope of $-\Delta H°/R$ and an intercept of $\Delta S°/R$ (Fig. 11–14). The equation works two ways. Besides giving equilibrium constants at various temperatures, it opens a route for calculating enthalpy and entropy changes from experimentally measured equilibrium constants. This is particularly useful for solution-phase reactions, in which entropies cannot be measured calorimetrically.

Moreover, if the value of K at one temperature and also $\Delta H°$ are known, the equation provides a means to calculate K at other temperatures. To see this, suppose that K_1 and K_2 are the equilibrium constants of a reaction at temperatures T_1 and

T_2, respectively. Then

$$\ln K_2 = -\frac{\Delta H^\circ}{RT_2} + \frac{\Delta S^\circ}{R}$$

$$\ln K_1 = -\frac{\Delta H^\circ}{RT_1} + \frac{\Delta S^\circ}{R}$$

Subtracting the second equation from the first gives

$$\ln K_2 - \ln K_1 = \ln \frac{K_2}{K_1} = \frac{\Delta H^\circ}{R}\left(\frac{1}{T_1} - \frac{1}{T_2}\right)$$

This is known as the **van't Hoff equation,** after the Dutch chemist Jacobus van't Hoff. Given ΔH° and K at one temperature, we use the equation to calculate K at another temperature, in the approximation that ΔH° and ΔS° are constant. Alternatively, we can use it to determine ΔH° without using a calorimeter if we know the values of K at two temperatures.

The van't Hoff equation shows that if ΔH° is negative (that is, if the reaction is exothermic, giving off heat), then an increase in temperature *reduces* K. If ΔH° is positive (the reaction is endothermic, taking up heat), then an increase in temperature *increases* K. These conclusions about the effect of temperature on equilibrium constants are the same as those obtained in Section 7–5 using Le Chatelier's principle.

EXAMPLE 11–10

An important step in the manufacture of sulfuric acid is the oxidation of sulfur dioxide to sulfur trioxide.

$$SO_2(g) + \tfrac{1}{2} O_2(g) \longrightarrow SO_3(g)$$

At 25°C, the equilibrium constant for this reaction is 2.6×10^{12}, but the reaction occurs very slowly. Calculate K for this equilibrium at 550°C, assuming ΔH° and ΔS° to be independent of temperature over the range from 25 to 550°C.

Solution

We apply Hess's law to the ΔH_f° data from Appendix D and calculate (just as in Example 10–8) that $\Delta H^\circ = -98.9$ kJ for the reaction given above. That is, $\Delta H^\circ = -98.9$ kJ per mole of reaction as written. From the van't Hoff equation, taking T_2 to be 823 K (550°C),

$$\ln\left(\frac{K_{823}}{K_{298}}\right) = \frac{\Delta H^\circ}{R}\left(\frac{1}{298 \text{ K}} - \frac{1}{823 \text{ K}}\right)$$

$$= \left(\frac{-98,900 \;\frac{\cancel{J}}{\cancel{mol}}}{8.315 \;\frac{\cancel{J}}{\text{K}\,\cancel{mol}}}\right)\left(\frac{1}{298\,\cancel{K}} - \frac{1}{823\,\cancel{K}}\right) = -25.46$$

$$\frac{K_{823}}{K_{298}} = e^{-25.46} = 8.8 \times 10^{-12}$$

$$K_{823} = (8.8 \times 10^{-12}) \times (2.6 \times 10^{12}) = 23$$

This result differs somewhat from the experimental K_{823} because the temperature dependence of ΔH° and ΔS° was neglected over quite a large range of temperature. Because the reaction is exothermic, an increase in temperature reduces the equilibrium constant.

Exercise

The reaction

$$2\ Al_3Cl_9(g) \longrightarrow 3\ Al_2Cl_6(g)$$

has an equilibrium constant of 8.8×10^3 at 443 K and a $\Delta H°$ at that temperature of 39.8 kJ. Estimate the equilibrium constant at a temperature of 600 K.

Answer: 1.5×10^5.

The Variation of Vapor Pressure with Temperature

Suppose a pure liquid such as water is in equilibrium with its vapor at temperature T.

$$H_2O(\ell) \rightleftharpoons H_2O(g) \qquad K = P_{H_2O}(g)$$

The temperature dependence of K (and therefore of the vapor pressure $P_{H_2O}(g)$) can be considered a special case of the temperature dependence of equilibrium constants. If ΔH_{vap} and ΔS_{vap} are approximately independent of temperature, then from the van't Hoff equation,

$$\ln\left(\frac{K_2}{K_1}\right) = \ln\left(\frac{P_2}{P_1}\right) = \left(\frac{\Delta H_{vap}}{R}\right)\left(\frac{1}{T_1} - \frac{1}{T_2}\right)$$

The vapor pressure at any given temperature can be estimated if its value is known at some other temperature and if the enthalpy of vaporization is also known. The above equation is known as the **Clausius–Clapeyron equation.**

At the normal boiling point of a substance T_b, the vapor pressure is 1 atm. Taking T_1 to correspond to T_b and T_2 to some other temperature T, we find

$$\ln P_T = \frac{\Delta H_{vap}}{R}\left(\frac{1}{T_b} - \frac{1}{T}\right)$$

where P_T is the vapor pressure expressed in atmospheres at the temperature T.

EXAMPLE 11–11

The molar enthalpy of vaporization of water is 40.66 kJ mol^{-1} at the normal boiling point $T_b = 373.15$ K. Assuming ΔH_{vap} and ΔS_{vap} are approximately independent of temperature from 50 to 100°C, estimate the vapor pressure of water at 50°C (323 K).

Solution

$$\ln P_{323} = \frac{40,660\ \dfrac{\cancel{J}}{\cancel{mol}}}{8.315\ \dfrac{\cancel{J}}{\cancel{K}\ \cancel{mol}}}\left(\frac{1}{373\ \cancel{K}} - \frac{1}{323\ \cancel{K}}\right) = -2.03$$

$$P_{323} = e^{-2.03} = 0.131\ atm$$

This differs by about 8% from the experimental value of 0.1217 atm because ΔH_{vap} *does* change with temperature, an effect that is neglected in the use of the Clausius–Clapeyron equation.

Exercise

The molar enthalpy of vaporization of butane is 21.93 kJ mol^{-1}, and its boiling point is -0.50°C. Estimate its vapor pressure at 25.0°C.

Answer: 2.29 atm.

CHEMISTRY IN YOUR LIFE

According to the second law, the entropy (and, therefore, in a loose sense, the disorder) of the universe is constantly increasing. How is this reconciled with the evolution and existence of highly sophisticated life forms that are clearly in states of considerable order?

The answer is implied early in this chapter: a decrease in entropy is possible in a system *if* it is coupled to another system in which the entropy increases by a larger amount. Life on earth, and the localized decrease in entropy that goes with it, is coupled to nuclear reactions in the sun that are strongly exothermic and cause increases in entropy that more than compensate. The means of coupling is sunlight.

The most direct use of sunlight on earth is photosynthesis by green plants to convert carbon dioxide and water to more complex chemical structures. Cellulose and starch are important plant constituents that are made up of chains of sugar molecules. One such sugar is glucose, which forms, after many steps, according to the overall equation

$$6\ CO_2(g) + 6\ H_2O(\ell) \longrightarrow C_6H_{12}O_6(aq) + 6\ O_2(g)$$

This reaction has a negative standard change in entropy ($\Delta S° = -207$ J K^{-1}) and a positive standard change in Gibbs function ($\Delta G° = +2872$ kJ at 298.15 K). Glucose forms in plants against these barriers only by the intervention of light energy. Once substances of high Gibbs function such as glucose are available, they can be decomposed (metabolized) to drive other, coupled chemical reactions. The overall scheme is outlined in Figure 11–A. Terrestrial organisms metabolize sugars and other substances to drive the production of a specific compound of high Gibbs function, adenosine triphosphate (ATP), from its precursor, adenosine diphosphate (ADP). Energy from subsequent reconversion of ATP to ADP is then harnessed to carry out processes that would otherwise not occur: the synthesis of proteins, the building of cell walls, and the replication of genetic material. The flow of energy from the sun drives the local build-up of order represented by living organisms on earth.

A similar analysis applies to the industrial production of materials of high Gibbs function. The metals are an important example. The change in the Gibbs function to produce a metal, let us say iron, from an oxide ore is positive:

$$Fe_2O_3(s) \longrightarrow 2\ Fe(s) + \tfrac{3}{2}\ O_2(g) \qquad \Delta G° = +742\ kJ$$

If it were possible, we would couple this reaction directly to the nuclear reactions in the sun, producing iron through the use of sunlight. Because we cannot yet do this, we drive it by coupling to another chemical reaction with a negative ΔG:

$$\tfrac{3}{2}\ C(graphite) + \tfrac{3}{2}\ O_2(g) \longrightarrow \tfrac{3}{2}\ CO_2(g)$$
$$\Delta G° = -592\ kJ$$

As the numbers show, even the coupling of this strongly spontaneous oxidation does not suffice to favor the overall process at 25°C. The reactants must be heated to a high temperature to produce metallic iron. The carbon used is in the form of coke, which is obtained from coal and ultimately from decayed plant matter, a resource of high G stored up from sunshine over many millions of years. A continuing major concern in the last years of the 20th century is the rate at which petroleum, coal, and other resources of high Gibbs function are being depleted.

SUMMARY

11–1 The first law of thermodynamics is a principle of constancy (the energy of the universe is always conserved). It says nothing about why **spontaneous** change (change taking place without outside intervention) proceeds in one direction only. The sign of ΔH, the change in enthalpy, does *not* predict the direction of spontaneous change in a process.

11–2 Entropy, a state property of chemical and physical systems, is related to the number of microscopic ways in which a particular state of a system can be attained. Higher entropy corresponds to a larger number of available microscopic states that

Life Does Not Violate the Second Law of Thermodynamics

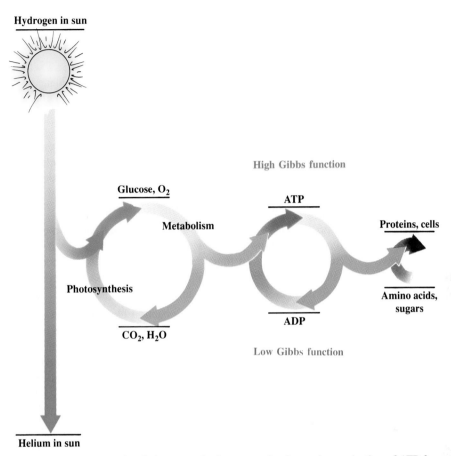

Figure 11-A The synthesis of glucose and other sugars in plants, the production of ATP from ADP, and the elaboration of proteins and other biological molecules are all processes in which the Gibbs function of the system must increase. They occur only through coupling to other processes in which the Gibbs function decreases by an even larger amount.

have identical macroscopic characteristics. When a constraint is removed from a system (as when a stopcock is opened and a gas is allowed to flow from its container into a vacuum), the entropy of the system increases because the number of accessible microscopic states (**microstates**) has increased. The direction of change is rooted in probability. The reverse of a spontaneous change is overwhelmingly unlikely to occur in the macroscopic systems dealt with in chemistry and in daily life.

11-3 According to the **third law of thermodynamics,** the entropy of any crystalline substance approaches zero as the absolute zero of temperature is approached. The absolute entropy of one mole of a substance under standard conditions is its

standard molar entropy. From tabulated standard molar entropies of the reactants and products, the standard entropy change in a chemical reaction can be calculated. Reactions in which the total chemical amount of gas increases almost invariably have positive entropy changes.

11–4 The entropy of the system may increase or decrease in a spontaneous process, but when the entropy of the surroundings is added, the total must always increase; this is a statement of the **second law of thermodynamics.**

11–5 The **Gibbs function** (G), is defined as $G = H - TS$. In constant-temperature changes, $\Delta G = \Delta H - T\,\Delta S$. For changes to occur spontaneously in a system at constant temperature and pressure, the Gibbs function of the system must *decrease* ($\Delta G < 0$). Below the freezing point T_{freez} of a liquid, ΔG for freezing is negative and freezing occurs spontaneously. Above T_{freez} (at atmospheric pressure), freezing is non-spontaneous because ΔG is positive, but melting (the reverse process) is spontaneous. Exactly at T_{freez} the liquid and solid are in equilibrium, and the difference in Gibbs function between them is zero. In the conversion between liquid and vapor, the entropy change (ΔS_{vap}) at the boiling point approximates 88 J K^{-1} mol^{-1} for many different substances. This is **Trouton's rule.**

11–6 Tabulated values of the **standard Gibbs function of formation** ($\Delta G_{\text{f}}^{\circ}$) of the substances taking part in a chemical reaction are used to calculate the standard change in the Gibbs function of the reaction. The method is the same one used to calculate the standard enthalpy change for a reaction from the standard enthalpies of formation of its reactants and products.

11–7 When ΔH° and ΔS° of a reaction have the same sign, a temperature exists at which the reaction changes from being non-spontaneous to spontaneous. When ΔH° and ΔS° have opposite signs, the reaction is either always spontaneous or always non-spontaneous. The ΔG of a reaction is related to ΔG° by the equation $\Delta G = \Delta G^{\circ} + RT \ln Q$. At equilibrium, $\Delta G = 0$, and this implies $\Delta G^{\circ} = -RT \ln K$. ΔG° and K can be estimated at temperatures other than 25°C by using ΔH° and ΔS° values at 25°C because the latter depend only weakly on temperature.

11–8 The **van't Hoff equation** relates ΔH° to K and T. In an exothermic reaction, the equilibrium constant decreases with increasing temperature; in an endothermic reaction, the equilibrium constant increases with increasing temperature. Specialization of the van't Hoff equation to the case of the liquid–vapor phase transition gives the vapor pressure curve of a substance based on its molar enthalpy of vaporization and its normal boiling point. This relationship is called the **Clausius–Clapeyron equation.**

PROBLEMS

Note: Answers to blue-numbered problems are given in Appendix F. Problems that are more challenging are indicated with asterisks.

Enthalpy and Spontaneous Change

1. Write a balanced equation for (a) a process that is exothermic and spontaneous, and (b) one that is endothermic and spontaneous.

2. Write a balanced equation for (a) a process that is exothermic and *not* spontaneous, and (b) one that is endothermic and *not* spontaneous.

3. According to the erroneous principle of Berthelot and Thomsen, spontaneous change occurs in systems in such a way that the maximum heat is released. Give three examples of spontaneous changes that absorb heat and therefore violate this principle (state the conditions of each change).

4. As the temperature is increased, does the principle of Berthelot and Thomsen become more nearly or less nearly correct?

Disorder and Entropy

5. (See Example 11–1).
   (a) How many "microstates" are there for the numbers that come up on a pair of dice?
   (b) What is the probability that a roll of a pair of dice shows two sixes?

6. (See Example 11–1.)
   (a) Suppose a volume is divided into *three* equal parts. How many microstates can be written for all possible ways of distributing four molecules among the three parts?
   (b) What is the probability that all four molecules are located in the left-most third of the volume at the same time?

7. (See Example 11–2.) Predict the sign of the entropy change of the system in each of the following processes.
   (a) Sodium chloride melts.
   (b) A building is demolished.
   (c) A volume of air is divided into three separate volumes of nitrogen, oxygen, and argon, each at the same pressure and temperature as the original air.

8. (See Example 11–2.) Predict the sign of the entropy change of the system in each of the following processes:
   (a) A computer is constructed from iron, copper, carbon, hydrogen, silicon, gallium, and arsenic.
   (b) A container holding a compressed gas develops a leak and the gas enters the atmosphere.
   (c) Solid carbon dioxide (Dry Ice) sublimes to gaseous carbon dioxide.

Absolute Entropies and Chemical Reactions

9. Without consulting Appendix D, predict the sign of the entropy change (ΔS) for the processes
   (a) $SF_4(g) + F_2(g) \longrightarrow SF_6(g)$
   (b) $H_2S(g) + NH_3(g) \longrightarrow NH_4HS(s)$
   (c) $O_2(g)$ (at 400 K) $\longrightarrow O_2(g)$ (at 800 K)
   (d) $CH_3OH(\ell) \longrightarrow CH_3OH(g)$

10. Without consulting Appendix D, predict the sign of the entropy change (ΔS) for the processes
    (a) $Ar(g)$ (at 600 K) $\longrightarrow Ar(g)$ (at 200 K)
    (b) $2 NH_4ClO_4(s) \longrightarrow$
       $N_2(g) + Cl_2(g) + 4 H_2O(g) + 2 O_2(g)$
    (c) $P_4(g) \longrightarrow 2 P_2(g)$
    (d) $C_6H_5CH_3(g) \longrightarrow C_6H_5CH_3(\ell)$

11. (a) Use data from Appendix D to calculate the standard entropy change at 25°C for the reaction

    $$N_2H_4(\ell) + 3 O_2(g) \longrightarrow 2 NO_2(g) + 2 H_2O(\ell)$$

    (b) Suppose that the hydrazine (N_2H_4) is in the gaseous, rather than the liquid, standard state. Is the standard entropy change for its reaction with oxygen higher or lower than that calculated in part (a)? (*Hint:* Entropies of reaction can be added when chemical equations are added, in the same way that Hess's law allows enthalpies to be added.)

12. (a) Use data from Appendix D to calculate the standard entropy change at 25°C for the reaction

    $$CH_3COOH(g) + NH_3(g) \longrightarrow$$
    $$CH_3NH_2(g) + CO_2(g) + H_2(g)$$

    (b) Suppose one mole of solid acetamide ($CH_3CONH_2(s)$) and one mole of water ($H_2O(\ell)$) react to give the same products. Is the standard entropy change larger or smaller than that calculated for the reaction in part (a)?

13. (See Example 11–3.) The alkali metals react with chlorine to give salts:

    $$2 Li(s) + Cl_2(g) \longrightarrow 2 LiCl(s)$$
    $$2 Na(s) + Cl_2(g) \longrightarrow 2 NaCl(s)$$
    $$2 K(s) + Cl_2(g) \longrightarrow 2 KCl(s)$$
    $$2 Rb(s) + Cl_2(g) \longrightarrow 2 RbCl(s)$$
    $$2 Cs(s) + Cl_2(g) \longrightarrow 2 CsCl(s)$$

    Using the data in Appendix D, compute $\Delta S°$ at 25°C for each reaction and identify a periodic trend, if any.

14. (See Example 11–3.) All of the halogens react directly with $H_2(g)$ to give binary compounds. The reactions are

    $$F_2(g) + H_2(g) \longrightarrow 2 HF(g)$$
    $$Cl_2(g) + H_2(g) \longrightarrow 2 HCl(g)$$
    $$Br_2(g) + H_2(g) \longrightarrow 2 HBr(g)$$
    $$I_2(g) + H_2(g) \longrightarrow 2 HI(g)$$

    Using the data in Appendix D, compute $\Delta S°$ at 25°C for each reaction and identify a periodic trend, if any.

15. (See Example 11–4.) Compute $\Delta S°$ at 25°C for the process

    $$Ba^{2+}(aq) + 2 Cl^-(aq) \longrightarrow BaCl_2(s)$$

16. (See Example 11–4.) Compute $\Delta S°$ for the process

    $$I^-(aq) + I_2(aq) \longrightarrow I_3^- (aq)$$

The Second Law of Thermodynamics

17. When $H_2O(\ell)$ and $D_2O(\ell)$ are mixed, the following reaction occurs spontaneously:

    $$H_2O(\ell) + D_2O(\ell) \longrightarrow 2 HOD(\ell)$$

    There is little difference between the enthalpy of an O—H and an O—D bond. What is the main driving force of this reaction?

18. The two gases $BF_3(g)$ and $BCl_3(g)$ are mixed in equal molar amounts. All B—F bonds have about the same bond enthalpy, as do all B—Cl bonds. Explain why the mixture tends to react to form $BF_2Cl(g)$ and $BCl_2F(g)$.

19. In a recent book about entropy, the author writes: "The second law, the Entropy Law, states that matter and energy can only be changed in one direction, that is, from usable to unusable, or from availabe to unavailable, or from ordered to disordered." This statement is glaringly wrong. Identify the errors.

20. The following advertisement is taken from a recent edition of a major metropolitan newspaper.

FREE ELECTRICITY EXTRACTED FROM THE AIR

We have discovered how to produce torque with a closed loop Rankine cycle turbine powered by free energy taken from the air with an advanced heat pump, even in sub-zero weather.

EXPLOSIVE INCOME

You can manufacture or sell this technology including an air conditioner that operates for free, a water heater that can do 3000 gallons of free hot water per day and torque for a variety of purposes.

Do the claims in this advertisement violate the first law of thermodynamics? Do the claims violate the second law of thermodynamics?

21. The dissolution of calcium chloride

$$CaCl_2(s) \longrightarrow Ca^{2+}(aq) + 2\ Cl^-\ (aq)$$

is a spontaneous process at 25°C, even though the standard entropy change of the reaction itself is negative ($\Delta S° = -44.7$ J K^{-1}). What conclusion can you draw about the change in entropy of the surroundings in this process?

22. Quartz ($SiO_2(s)$) does not spontaneously decompose to silicon and oxygen at 25°C in the reaction

$$SiO_2(s) \longrightarrow Si(s) + O_2(g)$$

even though the standard entropy change at this temperature of the reaction is large and positive ($\Delta S° = +182.02$ J K^{-1} mol^{-1}). Explain.

The Gibbs Function

23. (See Example 11–5.) Tungsten melts at 3410°C and has an enthalpy of fusion of 35.4 kJ mol^{-1}. Calculate the entropy of fusion of tungsten.

24. (See Example 11–5.) Tetraphenylgermane (($C_6H_5)_4$Ge) has a melting point of 232.5°C, and its enthalpy increases during fusion by 106.7 J g^{-1}. Calculate the molar enthalpy of fusion and the molar entropy of fusion of tetraphenylgermane.

25. (See Example 11–5.) The molar enthalpy of fusion of solid ammonia is 5.65 kJ mol^{-1}, and the molar entropy of fusion is 28.9 J K^{-1} mol^{-1}.
  (a) Calculate the change in the Gibbs function for the melting of 1.00 mol of ammonia at 170 K.
  (b) Calculate the change in the Gibbs function for the conversion of 3.60 mol of solid ammonia to liquid ammonia at 170 K.
  (c) Does ammonia melt spontaneously at 170 K?
  (d) At what temperature are solid and liquid ammonia in equilibrium at a pressure of 1 atm?

26. (See Example 11–5.) Solid tin exists in two forms: white and gray. For the transformation

$$Sn(white) \longrightarrow Sn(gray)$$

the enthalpy change is -2.1 kJ and the entropy change is -7.4 J K^{-1} at -30°C and 1 atm.
  (a) Calculate the change in the Gibbs function for the conversion of 1.00 mol of white tin to gray tin at -30°C.
  (b) Calculate the change in the Gibbs function for the conversion of 2.50 mol of white tin to gray tin at -30°C.
  (c) Does white tin convert spontaneously to gray tin at -30°C?
  (d) Estimate the temperature at which white and gray tin are in equilibrium at a pressure of 1 atm.

27. The normal boiling point of the solvent acetone is 56.2°C. Use Trouton's rule to estimate its molar enthalpy of vaporization.

28. At its boiling point, methanol has a molar enthalpy of vaporization of 35.28 kJ mol^{-1} and a molar entropy of vaporization of 104.5 J mol^{-1}. Use these data to determine the boiling point of methanol in degrees Celsius.

29. (See Example 11–6.) Use the data in Appendix D to calculate $\Delta G°$ at 25°C for the following reactions:
  (a) $SrCl_2(s) + Ca(s) \longrightarrow CaCl_2(s) + Sr(s)$
  (b) $Fe_2O_3(hematite) + 2\ Al \longrightarrow 2\ Fe(s) + Al_2O_3(s)$

30. (See Example 11–6.) Use the data in Appendix D to calculate $\Delta G°$ at 25°C for the following reactions:
  (a) $CaCO_3(s) + 2\ H_3O^+(aq) \longrightarrow$
  $$Ca^{2+}(aq) + CO_2(g) + 3\ H_2O(\ell)$$
  (b) $CaCO_3(s) + 2\ H^+(aq) \longrightarrow$
  $$Ca^{2+}(aq) + CO_2(g) + H_2O(\ell)$$

The Gibbs Function and Chemical Reactions

31. For $CoCl_2\cdot6H_2O(s)$ at 25°C, $\Delta H_f°$ is -2115.4 kJ mol^{-1}, $\Delta G_f°$ is -1725.2 kJ mol^{-1}, and $S°$ is 343 J K^{-1} mol^{-1}. Compute $\Delta H°$, $\Delta G°$, and $\Delta S°$ for the reaction

$$CoCl_2(s) + 6\ H_2O(\ell) \longrightarrow CoCl_2\cdot6H_2O(s)$$

at 25°C. Use Appendix D as necessary.

32. Use the data from the previous problem and Appendix D to compute $\Delta H°$, $\Delta G°$, and $\Delta S°$ at 25°C for the reaction

$$CoCl_2(s) + 6\ H_2O(g) \longrightarrow CoCl_2\cdot6H_2O(s)$$

33. Predict the sign of $\Delta G°$ for reactions at *low* temperature for which
  (a) $\Delta H°$ is negative and $\Delta S°$ is negative.
  (b) $\Delta H°$ is negative and $\Delta S°$ is positive.
  (c) $\Delta H°$ is positive and $\Delta S°$ is negative.
  (d) $\Delta H°$ is positive and $\Delta S°$ is positive.

34. Predict the sign of $\Delta G°$ for reactions at *high* temperature for which
  (a) $\Delta H°$ is negative and $\Delta S°$ is negative.
  (b) $\Delta H°$ is negative and $\Delta S°$ is positive.
  (c) $\Delta H°$ is positive and $\Delta S°$ is negative.
  (d) $\Delta H°$ is positive and $\Delta S°$ is positive.

35. (See Example 11–7.) A process at constant T and P can be described as spontaneous if $\Delta G° < 0$, and non-spontaneous if $\Delta G° > 0$. Over what range of temperatures is each of the following processes spontaneous? Assume that all gases are at a pressure of 1 atm. (*Hint:* Use Appendix D to calculate $\Delta H°$ and $\Delta S°$ (assumed independent of temperature) and then use the definition of $\Delta G°$.)

(a) The rusting of iron, a complex reaction that can be approximated as

$$4 \, Fe(s) + 3 \, O_2(g) \longrightarrow 2 \, Fe_2O_3(s)$$

(b) The preparation of $SO_3(g)$ from $SO_2(g)$, a step in the manufacture of sulfuric acid

$$SO_2(g) + \tfrac{1}{2} O_2(g) \longrightarrow SO_3(g)$$

(c) The production of the anesthetic dinitrogen oxide through the decomposition of ammonium nitrate

$$NH_4NO_3(s) \longrightarrow N_2O(g) + 2 \, H_2O(g)$$

36. (See Example 11–7.) Follow the procedure used in problem 35 to estimate the range of temperatures over which each of the following processes is spontaneous:

(a) The preparation of the gas phosgene ($COCl_2(g)$)

$$CO(g) + Cl_2(g) \longrightarrow COCl_2(g)$$

(b) The laboratory-scale production of oxygen from the decomposition of potassium chlorate

$$2 \, KClO_3(s) \longrightarrow 2 \, KCl(s) + 3 \, O_2(g)$$

(c) The reduction of iron(II) oxide (wüstite) by coke (carbon), a step in the production of iron in a blast furnace

$$FeO(s) + C(s) \longrightarrow Fe(s) + CO(g)$$

37. (See Example 11–7.) Explain why it is possible to reduce tungsten(VI) oxide (WO_3) to metal with hydrogen at an elevated temperature. At what temperature does $\Delta G°$ of the reaction taking place in this conversion become equal to zero? Use the data of Appendix D.

38. (See Example 11–7.) Tungsten(VI) oxide can also be reduced to tungsten by heating it with carbon in an electric furnace.

$$2 \, WO_3(s) + 3 \, C(s) \longrightarrow 2 \, W(s) + 3 \, CO_2(g)$$

(a) Calculate the standard change in the Gibbs function ($\Delta G°$) at 25°C for this reaction, and comment on the feasibility of the process at room conditions.

(b) What must be done to make the process thermodynamically feasible, assuming that $\Delta H°$ and $\Delta S°$ do not vary much with temperature?

39. The following data are gathered on the reaction of hydrogen with carbon dioxide to give formic acid at 298.15 K

	$\Delta H°$	$\Delta G°$
$H_2(g) + CO_2(g) \longrightarrow HCOOH(g)$	+30.9 kJ	+58.6 kJ
$H_2(g) + CO_2(g) \longrightarrow HCOOH(\ell)$	−31.2	+32.9

Compute $\Delta H°$, $\Delta G°$, and $\Delta S°$ in the vaporization of liquid formic acid at 298.15 K.

40. According to the data in problem 39, the reaction of $H_2(g)$ with $CO_2(g)$ to give gaseous formic acid is *endo*thermic and non-spontaneous, and the reaction to give liquid formic acid is *exo*thermic but also non-spontaneous. Explain these facts from the point of view of the making and breaking of bonds during the reaction and the relative entropies of the reactants and products.

The Gibbs Function and the Equilibrium Constant

41. (See Example 11–8.) Calculate $\Delta G°$ and the equilibrium constant K at 25°C for the reaction

$$2 \, NH_3(g) + \tfrac{7}{2} O_2(g) \rightleftharpoons NO_2(g) + 3 \, H_2O(g)$$

using data in Appendix D.

42. (See Example 11–8). Write a reaction for the dehydrogenation of gaseous ethane (C_2H_6) to acetylene (C_2H_2). Calculate $\Delta G°$ and the equilibrium constant for this reaction at 25°C, using data from Appendix D.

43. (See Example 11–8.) Use the thermodynamic data from Appendix D to calculate equilibrium constants at 25°C for the following reactions:
(a) $SO_2(g) + \tfrac{1}{2} O_2(g) \rightleftharpoons SO_3(g)$
(b) $3 \, Fe_2O_3(s) \rightleftharpoons 2 \, Fe_3O_4(s) + \tfrac{1}{2} O_2(g)$
(c) $CuCl_2(s) \rightleftharpoons Cu^{2+}(aq) + 2 \, Cl^-(aq)$
Write an equilibrium expression in each case

44. (See Example 11–8.) Use the thermodynamic data from Appendix D to calculate equilibrium constants at 25°C for the following reactions:
(a) $H_2(g) + N_2(g) + 2 \, O_2(g) \rightleftharpoons 2 \, HNO_2(g)$
(b) $Ca(OH)_2(s) \rightleftharpoons CaO(s) + H_2O(g)$
(c) $Zn^{2+}(aq) + 4 \, NH_3(aq) \rightleftharpoons Zn(NH_3)_4^{2+}(aq)$
Write an equilibrium expression in each case

45. (See Example 11–8.) Use the $\Delta G_f°$ values in Appendix D to compute the K_a at 25°C for the reaction

$$H_2PO_4^-(aq) + H_2O(\ell) \longrightarrow HPO_4^{2-}(aq) + H_3O^+(aq)$$

at 25°C. Compare the answer to the value for K_{a2} for $H_3PO_4(aq)$ in Table 8–2.

46. (See Example 11–8.) Use the $\Delta G_f°$ values in Appendix D to compute the K_a at 25°C for the reaction

$$H_3PO_4(aq) + H_2O(\ell) \longrightarrow H_2PO_4^-(aq) + H_3O^+(aq)$$

at 25°C. Compare the answer to the value for K_{a1} for $H_3PO_4(aq)$ in Table 8–2.

47. Although the process of solution of helium gas in water is energetically favored, helium is only very slightly soluble in water. What keeps this gas from dissolving in great quantities in water?

48. At 25°C the solubility of rhombic sulfur in water is slight, only 1.9×10^{-8} mol sulfur per liter of water.
(a) Compute $\Delta G°$ for the dissolution of sulfur in water.
(b) Despite the fact that the $\Delta G°$ computed in part (a) is positive, sulfur placed in water at 25°C still spontaneously dissolves to the extent given above. Explain.

The Temperature Dependence of Equilibrium Constants

49. For the reaction

$$3\ CuCl(s) + \tfrac{1}{2}\ Al_2Cl_6(g) \rightleftharpoons Cu_3AlCl_6(g)$$

$\Delta G°$ is $+29.4$ kJ per mole of $Cu_3AlCl_6(g)$ formed at 590 K. Calculate the equilibrium constant K for this reaction at 590 K.

50. The compounds $MgFe_2Cl_8$ and $CaFe_2Cl_8$ can be prepared by the reactions

$$MgCl_2(s) + 2\ FeCl_3(g) \rightleftharpoons MgFe_2Cl_8(g) \qquad K = 1.32$$
$$CaCl_2(s) + 2\ FeCl_3(g) \rightleftharpoons CaFe_2Cl_8(g) \qquad K = 0.0250$$

The equilibrium constants quoted above were measured at 873 K.
(a) Calculate $\Delta G°$ of each of these reactions at 873 K.
(b) Which compound is more stable at 873 K with respect to decomposition to $FeCl_3(g)$ and magnesium (or calcium) chloride?

51. Oxygen dissolves in water to an extent, and fish extract the dissolved O_2 in their gills to support their metabolism. Without dissolved oxygen in their water, all fish die. Heating the water in rivers and lakes has led on occasion to massive fish kills by suffocation. From these facts, determine the sign of $\Delta H°$ for the equilibrium

$$O_2(g) \rightleftharpoons O_2(aq)$$

52. (a) The reaction

$$Cl_2(g) \rightleftharpoons Cl_2(aq)$$

represents the dissolution of gaseous chlorine in water. The equilibrium constant for this reaction doubles when the temperature is increased from 10°C to 30°C. Is the production of $Cl_2(aq)$ by this reaction an endothermic or an exothermic process?
(b) Estimate the value of $\Delta H°$ of this reaction.

53. One way to manufacture ethanol is by the reaction

$$C_2H_4(g) + H_2O(g) \rightleftharpoons C_2H_5OH(g)$$

The $\Delta H_f°$ of $C_2H_4(g)$ is $+52.3$ kJ mol^{-1}; of $H_2O(g)$, -241.8 kJ mol^{-1}; and of $C_2H_5OH(g)$, -235.3 kJ mol^{-1}. Without doing detailed calculations, suggest the conditions of pressure and temperature that maximize the yield, at equilibrium, of ethanol.

54. Dimethyl ether (CH_3OCH_3) is a good substitute for environmentally harmful propellants in aerosol spray cans. It is produced by the dehydration of methanol:

$$2\ CH_3OH(g) \rightleftharpoons CH_3OCH_3(g) + H_2O(g)$$

Describe reaction conditions that favor the equilibrium production of this valuable chemical. As a basis for your answer, compute the $\Delta H°$ and $\Delta S°$ at 298.15 K of the reaction from the data in Appendix D.

55. (See Example 11–10.) For the synthesis of ammonia from its elements,

$$3\ H_2(g) + N_2(g) \rightleftharpoons 2\ NH_3(g)$$

calculate the equilibrium constant at 600 K, assuming no change in $\Delta H°$ and $\Delta S°$ between 298.15 K and 600 K. Use data from Appendix D.

56. (See Example 11–10.) The reaction of $Cu_2O(s)$ with $H_2(g)$ to yield $Cu(s)$ is represented by the chemical equation

$$Cu_2O(s) + H_2(g) \rightleftharpoons 2\ Cu(s) + H_2O(\ell)$$

Calculate the equilibrium constant for this reaction at 373 K, assuming $\Delta H°$ and $\Delta S°$ are approximately independent of temperature. Use data from Appendix D.

57. The equilibrium constant at 25°C for the reaction

$$2\ NO_2(g) \rightleftharpoons N_2O_4(g)$$

is 6.8. At 200°C, the equilibrium constant is 1.21×10^{-3}. Calculate the enthalpy change ($\Delta H°$) for this reaction assuming that $\Delta H°$ and $\Delta S°$ of the reaction are constant over the temperature range from 25 to 200°C.

58. The acidity constant of chloroacetic acid ($ClCH_2COOH$) in water is 1.528×10^{-3} at 0°C and 1.230×10^{-3} at 40°C. Calculate the enthalpy of dissociation of the acid in water, assuming that $\Delta H°$ and $\Delta S°$ are constant over this range of temperature.

59. Stearic acid dimerizes when dissolved in hexane:

$$2\ C_{17}H_{35}COOH(hexane) \rightleftharpoons (C_{17}H_{35}COOH)_2(hexane)$$

The equilibrium constant for this reaction is 2900 at 28°C, but it drops to 40 at 48°C. Estimate $\Delta H°$ and $\Delta S°$ of the reaction.

60. Stearic acid also dimerizes when dissolved in carbon tetrachloride:

$$2\ C_{17}H_{35}COOH(CCl_4) \rightleftharpoons (C_{17}H_{35}COOH)_2(CCl_4)$$

The equilibrium constant for this reaction is 2780 at 22°C, but it drops to 93.1 at 42°C. Estimate $\Delta H°$ and $\Delta S°$ of the reaction.

61. The equilibrium constant for the reaction

$$\tfrac{1}{2}\ Cl_2(g) + \tfrac{1}{2}\ F_2(g) \rightleftharpoons ClF(g)$$

is measured to be 9.3×10^9 at 298.15 K and 3.3×10^7 at 398.15 K.
(a) Calculate $\Delta G°$ at 298.15 K for the reaction.
(b) Calculate $\Delta H°$ and $\Delta S°$, assuming the enthalpy and entropy changes to be independent of temperature between 298.15 and 398.15 K.

62. The autoionization constant of water (K_w) is 1.139×10^{-15} at 0°C and 9.614×10^{-14} at 60°C.
(a) Calculate the enthalpy of autoionization of water

$$2\ H_2O(\ell) \rightleftharpoons H_3O^+(aq) + OH^-(aq)$$

(b) Calculate the entropy of autoionization of water.
(c) From these data, at what temperature is the pH of pure water 7.00?

63. The vapor pressure of ammonia at −50°C is 0.4034 atm, and at 0°C it is 4.2380 atm.
(a) Calculate the molar enthalpy of vaporization (ΔH_{vap}) of ammonia.
(b) Estimate the normal boiling temperature of $NH_3(\ell)$.

64. The vapor pressure of butyl alcohol (C_4H_9OH) at 70.1°C is 0.1316 atm; at 100.8°C, it is 0.5263 atm.
(a) Calculate the molar enthalpy of vaporization (ΔH_{vap}) of butyl alcohol.
(b) Estimate the normal boiling point of butyl alcohol.

65. Estimate ΔH_{vap} of a liquid if reducing the pressure on the liquid from 1.00 atm to 0.63 atm lowers its boiling point from 124°C to 107°C.

66. Estimate the boiling point of water at a pressure of 0.50 atm, assuming that ΔH_{vap} at this temperature is 40.66 kJ mol^{-1}.

67. Calculate the $\Delta H°$ of the reaction for the formation of 1.00 mol of aqueous glucose from gaseous carbon dioxide and liquid water, using the data available in Section 11–8.

68. Calculate the $\Delta H°$ of the reaction for the formation of aqueous glucose from its elements in their standard states, using the data available in Appendix D and the results of the previous problem.

Additional Problems

*69. A computer is programmed to select at random any whole number in the range 1 to 100 (this includes 1 and 100 themselves) and print it every time a key is pressed.
(a) What is the chance that 10 key strokes generate 6 even numbers, then 1 odd number, and then 3 even numbers?
(b) What is the chance that 10 key strokes generate 9 even numbers and 1 odd number irrespective of the order?

*70. The computer described in problem 69 is used again.
(a) Compute the probability that 10 key strokes generate 10 numbers less than or equal to 50.
(b) Compute the probability that pushing a key 1 billion (10^9) times generates 1 billion numbers less than or equal to 50.
(c) Compute the probability that pushing a key 10 times generates 10 numbers less than or equal to 99.
(d) Compute the probability that pushing a key 10^9 times generates 1 billion numbers less than or equal to 99.

71. The N_2O molecule has the structure N≡N—O. In an ordered crystal of N_2O, the molecules are lined up in a regular fashion, with the orientation of each determined by its position in the crystal. In a random crystal (formed on rapid freezing), each molecule has two equally likely orientations ("up" and "down").

(a) In which form is the entropy higher?
(b) What is the number of possible microstates available to a random crystal of N_0 (Avogadro's number) of molecules?

*72. Problem 96 in Chapter 5 described an optical atomic trap. In one experiment, a gas of 500 sodium atoms is confined in a volume of 1000 μm^3. The temperature of the system is 0.00024 K. Compute the probability that, by chance, these 500 slowly moving sodium atoms would all congregate in the left half of the available volume. Express your answer in scientific notation.

73. By examining the two graphs below, predict which element—copper or gold—has the higher absolute entropy at a temperature of 200 K.

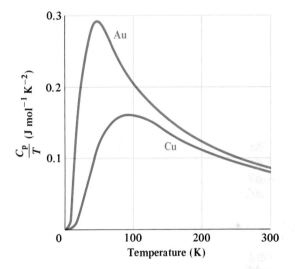

74. Without consulting Appendix D, predict the sign of $\Delta S°$ in each of the following reactions.
(a) $4\ NH_3(g) + 5\ O_2(g) \longrightarrow 4\ NO(g) + 6\ H_2O(g)$
(b) $4\ NH_3(g) + 5\ O_2(g) \longrightarrow 4\ NO(g) + 6\ H_2O(\ell)$

75. Comment on the truth or falsity of each of these statements. Revise all of the false statements so that they are true.
(a) Energy is created in power plants.
(b) Scientists have found only a few exceptions to the second law of thermodynamics.
(c) The laws of thermodynamics indicate that we will absolutely never be able to make perpetual-motion machines.
(d) The second law of thermodynamics allows chemists to predict the rates of reactions.
(e) Classical thermodynamics is derived from Dalton's atomic theory of matter.

76. A quantity of ice is mixed with a quantity of hot water in a sealed, rigid, insulated container. The insulation prevents any heat exchange between the ice-water mixture and the surroundings. The contents of the container soon reach equilibrium. Tell whether the total *energy* of the contents decreases, remains the same, or increases in this process. Make a similar statement about the total *entropy* of the contents. Explain your answers.

77. Sulfur vapors include both cyclic S_8 and cyclic S_6 molecules. The ΔH_f°'s of $S_8(g)$ and $S_6(g)$ are 101.55 kJ mol^{-1} and 103.72 kJ mol^{-1}, respectively.

(a) Determine which of the two forms is favored on the basis of energetics by calculating the ΔH° of the reaction

$$4\ S_6(g) \rightleftharpoons 3\ S_8(g)$$

(b) Which of the two forms do you expect to be favored on the basis of the entropy change in the reaction?

78. Two different crystalline forms of sulfur are the rhombic form and the monoclinic form. At atmospheric pressure, rhombic sulfur undergoes a transition to monoclinic when it is heated above 368.5 K.

$$S(rhombic) \longrightarrow S(monoclinic)$$

(a) What is the sign of the entropy change (ΔS) for this transition?

(b) $|\Delta H^\circ|$ for this transition is 400 J mol^{-1}. Calculate ΔS° for the transition.

79. Ethanol (CH_3CH_2OH) has a normal boiling point of 78.4°C and a measured molar enthalpy of vaporization of 38.74 kJ mol^{-1}. Calculate the molar entropy of vaporization of ethanol, and compare it with the prediction of Trouton's rule.

80. Use data from Appendix D to predict the normal boiling point of water (the temperature at which liquid water is in equilibrium with 1 atm of steam). Why does the answer differ from 100.00°C (373.15 K)?

81. Use data from Appendix D to predict the normal boiling point of carbon disulfide (CS_2).

*82. Use data from Appendix D to estimate the temperature at which $I_2(g)$ and $I_2(s)$ are in equilibrium at a pressure of 1 atm. Can this equilibrium actually be achieved? Iodine melts at 113.5°C and boils at 184.4°C.

83. Given that ΔG_f° of $Si_3N_4(s)$ is -642.6 kJ mol^{-1}, use the data in Appendix D to compute ΔG° of the reaction

$$3\ CO_2(g) + Si_3N_4(s) \longrightarrow$$
$$3\ SiO_2(quartz) + 2\ N_2(g) + 3\ C(graphite)$$

84. The compound $Pt(NH_3)_2I_2$ comes in two forms, the *cis* and the *trans*, which differ in their molecular structure. The following data are available:

	ΔH_f° (kJ mol^{-1})	ΔG_f° (kJ mol^{-1})
cis	-286.56	-130.25
trans	-316.94	-161.50

Combine these data with data from Appendix D to compute the standard entropy (S°) of both of these compounds at 25°C.

85. (a) Use data from Appendix D to calculate ΔH° and ΔS° at 25°C for the reaction

$$2\ CuCl_2(s) \rightleftharpoons 2\ CuCl(s) + Cl_2(g)$$

(b) Calculate ΔG° at 590 K, assuming that ΔH° and ΔS° are independent of temperature.

(c) Careful high-temperature measurements establish that at 590 K, ΔH_{590}° of this reaction is 158.36 kJ and ΔS_{590}° is 177.74 J K^{-1}. Use these facts to compute an improved value of ΔG_{590}° for this reaction. Determine the percentage error in ΔG_{590}° that comes from using the 298-K values in place of 590-K values in this case.

86. Calculate the equilibrium pressure (in atmospheres) of $O_2(g)$ over a sample of pure $NiO(s)$ in contact with pure $Ni(s)$ at 25°C. The $NiO(s)$ decomposes according to the equation

$$NiO(s) \longrightarrow Ni(s) + \tfrac{1}{2}\ O_2(g)$$

Use data from Appendix D.

87. Although iodine is not very soluble in pure water, it dissolves readily in water that contains $I^-(aq)$ ion, thanks to the reaction

$$I_2(aq) + I^-(aq) \rightleftharpoons I_3^-(aq)$$

The equilibrium constant of this reaction was measured as a function of temperature with these results:

T:	3.8°C	15.3°C	25.0°C	35.0°C	50.2°C
K:	1160	841	689	533	409

(a) Plot ln K against $1/T$, the reciprocal of the absolute temperature.

(b) Estimate the ΔH° of this reaction

88. Barium nitride vaporizes slightly at high temperature by means of the dissociation

$$Ba_3N_2(s) \rightleftharpoons 3\ Ba(g) + N_2(g)$$

At 1000 K, the equilibrium constant is 4.5×10^{-19}. At 1200 K, the equilibrium constant is 6.2×10^{-12}.

(a) Estimate ΔH° for this reaction.

(b) The equation is rewritten as

$$2\ Ba_3N_2(s) \rightleftharpoons 6\ Ba(g) + 2\ N_2(g)$$

Now the equilibrium constant is 2.0×10^{-37} at 1000 K and 3.8×10^{-23} at 1200 K. Estimate ΔH° of *this* reaction.

*89. The triple bond in the N_2 molecule is very strong, but at high enough temperatures, even it breaks down. At 5000 K, when the total pressure exerted by a sample of nitrogen is 1.00 atm, $N_2(g)$ is 0.65% dissociated at equilibrium.

$$N_2(g) \rightleftharpoons 2\ N(g)$$

At 6000 K with the same total pressure, the proportion of $N_2(g)$ dissociated at equilibrium rises to 11.6%. Use the van't Hoff equation to estimate the ΔH° of this reaction.

*90. The sublimation pressure of solid NbI_5 is the pressure of gaseous NbI_5 present in equilibrium with the solid. It is given by the empirical equation

$$\log_{10} P = -6762/T + 8.566$$

The vapor pressure of liquid NbI_5, on the other hand, is given by

$$\log_{10} P = -4653/T + 5.43$$

In these two equations, T is the absolute temperature in kelvins and P is the pressure in atmospheres.

(a) Determine the enthalpy and entropy of sublimation of $NbI_5(s)$.

(b) Determine the enthalpy and entropy of vaporization of $NbI_5(\ell)$.

(c) Calculate the normal boiling point of $NbI_5(\ell)$.

(d) Calculate the normal *melting* point of $NbI_5(s)$.

(*Hint:* At the normal melting point of a substance, the liquid and solid are in equilibrium and must have the same vapor pressure. If they did not, then vapor would continually escape from the phase with the higher vapor pressure and collect in the phase with the lower vapor pressure.)

91. "Equilibrium is like death." Explain and criticize this statement.

92. Consider the total entropy of the upper kilometer of the earth's crust, together with the oceans and atmosphere. Discuss ways in which human activity has served to increase or decrease the entropy of that portion of the earth. Consider such aspects as burning fossil fuels, reclaiming metals from their ores, air and water pollution, and building cities.

93. The standard change in the Gibbs function at 25°C and pH 7 for the conversion of 1.00 mol of ATP to ADP is -34.5 kJ. The standard enthalpy change of the conversion under these conditions is -19.7 kJ.

(a) Calculate the standard entropy change for this process.

(b) The metabolism of one mole of glucose is coupled to the formation of 38 mol of ATP from ADP. What fraction of the high Gibbs function of glucose is saved in ATP? Use data from *Chemistry in Your Life: Life Does Not Violate the Second Law of Thermodynamics*, in Section 11–8.

94. The reaction of nitrogen oxides (NO_x) with NH_3 to give N_2 has gained commercial acceptance as a means to reduce NO_x emissions from power plants.

(a) The reaction of NO_2 in this way is represented:

$$6\ NO_2(g) + 8\ NH_3(g) \longrightarrow 7\ N_2(g) + 12\ H_2O(g)$$

Calculate $\Delta G°$ and K for this reaction at 298.15 K.

(b) Write and balance similar equations for the conversion of NO and N_2O to $N_2(g)$ and $H_2O(g)$ by means of reaction with ammonia. Determine the equilibrium constants for these reactions at 298.15 K.

(c) Discuss the problem of finding the best temperature and pressure for running this reaction when dealing with a stream of gases that contains a mixture of NO, N_2O, and NO_2.

95. A report on the first synthesis of difluoromethanimine appeared recently. This compound exists in two isomeric forms:

$$\underset{H}{\overset{F}{\diagdown}}C{=}N\underset{F}{} \quad \text{and} \quad \underset{H}{\overset{F}{\diagdown}}C{=}N\overset{F}{}$$

When a sample of the isomer diagrammed on the left was dissolved in chloroform at room temperature (22°C), it slowly reacted to give the isomer in the diagram on the right. After 7 days, 95% of the original amount was converted. No further conversion occurred thereafter.

(a) Compute the equilibrium constant and $\Delta G°$ (at 22°C) for this *isomerization* reaction.

(b) Suppose that some of the isomer diagrammed on the right is prepared, purified, and then dissolved in chloroform at 22°C. Predict the relative amounts of the two isomers in this solution after 7 days.

CUMULATIVE PROBLEM

Purifying Nickel

Impure nickel, obtained from the smelting of nickel sulfide ores in a blast furnace, can be converted into metal of 99.90 to 99.99% purity by the Mond process, which relies on the equilibrium

$$Ni(s) + 4\ CO(g) \rightleftharpoons Ni(CO)_4(g)$$

The standard enthalpy of formation $\Delta H_f°$ of nickel tetracarbonyl ($Ni(CO)_4(g)$) is -602.9 kJ mol^{-1}, and its standard entropy $S°$ is 410.6 J K^{-1} mol^{-1} at 25°C.

(a) Predict (without referring to a table) whether the entropy change of the system (the reacting atoms and molecules) is positive or negative in this process.

(b) At a temperature for which this reaction is spontaneous, predict whether the entropy change of the surroundings during the reaction is positive or negative.

(c) Use data from Appendix D to calculate $\Delta H°$ and $\Delta S°$ for this reaction.

A huge reducer kiln at the heart of a nickel carbonyl refinery in Wales.

(d) At what temperature is $\Delta G^\circ = 0$ (and $K = 1$) for this reaction?

(e) The first step in the Mond process is the equilibration of impure nickel with CO and $Ni(CO)_4$ at about 50°C. In this step, the equilibrium constant should be as large as possible, to draw as much nickel as possible into the vapor-phase compound. Calculate the equilibrium constant for the above reaction at 50°C.

(f) In the second step of the Mond process, the mixture of gases is removed from the reaction chamber and heated to about 230°C. At high enough temperatures, the sign of ΔG° is reversed and the reaction occurs in the opposite direction, depositing pure nickel. In this step, the equilibrium constant for the forward reaction should be as small as possible. Calculate the equilibrium constant for the above reaction at 230°C.

(g) The Mond process relies on the volatility of $Ni(CO)_4$ for its success. Under room conditions, this compound is a liquid, but it boils at 42.2°C with an enthalpy of vaporization of 29.0 kJ mol^{-1}. Calculate the entropy of vaporization of $Ni(CO)_4$, and compare it with that predicted by Trouton's rule.

(h) A recently developed variation of the Mond process carries out the first step at higher pressures and at a temperature of 150°C. Estimate the maximum pressure of $Ni(CO)_4(g)$ that can be attained before the gas liquefies at this temperature (that is, estimate the vapor pressure of $Ni(CO)_4(\ell)$ at 150°C).

Redox Reactions and Electrochemistry

Copper ingots.

$\mathbf{T}$he burning of fuel for heat involves oxidation–reduction (redox) reactions, as do the winning of iron and other metals from ores, the growth of trees and crops, the contraction and extension of muscles, and, indeed, all metabolic changes. Redox reactions were introduced in Chapter 4 as one of three broad categories of chemical change. They are reactions that occur with the *transfer* of one or more electrons: when a chemical entity loses electrons, it is *oxidized;* when it gains electrons, it is *reduced.* In this chapter, we consider redox reactions in greater depth. We begin by showing how redox chemistry is exploited in practical processes for extracting and refining metals, taking copper and iron as examples. We then show how any redox reaction can be represented as the sum of simultaneous reduction and oxidation half-reactions. The concept of half-reactions proves helpful in balancing equations for redox reactions and in describing chemical events in electrochemical cells, which are defined and discussed. Finally, we outline ways in which electrochemical methods are used to isolate and purify metals such as aluminum, magnesium, and chromium.

12-1 METALS AND METALLURGY

The recovery of metals from their sources in the earth relies heavily on oxidation–reduction reactions. **Extractive metallurgy,** a blend of chemistry, physics, and engineering, is devoted to learning the most efficient and environmentally safe ways of winning essential metals from their ores. As a science, it is comparatively recent, but its beginnings, thought to have occurred in the Near East about 6000 years ago, marked the emergence of humanity from the Stone Age. The earliest known metals were undoubtedly gold, silver, and copper, because they could be found in their native (elemental) state (Fig. 12–1). Gold and silver were valued for their ornamental uses, but they were too soft to be made into tools. Iron was also found in elemental form, but only rarely, in meteorites.

Most metals are combined with other elements such as oxygen and sulfur in ores, and chemical processes are required to free them. As Table 12–1 shows, the Gibbs functions of formation of most metal oxides are negative. The reverse reactions, which yield the free metal and oxygen, therefore mostly have positive ΔG_f°'s; they are non-spontaneous. Non-spontaneous chemical reactions can be accomplished, but only by coupling them with a second, spontaneous reaction. The greater the rise in the Gibbs function, the more thermodynamically difficult it is to force a non-spontaneous reaction to take place. Thus, metals like silver or gold (at the bottom of Table 12–1) exist in nature as elements, and mercury can be released from its oxide or sulfide ore (cinnabar) simply by moderate heating (see Fig. 1–13). Winning copper, zinc, and iron requires more stringent conditions; the ores of these metals are reduced in chemical reactions at high temperatures, the collective term for which is **pyrometallurgy.** These reactions are carried out in huge furnaces, in which coke (coal from which volatile components have been expelled) serves both as the ultimate reducing agent and as fuel to heat the system to the required high temperature.

$$C(s) + O_2(g) \longrightarrow CO_2(g) \qquad \Delta G^\circ = -394 \text{ kJ mol}^{-1}$$

The process, called **smelting,** involves both chemical change and melting. Even smelting is not sufficient to win the metals at the top of Table 12–1, which have oxides (and also sulfides) with very negative ΔG_f°'s. Electrometallurgical methods, which we consider in Section 12–5, are needed.

The molar Gibbs functions of formation in Table 12–1 are divided by the total change in the oxidation number when the oxide of a metal is reduced to elemental

Figure 12–1 A specimen of native copper.

Table 12-1
Metal Oxides Arranged According to Ease of Reduction

Metal Oxide	Metal	n^a	$\Delta G_f^\circ / n^a$ (kJ mol^{-1})	A Method of Production of the Metal
MgO	Mg	2	-285	Electrolysis of $MgCl_2$
Al_2O_3	Al	6	-264	Electrolysis
TiO_2	Ti	4	-222	Reaction with Mg
Na_2O	Na	2	-188	Electrolysis of NaCl
Cr_2O_3	Cr	6	-176	Electrolysis, reduction by Al
ZnO	Zn	2	-159	Smelting of ZnS
SnO_2	Sn	4	-130	Smelting
Fe_2O_3	Fe	6	-124	Smelting
NiO	Ni	2	-106	Smelting of nickel sulfides
PbO	Pb	2	-94	Smelting of PbS
CuO	Cu	2	-65	Smelting of $CuFeS_2$
HgO	Hg	2	-29	Moderate heating of HgS
Ag_2O	Ag	2	-6	Found in elemental form
Au_2O_3	Au	6	$+13$	Found in elemental form

^{a} The standard Gibbs functions of formation of the metal oxides in kJ mol^{-1} are adjusted for fair comparison by dividing them by n, the total number of moles of electrons required to reduce the metal atoms combined in the oxide to oxidation states of zero. Thus, the reduction of 1 mole of Cr_2O_3 requires a total of 6 moles of electrons, so that $n = 6$.

form. This allows a fair comparison of the relative ease of reduction of two compounds that have the metals in different oxidation states. Reducing one mole of Fe_2O_3 to elemental iron, for example, requires oxidizing three times as much carbon (to CO_2) as the similar reduction of one mole of ZnO. Thus, the Gibbs function of formation of Fe_2O_3 is divided by 6 in the table, whereas the Gibbs function of formation of ZnO is divided only by 2. The change in the Gibbs function at 1 atm and 298.15 K for

$$Cr_2O_3(s) + 3\,Mg(s) \longrightarrow 2\,Cr(s) + 3\,MgO(s)$$

is

$$\Delta G^\circ = 3\,\Delta G_f^\circ\,(MgO) - \Delta G_f^\circ\,(Cr_2O_3)$$

which can be rewritten as

$$\Delta G^\circ = 6 \times [\tfrac{1}{2}\,\Delta G_f^\circ\,(MgO) - \tfrac{1}{6}\,\Delta G_f^\circ\,(Cr_2O_3)]$$

• Recall that the ΔG_f° of the most stable form of an element at 298.15 K is zero. This is why additional terms do not appear in the equation.

Because MgO lies above Cr_2O_3 in the table, the difference enclosed in brackets is negative, and this reaction is spontaneous. A similar computation linking any pair of lines in the table confirms that a free metal spontaneously reduces any oxide below it in the table when the reactants and products are in standard states at 298.15 K. Table 12–1 amounts to an activity series because the order of entries ranks the metals by their ability to displace each other from their compounds with oxygen. A comparison to Table 4–3 (page 155), which gives an activity series for the displacement of metal ions from aqueous solution, reveals strong general similarities,

• For example, Na(s) displaces
$Mg^{2+}(aq)$ from aqueous solution, but
Mg(s) displaces Na from $Na_2O(s)$.

but some reversals of order as well. The reversals occur because this series was constructed using ΔG_f°'s of metal oxides and the other used the ΔG_f°'s of metal ions in aqueous solution. The two sets of ΔG_f°'s differ because the solid-state and solution environments differ. Thus, generalizations such as "sodium is a better reducing agent than magnesium," although often helpful, can go wrong. Relative reactivity depends on the environment in which substances are used and on the products that form.

According to Table 12–1, magnesium should react with titanium(IV) oxide (TiO_2) to give elemental titanium and MgO, which it does, and aluminum should reduce iron(III) oxide to give elemental iron and Al_2O_3, a process that proceeds spectacularly once it is initiated (see Fig. 10–1). The metals at the top of the table are, in thermodynamic terms, the most difficult to produce, and they are the most reactive once made. It is no accident that, historically, the most reactive metallic elements were the last to be isolated in elemental form.

In the remainder of this section, we examine the production and uses of two of the most important metals produced by chemical reduction, copper and iron. In Section 12–5, we consider two other important metals, aluminum and magnesium, that require electrochemical processing.

Copper and Its Alloys

Copper occurs in a variety of minerals in the form of veins that are frequently exposed as outcrops at the earth's surface. Chalcopyrite ($CuFeS_2$), chalcocite (Cu_2S), and bornite (Cu_5FeS_4) are chief among the copper-bearing ores. On exposure to the weathering action of the atmosphere, they are oxidized to beautiful green or blue basic copper carbonates such as malachite ($Cu_2CO_3(OH)_2$) and azurite ($Cu_3(CO_3)_2(OH)_2$), both of which are semiprecious stones (Fig. 12–2). Metallurgy may well have begun when a potter introduced an ore such as malachite into a kiln, perhaps as a colored glaze, and found a globule of a red metal after heating the ore in the reducing atmosphere generated by burning charcoal (carbon):

$$Cu_2CO_3(OH)_2(s) + C(s) \longrightarrow 2\ Cu(s) + 2\ CO_2(g) + H_2O(g)$$

Figure 12–2 Several gems from copper minerals. Clockwise from top: malachite with azurite, malachite, azurite with chalcopyrite, and turquoise.

Pure copper is not hard enough to make serviceable cutting tools. Some copper daggers and axes that date to the third millennium B.C. contain a few percent of arsenic, which confers hardness upon copper. We cannot be sure whether it is present by design or from the accidental use of ores that contained arsenic as an impurity. But the presence of tin with copper in ancient tools, at first in trace amounts and increasing to about 10% as the centuries passed, surely resulted from intent and marked the start of Bronze Age civilization. Bronze (Fig. 12–3), an alloy that contains about 10% tin and 90% copper, is hard without being brittle, can be cast in a mold, and melts at a temperature (950°C) substantially below the melting point of pure copper (1084°C).

Brass, which combines copper with up to 42% of zinc, is a second important alloy of copper. This alloy was first discovered by the Romans and is prized to this day in the manufacture of precision instruments because it is hard, easily machined, and corrosion-resistant.

Modern demand for copper stems largely from its use in wires for conducting electricity. This has led to worldwide exploitation of copper ores, of which the most abundant is chalcopyrite ($CuFeS_2$). Open-pit mining operations on a mammoth scale (Fig. 12–4) remove and crush tons of rock that may contain less than 1% copper. **Froth flotation** is used to concentrate the copper ores (Fig. 12–5). In this process, crushed low-grade ore is mixed with water, a special oil, and a detergent. A stream of compressed air produces a foamy, frothy mixture in which oil-wetted particles of ore are buoyed up by clinging to air bubbles while water-wetted earth and rock sink to the bottom. The copper-rich froth is skimmed off, the oil is separated for reuse, and (after drying) a solid mass enriched in $CuFeS_2$ (typically to 25 to 30% copper) results.

Figure 12–3 This bronze of Greek origin is about 2500 years old.

Figure 12–4 An open-pit copper mine in Utah. The ore contains less than 1% copper, but the metal can nevertheless be extracted economically, depending on fluctuations in the world market price.

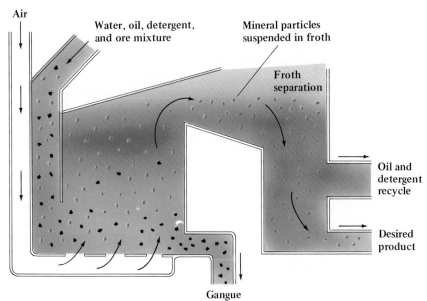

Figure 12–5 Froth-flotation for the enrichment of ore. Frothing agents are added to a water suspension of the pulverized ore. As compressed air is bubbled in, the desired mineral ($CuFeS_2$, in the case of copper production) is concentrated at the surface of a foam and borne upward while the unwanted rocky material settles to the bottom.

The next step is the **roasting** of this enriched ore, in which it reacts with air at high temperatures in a redox reaction that converts the iron to its oxide, while leaving the copper as a sulfide:

$$6 \ CuFeS_2(s) + 13 \ O_2(g) \longrightarrow 3 \ Cu_2S(s) + 2 \ Fe_3O_4(s) + 9 \ SO_2(g)$$

In the next stage, the mixture of Cu_2S and Fe_3O_4 is put in a furnace where the flame is reflected from the roof to play down on the material being heated. The temperature is raised to 1100°C, and limestone ($CaCO_3$) and silica (SiO_2) are added as a **flux** to remove the iron in the reaction

$$CaCO_3(s) + Fe_3O_4(s) + 2 \ SiO_2(s) \longrightarrow CO_2(g) + CaO{\cdot}FeO{\cdot}Fe_2O_3{\cdot}(SiO_2)_2(\ell)$$

slag, composition variable

• This slag could also be formulated as a mixture of $CaSiO_3(\ell)$ (equivalent to $CaO{\cdot}SiO_2$) and $Fe_2SiO_5(\ell)$ (equivalent to $Fe_2O_3{\cdot}SiO_2$) with FeO dissolved in it; additional amounts of all four oxides can be dissolved by such a mixture.

Because $Cu_2S(\ell)$ is not soluble in the calcium–iron slag and exceeds it in density, it settles to the bottom, and the molten slag layer can be poured off. The $Cu_2S(\ell)$ is run into another furnace (the converter), through which additional air is blown to cause the redox reaction:

$$Cu_2S(\ell) + O_2(g) \longrightarrow 2 \ Cu(\ell) + SO_2(g)$$

The liquid copper is then cooled and cast into large slabs of "blister copper" for further purification. The final refinement routinely uses an electrochemical method (see Section 12–5) that allows the recovery of the small amounts of gold and silver that accompany the copper to this stage.

Three major problems afflict the smelting of copper: the pollution that can be caused by the SO_2 by-product (Fig. 12–6), the disposal of the calcium–iron–silicate slag, which can contain toxic heavy-metal impurities such as cadmium, and the energy costs associated with operating high-temperature furnaces. Emission-control

Figure 12-6 The Ducktown region of Tennessee was laid waste in the 19th century by SO_2 from copper smelting. The copper is all mined out, and the denuded hillsides remain. Modern methods of copper smelting have virtually eliminated pollution from SO_2.

regulations have required operators to devise new processes in which nearly all of the SO_2 is captured and converted to commercially important sulfuric acid (see Chapter 22). Careful attention to the recycling of water in newer smelters has greatly reduced the amount of heavy metals leaching out from the slag. Finally, energy costs are reduced by new **hydrometallurgical** methods for separating the copper and iron in the ore. When an aqueous mixture of copper(II) chloride and iron(III) chloride is added to the ore, the copper-bearing minerals are reduced in the reactions

$$CuFeS_2(s) + 3\ CuCl_2(aq) \longrightarrow 4\ CuCl(s) + FeCl_2(aq) + 2\ S(s)$$
$$CuFeS_2(s) + 3\ FeCl_3(aq) \longrightarrow CuCl(s) + 4\ FeCl_2(aq) + 2\ S(s)$$

Excess aqueous sodium chloride is added to the solids that form to convert the $CuCl(s)$ into the soluble complex ion $[CuCl_2]^-(aq)$, which is separated from the sulfur and further converted to metallic copper and copper(II) chloride in the spontaneous reaction

$$2\ [CuCl_2]^-(aq) \longrightarrow Cu(s) + CuCl_2(aq) + 2\ Cl^-(aq)$$

The copper(II) chloride is recycled to reduce more ore.

Iron and Steel

The chief sources of iron are the minerals hematite (Fe_2O_3), magnetite (Fe_3O_4), goethite (FeO(OH)), limonite (FeO(OH)·nH$_2$O), and siderite ($FeCO_3$). The metallurgy of iron requires considerably higher temperatures than that of copper, and even though forced drafts were used to fan the fires, the molten iron produced in early times merely formed globules interspersed in a semiliquid slag. Repeated hammering squeezed the more fluid slag out of the red-hot mass to yield a product known as "wrought iron." Such iron was fairly soft and usually inferior to bronze until artisans learned that prolonged heating of iron in contact with charcoal caused it to absorb carbon. When iron containing several percent of dissolved carbon was quenched in water, it produced an extremely hard metal, a form of steel. This technology, developed by the Hittites around 1200 B.C., was further refined in Greco-Roman times. It was little changed until the **blast furnace** was devised in the late 14th and early 15th centuries.

Figure 12–7 illustrates the construction of a blast furnace for smelting iron ore. It consists of a cylindrical steel shell some 10 m in diameter at its widest part and about 30 m high, lined with a refractory "fire brick." Designed for continuous operation, it is typically run for a period of about five years, after which its lining must be torn out and replaced. A charge of iron ore, coke, and limestone ($CaCO_3$) or dolomite ($CaMg(CO_3)_2$) enters the top of the stack at intervals through a hopper and works its way downward as the reaction proceeds. Preheated air enters the stack through a number of nozzles called **tuyeres** above the **hearth,** where the molten iron collects.

High-grade iron ore is not pure Fe_2O_3 (approximately 70% Fe) but contains up to about 27% silica, alumina, clay, and minor impurities that must be converted into a fluid slag for the process to be continuous. For this reason limestone or dolomite is added with the charge. Either converts silica and alumina (melting points approximately 1700 and 2050°C, respectively) into lower-melting calcium aluminum silicates (melting point approximately 1550°C).

$$CaCO_3(s) \longrightarrow CaO(s) + CO_2(g)$$
$$CaO(s) + Al_2O_3(s) + 2\ SiO_2(s) \longrightarrow CaO{\cdot}Al_2O_3{\cdot}(SiO_2)_2(\ell)$$

composition variable

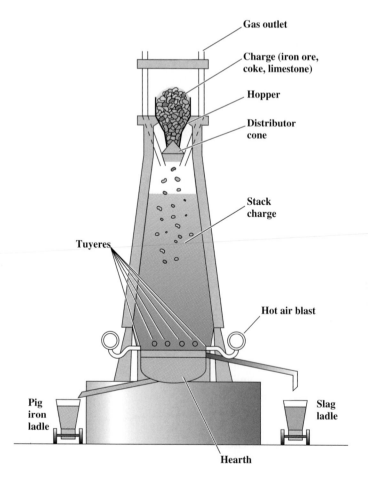

Figure 12–7 Pig iron flows from a blast furnace. The sparks mark the oxidation of splashed droplets of the iron by the air.

Hematite is not reduced by direct reaction with carbon itself but rather by carbon monoxide in the hot gases as they pass up through the charge under forced draft:

$$Fe_2O_3(s) + 3\ CO(g) \longrightarrow 2\ Fe(s) + 3\ CO_2(g)$$

The carbon monoxide is produced in reactions between hot coke and the preheated air that enters through the tuyeres. At this level in the blast furnace, the temperature is at a maximum (about 1875°C) because of the highly exothermic reaction

$$C(s) + O_2(g) \longrightarrow CO_2(g)$$

At this high temperature, carbon dioxide rapidly comes to equilibrium with carbon monoxide in the reaction

$$C(s) + CO_2(g) \rightleftharpoons 2\ CO(g)$$

The equilibrium constant of this reaction at 1875°C is readily calculated from data in Appendix D, if we assume temperature-independent enthalpies and entropies; it is large, about 1×10^5. The reduction of hematite to iron occurs as a gas–solid reaction fairly high in the stack, where the temperature is still well below the melting point of pure iron (1535°C). The result is a spongy form of iron interspersed with the other components of the charge. As the mass slowly descends through the stack, it encounters increasingly higher temperatures until, just above the tuyeres, both the iron and the slag components liquefy and trickle down into the hearth of the furnace. The density of the molten calcium aluminum silicate (slag) is less than that of molten iron, so the slag floats as a layer on the iron. Periodically, the two liquids are separately drawn off into ladle cars, the iron to be further refined and the slag to be solidified into a clinker (fused ash), which is ground up for use in portland cement or crushed for use as road ballast. The iron produced in the blast furnace, called **pig iron,** contains a total of 5 to 10% carbon, silicon, phosphorus, and other impurities, which are removed in a subsequent process. A modern blast furnace produces about 1500 metric tons of pig iron and 750 metric tons of slag per day while consuming about 1200 metric tons of coke.

Steel is iron mixed with a small amount (up to about 1.7% by mass) of carbon. The metallurgical properties of a steel depend on the exact percentage of carbon and on the levels and identities of other alloying elements as well. Steel is made from pig iron by the removal of silicon, phosphorus, manganese, sulfur, and other impurities, and the reduction of the carbon content to a target value that depends on the intended end use. Modern methods of conversion rely on the oxidation of the unwanted elements in the molten pig iron. The oxides escape as gases or react with a flux, which is added for this purpose, to form a slag phase that separates from the hot metal. At this time, the largest quantities of steel are produced in **basic oxygen converters.** These are typically barrel-shaped furnaces open at one end and lined with refractory bricks. They stand perhaps 10 m high, have a diameter of 5–6 m, and are tiltable to allow the tipping in of starting materials and pouring out of product. Such a converter is charged with some 150 to 250 tons of molten pig iron transferred in insulated ladles from a blast furnace (Fig. 12–8). A quantity of cold scrap steel is also often added. Then the converter is turned upright under a hood (to collect gases), and a high-velocity jet of pure oxygen is blown down onto the charge through nozzles in a water-cooled retractable lance positioned 1 or 2 m above the surface of the molten charge (Fig. 12–9). More oxygen is blown in through bottom-mounted tuyeres, which are protected from burning by injection of a small amount

Figure 12–8 Molten iron being charged into a basic oxygen furnace. The basic oxygen process for steelmaking is an *oxygen* process because it uses a blast of oxygen, rather than air, to oxidize impurities. It is a *basic* process because the flux is high in a base, lime, to capture the acidic oxides (such as SiO_2 and P_4O_{10}) of those impurities.

• The fate of slag is important. If it had no good use, it would present troublesome disposal problems. Its use in portland cement is treated in Section 21-4.

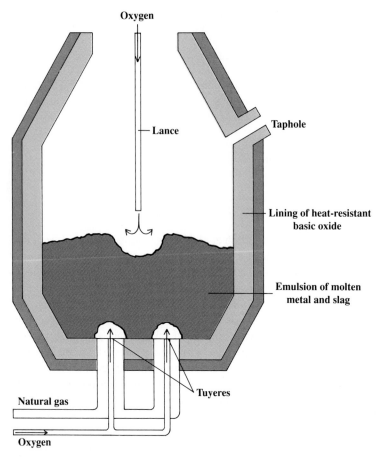

Figure 12–9 The operation of a basic oxygen converter. Blasts of pure oxygen from above (the lance) and below (the tuyeres) refine pig iron into steel. After processing, the converter is tipped on its side and the molten steel tapped off through the taphole. The impurity-containing slag floats atop the steel and is easily separated.

of propane gas through an annular space at their ends. The propane forms a sheath around the tuyeres. Its endothermic decomposition

$$C_3H_8 \longrightarrow 3\,C + 4\,H_2$$

in this region cools and protects the tuyeres.

Within seconds after blowing starts, strong oxidation commences in the pig iron. The flux, which is powdered burnt lime (CaO) or burnt dolomite (CaO·MgO), is blown into the mixture along with the oxygen. Some iron is oxidized to FeO and is rapidly distributed throughout the charge

$$2\,Fe + O_2 \longrightarrow 2\,FeO$$

Silicon burns directly to silicon dioxide, SiO_2, or reacts with FeO to give the same product

$$Si + O_2 \longrightarrow SiO_2$$
$$2\,FeO + Si \longrightarrow 2\,Fe + SiO_2$$

The acidic SiO_2 then reacts with the basic flux and forms a product that separates from the hot metal as the slag:

$$SiO_2 + CaO \longrightarrow CaO{\cdot}SiO_2(\ell)$$

Oxidation of the phosphorus and its removal into the slag proceed similarly, the overall equation being

$$5\ FeO + 2\ P + 3\ CaO \longrightarrow 5\ Fe + (CaO)_3{\cdot}P_2O_5(\ell)$$

The slag also takes up MnO as it forms by oxidation of manganese in the pig iron, and some FeO; both of these metal oxides are acidic. Carbon, the major impurity in the pig iron, is oxidized to gaseous carbon monoxide

$$C + FeO \longrightarrow Fe + CO$$
$$C + \tfrac{1}{2}\,O_2 \longrightarrow CO$$

Under the conditions in the converter, essentially no carbon dioxide ($CO_2(g)$) is produced. The escape of $CO(g)$ from the melt causes a boiling action that mixes the slag and metal phases thoroughly and hastens the transfer of oxides into the slag. All of the oxidation reactions are exothermic. The rising temperature in the converter melts the scrap steel and keeps the slag molten. Intermittently throughout the blow, a probe is plunged into the melt to measure the temperature (a good indicator of carbon content) and oxygen content of the steel. The data are used to adjust the blowing rate, the distance of the lance to the surface of the melt, the rate of addition of flux, and other operating conditions to reach the desired carbon content and temperature simultaneously. These adjustments are made under automatic computer control. The process is complete in 25 to 30 minutes.

Oxygen, hydrogen, and nitrogen are soluble in molten steel and must be removed before the steel can be cast. Otherwise, the escape of the dissolved gases would cause large void spaces (blowholes) in the casting. Dissolved oxygen also can react with alloying elements to form particles of oxides that remain in the finished steel as sources of internal and surface defects. The next stage of steel refining is therefore **vacuum degassing.** The pressure above the steel is reduced by large vacuum pumps as argon gas, which is not soluble in molten steel, is blown through the melt from the bottom. This circulates the liquid and speeds the escape of dissolved gases. Any desired alloying elements are also added during degassing. Additional oxygen is blown in briefly at the start of degassing to oxidize remaining carbon. During degassing, the low pressure above the melt shifts the equilibrium

$$C(\textit{dissolved}) + \tfrac{1}{2}\,O_2(\textit{dissolved}) \rightleftharpoons CO(g)$$

to the right and favors low levels of both carbon and oxygen in the steel. Toward the end of the degassing, aluminum is added. This further reduces the level of oxygen in the steel by the reactions

$$2\ Al + \tfrac{3}{2}\,O_2 \longrightarrow Al_2O_3$$
$$2\ Al + 3\ FeO \longrightarrow Al_2O_3 + 3\ Fe$$

The insoluble aluminum oxide forms an easily separated layer.

The carbon content of fully refined steel is far less than that of the original pig iron. For example, in steel intended for automobile body sheets, carbon is reduced to ultralow levels (0.002% C by mass) on a routine basis. Ultralow carbon steels have better press-forming characteristics than ordinary low-carbon or mild steel (0.1 to 0.4% C by mass). Steelmaking by the basic oxygen process allows the production of large batches of steel of controlled chemical make-up at short intervals. These are major advantages over other ways of making steel, particularly when the steel is taken directly for the continuous casting of sheets. The basic oxygen process in recent years has entirely supplanted the formerly dominant open-hearth process, in which pig iron was cooked in oxygen-enriched air with a basic flux for periods of several hours.

12–2 BALANCING REDOX EQUATIONS

In Section 2–3, we showed how to balance the chemical equation for burning butane in oxygen (a redox reaction) by logical reasoning from the fact that the chemical amounts of each element must be the same before and after the reaction. This approach works well for balancing a great many redox equations, including most of the gas- and solid-phase reactions characteristic of extractive metallurgy. Some redox equations are sufficiently complex, however, that an improved method for balancing becomes useful. These are most often equations for redox reactions occurring in either acidic or basic aqueous solution. In these reactions, water and H^+ (acidic solution) or OH^- (basic solution) may take part either as reactants or as products; therefore, the problem is to *complete* the equation (by including H_2O and H^+ or OH^- on the correct side of the equation) as well as to balance it.

> • For simplicity, we balance these equations using the symbol H^+ rather than $H_3O^+(aq)$. The latter form can always be obtained by replacing H^+ by H_3O^+ wherever it appears and adding a corresponding number of water molecules to the other side of the equation.

The best method to deal with this complication is to write and separately balance what are called **half-equations** to represent the two **half-reactions,** one of oxidation and the other of reduction, that make up a redox reaction. The balanced half-equations are then combined to give a balanced overall equation. A half-equation is a chemical equation in which electrons (e^-) are explicitly shown as either reactants or products. In a reduction half-equation, the electrons appear on the left side (are gained by the reactants); in an oxidation half-equation, the electrons appear on the right side (are lost by the reactants). The freedom to insert electrons on either side of the arrow makes it much easier to balance half-equations than "whole" equations. A correct balanced chemical equation may never include free electrons on either side, of course: the electrons in half-equations must always cancel out when the half-equations are combined to make a whole chemical equation. A step-by-step method for balancing redox equations by the use of half-equations follows. The steps are illustrated for the case of the dithionate ($S_2O_6^{2-}$) ion reacting with chlorous acid ($HClO_2$) in aqueous acidic solution. The equation to be completed and balanced is:

> • The concept of the half-equation and half-reaction also helps enormously in understanding redox reactions in electrochemical cells (see Section 12–3).

$$S_2O_6^{2-}(aq) + HClO_2(aq) \longrightarrow SO_4^{2-}(aq) + Cl_2(g)$$

Step 1. *Write two unbalanced half-equations, one for the species that is oxidized and its product and one for the species that is reduced and its product.*

Here, the half-reaction involving the dithionate ion is

$$S_2O_6^{2-} \longrightarrow SO_4^{2-}$$

and the half-reaction involving the chlorous acid is

$$HClO_2 \longrightarrow Cl_2$$

Step 2. *Insert coefficients to make the numbers of atoms of all elements* except *oxygen and hydrogen equal on the two sides of each half-equation.*

Balancing sulfur in the first half-equation and chlorine in the second gives

$$S_2O_6^{2-} \longrightarrow 2\ SO_4^{2-}$$

and

$$2\ HClO_2 \longrightarrow Cl_2$$

These are the only elements requiring attention in this step.

Step 3. *Balance oxygen by adding H_2O to the side deficient in O in each half-equation.*

There are 6 O's on the left of the first half-equation but 8 O's on the right. Hence 2 H_2O's are inserted on the left:

$$2\ H_2O + S_2O_6^{2-} \longrightarrow 2\ SO_4^{2-}$$

The second half-equation has the shortage of O on the right, so that is where the water goes:

$$2\ HClO_2 \longrightarrow Cl_2 + 4\ H_2O$$

Step 4. *Balance hydrogen. For a half-reaction in acidic solution, add H^+ ion to the side deficient in hydrogen. For a half-reaction in basic solution, add H_2O to the side that is deficient in hydrogen and an equal amount of OH^- to the other side.*

Note that this step does not disrupt the oxygen balance achieved in Step 3.

In this example (acidic solution), the first equation lacks a total of 4 H's on the right, so 4 H^+ ions are inserted on that side, giving

$$2\ H_2O + S_2O_6^{2-} \longrightarrow 2\ SO_4^{2-} + 4\ H^+$$

The shortage of hydrogen is on the left in the second half-equation. Hence 6 H^+ ions are inserted on the left

$$6\ H^+ + 2\ HClO_2 \longrightarrow Cl_2 + 4\ H_2O$$

Step 5. *Balance charge by inserting e^- (electrons) as a reactant or product in each half-equation.*

In the first half-equation, there is a total charge of -2 on the left and $2(-2) + 4(+1) = 0$ on the right. Correction of the imbalance requires 2 electrons on the right

$$2\ H_2O + S_2O_6^{2-} \longrightarrow 2\ SO_4^{2-} + 4\ H^+ + 2\ e^-$$

This half-equation represents an oxidation, because the electrons are among the products. In the second half-equation, the total charge equals $+6$ on the left and 0 on the right. The imbalance is removed by inserting 6 electrons on the left. Because the electrons are on the side of the reactants, this half-equation represents a reduction:

$$6\ H^+ + 2\ HClO_2 + 6\ e^- \longrightarrow Cl_2 + 4\ H_2O$$

We now have two balanced half-equations ready for combination to give the overall redox equation. Unless a mistake has been made, one half-equation represents a reduction and the other an oxidation.

Step 6. *Multiply the two half-equations by numbers chosen to make the number of electrons given off by the oxidation equal to the number taken*

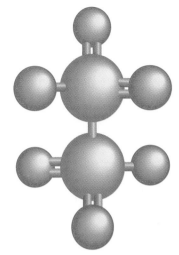

The dithionate ion ($S_2O_6^{2-}$). Note the sulfur–sulfur bond. Knowledge of oxidation numbers ($+5$ for the two sulfurs and -2 for the oxygens in this ion) is not needed to balance redox equations by the method presented here.

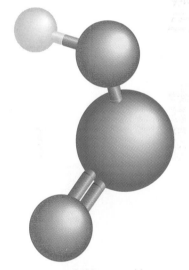

The structure of chlorous acid ($HClO_2$).

• The number of electrons appearing in a balanced half-equation equals in magnitude the total change in oxidation number experienced by all the atoms in the half-reaction.

up by the reduction. Then add the two half-equations and cancel out the electrons. If H^+ ion, OH^- ion, or H_2O appears on both sides of the final equation, cancel out the duplications.

Here the oxidation half-equation must be multiplied by 3, and the reduction half-equation by 1:

$$6\,H^+ + 2\,HClO_2 + 6\,e^- \longrightarrow Cl_2 + 4\,H_2O$$
$$6\,H_2O + 3\,S_2O_6^{2-} \longrightarrow 6\,SO_4^{2-} + 12\,H^+ + 6\,e^-$$

Addition of the balanced half-equations and cancellation of equal terms on the two sides gives

$$2\,H_2O(\ell) + 3\,S_2O_6^{2-}(aq) + 2\,HClO_2(aq) \longrightarrow$$
$$6\,SO_4^{2-}(aq) + 6\,H^+(aq) + Cl_2(g)$$

Electrons must *not* appear in the final equation.

Step 7. *Check for balance.*

In this case the final equation has 2 Cl, 6 S, 6 H, and 24 O atoms on each side. Also, the net charge on the left side is $3(-2) = -6$, which is equal to the net charge on the right side $6(-2) + 6 = -6$.

If the check fails, look first for omitted superscripts and subscripts. A particularly common error is writing H or H_2 when H^+ is intended.

EXAMPLE 12–1

Complete and balance the following equation, which represents the dissolution of copper(II) sulfide in aqueous nitric acid (Fig. 12–10):

$$CuS(s) + NO_3^-(aq) \longrightarrow Cu^{2+}(aq) + SO_4^{2-}(aq) + NO(g)$$

Solution

Step 1. Write two half-equations:

$$CuS \longrightarrow Cu^{2+} + SO_4^{2-}$$
$$NO_3^- \longrightarrow NO$$

Step 2. Balance elements other than H and O. This balance already exists.

Step 3. Balance oxygen by adding H_2O:

$$CuS + 4\,H_2O \longrightarrow Cu^{2+} + SO_4^{2-}$$
$$NO_3^- \longrightarrow NO + 2\,H_2O$$

Step 4. Balance hydrogen by adding H^+ (acidic solution):

$$CuS + 4\,H_2O \longrightarrow Cu^{2+} + SO_4^{2-} + 8\,H^+$$
$$NO_3^- + 4\,H^+ \longrightarrow NO + 2\,H_2O$$

Step 5. Balance charge by inserting electrons:

$$CuS + 4\,H_2O \longrightarrow Cu^{2+} + SO_4^{2-} + 8\,H^+ + 8\,e^- \quad \text{(oxidation)}$$
$$NO_3^- + 4\,H^+ + 3\,e^- \longrightarrow NO + 2\,H_2O \quad \text{(reduction)}$$

Step 6. Make the numbers of electrons represented in the two half-equations equal, add the half-equations, and cancel out duplicated terms. Here, if

Figure 12–10 Black particles of copper(II) sulfide react with concentrated nitric acid to liberate gaseous nitrogen monoxide and $SO_4^{2-}(aq)$ ion. The green color here comes from a complex of NO with Cu(II). It fades to blue as NO is lost, and the final product $Cu^{2+}(aq)$ forms.

the first half-equation is multiplied through by 3 and the second by 8, then each ends up with 24 electrons:

$$3\ CuS + 12\ H_2O \longrightarrow 3\ Cu^{2+} + 3\ SO_4^{2-} + 24\ H^+ + 24\ e^-$$
$$8\ NO_3^- + 32\ H^+ + 24\ e^- \longrightarrow 8\ NO + 16\ H_2O$$

Addition followed by cancellation of 24 e^-, 24 H^+, and 12 H_2O's gives

$$3\ CuS(s) + 8\ NO_3^-(aq) + 8\ H^+(aq) \longrightarrow$$
$$3\ Cu^{2+}(aq) + 3\ SO_4^{2-}(aq) + 8\ NO(g) + 4\ H_2O(\ell)$$

Step 7. Check. There are 3 Cu, 3 S, 8 N, 24 O, and 8 H atoms on each side; the net charge on each side is zero: $8(-1) + 8(+1) = 3(+2) + 3(-2)$.

Exercise

Complete and balance the following equation representing a reaction that takes place in acidic aqueous solution:

$$SO_2(aq) + Cr_2O_7^{2-}(aq) \longrightarrow Cr^{3+}(aq) + SO_4^{2-}(aq)$$

Answer: $3\ SO_2(aq) + Cr_2O_7^{2-}(aq) + 2\ H^+(aq) \longrightarrow$
$$2\ Cr^{3+}(aq) + 3\ SO_4^{2-}(aq) + H_2O(\ell).$$

When the reaction takes place in basic solution, remember to add H_2O and OH^-, rather than H^+ and H_2O, at step 4.

EXAMPLE 12–2

Complete and balance the following equation. It represents a reaction that takes place in basic aqueous solution.

$$Ag(s) + HS^-(aq) + CrO_4^{2-}(aq) \longrightarrow Ag_2S(s) + Cr(OH)_3(s)$$

Solution

Step 1. The two unbalanced half-equations are

$$Ag + HS^- \longrightarrow Ag_2S$$
$$CrO_4^{2-} \longrightarrow Cr(OH)_3$$

Step 2. Balance the elements other than hydrogen and oxygen. Here, only the first half-equation is affected. It becomes

$$2\ Ag + HS^- \longrightarrow Ag_2S$$

Step 3. Balance oxygen by adding H_2O. Here, all that is required is a single H_2O on the right side of the second half-equation. It becomes

$$CrO_4^{2-} \longrightarrow Cr(OH)_3 + H_2O$$

Step 4. Balance hydrogen by adding H_2O to the side that is deficient in hydrogen and an equal amount of OH^- to the other side (basic solution). The first half-equation becomes

$$2\ Ag + HS^- + OH^- \longrightarrow Ag_2S + H_2O$$

and the second becomes

$$CrO_4^{2-} + 4\,H_2O \longrightarrow Cr(OH)_3 + 5\,OH^-$$

after duplicated water molecules are canceled on the two sides.

Step 5. Balance charge by inserting electrons as necessary. The first half-reaction is revealed as the oxidation:

$$2\,Ag + HS^- + OH^- \longrightarrow Ag_2S + H_2O + 2\,e^- \qquad \text{(oxidation)}$$

and the second as the reduction:

$$CrO_4^{2-} + 4\,H_2O + 3\,e^- \longrightarrow Cr(OH)_3 + 5\,OH^- \qquad \text{(reduction)}$$

Step 6. Make the numbers of electrons represented in the two half-equations equal; then add the half-equations, and cancel out duplicated terms. The first equation is multiplied by 3 so that it liberates 6 electrons, and the second is multiplied by 2 so that it absorbs 6 electrons.

$$6\,Ag + 3\,HS^- + 3\,OH^- \longrightarrow 3\,Ag_2S + 3\,H_2O + 6\,e^- \qquad \text{(oxidation)}$$
$$2\,CrO_4^{2-} + 8\,H_2O + 6\,e^- \longrightarrow 2\,Cr(OH)_3 + 10\,OH^- \qquad \text{(reduction)}$$

Addition followed by cancellation of $6\,e^-$'s, $3\,OH^-$'s, and $3\,H_2O$'s gives

$$6\,Ag(s) + 3\,HS^-(aq) + 2\,CrO_4^{2-}(aq) + 5\,H_2O(\ell) \longrightarrow$$
$$3\,Ag_2S(s) + 2\,Cr(OH)_3(s) + 7\,OH^-(aq)$$

Step 7. Check. There are 6 Ag, 13 H, 3 S, 2 Cr, and 13 O atoms on each side. The net charge on the left is -7, which equals the net charge on the right.

Exercise

Complete and balance the following equation for a redox reaction that takes place in basic solution.

$$AsO_3^{3-}(aq) + Br_2(aq) \longrightarrow AsO_4^{3-}(aq) + Br^-(aq)$$

Answer: $AsO_3^{3-}(aq) + Br_2(aq) + 2\,OH^-(aq) \longrightarrow$
$$AsO_4^{3-}(aq) + 2\,Br^-(aq) + H_2O(aq)$$

Balancing Equations Representing Disproportionation

In a **disproportionation** reaction, the same chemical species is both oxidized and reduced; it reacts with itself (see Section 4–4). An example is the fate of chlorine dissolved in acidic solution:

$$Cl_2(aq) \longrightarrow ClO_3^-(aq) + Cl^-(aq) \qquad \text{(unbalanced)}$$

Balancing disproportionation equations is straightforward once it is realized that the same species may appear on the left in *both* half-equations. This gives for step 1 in this case:

$$Cl_2 \longrightarrow ClO_3^-$$
$$Cl_2 \longrightarrow Cl^-$$

CHEMISTRY IN YOUR LIFE

Redox Reactions Color Gems and Pigments

The pure, bright blue of sapphire comes from a simple redox reaction. If a small amount of titanium (mostly in the +4 oxidation state) is accidentally or deliberately doped into a crystal of corundum (Al_2O_3), no change in color results. Similarly, introduction of iron(II) or iron(III) impurities in corundum gives, at most, a very pale yellow color. When impurities of *both* titanium(IV) and iron(II) are present (even at the level of a few hundredths of 1%), however, the beautiful transparent blue of sapphire arises as electrons are transferred from iron-impurity sites to neighboring titanium-impurity sites

$$Ti^{4+} + Fe^{2+} \longrightarrow Ti^{3+} + Fe^{3+}$$

This reaction requires energy to make it occur (ΔE is positive). This energy can be supplied by red, orange, or yellow light. When ordinary white light (made up of all colors) passes through a crystal of sapphire, these colors are strongly absorbed as the redox reaction takes place, and only the blue light passes through, giving the color that we see (Fig. 12–A). The reverse reaction occurs quickly without emitting colored light, and so the blue color persists.

A redox reaction also causes the intense color of the artist's pigment called Prussian blue (see Fig. 12–B). Prussian blue is a hydrate of the compound iron(III)

Figure 12–A A synthetic sapphire.

Figure 12–B The dye "Prussian blue" forms upon mixing dilute solutions of $FeCl_3(aq)$ and $K_4Fe(CN)_6(aq)$.

hexacyanoferrate(II), with chemical formula $(Fe^{III})_4$ $[Fe^{II}(CN)_6]_3 \cdot xH_2O$, where the oxidation states of the iron atoms are explicitly indicated. The redox reaction in question transfers an electron from an iron(II) to an iron(III) atom, and can be written as

$$Fe(II) + Fe(III) \longrightarrow Fe(III) + Fe(II)$$

It seems from this equation as if nothing happens in the reaction, but the environments of the two types of iron atoms differ. The electron transfer oxidizes an iron atom within the $[Fe(CN)_6]^{4-}$ complex ion, converting it into $[Fe(CN)_6]^{3-}$, and simultaneously reduces an outer Fe^{III} ion to Fe^{II}. This reaction is induced by red, orange, and yellow light and makes the dye appear blue. Both oxidation states of iron must be present to give this reaction; neither iron(II) hexacyanoferrate(II) nor iron(III) hexacyanoferrate(III) is blue because no redox reaction can occur in these cases. The deep colors of other mixed-valence oxides, such as the black magnetite Fe_3O_4 and red Mn_3O_4, which can be formulated as $Fe^{II}(Fe^{III})_2O_4$ and $Mn^{II}(Mn^{III})_2O_4$, respectively, also result from redox reactions.

We then proceed with steps 2 to 5 to find

$$Cl_2 + 6\ H_2O \longrightarrow 2\ ClO_3^- + 12\ H^+ + 10\ e^- \quad \text{(oxidation)}$$
$$Cl_2 + 2\ e^- \longrightarrow 2\ Cl^- \quad \text{(reduction)}$$

Multiplying the reduction half-equation by 5 (to cancel the electrons in the final equation) and adding it to the oxidation half-equation gives

$$6\ Cl_2 + 6\ H_2O \longrightarrow 2\ ClO_3^- + 10\ Cl^- + 12\ H^+$$

which, when the coefficients are all divided by 2, gives the final result:

$$3\ Cl_2(aq) + 3\ H_2O(\ell) \longrightarrow ClO_3^-(aq) + 5\ Cl^-(aq) + 6\ H^+(aq)$$

12–3 ELECTROCHEMICAL CELLS

The separation of a redox reaction into two half-reactions is more than a trick for balancing equations. It can actually be achieved experimentally in a device called an **electrochemical cell.** As a first example, consider the oxidation–reduction reaction between metallic copper and an aqueous solution of silver nitrate:

$$Cu(s) + 2\ Ag^+(aq) \longrightarrow Cu^{2+}(aq) + 2\ Ag(s)$$

The nitrate ions are spectator ions. They are omitted in the preceding net ionic equation because they play no part in the reaction. Two electrons are transferred from each reacting copper atom to a pair of silver ions. The copper is thus oxidized, and the silver ions are reduced. This redox reaction can be carried out by simply putting a piece of copper in an aqueous solution of silver nitrate (Fig. 12–11a). Metallic silver immediately begins to plate out on the copper, the concentration of silver ion decreases, and blue $Cu^{2+}(aq)$ appears in solution and increases in concentration as time passes. The displacement reaction is clearly spontaneous. Hence, ΔG is less than zero. The reverse reaction does *not* take place spontaneously (Fig. 12–11b).

This same spontaneous reaction can be carried out in quite a different way without ever putting the two reactants in direct physical contact. We dip a copper strip in a solution of $Cu(NO_3)_2$ and a silver strip in a solution of $AgNO_3$, as Figures 12–12 and 12–13 illustrate. The two metallic strips are the **electrodes** of an electrochemical cell. Next, the two solutions are connected by a **salt bridge,** which is an inverted U-shaped tube containing a solution of a salt such as $NaNO_3$. The ends of the bridge are stuffed with porous plugs that prevent mixing of the two solutions but allow ions to pass between them. Finally, the two electrodes are connected by wires to an **ammeter,** an instrument that measures both the direction and the magnitude of electric current through it.

Figure 12–11 (a) When copper is placed in a solution of silver nitrate, silver forms in a tree-like structure, and the solution turns blue as Cu^{2+} ions form. (b) No reaction occurs between a silver spoon and a copper nitrate solution.

(a)

(b)

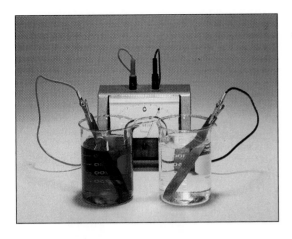

Figure 12-12 Metallic copper reacts with silver ions in a galvanic cell to produce metallic silver and a blue solution of copper(II) ions.

As copper is oxidized on the left side, Cu^{2+} ions enter the solution:

$$Cu(s) \longrightarrow Cu^{2+}(aq) + 2\ e^-$$

The electrons that are released pass through the external circuit from left to right, as shown by the deflection of the ammeter needle. The electrons flow into the silver strip, and at the metal–solution interface, they are picked up by Ag^+ ions, which plate out as atoms on the surface of the metal:

$$Ag^+(aq) + e^- \longrightarrow Ag(s)$$

By itself, this process would lead to an increase of positive charge in the left-hand beaker and an increase of negative charge in the right-hand beaker and therefore could not proceed. The salt bridge, however, permits a compensating flow of positive ions into the right-hand beaker and of negative ions into the left-hand beaker, preserving charge neutrality in each. In the cell diagrammed in Figure 12–13, electrochemical action continues until the copper electrode is corroded away at the waterline or essentially all the Ag^+ ion is plated out of solution.

The net chemical reaction that occurs in this simple electrochemical cell is the same as that which takes place when a copper rod is immersed in an aqueous solution of silver nitrate. What is different is that the half-reactions occur at sites remote from each other. The separation of the reaction components into two compartments (two **half-cells**) prevents the direct transfer of electrons from copper atoms to silver ions. If the reaction is to occur, the electrons must travel through a wire. The current of electrons through the wire can be used for a variety of purposes. If a light bulb is connected into the external circuit, the electron flow causes it to glow. The electrochemical cell thereby converts chemical energy into radiant energy. If the light bulb is replaced with a small electric motor, the energy change of the chemical reaction can be used to perform mechanical work.

The **electric current** just described is a direct current (DC) because it always flows in a single direction. The strength of a direct current is the amount of charge that flows past a point in a circuit per unit time. If Q is the magnitude of the charge in coulombs (abbreviation: C) and t is the time in seconds that it takes to pass by the test point in the circuit, then the current I is

$$I = \frac{Q}{t}$$

The SI unit for I is the ampere (abbreviation: A) or coulomb per second ($C\ s^{-1}$).

Figure 12–13 In the galvanic cell of Figure 12–12, charged particles move when the circuit is completed. Electrons flow from the copper to the silver electrode through the wire. In solution, anions migrate toward the copper electrode and cations toward the silver electrode.

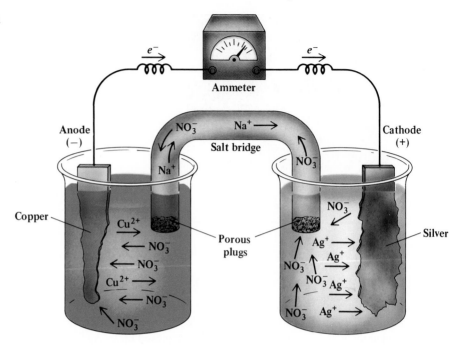

EXAMPLE 12–3

A battery delivers a steady current of 1.25 A for a period of 1.50 hours (h). Calculate the total charge (Q) in coulombs that passes through the circuit.

Solution

Because 1 h contains $60 \times 60 = 3600$ s,

$$1.50 \text{ h} \times \left(\frac{3600 \text{ s}}{1 \text{ h}}\right) = 5.40 \times 10^3 \text{ s}$$

The total charge is

$$Q = I \times t = 1.25 \frac{\text{C}}{\text{s}} \times 5.40 \times 10^3 \text{ s} = \boxed{6.75 \times 10^3 \text{ C}}$$

Exercise

Compute the time (in minutes) required for 12,600 C of electrical charge to pass through a light bulb in a DC circuit if the current in the circuit is a steady 0.850 A.

Answer: 247 minutes.

Galvanic and Electrolytic Cells

What causes an electric current to flow? There must be an electric **potential difference** ($\Delta \mathscr{E}$) between two points in the circuit to cause electrons to flow, just as a difference in gravitational potential between two points on a slope causes water to flow downhill. The potential difference developed by an electrochemical cell, or **cell voltage,** can be measured with an instrument called a **potentiometer.** A potentiometer is a variable voltage source that is put in the external circuit in such a way that its potential difference ($\Delta\mathcal{E}_{ext}$) *opposes* the intrinsic potential difference ($\Delta\mathcal{E}$) of the electrochemical cell. The total potential difference is then

$$\Delta\mathcal{E}_{total} = \Delta\mathcal{E} - \Delta\mathcal{E}_{ext}$$

$\Delta\mathcal{E}$ is measured by adjusting $\Delta\mathcal{E}_{ext}$ until $\Delta\mathcal{E}_{total}$ becomes zero, at which point the current through the circuit falls to zero as well, and the cell voltage is read from a dial.

If the opposing external voltage is made to *exceed* the potential difference developed by the cell, the flow of electrons is forced into reverse. In the copper–silver cell of Figures 12–12 and 12–13, copper ions in solution then accept electrons and deposit as metallic copper, and silver dissolves to furnish additional Ag^+ ions. The net chemical reaction that occurs is the reverse of the spontaneous reaction, namely,

$$2\ Ag(s) + Cu^{2+}(aq) \longrightarrow 2\ Ag^+(aq) + Cu(s)$$

An electrochemical cell that operates spontaneously is called a **galvanic cell.** Such a cell allows a net conversion of chemical energy into electrical energy, which can then be used to perform work. A cell in which an opposing external potential forces a reaction to occur in the reverse of its spontaneous direction is called an **electrolytic cell.** Such a cell uses electrical energy to carry out a chemical reaction that would otherwise not occur.

An oxidation–reduction reaction is the combination of two half-reactions. In the galvanic cell that we have been considering, the change that takes place spontaneously in the left-hand beaker of Figure 12–12 is the oxidation half-reaction

$$Cu(s) \longrightarrow Cu^{2+}(aq) + 2\ e^-$$

The change in the right-hand beaker is the reduction half-reaction

$$Ag^+(aq) + e^- \longrightarrow Ag(s)$$

When these are combined to give the overall balanced chemical equation, the electrons cancel out. In an electrochemical cell, the sites of the two half-reactions are remote from each other.

Chemists call the site at which oxidation occurs in an electrochemical cell the **anode.** The site at which reduction occurs is called the **cathode.**

• Remember this by noting that *anode* and *oxidation* both start with vowels, and *cathode* and *reduction* both start with consonants.

In our copper–silver galvanic cell, the copper electrode is the anode because it is oxidized; the silver electrode is the cathode because reduction occurs at its surface. When the same cell is forced by an external potential to run in reverse, as an electrolytic cell, the labels are exchanged. Oxidation then takes place at the silver electrode, which becomes the anode, and reduction takes place at the copper electrode, which becomes the cathode. In both galvanic and electrolytic cells, the flow of electrons in the external circuit always proceeds from anode to cathode. The solution, unlike the wire comprising the external circuit, permits both positive and negative particles to migrate under the influence of the difference in electrical potential. Negative ions (anions) move toward the anode, and positive ions (cations) move toward the cathode (see Fig. 12–13).

Rather than diagramming cells in full (as in Figure 12–13), chemists frequently resort to a more compact representation in which boundaries between phases are shown by a single vertical bar and a salt bridge by a double dashed bar. In these abbreviated diagrams, the anode is always shown on the left and the cathode on the right, so that electrons always flow through the external circuit from left to right in the diagram. The diagram of the copper–silver galvanic cell (Fig. 12–13) is accordingly compressed to

$$Cu(s)|Cu^{2+}(aq)\|Ag^+(aq)|Ag(s)$$

The same cell forced by an external potential difference to run in reverse is represented as

$$Ag(s)|Ag^+(aq)\|Cu^{2+}(aq)|Cu(s)$$

In many cells, the electrode itself does not react but serves only as a conduit to deliver electrons to or remove electrons from the solution, where a reaction involving other species takes place. Platinum and graphite are inert in most (but not all) electrochemical reactions. A platinum electrode might thus be used to remove electrons as one dissolved species is oxidized to give a second, as in the half-reaction

$$Fe^{2+}(aq) \longrightarrow Fe^{3+}(aq) + e^-$$

or to deliver electrons if this half-reaction is reversed, and remain itself unaffected. Its role in the above half-reaction would be diagrammed

$$Pt|Fe^{2+}(aq), Fe^{3+}(aq)$$

If a gas is bubbled over an immersed platinum electrode (or other inert electrode), half-reactions between that gas and dissolved ions can occur at the surface of the electrode. An example is

$$Cl_2(g) + 2\,e^- \longrightarrow 2\,Cl^-(aq)$$

Such a cathode half-cell would be diagrammed as $Cl_2(g)|Cl^-(aq)|Pt$. If this were combined with the previous half-cell through a suitable salt bridge, the cell

$$Pt|Fe^{2+}(aq), Fe^{3+}(aq)\|Cl_2(g)|Cl^-(aq)|Pt$$

would result.

• It was the unexpected dissolution of an "inert" platinum electrode in an electrochemical experiment that led to the discovery of the platinum-containing anticancer drug cisplatin.

EXAMPLE 12–4

The final step in the production of metallic magnesium from seawater (see Section 12–5) is the electrolysis of molten magnesium chloride in a large cell using a steel cathode and a graphite anode. The overall reaction is

$$Mg^{2+}(melt) + 2\,Cl^-(melt) \longrightarrow Mg(\ell) + Cl_2(g)$$

Write equations for the half-reactions occurring at the anode and at the cathode, and indicate the direction in which electrons flow through the external circuit. Diagram the cell.

Solution

The anode is the site at which oxidation takes place—that is, where electrons are given up. The anode half-reaction must be

$$2\,Cl^-(melt) \longrightarrow Cl_2(g) + 2\,e^-$$

At the cathode, reduction takes place, and the electrons are taken up. The half-reaction at the cathode is

$$Mg^{2+}(melt) + 2\,e^- \longrightarrow Mg(\ell)$$

Electrons move from the anode, where chlorine is liberated, through the external circuit to the cathode, where molten magnesium is produced. In a conventional cell diagram, the anode is always at the left.

$$graphite|Cl^-(melt),\ Cl_2(g),\ Mg^{2+}(melt),\ Mg(\ell)|steel$$

Exercise

The commercial production of fluorine relies on the electrolysis of a solution of potassium fluoride in liquid hydrogen fluoride using a steel cathode and a carbon anode. The overall reaction is

$$2\,HF(\ell) \longrightarrow F_2(g) + H_2(g)$$

Write equations for the half-reactions occurring at the anode and the cathode, and describe the direction of motion of the electrons in the external circuit. (*Hint*: Fluoride ion (F^-) is involved in the reaction.)

Answer: Anode: $2\,F^- \longrightarrow F_2(g) + 2\,e^-$. Cathode: $2\,HF + 2\,e^- \longrightarrow H_2(g) +$ $2\,F^-$. The electrons proceed from the anode, where fluorine is generated, to the cathode, where HF is reduced and H_2 is generated.

12-4   FARADAY'S LAWS

In both galvanic and electrolytic cells, electrons are liberated at the anode, pass through an external circuit, and are recaptured at the cathode. The number of electrons lost always equals the number of electrons gained; therefore, there must be a quantitative relationship between the amounts of substance that react at the two electrodes and the total electric charge that passes through the external circuit. In 1834, Michael Faraday reported this "doctrine of definite electrochemical action" on the basis of his experiments. It is summarized in modern terms in two statements:

1. The quantities of substance produced and consumed at the electrodes are directly proportional to the amount of electric charge passing through the cell.
2. When a given amount of electric charge passes through a cell, the quantity of a substance produced or consumed at an electrode is proportional to its molar mass divided by the number of moles of electrons required to produce or consume one mole of the substance.

These are **Faraday's laws** for the stoichiometry of electrochemical processes. Faraday's laws are best understood by thinking of electrons as specialized chemical reactants that cannot be stored and weighed out but must be generated on the spot and measured out with electrical meters. The charge e on a single electron has been very accurately determined to be

$$e = 1.6021773 \times 10^{-19}\ C$$

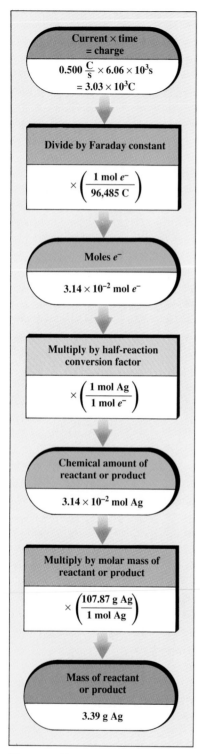

Figure 12–14 A flow chart for stoichiometric calculations in electrochemistry. Note the similarities to the flow chart in Figure 2–6.

This result allows the charge per mole of electrons to be calculated:

$$\left(1.6021773 \times 10^{-19}\ \frac{\mathrm{C}}{\text{electron}}\right) \times \left(6.022137 \times 10^{23}\ \frac{\text{electron}}{\mathrm{mol}}\right) =$$

$$9.648531 \times 10^{4}\ \frac{\mathrm{C}}{\mathrm{mol}}$$

The charge of one mole of electrons is the **Faraday constant** (symbol $\mathscr{F}$):

$$\mathscr{F} = 9.648531 \times 10^{4}\ \mathrm{C\ mol}^{-1} = 96{,}485.31\ \mathrm{C\ mol}^{-1}$$

A current of I amperes flowing steadily for t seconds causes $I \times t$ coulombs of charge to pass through a circuit. We use the Faraday constant to determine the chemical amount, in moles, of electrons to which this is equivalent:

$$(I \times t)\ \mathcal{C} \times \left(\frac{1\ \text{mol electrons}}{96{,}485.31\ \mathcal{C}}\right) = \frac{It}{96{,}485.31}\ \text{mol electrons}$$

From the number of moles of electrons that pass through a circuit, we can calculate the number of moles (and from that, the number of grams) of substances reacting at the electrodes in any electrochemical cell included in that circuit. Suppose that we connect up our copper–silver galvanic cell one more time. The anode half-reaction is

$$\mathrm{Cu}(s) \longrightarrow \mathrm{Cu}^{2+}(aq) + 2\ e^{-}$$

and the cathode half-reaction is

$$\mathrm{Ag}^{+}(aq) + e^{-} \longrightarrow \mathrm{Ag}(s)$$

Each mole of electrons that passes through the external circuit arises from the oxidation of one-half mole of Cu(s) (because each copper atom gives up two electrons) and reduces one mole of silver ions. From the molar masses of copper and silver and the n values in the half-reactions, we calculate that $63.55/2 = 31.77$ g of copper is dissolved at the anode and $107.87/1 = 107.87$ g of silver is deposited at the cathode for each mole of electrons passing through the circuit. The same relationships hold if the cell is operated as an electrolytic cell, but in that case silver is dissolved and copper is deposited. Figure 12–14 shows a "flow-chart" for Faraday's law calculations, one of which is illustrated in the following example.

EXAMPLE 12–5

An electrolytic cell is constructed in which the silver ions in silver chloride are reduced to silver at the cathode and zinc is oxidized to $\mathrm{Zn}^{2+}(aq)$ at the anode. A current of 0.500 A is passed through the cell for 101 min. Calculate the mass of zinc dissolved and the mass of silver deposited.

Solution

We convert the time of operation of the cell to seconds

$$t = 101\ \text{min} \times \left(\frac{60\ \mathrm{s}}{1\ \text{min}}\right) = 6.06 \times 10^{3}\ \mathrm{s}$$

and then use the definition of an ampere (1 C s^{-1}) to determine the charge passed through the cell first in coulombs and then (by use of the Faraday constant) in moles of electrons:

$$6.06 \times 10^3 \; \cancel{s} \times \left(\frac{0.500 \; \cancel{C}}{\cancel{s}}\right) \times \left(\frac{1 \; \text{mol}}{96{,}485 \; \cancel{C}}\right) = 3.14 \times 10^{-2} \; \text{mol}$$

The half-reactions are

$$AgCl(s) + e^- \longrightarrow Ag(s) + Cl^-(aq) \quad \text{(cathode)}$$
$$Zn(s) \longrightarrow Zn^{2+}(aq) + 2\,e^- \quad \text{(anode)}$$

so the passage of 3.14×10^{-2} mol of electrons brings about the deposition of 3.14×10^{-2} mol of silver and the dissolution of $(3.14 \times 10^{-2})/2$ mol of zinc.

$$m_{Ag} = (3.14 \times 10^{-2} \; \cancel{\text{mol Ag}}) \times \left(\frac{107.87 \; \text{g Ag}}{1 \; \cancel{\text{mol Ag}}}\right) = \boxed{3.39 \; \text{g Ag}}$$

$$m_{Zn} = \left(\frac{3.14 \times 10^{-2}}{2}\right) \cancel{\text{mol Zn}} \times \left(\frac{65.39 \; \text{g Zn}}{1 \; \cancel{\text{mol Zn}}}\right) = \boxed{1.03 \; \text{g Zn}}$$

Exercise
A galvanic cell generates an average current of 0.121 A for 15.6 min. The cathode half-reaction in the cell is $Pb^{2+}(aq) + 2\,e^- \longrightarrow Pb(s)$. What mass of lead is deposited at the cathode?

Answer: 0.122 g of lead.

12-5 ELECTROMETALLURGY

The methods of pyrometallurgy and hydrometallurgy (Section 12–1) work poorly to produce aluminum and the alkali and alkaline-earth elements as free metals. These metals have such high Gibbs functions relative to their ores that only **electrometallurgy,** or electrolytic production, can produce them in large quantity. Electrochemical cells are also used to purify these metals as well as metals that are produced by the techniques of pyrometallurgy.

Aluminum

Aluminum is the third most abundant element in the earth's crust (after oxygen and silicon), accounting for 8.2% of the total mass. The most important ore for the production of aluminum is bauxite, a hydrated aluminum oxide that contains 50 to 60% Al_2O_3; 1 to 20% Fe_2O_3; 1 to 10% silica; minor concentrations of transition metal oxides; and 20 to 30% water. Because bauxite is not pure hydrated alumina (Al_2O_3), it must be purified before reduction to elemental aluminum. This is accomplished by the **Bayer process,** which takes advantage of the fact that the amphoteric oxide, alumina, is soluble in strong bases, whereas iron(III) oxide is not. Bauxite is treated with aqueous sodium hydroxide, which dissolves the alumina

$$Al_2O_3(s) + 2\;OH^-(aq) + 3\;H_2O(\ell) \longrightarrow 2\;Al(OH)_4^-(aq)$$

and allows its separation from hydrated iron oxide and other insoluble impurities by filtration. Pure hydrated aluminum oxide is precipitated when the solution is cooled to supersaturation and seeded with crystals of the product

$$2\;Al(OH)_4^-(aq) \longrightarrow Al_2O_3 \cdot 3H_2O(s) + 2\;OH^-(aq)$$

The solid is separated, and the water of hydration is removed by heating to high temperature (1200°C).

Unlike copper, iron, gold, and lead, which were known in elemental form in antiquity, aluminum is a relative newcomer. Sir Humphrey Davy obtained it as an alloy of iron and proved its metallic nature in 1809. It was first prepared in relatively pure form by H. C. Oersted in 1825 by reduction of aluminum chloride with potassium dissolved in mercury,

$$AlCl_3(s) + 3\ K(Hg)_x(\ell) \longrightarrow 3\ KCl(s) + Al(Hg)_{3x}(\ell)$$

after which the mercury was removed by distillation. Aluminum remained largely a laboratory curiosity until 1886, when Charles Hall in the United States (then a 21-year-old recent graduate of Oberlin College) and Paul Héroult (a Frenchman of the same age) independently invented an efficient process for its production. In 1995, the worldwide annual production of aluminum by the **Hall–Héroult process** was approximately 1.9×10^7 metric tons.

• This is second only to the production of steel (about 7.6×10^8 metric tons).

The process involves the cathodic deposition of aluminum from molten cryolite (Na_3AlF_6) containing dissolved aluminum oxide in electrolysis cells. In industrial plants, these consist of rectangular steel boxes some 6 m long, 2 m wide, and 1 m high that serve as cathodes, and massive graphite anodes that extend through the roof of the cell into the molten cryolite bath (Fig. 12–15). Enormous currents (50,000 to 100,000 A) are passed through the cell, and as many as 100 such cells may be connected in series.

Molten cryolite, which is completely dissociated into Na^+ and AlF_6^{3-} ions, is an excellent solvent for aluminum oxide, giving rise to an equilibrium distribution of ions such as Al^{3+}, AlF^{2+}, AlF_2^+, ... , AlF_6^{3-}, and O^{2-} in the electrolyte. Cryolite melts at 1000°C, but its melting point is lowered by dissolved aluminum oxide, so the operating temperature of the cell is about 950°C. Compared to the melting point of pure Al_2O_3 (2050°C), this is a low temperature, and it is the reason the Hall–Héroult process is technologically successful. Molten metallic aluminum is somewhat denser than the melt at 950°C and insoluble in it. The molten product collects at the bottom of the cell, from which it is tapped periodically. Oxygen is the primary anode product, but it reacts with the graphite electrode to produce carbon dioxide. The overall cell reaction is

$$2\ Al_2O_3 + 3\ C \longrightarrow 4\ Al + 3\ CO_2$$

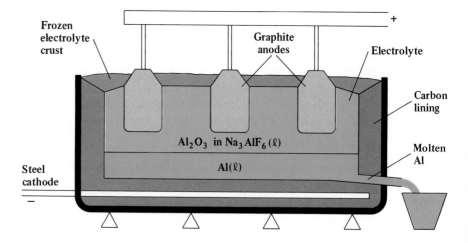

Figure 12–15 An electrolytic cell used in the Hall–Héroult process for making aluminum.

Molten aluminum is being tapped from an electrolytic cell.

Aluminum and its alloys have a tremendous variety of applications. Many of these make use of its low density (Table 12–2), an advantage over iron or steel in situations in which weight savings are desirable. This includes the transportation industry, in which aluminum is used in vehicles from automobiles to moon rockets. Its high electrical conductivity combines with its low density to make aluminum useful for electrical transmission lines. For structural and building applications, its resistance to corrosion is a positive feature, as is the fact that it becomes stronger at subzero temperatures that can make steel and iron brittle. Household applications such as aluminum foil, aluminum soft-drink cans, and cooking utensils are well known.

Magnesium

Like aluminum, magnesium is a very abundant element on the surface of the earth, but one that is difficult to prepare in elemental form. Although ores, such as dolomite ($CaMg(CO_3)_2$) and carnallite ($KCl \cdot MgCl_2 \cdot 6H_2O$) exist, the major commercial source of magnesium and its compounds is seawater. Magnesium forms the second most abundant positive ion in the sea, and Mg^{2+} is separated from the other cations in seawater (Na^+, Ca^{2+}, and K^+, in particular) by taking advantage of the fact that magnesium hydroxide is the least soluble hydroxide of the group. To recover magnesium economically requires a cheap base to treat large volumes of seawater and efficient methods for separating the $Mg(OH)_2(s)$ that precipitates from the solution. One base that is used in this way is calcined dolomite, which is prepared by heating dolomite to high temperatures to drive off carbon dioxide:

$$CaMg(CO_3)_2(s) \longrightarrow CaO \cdot MgO(s) + 2\,CO_2(g)$$

When $CaO \cdot MgO(s)$ is added to seawater, the greater solubility of calcium hydroxide ($K_{sp} = 5.5 \times 10^{-6}$) relative to magnesium hydroxide ($K_{sp} = 1.2 \times 10^{-11}$) leads to the reaction

$$CaO \cdot MgO(s) + Mg^{2+}(aq) + 2\,H_2O(\ell) \longrightarrow 2\,Mg(OH)_2(s) + Ca^{2+}(aq)$$

Table 12–2 Densities of Selected Metals	
Metal	**Density (g cm^{-3}) at Room Conditions**
Li	0.534
Na	0.971
Mg	1.738
Al	2.702
Ti	4.54
Zn	7.133
Fe	7.874
Ni	8.902
Cu	8.96
Ag	10.50
Pb	11.35
U	18.95
Au	19.32
Pt	21.45

CHEMISTRY IN YOUR LIFE

Redox Reactions in the Mouth

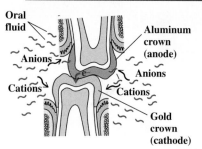

Figure 12–C A galvanic cell operates when aluminum touches gold in the mouth. Electrons (shown in red) pass directly from the aluminum to the gold. Aluminum is oxidized, and oxygen is reduced.

Dentists install temporary crowns so that patients can chew during the week or two it may take to fabricate a permanent crown. Aluminum is often used for this purpose because it is cheap and easily manipulated. It can however present problems. Aluminum is an active metal; it combines spontaneously with oxygen and displaces hydrogen from water at room conditions. Ordinarily, neither reaction proceeds because aluminum is protected by a coating of aluminum oxide that is so effective that aluminum can be melted in the air (at 660°C) without serious oxidation. The reactive tendency of aluminum is thwarted, but *not* removed.

The tendency can manifest itself if a tooth with an aluminum crown opposes one with a gold crown. The mouth then becomes the site of a galvanic cell:

$$Al(s) \longrightarrow Al^{3+}(aq) + 3\ e^- \qquad \text{(anode)}$$
$$O_2(aq) + 4\ H^+(aq) + 4\ e^- \longrightarrow 2\ H_2O(\ell) \qquad \text{(cathode)}$$

When the dissimilar metals touch, electrons pass from the aluminum to the gold as cations in the oral fluid move toward the gold and anions move toward the aluminum (Fig. 12–C). The electrical activity generally stimulates a nerve and causes pain. Moreover, the $Al^{3+}(aq)$ ion has an unpleasant metallic taste. The same effect occurs if a piece of aluminum foil from a candy wrapper lodges between two teeth and touches a gold crown or inlay.

Another dental scenario for unleashing aluminum's true reactivity involves mercury. Mercury forms alloys called *amalgams* with many metals, including aluminum, gold, copper, and silver. Because aluminum oxide does not adhere to an amalgamated surface, amalgamated aluminum reacts well with water or air. "Silver" fillings are prepared by "triturating" (grinding together) an alloy of silver, tin, copper and sometimes zinc, as well as liquid

mercury. The fresh mixture, which is called *dental amalgam*, is packed into the cavity in the tooth while still in a pasty semi-liquid state. It then hardens. If a dentist allows a temporary aluminum crown to come in contact with a new amalgam filling, mercury from the filling may amalgamate the aluminum. The protective coating of aluminum oxide then fails, and the aluminum crown starts to react with oxygen and water. The reactions are exothermic, and the heat can cause severe pain.

Dentists have to be cautious about another redox reaction when using a zinc-containing amalgam. If water contaminates the fresh amalgam during manipulation, it soon reacts with zinc

$$Zn(s) + H_2O(\ell) \longrightarrow Zn^{2+}(aq) + H_2(g) + 2\ OH^-(aq)$$

The trapped gas causes the filling to expand unduly as it hardens. The consequence is again pain and perhaps a cracked tooth.

An advantage of this process is that the magnesium hydroxide that is produced includes not only the magnesium from the seawater but also that from the dolomite.

An interesting alternative to dolomite as a base for magnesium production is employed in a process used on the coast of Texas (Fig. 12–16). In this process, oyster shells (composed largely of $CaCO_3$) are calcined to give lime (CaO), which is added to the seawater to yield magnesium hydroxide. The $Mg(OH)_2$ slurry (a suspension in water) is washed and filtered in huge nylon filters. The addition of hydrochloric acid neutralizes the magnesium hydroxide and yields aqueous magnesium chloride:

$$Mg(OH)_2(s) + 2\ HCl(aq) \longrightarrow MgCl_2(aq) + 2\ H_2O(\ell)$$

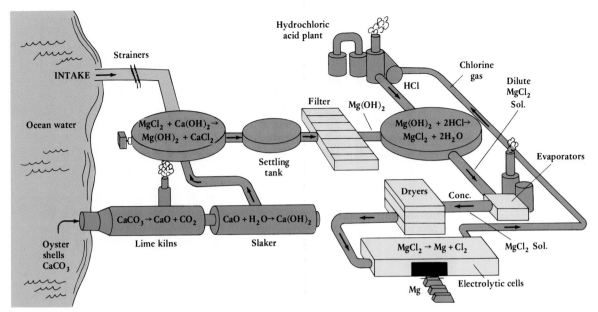

Figure 12–16 Magnesium hydroxide is produced starting with the addition of lime (CaO) to seawater. Reaction of the magnesium hydroxide with hydrochloric acid produces magnesium chloride, which, after drying, is electrolyzed to give metallic magnesium.

After the water is evaporated, the solid magnesium chloride can be melted (melting point 708°C) in a steel electrolysis cell that might hold as much as 10 tons of the molten salt. The steel in the cell acts as the cathode during electrolysis, with graphite anodes suspended from the top. The cell reaction is

$$MgCl_2(\ell) \longrightarrow Mg(\ell) + Cl_2(g)$$

The molten magnesium liberated at the cathode floats to the surface and is dipped out periodically, while the chlorine generated at the anodes is collected and reacted with steam at high temperatures to produce hydrochloric acid. This is recycled for further reaction with magnesium hydroxide.

Until 1918, the major use of elemental magnesium was in fireworks and flash-bulbs, which took advantage of its great reactivity with the oxygen in air and the bright light that reaction gives off. Since that time, many further uses for the metal and its alloys have been developed. Magnesium is even less dense than aluminum and is used in alloys with that metal to lower its density and improve its resistance to corrosion under basic conditions. As we discuss in Section 13–7, magnesium can be used as a "sacrificial anode" to prevent the oxidation of another metal with which it is in contact. It is also used as a reducing agent to produce other metals such as titanium, uranium, and beryllium from their compounds. Both of these uses reflect the ease with which magnesium can be oxidized.

Electrorefining and Electroplating

Metals such as copper, silver, nickel, and tin that have been produced by pyromet-allurgical methods are too impure for many purposes. The methods of **electrorefin-ing** are adopted to purify them further. Crude metallic copper, for example, is cast

Figure 12–17 In the electrolytic refining of copper, a large number of slabs of impure copper, which serve as anodes, alternate with thin sheets of pure copper (the cathodes). Both dip into a dilute acidic solution of copper sulfate. As the copper is oxidized from the impure anodes, it enters the solution and migrates to the cathodes, where it plates out in purer form.

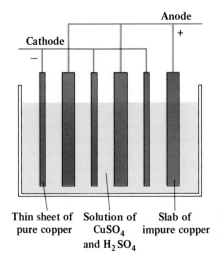

Thin sheet of   Solution of   Slab of
pure copper $CuSO_4$ impure copper
                and H_2SO_4

into slabs, which are used as anodes in electrolysis cells that contain a solution of $CuSO_4$ in aqueous H_2SO_4. Thin sheets of pure copper serve as cathodes. Copper that dissolves at the anodes passes through the solution to be deposited in purer form on the cathodes (Fig. 12–17). Impurities that are more easily oxidized than copper, such as nickel, go into solution along with the copper but then remain there; elements such as silver and gold that are less easily oxidized do not dissolve but fall away from the anode as a metallic slime. Periodically, the anode slime and the solution are removed and processed further to recover the valuable elements they contain.

A related process is **electroplating,** in which electrolysis is used to plate out a thin layer of a metal on top of another material, frequently a second metal. In chrome plating, for example, the piece of metal to be plated is placed in a hot bath of sul-

* About one quarter of the silver produced in the United States and one eighth of the gold are by-products of the electrolytic refining of copper.

Figure 12–18 Silver being plated onto tableware at the cathode of an electrolytic cell.

furic and chromic acid (H_2CrO_4) and is made the cathode in an electrolytic cell. As current passes through the cell, chromium is reduced from the +6 oxidation state in chromic acid to metallic chromium and plates out on the cathode. When such plating is used for decorative purposes, the chromium layer can be as thin as 2.5×10^{-5} cm (corresponding to 2 g of chromium to cover each square meter of surface). Thicker layers ranging up to 10^{-2} cm are found in hard chromium plate, which is prized for its resistance to wear. Electroplating is used with many other metals as well. Steel can be plated with cadmium to improve its resistance to corrosion in marine environments. Gold and silver plating are used both for decorative purposes (Fig. 12–18) and in electronic devices because of the low electrical resistance of these metals.

EXAMPLE 12–6

Suppose that a layer of metallic chromium of thickness 3.0×10^{-3} cm is to be plated onto an automobile bumper that has a surface area of 2.0×10^3 cm^2. If a current of 250 A is used, for how long must current be passed through the cell to achieve the desired thickness? The density of chromium is 7.2 g cm^{-3}.

Solution

This example requires a calculation that reverses the flow shown in Figure 12–14. The first step is to calculate the mass of chromium that must be deposited. The volume of the chromium is the product of the thickness of the layer and the area of the surface:

$$V_{Cr} = (3.0 \times 10^{-3} \text{ cm})(2.0 \times 10^3 \text{ cm}^2) = 6.0 \text{ cm}^3$$

The mass of chromium is the product of this volume and the density of the element:

$$m_{Cr} = (6.0 \text{ cm}^3) \times \left(\frac{7.2 \text{ g}}{\text{cm}^3}\right) = 43.2 \text{ g}$$

From this, we can calculate the chemical amount of chromium that must be reduced

$$n_{Cr} = 43.2 \text{ g Cr} \times \left(\frac{1 \text{ mol Cr}}{52.00 \text{ g Cr}}\right) = 0.831 \text{ mol Cr}$$

Next, we relate the number of moles of chromium to the number of moles of electrons that must pass through the cell. The half-reaction for the production of elemental chromium from H_2CrO_4 is

$$H_2CrO_4 + 6 \text{ H}^+ + 6 \text{ } e^- \longrightarrow Cr + 4 \text{ H}_2O$$

Hence 6 moles of electrons are required for each mole of chromium deposited. The number of moles of electrons is then

$$n_e = 0.831 \text{ mol Cr} \times \left(\frac{6 \text{ mol } e^-}{1 \text{ mol Cr}}\right) = 4.98 \text{ mol } e^-$$

The charge (in coulombs) is the product of this and the Faraday constant:

$$Q = (4.98 \text{ mol}) \times \left(\frac{9.6485 \times 10^4 \text{ C}}{\text{mol}}\right) = 4.81 \times 10^5 \text{ C}$$

The time required for this quantity of electric charge to pass through the cell is

$$\text{time} = \frac{4.81 \times 10^5 \text{ C}}{250 \text{ C s}^{-1}} = 1.9 \times 10^3 \text{ s} = \boxed{32 \text{ min}}$$

Exercise

A layer of gold 1.00 mm thick is to be plated from a solution that contains gold in the $+3$ oxidation state onto a specialized electronic part with a surface area of 4.50 cm^2. A steady current of 0.880 A is used. How long does the plating run last? The density of gold is 19.32 gm^{-3}.

Answer: 242 min.

SUMMARY

12–1 In **oxidation–reduction (redox) reactions,** one chemical species gains electrons (it is reduced) while another species loses electrons (it is oxidized). The extraction of metals from their ores provides a most important example of redox chemistry. Different methods are used for different metals, ranging from gentle heating to very energy-intensive electrochemical methods. Between the two limits are the techniques of **pyrometallurgy,** in which the desired non-spontaneous reaction to obtain the metal is coupled to a spontaneous oxidation reaction, typically the oxidation of carbon to carbon dioxide, in a **smelter** at high temperature. Frequently, a **flux** of limestone or other substances is added to the ores during smelting to remove impurities from the reaction mixture.

12–2 Chemical equations to represent redox reactions in aqueous solution can be both completed (with H^+, OH^-, and H_2O) and balanced by the half-equation method. The oxidation and reduction steps are conceived as **half-reactions** and represented by **half-equations** in which electrons appear explicitly. When the two half-equations are combined to give the overall balanced equation, the electrons must cancel out. **Disproportionation** reactions, in which different atoms of the same chemical element are oxidized and reduced in the course of a single reaction, can be balanced by the same method.

12–3 In an **electrochemical cell,** oxidation and reduction half-reactions are carried out at physically separated **electrodes.** The electrons released by oxidation at the **anode** pass through an external circuit and are taken up by reduction at the **cathode.** At the same time, ions migrate within the cell to maintain charge neutrality throughout. When a spontaneous reaction takes place in an electrochemical cell, the cell is a **galvanic cell;** the electric current produced by such a cell can be used to do work. Conversely, in an **electrolytic cell,** current from an outside source is forced through the cell and non-spontaneous reactions are caused to occur, allowing the generation of products possessing a higher Gibbs function than the reactants.

12–4 Faraday's laws state a quantitative relationship between the mass of a chemical species consumed or produced at an electrode in an electrochemical cell and the amount of charge passing through the circuit. They are best used by treating the electrons as a reactant or product in a stoichiometry calculation based on a half-equation.

12–5 Electrolytic cells are used in producing and purifying metals, especially those that have high Gibbs functions relative to their common ore sources. The **electrometallurgy** of aluminum and magnesium, which uses such cells, leads to metals and alloys that are particularly useful because of their low density and good electrical conductivity. Electrolytic cells are also used to purify metals and to plate a thin layer of one metal onto a second metal.

PROBLEMS

Note: Answers to blue-numbered problems are given in Appendix F. Problems that are more challenging are indicated with asterisks.

Metals and Metallurgy

1. The principal ore of manganese is pyrolusite (MnO_2), which can be reduced to elemental manganese with aluminum. Write a balanced equation for the reaction, and calculate the standard enthalpy and Gibbs function changes per mole of manganese produced from pyrolusite, using data in Appendix D.

2. Uranium can be prepared in elemental form by reducing U_3O_8 with calcium. Write a balanced equation for the reaction, and calculate the standard enthalpy change and standard Gibbs function change per mole of uranium produced from U_3O_8, using the following thermodynamic data:

	ΔH_f° (kJ mol^{-1})	S°(J K^{-1} mol^{-1})
$U_3O_8(s)$	-3575	282.6
$U(s)$	0	50.2
$CaO(s)$	-635	39.8
$Ca(s)$	0	41.4

3. As mentioned in the text, mercury(II) oxide (HgO) decomposes to mercury and oxygen when heated. Use data from Appendix D to estimate the temperature at which this change becomes spontaneous—that is, the temperature at which ΔG° becomes negative.

4. Use data from Appendix D to predict the temperature above which CuO decomposes spontaneously to Cu(s) and $\frac{1}{2} O_2(g)$ at atmospheric pressure.

5. Analysis of a copper-bearing seam of rock shows it to contain three copper-containing substances: 1.1% chalcopyrite ($CuFeS_2$), 0.42% covellite (CuS), and 0.51% bornite (Cu_5FeS_4) by mass. Calculate the total mass of copper present in 1.0 metric ton (1.0×10^3 kg) of this copper-bearing ore.

6. Calculate (a) the mass of copper and (b) the volume of sulfur dioxide at STP produced in the smelting of 1.0 metric ton (1.0×10^3 kg) of chalcopyrite ($CuFeS_2$).

7. (a) Write balanced equations for the reduction of hematite (Fe_2O_3), magnetite (Fe_3O_4), and siderite ($FeCO_3$) to iron with carbon monoxide.

    (b) Calculate ΔG° at 298.15 K per mole of iron produced from each of the three reactions in part (a). From your results, predict which iron compound should be easiest to reduce with CO(g) from the standpoint of thermodynamics.

8. Calculate ΔG° at 298.15 K for each of the reactions

$$Fe_2O_3(s) + \tfrac{3}{2} C(s) \longrightarrow 2 Fe(s) + \tfrac{3}{2} CO_2(g)$$
$$Fe_2O_3(s) + 3 C(s) \longrightarrow 2 Fe(s) + 3 CO(g)$$
$$Fe_2O_3(s) + 3 CO(g) \longrightarrow 2 Fe(s) + 3 CO_2(g)$$

From your results, decide which is the better reducing agent: C (being oxidized to CO or CO_2) or CO.

9. Set up an entry for CoO(s)/Co(s) for inclusion in Table 12–1; indicate where in the table it should be placed. Use data from Appendix D.

10. Calculate ΔG°'s at 298.15 K for the reactions

$$2 Na(s) + Mg^{2+}(aq) \longrightarrow Mg(s) + 2 Na^+(aq)$$
$$Na_2O(s) + Mg(s) \longrightarrow MgO(s) + 2 Na(s)$$

to confirm that in aqueous solution Na spontaneously reduces $Mg^{2+}(aq)$ but that metallic Mg spontaneously reduces sodium oxide.

Balancing Redox Equations

11. The following balanced equations represent reactions that occur in aqueous acid. Break them down into balanced oxidation and reduction half-equations.

    (a) $2 H^+(aq) + H_2O_2(aq) + 2 Fe^{2+}(aq) \longrightarrow$
    $\qquad\qquad\qquad\qquad 2 Fe^{3+}(aq) + 2 H_2O(\ell)$

    (b) $H^+(aq) + 2 H_2O(\ell) + 2 MnO_4^-(aq) + 5 SO_2(aq) \longrightarrow$
    $\qquad\qquad\qquad\qquad 2 Mn^{2+}(aq) + 5 HSO_4^-(aq)$

    (c) $5 ClO_2^-(aq) + 4 H^+(aq) \longrightarrow$
    $\qquad\qquad 4 ClO_2(g) + Cl^-(aq) + 2 H_2O(\ell)$

12. The following balanced equations represent reactions that occur in aqueous base. Break them down into balanced oxidation and reduction half-equations.

    (a) $4 PH_3(g) + 4 H_2O(\ell) + 4 CrO_4^{2-}(aq) \longrightarrow$
    $\qquad\qquad P_4(s) + 4 Cr(OH)_4^-(aq) + 4 OH^-(aq)$

    (b) $NiO_2(s) + 2 H_2O(\ell) + Fe(s) \longrightarrow$
    $\qquad\qquad\qquad Ni(OH)_2(s) + Fe(OH)_2(s)$

    (c) $2 OH^-(aq) + 2 NO_2(g) \longrightarrow$
    $\qquad\qquad NO_3^-(aq) + NO_2^-(aq) + H_2O(\ell)$

13. (See Example 12–1.) Complete and balance the following equations for reactions taking place in acidic solution:

    (a) $VO_2^+(aq) + SO_2(g) \longrightarrow VO^{2+}(aq) + SO_4^{2-}(aq)$
    (b) $Br_2(\ell) + SO_2(g) \longrightarrow Br^-(aq) + SO_4^{2-}(aq)$
    (c) $Cr_2O_7^{2-}(aq) + Np^{4+}(aq) \longrightarrow Cr^{3+}(aq) + NpO_2^{2+}(aq)$
    (d) $HCOOH(aq) + MnO_4^-(aq) \longrightarrow CO_2(g) + Mn^{2+}(aq)$
    (e) $Pb_3O_4(s) \longrightarrow Pb^{2+}(aq) + PbO_2(s)$
    (f) $Hg_2HPO_4(s) + Au(s) + Cl^-(aq) \longrightarrow$
    $\qquad\qquad Hg(\ell) + H_2PO_4^-(aq) + AuCl_4^-(aq)$

14. (See Example 12–1.) Complete and balance the following equations for reactions taking place in acidic solution:

    (a) $MnO_4^-(aq) + H_2S(aq) \longrightarrow Mn^{2+}(aq) + SO_4^{2-}(aq)$
    (b) $Zn(s) + NO_3^-(aq) \longrightarrow Zn^{2+}(aq) + NH_4^+(aq)$
    (c) $H_2O_2(aq) + MnO_4^-(aq) \longrightarrow O_2(g) + Mn^{2+}(aq)$
    (d) $Sn(s) + NO_3^-(aq) \longrightarrow Sn^{4+}(aq) + N_2O(g)$
    (e) $UO_2^{2+}(aq) + Te(s) \longrightarrow U^{4+}(aq) + TeO_4^{2-}(aq)$
    (f) $PbSO_4(s) \longrightarrow Pb(s) + PbO_2(s) + SO_4^{2-}(aq)$

15. (See Example 12–2.) Complete and balance the following equations for reactions taking place in basic solution:
    (a) $Cr(OH)_3(s) + Br_2(aq) \longrightarrow CrO_4^{2-}(aq) + Br^-(aq)$
    (b) $ZrO(OH)_2(s) + SO_3^{2-}(aq) \longrightarrow Zr(s) + SO_4^{2-}(aq)$
    (c) $HPbO_2^-(aq) + Re(s) \longrightarrow Pb(s) + ReO_4^-(aq)$
    (d) $HXeO_4^-(aq) \longrightarrow XeO_6^{4-}(aq) + Xe(g)$
    (e) $Ag_2S(s) + Cr(OH)_3(s) \longrightarrow$
    $$Ag(s) + HS^-(aq) + CrO_4^{2-}(aq)$$
    (f) $N_2H_4(aq) + CO_3^{2-}(aq) \longrightarrow N_2(g) + CO(g)$

16. (See Example 12–2.) Complete and balance the following equations for reactions taking place in basic solution:
    (a) $OCl^-(aq) + I^-(aq) \longrightarrow IO_3^-(aq) + Cl^-(aq)$
    (b) $SO_3^{2-}(aq) + Be(s) \longrightarrow S_2O_3^{2-}(aq) + Be_2O_3^{2-}(aq)$
    (c) $P_4(s) \longrightarrow HPO_3^{2-}(aq) + PH_3(g)$
    (d) $H_2BO_3^-(aq) + Al(s) \longrightarrow BH_4^-(aq) + H_2AlO_3^-(aq)$
    (e) $O_2(g) + Sb(s) \longrightarrow H_2O_2(aq) + SbO_2^-(aq)$
    (f) $Sn(OH)_6^{2-}(aq) + Si(s) \longrightarrow HSnO_2^-(aq) + SiO_3^{2-}(aq)$

17. "Doctor solution" for sweetening gasoline is prepared by the reaction
    $$PbS(s) + O_2(g) \longrightarrow PbO_2^{2-}(aq) + SO_4^{2-}(aq)$$
    in aqueous base. Complete and balance this equation.

18. Metallic copper is attacked by solutions of nitric acid and oxidized.
    (a) Complete and balance each of the following equations, which have been written to represent this reaction:
    $$Cu(s) + HNO_3(aq) \longrightarrow Cu^{2+}(aq) + NO_2(g)$$
    $$Cu(s) + HNO_3(aq) \longrightarrow Cu^{2+}(aq) + NO(g)$$
    (b) Recently, it was suggested that this reaction is actually
    $$Cu(s) + HNO_3(aq) \longrightarrow Cu^{2+}(aq) + HNO_2(aq)$$
    (The $HNO_2(aq)$ then reacts further.) Complete and balance this equation.

19. Write balanced chemical equations for the following reactions:
    (a) Solid calcium plus gaseous chlorine gives solid calcium chloride.
    (b) Iron(III) ion mixed with tin(II) ion in aqueous solution gives iron(II) ion and tin(IV) ion.
    (c) Solid iron(III) oxide reacts with gaseous hydrogen to give water vapor and solid iron.
    (d) Solid potassium reacts with aqueous hydrogen peroxide to give aqueous potassium hydroxide.

20. Write a balanced equation to represent the redox reaction of the mineral bornite (Cu_5FeS_4) with aqueous copper(II) chloride ($CuCl_2$) if the products of the reaction are the same as in the reaction of chalcopyrite with $CuCl_2$.

21. Write a balanced equation for the reaction of the mineral bornite (Cu_5FeS_4) with aqueous $FeCl_3$ if the products of the reaction are the same as in the reaction of chalcopyrite with $CuCl_2$.

22. Write balanced equations for the following reactions:
    (a) Solid barium reacts with gaseous oxygen to give solid barium peroxide.

    (b) Cerium(IV) ion plus iodide ion in aqueous solution gives cerium(III) ion and aqueous iodine.
    (c) Solid manganese(III) hydroxide reacts with solid chromium to give solid manganese(II) hydroxide and solid chromium(III) hydroxide.
    (d) Solid mercury(I) chloride reacts with solid cobalt to give liquid mercury and solid cobalt(II) chloride.

23. Nitrous acid (HNO_2) disproportionates in acidic solution to nitrate ion (NO_3^-) and nitrogen monoxide (NO). Write a balanced equation for this reaction.

24. Thiosulfate ion ($S_2O_3^{2-}$) disproportionates in acidic solution to give solid sulfur and aqueous hydrogen sulfite ion (HSO_3^-). Write a balanced equation for this reaction.

Electrochemical Cells

25. (See Example 12–3.) An automobile battery delivers a current of 45 A for a period of 1.2 s to start the engine. Calculate the quantity of charge (in coulombs) passing through the battery.

26. (See Example 12–3.) A light bulb in a flashlight draws a steady current of 3.00×10^{-3} A. Calculate the quantity of charge (in coulombs) passing through the bulb in 1.00 h.

27. A total of 6.5×10^3 C of charge passes through an electric iron in 18 min. Calculate the average current drawn by the iron in amperes.

28. Calculate the time required for 4.87×10^5 C to pass through an electric circuit if the current is 41.2 A.

29. (See Example 12–4.) Acidified water can be electrolyzed to produce hydrogen and oxygen (see Fig. 1–9). Write equations for the half-reactions that occur at the anode and at the cathode, and verify that they combine correctly to give the overall decomposition of water to its elements.

30. (See Example 12–4.) A galvanic cell operates to produce $Br^-(aq)$ and $Ni(OH)_2(s)$ from metallic nickel and bromate ion (BrO_3^-) in a basic aqueous solution. Write equations for the half-reactions that occur at the anode and at the cathode and for the overall reaction.

31. Diagram an electrochemical cell using the reaction
    $$2\,Mn^{3+} + H_2(aq) \longrightarrow 2\,Mn^{2+}(aq) + 2\,H^+(aq)$$

32. Diagram an electrochemical cell in which an external voltage source is used to produce nickel and iodine from a solution of nickel(II) iodide, according to the equation
    $$Ni^{2+}(aq) + 2\,I^-(aq) \longrightarrow Ni(s) + I_2(aq)$$

33. Sketch the following galvanic cell, indicating the direction of the flow of electrons in the external circuit and the motion of the ions in the salt bridge.
    $$Pt(s)|Cr^{2+}(aq),\ Cr^{3+}(aq)\|Cu^{2+}(aq)|Cu(s)$$
    Write a balanced equation for the overall reaction in this cell.

34. Sketch the following galvanic cell, indicating the direction of the flow of electrons in the external circuit and the motion of

ions in the salt bridge.

$$Ni(s)|Ni^{2+}(aq)\|HCl(aq)|H_2(g)|Pt(s)$$

Write a balanced equation for the overall reaction in this cell.

35. Sketch the following electrochemical cells. Indicate in the sketch the sites of oxidation and reduction and the direction of flow of electrons. Write a balanced equation to represent the reaction taking place in each cell.
   (a) $Pt(s)|Fe^{2+}(aq), Fe^{3+}(aq)\|MnO_4^-(aq), Mn^{2+}(aq)|Pt(s)$
   (b) $C(graphite)|I^-(aq), I_2(aq)\|Ag^+(aq)|Ag(s)$

36. Sketch the following electrochemical cells. Indicate in the sketch the sites of oxidation and reduction and the direction of flow of electrons.
   (a) $Pb(s)|Pb^{2+}(aq)\|Cu^{2+}(aq)|Cu(s)$
   (b) $Ag(s)|Ag^+(aq)\|I_2(aq), I^-(aq)|C(graphite)$

Faraday's Laws

37. An electric circuit from a battery carries a steady current of 2.5 A. How many electrons pass a given point in the wire per minute?

38. In a television set, the picture is created by an electron beam that sweeps over the screen, causing scintillations (flashes of light) where it strikes. A typical current for the electron beam is 60 microamperes (μA). How many electrons strike the screen per second?

39. How many coulombs of electricity is required for each of the following reductions?
   (a) 1.00 mol of $Ag^+(aq)$ to $Ag(s)$
   (b) 2.00 mol of $H_2O(\ell)$ to $H_2(g)$
   (c) 1.50 mol of $Mn^{3+}(aq)$ to $Mn^{2+}(aq)$
   (d) 3.00 mol of $ClO_4^-(aq)$ to $Cl^-(aq)$

40. What number of coulombs of electricity is required for each of the following oxidations?
   (a) 2.00 mol of $Zn(s)$ to $Zn^{2+}(aq)$
   (b) 1.00 mol of $H_2O_2(aq)$ to $O_2(g)$
   (c) 2.50 mol of $MnO_2(s)$ to $MnO_4^-(aq)$
   (d) 1.70 mol of $H_2S(aq)$ to 0.850 mol of $H_2S_2O_3(aq)$

41. (See Example 12–5.) A quantity of electricity equal to 6.95×10^4 C passes through an electrolytic cell that contains a solution of $Sn^{4+}(aq)$ ions. Compute the maximum chemical amount, in moles, of $Sn(s)$ that can be deposited at the cathode. What mass of Sn is this?

42. (See Example 12–5.) A quantity of electricity equal to 9.263×10^4 C passes through a galvanic cell that has an $Ni(s)$ anode. Compute the maximum chemical amount, in moles, of $Ni^{2+}(aq)$ that can be formed in solution. What maximum mass of $Ni(s)$ is lost from the anode?

43. A galvanic cell uses a zinc anode immersed in a $Zn(NO_3)_2$ solution and a platinum cathode immersed in a NaCl solution and in contact with $Cl_2(g)$ at 1 atm and 25°C.
   (a) Write a balanced equation for the cell reaction.
   (b) A steady current of 0.800 A is observed to flow for a period of 25.0 min. How much charge passes through the circuit during this time? How many moles of electrons is this charge equivalent to?

   (c) Calculate the change in mass of the zinc electrode.
   (d) Calculate the volume of gaseous chlorine consumed as a result of the reaction.

44. A galvanic cell uses a cadmium cathode immersed in a $CdSO_4$ solution and a zinc anode immersed in a $ZnSO_4$ solution.
   (a) Write a balanced equation for the cell reaction.
   (b) A current of 1.45 A is observed to flow for a period of 2.60 h. How much charge passes through the circuit during this time? How many moles of electrons is this charge equivalent to?
   (c) Calculate the change in mass of the zinc electrode.
   (d) Calculate the change in mass of the cadmium electrode.

45. Two electrolytic cells are connected in series. (In this arrangement, for every electron passing through the first cell, one electron also passes through the second.) In the first cell, $Ag^+(aq)$ is reduced to $Ag(s)$. In the second, $Cd^{2+}(aq)$ is reduced to $Cd(s)$. These are the only reduction reactions taking place in either cell. After a time, electrolysis is stopped, and it is found that 0.475 g of solid silver has been deposited in the first cell. Determine the mass of cadmium deposited in the second cell.

46. Two electrolytic cells are connected in series. In the first cell, H_2O is oxidized to $O_2(g)$, and in the second $Cl^-(aq)$ is oxidized to $Cl_2(g)$. These are the only oxidation reactions taking place in the cells. The gases are collected as they evolve. After a time, electrolysis is stopped, and it is found that 2.65 L of oxygen has been evolved from the first cell. Determine the volume of chlorine evolved from the second cell. The volume measurements are performed at the same temperature and pressure.

47. An acidic solution containing copper ions is electrolyzed, producing gaseous oxygen (from water) at the anode and metallic copper at the cathode. For every 16.0 g of oxygen that is generated, 63.5 g of copper plates out. What is the oxidation state of the copper in the solution?

48. Michael Faraday reported that passing electricity through one solution liberated 1 mass of hydrogen at the cathode and 8 masses of oxygen at the anode. The same quantity of electricity liberated 36 masses of chlorine at the anode and 58 masses of tin at the cathode from a second solution. What were the oxidation states of hydrogen, oxygen, chlorine, and tin in these solutions?

Electrometallurgy

49. In the Downs process, molten sodium chloride is electrolyzed to produce sodium and chlorine. Write equations representing the processes taking place at the anode and at the cathode in the Downs process.

50. The first element to be prepared by electrolysis was potassium. In 1807, Humphrey Davy passed an electric current through molten potassium hydroxide (KOH), obtaining liquid potassium at one electrode and water and oxygen at the other. Write equations representing the processes taking place at the anode and at the cathode.

51. A current of 55,000 A is passed through a series of 100 Hall–Héroult cells for a period of 24 hours. Calculate the maximum theoretical mass of aluminum that can be recovered.

52. A current of 75,000 A is passed through an electrolysis cell containing molten $MgCl_2$ for a period of 7.0 days. Calculate the maximum theoretical mass of magnesium that can be recovered.

53. An important use for magnesium is to make titanium. In the Kroll process, magnesium reduces titanium(IV) chloride to elemental titanium in a sealed vessel at 800°C. Write a balanced chemical equation for this reaction. What mass of magnesium is needed in theory to produce 100 kg of titanium from titanium(IV) chloride?

54. Calcium is used to reduce vanadium(V) oxide to elemental vanadium in a sealed steel vessel. Vanadium is used in vanadium steel alloys for jet engines, high-quality knives, and tools. Write a balanced chemical equation for this process. What mass of calcium is needed in theory to produce 20.0 kg of vanadium from vanadium(V) oxide?

55. (See Example 12–6.) Galvanized steel consists of steel with a thin coating of zinc to slow corrosion. The zinc can be deposited electrolytically by making the steel object the cathode and a block of zinc the anode in an electrochemical cell containing a dissolved zinc salt. Suppose a steel garbage can is to be galvanized and requires a total mass of 7.32 g of zinc to be coated to the required thickness. For how long should a current of 8.50 A be passed through the cell to achieve this?

56. (See Example 12–6.) In the electroplating of a silver spoon, the spoon acts as the cathode and a piece of pure silver as the anode. Both dip into a solution of silver cyanide (AgCN). Suppose that a current of 1.5 A is passed through such a cell for a period of 22 min, and that the spoon has a surface area of 16 cm^2. Calculate the average thickness of the silver layer deposited on the spoon, taking the density of silver to be 10.5 g cm^{-3}.

Additional Problems

*57. When pig iron is refined into steel, silicon and phosphorus impurities in the iron are oxidized to acidic oxides that are then trapped in the slag. Write balanced chemical equations for the oxidation of each of these two elements by oxygen. Write balanced equations for the reactions by which these products are trapped in the slag. Classify each reaction as a redox or an acid–base reaction.

58. Compute the equilibrium constant of the reaction

$$CO_2(g) + C(s) \rightleftharpoons 2\ CO(g)$$

at room temperature (25°C) and at 1875°C.

59. The principal ore of chromium is the mineral chromite $(FeCr_2O_4)$, for which ΔG_f° equals -1344 kJ mol^{-1} at 25°C. Because the main use of chromium is in chromium-containing steels, the ore is generally reduced directly to the alloy ferrochrome by the reaction

$$FeCr_2O_4(s) + 4\ C(s) \longrightarrow Fe(s) + 2\ Cr(s) + 4\ CO(g)$$

Calculate ΔG° for the production of ferrochrome according to this equation, using Appendix D for other information you may need. (*Note*: The ΔG° in mixing iron and chromium in the alloy is small and can be neglected.)

60. Compare and contrast the smelting of copper with that of iron. Include in your discussion the differences in starting materials, in temperatures and reducing agents employed, and in the by-products and their disposal.

61. The drain cleaner Drano consists of aluminum turnings mixed with sodium hydroxide. When it is added to water, the sodium hydroxide dissolves and releases heat. The aluminum reacts with the basic water to generate bubbles of hydrogen and aqueous $Al(OH)_4^-$ ions. Write a balanced net ionic equation for this reaction.

62. Sodium hypochlorite (NaOCl) is the major ingredient of household bleach. Suppose its aqueous solution reacts with an aqueous solution of sodium thiosulfate $(Na_2S_2O_3)$ under basic conditions, generating aqueous sodium sulfate and sodium chloride. Write a balanced net ionic equation for this reaction.

63. A current passed through inert electrodes immersed in an aqueous solution of sodium chloride produces chlorate ion $(ClO_3^-(aq))$ at the anode and gaseous hydrogen at the cathode. Given this fact, write a balanced equation for the chemical reaction if gaseous hydrogen and aqueous sodium chlorate are mixed and allowed to react spontaneously until they reach equilibrium.

64. Tarnish forms on silver spoons when airborne sulfur compounds react with the silver to produce a dark coating of silver sulfide (Ag_2S). Tarnish can be removed by putting the corroded spoons in an aluminum container (a pie plate works) containing a warm solution of sodium hydrogen carbonate $(NaHCO_3)$. The $NaHCO_3(aq)$ does not react but serves only as an electrolyte in an electrochemical cell in which the aluminum container is the anode. Write a balanced chemical equation for the tarnish-removing reaction.

*65. Balance the following redox equations. (*Note*: Some may be easier to balance by inspection than by following the rules for redox reactions.)
    (a) $Cl_2O(g) + N_2O_5(g) \longrightarrow NO_2Cl(g) + O_2(g)$
    (b) $Cl_2O(g) + N_2O_5(g) \longrightarrow NO_3Cl(g) + O_2(g)$
    (c) $N_2O_5(\ell) + Au(s) \longrightarrow (NO_2)Au(NO_3)_4(s) + NO_2(g)$
    (d) $TiCl_4(\ell) + NO_2Cl(g) \longrightarrow$
        $Cl_2(g) + TiOCl_2(s) + (NO)_2TiCl_6(s)$
    (e) $NHBr_2(s) \longrightarrow NH_4Br(s) + N_2(g) + Br_2(g)$
    (f) $NaN_3(s) + NO_2Cl(g) \longrightarrow$
        $N_2O(g) + NaNO_3(s) + Cl_2(g)$
    (g) $KSCN(s) + NOCl(g) \longrightarrow$
        $(SCN)_2(g) + NO(g) + KCl(s)$
    (h) $K_2MnF_6(s) + SbF_5(s) \longrightarrow$
        $KSbF_6(s) + MnF_3(s) + F_2(g)$
    (i) $NF_4SbF_6(s) + H_2O(\ell) \longrightarrow$
        $NF_3(g) + HSb(OH)_6(s) + O_2(g) + HF(g)$

*66. The following reaction occurs in aqueous base:

$$2\ NO(g) + 2\ NH_2OH(aq) \longrightarrow$$
$$N_2O(g) + N_2(g) + 3\ H_2O(g)$$

Break its balanced equation down into balanced oxidation and reduction half-equations in two different ways.

67. Determine the electric charge on a single electron by dividing the Faraday constant by Avogadro's number.

68. A newly discovered bacterium is able to reduce selenate ion ($SeO_4^{2-}(aq)$) to elemental selenium ($Se(s)$) in reservoirs. This is significant because the soluble selenate ion is potentially toxic; elemental selenium, however, is insoluble and harmless. Assume that water is oxidized to oxygen as the selenate ion is reduced. Compute the mass of oxygen produced if all the selenate in a 10^{12}-L reservoir contaminated with 100 mg L^{-1} of selenate ion is reduced to selenium.

69. Thomas Edison invented an electric meter that was nothing more than a simple coulometer, a device to measure the amount of electricity passing through a circuit. In this meter, a small, fixed fraction of the total current supplied to a household was passed through an electrolytic cell, plating out zinc at the cathode. Each month, the cathode could then be removed and weighed to determine the amount of electricity used. If 0.25% of a household's electricity passed through such a coulometer, and the cathode increased in mass by 1.83 g in a month, how many coulombs of electricity were used during that month?

*70. The chief chemist of the Brite-Metal Electroplating Co. is required to certify that the rinse solutions that are discharged from its tin-plating process into the municipal sewer system contain no more than 10 ppm (parts per million) by mass of Sn^{2+}. The chemist devises the following analytical procedure to determine the concentration. At regular intervals, a 100-mL (= 100 g) sample is withdrawn from the waste stream and acidified to pH = 1.0. A starch solution and 10 mL of 0.1 M potassium iodide are added, and a 25.0-milliampere current is passed through the solution between platinum electrodes. Iodine appears as a product of electrolysis at the anode when the oxidation of Sn^{2+} to Sn^{4+} is practically complete and signals its appearance by the deep blue color of a complex formed with starch. What is the maximum time of electrolysis to the appearance of the blue color that ensures that the concentration of Sn^{2+} does not exceed 10 ppm?

71. The galvanic cell $Zn(s)|Zn^{2+}(aq)\|Ni^{2+}(aq)|Ni(s)$ is constructed using a completely immersed zinc electrode that weighs 32.68 g and a nickel electrode immersed in 575 mL of 1.00 M $Ni^{2+}(aq)$ solution. A steady current of 0.0715 A is drawn from the cell (the electrons move from the zinc electrode to the nickel electrode).
(a) Which reactant is the limiting reactant in this cell?
(b) How long does it take for the cell to be completely discharged?
(c) How much mass has the nickel electrode gained when the cell is completely discharged?
(d) What is the concentration of the $Ni^{2+}(aq)$ when the cell is completely discharged?

72. Use data from Appendix D to predict whether elemental magnesium should react with water on purely thermodynamic grounds. What is observed in practice? Explain.

73. Compare and contrast the production of aluminum with that of magnesium. Include in your discussion the differences in starting materials, in reactions preparatory to the electrolysis, and in reaction by-products.

74. A 55.5-kg slab of crude copper from a smelter has a copper content of 98.3%. Estimate the length of time it takes to purify it electrochemically, if it is used as the anode in a cell that has acidic copper(II) sulfate as its electrolyte and a current of 2.00×10^3 A is passed through the cell.

*75. Chromium is separated from iron in the ore chromite ($FeCr_2O_4(s)$) by fusing the ore with sodium carbonate.

$$FeCr_2O_4(s) + Na_2CO_3(\ell) + O_2(g) \longrightarrow$$
$$Fe_2O_3(s) + Na_2CrO_4(\ell) + CO_2(g)$$

(a) Balance the preceding chemical equation.
(b) Calculate the standard enthalpy change of the reaction, assuming that the enthalpies of fusion of Na_2CO_3 and Na_2CrO_4 are approximately equal. The standard molar enthalpy of formation of $FeCr_2O_4(s)$ is -1445 kJ mol^{-1} and that of $Na_2CrO_4(s)$ is -1342 kJ mol^{-1}. Use Appendix D for the additional data you require.
(c) The sodium chromate produced in the reaction is soluble in water and can be leached from the solidified melt. Suggest a procedure for converting it to elemental chromium.

*76. A steady current of 2.0 A is passed through a molten salt for 12.1 min, resulting in the deposition of 260 mg of a metallic element at the cathode. Name the two elements that could produce this behavior.

77. Seawater furnishes the "ore" for the preparation of metallic magnesium, and the cost of the product is accordingly almost completely determined by the cost of the energy used in the electrolytic reduction process. An engineer suggests reducing MgO directly with carbon (from coal).
(a) Estimate the temperature at which K equals 1 for the reaction

$$2\ MgO(s) + C(s) \rightleftharpoons CO_2(g) + 2\ Mg(\ell)$$

(b) Tests of this reaction establish that a practical process would have to be run at at least 2500°C, a temperature that is difficult to reach and maintain. Moreover, it is difficult to separate the two products, and the back-reaction between them proves hazardously rapid. To reduce the possibility of back-reaction, the process is redesigned in such a way that that the product Mg is produced in solution in liquid antimony.

$$2\ MgO(s) + C(s) \rightleftharpoons CO_2(g) + 2\ Mg(Sb)$$

Predict whether this change raises or lowers the temperature required for a practical process. Explain.

78. The following "balanced" equation appeared recently in the pages of the *Journal of the American Chemical Society*

$$[Co(H_2O)_6]^{2+} + HSO_5^- \longrightarrow$$
$$[Co(H_2O)_5(OH)]^{2+} + 2\,H^+ + SO_4^-$$

In what respect is this equation not balanced? Balance the equation. (*Note:* The formulas HSO_5^- and SO_4^- are correct.)

79. A scientific paper weighing various possibilities for the chemical removal of carbon dioxide from the atmosphere points out that the reaction

$$4\,H_2(g) + CO_2(g) \longrightarrow 2\,H_2O(\ell) + CH_4(g)$$

has both a negative $\Delta H°$ (-253 kJ) and a negative $\Delta G°$ (-131 kJ). The paper then goes on to conclude that this reaction is *not* a good candidate for use in removing CO_2. Why?

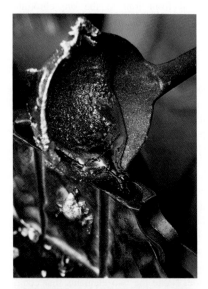

Molten lead being poured to make lead soldiers.

CUMULATIVE PROBLEM

Lead

The metallurgy of lead parallels that of iron and copper in many respects. Lead ores generally contain lead sulfide (PbS) in some form, sometimes as the mineral galena.

(a) Balance the following equations, which represent reactions occurring in the smelting of lead ores:

    1. $PbS(s) + O_2(g) \longrightarrow PbSO_4(s)$

    2. $PbS(s) + PbO(s) \longrightarrow Pb(s) + SO_2(g)$

    3. $PbS(s) + PbSO_4(s) \longrightarrow Pb(s) + SO_2(g)$

(b) Metallic lead is sometimes prepared by heating ores containing lead sulfide in a furnace with scrap iron. Write a balanced equation for this reaction. Determine whether this is a spontaneous reaction at 298.15 K.

(c) Lead can be purified by passing an electric current through an aqueous solution of $PbSiF_6$ (which contains Pb^{2+} and SiF_6^{2-} ions) using the crude lead as the anode and a piece of pure lead as the cathode. Write equations for the half-reactions that take place at the anode and at the cathode.

(d) Suppose that, by mistake, the source of electric current (in the purification cell of the previous part) is connected so that the crude lead is the cathode and the pure lead is the anode. Does an electric current flow? Does the electrochemical purification work? Explain why or why not.

(e) Some lead ores are high in bismuth(III) oxide (Bi_2O_3), which has a standard Gibbs function of formation of -497 kJ mol^{-1}. Between what two entries would bismuth(III) oxide be placed in Table 12–1?

(f) A major use of lead is in the manufacture of lead–sulfuric acid storage batteries for automobiles. The reaction occurring when such a battery is discharged is

$$Pb(s) + PbO_2(s) + SO_4^{2-}(aq) \longrightarrow PbSO_4(s)$$

Complete and balance this equation for acidic conditions.

(g) A lead–acid storage battery delivers an average current of 0.705 A for a period of 17.4 min. Determine the mass of $PbSO_4$ that is formed inside the battery.

Electrochemistry and Cell Voltage

Lead-acid batteries connected in series in an electric vehicle.

Electrochemistry combines the chemistry of oxidation–reduction reactions with the physics of the flow of electric currents. Galvanic cells use the decrease in Gibbs function during a spontaneous chemical reaction to push electrons through a wire from anode to cathode. The voltage (potential difference) developed by such cells is the innate driving force for this transfer. In electrolytic cells, the chemical reactions are non-spontaneous, but controlled intervention with an outside voltage forces them along to give valuable compounds with high Gibbs functions.

Electrochemistry is a supremely practical field, dealing with the storage of energy and its efficient conversion from readily available sources (such as solar or chemical energy) into forms that are useful for technological applications. Although the thermodynamic considerations governing electrochemical transformations have been thoroughly worked out, the design of efficient electrochemical cells must contend with such a multitude of other, conflicting factors that it remains very much an art as well as a science. In this chapter, we present the fundamental theory and show its application in cases in which electrochemical action is desired (batteries and electrolysis cells) as well as in cases in which it is not desired (corrosion).

13–1 THE GIBBS FUNCTION AND CELL VOLTAGE

The driving force for current in any electrical device is a difference in electric potential ($\Delta \mathscr{E}$). If two regions of different potential are physically separated, no current flows unless a path (a copper wire, for example) is in place to conduct the flow of electrons. Without a path, the potential for change simply abides, unfulfilled; hence, the word "potential" is well chosen.

- An electric potential difference is measured with a potentiometer. Zero electric current flows during such a measurement.

As it flows between two regions of different potential, an electric current can perform a type of work not previously discussed, **electrical work.** Recall that work is one means (the other is heat) by which energy is transferred between a system and its surroundings. The transfer of pressure–volume work, considered in Chapters 10 and 11, requires a mechanical connection between a system and its surroundings. In exactly the same way, the transfer of electrical work requires an electrical connection between system and surroundings. The apparatus of electrodes, salt bridge, and wires discussed in Section 12–3 is such a connection. If an amount of charge (Q) moves through an electrical connection, driven by a potential difference ($\Delta \mathscr{E}$), the electrical work is

$$w_{\text{elec}} = -Q \, \Delta \mathscr{E}$$

The minus sign appears in this equation because of the convention (adopted in Chapters 10 and 11) that positive work is work that is done *on* a system. If the system is a galvanic cell, the potential difference $\Delta \mathscr{E}$ is positive and w_{elec} is negative; the cell performs electrical work in the surroundings (acts as a battery). If the system is an electrolytic cell, $\Delta \mathscr{E}$ is negative and w_{elec} is positive; an energy source in the surroundings, such as an electrical generator, performs work on the cell.

- It is important not to confuse $\mathscr{E}$ and $\Delta \mathscr{E}$, here signifying the potential and potential difference, with E and ΔE, used in Chapters 10 and 11 to mean internal energy and difference in internal energy.

If the preceding equation is used with SI units, then w_{elec} is measured in joules and Q in coulombs. It follows that $\Delta \mathscr{E}$ has units of joules per coulomb. One joule per coulomb is defined as one **volt.** The volt (and derivatives such as the millivolt or kilovolt) is the only unit used to measure potential difference, which accordingly is often called **voltage.** The total charge (Q) is the average current (I) multiplied by the time (t) during which the current is flowing, so the equation for the electrical work can also be written

$$w_{elec} = -I \, t \, \Delta\mathscr{E}$$

One further unit for electrical work or energy that is in common use is the kilowatt-hour (kW-h) The **watt** is a unit of power, or the rate of transfer of energy. It is defined as one joule per second. The kilowatt is 1000 watts, or 10^3 J s^{-1}. The kilowatt-hour, then, is the total energy transferred (or work done) in 1 h at a rate of 10^3 J per second. It is equal to

$$1 \text{ kW-h} = \left(\frac{10^3 \text{ J}}{\cancel{s}}\right)(3600 \, \cancel{s}) = 3.6 \times 10^6 \text{ J} \quad \text{(exactly)}$$

because 1 h has exactly 3600 s.

EXAMPLE 13-1

The battery in Example 12–3 transferred 6750 C of charge. Suppose that it maintained a constant voltage of 6.00 V during this process. Calculate the total work done *by* the battery in joules and in kilowatt-hours.

Solution

The electrical work is the product of the charge transferred (6750 C) and the cell voltage, with a minus sign:

$$w_{elec} = -Q \, \Delta\mathscr{E} = -(6750 \text{ C})(6.00 \text{ V}) = -40{,}500 \text{ C V} = -4.05 \times 10^4 \text{ J}$$

This is the work done *on* the battery. The work performed *by* the battery is the negative of this, or $+4.05 \times 10^4$ J. Converting to kilowatt-hours gives

$$\text{work done by battery} = +4.05 \times 10^4 \cancel{J} \times \left(\frac{1 \text{ kW-h}}{3.600 \times 10^6 \cancel{J}}\right) = 0.0112 \text{ kW-h}$$

Exercise

Calculate the total electrical work done on an electrolytic cell if a current of 0.741 A is run through it at a voltage of 2.53 V for 4150 s.

Answer: 7.78×10^3 J.

Let us consider a galvanic cell at constant temperature and pressure that is ready to operate but is disconnected so that no electric current is actually flowing. Thermodynamics provides a fundamental relationship between the change in the Gibbs function (ΔG) of the chemical reaction poised to take place within this cell and the maximum electrical work that can be transferred:

$$\Delta G = w_{elec, \, max} \quad \text{(at constant } T \text{ and } P\text{)}$$

According to this equation, if the cell is connected for a while, the change in the Gibbs function of the reaction equals the maximum amount of energy that can appear in the surroundings as electrical work.

How does the transfer occur? The decrease in the Gibbs function in the cell first manifests itself as an increase in the Gibbs function of electrons as they are caused to move through the wire. This extra G can then be "harvested" by connecting the wire to a motor that is set up to do work, or it can be "wasted" by simply allowing electrical resistance to dissipate it in the form of heat.

The subscript "max" appears in the preceding equation because real galvanic cells inevitably waste some of their decrease in Gibbs function as heat when the circuit is closed and a measurable current flows. If the current flows very slowly, essentially one electron at a time, then this wastage is negligible and

$$w_{\text{elec, max}} = -Q\Delta\mathscr{E}$$

although the transfer of the charge would then take a prohibitively long time. Combining this equation with the previous one gives

$$\Delta G = -Q\Delta\mathscr{E}$$
(at constant T and P)

This equation shows that the potential difference of the cell, an easily measured property at any stage in the progress of the redox reaction, is proportional to the change in the Gibbs function of the reaction. When $\Delta\mathscr{E}$ is positive, then ΔG is negative; the chemical reaction is spontaneous and can produce electrical work in the surroundings. This is the case of a galvanic cell, which we have been discussing. Conversely, a negative $\Delta\mathscr{E}$ means that ΔG is positive; the chemical reaction is non-spontaneous and requires the input of electrical work from an external source of voltage before it occurs.

If n moles of electrons (or $n\mathscr{F}$ coulombs of charge, where $\mathscr{F}$ is the Faraday constant, 96,485 C mol^{-1}) pass through the external circuit of the cell, then the last equation becomes

$$\Delta G = -Q\Delta\mathscr{E} = -n\mathscr{F}\Delta\mathscr{E} \quad \text{(at constant } T \text{ and } P\text{)}$$

Because differences in electric potential are very easily measured, this relationship provides a most attractive, direct way to determine changes in the Gibbs function. Dividing both sides by n gives $\Delta G/n = -\mathscr{F}\,\Delta\mathscr{E}$. This form highlights an important fact: the potential difference of an electrochemical cell does not depend on the size of the cell. If the chemical amount of reactants that is converted to products is, for example, doubled, then ΔG is doubled, but the number of moles of electrons transferred in the conversion is also doubled, and $\Delta\mathscr{E}$ remains unchanged.

Standard States and Cell Voltages

Recall from Section 10–4 that the standard state of a liquid or solid substance is its state at a pressure of 1 atm, that the standard state of a gas is its state under a pressure of 1 atm and exhibiting ideal-gas behavior, and that the standard state of a solute is a 1 M solution under a pressure of 1 atm and exhibiting ideal-solution behavior. Although any temperature can be specified, we are, as before, almost exclusively concerned with standard states at 25°C. The standard change in the Gibbs function ($\Delta G°$) applies to a reaction in which all reactants and products are in standard states. It can be calculated from the standard Gibbs functions of formation (the $\Delta G_f°$'s) of the substances taking part in the reaction. Tables of $\Delta G_f°$ are widely available (see Appendix D). For redox reactions, all of which can in principle be carried out in electrochemical cells, the standard change in the Gibbs function ($\Delta G°$) is related to a **standard cell voltage** ($\Delta\mathscr{E}°$) by the previous equation, with added superscript zeros ("naughts") to indicate that standard states are specified:

$$\Delta G° = n\mathscr{F}\Delta\mathscr{E}°$$

The $\Delta\mathscr{E}°$ is the potential difference or voltage of a galvanic cell in which all reactants and products are in standard states. A standard cell voltage is an intrinsic

• Review the computation of the $\Delta G°$ of a reaction from $\Delta G_f°$ values (see Section 11-6) and the computation of $\Delta H°$ of a reaction from values of $\Delta H_f°$ (see Section 10-4).

• This equation forges a deep link between calorimetry (the results of which are summarized in tables of $\Delta G_f°$) and electrochemistry. In their outward appearance, the two kinds of experiment are unrelated.

electrical property of a redox reaction that can be calculated from the standard change in the Gibbs function ($\Delta G°$) of the reaction by means of the preceding equation. The reverse is also true. The $\Delta G°$ (and from it the equilibrium constant) of any redox reaction can be determined by a single voltage measurement, whenever a cell can be constructed that uses the reaction and has the reactants and products in standard states.

Figure 13-1 The reduction of $Cu^{2+}(aq)$ ion by nickel (see Example 13–2) occurs spontaneously when a bar of metallic nickel is put into a solution containing $Cu^{2+}(aq)$ ions, which are blue. Copper plates out, and the solution turns green as $Ni^{2+}(aq)$ forms. In this set-up, the change in Gibbs function is wasted as heat. In an electrochemical cell, it can be converted to electrical work.

EXAMPLE 13–2

A $Ni^{2+}|Ni$ half-cell is connected to a $Cu^{2+}|Cu$ half-cell to make a galvanic cell in which $[Ni^{2+}] = [Cu^{2+}] = 1.00$ M. The cell voltage at 25°C when no current is flowing is measured to be $\Delta\mathscr{E}° = 0.57$ V, and copper is observed to plate out spontaneously when the reaction is allowed to proceed. Calculate $\Delta G°$ for the chemical reaction that takes place in the cell, for one mole of nickel dissolved.

Solution

The redox reaction is evidently

$$Ni(s) + Cu^{2+}(aq) \longrightarrow Ni^{2+}(aq) + Cu(s)$$

because metallic Cu is a product (Fig. 13–1).

For the reaction as written, if one mole of $Ni(s)$ reacts, two moles of electrons pass through the external circuit, and so $n = 2$ mol e^-. Insert this value into the equation, together with the $\Delta\mathscr{E}°$:

$$\Delta G° = n\mathscr{F}\,\Delta\mathscr{E}°$$

$$= -(2.00 \text{ mol } e^-) \times \left(\frac{96{,}485 \text{ C}}{1 \text{ mol } e^-}\right) \times (0.57 \text{ V}) = -1.10 \times 10^5 \text{ C V}$$

$$\Delta G° = -1.10 \times 10^5 \text{ C V} \times \left(\frac{1 \text{ J}}{1 \text{ C V}}\right) \times \left(\frac{1 \text{ kJ}}{1000 \text{ J}}\right) = \boxed{-110 \text{ kJ}}$$

Exercise

Suppose that an enlarged version of the above cell, in which the spontaneous reaction is $Ni(s) + Cu^{2+}(aq) \longrightarrow Ni^{2+}(aq) + Cu(s)$, is constructed. Every part of the new cell (the electrodes, salt bridge, and so forth) is dozens of times bigger, but the concentrations of the $Cu^{2+}(aq)$ and $Ni^{2+}(aq)$ are still 1.00 M. (a) Predict the voltage that would be measured (at 25°C) from this cell. (b) Calculate $\Delta G°$ for the chemical reaction if 10.00 mol of nickel dissolves.

Answer: (a) 0.57 V. (b) -1100 kJ.

13-2   HALF-CELL POTENTIALS

We could tabulate all the conceivable galvanic cells and their standard potential differences, but the list would be very long. To avoid this, we introduce the concept of a **standard half-cell reduction potential.**

Consider the cell described in Example 13–2, which is made up of $Ni^{2+}|Ni$ and $Cu^{2+}|Cu$ half-cells; it generates a voltage of 0.57 V and plates out copper. We write each half-cell reaction as a reduction:

$$Ni^{2+}(aq) + 2\,e^- \longrightarrow Ni(s)$$
$$Cu^{2+}(aq) + 2\,e^- \longrightarrow Cu(s)$$

Each of these half-reactions has a certain tendency to occur. We associate with each a numerical reduction potential. The reduction potential $\mathscr{E}°(Ni^{2+}|Ni)$ expresses the tendency of the first half-reaction to occur as written, and $\mathscr{E}°(Cu^{2+}|Cu)$ does the same for the second. The more positive the reduction potential, the greater is the innate tendency of a half-reaction to take place as a reduction. Of course, reduction can never take place in isolation because the electrons must come from somewhere. The source can only be another half-cell. When two half-cells are combined, the one with the more positive (or algebraically greater) reduction potential wins out and occurs as a reduction. The one with the less positive reduction potential runs as an oxidation to supply electrons. The potential difference observed between the two electrodes of the cell (the cell voltage) is then the *difference* between the two half-cell reduction potentials. In the preceding example, new copper metal plates out as a coating on the cathode, so that the $Cu^{2+}(1 \text{ M})|Cu$ reduction potential is evidently more positive than the $Ni^{2+}(1 \text{ M})|Ni$ reduction potential.

$$\Delta\mathscr{E}° = \mathscr{E}°(Cu^{2+}|Cu) - \mathscr{E}°(Ni^{2+}|Ni) = 0.57 \text{ V}$$

• In algebraic comparisons, the signs of the numbers are considered. The number 2 is algebraically greater than −3, although smaller than −3 in absolute value.

When two half-cells are combined to make a galvanic cell, reduction occurs in the half-cell having the algebraically greater reduction potential (making it the cathode). Oxidation occurs in the other half-cell (making it the anode). Therefore,

$$\Delta\mathscr{E}° = \mathscr{E}°(\text{cathode}) - \mathscr{E}°(\text{anode})$$

for a galvanic cell.

In thermochemistry, only *differences* in enthalpy (ΔH) are directly measurable; absolute enthalpies are not. The same is true of the internal energy and the Gibbs function. Consequently, the enthalpies (and internal energies and Gibbs functions) of certain reference substances (namely, the elements in their most stable standard states at 25°C) are arbitrarily taken to equal zero (see Section 10–4) so that data can be listed conveniently. The same tactic is necessary with reduction potentials. Because only differences in electric potential are directly measurable, the reduction potential for a particular half-cell is assigned arbitrarily and other half-cell potentials are tabulated relative to it. By international convention, the $\mathscr{E}°$ for the reduction of $H^+(aq)$ to give $H_2(g)$ is set equal to zero at all temperatures, when the pressure of gaseous H_2 at the electrode is 1 atm and the $H^+(aq)$ concentration in solution is 1 M:

$$2 \text{ H}^+(aq) + 2 \text{ } e^- \longrightarrow H_2(g) \qquad \mathscr{E}° = 0 \text{ V} \quad \text{(by definition)}$$

All other standard reduction potentials can then be determined relative to this reference potential by building galvanic cells that combine their standard half-cells with the standard $H^+|H_2$ half-cell and measuring the cell voltage. Although it seems strange to have a gas as a cathode or anode, it is fairly easy to arrange (Fig. 13–2). The experimental voltages (the $\Delta\mathscr{E}°$'s) that come from such devices are then substituted in the equation

$$\Delta\mathscr{E}° = \mathscr{E}°(\text{cathode}) - \mathscr{E}°(\text{anode})$$

to compute the standard reduction potential of the half-cell in question. The $\mathscr{E}°$ of the standard $H^+|H_2$ half-cell is zero in this equation, whether it serves as the anode or cathode. Consequently, if a half-cell is the anode (the site of oxidation) in such an experimental galvanic cell, its $\mathscr{E}°$ is negative. If it is the cathode (the site of

Gaseous hydrogen evolves from an illuminated photoelectrode.

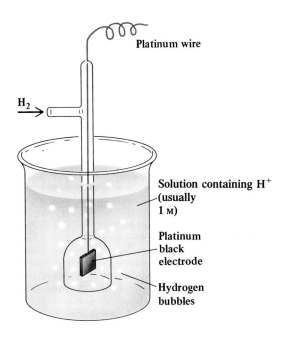

Platinum wire

H₂ →

Solution containing H⁺
(usually
1 M)

Platinum
black
electrode

Hydrogen
bubbles

Figure 13-2 Hydrogen is a gas at room conditions, and electrodes cannot be constructed from it directly. In this hydrogen half-cell, a foil of platinum covered with a coating of finely divided platinum is dipped into the solution, and a stream of hydrogen is passed over the surface. The platinum itself does not react but provides a support surface for the reaction of the H_2 to give H^+, or the reverse. It also provides the necessary electrical connection to carry away or supply electrons.

reduction), then its $\mathscr{E}°$ is positive. A positive $\mathscr{E}°$ signals that the standard reduction half-reaction has a greater innate tendency to occur than the reduction of 1 M $H^+(aq)$ to $H_2(g)$ at 1 atm. A negative $\mathscr{E}°$ signals that it has a lesser tendency.

For example, if a Cu^{2+}(1 M)|Cu half-cell is connected to the H^+(1 M)|H_2 half-cell, metallic copper is observed to plate out, showing that the copper half-cell is the cathode and the hydrogen half-cell is the anode. The observed cell voltage is 0.34 V, so

$$\Delta\mathscr{E}° = \mathscr{E}° \text{ (cathode)} - \mathscr{E}°\text{(anode)}$$
$$= \mathscr{E}°(Cu^{2+}|Cu) - \mathscr{E}°(H^+|H_2)$$
$$0.34 \text{ V} = \mathscr{E}°(Cu^{2+}|Cu) - 0$$
$$\mathscr{E}°(Cu^{2+}|Cu) = 0.34 \text{ V}$$

This means that the standard Cu^{2+}|Cu half-cell potential is 0.34 V on a scale in which the standard H^+|H_2 half-cell potential is zero:

$$Cu^{2+}(1 \text{ M}) + 2 e^- \longrightarrow Cu(s) \qquad \mathscr{E}° = 0.34 \text{ V}$$

When a Ni^{2+}(1 M)|Ni half-cell is connected to the standard hydrogen half-cell, nickel dissolves. The nickel half-cell is the anode because oxidation occurs in it. The measured cell voltage is 0.23 V, so that

$$\Delta\mathscr{E}° = 0.23 \text{ V} = \mathscr{E}°\text{(cathode)} - \mathscr{E}°\text{(anode)}$$
$$0.23 \text{ V} = 0 - \mathscr{E}°(Ni^{2+}|Ni)$$
$$\mathscr{E}°(Ni^{2+}|Ni) = -0.23 \text{ V}$$

The Ni^{2+}(1 M)|Ni standard half-cell potential is −0.23 V:

$$Ni^{2+}(1 \text{ M}) + 2 e^- \longrightarrow Ni(s) \qquad \mathscr{E}° = -0.23 \text{ V}$$

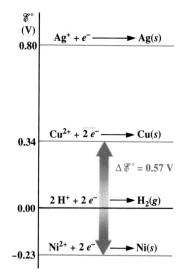

Figure 13–3 Sometimes standard reduction potentials are represented to scale. In this diagram, the distance between a pair of standard half-cell potentials is proportional to the $\Delta\mathscr{E}°$ generated by the electrochemical cell that combines the two. The half-reaction with the higher reduction potential forces the other half-reaction to occur in reverse.

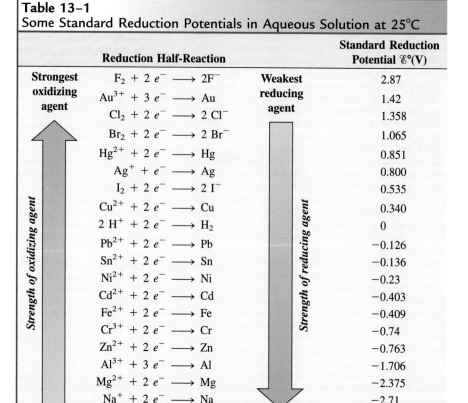

Table 13–1
Some Standard Reduction Potentials in Aqueous Solution at 25°C

	Reduction Half-Reaction		Standard Reduction Potential $\mathscr{E}°$(V)
Strongest oxidizing agent	$F_2 + 2\,e^- \longrightarrow 2F^-$	**Weakest reducing agent**	2.87
	$Au^{3+} + 3\,e^- \longrightarrow Au$		1.42
	$Cl_2 + 2\,e^- \longrightarrow 2\,Cl^-$		1.358
	$Br_2 + 2\,e^- \longrightarrow 2\,Br^-$		1.065
	$Hg^{2+} + 2\,e^- \longrightarrow Hg$		0.851
	$Ag^+ + e^- \longrightarrow Ag$		0.800
	$I_2 + 2\,e^- \longrightarrow 2\,I^-$		0.535
	$Cu^{2+} + 2\,e^- \longrightarrow Cu$		0.340
	$2\,H^+ + 2\,e^- \longrightarrow H_2$		0
	$Pb^{2+} + 2\,e^- \longrightarrow Pb$		−0.126
	$Sn^{2+} + 2\,e^- \longrightarrow Sn$		−0.136
	$Ni^{2+} + 2\,e^- \longrightarrow Ni$		−0.23
	$Cd^{2+} + 2\,e^- \longrightarrow Cd$		−0.403
	$Fe^{2+} + 2\,e^- \longrightarrow Fe$		−0.409
	$Cr^{3+} + 2\,e^- \longrightarrow Cr$		−0.74
	$Zn^{2+} + 2\,e^- \longrightarrow Zn$		−0.763
	$Al^{3+} + 3\,e^- \longrightarrow Al$		−1.706
	$Mg^{2+} + 2\,e^- \longrightarrow Mg$		−2.375
	$Na^+ + 2\,e^- \longrightarrow Na$		−2.71
	$Ca^{2+} + 2\,e^- \longrightarrow Ca$		−2.76
Weakest oxidizing agent	$K^+ + e^- \longrightarrow K$	**Strongest reducing agent**	−2.92
	$Li^+ + e^- \longrightarrow Li$		−3.04

If now a standard nickel and a standard copper half-cell are combined, the copper half-cell is the cathode because it has the more positive reduction potential. The voltage of the resulting galvanic cell should be (Fig. 13–3)

$$\Delta\mathscr{E}° = \mathscr{E}°(\text{cathode}) - \mathscr{E}°(\text{anode}) = 0.34\ \text{V} - (-0.23\ \text{V}) = 0.57\ \text{V}$$

This prediction agrees with the experimentally measured value.

The number of electrons transferred in a redox reaction affects the Gibbs function change of the reaction system but not the potential difference the reaction develops in an electrochemical cell. Consequently, the numbers of electrons appearing in half-equations are ignored in the computation of the potential difference of a cell. For example, when a standard $Ag^+(1\ \text{M})|Ag$ half-cell is connected to a standard $H^+(1\ \text{M})|H_2$ half-cell, metallic silver plates out and the measured voltage is 0.80 V. This means that the silver is the cathode, at which reduction occurs:

$$Ag^+(aq) + e^- \longrightarrow Ag(s)$$

and the hydrogen electrode is the anode, where oxidation occurs:

$$H_2(g) \longrightarrow 2\,H^+(aq) + 2\,e^-$$

To obtain the overall reaction, the first half-equation must be multiplied by 2 before being added to the second, in order that the electrons cancel out. However, a parallel multiplication is *not* performed in computing the potential difference. The half-cell reduction potentials are simply subtracted:

$$\Delta\mathscr{E}° = \mathscr{E}°(Ag^+|Ag) - \mathscr{E}°(H^+|H_2) = 0.80 \text{ V}$$

from which $\mathscr{E}°(Ag^+|Ag)$ is 0.80 V.

The standard reduction potentials for a large number of half-reactions are given in Appendix E. An abbreviated listing appears in Table 13–1. The half-reactions in the tables are arranged in order of decreasing reduction potential. The most easily reduced species has the most positive reduction potential and is at the top. In any galvanic cell that is constructed by combining half-cells that have their reactants and products in standard states at 25°C, the half-reaction that is listed higher in the table proceeds as written at the cathode, and the half-reaction that is listed lower proceeds in reverse at the anode.

EXAMPLE 13–3

An aqueous solution of potassium permanganate ($KMnO_4$) has a deep purple color. In aqueous acidic solution, the permanganate ion can be reduced to the pale pink manganese(II) ion (Mn^{2+}). The standard reduction potential of a $MnO_4^-|Mn^{2+}$ half-cell is $\mathscr{E}° = 1.49$ V at 25°C. Suppose that this half-cell is combined with a $Zn^{2+}|Zn$ half-cell in a galvanic cell, with $[Zn^{2+}] = [MnO_4^-] = [Mn^{2+}] = [H^+] = 1$ M.
(a) Write equations for the half-reactions at the anode and the cathode.
(b) Write a balanced equation for the overall cell reaction.
(c) Calculate the standard potential difference ($\Delta\mathscr{E}°$) of the cell at 25°C.

Solution

(a) Because the potential $\mathscr{E}°(MnO_4^-|Mn^{2+}) = 1.49$ V is algebraically greater than $\mathscr{E}°(Zn^{2+}|Zn) = -0.76$ V, permanganate ions are reduced at the cathode. The balanced half-cell reaction requires the presence of H^+ ions and water, as discussed in Section 12–2. Inserting four moles of water on the right balances oxygen:

$$MnO_4^- \longrightarrow Mn^{2+} + 4 H_2O \quad \text{(unbalanced)}$$

and eight moles of H^+ ions on the left balances hydrogen:

$$MnO_4^- + 8 H^+ \longrightarrow Mn^{2+} + 4 H_2O \quad \text{(unbalanced)}$$

Finally, five moles of electrons on the left balances the charge:

$$MnO_4^-(aq) + 8 H^+(aq) + 5 e^- \longrightarrow Mn^{2+}(aq) + 4 H_2O(\ell)$$

Balancing the half-equation for the oxidation of zinc at the anode takes only one step:

$$Zn(s) \longrightarrow Zn^{2+}(aq) + 2 e^-$$

(b) In the overall reaction, the number of electrons taken up from the cathode must equal the number released to the anode, so the first half-equation is multiplied by 2 and the second by 5. Adding the two gives the overall equation

$$2 MnO_4^-(aq) + 5 Zn(s) + 16 H^+(aq) \longrightarrow$$
$$2 Mn^{2+}(aq) + 5 Zn^{2+}(aq) + 8 H_2O(\ell)$$

(c) The standard cell potential is the difference between the standard reduction potential for permanganate (at the cathode) and the standard reduction potential for zinc (at the anode):

$$\Delta\mathscr{E}^° = \mathscr{E}^°(MnO_4^-|Mn^{2+}) - \mathscr{E}^°(Zn^{2+}|Zn) = 1.49 - (-0.76) = \boxed{2.25 \text{ V}}$$

Note that the reduction potentials are *not* multiplied by the coefficients in their associated half-equations (2 and 5) before the subtraction.

Exercise

A galvanic cell is constructed using the standard half-cells $Cr^{3+}|Cr$ and $Zn^{2+}|Zn$. (a) Write half-equations for the half-reactions at the anode and the cathode. (b) Write a balanced equation for the overall cell reaction. (c) Calculate the standard potential difference ($\Delta\mathscr{E}^°$) (consult Appendix E for standard half-cell potentials).

Answer: (a) Cathode: $Cr^{3+}(aq) + 3e^- \longrightarrow Cr(s)$. Anode: $Zn(s) \longrightarrow Zn^{2+}(aq) + 2\ e^-$. (b) $2\ Cr^{3+}(aq) + 3\ Zn(s) \longrightarrow 2\ Cr(s) + 3\ Zn^{2+}(aq)$. (c) 0.02 V.

Disproportionation

Recall from Sections 4–4 and 12–2 that *disproportionation* is a process in which a single chemical species is both oxidized and reduced. Reduction potentials allow a straightforward determination of which species are and which are not stable with respect to disproportionation. For a species to be susceptible to disproportionation, it must be in an intermediate oxidation state, so that it can both give up electrons (increasing its oxidation number) *and* accept them (decreasing its oxidation number). In a listing of standard reduction potentials, a species that disproportionates must appear on the right side of one half-reaction and on the left side of another.

This alone is not enough, however. For a species to disproportionate spontaneously, the half-reaction in which it is reduced must lie higher in the table (have a higher reduction potential) than the half-reaction in which it is oxidized. If this is the case, the first half-reaction drives the second to go in reverse, and spontaneous disproportionation occurs. As an example, consider two half-reactions involving copper in oxidation states 0, +1, and +2:

$$Cu^+(aq) + e^- \longrightarrow Cu(s) \qquad\qquad \mathscr{E}^° = 0.522 \text{ V}$$
$$Cu^{2+}(aq) + e^- \longrightarrow Cu^+(aq) \qquad\qquad \mathscr{E}^° = 0.158 \text{ V}$$

The first reaction occurs as a reduction and drives the second as an oxidation, giving the net reaction

$$2\ Cu^+(aq) \longrightarrow Cu^{2+}(aq) + Cu(s) \qquad \Delta\mathscr{E}^° = 0.522 - 0.158 = 0.364 \text{ V}$$

$Cu^+(aq)$ disproportionates spontaneously to $Cu^{2+}(aq)$ and $Cu(s)$ in solution. Whenever $\Delta\mathscr{E}^° > 0$, we have $\Delta G^° < 0$ and the reaction (if reactants and products are in standard states) occurs spontaneously, although in some cases the rate may be very slow.

EXAMPLE 13–4

Use data in Appendix E to decide whether $Fe^{2+}(aq)$ ion in its standard state at 25°C is unstable with respect to disproportionation to products in standard states at the same temperature.

Solution

Appendix E lists two relevant half-cell potentials:

$$Fe^{3+} + e^- \longrightarrow Fe^{2+} \qquad \mathscr{E}_1^\circ = 0.770 \text{ V}$$
$$Fe^{2+} + 2\,e^- \longrightarrow Fe(s) \qquad \mathscr{E}_2^\circ = -0.409 \text{ V}$$

The Fe^{2+} appears on the left in one and on the right in the other half-reaction. For Fe^{2+} to disproportionate, however, the *top* reaction would have to be driven backward as an oxidation. This is thermodynamically not favored (it would give $\Delta\mathscr{E}^\circ < 0$), so the Fe^{2+} ion is stable against disproportionation under these conditions.

- Aqueous solutions of Fe(II) are still notoriously contaminated by Fe(III). The problem is not that the Fe(II) disproportionates but that it is easily oxidized by oxygen (from the air).

Exercise

Use data in Appendix E to decide whether the $Hg_2^{2+}(aq)$ ion in the standard state at 25°C is unstable with respect to disproportionation to products in standard states at 25°C.

Answer: The Hg_2^{2+} ion is (barely) stable against disproportionation to $Hg(\ell)$ and $Hg^{2+}(aq)$.

13-3 OXIDIZING AND REDUCING AGENTS

A strong **oxidizing agent** is a chemical species that is itself easily reduced. Strong oxidizing agents have large positive reduction potentials. They appear as reactants in half-equations toward the top of Table 13–1 (and Appendix E). Fluorine has the largest reduction potential listed, and fluorine molecules indeed avidly accept electrons to produce fluoride ions. Other strong oxidizing agents are hydrogen peroxide (H_2O_2) and the permanganate ion (MnO_4^-) in solution.

- The chemistry of fluorine is discussed in some detail in Chapter 23.

A strong **reducing agent,** on the other hand, is itself very easily oxidized. Strong reducing agents are *products* in half-reactions having large *negative* reduction potentials. The best reducing agents appear on the right-hand side of half-equations toward the bottom of Table 13–1 and Appendix E. Metallic lithium and potassium (not Li^+ ion or K^+ ion) are potent reducing agents in aqueous systems. Arranging the products of the half-reactions in Table 13–1 from the first to the last generates an activity series of reducing agents ranked from weakest to strongest. The order of the metals in this activity series is the same as in Table 4–3, which also deals with relative reducing power in aqueous systems. Table 13–1 and Appendix E, however, include both metals and non-metals. Figure 13–4 shows some standard reduction potentials of the elements in aqueous solution. The elements to the left in the periodic table (the metals) are good reducing agents; the elements to the right are good oxidizing agents.

Oxygen as an Oxidizing Agent

The most important oxidizing agent for terrestrial life is the element oxygen itself. Gaseous O_2 has a fairly high reduction potential in standard acidic solution (in which $[H^+] = 1$ M and the pH is 0)

$$O_2(g) + 4\,H^+(aq) + 4\,e^- \longrightarrow 2\,H_2O(\ell) \qquad \mathscr{E}^\circ = 1.229 \text{ V}$$

and is accordingly a good oxidizing agent. Ozone (O_3) is even better, as is shown

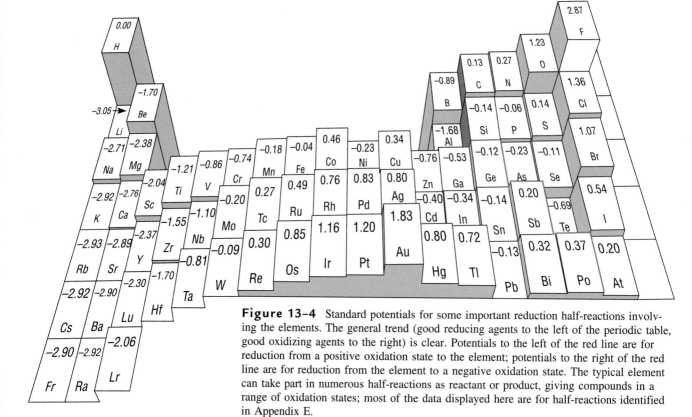

Figure 13-4 Standard potentials for some important reduction half-reactions involving the elements. The general trend (good reducing agents to the left of the periodic table, good oxidizing agents to the right) is clear. Potentials to the left of the red line are for reduction from a positive oxidation state to the element; potentials to the right of the red line are for reduction from the element to a negative oxidation state. The typical element can take part in numerous half-reactions as reactant or product, giving compounds in a range of oxidation states; most of the data displayed here are for half-reactions identified in Appendix E.

by the larger standard reduction potential in aqueous acidic solution for the half-reaction

$$O_3(g) + 2\ H^+(aq) + 2\ e^- \longrightarrow O_2(g) + H_2O(\ell) \qquad \mathscr{E}° = 2.07\ V$$

Because of its great oxidizing power, ozone is used commercially as a bleach for wood pulp and as a disinfectant and sterilizing agent for water, where it oxidizes algae and organic impurities but leaves no undesirable residue.

Oxygen in hydrogen peroxide has the intermediate oxidation state -1, so that it can be either reduced to water (oxidation state -2) or oxidized to molecular oxygen (oxidation state 0). These processes are represented by the reduction half-equations (if the reactions take place in acidic solution):

$$H_2O_2(aq) + 2\ H^+(aq) + 2\ e^- \longrightarrow 2\ H_2O(\ell) \qquad \mathscr{E}° = 1.77\ V$$
$$O_2(g) + 2\ H^+(aq) + 2\ e^- \longrightarrow H_2O_2(aq) \qquad \mathscr{E}° = 0.68\ V$$

Hydrogen peroxide can act as either an oxidizing agent or a reducing agent in acidic solution, although it is far stronger as an oxidizing agent. This oxidizing power makes hydrogen peroxide useful as a bleach and germicide. As a reducing agent, it reduces only chemical species with reduction potentials greater than 0.68 V. The preceding pair of reduction potentials also shows that acidic solutions of H_2O_2 disproportionate spontaneously into O_2 and water. This reaction is also spontaneous in neutral solution, but it is slow enough in that case that aqueous solutions of hydrogen peroxide can be stored for long periods of time as long as they are kept out of the light.

In basic aqueous solution, the half-reactions of oxygen-containing species involve hydroxide ions instead of hydrogen ions, and the reduction potentials change considerably. At pH 14, the pH at which $OH^-(aq)$ is in the standard state ($[OH^-]$ = 1 M),

$$O_2(g) + 2\,H_2O(\ell) + 4\,e^- \longrightarrow 4\,OH^-(aq) \qquad \mathscr{E}° = 0.401\text{ V}$$

$$O_3(g) + H_2O(\ell) + 2\,e^- \longrightarrow O_2(g) + 2\,OH^-(aq) \quad \mathscr{E}° = 1.24\text{ V}$$

Both oxygen and ozone are less effective oxidizing agents in basic solution than in acidic solution. The same is true of hydrogen peroxide (Fig. 13–5).

Nitrogen, Sulfur, and Phosphorus

Figure 13–6 shows standard reduction potentials for several half-reactions involving the oxoacids of nitrogen, sulfur, and phosphorus, and their conjugate bases. The species listed are those that predominate at the pH given (0 or 14). In acidic solution nitrous acid exists as HNO_2 molecules, but in basic solution almost all these molecules have reacted—to form NO_2^- ions, for example. Nitric acid, on the other hand, is a strong acid and is almost fully ionized to nitrate ion and hydrogen ion, even at pH 0. Because of the participation of hydrogen ion or hydroxide ion in the reduction half-reactions, the half-cell potentials depend strongly on pH, as seen by comparing the left side of the figure with the right side.

Comparison of the reduction potentials of the nitrogen, sulfur, and phosphorus oxoacids on the left side of the figure shows immediately that the phosphorus oxoacids are the most difficult to reduce and therefore are the least effective oxidizing agents. The oxoacids of sulfur are intermediate in this regard, and nitrate ion in acidic solution (nitric acid) and nitrous acid are the strongest oxidizing acids

Figure 13-5 Hydrogen peroxide is a reasonably strong oxidizing agent even in basic solution. Here, a solution containing pink $Cr^{3+}(aq)$ ion (*left*) is made basic (*center*). When H_2O_2 is added, it still oxidizes the $Cr^{3+}(aq)$ to the orange-yellow chromate ($CrO_4^{2-}(aq)$) ion (*right*).

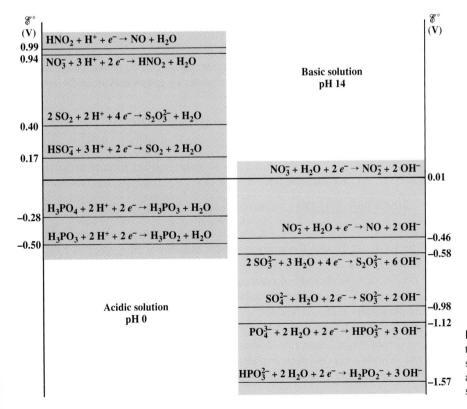

Figure 13-6 The standard reduction potentials for oxides of nitrogen, sulfur, and phosphorus are higher in acidic solution (left, pH 0) than in basic solution (right, pH 14).

Figure 13–7 When a copper wire is inserted into concentrated nitric acid, copper ions are produced. The brown fumes are $NO_2(g)$, one of the products of reduction of HNO_3.

listed. In concentrated form, nitric acid attacks and dissolves many metals, such as copper (Fig. 13–7), which phosphoric acid does not attack. In applications for which an acid is needed that is *not* a strong oxidizing agent, phosphoric acid is preferable to nitric or sulfuric acid (Fig. 13–8).

The right side of Figure 13–6 shows that the oxidizing strengths of all of these oxoacids are drastically curtailed when they are replaced by their basic forms at pH 14. The highest reduction potential (that for nitrate ion) is only +0.01 V at pH 14, showing that basic conditions should be avoided if oxidation is desired. On the other hand, the lower oxidation states of the sulfur and phosphorus compounds can be excellent *reducing* agents at pH 14. Both sulfite (SO_3^{2-}) and thiosulfate ($S_2O_3^{2-}$) ions are fairly strong reducing agents in basic solution, as shown by their appearance on the right side of half-equations with negative reduction potentials (−0.98 and −0.58 V, respectively). Two conjugate bases of the lower oxoacids of phosphorus, $H_2PO_2^-$ and HPO_3^{2-}, are even stronger reducing agents.

From Figure 13–6, we can immediately identify which species of intermediate oxidation state are unstable with respect to disproportionation under the conditions given. Nitrous acid, for example, disproportionates spontaneously in acidic solution according to

$$3\ HNO_2(aq) \longrightarrow NO_3^-(aq) + 2\ NO(g) + H^+(aq) + H_2O(\ell)$$
$$\Delta\mathscr{E}° + 0.99 - 0.94 = +0.05\ V$$

In basic solution, however, the nitrite ion is stable against disproportionation to NO and NO_3^- because the reaction

$$3\ NO_2^-(aq) + H_2O(\ell) \longrightarrow NO_3^-(aq) + 2\ NO(g) + 2\ OH^-(aq)$$

has $\Delta\mathscr{E}° = -0.46 - 0.01 = -0.47\ V < 0$, so that equilibrium lies very far to the left. The NO_2^- ion is still thermodynamically unstable with respect to NO_3^- and N_2O or N_2 (not shown in the table), but these reactions occur so slowly that NO_2^- can be kept indefinitely in basic solution.

It is frequently supposed that containing an element in a very high oxidation state automatically makes a species a strong oxidizing agent. It is certainly true that aqueous permanganate ion, which contains Mn in the +7 oxidation state, has great oxidizing power. Figure 13–6 shows however that a general correlation does not exist. Nitrous acid (nitrogen oxidation state +3) would be a stronger oxidizing agent than nitric acid (+5) at pH 0 if it were stable against disproportionation.

• The identification is immediate because we simply look for a pattern of the same species appearing "above and on the left" and "below and on the right" in the figure.

Aqueous sulfur dioxide (sulfur oxidation state +4) is a stronger oxidizing agent than hydrogen sulfate ion (sulfur oxidation state +6). The decisive factor is not the magnitude of the oxidation number but rather the relative stabilities of the redox pair of chemical species. Nitrous acid is a strong oxidizing agent precisely because it is much less stable than its reduced form, nitrogen monoxide. Thermodynamic stability changes with oxidation state in an irregular fashion.

EXAMPLE 13–5

By referring to Figure 13–6, determine (a) whether H_3PO_3 and SO_2 are stable with respect to disproportionation in acidic solution and (b) which is the stronger reducing agent at pH 0.

Solution

(a) H_3PO_3 is stable because the reaction

$$2\ H_3PO_3(aq) \longrightarrow H_3PO_4(aq) + H_3PO_2(aq)$$

has $\Delta\mathscr{E}° = -0.50 - (-0.28) = -0.22$ V and thus is not spontaneous. On the other hand, SO_2 is unstable because the reaction

$$4\ SO_2(aq) + 3\ H_2O(\ell) \longrightarrow 2\ HSO_4^-(aq) + S_2O_3^{2-}(aq) + 4\ H^+(aq)$$

has $\Delta\mathscr{E}° = 0.40 - 0.17 = +0.23$ V and is spontaneous.

(b) H_3PO_3 is the stronger reducing agent because the half-reaction in which it appears on the right side

$$H_3PO_4(aq) + 2\ H^+(aq) + 2\ e^- \longrightarrow$$
$$H_3PO_3(aq) + H_2O(\ell) \qquad \mathscr{E}° = -0.28 \text{ V}$$

has a more negative reduction potential (lies below) the half-reaction involving SO_2:

$$HSO_4^-(aq) + 3\ H^+(aq) + 2\ e^- \longrightarrow SO_2(aq) + 2\ H_2O(\ell) \qquad \mathscr{E}° = 0.17 \text{ V}$$

Exercise

By referring to Figure 13–6, determine (a) whether HPO_3^{2-} and SO_3^{2-} ions are stable with respect to disproportionation in basic solution and (b) which is the stronger oxidizing agent at pH 14.

Answer: (a) HPO_3^{2-} is stable, but SO_3^{2-} is not. (b) SO_3^{2-} is the stronger oxidizing agent (although neither one is very strong).

Figure 13-8 Concentrated nitric acid reacts with sugar after some time (*left*), and sulfuric acid (*center*) reacts immediately. By contrast, no reaction is seen with phosphoric acid (*right*). Soft drinks contain both phosphoric acid and dissolved sugar.

13-4   CONCENTRATIONS AND THE NERNST EQUATION

Electrochemical cells in which the concentrations of solutes or the partial pressures of gases differ from 1 M and 1 atm, their standard-state values, are both common and important. It is fortunately not difficult to extend the thermodynamic principles of Chapter 11 to explain the effects on cell voltage of changes in concentration or pressure. From Chapter 11 the change in the Gibbs function of a chemical reaction is related to the reaction quotient (Q) through

$$\Delta G = \Delta G° + RT \ln Q$$

Combining this equation with

$$\Delta G° = -n\mathscr{F}\Delta\mathscr{E}°$$

and

$$\Delta G = -n\mathscr{F}\Delta\mathscr{E}$$

gives

$$-n\mathscr{F}\Delta\mathscr{E} = -n\mathscr{F}\Delta\mathscr{E}° + RT \ln Q$$

Division by $-n\mathscr{F}$ then gives

$$\Delta\mathscr{E} = \Delta\mathscr{E}° - \left(\frac{RT}{n\mathscr{F}}\right) \ln Q$$

This relationship, known as the **Nernst equation,** gives the voltage of a cell as a function of the concentrations and pressures entering into the reaction quotient (Q). It is assumed that the standard cell voltage ($\Delta\mathscr{E}°$) is known or can be calculated from tabulated values, as in Example 13–3.

We can rewrite the Nernst equation using base-10 logarithms rather than natural logarithms as

$$\Delta\mathscr{E} = \Delta\mathscr{E}° - \frac{2.303 \, RT}{n\mathscr{F}} \log_{10} Q$$

At 25°C (298.15 K), the combination of constants $2.303 \, RT/\mathscr{F}$ becomes

$$2.303 \, \frac{RT}{\mathscr{F}} = (2.303)\frac{(8.315 \text{ J K}^{-1} \text{ mol}^{-1})(298.15 \text{ K})}{96,485 \text{ C mol}^{-1}}$$

$$= 0.0592 \text{ J C}^{-1} = 0.0592 \text{ V}$$

because one joule per coulomb is one volt. The Nernst equation then takes on the handy form

$$\Delta\mathscr{E} = \Delta\mathscr{E}° - \frac{0.0592 \text{ V}}{n} \log_{10} Q \qquad (T = 298.15 \text{ K only})$$

Here n is the number of moles of electrons transferred in the overall chemical reaction as written. In a galvanic cell made from zinc, aluminum, and their ions, for example,

$$\begin{array}{ll} 3 \text{ Zn}^{2+}(aq) + 6 \, e^- \longrightarrow 3 \text{ Zn}(s) \\ \underline{2 \text{ Al}(s) \hspace{3cm} \longrightarrow 2 \text{ Al}^{3+}(aq) + 6 \, e^-} \\ 3 \text{ Zn}^{2+}(aq) + 2 \text{ Al}(s) \longrightarrow 2 \text{ Al}^{3+}(aq) + 3 \text{ Zn}(s) \end{array}$$

there is a net transfer of 6 mol of electrons through the external circuit, so $n = 6$.

The Nernst equation applies to half-cells in exactly the same way that it does to complete electrochemical cells. For any half-reaction at 25°C, the reduction potential is

$$\mathscr{E} = \mathscr{E}° - \frac{0.0592 \text{ V}}{n_{hc}} \log_{10} Q_{hc}$$

where n_{hc} is the number of electrons appearing in the half-equation and Q_{hc} is the reaction quotient for the half-cell reaction written as a reduction. Naturally, the electrons themselves do not appear in Q_{hc}. In the case of $\text{Zn}^{2+}|\text{Zn}$, the half-reaction written as a reduction is

$$Zn^{2+}(aq) + 2\ e^- \longrightarrow Zn(s)$$

so that $n_{hc} = 2$ and $Q_{hc} = 1/[Zn^{2+}]$.

If two half-cells in which reactants and products are not in standard states are combined, the one with the more positive value of $\mathscr{E}$ is the cathode, where reduction takes place, and the cell voltage is

$$\Delta\mathscr{E} = \mathscr{E}(\text{cathode}) - \mathscr{E}(\text{anode})$$

EXAMPLE 13-6

Suppose that the $Zn|Zn^{2+}\|MnO_4^-|Mn^{2+}|Pt$ cell (the same cell used in Example 13–3) is operated at pH 2.00 with $[MnO_4^-] = 0.12$ M, $[Mn^{2+}] = 0.0010$ M, and $[Zn^{2+}] = 0.015$ M. Calculate the cell voltage ($\Delta\mathscr{E}$) at 25°C.

Solution

The reaction in this cell is

$$2\ MnO_4^-(aq) + 5\ Zn(s) + 16\ H^+(aq) \longrightarrow$$
$$2\ Mn^{2+}(aq) + 5\ Zn^{2+}(aq) + 8\ H_2O(\ell)$$

At 25°C the Nernst equation reads:

$$\Delta\mathscr{E} = \Delta\mathscr{E}° - \frac{0.0592\ V}{n} \log_{10} Q$$

From Example 13–3, for every 5 mol of zinc oxidized (or 2 mol of MnO_4^- reduced), 10 mol of electrons passes through the external circuit, so $n = 10$. A pH of 2.00 corresponds to a hydrogen-ion concentration of 0.010 M. Putting this and the other concentrations into the expression for the reaction quotient and substituting both Q and the $\Delta\mathscr{E}°$ in Example 13–3 into the Nernst equation gives

$$\Delta\mathscr{E} = 2.25\ V - \frac{0.0592\ V}{10} \log_{10} \frac{[Mn^{2+}]^2[Zn^{2+}]^5}{[MnO_4^-]^2[H^+]^{16}}$$

$$= 2.25\ V - \frac{0.0592\ V}{10} \log_{10} \frac{(0.0010)^2(0.015)^5}{(0.12)^2(0.010)^{16}}$$

$$= 2.25\ V - \frac{0.0592\ V}{10} \log_{10} (5.3 \times 10^{18}) = 2.14\ V$$

Exercise

Suppose that the $Zn|Zn^{2+}\|Cr^{3+}|Cr$ cell from the exercise in Example 13–3 were set up with $[Zn^{2+}] = 0.78$ M and $[Cr^{3+}] = 0.00011$ M. Calculate the cell voltage ($\Delta\mathscr{E}$) at 25°C.

Answer: $\Delta\mathscr{E} = -0.09$ V. The negative sign means the cell reaction $3\ Zn(s) + 2\ Cr^{3+}(aq) \longrightarrow 3\ Zn^{2+}(aq) + 2\ Cr(s)$, which is spontaneous when the reactants and products are in standard states, is *not* spontaneous with these concentrations. An external voltage would be needed to drive it. At these concentrations, the reverse reaction *is* spontaneous.

Measuring Concentrations with Electrochemical Cells

In the first part of this section, we used the Nernst equation to calculate the potential difference that is generated between the electrodes of an electrochemical cell in which the concentrations differ from standard-state values. We now turn this procedure around and show how a measured voltage provides a basis to calculate the concentration of a dissolved species in a cell. Suppose we know the half-reactions taking place (and, thus, the standard cell voltage $\Delta\mathscr{E}°$ by the use of listings such as that in Appendix E). If we then measure the actual $\Delta\mathscr{E}$ we can calculate the reaction quotient Q from the Nernst equation. This permits calculation of an unknown concentration when the concentrations of all the other species taking part in the reaction are known. The procedure is illustrated in the following example.

EXAMPLE 13–7

An electrochemical cell is set up at 25°C. One half-cell consists of a zinc anode immersed in a 1.00 M solution of $Zn(NO_3)_2$. The second half-cell consists of a platinum cathode that has gaseous hydrogen bubbling over it at a pressure of 1.00 atm in a solution of unknown hydrogen-ion concentration. The observed cell voltage is 0.472 V.
(a) Calculate the reaction quotient Q.
(b) Calculate the concentration of hydrogen ion in the second half-cell.

Solution

(a) The half-reactions are

$$\text{anode: } Zn(s) \longrightarrow Zn^{2+}(aq) + 2\ e^- \quad \text{(oxidation)}$$

$$\text{cathode: } 2\ H^+(aq) + 2\ e^- \longrightarrow H_2(g) \quad \text{(reduction)}$$

and the overall reaction is

$$Zn(s) + 2\ H^+(aq) \longrightarrow Zn^{2+}(aq) + H_2(g)$$

with $n = 2$. The voltage for the standard cell is

$$\Delta\mathscr{E}° = \Delta\mathscr{E}\ \text{(cathode)} - \mathscr{E}°\ \text{(anode)} = 0.000 - (-0.763) = +0.763\ \text{V}$$

However, the measured cell voltage is only 0.472 V, so that the reaction quotient must differ from 1. We can calculate it from the Nernst equation at 25°C:

$$\Delta\mathscr{E} = \Delta\mathscr{E}° - \frac{0.0592\ \text{V}}{n} \log_{10} Q$$

$$\log_{10} Q = \frac{n}{0.0592\ \text{V}} \times (\Delta\mathscr{E}° - \Delta\mathscr{E}) = \frac{2}{0.0592\ \cancel{\text{V}}} \times (0.763 - 0.472\ \cancel{\text{V}})$$

$$= 9.83$$

$$Q = 10^{9.83} = 6.8 \times 10^9$$

(b) The expression for the reaction quotient for this heterogeneous equilibrium is:

$$Q = \frac{[Zn^{2+}]P_{H_2}}{[H^+]^2}$$

$$6.8 \times 10^9 = \frac{(1.00)(1.00)}{[H^+]^2}$$

Solving for $[H^+]$ gives

$$[H^+] = 1.2 \times 10^{-5}\ \text{M}$$

Exercise

Suppose that in a cell of the type described in Example 13–7, it is the *zinc* concentration that is unknown. If the concentration of H^+ is 1.0×10^{-3} M and the partial pressure of $H_2(g)$ is 1.00 atm, determine $[Zn^{2+}]$ if the voltage is 0.851 V.

Answer: 1.1×10^{-9} M

The preceding example and exercise show the extraordinary sensitivity that the potential difference of an electrochemical cell can have to small concentrations of dissolved species. No other analytical technique could so easily measure a concentration of $Zn^{2+}(aq)$ on the order of 10^{-9} M.

pH Meters

The electrochemical cell described in Example 13–7 is a **pH meter,** a device to measure the H^+ concentration (or pH) of an unknown sample. Commercial pH meters (see Fig. 8–2) are constructed somewhat differently, however. A $Zn^{2+}|Zn$ electrode is not used as the reference electrode, and it is a nuisance to have to bubble hydrogen through the half-cell containing H^+ at the unknown concentration. Instead, a more portable and miniaturized pair of electrodes is used. In a typical pH meter, two electrodes are dipped into the solution of unknown pH. One of these, the **glass electrode,** usually consists of an AgCl-coated silver wire in contact with a solution of HCl of known (for example, 1.000 M) concentration in a thin-walled glass bulb. A pH-dependent potential develops across this thin glass membrane when the glass electrode is immersed in a solution of different, unknown $[H^+]$. The second half-cell is frequently a **saturated calomel electrode** that consists of a platinum wire in electrical contact with a paste of liquid mercury, calomel ($Hg_2Cl_2(s)$), and a saturated solution of potassium chloride. Figure 13–9 diagrams

• Recall that $H^+(aq)$ and $H_3O^+(aq)$ refer to the same thing. See Section 8-1.

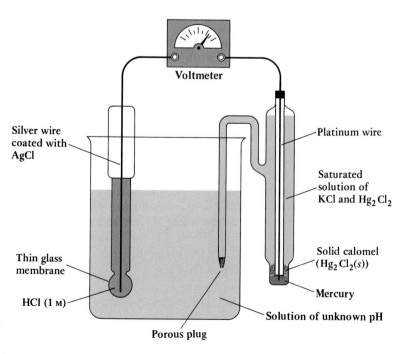

Figure 13-9 A pH meter consists of a glass electrode (*left*) and a calomel electrode (*right*), here shown dipping into a solution of unknown hydrogen ion concentration.

a glass electrode and a calomel electrode inserted in a solution of unknown pH. A pH meter is calibrated by measuring voltages for solutions buffered at known pH's.

The glass electrode has a number of advantages. It responds only to changes in $[H^+]$ and does so over a wide range of pH. It is unaffected by substances that may be present (such as strong oxidizing agents) that make a hydrogen electrode unreliable. Strongly colored solutions that obscure the color changes of acid–base indicators do not interfere with the glass electrode. Finally, the glass electrode can be miniaturized to an astonishing extent, enough to permit insertion into individual living cells, and finds wide use in biology.

Other types of electrodes have been designed that are sensitive to the concentrations of ions other than H^+. The simplest example of such an **ion-selective electrode** is a metal wire that can be used to detect the concentration of the corresponding metal ion in solution. Silver and copper wires can be used in this way to determine the concentrations of Ag^+ and Cu^{2+} in solutions that do not contain interfering ions. Many other electrodes have been developed for detecting and determining the concentrations of specific ions. Glass of chemically modified composition, for example, can be used to construct membrane electrodes to determine potassium, sodium, or halide ions.

13–5 EQUILIBRIUM CONSTANTS FROM ELECTROCHEMISTRY

We have seen that for a reaction taking place in an electrochemical cell

$$\Delta G° = -n\mathscr{F}\Delta\mathscr{E}°$$

In addition, $\Delta G°$ is related to the equilibrium constant (K) through

$$\Delta G° = -RT \ln K$$

so we conclude that

$$RT \ln K = n\mathscr{F}\Delta\mathscr{E}°$$

$$\ln K = \left(\frac{n\mathscr{F}}{RT}\right)\Delta\mathscr{E}°$$

$$\log_{10} K = \frac{n}{0.0592 \text{ V}}\Delta\mathscr{E}° \qquad (T = 298.15 \text{ K only})$$

The same result can be obtained in a slightly different way. We return in the Nernst equation, which reads (at 298.15 K).

$$\Delta\mathscr{E} = \Delta\mathscr{E}° - \frac{0.0592 \text{ V}}{n}\log_{10} Q$$

Suppose that we begin with all species present in standard states. Then, initially, $Q = 1$ and the observed cell voltage ($\Delta\mathscr{E}$) is the standard cell voltage ($\Delta\mathscr{E}°$). As the reaction proceeds, Q increases and $\Delta\mathscr{E}$ decreases. As reactants are used up and products are formed, the cell voltage drops. Eventually, equilibrium is reached. At this point, the Gibbs function of the cell is a minimum, the electric current ceases to flow, and $\Delta\mathscr{E}$ becomes zero. The cell is dead. Then

$$\Delta\mathscr{E}° = \frac{0.0592 \text{ V}}{n}\log_{10} K \qquad (T = 298.15 \text{ K only})$$

which is the same relationship obtained above.

EXAMPLE 13-8

Calculate the equilibrium constant of the redox reaction

$$2\ MnO_4^-(aq) + 5\ Zn(s) + 16\ H^+(aq) \longrightarrow$$
$$2\ Mn^{2+}(aq) + 5\ Zn^{2+}(aq) + 8\ H_2O(\ell)$$

at 25°C using the standard potential difference established in Example 13–3.

Solution

From Example 13–3, $\Delta\mathscr{E}° = 2.25$ V, and from the analysis there and in Example 13–6, $n = 10$ for the reaction written above. Therefore,

$$\log_{10} K = \frac{n}{0.0592\ \text{V}}\ \Delta\mathscr{E} = \frac{10}{0.0592\ \text{V}}\ (2.25\ \text{V}) = 380$$
$$K = 10^{380}$$

This overwhelmingly large equilibrium constant reflects the great strength of permanganate ion as an oxidizing agent and the great strength of zinc as a reducing agent. It means that for all practical purposes, no MnO_4^- ions are present at equilibrium. Such a large equilibrium constant is impossible to determine from direct measurements of concentration, but it is easy to measure using an electrochemical cell.

• When the value of n is not obvious upon inspection of a redox equation, rewrite the equation as the sum of two half-equations.

Exercise

Calculate the equilibrium constant for the redox reaction

$$2\ Cr^{3+}(aq) + 3\ Zn(s) \longrightarrow 2\ Cr(s) + 3\ Zn^{2+}(aq)$$

at 25°C using the standard potential difference established in Exercise 13–3.

Answer: $K = 10^2$.

Acid–Base and Solubility Equilibria

The example just presented illustrates how equilibrium constants for overall cell reactions can be determined electrochemically. It happened to have been a redox equilibrium, but related procedures can be used to measure acidity and basicity constants and solubility-product constants. Electrochemical measurements provide many equilibrium constants that would be difficult to determine by other means.

It is straightforward to determine the K_a of a weak acid electrochemically with a pH meter. We prepare a buffer solution (see Section 8–5) in which the concentrations of both the acid [HA] and its conjugate base [A$^-$] are known. A measurement with a pH meter determines [H$^+$], and we can then calculate the acidity constant by substituting the three concentrations into the expression

$$K_a = \frac{[H^+][A^-]}{[HA]}$$

Suppose that now we wish to determine a solubility-product constant, such as that of silver chloride, by means of an electrochemical cell. Here we can take advantage of the common-ion effect (see Section 9–3) by constructing a cell in which one half-cell contains solid AgCl in equilibrium with a known concentration of Cl$^-$(aq) (established with 0.00100 M NaCl, for example) and an unknown concentration of Ag$^+$(aq). A silver electrode is used so that the half-reaction involved

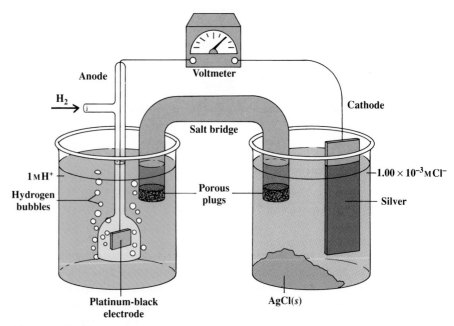

Figure 13–10 An electrochemical cell for measuring the solubility product of AgCl. An $Ag^+|Ag(s)$ half-cell containing $AgCl(s)$ in contact with $Cl^-(aq)$ at a known concentration forms the cathode; the anode is a standard $H^+(aq)|H_2(g)$ half-cell.

is either the reduction of $Ag^+(aq)$ or the oxidation of $Ag(s)$. This is, in effect, an $Ag^+|Ag$ half-cell with a reduction potential to be determined. The other half-cell can use any half-reaction at all, as long as its reduction potential is accurately known. In the following example, the second half-cell is a standard $H^+|H_2$ half-cell (Fig. 13–10).

EXAMPLE 13–9

A galvanic cell consists of a standard hydrogen half-cell (with platinum electrode) operating as the anode, and a silver half-cell:

$$Pt|H_2(1\ atm)|H^+(1\ \text{M})\|Ag^+|Ag$$

The $Ag^+(aq)$ in the cathode compartment is in equilibrium with some solid AgCl and $Cl^-(aq)$; the concentration of the $Cl^-(aq)$ is 0.00100 M. The measured cell voltage is $\Delta\mathscr{E} = 0.397$ V. Calculate the silver-ion concentration in the cell and the K_{sp} of silver chloride at 25°C.

Solution

The half-cell reactions are

$$\text{anode: } H_2(g) \longrightarrow 2\ H^+(aq) + 2\ e^-$$
$$\text{cathode: } 2\ Ag^+(aq) + 2\ e^- \longrightarrow 2\ Ag(s)$$
$$\Delta\mathscr{E}° = \mathscr{E}°(\text{cathode}) - \mathscr{E}°(\text{anode}) = 0.800 - 0.000\ \text{V} = 0.800\ \text{V}$$

Note that $n = 2$ for the overall cell reaction. The Nernst equation at 298.15 K becomes, for this cell,

$$\Delta\mathscr{E} = \Delta\mathscr{E}° - \frac{0.0592 \text{ V}}{n} \log_{10} Q$$

$$\log_{10} Q = \frac{n}{0.0592 \text{ V}} \times (\Delta\mathscr{E}° - \Delta\mathscr{E})$$

$$= \frac{2}{0.0592 \text{ V}} \times (0.800 - 0.397 \text{ V})$$

$$= 13.6$$

$$Q = 10^{13.6} = 4 \times 10^{13}$$

Because $[H^+] = 1$ M and $P_{H_2} = 1$ atm, the reaction quotient simplifies to

$$Q = \frac{[H^+]^2}{[Ag^+]^2 P_{H_2}} = \frac{1}{[Ag^+]^2}$$

Substituting the value 4×10^{13} for Q and solving for the concentration of silver ion gives

$$[Ag^+] = 1.6 \times 10^{-7} \text{ M}$$

so that

$$[Ag^+][Cl^-] = (1.6 \times 10^{-7})(1.00 \times 10^{-3})$$

$$K_{sp} = 1.6 \times 10^{-10}$$

Exercise

A half-cell is constructed in which a liquid-mercury cathode is in contact with a solution in which the concentration of $Br^-(aq)$ is known to be 1.0×10^{-4} M. The solution is also in contact with some solid mercury(I) bromide, so the equilibrium

$$Hg_2Br_2(s) \rightleftharpoons Hg_2^{2+}(aq) + 2 Br^-(aq)$$

is established. This half-cell is connected to a $Zn|Zn^{2+}$ reference half-cell in which the Zn^{2+} concentration is 1.00 M. The measured cell voltage is 1.178 V at 25°C. Calculate the concentration of Hg_2^{2+} and the solubility-product constant K_{sp} for Hg_2Br_2.

Answer: $[Hg_2^{2+}] = 1.3 \times 10^{-13}$ M; $K_{sp} = 1.3 \times 10^{-21}$.

13–6 BATTERIES AND FUEL CELLS

Electrochemical cells have numerous applications in industry and in everyday life. In this section, we focus on two of the most important: batteries to store energy and fuel cells to convert chemical energy into electrical energy.

Batteries

The origin of the **battery** is lost in history, but it is believed that Persian artisans must have used some form of battery to gold-plate jewelry as long ago as the second century B.C. The modern development of the battery began with Alessandro Volta in 1800. Volta constructed a "pile" that consisted of a stack of alternating zinc and silver disks, between pairs of which were inserted disks of paper moistened with an acid solution. This device was found to generate a potential difference across its ends that became larger as more pairs of disks were added. It could deliver a severe shock if the stack contained a great many disks.

Batteries vary in size and chemistry.

Figure 13-11 Much of the zinc used in the ordinary flashlight battery is produced by electrolysis of aqueous zinc sulfate solutions. In this cell house, there are hundreds of cells, each containing a series of cathodes where zinc plates out.

In Volta's invention, a large number of galvanic cells were arranged in **series** (cathode to anode) so that their individual voltages added together. Such an arrangement is, strictly speaking, a **battery of cells,** but modern usage makes no distinction between a single cell and such a combination, and the word "battery" has come to mean the cell itself. A distinction is usually made between cells that are discarded when their electrical energy has been spent and those that can be recharged. The former are called **primary** cells and the latter are called **secondary** cells.

The most familiar primary cell is the **Leclanché cell** (also called a "zinc–carbon dry cell"), which is used for flashlights, portable radios, and a host of other purposes. Each year, more than 5 billion such dry cells are used worldwide, and estimates place the quantity of zinc reacted in dry cells at over 30 metric tons per day (Fig. 13–11). The "dry cell" is not really dry at all; rather, its electrolyte is a moist powder that contains ammonium chloride and zinc chloride. Figure 13–12 is a cutaway illustration of a dry cell, which consists of a zinc shell for an anode (negative electrode) and an axial graphite rod for a cathode (positive electrode); the rod is surrounded by a densely packed layer of graphite and manganese dioxide. Each of these components performs an interesting and essential function. At the zinc anode, oxidation takes place:

$$\text{anode: } Zn(s) \longrightarrow Zn^{2+}(aq) + 2\ e^-$$

The moist salt mixture permits ions to transport net charge through the cell equal to that carried by electrons in the external circuit, just as a salt bridge does in the galvanic cells considered up to this point. Manganese dioxide is the ultimate electron acceptor and is reduced to Mn_2O_3 by electrons that pass to it from the graphite rod through the graphite particles:

$$\text{cathode: } 2\ MnO_2(s) + 2\ NH_4^+(aq) + 2\ e^- \longrightarrow$$
$$Mn_2O_3(s) + 2\ NH_3(aq) + H_2O(\ell)$$

In addition, some ammonium ion is reduced to gaseous ammonia and hydrogen

$$\text{cathode: } 2\ NH_4^+(aq) + 2e^- \longrightarrow 2\ NH_3(g) + H_2(g)$$

A build-up of gases is, however, prevented by formation of a complex ion

$$Zn^{2+}(aq) + 2\ NH_3(g) \longrightarrow [Zn(NH_3)_2]^{2+}(aq)$$

and the re-oxidation of the hydrogen

$$2\ MnO_2(s) + H_2(g) \longrightarrow Mn_2O_3(s) + H_2O(\ell)$$

The ingenious device of mixing powdered graphite with powdered MnO_2 greatly increases the effective surface area of the cathode and enables currents of several amperes to flow. The overall cell reaction is

$$Zn(s) + 2\ MnO_2(s) + 2\ NH_4^+(aq) \longrightarrow$$
$$[Zn(NH_3)_2]^{2+}(aq) + Mn_2O_3(s) + H_2O(\ell)$$

The cell components are hermetically sealed in a steel shell that, because it is in contact with the zinc, is the negative terminal of the battery. A fresh Leclanché dry cell generates a potential difference of 1.5 V.

The Leclanché cell has several disadvantages. Concentrations change as the cell reaction progresses, so the cell voltage falls with use. Rapid withdrawal of energy may cause the production of ammonia and hydrogen to outpace their removal, leading to a sharp drop in voltage (a cell in this condition recovers on standing). Unused cells deteriorate because of a slow reaction between $Zn(s)$ and $NH_4^+(aq)$; thus, they can go dead while sitting on the shelf. Finally, the cells are not rechargeable because the $Zn^{2+}(aq)$ ions migrate away from the anode and are trapped in the complex ion.

In an **alkaline dry cell,** the ammonium chloride is replaced by potassium hydroxide. Here the half-cell reactions become

anode: $Zn(s) + 2\ OH^-(aq) \longrightarrow Zn(OH)_2(s) + 2\ e^-$

cathode: $2\ MnO_2(s) + H_2O(\ell) + 2\ e^- \longrightarrow Mn_2O_3(s) + 2\ OH^-(aq)$

The overall reaction is then

$$Zn(s) + 2\ MnO_2(s) + H_2O(\ell) \longrightarrow Zn(OH)_2(s) + Mn_2O_3(s)$$

• Most reactions are slowed by lowering the temperature. The shelf life of a dry cell can be increased by 100% to 200% by storage in a freezer or refrigerator.

• Recall that NH_4Cl is acidic in aqueous solution, and KOH is basic (alkaline).

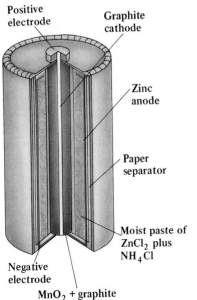

Positive electrode

Graphite cathode

Zinc anode

Paper separator

Moist paste of $ZnCl_2$ plus NH_4Cl

Negative electrode

MnO_2 + graphite

Figure 13–12 In a Leclanché dry cell, electrons are released to an external circuit at the anode and enter the cell again at the cathode, where the reduction of MnO_2 occurs.

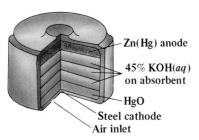

Figure 13–13 A zinc–mercuric oxide dry cell that is used in electric watches and cameras.

Figure 13–14 A rechargeable nickel–cadmium battery.

• A secondary battery is converted from a galvanic cell into an electrolytic cell during recharging. The terms "discharge" and "recharge" are misleading because the actual quantity of electrical charge in the battery does not change in either process. Perhaps "de-energize" and "re-energize" would be better.

Because no dissolved species take part in the overall reaction, the reaction quotient Q does not change as the reaction progresses, and a steadier voltage results. The standard voltage of "alkaline batteries" is 1.54 V; they replace Leclanché cells without problems.

A third primary dry cell is the **zinc–mercuric oxide cell,** or **mercury battery,** depicted in Figure 13–13. It frequently is given the shape of a small button and is used in automatic cameras, hearing aids, digital calculators, and quartz-electric watches. The anode in this cell is a mixture of mercury and zinc; the cathode is steel in contact with solid mercury(II) oxide (HgO). The electrolyte is a 45% potassium hydroxide solution that saturates an absorbent material. The anode half-reaction is the same as that in an alkaline dry cell:

$$\text{anode: } Zn(s) + 2\ OH^-(aq) \longrightarrow Zn(OH)_2(s) + 2\ e^-$$

The cathode half-reaction is now

$$\text{cathode: } HgO(s) + H_2O(\ell) + 2\ e^- \longrightarrow Hg(\ell) + 2\ OH^-(aq)$$

The overall reaction is

$$Zn(s) + HgO(s) + H_2O(\ell) \longrightarrow Zn(OH)_2(s) + Hg(\ell)$$

In the mercury cell, mercury serves also to alloy or amalgamate the zinc (alloys of mercury with other metals are called "amalgams"). This cell has a very stable output of 1.34 V, a fact that makes it especially valuable for use in communication equipment and scientific instruments. A disadvantage is that mercury and mercury compounds are toxic. Careless disposal of mercury batteries leads to environmental contamination.

Rechargeable Batteries

Chemical change at the electrodes always accompanies the extraction of energy from a battery. When the redox reaction in a battery reaches equilibrium, this change has gone to its fullest extent: the battery is completely discharged. In some battery designs, the electrodes can be regenerated to their original form (or near it) by imposing an external potential difference to force current through the battery in a direction opposite to the way it flowed during discharge. Such batteries are **secondary batteries,** and the process of reconstituting them to their original state is called "recharging." To recharge a run-down secondary battery, the voltage of the external source must be larger than that of the battery in its original state and, of course, opposite in polarity.

The **nickel–cadmium cell** (or nicad battery, Fig. 13–14) is used in hand-held electronic calculators, cordless electric shavers, camcorders, and portable radios. Its half-cell reactions during discharge are

anode: $Cd(s) + 2\ OH^-(aq) \longrightarrow Cd(OH)_2(s) + 2\ e^-$
cathode: $2\ NiO(OH)(s) + 2\ H_2O(\ell) + 2\ e^- \longrightarrow 2\ Ni(OH)_2(s) + 2\ OH^-(aq)$

total: $Cd(s) + 2\ NiO(OH)(s) + 2\ H_2O(\ell) \longrightarrow Cd(OH)_2(s) + 2\ Ni(OH)_2(s)$

This battery gives a fairly constant voltage of 1.4 V. When the battery is connected to an external voltage source, the preceding reactions are reversed as the battery is recharged. Unfortunately, the nicad battery suffers from "discharge memory." If the battery is frequently recharged after only partial discharge, it starts to require recharging on that basis. In addition, cadmium, like mercury, is a toxic substance.

Perhaps the best-known secondary battery is the **lead-acid battery,** which is universally used in automobiles. A 12-V lead storage battery consists of six cells (Fig. 13–15) that are connected in series (cathode to anode) by an internal lead linkage and housed together in a hard rubber or plastic case. When cells are connected in series, their voltages add, so the six cells in this battery generate 2 V each. In each cell the anode consists of porous metallic lead. The sponge-like texture of this electrode maximizes its area of contact with the electrolyte. The cathode is of similar design, but its lead has been converted to lead dioxide. A sulfuric acid solution (37% by mass) serves as the electrolyte.

When the external circuit is completed, electrons are released from the anode. The resulting Pb^{2+} ions immediately react with SO_4^{2-} ions to precipitate a layer of insoluble lead sulfate on the surface of the electrode. At the cathode, the electrons from the external circuit reduce PbO_2 to water and Pb^{2+} ions, which also react immediately with the sulfate ions in the electrolyte to precipitate $PbSO_4$ on the electrode. The half-cell reactions are

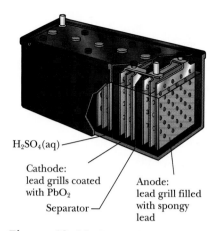

$H_2SO_4(aq)$

Cathode: lead grills coated with PbO_2

Separator

Anode: lead grill filled with spongy lead

Figure 13–15 In a lead–acid storage battery, anodes made of Pb alternate with cathodes of Pb coated with PbO_2; the electrolyte is sulfuric acid (sometimes therefore called "battery acid").

anode: $Pb(s) + SO_4^{2-}(aq) \longrightarrow PbSO_4(s) + 2\ e^-$

cathode: $PbO_2(s) + SO_4^{2-}(aq) + 4\ H^+(aq) + 2\ e^- \longrightarrow PbSO_4(s) + 2\ H_2O(\ell)$

total: $Pb(s) + PbO_2(s) + 2\ SO_4^{2-}(aq) + 4\ H^+(aq) \longrightarrow$
$2\ PbSO_4(s) + 2\ H_2O(\ell)$

The anode and cathode are both largely converted to $PbSO_4(s)$ when the storage battery is fully discharged. Because sulfuric acid is a reactant, its concentration falls as the battery is discharged. Measuring the density of the electrolyte provides a quick way of estimating the state of charge of the battery.

When a voltage exceeding 2 V is applied across the terminals of a single cell (or 12 V for the entire battery) in such a direction as to make the electrode that was originally the cathode into the anode and the electrode that was originally the anode into the cathode, the half-cell reactions are reversed. Continued charging of the cell restores it to its initial state, ready for another work-producing discharge half-cycle. Lead storage batteries endure many thousands of cycles of discharge and charge before they finally fail because $PbSO_4$ flakes off from the electrodes or internal short circuits develop. These short circuits occur when elongated crystals of lead and lead oxide, formed during recharging, finally connect the electrodes. In automobiles, lead-acid batteries are usually not designed to undergo complete discharge–recharge cycles. Instead, a DC generator converts some of the kinetic energy of the vehicle into electrical energy for continuous or intermittent charging. About 1.8×10^7 J of energy can be obtained in the discharge of an average automobile battery, enough to light a 100-watt light bulb for 50 hours. Currents as large as 100 A are drawn from the battery for the short time needed to start the engine. Of course, when such huge currents flow, the operation of a battery is very far from equilibrium, and the potential difference across its electrodes is much smaller than the theoretical potential difference. The Nernst equation is of no value in calculating the voltage of a battery during this nonequilibrium process.

• When a dead battery is to be recharged, the wire from the charger that is marked "plus" must always connect to the battery terminal marked "plus," and "minus" to "minus." In this way, the external source opposes the normal flow in the battery during discharge.

Fuel Cells

A battery is a closed system that delivers electrical energy by electrochemical reactions. Once the chemicals originally present are consumed, the battery must be either recharged or discarded. In contrast, a **fuel cell** is designed for continuous operation, with reactants (fuel) being supplied and products removed continuously.

Figure 13-16 This hydrogen–oxygen fuel cell was used on U. S. space missions.

It is an energy converter, transforming chemical energy into electrical energy. Fuel cells based on the reaction

$$2 \text{ H}_2(g) + \text{O}_2(g) \longrightarrow 2 \text{ H}_2\text{O}(\ell)$$

are used on space vehicles (Fig. 13–16).

The hydrogen–oxygen fuel cell is schematically represented in Figure 13–17. The electrodes can be any nonreactive conductor (graphite, for example). Their function is merely to conduct electrons into and out of the cell and to facilitate the exchange of electrons between the gases and the ions in solution. The electrolyte transports charge through the cell and the ions dissolved in it participate in the half-reactions at each electrode. Acidic solutions present corrosion problems, so an alkaline solution is preferable (1 M KOH, for instance). The anode half-reaction is then

$$\text{H}_2(g) + 2 \text{ OH}^-(aq) \longrightarrow 2 \text{ H}_2\text{O}(\ell) + 2 \ e^-$$

Figure 13-17 In the hydrogen–oxygen fuel cell, the two gases are fed in separately and are oxidized or reduced at the electrodes. A hot solution of potassium hydroxide between the electrodes completes the circuit, and the steam produced in the reaction evaporates from the cell continuously.

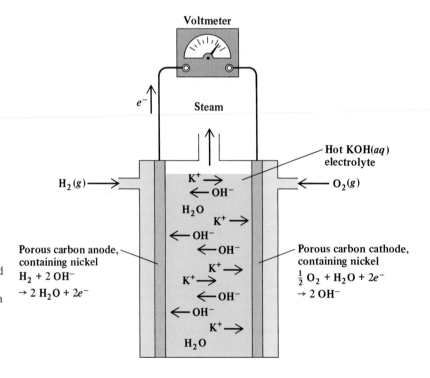

and the cathode half-reaction is

$$\tfrac{1}{2} O_2(g) + H_2O(\ell) + 2\ e^- \longrightarrow 2\ OH^-(aq)$$

The standard reduction potentials at 25°C (1 M concentrations, 1 atm pressure) are

$$\mathscr{E}°(H_2|H_2O) = -0.828\ V \quad \text{and} \quad \mathscr{E}°(O_2|OH^-) = 0.401\ V$$

and the overall cell reaction is the production of water

$$H_2(g) + \tfrac{1}{2} O_2(g) \longrightarrow H_2O(\ell)$$

The standard cell voltage is therefore

$$\Delta\mathscr{E}° = \mathscr{E}°(\text{cathode}) - \mathscr{E}°(\text{anode}) = 0.401 - (-0.828) = 1.229\ V$$

Another practical fuel cell employs the overall reaction

$$CO(g) + \tfrac{1}{2} O_2(g) \longrightarrow CO_2(g)$$

for which the half-reactions are

anode: $CO(g) + H_2O(\ell) \longrightarrow CO_2(g) + 2\ H^+(aq) + 2\ e^-$
cathode: $\tfrac{1}{2} O_2(g) + 2\ H^+(aq) + 2\ e^- \longrightarrow H_2O(\ell)$

The electrolyte is usually concentrated phosphoric acid, and the operating temperature is between 100°C and 200°C. Platinum is the electrode material of choice.

Natural gas (largely CH_4) and even fuel oil can be "burned" electrochemically to provide electrical energy by either of two approaches. They can be converted into CO and H_2 or to CO_2 and H_2 prior to use in a fuel cell by so-called reforming reactions with steam

$$CH_4(g) + H_2O(g) \longrightarrow CO(g) + 3\ H_2(g)$$
$$CO(g) + H_2O(g) \longrightarrow CO_2(g) + H_2(g)$$

at temperatures of about 500°C. Alternatively, they can be used directly in a fuel cell at higher temperatures (up to 750°C) with a molten alkali-metal carbonate as the electrolyte. Either type of fuel cell is attractive as an electrochemical energy converter in areas where hydrocarbon fuels are readily available but large-scale power plants (conventional or nuclear) are remote.

13-7   CORROSION AND ITS PREVENTION

The **corrosion** of metals (Fig. 13–18) is a significant problem. It has been estimated that, in the United States alone, the annual cost of corrosion is on the order of tens of *billions* of dollars. The effects of corrosion are both visible (the formation of rust on exposed iron and steel surfaces) and invisible (the cracking and resulting loss of strength of metal beneath the surface). The corrosion of the various structural metals under ordinary terrestrial conditions is thermodynamically spontaneous and therefore cannot, in the long run, be stopped. Iron, steel, and aluminum eventually corrode. What really matters, in human terms, is how long such processes take and how they can be prevented from happening prematurely and catastrophically. Before ways to achieve this kind of prevention can be developed, the mechanisms of corrosion must be studied and understood.

Figure 13-18 The corrosion of iron and steel is a major problem for industrial society, ranging in scale from the rusting of nails and screws to the structural weakening of girders and bridges.

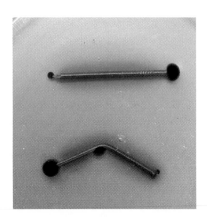

Figure 13-19 Nails corroding in water in the absence of oxygen. Blue-green indicates anodic regions where $Fe(s)$ has been oxidized to $Fe^{2+}(aq)$ ion. Red indicates cathodic regions where water has been reduced to $OH^-(aq)$ ion. The colors appear because the water contains $[Fe(CN)_6]^{3+}$ ion, which reacts with Fe^{2+} to give a blue-green complex, and phenolphthalein, which turns red in the presence of OH^- ion; the solution also contains a thickening gel to slow diffusion. Note how oxidation occurs at the bend and at the ends of the nails; these are regions of stress.

Although corrosion is a serious problem for many metals, we focus on the spontaneous electrochemical reactions of iron. The corrosion of iron can be pictured as resulting from the operation of a galvanic cell, in which some regions of the metal surface act as cathodes and others as anodes, and the electrical circuit is completed by electron flow through the iron itself. The electrochemical half-cells appear in parts of the metal containing impurities or in regions that are subject to stress. The anode reaction is

$$Fe(s) \longrightarrow Fe^{2+}(aq) + 2\ e^-$$

Various cathode reactions are possible. In the absence of oxygen (for example, at the bottom of a lake), the corrosion reactions are

anode: $Fe(s) \longrightarrow Fe^{2+}(aq) + 2\ e^-$
cathode: $2\ H_2O(\ell) + 2\ e^- \longrightarrow 2\ OH^-(aq) + H_2(g)$
$$\overline{\text{total:}\quad Fe(s) + 2\ H_2O(\ell) \longrightarrow Fe^{2+}(aq) + 2\ OH^-(aq) + H_2(g)}$$

However, these (and similar reactions) are generally slow and do not cause serious amounts of corrosion unless accelerated by the intervention of bacteria (see later). In Figure 13-19, indicators make it possible to distinguish cathodic and anodic regions in nails corroding in the absence of oxygen. More extensive corrosion takes place when the iron is in contact with both oxygen and water. In this case, the cathode reaction is

$$\tfrac{1}{2} O_2(g) + 2\ H^+(aq) + 2\ e^- \longrightarrow H_2O(\ell)$$

The Fe^{2+} ions formed simultaneously at the anode can migrate to the cathode, where they are further oxidized by O_2 to the $+3$ oxidation state to form rust ($Fe_2O_3 \cdot xH_2O$), a hydrated form of iron(III) oxide:

$$2\ Fe^{2+}(aq) + \tfrac{1}{2} O_2(g) + (2 + x)\ H_2O(\ell) \longrightarrow Fe_2O_3 \cdot xH_2O + 4\ H^+(aq)$$

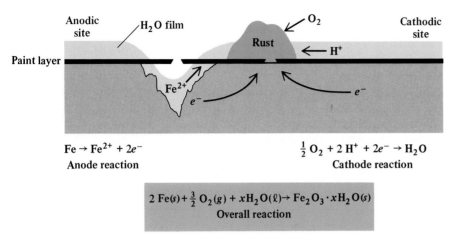

Figure 13-20 The corrosion of iron. Note that pitting occurs at the anodic region and rust appears at the cathodic region.

The hydrogen ions produced in this reaction allow the corrosion cycle to continue.

The nature of the cathodic and anodic regions can be illustrated by the way in which a piece of iron or steel corrodes when a portion of the paint that protects it is chipped off (Fig. 13–20). The exposed area acts as the cathode because it is open to the atmosphere (air and water) and is therefore rich in oxygen, whereas oxygen-poor areas *underneath* the paint act as anodes. The result is that rust forms on the cathode (the visible, exposed region) and pitting occurs at the anode (loss of metal through oxidation of iron and flow of metal ions to the cathode). This pitting can lead to loss of structural strength in girders and other supports. The most serious harm done by corrosion is not the visible rusting but the damage done underneath the painted surface.

A number of factors speed up corrosion. Dissolved salt improves the flow of charge through solution; a well-known example is the more rapid rusting of cars in areas where salt is spread to combat icy roads. Higher acidity also favors corrosion (note that H^+ is a reactant in the half-reaction at the cathode). Acidity is increased by dissolved CO_2 (which produces H^+ and HCO_3^- ions) and, more seriously, by dissolved oxides of sulfur from air pollution brought down in acid rain.

Biocorrosion

Although corrosion is usually thought of as occurring when a metal is in contact with air and water (*aerobic* conditions), significant corrosion also occurs under conditions that completely exclude oxygen from the metal (*anaerobic* conditions). Quite often, microorganisms accelerate these processes in a mechanism referred to as **biocorrosion.**

One major group of agents for biocorrosion are the sulfate-reducing bacteria. These organisms greatly speed up the corrosion of iron in pipes buried in clay soils or covered by polluted water. Such environments contain sulfate ion ($SO_4^{2-}(aq)$) but no $O_2(g)$ or $O_2(aq)$, so that corrosion by the reactions just described cannot occur; however, the iron is oxidized, and the ultimate oxidizing agent is the sulfate ion. Again, the process is best viewed as a short-circuited galvanic cell with the half-reactions taking place at separate locations on the metal surface. The half-reaction at the anodic sites is

$$4\ Fe(s) + HS^-(aq) + 7\ OH^-(aq) \longrightarrow$$
$$3\ Fe(OH)_2(s) + FeS(s) + H_2O(\ell) + 8\ e^-$$

At the cathodic sites, the sulfate ion is reduced:

$$SO_4^{2-}(aq) + 5\ H_2O(\ell) + 8\ e^- \longrightarrow HS^-(aq) + 9\ OH^-(aq)$$

The overall reaction

$$4\ Fe(s) + SO_4^{2-}(aq) + 4\ H_2O(\ell) \longrightarrow 3\ Fe(OH)_2(s) + FeS(s) + 2\ OH^-(aq)$$

is thermodynamically spontaneous. In the absence of the bacteria, however, this reaction does not take place at a significant rate. The bacteria provide the necessary pathway for the reaction to occur.

Events in the cathodic regions, where the bacteria live, take place in two steps:

$$8\ H_2O(\ell) + 8\ e^- \longrightarrow 4\ H_2(g) + 8\ OH^-(aq) \qquad \text{(reduction of water to } H_2)$$
$$SO_4^{2-}(aq) + 4\ H_2(g) \longrightarrow$$
$$HS^-(aq) + OH^-(aq) + 3\ H_2O(\ell) \qquad \text{(reduction of } SO_4^{2-} \text{ by } H_2)$$

In the absence of the bacteria, there is no rapid way for the second step to occur. Corrosion quickly stalls with the build-up of $H_2(g)$ generated in the first step. The bacteria accelerate the second step by combining the $H_2(g)$ with $SO_4^{2-}(aq)$ as it forms. The overall corrosive reaction then runs smoothly. Although the bacteria do not incorporate the iron into their own organisms, they still feed on the pipes in the sense that they use them as an energetically accessible source of electrons to reduce sulfate. Iron pipes corroded in this way become covered with black blotches of iron(II) sulfide. If this soft $FeS(s)$ is wiped away, an anodic pit is revealed, and within the pit shines a bright surface of metallic iron.

A second example of biocorrosion involves a group of anaerobic microorganisms that live by oxidizing hydrogen with carbon dioxide to generate water and methane:

$$4\ H_2(g) + CO_2(g) \longrightarrow CH_4(g) + 2\ H_2O(\ell)$$

* For this reason, methane is sometimes called "swamp gas."

Such *methanogens* (methane-generating bacteria) thrive in the airless muck at the bottom of swamps, generating considerable amounts of methane, which collects in bubbles and eventually rises to the surface. The preceding reaction of hydrogen with carbon dioxide has a $\Delta G°$ of -131 kJ, which fuels the organisms' growth. What is significant for corrosion is that some methanogens also grow in the complete absence of H_2, if $Fe(s)$ is present. Energy for their growth comes from the reaction

$$8\ H^+(aq) + 4\ Fe(s) + CO_2(g) \longrightarrow$$
$$CH_4(g) + 4\ Fe^{2+}(aq) + 2\ H_2O(\ell) \qquad \Delta G° = -136\ kJ$$

Observe that the $\Delta G°$ of this reaction is nearly the same as in the reaction in which H_2 is oxidized. In a sense, the bacteria can eat iron as a substitute for H_2. The iron corrodes as the microorganism uses it as a source of electrons to reduce carbon dioxide.

Inhibition of Corrosion

Corrosion of metals can be inhibited in a number of ways. Coatings of paint or plastics obviously protect the metal, but they can crack or suffer other damage,

CHEMISTRY IN PROGRESS

Turning Corrosion to Good Use

Chlorinated hydrocarbons are the products, by-products, or wastes of numerous industrial processes. Some are quite noxious and have unfortunately escaped into the air (see Section 18–5) or the ground, where they resist decomposition and present long-term hazards. These chemicals have made the term "toxic waste" part of everyday language.

Dozens of schemes for the remediation of spills of liquid and solid chlorinated hydrocarbons have been suggested. The standard procedure is "pump and treat." Wells are drilled. Contaminated groundwater is brought up, cleansed chemically in a special plant, and pumped back into the ground. Other concepts range from *bioremediation* (encouraging existing microorganisms to consume specific pollutants or even breeding new ones) to a Star Wars approach in which electron-beam accelerators originally designed for weapons research blast the pollutants as the contaminated water flows by.

One new idea is stunning in its simplicity and apparent success. Tons of iron filings are buried in underground walls positioned to intercept the movement of contaminated groundwater. Chlorinated hydrocarbons in the groundwater percolate through the barrier and, over a period of several days, corrode the iron(0), oxidizing it to iron(II) or iron(III). They themselves are reduced to inorganic chloride (the aqueous chloride ion, which is innocuous in the environment) and a nontoxic or less toxic hydrocarbon.

Clean-up by corrosion is working right now to destroy spilled trichloroethylene (C_2HCl_3) at a site in Sunnyvale, California, where a slow, arduous course of remediation by the standard method had been anticipated. The water seeping through the barrier meets the standards of the U. S. Environmental Protection Agency, its operators say.

In clean-up by corrosion, the high Gibbs function of elemental iron relative to its oxides and chlorides drives the reactions that consume the pollutants. The op-

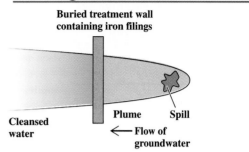

An underground wall of iron filings intercepts and destroys pollutants in groundwater.

eration consumes no energy once the iron is smelted. Although the land and its patterns of water flow must be surveyed very carefully to ensure that the flow of contaminated groundwater does not somehow miss the barrier containing the iron, researching a site and building an underground wall cost no more than constructing a water-treatment plant. The underground walls require little maintenance, cleanse the water more rapidly than a treatment plant, waste less water, and, it is thought, will remain effective for many years. Finally, the land above can be used for other purposes than a waste-treatment facility.

The planned corrosion of iron may help eliminate pollutants other than chlorinated hydrocarbons. Iron seems capable of degrading dye wastes from the textile industry. It also reduces chromium in the +6 oxidation state to chromium in the +3 oxidation state. This is particularly useful because many highly toxic compounds in which chromium is in the +6 state are water soluble and disperse rapidly in groundwater when spilled. Compounds in which chromium is in the +3 state are generally insoluble and stick on the iron barrier. Iron may also produce immobile compounds of technetium, a radioactive element that contaminates several sites operated by the Department of Energy.

thereby localizing and accentuating the process. A very important method of protecting metals arises from the phenomenon of **passivation,** in which a thin metal-oxide layer forms on the surface and prevents further electrochemical reactions. Some metals become passivated spontaneously upon exposure to air; aluminum, for example, reacts with oxygen to form a thin protective layer of Al_2O_3. A tightly adhering oxide layer less than 20 atoms deep protects magnesium, commonly used for transmission cases in automobiles and in other structural applications, from corroding in air. Special paints designed for rust prevention contain potassium

Figure 13–21 The iron of this ship's hull is protected against corrosion by the blocks of zinc bolted to it. In salt water, the zinc acts as a sacrificial anode, suffering oxidation in place of the ship's hull.

dichromate ($K_2Cr_2O_7$) and lead oxide (Pb_3O_4), which cause the surface of iron to be oxidized and passivated. The presence of small quantities of alloying materials can greatly influence the amount of protection that a film provides. Stainless steel is an alloy of iron with chromium in which the chromium leads to passivation and prevents rusting. The integrity of a surface film is constantly assaulted by routine day-to-day events; the metal may be bent, shocked by blows, and scoured by abrasive particles. Protection against corrosion requires a surface film that can heal itself as fast as it is broken down.

A different way of preventing the corrosion of iron is to use a **sacrificial anode.** According to the standard reduction potentials of iron and magnesium

$$Fe^{2+}(aq) + 2\ e^- \longrightarrow Fe(s) \qquad \mathscr{E}° = -0.41\ V$$
$$Mg^{2+}(aq) + 2\ e^- \longrightarrow Mg(s) \qquad \mathscr{E}° = -2.39\ V$$

$Mg^{2+}(aq)$ is much harder to reduce than $Fe^{2+}(aq)$. It follows that $Mg(s)$ is more easily oxidized than $Fe(s)$. A piece of magnesium in electrical contact with iron is oxidized in preference to the iron, and the iron is thereby protected. The magnesium is the sacrificial anode: once it is consumed by oxidation, it must be replaced. Sacrificial anodes are used to protect iron ship hulls, bridges, and water pipes from corrosion (Fig. 13–21). Magnesium plates are attached at regular intervals along a piece of buried pipe. It is far easier to replace them periodically than to have to replace the entire pipe.

SUMMARY

13–1 Thermodynamics demonstrates a direct connection between work, the Gibbs function, and cell voltage in an electrochemical cell. In a cell operated at constant temperature and pressure, the maximum **electrical work** done is equal to the change in the Gibbs function of the redox reaction and to the product (with a minus sign) of the total charge passing through the external circuit and the cell voltage. A galvanic cell can produce useful work from a spontaneous reaction; an electrolytic cell requires the input of work, from an outside source of electrical energy, to force a non-spontaneous chemical reaction to occur. The **standard cell voltage** is the voltage measured in an electrochemical cell in which all reactants and products are in thermodynamic standard states.

13–2 Half-cell reduction potentials are tabulated using the convention that the reduction potential of a standard $H^+(aq)|H_2(g)$ half-cell is exactly zero. The reduction potential of any other half-cell can be measured by setting it up in combination with a half-cell that has a reduction potential that is already known relative to this standard and measuring the potential difference (the cell voltage).

13–3 The relative strengths of oxidizing and reducing agents depend on the products they form; in aqueous systems, the pH of the solution is a particularly important influence. Tables of standard reduction potentials show which species are unstable with respect to disproportionation and which are strong oxidizing or reducing agents.

13–4 The **Nernst equation** describes the effect of changes of concentrations or pressures on cell voltages. It can be used to compute the concentration of a dissolved species taking part in a redox reaction from the cell voltage, provided that the concentrations of other species being consumed and produced are all known. The **pH meter** is an application of the equation.

13-5 As a reaction in a galvanic cell proceeds toward equilibrium, the cell voltage falls toward zero. This fact allows the equilibrium constant of a redox reaction to be calculated from its standard cell voltage. Electrochemical cells can also be used to measure K_a's, K_b's, and K_{sp}'s.

13-6 Electrochemical cells are used to generate and store electricity. In **batteries,** spontaneous reactions generate an electric current. In a **primary battery,** the cells are discarded after the reactants are consumed and the voltage drops, but in a **secondary battery,** the original reactants can be regenerated by recharging. A **fuel cell** allows the continuous conversion of chemical to electrical energy as reactants are fed in and caused to react in an electrochemical cell.

13-7 Corrosion in metals can often be understood as the operation of short-circuited electrochemical cells, in which stress or exposure to water and air creates cathodic and anodic regions in a single material and permits the oxidation and dissolution of the metal. Biological organisms play a major role in accelerating corrosion in some settings. The effects of corrosion are reduced through **passivation** (the use of oxidized surface layers) and the provision of **sacrificial anodes** that are corroded away in preference to the metal that is being protected.

PROBLEMS

Note: Answers to blue-numbered problems are given in Appendix F. Problems that are more challenging are indicated with asterisks.

The Gibbs Function and Cell Voltage

1. Electricity is usually sold by the kilowatt-hour (kW-h). If 1 kW-h of electrical energy costs 10 cents, compute the cost of leaving a 150-W porch light burning overnight (a period of 10 h).

2. (See Example 13–1.) A flashlight, in which two batteries generate an average current of 1.1 A across a voltage of 2.9 V, is left on for 10 h. How many kW-h of energy are expended? Use this answer and the answer to problem 1 to comment on the relative cost of producing electricity from a battery and from a power station, if the batteries cost $1.00 each and are essentially dead after the 10 h of use.

3. How long does it take a 1.34-V battery, from which a steady current of 0.800 A is drawn, to perform 1.00 J of electrical work in the surroundings?

4. A galvanic cell is operated extremely slowly, in order to extract the maximum theoretical electrical work. A current of 1.60×10^{-18} A (equal to 10.0 electrons per second) flows across a potential difference of 2.00 V. How long does it take to perform 1.00 J of electrical work in the surroundings?

5. (See Example 13–2.) Compute $\Delta \mathscr{E}°$ and $\Delta G°$ at 25°C for the chemical reaction taking place in each of the following electrochemical cells. Use data from Appendix E.
   (a) $Pb(s)|Pb^{2+}\|Cu^{2+}|Cu(s)$
   (b) $Pt(s)|Fe^{2+}, Fe^{3+}\|MnO_4^-, Mn^{2+}, H^+|Pt(s)$
   (c) $C(graphite)|I^-, I_2(s)\|Ag^+|Ag(s)$

6. (See Example 13–2.) Compute $\Delta \mathscr{E}°$ and $\Delta G°$ at 25°C for the chemical reaction taking place in each of the following electrochemical cells. Use data from Appendix E.
   (a) $Mn(s)|Mn^{2+}\|Cr^{2+}|Cr(s)$
   (b) $Pt(s)|Mn(s), Mn^{2+}\|Mn^{3+}, Mn^{2+}|Pt(s)$
   (c) $Ag(s)|Ag^+\|I_2(s), I^-|C(graphite)$

7. A $Ni|Ni^{2+}\|Ag^+|Ag$ galvanic cell is constructed in which the standard cell voltage is 1.03 V. Calculate the Gibbs function change at 25°C when 1.00 *gram* of silver plates out, if all concentrations remain at their standard-state values of 1 M throughout the process. What is the maximum electrical work done *by* the cell on its surroundings during this experiment?

8. A $Zn|Zn^{2+}\|Co^{2+}|Co$ galvanic cell is constructed in which the standard cell voltage is 0.48 V. Calculate the Gibbs function change at 25°C per gram of zinc lost at the anode, if all concentrations remain at their standard-state values of 1 M throughout the process. What is the maximum electrical work done *by* the cell on its surroundings during this experiment?

Half-Cell Potentials

9. (See Example 13–3.) A galvanic cell is constructed in which a $Br_2(\ell)|Br^-$ half-cell is connected to a $Co^{2+}|Co(s)$ half-cell.
   (a) By referring to Appendix E, write balanced chemical

equations for the half-reactions at the anode and the cathode and for the overall cell reaction.

(b) Calculate the cell voltage, assuming that all reactants and products are in standard states at 298.15 K.

10. (See Example 13–3.) A galvanic cell is constructed in which a $Pt|Fe^{2+}, Fe^{3+}$ half-cell is connected to a $Cd^{2+}|Cd$ half-cell.

(a) By referring to Appendix E, write balanced chemical equations for the half-reactions at the anode and the cathode and for the overall cell reaction.

(b) Calculate the cell voltage, assuming that all reactants and products are in standard states at 298.15 K.

11. In a galvanic cell, one half-cell consists of a zinc strip dipped into a 1.00 M solution of $Zn(NO_3)_2$. In the second half-cell, solid indium adsorbed on graphite is in contact with a 1.00 M solution of $In(NO_3)_3$. Indium is observed to plate out as the galvanic cell operates, and the initial cell voltage is measured to be 0.425 V at 25°C.

(a) Write balanced equations for the half-reaction at the anode and the half-reaction at the cathode.

(b) Calculate the standard reduction potential of an $In^{3+}|In$ half-cell. Consult Appendix E for the reduction potential of the $Zn^{2+}|Zn$ half-cell.

12. In a galvanic cell, one half-cell consists of gaseous chlorine bubbled over a platinum electrode at a pressure of 1.00 atm into a 1.00 M solution of NaCl. The second half-cell has a strip of solid gallium immersed in a 1.00 M $Ga(NO_3)_3$ solution. The initial cell voltage is measured to be 1.918 V at 25°C, and, as the cell operates, the concentration of chloride ion is observed to increase.

(a) Write balanced equations for the half-reaction at the anode and the half-reaction at the cathode.

(b) Calculate the standard reduction potential of a $Ga^{3+}|Ga$ half-cell. Consult Appendix E for the standard reduction potential of the $Cl_2|Cl^-$ electrode.

13. (See Example 13–4.) By referring to Appendix E, decide whether $Mn^{2+}(aq)$ in its standard state disproportionates spontaneously according to

$$3 \ Mn^{2+}(aq) \longrightarrow Mn(s) + 2 \ Mn^{3+}(aq)$$

Explain your answer.

14. (See Example 13–4.) The following reduction potentials have been measured for the oxidation states of thallium:

$$Tl^{3+} + e^- \longrightarrow Tl^{2+} \qquad \mathscr{E}° = -0.37 \text{ V}$$
$$Tl^{2+} + e^- \longrightarrow Tl^+ \qquad \mathscr{E}° = +2.87 \text{ V}$$

Does Tl^{2+} disproportionate spontaneously to Tl^{3+} and Tl^+? Explain.

15. One method to reduce the concentration of unwanted $Fe^{3+}(aq)$ ion in solutions of $Fe^{2+}(aq)$ is to drop a piece of metallic iron into the storage container. Write the reaction that removes the $Fe^{3+}(aq)$, and compute its standard potential difference.

16. By referring to problems 13 and 15, suggest a way to remove unwanted $Mn^{3+}(aq)$ ions from solutions of $Mn^{2+}(aq)$. Com-

pute the standard potential difference for the reaction proposed for this use.

Oxidizing and Reducing Agents

17. Would you expect powdered solid aluminum to act as an oxidizing agent or a reducing agent?

18. Would you expect sodium perchlorate ($NaClO_4(aq)$) in a concentrated acidic solution to act as an oxidizing agent or a reducing agent?

19. Bromine is sometimes used in place of chlorine as a disinfectant in swimming pools. If the effectiveness of a chemical as a disinfectant depends solely on its strength as an oxidizing agent, do you expect bromine to be better or worse than chlorine as a disinfectant, at a given concentration?

20. Many bleaches, including chlorine and its oxides, oxidize dye compounds in cloth. Predict which of the following is the strongest bleach at a given concentration and pH 0: $NaClO_3(aq)$, $NaClO(aq)$, $Cl_2(aq)$. How does the strongest chlorine-containing bleach compare in strength with ozone ($O_3(g)$)?

21. Find in Appendix E the standard potential $\mathscr{E}°$ for the reduction in acidic solution of the permanganate ion (MnO_4^-) to manganese(II) ion (Mn^{2+}). Use this potential and standard potentials from Appendix E to decide which of the following tend to be oxidized by MnO_4^- in acidic solution with the reactants and products in standard states at 298.15 K.

(a) $F^-(aq)$ (b) $Cl^-(aq)$ (c) $Cr^{3+}(aq)$ (d) $ClO_4^-(aq)$

22. The standard potential $\mathscr{E}°$ for the reduction in basic solution of platinum(II) hydroxide ($Pt(OH)_2(s)$) to elemental platinum ($Pt(s)$) is 0.16 V. Use this potential and standard potentials from Appendix E to decide which of the following tend to be oxidized by $Pt(OH)_2(s)$ in basic solution when all reactants and products are in standard states at 298.15 K.

(a) $Ni(s)$ (b) $OH^-(aq)$ (c) $Pb(s)$ (d) $F^-(aq)$

23. (See Example 13–5.) Figure 13–6 can be extended to include species in which the oxidation states of nitrogen, sulfur, and phosphorus are zero or negative. In basic solution, for example, the following additional reduction potentials are measured at pH 14:

$$P_4(s) + 12 \ H_2O(\ell) + 12 \ e^- \longrightarrow 4 \ PH_3(g) + 12 \ OH^-$$
$$\mathscr{E}° = -0.89 \text{ V}$$
$$4 \ H_2PO_2^- + 4 \ e^- \longrightarrow P_4(s) + 8 \ OH^- \qquad \mathscr{E}° = -2.05 \text{ V}$$

(a) Does $P_4(s)$ disproportionate spontaneously in basic solution (at pH 14)?

(b) Which is the stronger reducing agent at pH 14: $P_4(s)$ or $PH_3(g)$?

24. (See Example 13–5.) Extending Figure 13–6, the following additional reduction potentials are measured at pH 14:

$$S(s) + H_2O(\ell) + 2 \ e^- \longrightarrow$$
$$HS^- + OH^- \qquad \mathscr{E}° = -0.51 \text{ V}$$
$$S_2O_3^{2-} + 3 \ H_2O(\ell) + 4 \ e^- \longrightarrow$$
$$2 \ S(s) + 6 \ OH^- \qquad \mathscr{E}° = -0.74 \text{ V}$$

(a) Does sulfur disproportionate spontaneously under standard basic conditions?

(b) Which is the stronger reducing agent: $S(s)$ or $HS^-(aq)$?

Concentrations and the Nernst Equation

25. (See Example 13–6.) A galvanic cell is constructed that employs the reaction

$$Pb^{2+}(aq) + 2\ Cr^{2+}(aq) \longrightarrow Pb(s) + 2\ Cr^{3+}(aq)$$

If the initial concentration of $Pb^{2+}(aq)$ is 0.15 M, that of $Cr^{2+}(aq)$ is 0.20 M, and that of $Cr^{3+}(aq)$ is 0.0030 M, calculate the initial voltage generated by the cell at 25°C.

26. (See Example 13–6.) A galvanic cell is constructed that employs the reaction

$$2\ Ag(s) + Cl_2(g) \longrightarrow 2\ Ag^+(aq) + 2\ Cl^-(aq)$$

If the partial pressure of $Cl_2(g)$ is 1.00 atm, the initial concentration of $Ag^+(aq)$ is 0.25 M, and that of $Cl^-(aq)$ is 0.016 M, calculate the initial voltage generated by the cell at 25°C.

27. (See Example 13–6.) Use the Nernst equation to determine the potential difference at 25°C in the following cells when the dissolved ions have the concentrations indicated.

(a) $Pb(s)|Pb^{2+}(0.002\ \text{M})\|Cu^{2+}(0.075\ \text{M})|Cu(s)$

(b) $Pt(s)|Fe^{2+}(0.020\ \text{M}),\ Fe^{3+}(0.010\ \text{M})\|MnO_4^-(0.050\ \text{M}),$
$Mn^{2+}(0.95\ \text{M}),\ H^+(0.10\ \text{M})|Pt(s)$

(c) $C(graphite)|I^-(0.10\ \text{M}),\ I_2(s)\|Ag^+(0.10\ \text{M})|Ag(s)$

28. (See Example 13–6.) Use the Nernst equation to determine the potential difference at 25°C in the following cells when the dissolved ions have the concentrations indicated.

(a) $Mn(s)|Mn^{2+}(0.0020\ \text{M})\|Cr^{2+}(0.075\ \text{M})|Cr(s)$

(b) $Pt(s)|Mn(s),\ Mn^{2+}(0.010\ \text{M})\|Mn^{3+}(0.050\ \text{M}),$
$Mn^{2+}(0.95\ \text{M})|Pt(s)$

(c) $Ag(s)|Ag^+(0.10\ \text{M})\|I_2(s),\ I^-(0.075\ \text{M})|C(graphite)$

29. Calculate the reduction potential at 25°C for a $Pt|Cr^{3+}$, Cr^{2+} half-cell in which $[Cr^{3+}]$ is 0.15 M and $[Cr^{2+}]$ is 0.0019 M.

30. Calculate the reduction potential at 25°C for an $I_2(s)|I^-$ half-cell in which $[I^-]$ is 1.5×10^{-6} M.

31. If the half-cell from problem 29 is connected to a standard $Cd^{2+}|Cd$ half-cell to make a galvanic cell, which half-cell acts as the anode?

32. If the half-cell from problem 30 is connected to a standard $Pt|Fe^{3+}$, Fe^{2+} half-cell to make a galvanic cell, which half-cell acts as the cathode?

33. (See Example 13–7.) A standard $I_2(s)|I^-$ half-cell is connected to a $Pt|H^+|H_2(1\ \text{atm})$ half-cell in which the concentration of H^+ is unknown. The measured cell voltage is 0.841 V and the $I_2(s)|I^-$ half-cell is the cathode. What is the pH in the $PtH^+|H_2$ half-cell?

34. (See Example 13–7.) A standard $Cu^{2+}|Cu$ half-cell is connected to a $Br_2(aq)|Br^-$ half-cell in which the concentration of bromide ion is unknown. The measured cell voltage is 0.963 V and the Cu is the anode. What is the bromide ion concentration in the $Br_2(aq)|Br^-$ half-cell?

35. (See Example 13–7.) The following reaction occurs in an electrochemical cell:

$$3\ HClO_2(aq) + 2\ Cr^{3+}(aq) + 4\ H_2O)(\ell) \longrightarrow$$
$$3\ HClO(aq) + Cr_2O_7^{2-}(aq) + 8\ H^+(aq)$$

(a) Calculate $\Delta\mathscr{E}°$ for this cell.

(b) At pH 0, with $[Cr_2O_7^{2-}] = 0.80$ M $[HClO_2] = 0.15$ M, and $[HClO] = 0.20$ M, the cell voltage is found to equal 0.15 V. Calculate the concentration of $Cr^{3+}(aq)$ in the cell.

36. (See Example 13–7.) A galvanic cell is constructed in which the overall reaction is

$$Cr_2O_7^{2-}(aq) + 14\ H^+(aq) + 6\ I^-(aq) \longrightarrow$$
$$2\ Cr^{3+}(aq) + 3\ I_2(s) + 7\ H_2O(\ell)$$

(a) Calculate $\Delta\mathscr{E}°$ for this cell.

(b) At pH 0, with $[Cr_2O_7^{2-}] = 1.5$ M and $[I^-] = 0.40$ M the cell voltage is found to equal 0.870 V. Calculate the concentration of $Cr^{3+}(aq)$ in the cell.

Equilibrium Constants from Electrochemistry

37. A galvanic cell consists of a silver electrode dipping into a 0.10 M $AgNO_3$ solution connected through a salt bridge and an external circuit to a nickel electrode dipping into a 0.10 M $Ni(NO_3)_2$ solution.

(a) Write balanced chemical equations for the half-reaction occurring at the anode, the half-reaction occurring at the cathode, and the overall cell reaction.

(b) Calculate the initial cell voltage.

(c) Calculate the equilibrium constant for the overall reaction given in part (a).

38. A galvanic cell operating at 25°C is composed of two half-cells. The first consists of a Pt electrode immersed in a solution of MnO_4^- and Mn^{2+} ions at pH = 0, and the second of a Pt electrode immersed in a solution of Fe^{3+} and Fe^{2+} ions at pH = 0. The concentrations of all these ions are 0.010 M.

(a) Write balanced chemical equations for the half-reaction occurring at the anode, the half-reaction occurring at the cathode, and the overall cell reaction.

(b) Calculate the initial cell voltage.

(c) Calculate the equilibrium constant for the overall reaction given in part (a).

39. (See Example 13–8.) By using the reduction potentials in Appendix E, calculate the equilibrium constant at 25°C for the reaction in problem 35. Dichromate ion ($Cr_2O_7^{2-}$) is orange, and Cr^{3+} ion is light green in aqueous solution. If 2.00 L of 1.00 M $HClO_2$ solution is added to 2.00 L of 0.50 M $Cr(NO_3)_3$ solution, what color does the resulting solution have?

40. (See Example 13–8.) By using the reduction potentials in Appendix E, calculate the equilibrium constant at 25°C for the reaction

$$6 \text{ Hg}^{2+}(aq) + 2 \text{ Au}(s) \longrightarrow 3 \text{ Hg}_2^{2+}(aq) + 2 \text{ Au}^{3+}(aq)$$

If 1.00 L of a 1.00 M $\text{Au(NO}_3)_3$ solution is added to 1.00 L of a 1.00 M $\text{Hg}_2(\text{NO}_3)_2$ solution, calculate the concentrations of Hg^{2+}, Hg_2^{2+}, and Au^{3+} at equilibrium at 25°C.

41. The following standard reduction potentials at 25°C have been determined for the aqueous chemistry of indium:

$$\text{In}^{3+}(aq) + 2 \ e^- \longrightarrow \text{In}^+(aq) \qquad \mathscr{E}° = -0.40 \text{ V}$$
$$\text{In}^+(aq) + e^- \longrightarrow \text{In}(s) \qquad \mathscr{E}° = -0.21 \text{ V}$$

Calculate the equilibrium constant (K) for the disproportionation of $\text{In}^+(aq)$ at this temperature.

$$3 \text{ In}^+(aq) \longrightarrow \text{In}(s) + \text{In}^{3+}(aq)$$

42. (a) Use data from Appendix E to compute the equilibrium constant at 25°C for the reaction

$$\text{Hg}^{2+}(aq) + \text{Hg}(\ell) \rightleftharpoons \text{Hg}_2^{2+}(aq)$$

(b) Ammonia reacts with an aqueous solution of Hg_2Cl_2 to form the white, very insoluble compound HgNH_2Cl (see Section 9–6). Apply Le Chatelier's principle to the above equilibrium to explain why the addition of ammonia to such a solution also always produces black elemental mercury that mixes with the white HgNH_2Cl to give a gray precipitate.

43. A galvanic cell is made up of a standard $\text{Pt}|\text{H}^+|\text{H}_2(g)$ cathode connected to a nonstandard $\text{Pt}|\text{H}^+(aq)|\text{H}_2(g)$ anode in which the concentration of H^+ is unknown but is kept constant by the action of a buffer consisting of a weak acid HA (0.10 M) mixed with its conjugate base A^-(0.10 M). The measured cell voltage is $\Delta\mathscr{E} = 0.150$ V at 25°C, with a hydrogen pressure of 1.00 atm at both electrodes. Calculate the pH in the buffer solution, and from it, determine the K_a of the weak acid. (*Hint:* Because the same species are taking part in the reactions at anode and cathode, the standard voltage $\Delta\mathscr{E}°$ is 0.)

44. In a galvanic cell, the reduction occurs in a standard $\text{Ag}^+|\text{Ag}$ half-cell and oxidation occurs at a platinum wire, with hydrogen bubbling over it at 1.00-atm pressure, that is immersed in a buffer solution containing benzoic acid and sodium benzoate. The concentration of benzoic acid ($\text{C}_6\text{H}_5\text{COOH}$) is 0.10 M, and that of benzoate ion ($\text{C}_6\text{H}_5\text{COO}^-$) is 0.05 M. The overall cell reaction is then

$$\text{Ag}^+(aq) + \tfrac{1}{2} \text{ H}_2(g) \longrightarrow \text{Ag}(s) + \text{H}^+(aq)$$

and the measured cell voltage is 1.030 V. Calculate the pH in the buffer solution and determine the K_a of benzoic acid.

45. (See Example 13–9.) A galvanic cell is constructed in which the overall reaction is

$$\text{Br}_2(aq) + \text{H}_2(g) \longrightarrow 2 \text{ Br}^-(aq) + 2 \text{ H}^+(aq)$$

(a) Calculate $\Delta\mathscr{E}°$ for this cell at 25°C.
(b) Silver ions are added until AgBr precipitates in the cathode compartment and $[\text{Ag}^+]$ reaches 0.060 M. At this point, the cell voltage is 1.710 V, the pH is 0, and $P_{\text{H}_2} = 1.0$ atm. Calculate $[\text{Br}^-]$ under these conditions.
(c) Calculate the solubility-product constant K_{sp} for AgBr.

46. (See Example 13–9.) A galvanic cell is constructed in which the overall reaction is

$$\text{Pb}(s) + 2 \text{ H}^+(aq) \longrightarrow \text{Pb}^{2+}(aq) + \text{H}_2(g)$$

(a) Calculate $\Delta\mathscr{E}°$ for this cell at 25°C.
(b) Chloride ions are added until PbCl_2 precipitates in the anode compartment and $[\text{Cl}^-]$ reaches 0.15 M. At that point, the cell voltage is 0.22 V, the pH is 0, and $P_{\text{H}_2} = 1.0$ atm. Calculate $[\text{Pb}^{2+}]$ under these conditions.
(c) Calculate the solubility-product constant K_{sp} of PbCl_2.

Batteries and Fuel Cells

47. Calculate the standard voltage $\Delta\mathscr{E}°$ of a lead–acid cell at 25°C. What is the voltage if six such cells are connected in series?

48. Calculate the standard voltage of the zinc–mercuric oxide cell shown in Figure 13–13. Take $\Delta G_f°(\text{Zn(OH)}_2(s)) = -553.5$ kJ mol^{-1}. (*Hint:* The easiest way to proceed is to calculate $\Delta G°$ for the corresponding overall reaction, and then to find $\Delta\mathscr{E}°$ from it.)

49. What quantity of charge (in coulombs) is a fully charged 2.04-V lead–acid cell theoretically capable of furnishing if the spongy lead available for reaction at the anodes weighs 10 kg and there is excess PbO_2?

50. What quantity of charge (in coulombs) is a fully charged 1.34-V zinc–mercuric oxide watch battery theoretically capable of furnishing if the mass of HgO in the battery is 0.50 g?

51. What is the theoretical maximum amount of work (in joules) that can be obtained from the lead–acid storage cell in problem 49?

52. What is the theoretical maximum amount of work (in joules) that can be obtained from the watch battery in problem 50?

53. The concentration of the electrolyte, sulfuric acid, in a lead–acid storage battery diminishes as it is discharged. Is a discharged battery recharged by replacing the dilute H_2SO_4 with fresh, concentrated H_2SO_4? Explain.

54. One cold winter morning the temperature is well below 0°F. In trying to start your car, you run the battery down completely. Several hours later, you return to replace your fouled spark plugs and discover that the liquid in the battery has now frozen even through the temperature is actually a bit higher than it was in the morning. Explain how this can happen.

55. Consider the fuel cell that accomplishes the overall reaction

$$\text{H}_2(g) + \tfrac{1}{2} \text{ O}_2(g) \longrightarrow \text{H}_2\text{O}(\ell)$$

If the fuel cell operates with 60% efficiency, calculate the amount of electrical work generated per gram of water produced. The gas pressures are constant at 1 atm and the temperature is 25°C.

56. Consider the fuel cell that accomplishes the overall reaction

$$CO(g) + \tfrac{1}{2} O_2(g) \longrightarrow CO_2(g)$$

Calculate the maximum electrical work that could be obtained from the conversion of 1.00 mol of $CO(g)$ to $CO_2(g)$ in such a fuel cell operated with 100% efficiency at 25°C and with the pressure of each gas equal to 1 atm. (*Hint*: In the absence of half-cell voltages from Appendix E, use Appendix D to calculate $\Delta G°$ of the reaction and relate that to the work.)

Corrosion and Its Prevention

57. Two half-reactions proposed for the corrosion of iron in the absence of oxygen are

$$Fe(s) \longrightarrow Fe^{2+}(aq) + 2\ e^-$$
$$2\ H_2O(\ell) + 2\ e^- \longrightarrow 2\ OH^-(aq) + H_2(g)$$

Calculate the standard voltage generated by a galvanic cell running this pair of half-reactions. Is the overall reaction spontaneous under standard conditions? As the pH falls from 14, does the reaction become spontaneous?

58. In the presence of oxygen, the cathode half-reaction written in the preceding problem is replaced by

$$\tfrac{1}{2} O_2(g) + 2\ H^+(aq) + 2\ e^- \longrightarrow H_2O(\ell)$$

but the anode half-reaction is unchanged. Calculate the standard cell voltage for *this* pair of half-reactions running in a galvanic cell. Is the overall reaction spontaneous when the reactants and products are present in standard states at 25°C? As the water becomes more acidic, does the driving force for the rusting of iron increase or decrease?

59. Could sodium be used as a sacrificial anode to protect the iron hull of a ship? Explain.

60. If it is shown that titanium can be used as a sacrificial anode to protect iron, what conclusion can be drawn about the standard reduction potential of its half-reaction?

$$Ti^{3+}(aq) + 3\ e^- \longrightarrow Ti(s)$$

Additional Problems

61. Estimate the cost of the electrical energy needed to produce 1.9×10^{10} kg (a year's supply for the world) of aluminum from $Al_2O_3(s)$ if electrical energy costs 10 cents per kilowatt-hour (1 kW-h = 3.6 MJ = 3.6×10^6 J) and if the cell voltage is 5 V.

*62. Sheet iron can be galvanized by passing a direct current through a cell containing a solution of zinc sulfate between a graphite anode and the iron sheet. Zinc plates out on the iron. The process can be made continuous if the iron sheet

is a coil that unwinds as it passes through the electrolysis cell and coils up again after it emerges from a rinse bath. Calculate the cost of the electricity needed to deposit a 0.250-mm-thick layer of zinc on both sides of an iron sheet that is 1.00 m wide and 100 m long, if a current of 25 A at a voltage of 3.5 V is used and the energy efficiency of the process is 90%. The cost of electricity is 10 cents per kilowatt-hour (1 kW-h = 3.6 MJ), and the density of zinc is 7.133 g cm^{-3}.

63. A half-cell has a graphite electrode immersed in an acidic solution (pH 0) of Mn^{2+} (concentration 1.00 M) in contact with solid MnO_2. A second half-cell has an acidic solution (pH 0) of H_2O_2 (concentration 1.00 M) in contact with a platinum electrode past which gaseous oxygen at a pressure of 1.00 atm is bubbled. The two half-cells are connected to form a galvanic cell.
(a) By referring to Appendix E, write balanced chemical equations for the half-reactions at the anode and the cathode, and for the overall cell reaction.
(b) Calculate the cell voltage.

64. (a) Based only on the standard reduction potentials for the $Cu^{2+}|Cu^+$ and the $I_2(s)|I^-$ half-reactions, would you expect $Cu^{2+}(aq)$ to be reduced to $Cu^+(aq)$ by $I^-(aq)$?
(b) The formation of solid CuI plays a role in the interaction between $Cu^{2+}(aq)$ and $I^-(aq)$.

$$Cu^{2+}(aq) + I^-(aq) + e^- \longrightarrow CuI(s) \quad \mathscr{E}° = 0.86\ V$$

Taking into account this added information, do you expect $Cu^{2+}(aq)$ to be reduced by aqueous iodide ion?

65. Calculate the minimum external voltage that must be applied to produce $Ni(s)$ and $I_2(s)$ in an electrolytic cell that is 1.00 M in $NiSO_4$ and 1.00 M in NaI. Write balanced chemical equations for the half-reactions that take place at the anode and at the cathode.

*66. In an old European church, the stained-glass windows have experienced such extreme darkening from corrosion that hardly any light comes through. A microprobe analysis shows that tiny cracks and defects on the glass surface are enriched in insoluble Mn(III) and Mn(IV) compounds. From Appendix E, suggest a reducing agent and conditions that might successfully convert these compounds to soluble Mn(II) without simultaneously reducing Fe(III), which colors the glass, to Fe(II). Take MnO_2 as representative of the insoluble Mn(III) and Mn(IV) compounds.

67. By referring to the half-equations in problems 23 and 24, decide whether $PH_3(g)$ or $HS^-(aq)$ is the stronger reducing agent in basic aqueous solution.

*68. (a) Calculate the half-cell potential for the reaction

$$O_2(g) + 4\ H^+(aq) + 4\ e^- \longrightarrow 2\ H_2O(\ell)$$

at pH 7 with the oxygen pressure at 1 atm.
(b) Explain why aeration of solutions of I^- leads to oxidation of the I^- ion. Write a balanced equation for the redox reaction that occurs.

(c) Does the same problem arise with solutions containing Br^- or Cl^-? Explain.

(d) Is oxidation of the halide ions favored or opposed by increasing acidity?

69. An engineer needs to prepare a galvanic cell that uses the reaction

$$2 Ag^+(aq) + Zn(s) \longrightarrow Zn^{2+}(aq) + 2 Ag(s)$$

and generates an initial voltage of 1.50 V. She has 0.010 M $AgNO_3(aq)$ and 0.100 M $Zn(NO_3)_2(aq)$ solutions as well as electrodes of zinc and silver, wires, containers, water, and a KNO_3 salt bridge. Sketch the cell. Clearly indicate the concentrations of all solutions.

70. Consider a galvanic cell for which the anode reaction is

$$Pb(s) \longrightarrow Pb^{2+}(1.0 \times 10^{-2} \text{ M}) + 2 e^-$$

and the cathode reaction is

$$VO^{2+}(0.10 \text{ M}) + 2 H^+(0.10 \text{ M}) + e^- \longrightarrow V^{3+}(1.0 \times 10^{-5} \text{ M}) + H_2O(\ell)$$

The measured cell voltage is 0.640 V.

(a) Calculate $\mathscr{E}°$ for the $VO^{2+}|V^{3+}$ half-reaction, using $\mathscr{E}°$ $(Pb^{2+}|Pb)$ from Appendix E.

(b) Calculate the equilibrium constant (K) at 25°C for the reaction

$$Pb(s) + 2 VO^{2+}(aq) + 4 H^+(aq) \longrightarrow Pb^{2+}(aq) + 2 V^{3+}(aq) + 2 H_2O(\ell)$$

71. A galvanic cell operating at 25°C is made up of two half-cells. The first half-cell contains a Pb rod immersed in a 0.10 M $Pb(NO_3)_2$ solution. The second half-cell contains a Pb rod immersed in a 0.010 M $Pb(NO_3)_2$ solution. A salt bridge connects the two half-cells.

(a) Calculate the cell voltage generated by the cell at 25°C.

(b) Concentrated H_2SO_4 is added to the first half-cell until the equilibrium concentration of SO_4^{2-} is 0.10 M. (Lead sulfate is insoluble and precipitates: $K_{sp}(25°C) = 1.1 \times 10^{-8}$.) Calculate the cell voltage generated by the cell at room temperature, after the addition of the H_2SO_4.

72. Write a balanced equation for the reduction of Cu^{2+} by I^- ion, based on the discussion in problem 64. Calculate the equilibrium constant for this reaction at 25°C.

73. A wire is fastened across the terminals of the Leclanché cell in Figure 13–12. Indicate in which direction the electrons flow in the wire.

74. The zinc–silver cell is used in miniature "button" batteries for a variety of specialized uses. The cell is represented

$$Zn(s)|ZnO(s)|KOH(satd)|Ag_2O(s)|Ag(s)$$

The KOH electrolyte is held on an absorbent disc between the two electrodes (as shown in Figure 13–13); hence, no salt bridge is needed in this cell.

(a) Write half-equations for the oxidation and reduction taking place in this cell.

(b) What is the overall reaction taking place in this cell?

(c) The $\Delta G_f°$ of $ZnO(s)$ is -318.32 kJ mol^{-1}, and the $\Delta G_f°$ of $Ag_2O(s)$ is -10.84 kJ mol^{-1} at 25°C. Calculate $\Delta\mathscr{E}°$ for the zinc–silver cell at this temperature.

75. As a junior executive trainee in a large battery-making company, you are assigned to lead a group of important visitors on a tour of the factory. After seeing every stage of the fabrication process, one visitor asks, "Why is there no provision for charging up the batteries before they are shipped? When do you put the electricity into them?" Write a brief statement explaining to the visitor exactly why such a procedure is unnecessary and telling where the energy in your firm's batteries comes from.

76. Overcharging a lead–acid storage battery can generate hydrogen. Write a balanced equation to represent the reaction taking place.

77. The rights to manufacture a new fuel cell in which hydrazine is oxidized to nitrogen and water according to the equation

$$N_2H_4(aq) + O_2(g) \longrightarrow N_2(g) + 2 H_2O(\ell)$$

are for sale. The sellers point out that the products of the reaction are completely innocuous in the environment. Your firm asks you to evaluate this new invention. As a start, check whether the cell is thermodynamically feasible. Do this by using data from Appendix D to compute its $\Delta\mathscr{E}°$.

*78. The following reaction is carried out in a fuel cell at 1200 K:

$$CO(g) + \tfrac{1}{2} O_2(g) \longrightarrow CO_2(g)$$

The electrolyte is molten Na_2CO_3, and the reactant and product gases are all at 1.0 atm pressure. What is the change in the standard Gibbs function ($\Delta G°$) for the reaction at this temperature (use Appendix D, and assume $\Delta H°$ and $\Delta S°$ are independent of temperature)? What voltage should the fuel cell produce?

79. Iron or steel is often covered by a thin layer of a second metal to prevent rusting: tins cans consist of steel covered with tin, and galvanized iron is made by coating iron with a layer of zinc. If the protective layer is broken, however, iron rusts more readily in a tin can than in galvanized iron. Explain this observation by comparing the reduction potentials of iron, tin, and zinc.

80. The Bobay process is used to produce $Cl_2(g)$ by the direct electrolysis of HCl dissolved in water. This reaction requires a potential difference of 1.4 V when the concentration of the HCl is about 1 M. A new process replaces water as the solvent with a molten mixture of KCl (59 mol %) and LiCl (41 mol %) at 400°C. Hydrogen chloride dissolves without dissociation in this mixture, and applying a potential difference of about 1 V generates gaseous Cl_2 and H_2 at the electrodes.

(a) Write half-equations for the chemical events at the anode and at the cathode in both the original and the new process. Take care to show the state of dissolution of each reactant and product.

(b) Estimate the energy savings per mole of Cl_2.

(c) A savings in energy is one reason the new process is deemed an improvement, despite requiring an elevated temperature. Suggest at least one other reason.

(d) Hydrogen chloride is soluble in both molten lithium chloride ($LiCl(\ell)$) and potassium chloride ($KCl(\ell)$). Explain why a mixture is used.

CUMULATIVE PROBLEM

Manganese

Manganese is the 12th most abundant element on the earth's surface. Its most important ore is pyrolusite (MnO_2). The preparation and uses of manganese and its compounds are intimately bound up with electrochemistry.

(a) Elemental manganese in a state of high purity can be prepared by electrolyzing aqueous solutions of Mn^{2+}. At which electrode (anode or cathode) does the manganese appear? Electrolysis is also used to make MnO_2 in high purity from Mn^{2+} solutions. At which electrode does the MnO_2 appear?

(b) The Winkler method is an analytical procedure for determining the amount of oxygen dissolved in water. In the first step, $Mn(OH)_2(s)$ is oxidized by dissolved oxygen to $Mn(OH)_3(s)$ in basic aqueous solution. Write the oxidation and reduction half-equations for this step, and write the balanced overall equation. Then use Appendix E to calculate the standard voltage that would be measured if this reaction were carried out in an electrochemical cell at 25°C.

Black crystals of manganite, MnOOH, an ore of manganese.

(c) Calculate the equilibrium constant at 25°C for the reaction in part (b).

(d) In the second step of the Winkler method, the $Mn(OH)_3$ is acidified to give Mn^{3+}, and iodide ion is added. Does Mn^{3+} spontaneously oxidize I^-? Write a balanced equation for its reaction with I^-, and use data from Appendix E to calculate its equilibrium constant at 25°C. Titration of the I_2 produced completes the use of the Winkler method.

(e) Manganese(IV) oxide is an even stronger oxidizing agent than $Mn(OH)_3$. It oxidizes zinc in the "dry cell" or flashlight battery found in every home. Such a battery has a cell voltage of 1.5 V. Calculate the electrical work done by a dry cell in 1.00 h if it produces a steady current of 0.70 A at 1.5 V. Express your answer in joules and in kilowatt-hours.

(f) The reduction of permanganate ion (in which Mn has the $+7$ oxidation state) in acidic aqueous solution is represented

$$MnO_4^-(aq) + 8\ H^+(aq) + 5\ e^- \longrightarrow Mn^{2+}(aq) + 4\ H_2O(\ell) \quad \mathscr{E}° = 1.491\ V$$

The reduction of the analogous fifth period species, pertechnetate ion, follows a similar course:

$$TcO_4^-(aq) + 8\ H^+(aq) + 5\ e^- \longrightarrow Tc^{2+}(aq) + 4\ H_2O(\ell) \quad \mathscr{E}° = 0.500\ V$$

Which is the stronger oxidizing agent: permanganate ion or pertechnetate ion?

(g) A galvanic cell is made from two half-cells. In the first, a platinum electrode is immersed in a solution at pH 2.00 that is 0.100 M in both MnO_4^- and Mn^{2+}. In the second, a zinc electrode is immersed in a 0.0100 M solution of $Zn(NO_3)_2$. Calculate the theoretical cell voltage at 25°C.

14 Chemical Kinetics

Magnesium ribbon burning in air.

Why do some chemical reactions proceed with lightning-like speed, whereas others require days, months, or years to produce a detectable amount of product? How do catalysts increase the rates of chemical reactions? What does a study of the rate of a chemical reaction tell us about the way in which reactants combine to form products? These questions are the province of chemical kinetics.

Chemical kinetics is the study of how rapidly reactions proceed and of the detailed events along the way as reactants are transformed to products. Although thermodynamics can accurately predict the position of the equilibrium in a chemical reaction, it reveals nothing about the rate at which that equilibrium is achieved. The speeds of chemical reactions range from explosive rapidity to glacial sluggishness. A mixture of hydrogen and oxygen at room temperature and pressure exists unchanged almost indefinitely in a closed container, despite the inherent tendency of the two gases to react to form water. The introduction of a spark or a tiny piece of platinum then sets off the reaction, which goes explosively. It is of enormous practical importance to know the rates of reactions and how conditions affect them. In the chemical industry we want fast reactions, but we need them under control. In atmospheric chemistry, we urgently seek to understand the rates at which ozone-depleting species form, react, and are converted into other species.

Closely linked to the practical knowledge of which reactions go rapidly and which go slowly under different conditions is the study of the **mechanisms** of reactions. These are the detailed pathways by which the reactants combine. For example, the redox reaction

$$5\ Fe^{2+}(aq) + MnO_4^-(aq) + 8\ H^+(aq) \longrightarrow 5\ Fe^{3+}(aq) + Mn^{2+}(aq) + 4\ H_2O(\ell)$$

certainly does *not* involve the simultaneous collision of five $Fe^{2+}(aq)$ ions and one $MnO_4^-(aq)$ ion with eight $H^+(aq)$ ions because such a collision would be exceedingly rare. Instead, it proceeds through a series of elementary steps, each of which involves a collision of two or, at most, three species. Early steps in the series give intermediates that later react further to generate the final products. A major goal of chemical kinetics is to deduce the mechanisms of reactions from experimental knowledge of their rates under different conditions.

14-1 RATES OF CHEMICAL REACTIONS

The speed, or rate, of a reaction depends on many factors. Increased concentrations of reactants often speed up a reaction (Fig. 14–1) but in another case may have no effect or may even slow it down. Higher concentrations of *nonreactants* (including reaction products) also sometimes increase reaction rates, sometimes decrease them, and sometimes leave them unchanged. Reaction rates in general are strongly affected by changes in temperature, a fact that makes careful monitoring and control of the temperature critical for quantitative measurements in chemical kinetics. Finally, the physical form of the reactants is often crucial to the observed rate. An iron nail oxidizes only very slowly in dry air to iron oxide, but steel wool burns spectacularly in oxygen (Fig. 14–2). The quantitative study of such heterogeneous reactions (those involving two or more phases, such as a solid and a gas) is difficult, so we begin with homogeneous reactions (those that take place entirely within the gas phase or solution). In Section 14–7, we return briefly to some important aspects of heterogeneous reactions.

Figure 14–1 The rate of reaction of zinc with aqueous sulfuric acid depends on the concentration of the acid. The dilute solution reacts slowly (*left*), and the more concentrated solution reacts rapidly (*right*).

Figure 14–2 Steel wool burning in oxygen.

Measuring Reaction Rates

A kinetics experiment measures the rate at which the concentration of a substance taking part in a chemical reaction changes with time. There are many ways to monitor a changing concentration, and there is much scope for ingenuity in inventing new ones. If the reaction is slow enough, we can allow it to proceed for a measured length of time and then abruptly stop it ("quench" it) by cooling the reaction mixture rapidly to such a low temperature that the reaction effectively stops. At that low temperature, leisurely chemical analysis for the concentration of a particular reactant or product is possible. This procedure is not useful for rapid reactions, especially those involving gas mixtures, which are difficult to cool quickly. An alternative is to use the absorption of light as a probe. As we shall see in Chapters 17 and 18, substances differ in the wavelengths of light they absorb. If a wavelength is absorbed by only one particular reactant or product, then measurements of the amount of light absorbed at that wavelength track the concentration of the absorbing species as time progresses.

Determining the average rate of a reaction closely resembles determining the average speed of a traveling car. On a road trip

$$\text{average speed} = \frac{\text{distance traveled}}{\text{elapsed time}} = \frac{\text{change in location}}{\text{change in time}}$$

The average speed has units of miles per hour, meters per second, or any other convenient combination of distance units divided by time units. The **average reaction rate** is obtained by dividing the change in concentration of a reactant or product by the time interval over which the change occurs:

$$\text{average reaction rate} = \frac{\text{change in concentration}}{\text{change in time}}$$

If concentration is measured in mol L^{-1} and time in seconds, then the rate of a reaction has units of

$$\frac{\text{mol } L^{-1}}{s} = \frac{\text{mol}}{L\ s} = \text{mol } L^{-1}\ s^{-1}$$

Consider a specific example. In the gas-phase reaction

$$NO_2(g) + CO(g) \longrightarrow NO(g) + CO_2(g)$$

$NO_2(g)$ and $CO(g)$ are consumed as $NO(g)$ and $CO_2(g)$ are produced. Suppose that a probe that measures the concentration of $NO(g)$ is used. The average rate of the reaction over any interval of time then equals the change in the concentration of $NO(g)$ (symbolized $\Delta[NO]$) divided by the duration of the time interval (Δt):

$$\text{average rate} = \frac{[NO]_{final} - [NO]_{initial}}{t_{final} - t_{initial}} = \frac{\Delta[NO]}{\Delta t}$$

This average rate depends on which time interval (Δt) is selected, because the rate at which NO is produced changes with time, just as the average speed of a car on a cross-country trip certainly changes with time. Results like these are obtained when NO_2 and CO are mixed and the concentration of NO is monitored over time:

Time (s)	[NO] (mol L^{-1})
0	0
50	0.0160
100	0.0240
150	0.0288
200	0.0320

The average rate of reaction during the first 50 s is

$$\text{average rate} = \frac{\Delta[NO]}{\Delta t} = \frac{(0.0160 - 0)\,\dfrac{mol}{L}}{(50 - 0)\,s} = 3.2 \times 10^{-4}\,\frac{mol}{L\,s}$$

The average rate during the second 50 s is 1.6×10^{-4} mol L^{-1} s^{-1}, and the average rate during the third 50 s is 9.6×10^{-5} mol L^{-1} s^{-1}. Clearly, this reaction slows down as it progresses. A graphical method for determining average rates is shown in Figure 14–3. The average rate is the slope of the straight (green) line connecting the concentrations at the initial and final points of a time interval.

The **instantaneous rate** of a reaction is its rate at a particular moment in its course. An instantaneous rate is found by taking the average rate over shorter and shorter time intervals (smaller Δt's) running from just before to just after the instant of time (t) at which we wish to know the rate, and following the trend to an infinitely brief Δt. Graphically, this is done by drawing a line tangent to the curve at time t (see the red line in Fig. 14–3). The slope of the tangent line is the instantaneous rate at that moment. At time $t = 150$ s, for example, the slope of the tangent line from the figure is

$$\text{instantaneous rate} = \frac{(0.0326 - 0.0249)\,mol\,L^{-1}}{(200 - 100)\,s} = 7.7 \times 10^{-5}\,mol\,L^{-1}\,s^{-1}$$

The concentrations used here were read from the graph from the points where the red line intersects the $t = 100$ and $t = 200$ vertical lines; any pair of points on the line could have been used (see Appendix C).

Chemical kinetics uses mainly instantaneous rates. Therefore, the instantaneous rate is referred to from now on simply as the *rate*. The rate of a reaction at the mo-

• A traveler may drive 480 mi in 8 h, for an average speed of 60 miles per hour (mph). An 8-h stop then lowers the average speed over the 16-h trip to 30 mph. Another 480 mi in 8 h raises the average speed on the trip to 40 mph (960 mi in 24 h).

• Most reactions slow down as they progress, but such behavior is by no means a rule. Numerous reactions start off slowly and pick up speed as they go along.

• A tangent line is a straight line that touches a curve at one point without crossing it.

Figure 14–3 The blue curve is the concentration of NO against time in the reaction $NO_2 + CO \longrightarrow NO + CO_2$. The slope of the green line gives the *average rate* of the reaction over the time interval from 50 to 150 s. It equals the change in NO concentration divided by the duration of the interval, as calculated in the green box. The red line is tangent to the blue curve at $t = 150$ s. Its slope, which is calculated in the red box, is the *instantaneous rate* 150 s after the start of the reaction. An instantaneous rate exists at any moment in the progress of the reaction; the slope of the brown line gives the *initial rate*, which is the rate at the instant the reaction starts.

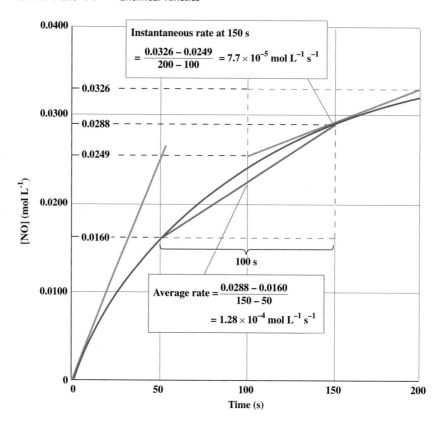

ment that it begins (at $t = 0$) is often of special interest. It is the (instantaneous) **initial rate** of the reaction.

The rate of this example reaction could just as well have been measured by monitoring changes in the concentrations of $CO_2(g)$, $NO_2(g)$, or $CO(g)$, instead of changes in the concentration of $NO(g)$. Because every molecule of NO produced is accompanied by one molecule of CO_2, the rate of increase of the concentration of CO_2 must equal that of the NO. The concentrations of the two reactants, NO_2 and CO, *decrease* at the same rate that the concentrations of the products increase because their coefficients in the balanced equation are also both equal to 1. This is summarized as

$$\text{rate} = -\frac{\Delta[NO_2]}{\Delta t} = -\frac{\Delta[CO]}{\Delta t} = \frac{\Delta[NO]}{\Delta t} = \frac{\Delta[CO_2]}{\Delta t}$$

EXAMPLE 14–1

Using data from the table on page 569 and Figure 14–3, calculate the following:
(a) The average rate of disappearance of CO between $t = 100$ and $t = 200$ s.
(b) The instantaneous rate of appearance of NO at $t = 0$ (the initial rate of the reaction).

Solution

(a) The average rate of the *appearance* of NO between 100 and 200 s is

$$\text{average rate} = \frac{\Delta[NO]}{\Delta t} = \frac{(0.0320 - 0.0240) \text{ mol L}^{-1}}{(200 - 100) \text{ s}}$$
$$= 8.0 \times 10^{-5} \text{ mol L}^{-1} \text{ s}^{-1}$$

The rate of *disappearance* of CO over this time interval must be equal to the rate of appearance of NO and is therefore 8.0×10^{-5} mol L^{-1} s^{-1}.
(b) The initial rate equals the slope of the line tangent to the blue curve in Figure 14–3 at $t = 0$. This line is drawn in brown in the figure. Using the points at which it intersects the $t = 0$ and $t = 50$ vertical lines to compute its slope gives

$$\text{initial rate} = \frac{(0.0240 - 0) \text{ mol } L^{-1}}{(50 - 0) \text{ s}} = 4.8 \times 10^{-4} \frac{\text{mol}}{\text{L s}}$$

Exercise
(a) Using the same data, calculate the average rate of appearance of CO_2 between $t = 50$ s and $t = 150$ s. (b) Graphically estimate the instantaneous rate of the reaction at $t = 50$ s.

Answer: (a) 1.3×10^{-4} mol L^{-1} s^{-1}. (b) 2.1×10^{-4} mol L^{-1} s^{-1}.

Another gas-phase reaction is

$$2 NO_2(g) + F_2(g) \longrightarrow 2 NO_2F(g)$$

This equation states that two moles of NO_2 disappear and two of NO_2F appear for each mole of F_2 that reacts. Thus, the concentration of NO_2 changes twice as fast as the concentration of F_2. The concentration of NO_2F also changes twice as fast as that of F_2, and the change has the opposite sign. We can write the rate in this case as

$$\text{rate} = -\left(\frac{1}{2}\right)\frac{\Delta[NO_2]}{\Delta t} = -\frac{\Delta[F_2]}{\Delta t} = \left(\frac{1}{2}\right)\frac{\Delta[NO_2F]}{\Delta t}$$

In this equation, the rate of change of the concentration of each species is divided by its coefficient in the balanced chemical equation. Rates of change of reactants appear with negative signs, and those of products with positive signs. If these conventions are used, the rate of a reaction comes out the same regardless of which reactant or product is being monitored. For the general reaction

$$a\text{A} + b\text{B} \longrightarrow c\text{C} + d\text{D}$$

the rate is

$$\text{rate} = -\frac{1}{a}\frac{\Delta[\text{A}]}{\Delta t} = -\frac{1}{b}\frac{\Delta[\text{B}]}{\Delta t} = \frac{1}{c}\frac{\Delta[\text{C}]}{\Delta t} = \frac{1}{d}\frac{\Delta[\text{D}]}{\Delta t}$$

These relations hold true provided that there are no transient intermediate species or, if there are intermediates, their concentrations are independent of time for most of the reaction period.

14-2   REACTION RATES AND CONCENTRATIONS

In discussing chemical equilibrium, we stressed that both forward and reverse reactions can occur: once products are formed, they can react back to give the original reactants. The **net rate** is the difference between the forward and reverse rates:

$$\text{net rate} = \text{forward rate} - \text{reverse rate}$$

Strictly speaking, measurements of concentration versus time give the net rate rather than simply the forward rate. Near the beginning of a reaction that starts from pure reactants, however, the concentrations of reactants are far higher than those of products, and the reverse rate can be neglected. Furthermore, many reactions go to "completion" (have very large equilibrium constants) and so have a measurable rate only in the forward direction, or else the experiment can be arranged in such a way that the products are removed as fast as they are formed. In this section, the focus is on forward rates exclusively.

Order of a Reaction

• The forward rate may depend on the concentrations of nonreactants as well.

The forward rate of a homogeneous chemical reaction nearly always depends on the concentrations of one (or more) of the reactants. As an example, consider the decomposition of dinitrogen pentaoxide (N_2O_5). This compound is a white solid that is stable below 0°C but decomposes when vaporized:

$$N_2O_5(g) \longrightarrow 2\,NO_2(g) + \tfrac{1}{2}\,O_2(g)$$

The rate of the reaction depends on the concentration of $N_2O_5(g)$. In fact, as Figure 14–4 shows, a graph of the rate versus the concentration is a straight line that passes through the origin when extrapolated to zero concentration. The equation of the straight line is

$$\text{rate} = k\,[N_2O_5]$$

• The distinction between the *rate* of a reaction and the *rate constant* for that reaction is important.

Such a relation between the rate of a reaction and concentration is called a **rate expression,** or **rate law,** and the proportionality constant k is called the **rate constant** for the reaction. Rate constants, like equilibrium constants, are independent of concentration but depend strongly on temperature.

For many (but not all) reactions with a single reactant, the rate is proportional to the concentration of that reactant raised to a power; that is, the rate expression for

$$a\,A \longrightarrow \text{products}$$

frequently has the form

$$\text{rate} = k\,[A]^n$$

The power n in such a rate expression is *not*, in general, equal to the coefficient a in the balanced chemical equation. For example, for the decomposition of ethane at high temperatures and low pressures

$$C_2H_6(g) \longrightarrow 2\,CH_3(g)$$

the rate expression has the form

$$\text{rate} = k\,[C_2H_6]^2$$

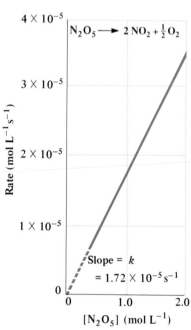

so that $n = 2$ even though the coefficient of C_2H_6 in the chemical equation is 1.

The power to which the concentration is raised is called the **order** of the reaction with respect to that reactant. Thus, the decomposition of N_2O_5 is **first order,** whereas that of C_2H_6 is **second order.** Some processes are **zeroth order** over a range of concentrations. Because $[A]^0 = 1$, such reactions have rates that are independent of concentration.

$$\text{rate} = k \quad \text{(zeroth-order kinetics)}$$

Figure 14–4 The rate of decomposition of $N_2O_5(g)$ at 25°C is proportional to its concentration. The slope of this line is equal to the rate constant k for the reaction.

The order of a reaction does not have to be an integer; fractional powers are sometimes found. At 450 K, the decomposition of acetaldehyde (CH_3CHO) is described by the rate expression

$$\text{rate} = k\,[CH_3CHO]^{3/2}$$

The lesson from all this is that reaction order is an experimental property that cannot be predicted from the form of the chemical equation.

The following example illustrates the way that the order of a reaction with respect to a reactant is deduced from experimental data.

EXAMPLE 14-2

At elevated temperatures, HI reacts according to the chemical equation

$$2\,HI(g) \longrightarrow H_2(g) + I_2(g)$$

At 443°C, the rate of the reaction increases with concentration of HI, as shown in the following table:

Data Point	1	2	3
[HI] (mol L^{-1})	0.0050	0.010	0.020
Rate (mol L^{-1} s^{-1})	7.5×10^{-4}	3.0×10^{-3}	1.2×10^{-2}

(a) Determine the order of the reaction, and write the rate expression.
(b) Calculate the rate constant, and give its units.
(c) Calculate the reaction rate for a concentration of HI equal to 0.0020 M.

Solution

(a) Write the rate expression for each of two different concentrations, [HI]$_1$ and [HI]$_2$

$$\text{rate}_1 = k\,([HI]_1)^n$$
$$\text{rate}_2 = k\,([HI]_2)^n$$

If the second equation is now divided by the first, the rate constant k drops out, leaving the reaction order n as the only unknown quantity.

$$\frac{\text{rate}_2}{\text{rate}_1} = \left(\frac{[HI]_2}{[HI]_1}\right)^n$$

Putting the numbers for any pair of data points into this equation allows solution for n. Choosing the first two gives

$$\frac{3.0 \times 10^{-3}}{7.5 \times 10^{-4}} = \left(\frac{0.010}{0.0050}\right)^n$$

which can be simplified to

$$4 = (2)^n$$

The solution is clearly $n = 2$, so the reaction is second order in HI. The rate expression has the form

$$\text{rate} = k\,[HI]^2$$

(b) The rate constant k is calculated by inserting any of the data points into the rate expression. Taking the first gives

$$7.5 \times 10^{-4}\ \text{mol L}^{-1}\ \text{s}^{-1} = k\,(0.0050\ \text{mol L}^{-1})^2$$
$$k = 30\ \text{L mol}^{-1}\ \text{s}^{-1}$$

(c) Once k and the rate expression are known, the rate at any other concentration of HI can be calculated by putting that concentration into the expression. At [HI] = 0.0020 M,

$$\text{rate} = k\,[\text{HI}]^2 = (30 \text{ L mol}^{-1} \text{ s}^{-1})(0.0020 \text{ mol L}^{-1})^2$$
$$= 1.2 \times 10^{-4} \text{ mol L}^{-1} \text{ s}^{-1}$$

Exercise

Ammonium cyanate (NH_4CNO) reacts to form urea (NH_2CONH_2) according to the equation $NH_4CNO(aq) \longrightarrow NH_2CONH_2(aq)$. The initial rate of the reaction is measured at 65°C for several initial concentrations of ammonium cyanate:

[NH4CNO] (mol L^{-1})	0.100	0.200	0.400
Rate (mol L^{-1} s^{-1})	3.60×10^{-2}	1.44×10^{-1}	5.76×10^{-1}

(a) Determine the order of the reaction with respect to ammonium cyanate and write the rate expression. (b) Calculate the rate constant, and give its units. (c) Calculate the initial reaction rate if the concentration of NH_4CNO is 0.0500 M.

Answer: (a) Second order, rate = $k\,[NH_4CNO]^2$. (b) $k = 3.60$ L mol^{-1} s^{-1}. (c) 9.00×10^{-3} mol L^{-1} s^{-1}.

• This is a very famous reaction. In 1828 Wohler showed that urea prepared in this way is identical to urea from urine. This helped to overthrow "vitalism," the doctrine holding that compounds produced by living organisms differ from synthetic compounds by possessing a "vital force."

Rates That Depend on Two or More Concentrations

So far, each reaction rate has depended on only a single concentration. In some reactions, however, the rate depends on the concentrations of two or more different chemical species. When this is so, the rate expression has the form

$$\text{rate} = k\,[\text{A}]^m[\text{B}]^n$$

with an additional factor put in for every additional substance found to affect the rate. The exponents m and n do *not* derive from the coefficients that A and B might have in the balanced equation for the reaction. Indeed, A and B are not necessarily even reactants or products in the reaction. Any substance might influence the rate of a particular reaction. Discovering the species that affect the rate and then determining the order of the reaction with respect to each such species require experimentation.

The exponents m, n, ... , which are usually integers or half-integers, give the order of the reaction, just as in the simpler case in which only one concentration appeared in the rate expression. For example, the preceding reaction is "mth order in A." This means that a change in the concentration of A by a certain factor leads to a change in the rate equal to that factor raised to the mth power. The reaction is nth order in B, and its **overall reaction order** is $m + n$. For the reaction

$$H_2PO_2^-(aq) + OH^-(aq) \longrightarrow HPO_3^{2-}(aq) + H_2(g)$$

the experimentally determined rate expression is

$$\text{rate} = k\,[H_2PO_2^-][OH^-]^2$$

so that the reaction is first order in $H_2PO_2^-(aq)$ and second order in $OH^-(aq)$, with an overall reaction order of 3. The units of the rate constant k depend on the overall reaction order. If all concentrations are expressed in mol L^{-1} and if $p = m + n + ...$ is the overall reaction order, then k has units of mol$^{-(p-1)}$ L$^{(p-1)}$ s^{-1}.

EXAMPLE 14-3

Use the preceding rate equation to determine the effect of the following changes on the rate of decomposition of $H_2PO_2^-(aq)$:
(a) Tripling the concentration of $H_2PO_2^-(aq)$ at constant pH.
(b) Changing the pH from 13 to 14 at a constant concentration of $H_2PO_2^-(aq)$.

Solution

(a) Because the reaction is first order in $H_2PO_2^-(aq)$, tripling this concentration leads to a tripling of the reaction rate.
(b) A change in pH from 13 to 14 corresponds to an increase in $OH^-(aq)$ concentration by a factor of 10. Because the reaction is second order in $OH^-(aq)$ (that is, this term is squared in the reaction rate equation), this leads to an increase in reaction rate by a factor of 10^2, or 100.

Exercise
Sucrose breaks down to a mixture of fructose and glucose

$$\text{sucrose}(aq) + H_2O(\ell) \longrightarrow \text{fructose}(aq) + \text{glucose}(aq)$$

in acidic solution. This reaction is important to the food and beverage industry because the product sugars (fructose and glucose) are sweeter than sucrose. The rate of this reaction is

$$\text{rate} = k\,[H^+][\text{sucrose}]$$

• The concentration of the nonreactant H^+ ion exerts a strong influence on the rate of this reaction.

What is the effect on the rate of: (a) cutting the concentration of sucrose in half at constant pH; (b) keeping the concentration of sucrose the same but lowering the pH from 1.5 to 0.5?

Answer: (a) Rate is cut in half. (b) Rate increases by a factor of 10.

Rate expressions that depend on more than one concentration are more difficult to obtain experimentally than those that depend on only one. One way to proceed is to find the initial rate of the reaction for several values of one of the concentrations while holding the other initial concentrations fixed from one run to the next. Then the experiment is repeated, changing one of the other concentrations. The following example illustrates this procedure.

EXAMPLE 14-4

The reaction of $NO(g)$ with $O_2(g)$ gives $NO_2(g)$:

$$2\,NO(g) + O_2(g) \longrightarrow 2\,NO_2(g)$$

From the dependence of the initial rate $(-\frac{1}{2}\Delta[NO]/\Delta t)$ on the initial concentrations of NO and O_2, determine the rate expression and the value of the rate constant.

[NO] (mol L^{-1})	[O_2] (mol L^{-1})	Initial Rate (mol $L^{-1}\,s^{-1}$)
1.0×10^{-4}	1.0×10^{-4}	2.8×10^{-6}
1.0×10^{-4}	3.0×10^{-4}	8.4×10^{-6}
2.0×10^{-4}	3.0×10^{-4}	3.4×10^{-5}

Solution

When $[O_2]$ is multiplied by 3 (with $[NO]$ constant), the rate is also multiplied by 3 (from 2.8×10^{-6} to 8.4×10^{-6} mol L^{-1} s^{-1}), so the reaction is first order in O_2. When $[NO]$ is multiplied by 2 (with $[O_2]$ constant), the rate is multiplied by

$$\frac{3.4 \times 10^{-5}}{8.4 \times 10^{-6}} \approx 4 = 2^2$$

so the reaction is second order in NO. The form of the rate expression is

$$\text{rate} = k\,[O_2][NO]^2$$

To obtain a value for k, insert the data on any of the three lines of the table into the rate expression. Data from the first line give

$$2.8 \times 10^{-6} \text{ mol } L^{-1} \text{ s}^{-1} = k\,(1.0 \times 10^{-4} \text{ mol } L^{-1})(1.0 \times 10^{-4} \text{ mol } L^{-1})^2$$
$$k = 2.8 \times 10^6 \text{ L}^2 \text{ mol}^{-2} \text{ s}^{-1}$$

Exercise

Ozone (O_3) also reacts with $NO(g)$:

$$NO(g) + O_3(g) \longrightarrow NO_2(g) + O_2(g)$$

The table gives the initial rate observed at different initial concentrations of NO and O_3:

[NO] (mol L^{-1})	[O_3] (mol L^{-1})	Initial Rate (mol L^{-1} s^{-1})
1.10×10^{-5}	1.10×10^{-5}	1.32×10^{-3}
1.10×10^{-5}	3.40×10^{-5}	4.08×10^{-3}
6.00×10^{-5}	3.40×10^{-5}	2.22×10^{-2}

Determine the rate expression and the value of the rate constant.

Answer: Rate $= k\,[NO][O_3]$ and $k = 1.09 \times 10^7$ L mol^{-1} s^{-1}.

14–3 THE DEPENDENCE OF CONCENTRATIONS ON TIME

Measuring the initial rate or other instantaneous rate of a chemical reaction involves determining small changes in concentration ($\Delta[A]$) occurring during a short time interval (Δt). Sometimes it can be difficult to obtain sufficiently precise experimental data for these small changes. An alternative is to fit all the data over a longer time interval with an equation that expresses the concentration of a species directly in terms of the elapsed time. For any given simple rate expression, a corresponding equation for the dependence of concentration on time can be obtained.

First-Order Reactions

Consider the reaction

$$N_2O_5(g) \longrightarrow 2\,NO_2(g) + \tfrac{1}{2}\,O_2(g)$$

The experimental rate law is

$$\text{rate} = -\frac{\Delta[N_2O_5]}{\Delta t} = k\,[N_2O_5]$$

This is a first-order reaction. The concentration of N_2O_5 dwindles from its original value, which we symbolize $[N_2O_5]_0$, as time passes. But by how much? To answer this question, we need an explicit equation giving $[N_2O_5]$ in terms of the time (t) and $[N_2O_5]_0$. This is arrived at using the methods of calculus, and we give only the result here. It is

$$[N_2O_5] = [N_2O_5]_0\,e^{-kt}$$

where e^{-kt} is the exponential (exp function on a calculator) of $-kt$. We now take the natural logarithm (the "ln") of both sides of this equation. Following the rules for the manipulation of logarithms reviewed in Appendix C then gives

$$\ln[N_2O_5] = \ln[N_2O_5]_0 - kt$$

With the rate law in either of these forms (and knowing the value of k), we can determine the remaining concentration of N_2O_5 at any time after pure N_2O_5 starts to react, or we can compute how long it takes the concentration of N_2O_5 to fall from its original value to any specified value. Similar equations can be written for any first-order reaction:

$$[A] = [A]_0\,e^{-kt}$$

which is equivalent to

$$\ln[A] = \ln[A]_0 - kt$$

A plot of the natural logarithm of the concentration [A] against time (t) is then a straight line with slope $-k$ and intercept $\ln[A]_0$ (Fig. 14–5).

A useful concept in discussions of first-order reactions is the **half-life ($t_{1/2}$)**— the time it takes for the concentration of a reactant A to fall to one half of its original value. This is, of course, $[A]_0/2$. The preceding equation can be rewritten as

• We call this form of the rate law, in which time appears explicitly, an integrated rate law. The "ln" stands for natural logarithm, 2.303 times larger than the common logarithm, or base-10 logarithm. See Appendix C.

• The general form of a linear equation is $y = mx + b$ (see Appendix C). Here, $\ln[A]$ plays the part of y, $-k$ the part of m, t the part of x, and $\ln[A]_0$ the part of b.

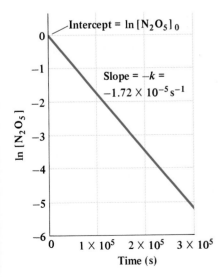

Figure 14–5 In a first-order reaction such as the decomposition of N_2O_5, a graph of the natural logarithm of the concentration against time is a straight line, the slope of which gives the rate constant for the reaction.

$$\ln [A] - \ln [A]_0 = -kt$$

$$\frac{\ln [A]}{[A]_0} = -kt$$

At the moment that t becomes equal to $t_{1/2}$, the concentration [A] has fallen to $[A]_0/2$. Substituting these values:

$$\ln \frac{\left(\dfrac{[A]_0}{2}\right)}{[A]_0} = \ln \tfrac{1}{2} = -\ln 2 = -kt_{1/2}$$

$$t_{1/2} = \frac{\ln 2}{k} = \frac{0.6931}{k}$$

• The concept of a half-life proves useful in our discussion of nuclear decay in Chapter 15.

If the rate constant of a reaction is given in s^{-1}, then the half-life comes out in seconds. During the first half-life of the reaction, the concentration of A falls to one half of its initial value; in two half-lives, it falls to one quarter (which is $\tfrac{1}{2} \times \tfrac{1}{2}$) of its initial value; in three half-lives, it falls to one eighth (which is $\tfrac{1}{2} \times \tfrac{1}{2} \times \tfrac{1}{2}$), and so forth (Fig. 14–6).

EXAMPLE 14–5

(a) What is the rate constant k for the first-order decomposition of $N_2O_5(g)$ at 25°C if the half-life of $N_2O_5(g)$ at that temperature is 4.03×10^4 s?
(b) Under these conditions, what percent of the N_2O_5 molecules have *not* reacted after one day?

Solution

(a) We have derived the relationship between the first-order rate constant and the half-life:

$$t_{1/2} = \frac{0.6931}{k} = 4.03 \times 10^4 \text{ s}$$

Solving for the rate constant k gives

$$k = \frac{0.6931}{t_{1/2}} = \frac{0.6931}{4.03 \times 10^4 \text{ s}} = 1.72 \times 10^{-5} \text{ s}^{-1}$$

(b) Because this is a first-order reaction, the explicit equation for the concentration of the reactant as a function of time is

$$[N_2O_5] = [N_2O_5]_0 \, e^{-kt}$$

Putting in the value for k and substituting $t = 8.64 \times 10^4$ s (one day has 86,400 s) gives

$$[N_2O_5] = [N_2O_5]_0 \exp[-(1.72 \times 10^{-5} \text{ s}^{-1})(8.64 \times 10^4 \text{ s})] = [N_2O_5]_0 \, e^{-1.486}$$

All scientific calculators provide for raising e to a power: simply put the power in the display and push the "e^x" button. Hence

$$\frac{[N_2O_5]}{[N_2O_5]_0} = e^{-1.486} = 0.226$$

Therefore, 22.6% of the N_2O_5 molecules have *not* reacted after one day at 25°C.

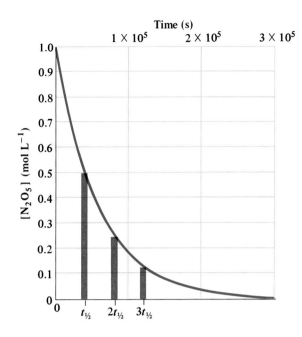

Figure 14-6 The same data as in Figure 14-5 are graphed in a concentration-versus-time picture. The half-life ($t_{1/2}$) is the time it takes for the initial concentration to be reduced to one half of its original value. In two half-lives, the concentration falls to one quarter of its initial value.

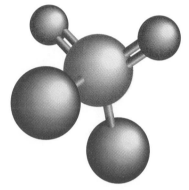

The structure of sulfuryl chloride SO_2Cl_2. This compound is used as a chlorinating agent.

Exercise

Sulfuryl chloride ($SO_2Cl_2(g)$) decomposes at an elevated temperature to $SO_2(g)$ and $Cl_2(g)$. The reaction is first order in SO_2Cl_2 with a half-life of 1.48×10^5 s. (a) Determine the rate constant of this reaction. (b) If 1.00 mol of pure SO_2Cl_2 is sealed in a container at this temperature, how much remains after 1.00 h? Assume that there is no recombination of SO_2 with Cl_2.

Answer: (a) $k = 4.68 \times 10^{-5}$ s^{-1}. (b) 0.845 mol of SO_2Cl_2.

Second-Order Reactions

Explicit equations giving the concentration of a reactant as time passes can be derived for second-order reactions by the same method (calculus) used for first-order reactions. An example is the reaction

$$2\ NO_2(g) \longrightarrow 2\ NO(g) + O_2(g)$$

for which the observed rate law is

$$\text{rate} = -\tfrac{1}{2}\frac{\Delta[NO_2]}{\Delta t} = k\,[NO_2]^2$$

The desired explicit equation is

$$\frac{1}{[NO_2]} = 2kt + \frac{1}{[NO_2]_0}$$

In this case, a plot of the reciprocal of the concentration of NO_2 against t is linear. The straight line has a slope of $2k$ and an intercept of 1 divided by the initial con-

Figure 14–7 (a) For a second-order reaction such as 2 NO$_2$ $\longrightarrow$ 2 NO + O$_2$, a graph of the reciprocal of the concentration against time is a straight line with slope 2k; k in this case equals 0.050 L mol^{-1}s^{-1}. (b) The same data graphed in a concentration-versus-time picture. The data are more difficult to interpret when graphed directly than when graphed as in part (a).

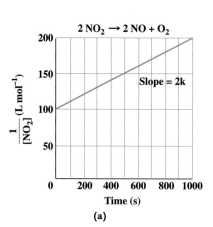

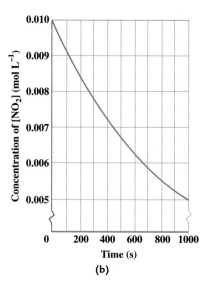

centration (Fig. 14–7). The factor of 2 in this equation comes from the factor of 1/2 in the experimental rate law.

The concept of half-life has little use for second-order reactions. If [NO$_2$] is set equal to [NO$_2$]$_0$/2 in the preceding equation, the result of solving for t is

$$\frac{2}{[NO_2]_0} = 2kt_{1/2} + \frac{1}{[NO_2]_0}$$

$$t_{1/2} = \frac{1}{2k\,[NO_2]_0}$$

The second-order half-life thus depends on the initial concentration, rather than behaving as a constant. The half-life is not a property of the reaction but of the particular system under study.

EXAMPLE 14–6

The dimerization of tetrafluoroethylene (C$_2$F$_4$) to octafluorocyclobutane (C$_4$F$_8$) is second order in the reactant C$_2$F$_4$ and second order overall. At 450 K, the rate constant is $k = 0.0448$ L mol^{-1} s^{-1}. If the initial concentration of C$_2$F$_4$ is 0.100 mol L^{-1}, what is its concentration after 203 s?

Solution

The rate expression for the reaction is

$$\text{rate} = -\frac{1}{2}\frac{\Delta[C_2F_4]}{\Delta t} = k\,[C_2F_4]^2$$

This second-order expression is just like the one previously given for the decomposition of NO$_2$. By comparison with that expression, the concentration of C$_2$F$_4$ changes with time according to

$$\frac{1}{[C_2F_4]} = 2kt + \frac{1}{[C_2F_4]_0}$$

Inserting $t = 203$ s and $[C_2F_4]_0 = 0.0100$ mol L^{-1} gives

$$\frac{1}{[C_2F_4]} = 2\left(\frac{0.0448 \text{ L}}{\text{mol s}}\right)(203 \text{ s}) + \frac{1}{0.100 \frac{\text{mol}}{\text{L}}}$$

$$= 18.2 \frac{\text{L}}{\text{mol}} + 10.0 \frac{\text{L}}{\text{mol}} = 28.2 \frac{\text{L}}{\text{mol}}$$

Solving for $[C_2F_4]$ gives

$$[C_2F_4] = 3.55 \times 10^{-2} \text{ mol L}^{-1}$$

Exercise

In another dimerization experiment with C_2F_4, all conditions are the same except that the initial concentration of C_2F_4 is 0.200 mol L^{-1}. Compute the concentration of C_2F_4 203 s after the reaction starts.

Answer: 4.31×10^{-2} mol L^{-1}.

In a real experimental study, the reaction order may be unknown. A standard procedure is to plot first the logarithm of the concentration against time, and then the reciprocal of the concentration against time (as in Fig. 14–8). A straight line in the first plot indicates first-order kinetics; a straight line in the second indicates second-order kinetics. If neither plot gives a straight line, the kinetics are more complex.

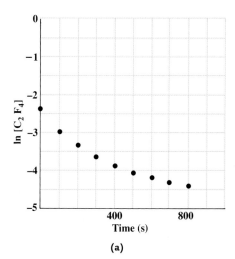
(a)

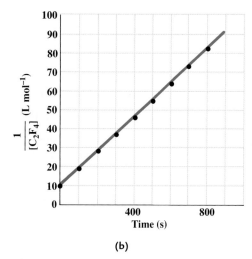
(b)

Figure 14–8   Suppose that the rate of reaction $2\,C_2F_4 \longrightarrow C_4F_8$ is being studied for the first time, under the conditions given in Example 14–6. Plotting the logarithm of the concentration of C_2F_4 against time (graph a) tests whether the kinetics are first order in C_2F_4; the curving line shows they are not. Plotting the reciprocal of the concentration of C_2F_4 against time (graph b) tests whether the kinetics are second order in C_2F_4. The nearly perfect straight-line relationship strongly suggests that they are.

14-4   REACTION MECHANISMS

Many reactions do not occur in a single step but proceed through a series of steps. Each step is called an **elementary step** and is directly caused by the collisions of atoms, ions, or molecules. The rate expression for an overall reaction cannot be derived from the stoichiometry of the balanced equation, as we have emphasized, but must be determined experimentally. The rate of an elementary step, on the other hand, *is* directly proportional to the product of the concentrations of the reacting species, each raised to a power equal to its coefficient in the balanced equation that represents the step. The **molecularity** of an elementary step equals the number of molecules that take part in it.

Types of Elementary Reactions

• Chemists call the process by which a molecule gains energy an "excitation" and the result an "excited state"; the loss of energy is a "relaxation."

A **unimolecular** elementary reaction involves only a single reactant molecule. An example is the dissociation of energized or "excited" molecules of N_2O_5 in the gas phase:

$$N_2O_5^* \longrightarrow NO_2 + NO_3$$

The asterisk indicates that the reacting molecules are in excited states, usually as the result of prior collisions with other molecules. The rate expression for the step is

$$\text{rate} = k\,[N_2O_5^*]$$

The most common type of elementary step involves the collision of two atoms, ions, or molecules and is called **bimolecular.** An example is the reaction

$$NO(g) + O_3(g) \longrightarrow NO_2(g) + O_2(g)$$

Because the rate of collisions is proportional to the product of the concentrations of the two reacting species, the rate expression for this bimolecular elementary step is

$$\text{rate} = k\,[NO][O_3]$$

• The standard energy change ($\Delta E°$) for the reaction $2\,I(g) \longrightarrow I_2(g)$ is -149 kJ mol^{-1}.

A **termolecular** elementary step involves the simultaneous collision of three reacting species, which is a much less likely event. An example is the recombination of iodine atoms in the gas phase to form iodine molecules. So much energy is released in forming the I—I bond that the molecule would simply fly apart as soon as it was formed if the event were a binary collision. A third atom or molecule is necessary to take away some of the excess energy. In the presence of an inert gas such as argon, such termolecular elementary steps as

$$I + I + Ar \longrightarrow I_2 + Ar$$

occur in which the argon atom leaves with more kinetic energy than it had initially. The rate law for this termolecular reaction is

$$\text{rate} = k\,[I]^2[Ar]$$

Elementary steps involving collisions of four or more molecules are not observed, and even termolecular collisions are rare, particularly if other pathways are possible.

Elementary steps in liquid solvents involve encounters of solute species with one another. If the solution is ideal, the rates of these processes are proportional to the product of the concentrations of the solute species involved. Solvent molecules are always present and may affect the reaction, even though they do not appear in the rate expression because the concentration of the solvent cannot be varied appreciably. A reaction such as the recombination of iodine atoms occurs readily in solution. It appears to be second order with the rate expression

• This invariance of solvent concentration is the reason that the concentrations of solvents can be replaced by a 1 in equilibrium expressions (see Section 7-6).

$$\text{rate} = k\,[I]^2$$

only because the third body involved is a solvent molecule. In the same way, a reaction between a solvent and a solute appears to be unimolecular, and only the concentration of solute molecules enters the rate expressions.

Reaction Mechanisms

A **reaction mechanism** is a detailed series of elementary steps, together with their rates, which are combined to yield the overall reaction. It is generally possible to write several reaction mechanisms, each of which is consistent with a given overall reaction. One of the goals of chemical kinetics is to use the observed rate of a reaction to choose among the many conceivable reaction mechanisms.

The gas-phase reaction of nitrogen dioxide with carbon monoxide provides a good example of a reaction mechanism. The mechanism that is generally accepted has two steps, both bimolecular:

$$NO_2 + NO_2 \longrightarrow NO_3 + NO \quad \text{(slow)}$$
$$NO_3 + CO \longrightarrow NO_2 + CO_2 \quad \text{(fast)}$$

For any reaction mechanism, combining the steps must give the overall reaction. When each elementary step occurs the same number of times in the course of the reaction, the equations can simply be added. (If one step occurs, say, twice as often as the others, it must be multiplied by 2 before the elementary reactions are added.) In this case, we add the two chemical equations to get

$$2\,NO_2 + NO_3 + CO \longrightarrow NO_3 + NO + NO_2 + CO_2$$

Canceling out the NO_3 and one molecule of NO_2 from each side leads to

$$NO_2 + CO \longrightarrow NO + CO_2$$

A **reaction intermediate** (here, NO_3) is a chemical species that is formed and consumed in the reaction but does not appear in the overall balanced chemical equation. One of the major challenges in chemical kinetics is to identify intermediates, which are frequently so short-lived that they are difficult to detect directly.

EXAMPLE 14–7

Consider the following reaction mechanism:

$$Cl_2 \longrightarrow 2\,Cl$$
$$Cl + CHCl_3 \longrightarrow HCl + CCl_3$$
$$CCl_3 + Cl \longrightarrow CCl_4$$

(a) What is the molecularity of each elementary step?
(b) Write the equation for the overall reaction.
(c) Identify the reaction intermediate(s).

Solution

(a) The first step is unimolecular, and the other two are bimolecular.
(b) Adding the three steps gives

$$Cl_2 + 2\,Cl + CHCl_3 + CCl_3 \longrightarrow 2\,Cl + HCl + CCl_3 + CCl_4$$

The two species that appear in equal amounts on both sides cancel out to leave

$$Cl_2 + CHCl_3 \longrightarrow HCl + CCl_4$$

(c) The two reaction intermediates are Cl and CCl_3.

Exercise

Consider the mechanism

$$H_2 + 2\,NO \longrightarrow N_2O + H_2O$$
$$N_2O + H_2 \longrightarrow N_2 + H_2O$$

(a) What is the molecularity of each elementary step? (b) Write the equation for the overall reaction. (c) Identify the reaction intermediate(s).

Answer: (a) The first step is termolecular and the second is bimolecular. (b) The overall reaction is $2\,H_2 + 2\,NO \longrightarrow N_2 + 2\,H_2O$. (c) The only intermediate is N_2O.

Kinetics and Chemical Equilibrium

If a reaction occurs in a single elementary step, a direct connection exists between the rate constants of the forward and reverse reactions and the equilibrium constant for the reaction. Consider the equilibrium

$$NO + O_3 \rightleftharpoons NO_2 + O_2$$

which takes place in a single elementary step. Here, the rate of the forward reaction is

$$\text{forward rate} = k_1\,[NO][O_3]$$

and the rate of the reverse reaction (in which collisions of NO_2 and O_2 molecules cause NO and O_3 to form) is

$$\text{reverse rate} = k_{-1}\,[NO_2][O_2]$$

For equilibrium to exist, the forward and reverse rates must be equal; this means that

$$k_1\,[NO]_{eq}[O_3]_{eq} = k_{-1}\,[NO_2]_{eq}[O_2]_{eq}$$

This equation can be rearranged into the form

$$\frac{[NO_2]_{eq}[O_2]_{eq}}{[NO]_{eq}[O_3]_{eq}} = \frac{k_1}{k_{-1}} = K_1$$

where K_1 is the equilibrium constant.[1] For an elementary step, the equilibrium constant is the ratio of the forward to the reverse rate constant.

Now, consider an equilibrium that is reached by a series of elementary steps, such as

$$2\,NO(g) + 2\,H_2(g) \rightleftharpoons N_2(g) + 2\,H_2O(g)$$

This reaction is believed to occur by a three-step mechanism involving N_2O_2 and N_2O as intermediates:

$$NO + NO \underset{k_{-1}}{\overset{k_1}{\rightleftharpoons}} N_2O_2$$

---

[1]Gas-phase equilibrium constants involve partial pressures expressed in atmospheres, whereas the convention in chemical kinetics is to use concentrations even for gaseous species. Therefore, the constants such as K_1, which we introduce here, are related to true dimensionless equilibrium constants by multiplicative factors of the form (RT/P_{ref}) raised to a power, as discussed in Section 7–4. Nevertheless, we refer to constants such as K_1 in this section as equilibrium constants.

$$N_2O_2 + H_2 \underset{k_{-2}}{\overset{k_2}{\rightleftharpoons}} N_2O + H_2O$$

$$N_2O + H_2 \underset{k_{-3}}{\overset{k_3}{\rightleftharpoons}} N_2 + H_2O$$

The reverse reactions (from products to reactants) are included here as well; k_1, k_2, and k_3 are the rate constants for the three forward elementary steps, and k_{-1}, k_{-2}, and k_{-3} are the rate constants for the corresponding reverse reactions.

We now invoke the **principle of detailed balance,** which states that, at equilibrium, the rate of *each* elementary step is balanced by (equal to) the rate of its reverse. For the above mechanism, we conclude that

$$k_1 [NO]^2_{eq} = k_{-1}[N_2O_2]_{eq}$$
$$k_2 [N_2O_2]_{eq}[H_2]_{eq} = k_{-2} [N_2O]_{eq}[H_2O]_{eq}$$
$$k_3 [N_2O]_{eq}[H_2]_{eq} = k_{-3}[N_2]_{eq}[H_2O]_{eq}$$

We can define equilibrium constants K_1 and K_2, and K_3 for the three elementary steps

$$K_1 = \frac{[N_2O_2]_{eq}}{[NO]^2_{eq}} = \frac{k_1}{k_{-1}}$$

$$K_2 = \frac{[N_2O]_{eq}[H_2O]_{eq}}{[N_2O_2]_{eq}[H_2]_{eq}} = \frac{k_2}{k_{-2}}$$

$$K_3 = \frac{[N_2]_{eq}[H_2O]_{eq}}{[N_2O]_{eq}[H_2]_{eq}} = \frac{k_3}{k_{-3}}$$

Each equilibrium constant is the ratio of the forward rate constant for that step to the reverse rate constant.

Adding together the three steps of the mechanism gives the overall reaction. It is known (see Section 7–2) that when reactions are added, their equilibrium constants are multiplied. Therefore, the equilibrium constant for the overall reaction is

$$K = K_1K_2K_3 = \frac{k_1k_2k_3}{k_{-1}k_{-2}k_{-3}} = \frac{[N_2O_2]_{eq}[N_2O]_{eq}[H_2O]_{eq}[N_2]_{eq}[H_2O]_{eq}}{[NO]^2_{eq}[N_2O_2]_{eq}[H_2]_{eq}[N_2O]_{eq}[H_2]_{eq}}$$
$$= \frac{[H_2O]^2_{eq}[N_2]_{eq}}{[NO]^2_{eq}[H_2]^2_{eq}}$$

The concentrations of the intermediates N_2O_2 and N_2O cancel out, giving the usual mass-action expression.

This result can be generalized to any reaction mechanism. The product of the forward rate constants for the elementary steps divided by the product of the reverse rate constants is always equal to the equilibrium constant of the overall reaction. Should there be other possible mechanisms for a given reaction (which might involve intermediates other than N_2O_2 and N_2O in our example), the forward and reverse rate constants of the various steps are still consistent in this way with the equilibrium constant of the reaction.

14-5 REACTION MECHANISM AND RATE

In many reaction mechanisms, one step is significantly slower than all the others. Such a step is called the **rate-determining step** because an overall reaction can occur only as fast as its slowest step, just as diners go through a cafeteria line only as fast as they pass the "bottleneck" of a slow cashier. Figure 14–9 shows passage through a literal bottleneck as the rate-determining step in the three-step "reaction" in which water runs from one soft-drink bottle to another through a connector.

If the rate-determining step is the first one, the analysis is particularly simple. An example is the reaction

$$2\,NO_2 + F_2 \longrightarrow 2\,NO_2F$$

for which the experimental rate expression is

$$\text{rate} = k_{obs}\,[NO_2][F_2]$$

A possible mechanism for the reaction is

$$NO_2 + F_2 \xrightarrow{k_1} NO_2F + F \quad \text{(slow)}$$
$$NO_2 + F \xrightarrow{k_2} NO_2F \quad \text{(fast)}$$

The first step is slow and determines the rate

$$\text{rate} = k_1\,[NO_2][F_2]$$

The subsequent fast step does not affect the reaction rate, because fluorine atoms join with NO_2 molecules as soon as they are produced.

Mechanisms in which the rate-determining step occurs after one or more fast steps are often signaled by a reaction order greater than two, by a nonintegral reaction order, or by an inverse dependence of the rate on the concentration of a species taking part in the reaction. An example is the reaction

$$2\,NO + O_2 \longrightarrow 2\,NO_2$$

for which the experimental rate expression is

$$\text{rate} = k_{obs}\,[NO]^2[O_2]$$

One possible mechanism would be a single-step termolecular reaction of two NO molecules with one O_2 molecule. This would be consistent with the form of the rate expression, but three-way collisions are quite rare; if there is an alternative pathway, it is usually followed.

One such alternative is the two-step mechanism:

$$NO + NO \underset{k_{-1}}{\overset{k_1}{\rightleftharpoons}} N_2O_2 \quad \text{(fast equilibrium)}$$

$$N_2O_2 + O_2 \xrightarrow{k_2} 2\,NO_2 \quad \text{(slow)}$$

• A three-way collision requires a two-way collision *plus* the simultaneous arrival of a third particle.

(a)   (b)

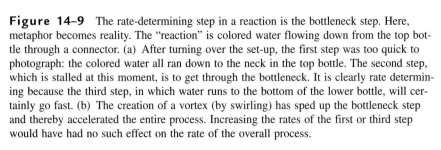

Figure 14–9 The rate-determining step in a reaction is the bottleneck step. Here, metaphor becomes reality. The "reaction" is colored water flowing down from the top bottle through a connector. (a) After turning over the set-up, the first step was too quick to photograph: the colored water all ran down to the neck in the top bottle. The second step, which is stalled at this moment, is to get through the bottleneck. It is clearly rate determining because the third step, in which water runs to the bottom of the lower bottle, will certainly go fast. (b) The creation of a vortex (by swirling) has sped up the bottleneck step and thereby accelerated the entire process. Increasing the rates of the first or third step would have had no such effect on the rate of the overall process.

Because the slow step determines the overall rate, we can write

$$\text{rate} = k_2\,[N_2O_2][O_2]$$

The concentration of a reactive intermediate such as N_2O_2 cannot be varied at will, however. Because the N_2O_2 reacts only slowly with O_2, the reverse reaction (to 2 NO) is possible and must be taken into account. In fact, it is reasonable to assume that all of the elementary steps that occur *before* the rate-determining step are in equilibrium, with the forward and reverse reactions occurring at the same rate. In this case, we have

$$\frac{[N_2O_2]}{[NO]^2} = \frac{k_1}{k_{-1}} = K_1$$
$$[N_2O_2] = K_1\,[NO]^2$$

Substituting this relationship into the rate expression gives

$$\text{rate} = k_2K_1\,[NO]^2[O_2]$$

This result is consistent with the observed reaction order, with $k_2K_1 = k_{obs}$.

EXAMPLE 14-8

In basic aqueous solution, the reaction

$$I^- + OCl^- \longrightarrow Cl^- + OI^-$$

is generally agreed to occur according to the following mechanism:

$$OCl^-(aq) + H_2O(\ell) \underset{k_{-1}}{\overset{k_1}{\rightleftharpoons}} HOCl(aq) + OH^-(aq) \qquad \text{(fast equilibrium)}$$

$$I^-(aq) + HOCl(aq) \overset{k_2}{\longrightarrow} HOI(aq) + Cl^-(aq) \qquad \text{(slow)}$$

$$OH^-(aq) + HOI(aq) \overset{k_3}{\longrightarrow} H_2O(\ell) + OI^-(aq) \qquad \text{(fast)}$$

What rate expression is predicted by this mechanism?

Solution

The rate is determined by the slowest elementary step, the second one:

$$\text{rate} = k_2\,[I^-][HOCl]$$

However, the HOCl is in equilibrium with OCl^- and OH^- due to the first step:

$$\frac{[HOCl][OH^-]}{[OCl^-]} = K_1 = \frac{k_1}{k_{-1}}$$

Solving this equation for [HOCl] and inserting it into the previous expression gives the prediction

$$\text{rate} = k_2K_1\,\frac{[I^-][OCl^-]}{[OH^-]}$$

which is, in fact, the experimentally observed rate expression.

• The hypochlorite ion OCl^- in basic solution is the active ingredient in household bleach. Here it oxidizes the iodide ion.

Exercise

Suppose the mechanism for the reaction given in the preceding example were

$$OCl^-(aq) + H_2O(\ell) \underset{k_{-1}}{\overset{k_1}{\rightleftharpoons}} HOCl(aq) + OH^-(aq) \quad \text{(fast equilibrium)}$$

$$I^-(aq) + HOCl(aq) \underset{k_{-2}}{\overset{k_2}{\rightleftharpoons}} HOI(aq) + Cl^-(aq) \quad \text{(fast equilibrium)}$$

$$OH^-(aq) + HOI(aq) \overset{k_3}{\longrightarrow} H_2O(\ell) + OI^-(aq) \quad \text{(slow)}$$

What rate expression is now predicted?

Answer: rate $= k_3 K_2 K_1 \dfrac{[I^-][OCl^-]}{[Cl^-]}$

The rate of the reaction between I^- and OCl^- (in Example 14–8) depends inversely on the concentration of OH^- ion. Such a form is often a clue that a rapid equilibrium occurs in the first steps of a reaction, preceding the rate-determining step. Fractional orders of reaction provide a similar clue, as in the reaction of H_2 with Br_2 to form HBr,

$$H_2 + Br_2 \longrightarrow 2\,HBr$$

for which the initial reaction rate (before very much HBr builds up) follows the rate expression

$$\text{rate} = k_{\text{obs}}\,[H_2][Br_2]^{1/2}$$

How can such a fractional power appear? One reaction mechanism that predicts this rate expression is

$$Br_2 + M \underset{k_{-1}}{\overset{k_1}{\rightleftharpoons}} Br + Br + M \quad \text{(fast equilibrium)}$$

$$Br + H_2 \overset{k_2}{\longrightarrow} HBr + H \quad \text{(slow)}$$

$$H + Br_2 \overset{k_3}{\longrightarrow} HBr + Br \quad \text{(fast)}$$

Here M stands for a second molecule that does not enter into combination with the reactant but supplies the energy to break up the bromine molecules. The reaction rate is determined by the slow step:

$$\text{rate} = k_2\,[Br][H_2]$$

However, [Br] is fixed by the establishment of equilibrium in the first reaction:

$$\frac{[Br]^2}{[Br_2]} = K_1 = \frac{k_1}{k_{-1}}$$

so that

$$[Br] = K_1^{1/2}\,[Br_2]^{1/2}$$

The rate expression predicted by this mechanism is thus

$$\text{rate} = k_2 K_1^{1/2}\,[H_2][Br_2]^{1/2}$$

This is in accord with the observed fractional power in the rate expression. On the other hand, the simple bimolecular mechanism

$$H_2 + Br_2 \xrightarrow{k_1} 2\ HBr \quad \text{(slow)}$$

predicts the rate expression

$$\text{rate} = k_1\ [H_2][Br_2]$$

This disagrees with the observed rate expression, so the simple bimolecular mechanism must be ruled out as the major contributor to the measured rate.

It should be clear by now that deducing a rate expression from a proposed mechanism is relatively straightforward but that doing the reverse is much harder. In fact, several competing mechanisms often give rise to the same rate expression, and only some independent type of measurement enables a choice between mechanisms.

A reaction mechanism can never be proved from an experimental rate law; it can only be disproved if it is inconsistent with the experimental behavior. A classic example is the combination reaction

$$H_2(g) + I_2(g) \longrightarrow 2\ HI(g)$$

for which the observed rate expression is

$$\text{rate} = k_{obs}\ [H_2][I_2]$$

(contrast this with the rate expression given above for the combination of H_2 with Br_2). This is one of the earliest and most extensively studied reactions in chemical kinetics, and until 1967 it was universally believed to occur as a one-step elementary reaction. At that time, J. H. Sullivan investigated the effect of illuminating the reacting sample with light, which splits some of the I_2 molecules into iodine atoms. If the mechanism proposed above is correct, the effect of the light on the reaction should be small because it leads only to a small decrease in the concentration of I_2.

Instead, Sullivan observed a dramatic *increase* in the rate of reaction under illumination. This increase could be explained only by the participation of iodine *atoms* in the reaction mechanism. One such mechanism is

$$I_2 + M \underset{k_{-1}}{\overset{k_1}{\rightleftharpoons}} I + I + M \quad \text{(fast equilibrium)}$$

$$H_2 + I \underset{k_{-2}}{\overset{k_2}{\rightleftharpoons}} H_2I \quad \text{(fast equilibrium)}$$

$$H_2I + I \xrightarrow{k_3} 2\ HI \quad \text{(slow)}$$

for which the rate expression is

$$\begin{aligned}
\text{rate} &= k_3\ [H_2I][I] \\
&= k_3 K_2\ [H_2][I]^2 \\
&= k_3 K_2 K_1\ [H_2][I_2]
\end{aligned}$$

This mechanism gives the same rate expression that is observed experimentally but also explains the effect of light on the reaction. The first reaction mechanism may *also* occur, but, at least in the presence of light, it appears to be less important.

This example illustrates the hazards of trying to determine reaction mechanisms from rate laws: several mechanisms can fit any given empirical rate law, and it is always possible that a new piece of information will come along to suggest a different mechanism. The problem is that, under ordinary conditions, the reaction intermediates can only be postulated and not isolated and studied like the reactants and products. This state of affairs is changing, however, with the development of experimental techniques that allow direct study of the transient species that form in small concentrations during the course of a chemical reaction.

Chain Reactions

A **chain reaction** is one that proceeds through a series of elementary steps, some of which are repeated many times. Chain reactions consist of three stages: first, **initiation,** in which two or more reactive intermediates are generated; second, **propagation,** in which products are formed but reactive intermediates are continuously regenerated; and third, **termination,** in which two intermediates combine to give a stable product.

An example of a chain reaction is the reaction of methane with fluorine to give CH_3F and HF:

$$CH_4(g) + F_2(g) \longrightarrow CH_3F(g) + HF(g)$$

Although in principle this reaction could occur in a one-step bimolecular process, that route turns out to be too slow to contribute significantly under normal conditions. Instead, the main mechanism involves a chain reaction:

$$CH_4 + F_2 \longrightarrow \cdot CH_3 + HF + F\cdot \quad \text{(initiation)}$$
$$\cdot CH_3 + F_2 \longrightarrow CH_3F + F\cdot \quad \text{(propagation)}$$
$$CH_4 + F\cdot \longrightarrow \cdot CH_3 + HF \quad \text{(propagation)}$$
$$\cdot CH_3 + F\cdot + M \longrightarrow CH_3F + M \quad \text{(termination)}$$

In the initiation step, two reactive intermediates ($\cdot CH_3$ and $\cdot F$) are produced. These species are radicals, species in which one (or more) valence electron is unpaired. Dots in the formulas of radicals represent unpaired electrons. Radicals tend to transfer or share electrons to attain valence octets. Consequently, they are usually highly reactive. During the propagation steps, these intermediates react to form products but are both also regenerated. The two propagation steps occur again and again, churning out quantities of products (CH_3F and HF) as reactants (CH_4 and F_2) are consumed. Eventually, two reactive intermediates come together in a termination step. Chain reactions are important in building up the long-chain molecules called polymers, as we shall see in Chapter 25.

• In a chain reaction, the overall reaction is the sum of the propagation steps only. These steps occur far more times than the initiation or termination steps, so they dominate the overall stoichiometry of the reaction.

The chain reaction between CH_4 and F_2 proceeds at a constant rate because each propagation step both consumes and produces a reactive intermediate. The concentrations of the reactive intermediates remain approximately constant and are determined by the rates of chain initiation and termination. Another type of chain reaction is possible in which the number of reactive intermediates increases during one or more propagation steps. This is called a **branching chain reaction.** An example is the reaction of oxygen with hydrogen. The mechanism is complex and can be initiated in various ways, causing several reactive intermediates such as O, H, and OH to form. Some propagation steps are of the type we have seen already for CH_4 and F_2, such as

$$\cdot OH + H_2 \longrightarrow H_2O + H\cdot$$

in which one reactive intermediate ($\cdot OH$) is used up and one ($H\cdot$) is produced. Other propagation steps are branching, however:

$$\cdot H + O_2 \longrightarrow \cdot OH + \cdot O\cdot$$
$$\cdot O\cdot + H_2 \longrightarrow \cdot OH + H\cdot$$

In these steps, each reactive intermediate that is used up leads to the generation of two others. This leads to a rapid growth in the number of reactive species, speeding the rate further and possibly causing an explosion.

14-6   EFFECT OF TEMPERATURE ON REACTION RATES

The first five sections of this chapter discuss the experimental determination of rate laws and their relation to assumed mechanisms for chemical reactions. Little is said, however, about the actual magnitudes of rate constants (either for elementary reactions or for multistep reaction), nor is the effect of temperature on reactions considered. Before these matters are taken up, it is necessary to establish the connection between the rates of molecular collisions and the rates of chemical reactions. Such a connection is fairly easy to make for gas-phase reactions because the simple kinetic theory of Chapter 5 applies. Its results are of critical importance in studying the chemistry of the atmosphere, a matter of highest priority in environmental affairs.

Rate Constants for Gas-Phase Reactions

The kinetic theory of gases (see Section 5–7) allows an estimate of the frequency of collisions between a given molecule and others in a gas at ordinary conditions. A typical small molecule (one of oxygen, for example) in a gas confined at ordinary room conditions strikes other molecules with enormous frequency, undergoing on the order of 10^{10} collisions per second. If every collision led to reaction, the reaction would be practically complete in a period of about 10^{-9} second. Some reactions do proceed at almost this rate. An example is the bimolecular reaction between two molecules of ClO:

$$2\ ClO \longrightarrow Cl_2 + O_2$$

for which the observed rate constant at 273 K is 6×10^{10} L mol^{-1} s^{-1}. If the initial pressure of ClO could be raised as high as 1 atm, then at 273 K the initial concentration of ClO would be about 0.04 M. According to the second-order rate law (see Section 14–3), after one billionth of a second (10^{-9} s), the concentration of ClO would drop to 0.007 M, and after two billionths of a second, it would be 0.004 M! Most reactions proceed far more slowly than this. Indeed, rates can be smaller by factors of 10^{12} or more. The naive idea that "to collide is to react" clearly must be modified if such rates are to be understood.

The relative orientation of the colliding molecules must certainly play a role in determining whether a particular collision results in a reaction. In a biomolecular reaction such as

$$2\ NOCl \longrightarrow 2\ NO + Cl_2$$

it seems obvious that the two NOCl molecules must approach each other in such a way that the chlorine atoms are close together in order for a Cl_2 molecule to split off (Fig. 14–10). The collision frequency must therefore be multiplied by a **steric factor** (less than unity) to account for the fact that only a fraction of the collisions occurs with the proper orientation to lead to reaction. For small molecules, however, such a steric factor could hardly reduce the reaction rate by much more than one order of magnitude, so some further explanation must be sought for the slow rates that are frequently observed.

A clue is found in the observed temperature dependence of rate constants. The rates of many reactions increase extremely rapidly with increases in temperature;

• The key role played by chlorine monoxide, ClO, in the depletion of ozone in the outer atmosphere is discussed in Section 18–5.

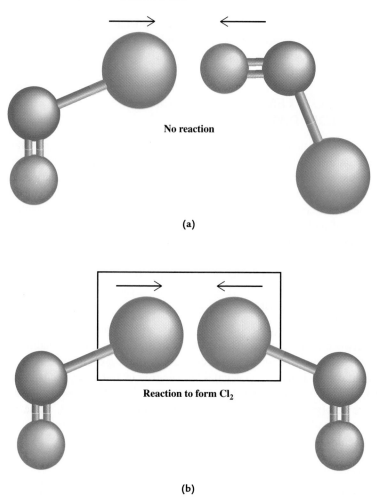

Figure 14–10 The steric effect. (a) This collision of NOCl molecules, although energetic, produces no Cl_2; the Cl atoms (green) are too far away from each other. (b) When two fast-moving NOCl molecules collide with two chlorine atoms close together, Cl_2 forms (along with NO(g)).

typically, the rate may double for a 10°C rise in temperature. In 1887, Svante Arrhenius suggested that rate constants vary exponentially with the reciprocal of the absolute temperature,

$$k = Ae^{-E_a/RT} = A \exp{(-E_a/RT)}$$

where the constant E_a has units of energy per mole, and the pre-exponential factor A is a constant having the same units as k. This relationship is now known at the **Arrhenius equation.** Taking the natural logarithm of both sides of the equation gives

$$\ln k = \ln A - \frac{E_a}{RT}$$

so that a plot of $\ln k$ against $1/T$ should be a straight line with slope $-E_a/R$ and intercept $\ln A$. Many rate constants do show just this kind of temperature dependence (Fig. 14–11).

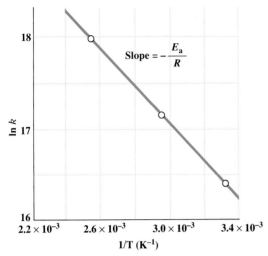

Figure 14-11 An Arrhenius plot of ln k against $1/T$ for the reaction of benzene vapor with oxygen atoms. A considerable extrapolation to $1/T = 0$ would be necessary to determine the constant ln A from the intercept of this line.

EXAMPLE 14-9

The decomposition of hydroxylamine (NH_2OH) in the presence of oxygen follows the rate expression

$$\text{rate} = k_{\text{obs}} [NH_2OH][O_2]$$

where k_{obs} is 0.237×10^{-4} L mol^{-1} s^{-1} at 0°C and rises to 2.64×10^{-4} L mol^{-1} s^{-1} at 25°C. Calculate E_a and A for this reaction.

Solution

Let us write the Arrhenius equation at two different temperatures T_1 and T_2:

$$\ln k_1 = \ln A - \frac{E_a}{RT_1} \qquad \text{and} \qquad \ln k_2 = \ln A - \frac{E_a}{RT_2}$$

If the first equation is subtracted from the second, the term ln A cancels out, leaving

$$\ln k_2 - \ln k_1 = \ln \frac{k_2}{k_1} = -\frac{E_a}{R}\left(\frac{1}{T_2} - \frac{1}{T_1}\right)$$

which can be solved for E_a. In the present case, $T_1 = 273.15$ K and $T_2 = 298.15$ K so that

> • Notice the similarity in form between this equation and the van't Hoff equation (see Section 11-8), which describes the temperature dependence of *equilibrium* constants. Also, compare Figure 14-11 with Figure 11-14.

$$\ln\left(\frac{2.64 \times 10^{-4}}{0.237 \times 10^{-4}}\right) = -\frac{E_a}{8.315 \text{ J K}^{-1}\text{ mol}^{-1}}\left(\frac{1}{298.15 \text{ K}} - \frac{1}{273.15 \text{ K}}\right)$$

$$2.410 = -\frac{E_a}{8.315 \text{ J K}^{-1}\text{ mol}^{-1}}(-3.07 \times 10^{-4} \text{ K}^{-1})$$

$$E_a = 6.53 \times 10^4 \text{ J mol}^{-1} = \boxed{65.3 \text{ kJ mol}^{-1}}$$

Now that E_a is known, the constant A can be calculated by using data at either temperature. At 273.15 K

$$\ln A = \ln k_1 + \frac{E_a}{RT}$$

$$= \ln (0.237 \times 10^{-4}) + \frac{65.3 \times 10^3 \; J \; mol^{-1}}{8.315 \; J \; K^{-1} \; mol^{-1} \; (273.15 \; K)}$$

$$= -10.65 + 28.75 = 18.10$$

$$A = e^{18.10} = 7.3 \times 10^7 \; L \; mol^{-1} \; s^{-1}$$

A more accurate determination of E_a and A would use measurements at a series of temperatures rather than at only two, with a fit to a plot such as that in Figure 14–11.

Exercise

The reaction

$$N(SO_3)_3^{3-}(aq) + H_2O(\ell) \longrightarrow HN(SO_3)_2^{2-}(aq) + HSO_4^{-}(aq)$$

follows the rate law

$$\text{rate} = k_{obs} \, [N(SO_3)_3^{3-}][H^+]$$

The rate increases by a factor of 23.1 when the temperature is raised from 283 K to 313 K. Compute E_a for the reaction.

Answer: 77.1 kJ mol^{-1}.

Arrhenius believed that to react upon collision, molecules must become "activated," and the parameter E_a became known as the **activation energy.** His ideas were refined by scientists who followed. In 1915, Marcelin pointed out that although molecules make many collisions, not all collisions are reactive. Only those collisions in which the collision energy (that is, the relative translational kinetic energy of the colliding molecules) exceeds some critical minimum can result in reaction. Thus, Marcelin gave a dynamic interpretation to the activation energy inferred from reaction rates.

The strong temperature dependence of rate constants described by the Arrhenius equation can be related to the Maxwell–Boltzmann distribution of molecular kinetic energies (Fig. 14–12). When the temperature is sufficiently low, only a small fraction of all pairs of colliding molecules possesses the minimum collision energy required for reaction. This fraction corresponds to the area under the Maxwell–Boltzmann distribution curve between E_a and infinity. As the temperature is increased, the distribution curve spreads out toward higher energies. The fraction of molecules that have energies exceeding the critical energy E_a increases exponentially with temperature as $e^{-E_a/RT}$, as the Arrhenius law and experiment require. The reaction rate is then proportional to $e^{-E_a/RT}$, and both the strong temperature dependence and the order of magnitude of the experimental rate constants can be understood.

The Activated Complex

Why should there be a critical collision energy E_a for the reaction between two molecules? To understand this, consider the analogy in Figure 14–13. In a landslide, a boulder freed from one slope may actually roll up a neighboring hill. If it does, its potential energy increases and its kinetic energy decreases (it gets higher and slows down). If it can reach the top of the hill, it can topple over to the other side, with its kinetic energy then increasing and its potential energy decreasing. Not every boul-

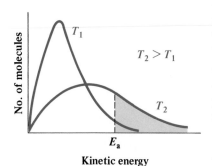

Figure 14–12 The Maxwell–Boltzmann distribution of molecular kinetic energies at two different temperatures. A relatively small temperature change has a large effect on the fraction of molecules having enough kinetic energy to react.

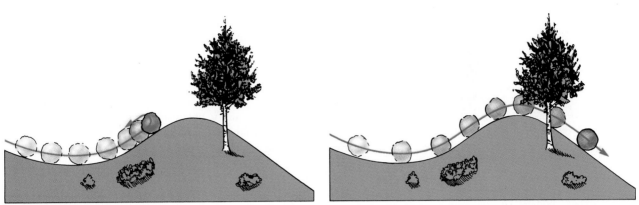

Figure 14-13 A landslide has set some large boulders rolling, Their fates are analogous to those of molecules in a chemical reaction. The blue boulder rolls up a hill and then back. It does not have enough energy to make it over. The red boulder has more kinetic energy. It makes it over the hill to reach the valley beyond.

der makes it over the hill, however. If its initial speed (and therefore its kinetic energy) is too small, a boulder rolls part of the way up and then falls back. Only a boulder with initial kinetic energy higher than a critical threshold passes over the barrier.

We can translate this physical model into a description of molecular collisions and reactions. As two reactant molecules, atoms, or ions approach each other along a **reaction path,** their potential energy increases as the bonds within them distort. At some critical potential energy, they are associated in an unstable entity called an **activated complex.** The activated complex is the crossover stage or **transition state,** in which the smooth ascent in potential energy as the reactants come together passes through a maximum and becomes a smooth descent as the product molecules separate. Just as a boulder can fail to roll over a hill, some pairs of reactant bodies fail to react. Only those pairs with sufficient kinetic energy can stretch bonds and rearrange atoms sufficiently to reach the transition state that separates reactants from products. If the barrier is too high, almost all colliding pairs of reactant molecules separate from each other without reacting. The height of the barrier is the activation energy for the reaction.

Figure 14–14 is a graph of potential energy versus position along the reaction path for the reaction

$$NO_2(g) + CO(g) \longrightarrow NO(g) + CO_2(g)$$

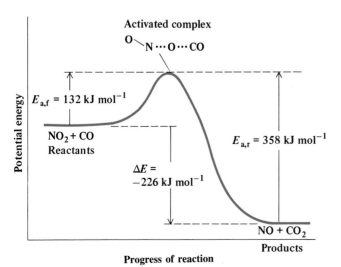

Figure 14-14 A transition state diagram for the reaction $NO_2 + CO \longrightarrow NO + CO_2$.

Figure 14–15 The decomposition of hydrogen peroxide (H_2O_2), the liquid in the dropper, to water and oxygen is catalyzed by solid MnO_2. Here the water evolves as steam because of the heat generated by the reaction.

The mechanism of this reaction is a simple bimolecular collision. Two activation energies are shown: $E_{a,f}$ for the forward reaction, and $E_{a,r}$ for the reverse reaction, in which NO takes an oxygen atom from CO_2 to form NO_2 and CO. The difference between the two is ΔE, the change in internal energy of the chemical reaction:

$$\Delta E = E_{a,f} - E_{a,r}$$

Although ΔE is a thermodynamic quantity, obtainable from calorimetric measurements, $E_{a,f}$ and $E_{a,r}$ must be found from the temperature dependence of the rate constants for the forward and reverse reactions. In this reaction, the forward and reverse activation energies are 132 and 358 kJ mol^{-1}, respectively, and ΔE from calorimetry is -226 kJ mol^{-1}.

Some elementary reactions are *barrierless* and so have an activation energy of zero and no well-defined transition state. In a reaction between an ion and a molecule, the attractive ion-dipole or ion-induced dipole force (see Section 6–1) may more than counteract any increase in potential energy due to bond reorganization. Some reactions between neutral particles are also barrierless. For example, the rate constant for the bimolecular elementary step

$$H(g) + NO_2(g) \longrightarrow OH(g) + NO(g)$$

does not change over the temperature range from 195 to 2000 K; this implies that E_a equals zero.

Remarkably, multistep reactions sometimes even have *negative* activation energies, reflecting the fact that the overall rate of reaction slows at higher temperature. How can this be? Let us examine a specific example: the reaction of NO with oxygen

$$2\,NO(g) + O_2(g) \longrightarrow 2\,NO_2(g)$$

for which the observed rate expression is

$$\text{rate} = k_{\text{obs}}\,[NO]^2[O_2]$$

where k_{obs} *decreases* with increasing temperature. In Section 14–5, we show that this rate expression can be accounted for with a two-step mechanism. The first step is a rapid equilibrium (with equilibrium constant K_1) of two NO molecules with their dimer, N_2O_2. The second step is the slow reaction (with rate constant k_2) of N_2O_2 with O_2 to form products. The observed rate constant is therefore the product of k_2 and K_1. Although k_2 increases with increasing temperature, K_1 is an equilibrium constant. Provided that the corresponding reaction is sufficiently exothermic (as it is in this case), K_1 decreases so rapidly with increasing temperature that the product k_2K_1 decreases as well.

14–7 KINETICS OF CATALYSIS

A **catalyst** is a substance that takes part in a chemical reaction and speeds it up but itself undergoes no permanent chemical change. Catalysts therefore do not appear in overall balanced chemical equations, but their presence very much affects the rate expression, modifying and speeding existing pathways or, more commonly, providing completely new paths by which a reaction can occur (Fig. 14–15). Catalysts exert significant effects on reaction rates even when they are present in very small amounts. Great effort in industrial chemistry is devoted to finding catalysts that accelerate particular desired reactions without increasing the production of undesired products.

Catalysis can be classified into two types: homogeneous and heterogeneous. In **homogeneous catalysis,** the catalyst is present in the same phase as the reactants, as when a gas-phase catalyst speeds up a gas-phase reaction or a species dissolved in solution speeds up a reaction in solution. An example of homogeneous catalysis is the effect of silver ions on the oxidation–reduction reaction:

$$Tl^+(aq) + 2\,Ce^{4+}(aq) \longrightarrow Tl^{3+}(aq) + 2\,Ce^{3+}(aq)$$

The direct reaction of Tl^+ with a single Ce^{4+} ion to give Tl^{2+} as an intermediate is slow. The reaction can be speeded by the addition of Ag^+ ion, which takes part in a reaction mechanism of the form

$$Ag^+ + Ce^{4+} \underset{k_{-1}}{\overset{k_1}{\rightleftharpoons}} Ag^{2+} + Ce^{3+} \quad \text{(fast equilibrium)}$$

$$Tl^+ + Ag^{2+} \overset{k_2}{\longrightarrow} Tl^{2+} + Ag^+ \quad \text{(slow)}$$

$$Tl^{2+} + Ce^{4+} \overset{k_3}{\longrightarrow} Tl^{3+} + Ce^{3+} \quad \text{(fast)}$$

• Although this second step is "slow," it is much faster than the corresponding step in the uncatalyzed mechanism: $Tl^+ + Ce^{4+} \longrightarrow Tl^{2+} + Ce^{3+}$.

The Ag^+ ions are not permanently transformed by this reaction because those used up in the first step are regenerated in the second; they play the role of catalyst in significantly speeding the rate of the overall reaction. The $H^+(aq)$ and $OH^-(aq)$ ions often act as homogeneous catalysts, especially in biochemistry. Recall from Exercise 14–3 that $H^+(aq)$ speeds the conversion of sucrose to fructose and glucose; it can play a similar role in breaking proteins into smaller molecules.

In **heterogeneous catalysis,** the catalyst is present as a distinct phase. The most important case is the catalytic action of certain solid surfaces (Fig. 14–16) on gas-phase and solution-phase reactions. A critical step in the production of sulfuric acid, for example, involves the use of a solid oxide of vanadium (V_2O_5) as catalyst. Many other solid catalysts are used in industrial processes. One of the best-studied of such reactions is the addition of hydrogen to ethylene (C_2H_4) to form ethane (C_2H_6):

• Sulfuric acid production is examined further in Section 22-2.

$$C_2H_4(g) + H_2(g) \longrightarrow C_2H_6(g)$$

The process occurs extremely slowly in the gas phase but is catalyzed by the presence of a platinum surface. Platinum speeds the reaction by causing H_2 molecules to dissociate to hydrogen atoms attached to its surface (Fig. 14–17). The atoms can then react sequentially with ethylene molecules to form ethane. Heterogeneous catalysis is also important in the atmosphere, where small solid particles (aerosols) can speed up reactions that would occur slowly in the gas phase.

A catalyst speeds up a reaction by increasing the pre-exponential factor A in the Arrhenius equation or, more often, by providing an alternative mechanism with a lower activation energy E_a. It lowers E_a by providing a new activated complex of lower potential energy (Fig. 14–18). A corollary follows: the same catalyst speeds both the forward *and* reverse reactions, because it lowers both the forward and reverse activation energies equally. However:

A catalyst has no effect on the thermodynamics of the overall reaction.

The change in the Gibbs function (ΔG) is independent of the path followed; therefore, the equilibrium constant is never changed by catalytic action. Reaction products that are not favored thermodynamically cannot be caused to form by catalysis.

Figure 14–16 Several solid catalysts used by the petrochemical industry.

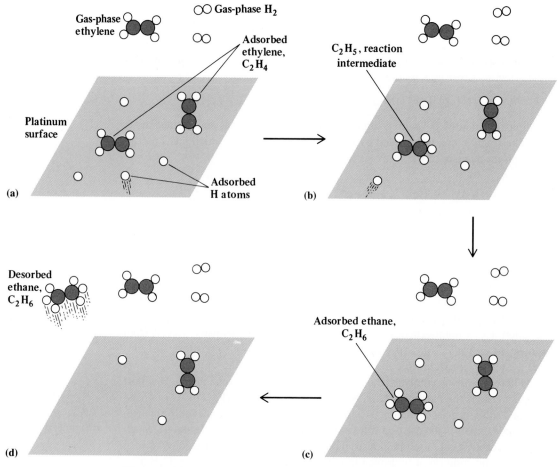

Figure 14–17 Platinum catalyzes the reaction $H_2 + C_2H_4$ by providing a surface that promotes the dissociation of H_2 to H atoms, which can then add stepwise to the C_2H_4 to give ethane (C_2H_6).

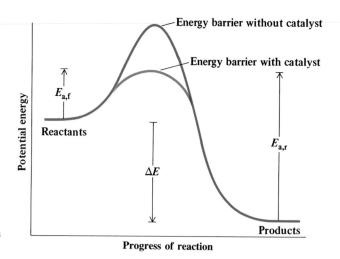

Figure 14–18 The most important way in which catalysts speed up reactions is by reducing the activation energy.

CHEMISTRY IN YOUR LIFE

Catalytic Converters

One of the most widely used chemical catalysts in everyday life is found in the **catalytic converters** installed in the exhaust manifolds of automobiles. The combustion of gasoline in the engine of a car produces carbon dioxide and water as primary products, but some incomplete combustion always occurs as well. The result is a mixture of carbon monoxide (CO) and hydrocarbon fragments that can be represented as C_mH_n. In addition, the high temperatures in the engine cause the formation of oxides of nitrogen (NO and NO_2) from the nitrogen and oxygen in the air. If released in quantity into the air, these gases are further converted into photochemical smog, the notorious and unpleasant haze that obscures many cities and can impair respiratory health.

A number of changes in automobile design have been used to reduce this pollution. One major step was the introduction of catalysts into the exhaust stream of the car, to convert the pollutant gases into less harmful forms before releasing them into the air. These catalysts have a dual mission. First, they must speed up the *oxidation* of unburned hydrocarbons to carbon dioxide and water:

$$CO, C_mH_n, O_2 \xrightarrow{\text{catalyst}} CO_2, H_2O$$

Second, they must accelerate the *reduction* of nitrogen oxides to nitrogen and oxygen:

$$NO, NO_2 \xrightarrow{\text{catalyst}} N_2, O_2$$

Both these processes are favored by thermodynamics at low temperatures but are far too slow to take place before the gases leave the tailpipe.

Figure 14-A The catalytic converter of an automobile, cut open to expose the pellets that contain the platinum/palladium/rhodium catalyst.

Catalysts that are good at speeding up oxidation reactions are usually poor at speeding the reduction of nitrogen oxides. A combination of two catalysts is therefore used. The noble metals, although expensive, are particularly useful catalysts. Typically, platinum and rhodium are deposited on a fine honeycomb mesh of alumina (Al_2O_3), giving a large surface area that increases the contact time of the exhaust gas with the catalysts (Fig. 14–A). The platinum serves primarily as an oxidation catalyst and the rhodium as a reduction catalyst. Catalytic converters can be poisoned with certain metals that block their active sites and reduce their effectiveness. Lead is one of the most serious of such poisons; as a result, unleaded fuel must be used in automobiles with catalytic converters.

The role of catalysts is to speed up the rate of production of products that *are* allowed by thermodynamics.

An **inhibitor** is a negative catalyst. It slows the rate of a reaction, frequently by barring access to a path of low E_a and thereby forcing the reaction to proceed by a path of higher E_a. Inhibitors are also important industrially because they can be used to reduce the rates of undesirable side-reactions, allowing desired products to be removed in greater yield.

Many chemical reactions in living systems are accelerated by catalysts called enzymes. An **enzyme** is a large protein molecule (typically of molar mass $20,000 \text{ g mol}^{-1}$ or more) that is both very effective and selective in its catalytic action. One or more reactant molecules (called **substrates**) bind to an enzyme at its **active sites.** These are

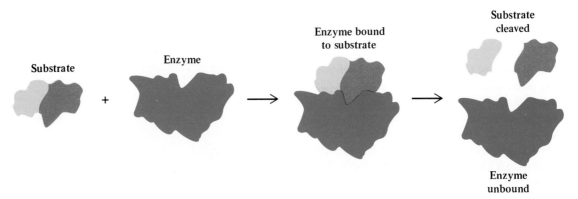

Figure 14–19 The binding of a substrate to an enzyme, and the subsequent reaction of the substrate.

• The ending *-ase* in the name of a substance signals that the substance is an enzyme.

regions of the enzyme molecule where local structure favors quick chemical transformation to a specific product, followed by release of that product (Fig. 14–19). The molecules of most other substances are not accommodated at the active sites. The enzyme urease, for example, catalyzes only the hydrolysis of urea, $(NH_2)_2CO$:

$$H^+(aq) + (NH_2)_2CO(aq) + 2\,H_2O(\ell) \underset{\text{urease}}{\longrightarrow} 2\,NH_4^+(aq) + HCO_3^-(aq)$$

It does not bind most other kinds of molecules, even those of very similar structure. In some cases, a second species *does* bind to the enzyme and can act as an inhibitor, preventing the enzyme from carrying out its usual role as catalyst. Much modern chemical research is aimed at designing catalysts that are as powerful as enzymes in speeding up selected chemical reactions.

SUMMARY

14–1 The **average rate** of a chemical reaction is given by the change in the concentration of a reactant or product divided by the elapsed time. The **instantaneous rate** (or simply the rate) is obtained graphically from the slope of the line tangent to a curve obtained by plotting the concentration of a reactant or product against time. Such slopes are divided by the coefficient of the chemical species in the balanced equation, with a minus sign for reactants and a plus sign for products. In general, the rates of reactions change from moment to moment. An **initial rate** is the rate at the instant a reaction begins.

14–2 An **experimental rate expression** or **rate law** gives the dependence of the observed rate of a reaction on the concentrations of various species. In a rate expression, the rate equals a **rate constant** k multiplied by the concentrations of reactant, product, or other species raised to integral or fractional powers. The powers are the **orders** of the reaction with respect to each chemical species, and their sum is the **overall reaction order.**

14–3 The **half-life** of a reaction is the time it takes for the concentration of a reacting species to fall to half its initial value. This concept is most useful for first-order reactions. From a rate expression, explicit equations giving the dependence of reactant or product concentrations on time can be devised for zero-order and (using calculus) for first-order and second-order reactions. By graphing appropriate functions of the concentration (the concentration itself for zero-order reactions, the logarithm for first-order, the reciprocal for second-order) against time, the order of the reaction can be deduced and the rate constant evaluated.

14–4 An **elementary step** involves a direct collision of atoms, ions, or molecules. Such steps may be **unimolecular, bimolecular,** or **termolecular,** according to whether one, two, or three colliding bodies take part. A reaction **mechanism** consists of a series of elementary steps that add to give the overall reaction and that detail the progress from reactants to products. The **principle of detailed balance** states that, at equilibrium, the rate of every elementary step in a reaction mechanism equals the rate of its reverse. Therefore, the product of the forward rate constants in a reaction mechanism divided by the product of the reverse rate constants gives the equilibrium constant for the reaction; the reaction **intermediates** cancel out of the final expression.

14–5 A major goal of chemical kinetics is to determine reaction mechanisms from the experimental rate expressions for the reactions. If one step is much slower than the others, it is possible to write the form of the rate expression for any given mechanism and compare it with an experiment. A particularly important type of reaction is the **chain reaction,** the mechanism of which involves three types of steps: **initiation** (in which highly reactive intermediates are produced), **propagation** (in which reactants are converted to products as reactive intermediates are regenerated), and **termination** (in which the reactive intermediates combine to form stable species and the reaction ends).

14–6 Reaction rates usually depend strongly on temperature. In a reaction, the reactants come together to form a high-energy **activated complex.** In this **transition state,** bonds are distorted to make an entity that is intermediate between reactants and products. Formation of an activated complex requires both sufficient energy in the collision of the reacting particles and proper orientation (the **steric factor**). The difference between the energy of this activated complex and the energy of the reactants is the **activation energy** for the forward reaction. The activation energy and the temperature are related to the rate constant by the **Arrhenius equation.**

14–7 A **catalyst** speeds up a reaction by increasing the pre-exponential factor A in the Arrhenius equation or, more often, by providing an alternative mechanism with a lower activation energy E_a. **Inhibitors** have the opposite effect. Catalysis can be **homogeneous** (occurring within a single phase) or **heterogeneous** (occurring at the boundary between two phases, as on the surface of a solid). An especially important group of catalysts are **enzymes,** which selectively speed up certain reactions in living systems. Enzymes are large, complex molecules that are capable of highly specific interactions (at their **active sites**) with the reacting **substrate** molecules. The interactions promote specific reaction pathways.

PROBLEMS

Note: Answers to blue-numbered problems are given in Appendix F. Problems that are more challenging are indicated with asterisks.

Rates of Chemical Reactions

1. A car going north on an interstate highway passes mile marker 11.5 at 11:46 A.M. and mile marker 30.0 at 12:02:24 P.M. Determine its average rate of speed in miles per minute and in miles per hour.

2. (See Example 14–1.) A reaction generates carbon dioxide in a vessel of constant volume at a temperature of 25°C. The partial pressure of the $CO_2(g)$ is 0.520 atm at 11:00:00 A.M. and 0.645 atm at 11:30:00 A.M. the same day. Compute the average rate of appearance of $CO_2(g)$ during this period of time (in mol $L^{-1} s^{-1}$).

3. A cross-country hitchhiker gets a ride in New York that takes him 550 mi west in 11 h and lets him out at a truck stop. It takes 2 h before he gets a new ride that takes him another 300 mi in 5.5 h. Determine his average rate of travel after 11 h, after 12 h, after 13 h, and after 18.5 h.

4. (See Example 14–1.) A chemical reaction generates 1.0 mol L^{-1} of product in the first minute after the reactants are mixed. In the second minute, it generates 0.5 mol L^{-1} of the product, and in the third minute, it generates 0.25 mol L^{-1} of product. Determine the average rate of the reaction in mol $L^{-1} s^{-1}$ during the first 60 s, during the first 120 s, and during the last 120 s.

5. (See Example 14–1.) Use Figure 14–3 to estimate graphically the instantaneous rate of production of NO at $t = 200$ s.

6. (See Example 14–1.) Use Figure 14–3 to estimate graphically the instantaneous rate of production of NO at $t = 100$ s.

7. Give three related expressions for the rate of the reaction

$$N_2(g) + 3 H_2(g) \longrightarrow 2 NH_3(g)$$

assuming that the concentrations of any intermediates are constant and that the volume of the reaction vessel does not change.

8. Give four related expressions for the rate of the reaction

$$2 H_2CO(g) + O_2(g) \longrightarrow 2 CO(g) + 2 H_2O(g)$$

assuming that the concentrations of any intermediates are constant and that the volume of the reaction vessel does not change.

9. At one moment in the course of the reaction $N_2(g) + 3 H_2(g) \longrightarrow 2 NH_3(g)$, the rate of appearance of $NH_3(g)$ is 0.150 mol $L^{-1} s^{-1}$.
   (a) What is the rate of disappearance of $H_2(g)$ at that instant?
   (b) What is the rate of disappearance of $N_2(g)$ at that instant?

10. In the reaction $SO_2(g) + \frac{1}{3} O_3(g) \longrightarrow SO_3(g)$, the rate of disappearance of the reactant $SO_2(g)$ at a certain instant is 0.100 mol $L^{-1} s^{-1}$.
   (a) What is the rate of disappearance of $O_3(g)$ at that moment?
   (b) What is the rate of appearance of $SO_3(g)$ at that instant?

Reaction Rates and Concentrations

11. Determine the overall order of the reaction, given the rate law:
   (a) Rate = $k [NO_2]^2$.
   (b) Rate = $k [I_2]$.

12. Determine the overall order of the reaction, given the rate law:
   (a) Rate = $k [CH_3CHO]^{3/2}$.
   (b) Rate = k.

13. Give the units for the rate constant k of a third-order reaction, if concentrations are measured in mol L^{-1} and time is measured in seconds.

14. Give the units for the rate constant k of a one-half–order reaction, if concentrations are measured in mol L^{-1} and time is measured in seconds.

15. (See Example 14–2.) The following data were obtained for the reaction

$$2 NOBr(g) \longrightarrow 2 NO(g) + Br_2(g):$$

[NOBr] (mol L^{-1})	Rate (mol $L^{-1} s^{-1}$)
0.0130	1.35×10^{-4}
0.0088	6.2×10^{-5}
0.0049	1.92×10^{-5}

   (a) Write the rate expression for this reaction.
   (b) Calculate the rate constant.

16. (See Example 14–2.) The following data were obtained for the reaction $C_4H_8(g) \longrightarrow$ products at 750 K:

[C_4H_8] (mol L^{-1})	Rate (mol $L^{-1} s^{-1}$)
0.0095	3.42×10^{-5}
0.0070	2.52×10^{-5}
0.0025	9.0×10^{-6}

   (a) Write the rate expression for this reaction.
   (b) Calculate the rate constant.

17. (See Example 14–3.) At a fixed pH of 2.0, the rate expression for the reaction $2 H^+(aq) + 2 I^-(aq) + H_2O_2(aq) \longrightarrow I_2(aq) + 2 H_2O(\ell)$ is

$$\text{rate} = k [H_2O_2][I^-]$$

State the effect on the rate of:
   (a) increasing the concentration of I^- by a factor of 10.
   (b) decreasing the concentration of H_2O_2 by a factor of 4.

18. (See Example 14–3.) The rate expression for the reaction $2 ClO_2(aq) + 2 OH^-(aq) \longrightarrow ClO_3^-(aq) + ClO_2^-(aq) + H_2O(\ell)$ is

$$\text{rate} = k [ClO_2]^2[OH^-]$$

State the effect on the rate of:
(a) increasing the concentration of ClO_2 by a factor of 3.
(b) changing the pH from 12.0 to 11.0.

19. Nitrogen monooxide reacts with hydrogen at elevated temperatures according to the following chemical equation:

$$2 NO(g) + 2 H_2(g) \longrightarrow N_2(g) + 2 H_2O(g)$$

It is observed that when the concentration of H_2 is cut in half, the rate of the reaction is also cut in half. When the concentration of NO is multiplied by 10, the rate of the reaction increases by a factor of 100.
(a) Write the rate expression for this reaction, and give the units of the rate constant k.
(b) If [NO] were multiplied by 3 and [H_2] by 2, what change in the rate would be observed?

20. In the presence of vanadium(V) oxide, $SO_2(g)$ reacts with an excess of oxygen to give $SO_3(g)$:

$$SO_2(g) + \tfrac{1}{2} O_2(g) \xrightarrow[V_2O_5]{} SO_3(g)$$

This reaction is an important step in the manufacture of sulfuric acid. It is observed that tripling the SO_2 concentration increases the rate by a factor of three, but tripling the SO_3 concentration *decreases* the rate by a factor of $1.7 \approx \sqrt{3}$. The rate is insensitive to the O_2 concentration as long as an excess of oxygen is present.
(a) Write the rate expression for this reaction, and give the units of the rate constant k.
(b) If [SO_2] is multiplied by 2 and [SO_3] by 4 but all other conditions are unchanged, what change in the rate is observed?

21. (See Example 14–4.) In a study of the reaction of pyridine (C_5H_5N) with methyl iodide (CH_3I) in a benzene solution, the following set of initial reaction rates was measured at 25°C for different initial concentrations of the two reactants:

$[C_5H_5N]$ (mol L^{-1})	$[CH_3I]$ (mol L^{-1})	Rate (mol L^{-1} s^{-1})
1.00×10^{-4}	1.00×10^{-4}	7.5×10^{-7}
2.00×10^{-4}	2.00×10^{-4}	3.0×10^{-6}
2.00×10^{-4}	4.00×10^{-4}	6.0×10^{-6}

(a) Write the rate expression for this reaction.
(b) Calculate the rate constant k, and give its units.
(c) Predict the initial reaction rate that would be seen in a solution in which [C_5H_5N] is 5.0×10^{-5} M and [CH_3I] is 2.0×10^{-5} M.

22. (See Example 14–4.) The initial rate for the oxidation of iron(II) by cerium(IV)

$$Ce^{4+}(aq) + Fe^{2+}(aq) \longrightarrow Ce^{3+}(aq) + Fe^{3+}(aq)$$

is measured at several different initial concentrations of the two reactants:

$[Ce^{4+}]$ (mol L^{-1})	$[Fe^{2+}]$ (mol L^{-1})	Rate (mol L^{-1} s^{-1})
1.1×10^{-5}	1.8×10^{-5}	2.0×10^{-7}
1.1×10^{-5}	2.8×10^{-5}	3.1×10^{-7}
3.4×10^{-5}	2.8×10^{-5}	9.5×10^{-7}

(a) Write the rate expression for this reaction.
(b) Calculate the rate constant k, and give its units.
(c) Predict the initial reaction rate for a solution in which [Ce^{4+}] is 2.6×10^{-5} M and [Fe^{2+}] is 1.3×10^{-5} M.

23. (See Example 14–4.) For the reaction $Cr(H_2O)_6^{3+}(aq) + SCN^-(aq) \longrightarrow Cr(H_2O)_5SCN^{2+}(aq) + H_2O(\ell)$ the following data were obtained:

$[Cr(H_2O)_6^{3+}]$ (mol L^{-1})	$[SCN^-]$ (mol L^{-1})	Rate (mol^{-1} s^{-1})
0.021	0.045	1.9×10^{-9}
0.021	0.084	3.5×10^{-9}
0.053	0.084	8.9×10^{-9}

(a) Determine the rate expression for this reaction.
(b) Calculate the rate constant k, and give its units.

24. (See Example 14–4.) For the reaction $2 HgCl_2(aq) + C_2O_4^{2-}(aq) \longrightarrow Hg_2Cl_2(s) + 2 Cl^-(aq) + 2 CO_2(g)$ the following data were obtained:

$[HgCl_2]$ (mol L^{-1})	$[C_2O_4^{2-}]$ (mol L^{-1})	Rate (mol L^{-1} s^{-1})
0.096	0.13	2.1×10^{-7}
0.096	0.21	5.5×10^{-7}
0.171	0.21	9.8×10^{-7}

(a) Determine the rate expression for this reaction.
(b) Calculate the rate constant k, and give its units.

The Dependence of Concentrations on Time

25. (See Example 14–5.) The reaction $SO_2Cl_2(g) \longrightarrow SO_2(g) + Cl_2(g)$ is first order, with a half-life of 4.5×10^4 s at 320°C.
(a) Calculate the rate constant k.
(b) What fraction of the SO_2Cl_2 initially present has *not* reacted after 16.2 h?

26. (See Example 14–5.) The reaction $FClO_2(g) \longrightarrow FClO(g) + O(g)$ is first order, with a half-life of 1.48×10^3 s at 322°C.
(a) Calculate the rate constant k.
(b) What fraction of the $FClO_2$ initially present has reacted after 2.30 h?

27. Chloroethane decomposes at elevated temperatures according to the equation

$$C_2H_5Cl(g) \longrightarrow C_2H_4(g) + HCl(g)$$

This reaction obeys first-order kinetics. After 340 s at 800 K, a measurement shows that the concentration of C_2H_5Cl has decreased from 0.0098 mol L^{-1} to 0.0016 mol L^{-1}. Calculate the rate constant k at 800 K.

28. The isomerization reaction

$$CH_3NC \longrightarrow CH_3CN$$

obeys the first-order rate law

$$\text{rate} = -k\,[CH_3NC]$$

in the presence of an excess of argon. Measurements at 500 K reveal that in 520 s the concentration of CH_3NC decreases to 71% of its original value. Calculate the rate constant k of the reaction at 500 K.

29. The first-order reaction A $\longrightarrow$ products has the following rate law: rate $= k\,[A]$ with $k = 2.0\ s^{-1}$. Indicate which of the following statements about this reaction are true and which are false. Explain your reasoning.
(a) The reaction slows down as time goes on.
(b) After 2.0 s the reaction is over.
(c) The rate of the reaction doubles if the concentration of A is doubled while all other conditions remain unchanged.
(d) The half-life of the reaction is 2.0 s.
(e) The half-life of the reaction is 0.50 s.

30. The second-order reaction 2 B $\longrightarrow$ products has the following rate law: rate $= k\,[B]^2$ with $k = 2.0\ L\ mol^{-1}\ s^{-1}$. Indicate which of the following statements about this reaction are true and which are false. Explain your reasoning.
(a) The concentration of B increases as time goes on.
(b) After 1.0 s the reaction is half over.
(c) The rate of the reaction quadruples if the concentration of B is doubled while all other conditions remain unchanged.
(d) A plot of [B] versus time gives a straight line.
(e) A plot of 1/[B] versus time gives a straight line.

31. (See Example 14–6.) At 25°C in CCl_4 solution, the reaction

$$I + I \longrightarrow I_2$$

is second order in the concentration of the iodine atoms. The rate constant k has been measured as $8.2 \times 10^9\ L\ mol^{-1}\ s^{-1}$. Suppose that the initial concentration of I atoms is 1.00×10^{-4} mol L^{-1}. Calculate their concentration after 2.0×10^{-6} s.

32. (See Example 14–6.) HO_2 is a highly reactive chemical species that plays a role in atmospheric chemistry. The rate of the gas-phase reaction

$$HO_2(g) + HO_2(g) \longrightarrow H_2O_2(g) + O_2(g)$$

is second order in $[HO_2]$, with a rate constant at 25°C of $1.4 \times 10^9\ L\ mol^{-1}\ s^{-1}$. Suppose some HO_2 could be confined at an initial concentration of 2.0×10^{-18} M at 25°C. Calculate the concentration that would remain after 1.0 s, assuming that no other reactions take place.

33. The decomposition of N_2O_5 dissolved in CCl_4 has a rate constant of $5.5 \times 10^{-4}\ s^{-1}$ at a certain temperature. A 0.200 M solution of N_2O_5 is prepared at this temperature. Plot the concentration of N_2O_5 as a function of time for the first hour

(3600 s) after mixing, assuming that no back-reaction of the products occurs.

34. At a certain temperature, the dimerization of $C_2F_4(g)$ to $C_4F_8(g)$ has a rate constant of 0.0600 $L\ mol^{-1}\ s^{-1}$. A container is prepared in which the initial concentration of C_2F_4 is 0.200 mol L^{-1}.
(a) Plot the concentration of $C_2F_4(g)$ as a function of time for the three minutes (180 s) after the reaction starts, assuming that no back-reaction occurs.
(b) On the same graph paper, plot the concentration of $C_4F_8(g)$ as a function of time over the same period, making the same assumption.

Reaction Mechanisms

35. Identify each of the following elementary reactions as unimolecular, bimolecular, or termolecular, and write the rate expression:
(a) $HCO + O_2 \longrightarrow HO_2 + CO$
(b) $CH_3 + O_2 + N_2 \longrightarrow CH_3O_2 + N_2$
(c) $HO_2NO_2 \longrightarrow HO_2 + NO_2$

36. Identify each of the following elementary reactions as unimolecular, bimolecular, or termolecular, and write the rate expression.
(a) $BrONO_2 \longrightarrow BrO + NO_2$
(b) $HO + NO_2 + Ar \longrightarrow HNO_3 + Ar$
(c) $O + H_2S \longrightarrow HO + HS$

37. (See Example 14–7.) Consider the following reaction mechanism:

$$H_2O_2 \longrightarrow H_2O + O$$

$$O + CF_2Cl_2 \longrightarrow ClO + CF_2Cl]$$

$$ClO + O_3 \longrightarrow Cl + 2\ O_2$$

$$Cl + CF_2Cl \longrightarrow CF_2Cl_2$$

(a) What is the molecularity of each elementary step?
(b) Write the overall equation for the reaction.
(c) Identify the reaction intermediate(s).

38. (See Example 14–7.) Consider the following reaction mechanism:

$$NO_2Cl \longrightarrow NO_2 + Cl$$

$$Cl + H_2O \longrightarrow HCl + OH$$

$$OH + NO_2 + N_2 \longrightarrow HNO_3 + N_2$$

(a) What is the molecularity of each elementary step?
(b) Write the overall equation for the reaction.
(c) Identify the reaction intermediate(s).

39. The rate constant of the elementary reaction

$$BrO(g) + NO(g) \longrightarrow Br(g) + NO_2(g)$$

is $1.3 \times 10^{10}\ L\ mol^{-1}\ s^{-1}$ at 25°C, and its equilibrium constant is 5.0×10^{10} at this temperature. Calculate the rate constant at 25°C of the elementary reaction

$$Br(g) + NO_2(g) \longrightarrow BrO(g) + NO(g)$$

40. The compound $IrH_3(CO)(P(C_6H_5)_3)_2$ exists in two forms, the meridional ("mer") and the facial ("fac"). At 25°C in a non-aqueous solvent, the reaction mer $\longrightarrow$ fac has a rate constant of 2.33 s^{-1}, and the reaction fac $\longrightarrow$ mer has a rate constant of 2.10 s^{-1}. What is the equilibrium constant of the mer-to-fac reaction at 25°C?

Reaction Mechanism and Rate

41. (See Example 14–8.) Write the overall reaction and the rate expressions that correspond to the following reaction mechanisms. Be sure to eliminate intermediates from the answers:

(a) $A + B \underset{k_{-1}}{\overset{k_1}{\rightleftharpoons}} C + D$ (fast equilibrium)

    $C + E \overset{k_2}{\longrightarrow} F$ (slow)

(b) $A \underset{k_{-1}}{\overset{k_1}{\rightleftharpoons}} B + C$ (fast equilibrium)

    $C + D \underset{k_{-2}}{\overset{k_2}{\rightleftharpoons}} E$ (fast equilibrium)

    $E \overset{k_3}{\longrightarrow} F$ (slow)

42. (See Example 14–8.) Write the overall reaction and the rate expressions that correspond to the following mechanisms. Be sure to eliminate intermediates from the answers:

(a) $2A + B \underset{k_{-1}}{\overset{k_1}{\rightleftharpoons}} Dt$ (fast equilibrium)

    $D + B \overset{k_2}{\longrightarrow} E + F$ (slow)

    $F \overset{k_3}{\longrightarrow} G$ (fast)

(b) $A + B \underset{k_{-1}}{\overset{k_1}{\rightleftharpoons}} C$ (fast equilibrium)

    $C + D \underset{k_{-2}}{\overset{k_2}{\rightleftharpoons}} F$ (fast equilibrium)

    $F \overset{k_3}{\longrightarrow} G$ (slow)

43. HCl reacts with propene (CH_3CHCH_2) in the gas phase according to the overall reaction

$$HCl + CH_3CHCH_2 \longrightarrow CH_3CHClCH_3$$

The experimental rate expression is

$$\text{rate} = k \, [HCl]^3[CH_3CHCH_2]$$

Which, if any, of the following mechanisms are consistent with the observed rate expression?

(a) $HCl + HCl \rightleftharpoons H + HCl_2$ (fast equilibrium)

    $H + CH_3CHCH_2 \longrightarrow CH_3CHCH_3$ (slow)

    $HCl_2 + CH_3CHCH_3 \longrightarrow CH_3CHClCH_3 + HCl$ (fast)

(b) $HCl + HCl \rightleftharpoons H_2Cl_2$ (fast equilibrium)

    $HCl + CH_3CHCH_2 \rightleftharpoons CH_3CHClCH_3^*$ (fast equilibrium)

    $CH_3CHClCH_3^* + H_2Cl_2 \longrightarrow CH_3CHClCH_3 + 2 \; HCl$ (slow)

(c) $HCl + CH_3CHCH_2 \rightleftharpoons H + CH_3CHClCH_2$ (fast equilibrium)

    $H + HCl \rightleftharpoons H_2Cl$ (fast equilibrium)

$$H_2Cl + CH_3CHClCH_2 \longrightarrow HCl + CH_3CHClCH_3$$
         (slow)

44. Chlorine reacts with hydrogen sulfide in aqueous solution

$$Cl_2(aq) + H_2S(aq) \longrightarrow S(s) + 2 \, H^+(aq) + 2 \, Cl^-(aq)$$

in a second-order reaction that follows the rate expression

$$\text{rate} = k \, [Cl_2][H_2S]$$

Which, if any, of the following mechanisms are consistent with the observed rate expression?

(a) $Cl_2 + H_2S \longrightarrow H^+ + Cl^- + Cl^+ + HS^-$ (slow)

    $Cl^+ + HS^- \longrightarrow H^+ + Cl^- + S$ (fast)

(b) $H_2S \rightleftharpoons HS^- + H^+$ (fast equilibrium)

    $HS^- + Cl_2 \longrightarrow 2 \, Cl^- + S + H^+$ (slow)

(c) $H_2S \rightleftharpoons HS^- + H^+$ (fast equilibrium)

    $H^+ + Cl_2 \rightleftharpoons H^+ + Cl^- + Cl^+$ (fast equilibrium)

    $Cl^+ + HS^- \longrightarrow H^+ + Cl^- + S$ (slow)

45. Nitryl chloride is a reactive gas with a normal boiling point of $-16°C$. Its decomposition to nitrogen dioxide and chlorine is described by the equation

$$2 \, NO_2Cl \longrightarrow 2 \, NO_2 + Cl_2$$

The rate expression for this reaction has the form

$$\text{rate} = k \, [NO_2Cl]$$

Which, if any, of the following mechanisms are consistent with the observed rate expression?

(a) $NO_2Cl \longrightarrow NO_2 + Cl$ (slow)

    $Cl + NO_2Cl \longrightarrow NO_2 + Cl_2$ (fast)

(b) $2 \, NO_2Cl \rightleftharpoons N_2O_4 + Cl_2$ (fast equilibrium)

    $N_2O_4 \longrightarrow 2 \, NO_2$ (slow)

(c) $2 \, NO_2Cl \rightleftharpoons ClO_2 + N_2O + ClO$ (fast equilibrium)

    $N_2O + ClO_2 \rightleftharpoons NO_2 + NOCl$ (fast equilibrium)

    $NOCl + ClO \longrightarrow NO_2 + Cl_2$ (slow)

46. Ozone in the upper atmosphere is decomposed by nitrogen monoxide through the reaction

$$O_3 + NO \longrightarrow O_2 + NO_2$$

The experimental rate expression for this reaction is

$$\text{rate} = k \, [O_3][NO]$$

Which, if any, of the following mechanisms are consistent with the observed rate expression?

(a) $O_3 + NO \longrightarrow O + NO_3$ (slow)

    $O + O_3 \longrightarrow 2 \, O_2$ (fast)

    $NO_3 + NO \longrightarrow 2 \, NO_2$ (fast)

(b) $O_3 + NO \longrightarrow O_2 + NO_2$ (slow)

(c) $NO + NO \rightleftharpoons N_2O_2$ (fast equilibrium)

    $N_2O_2 + O_3 \longrightarrow NO + NO_2 + O_2$ (slow)

Effect of Temperature on Reaction Rates

47. (See Example 14–9.) The rate of the elementary reaction

$$Ar + O_2 \longrightarrow Ar + O + O$$

has been studied as a function of temperature between 5000 and 18,000 K. The following data were obtained for the rate constant k:

Temperature (K)	k (L mol^{-1} s^{-1})
5,000	5.49×10^6
10,000	9.86×10^8
15,000	5.09×10^9
18,000	8.60×10^9

(a) Calculate the activation energy of this reaction.
(b) Calculate the factor A in the Arrhenius equation for the temperature dependence of the rate constant.

48. (See Example 14–9.) The gas-phase reaction

$$H + D_2 \longrightarrow HD + D$$

is the exchange of isotopes of hydrogen of mass number 1 (H) and 2 (D, deuterium). The following data were obtained for the rate constant k of this reaction:

Temperature (K)	k (L mol^{-1} s^{-1})
299	1.56×10^4
327	3.77×10^4
346	7.6×10^4
440	1.07×10^6
549	8.7×10^6
745	8.7×10^7

(a) Calculate the activation energy of this reaction.
(b) Calculate the factor A in the Arrhenius equation for the temperature dependence of the rate constant.

49. The rate constant of the elementary reaction

$$BH_4^-(aq) + NH_4^+(aq) \longrightarrow BH_3NH_3(aq) + H_2(g)$$

is $k = 1.94 \times 10^{-4}$ L mol^{-1} s^{-1} at 30.0°C, and the reaction has an activation energy of 161 kJ mol^{-1}
(a) Compute the rate constant of the reaction at a temperature of 40.0°C.
(b) After equal concentrations of $BH_4^-(aq)$ and $NH_4^+(aq)$ are mixed at 30.0°C, it requires 1.00×10^4 s for half of them to be consumed. How long does it take to consume half of the reactants if an identical experiment is performed at 40.0°C?

50. Dinitrogen tetraoxide (N_2O_4) decomposes spontaneously at room temperature in the gas phase:

$$N_2O_4(g) \longrightarrow 2 NO_2(g)$$

At 30°C, $k = 5.1 \times 10^6$ s^{-1}, and the activation energy of the reaction is 54.0 kJ mol^{-1}.
(a) Calculate the time it takes, in seconds, for the partial pressure of $N_2O_4(g)$ to decrease from 0.100 atm to 0.099 atm at 30°C.

(b) Repeat the calculation of part (a) for the reaction at 60°C.

51. The activation energy of the isomerization reaction of CH_3NC in problem 28 is 161 kJ mol^{-1}, and the reaction rate constant at 600 K is 0.41 s^{-1}.
(a) Calculate the Arrhenius factor A for this reaction.
(b) Calculate the rate constant for this reaction at 1000 K.

52. Cyclopropane isomerizes to propylene according to a first-order reaction:

$$\text{cyclopropane} \longrightarrow \text{propylene}$$

The activation energy is $E_a = 272$ kJ mol^{-1}. At 500°C, the reaction rate constant is 6.1×10^{-4} s^{-1}.
(a) Calculate the Arrhenius factor A for this reaction.
(b) Calculate the rate constant for this reaction at 25°C.

53. The activation energy of the gas-phase reaction

$$OH(g) + HCl(g) \longrightarrow H_2O(g) + Cl(g)$$

is 3.5 kJ mol^{-1} and the change in the internal energy in the reaction is $\Delta E = -66.8$ kJ mol^{-1}. Calculate the activation energy of the reaction

$$H_2O(g) + Cl(g) \longrightarrow OH(g) + HCl(g)$$

54. The compound HOCl is known, but the related compound HClO, with a different order for the atoms in the molecule, is not known. Calculations suggest that the activation energy of the conversion HOCl $\longrightarrow$ HClO is 311 kJ mol^{-1} and that of the conversion HClO $\longrightarrow$ HOCl is 31 kJ mol^{-1}. Estimate ΔE for the reaction HOCl $\longrightarrow$ HClO.

Kinetics of Catalysis

55. How would you describe the role of the CF_2Cl_2 in the reaction mechanism of problem 37?

56. Compare homogeneous catalysis with heterogeneous catalysis, and give an example of each.

Additional Problems

57. A freight train traveling west starts up a long mountain slope at a speed of 60.0 miles per hour (mph). By the end of the first minute, its speed has slowed to 59.0 mph, and it slows down by 1.0 mph every minute thereafter, because of the steepness of the grade. Compute its average speed during the first minute, during the first 3 min, during the second 3 min, and during the first 20 min.

58. Jones writes a chemical reaction as $C_2H_6 \longrightarrow C_2H_4 + H_2$ and concludes that it is a first-order reaction. Smith writes it as $2\,C_2H_6 \longrightarrow 2\,C_2H_4 + 2\,H_2$ and concludes that it is a second-order reaction. Both are amazed to learn that the reaction is 3/2 order. Explain the misunderstanding.

59. Hemoglobin (Hb) molecules in blood bind oxygen and carry it to cells, where it takes part in metabolism. The first step in the binding of oxygen

$$Hb(aq) + O_2(aq) \longrightarrow (HbO_2)(aq)$$

is first order in hemoglobin and first order in dissolved oxygen, with a rate constant of 4×10^7 L mol^{-1} s^{-1}. Calculate the initial rate at which oxygen is bound to hemoglobin if the concentration of hemoglobin is 2×10^{-9} mol L^{-1} and that of oxygen is 5×10^{-5} mol L^{-1}.

*60. A compound called di-*t*-butyl peroxide (abbreviation: DTBP, formula: $(CH_3)_3COOC(CH_3)_3$) decomposes to give acetone $((CH_3)_2CO)$ and ethane (C_2H_6):

$$(CH_3)_3COOC(CH_3)_3(g) \longrightarrow 2\,(CH_3)_2CO(g) + C_2H_6(g)$$

The *total* pressure of the reaction mixture changes with time as shown by the following data at 147.2°C:

Time (min)	P_{tot} (atm)	Time (min)	P_{tot} (atm)
0	0.2362	26	0.3322
2	0.2466	30	0.3449
6	0.2613	34	0.3570
10	0.2770	38	0.3687
14	0.2911	40	0.3749
18	0.3051	42	0.3801
20	0.3122	46	0.3909
22	0.3188		

(a) Calculate the partial pressure of DTBP at each time from these data. Assume that at time 0, DTBP is the only gas present. Recall that each decrease by x atm in the partial pressure of DTBP is accompanied by an increase of $2x$ in the total pressure because every gas molecule that dissociates gives three new gas molecules.

(b) Are the data better described by a first-order or a second-order rate expression with respect to DTBP concentration?

61. 8.23×10^{-3} mol of InCl(s) is placed in 1.00 L of 0.010 M HCl(aq) at 75°C. The InCl(s) dissolves quite quickly, and then the following reaction occurs

$$3\,In^+(aq) \longrightarrow 2\,In(s) + In^{3+}(aq)$$

As this disproportionation proceeds, the solution is analyzed at intervals to determine the concentration of $In^+(aq)$ that remains.

Time (s)	$[In^+]$ (mol L^{-1})
0	8.23×10^{-3}
240	6.41×10^{-3}
480	5.00×10^{-3}
720	3.89×10^{-3}
1000	3.03×10^{-3}
1200	3.03×10^{-3}
10,000	3.03×10^{-3}

(a) Plot ln $[In^+]$ versus time, and determine the apparent rate constant for this first-order reaction.

(b) Determine the half-life of this reaction.

(c) Determine the equilibrium constant K for the reaction under the experimental conditions.

*62. Carbon dioxide reacts with ammonia to give ammonium carbamate, a solid. The reverse reaction also occurs:

$$CO_2(g) + 2\,NH_3(g) \rightleftharpoons NH_4OCONH_2(s)$$

The forward reaction is first order in $CO_2(g)$ and second order in $NH_3(g)$. Its rate constant is 0.238 atm^{-2} s^{-1} at 0.0°C (expressed in terms of partial pressures rather than concentrations). The reaction in the reverse direction is zero order, and its rate constant, at the same temperature, is 1.60×10^{-7} atm s^{-1}. Experimental studies show that, at all stages in the progress of this reaction, the net rate is equal to the forward rate minus the reverse rate. Compute the equilibrium constant of this reaction at 0.0°C.

63. The reaction of OH$^-$ with HCN in aqueous solution at 25°C has a forward rate constant k_f of 3.7×10^9 L mol^{-1} s^{-1}. Using this information and the measured acidity constant of HCN (see Table 8–2), calculate the rate constant k_r in the first-order rate law

$$\text{rate} = k_r[CN^-]$$

for the transfer of hydrogen ions to CN$^-$ from surrounding water molecules,

$$H_2O(\ell) + CN^-(aq) \longrightarrow OH^-(aq) + HCN(aq)$$

*64. When pure solid magnesium perchlorate $(Mg(ClO_4)_2)$ is heated to 440°C, it first breaks down to give magnesium chloride $(MgCl_2)$ and oxygen. Then, as the product magnesium chloride builds up, it starts to react with the magnesium perchlorate to generate magnesium oxide, chlorine, and oxygen. Pure magnesium chloride undergoes no changes at 440°C.

(a) Write balanced chemical equations to represent the two processes.

(b) There is always some magnesium chloride mixed with the magnesium oxide residue from the decomposition of magnesium perchlorate at 440°C. On the basis of this fact, determine which of the two reactions in part (a) goes faster.

65. The gas-phase decomposition of acetaldehyde can be represented by the overall chemical equation

$$CH_3CHO \longrightarrow CH_4 + CO$$

It is thought to occur through the sequence of reactions

$$CH_3CHO \longrightarrow CH_3 + CHO$$
$$CH_3 + CH_3CHO \longrightarrow CH_4 + CH_2CHO$$
$$CH_2CHO \longrightarrow CO + CH_3$$
$$CH_3 + CH_3 \longrightarrow CH_3CH_3$$

Show that this reaction mechanism corresponds to a chain reaction, and identify the initiation, propagation, and termination steps.

66. Lanthanum(III) phosphate crystallizes as a hemihydrate, $LaPO_4 \cdot \frac{1}{2} H_2O$. When it is heated, it loses water to give anhydrous lanthanum(III) phosphate.

$$2(LaPO_4 \cdot \tfrac{1}{2}H_2O(s)) \longrightarrow 2\, LaPO_4(s) + H_2O(g)$$

This reaction is first order in the chemical amount of the reactant. The rate constant varies with temperature as follows

Temperature (°C)	k (s^{-1})
205	2.3×10^{-4}
219	3.69×10^{-4}
246	7.75×10^{-4}
260	12.3×10^{-4}

Compute the activation energy of this reaction.

67. The water in a pressure cooker boils at a temperature greater than 100°C because it is under pressure. At this higher temperature, the chemical reactions associated with the cooking of food take place at a greater rate.

  (a) Some food cooks fully in 5 minutes in a pressure cooker at 112°C and in 10 minutes in an open pot at 100°C. Calculate the average activation energy for the reactions associated with the cooking of this food.

  (b) How long does the same food take to cook in an open pot of boiling water in Denver, where the average atmospheric pressure is 0.818 atm and the boiling point of water is 94.4°C?

68. The figure below shows the decrease in the concentration of A as it reacts from an initial concentration of 1.00 mol L^{-1} according to three rate laws: one zeroth-order, one first-order in A, and one second-order in A. The numerical values of the rate constants are the same in each case.

  (a) From the graph, estimate the first half-life of the reaction

according to each rate law. This equals the time it takes for the concentration of A to reach half of its original value. Estimate the second half-life of each reaction (the time it takes for the concentration of A to reach to half the value it had after the first half-life). Use these estimates to predict the third half-life of the reaction according to each rate law.

  (b) Determine the rate constant for each rate law. Take care to include the proper units.

  (c) Determine the rate of the reaction according to each rate law after 300 s has elapsed.

  (d) Explain on the basis of molecular collisions why the zeroth-order reaction does not start fastest but finishes first and why the second-order reaction starts out fastest but slows down the most.

*69. A stream of gaseous H_2 is directed onto finely divided platinum powder in the open air. The metal immediately glows white-hot and continues to do so as long as the stream continues. Explain.

70. (a) A certain first-order reaction has an activation energy of 53 kJ mol^{-1}. It is run twice, first at 298 K and then at 308 K (10°C higher). All other conditions are identical. Show that, in the second run, the reaction occurs at double its rate in the first run.

  (b) The same reaction is run twice more, but these runs take place at 398 K and 408 K. Show that the reaction goes 1.5 times faster at 408 K than it does at 398 K.

*71. The gas-phase reaction between hydrogen and iodine

$$H_2(g) + I_2(g) \underset{k_r}{\overset{k_f}{\rightleftharpoons}} 2\, HI(g)$$

proceeds with a forward rate constant at 1000 K of $k_f = 240$ L mol^{-1} s^{-1} and an activation energy of 165 kJ mol^{-1}. By using this information and data from Appendix D, calculate

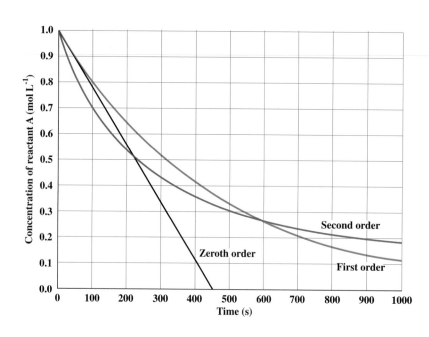

the activation energy for the reverse reaction and the value of k_r at 1000 K. Assume that $\Delta H°$ and $\Delta S°$ for the reaction are independent of temperature between 298.15 and 1000 K.

72. The following reaction mechanism has been proposed for a chemical reaction:

$$A_2 \underset{k_{-1}}{\overset{k_1}{\rightleftharpoons}} A + A \qquad \text{(fast equilibrium)}$$

$$A + B \underset{k_{-2}}{\overset{k_2}{\rightleftharpoons}} AB \qquad \text{(fast equilibrium)}$$

$$AB + CD \overset{k_3}{\longrightarrow} AC + BD \qquad \text{(slow)}$$

(a) Write a balanced equation for the overall reaction.
(b) Write the rate expression that corresponds to the preceding mechanism. Express the rate in terms of concentrations of reactants only (A_2, B, CD).
(c) Suppose that the first two steps in the preceding mechanism are endothermic and the third one is exothermic. Does an increase in temperature *increase* the reaction rate constant, *decrease* it, or cause *no change?* Explain.

*73. The rate of the oxidation of glycine (NH_2CH_2COOH) by potassium permanganate was studied in a buffer solution at a pH of 6.78. The balanced equation for this chemical change is

$$5\,H^+(aq) + 3\,NH_2CH_2COOH(aq) + 2\,MnO_4^-(aq) \longrightarrow$$
$$2\,MnO_2(s) + 3\,CH_2O(aq) + 3\,CO_2(g) +$$
$$3\,NH_4^+(aq) + H_2O(\ell)$$

The graph below shows that the rate of this reaction (at 25°C) actually *increases* during the first 2500 seconds, reaches a maximum, and then falls off toward zero.

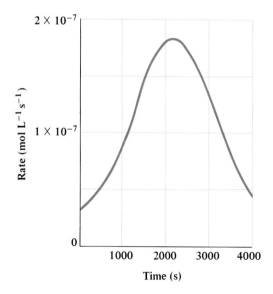

Usually reactions slow down as they proceed. Suggest a reason why this reaction speeds up for a while before it slows down.

74. The nerve gas Soman is a cholinesterase inhibitor. Antidotes to poisoning by Soman are cholinesterase reactivators. What kind of substance is cholinesterase? Suggest what happens to it when it is inhibited by Soman.

75. You and your lab partner measure the rate constant for the reaction

$$S_2O_8^{2-}(aq) + 2\,I^-(aq) \longrightarrow 2\,SO_4^{2-}(aq) + I_2(aq)$$

at 12 different temperatures ranging from 15 to 40°C. You prepare an Arrhenius plot (a plot of $\ln k$ versus $1/T$ as in Figure 14–11) and calculate the activation energy of the reaction, which comes out to 54 kJ mol^{-1}. You then repeat the 12 runs under identical conditions except that you add a small amount of $Fe^{2+}(aq)$ ion as a catalyst. The reaction runs faster in every case. A plot of $\ln k$ versus $1/T$ gives an activation energy of 63 kJ mol^{-1}. Your lab partner says, "The catalyzed reaction must have a lower activation energy, not a higher one because catalysts work by lowering the activation energy of reactions." He accuses you of doing the calculations wrong. You accuse him of blunders in setting up the experiments! In fact, neither the experiments nor the calculations contain any errors. Explain.

*76. The burning of fossil fuels increases the concentration of CO_2 in the atmosphere and may lead to global warming (see Chapter 18). Alarm over this prospect has prompted research into the possible disposal of CO_2 on the deep-ocean floor. In this concept, the pressure of the overlying water maintains large lakes of liquid CO_2 safely on the bottom. CO_2 might, however, leak out of the lakes by the reaction

$$CO_2(\ell) \longrightarrow CO_2(aq)$$

with unknown consequences for the benthic environment. Accordingly, the rate of the preceding reaction was studied. Droplets of liquid CO_2 were found to shrink slowly when held under water at high pressure:

Time (hr)	Diameter (mm)	
	Droplet 1 (at 28 MPa)	*Droplet 2 (at 35 MPa)*
0	14.4	13.0
1	12.2	10.5
2	10.1	9.0
3	9.0	7.3
4	7.5	5.9
5	4.6	3.0
6	3.0	—

(a) Plot the diameters of the two droplets as a function of time. Does the rate of dissolution of the liquid CO_2 depend on the pressure in these experiments?
(b) Determine the order of the dissolution reaction in these experiments.
(c) Compute the rate constant for this dissolution. Explain the choice of units in your answer.

CUMULATIVE PROBLEM

Sulfite and Sulfate Kinetics

The pollutant sulfur dioxide dissolves in water droplets (fog, clouds, and rain) in the atmosphere and reacts according to the equation

$$SO_2(aq) + H_2O(\ell) \longrightarrow HSO_3^-(aq) + H^+(aq)$$

The $HSO_3^-(aq)$ is then, much more slowly, oxidized by dissolved oxygen that is also in the droplets:

$$2\,HSO_3^-(aq) + O_2(aq) \longrightarrow 2\,SO_4^{2-}(aq) + 2\,H^+(aq)$$

Although the second reaction has been studied for many years, only recently was it discovered that it can proceed by the steps

$$2\,HSO_3^-(aq) + O_2(aq) \longrightarrow S_2O_7^{2-}(aq) + H_2O(\ell) \qquad \text{(fast)}$$
$$S_2O_7^{2-}(aq) + H_2O(\ell) \longrightarrow 2\,SO_4^{2-}(aq) + 2\,H^+(aq) \qquad \text{(slow)}$$

The previously undetected intermediate, $S_2O_7^{2-}(aq)$, is well known in other reactions. It is the *disulfate* ion.

In an experiment at 25°C, a solution was mixed with the realistic initial concentrations of 0.270 M $HSO_3^-(aq)$ and 0.0135 M $O_2(aq)$. The initial pH was 3.90. The following table tells what happened in the solution, beginning from the moment of mixing.

Time (s)	$[HSO_3^-]$ (M)	$[O_2]$ (M)	$[S_2O_7^{2-}]$ (M)	$[HSO_4^-] + [SO_4^{2-}]$ (M)
0.000	0.270	0.0135	0.000	0.000
0.010	0.243	0.000	13.5×10^{-3}	0.000
10.0	0.243	0.000	11.8×10^{-3}	3.40×10^{-3}
45.0	0.243	0.000	7.42×10^{-3}	12.2×10^{-3}
90.0	0.243	0.000	4.08×10^{-3}	18.8×10^{-3}
150.0	0.243	0.000	1.84×10^{-3}	23.3×10^{-3}
450.0	0.243	0.000	0.034×10^{-3}	26.9×10^{-3}
600.0	0.243	0.000	0.005×10^{-3}	27.0×10^{-3}

(a) Determine the average rate of increase of the total of the concentrations of the sulfate plus hydrogen sulfate ions during the first 10 s of the experiment.

(b) Determine the average rate of disappearance of hydrogen sulfite ion during the first 0.010 s of the experiment.

(c) Explain why the hydrogen sulfite ion (HSO_3^-) stops disappearing after 0.010 s.

(d) Plot the concentration of disulfate ion versus time on a piece of graph paper. Use the graph to estimate the instantaneous rate of disappearance of disulfate ion 90.0 s after the reaction starts.

(e) Plot the total of the concentrations of sulfate plus hydrogen sulfate ions versus time, and estimate the rate of increase of this total 90.0 s after the reaction starts.

(f) Use the graphs of the previous two parts to estimate the rate of the reaction 150 s after the reactants are mixed.

Sulfur burns readily in oxygen to generate SO_2, which is further oxidized in the atmosphere to generate H_2SO_4. The Cumulative Problem turns on recent findings on the mechanism of this oxidation.

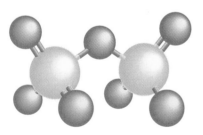

The structure of the disulfate ion, $S_2O_7^{2-}$.

(g) Determine the order with respect to the disulfate ion of the second step of the conversion and also the rate constant of the second step.

(h) Determine the half-life of the second step of the conversion process.

(i) Explain why nothing definite can be concluded from this experiment about the order and rate constant of the first step of the conversion.

(j) At 15°C, the rate constant of the second step of the conversion is only 62% of its value at 25°C. Compute the activation energy of the second step.

(k) The first step of the conversion occurs much faster when 1.0×10^{-6} M $Fe^{2+}(aq)$ ion is added. What role does Fe^{2+} play?

(l) Write a balanced equation for the overall reaction that gives sulfuric acid from SO_2 dissolved in water droplets in the air.

CHAPTER OUTLINE

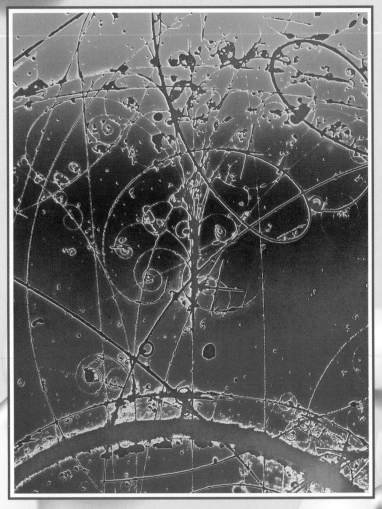

The tracks of subatomic particles in a bubble chamber.

O ur approach to chemistry has focused so far on the bulk properties of matter: the masses of reactants and yields of products; melting, boiling, and sublimation; and the drift of reaction systems toward equilibrium. Until the end of the 19th century, chemistry was concerned almost exclusively with such *macroscopic* properties, and it was appropriate to begin with such a viewpoint. Of course, some *microscopic* concepts have appeared. The laws of combining masses and volumes, macroscopic observations that they are, demanded interpretation in terms of atoms; the relationship between the energy and temperature of ideal gases was clarified through the kinetic–molecular theory. Such excursions into the microscopic world in the first 14 chapters were aimed largely at clarifying macroscopic chemical phenomena, however, rather than at discovering their ultimate causes.

We now shift our point of view. In this and the following six chapters, we examine matter and chemical phenomena from a microscopic point of view, considering the ways in which the building blocks introduced in Section 1–5 (the proton, neutron, and electron) fit together to give matter its properties. We start with the properties and reactions of the atomic nucleus. In subsequent chapters, we move on to atoms, molecules, and the aggregates of molecules that form liquids and solids.

15–1 MASS–ENERGY RELATIONSHIPS IN NUCLEI

Recall from Section 1–5 that almost all the mass of an atom derives from protons and neutrons contained in a very small volume in its central kernel, or nucleus. The small size of the nucleus (which occupies less than one *trillionth* of the space in the atom) and the strength of the forces that hold it together largely isolate its behavior from the outside world of electrons and other nuclei. This allows meaningful discussion of single nuclei without being concerned about the atoms, ions, or molecules in which they might be found.

The term **nuclide** is applied to atoms when the exact constitution of their nuclei is known or is important. A nuclide is characterized by the number of protons, Z, and the number of neutrons, N, that its nucleus contains. The atomic number, Z, determines the charge $+Ze$ on the nucleus and controls the chemical identity of the nuclide; the sum $Z + N = A$ is the mass number of the nuclide and equals the integer closest to the relative atomic mass of the nuclide. Symbols of the form $^A_Z E$, where E is the chemical symbol for the atom, represent nuclides fully.

• The term "isotope" should not be used in place of "nuclide." Two nuclides are isotopes of each other only if their Z's are the same and their N's are different.

As pointed out in Section 1–7, the gram and kilogram are inconveniently large units for measuring the masses of single atoms (single nuclides) and the elementary particles that make them up. The atomic mass unit (u) is better suited for measurements on this scale. The definition of this tiny unit of mass has been deliberately arranged so that Avogadro's number of atomic mass units equals 1 gram of mass

$$6.022137 \times 10^{23} \text{ u} = 1 \text{ g}$$

As a consequence, the mass of a single atom in atomic mass units is numerically equal to the mass of 1 mol of such atoms in grams. Table 15–1 gives the masses of elementary particles and of selected nuclides in atomic mass units. Each nuclidic mass includes the contribution from the surrounding electrons in the atom as well as the nucleus. It is instructive to compare these nuclidic masses to the chemical relative atomic masses of the elements (listed in the table on the inside back cover of this book). The numbers are identical when the nuclide listed is the only one that occurs in nature (the case with sodium, fluorine, and iodine), but they differ when isotopes contribute to a weighted average atomic mass (the case with hydrogen, carbon, and uranium).

• It follows that 1 u = 1.660540 × 10^{-24} g = 1.660540 × 10^{-27} kg.

• The data, which are quite precise, come from mass spectrometry experiments (see Fig. 1–22).

Table 15–1
Masses of Selected Elementary Particles and Atoms

Elementary Particle	Symbol	Mass (u)	Mass (kg)
Electron (beta particle)	$_{-1}^{0}e^{-}$, $_{-1}^{0}\beta^{-}$	0.00054857990	9.109390×10^{-31}
Positron	$_{-1}^{0}e^{+}$, $_{-1}^{0}\beta^{+}$	0.00054857990	9.109390×10^{-31}
Proton	$_{1}^{1}p^{+}$	1.00727647	1.672623×10^{-27}
Neutron	$_{0}^{1}n$	1.00866490	1.674929×10^{-27}

Nuclide	Mass (u)	Nuclide	Mass (u)
$_{1}^{1}\text{H}$	1.00782504	$_{12}^{24}\text{Mg}$	23.985042
$_{1}^{2}\text{H}$	2.01410178	$_{14}^{30}\text{Si}$	29.973771
$_{1}^{3}\text{H}$	3.01604927	$_{15}^{30}\text{P}$	29.9783138
$_{2}^{3}\text{He}$	3.01602931	$_{16}^{32}\text{S}$	31.9720707
$_{2}^{4}\text{He}$	4.00260324	$_{17}^{35}\text{Cl}$	34.96885272
$_{3}^{7}\text{Li}$	7.016003	$_{19}^{40}\text{K}$	39.963999
$_{4}^{8}\text{Be}$	8.0053052	$_{20}^{40}\text{Ca}$	39.962591
$_{4}^{9}\text{Be}$	9.0121822	$_{22}^{49}\text{Ti}$	48.947871
$_{4}^{10}\text{Be}$	10.013535	$_{35}^{81}\text{Br}$	80.916289
$_{5}^{8}\text{B}$	8.024607	$_{37}^{87}\text{Rb}$	86.909187
$_{5}^{10}\text{B}$	10.0129369	$_{38}^{87}\text{Sr}$	86.908884
$_{5}^{11}\text{B}$	11.0093054	$_{53}^{127}\text{I}$	126.904473
$_{6}^{11}\text{C}$	11.011433	$_{88}^{226}\text{Ra}$	226.025403
$_{6}^{12}\text{C}$	12 exactly	$_{88}^{228}\text{Ra}$	228.031064
$_{6}^{13}\text{C}$	13.00335483	$_{89}^{228}\text{Ac}$	228.031015
$_{6}^{14}\text{C}$	14.00324198	$_{90}^{232}\text{Th}$	232.038051
$_{7}^{14}\text{N}$	14.00307400	$_{90}^{234}\text{Th}$	234.0436
$_{8}^{16}\text{O}$	15.99491463	$_{91}^{231}\text{Pa}$	231.035879
$_{8}^{17}\text{O}$	16.991312	$_{92}^{231}\text{U}$	231.036289
$_{8}^{18}\text{O}$	17.999160	$_{92}^{234}\text{U}$	234.040956
$_{9}^{19}\text{F}$	18.9984032	$_{92}^{235}\text{U}$	235.043925
$_{11}^{23}\text{Na}$	22.989768	$_{92}^{238}\text{U}$	238.050786

The Einstein Mass–Energy Relationship

Some nuclides exist indefinitely, but others are **radioactive** and decay at detectable rates by processes considered in Section 15–2. No nuclide (except $_{1}^{1}\text{H}$) exists without one or more neutrons in the nucleus. How many are required? Figure 15–1 shows a plot of neutron number N ($= A - Z$) against atomic number Z (protons in the nucleus) for the known nuclides of the elements. Stable nuclides are shown in black so that a *belt of stability* is visible in black. For the light elements up to about $Z = 20$, $N/Z \approx 1$ for stable nuclides; nearly equal numbers of protons and neutrons are required. For the heavier elements, $N/Z > 1$, and progressively more neutrons than protons are required in the nucleus as Z increases. The belt of stability ends at $_{83}^{209}\text{Bi}$, the heaviest stable nuclide. It appears that holding more than 83 protons in the same nucleus indefinitely is just not possible, no matter how many neutrons are present.

Although neutrons lend stability to the nuclei of atoms, free neutrons are not stable. They decay with a half-life of about 14.6 minutes into a proton and an elec-

• Some of the heavier nuclides are *nearly* stable: the half-life of ^{238}U is 4.5 billion years.

tron. The change is represented by the following balanced **nuclear equation**

$$_0^1 n \longrightarrow {}_1^1 p^+ + {}_{-1}^0 e^-$$

Like chemical equations, nuclear equations must be balanced. The criteria for balance in nuclear reactions differ because nuclear reactions create, destroy, and transmute atoms. In balanced nuclear equations:

- The total mass number (the sum of the A's, the left superscripts) of the particles on the two sides must be equal.
- The total atomic number (the sum of the Z's, the left subscripts) of the particles on the two sides must be equal.
- The overall charge on the two sides (the sum of the right superscripts) on the two sides must be equal.

Mass is *not* conserved in nuclear reactions. This is evident in the decay of a free neutron: subtracting the exact mass of the reactant (a neutron) from the exact

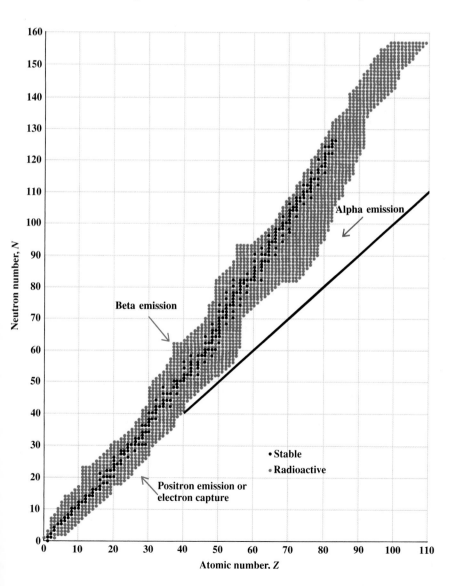

Figure 15–1 A plot of N versus Z for all stable (marked in black) and many unstable (marked in red) nuclides. The transforming effects of decay of unstable nuclides via alpha emission, beta emission, or positron emission and electron capture are shown by the colored arrows.

masses of the products (a proton and an electron) gives

$$\Delta m = \underbrace{1.00727647}_{\text{proton mass}} + \underbrace{0.0005485799}_{\text{electron mass}} - \underbrace{1.00866490}_{\text{neutron mass}} = -8.3985 \times 10^{-4}\ \text{u}$$

$$\Delta m = -8.3985 \times 10^{-4}\ \text{u} \times \left(\frac{1\ \text{g}}{6.022037 \times 10^{23}\ \text{u}}\right) = -1.395 \times 10^{-27}\ \text{g}$$

$$= -1.395 \times 10^{-30}\ \text{kg}$$

If mass were conserved, Δm would equal zero. The negative sign means that the mass of the system decreases. The failure of the law of conservation of mass is explained by appealing to the special theory of relativity, which was formulated by Albert Einstein in 1905. Einstein assumed as an axiom that energy is conserved. He then showed that the transfer of energy by any means (heat, work, emission of radiation, and so on) implies the transfer of a proportional amount of mass. According to the special theory of relativity, energy and mass are merely different manifestations of the same quantity and are related by the equation

$$\boxed{\Delta E = c^2\ \Delta m}$$

where c is the speed of light in a vacuum. Any process (including a chemical reaction) that releases energy from a system (ΔE negative) is thus accompanied by a decrease in the mass of the system (Δm negative). The loss of mass when a neutron decays therefore simply means that the decay releases energy. The release appears as kinetic energy of the particles that are produced. If this energy is to be computed in joules, c must be in meters per second and Δm in kilograms

$$\Delta E = c^2\ \Delta m = \left(2.9979 \times 10^8\ \frac{\text{m}}{\text{s}}\right)^2 (-1.395 \times 10^{-30}\ \text{kg}) = -1.254 \times 10^{-13}\ \text{J}$$

The energy released in the decay of one mole of neutrons is obtained by multiplying by Avogadro's number

$$\Delta E = \left(\frac{-1.254 \times 10^{-13}\ \text{J}}{\text{neutron}}\right) \times \left(\frac{6.022 \times 10^{23}\ \text{neutron}}{1\ \text{mol}}\right) = -7.552 \times 10^{10}\ \text{J mol}^{-1}$$

$$\Delta E = -7.552 \times 10^{7}\ \text{kJ mol}^{-1}$$

This ΔE is hundreds of thousands of times larger than the ΔE's of chemical reactions. For example, the energy release in the combustion of carbon to CO_2 is only about 400 kJ mol^{-1}. Nuclear reactions without exception involve energy changes orders of magnitude larger than those of chemical reactions (Fig. 15–2).

Changes in energy in nuclear reactions are nearly always expressed in more convenient energy units than the joule—namely, the **electron volt (eV)** and the **megaelectron volt (MeV)**. The electron volt is defined as the energy given to an electron when it is accelerated through a potential difference of exactly one volt:

$$\Delta E(\text{joules}) = Q(\text{coulombs}) \times \Delta\mathscr{E}(\text{volts})$$

$$1\ \text{eV} = 1.602177 \times 10^{-19}\ \text{C} \times 1\ \text{V} = 1.602177 \times 10^{-19}\ \text{J}$$

The **megaelectron volt (MeV)** is one million times larger than an electron volt:

$$1\ \text{MeV} = 10^6\ \text{eV} = 1.602177 \times 10^{-13}\ \text{J}$$

Let us use Einstein's equation to compute in MeV the change in energy associated with a change in mass of 1 atomic mass unit. This value provides a useful stepping stone in later computations of mass–energy equivalencies. We substitute

$\Delta m = 1.66054 \times 10^{-27}$ kg (1 u) into Einstein's equation

$$\Delta E = c^2\, \Delta m = (2.99792 \times 10^8 \text{ m s}^{-1})^2(1.66054 \times 10^{-27} \text{ kg})$$
$$= 1.492419 \times 10^{-10} \text{ J}$$

and then convert to MeV:

$$1.492419 \times 10^{-10} \text{ J} \times \left(\frac{1 \text{ MeV}}{1.602177 \times 10^{-13} \text{ J}} \right) = 931.494 \text{ MeV}$$

That is, the **energy equivalent** of 1 u of mass is 931.494 MeV. In the decay of a neutron

$$\Delta E = (-8.33985 \times 10^{-4} \text{ u}) \times \left(\frac{931.494 \text{ MeV}}{\text{u}} \right) = -0.782 \text{ MeV}$$

Binding Energies of Nuclei

The **binding energy** E_B of a nucleus is defined as the negative of the energy change ΔE that occurs when the nucleus is formed from its component protons and neutrons. For the ^4_2He nucleus, for example

$$2\, ^1_1p^+ + 2\, ^1_0n \longrightarrow\, ^4_2\text{He}^{2+} \qquad E_B = -\Delta E = \text{?}$$

Binding energies of nuclei are calculated using Einstein's equation and accurate mass data from a mass spectrometer. Most measurements, however, determine the mass of the atom with most or all of its electrons, not the mass of the bare nucleus. For this reason, we take the binding energy E_B of a nucleus to equal the negative of the energy change ΔE for the formation of the neutral *atom* from hydrogen *atoms* and neutrons. For the ^4_2He nucleus, this corresponds to the nuclear reaction

$$2\, ^1_1\text{H} + 2\, ^1_0n \longrightarrow\, ^4_2\text{He}$$

The correction for the differences in binding energies of the electrons that are present in the hydrogen atoms and the helium atom is very small and can be neglected.

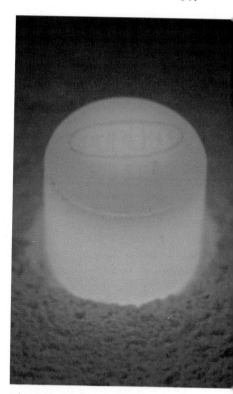

Figure 15–2 Self-heating in a radioactive metal. Relatively few atoms of plutonium (about 2.5 per second out of every 10 billion) are taking part in nuclear reactions in this pellet of nearly pure plutonium-238, but the energy released is enough to make it glow red-hot. The pellet was manufactured as a compact power source for a space probe.

EXAMPLE 15–1

Calculate the binding energy of ^4_2He from the data in Table 15–1, and express it in joules, in megaelectron volts (MeV), and in kilojoules per mole.

Solution

The mass change in forming one atom of ^4_2He, first in u and then in kg, is

$$\Delta m = m[^4_2\text{He}] - 2\, m[^1_1\text{H}] - 2\, m[^1_0n]$$
$$\Delta m = 4.00260324 - 2(1.00782504) - 2(1.00866490) = -0.03037664 \text{ u}$$
$$\Delta m = -0.03037664 \text{ u} \times \left(\frac{1 \text{ g}}{6.022137 \times 10^{23} \text{ u}} \right) \times \left(\frac{1 \text{ kg}}{1000 \text{ g}} \right)$$
$$= -5.044163 \times 10^{-29} \text{ kg}$$

Substitution of the mass change into Einstein's equation gives

$$\Delta E = c^2\, \Delta m = (2.9979246 \times 10^8 \text{ m s}^{-1})^2 (-5.044163 \times 10^{-29} \text{ kg})$$
$$= -4.5334676 \times 10^{-12} \text{ J}$$
$$E_B = -\Delta E = +4.5334676 \times 10^{-12} \text{ J}$$

The easiest way to compute the binding energy in MeV is to use the energy equivalent of 1 atomic mass unit that was just established

$$E_B = -\Delta E = -(-0.03037664 \text{ u}) \times \left(\frac{931.494 \text{ MeV}}{1 \text{ u}} \right) = 28.2956 \text{ MeV}$$

The binding energy of 1 mol of $_2^4$He atoms exceeds the binding energy of a single atom by a factor of Avogadro's number

$$E_B = \left(\frac{+4.5334676 \times 10^{-12} \text{ J}}{_2^4 \text{He atom}} \right) \times \left(\frac{6.022137 \times 10^{23} \, _2^4 \text{He atom}}{\text{mol } _2^4 \text{He}} \right) \times \left(\frac{1 \text{ kJ}}{1000 \text{ J}} \right)$$

$$= 2.730116 \times 10^9 \text{ kJ mol}^{-1}$$

Exercise

Calculate the binding energy of $_2^3$He in joules, in megaelectron volts, and in kilojoules per mole.

Answer: 1.2366×10^{-12} J; 7.7181 MeV; 7.4468×10^8 kJ mol^{-1}.

There are four **nucleons** (protons plus neutrons) in the $_2^4$He nucleus, so the **binding energy per nucleon** is

$$\frac{E_B}{4} = \frac{28.2956 \text{ MeV}}{4} = 7.07390 \text{ MeV}$$

The binding energy per nucleon is a direct measure of the stability of the nucleus. It varies with the mass number among the nuclides, increasing to a maximum of about 8.8 MeV per nucleon for iron and nickel, as Figure 15–3 indicates.

15–2 NUCLEAR DECAY PROCESSES

Why are some nuclides radioactive but others stable? The thermodynamic criterion for spontaneity in a process at constant T and P is that $\Delta G < 0$. For nuclear reactions, the change in the Gibbs function is essentially equal to the energy change (ΔE). Consequently, the thermodynamic criterion for a spontaneous nuclear reaction simplifies to

$$\Delta E < 0 \qquad \text{or equivalently} \qquad \Delta m < 0$$

The equivalence of these criteria follows from Einstein's equation.

Nuclear reactions with $\Delta m < 0$ always increase the binding energy per nucleon among the products. In terms of Figure 15–3, this means that reactions producing nuclides positioned higher up the page are thermodynamically spontaneous. These include reactions that combine the nuclei of light elements to produce heavier elements (fusion reactions) and well as those that split apart the nuclei of the heaviest elements (fission reactions). Thus, nearly all nuclides are thermodynamically unstable with respect to either fusion or fission. Fortunately, most nuclear reactions have exceedingly large activation energies and proceed at zero rates except at extremely high temperatures (see Section 15–6). Otherwise, all matter would long since have reacted to give the most stable nuclides in Figure 15–3. The nuclear reactions that *do* occur with detectable rates under ordinary conditions have been the

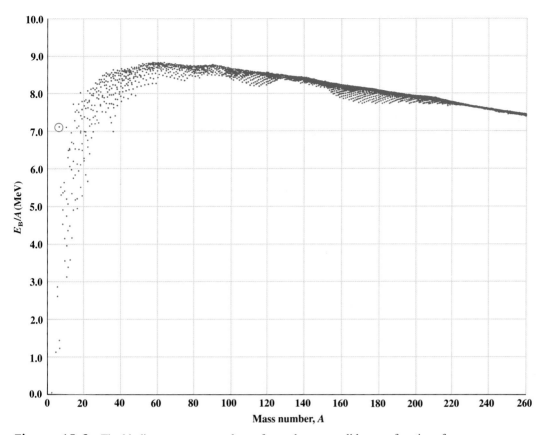

Figure 15–3 The binding energy per nucleon of most known nuclides as a function of the mass number. A total of 2212 nuclides are shown. Of these only 274 are stable; the rest undergo some type of radioactive decay. Maximum stability occurs in the vicinity of ^{56}Fe. Note that ^{4}He (marked with a red circle) is unusually stable for its mass number.

subject of intense experimental study since the discovery of radioactivity in 1896. They proceed along relatively few pathways.

Before discussing these pathways, it is necessary to mention **antiparticles.** Each subatomic particle is thought to have an antiparticle of the same mass but opposite charge. Thus, the antiparticle of an electron is a **positron** (e^+) and the antiparticle of a proton is a negatively charged particle called an **antiproton.** Antiparticles have only transient existences because matter and antimatter quickly annihilate one another when they come close together, emitting radiation carrying an equivalent quantity of energy. In the case of an electron and a positron, two oppositely directed **gamma rays** (high-energy electromagnetic waves), called **annihilation radiation,** are emitted. Another important particle–antiparticle pair is the neutrino ν and antineutrino $\bar{\nu}$, both of which are uncharged and appear to have no mass. They carry angular momentum (see Section 16–3) and interact only weakly with ordinary matter.

• Nuclear reactions in the sun emit large numbers of neutrinos, but almost all of those that strike the earth pass completely through with no effect at all. Elaborate detectors are required to record the extremely rare events induced by neutrinos in certain nuclides.

Pathways for Decay

If an unstable nuclide contains fewer protons than do stable nuclides of the same mass number, it is called "proton deficient." Such a nuclide may decay by transforming one of its neutrons into a proton and emitting a high-energy electron, also called a **beta particle.** The transformation simultaneously emits an antineutrino $\tilde{\nu}$. The daughter nuclide has the same mass number A as the parent, but its atomic number Z has been increased by one, corresponding to the conversion of a neutron to a proton. This is **beta decay,** or *beta emission.* An electron (beta particle) is symbolized $_{-1}^{0}e^-$ (or $_{-1}^{0}\beta^-$) in equations for nuclear reactions. The superscript 0 indicates a mass number of 0, which is in fact the integer closest to the atomic mass of the electron. The subscript -1 is not a true atomic number but accounts for the change in the charge of a nucleus when an electron leaves or joins it. Finally, as in chemical equations, the superscript $-$ gives the overall charge of the species. Some equations for the beta decay of unstable nuclei are

• Antineutrinos and neutrinos affect none of these aspects of balance.

$$_{6}^{14}C \longrightarrow {}_{7}^{14}N^+ + {}_{-1}^{0}e^- + \tilde{\nu} \qquad {}_{19}^{40}K \longrightarrow {}_{20}^{40}Ca^+ + {}_{-1}^{0}e^- + \tilde{\nu}$$

These equations are balanced as to mass number (left superscript), atomic number (left subscript), and overall charge (right superscript).

When a nuclide has too *many* protons for stability relative to the number of neutrons it contains, it may decay by emitting a positron, symbolized $_{1}^{0}e^+$ (or $_{1}^{0}\beta^+$). Like its antiparticle the electron, a positron has a mass number of zero, but its charge is $+1$ instead of -1. In **positron emission,** a proton is converted into a neutron, which is retained in the nucleus, and a high-energy positron is emitted together with a neutrino (ν). The mass number of the nuclide remains the same, but the atomic number Z *decreases* by one. Examples of positron emission are

$$_{10}^{19}Ne \longrightarrow {}_{9}^{19}F^- + {}_{1}^{0}e^+ + \nu \qquad {}_{29}^{64}Cu \longrightarrow {}_{28}^{64}Ni^- + {}_{1}^{0}e^+ + \nu$$

EXAMPLE 15-2

Write balanced nuclear equations for (a) beta decay of $_{11}^{24}Na$ and (b) positron emission by $_{11}^{21}Na$.

Solution

(a) In beta decay, the mass number A (here, 24) is unchanged, but one neutron is converted into a proton. After the reaction, then, $11 + 1 = 12$ protons are present. The element with atomic number $Z = 12$ is magnesium, so the balanced equation for this nuclear reaction is

$$_{11}^{24}Na \longrightarrow {}_{12}^{24}Mg^+ + {}_{-1}^{0}e^- + \tilde{\nu}$$

We verify that the sums of the mass numbers are the same on the two sides, that the sums of the nuclear charges are the same on the two sides, and that the overall charge is the same on the two sides.

(b) In positron emission, the mass number (21 in this case) is unchanged, but one proton is converted into a neutron, decreasing Z by one (from 11 to 10 in this case). This implies the formation of a Ne nucleus, so the balanced nuclear equation is

$$_{11}^{21}Na \longrightarrow {}_{10}^{21}Ne^- + {}_{1}^{0}e^+ + \nu$$

Exercise

Write balanced nuclear equations for (a) beta decay of $_{4}^{10}Be$ and (b) positron emission by $_{5}^{8}B$.

Answer: (a) $_{4}^{10}Be \longrightarrow {}_{5}^{10}B^+ + {}_{-1}^{0}e^- + \tilde{\nu}$. (b) $_{5}^{8}B \longrightarrow {}_{4}^{8}Be^- + {}_{1}^{0}e^+ + \nu$.

There is another pathway by which a nuclide with an excess of protons can diminish that excess and seek stability. Its nucleus may capture one of its own surrounding electrons and use it to convert a proton to a neutron. As in positron emission, the mass number of the parent and daughter atoms are the same and the atomic number decreases by one. In this case, however, only a massless neutrino is emitted. Examples of **electron capture** (E.C.) are

$$^{26}_{13}\text{Al} \xrightarrow{\text{E.C.}} {}^{26}_{12}\text{Mg} + \nu \qquad {}^{36}_{17}\text{Cl} \xrightarrow{\text{E.C.}} {}^{36}_{16}\text{S} + \nu$$

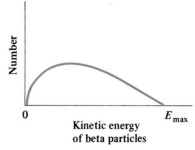

Figure 15–4 Emitted electrons (beta particles) in beta decay have a distribution of kinetic energies up to a cut-off value of E_{max}.

These nuclear equations are not balanced. Balancing them requires showing the captured electrons explicitly

$$(^{26}_{13}\text{Al}^+ + {}^{0}_{-1}e^-) \longrightarrow {}^{26}_{12}\text{Mg} + \nu \qquad (^{36}_{17}\text{Cl}^+ + {}^{0}_{-1}e^-) \longrightarrow {}^{36}_{16}\text{S} + \nu$$

where the parentheses emphasize that the electrons that are captured start as part of the atom whose nucleus absorbs them. Electron capture is quite common for heavier, neutron-deficient nuclei.

The three paths for nuclear decay discussed so far lead to changes in the atomic number (Z) and neutron number (N) but not in the mass number (A). They are indicated by two blue arrows running "northwest–southeast" in Figure 15–1. By contrast, the process of **alpha decay** involves the emission of an alpha particle (a $^4_2\text{He}^{2+}$ ion, also written as $^4_2\alpha^{2+}$) and a decrease in mass number A by 4 (the atomic number Z and the neutron number N each decrease by 2). Alpha decay is indicated in Figure 15–1 by the green arrow running "northeast–southwest." Examples of alpha decay are

$$^{226}_{88}\text{Ra} \longrightarrow {}^{222}_{86}\text{Rn} + {}^4_2\text{He} \qquad {}^{238}_{92}\text{U} \longrightarrow {}^{234}_{90}\text{Th} + {}^4_2\text{He}$$

Alpha decay occurs chiefly among the heavier elements, with atomic numbers in the region of unstable nuclei beyond bismuth ($Z = 83$) in the periodic table.

When nuclei are very proton deficient (have a large excess of neutrons) or very neutron deficient (have a large excess of protons), an excess particle may "boil off," that is, may be ejected directly from the nucleus. These decay modes are called **neutron emission** and **proton emission,** respectively. Emission of a neutron reduces N but not Z and produces daughter nuclides lying one step lower than the parent on the N axis in Figure 15–1. Proton emission reduces Z but not N and produces daughter nuclides lying to the left of the parent in the figure. Finally, certain nuclides of high A are known to spontaneously fission, that is, to split into two chunks of roughly equal size.

• Fission reactions are discussed more extensively in Section 15–5.

Mass Changes in Nuclear Decay

In nuclear reactions, a negative Δm corresponds to a negative ΔE, which corresponds to the liberation of energy. In beta decay, for example, this liberated energy appears as kinetic energy of the beta particle (emitted election), the antineutrino, and the daughter nuclide. The daughter nuclide is massive compared to the others, and the recoil kinetic energy it carries away is small and can be neglected. The kinetic energy of the beta particle plus that of the antineutrino therefore equals $-\Delta E$. The antineutrino is not easily observed, but the kinetic energy of the emitted beta particle varies in a continuous range from zero up to a maximum, as shown in Figure 15–4, reflecting all the ways that the kinetic energy can be parceled out between it and the antineutrino.

EXAMPLE 15-3

Calculate the maximum kinetic energy (in MeV) of the beta particle in the decay

$$^{24}_{11}\text{Na} \longrightarrow {}^{24}_{12}\text{Mg}^+ + {}^{0}_{-1}e^- + \tilde{\nu}$$

considered in Example 15–2. The masses of *neutral atoms* of ^{24}Na and ^{24}Mg are 23.99096 and 23.98504 u, respectively.

Solution

The maximum kinetic energy of the electron equals $-\Delta E$ for the reaction. Computing ΔE (by use of Einstein's equation) requires Δm, the mass of the products minus the mass of the reactants. The balanced equation shows the daughter nuclide as a $+1$ ion in order to maintain balance as to overall charge: $^{24}_{12}\text{Mg}$ has 12 electrons, but $^{24}_{11}\text{Na}$ has only 11. Therefore, the mass of the $^{24}_{12}\text{Mg}^+$ ion is used.

$$\Delta m = \text{mass of } {}^{24}_{12}\text{Mg}^+ \text{ ion} + \text{mass of } {}^{0}_{-1}e^- - \text{mass of } {}^{24}_{11}\text{Na atom}$$
$$\Delta m = (23.98504 - 0.00054857990) + 0.00054857990 - 23.99096$$
$$= 23.98504 - 23.99096 = -0.00592 \text{ u}$$

Note that the electron mass (taken from Table 15–1) cancels out and is actually not required. The energy change is

$$\Delta E = (-0.00592 \text{ u}) \times \left(\frac{931.494 \text{ MeV}}{1 \text{ u}}\right) = -5.51 \text{ MeV}$$

The maximum kinetic energy of the beta particle is $+5.51$ MeV.

Exercise

One atom of the nuclide $^{10}_4\text{Be}$ undergoes beta decay (Exercise 15–2). Using data from Table 15–1, calculate the maximum kinetic energy (in MeV) of the emitted electron.

Answer: 0.557 MeV.

Positron emission is the "anti-process" of beta decay, a symmetry that suggests that emitted positrons should, like emitted electrons in beta decay, have kinetic energies that range from zero to a maximum of ΔE. This is exactly what is found. However, positron emission lowers Z, whereas beta decay raises it. This causes the computation of the maximum kinetic energy of an emitted positron to proceed slightly differently.

EXAMPLE 15-4

Calculate the maximum kinetic energy of the positron emitted in the decay

$$^{11}_{6}\text{C} \longrightarrow {}^{11}_{5}\text{B}^- + {}^{0}_{1}e^+ + \nu$$

The masses of *neutral atoms* of ^{11}C and ^{11}B are 11.011433 and 11.009305 u, respectively.

Solution

The daughter nuclide appears as a negative ion in the balanced nuclear equation because otherwise the nuclear equation would represent six electrons on the left and five on the right and not be balanced as to overall charge. The mass of the electron that gives the required charge to $^{11}_5\text{B}$ must be added to the mass of the neutral atom

$$\Delta m = \text{mass of } ^{11}_5\text{B}^- \text{ ion} + \text{mass of } _{-1}^0 e^+ - \text{mass of } ^{11}_6\text{C atom}$$

$$\Delta m = (11.009305 + 0.00054857990) + 0.00054857990 - 11.011433$$
$$= -0.001031 \text{ u}$$

Note how the (identical) masses of the electron and positron add together. The energy change is

$$\Delta E = (-0.001031 \text{ u}) \times \left(\frac{931.494 \text{ MeV}}{1 \text{ u}} \right) = -0.960 \text{ MeV}$$

The maximum kinetic energy of the positron is $+0.960$ MeV.

Exercise

The nuclide ^8_5B decays by positron emission (see Exercise 15–2). Use data from Table 15–1 to calculate the maximum kinetic energy (in MeV) of the emitted positron.

Answer: 16.96 MeV.

Detecting and Measuring Radioactivity

Many methods have been developed to detect, identify, and quantitatively measure the products formed in nuclear reactions. Some are quite simple, but others require complex electronic instrumentation. Perhaps the simplest radiation detector is the photographic emulsion, which was first used by Henri Becquerel, the discoverer of radioactivity. In 1896, Becquerel reported his observation that potassium uranyl sulfate $(\text{K}_2\text{UO}_2(\text{SO}_4)_2 \cdot 2\text{H}_2\text{O})$ could expose a photographic plate even in the dark. Such detectors are still in use in the form of film badges, worn to monitor personal exposure to radiation. The degree of darkening of the film is proportional to the quantity of radiation received.

Rutherford and his students used a screen coated with zinc sulfide to detect the location of alpha particles by the pinpoint flashes or scintillations of light they produce (see Section 1–5). The modern **scintillation counter** is an outgrowth of that technique. The zinc sulfide scintillation screen is replaced by a sodium bromide crystal in which Tl^+ ions replace a small fraction of the Na^+ ions. Such a crystal emits a pulse of light when it absorbs a beta particle or a gamma ray. A photomultiplier tube detects and counts the light pulses.

The Geiger counter (Fig. 15–5) measures radiation by a different method. A cylindrical tube, usually of glass, is coated internally with metal to provide a negative electrode, and a wire is run down the center of the tube to serve as a positive electrode. The tube is filled to a pressure of about 0.1 atm with a mixture of 90% argon and 10% ethyl alcohol vapor. Then a potential difference of about 1000 V is applied across the electrodes. When a high-energy beta particle enters the tube, it produces ion pairs (electrons and positive ions). The electrons, being much lighter than the positive ions, are quickly accelerated toward the wire, which has a positive charge. As they advance, they encounter and ionize other neutral atoms. An avalanche of electrons builds up, and a large electron current flows into the central wire. The sudden increase in the electrical conductance of the tube causes a drop in the potential difference, which is recorded, and the multiplicative electron discharge is

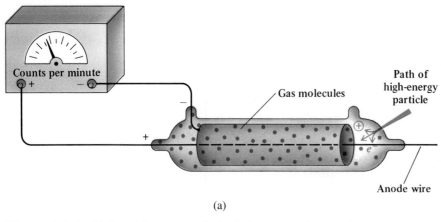

Figure 15-5 (a) In a Geiger tube, radiation ionizes gas in the tube, freeing electrons that are accelerated to the anode wire in a cascade. Their arrival creates an electrical pulse. (b) A Geiger counter measuring the emissions from a sample of carnotite, a uranium-containing mineral.

quenched by the alcohol molecules. In this way, single beta particles produce electrical pulses that can be amplified and counted.

15-3 KINETICS OF RADIOACTIVE DECAY

Nuclear decay occurs at random, making it impossible to predict exactly when a particular nucleus among a group of identical unstable nuclei will disintegrate. Nevertheless, the laws of probability assure us that if the group is large, a definite fraction will decay within a given period of time. The process of nuclear decay follows first-order kinetics and is described by the same equations developed in Chapter 14 for the rates of first-order reactions. The first-order rate law has the form

$$N = N_i \, e^{-kt}$$

• An analogy: insurance companies cannot predict which policyholders in a large group will die by a given age but can establish, from experience, the *fraction* who will die.

where N_i is the number of nuclei initially present at $t = 0$, and N is the number that remain after a time, t. The first-order rate constant (k), which is now called a **decay constant,** is related to the half-life ($t_{1/2}$) through

$$t_{1/2} = \frac{\ln 2}{k} = \frac{0.6931}{k}$$

just as in first-order chemical kinetics. The half-life is the time required for one half of the nuclei in a sample to decay. Known half-lives of unstable nuclei range from 10^{-21} s to more than 10^{24} years. Table 15-2 lists the half-lives and decay modes for a number of unstable nuclides.

There is one important practical difference between chemical kinetics and nuclear kinetics. In chemical kinetics the concentration of a reactant or product is monitored over time, and the rate of the reaction is then found from the rate of change of that concentration. In nuclear kinetics, the rate of occurrence of decay events, $-\Delta N/\Delta t$, is measured with a Geiger counter or other radiation detector. This decay rate is the **activity,** the average disintegration rate, in numbers of nuclei per unit of time.

$$\text{activity} = A = \frac{-\Delta N}{\Delta t} = kN$$

Because the activity is proportional to the number of nuclei (N), the activity also falls off exponentially with time:

$$A = A_i\, e^{-kt}$$

A plot of $\ln A$ against t is linear, with slope $-k = -(\ln 2)/t_{1/2}$, as Figure 15–6 shows. The activity A is reduced to half its initial value in a time $t_{1/2}$. Once A and k of a particular sample are known, the number of nuclei (N) in the sample can be calculated:

$$N = \frac{A}{k} = \frac{A\, t_{1/2}}{\ln 2} = \frac{A\, t_{1/2}}{0.6931}$$

The SI unit of activity is the **becquerel** (Bq), which equals 1 radioactive disintegration per second (1 Bq = 1 s^{-1}). The becquerel supersedes an older and much larger unit of activity called the **curie** (Ci), which is defined as 3.7×10^{10} disintegrations per second.

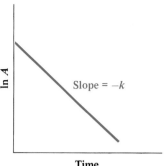

Figure 15–6 A graph of the logarithm of the activity of a radioactive nuclide against time is a straight line with slope $-k$.

Table 15–2
Decay Characteristics of Some Radioactive Nuclides

Nuclide	$t_{1/2}$	Decay Mode	Daughter
$^{3}_{1}$H (tritium)	12.26 yr	beta	$^{3}_{2}$He
$^{8}_{4}$Be	$\approx 1 \times 10^{-16}$ s	alpha	$^{4}_{2}$He
$^{14}_{6}$C	5730 yr	beta	$^{14}_{7}$N
$^{22}_{11}$Na	2.601 yr	positron emission	$^{22}_{10}$Ne
$^{24}_{11}$Na	15.02 h	beta	$^{24}_{12}$Mg
$^{32}_{15}$P	14.28 d	beta	$^{32}_{16}$S
$^{35}_{16}$S	87.2 d	beta	$^{35}_{17}$Cl
$^{36}_{17}$Cl	3.01×10^{5} yr	beta	$^{36}_{18}$Ar
$^{40}_{19}$K	1.28×10^{9} yr	beta (89.3%) / electron capture (10.7%)	$^{40}_{20}$Ca / $^{40}_{18}$Ar
$^{59}_{26}$Fe	44.6 d	beta	$^{59}_{27}$Co
$^{60}_{27}$Co	5.27 yr	beta	$^{60}_{28}$Ni
$^{90}_{38}$Sr	29 yr	beta	$^{90}_{39}$Y
$^{109}_{48}$Cd	453 d	electron capture	$^{109}_{47}$Ag
$^{125}_{53}$I	59.7 d	electron capture	$^{125}_{52}$Te
$^{131}_{53}$I	8.021 d	beta	$^{131}_{54}$Xe
$^{127}_{54}$Xe	36.41 d	electron capture	$^{127}_{53}$I
$^{137}_{57}$La	$\approx 6 \times 10^{4}$ yr	electron capture	$^{137}_{56}$Ba
$^{222}_{86}$Rn	3.824 d	alpha	$^{218}_{84}$Po
$^{226}_{88}$Ra	1600 yr	alpha	$^{222}_{86}$Rn
$^{232}_{90}$Th	1.40×10^{10} yr	alpha	$^{228}_{88}$Ra
$^{235}_{92}$U	7.04×10^{8} yr	alpha	$^{231}_{90}$Th
$^{238}_{92}$U	4.468×10^{9} yr	alpha	$^{234}_{90}$Th
$^{239}_{93}$Np	2.350 d	beta	$^{239}_{94}$Pu
$^{239}_{94}$Pu	2.411×10^{4} yr	alpha	$^{235}_{92}$U

EXAMPLE 15–5

Tritium (^{3}H) decays by beta emission to ^{3}He with a half-life of 12.26 years. A sample of a tritium-containing compound has an initial activity of 0.833 Bq. Calculate (a) the number of tritium nuclei present in the sample initially (N_i), (b) the decay constant k, and (c) the activity after 2.50 years.

Solution

(a) The number of nuclei of tritium present at any moment is given by

$$N = \frac{A\, t_{1/2}}{\ln 2}$$

where A is the activity of the tritium at that moment and $t_{1/2}$ is its half-life. Because the activity is given in Bq, which are equivalent to s^{-1}, convert the half-life from years to seconds

$$t_{1/2} = 12.26 \text{ yr} \times \left(\frac{365 \text{ days}}{1 \text{ yr}}\right) \times \left(\frac{24 \text{ hr}}{1 \text{ day}}\right) \times \left(\frac{3600 \text{ s}}{1 \text{ hr}}\right) = 3.866 \times 10^8 \text{ s}$$

Substitution then gives

$$N_i = \frac{(0.833 \text{ s}^{-1})(3.866 \times 10^8 \text{ s})}{0.6931} = 4.65 \times 10^8 \,^3\text{H nuclei}$$

(b) The decay constant is computed directly from the half-life:

$$k = \frac{\ln 2}{t_{1/2}} = \frac{0.6931}{3.866 \times 10^8 \text{ s}} = 1.793 \times 10^{-9} \text{ s}^{-1}$$

(c) The activity falls off with time as the number of radioactive atoms diminishes. The equation for this decay in activity is

$$A = A_i\, e^{-kt}$$

The k and A_i are known, so substitution into this equation gives the activity after $t = 2.50$ years. First, however, the 2.50 years must be converted to seconds so that the units in the exponent cancel out. The answer, computed as in part (a), is $t = 7.884 \times 10^7$ s. Then

$$A = (0.833 \text{ Bq})\, e^{-(1.793 \times 10^{-9} \text{ s}^{-1})(7.884 \times 10^7 \text{ s})} = 0.723 \text{ Bq}$$

Exercise

The ^{60}Co used in cancer therapy "cobalt treatments" emits gamma rays that kill cells. Cobalt-60 decays with a half-life of 5.27 years. A 0.19-g sample of cobalt contains ^{60}Co mixed with stable ^{59}Co and has an activity of 4.0×10^{11} Bq. What percentage by mass of the sample is ^{60}Co? Take the relative atomic mass of ^{60}Co to be 60 and that of ^{59}Co to be 59.

Answer: 5.0%.

Radioactive Dating

The decay of radioactive nuclides with known half-lives enables geochemists to measure the ages of rocks from their isotopic composition. Suppose that a uranium-bearing mineral was deposited in the earth 2 billion (2.00×10^9) years ago and that

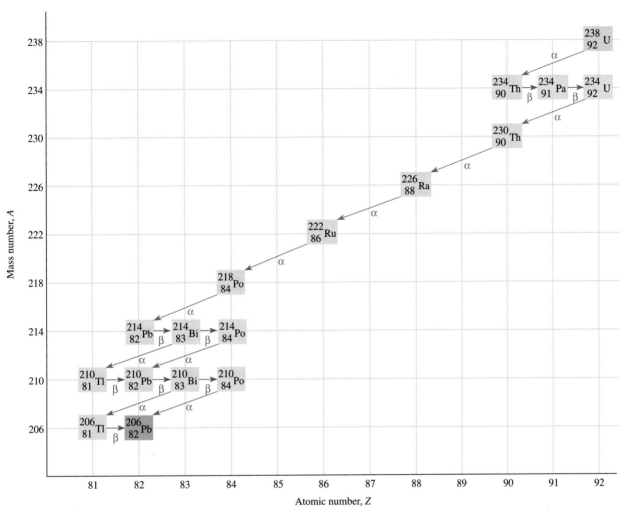

Figure 15-7   The radioactive nuclide ^{238}U decays by a series of alpha (α) and beta (β) emissions to the stable nuclide ^{206}Pb.

it has remained geologically unaltered. The ^{238}U atoms in the mineral decay with a half-life of 4.468×10^9 years to form a series of short-lived intermediates, ending in the stable isotope of lead ^{206}Pb (Fig. 15-7). The fraction of ^{238}U remaining after 2.00×10^9 years should be

$$\frac{N}{N_i} = e^{-kt} = \exp\left(\frac{-0.6931}{4.468 \times 10^9 \text{ yr}} \times 2.00 \times 10^9 \text{ yr}\right) = 0.733$$

One ^{206}Pb atom forms for each ^{238}U that decays, so the number of ^{206}Pb atoms should be $1.00 - 0.733$, or 0.267 times the original number of ^{238}U atoms. The present-day ratio of the abundances of the lead and uranium isotopes is then $0.267{:}0.733$, which equals 0.364. This ratio is fixed by the time elapsed since the mineral was formed. Of course, to calculate the age of a mineral we work backward from the measured $N(^{206}\text{Pb}){:}N(^{238}\text{U})$ ratio.

CHEMISTRY IN PROGRESS

Element 112

The periodic table was extended to 112 elements in 1996 with the synthesis of the nuclide $^{277}112$ by researchers at the Society for Heavy Ion Research in Darmstadt, Germany. Two atoms of this element were prepared by irradiation of a target of ^{208}Pb with a beam of $^{70}_{30}$Zn^{10+} ions accelerated to an energy of 343.8 MeV:

$$^{70}_{30}\text{Zn} + ^{208}_{82}\text{Pb} \longrightarrow ^{277}_{112}112 + ^{1}_{0}n$$

Figure 15-A diagrams the chain of decay events that followed synthesis of the second atom. After 280 μs, the nucleus emitted an alpha particle (in the step labeled α_1 in the diagram) to give element 110

$$^{277}_{112}112 \longrightarrow ^{273}_{110}110 + ^{4}_{2}\text{He}$$

The 110 atom then decayed further, transforming in five successive alpha decays over a period of 47 s to hassium-269, seaborgium-265, rutherfordium-261 (element 104), nobelium-257, and then fermium-253. Although fermium-253 is also unstable, it is long-lived (half-life = 3 days), and additional decay steps were not observed. The conclusion that an atom of element 112 was present emerges unambiguously from analysis of the energies, timing, and location of the decay events, and comparison to the known modes and energies of radioactive decay of the nuclides in the chain.

The diagram gives lifetimes of specific atoms and not half-lives, which are statistical quantities. Thus, the *other* atom of element $^{77}_{112}112$, which was created on February 1, 1996, survived 400 μs before decaying.

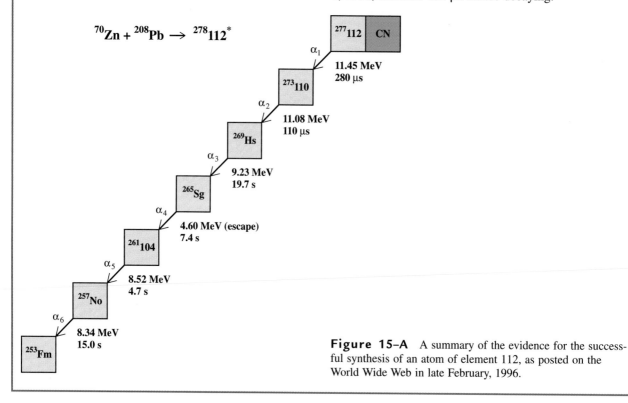

Figure 15–A   A summary of the evidence for the successful synthesis of an atom of element 112, as posted on the World Wide Web in late February, 1996.

This method of dating assumes that the stable nuclide generated at the end of the decay series (^{206}Pb in this example) derives exclusively from the parent species (^{238}U in this case) and that neither it nor the parent has entered or left the rock over the course of geological time. Whenever possible, geochemists determine the ratios of several different parent–daughter pairs of nuclides from the same rock sample.

For example, ^{40}K decays to ^{40}Ar (a gas, but trapped in the rock matrix) with a half-life of 1.28×10^9 years, and ^{87}Rb decays to ^{87}Sr with a half-life of 4.9×10^{11} years. Each pair should ideally yield the same age. Analysis of a large number of rock samples suggests that the oldest surface rocks on earth are about 3.8 billion years old. An estimate of 4.55 billion years for the age of the earth and the solar system comes from isotopic analysis of meteorites, which are believed to have formed at the same time.

Another type of radioactive dating, one that has been of particular value to archaeologists and anthropologists, relies on the decay of carbon-14. This nuclide has a half-life of 5730 years and is continuously produced in the earth's atmosphere by ordinary ^{14}N reacting with high-energy neutrons generated by cosmic rays.

$$^{14}_{7}N + {}^{1}_{0}n \longrightarrow {}^{14}_{6}C + {}^{1}_{1}H$$

The radioactive carbon produced in this way enters the carbon reservoir on the earth's surface, mixing with stable ^{12}C as dissolved $H^{14}CO_3^-$ in the oceans, as $^{14}CO_2$ in the atmosphere, and in the tissues of plants and animals. Right now, the ^{14}C in the environment gives a specific activity of 15.3 disintegrations per minute per gram of total carbon, that is, $0.255\ Bq\ g^{-1}$. This corresponds to one ^{14}C atom for every 7.54×10^{11} atoms of ^{12}C. When a plant or animal dies, the exchange of its carbon atoms with those of the environment stops, and the amount of ^{14}C in it then begins to decrease exponentially. By measuring the remaining activity of ^{14}C in an archaeological sample, we can estimate its age. The method depends upon the constancy of the cosmic-ray flux through the earth's atmosphere, of course. Comparisons of ^{14}C dates against those obtained by other means (such as counting the annual growth rings of the long-lived bristlecone pine or studying the written records that may accompany a carbon-containing artifact) show that the technique is reliable over the time span for which it can be checked. The burning of fossil fuels (coal and oil) over the last 100 years has increased the proportion of ^{12}C in living organisms and will cause difficulty in applying ^{14}C dating in the future.

EXAMPLE 15-6

An ancient wooden shovel has a specific activity of ^{14}C of $0.195\ Bq\ g^{-1}$. Estimate the age of the shovel.

Solution

The decay constant for ^{14}C is

$$k = \frac{0.6931}{5730\ \text{yr}} = 1.21 \times 10^{-4}\ \text{yr}^{-1}$$

Initially the specific activity of the shovel was $0.255\ Bq\ g^{-1}$. This has fallen to $0.195\ Bq\ g^{-1}$. These values are on a per gram basis, but the mass of the shovel does not change, so they can be used for A_i and A. Substitute them into the equation for the decay of the activity in the form

$$\frac{A}{A_i} = e^{-kt}$$

$$\frac{0.195\ \cancel{Bq\ g^{-1}}}{0.255\ \cancel{Bq\ g^{-1}}} = e^{-kt} = e^{-(1.21 \times 10^{-4}\ \text{yr}^{-1})\,t}$$

Taking the natural logarithm (ln) of both sides gives

$$\ln\!\left(\frac{0.195}{0.255}\right) = -(1.21 \times 10^{-4}\ \text{yr}^{-1})\, t$$

$$t = \boxed{2200\ \text{yr}}$$

The shovel was made with wood from a tree that died approximately 2200 years ago.

Exercise

Linen cloth from a religious relic contains one ^{14}C atom for every 8.25×10^{11} atoms of ^{12}C. Estimate the age of the cloth.

Answer: 740 years.

15–4   RADIATION IN BIOLOGY AND MEDICINE

Radiation has both harmful and beneficial effects for living organisms. All forms of radiation cause damage in proportion to the amount of energy they deposit in cells and tissues. The damage takes the form of chemical changes in cellular molecules that alter their functions and may lead either to uncontrolled multiplication and growth of cells or to their death. Alpha particles lose their kinetic energy over very short distances in matter (typically 10 cm in air and 0.05 cm in water and tissues), producing intense ionization in their wakes until they accept electrons and become harmless helium atoms. Radium, for example, is an alpha emitter that substitutes for calcium in bone tissue and destroys its capacity to produce both red and white blood cells. Beta particles and gamma rays have greater penetrating power than alpha particles and so present a radiation hazard even when their source is well outside an organism.

> • The rad is more widely used in the United States than the SI unit of absorbed dose, the gray (Gy). One gray deposits 1 joule of energy per kilogram of absorbing material; 1 Gy = 100 rad.

The amount of damage produced in tissue by radiation is proportional to the quantity of energy deposited in the tissue. A dose of all types of radiation that actually deposits 10^{-2} J of energy per kilogram of tissue is 1 **rad** (radiation absorbed dose). Two 1-rad doses of different kinds of radiation may differ in toxicity depending on the actual volume of affected tissue, the type of tissue (some organs are more susceptible than others), and how fast the doses are administered (a slow dose does not do the same harm as a sudden one). To take all of these factors into account, the **rem** (**r**oentgen **e**quivalent in **m**an) has been defined to measure the effective dosages of radiation received by humans. A physical dose of 1 rad of beta or gamma radiation converts into a human dose of 1 rem. Alpha radiation is more toxic; a physical dose of 1 rad of alpha radiation equals about 10 rem.

Exposure to radiation is unavoidable. The average person in the United States receives an annual radiation dose of about 11 mrem from natural sources, which include cosmic radiation and natural radioactive nuclides like ^{14}C, ^{40}K, and ^{222}Rn. Another 50 to 100 mrem come from human activities (including dental and medical x-rays and airplane flights, which cause greater exposure to cosmic rays higher in the atmosphere). The safe level of exposure to radiation is a controversial issue among biologists because the consequences of exposure may become apparent only after many years. Also, it is necessary to distinguish between tissue damage experienced by the exposed individual and genetic damage, which may not become apparent for several generations. It is much easier to define the radiation level that will, with high probability, cause death to an exposed person. The LD_{50} level in human beings (50% probability that death will result in 30 days after a single exposure) is 500 rad.

EXAMPLE 15-7

The beta decay of the potassium-40 (^{40}K) that is a natural part of the body makes all human beings slightly radioactive. The relative natural abundance of ^{40}K is 0.0118%, its half-life is 1.28×10^9 years, and its beta particles have an average kinetic energy of 0.55 MeV. A 70.0-kg person contains about 170 g of potassium. (a) Calculate the total activity of ^{40}K in this person. (b) Determine (in rad per year) the annual radiation absorbed dose arising from this person's internal ^{40}K.

Solution

(a) The activity (A) equals the number of ^{40}K atoms present multiplied by their decay constant. First calculate the number of atoms of ^{40}K present:

$$170 \text{ g K} \times \left(\frac{1 \text{ mol K}}{40.0 \text{ g K}}\right) \times \left(\frac{6.022 \times 10^{23} \text{ atoms K}}{1 \text{ mol K}}\right) \times \left(\frac{0.0118 \text{ atoms }^{40}K}{100 \text{ atoms K}}\right)$$
$$= 3.02 \times 10^{20} \text{ atoms of }^{40}K$$

Then determine the first-order decay constant from the half-life. We need the decay constant in s^{-1}, so some conversions of the time units are necessary.

$$k = \frac{0.6931}{t_{1/2}} = \frac{0.6931}{1.28 \times 10^9 \text{ yr} \times \left(\frac{365 \text{ day}}{1 \text{ yr}}\right) \times \left(\frac{24 \text{ hr}}{1 \text{ day}}\right) \times \left(\frac{3600 \text{ s}}{1 \text{ hr}}\right)}$$
$$= 1.72 \times 10^{-17} \text{ s}^{-1}$$

Because $A = k\,N$, we have

$$A = (1.72 \times 10^{-17} \text{ s}^{-1})(3.02 \times 10^{20}) = 5.19 \times 10^3 \text{ s}^{-1} = 5.19 \times 10^3 \text{ Bq}$$

This means that about 5190 atoms of ^{40}K disintegrate within the person every second. (b) A disintegration of radioactive ^{40}K emits a beta particle with an average of 0.55 MeV of energy, and all of this energy is presumably deposited within the person. From part (a) there are 5.19×10^3 disintegrations per second. Therefore, the total energy deposited per year is

$$\frac{5.19 \times 10^3 \text{ disint.}}{\text{s}} \times \left(\frac{0.55 \text{ MeV}}{\text{disint.}}\right) \times \left(\frac{3600 \text{ s}}{1 \text{ hr}}\right) \times \left(\frac{24 \text{ hr}}{1 \text{ day}}\right) \times$$
$$\left(\frac{365 \text{ day}}{1 \text{ yr}}\right) = 9.00 \times 10^{10} \frac{\text{MeV}}{\text{yr}}$$

Because a rad is a centijoule (cJ) per kilogram of tissue, express this answer in centijoules per year:

$$9.00 \times 10^{10} \frac{\text{MeV}}{\text{yr}} \times \left(\frac{1.602 \times 10^{-13} \text{ J}}{1 \text{ MeV}}\right) \times \left(\frac{10^2 \text{ cJ}}{1 \text{ J}}\right) = 1.44 \frac{\text{cJ}}{\text{yr}}$$

Each kilogram of body tissue receives 1/70.0 of this amount of energy per year because the person weighs 70.0 kg. The dose is thus 21×10^{-3} cJ kg^{-1} yr^{-1}, which is equivalent to 21 mrad yr^{-1}. This is about one fifth of the annual background dosage received by a person.

An advanced generator for ^{99m}Tc, an isotope used in nuclear medicine to study heart tissue. The m stands for metastable and refers to a short-lived excited nuclear state.

Figure 15–8 With her head centered in a ring of radiation detectors, a patient prepares to undergo a PET (positron emission tomography) scan.

Exercise

Strontium-90 decays by beta emission with a decay energy of 0.546 MeV and a half-life of 28.1 years. How much ^{90}Sr (in grams) would inflict a 125-mrad dose of radiation in a 65.0-kg man in 60.0 s if all the decay energy were absorbed by the man? The molar mass of ^{90}Sr is 89.9073 g mol^{-1}.

Answer: 0.00295 g.

Although radiation can do great harm, it confers great benefits in medical applications. The diagnostic importance of x-ray imaging hardly needs to be mentioned. Both x-rays and gamma rays are used selectively to destroy malignant cells in cancer therapy. The beta-emitting ^{131}I nuclide finds use in the treatment of cancer of the thyroid because iodine is taken up preferentially by the thyroid gland. Heart pacemakers use the energy generated by the decay of tiny amounts of radioactive ^{238}Pu (see Fig. 15–2), converted into electrical energy. The element technetium has an important radioactive isotope, ^{99m}Tc, an excited state of ^{99}Tc that decays to the nuclear ground state by emission of gamma rays with a half-life of 6 hours. It is taken up by healthy heart tissue, allowing a gamma-ray detector to provide an image of the heart that is useful diagnostically.

Positron emission tomography (PET) is another important diagnostic technique using radiation (Fig. 15–8). It uses radioactive nuclides such as ^{11}C (half-life, 20.4 minutes) and ^{15}O (half-life, 2.0 minutes) that emit positrons when they decay. These nuclides are incorporated (quickly, because of their short half-lives) into substances such as glucose, which are injected into the patient. By following the pattern of annihilation radiation issuing from the body (generated as positrons encounter normal matter), researchers can study blood flow and glucose metabolism in healthy and diseased individuals. Computer-constructed pictures of positron emissions are useful in understanding how the brain works; for example, the location and level of glucose metabolism in the brain change as Alzheimer's disease progresses (Fig. 15–9).

Less direct benefits are also numerous. An example is the study of the mechanism of photosynthesis, in which light drives the combination of carbon dioxide and water to form glucose in the green leaves of plants.

$$6 \; CO_2(g) + 6 \; H_2O(\ell) \longrightarrow C_6H_{12}O_6(s) + 6 \; O_2(g)$$

Exposure of plants to carbon dioxide containing an artificially increased proportion of ^{14}C as a tracer permits the mechanism of the reaction to be followed. At intervals, the plants are analyzed to find what compounds have incorporated ^{14}C in their molecules and thereby to identify the intermediates in photosynthesis. Radioactive tracers are also widely used in medical diagnosis. The *radioimmunoassay* technique, invented by Nobel laureate Rosalyn Yalow, allows determination of very low levels of drugs and hormones in body fluids (see Chemistry in Progress: Radioimmunoassay).

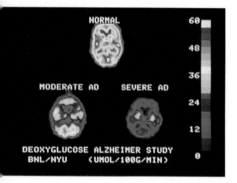

Figure 15–9 Positron emission tomographs of the brains of a healthy older person and of patients suffering from Alzheimer's disease. Lighter colors indicate regions having higher concentrations of radioactive glucose and therefore more active metabolism.

CHEMISTRY IN PROGRESS

Radioimmunoassay

One of the most dangerous radioactive nuclides in the fallout from the 1986 meltdown and fire at Chernobyl was iodine-131, which undergoes beta decay with a half-life of 8.021 days. Microgram quantities of ^{131}I deliver a lethal whole-body dose of radiation (see Problem 39 in this chapter). The much smaller amounts in the fallout hundreds of miles downwind from Chernobyl posed a hazard because the thyroid gland concentrates iodine in the course of synthesizing the hormone thyroxin, $C_{15}H_{11}I_4NO_4$. People in the path of the fallout were told to take potassium iodide pills to protect their thyroids against radiation damage. The basis for the intervention was chemical: binding sites in the thyroid gland would pick up very little outside iodine, radioactive or not, against the flood of competition offered by the iodine in the pills.

The concept of competition for binding sites underlies the important technique called **radioimmunoassay (RIA).** This method permits the measurement of the concentration of substances of biological interest, such as hormones, at concentrations as low as 10^{-12} M. In RIA, the binding sites are on *antibodies*, which are proteins produced by the immune system in response to the presence of a foreign protein. Antibodies bind quite specifically to foreign proteins as part of the body's defense against disease. Suppose that we wish to assay for protein H, a hormone, in a sample. We add H*, a radioisotopically labeled form of H, together with an H-binding antibody. This antibody naturally also binds H*, which is chemically equivalent to H. If any H is in the sample, it prevents some H* from binding to the antibody because it competes for binding sites. The radioactive emissions from the H*–antibody complex are then lower than if no H were in the sample. By comparing the rate of decay events in H*–antibody complex obtained from the sample with the rate in H*–antibody complex obtained from a series of standards that were prepared with known amounts of H, the amount of H in the sample can be inferred. The method is sensitive, highly specific, and easily automated.

RIA can be adapted to analyze for non-hormonal substances. Suppose that a sample of urine is being tested for cocaine. A known amount of radioisotopically labeled cocaine is complexed to a carrier protein. This complex is added to the sample, and the mixture is allowed to equilibrate. Cocaine, if present in the sample, replaces some of the radioactive cocaine in the complex. The complex is then separated, and its radioactivity is measured. The observed count rate is inversely related to the concentration of cocaine in the sample.

15-5   NUCLEAR FISSION

The nuclear reactions considered so far have been spontaneous first-order decays of unstable nuclides. By the 1920s, physicists and chemists began using particle accelerators to bombard samples with high-energy particles in order to *induce* nuclear reactions. One of the first results of this program was the 1932 discovery by James Chadwick of the neutron as a product of the reaction of alpha particles with light nuclides such as $^9_4$Be:

$$^4_2\text{He} + ^9_4\text{Be} \longrightarrow ^1_0n + ^{12}_6\text{C}$$

Shortly after Chadwick's discovery, a group of physicists in Rome, led by Enrico Fermi, began to study the interaction of neutrons with nuclei of various elements. The experiments produced a number of radioactive species, and it was evident that the absorption of a neutron increased the *N:Z* ratio in target nuclei above the stability line (see Fig. 15–1). One of the targets used was uranium, the heaviest naturally occurring element. Several radioactive products resulted, but none had chemical properties resembling those of the elements lying between *Z* = 86 (radon) and *Z* = 92 (uranium). It appeared to the Italian scientists in 1934 that several new transuranic elements (*Z* > 92) had been synthesized, and an active period of investigation followed.

In Berlin in 1938, Otto Hahn and Fritz Strassmann sought to characterize the supposed transuranic elements. To their bewilderment, they found that barium ($Z = 56$) appeared among the products of the bombardment of uranium by neutrons. Hahn informed his former colleague Lise Meitner, and she suggested that the products of the bombardment of uranium by neutrons were not transuranic elements but fragments of uranium atoms resulting from a process she termed **fission.** The implication of this idea—the possible release of enormous amounts of energy from the splitting of the nucleus—was instantly apparent. In August, 1939, with the outbreak of World War II imminent, Albert Einstein wrote to U.S. President Franklin Roosevelt to inform him of the possible military uses of nuclear fission and his concern that Germany might develop a nuclear explosive. As a result, in 1942 President Roosevelt authorized the Manhattan District Project, an intense coordinated effort by a large number of physicists, chemists, and engineers to make a fission bomb of unprecedented destructive power.

The operation of the first atomic bomb hinged on the fission of uranium in a chain reaction induced by absorption of neutrons. The two most abundant isotopes of uranium are ^{235}U and ^{238}U, which have natural abundances of 0.720% and 99.275%, respectively. Both isotopes undergo fission upon absorbing a neutron, ^{238}U only with "fast" neutrons but ^{235}U with both "fast" and "slow" neutrons. In the early days of neutron research, it was not known that the absorption of neutrons by nuclei depends strongly upon neutron velocity. Fermi and his co-workers observed that neutron absorption experiments conducted on a wooden table led to a much higher yield of radioactive products than those performed on a marble-topped table. Fermi then repeated the irradiation experiments with a block of paraffin wax interposed between the radium–beryllium neutron source and the target, with the startling result that the level of induced radioactivity was greatly enhanced. Within hours Fermi had found the explanation: the high-energy (fast) neutrons emitted from the radium–beryllium source were slowed down by collision with the nuclei of the paraffin molecules. Because of their lower kinetic energies, their probability of reaction with ^{235}U was greater and a higher yield was achieved. Hydrogen nuclei and the nuclei of other light elements such as ^{12}C (in graphite) are very effective in slowing down high-velocity neutrons and are called **moderators.**

In the electrorefining of uranium for nuclear-reactor fuel, uranium(III) chloride is deposited on the anode of an electrolytic cell. Here, it comprises 3% of the mass in a matrix of LiCl and KCl and causes the amethyst color of the deposit.

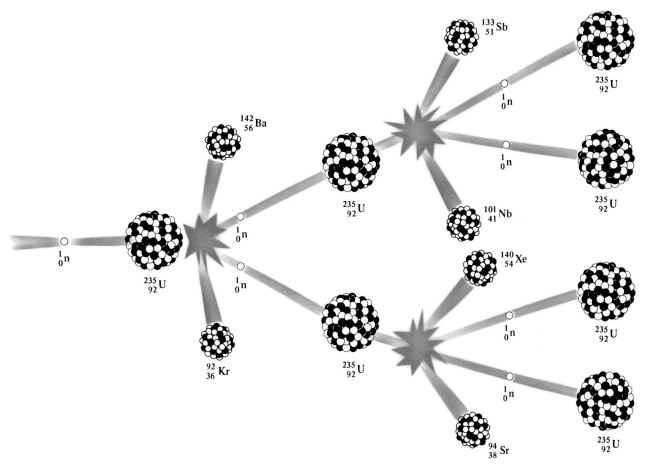

Figure 15–10   In a self-propagating nuclear chain reaction, the number of neutrons grows exponentially during fission.

The fission of ^{235}U follows many different patterns, and some 34 elements have been detected among the products. In any single fission event, two particular nuclides are produced, together with two or three neutrons; collectively, they carry away about 200 MeV of kinetic energy. Three of the many pathways are

$$_0^1n + {}_{92}^{235}U \longrightarrow \begin{array}{l} {}_{30}^{72}Zn + {}_{62}^{162}Sm + 2\,{}_0^1n \\ {}_{38}^{80}Sr + {}_{54}^{153}Xe + 3\,{}_0^1n \\ {}_{36}^{94}Kr + {}_{56}^{139}Ba + 3\,{}_0^1n \end{array}$$

The fission process is a branching chain reaction because two or three new neutrons are produced for each neutron absorbed. Figure 15–10 illustrates the self-propagating nuclear chain reaction and its accompanying exponential growth of numbers of neutrons.

Such growth results in an explosion if the rate of production of neutrons is not soon matched by the rate of their loss. Fermi and his associates demonstrated on December 2, 1942, that a controlled self-sustaining neutron chain reaction occurred in a uranium "pile" that used cadmium control rods to absorb neutrons, thereby balancing the rates of neutron generation and loss. This proved the feasibility of having nuclear reactors that operate at a controlled and steady neutron concentration and, hence, a constant rate of energy production.

Although Enrico Fermi and his associates were the first scientists to demonstrate a self-sustaining nuclear chain reaction, a natural uranium fission reactor "went critical" about 1.8 billion years ago in a place now called Oklo in the Gabon Republic of equatorial Africa. The clue that this had happened came when French scientists discovered that the ^{235}U content of ore from a site in the open-pit mine at Oklo was only 0.7171%; the normal ^{235}U content in other areas of the mine was 0.7207%. Although this was not a large deviation, it was significant, and investigation revealed that other elements were also present in the ore in the exact proportions expected if nuclear fission had occurred. This established beyond all doubt that a self-sustaining nuclear reaction had occurred there. From the size of the active ore mass and the depletion of ^{235}U, it is estimated that the reactor generated about 15,000 megawatt-years (5×10^{17} J) of energy over a period of about 100,000 years.

Nuclear Power Reactors

Most **nuclear power reactors** in the United States use rods containing oxides of uranium as fuel. The uranium is primarily ^{238}U, but the amount of ^{235}U is enriched above natural abundance to a level of about 3%. The moderator used to slow the neutrons (to increase the efficiency of the fission) is ordinary water in most cases, so these reactors are called "light water" reactors. The controlled release of energy by nuclear fission in power reactors demands a delicate balance between neutron generation and neutron loss. As mentioned earlier, this is accomplished by means of steel control rods containing ^{112}Cd or ^{10}B, nuclides that have a very large neutron-capture probability. These rods are automatically inserted into or withdrawn from the fissioning system in response to changes in the neutron flux. As the nuclear reaction proceeds, the moderator (water) transfers the heat that it releases to a steam generator. The steam then drives turbines that generate electricity (Fig. 15–11).

The risks associated with the operation of fission reactors are small but not negligible, as the accident at the Three Mile Island reactor in March 1979 and the dis-

• A different reactor design employs "heavy water," in which the protons of ordinary water are replaced by deuterons, giving $(^2H)_2O$ (or D_2O). Heavy water is an efficient moderator, but expensive.

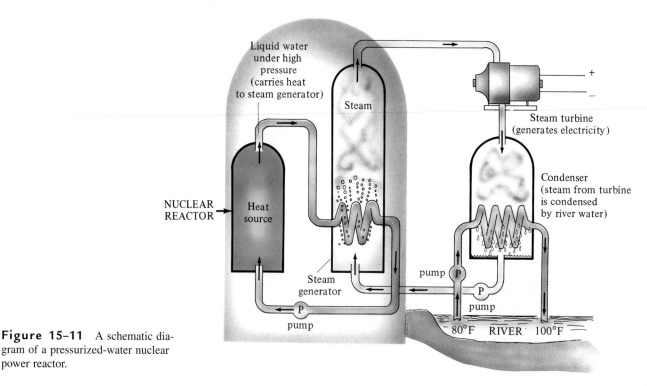

Figure 15–11 A schematic diagram of a pressurized-water nuclear power reactor.

Inside a containment building. The reactor core sits under a pool of water. A blue glow emanates as high-energy particles from the core are slowed by the water.

Figure 15–12 An aerial view of the Three Mile Island nuclear power plant in Pennsylvania. The containment buildings for the two power reactors are the smaller cylindrical buildings to the right. The four large structures are cooling towers.

aster at Chernobyl in April 1986 attest. If a reactor has to be shut down quickly, there is danger of a meltdown, in which the heat from the continuing fission processes melts the uranium fuel. Coolant must be circulated until heat from the decay of short-lived unstable nuclides has been dissipated. The Three Mile Island accident resulted in a partial meltdown because some water-coolant pumps were inoperative and others were shut down too soon, causing damage to the core and slight release of radioactivity in the environment. The Chernobyl disaster was caused by a failure of the water-cooling system and a meltdown. The rapid and uncontrolled nuclear reaction that took place set the graphite moderator on fire and burst through the reactor building, spreading radioactive nuclides with an activity estimated at 2×10^{20} Bq into the environment. A major part of the problem was that the reactor at Chernobyl, unlike those in the United States, was not in a massive containment building (Fig. 15–12).

The safe disposal of the radioactive wastes from nuclear reactors is also an important but controversial matter that has not yet been resolved. A variety of proposals have been advanced, including burial of radioactive waste in deep mines on either a recoverable or permanent basis, burial at sea, and launching the waste into outer space. The first alternative is the only one that appears credible. The essential requirement is that the disposal site(s) be stable with respect to possible earthquakes and invasion by underground water movements. For a nuclide such as plutonium-239, which has a half-life of 24,000 years, a storage site that is stable over 240,000 years is needed before the activity drops to 0.1% of its original value. Some shorter-lived nuclides are more hazardous over short periods of time, but the threat diminishes more quickly.

15–6 NUCLEAR FUSION AND NUCLEOSYNTHESIS

Nuclear fusion is the union of two light nuclides to form a heavier nuclide with the release of energy. Fusion reactions are often called **thermonuclear** reactions because the colliding particles must possess very high kinetic energies, corresponding to temperatures of millions of degrees, before they are initiated. They are the processes that occur in the sun and other stars. In 1939, Hans Bethe (and, independently, Carl von Weizsäcker) proposed that in normal stars the following reactions take place:

$$
{}^1_1p^+ + {}^1_1p^+ \longrightarrow {}^2_1H^+ + {}^0_1e^+ + \nu
$$
$$
{}^2_1H^+ + {}^1_1p^+ \longrightarrow {}^3_2He^{2+} + \gamma
$$
$$
{}^3_2He^{2+} + {}^3_2He^{2+} \longrightarrow {}^4_2He^{2+} + 2\,{}^1_1p^+
$$

In the first reaction, two high-velocity protons fuse to form a deuteron ($^2H^+$), with the emission of a positron and a neutrino that have a total of 1.026 MeV of kinetic energy. In the second reaction, a high-energy deuteron combines with a high-velocity proton to form a helium nucleus of mass 3, together with a gamma ray. The third reaction completes the cycle with formation of a normal helium nucleus and regeneration of the two protons. Each of these reactions is exothermic, but up to 1.25 MeV is required to overcome the electrostatic repulsion between the positively charged nuclei. The overall result of the cycle is to convert hydrogen nuclei into helium nuclei, and the process is called **hydrogen burning.**

As such a star ages and accumulates helium, it begins to contract under the influence of its immense gravity. The star's helium core begins to heat up. When the

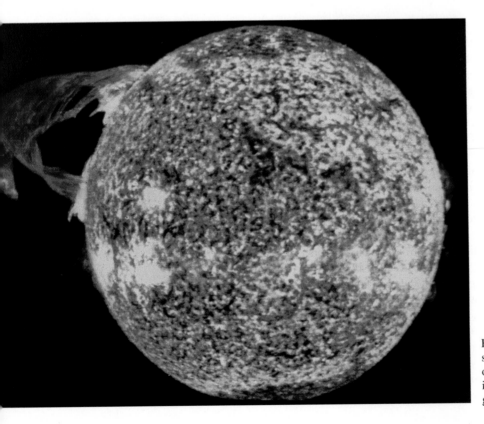

Fusion reactions in the interior of the sun are the ultimate source of most of our energy resources on earth. Our sun is still in its hydrogen-burning period, generating helium from hydrogen.

core reaches a temperature of about 10^8 K, another cycle of **nucleosynthesis** begins, the first reaction of which is

$$2\,{}_{2}^{4}\mathrm{He}^{2+} \rightleftharpoons {}_{4}^{8}\mathrm{Be}^{4+}$$

This reaction is written as an equilibrium because the ^{8}Be nucleus reverts to helium nuclei with a half-life of only 2×10^{-16} s. Even with this short half-life, the ^{8}Be nuclei are believed to react with alpha particles to form ^{12}C in the exothermic reaction

$$ {}_{4}^{8}\mathrm{Be}^{4+} + {}_{2}^{4}\mathrm{He}^{2+} \longrightarrow {}_{6}^{12}\mathrm{C}^{6+}$$

The product of this set of reactions is the stable ^{12}C nucleus. The density of the core of a star that is burning helium in this way is on the order of 10^5 g cm^{-3}.

This process of **nucleosynthesis** continues beyond the formation of ^{12}C to produce additional, heavier nuclides: ^{13}N, ^{13}C, ^{14}N, ^{15}O, ^{15}N, and ^{16}O. Stars in this stage have moved away from the main sequence that characterizes the younger stars and are classified as red giants. Even here, the process does not end. Similar cycles occur until the temperature of the core is about 4×10^9 K, its density is about 3×10^6 g cm^{-3}, and the nuclei are ^{56}Fe, ^{59}Co, and ^{60}Ni. These nuclei have the maximum binding energy per nucleon (see Fig. 15–3). It is thought that the synthesis of still heavier nuclei occurs in the immense explosions of supernovas.

Heavy elements can also be produced in **particle accelerators,** which accelerate electrons, protons, and other ions to high speeds. The resulting collisions can cause new elements to form. Technetium, for example, is not found in nature but was first produced in 1937 by directing high-energy deuterons at a molybdenum target:

$$ {}_{42}^{96}\mathrm{Mo} + {}_{1}^{2}\mathrm{H} \longrightarrow {}_{43}^{97}\mathrm{Tc} + {}_{0}^{1}n$$

The first **transuranic element** was produced in 1940. Neptunium ($Z = 93$) results from the capture of a neutron by ^{238}U, followed by beta decay. Subsequent work by Glenn Seaborg and co-workers led to the production of plutonium ($Z = 94$) and heavier elements. In recent years, nuclides with Z as high as 112 have been made, but in tiny quantities. These nuclides have very short half-lives.

Major efforts are now under way to achieve controlled nuclear fusion as a source of energy. The most suitable reaction is

$$ {}_{1}^{2}\mathrm{H}^{+} + {}_{1}^{3}\mathrm{H}^{+} \longrightarrow {}_{2}^{4}\mathrm{He}^{2+} + {}_{0}^{1}n$$

with a predicted energy release of 17.6 MeV. This reaction is thermodynamically favored, but its activation energy is very large because the electrostatic repulsion of the nuclei must be overcome to bring them close enough together to fuse. When the repulsion is overcome, fusion ignition occurs. The neutrons that are produced in this reaction can be used to produce the tritium needed by means of the reaction

$$ {}_{3}^{6}\mathrm{Li} + {}_{0}^{1}n \longrightarrow {}_{1}^{3}\mathrm{H} + {}_{2}^{4}\mathrm{He}$$

in a lithium "blanket" that surrounds the region of the reaction. The temperatures required for initiation of fusion are on the order of 10^8 K, conditions under which atoms are stripped of all their electrons, and the system consists of a **plasma** of nuclei, electrons, and neutrons. The major problem is confining the plasma at this high temperature. Two methods are used. In **inertial confinement,** pellets of deuterium–tritium fuel are raised to the thermonuclear ignition temperature by sudden bursts of light fired from lasers (Fig. 15–13). The targets do not have time to move before ignition is achieved. Fusion ignition under inertial confinement is anticipated at the National Ignition Facility, now in the planning stages in California, early in the next century. This facility will use a very powerful neodymium glass laser sys-

(a)   (b)

Figure 15–13 (a) The target chamber of Nova, an inertial confinement fusion experiment, is a 16-foot-diameter aluminum sphere inside of which ten powerful laser beams converge. (b) The target itself is a tiny capsule (1 mm in diameter) of deuterium mixed with tritium. A laser pulse 1 billionth of a second in duration heats the target hotter than the sun's core, raising its pressure to more than 100 million atmospheres, to initiate nuclear fusion.

tem (50 times more powerful than the one at the heart of the Nova system in Fig. 15–13). In magnetic confinement, a magnetic field confines the plasma, which is much less dense than that in inertial confinement but which lasts much longer. Numerous designs for "magnetic bottles" are possible. The most promising are toroidal (ring-shaped) and, of these, the most advanced is the Tokamak. Figure 15–14 shows a plasma under magnetic confinement in a Tokamak.

Research reactors using magnetic confinement have made steady progress toward the breakeven point (at which the energy generated by the fusing nuclei equals the energy required to start the process). Formidable problems still must be solved, including the development of materials for containment vessels that can withstand the corrosive effects of the intense x-ray and neutron radiation that is present.

The rewards of a workable nuclear fusion process would be great. Fusion produces neither the long-lived radioactive nuclides that accompany nuclear fission nor the environmental pollutants that the burning of fossil fuels releases. Although deuterium is present to the extent of only 1/6000 of the abundance of ordinary hydrogen, its separation from the latter by the electrolysis of water is readily accomplished, and the oceans contain a virtually unlimited quantity of deuterium.

Figure 15–14 Plasma in the START (Small Tight Aspect Ratio Tokamak) device at Culham in the United Kingdom. The "tight aspect ratio" in this advanced Tokamak design means that the plasma is nearly spherical. The center column supplies the magnetic field for magnetic confinement of the plasma.

SUMMARY

15–1 The term **nuclide** is applied to atoms when the exact constitution of their nuclei is known or is important. A particular nuclide differs from all others in atomic number Z, the number of protons, or mass number A (the number of **nucleons** (protons plus neutrons)), or both. The masses of most nuclides are known to high precision and are usually given in atomic mass units. Some nuclides undergo nuclear reactions (changes in Z or A or both) at measurable rates; others are indefinitely stable. The criteria for balance in nuclear equations are: total A stays constant, total Z stays constant, and total charge stays constant. Mass and the chemical identity of nuclides are *not* conserved in nuclear reactions. Any change in energy causes a proportional change in mass according to $\Delta E = c^2 \Delta m$. The **electron volt (eV),** which is equal to 1.60218×10^{-19} J, and the **megaelectron volt (MeV)** are useful units for stating the energy of nuclear reactions. The **binding energy** of a nucleus is the negative of the energy change that would occur if the nucleus were to be formed directly from its component protons and neutrons. The most stable nuclides have the largest **binding energy per nucleon.**

15–2 In all spontaneous nuclear reactions, the mass of the products is less than the mass of the starting materials. If an unstable nuclide is deficient in protons, it tends to undergo **beta decay,** emission of a **beta particle** (an electron) plus an antineutrino, and so increase its atomic number by 1. If an unstable nuclide has an excess of protons, it tends to decay by emission of a **positron** plus a neutrino (**positron emission**) and so decrease its atomic number by 1. Another pathway for the decay of proton-rich nuclides is **electron capture,** in which the atomic number decreases by 1, but only a neutrino is emitted. Heavier unstable nuclides often decay by emission of an alpha particle (**alpha decay**), which reduces their atomic number by 2 and their mass number by 4. The energy change in any of these reactions can be computed given a balanced nuclear equation and a table of masses of the nuclides: Δm is computed and Einstein's equation applied. This energy appears as kinetic energy of the product particles. **Neutron emission** and **proton emission,** in which neutrons or protons "boil off" of a nucleus, are additional modes of decay. Important means of detecting emissions from radioactive materials include the photographic emulsion, the scintillation counter, and the Geiger counter.

15–3 Processes of radioactive decay follow first-order kinetics, and the **half-lives** of unstable nuclides can be measured. The rate of nuclear disintegration in a sample of radioactive substance is its **activity,** measured in disintegrations per second (**becquerels**). The activity is proportional to the number of atoms of radioactive substance times the rate constant for the decay. The activity itself decays according to first-order kinetics. If the half-lives of the unstable nuclides are known, then measurements of nuclidic composition or activity provide the age of specimens containing the nuclides. An especially important example is **carbon-14 dating.**

15–4 Penetrating radiation can inflict damage on living tissue. The **rad** is defined as the quantity of radiation that deposits 10^{-2} J of energy in 1 kg of tissue. It is closely related to the **rem,** the unit of effective radiation dosage in human beings.

15–5 In nuclear **fission,** heavy nuclides split apart. The fission of several elements, in particular uranium, can be induced by the absorption of neutrons. When a fission reaction generates more neutrons than it consumes, a self-propagating chain reaction is possible. Uncontrolled exponential growth in the number of neutrons results in a nuclear explosion. When the reaction is moderated so that neutron production and loss are in balance, it can serve as a source of power.

15–6 In nuclear **fusion,** light nuclides join together. Nuclear fusion reactions, or **thermonuclear** reactions, provide the energy emitted by the sun and other stars. In stars, helium is formed from hydrogen. This **hydrogen burning** releases much energy. Additional processes of **nucleosynthesis** produce heavier nuclides.

PROBLEMS

Note: Answers to blue-numbered problems are given in Appendix F.

Mass–Energy Relationships in Nuclei

1. Complete and balance the following equations, which represent nuclear reactions thought to take place in stars:
   (a) $2 \, {}^{12}_{6}\text{C} \longrightarrow \underline{?} + {}^{1}_{0}n$
   (b) $\underline{?} + {}^{1}_{1}\text{H} \longrightarrow {}^{12}_{6}\text{C} + {}^{4}_{2}\text{He}$
   (c) $2 \, {}^{3}_{2}\text{He} \longrightarrow \underline{?} + 2 \, {}^{1}_{1}\text{H}$

2. Complete and balance the following equations, which represent reactions used in particle accelerators to synthesize elements beyond uranium:
   (a) ${}^{4}_{2}\text{He} + {}^{235}_{99}\text{Es} \longrightarrow \underline{?} + 2 \, {}^{1}_{0}n$
   (b) ${}^{249}_{98}\text{Cf} + \underline{?} \longrightarrow {}^{257}_{103}\text{Lr} + 2 \, {}^{1}_{0}n$
   (c) ${}^{238}_{92}\text{U} + {}^{12}_{6}\text{C} \longrightarrow {}^{224}_{98}\text{Cf} + \underline{?}$

3. (See Example 15–1.) Calculate the total binding energy, in both kilojoules per mole and MeV per atom, and the binding energy per nucleon of the following nuclides, using the data from Table 15–1.
   (a) ${}^{40}_{20}\text{Ca}$
   (b) ${}^{87}_{37}\text{Rb}$
   (c) ${}^{238}_{92}\text{U}$

4. (See Example 15–1.) Calculate the total binding energy, in both kilojoules per mole and MeV per atom, and the binding energy per nucleon of the following nuclides, using the data from Table 15–1.
   (a) ${}^{10}_{4}\text{Be}$
   (b) ${}^{35}_{17}\text{Cl}$
   (c) ${}^{49}_{22}\text{Ti}$

5. The nuclide hassium-269, first synthesized in 1994, decays rapidly according to the equation

   $${}^{269}_{108}\text{Hs} \longrightarrow {}^{265}_{106}\text{Sg} + {}^{4}_{2}\text{He}$$

   An experiment establishes that $\Delta E = -9.23$ MeV for this reaction. Compute the difference in mass (in atomic mass units) between an atom of hassium-269 and one of seaborgium-265.

6. Nobelium-257 has a nuclidic mass of 257.096853 u. It reacts to give fermium-253 according to the equation

   $${}^{257}_{102}\text{No} \longrightarrow {}^{253}_{100}\text{Fm} + {}^{4}_{2}\text{He}$$

   The energy change of the system is -8.45 MeV. Compute the mass (in u) of an atom of fermium-253.

7. Use data from Table 15–1 to predict which is more stable: four protons, four neutrons, and four electrons organized as two ${}^{4}\text{He}$ atoms or as one ${}^{8}\text{Be}$ atom. What is the mass difference?

8. (a) Use data from Table 15–1 to predict which is more stable: 16 protons, 16 neutrons, and 16 electrons organized as two ${}^{16}\text{O}$ atoms or as one ${}^{32}\text{S}$ atom. What is the mass difference?
   (b) If the answer is that a ${}^{32}\text{S}$ atom is more stable, why do oxygen molecules in the atmosphere (${}^{16}\text{O}={}^{16}\text{O}$) not combine into sulfur atoms? If the answer is that two ${}^{16}\text{O}$ atoms are more stable, why do ${}^{32}\text{S}$ atoms in rocks and minerals not break down into ${}^{16}\text{O}$ atoms?

Nuclear Decay Processes

9. (See Example 15–2.) Write balanced equations that represent the following nuclear reactions:
   (a) Beta emission by ${}^{39}_{17}\text{Cl}$
   (b) Positron emission by ${}^{22}_{11}\text{Na}$
   (c) Alpha emission by ${}^{224}_{88}\text{Ra}$
   (d) Electron capture by ${}^{82}_{38}\text{Sr}$

10. (See Example 15–2.) Write balanced equations that represent the following nuclear reactions:
    (a) Alpha emission by ${}^{155}_{70}\text{Yb}$
    (b) Positron emission by ${}^{26}_{14}\text{Si}$
    (c) Electron capture by ${}^{65}_{30}\text{Zn}$
    (d) Beta emission by ${}^{100}_{41}\text{Nb}$

11. The natural abundance of ${}^{30}\text{Si}$ is 3.1%. Upon irradiation with neutrons, this nuclide is converted into ${}^{31}\text{Si}$, which decays to form the stable nuclide ${}^{31}\text{P}$. This provides a way of doping silicon with phosphorus in a much more uniform fashion than is possible by ordinary mixing of silicon and phosphorus and gives semiconductor devices capable of handling much higher levels of power. Write balanced nuclear equations for the two steps in the preparation of ${}^{31}\text{P}$ from ${}^{30}\text{Si}$.

12. The most convenient way to prepare the element polonium is to expose bismuth (which is 100% ${}^{209}\text{Bi}$) to neutrons. Write balanced nuclear equations for the two steps in the preparation of polonium.

13. (See Example 15–3.) The nuclide ${}^{40}_{19}\text{K}$ undergoes spontaneous decay to ${}^{40}_{20}\text{Ca}$ with emission of a beta particle and an antineutrino. Calculate the maximum kinetic energy of the beta particle (in MeV), using nuclidic masses from Table 15–1.

14. (See Example 15–3.) The nuclide ${}^{14}_{6}\text{C}$ undergoes spontaneous decay to ${}^{14}_{7}\text{N}$ with emission of a beta particle and an antineutrino. Calculate the maximum kinetic energy of the beta particle (in MeV), using nuclidic masses from Table 15–1.

15. (See Example 15–4.) Find the maximum kinetic energy (in MeV) of the positron in the decay of neon-19 (mass of neutral atom 19.0018798 u) to fluorine-19 (mass of neutral atom 18.9984032 u)

    $${}^{19}_{10}\text{Ne} \longrightarrow {}^{19}_{9}\text{F}^{-} + {}^{0}_{1}e^{+} + \nu$$

16. (See Example 15–4.) Find the mass in atomic mass units of a neutral atom of oxygen-15 if the maximum kinetic energy of the emitted positron in the reaction

    $${}^{15}_{8}\text{O} \longrightarrow {}^{15}_{7}\text{N}^{-} + {}^{0}_{1}e^{+} + \nu$$

    is 1.732 MeV, and the mass of a neutral atom of nitrogen-15 equals 15.0001089 u.

17. Compute ΔE (in MeV) for the following electron capture

$$^{231}_{92}\text{U} \xrightarrow{\text{E.C.}} {}^{231}_{91}\text{Pa} + \nu$$

The masses of the neutral atoms are given in Table 15–1.

18. Most $^{231}_{92}\text{U}$ atoms decay by electron capture (see the previous problem). However, some decay by alpha emission

$$^{231}_{92}\text{U} \longrightarrow {}^{227}_{89}\text{Ac} + {}^{4}_{2}\text{He}$$

Compute ΔE (in MeV) if the mass of a neutral atom of actinium-227 equals 221.015576 u. Consult Table 15–1 for other nuclidic masses.

Kinetics of Radioactive Decay

19. How many radioactive disintegrations occur per minute in a 0.0010-g sample of ^{209}Po that has been freshly separated from its decay products? The half-life of ^{209}Po is 103 years.

20. How many alpha particles are emitted per minute by a 0.0010-g sample of ^{238}U that has been freshly separated from its decay products? Assume that each decay emits one alpha particle. The half-life of ^{238}U is 4.47×10^9 years.

21. (See Example 15–5.) The nuclide ^{19}O, prepared by neutron irradiation of ^{19}F, has a half-life of 29 s.
    (a) How many ^{19}O atoms are in a freshly prepared sample if its activity is 2.5×10^4 Bq?
    (b) After 2.00 min, how many ^{19}O atoms remain?

22. (See Example 15–5.) The nuclide ^{35}S decays by beta emission with a half-life of 87.1 days.
    (a) How many grams of ^{35}S are in a sample that has a beta emission activity from that nuclide of 3.70×10^2 Bq?
    (b) After 365 days, how many grams of ^{35}S remain?

23. The activity of a 1.00-mg sample of pure ^{137}Cs is 3.19×10^9 Bq. Determine the half-life of ^{137}Cs in years. Take the atomic mass of this nuclide to equal 137 g mol^{-1}.

24. A sample of mixed metals weighing 125.3 mg is found to be radioactive with an activity of 3.0×10^7 Bq. A study of the energy of the emitted beta particles establishes that the radioactivity is due entirely to ^{59}Fe. What percentage by mass of the sample is ^{59}Fe, assuming that its atomic mass is 59.0 g mol^{-1}? The half-life of ^{59}Fe is 44.6 days.

25. The half-life of a free neutron was thought to be about 1100 s until a new experiment established it as 876 ± 21 s. Suppose that a procedure starts with 1.00 mol of free neutrons. Calculate how much time is required for the quantity of neutrons to be reduced to 0.90 mol, according to the old half-life and the new half-life.

26. Gallium citrate, which contains the radioactive nuclide ^{67}Ga, is used in medicine as a tumor-seeking agent. Gallium-67 decays with a half-life of 77.9 h. How much time (in hours) is required for the activity to decay to 5.0% of its initial activity?

27. Astatine is the rarest naturally occurring element, with ^{219}At appearing as the product of a very minor side-branch in the decay of ^{235}U (itself not a very abundant nuclide). It is estimated that the mass of all the naturally occurring ^{219}At in the upper kilometer of the earth's surface has a steady value of only 44 mg. Calculate the total activity (in becquerels) of all the naturally occurring astatine in this part of the earth. The half-life of ^{219}At is 54 s, and its nuclidic mass is 219.01 u.

28. Technetium has not been found in nature. It can be obtained readily as a product of uranium fission in nuclear power plants, however, and is now produced in quantities of many kilograms per year. One use in medicine relies on the tendency of ^{99m}Tc (an excited nuclear state of ^{99}Tc) to concentrate in abnormal heart tissue. Calculate the total activity (in becquerels) of 1.0 μg of ^{99m}Tc, which has a half-life of 6.0 h.

29. Uranium-238 decays over several steps to lead-206, as diagrammed in Figure 15–7. An experiment on a uranium-bearing rock establishes that the ratio of the number of atoms of ^{206}Pb to the number of atoms of ^{238}U equals 0.333. All the ^{206}Pb in the rock comes from the decay of the ^{238}U, and ^{238}U is lost from the rock only by its radioactive decay.
    (a) Determine the fraction of the original amount of ^{238}U that has decayed.
    (b) The half-life of ^{238}U is 4.468×10^9 years. How old is the rock?

30. The nuclide ^{87}Rb decays to ^{87}Sr with a half-life of 4.9×10^{10} years. A mineral that contained ^{87}Rb but no ^{87}Sr when it was originally formed is now found to contain one ^{87}Sr atom for each 120 ^{87}Rb atoms. Estimate the age of the mineral, in years.

31. (See Example 15–6.) The specific activity of ^{14}C in the biosphere is 0.255 Bq g^{-1}. What is the age of a piece of papyrus from an Egyptian tomb if its beta counting rate is 0.153 Bq g^{-1}? The half-life of ^{14}C is 5730 years.

32. (See Example 15–6.) The charcoal residues at an ancient hearth have only 0.125 of the carbon-14 activity of growing wood. Estimate the age of the residues.

33. (See Example 15–6.) Detecting carbon-14 activity at levels lower than 5×10^{-4} Bq g^{-1} is very difficult experimentally. If 5×10^{-4} Bq g^{-1} is the lowest activity that can be measured, determine the maximum age that can be assigned to an object by carbon-14 methods.

34. (See Example 15–6.) The ^{14}C activity of carbon extracted from the wood of an ancient bow is counted with great difficulty and found to equal $5 \pm 2 \times 10^{-4}$ Bq g^{-1}. Determine the maximum age of the bow (corresponding to the lower end of the range of activity) and the minimum age of the bow (corresponding to the higher end of the range of activity).

Radiation in Biology and Medicine

35. Write balanced equations for the decays of ^{11}C and ^{15}O, both of which are used in positron emission tomography to scan the uptake of glucose in the body.

36. Write balanced equations for the decays of ^{13}N and ^{18}F, two other radioactive nuclides that are also used in positron emission tomography. What is the ultimate fate of the positrons?

37. The positrons emitted by ^{11}C have a maximum kinetic energy of 0.99 MeV, and those emitted by ^{15}O have a maximum kinetic energy of 1.72 MeV. Calculate the ratio of the number of millirems of radiation exposure caused by in-

gesting a given fixed chemical amount (equal numbers of atoms) of each of these radioactive nuclides.

38. Compare the relative health risks of contact with a given amount of ^{226}Ra, which decays with a half-life of 1622 years and emits 4.78 MeV alpha particles, with contact with the same chemical amount of ^{14}C, which decays with a half-life of 5730 years and emits beta particles with energies of up to 0.155 MeV.

39. (See Example 15–7.) The radioactive nuclide ^{131}I undergoes beta decay with a half-life of 8.021 days. Quantities of this nuclide, which is a product of nuclear fission, were released into the environment in the Chernobyl accident. A victim of radiation poisoning has absorbed 5.0×10^{-6} g (5.0 μg) of ^{131}I.
   (a) Compute the activity in becquerels of the ^{131}I in this person, taking the atomic mass of the nuclide to equal 131 g mol^{-1}.
   (b) Compute the radiation absorbed dose in millirads caused by this nuclide during the first *second* after its ingestion. Assume that beta particles emitted by ^{131}I have an average kinetic energy of 0.40 MeV, that all of this energy is deposited within the victim's body, and that the victim weighs 60 kg.
   (c) Is this poisoning incident likely to be fatal? Remember that the ^{131}I diminishes its activity as it decays.

40. (See Example 15–7.) The nuclide ^{239}Pu undergoes alpha decay with a half-life of 2.411×10^4 years. An atomic energy worker breathes in 5.0×10^{-6} g (5.0 μg) of ^{239}Pu, which lodges permanently in a lung.
   (a) Compute the activity in becquerels of the ^{239}Pu ingested, taking the atomic mass of the nuclide to equal 239 g mol^{-1}.
   (b) Determine the radiation absorbed dose in millirads during the first *year* after its ingestion. Assume that alpha particles emitted by ^{239}Pu have an average kinetic energy of 5.24 MeV, that all of this energy is deposited within the worker's body, and that the worker weighs 60 kg.
   (c) Is this dose likely to be lethal?

Nuclear Fission and Fusion

41. Strontium-90 is one of the most hazardous products of atomic-weapons testing because of its long half-life ($t_{1/2} = 28.1$ years) and its chemical tendency to accumulate in the bones.
   (a) Write nuclear equations for the decay of ^{90}Sr via the successive emission of two beta particles.
   (b) The nuclidic mass of ^{90}Sr is 89.9073 u and that of ^{90}Zr is 89.9043 u. Calculate the energy released per ^{90}Sr atom, in MeV, in decaying to ^{90}Zr.
   (c) What is the initial activity of 1.00 g of ^{90}Sr released into the environment, in becquerels?
   (d) What activity would the material from part (c) show after 100 years?

42. Plutonium-239 is the fissionable nuclide produced in breeder reactors; it is also produced in ordinary nuclear plants and in weapons tests. It is an extremely poisonous substance with a half-life of 24,100 years.

(a) Write an equation for the decay of ^{239}Pu via alpha emission.
(b) The nuclidic mass of ^{239}Pu is 239.05216 u and that of ^{235}U is 235.043925 u. Calculate the energy released per ^{239}Pu atom decaying via alpha emission, in MeV.
(c) What is the initial activity, in becquerels, of 1.00 g of ^{239}Pu buried in a disposal site for radioactive wastes?
(d) What activity would the material from part (c) show after 100,000 years?

43. The three naturally occurring isotopes of uranium are ^{234}U (half-life 2.5×10^5 years), ^{235}U (half-life 7.0×10^8 years), and ^{238}U (half-life 4.5×10^9 years). As time passes, does the average atomic mass of the uranium in a sample taken from nature increase, decrease, or remain constant? Explain.

44. Natural lithium consists of 7.42% ^{6}Li and 92.58% ^{7}Li. Much of the tritium ($^3_1$H) used in experiments with fusion reactions is made by the capture of neutrons by ^{6}Li atoms.
   (a) Write a balanced nuclear equation for the process. What is the other particle produced?
   (b) After ^{6}Li is removed from natural lithium, the remainder is sold for other uses. Is its molar mass greater or smaller than that of natural lithium?

45. Calculate the amount of energy released, in kilojoules per *gram* of uranium, in the fission reaction

$$^{235}_{92}\text{U} + ^1_0n \longrightarrow ^{94}_{36}\text{Kr} + ^{139}_{56}\text{Ba} + 3\,^1_0n$$

Use the nuclidic masses given in Table 15–1. The mass of ^{94}Kr is 93.919 u and that of ^{139}Ba is 138.909 u.

46. Calculate the amount of energy released, in kilojoules per *gram* of deuterium (^{2}H), for the fusion reaction

$$^2_1\text{H} + ^2_1\text{H} \longrightarrow ^4_2\text{He}$$

Use the nuclidic masses given in Table 15–1. Compare your answer with that from the preceding problem.

Additional Problems

47. Tritium is $^3_1$H, an isotope of hydrogen. It is radioactive, emitting beta particles with a half-life of 12.26 years. Iodine-131 is one of many radioactive isotopes of iodine. It also decays by beta emission and has a half-life of 8.021 days. Ingestion of iodine-131 is far more dangerous than ingestion of tritium. Explain why.

48. A recent report states that the tau neutrino has (with 95% confidence) a maximum mass corresponding to 76 eV. Compute this upper limit in kilograms and in atomic mass units.

49. Hydrazine ($N_2H_4(\ell)$) reacts with oxygen in a rocket engine to form nitrogen and water vapor:

$$N_2H_4(\ell) + O_2(g) \longrightarrow N_2(g) + 2\,H_2O(g)$$

(a) Calculate ΔH for this highly exothermic reaction at 25°C using data from Appendix D.
(b) Calculate ΔE of this reaction at 25°C.
(c) Calculate the loss in mass (in grams) during the reaction of one mole of hydrazine.

50. Working in Rutherford's laboratory in 1932, Cockcroft and Walton bombarded a lithium target with 0.70-MeV protons and observed the following reaction:

$$^{7}_{3}\text{Li} + ^{1}_{1}\text{H} \longrightarrow 2\,^{4}_{2}\text{He}$$

Each of the alpha particles was found to have a kinetic energy of 8.5 MeV. This research provided the first experimental confirmation of Einstein's $\Delta E = c^2\,\Delta m$ relationship. Discuss. Using nuclidic masses from Table 15–1, calculate the value of c needed to account for this result.

51. The radioactive nuclide $^{64}_{29}\text{Cu}$ decays by emission of positrons to $^{64}_{28}\text{Ni}$. The mass of the neutral $^{64}_{29}\text{Cu}$ atom is 63.92976 u, and that of the neutral $^{64}_{28}\text{Ni}$ atom is 63.92796 u. Calculate the maximum kinetic energy of the positrons emitted (in MeV).

52. The nuclide ^{64}Cu (see previous problem) sometimes decays by emission of beta particles having a maximum kinetic energy of 0.573 MeV. Identify the daughter nuclide in this decay, and compute its mass in atomic mass units.

53. The nuclide $^{231}_{92}\text{U}$ converts spontaneously to $^{231}_{91}\text{Pa}$.
    (a) Write two balanced nuclear equations, one for this conversion proceeding by electron capture and the other for it proceeding by positron emission.
    (b) Using the nuclidic masses given in Table 15–1, calculate the change in mass for each process. Explain why in this case electron capture can occur spontaneously, but positron emission cannot.

54. Selenium-82 undergoes *double* beta decay.

$$^{82}_{34}\text{Se} \longrightarrow ^{82}_{36}\text{Kr}^{2+} + 2\,^{0}_{-1}e^- + 2\,\tilde{\nu}$$

This low-probability process occurs with a half-life of 3.5×10^{27} s, one of the longest half-lives ever measured. Estimate the activity in an 82.0-g (1.00 mol) sample of this nuclide. How many ^{82}Se nuclei decay in a day?

55. The half-life of ^{14}C is $t_{1/2} = 5730$ years, and carbon separated from modern wood charcoal has a specific activity of 0.255 Bq g^{-1}.
    (a) Calculate the number of ^{14}C atoms per gram of carbon in modern wood charcoal.
    (b) Calculate the fraction of carbon atoms in the biosphere that are ^{14}C.

56. Carbon-14 is produced in the upper atmosphere by the reaction

$$^{14}_{7}\text{N} + ^{1}_{0}n \longrightarrow ^{14}_{6}\text{C} + ^{1}_{1}\text{H}$$

where the neutrons come from nuclear processes induced by cosmic rays. It is estimated that the steady-state ^{14}C activity in the biosphere is 1.1×10^{19} Bq.
    (a) Estimate the total mass of carbon in the biosphere, using the data in problem 55.
    (b) The earth's crust has an average carbon content of 250 parts per million by mass, and the total crustal mass is 2.9×10^{25} g. Estimate the fraction of the carbon in the earth's crust that is part of the biosphere. Speculate on the whereabouts of the rest of the carbon in the earth's crust.

57. Over geologic time scales, an atom of ^{238}U decays to a stable ^{206}Pb atom in a series of eight alpha emissions, each of which leads to the formation of one helium atom. A geochemist analyzes a rock and finds that it contains 9.0×10^{-5} cm^3 of helium (at STP) per gram and 2.0×10^{-7} g ^{238}U per gram. Estimate the age of the mineral, given that $t_{1/2}$ of ^{238}U is 4.468×10^9 years.

58. The half-lives of ^{235}U and ^{238}U are 7.04×10^8 years and 4.468×10^9 years, respectively, and the present abundance ratio is $^{238}\text{U}:^{235}\text{U} = 137.7$. It is thought that their abundance ratio was 1 at some time before our earth and solar system were formed about 4.5×10^9 years ago. Estimate how long ago a hypothetical supernova occurred that produced all the uranium isotopes in equal abundance, including the two longest-lived isotopes, ^{238}U and ^{235}U.

59. Cobalt-60 and iodine-131 are used in treatments of different types of cancer. Cobalt-60 decays with a half-life of 5.27 years, emitting beta particles with a maximum energy of 0.32 MeV. Iodine-131 decays with a half-life of 8.021 days, emitting beta particles with a maximum energy of 0.60 MeV.
    (a) Suppose a fixed small number of moles of each of these nuclides were to be ingested and were to remain in the body indefinitely. What is the *ratio* of the number of millirems of total lifetime radiation exposure that would be caused by the two radioactive nuclides?
    (b) Now suppose that the contact with each of these nuclides is for a fixed short period of time, such as 1 hour. What is the ratio of millirems of radiation exposure for the two in this case?

60. The caption of Figure 15–2 states that plutonium-238 self-heats (to red heat!) from the disintegration of about 2.5 atoms per second out of every 10 billion. Estimate the half-life of ^{238}Pu (in years).

61. Boron is used in control rods in nuclear power reactors because it is a good neutron absorber. When the nuclide ^{10}B captures a neutron, an alpha particle (helium nucleus) is emitted. What other atom is formed? Write a balanced equation.

62. A puzzling observation that led to the discovery of isotopes was the fact that lead obtained from uranium-containing ores had an atomic mass lower by two full atomic mass units than lead obtained from thorium-containing ores. Explain this result, using the fact that decay of radioactive uranium and thorium to stable lead occurs via alpha and beta emission.

63. By 1913, the elements radium, actinium, thorium, and uranium had all been discovered, but element 91, between thorium and uranium in the periodic table, was not yet known. The approach used by Meitner and Hahn was to look for the parent that decays to form actinium. Alpha and beta emission are the most important decay pathways among the heavy radioactive elements. What elements would decay to actinium by each of these two pathways? If radium salts show no sign of actinium, what does this suggest about the parent of actinium? What is the origin of the name of element 91, discovered by Meitner and Hahn in 1918?

64. The average energy released in the fission of a ^{235}U nucleus

is about 200 MeV. Suppose that the conversion of this energy into electrical energy is 40% efficient. What mass of ^{235}U is converted into its fission products in a year's operation of a 1000-megawatt nuclear power station? Recall that 1 watt is 1 joule per second.

65. The radiant energy received by the earth from the sun is approximately 3.4×10^{17} J s^{-1}. This energy represents approximately 4.5×10^{-10} times the total energy put out by the sun.

(a) Calculate the change in mass of the sun per second.

(b) Assuming that the overall hydrogen-burning reaction in the sun is the conversion of four ^1_1H atoms to one ^4_2He atom and that the entire energy change appears as radiant energy, calculate the mass of hydrogen reacting per second in the sun. Use the data of Table 15–1.

66. Compare fission and fusion nuclear reactions with regard to the nature of the starting materials, their use to generate electrical energy, the technical problems that need to be overcome in their use, and their by-products.

CUMULATIVE PROBLEM

Radon

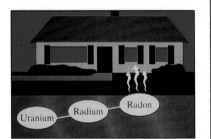

Radon most commonly enters houses by seeping in from the ground through foundations or basement walls.

Radioactive ^{222}Rn and ^{220}Rn constantly form from the decay of uranium and thorium in rocks and soil and, being gaseous, seep out of the ground. The radon isotopes decay fairly quickly, but their products, which are also radioactive, are then in the air and attach themselves to dust particles. Thus, airborne radioactivity can accumulate to worrisome levels in poorly ventilated basements dug in ground that is rich in uranium and thorium.

(a) Describe the composition of an atom of ^{222}Rn, and compare it with that of an atom of ^{220}Rn.

(b) Although ^{222}Rn is a decay product of ^{238}U, ^{220}Rn comes from ^{232}Th. How many alpha particles are emitted in forming these radon isotopes from their uranium or thorium starting points? (*Hint*: Alpha decay changes the mass number (A), but other decay processes do not.)

(c) Can alpha decay alone explain the formation of these radon isotopes from ^{238}U and ^{232}Th? If not, state what other types of decay must occur.

(d) Can $^{222}_{86}Rn$ and $^{220}_{86}Rn$ decay by alpha-particle emission? Write balanced nuclear equations for these two decay processes, and calculate the change in mass that would result. The nuclidic masses of ^{222}Rn and ^{220}Rn are 222.0175 and 220.01140 u, respectively; those of ^{218}Po and ^{216}Po are 218.0089 and 216.00192 u, respectively.

(e) Calculate the change in energy in the alpha decay of one ^{220}Rn nucleus, in MeV and in joules.

(f) The half-life of ^{222}Rn is 3.82 days. Calculate the initial activity of 2.00×10^{-8} g of ^{22}Rn, in becquerels.

(g) What is the activity of the ^{222}Rn from part (f) after 14 days?

(h) The half-life of ^{220}Rn is 54 s. Are the health risks for exposure to a given amount of radon for a given short period of time greater or smaller for ^{220}Rn than for ^{222}Rn? Explain.

Quantum Mechanics and the Hydrogen Atom

Traveling water waves exhibit interference and diffraction, just as do waves associated with light and matter.

Mechanics is the branch of physics that deals with the interactions and motions of particles and the collections of particles that make up working mechanisms. It uses concepts such as mass, velocity, and force. The 200 years that followed the seminal work of Isaac Newton were the classical period in the study of mechanics (and of physics in general). By the end of this period (about 1900), physicists had achieved a deep and fruitful understanding that successfully dealt with problems ranging from the motions of the planets in their orbits to the design of a bicycle. These attainments now form the subject called *classical mechanics* and are eminently useful. In classical mechanics, a fundamental role is played by energy, which consists of two parts: kinetic energy, arising from the motion of particles, and potential energy, arising from the interactions of particles with each other or with an external field (such as gravity), and depending on the *positions* of the particles. In classical mechanics, one may in principle predict the future positions of a group of particles from a knowledge of their present positions and momenta and of the forces between them. A major triumph of classical mechanics was the kinetic theory of gases, which is outlined in Section 5–7.

At the end of the 19th century, it was naturally thought that the motion of elementary particles (such as the recently discovered electron) could be described by classical mechanics and that once the correct laws of force were discovered, the properties of atoms and molecules could be predicted to any desired degree of accuracy by solving Newton's equations of motion. There was a feeling that all the fundamental laws of physics had been discovered. At the dedication of the Ryerson Physics Laboratory at the University of Chicago in 1894, A. A. Michelson said that "our future discoveries must be looked for in the sixth decimal place." Little did he realize the far-reaching changes that would shake physics and chemistry during the subsequent 30 years. Central to those changes was the discovery that all particles have wave-like properties, the effects of which are most pronounced for low-mass particles such as electrons. The incorporation of wave and particle aspects of matter into a single comprehensive theory is the achievement of *quantum mechanics,* the great 20th-century extension of classical mechanics. In this chapter, we discuss the origins of the quantum theory and the implications of that theory for the structure of the hydrogen atom. In subsequent chapters, we consider the consequences of the quantum nature of matter for atoms containing two or more electrons, for chemical bonds, and for the liquid and solid states.

• The nucleus of a hydrogen atom consists of a single proton. Addition of a single electron forms a neutral atom of hydrogen, $_1^1H$. No chemical element has a simpler atomic structure.

16–1 WAVES AND LIGHT

Many kinds of waves appear in physics and chemistry. The waves in water are a familiar example, whether they are stirred up by the winds over the oceans, set off by a stone dropped into a quiet pool, or created by a laboratory water-wave ma-

Table 16–1
Kinds of Waves

Wave	Oscillating Quantity
Water	Height of water surface
Sound	Density of air
Light	Electric and magnetic fields

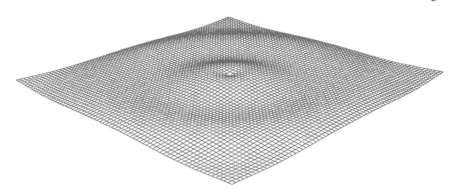

Figure 16-1 A traveling wave moves out from a central point.

chine. Sound waves are another example: periodic compressions of the air that move from a source to a detector such as the human ear. Light waves, as we shall see, are the consequence of oscillating electric and magnetic fields moving through space. Even some chemical reactions occur in such a way that waves of color move through the sample as the reaction proceeds. Common to all these wave phenomena is the oscillatory variation with time of some property (Table 16–1) at a fixed location in space.

This is visualized most easily for a wave in water. A snapshot of such a wave (Fig. 16–1) records the peaks and troughs present at some instant in time. The **amplitude** of the wave at any point is the height of the water surface over the level of the water when it is undisturbed. The distance between two successive peaks (or troughs) is called the **wavelength** of the wave, provided that this distance is reproducible from peak to peak (Fig. 16–2). Wavelengths are usually given the symbol λ (Greek "lambda"). The **frequency** of a water wave is measured by counting the number of peaks that pass a fixed point in space per second. The frequency has units of waves (or cycles) per second, or simply s^{-1}. If 12 water-wave peaks pass a certain point in 30 seconds, for example, the frequency is

$$\text{frequency} = \nu = \frac{12}{30 \text{ s}} = 0.40 \text{ s}^{-1}$$

where the Greek letter ν ("nu") represents the frequency. If the wavelength and frequency of a wave are both known, its speed, which is the rate at which a particular wave crest moves along through the medium, is readily calculated. As Figure 16–2 shows, in a time interval $\Delta t = \nu^{-1}$ the wave moves through one

• The "per second" (s^{-1}) in the context of the frequency of waves is often called a hertz (Hz). In Chapter 15 the same unit (s^{-1}) was used to measure the frequency of radioactive disintegrations and was, in that context, called a becquerel (Bq).

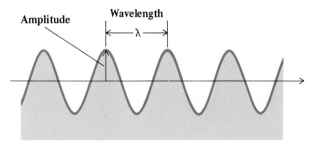

$$\text{Speed} = c = \frac{\text{distance}}{\text{time}}$$

Figure 16-2 As a water wave moves across an otherwise calm tank, both its amplitude and its wavelength can be easily determined. Its speed is found by taking the distance traveled by a particular wave crest and dividing by the time elapsed.

wavelength, so the speed (the distance traveled divided by the time elapsed) is

$$\text{speed} = \frac{\text{distance traveled}}{\text{time elapsed}} = \frac{\lambda}{\nu^{-1}} = \lambda\nu$$

The speed of a wave is the product of its wavelength and its frequency.

James Clerk Maxwell had demonstrated by 1865 that light is **electromagnetic radiation.** A beam of light emitted by a laser (Fig. 16–3) consists of oscillating electric and magnetic fields perpendicular to the direction in which the light is propagating (Fig. 16–4). As in the case of a water wave, the wavelength is called λ and the frequency ν. The speed of light passing through a vacuum (c) is equal to the product $\lambda\nu$.

$$c = \lambda\nu = 2.99792458 \times 10^8 \text{ m s}^{-1}$$

- The constancy of c allows definition of the meter in terms of the second. Officially, one meter equals "the length of path traveled by light in a vacuum during 1/299,792,458 of a second."

- We often use the nanometer (abbreviation: nm) as a unit of measure for the wavelength of light. Recall that 1 nm is 10^{-9} m.

The speed of light in a vacuum is a universal constant, the same for all types of light, although the wavelength and frequency depend on the color of the light. Green light has a range of frequencies near $5.7 \times 10^{14} \text{ s}^{-1}$ and wavelengths near 5.3×10^{-7} m (530 nm). Red light has a lower frequency and a longer wavelength than green, and violet light a higher frequency and shorter wavelength than green (Fig. 16–5). A laser such as that shown in Figure 16–3 emits nearly monochromatic light (light with a single frequency and wavelength), but white light or simple daylight contains the full range of visible wavelengths. White light can be resolved into its component wavelengths with a prism or a diffraction grating.

The light visible to the eye is only a very small part of the entire electromagnetic spectrum (see Fig. 16–5). Light can have wavelengths longer than red light or shorter than violet. Such radiation is not visible to human eyes but can be detected by other means. The warmth felt radiating from a stone pulled out of a fire consists largely of **infrared** radiation, which has wavelengths that are longer than those of visible light. Microwave ovens use radiation of longer wavelength than infrared, and still longer waves are used in radio communication. The position of a radio station on the dial is given by its frequency in hertz (Hz), where one hertz is one cycle per second. Thus, FM stations typically broadcast at frequencies in the region of tens to hundreds of megahertz (1 MHz = 10^6 s^{-1}); AM stations have much lower broadcast frequencies, from hundreds to thousands of kilohertz

Figure 16–3 A laser emits a sharply focused beam of light with a very narrow range of wavelengths. The direction of motion of a laser beam can be manipulated by inserting mirrors in its path.

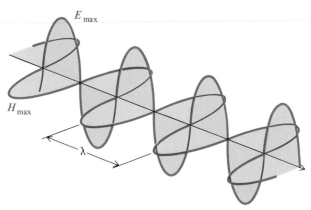

Figure 16–4 Light consists of waves of oscillating electric (E) and magnetic (H) fields that are perpendicular to one another and to the direction of propagation of the light.

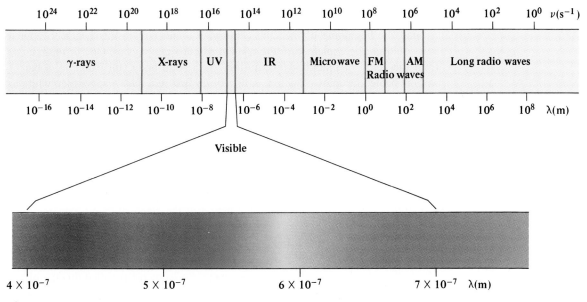

Figure 16–5 The electromagnetic spectrum. The frequency doubles 80 times from right to left in the diagram; the wavelength doubles 80 times from left to right. Visible light is confined within a single doubling near the middle.

(1 kHz $= 10^3$ s^{-1}). Types of light with shorter wavelengths than those of visible light include ultraviolet light, x-rays, and gamma rays. The latter two kinds of radiation, with their high frequencies and short wavelengths, are more penetrating than visible or infrared radiation and so can be used to image bones or organs beneath the skin. The high energy and penetrating power of x-rays and gamma rays also make them hazardous, so exposure must be monitored carefully.

EXAMPLE 16–1

Cellular telephones in the United States operate in the frequency range 869 to 894 MHz. Compute the wavelength of the radiation from a cellular telephone emitting at a frequency of 879 MHz.

Solution

The wavelength is related to the frequency through

$$\lambda = \frac{c}{\nu} = \frac{3.00 \times 10^8 \text{ m s}^{-1}}{879 \times 10^6 \text{ s}^{-1}} = 0.341 \text{ m}$$

so the wavelength is 34.1 cm (about $13\frac{1}{2}$ inches).

Exercise

An AM radio station calls itself "Radio-89" because it broadcasts at a frequency of 890 kHz (8.90×10^5 s^{-1}). Compute the wavelength of the wave broadcast from this station.

Answer: 337 m (about 0.2 mi).

16–2 PARADOXES IN CLASSICAL PHYSICS

Around the beginning of the 20th century, a number of results began to appear that could not be explained in the context of classical physics. One of these arose from the study of **blackbody radiation.** As a solid body (for example, a bar of iron) is heated, it emits radiation and becomes first red, then orange, then white as its temperature increases. The distribution of frequencies of the radiated light changes with the temperature of the body. It is exactly this effect that gives stars of different temperatures their different colors. A blackbody is an idealized version of the iron bar. It consists of an enclosure made of perfectly absorbing (perfectly black) material but having a small hole from which radiation can escape.

What is the intensity of the light emitted through the hole at each wavelength λ for a blackbody at temperature T? Two typical experimental curves are shown in Figure 16–6. Also shown is the prediction of the classical theory of radiation,

$$\text{intensity} = \left(\frac{8\pi R}{N_0}\right)\left(\frac{T}{\lambda^4}\right)$$

which disagrees disastrously with experiment. Some theorists called this the **ultraviolet catastrophe:** a heated body does *not* emit radiation approaching infinite intensity at short wavelengths (the violet end of the spectrum), yet classical theory predicts that it does. The paradox could not be resolved within classical physics.

A second paradox arose from the discovery of the **photoelectric effect.** In this effect, a beam of light falling on a metal surface in an evacuated space ejects electrons from the surface, causing an electric current (called a *photocurrent*) to flow. This fact was in itself not difficult to understand, because electromagnetic radiation carries energy. The problem came in explaining the observed dependence of the effect on the frequency of the light. It was found that for frequencies less than a certain threshold frequency ν_0, no electrons were ejected; once the frequency exceeded the threshold, the photocurrent increased rapidly. According to classical theory, the energy associated with electromagnetic radiation depends only on the *intensity* (or square of the magnitude of the electric field), and not on the frequency.

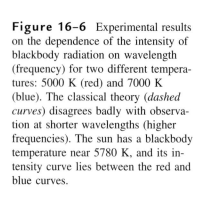

Figure 16–6 Experimental results on the dependence of the intensity of blackbody radiation on wavelength (frequency) for two different temperatures: 5000 K (red) and 7000 K (blue). The classical theory (*dashed curves*) disagrees badly with observation at shorter wavelengths (higher frequencies). The sun has a blackbody temperature near 5780 K, and its intensity curve lies between the red and blue curves.

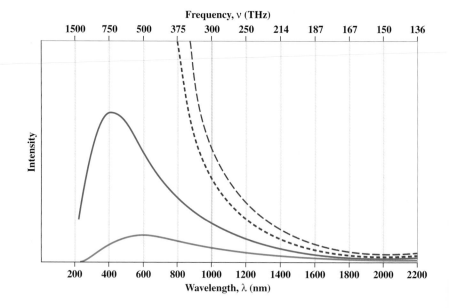

• Red-hot is surely too hot to touch, but white-hot is hotter still.

CHEMISTRY IN YOUR LIFE

Incandescent Light

The production of light by a hot object is called *incandescence*. Many hot objects behave much like ideal blackbodies in their emission of light at different wavelengths. Light from the sun, for example, closely follows the blackbody radiation curve for its surface temperature of 5780 K. The peak of this curve lies near the center of the visible region of the spectrum. Human vision in fact evolved toward efficient detection of this, the most abundant light present on earth. Stars other than the sun with different surface temperatures, vary in color from cooler red giants to hotter blue-white stars.

Wood fires represent, after the sun, our most ancient source of light. The red glow of the coals comes from a blackbody temperature of about 1200°C, and the yellow-orange of the flames results from small particles of soot (carbon) inside the flame emitting at a temperature of about 1500°C. Most modern incandescent lights use tungsten filaments heated to about 2500°C; tungsten is the metal of choice because of its high melting point and small rate of evaporation. In the quartz–halogen light bulb, a quartz (SiO_2) bulb permits still higher temperatures. The filament is again made of tungsten, and bromine or iodine is put inside the bulb as well. These halogens react with evaporated tungsten atoms and redeposit them on the filament. A whiter and more natural light results from the higher temperature.

An *optical pyrometer* is a device that measures the intensity of radiated light. Measurements at two or more wavelengths permit the temperature of the radiating object to be estimated. Such instruments are needed at high

Figure 16–A   A candle flame.

temperature where ordinary thermometers cannot be used. An experienced observer can estimate the temperature of a glowing object to within 50 to 100°C just from the color of the light it emits.

How, then, could a very weak beam of blue light eject electrons from sodium when an intense red beam had no effect (Fig. 16–7a)? Further experiments measured the kinetic energy of the ejected electrons and revealed a linear dependence of the maximum kinetic energy on frequency, as shown in Figure 16–7b.

$$\text{maximum kinetic energy} = \tfrac{1}{2}m_e v^2 = h(\nu - \nu_0)$$

where m_e is the mass of the electron and h is a constant. This behavior also was inexplicable in classical physics.

The third, and in some ways most fundamental, paradox was the stability of the atom. Rutherford's experiments (see Section 1–5) had shown that the atom consists of a very small, massive nucleus surrounded by electrons. If the negatively charged electrons are stationary, what keeps them from falling into the nucleus in response to the electrostatic tug of the positively charged nucleus? If the electrons move in orbits around the nucleus like planets in the solar system, why do they

Figure 16–7 (a) The photoelectric effect. Blue light is effective in ejecting electrons from the surface of this metal, but red light is not. (b) The maximum kinetic energy of the ejected electrons varies linearly with the frequency of light used.

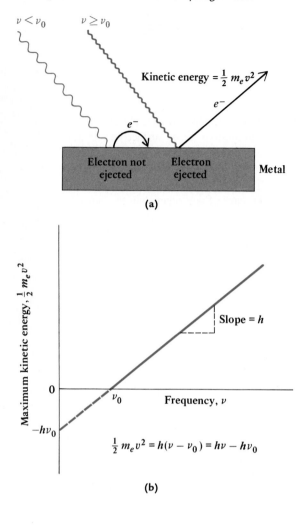

(a)

(b)

$$\tfrac{1}{2} m_e v^2 = h(v - v_0) = hv - hv_0$$

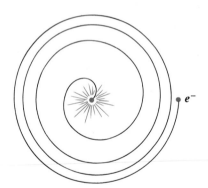

Figure 16–8 In classical theory, atoms constructed according to the Rutherford model are not stable; the electron would quickly spiral into the nucleus.

• The detector in this arrangement may be a human eye. An observer sees a series of differently colored lines if the light consists of only certain frequencies. If the light is white light, the observer sees a rainbow.

not emit radiation continuously as they lose energy and spiral into the nucleus (Fig. 16–8), as is predicted by classical theory?

Atomic Spectra

Closely related to the question of atomic stability were the results of some experiments on **atomic spectra.** Light consisting of more than one frequency (Fig. 16–9) can be resolved according to frequency by passing it through a glass prism. Components of higher frequency (blue light, for example) interact differently with the glass than those of lower frequency (red light, for example) and are bent to greater angles by the prism. One instrument used to separate light into its component frequencies is called a **spectrograph** (Fig. 16–10). It consists of a narrow vertical slit through which the light enters, the prism, and a photographic plate or other detector. If the light consists of several discrete frequencies, the detector records an array of lines (images of the slit), each having a different color; these are **spectroscopic lines.** If the incoming light contains all frequencies, then the spectrum is a continuous band, made up of overlapping spectroscopic lines. Early spectrographic experiments quickly showed that atoms do not emit a continuous dis-

text continued on p. 656

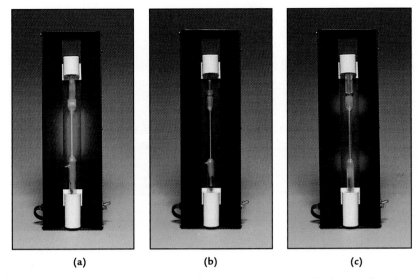

(a)   (b)   (c)

Figure 16-9   When a gas is excited in an electrical discharge, it glows as it emits light. Here the colors of the light emitted by three gases are shown: (a) neon, (b) argon, and (c) mercury. Each emission consists of several frequencies (wavelengths) of light, and the perceived color depends on which predominate.

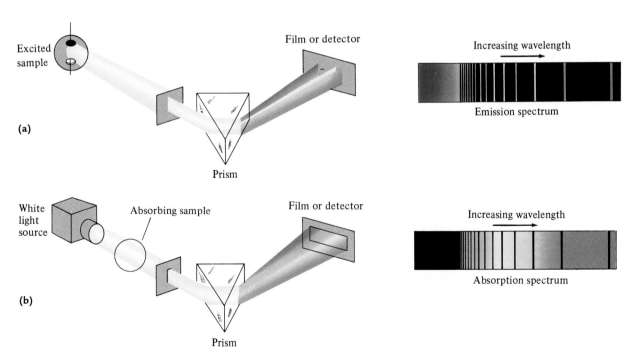

Figure 16-10   (a) The emission spectrum of atoms or molecules is measured by passing the light emitted from an excited sample through a prism to separate it according to wavelength and then recording the separated light on photographic film or with another detector. (b) In absorption spectroscopy, white light from a source passes through the un-excited sample, which absorbs certain discrete wavelengths of light from it. The result is the appearance of dark lines superimposed on a continuous bright background.

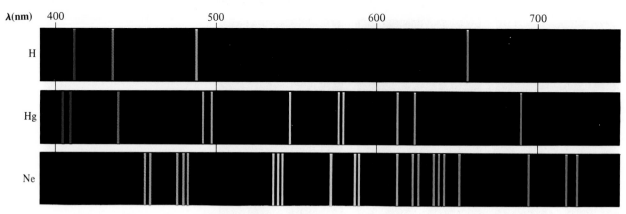

Figure 16–11 Atoms of hydrogen, mercury, and neon emit light at discrete wavelengths. The pattern seen is characteristic of the element under study. (1 Å = 10^{-10} m; see page 659.)

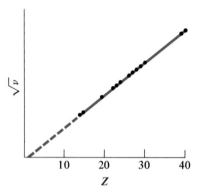

Figure 16–12 A graph of data from Moseley's experiments. The square root of the frequency of the characteristic x-rays emitted by the elements under a bombardment of electrons is a straight-line function of their atomic number.

• The Moseley experiment is still routinely performed under the name "x-ray fluorescence spectroscopy." The characteristic frequencies of the x-rays emitted by a sample of a solid unknown bombarded by an electron beam reveal the identities of the elements present.

tribution of frequencies. Instead, single atoms emit light at discrete frequencies that are characteristic of their chemical identity (Fig. 16–11).

Any theory explaining the structure and stability of atoms had to explain such observations. Moreover, the frequencies of the spectroscopic lines followed patterns that could be fitted to empirical equations. In 1885, for example, J. J. Balmer discovered that when hydrogen atoms are excited by the passage of an electric current, they emit a series of lines in the visible region with frequencies given by the simple formula

$$\nu = \left(\frac{1}{4} - \frac{1}{n^2} \right) \times (3.29 \times 10^{15} \text{ s}^{-1}) \qquad n = 3, 4, 5, \ldots$$

A successful theory of atomic structure would predict Balmer's formula in a natural way.

Line spectra from atoms are observed in all regions of the electromagnetic spectrum. In 1913 Henry Moseley measured the frequencies of the x-rays emitted by different elements when bombarded with high-energy electrons. He found that each element emits a characteristic frequency ν and that the frequencies followed the particularly simple relationship

$$\sqrt{\nu} = a(Z - b)$$

where a and b are constants and Z is an integer having a different value for each element (Fig. 16–12). These results offered strong early support for Rutherford's nuclear model of the atom because it was clear that the Z in Moseley's equation was the same as Rutherford's Z, the atomic number. In this way, Moseley's work strengthened the challenge that atomic structure posed to theorists in the early part of the 20th century.

16–3 PLANCK, EINSTEIN, AND BOHR

One of the fundamental assumptions of early science was that nature is continuous: nature does not make jumps. On a macroscopic scale, this seems true enough. We can measure out an amount of graphite (carbon) of mass 9 kg, or 8.23 kg, or 6.4257 kg, and it appears that the mass can have any value, to any number of decimal places, provided that our balance is sufficiently accurate.

On an atomic scale, however, this apparent seamless continuity breaks down. An analogy may be useful here: when viewed from a distance, a sandy beach appears smooth and unbroken, but up close individual grains of sand are distinguishable. This same "graininess" is found when matter is observed on an atomic scale. Although on a macroscopic scale it appears that we can measure out an arbitrary mass of graphite, if we try to extend this to the atomic level, we find that the mass of carbon (^{12}C) come in "packets," each of which weighs 1.99265×10^{-26} kg. Although we can, in principle, "weigh out" two, three, or any integral number of such packets, we cannot obtain $1\frac{1}{2}$ packets. Carbon is not a continuous material but comes in chunks, each of which contains the minimum measurable mass of carbon. The chunks of carbon are in fact its atoms. Similarly, electric charge comes in packets of size e, called electrons or protons. We cannot obtain nonintegral numbers of such packets in experiments involving chemical energies.

The central idea of quantum theory is that energy also is not continuous but comes in discrete packets. A packet of energy is called a **quantum** of energy. The quantization of the energy of electromagnetic radiation and of matter leads to a resolution of the paradoxes presented in the last section.

• At very high energies (high enough to probe the structure of the proton itself), there is evidence for the existence of *quarks*, packets in which the minimum charge is $\pm\frac{1}{3}e$ rather than e.

Planck's Constant and the Photon

The paradox of blackbody radiation was resolved by Max Planck in 1900. In previous theory, radiation from a blackbody or other solid object was thought to arise from the oscillations of groups of atoms on the surface of the body: if the atoms oscillated with a frequency ν, then radiation of frequency ν would be emitted. In this theory, oscillators with high frequencies greatly outnumber those with low frequencies. Even a small amount of energy placed in these oscillators (by a rise in temperature) shifts the radiation from the blackbody drastically to high frequencies, leading to the ultraviolet catastrophe. Planck made the daring hypothesis that it was *not* possible to put an arbitrarily small amount of energy into an oscillator of frequency ν. Instead, he suggested that the energy must come in "packets" or *quanta* of magnitude $h\nu$, so that the energy in an oscillator must be an integral multiple of $h\nu$—that is, it must go up or down in steps: $1 \times (h\nu)$, $2 \times (h\nu)$, and so forth. In this scheme, most of the high-frequency oscillators can gain no energy, because packets of energy that are large enough to be accepted by them are scarce. Furthermore, an oscillator with its energy at the allowed minimum cannot lose energy by emitting radiation. The mathematical development of Planck's hypothesis predicts a fall-off of the spectrum at high frequencies and describes the experimental observations quite accurately. The constant h in the relation

$$E = h\nu$$

has since been measured to very high precision. It is referred to as **Planck's constant** and has the value

$$h = 6.62608 \times 10^{-34} \text{ J s} = 6.62608 \times 10^{-34} \text{ kg m}^2 \text{ s}^{-1}$$

• Planck's constant is a fundamental constant of nature, occurring again and again in quantum theory.

In 1905, Einstein used Planck's quantum hypothesis to explain the photoelectric effect. He suggested that light consists of a series of "packets" of energy called **photons,** each of which carries energy $E = h\nu$. A photon of low-frequency light that strikes an electron on a metal surface does not deliver enough energy to eject the electron, but a photon of high-frequency light does. If a threshold energy of $h\nu_0$ is required to overcome the binding of the electron to the atoms of the surface, then, from the law of conservation of energy, any excess energy $h\nu - h\nu_0$ should appear as kinetic energy of the outgoing electron:

$$\text{maximum kinetic energy} = \tfrac{1}{2}m_e v^2 = h(\nu - \nu_0) = h\nu - \Phi$$

That is what is observed experimentally (see Fig. 16–7b). The constant h obtained from this experiment is Planck's constant, identical to that from the blackbody radiation experiment; $h\nu_0$ is equal to the energy Φ binding the ejected electron to the surface and called the *work function* of the material.

• The symbol Φ is an uppercase Greek "phi."

EXAMPLE 16–2

Light with a wavelength of 400 nm strikes the surface of metallic cesium in a photoelectric cell, and the maximum kinetic energy of the electrons ejected is 1.54×10^{-19} J. Calculate the work function of cesium and the longest wavelength of light that is capable of ejecting electrons from cesium.

Solution

The frequency of the light is

$$\nu = \frac{c}{\lambda} = \frac{3.00 \times 10^8 \text{ m s}^{-1}}{4.00 \times 10^{-7} \text{ m}} = 7.50 \times 10^{14} \text{ s}^{-1}$$

The work function Φ can be calculated from Einstein's formula:

$$\left(\tfrac{1}{2}m_e v^2\right)_{\text{max}} = h\nu - \Phi$$

$$1.54 \times 10^{-19} \text{ J} = (6.626 \times 10^{-34} \text{ J s})(7.50 \times 10^{14} \text{ s}^{-1}) - \Phi$$

$$= 4.97 \times 10^{-19} \text{ J} - \Phi$$

$$\Phi = h\nu_0 = (4.97 - 1.54) \times 10^{-19} \text{ J} = \boxed{3.43 \times 10^{-19} \text{ J}}$$

The minimum frequency ν_0 for the light to eject electrons is the frequency that supplies just enough energy to shake electrons loose from the surface:

$$h\nu_0 = \Phi$$

$$\nu_0 = \frac{3.43 \times 10^{-19} \text{ J}}{6.626 \times 10^{-34} \text{ J s}} = 5.18 \times 10^{14} \text{ s}^{-1}$$

From this, the maximum wavelength λ_0 can be calculated:

$$\lambda_0 = \frac{c}{\nu_0} = \frac{3.00 \times 10^8 \text{ m s}^{-1}}{5.18 \times 10^{14} \text{ s}^{-1}} = 5.79 \times 10^{-7} \text{ m} = \boxed{579 \text{ nm}}$$

Exercise

It requires a photon with a minimum of 6.94×10^{-19} J of energy to eject an electron from a polished zinc surface. (a) Does a photon with a wavelength of 210 nm suffice to do this? (b) If so, what is the maximum kinetic energy of the ejected electron?

Answer: (a) Yes. (b) 2.53×10^{-19} J.

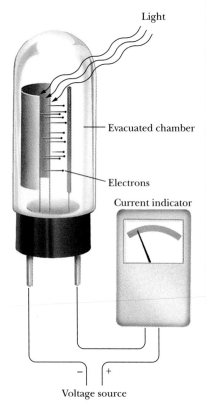

Light

Evacuated chamber

Electrons

Current indicator

– +

Voltage source

In a photoelectric cell (photocell), light strikes a metal surface and ejects electrons. The electrons are attracted to a positively charged collector, and a current flows through the cell. If the light stops or if its frequency falls too low, this photocurrent stops. Photocells are widely used for detecting and measuring light.

Planck's and Einstein's results suggest that, in some circumstances, light displays wave-like properties, whereas in others it behaves like a stream of particles (photons). This **wave–particle duality** is not a contradiction in terms but rather part of the fundamental nature of light and of matter as well.

The Bohr Atom

In 1913, Niels Bohr proposed a model for the atom that accounted for the striking regularities seen in the spectrum of the hydrogen atom and one-electron ions such as He^+, Li^{2+}, and Be^{3+}. His model is sometimes referred to as the planetary model because in it an electron of mass m_e is assumed to move in a circular orbit of radius r around a fixed nucleus, like the earth in its orbit around the sun. As mentioned earlier, classical physics predicts that an electron moving in this way emits light of continuously increasing frequency and spirals in toward the nucleus. Bohr avoided this difficulty by simply taking it as a postulate that a set of stable, discrete orbits (characterized by radius r_n and energy E_n) occurs, and that light is emitted or absorbed only when the electron "jumps" from one stable orbit to another.

The *linear* momentum of a particle is the product of its mass and its velocity; this is $m_e v$ for an electron. The *angular* momentum is a different quantity that describes turning motion around an axis. For the circular paths of the Bohr model, the angular momentum of the electron is the product of its mass, its velocity, and the radius of the orbit ($m_e vr$). Bohr assumed that the angular momentum of the electron was quantized and equal to some integral multiple of $h/2\pi$, where h is Planck's constant:

$$\text{angular momentum} = m_e vr = n\frac{h}{2\pi} \qquad n = 1, 2, 3, \ldots$$

This hypothesis cannot be derived, and in fact it is likely that Bohr adopted it after working backward from the known spectrum of hydrogen. Having assumed this condition of quantization, Bohr used the classical equations of circular motion to calculate the allowed radius r_n for each integral value of n. He found

• This is frequently the way in which new theories are developed.

$$r_n = \frac{n^2}{Z} a_0$$

where Z is the positive charge on the nucleus (1 for a hydrogen atom, 2 for a He^+ ion, and so forth), and a_0 is the **Bohr radius,** a constant with the dimensions of length. Bohr was able to calculate its value as a combination of Planck's constant, the charge of the electron, and the mass of the electron:

$$a_0 = 5.29177 \times 10^{-11} \text{ m}$$

Chemists often use the **ångström** (abbreviation: Å), a unit of length equal to 10^{-10} m, in discussions of atomic size. The Bohr radius, 0.529 Å, is the predicted distance of the electron from the nucleus in the state $n = 1$ of the hydrogen atom. In hydrogen atoms having higher values of n, the electron is farther away from the nucleus: when $n = 2$, it is (4×0.529) Å away; when $n = 3$, it is (9×0.529) Å away, and so forth. In the He^+ ion, which like H has just one electron, the nuclear charge is larger ($Z = 2$). Its stronger attraction contracts the entire pattern of distances by a factor of 2.

• The Bohr radius (also called the bohr) is the natural unit of length on the atomic scale. As such, it agrees well in magnitude with the approximate atomic dimensions estimated in Section 1–8.

Bohr also calculated the energy of the electron (the sum of its kinetic and potential energies) in the stable orbits. It equals

$$E_n = -\frac{Z^2}{n^2}\left(\frac{h^2}{8\pi^2 m_e a_0^2}\right) \qquad n = 1, 2, 3, \ldots$$

The combination in parentheses of the Bohr radius, Planck's constant, the mass of the electron, and other constants has units of energy. It occurs so frequently that it is defined as a (non-SI) unit of energy called the **rydberg (Ry).** Substitution of the various constants shows that 1 Ry = 2.17987×10^{-18} J. Hence

$$E_n = -\frac{Z^2}{n^2} Ry = -\frac{Z^2}{n^2} (2.18 \times 10^{-18} \text{ J})$$

The allowed values of the energy are shown in Figure 16–13 for hydrogen, for which $Z = 1$. The state with $n = 1$ is called the **ground state** because it is the state of lowest energy for the system of nucleus plus electron. An atom of hydrogen in this state has an energy of -1 Ry, or -2.18×10^{-18} J. States with higher values of n are called **excited states** and have higher energy. As n approaches infinity, it is found from the preceding equations that the radius becomes infinite and the energy rises toward zero. As n becomes very large, the orbit of the electron is farther and farther from the nucleus. The energy reaches zero when the electron and proton are at rest and separated by an infinite distance. It is negative for all smaller distances (Fig. 16–14). It may seem strange to speak of a "negative energy," but the negative sign is solely a result of choosing the zero of energy to correspond to the infinitely separated electron and nucleus.

• This convention was previously used in Figure 6-1, which displays potential energy diagrams for chemical bonding.

The **ionization energy** of an atom is the minimum energy needed to remove an electron from the atom when it is in its ground state. Removing an electron from hydrogen in its ground state involves a change from the $n = 1$ to the $n = \infty$ state with $Z = 1$. The energy change is

$$\Delta E = E_{\text{final}} - E_{\text{initial}} = \left(-\frac{1^2}{\infty^2}\right) - \left(-\frac{1^2}{1^2}\right) Ry = 0 - (-1 \text{ Ry})$$

$$= 1 \text{ Ry} = 2.18 \times 10^{-18} \text{ J}$$

This is the ionization energy of a single hydrogen atom. Multiplying it by Avo-

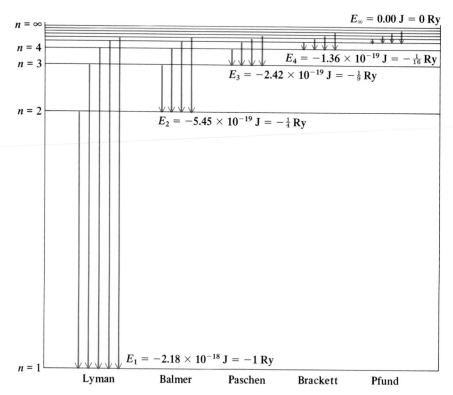

Figure 16–13 In the energy levels of the hydrogen atom the separated electron and proton are assigned zero energy, and all other energies are more negative than that. Atoms emit light in falling from higher to lower energy levels (*blue arrows*). Each series of related transitions is named after its discoverer.

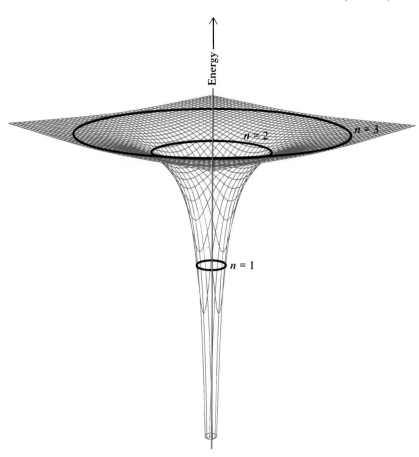

Figure 16–14 The interaction between the electron and nucleus has its lowest energy when the electron is closest to the nucleus. Moving the electron away from the nucleus can be seen as moving it up the sides of a steep potential energy well. In the Bohr theory, it can "catch" and stick on the sides only at certain allowed values of r, the radius, and of E, the energy. The first three of these are shown by rings.

gadro's number gives the ionization energy (IE) per mole of hydrogen atoms:

$$IE = (6.022 \times 10^{23} \text{ mol}^{-1})(2.18 \times 10^{-18} \text{ J})$$
$$= 1.31 \times 10^{6} \text{ J mol}^{-1} = 1310 \text{ kJ mol}^{-1}$$

This value is in agreement with the experimentally observed ionization energy of hydrogen atoms.

EXAMPLE 16–3

Consider the $n = 2$ state of the Li^{2+} ion. Using the Bohr model, calculate the radius of the electron orbit and the energy of the ion relative to the separated nucleus and electron.

Solution

Because $Z = 3$ for a Li^{2+} ion (the nuclear charge is $+3e$) and because $n = 2$ in this case, the radius is

$$r_2 = \frac{n^2}{Z} a_0 = \frac{4}{3} a_0 = \frac{4}{3} (0.529 \text{ Å}) = 0.705 \text{ Å}$$

The energy is

$$E_2 = -\frac{Z^2}{n^2} \text{Ry} = -\frac{(3)^2}{(2)^2} \text{Ry} = -\frac{9}{4} (2.18 \times 10^{-18} \text{ J}) = -4.90 \times 10^{-18} \text{ J}$$

> **Exercise**
>
> A certain one-electron ion has a radius of 2.12 Å and an energy of -2.18×10^{-18} J. Name the ion and tell what quantum state it is in.
>
> **Answer:**   The ion is beryllium(III), Be^{3+}. It is in the $n = 4$ state.

Atomic Spectra

When a one-electron atom or ion undergoes a transition from a state with quantum number $n_{initial}$ to one lower in energy with quantum number n_{final} ($n_{initial} > n_{final}$), a photon is emitted to carry off the energy lost by the atom. If the frequency of the photon is ν, it carries energy $h\nu$, and by the law of conservation of energy,

$$h\nu = -Z^2 \left(\frac{1}{n_{initial}^2} - \frac{1}{n_{final}^2} \right) Ry$$

Lines are seen in the emission spectrum with frequencies

$$\nu = Z^2 \left(\frac{1}{n_{final}^2} - \frac{1}{n_{initial}^2} \right) 3.29 \times 10^{15} \ s^{-1}$$

$$n_{initial} > n_{final} = 1, 2, 3, \ldots \text{(emission)}$$

Thus, each arrow in Figure 16–13 links a higher energy level to a lower and corresponds to one line in the emission spectrum of hydrogen. On the other hand, an atom or ion can *absorb* energy $h\nu$ from a photon as it undergoes a transition to a *higher* energy state ($n_{final} > n_{initial}$). The absorption spectrum of a one-electron atom or ion shows a series of lines at frequencies

$$\nu = Z^2 \left(\frac{1}{n_{initial}^2} - \frac{1}{n_{final}^2} \right) 3.29 \times 10^{15} \ s^{-1}$$

$$n_{final} > n_{initial} = 1, 2, 3, \ldots \text{(absorption)}$$

For hydrogen, which has $Z = 1$, the predicted emission spectrum with $n_{final} = 2$ corresponds exactly to the series of lines in the visible region discovered by Balmer and mentioned in Section 16–2. A series of lines at higher frequencies (in the ultraviolet region) is predicted for $n_{final} = 1$ (the Lyman series), and other series are predicted at lower frequencies (in the infrared region) for $n_{final} = 3, 4, \ldots$. In fact, the predicted and observed spectra of the hydrogen atom are in excellent agreement, a major triumph of the Bohr theory. Similar close agreement is found between Bohr theory and experiment for the various one-electron ions.

The Bohr theory has a number of shortcomings, however. The most important is that it is incapable of predicting the energy levels and spectra of atoms and ions with more than one electron. The modern theory of quantum mechanics that was subsequently developed to overcome this deficiency retains certain features of the Bohr theory and replaces others. The concept of discrete, stationary states and electrons moving between them is retained, and so is the quantization of angular momentum, although in a form slightly different from that advanced by Bohr. On the other hand, the circular orbits of the Bohr theory do not appear in the current quantum mechanics.

16–4 WAVES, PARTICLES, AND THE SCHRÖDINGER EQUATION

In his theory of the photoelectric effect, Einstein showed that light has particle-like as well as wave-like properties. In 1924, the young French physicist Louis de Broglie posed the following question: If light, which we think of as a wave, can have particle-like properties, then why cannot particles of matter have wave-like properties? He demonstrated that Bohr's assumption about the quantization of angular momentum in the hydrogen atom could be rationalized by ascribing such wave-like properties to the electron.

Up to this point we have considered only one type of wave: a **traveling wave.** Electromagnetic radiation (light, x-rays, and gamma rays) is described by such traveling waves, which move through space at speed c. Another type of wave that arises in physical situations is a **standing wave.** The vibrations set up by plucking a guitar string stretched taut between two fixed pegs are standing waves. A plucked string starts to vibrate, but not all oscillations are possible. Because the ends are fixed, the only oscillations that can persist are those in which an integral number of half-wavelengths fits into the length of the string L (Fig. 16–15). The condition on the allowed wavelengths is

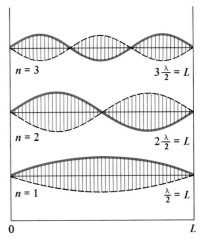

Figure 16–15 A string of length L with fixed ends can vibrate in only a restricted set of ways. The positions of largest amplitude of vibration are shown here for the first three of these harmonics. In a standing wave like this, the whole string is in motion except at its ends and at the nodes.

$$\frac{n\lambda}{2} = L \qquad n = 1, 2, 3, \ldots$$

It is impossible to create a wave with any other value of λ if the ends of the string are fixed. The oscillation with $n = 1$ is called the **fundamental** or first harmonic, and the higher values of n correspond to higher harmonics; for example, the oscillation with $n = 2$ is the second harmonic. In the higher ($n > 1$) harmonics, the points on the string at which half-wavelength sections meet are special. The wave moves up and down on the two sides of these points, but no vibration occurs at the points themselves. Regions of no vibration in a standing wave are called **nodes.** (The fixed ends do not count as nodes.) The study of standing waves allows some important conclusions:

> As the number of the harmonic increases, the number of nodes increases as well. The standing wave then has a shorter wavelength, higher frequency, and higher energy.

A simple experiment with a guitar confirms the last point. Plucking a string midway along its length excites mainly the first harmonic and requires minimal effort. Plucking near one end excites higher harmonics and requires distinctly more effort. The presence of higher frequencies in the second case is evident from the harsh timbre of the note.

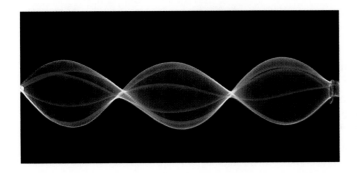

The third harmonic ($n = 3$) of vibration in a stretched rubber tube fastened at its ends. This standing wave has two nodes and a wavelength equal to two thirds of the tube's length.

Figure 16–16 A circular standing wave on a closed loop. The state shown has $n = 7$, with seven full wavelengths around the circle.

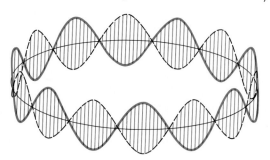

De Broglie realized that standing waves are examples of quantization: only certain discrete states of vibration are allowed for the vibrating string, characterized by the "quantum number," n. He suggested that the quantization of a one-electron atom might have the same origin and that the electron might be associated with a standing wave—in this case, a *circular* standing wave about the nucleus of the atom (Fig. 16–16). Standing waves in guitar strings encounter friction and tend to die away with time. If a standing wave is set up and there is no damping to make it fade away, then it persists indefinitely. Such situations are called **stationary states.** If electrons are wave-like, then they must exist in stable atoms as standing "matter-waves," stationary states that form a pattern about the nucleus. The admittedly uncomfortable notion of an electron in an atom as a motionless standing matter-wave succeeds in explaining why electrons in atoms do not continuously radiate energy. The electrons do *not* move in curved paths, as in the Bohr planetary model.

For a circular standing wave to persist, a whole number of wavelengths must fit into the circumference of the circle $2\pi r$. The condition on the allowed wavelengths is

$$2\pi r = n\lambda \qquad n = 1, 2, 3, \ldots$$

Bohr's assumption about quantization of the angular momentum of the electron is

$$m_e v r = n\frac{h}{2\pi}$$

which can be rewritten as

$$2\pi r = n\left(\frac{h}{m_e v}\right)$$

• In this case, there is an integral number of wavelengths rather than of half-wavelengths, as for the guitar string. The reason is that the circular loop comes back on itself, rather than reaching a fixed end.

Setting the two quantities that are equal to $2\pi r$ equal to each other (and canceling out the n) shows that the wavelength of the standing wave is related to the linear momentum $p = m_e v$ of the electron by the formula

$$\lambda = \frac{h}{m_e v} = \frac{h}{p}$$

De Broglie showed from the theory of relativity that just this relationship exists between the wavelength and the momentum of a *photon*. He therefore proposed as a generalization that any particle moving with linear momentum p has associated with it wave-like properties and a wavelength λ given by h/p.

Under what circumstances does the wave-like nature of particles become apparent? When "real" waves from two sources pass through the same region of space, they *interfere* with each other. Consider two water waves as an example.

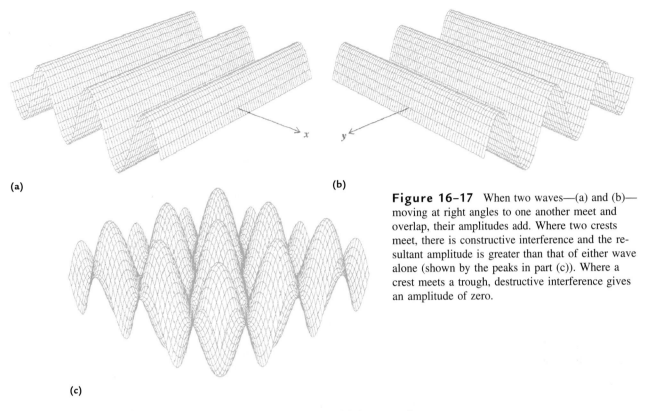

(a)

(b)

(c)

Figure 16-17 When two waves—(a) and (b)—moving at right angles to one another meet and overlap, their amplitudes add. Where two crests meet, there is constructive interference and the resultant amplitude is greater than that of either wave alone (shown by the peaks in part (c)). Where a crest meets a trough, destructive interference gives an amplitude of zero.

When two crests meet, **constructive interference** occurs and a higher crest (larger amplitude) appears; where a crest of one wave meets a trough of the other, **destructive interference** (smaller amplitude) occurs (Fig. 16–17). When x-rays are sent into a crystal, they are scattered by the atoms of the crystal (picture a water wave encountering a vertical post embedded in the bottom of a lake). The x-rays scattered from each atom in the crystal interfere with each other to give a characteristic x-ray **diffraction pattern.** According to de Broglie's generalization, such diffraction should also be observed when a beam of *electrons* is sent into a crystal. In 1927, C. Davisson and L. H. Germer showed that electrons in fact are diffracted by a crystal and that de Broglie's relationship correctly predicts their wavelength. The experiment provided a brilliant confirmation of de Broglie's hypothesis that electrons have wave-like properties. "Waves" and "particles" are idealized limits, and protons, electrons, and photons all possess both wave and particle aspects.

• We discuss x-ray diffraction further in Chapter 20.

EXAMPLE 16-4

Calculate the wavelength of an electron moving with velocity 1.0×10^6 m s^{-1}.

Solution

$$\lambda = \frac{h}{p} = \frac{h}{m_e v} = \frac{6.626 \times 10^{-34} \text{ kg m}^2 \text{ s}^{-1}}{(9.11 \times 10^{-31} \text{ kg})(1.0 \times 10^6 \text{ m s}^{-1})}$$

$$= 7.3 \times 10^{-10} \text{ m} = \boxed{7.3 \text{ Å}}$$

This wavelength is only about a dozen times larger than the radius of a ground-state hydrogen atom. Clearly, the wave-like properties of electrons are essential to an understanding of atomic structure.

Exercise

Calculate the de Broglie wavelength of a baseball of mass 0.17 kg that is thrown with a velocity of 30 m s^{-1}.

Answer: 1.3×10^{-34} m $= 1.3 \times 10^{-24}$ Å. This wavelength is far too small to be observed and need not be considered in studying the motion of a thrown baseball.

The Heisenberg Indeterminacy Principle

In classical physics, particles in motion have well-defined positions and momenta and follow precise trajectories. On an atomic scale, however, positions and momenta cannot be precisely determined because of a fundamental principle first stated by Werner Heisenberg in 1927: the **indeterminacy principle.**

• This is also known as the **uncertainty principle.** We use the term "indeterminacy" because it more accurately describes the true meaning of the principle and is also a better translation of the German word used originally by Heisenberg.

The indeterminacy principle is directly connected with the process of measurement. How do we determine the position and momentum of a macroscopic object like a baseball in motion? The simplest way is to take a series of snapshots of it at different times. Each picture is a record of the light (photons) scattered by the baseball. To observe an object, we must scatter something (such as a photon) from it (Fig. 16–18). Bouncing photons off a moving baseball does not change the momentum and trajectory of the baseball very much. If a proton or electron replaced the ball, however, the same photons would strongly affect it because their momenta would be comparable to that of the subatomic particle. If we sought to avoid this by using low-momentum photons to observe subatomic particles, then the momentum of the subatomic particle would not be altered much in an observation, but our knowledge of its position would suffer: to locate an object accurately we must use light that has a wavelength comparable to, or shorter than, the size of the object, and by de Broglie's relationship, the smaller the momentum, the longer the wavelength of a photon. We cannot have it both ways. There is a fundamental limit to our ability to measure simultaneously both the position and momentum of an object of atomic or subatomic dimensions.

A rough estimate takes the minimum indeterminacy of the position measurement (Δx) to be on the order of the wavelength of the light (λ) and the minimum indeterminacy of the momentum measurement (Δp) to be on the order of h/λ (because that is the momentum carried by the photon). The product of the two indeterminacies is then $(\Delta x)(\Delta p) \geq h$. A fuller analysis gives the result

$$(\Delta p)(\Delta x) \geq h/4\pi$$

• To speak of trajectories of bound electrons, such as the planetary orbits of the Bohr model, is not meaningful.

The indeterminacy principle places a fundamental limitation on the degree of accuracy to which the position and momentum of a particle can simultaneously be known.

EXAMPLE 16–5

Suppose that photons of green light (wavelength 5.3×10^{-7} m) are used to locate the position of the baseball from Exercise 16–4 to an accuracy of one wavelength. Calculate the minimum indeterminacy in the *speed* of the baseball.

Solution

The Heisenberg relation

$$(\Delta x)(\Delta p) \geq h/4\pi$$

gives

$$\Delta p \geq \frac{h}{4\pi \, \Delta x} = \frac{6.626 \times 10^{-34} \text{ J s}}{4\pi (5.3 \times 10^{-7} \text{ m})} = 9.9 \times 10^{-29} \text{ kg m s}^{-1}$$

Because the momentum is just the mass (a constant) times the speed, the indeterminacy in the speed is

$$\Delta v = \frac{\Delta p}{m} \geq \frac{9.9 \times 10^{-29} \text{ kg m s}^{-1}}{0.17 \text{ kg}} = 5.8 \times 10^{-28} \text{ m s}^{-1}$$

This is such a tiny fraction of the speed of the baseball (30 m s^{-1}) that the indeterminacy principle plays a negligible role in the measurement of its motion. Such is the case for all macroscopic objects.

Exercise

Suppose that the location of an electron is determined to within 0.53 Å (the Bohr radius). Calculate the minimum indeterminacy in its speed.

Answer: 1.1×10^6 m s^{-1}.

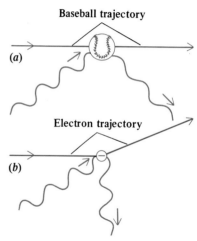

Figure 16–18 A photon that has a negligible effect on the trajectory of a baseball (a) significantly perturbs the trajectory of the far less massive electron (b).

The Schrödinger Equation

De Broglie's work demonstrates that electrons in atoms have wave-like properties, and the indeterminacy principle shows that detailed trajectories of electrons cannot be defined: we must instead deal in terms of the *probability* of electrons having certain positions and momenta. These ideas are combined in the fundamental equation of quantum mechanics, the **Schrödinger equation,** which was discovered by the Austrian physicist Erwin Schrödinger in 1925. Schrödinger reasoned that if an electron (or any other particle) has wave-like properties, it should be described by a **wave function** that has an amplitude at each position in space (just as for a water wave or a classical electromagnetic wave). This wave function is symbolized by the Greek letter ψ ("psi"), and $\psi(x, y, z)$ is the amplitude of the wave at the point in space defined by the Cartesian coordinates (x, y, z). According to quantum mechanics, only wave functions ψ that are solutions to the Schrödinger equation are allowed and have meaning as descriptions of atoms. Such a wave function for an atom tells all that can be known about the physical state of the electrons bound in that atom.

Whole families of allowed wave functions arise in every kind of atom but do so only for certain discrete values of the energy of the atom. For other energies, allowed wave functions simply do not exist. Hence, quantization of energy is a natural consequence of the Schrödinger equation. The ψ that corresponds to the lowest value of the energy of an atom is called the ground state (just as in the Bohr model), and ψ's that correspond to higher energies are called excited states. A wave function, like the amplitude of an electromagnetic wave, can take positive values in some regions of space (it is said to have *positive phase* in these regions) and negative values in other regions (where it has *negative phase*). The points at which the wave function passes through zero and changes sign are called *nodes,* just as in the guitar-string model for standing waves.

What is the physical meaning of the wave function ψ? There is no way to measure ψ directly, just as in classical wave optics there is no direct way to measure the amplitudes of the electric and magnetic fields that constitute the light wave (see Fig. 16–4). What *can* be measured in the latter case is the *intensity* of the

* In Chapters 17 and 18, the phase of the wave function proves important. Wave functions must have the same phase (that is, constructively reinforce one another) if they are to combine to form a chemical bond between atoms.

light wave, which is proportional to the *square* of the amplitude of the electric field. This intensity is in turn proportional to the density of photons at that point in space, or, in other words, to the probability density of finding a photon there.

By analogy, we propose that the square of the wave function for a particle (ψ^2) is a probability density for that particle. In other words, $\psi^2(x, y, z)\Delta x\Delta y\Delta z$ is the probability that the particle will be found in a small volume $\Delta x\Delta y\Delta z$ around the point (x, y, z). This probabilistic interpretation of the wave function, first suggested by the German physicist Max Born, is now generally accepted because it provides a consistent picture of particle motion on a microscopic scale. We are obliged to acknowledge that our information about the location of a particle is statistical.

• The "Δx" in this expression stands for one edge of a small box centered at the point (x, y, z). "Δy" and "Δz" refer to the other two edges of the box. Hence, $\Delta x\Delta y\Delta z$ is the volume of the box.

16–5 THE HYDROGEN ATOM

A hydrogen atom consists of a single electron interacting with a nucleus of charge $+1e$. The solutions of the Schrödinger equation for this arrangement of particles have great importance in chemistry. Recall that the hydrogen atom is the simplest example of a one-electron atom or ion; others are He^+, Li^{2+}, and all other ions in which all but one electron have been stripped off. They differ in the charge (Ze) on the nucleus and therefore in the attractive force felt by the electron. Mathematical analysis shows that wave functions that satisfy the Schrödinger equation for hydrogen atoms and one-electron ions exist only for these values of the energy:

$$E_n = -\frac{Z^2}{n^2}\,\text{Ry} \qquad n = 1, 2, 3, \ldots$$

which are exactly the energies found in the Bohr atom. Here, however, quantization of the energy comes about naturally from the Schrödinger description of the electron as a standing wave rather than through an arbitrary assumption about the angular momentum.

The energy of one-electron atoms or ions depends only on the quantum number n, which is the **principal quantum number.** Two more quantum numbers appear when the functions that satisfy the one-electron Schrödinger equation are systematically investigated. Each is associated with an observable property of the atom. The first is the **angular momentum quantum number** (ℓ) which may take on any integral value from 0 up to and including $n - 1$. The second is the **magnetic quantum number** (m_ℓ), which may take on any integral value (including 0) from $-\ell$ to ℓ. It is called the "magnetic quantum number" because its value governs the behavior of the atom in an external magnetic field. For $n = 1$ (the ground state), the only allowed values of ℓ and m_ℓ are ($\ell = 0$, $m_\ell = 0$). For $n = 2$, there are $n^2 = 4$ sets of allowed values:

$$(\ell = 0, m_\ell = 0), (\ell = 1, m_\ell = 1), (\ell = 1, m_\ell = 0), (\ell = 1, m_\ell = -1)$$

The restrictions on ℓ and m_ℓ cause exactly n^2 solutions of the Schrödinger equation to exist for every value of n.

A wave function ψ that satisfies the one-electron Schrödinger equation with quantum numbers n, ℓ, and m_ℓ is called an **orbital.** Although the word "orbital" recalls the orbits of the Bohr atom, there is no real resemblance. An orbital is *not* some trajectory traced by an individual electron. Instead, it refers to a wave function ψ that gives the amplitude of the standing wave corresponding to the electron. The value of this function changes from one point in space to the next. The square of

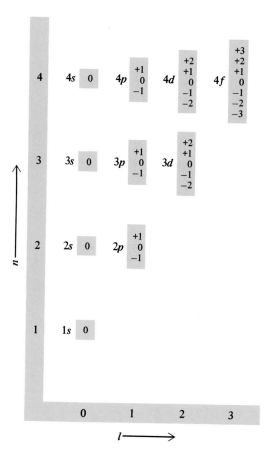

Figure 16-19 The allowed sets of quantum numbers for a one-electron atom or ion. The allowed values of the m_ℓ quantum number are shown to the right of each (n, ℓ) symbol.

this value tells the probability of finding the electron at each different point in space.

An orbital is identified by its three quantum numbers—n, ℓ, and m_ℓ. The usual convention is to replace the angular momentum quantum number by a letter:

$$\ell = 0 \text{ is signified by } s$$
$$\ell = 1 \text{ is signified by } p$$
$$\ell = 2 \text{ is signified by } d$$
$$\ell = 3 \text{ is signified by } f$$
$$\ell = 4 \text{ is signified by } g$$

• The letters s, p, d, f come from early (pre-quantum mechanics) spectroscopy, in which certain spectroscopic lines were referred to as "sharp," "principal," "diffuse," and "fundamental."

and then on down the alphabet. A wave function with $n = 1$ and $\ell = 0$ is thus called a 1s-orbital. One with $n = 3$ and $\ell = 1$ is a 3p-orbital, one with $n = 4$ and $\ell = 3$ is a 4f-orbital, and so forth. Figure 16-19 summarizes the allowed combinations of quantum numbers.

EXAMPLE 16-6

Give the labels of all the groups of orbitals with $n = 4$, and state how many m_ℓ values are present in each group.

Solution

The quantum number ℓ may range from 0 to $n - 1$, so its allowed values in this case ($n = 4$) are 0, 1, 2, and 3. The labels for the groups of orbitals are then

$$\ell = 0 \quad 4s$$

$$\ell = 1 \quad 4p$$

$$\ell = 2 \quad 4d$$

$$\ell = 3 \quad 4f$$

The quantum number m_ℓ ranges from -1 to $+1$, so the number of m_ℓ values is $2\ell + 1$. This gives

one 4s-orbital

three 4p-orbitals

five 4d-orbitals

seven 4f-orbitals

for a total of $16 = 4^2 = n^2$ orbitals with $n = 4$. All have the same energy but differ in their shapes and orientations.

Exercise

(a) How many orbitals are there with $n = 5$? (b) How many of these are 5g-orbitals? (c) How many are 5h-orbitals?

Answer: (a) 25 orbitals. (b) 9 orbitals. (c) Zero, because 5h-orbitals are not possible.

Sizes and Shapes of Orbitals

The shapes and sizes of the hydrogen-atom orbitals are important in chemistry. All **s-orbitals** have $\ell = 0$ (therefore, $m_\ell = 0$ as well) and are spherically symmetrical about the nucleus. This means that the amplitude ψ of an s-orbital (and, therefore, also the probability of finding the electron at or near some point in space) depends only on the distance r of the point from the nucleus. There are several ways to visualize the radial variation of the ns-orbitals with $n = 1, 2, 3, \ldots$ One way (shown in Fig. 16–20a) is to make a "three-dimensional" picture, in which the shading is heaviest where ψ^2 is largest and lighter where ψ^2 is smaller. A second way (Fig. 16–20b) is to plot the wave functions themselves (ψ_{1s}, ψ_{2s}, and ψ_{3s}) versus r, the distance from the nucleus. A third way, perhaps the most useful in chemistry, is to plot a **radial probability distribution** $r^2\psi^2$ (Fig. 16–20c). Again, ψ^2 gives the probability of finding the electron in a small volume $\Delta x \Delta y \Delta z$ around a point (x, y, z) in space. Multiplying it by r^2 accounts for the larger number of points that become available as the distance from the origin increases. These are all the points located on the surface of a sphere of radius r. Although ψ^2 of all s-orbitals goes to a maximum within the nucleus, where r is nearly zero, the nucleus offers very few small volumes $\Delta x \Delta y \Delta z$ for an electron to occupy. This greatly reduces the chance of finding an s-electron within the nucleus. At larger r, more points meet the distance specification so that more small volumes $\Delta x \Delta y \Delta z$ are available. Consequently, the probability of finding the electron at the distance r, as given by $r^2\psi^2$, is larger even though ψ^2 is less.

What is meant by the *size* of an orbital? The wave function of an electron in an atom (and its square) stretches out to infinity. It follows that an atom has no

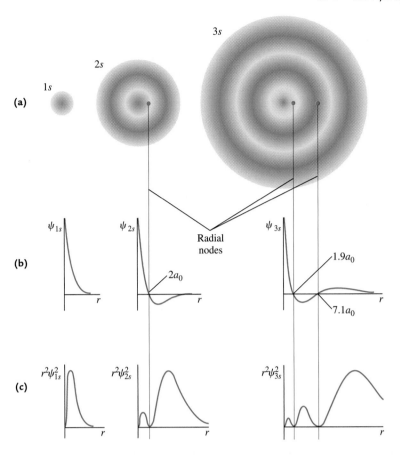

Figure 16–20   Three representations of hydrogen *s*-orbitals. (a) An electron-density representation of a hydrogen atom in its 1*s*, 2*s*, and 3*s* states. The spheres are cut off at a radius that encloses 90% of the total electron density. (b) The wave functions graphed against distance from the nucleus *r*. (c) The radial probability distribution, equal to $r^2\psi^2$. The distance a_0 is the Bohr radius (0.529 Å).

"natural" boundary. We must set some arbitrary limit upon the extension of its wave function. We might define the size of an atom as that of the contour at which ψ^2 has fallen off to some particular numerical value, or we could define the size as the extent of a "balloon skin," inside which the electron has a definite probability (for example, 90% or 99%) of being found. We have adopted the latter practice in drawing pictures of orbitals. Figure 16–20a shows spheres that enclose 90% of the electron probability for the 1*s*-, 2*s*-, and 3*s*-orbitals for hydrogen.

Note two features of *s*-orbitals. First, the size of an orbital increases with increasing quantum number *n*, meaning that the electron is most likely to be found farther from the nucleus for larger *n*. A 3*s*-orbital is larger than a 2*s*-orbital, which in turn is larger than a 1*s*-orbital. This is the quantum-mechanical analog of the increase in radius of the Bohr orbits with *n*. Second, an *ns*-orbital has $n - 1$ **radial nodes** (a radial node for an *ns*-orbital is a spherical shell around the nucleus on which ψ and ψ^2 are zero). The more numerous the nodes, the higher the energy of the orbital. This recalls the case of standing waves on a guitar string, in which

• Thus, a 2*s*-electron has one radial node. The electron has a non-zero probability of being within this closed nodal surface and also a non-zero probability of being outside of it, but its probability density on the nodal surface is zero.

higher frequency (and, thus, higher energy) vibrations have more nodes. The radial nodes of Figure 16–20a lie inside the outer spherical surface shown, close to the nucleus. As we shall see, the chemical behavior of atoms is dominated by the outer regions of electron density rather than the inner nodal structure.

It is more difficult to draw pictures showing the behavior of orbitals with angular quantum numbers ℓ different from zero, because these orbitals are not spherically symmetrical. For the case $\ell = 1$ (**p-orbitals**), there are three such orbitals for each value of n. The shapes of these p-orbitals are shown in Figure 16–21 (for $n = 2$). The wave function for the p_z-orbital has maximum amplitude along the z-axis and is zero in the xy-plane. Therefore, it points along the z-axis, with one lobe on one side of the xy-plane and the other lobe on the other side. The p_x-orbital has the same shape, but it points along the x-axis, and the p_y-orbital points along the y-axis. If the electron is in a p_z-orbital, for example, it has the highest probability of being found along the direction of the z-axis and has zero probability of being found in the xy-plane. This plane is a nodal plane, or, more generally, an **angular node** across which the wave function changes sign.

The p-orbitals, like the s-orbitals, may also have radial nodes, in which the electron density vanishes at certain distances from the nucleus irrespective of direction. An np wave function has $n - 2$ radial nodes: thus, a $2p$ wave function has no such nodes, a $3p$ wave function has one, and a $4p$ wave function has two. Because an np wave function always has exactly one angular node, it has a total of $n - 1$ nodes ($n - 2$ radial and 1 angular), which is the same number as an s orbital that has the same principal quantum number. For one-electron atoms, the energy depends only on the total number of nodes $n - 1$ and not on the type of node (radial or angular). As in the case of s-orbitals, the radial nodes in p-orbitals occur so close to the nucleus that they have little effect on the chemical bonding of atoms. For the purposes of chemistry, we may therefore represent all p-orbitals (not just the $2p$-orbitals) by pictures like those in Figure 16–21.

Because p-orbitals have angular nodes that pass through the nucleus, the electron density at the nucleus equals zero. This is true, in fact, of all types of orbitals except s-orbitals, and it means that the electron is never at the nucleus in wave functions with $\ell > 0$ ($p, d, f \ldots$).

The spatial distribution of the five **d-orbitals** is somewhat more complicated. Their shapes are shown in Figure 16–22. The d_{xy}-, d_{yz}-, d_{xz}-, and $d_{x^2-y^2}$-orbitals all have the same shape but different orientations with respect to the Cartesian axes. A d_{xy}-orbital, for example, has four lobes; the maximum amplitude is at 45 degrees to the x- and y-axes. The $d_{x^2-y^2}$-orbital has maximum amplitude along the x- and y-axes. The fifth orbital, d_{z^2}, has a different shape, with maximum amplitude along the z-axis and a little "doughnut" in the xy-plane. Each of these d-orbitals has exactly two angular nodes (for example, for the d_{xy}-orbital the xz- and yz-planes are nodal surfaces). Their wave functions also have $n - 3$ radial nodes, giving once again $n - 1$ total nodes.

The wave functions for f-orbitals and orbitals of higher ℓ can be calculated, but they are hard to draw or to visualize. They play a smaller role in chemistry than the s-, p-, and d-orbitals. We therefore pause at this point and summarize the important features of orbital shapes and sizes:

$2p_x$

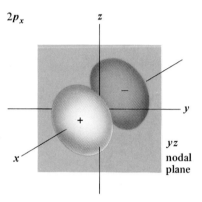

$2p_y$

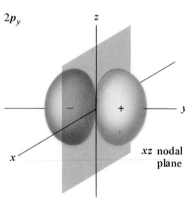

$2p_z$

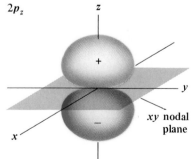

Figure 16–21 The distribution of electron density in the three $2p$-orbitals, with the relative phase of the wave function and the nodal planes indicated.

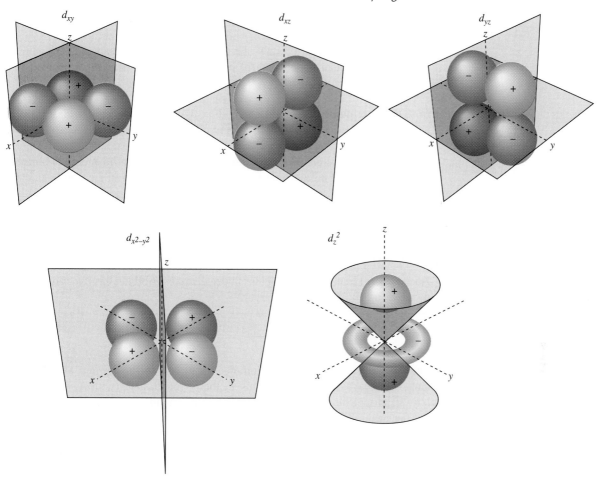

Figure 16–22 The shapes of the five $3d$-orbitals, with relative phases and nodal surfaces indicated.

1. For a given value of ℓ, an increase in n leads to an increase in the average distance of the electron from the nucleus and therefore in the size of the orbital. A $2s$-orbital, for example, is larger than a $1s$-orbital, and a $3s$-orbital is larger than a $2s$ (see Fig. 16–20).
2. An orbital with quantum numbers n and ℓ has ℓ angular nodes and $n - \ell - 1$ radial nodes, giving a total of $n - 1$ nodes. For a one-electron atom or ion, the energy depends only on the number of nodes; that is, on n but not on ℓ or m_ℓ.
3. As r approaches 0, the wave function vanishes for all orbitals except the s-orbitals; thus, only an electron in an s-orbital can "penetrate to the nucleus"—that is, have a finite probability of being found at the nucleus.

These general statements will prove helpful in studying the electronic structure of many-electron atoms, even though they are deduced from the one-electron case.

EXAMPLE 16–7

Compare the 3p- and 4d-orbitals of a hydrogen atom with respect to (a) number of radial and angular nodes, and (b) energy of the corresponding atom.

Solution

(a) The 3p-orbital has a total of $n - 1 = 3 - 1 = 2$ nodes. Of these, one is angular ($\ell = 1$) and one is radial. The 4d-orbital has $4 - 1 = 3$ nodes. Of these, two are angular ($\ell = 2$) and one is radial.

(b) The energy of a one-electron atom depends only on n. The energy of an atom with an electron in a 4d-orbital is then higher than that of an atom with an electron in a 3p-orbital because $4 > 3$.

Exercise

Compare the 3d- and 5d-orbitals of a He$^+$ ion with respect to (a) number of radial and angular nodes and (b) size of the orbital.

Answer: (a) Each has two angular nodes, but the 5d-orbital has two radial nodes as well. (b) The 5d-orbital is larger.

Electron Spin

Our discussion of one-electron atoms and ions has so far omitted an important point. If a beam of hydrogen atoms in their ground state (with $n = 1$, $\ell = 0$, $m_\ell = 0$) is sent through a magnetic field, the strength of which varies with position, the beam splits into two beams, each containing half of the atoms (Fig. 16–23). The results of this experiment are explained by introducing a *fourth* quantum number to complete the description of the wave function of the electron. This quantum number, m_s, may take on only two values, conventionally chosen to be $\frac{1}{2}$ and $-\frac{1}{2}$. Because the behavior in the magnetic field is what would be predicted by classical theory if electrons were balls of charge spinning around their own axes, the fourth quantum number is referred to as the **spin quantum number.** When $m_s = \frac{1}{2}$, the electron spin is in one direction; when $m_s = -\frac{1}{2}$, the spin is in the other. We should not take the classical picture of a spinning ball of charge literally, however. The spin quantum number comes from relativistic effects that are not included in the Schrödinger equation. For most purposes in chemistry, it is sufficient simply to solve the Schrödinger equation and then associate with each electron a spin quantum number $m_s = +\frac{1}{2}$ or $m_s = -\frac{1}{2}$, which does not affect the spatial probability density of the electron. The spin does affect the number of

Figure 16–23 A beam of hydrogen atoms is split into two beams when traversing a nonuniform magnetic field. Atoms with spin quantum number $m_s = +\frac{1}{2}$ comprise one beam, and those with $m_s = -\frac{1}{2}$ the other.

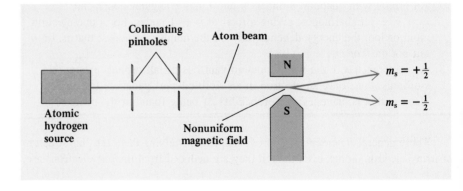

different quantum states with energy E_n, however, doubling it from n^2 to $2n^2$. This fact assumes considerable importance when we consider many-electron atoms in Chapter 17.

The single quantum number n of the Bohr theory is thus superseded by a set of four quantum numbers (n, ℓ, m_ℓ, and m_s) in modern quantum mechanics. For a one-electron atom or ion in free space, only n affects the energy, and only ℓ and m_ℓ affect the orbital shape. All four quantum numbers must be specified to give a complete description of a one-electron system.

SUMMARY

16-1 A traveling wave has **amplitude** (the extent of variation of the quantity that oscillates), **wavelength** λ (the distance that the oscillating quantity covers before repeating itself), **frequency** ν (the number of cycles passing a fixed point per second), and speed (the product of frequency and wavelength). **Electromagnetic waves** range in wavelength from the meter or kilometer scale of radio waves down through the micrometer scale of visible light to the nanometer scale of x-rays and beyond. All electromagnetic radiation propagates at the same speed c in a vacuum (approximately 3.00×10^8 m s^{-1}).

16-2 Classical physics fails to explain how the intensity of electromagnetic radiation emitted by a heated object (idealized as a **blackbody**) varies with wavelength. Instead it predicts the **ultraviolet catastrophe.** Classical physics also fails to explain the existence of a threshold frequency (in which light of a certain minimum frequency is required to eject electrons from a metallic surface) in the **photoelectric effect.** Finally, classical physics does not explain the observation of discrete frequencies in the emission spectra of single atoms and in fact predicts that the nuclear atom is not stable.

16-3 Light has particle properties as well as wave properties. This **wave-particle duality** means that the energy of electromagnetic waves comes in quantized packets called **photons.** Quantum theory allowed explanations of blackbody radiation (by Planck), the photoelectric effect (by Einstein), and the stability and characteristic spectra of atoms (by Bohr). In Bohr's planetary model of the hydrogen atom, only certain orbital radii and energies are allowed, and transitions between orbits are accompanied by the emission or absorption of photons of light having the frequency $\nu = \Delta E/h$, where ΔE equals the difference in the energies of the allowed orbits and h is **Planck's constant.** The Bohr model predicts the observed spectrum of hydrogen atoms and one-electron ions but fails for atoms or ions with more than one electron.

16-4 Louis de Broglie proposed in 1924 that the wavelength of a moving particle of matter is inversely proportional to its momentum, with the proportionality constant (Planck's constant) so small that macroscopic objects have negligible wavelengths. For electrons and other particles on the atomic scale, wave properties play an essential role. According to the **Heisenberg indeterminacy principle,** the position and momentum of small particles cannot both be specified to arbitrary accuracy. There is an inevitable degree of indeterminacy in position and momentum due to the act of observation itself. The **Schrödinger wave equation** associates a **wave function** (symbolized by ψ) with the electrons in an atom. The square of the wave function gives the electron density at a point in space, or, equivalently, the probability of finding an electron there. The Schrödinger equation provides the starting point for modern quantitative prediction of the properties of atoms and molecules.

16–5 Solutions to the Schrödinger equation for one-electron atoms and ions correspond to the same energies as the Bohr model but replace circular orbits by **orbitals,** which are standing waves by which the electron density is distributed throughout the atom. The energies, shapes, and sizes of orbitals are characterized by three **quantum numbers.** A fourth quantum number (the spin quantum number m_s) describes the inherent magnetic properties of the electron. The energy of an electron in a one-electron atom or ion depends only on the first quantum number (the principal quantum number) n; the second and third quantum numbers, ℓ and m_ℓ, determine the shapes and orientations of the orbitals and, therefore, the electron-density distribution. The letters s, p, d, and f are traditionally associated with values of ℓ from 0 to 3, respectively. Only s-orbitals are spherically symmetrical; all others describe electron-density distributions that depend on direction as well as distance from the nucleus. The overall number of **nodes** in a wave function is $n - 1$, and the number of **angular nodes** is ℓ. As a result, the number of lobes in orbitals increases with increasing ℓ. Sizes and shapes can be associated with orbitals most easily by identifying "balloon skins," inside which the electron has a 90% probability of being found.

PROBLEMS

Note: Answers to blue-numbered problems are given in Appendix F. Problems that are more challenging are indicated with asterisks.

Waves and Light

1. Some water waves reach the beach at a rate of one every 3.2 s, and the distance between their crests is 2.1 m. Calculate the speed of these waves.

2. The spacing between bands of color in a chemical wave from an oscillating reaction is measured to be 1.2 cm, and a new wave appears every 42 s. Calculate the speed of propagation of the chemical waves.

3. (See Example 16–1.) An FM radio station broadcasts at a frequency of 9.86×10^7 s^{-1} (98.6 MHz). Calculate the wavelength of the radio waves.

4. (See Example 16–1.) The gamma rays emitted by ^{60}Co are used in radiation treatment of cancer. These gamma rays have a frequency of 2.83×10^{20} s^{-1}. Calculate their wavelength, expressing your answer in meters and in ångströms.

5. Radio waves of wavelength 6.00×10^2 m can be used to communicate with spacecraft over large distances.
   (a) Calculate the frequency of these radio waves.
   (b) Suppose that a radio message is sent home by the astronauts in a spaceship approaching the planet Mars at a distance of 8.0×10^{10} m from earth. How long (in minutes) does it take for the message to travel from the spaceship to earth?

6. An argon-ion laser emits light of wavelength 488 nm.
   (a) Calculate the frequency of the light.
   (b) Suppose that a pulse of light from this laser is sent from earth, is reflected from a mirror on the moon, and returns to its starting point. Calculate the time elapsed for the round trip, taking the distance from earth to moon to be 3.8×10^5 km.

7. The speed of sound in dry air at 20°C is 343.5 m s^{-1}, and the frequency of the sound from the middle C note on a piano is 261.6 s^{-1}, according to the American standard-pitch scale. Calculate the wavelength of this sound and the time it takes to travel 30.0 m across a concert hall.

8. Ultrasonic waves have frequencies too high to be detected by the human ear but can be produced and detected by vibrating crystals. Calculate the wavelength of an ultrasonic wave of frequency 5.0×10^4 s^{-1} that is propagating through a sample of water at a speed of 1.5×10^3 m s^{-1}. Explain why ultrasound can be used to probe the size and position of the fetus inside the abdomen of the mother. Could audible sound with a frequency of 8000 s^{-1} be used for this purpose?

Paradoxes in Classical Physics

9. Both blue and green light eject electrons from the surface of potassium. In which case do the ejected electrons have the higher average kinetic energy?

10. When an intense beam of green light is directed on a copper surface, no electrons are ejected. What happens if the green light is replaced with red light?

11. Excited lithium atoms emit light strongly at a wavelength of 671 nm. This emission predominates when Li atoms are excited in a flame. Predict the color of the flame.

12. Excited mercury atoms emit light strongly at a wavelength of 454 nm. This emission predominates when Hg atoms are excited in a flame. Predict the color of the flame.

Planck, Einstein, and Bohr

13. Barium atoms in a flame emit light as they undergo transitions from one energy level to another one that lies 3.6×10^{-19} J lower in energy. Calculate the wavelength of light emitted and, by referring to Figure 16–5, predict the color visible in the flame.

14. Potassium atoms in a flame emit light as they undergo transitions from one energy level to another one that lies 4.9×10^{-19} J lower in energy. Calculate the wavelength of light emitted and, by referring to Figure 16–5, predict the color visible in the flame.

15. The sodium D-line is actually a pair of closely spaced spectroscopic lines seen in the emission spectrum of sodium atoms. The wavelengths are centered at 589.3 nm. The intensity of this emission makes it the major source of light in the sodium arc light and causes the yellow color of that light.
    (a) Calculate the energy change per sodium atom emitting a photon at the D-line wavelength.
    (b) Calculate the energy change per mole of sodium atoms emitting photons at the D-line wavelength.
    (c) If a sodium arc light produces 1.000 kilowatt (1000 J s^{-1}) of radiant energy at this wavelength, how many moles of sodium atoms emit photons per second?

16. The power output of a laser is measured by its wattage, the number of joules of energy it radiates per second (1 watt = 1 J s^{-1}). A 10-watt laser produces a beam of green light with a wavelength of 520 nm (5.2×10^{-7} m).
    (a) Calculate the energy carried by each photon.
    (b) Calculate the number of photons emitted by the laser per second.

17. (See Example 16–2.) Cesium is frequently used in photocells because its work function (3.43×10^{-19} J) is the lowest of all the elements. Such photocells are efficient because the broadest range of wavelengths of light can eject electrons. What colors of light eject electrons from cesium? What colors of light eject electrons from selenium, which has a work function of 9.5×10^{-19} J?

18. (See Example 16–2.) Some alarm systems use the photoelectric effect. A beam of light strikes a piece of metal in a photocell, ejecting electrons continuously and causing a small electric current to flow. When someone steps into the light beam, the current is interrupted and the alarm is triggered. What is the maximum wavelength of light that can be used in such an alarm system if the photocell metal is sodium, with a work function of 4.41×10^{-19} J?

19. Light having a wavelength of 2.50×10^{-7} m falls upon the surface of a piece of chromium in an evacuated glass tube. If the work function of the chromium is 7.21×10^{-19} J, determine the following:
    (a) The maximum kinetic energy of the photoelectrons emitted from the chromium.
    (b) The speed of photoelectrons having this maximum kinetic energy.

20. Calculate the maximum wavelength of electromagnetic radiation if it is to cause detachment of electrons from the surface of metallic tungsten, which has a work function of 7.29×10^{-19} J. If the maximum speed of the emitted photoelectrons is to be 2.00×10^{6} m s^{-1}, what should be the wavelength of the radiation?

21. (See Example 16–3.) Consider three hydrogen atoms in the quantum states $n = 3$, $n = 6$, and $n = 8$.
    (a) Compute the energies of these atoms. Express the answers both in joules (J) and in rydbergs (Ry). State which atom is in the highest energy state.
    (b) Compute the radii of these atoms. Express the answers both as multiples of a_0 and in meters. State which of the atoms is largest.

22. (See Example 16–3.) Consider three He$^+$ ions in the quantum states $n = 2$, $n = 4$, and $n = 7$.
    (a) Compute the energies of these ions in joules (J) and in rydbergs (Ry).
    (b) Compute the radii of these atoms as multiples of a_0 and in meters.

23. A hydrogen atom undergoes the following transitions. Determine the amount of energy lost by the atom in each case, and express it both in rydbergs and in joules.
    (a) $n = 7$ to $n = 4$
    (b) $n = 4$ to $n = 3$
    (c) $n = 3$ to $n = 4$
    (*Hint*: Recall that the loss of a negative is a gain.)

24. A Li^{2+} ion undergoes the following transitions. Determine the amount of energy lost by the ion in each case and express it both in rydbergs and in joules.
    (a) $n = 3$ to $n = 1$
    (b) $n = 6$ to $n = 2$
    (c) $n = 3$ to $n = 4$

25. A single photon is absorbed or emitted in each transition in problem 23. Calculate its wavelength.

26. A single photon is absorbed or emitted in each transition in problem 24. Calculate its wavelength.

27. Use the Bohr model to calculate the radius and the energy of the B^{4+} ion in the $n = 3$ state. How much energy is required to remove the electrons from one mole of B^{4+} ions

in this state? What frequency and wavelength of light are emitted in a transition from the $n = 3$ to the $n = 2$ state of this ion? Express all results in SI units.

28. He$^+$ ions are observed in stellar atmospheres. Use the Bohr model to calculate the radius and the energy of the He$^+$ ion in the $n = 5$ state. How much energy is required to remove the electrons from one mole of He$^+$ in this state? What frequency and wavelength of light are emitted in a transition from the $n = 5$ to the $n = 3$ state of this ion? Express all results in SI units.

29. The radiation emitted in the transition from $n = 3$ to $n = 2$ in a neutral hydrogen atom has a wavelength equal to 656.1 nm. What is the wavelength of radiation emitted from a doubly ionized lithium atom (Li^{2+}) if a transition occurs from $n = 3$ to $n = 2$? In what region of the spectrum does this radiation lie?

30. The Be^{3+} ion has a single electron. Calculate the frequencies and wavelengths of light seen in its emission spectrum for the first three lines of each of the series analogous to the Lyman and the Balmer series of neutral hydrogen. In what region of the spectrum does this radiation lie?

Waves, Particles, and the Schrödinger Equation

31. A guitar string with fixed ends has a length of 50 cm.
    (a) Calculate the wavelength of its fundamental mode of vibration (i.e., its first harmonic) and of its third harmonic.
    (b) How many nodes does the third harmonic have?

32. Suppose we picture an electron in a chemical bond as being a wave with fixed ends. Take the length of the bond to be 1.0 Å.
    (a) Calculate the wavelength of the electron wave in its ground state and in its first excited state.
    (b) How many nodes does the first excited state have?

33. (See Example 16–4.) Calculate the de Broglie wavelength of the following:
    (a) An electron moving at a speed of 1.00×10^3 m s^{-1}.
    (b) A proton moving at a speed of 1.00×10^3 m s^{-1}.
    (c) A baseball having a mass of 170.0 g and moving at a speed of 95 km hr^{-1}.

34. (See Example 16–4.) Calculate the de Broglie wavelength of the following:
    (a) An electron that has been accelerated to a kinetic energy equivalent to 1.20×10^7 J mol^{-1}.
    (b) A helium atom moving at a speed of 353 m s^{-1}. This is the root-mean-square speed of helium atoms at 20 K.
    (c) A krypton atom moving at a speed of 299 m s^{-1}. This is the root-mean-square speed of krypton atoms at 300 K.

35. (See Example 16–5.)
    (a) The position of an electron is known to within 10 Å (1.0×10^{-9} m). What is the minimum indeterminacy in its velocity?
    (b) Repeat the calculation of part (a) for a helium atom.

36. No object can travel faster than the speed of light, so it is evident that the indeterminacy in the speed of any object is at most 3×10^8 m s^{-1}.
    (a) What is the minimum indeterminacy in the position of an electron, given that we know nothing about its speed except that it is less than the speed of light?
    (b) Repeat the calculation of part (a) for the position of a helium atom.

The Hydrogen Atom

37. Give labels for the orbitals described by each of the following sets of quantum numbers:
    (a) $n = 4$, $\ell = 1$
    (b) $n = 2$, $\ell = 0$
    (c) $n = 6$, $\ell = 3$

38. Give labels for the orbitals described by each of the following sets of quantum numbers:
    (a) $n = 3$, $\ell = 2$
    (b) $n = 7$, $\ell = 4$
    (c) $n = 5$, $\ell = 1$

39. (See Example 16–7.) How many radial nodes and how many angular nodes does each of the orbitals in problem 37 have?

40. (See Example 16–7.) How many radial nodes and how many angular nodes does each of the orbitals in problem 38 have?

41. Where is an electron in a $3p_y$-orbital most likely to be found? Where does the probability of finding it vanish?

42. Where is an electron in a $3d_{x^2-y^2}$-orbital most likely to be found? Where does the probability of finding it vanish?

43. What is a quantum number? How many quantum numbers are needed to specify the state of the electron in a Cl^{16+} ion?

44. What is an atomic orbital? How does it resemble and how does it differ from the Bohr planetary orbits?

45. Complete the following table:

n	ℓ	Label	No. of Orbitals
2	1	___	___
___	___	$3d$	___
4	___	___	7

46. Complete the following table:

		No. of Nodes	
n	ℓ	Radial	Angular
1	0	___	___
2	___	1	___
___	___	3	1
___	___	1	3

47. Which of the following combinations of quantum numbers are allowed for an electron in a one-electron atom and which are not?
    (a) $n = 2$, $\ell = 2$, $m_\ell = 1$, $m_s = \frac{1}{2}$
    (b) $n = 3$, $\ell = 1$, $m_\ell = 0$, $m_s = -\frac{1}{2}$

(c) $n = 5$, $\ell = 1$, $m_\ell = 2$, $m_s = \frac{1}{2}$

(d) $n = 4$, $\ell = -1$, $m_\ell = 0$, $m_s = \frac{1}{2}$

48. Which of the following combinations of quantum numbers are allowed for an electron in a one-electron atom and which are not?

(a) $n = 3$, $\ell = 2$, $m_\ell = 1$, $m_s = 0$

(b) $n = 2$, $\ell = 0$, $m_\ell = 0$, $m_s = -\frac{1}{2}$

(c) $n = 7$, $\ell = 2$, $m_\ell = -2$, $m_s = \frac{1}{2}$

(d) $n = 3$, $\ell = -3$, $m_\ell = -2$, $m_s = -\frac{1}{2}$

Additional Problems

49. A piano tuner uses a tuning fork that emits sound with a frequency of 440 s^{-1}. Calculate the wavelength of the sound from this tuning fork and the time the sound takes to travel 10.0 m across a large room. Take the speed of sound in air to be 343 m s^{-1}.

50. The distant galaxy called Cygnus A is one of the strongest sources of radio waves reaching the earth. The distance of this galaxy from the earth is 3×10^{24} m. How long (in years) does it take a radio wave from Cygnus A to reach the earth? If the wavelength of the radio wave is 10 m, what is its frequency?

51. Hot objects can emit blackbody radiation that appears red, orange, white, or bluish white, but never green. Explain.

52. Describe briefly the ways in which the classical physics at the end of the 19th century was unable to account for several important observations about the properties of light, electrons, and atoms.

53. Compare the energy (in joules) carried by an x-ray photon (wavelength $\lambda = 0.20$ nm) with that carried by an AM radio wave photon ($\lambda = 200$ m). Calculate the energy of one mole of each type of photon. What effect do you expect each type of radiation to have in inducing chemical reactions in substances through which it passes?

54. When ultraviolet light of wavelength 131 nm strikes a polished nickel surface, the maximum kinetic energy of ejected electrons is measured to be 7.04×10^{-19} J. Calculate the work function of nickel.

55. Photons are emitted in the Lyman series, as hydrogen atoms undergo transitions from various excited states to the ground state. If ground-state He^+ ions are present in the same gas (near stars, for example), can they absorb these photons? Explain.

* 56. Name a transition in the C^{5+} ion that leads to the absorption of green light.

57. In two-photon ionization spectroscopy, two photons are absorbed by an atom, ion, or molecule to remove an electron from it. Suppose two photons of the same wavelength are used to remove the single electron from a He^+ ion in its $2s$ state. Calculate the maximum wavelength that can be used.

58. Frequency-doubling crystals are used in optics to double the frequency of light that passes through them. Suppose that the output from such a crystal is to be used to excite a hydrogen atom from the $n = 2$ to the $n = 5$ state. What wavelength of input light should be used?

* 59. The energies of macroscopic objects as well as those of microscopic ones are quantized, but the effects of the quantization are not seen because the difference in energy between adjacent states is so small. Apply Bohr's quantization of angular momentum to the revolution of the earth (mass 6.0×10^{24} kg), which moves with a velocity of 3.0×10^4 m s^{-1} in a circular orbit (radius 1.5×10^{11} m) around the sun, which can be treated as fixed. Calculate the value of the quantum number n for the present state of the earth-sun system. What would be the effect of an increase in n by 1?

* 60. Sound waves, like light waves, can interfere with each other, giving maximum and minimum levels of sound. Suppose that a listener standing directly between two loudspeakers hears the same tone being emitted from both. This listener observes that, upon moving one of the speakers a distance of 0.16 m farther away, the perceived intensity of the tone decreases from a maximum to a minimum.

(a) Calculate the wavelength of the sound.

(b) Calculate its frequency, using 343 m s^{-1} as the speed of sound.

61. (a) If the kinetic energy of an electron is known to lie between 1.59×10^{-19} J and 1.61×10^{-19} J, what is the smallest distance within which it can be known to lie?

(b) Repeat the calculation of part (a) for a helium atom instead of an electron.

* 62. It is interesting to speculate about the properties of a universe with different values for the fundamental constants.

(a) In a universe in which Planck's constant had the value $h = 1$ J s, what would be the de Broglie wavelength of a baseball of mass 170 g moving at a speed of 30 m s^{-1}?

(b) Suppose that the velocity of the ball from part (a) is known to lie between 29 and 31 m s^{-1}. What is the smallest distance within which it can be known to lie?

63. How does the $3d_{xy}$-orbital of an electron in an O^{7+} ion resemble, and how does it differ from, the $3d_{xy}$-orbital of an electron in a hydrogen atom?

64. What is the difference between the amplitude of its wave function and the average electron density in a hydrogen atom?

CUMULATIVE PROBLEM

Interstellar Space

The vast stretches of space between the stars are by no means empty. They contain both gases and dust particles at very low concentrations. These low-density contents can significantly affect the electromagnetic radiation that telescopes detect arriving from distant stars and other sources. The gas in interstellar space consists primarily of hydrogen (either neutral or ionized) at a concentration on the order of one atom per cubic centimeter. The dust (thought to be mostly solid water, methane, or ammonia) is even less concentrated, with typically only a few dust particles (10^{-4} to 10^{-5} cm in radius) per cubic *kilometer*.

(a) The hydrogen in interstellar space near a star is largely ionized by the high-energy photons from the star. Such regions are called "H II regions." Calculate the longest wavelength of light that ionizes a ground-state hydrogen atom.

(b) Suppose a ground-state hydrogen atom absorbs a photon with a wavelength of 65 nm. Calculate the kinetic energy of the electron ejected. (*Note*: This is the gas-phase analog of the photoelectric effect for solids.)

(c) What is the de Broglie wavelength of the electron from part (b)?

(d) Ionized electrons in H II regions can be recaptured by hydrogen nuclei. In such events, the atom emits a series of photons of increasing energy as the electrons cascade down through the quantum states of the hydrogen atom. Extremely high quantum states can be detected. In particular, the transition from the state $n = 100$ to $n = 109$ for the hydrogen atom has been detected. What is the Bohr radius of an electron in the state $n = 110$ for hydrogen?

(e) Calculate the wavelength of light emitted as an electron in a hydrogen atom undergoes a transition from level $n = 110$ to $n = 109$. In what region of the electromagnetic spectrum does this lie?

(f) The regions farther from stars are called "H I regions." In these regions, almost all of the hydrogen is neutral rather than ionized and is in its ground state. Do such hydrogen atoms absorb light in the Balmer series emitted from the atoms in the H II regions?

(g) Determine the number of angular nodes and the number of radial nodes in the $109g$ wave function.

(h) According to Section 16–5, the energy of the hydrogen atom depends only on the quantum number n. In fact, this is not quite true. The electron spin (m_s quantum number) couples very weakly with the spin of the nucleus, making the ground state split into two states with slightly different energies. The radiation emitted in a transition from the upper to the lower of these levels has a wavelength of 21.2 cm and is of great importance in astronomy because it allows the H I regions to be studied. What is the energy difference between these two levels, both for a single atom and for a mole of atoms?

(i) The gas and dust particles between a star and the earth absorb the star's light more strongly in the blue region of the spectrum than in the red. As a result, stars appear slightly redder than they actually are. Does an estimate of the temperature of a star based on its apparent color give too high or too low a number?

A cloud of luminous ionized hydrogen atoms surrounds a compact cluster of hot stars.

Many-Electron Atoms and Chemical Bonding

Types of compounds: gaseous covalent nitrogen dioxide (NO_2) and solid ionic calcium fluoride (CaF_2).

Modern quantum mechanics embraces the disquieting notion that matter and energy have a dual nature, part particle and part wave, and from it correctly predicts the atomic spectra of one-electron atoms and ions. This is a substantial achievement, but if it were all that quantum mechanics could do, it would matter little to chemists, because chemistry arises from the interactions of many electrons and nuclei in atoms and molecules. In fact, however, quantum mechanics goes far beyond the Bohr model, making quantitative, verifiable predictions about the properties of larger atoms and their interactions with one another. Equally important for chemistry, quantum mechanics provides a qualitative structure for describing the nature of atoms throughout the periodic table: their sizes, energies, oxidation states, and bonding properties. In this way, it gives a deep underpinning to the simple ideas of bonding used in the Lewis electron-dot model.

At the beginning of Chapter 3, we asked why some combinations of atoms form molecules of great stability, but others join together only for fleeting moments or not at all. We now return to this question, using the insights provided by quantum mechanics. Our first step is to understand the properties of electrons in isolated atoms; as we shall show, the behavior of single atoms provides important clues to the ways in which the atoms combine to form molecules.

17–1 MANY-ELECTRON ATOMS AND THE PERIODIC TABLE

In principle, mathematical solution of the Schrödinger equation gives a complete picture of the distribution of the electrons in any atom, no matter how many electrons there may be. Although the Schrödinger equation can be set up for any atom, solutions in explicit form are, unfortunately, not available, even for the simplest many-electron atom, the helium atom. Despite this, much of what is known about the properties of one-electron wave functions carries over to many-electron atoms by use of the **orbital approximation:**

> The electron density of an isolated many-electron atom is approximately the sum of the electron densities of each of the individual electrons taken separately.

Thus, an electron in a many-electron atom dwells in an atomic orbital with a characteristic size, shape, and energy and has a spin (up or down). These orbitals derive from the atomic orbitals of hydrogen and resemble them in many ways. Their angular dependence is identical, so it is straightforward to associate quantum numbers ℓ and m_ℓ with each atomic orbital. The radial dependence of the orbitals in many-electron atoms differs from that found for one-electron orbitals, but a principal quantum number n can still be defined. The lowest-energy orbital is a $1s$-orbital (and has no radial nodes), the next lowest s-orbital is a $2s$-orbital (with one radial node), and so forth. Each electron in an atom has associated with it a set of four quantum numbers (n, ℓ, m_ℓ, m_s) that describe its spatial distribution and spin state. The allowed quantum numbers follow the same pattern as those for the hydrogen atom, but the relative energies of the orbitals differ.

The Aufbau Principle

The electronic structure of atoms is governed by the **Pauli exclusion principle:**

No two electrons in an atom can have the same set of four quantum numbers (n, ℓ, m_ℓ, m_s).

An equivalent statement of this principle is that each atomic orbital (characterized by a set of three quantum numbers: n, ℓ, and m_ℓ) holds at most two electrons, one with m_s equal to $+\frac{1}{2}$ and the other with m_s equal to $-\frac{1}{2}$. The ground-state electronic structures of atoms are then obtained by arranging the atomic orbitals in order of increasing energy and filling them, one electron at a time, starting with the lowest-energy orbital and heeding the Pauli principle at all times. Starting at the energetic bottom and working up embodies what is called the **aufbau** principle (*Aufbau* is German for "building-up"). Often the atomic orbitals that are to be filled fall into groups that have nearly or exactly equal energies. A group of orbitals with the same energy (such as the three different $2p$-orbitals) comprise a **subshell,** and a group of subshells with similar energies make up a **shell.** The energy ordering of the orbitals is not absolutely fixed, however, but can vary from one kind of atom to another or from an atom to its own ions.

• Two such electrons are said to be "spin-paired."

Building Up from Helium to Argon

Let us see how the building-up procedure works for the atoms from helium through argon. The lowest-energy orbital is always the $1s$-orbital, so ground-state helium has two electrons (with opposite spins) in that orbital. The ground-state **electron configuration,** which shows the orbitals that are occupied in a given atom, is symbolized by $1s^2$ for the helium atom, and is conveniently shown by an orbital diagram, as in Figure 17–1. The $1s$-orbital in the helium atom differs from the $1s$-orbital in the helium *ion*, He^+, by being somewhat larger. In the ion, the electron feels the full nuclear charge, which is $+2e$, but in the atom, each electron partially shields the other electron from the nuclear charge. The orbital in the helium atom can be approximately described in terms of an "effective" nuclear charge Z_{eff} of 1.7, which lies between $+1$ (the value for complete shielding by the other electron) and $+2$ (no shielding).

• In the notation "$1s^2$," the superscript "2" is not an exponent, but tells how many electrons occupy the $1s$-orbital. It is read "one s two," not "one s squared" and not "one s to the second power."

An atom of lithium has three electrons. The third electron does not join the first two in the $1s$-orbital. It occupies a different orbital because, by the Pauli principle, a maximum of two electrons occupies any orbital. Does the third electron go into the $2s$-orbital or the $2p$-orbital? In a one-electron atom, these two orbitals have the same energy because the energy in one-electron atoms depends only on the principal quantum number n. In many-electron atoms, things change: the energies of orbitals depend on ℓ as well as n (but not on m_ℓ). An s-orbital penetrates more toward the nucleus than a p-orbital, so an electron in an s-orbital experiences (at least part of the time) an almost unshielded nuclear attraction. This lowers its energy relative to a p-orbital, which experiences the smaller attraction of a more highly shielded nucleus. The ground state of the lithium atom is therefore $1s^22s^1$, and $1s^22p^1$ is an excited state.

• These factors, which are responsible for the difference in energies, are discussed in Section 16–5.

The next two elements present no difficulties. Beryllium has the ground-state configuration $1s^22s^2$. The $2s$-orbital is now filled, and in boron, with five electrons, the latest addition occupies a $2p$-orbital; boron has the ground-state configuration $1s^22s^22p^1$. Because the three $2p$-orbitals of boron (designated the $2p_x$, $2p_y$, and $2p_z$) have the same energy, there is an equal chance for the electron to be in each one. It is not necessary to specify which of these p-orbitals is occupied in writing the electron configuration.

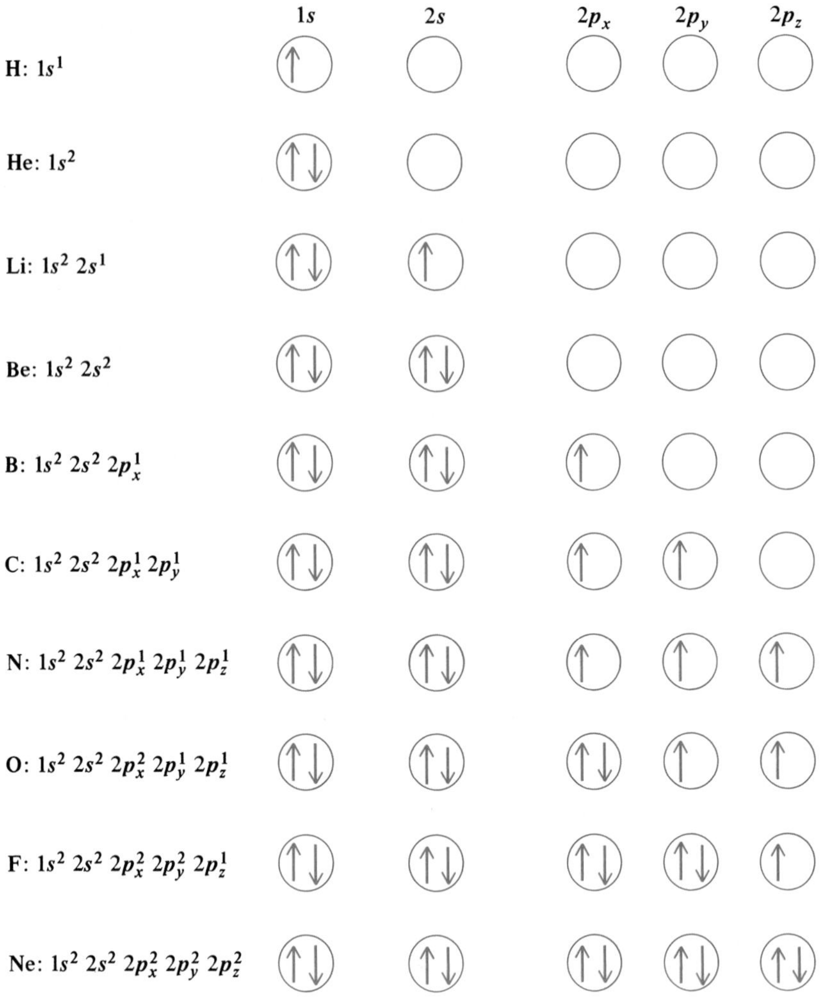

Figure 17–1 The ground-state electron configurations of first- and second-period atoms. The red arrows represent electrons with spin quantum numbers $m_s = +\frac{1}{2}$ or $m_s = -\frac{1}{2}$; each blue circle represents an orbital.

	$1s$	$2s$	$2p_x$	$2p_y$	$2p_z$
H: $1s^1$					
He: $1s^2$					
Li: $1s^2\,2s^1$					
Be: $1s^2\,2s^2$					
B: $1s^2\,2s^2\,2p_x^1$					
C: $1s^2\,2s^2\,2p_x^1\,2p_y^1$					
N: $1s^2\,2s^2\,2p_x^1\,2p_y^1\,2p_z^1$					
O: $1s^2\,2s^2\,2p_x^2\,2p_y^1\,2p_z^1$					
F: $1s^2\,2s^2\,2p_x^2\,2p_y^2\,2p_z^1$					
Ne: $1s^2\,2s^2\,2p_x^2\,2p_y^2\,2p_z^2$					

With carbon, the sixth element in the periodic table, a new question arises. Does the sixth electron go into the same p-orbital as the fifth (for example, both in a $2p_x$-orbital with opposite spins), or does it go into a fresh orbital of equal energy, such as the $2p_y$-orbital, with the same spin? The answer is found in the observation that two electrons occupying different atomic orbitals in the same subshell are more strongly attracted by the positive charge of the nucleus than if the two occupy the same orbital, because the separately occupied orbitals are somewhat smaller. Thus, putting the last two electrons of carbon into two different p-orbitals, say, the $2p_x$ and $2p_y$, leads to lower energy than putting them into the same p-orbital. This is formalized in **Hund's rules:**

> When electrons are added to orbitals of equal energy, a single electron enters each orbital before a second electron enters any orbital. In addition, the spins remain parallel, if possible.

• Two spins are parallel if both are "up" or both are "down."

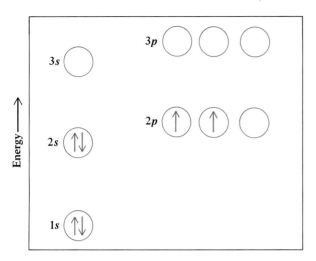

Figure 17–2   For a many-electron atom such as carbon, orbitals with different ℓ-values (such as the 2s and 2p) have different energies. When two or more orbitals have the *same* energy (such as the three 2p-orbitals here), electrons occupy different orbitals with spins parallel.

The lowest-energy electron configuration of carbon is then $1s^2 2s^2 2p_x^1 2p_y^1$ or, more simply, $1s^2 2s^2 2p^2$, with the understanding that two different p-orbitals are intended. The ground-state configuration of carbon is shown in Figure 17–2.

The presence of unpaired electrons in atoms of carbon has immediate implications for the physical behavior of the element. A substance is **paramagnetic** if it is attracted into a magnetic field. All substances that have one or more electrons with unpaired spins in the atoms, molecules, or ions that compose them are paramagnetic. Thus, the ground-state carbon atom is paramagnetic. Substances in which all the electrons are paired are weakly **diamagnetic:** they are pushed *out* of a magnetic field, although the force they experience is orders of magnitude smaller than the force that pulls a typical paramagnetic substance *into* a magnetic field. Of the atoms examined so far, H, Li, B, and C are paramagnetic, but He and Be have all of their electron spins paired and so are diamagnetic.

The electron configurations from nitrogen through neon follow from the stepwise filling of the 2p-orbitals. The six elements from boron to neon are ***p*-block** elements because they appear in the building-up process during the filling of a set of p-orbitals. The two elements that precede them (lithium and beryllium) are ***s*-block** elements, as are hydrogen and helium.

The build-up of the third period, from sodium to argon, is an echo of the second; first the one 3s-orbital is filled and then the three 3p-orbitals. As the number of electrons in an atom reaches 15 or 20, writing full-length electron configurations becomes tedious. It is frequently the practice to include explicitly only those electrons added in the building-up beyond the last preceding noble-gas element. The configuration of that noble gas is then represented by its chemical symbol enclosed in brackets. The ground-state configuration of silicon, for example, is written as $[Ne]3s^2 3p^2$ under this system.

• Effects of magnetic fields on substances are particularly important among compounds of the transition elements and are discussed more extensively in Section 19–3.

EXAMPLE 17–1

Write the ground-state electron configurations of magnesium and sulfur. Are the gaseous atoms of these elements paramagnetic or diamagnetic?

Solution

The noble-gas element preceding magnesium and sulfur is neon. Magnesium has two electrons beyond the neon core. These electrons go into the 3s-orbital, the next lowest in energy, to give the ground-state configuration [Ne]$3s^2$. An atom of magnesium is diamagnetic because all its electrons are paired in orbitals.

Sulfur has six electrons beyond the neon core; the first two are in the 3s-orbital, the next four in the 3p-orbital. Its ground-state configuration is [Ne]$3s^2 3p^4$. In putting four electrons into three p-orbitals, two electrons must occupy one of the orbitals, but the other two occupy different orbitals to reduce electron–electron repulsion. By Hund's rules, their spins are parallel. The sulfur atom is paramagnetic.

Exercise

The ground-state electron configuration of an atom is $1s^2 2s^2 2p^6 3s^2 3p^3$. Name the atom. Is it paramagnetic? If so, state how many unpaired electrons it has.

Answer: The atom is phosphorus, and it is paramagnetic with three unpaired electrons.

Transition-Metal Elements and Beyond

The natural inclination after filling up the three 3p-orbitals with six electrons is to continue the building-up process using the 3d-subshell. The aufbau principle, however, requires that orbitals be filled in order of their increasing energy. For neutral atoms at the beginning of the fourth period, it is an experimental fact that the 4s-orbital ($n = 4$, $\ell = 0$) actually has a lower energy than the 3d-orbital ($n = 3$, $\ell = 2$), although the two energies are very close; the 4s-orbital is therefore filled before the 3d-orbital. The resulting ground-state configurations are [Ar]$4s^1$ for potassium and [Ar]$4s^2$ for calcium. The energy of the 3d-orbitals declines relative to the 4s-orbital with increasing nuclear charge, however. In scandium, the next element after calcium, the 3d-orbital has a lower energy than the 4s-orbital, so an electron is removed more easily from the 4s than the 3d orbital in scandium. Accordingly, the ground state of the Sc^+ ion has the configuration [Ar]$3d^1 4s^1$ rather than [Ar]$4s^2$, and Sc^{2+} has the configuration [Ar]$3d^1$. The drop in the energy of the 3d-orbitals relative to the 4s continues throughout the transition-metal series from scandium to zinc, with the two 4s-orbital electrons always more easily removed than the 3d-electrons. The energy drop has important chemical consequences: the elements copper and zinc rarely have oxidation states higher than +2, and in the elements beyond zinc, the 3d-electrons are held so tightly that they play no chemical role. The ten elements from scandium to zinc are ***d*-block** elements because they appear with filling of d-orbitals in the building-up process.

Experimental evidence shows that two of the transition-metal elements in the fourth period have ground-state electron configurations that are contrary to expectation, based on the configurations that precede them. In its ground state, chromium has the configuration [Ar]$3d^5 4s^1$ rather than [Ar]$3d^4 4s^2$, and copper has the configuration [Ar]$3d^{10} 4s^1$ rather than [Ar]$3d^9 4s^2$. Similar reversals occur in the transition-metal atoms of the fifth period, and some other reversals occur there as well, such as the ground-state configuration [Kr]$4d^7 5s^1$ that is observed for ruthenium in place of the expected [Kr]$4d^6 5s^2$. These anomalous configurations also arise be-

• Once again, the fact that an s-electron (here, 4s) can penetrate to the nucleus lowers its energy relative to that of a d-electron.

• The fifth-period transition metals are also called d-block elements.

1s																	1s
H																	**He**

2s–filling										2p–filling					
Li	**Be**									**B**	**C**	**N**	**O**	**F**	**Ne**

3s–filling										3p–filling					
Na	**Mg**									**Al**	**Si**	**P**	**S**	**Cl**	**Ar**

4s–filling		3d–filling										4p–filling					
K	**Ca**	**Sc**	**Ti**	**V**	**Cr** $3d^5 4s^1$	**Mn**	**Fe**	**Co**	**Ni**	**Cu** $3d^{10}4s^1$	**Zn**	**Ga**	**Ge**	**As**	**Se**	**Br**	**Kr**

5s–filling		4d–filling										5p–filling					
Rb	**Sr**	**Y**	**Zr**	**Nb** $4d^4 5s^1$	**Mo** $4d^5 5s^1$	**Tc**	**Ru** $4d^7 5s^1$	**Rh** $4d^8 5s^1$	**Pd** $4d^{10}$	**Ag** $4d^{10}5s^1$	**Cd**	**In**	**Sn**	**Sb**	**Te**	**I**	**Xe**

6s–filling		5d–filling										6p–filling					
Cs	**Ba**	**Lu**	**Hf**	**Ta**	**W**	**Re**	**Os**	**Ir**	**Pt** $5d^9 6s^1$	**Au** $5d^{10}6s^1$	**Hg**	**Tl**	**Pb**	**Bi**	**Po**	**At**	**Rn**

7s–filling		6d–filling									
Fr	**Ra**	**Lw**	**Rf**	**Ha**	**Sg**	**Bh**	**Hs**	**Mt**	**110**	**111**	**112**

4f–filling													
La $5d^1 6s^2$	**Ce** $4f^2 6s^2$	**Pr**	**Nd**	**Pm**	**Sm**	**Eu**	**Gd** $4f^7 5d^1 6s^2$	**Tb**	**Dy**	**Ho**	**Er**	**Tm**	**Yb**

5f–filling													
Ac $6d^1 7s^2$	**Th** $6d^2 7s^2$	**Pa** $5f^2 6d^1 7s^2$	**U** $5f^3 6d^1 7s^2$	**Np** $5f^4 d^1 7s^2$	**Pu**	**Am**	**Cm** $5f^7 6d^1 7s^2$	**Bk**	**Cf**	**Es**	**Fm**	**Md**	**No**

Figure 17–3 The filling of subshells and the structure of the periodic table. Anomalous ground-state electron configurations are indicated explicitly; all others can be derived from the position of the element in the table.

cause the energy of the *d*-orbitals decreases relative to that of the *s*-orbitals as the charge on the nucleus increases across the period.

In the sixth period, the filling of the 4*f*-orbitals (and the generation of the **f-block** elements) begins as the rare-earth (lanthanide) elements from lanthanum to ytterbium are reached. The energies of the 4*f*-, 5*d*-, and 6*s*-orbitals are comparable over much of the sixth period, so that the order of filling does not follow any simple rules. The sixth period ends with the radioactive noble gas radon. All of the elements in the seventh period are radioactive, including the actinides, the second set of *f*-block elements. The building-up of the elements in this period starts with the 7*s*-orbital and continues with the 5*f*- and 6*d*-orbitals. We would expect the period to conclude with the filling of the 7*p*-orbitals when element 118 is reached. Very heavy nuclei are unstable, however, and no nucleus with an atomic number exceeding 112 has been discovered or synthesized to date. Hence, the expectation remains untested.

The periodic table of Figure 17–3 explicitly indicates the order by which subshells are filled as building up proceeds from left to right across the periods. The ground-state electron configurations of the majority of the elements can be written by locating them in the periodic table and then tracing the filling of the subshells to come up to that location. The correct ground-state configurations of the elements for which this procedure fails are given in Figure 17–3.

Shells and the Periodic Table

A shell is defined as a set of orbitals that have similar energies. A shell structure in atoms is real, as disclosed by the electron densities of atoms, obtained from either experiment or calculation. The electron density of argon (Fig. 17–4) shows clearly the different regions of space occupied by the first three shells, corresponding to the quantum numbers $n = 1$, 2, and 3. There is some spreading out of electron density to the regions between shells, but each peak can clearly be assigned to a shell.

The **valence electrons** are those that occupy the outermost shell, beyond the immediately preceding noble-gas configuration. The nonvalence electrons in inner shells are close to the nucleus and lower in energy than valence electrons; they are called **core electrons.** A **closed-shell atom** is one in which the outermost shell is filled with electrons, so that any excitation (promotion of an electron to a higher-energy orbital) requires a large amount of energy; the noble-gas atoms have closed shells, and the next higher unoccupied orbitals are separated from the occupied orbitals by a significant energy gap.

Two elements in the same group (column) of the periodic table have related valence-electron configurations. An atom's bonding properties and therefore its chemical behavior stem from the configuration of its valence electrons. Sodium (configuration $[Ne]3s^1$) and potassium (configuration $[Ar]4s^1$), for example, both have a single valence electron in an s-orbital outside a closed shell; consequently, the two elements resemble each other closely in their chemical properties. A major triumph of quantum mechanics is its ability to account for such relationships. Mendeleev and others (see Section 3–2) developed the periodic table of the elements purely on the basis of parallel chemical and physical properties. Now we see that the structure of the periodic table can be derived from quantum mechanics. The ubiquitous octets in the Lewis electron-dot structures of second- and third-period atoms and ions arise from the eight available sites for electrons in one s-orbital and three p-orbitals of the valence shell. The special properties of the transition-metal elements can be ascribed to the partial filling of their d-orbitals, a feature to which we return in Chapter 19.

• These terms were previously defined in Section 3-3.

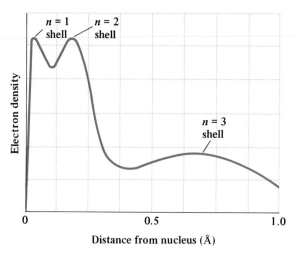

Figure 17–4 The shell structure of the distribution of electrons in an argon atom is clearly visible in the three peaks in its electron density.

17-2 EXPERIMENTAL MEASURES OF ORBITAL ENERGIES

The order of the energies of atomic orbitals determines the building up of electron configurations. How are these energies obtained? An important experimental technique called **photoelectron spectroscopy** allows orbital energies to be directly determined. The technique is simply the photoelectric effect discussed in Section 16–2 applied not to metals but to free atoms or molecules. If radiation of sufficiently high frequency ν strikes an atom, an electron is ejected with kinetic energy $\frac{1}{2} m_e v^2$. If the binding energy of an electron is Φ, then, by the principle of conservation of energy,

$$h\nu = \Phi + \tfrac{1}{2} m_e v^2$$

as shown in Section 16–3. Experiments in photoelectron spectroscopy use high-energy radiation of a fixed frequency (in the ultraviolet or x-ray region of the spectrum), and the kinetic energies of the emitted electrons are observed. From these data, the binding energies of the electrons in the atom are evaluated.

Figure 17–5 shows an experimental spectrum. The binding energy of the most weakly bound electron is called the **ionization energy,** the minimum amount of energy necessary to detach an electron from a ground-state atom. Clearly, no signal in the photoelectron spectrum for binding energies can be less than this. At greater binding energies, the spectrum consists of a series of peaks, each of which corresponds to the removal of an electron from a different orbital in the atom. Neon, for example, has a peak at the binding energy of the $2p$-orbital electrons, then one for the $2s$-orbital electrons, and, finally, at the highest binding energy, one for the $1s$-orbital electrons. Figure 17–6 shows measured orbital energies for electrons in neutral atoms.

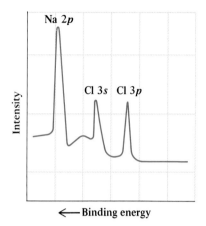

Figure 17–5 Sodium chloride consists of Na^+ ions (with the $[He]2s^2 2p^6$ electron configuration) and Cl^- ions (with the $[Ne]3s^2 3p^6$ configuration). The portion of the photoelectron spectrum of NaCl that covers the binding energies of the outer electrons is shown here. The binding energy increases from right to left. Thus, the core electrons on the two ions, which are more tightly held than the outer electrons, give peaks lying farther to the left. The Na $2s$ electrons lie well off the figure to the left.

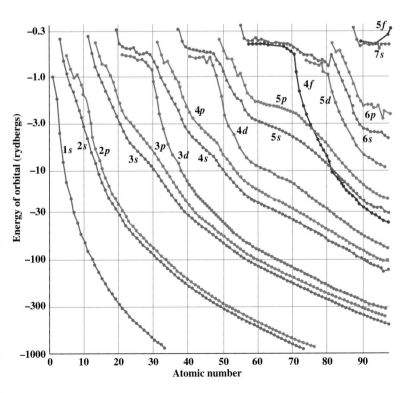

Figure 17–6 The orbital energies of the first 97 elements, based on photoelectron spectroscopy experiments. Orbitals of the same principal quantum number n, such as the $2s$ and $2p$ orbitals, have similar energies and are well separated from orbitals of different n. This confirms the shell structure of many-electron atoms. Note, however, the $3d$ orbitals. In elements 21 through 30 (scandium through zinc), they lie well above the $3s$ and $3p$ orbitals and only slightly below the $4s$ orbitals. They then rapidly become more tightly bound as Z increases further. The $4d$-, $5d$-, and $4f$-orbitals behave similarly.

Figure 17–7 First and second ionization energies of atoms of the first three periods.

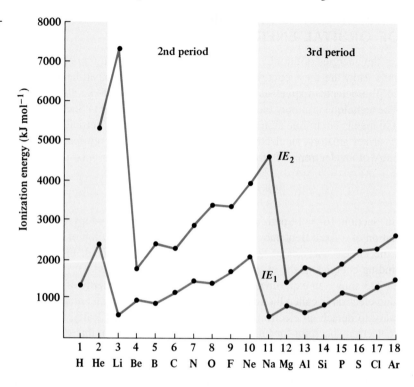

Periodic Trends in Ionization Energies

The ionization energy of an atom has just been defined as the minimum energy necessary to detach an electron from the neutral gaseous atom in its ground state. It is the change in energy (ΔE) for the process

$$X(g) \longrightarrow X^+(g) + e^- \qquad \Delta E = IE_1$$

Ionization energies are always positive because ground-state atoms are stable systems. It requires an input of energy to detach an electron from an atom. The abbreviation "IE" in the equation is subscripted with a 1 because this is the first ionization energy of the atom. The second ionization energy (IE_2) is defined as the minimum energy to remove a second electron, or ΔE for the process

$$X^+(g) \longrightarrow X^{2+}(g) + e^- \qquad \Delta E = IE_2$$

• There are as many ionization energies as there are electrons. The element krypton ($Z = 36$) has IE_1 through IE_{36}, all of which have been measured.

The third, fourth, and higher ionization energies are defined in an analogous fashion. Ionization energies differ from the series of photoelectron binding energies. The photoelectron spectrum of an atomic gas measures the energy required to eject the different electrons from a neutral atom. In contrast, the ionization energies (except for IE_1) are for the removal of electrons from ions, not neutral atoms. Figure 17–7 shows the periodic variation of the first and second ionization energies of the first 18 elements.

The ionization energy influences how an atom interacts with other atoms to form chemical bonds. It is a measure of the stability of the outer-electron configuration: the larger it is, the more energy it takes to remove an outer electron and therefore the more stable is the electronic structure. Let us examine the periodic trends in the first ionization energy (Figs. 17–7 and 17–8) to learn about the stability of the various electron configurations. There is a large drop in IE_1 from helium to

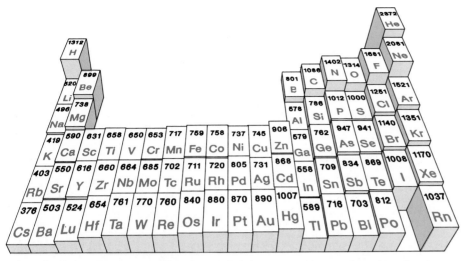

Figure 17–8 The variation of the first ionization energies of atoms throughout the periodic table. The energies are given in kJ mol.$^{-1}$.

lithium. There are two reasons for this decrease: first, a 2s-electron is much farther from the nucleus than a 1s-electron; second, the 1s-electrons screen the nucleus in lithium so effectively that the 2s-electron "sees" a net positive charge close to +1 rather than the larger charge seen by electrons in helium. Beryllium shows an increase in IE_1 because the nuclear charge has increased, but the electron being removed is still from a 2s-orbital.

The IE_1 of boron is somewhat less than that of beryllium because the fifth electron is in a higher-energy (and, therefore, less tightly bound) 2p-orbital. In carbon and nitrogen, the additional electrons go into 2p-orbitals as the nuclear charge increases to hold the outer electrons more tightly; hence, IE_1 increases. The nuclear charge is higher in oxygen than in nitrogen, which would give it a higher ionization energy; however, this is more than compensated for by the fact that ground-state oxygen must accommodate two electrons in a single 2p-orbital, leading to expansion of the p-orbital and weaker electron–nucleus attractions. Oxygen has a lower IE_1 than nitrogen. Fluorine and neon show successively higher first ionization energies as the nuclear charge increases. The general trends of increasing ionization energy across a given period, as well as the dips that occur at certain points, can thus be understood through the orbital description of many-electron atoms.

The ionization energy tends to decrease for the successively heavier elements in a group in the periodic table (for example, from lithium to sodium to potassium). As the principal quantum number increases, so does the distance of the outer electrons from the nucleus. There are exceptions to this trend, however, especially for the heavier elements in the middle of the periodic table. The first ionization energy of gold, for example, is larger than that of silver or copper. This fact is crucial in making gold a "noble metal," one that resists attack by oxygen (Fig. 17–9) and most other oxidizing agents.

Similar trends are observed in the second ionization energy but are shifted higher in atomic number by one unit (see Fig. 17–7). Thus, IE_2 is very high for Li (because Li$^+$ has a filled $1s^2$ shell), but it is relatively low for Be because Be$^+$ has a single electron in the outermost 2s-orbital.

• The ability of the positively charged nucleus to bind electrons is reduced by the repulsions among the electrons themselves. This is screening.

Figure 17–9 Copper is easily oxidized and silver tarnishes, but gold resists oxidation.

Figure 17–10   The electron affinities (in kJ mol^{-1}) of gaseous atoms of the elements. Some of the elements have negative electron affinities; these include the noble gases, which are not visible at the extreme right.

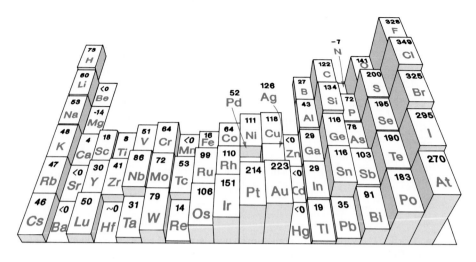

Electron Affinity

• This process is *not* the same as either the process that defines IE_1 or the reverse of that process.

The **electron attachment energy** of an atom is the energy change when the atom in its ground state *gains* a single electron. It equals ΔE for the reaction

$$X(g) + e^- \longrightarrow X^-(g) \qquad \Delta E = \text{electron attachment energy}$$

Electron attachment energies are either positive or negative, depending on the element involved. A negative electron attachment energy means that energy is released as the atom gains an electron and that the resulting anion is more stable than the atom from which it came. A positive electron attachment energy means that the electron must be forced to "stick" and spontaneously flies away from the negative ion if it is in free space. Seventy-five percent of the elements have negative electron attachment energies. The gaseous atoms of these elements "prefer" to gain an electron to become anions; they have an *affinity* for an additional electron. The **electron affinity** (**EA**) is defined as the negative of the electron attachment energy:

$$\text{electron affinity} = -\Delta E \text{ (electron attachment)}$$

• In general, electron affinities decrease in moving down through a group in the periodic table. However, the third-period elements in several cases have higher electron affinities than the corresponding second-period elements. The atom with the largest electron affinity is chlorine rather than fluorine, for example.

The periodic trends in the electron affinity (Fig. 17–10) parallel those in the ionization energy for the most part, except that they are shifted one unit *lower* in atomic number. The reason is clear in the following examples. Attaching an electron to F gives F$^-$, with the configuration $1s^2 2s^2 2p^6$, the same as that of Ne. Fluorine has a large affinity for electrons because the nuclear charge increases steadily as the 2p-orbitals are filled through the second period. This stability is, of course, the same reason that neon has a large ionization energy. Similarly, the noble gases have low (actually negative) electron affinities for the same reason that the alkali metals have small ionization energies: the last electron in Ne$^-$ or Xe$^-$ resides in a new shell far from the nucleus and is almost totally screened from the nuclear charge.

No atom has a positive electron affinity for a *second* electron. A gaseous ion with a net charge of $-2e$ is always unstable with respect to ionization. Attaching a second electron means bringing it close to a species that is already negatively charged. The two repel each other, and the energy rises. Doubly negative ions such as O^{2-}, however, can sometimes be stabilized in crystalline environments by electrostatic interaction with neighboring positive ions, as in CaO(s).

CHEMISTRY IN YOUR LIFE

Fireworks and Neon Signs

The shower of colors in a fireworks display (Figure 17–A) and the cold glare of a neon sign have a common origin: the emission of light from excited states of atoms and ions. In fireworks, strongly exothermic reactions such as the oxidation of magnesium

$$Mg(s) + \tfrac{1}{2} O_2(g) \longrightarrow MgO(s) \qquad \Delta H^\circ = -602 \text{ kJ}$$

not only provide brilliant flashes of light (see Fig. 4–11) but also generate enough heat to raise metal atoms from their ground states to excited states. These excited atoms then emit light as they return to their ground states. The same process occurs when metal atoms are placed in a flame (see Fig. 9–16). Sodium is excited from its ground state (configuration [Ne]3s) to a state with configuration [Ne]3p, in which the valence electron is in a higher-energy 3p-orbital. The transition back to the ground state is accompanied by the emission of yellow light at a wavelength of 589 nm. Red colors in fireworks are produced by excited strontium atoms, violet by potassium, and green by barium, just as in the corresponding flame tests. Blue colors are more difficult to achieve.

Gas discharges are used in many types of lights. The neon sign is the direct descendant of the gas-discharge tubes that figured in the discovery of the electron (see Section 1–5). When an electric current is passed through a tube that has been evacuated to a low gas pressure, ionization occurs, and the atoms or molecules of the gas are excited to higher-energy states. The light emitted by neon is red, and the light emitted by argon is violet (see Fig. 16–9). Other colors are achieved by using colored glass, because other gases are either too expensive or react too readily with the electrodes. In a sodium-vapor light, a small amount of neon is used to start the electrical dis-

Figure 17–A Fireworks.

charge because sodium, the main emitter, is solid at room temperature. As the temperature rises, the sodium is vaporized and excited to give the characteristic yellow light of the [Ne]3$p \longrightarrow$ [Ne]3s transition as its atoms return to the ground state. Similar principles apply to the bluish white light produced by mercury-vapor lamps. Vapor-discharge lamps are more efficient sources of illumination than incandescent filaments (like those in common light bulbs) because more of their output lies in the visible region of the spectrum. Their light is further from true white, however.

Excited states of atoms also figure in the production of light in lasers. In a helium–neon laser, light is emitted as neon atoms relax from one excited state ($1s^2 2s^2 2p^5 5s^1$) to a second, lower state ($1s^2 2s^2 2p^5 3p^1$). In a properly engineered laser, many excited atoms are made to emit light at the same moment, giving an intense coherent beam of light—a laser beam.

EXAMPLE 17–2

Consider the elements selenium (Se) and bromine (Br). Which should have the higher first ionization energy and which the higher electron affinity?

Solution

These two elements are adjacent to one another in the periodic table. Bromine has one more electron in the 4p-subshell, and this electron should be more tightly bound than the 4p-electrons in Se because of the extra unit of positive charge on the nucleus. Thus, IE_1 should be greater for Br.

Bromine has a larger electron affinity than selenium. The additional electron is still in the 4p-subshell, but the nuclear charge is higher for Br, stabilizing the negative ion.

Exercise
Which should have the larger electron affinity—Mg or Al? Explain.

Answer: Al. The Mg atom has a [Ne]$2s^2$ electron configuration. The additional electron must go into a new subshell. In the Al atom, which follows Mg in the periodic table, this next subshell has already been opened.

17-3   SIZES OF ATOMS AND IONS

The sizes of atoms and ions play a major role in determining how they interact in chemical compounds. Size can be estimated in several ways. If the electron density is known from theory or experiment, a contour surface of fixed electron density can be defined, as illustrated in Section 16–5 for one-electron atoms. Alternatively, the packing of atoms and ions in crystals can be examined. If the atoms in a crystal are assumed to be in contact with one another, the measured distances between the centers of the atoms define a size (see Section 20–3). These and other measures of size are reasonably consistent with each other. Figure 17–11 gives the sizes (the atomic

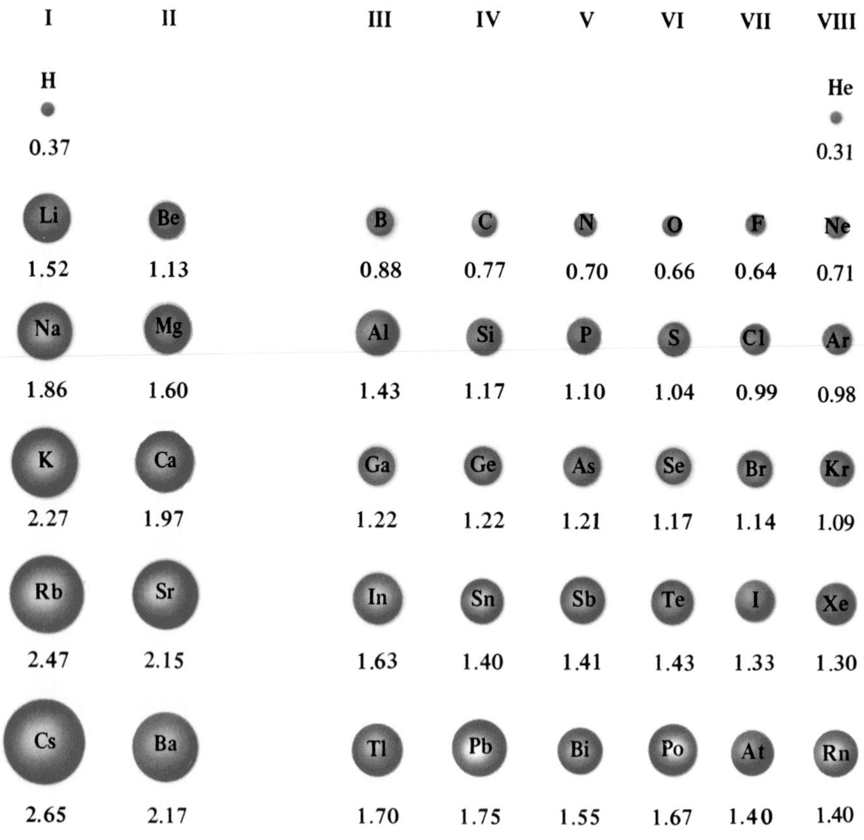

I	II	III	IV	V	VI	VII	VIII
H							He
0.37							0.31
Li	Be	B	C	N	O	F	Ne
1.52	1.13	0.88	0.77	0.70	0.66	0.64	0.71
Na	Mg	Al	Si	P	S	Cl	Ar
1.86	1.60	1.43	1.17	1.10	1.04	0.99	0.98
K	Ca	Ga	Ge	As	Se	Br	Kr
2.27	1.97	1.22	1.22	1.21	1.17	1.14	1.09
Rb	Sr	In	Sn	Sb	Te	I	Xe
2.47	2.15	1.63	1.40	1.41	1.43	1.33	1.30
Cs	Ba	Tl	Pb	Bi	Po	At	Rn
2.65	2.17	1.70	1.75	1.55	1.67	1.40	1.40

Figure 17–11   Radii of atoms of the main-group elements, in ångströms. Atomic radii increase from right to left and from top to bottom in the periodic table. 1 Å = 1×10^{-10} m.

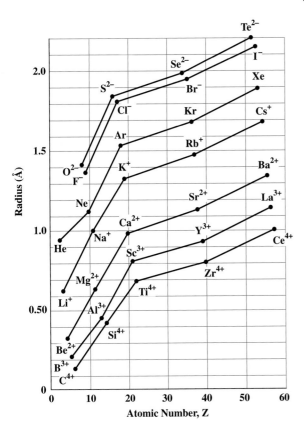

Figure 17–12 Ionic and atomic radii plotted versus atomic number. Each line connects a set of atoms or ions having the same charge; all species have noble gas electron configurations. (The noble gas radii here were estimated on a different basis from those in Figure 17–11; the trend displayed is the same with either set of values.)

radii) of the main-group elements. The sizes of *ions* naturally differ from the sizes of their parent atoms. Cations are always smaller than the atoms from which they derive because fewer electrons take up less space. Anions are always larger because added electrons require additional space.

Certain systematic trends appear among atomic and ionic radii. For a series of atoms (or ions of the same charge) in the same group (column) of the periodic table, the radius usually increases with increasing atomic number. Added electrons cannot go into the close-in orbitals already filled by the core electrons but instead must occupy more distant electron shells, which makes for an increase in size. On the other hand, Coulomb (electrostatic) forces cause the radii of atoms to *decrease* with increasing atomic number across a period. As the nuclear charge increases steadily, electrons successively join the same shell, in which they have about the same distance from the nucleus and are ineffective in shielding each other from its attraction.

Superimposed upon these broad trends are some more subtle effects that have significant consequences for chemistry. One such effect is shown dramatically in Figure 17–12, in which several sets of ionic and atomic radii are shown. The radii increase with atomic number in a given set, as explained earlier, but the *rate* of this increase changes considerably when argon and the ions containing the same number of electrons as argon are reached. The change in size per step from Li^+ to Na^+ to K^+ is quite large, for example, but the subsequent changes, to Rb^+ and Cs^+, are significantly smaller. The filling of the *d*-orbitals, which begins in the fourth period, causes this. Because both atomic and ionic sizes decrease from left to right across a series of transition-metal elements (owing to increased nuclear charge), the radius

• These ions are S^{2-}, Cl^-, Ar, K^+, Ca^{2+}, Sc^{3+}, and Ti^{4+}.

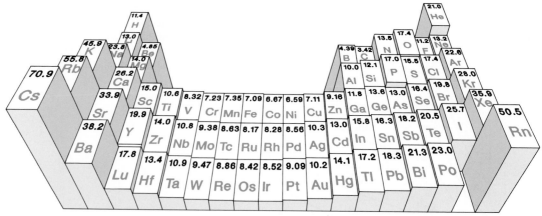

Figure 17–13 The molar volume (in cm^3 mol^{-1}) of some elements in the solid state. Note the high peaks corresponding to the alkali metals. (For higher-melting elements, the data are for room temperature; for lower-melting elements, they are for the highest temperature within the solid range.)

of a main-group element when it is ultimately reached is smaller than it would have been had the transition series not intervened.

A similar effect occurs after the filling of the 4f-orbitals in the lanthanide series and is called the *lanthanide contraction*. It causes the atomic and ionic radii of the post-lanthanide metals of the sixth period to be quite close to those of the metals of the fifth period. Hafnium in the sixth period has an atomic radius that is almost the same as that of zirconium in the fifth period. Because these elements are similar both in valence electron configuration *and* in size, they have quite similar properties and are difficult to separate from one another.

Another measure of atomic size is the volume occupied by a mole of atoms of the element in the solid state. The pronounced periodicity of the molar volume is shown in Figure 17–13, with maxima occurring for the alkali metals. Two factors affect the experimentally measured molar volumes: the inherent size of the atoms and the geometry of the bonding that connects them. The very large molar volumes of the alkali metals stem both from the large size of the atoms and their organization in rather open, loosely packed structures in the solid.

EXAMPLE 17–3

Predict which atom or ion in each of the following pairs should be larger:
(a) Kr or Rb
(b) Y or Cd
(c) F$^-$ or Br$^-$

Solution

(a) Rb should be larger because of its extra electron in a 5s-orbital beyond the Kr closed shell.
(b) Y should be larger because nuclear charge increases through the transition series from Y to Cd.
(c) Br$^-$ should be larger because its additional electrons occupy more distant electron shells.

Exercise

Predict which atom or ion in each of the following pairs should be larger:
(a) Lu or Ir. (b) Zn^{2+} or Cd^{2+}. (c) Cs or Kr.

Answer: (a) Lu. (b) Cd^{2+}. (c) Cs.

17-4 PROPERTIES OF THE CHEMICAL BOND

The noble gas atoms, with their filled valence shells, are quite unreactive because of the high cost in energy to remove or add electrons to form chemical bonds. By contrast, atoms with partially filled valence shells (**open-shell** atoms) can be quite reactive, attaining relative stability by exchanging or sharing electrons with other atoms to form either single molecules or large-scale networks of interacting atoms. Two properties are crucial in characterizing bonds: length and energy. These two properties are related, in turn, to another property: the bond order.

Bond Lengths

The first question to ask about a molecule is: what is its structure? For a diatomic molecule, the only relevant parameter is the bond length, the distance between the nuclei of the two atoms. For a polyatomic molecule, the structure is given by two or more bond lengths and one or more bond angles. Of course, molecules are not rigid structures with unchanging bond lengths and angles; their atoms can vibrate, usually with a small amplitude, around an average position. Many average bond lengths and bond angles are known. They are measured by spectroscopic techniques (discussed in Section 18–4) and by x-ray diffraction methods (introduced in Section 20–1).

Table 17–1 lists the bond lengths of a number of diatomic molecules. Some trends are immediately obvious. Within a group in the periodic table, bond lengths usually increase with increasing atomic number. Thus, the bond length increases in the series F_2 to Cl_2 to Br_2 to I_2. Similarly, Cl—F has a shorter bond length than Cl—Br. These trends are directly related to the corresponding trends in atomic sizes, which usually increase with increasing atomic number in a given group of the periodic table. This qualitative connection cannot be made quantitative, however, be-

Table 17–1
Properties of Some Diatomic Molecules

	Average Bond Length (Å)	Bond Enthalpy (kJ mol^{-1})
H_2	0.751	436
N_2	1.100	945
O_2	1.211	498
F_2	1.417	158
Cl_2	1.991	243
Br_2	2.286	193
I_2	2.669	151
HF	0.926	568
HCl	1.284	432
HBr	1.424	366
HI	1.620	298
ClF	1.632	255
BrF	1.759	285
BrCl	2.139	219
ICl	2.324	211
NO	1.154	632
CO	1.131	1076

Table 17–2
Reproducibility of Bond Lengths

Bond	Molecule	Bond Length (Å)
O—H	H_2O	0.958
	H_2O_2	0.960
	HCOOH	0.95
	CH_2OH	0.956
C—C	diamond	1.5445
	C_2H_6	1.536
	CH_3CHF_2	1.540
	CH_3CHO	1.50
C—H	CH_4	1.091
	C_2H_6	1.107
	C_2H_4	1.087
	C_6H_6	1.084
	CH_3Cl	1.11
	CH_2O	1.06

• This stability in bond length allows great simplification in thinking about chemical bonding. One must be alert for exceptions, however, as in multiple bonding (see later).

cause bond lengths are not simply sums of atomic radii. Too much rearrangement of electron density occurs during bond formation to allow any such approximation.

A striking result from the extensive experimental studies of bond lengths is that a given type of bond changes very little in length from one substance to the next. A C—H bond has almost the same length in acetylsalicylic acid (aspirin, $C_9H_8O_4$) as in methane (CH_4). Table 17–2 shows the lengths for O—H, C—C, and C—H bonds in a number of molecules. They are constant to within a few percent. The shortest known chemical bond is the one between the two hydrogen atoms in the hydrogen molecule. Its length is 0.751 Å. The lengths of the longest chemical bonds range upward to 3.0 Å or more.

Bond Enthalpies

Energetics are crucial in the stability of molecules. A chemical bond forms because the involved atoms have a lower energy when they are close to each other than when they are far apart. The observed molecular geometry is the one that gives the molecule the lowest energy. The energy change ΔE_d when a bond is broken ("d" stands for *dissociation* here) reveals the strength of that bond. It is more common to tabulate the enthalpy change ΔH_d for breaking the bond (that is, the heat absorbed at constant pressure) because ΔH_d is measured directly in calorimetric experiments at constant pressure. This is the **bond enthalpy.** The ΔE_d and ΔH_d per mole of bonds broken are related through

• At room temperature, RT is only 2–3 kJ mol^{-1}.

$$\Delta E_d = \Delta H_d - RT$$

where R is the gas constant, 8.315 J K^{-1} mol^{-1}.

Table 17–1 lists bond enthalpies for selected diatomic molecules. Again, certain systematic trends with changes in atomic number are evident. Bonds generally grow weaker with increasing atomic number, as shown by the decrease in the bond enthalpies of the hydrogen halides in the order HF > HCl > HBr > HI. Note, however, the unusual weakness of the bond in the fluorine molecule F_2 (its bond enthalpy is significantly *smaller* than that of Cl_2 and is comparable to that of I_2). Bond strength decreases dramatically in the diatomic molecules moving from N_2 (945 kJ mol^{-1}) to O_2 (498 kJ mol^{-1}) to F_2 (158 kJ mol^{-1}). What accounts for this behavior? A successful theory of bonding must explain both general trends and the reasons for particular exceptions. We examine these questions further in Chapter 18.

Bond enthalpies, like bond lengths, are fairly reproducible (within about 10%) from one compound to another. It is therefore possible to tabulate *average* bond enthalpies from measurements on a series of compounds (see Table 10–3). For any given bond, the bond enthalpy differs somewhat from that tabulated, but in most cases the deviations are small.

• Average bond enthalpies can be used to estimate the standard enthalpies of formation of compounds, as explained in Section 10-5.

Chemical bonds come in many different kinds and exhibit a considerable range of strength. A rough categorization of bonds by strength is

Weak chemical bonds	Up to approximately 200 kJ mol^{-1}
Average chemical bonds	Centered at 500 kJ mol^{-1}
Strong chemical bonds	Greater than 800 kJ mol^{-1}

One of the very strongest chemical bonds is that between carbon and oxygen in carbon monoxide (1076 kJ mol^{-1}). No exact lower limit to the strength of chemical bonds exists. It is a matter of judgment and circumstances when an interaction between two atoms becomes weak enough to demote it from a chemical bond to a "nonbonded attraction."

Bond Order

Sometimes the length and enthalpy of the bond between two specific kinds of atoms are *not* reproduced from one chemical compound to another. Table 17–3 shows the bond lengths and bond enthalpies of carbon–carbon bonds in ethane (H_3CCH_3), ethylene (H_2CCH_2), and acetylene (HCCH). The differences are great. Carbon–carbon bonds from many other molecules fall into one of the three classes given in the table (that is, some carbon–carbon bond lengths are close to 1.54 Å, others are close to 1.34 Å, and still others are close to 1.20 Å). This confirms the existence of not one, but three types of carbon–carbon bonds. The weakest and longest (as in ethane) is a single bond and is indicated by C—C; that of intermediate strength (as in ethylene) is a double bond, C=C; and the strongest and shortest (as in acetylene) is a triple bond, C≡C. The **bond order** is defined as the number of shared electron pairs, so the bond orders of these three types of bonds are 1, 2, and 3.

• As shown in Section 3-3, the different carbon-carbon bonds have Lewis structures with different numbers of electron pairs shared between the atoms.

Even these three types do not cover all the carbon–carbon bonds found in nature, however. In benzene (C_6H_6), the experimental carbon–carbon bond length is 1.397 Å, and the bond enthalpy is 505 kJ mol^{-1}. This bond is intermediate between

Table 17–3
Three Types of Carbon-Carbon Bonds

Bond	Molecule	Bond Length (Å)	Bond Enthalpy (kJ mol^{-1})
C—C	C_2H_6 (H_3CCH_3)	1.536	348
C=C	C_2H_4 (H_2CCH_2)	1.337	615
C≡C	C_2H_2 (HCCH)	1.204	812

a single and a double bond (bond order $1\frac{1}{2}$). In fact, as is discussed in Section 18–3, the bonding in benzene differs from that in many other compounds. Most bonds have properties that depend almost entirely on the two atoms forming the bond (and thus are similar from one compound to another). In benzene and certain related molecules the bonding depends on the nature of the whole molecule. In sum, the properties of the bond between specific elements are usually quite reproducible from one compound to another. Exceptions often signal a different class of bonding.

17–5   IONIC AND COVALENT BONDS

Chemical bonds arise from the sharing or transfer of electrons between two or more atoms. When electron density is mainly shared between two atoms, then the bond between them is **covalent.** When electron density is mainly transferred from one atom to another, the resulting chemical bond is **ionic.** Ionic and covalent bonds are idealized limits. Most real bonds are neither fully ionic nor fully covalent but instead possess a mixture of ionic and covalent character. Bonds in which a partial transfer of charge occurs are **polar covalent.** In this section we present a classical picture of chemical bonding, deferring a quantum-mechanical description to Chapter 18. To start, we re-introduce a most useful concept—the electronegativity. We used considerations of electronegativity in Chapter 3 as a rough guide to distinguish between ionic and covalent bonds. Now we establish the electronegativity as an atomic property on a more exact basis.

Electronegativity

The electronegativity of an atom is a measure of its power in a molecule to attract electrons to itself. Elements toward the lower left corner of the periodic table have low ionization energies and small electron affinities. This means that they give up electrons readily and accept electrons poorly. They tend to act as electron *donors* in interactions with other elements. In contrast, elements in the upper right corner of the periodic table have high ionization energies and also (except for the noble gases) large electron affinities. As a result, these elements accept electrons easily but give them up only reluctantly: such atoms act as electron *acceptors.*

These facts suggested to chemist Robert Mulliken in 1934 a simple way to assign numerical electronegativities: make them proportional to the average of the ionization energy and the electron affinity of the atom:

$$\text{electronegativity (Mulliken)} \propto \tfrac{1}{2}\,(IE_1 + EA)$$

* The electronegativity of an atom is sometimes confused with its electron affinity. This equation makes it clear that although the electronegativity is related to the electron affinity, the two properties are *not* the same.

Electron acceptors (like the halogens) have high ionization energies and electron affinities and are thus highly electronegative. Electron donors (like the alkali metals) have low ionization energies and electron affinities and therefore low electronegativities; they are electropositive.

By the Mulliken formula, the electronegativity is an invariant atomic property, as accurate as the measured ionization energies and electron affinities from which it is calculated. The "power of an atom in a molecule to attract electrons to itself" in fact depends upon the number and kind of the atom's neighbors, its oxidation state, and other factors. The Mulliken formula can be fine-tuned to deal with some of these factors, but detailed information about atomic environments is often lacking. Tables of *average* electronegativities, in which some kind of typical environment is assumed for each type of atom, suffice for most purposes. The most widely used set of average electronegativities (Fig. 17–14) was derived using a method devised by Linus Pauling. On the Pauling scale, fluorine, the most electronegative el-

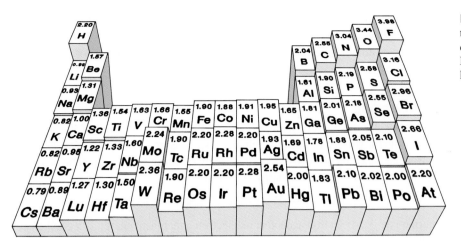

Figure 17–14 Average electronegativities of atoms, computed according to the method developed by Linus Pauling. Electronegativity values have no units.

ement, is assigned a value of 3.98, and cesium, the least electronegative element, has an electronegativity of 0.79. Average electronegativities are useful in exploring periodic trends and making semiquantitative comparisons. They should not be treated as high-precision experimental results.

The periodic trends in the electronegativity, as shown in Figure 17–14, are quite interesting. High electronegativity is favored by small size and a large effective nuclear charge felt by the outer electrons. The electronegativity thus decreases going down the periodic table in the groups at the left and right sides of the table, such as the alkali metals and the halogens. In these groups, atomic size increases sufficiently rapidly with atomic number to make it the predominant factor and to cause the decrease in electronegativity. For a group of late-transition elements such as copper, silver, and gold, however, the size changes only slowly. The rise in effective nuclear charge then predominates so that the electronegativity *increases* going down the group's column in the table.

The absolute value of the difference in the electronegativity of two bonded atoms tells the degree of polarity to be found in their bond. A large difference means that the bond is ionic, and electrons transfer nearly completely to the more electronegative atom. A small difference means that the bond is covalent, so the electrons in the bond are fairly evenly shared. Intermediate values of the difference signify intermediate character in the bond.

• The "late" transition metals are found toward the end of each transition series; the "early" transition metals are found toward the beginning.

EXAMPLE 17–4

Using Figure 17–15, arrange the following bonds in order of decreasing polarity: H—C, O—O, H—F, I—Cl, Cs—Au.

Solution

The differences in electronegativity between the five pairs of atoms (without regard to sign) are 0.35, 0.00, 1.78, 0.50, and 1.75, respectively. The order of decreasing polarity is the order of decrease in this difference. H—F, Cs—Au, I—Cl, H—C, and O—O. The last of these is nonpolar.

Exercise

Which bond has the greatest ionic character: H—Li, N—P, N—S?

Answer: The H—Li bond.

The noble gases participate only rarely in chemical bonding. Their high ionization energies and negative electron affinities mean that they are reluctant either to give up or to accept electrons. Electronegativities are generally not assigned to the noble gases.

Ionic Bonding

Let us consider the energy changes that accompany the formation of ionic bonds. Suppose that a very electropositive element, such as potassium, reacts with a very electronegative element, such as fluorine. Being electropositive, potassium is a good electron donor; its ionization energy is relatively low:

$$K(g) \longrightarrow K^+(g) + e^- \qquad \Delta E = +419 \text{ kJ}$$

Fluorine is a good electron acceptor; its electron attachment energy is a good-sized, negative number:

$$F(g) + e^- \longrightarrow F^-(g) \qquad \Delta E = -328 \text{ kJ}$$

Adding these two equations gives an equation representing the transfer of an electron from potassium to fluorine with the two atoms held far apart. The energy change for the transfer is the sum of the ΔE's:

$$K(g) + F(g) \longrightarrow K^+(g) + F^-(g) \qquad \Delta E_\infty = 419 \text{ kJ} + (-328 \text{ kJ}) = +91 \text{ kJ}$$

The subscript "∞" on ΔE emphasizes that the reactant atoms are infinitely distant from each other. They are "free atoms," just as the product ions are free ions. Even in this favorable case (with very different electronegativities), it *costs* energy to transfer an electron from a potassium atom to a fluorine atom. Examining the ionization energies and electron attachment energies of the elements shows that this is *always* the case. For atoms separated by large distances in the gas phase, electron transfer to form ions is *never* favored energetically.

• The smallest ionization energy (for cesium, 376 kJ mol^{-1}) and the most negative electron attachment energy (for chlorine, -349 kJ mol^{-1}) still add up to a positive ΔE.

How, then, does an ionic bond form? As the two ions approach each other, the attractive Coulomb force between them gives rise to an interaction energy proportional to the product of the charges divided by the separation R of the nuclei:

$$\Delta E_{\text{Coulomb}} = k \frac{Q_1 Q_2}{R}$$

where k is a constant of proportionality. Because Q_1 and Q_2 have opposite signs in this case, this Coulomb energy is negative, and the total energy is *lowered* as the ions move toward each other. At short enough distances, the Coulomb attraction more than compensates for the energy required to transfer the electron, and an ionic bond forms. Figure 17–15 shows the variation of ionic and neutral-state energies with separation R for the case of potassium fluoride. For large separations, a pair of neutral atoms is more stable, but at shorter distances the ionic species may be favored because of the Coulomb attraction. At very short distances, the electron clouds of the two atoms or ions begin to repel each other and the energy rises steeply. The length of an ionic bond is determined by a balance of attractive and repulsive forces. In KF(g) this is at $R = 2.13$ Å. It requires 489 kJ mol^{-1} to break the ionic bond in KF(g).

• Potential energy curves of this type also appear in Section 6–1.

Gaseous molecules of potassium fluoride exist only at high temperatures. Ordinarily, a collection of gaseous molecules of KF interacts further to form an ionic solid. The exothermic change

$$KF(g) \longrightarrow KF(s) \qquad \Delta E^\circ = -241 \text{ kJ mol}^{-1}$$

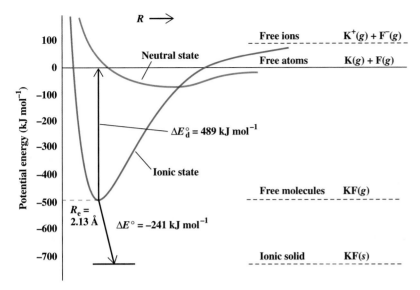

Figure 17–15   The energetics of ionic bonding in potassium fluoride. The curves show the potential energies of the ionic and neutral states of K + F as a function of the distance R between the nuclei. A pair of neutral atoms has lower energy at large separations, but a pair of ions has lower energy at small separations. The lowest energy is for *many* pairs of ions interacting in an ionic solid.

occurs as each positive ion is surrounded by several negative ions, and vice versa. The solid product has a three-dimensional structure in which K^+ and F^- ions occupy alternate adjacent sites. The ionic solid KF(s) has the lowest energy in Figure 17–15.

• The structures of ionic solids are examined in Section 20–3.

Covalent Bonding

Charge transfer followed by Coulomb attraction can account for the formation of ionic bonds. What happens when two atoms with similar or equal electronegativities interact? The H_2 molecule is quite stable, for example, with a bond dissociation energy of 432 kJ mol^{-1}, yet it consists of two identical atoms. There is no possibility of a net charge transfer from one to the other to form an ionic bond. The explanation of the stability of H_2 comes from a different type of bonding, *covalent bonding*, which arises from the sharing of electrons between atoms.

Consider the bonding of the H_2^+ molecular ion. This, the simplest of all possible polyatomic species, is stable but rather reactive. Its properties have been studied both experimentally and theoretically; the H—H bond length R_e is 1.06 Å, and the dissociation energy ΔE_d is 255.5 kJ mol^{-1}. To explain the existence of H_2^+, we use a simple classical model for chemical bonding. The electron is treated as a point negative charge interacting with the two hydrogen nuclei that are separated by a distance R (Fig. 17–16). The electron exerts attractive Coulomb forces on the two nuclei. If the electron lies *between* the two (Fig. 17–16a), then these forces pull the two nuclei together, leading to a bond. The electron screens the positive-to-positive repulsion of the nuclei by interposing its negative charge, to which both are attracted. If the electron in H_2^+ lies *outside* the region between the nuclei (Fig. 17–16b), then the Coulomb attractions between it and the nuclei push the two nuclei apart.

This picture of bonding can be extended to other molecules. Covalent bonds arise when electrons spend most of their time located in the regions between nuclei, that is, when electrons are shared between nuclei. Electrons that spend most of their time outside of such regions actively oppose bonding because they add to the repulsion between the nuclei. The shared valence-electron pairs of the Lewis model (see

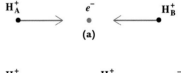

Figure 17–16   Electrons attract hydrogen nuclei because the two have opposite charges. (a) An electron between two hydrogen nuclei causes a net force pulling the nuclei together. (b) An electron outside the internuclear region attracts the nearer nucleus more strongly than the farther, causing a net force that pulls the two nuclei apart.

Chapter 3) are located between two nuclei, where they experience net attractive interactions with each nucleus and strengthen the bond through the mechanism illustrated in Figure 17–16). This simple classical model is developed further in Chapter 18 to a quantum-mechanical description in terms of electrons in molecular orbitals. The fundamental conclusion reached there is the same: covalent bonding arises from electrons shared in the region between two nuclei.

Percent Ionic Character

In a bond that is almost purely ionic, such as that of KF, nearly complete transfer of an electron from the electropositive to the electronegative species takes place. Hence, KF can be represented fairly accurately as K^+F^-, with a charge $+e$ on the positive ion and $-e$ on the negative ion. The charge distribution for a molecule such as HF, with significant covalent character, is more complex. If charges are assigned to its two atoms, then it is best described as $H^{\delta+}F^{\delta-}$, a notation that means that some fraction δ of the full charge $\pm e$ is located at each atom.

- Here δ is a lowercase Greek "delta."

Dipole moments (see Section 3–6) provide a useful measure of ionic character and of electronegativity differences, especially for bonds in diatomic molecules. If two charges of equal magnitude and opposite sign, $+e\delta$ and $-e\delta$ are separated by a distance R, the dipole moment μ (Greek "mu") of that charge distribution is the product

$$\mu = (e\delta)R$$

Dipole moments can be measured experimentally by electrical and spectroscopic methods. In HF, for example, the value of δ calculated from the measured dipole

Table 17–4
Ionic Character of Diatomic Molecules

Molecule	Percent Ionic Character ($100\ \delta$)
H_2	0
CO	2
NO	3
HI	6
ClF	11
HBr	12
HCl	18
HF	41
CsF	70
LiCl	73
LiH	76
KBr	78
NaCl	79
KCl	82
KF	82
LiF	84
NaF	88

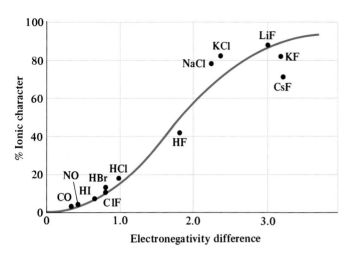

Figure 17–17   Two measures of ionic character for diatomic molecules are the electronegativity difference (From Fig. 17–14) and the percent ionic character 100δ (calculated from the observed dipole moments and bond lengths). The black points represent the actual data, and the blue curve is drawn to guide the eye. The two measures correlate approximately.

moment is 0.41, substantially less than the value of 1 for a purely ionic bond. We convert to a "percent ionic character" by multiplying by 100%, and say that the bond in HF is 41% ionic. Less than 100% ionic bonding occurs for two reasons: (1) incomplete transfer of electrons between atoms, which is to say partially covalent bonding, and (2) distortion of the electron charge distribution around one ion by the electric field of the other ion. This distortion, or **polarization,** of the electron charge alters the dipole moment of the molecule. When polarization is extreme, it is no longer a good approximation to regard the ions as point charges, and a more accurate description of the distribution of electron charge is necessary.

Table 17–4 gives a scale of ionic character for diatomic molecules, based on the definition of δ. For the most part, the proportion of ionic character computed from observed dipole moments and bond lengths parallels the difference in Pauling electronegativity (Fig. 17–17): high ionic character usually corresponds to large differences in electronegativity, with the more electropositive atom carrying the charge $+\delta$. There are exceptions to this general trend, however. Carbon is less electronegative than oxygen, so a charge distribution of $C^{\delta+}O^{\delta-}$ would be predicted for the CO molecule. In fact, the measured dipole moment is quite small in magnitude, and in the other direction: $C^{\delta-}O^{\delta+}$, with $\delta = 0.02$. This discrepancy arises because of the high electron density of the lone pair on the carbon atom. Interestingly, it is forecast by the formal charge of -1 assigned to C in the Lewis structure of CO in Section 3–4. Clearly, tabulated average electronegativities are not appropriate for every possible environment that an atom may have.

EXAMPLE 17–5

The OH radical is a highly reactive species found in flames. Use tabulated electronegativities to predict whether the oxygen or the hydrogen atom should carry a fractional positive charge.

Solution

The hydrogen should carry the positive charge $+\delta$ because its electronegativity (2.20) is smaller than that of oxygen (3.44).

Exercise

Another radical found in flames is CH. (a) Predict which atom in CH should carry a fractional positive charge. (b) Does this radical have more ionic character than the OH radical in Example 17–5, or less?

Answer: (a) Hydrogen should carry the positive charge. (b) Less ionic character than OH.

17–6   OXIDATION STATES AND CHEMICAL BONDING

- A brief refresher definition of oxidation number: the electrical charge that an atom in a molecule would have if all shared electrons were transferred completely to the more electronegative element.

Oxidation states, or numbers (see Section 3–8), are useful guides to systematizing data on the chemical bonds formed by the elements. Figure 17–18 shows the most common oxidation numbers of the elements of the main groups and transition metals. The oxidation numbers range up to $+8$ and down to -3, although oxidation numbers of $-1, -2, -3$ are seen only for the more electronegative elements on the right side of the periodic table. The diagonal red lines in Figure 17–18 emphasize the tendency of the main-group elements to achieve an oxidation state that gives them a closed-shell outer-electron configuration of ns^2np^6 or nd^{10}. For example, the most important oxidation states of sulfur are -2 and $+6$, corresponding to electron configurations of $[Ne]3s^23p^6$ and $[He]2s^22p^6$, respectively. The compounds H_2S and SF_6 have sulfur in these two oxidation states. (It should again be stressed that oxidation state is *not* an indication of the true charge on an atom because, except for the most electropositive and electronegative elements, there is substantial covalent bonding and electron sharing.)

Careful study of Figure 17–18a reveals a second interesting trend: the common positive oxidation state for compounds of the heavier elements in Groups III, IV, V, VI, and VII is often two lower than the maximum. Thus, the chemistry of carbon, silicon, and germanium is dominated by the $+4$ oxidation state, but the $+2$ oxidation state is more important for the later members of the group, tin and lead. In the $+2$ state of these elements, the outer-shell *p*-electrons participate in bonding, but the outer-shell *s*-electrons do not. The halides of the Group IV elements show the point well. The tetrachlorides CCl_4, $SiCl_4$, and $GeCl_4$ are all largely covalently bonded compounds that are quite volatile at room temperature. The next heavier element, tin, forms both $SnCl_2$ and $SnCl_4$. The former, with its $+2$ oxidation state, has mainly ionic bonding and is a solid at room temperature. The latter ($+4$ oxidation state) has largely covalent bonding and is a volatile liquid. Lead, the last element of the group, forms lead(II) chloride, a fairly soluble (1 g/100 g H_2O) ionic compound, but lead(IV) chloride does not exist. A related trend occurs among the Group II elements. Boron, aluminum, and gallium show few signs of a $+1$ oxidation state, but this state is known in indium and is important in the chemistry of thallium, as in thallium(I) chloride (TlCl), a slightly soluble ionic compound.

- A familiar tin(II) compound is stannous fluoride (systematic name tin(II) fluoride, SnF_2), which was formerly added to toothpaste to prevent tooth decay. It has now been replaced by sodium fluoride or sodium monofluorophosphate in most brands.

The transition metals have a great variety of oxidation states in their compounds (Fig. 17–18b). The early members of each period show maximum oxidation numbers that correspond to the participation of all the outer *s*-orbital and *d*-orbital electrons in ionic or covalent bonds. Thus, elements in the scandium group have the valence electron configuration $ns^2(n-1)d^1$ and show only the $+3$ oxidation state. Manganese has the valence electron configuration $3d^54s^2$ and has a maximum oxidation state of $+7$, as do the other elements in its group; the group forms compounds like perrhenic acid (HReO$_4$) and the dark-red liquid manganese(VII) oxide (Mn$_2$O$_7$). Lower oxidation states occur as well, of course (Fig. 17–19). Other ox-

(Text continues on p. 709.)

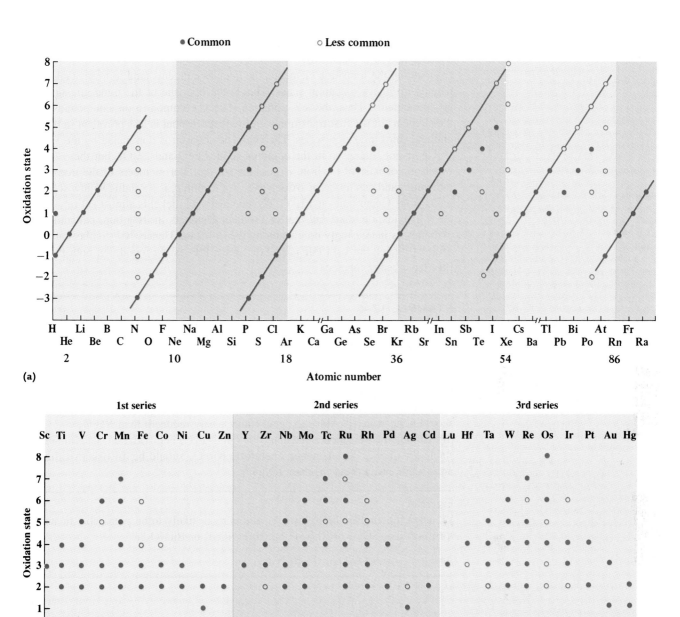

(a)

(b)

Figure 17–18 Some of the states found in compounds of (a) the main-group elements and (b) the transition-metal elements. All elements have oxidation states of zero in their elemental form, but these states are omitted except for the noble gases.

Figure 17–19 Several oxides of manganese. They are arranged in order of increasing oxidation number of the Mn, with MnO at the bottom; Mn_2O_3 and Mn_3O_4 at the left and right, respectively; and MnO_2 at the top. A compound of still higher oxidation state, Mn_2O_7, is a dark red liquid that explodes easily.

ides of manganese include the green solid MnO ($+2$ oxidation state) and the black solids Mn_2O_3 ($+3$) and MnO_2 ($+4$).

Among the transition elements, the *heavier* elements tend to form stable compounds in *higher* oxidation states. This is the opposite of the trend among main-group elements. Thus, the $+2$ oxidation state is common in the compounds of all the elements of the first transition series (except Sc) but is less prevalent in the second and third transition series. The chemistry of iron is dominated by the $+2$ and $+3$ oxidation states, as in the common oxides FeO and Fe_2O_3, but the $+8$ state, which is nonexistent for iron, is possible for the later members of the iron group, ruthenium and osmium. The oxide OsO_4, for example, is a volatile yellow solid that melts at $41°C$ and boils at $131°C$. The chemistry of nickel is almost entirely that of the $+2$ oxidation state, but that of the later elements in its group, palladium and platinum, is increasingly dominated by the $+4$ state. Hence, NiF_2 is the stable fluoride of nickel, but both PdF_2 and PdF_4 exist; PtF_2 is *not* found, but both PtF_4 and PtF_6 have been prepared.

EXAMPLE 17–6

Predict which compound in each pair is likely to be the stronger oxidizing agent: (a) $NaBiO_3$ or $NaPO_3$. (b) $KMnO_4$ or $KReO_4$.

Solution

(a) Among main-group elements, compounds in higher oxidation states are *less* stable lower down the periodic table. Bi lies below P in Group V. Hence $NaBiO_3$ should be a stronger oxidizing agent (more easily reduced) than $NaPO_3$.

(b) Among the transition elements, compounds in higher oxidation states are *more* stable for the heavier elements. Therefore, $KReO_4$ should be a weaker oxidizing agent (less easily reduced) than $KMnO_4$.

Exercise

Which compound in the following pairs is more likely to be preparable in stable form: (a) $Co_2(SO_4)_3$ or $Rh_2(SO_4)_3$; (b) $P(NO_3)_3$ or $Bi(NO_3)_3$.

Answer: (a) $Rh_2(SO_4)_3$ (b) $Bi(NO_3)_3$.

The Strengths of Oxoacids

Oxidation states, electronegativities, and bond polarities are interrelated properties. When the electronegativity difference between two atoms becomes small, the bond between them becomes more covalent, and the bond dipole moment is correspondingly reduced. Also, as the oxidation state of a chemical element increases, its bonds with an electronegative element such as oxygen become increasingly covalent. The two observations together imply that an element in a high oxidation state is effectively more electronegative than the same element in a low oxidation state.

These general statements have practical chemical consequences, one of which concerns the acidity constants K_a of oxoacids. Recall from Chapter 8 that the larger the value of K_a, the stronger the acid. An oxoacid is one in which the acidic hydrogen atoms are bonded to an oxygen atom. An example is sulfuric acid, for which one resonance Lewis structure is

$$\overset{\displaystyle :\overset{..}{O}:H}{\underset{\displaystyle :\overset{..}{O}:H}{:\overset{..}{O}=S=O:}}$$

In this oxoacid, two hydrogen atoms are bonded to oxygen atoms that in turn are bonded to the central sulfur atom. Each —OH group donates a H^+ to a water molecule, yielding HSO_4^- and then SO_4^{2-}. Other important oxoacids include nitric acid (HNO_3) and phosphoric acid (H_3PO_4).

Let us examine how the strength of an oxoacid depends on the electronegativity and oxidation state of the central atom. If this atom is given the symbol "X," then the oxoacid has the structure

$$—X—O—H$$

where X can be bonded to additional —OH groups, to oxygen atoms, or to hydrogen atoms. How does the strength of the oxoacid change with the electronegativity of X? Consider first the extreme case in which X is a highly electropositive element, such as the alkali metal sodium. Now, NaOH is not an acid at all, but a base. The sodium atoms in Na—O—H each give up a full electron to make Na^+ and OH^- ions. Because the X—O bond here is almost completely ionic, the OH^- group has a net negative charge that holds the H^+ tightly to the oxygen and prevents formation of H^+ ions. The alkaline earth elements behave similarly. They form hydroxides (such as $Mg(OH)_2$) that are somewhat weaker bases than NaOH, but they do not act at all as acids.

What happens as the central atom X becomes more electronegative, reaching values between 2 and 3, as in the oxoacids of the elements B, C, P, As, S, Se, Br, and I? As X gets better at withdrawing electron density from the oxygen atom, the X—O bond becomes more covalent. This leaves less negative charge on the oxygen atom, and consequently the oxoacid releases H^+ more readily (Fig. 17–20). Other things being equal, the acid strength should increase with the electronegativity of the central atom. This is observed and is illustrated by the data in Table 17–5 for series of oxoacids with the same structures. To summarize:

Oxoacids of the same structure show increasing acid strength as the electronegativity of the central atom increases.

Now suppose that several oxoacids have the same central atom but the oxidation state of that atom changes. The higher oxidation states correspond to more co-

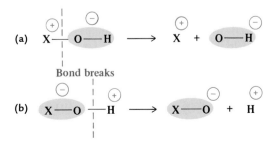

(a) **Bond breaks**

(b)

Figure 17-20 In (a) the atom X is electropositive, so extra electron density (shown in blue) accumulates on the OH group. The X—O bond then can break easily, making the compound a base. In (b), the atom X is electronegative, so electron density is withdrawn from the H atom to the X—O bond. In this case, it is the O—H bond that breaks easily, and the compound is an acid.

Table 17–5
Acid Ionization Constants for Oxoacids of the Non-metals

$X(OH)_m$ Very Weak	K_a	$XO(OH)_m$ Weak	K_a	$XO_2(OH)_m$ Strong	K_a	$XO_3OH)_m$ Very Strong	K_a
$Cl(OH)$	3×10^{-8}	$H_2PO(OH)$	8×10^{-2}	$SeO_2(OH)_2$	1×10^3	$ClO_3(OH)$	2×10^7
$Te(OH)_6$	2×10^{-8}	$IO(OH)_5$	2×10^{-2}	$ClO_2(OH)$	5×10^2		
$Br(OH)$	2×10^{-9}	$SO(OH)_2$	2×10^{-2}	$SO_2(OH)_2$	1×10^2		
$As(OH)_3$	6×10^{-10}	$ClO(OH)$	1×10^{-2}	$NO_2(OH)$	20		
$B(OH)_3$	6×10^{-10}	$HPO(OH)_2$	1×10^{-2}	$IO_2(OH)$	1.6×10^{-1}		
$Ge(OH)_4$	4×10^{-10}	$PO(OH)_3$	8×10^{-3}				
$Si(OH)_4$	2×10^{-10}	$AsO(OH)_3$	5×10^{-3}				
$I(OH)$	4×10^{-11}	$SeO(OH)_2$	3×10^{-3}				
		$TeO(OH)_2$	3×10^{-3}				
		$NO(OH)$	5×10^{-4}				

valent X—O bonds and thus should show higher acidity constants. An even better correlation than oxidation state, however, is a related quantity: the number of lone oxygen atoms attached to the central atom. We can state the following rule:

> The acid strength of oxoacids with a given central element increases with the number of lone oxygen atoms attached to the central atom.

If the formula of these acids is rewritten as $XO_n(OH)_m$, then the acid strengths fall into distinct classes, according to the value of n, which is the number of lone oxygen atoms (see Table 17–5). Each increase by 1 in n leads to an increase in the acidity constant K_a by a factor of about 10^5. Another way to describe this effect is to focus on the stability of the conjugate base, $XO_{n+1}(OH)^-_{m-1}$, of the oxoacid. The larger the number of lone oxygen atoms attached to the central atom, the more easily the net negative charge can be spread out over the ion, and, therefore, the more stable the base is. This leads to a larger K_a.

• Note that Table 17–5 includes many familiar acids written in unfamiliar forms. Thus, H_2SO_4 is listed as $SO_2(OH)_2$, HNO_3 is $NO_2(OH)$, and so forth.

EXAMPLE 17–7

Identify the stronger acid in each of the following pairs:
(a) $HClO_2$, HIO_2
(b) $HClO_3$, $HClO$

Solution

(a) Because chlorine is more electronegative than iodine, $HClO_2$ should be a stronger acid than HIO_2.
(b) $HClO_3$ can be written as $ClO_2(OH)$, with two lone oxygen atoms on the chlorine atom. It is a stronger acid than $Cl(OH)$, in which there are no lone oxygen atoms.

Exercise
Identify the stronger acid in each pair: (a) H_2MoO_4, H_2CrO_4. (b) $HMnO_4$, H_2MnO_4.

Answer: (a) H_2MoO_4. (b) $HMnO_4$.

One unusual and interesting structural result can be obtained from Table 17–5. The simplest Lewis structure for the oxoacid with formula H_3PO_3 is shown in Figure 17–21a. Note that each atom in it achieves an octet configuration. Such a structure could be formulated $P(OH)_3$ and is analogous to $As(OH)_3$, which has no lone oxygen atoms bonded to the central atom ($n = 0$). The expected value of K_a for $P(OH)_3$ based on this analogy is on the order of 10^{-9} (a very weak acid). In fact, however, H_3PO_3 is a much stronger acid ($K_a = 1 \times 10^{-2}$). It fits better into the class of acids with one lone oxygen atom bonded to the central atom. This analysis based on chemical properties is shown to be correct by x-ray diffraction measurements; the true structure of H_3PO_3 is best represented by Figure 17–21b! This corresponds either to a Lewis structure with more than eight electrons around the central phosphorus atom or to one with formal charges on the central phosphorus and lone oxygen atoms. The formula of this acid is written as $HPO(OH)_2$ in Table 17–5. Unlike phosphoric acid (H_3PO_4), which is a triprotic acid, H_3PO_3 is a diprotic acid. The third hydrogen atom, the one bonded directly to the phosphorus atom, is not lost even in strongly basic aqueous solution.

Further Comparisons of Acid Strength

Consider the trends in the acidity constants of the binary hydrides in Groups V through VII of the periodic table (these are not oxoacids, but compounds like NH_3, H_2O, and HF). The K_a's increase from left to right in the periodic table and also increase from top to bottom. The horizontal trend is understood by noting that a central atom of higher electronegativity withdraws more electron density from the vicinity of the hydrogen atom, making it easier to donate as an H^+ ion. However, the vertical trend in the K_a's does *not* match the decrease in electronegativity going down the columns in this part of the table: H_2S is *more* acidic than H_2O, although S is *less* electronegative than O. Similarly in Group VII, the K_a's increase in the order $HF < HCl < HBr < HI$ but the electronegativity of the halogen decreases. The unconsidered factor is the greater **polarizability** (susceptibility to polarization) of larger atoms. In large atoms, the distribution of the electrons is more easily distorted by approaching electric charges or dipoles than it is in small atoms. Thus, the bonding electrons in HI are more readily pushed back by negative charge on an approaching base than are the bonding electrons in HF. The result is that H^+ ion is more easily extracted, and K_a is larger.

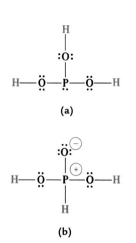

Figure 17–21   (a) The simplest Lewis diagram that can be drawn for H_3PO_3 is plausible but not in accord with the facts of the substance's chemical behavior. This acid would be triprotic, like H_3PO_4. (b) The observed structure of H_3PO_3. The hydrogen atom attached to the phosphorus atom is not released into acidic solution, so the acid is diprotic.

SUMMARY

17–1 According to the **orbital approximation,** electrons in many-electron atoms occupy orbitals resembling those of the hydrogen atom in their shapes and the number of their nodes. The **Pauli exclusion principle** states that at most two electrons (with opposite spin) can occupy any such atomic orbital. As the number of electrons increases with increasing atomic number in the periodic table, the ground-state electron configurations of successive atoms are built up (according to the **aufbau principle**) by the systematic filling of **subshells** and **shells** of orbitals, starting with those of lowest energy. **Closed-shell** ions or atoms (such as the noble gases) are those for which the next atomic orbitals available for occupancy by electrons are separated by a large energy gap from the highest occupied orbitals. Such species are comparatively unreactive. **Open-shell** ions and atoms tend to interact chemically to attain closed-shell configurations.

17–2 Orbital energies can be measured by **photoelectron spectroscopy,** in which the kinetic energies of electrons ejected by light are determined. The minimum energy to eject an electron is the **first ionization energy,** which generally increases across a period and decreases through a group in the periodic table. The **electron attachment energy** of an atom is the energy change that occurs when the atom gains an electron. The **electron affinity** is the negative of the electron attachment energy. Periodic trends in the electron affinity generally parallel those in the first ionization energy.

17–3 The sizes of atoms and ions generally increase moving down columns in the periodic table and decrease moving from left to right across rows. The interpolation of d-block and f-block elements leads to a relative reduction in size of the atoms or ions that follow them in a row of the periodic table.

17–4 A **bond length** is an internuclear distance between bonded atoms; a **bond enthalpy** is the enthalpy change in rupturing a bond. Bond lengths and enthalpies involving a given pair of elements stay constant to within a few percent. Multiple bonds cause exceptions, but even then, a similar constancy prevails within the set of doubly or triply bonded atoms.

17–5 The electronegativity of an atom is a measure of its power in a molecule to attract electrons to itself. Atoms of very different electronegativities form ionic bonds in which electrons are transferred and stability is achieved from Coulomb attraction; atoms of similar electronegativities form polar covalent bonds in which the sharing of electrons between the two nuclei is responsible for stability. The percent ionic character in a polar covalent bond is related to difference in electronegativity of the bonded atoms.

17–6 The most important oxidation states of the elements vary systematically through the periodic table. Among main-group elements, lower oxidation states predominate among the heavier elements, but the reverse is true for transition-metal elements. Trends in electronegativity and oxidation states can account for many relationships in chemical behavior. An example is oxoacid strength, which increases with increasing electronegativity of the central atom and with an increasing number of lone oxygen atoms attached to the central atom. High **polarizability** in the atom to which a hydrogen atom is bonded increases the acidity of that hydrogen.

PROBLEMS

Note: Answers to blue-numbered problems are given in Appendix F. Problems that are more challenging are indicated with asterisks.

Many-Electron Atoms and the Periodic Table

1. (See Example 17–1.) Give the ground-state electron configurations of the following elements:
   (a) Si   (b) S   (c) Co

2. (See Example 17–1.) Give the ground-state electron configurations of the following elements:
   (a) N   (b) Mn   (c) Eu

3. Write ground-state electron configurations for the following ions: Be^+, C^-, Ne^{2+}, Mg^+, P^{2+}, Cl^-, As^+, I^-. Which are paramagnetic due to the presence of unpaired electrons?

4. Write ground-state electron configurations for the following ions: Li^-, B^+, F^-, Al^{3+}, S^-, Ar^+, Br^+, Te^-. Which are paramagnetic due to the presence of unpaired electrons?

5. Identify the atom or ion corresponding to each of the following descriptions:
   (a) An atom with ground-state electron configuration $[Kr]4d^{10}5s^25p^1$.
   (b) An ion with charge -3 and ground-state electron configuration $[Ne]3s^23p^6$.
   (c) An ion with charge $+2$ and ground-state electron configuration $[Ar]3d^3$.

6. Identify the atom or ion corresponding to each of the following descriptions:
   (a) An atom with ground-state electron configuration $[Xe]4f^{14}5d^66s^2$.
   (b) An ion with charge -2 and ground-state electron configuration $[He]2s^22p^6$.
   (c) An ion with charge $+3$ and ground-state electron configuration $[Kr]4d^6$

7. Predict the atomic number of the (as yet undiscovered) element in the seventh period that is a halogen.

8. Predict the atomic number of the (as yet undiscovered) alkaline earth element in the eighth period. Suppose that the eighth-period alkaline earth element is discovered and turns out to have atomic number 138. Explain. (*Hint*: Recall that the atomic number of radium is only 88.)

9. Suppose that the spin quantum number did not exist, so that only one electron could occupy each orbital of a many-electron atom. Give the atomic numbers of the first three noble-gas atoms in this case.

10. Suppose that the spin quantum number had three allowed values ($m_s = 0, +\frac{1}{2}, -\frac{1}{2}$). Give the atomic numbers of the first three noble-gas atoms in this case.

Experimental Measures of Orbital Energies

11. (See Example 17–2.) For each of the following pairs of atoms, state which you expect to have the higher first ionization energy:
    (a) Rb or Sr   (c) Xe or Cs
    (b) Po or Rn   (d) Ba or Sr

12. (See Example 17–2.) For each of the following pairs of atoms, state which you expect to have the higher first ionization energy:
    (a) Bi or Xe   (c) Rb or Y
    (b) Se or Te   (d) K or Ne

13. (See Example 17–2.) For each of the following pairs of atoms, state which you expect to have the greater electron affinity:
    (a) Xe or Cs   (c) Ca or K
    (b) Pm or F   (d) Po or At

14. (See Example 17–2.) For each of the following pairs of atoms, state which you expect to have the greater electron affinity:
    (a) Rb or Sr   (c) Ba or Te
    (b) I or Rn   (d) Bi or Cl

15. Suppose that the spin quantum number did not exist (see problem 9). Sketch the variation of the first ionization energy with atomic number Z for values up to $Z = 8$ in this case.

16. Suppose that the spin quantum number had three allowed values ($m_s = 0, +\frac{1}{2}, -\frac{1}{2}$; see problem 10). Sketch the variation of the first ionization energy with atomic number Z for values up to $Z = 12$ in this case.

17. The cesium atom has the lowest ionization energy of all the neutral atoms in the periodic table, 375.5 kJ mol^{-1}. What is the longest wavelength of light that can ionize a cesium atom? In which region of the electromagnetic spectrum does this fall?

18. Until recently, it was thought that the Ca$^-$ ion was unstable, so that the calcium atom had a negative electron affinity. Some new experiments have now measured an electron affinity of $+4.1$ kJ mol^{-1} for Ca. What is the longest wavelength of light that can remove an electron from a Ca$^-$ ion? In which region of the electromagnetic spectrum does this fall?

Sizes of Atoms and Ions

19. Why do atomic radii increase from top to bottom within a group in the periodic table?

20. Why do atomic radii in general decrease from left to right within a period in the periodic table?

21. (See Example 17–3.) For each of the following pairs of atoms or ions, state which you expect to have the larger radius:
    (a) Li or Rb   (d) K or Ca
    (b) K or K$^+$ (e) Ne or O^{2-}
    (c) Rb$^+$ or Kr

22. (See Example 17–3.) For each of the following pairs of atoms or ions, state which you expect to have the larger radius:
    (a) Mn or Mn^{2+} (d) Ge or As
    (b) Mg or Ca   (e) Ba$^+$ or Cs
    (c) I$^-$ or Xe

23. (See Example 17–3.) Predict the larger ion in each of the following pairs:
    (a) O$^-$, S^{2-} (c) Mn^{2+}, Mn^{4+}
    (b) Co^{2+}, Ti^{2+} (d) Ca^{2+}, Sr^{2+}
    Give reasons for your answers.

24. (See Example 17–3.) Predict the larger ion in each of the following pairs:
    (a) S^{2-}, Cl$^-$ (c) Ce^{3+}, Dy^{3+}
    (b) Tl$^+$, Tl^{3+} (d) S$^-$, I$^-$
    Give reasons for your answers.

Properties of the Chemical Bond

25. Predict how the bond length and bond enthalpy vary in a series of diatomic compounds involving one element in combination with successive elements that are members of a group in the periodic table. Are there exceptions to these general trends?

26. Describe how the bond length and bond enthalpy vary through a series of compounds in which the bonded atoms remain the same but the bond order increases.

27. The bond lengths of the X—H bond in NH$_3$, PH$_3$, and SbH$_3$ are 1.02, 1.42, and 1.71 Å, respectively. Estimate the length of the As—H bond in AsH$_3$, the gaseous compound that decomposes on a heated glass surface in Marsh's test for arsenic. Which of these four hydrides has the weakest X—H bond?

28. Arrange the following covalent diatomic molecules in order of the length of the bond: BrCl, ClF, IBr. Which of the three has the weakest bond (the smallest bond enthalpy)?

CHAPTER 17

Ionic and Covalent Bonds

29. Ionic compounds tend to have higher melting and boiling points and to be less volatile (that is, have lower vapor pressures) than covalent compounds. Use electronegativity differences to predict which of the following pairs of compounds has the higher vapor pressure at room temperature:
    (a) CI_4 and KI
    (b) BaF_2 and OF_2
    (c) SiH_4 and NaH

30. Use electronegativity differences to predict which compound in each of the following pairs has the higher boiling point:
    (a) $MgBr_2$ and PBr_3
    (b) OsO_4 and SrO
    (c) Cl_2O and Al_2O_3

31. Use the data in Figures 17–8 and 17–10 to compute the energy changes (ΔE) of the following pairs of reactions:
    (a) $K(g) + Cl(g) \longrightarrow K^+(g) + Cl^-(g)$ and
        $K(g) + Cl(g) \longrightarrow K^-(g) + Cl^+(g)$
    (b) $Na(g) + Cl(g) \longrightarrow Na^+(g) + Cl^-(g)$ and
        $Na(g) + Cl(g) \longrightarrow Na^-(g) + Cl^+(g)$

32. Use the data in Figures 17–8 and 17–10 to compute the energy changes (ΔE) of the following pairs of reactions:
    (a) $Na(g) + I(g) \longrightarrow Na^+(g) + I^-(g)$ and
        $Na(g) + I(g) \longrightarrow Na^-(g) + I^+(g)$
    (b) $K(g) + Cl(g) \longrightarrow K^+(g) + Cl^-(g)$ and
        $K(g) + Cl(g) \longrightarrow K^-(g) + Cl^+(g)$
    Explain why Na^+I^- and K^+Cl^- form in preference to Na^-I^+ and K^-Cl^+.

33. The percent ionic character of a bond can be approximated by the formula $16\Delta + 3.5\Delta^2$, where Δ is the magnitude of the difference in the electronegativities of the atoms (see Fig. 17–14). Calculate the percent ionic character of HF, HCl, HBr, HI, and CsF, and compare the results with those of Table 17–4.

34. The percent ionic character of the bonds in several interhalogen molecules (as estimated from their measured dipole moments and bond lengths) are ClF (11%), BrF (15%), BrCl (5.6%), ICl (5.8%), and IBr (10%). Estimate the percent ionic characters for each of these molecules using the equation in the preceding problem, and compare them with the given values.

35. (See Examples 17–4 and 17–5.)
    (a) Use electronegativities to arrange the following bonds in order of decreasing polarity: N—O, N—N, N—P, and C—N.
    (b) Predict the atom that carries the fractional positive charge in each case.

36. (See Examples 17–4 and 17–5.) Among the diatomic molecules formed by the halogen atoms are IF, ICl, ClF, BrCl, and Cl_2.
    (a) Use electronegatitivies from Figure 17–14 to rank the bonds in these compounds from least ionic to most ionic in character.
    (b) Predict the atom that carries the fractional positive charge in each case.

Oxidation States and Chemical Bonding

37. Predict the highest oxidation state for the element in question in the compounds of each of the following elements: V, P, I, Sr.

38. Predict the highest oxidation state for the element in question in the compounds of each of the following elements: W, At, Xe, Fr.

39. Suggest why the $6s$ electrons in lead tend not to participate in bonding (making the $+2$ oxidation state more prevalent than the $+4$ state), but the $2s$ electrons in carbon, which is in the same column of the periodic table, generally do participate in bonding.

40. Iron almost invariably occurs in its compounds in the $+2$ or $+3$ oxidation state, but osmium does exhibit the $+8$ oxidation state in its compounds. Why is it easier for osmium to share its $6s$ and $5d$ electrons than it is for iron to share its $4s$ and $3d$ electrons?

41. Explain why Mn_2O_7, RuO_4, and OsO_4 have quite low melting points (6°C, 25°C, and 40°C, respectively) compared to the melting points of most transition-metal oxides.

42. Tin(IV) chloride is a liquid at room temperature and pressure, whereas tin(II) chloride is a solid under the same conditions. Explain why.

43. In formic acid (HCOOH), one hydrogen atom and both oxygen atoms are bonded directly to the central carbon atom. Predict a range of K_{a1} for this acid, based on the correlations shown in Table 17–5. Compare your prediction with the measured value from Table 8–2. At pH 14, do you expect HCOOH (aq), $HCOO^-(aq)$, or $COO^{2-}(aq)$ to be the predominant species present? Explain.

44. In carbonic acid (H_2CO_3), both hydrogen atoms are bonded to oxygen atoms, and all three oxygen atoms are directly bonded to the central carbon atom. In what range do you predict K_{a1} for this acid to occur, based on the correlations shown in Table 17–5? The measured value of 4.3×10^{-7} quoted in Table 8–2 actually applies to the equilibrium

$$CO_2(aq) + 2\ H_2O(\ell) \rightleftharpoons HCO_3^-(aq) + H_3O^+(aq)$$

Explain how this information provides evidence that only a small fraction of the dissolved CO_2 is present as $H_2CO_3(aq)$.

45. Oxoacids can be formed that involve several central atoms of the same chemical element. An example is $H_2B_4O_7$, which can be written as $B_4O_5(OH)_2$. In such a case, we expect acid strength to correlate approximately with the *ratio* of the number of lone oxygen atoms to the number of central atoms (this ratio is 5:4 for $H_2B_4O_7$, for example). Rank the following in order of increasing acid strength: $H_2B_4O_7$, H_3BO_3, $H_5B_3O_7$, and $H_6B_4O_9$.

46. Use the approach suggested in the previous problem to rank the following in order of increasing acid strength: H_3PO_4, $H_3P_3O_9$, $H_4P_2O_6$, $H_4P_2O_7$, $H_5P_3O_{10}$. Assume that no hydrogen atoms are directly bonded to phosphorus in these compounds. Sodium salts of these polyphosphoric acids are used as "builders" in detergents to improve cleaning power.

47. Draw Lewis structures that satisfy the octet rule for H_2SO_3 and H_2SO_4. Which acid is stronger, and why?

48. Oxoacids of the transition-metal elements fall into classes similar to those shown in Table 17–5 for the non-metal oxoacids, although the values of K_a are smaller by about a factor of 10^4 for the compounds of the transition metals. Use this information to estimate the first acidity constant K_a for H_2MnO_4, H_4ZrO_4, $HMnO_4$, and H_3MnO_4. In none of these oxoacids is a hydrogen atom directly bonded to the central metal atom.

Additional Problems

49. An atom of sodium has the electron configuration $[Ne]6s^1$. Explain how this is possible.

50. An atom or ion in which each electron shell is filled or half-filled has a total electron density that is spherically symmetrical (that is, the electron density varies with distance from the nucleus but not with direction). Identify the atoms and ions in the following list that are spherically symmetrical in their ground states: F^-, Na, Si, S^{2-}, Ar^+, Ni, Cu, Mo, Rh, Sb, W, Au.

51. Chromium(IV) oxide is used in making magnetic recording tapes because of its paramagnetic properties. It can be described as a solid made up of Cr^{4+} and O^{2-} ions. Give the electron configuration of Cr^{4+} in CrO_2, and determine the number of unpaired electrons on each chromium ion.

* 52. Which is higher, the third ionization energy of lithium or the energy required to eject a $1s$-electron from a lithium atom in a photoelectron spectroscopy experiment? Explain.

53. Consider the elements Sr, Li, P, and Si.
    (a) Which has the greatest difference between its first and second ionization energy? Explain.
    (b) Which has the greatest difference between its second and third ionization energy?

54. The radii of the ions N^{3-}, O^{2-}, and F^- equal 1.71, 1.40, and 1.36 Å, respectively. The contraction is explained as the effect of increased nuclear charge from nitrogen ($Z = 7$) to fluorine ($Z = 9$) on the same electron configuration. Suggest why the contraction is much larger from N^{3-} to O^{2-} than from O^{2-} to F^-.

55. Arrange the following six atoms or ions in order of size, from the smallest to the largest: K, F^+, Rb, Co^{25+}, Br, F, Rb^-.

56. The C—O bond length in free carbon monoxide is 1.13 Å. When four CO molecules bond to a central nickel atom to form $Ni(CO)_4$, the C—O bond length increases to 1.15 Å. (Recall from the Cumulative Problem of Chapter 11 that this compound is used to purify nickel.) Is the C—O bond stronger or weaker in the complex than it is in a free CO molecule?

57. Consult Figure 17–14, and compute the difference in electronegativity between the atoms in LiCl and in HF. Based on their physical properties (see below), are the two similar or different in their bonding?

	LiCl	HF
Melting point	605°C	−83.1°C
Boiling point	1350°C	19.5°C

58. Ordinarily two metals, when mixed, form alloys that maintain their metallic character. If the two metals differ sufficiently in electronegativity, they can form compounds with significant ionic character. Consider the solid produced by mixing equal chemical amounts of Cs and Rb in comparison with that produced by mixing Cs and Au. Compute the electronegativity difference in each case, and determine whether either has significant ionic character. If either compound is ionic or partially ionic, which atom carries the net negative charge? Are there alkali halides with similar or smaller electronegativity differences?

* 59. The electronegativities of the elements in Group IV of the periodic table are as follows: C, 2.55; Si, 1.90; Ge, 2.01; Sn, 1.88; Pb, 2.10. Explain the lack of a clear-cut trend. Is any trend seen in Group III? (*Hint*: Consider the relative influence of the size and the nuclear charge of the elements on the electronegativity.)

* 60. A stable triatomic molecule can be formed containing one atom each of nitrogen, sulfur, and fluorine. Three bonding structures are possible, depending on which is the central atom: NSF, SNF, and SFN.
    (a) Write a Lewis structure for each of these molecules, indicating the formal charges on each atom.
    (b) Frequently, the structure with the least separation of formal charge is the most stable. Is this statement consistent with the observed structure for this molecule, namely NSF, with a central sulfur atom?
    (c) Does consideration of the electronegativities of N, S, and F from Figure 17–14 help to rationalize this observed structure? Explain.

61. Compare the mechanism of covalent bond formation with that of ionic bond formation. Can both mechanisms contribute simultaneously to the strength of a bond?

62. A reference book tabulates five different values of the electronegativity of molybdenum, a different one for each oxidation state from +2 to +6. Predict which electronegativity is highest and which is lowest.

63. There have been some predictions that element 114 (not yet discovered) will be relatively stable in comparison with many other elements beyond uranium in the periodic table. Predict the maximum oxidation state of this element. Based on the trends in the oxidation states of other members of its group, is it likely that this oxidation state will be the dominant one?

* 64. H_3PO_2 is an oxoacid that has a first acidity constant of 8×10^{-2}. What molecular structure do you predict for H_3PO_2? Is this acid monoprotic, diprotic, or triprotic in aqueous solution?

65. Fluorine chemists sometimes say, "Fluorine gives wings to the metals," meaning that metal fluorides, particularly those of transition elements in high oxidation states, are volatile. Explain why this is so.

Iodine sublimes from the bottom and recondenses at the top of this flask.

CUMULATIVE PROBLEM

Iodine

The shiny purple-black crystals of elemental iodine were first prepared in 1811 from the ashes of seaweed. Several species of seaweed concentrate the iodine that is present in very low proportions in seawater. For many years, seaweed continued as the major practical source of this element. Most iodine is now produced from natural brines via oxidation of iodide ion with chlorine.

(a) Write the ground-state electron configuration for the iodine atom.

(b) Is the first ionization energy of an iodine atom larger or smaller than the first ionization energy of its immediate neighbors in the periodic table, tellurium and xenon? Make the same comparison for the electron affinity.

(c) Iodine is an essential trace element in the human diet, and iodine deficiency causes goiter, the enlargement of the thyroid gland. To prevent goiter, much salt intended for human consumption is "iodized" by the addition of small quantities of sodium iodide. Calculate the electronegativity difference between sodium and iodine. Is sodium iodide an ionic or a covalent compound? What is its chemical formula?

(d) Iodine is an important reactant in synthetic organic chemistry because bonds form readily between carbon and iodine. Use electronegativities to determine whether the C—I bond is ionic, purely covalent, or polar covalent in character.

(e) The highest oxidation state observed for iodine is +7. A synthetic route to this state begins with the disproportionation, by heating, of barium iodate (in which iodine is in the +5 oxidation state):

$$5\ Ba(IO_3)_2 \longrightarrow Ba_5(IO_6)_2 + 4\ I_2 + 9\ O_2$$

Treatment of $Ba_5(IO_6)_2$ with concentrated nitric acid yields white crystals of orthoperiodic acid (H_5IO_6). Dehydration of this acid gives periodic acid (HIO_4) but not the acid anhydride (I_2O_7). Why is the highest oxidation state of iodine +7 and not a higher or lower number?

(f) Rank the three oxoacids HIO_3, H_5IO_6, and HIO_4 according to their expected acid strength.

Molecular Orbitals, Spectroscopy, and Atmospheric Chemistry

18

In-phase water waves in a wave tank. Standing waves in water can be used to model the combination of atomic orbitals to give molecular orbitals.

Nitrogen and oxygen make up 99% of the earth's atmosphere. These two gases resemble each other superficially; both are diatomic molecular substances of low molar mass that liquefy only below 100 K. They differ in fundamental respects, however. The chemical bond in nitrogen is significantly stronger than that in oxygen; oxygen is paramagnetic, but nitrogen is not; oxygen reacts far more readily than nitrogen. What causes these differences, and how do they shape the roles the two gases play in the structure and dynamics of the atmosphere?

In this chapter, we answer these questions by broadening the ideas presented in Chapters 16 and 17. We show how *molecular* orbitals, a natural extension of atomic orbitals, help to explain the stability of molecules and the nature of their bonds. We explore the electron configurations of molecules and present an aufbau principle for molecules that is similar to the one for atoms. We examine the rearrangement of electrons in polyatomic molecules that leads to the valence shell electron-pair repulsion (VSEPR) model for molecular geometry. Finally, we consider the interactions between light (radiant energy) and molecules. Light provides our most important probe for measuring the properties of molecules: their geometries, bond lengths, and bond energies. Light can also play a crucial role in changing molecules, causing bonds to break or rearrange, and inducing chemical reactions both in the laboratory and in the environment.

18-1 DIATOMIC MOLECULES

The role of valence electrons in covalent chemical bonds is touched on at several points in this book. In the Lewis model (see Section 3–4), covalent bonding is introduced as arising quite simply from the tendency of atoms to achieve noble-gas electron configurations by sharing electrons. Section 17–5 points out that electron pairs located between nuclei exert electrostatic forces to pull the nuclei together, and Chapter 16 establishes the part-wave, part-particle nature of electrons, which are never correctly pictured as simply "at" some point in space. Instead, an electron is best described by a characteristic wave function, the square of which gives the *probability* of finding the electron at every point (the electron probability density). This applies to electrons in molecules as well as in atoms. In molecules, the wave functions, or orbitals, may be spread out, or *delocalized*, over more than one atom. Such orbitals are called **molecular orbitals,** in distinction to atomic orbitals, which are localized on single atoms. In this section we discuss the molecular orbitals of diatomic molecules. Polyatomic molecules are considered in the next section.

• Our molecular orbital model is a simplified quantum-mechanical description that contains most, although not all, of the physical factors that lead to the formation of chemical bonds.

Molecular Orbitals and Covalent Bonding

The simplest possible bonded species is the one-electron molecular ion H_2^+. The wave functions of the electron in H_2^+ are the solutions of the Schrödinger equation for an electron bound in the electrostatic field of *two* nuclei of charge $+e$ separated by a distance R. A full calculation of the ground-state and higher-energy molecular orbitals in H_2^+ has been carried out numerically, but the answer is complex. For many purposes, a pictorial description is sufficient and provides useful insight. In this description, the molecular orbitals in H_2^+ are regarded as constructed by *superposition* of the atomic orbitals centered on each of the two nuclei. As mentioned previously (and shown in Fig. 16–17), the overlap (superposition) of waves leads to two kinds of interference: if the waves are *in phase*, interference is constructive,

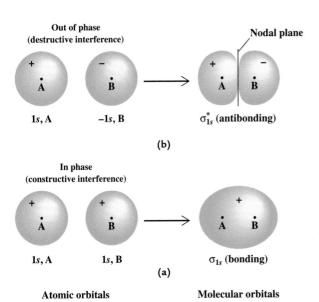

Figure 18-1 The $1s$ atomic orbitals on nuclei A and B combine in two ways. (a) If both have positive phase or both have negative phase, the two orbitals are *in phase* and interfere constructively to give a σ_{1s} bonding molecular orbital that has *increased* electron density between the nuclei. (b) If *out of phase* (one having positive phase and the other negative), the two interfere destructively to give a σ_{1s}^* antibonding molecular orbital that has *decreased* electron density (a nodal plane) between the nuclei.

and the amplitudes add to give a larger total amplitude; if the waves are *out of phase*, interference is destructive, and the amplitudes cancel to give smaller amplitude or zero amplitude (a node is formed). Thus, when two $1s$ atomic orbitals (wave functions) on neighboring centers are superposed, two different molecular orbitals form. In the first, constructive interference gives a higher amplitude (and therefore a greater average electron density) between the two nuclei. In the second, destructive interference gives a lower amplitude, and a node appears between the nuclei.

The two molecular orbitals arising from overlap of two $1s$ atomic orbitals are designated as σ_{1s} (constructive interference) and σ_{1s}^* (destructive interference), using the Greek letter σ (sigma) to indicate that the electron probability density is distributed symmetrically around the bond axis. The subscripts indicate which atomic orbitals are combined, that is, the parentage of the molecular orbitals. The two modes of combination and the electron probability densities in the resulting molecular orbitals are shown in Figure 18–1. Comparison reveals that an electron in a σ-orbital has an enhanced probability of being found where the two atomic orbitals overlap (between the nuclei), so σ is a **bonding orbital.** In contrast, an electron in a σ^*-orbital has a *reduced* probability of being found between the nuclei, so σ^* is an **antibonding orbital.** The antibonding orbital has a higher energy because it has a node. Therefore, the electron occupies the σ_{1s} molecular orbital in the ground state of H_2^+.

• The name "antibonding" is well chosen. An electron in an antibonding orbital in a molecule diminishes the stability of the molecule.

A molecular orbital, just like an atomic orbital, can hold two electrons if they are spin-paired (one with spin up and the other with spin down). The H_2 molecule has the same molecular orbitals as the H_2^+ ion but has two electrons. In its ground state, H_2 accommodates the two with opposing spins in a σ_{1s} bonding molecular orbital. Quite generally, covalent bonding arises from the sharing of electrons (most often electron pairs with opposite spins) in bonding molecular orbitals. In these orbitals, the electron probability density is largest between the nuclei and tends to pull the nuclei together. Electron sharing *alone* is not sufficient for chemical-bond formation. Electrons that are shared in an *antibonding* molecular orbital tend to force the nuclei apart, reducing the bond strength.

Figure 18–2 A correlation diagram for first-period diatomic molecules. The red arrows represent electrons. In the ground-state H_2 molecule (shown here) both electrons occupy the σ_{1s} molecular orbital (*center*).

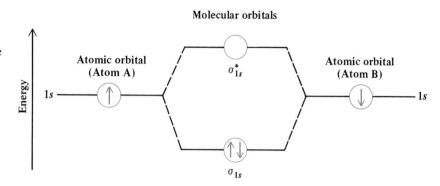

The relative energies and the parentage of molecular orbitals are often displayed in **correlation diagrams,** of which Figure 18–2 is an example. This diagram shows how two $1s$ atomic orbitals of equal energy on different centers overlap to give a lower-energy σ_{1s} molecular orbital and a higher-energy σ_{1s}^* molecular orbital. An aufbau principle analogous to that for atomic orbitals is then used in conjunction with such diagrams. Available electrons are "fed in" to the molecular orbitals, starting with the one of lowest energy. A maximum of two electrons occupies each molecular orbital. The correlation diagram in Figure 18–2 thus works for both H_2^+ and H_2 (and other first-period diatomic species as well). The ground-state electron configuration of H_2^+ is represented by $(\sigma_{1s})^1$, and the ground-state electron configuration of H_2 is $(\sigma_{1s})^2$.

In the Lewis theory of chemical bonding, a shared pair of electrons corresponds to a single bond, two shared pairs to a double bond, and so forth (see Section 17–4). In the molecular-orbital theory, electrons can be shared in antibonding orbitals as well as bonding orbitals. Antibonding electrons reduce the bond strength, however, and also the bond order. The definition of bond order is therefore expanded as follows:

bond order $= \frac{1}{2}$ (number of electrons in bonding molecular orbitals $-$

number of electrons in antibonding molecular orbitals)

EXAMPLE 18–1

Give the ground-state electron configuration and the bond order of the He_2^+ molecular ion.

Solution

The He_2^+ ion has three electrons, which can be placed into the molecular orbitals shown in Figure 18–2 to give the ground-state configuration $(\sigma_{1s})^2 (\sigma_{1s}^*)^1$, indicating that it has a doubly occupied σ_{1s} orbital (bonding) and a singly occupied σ_{1s}^* orbital (antibonding). The bond order is

$$\text{bond order} = \tfrac{1}{2}(2 \text{ electrons in } \sigma_{1s} - 1 \text{ electron in } \sigma_{1s}^*) = \tfrac{1}{2}$$

This should be a weaker bond than that in H_2.

Exercise

Give the ground-state electron configuration and the bond order of the H_2^{2-} molecular ion.

Answer: Electron configuration: $(\sigma_{1s})^2(\sigma_{1s}^*)^2$. The bond order is zero.

Homonuclear Diatomic Molecules

The molecular orbital configurations of **homonuclear** (that is, having identical nuclei) diatomic molecules and molecular ions made from first-period elements are shown in Table 18–1. These configurations are simply a listing of the occupied molecular orbitals in order of increasing energy, together with the number of electrons in each orbital. The observed bond lengths and bond energies are also given. Higher bond order corresponds to larger bond energies and shorter bond lengths. The species He_2 has a bond order of zero, and it does not form a true chemical bond.*

A general prescription for obtaining a molecular orbital description of the bonding in molecules can now be written:

* There is a very weak attractive interaction between two helium atoms due to dispersion forces. It was mentioned in Section 6-1.

> **1.** Combine atomic orbitals to form molecular orbitals. The total number of molecular orbitals formed in this way must equal the number of atomic orbitals used.
> **2.** Arrange the molecular orbitals in order from lowest to highest energy.
> **3.** Put in electrons (at most two per molecular orbital), starting from the orbital of lowest energy. Apply Hund's rules where appropriate.

This procedure is readily applied to second-period homonuclear diatomic molecules. The $2s$ atomic orbitals of the two atoms are combined in the same fashion as $1s$ orbitals, giving one σ_{2s} bonding orbital and one σ_{2s}^* antibonding orbital. The $2p$-orbitals form different kinds of molecular orbitals, depending on whether they are parallel to or perpendicular to the internuclear (bond) axis. The z-axis is customarily taken to lie along the bond. Then, the $2p_z$-orbitals of the two atoms combine "end to end" to give a bonding σ_{2p_z} molecular orbital and an anti-

Table 18–1
Configurations and Bond Orders for First-Row Homonuclear Diatomic Molecules

Species	Electron Configuration	Bond Order	Bond Energy (kJ mol^{-1})	Bond Length (Å)
H_2^+	$(\sigma_{1s})^1$	$\frac{1}{2}$	255	1.06
H_2	$(\sigma_{1s})^2$	1	431	0.74
He_2^+	$(\sigma_{1s})^2(\sigma_{1s}^*)^1$	$\frac{1}{2}$	251	1.08
He_2	$(\sigma_{1s})^2(\sigma_{1s}^*)^2$	0	Not observed	

Figure 18-3 The overlap of $2p_z$-orbitals on neighboring atoms A and B. (a) Constructive (in-phase) interference gives a σ_{2p_z} bonding molecular orbital. (b) Destructive (out-of-phase) interference gives a $\sigma_{2p_z}^*$ antibonding molecular orbital. The signs on the orbitals give relative phase, not electric charge.

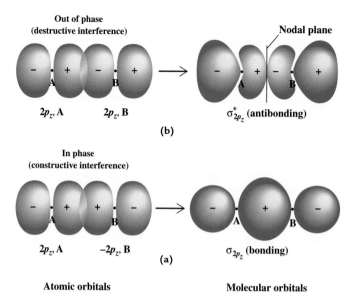

Out of phase
(destructive interference)

Nodal plane

$2p_z$, A $2p_z$, B

$\sigma_{2p_z}^*$ (antibonding)

(b)

In phase
(constructive interference)

$2p_z$, A $-2p_z$, B

σ_{2p_z} (bonding)

(a)

Atomic orbitals   Molecular orbitals

bonding $\sigma_{2p_z}^*$ molecular orbital (Fig. 18–3). These orbitals are designated σ molecular orbitals because, like the σ_{1s}-orbital and the σ_{1s}^*-orbital, their electron density is symmetrically distributed around the internuclear axis.

The two $2p_x$-orbitals, which are perpendicular to the bond axis and oriented "side by side," also combine to form a bonding and an antibonding molecular orbital (Fig. 18–4). These orbitals give maximum electron density on either side of the internuclear axis, with that axis lying in a nodal plane (in this case, the yz-plane). They are designated by the Greek letter π rather than σ. The π_{2p_x}-orbital is bonding, and the $\pi_{2p_x}^*$ is antibonding. In the same way, π_{2p_y}-orbitals and $\pi_{2p_y}^*$-orbitals can form from the $2p_y$ atomic orbitals. Their lobes project above and below the xz nodal plane, which is the plane of the page in Figure 18–4.

• Atomic orbitals are designated s (no angular nodes around the nucleus), p (one angular node), d (two angular nodes), and so forth. Molecular orbitals use the Greek equivalents: σ (no angular nodes about the internuclear axis), π (one angular node), δ (two angular nodes).

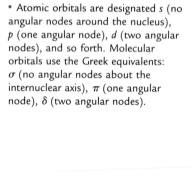

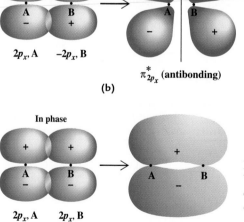

Out of phase

Nodal plane

$2p_x$, A $-2p_x$, B

$\pi_{2p_x}^*$ (antibonding)

(b)

In phase

$2p_x$, A $2p_x$, B

π_{2p_x} (bonding)

(a)

Figure 18–4 The overlap of $2p_x$-orbitals on neighboring atoms A and B. These orbitals lie "side by side" but still interact. (a) In-phase interference gives a π_{2p_x} bonding molecular orbital. (b) Out-of-phase interference gives a $\pi_{2p_x}^*$ antibonding molecular orbital. The signs on the orbitals give relative phase, not electric charge.

It is important to remember that the + and − signs in Figures 18–1, 18–3, and 18–4 refer not to positive or negative *charges* but to the phase of the wave function in the various regions of space. The relative phase of the two atomic orbitals as they combine determines whether the resulting molecular orbital is bonding or antibonding. Bonding orbitals form from the overlap of wave functions having the same phase in the same region; antibonding orbitals form from the overlap of wave functions having opposite phase in the same region.

The next step is to determine the energy ordering of the molecular orbitals. In general, this requires a calculation that is beyond the scope of this text, but some simple qualitative rules can be given:

---

1. The *average* energy of a bonding–antibonding pair of molecular orbitals lies approximately at the energy of the original atomic orbitals.
2. The difference in energy between members of a bonding–antibonding pair increases as the overlap of the atomic orbitals increases.

In second-period diatomic molecules, the σ_{1s} bonding orbitals and σ_{1s}^* antibonding orbitals are equally occupied (by two electrons each). Therefore, these core electrons have little net effect on bonding properties and need not be considered. Once again, we see that chemical bonding is dominated by valence-electron behavior. Correlation diagrams for the molecular orbitals formed from the 2s-orbital and 2p-orbital appear in Figure 18–5. There are two different energy orderings for diatomic molecules formed from second-period elements. The first (Fig. 18–5a) applies to the molecules with atoms lithium through nitrogen (that is, the first part of the

Figure 18–5 Correlation diagrams for second-period diatomic molecules. In each case, the atomic orbitals at the sides combine to give the molecular orbitals in the center. The red arrows represent the valence electrons in B_2 and O_2. Each of these two molecules is paramagnetic, having two unpaired electrons.

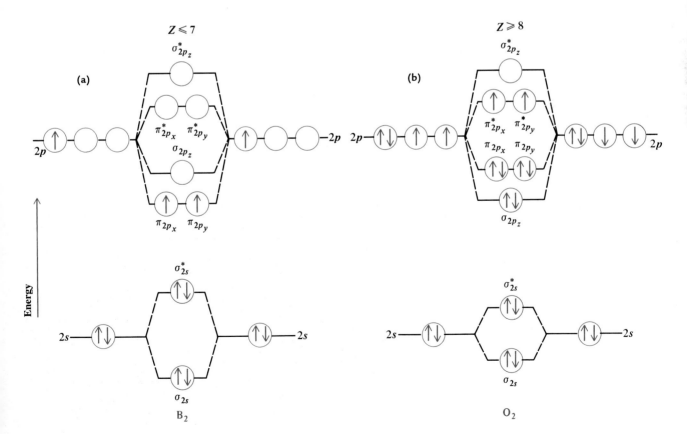

period) and their positive and negative molecular ions. The second (Fig. 18–5b) applies to molecules of the later elements oxygen, fluorine, and neon and their positive and negative molecular ions. The distinction between the two is with the π_{2p} and σ_{2p} orbitals, which are quite close in energy and reverse their order once oxygen is reached.

Two valence electrons, at most, are now put into each molecular orbital, starting from the orbital of lowest energy and building up according to the patterns given in Figure 18–5. Placement of all available valence electrons gives the ground-state molecular orbital configuration. The overall bond orders can be calculated as before or expressed in terms of **σ-bonds** and **π-bonds**:

* It is important in counting electrons at this stage to include *only* valence electrons.

$$\text{No. of } \sigma\text{-bonds} = \tfrac{1}{2}(\text{number of electrons in } \sigma \text{ molecular orbitals} -$$
$$\text{number of electrons in } \sigma^* \text{ molecular orbitals})$$

$$\text{No. of } \pi\text{-bonds} = \tfrac{1}{2}(\text{number of electrons in } \pi \text{ molecular orbitals} -$$
$$\text{number of electrons in } \pi^* \text{ molecular orbitals})$$

Overall bond order = number of σ-bonds + number of π-bonds

EXAMPLE 18–2

Determine the ground-state electron configuration, bond order, and the number of σ- and π-bonds in the F_2 molecule.

Solution

Each atom of fluorine has 7 valence electrons, so 14 electrons must be placed into molecular orbitals to represent bonding in the F_2 molecule. Using the correlation diagram of Figure 18–5 leads to the electron configuration

$$(\sigma_{2s})^2(\sigma_{2s}^*)^2(\sigma_{2p_z})^2(\pi_{2p})^4(\pi_{2p}^*)^4$$

Because there are eight valence electrons in bonding orbitals and six in antibonding orbitals, the bond order is

$$\text{bond order} = \tfrac{1}{2}(8 - 6) = 1$$

and the F_2 molecule has a single bond. This bond arises from electrons in a σ molecular orbital and is therefore a σ-bond; there are no π-bonds.

Exercise

Determine the ground-state valence electron configuration, bond order, and the number of σ- and π-bonds in the B_2^- molecular ion.

Answer: Ground-state valence electron configuration: $(\sigma_{2s})^2(\sigma_{2s}^*)^2(\pi_{2p})^3$. Bond order = $\tfrac{1}{2}(5 - 2) = \tfrac{3}{2}$. Zero σ-bonds and $\tfrac{3}{2}$ π-bonds.

Table 18–2 summarizes the bonding formalism for and the properties of first- and second-period homonuclear diatomic molecules. Note the strong relationship between bond order, bond length, and bond energy. Also, the bond orders obtained from molecular-orbital theory agree completely with the results of the Lewis electron-dot model.

An important prediction comes out of the two correlation diagrams in Figure 18–5. In the ground states of both B_2 and O_2, the last two electrons should follow Hund's rules and have parallel spins. Thus, the molecular-orbital approach predicts B_2 and O_2 to be paramagnetic because they contain unpaired electrons. This is

Table 18–2
Molecular Orbitals of Homonuclear Diatomic Molecules

Species	Number of Valence Electrons	Ground-State Valence-Electron Configuration	Bond Order	Bond Length (Å)	Bond Energy (kJ mol^{-1})
H_2	2	$(\sigma_{1s})^2$	1	0.74	431
He_2	4	$(\sigma_{1s})^2(\sigma_{1s}^*)^2$	0	Not observed	
Li_2	2	$(\sigma_{2s})^2$	1	2.67	105
Be_2	4	$(\sigma_{2s})^2(\sigma_{2s}^*)^2$	0	2.45	9
B_2	6	$(\sigma_{2s})^2(\sigma_{2s}^*)^2(\pi_{2p})^2$	1	1.59	289
C_2	8	$(\sigma_{2s})^2(\sigma_{2s}^*)^2(\pi_{2p})^4$	2	1.24	599
N_2	10	$(\sigma_{2s})^2(\sigma_{2s}^*)^2(\pi_{2p})^4(\sigma_{2p_z})^2$	3	1.10	942
O_2	12	$(\sigma_{2s})^2(\sigma_{2s}^*)^2(\sigma_{2p_z})^2(\pi_{2p})^4(\pi_{2p}^*)^2$	2	1.21	494
F_2	14	$(\sigma_{2s})^2(\sigma_{2s}^*)^2(\sigma_{2p_z})^2(\pi_{2p})^4(\pi_{2p}^*)^4$	1	1.41	154
Ne_2	16	$(\sigma_{2s})^2(\sigma_{2s}^*)^2(\sigma_{2p_z})^2(\pi_{2p})^4(\pi_{2p}^*)^4(\sigma_{2p_z}^*)^2$	0	Not observed	

exactly what is found experimentally (Fig. 18–6). In contrast, in the Lewis electron-dot structure for O_2

$$:\overset{..}{O}::\overset{..}{O}:$$

all the electrons appear paired. Moreover, the high chemical reactivity of molecular oxygen can be rationalized as resulting from the readiness of the two π^* electrons, unpaired and located in different regions of space, to find additional bonding partners in other molecules.

* Hund's rules state that when two orbitals of equal energy are available for two electrons, one electron goes into each orbital and the electrons are unpaired. These rules were introduced in Chapter 17.

Heteronuclear Diatomic Molecules

Diatomic molecules such as carbon monoxide and nitrogen monoxide, formed from atoms of two different elements, are called **heteronuclear.** Molecular orbitals can

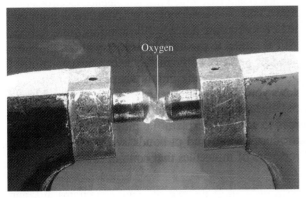

Figure 18–6 (a) Oxygen is paramagnetic; liquid oxygen poured between the pole faces of a magnet is attracted and held there. (b) When the experiment is repeated with liquid nitrogen, which is diamagnetic, the liquid pours straight through.

Figure 18–7 Correlation diagram for second-period heteronuclear diatomic molecules of the general formula AB. The atomic orbitals for the more electronegative atom (B) are displaced downward because they have lower energies than those of the less electronegative atom (A). The red arrows show the orbital filling for CO in its ground state.

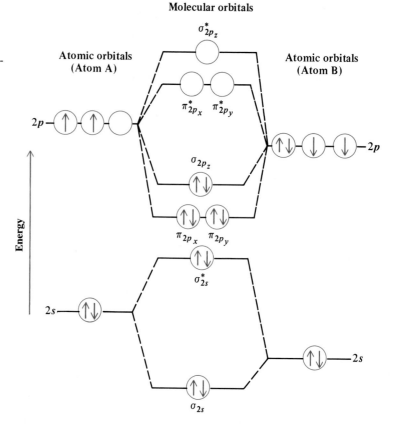

be constructed from their atomic orbitals, just as in the homonuclear case. In the correlation diagrams, the energies of the starting sets of atomic orbitals are no longer the same because the atoms are no longer the same. The energy levels of the more electronegative atom are displaced *downward* because it attracts valence electrons more strongly. For many heteronuclear diatomic molecules of second-period elements (those in which the electronegativity difference is not too large), the correlation diagram shown in Figure 18–7 results. It resembles Figure 18–5a, except for the difference just mentioned. This diagram guides determination of the electron configurations of heteronuclear diatomic molecules and molecular ions of the second period. As an example, take CO, which has ten valence electrons (four from C, six from O). Its ground-state configuration is

$$(\sigma_{2s})^2(\sigma_{2s}^*)^2(\pi_{2p})^4(\sigma_{2p_z})^2$$

From the fact that CO has eight electrons in bonding orbitals and two in antibonding orbitals, we compute a bond order of $\frac{1}{2}(8 - 2) = 3$ (one σ-bond and two π-bonds). Carbon monoxide is indeed tightly bound; it has the largest bond energy (1074 kJ mol^{-1}) of all second-period diatomic molecules. Nitrogen monoxide resembles carbon monoxide somewhat but possesses an additional electron in an antibonding orbital (a π_{2p}^*-orbital), giving the ground-state configuration.

$$(\sigma_{2s})^2(\sigma_{2s}^*)^2(\pi_{2p})^4(\sigma_{2p_z})^2(\pi_{2p}^*)^1$$

Hence, NO is paramagnetic and has a bond order of only $2\frac{1}{2}$. Its bond energy is predicted to be (and is) less than the bond energy of CO.

18-2 POLYATOMIC MOLECULES

There are two ways to describe bonding in polyatomic molecules. The first uses *delocalized* molecular orbitals, with electron wave functions that are spread out over the entire molecule. These molecular orbitals are first constructed from valence atomic orbitals, then placed in order of increasing energy, and finally filled with the available valence electrons. Complete delocalization encounters a fundamentally *chemical* objection. Bonds have fairly universal characteristics such as energy and length (see Section 17–4). If electrons are described by molecular orbitals that spread over the entire molecule, then why do the properties of a given type of bond depend only weakly on the particular molecule in which it finds itself? Another approach to bonding develops from this consideration—one that uses molecular orbitals that are *localized*, either as bonding orbitals between particular pairs of atoms or as lone-pair orbitals on individual atoms. This approach, **valence-bond theory,** is related to the qualitative Lewis theory, in which a similar localization of electrons is assumed. Which description is better, the delocalized molecular-orbital picture or the localized valence-bond approach? Each has its advantages and offers insights into the nature of chemical bonding. They differ primarily in their treatment of the *correlation* of the positions of different electrons. In the delocalized molecular-orbital picture, the electrons are considered to move independently of each other, so that there is a reasonable probability that several electrons will be found in the same region of space. The valence-bond approach prevents this by localizing electron distributions in individual bonds or on individual atoms. In most cases, the delocalized picture underestimates the importance of electron correlation, but the valence-bond approach overestimates it. The truth lies somewhere in between, and accurate computational techniques for predicting the properties of small molecules have been developed using each model as a starting point.

In this section, localized orbitals are used for σ-bonds and, wherever possible, for π-bonds. Delocalized molecular orbitals are brought in to describe π-bonds as required in certain bonding situations. This approach gives physical insight with few mathematical complications.

Hybridization

We begin with beryllium hydride (BeH_2). This molecule has four valence electrons (two from the Be and one from each H atom). Valence shell electron-pair repulsion (VSEPR) theory (see Section 3–6) assigns a steric number of 2 to the beryllium atom and so predicts that the molecule is linear, with the hydrogen atoms diametrically opposed about the central beryllium atom. The ground-state electron configuration of Be is $1s^2 2s^2$, with all electrons paired. How can it then form bonds by sharing electrons with hydrogen atoms? The answer depends on the fact that the $2p$-subshell lies close above the $2s$-subshell in energy. It costs only a small amount of energy to promote one electron from the $2s$-subshell to the $2p$-subshell, leaving two unpaired electrons to form bonds. In the valence-bond method, these half-occupied atomic orbitals are *mixed* to form new, singly occupied *atomic* orbitals, each of which can overlap with a singly occupied orbital of another atom to form a σ-bond along the internuclear axis. In BeH_2 we mix the $2s$ and $2p_z$ orbitals of Be to form two *sp* **hybrid atomic orbitals** on the central Be atom (Fig. 18–8a). An electron in one of these orbitals is localized, in that it is much more likely to be found on the left side of the nucleus than on the right; the reverse is

• An *sp* hybrid orbital is a "daughter" of an *s*-orbital and a *p*-orbital and is named after them.

Figure 18–8 (a) The two *sp* hybrid orbitals formed from the 2*s*- and 2*p_z*-orbitals on the Be atom. (b) The two *σ*-bonds that form from overlap of these orbitals with the hydrogen 1*s*-orbitals, giving two single bonds in the BeH$_2$ molecule.

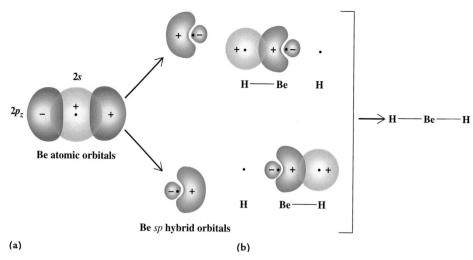

true of an electron in the other hybrid orbital. Once the hybrid atomic orbitals of the central atom are formed, the two hydrogen atoms can be brought near. The 1*s* orbital of the first H atom then overlaps with one of the *sp* hybrid atomic orbitals on the Be atom, and the 1*s* orbital of the second H atom overlaps with the second. The electrons in these orbitals are thereby shared in two *σ*-bonds (Fig. 18–8b). The sharing releases more than enough energy to make up for the investment in promoting one electron in the Be atom from the 2*s*-orbital to the 2*p*-orbital.

The formation of chemical bonds with hybrid orbitals (as in BeH$_2$) involves two steps.

1. Atomic orbitals are mixed to obtain new hybrid atomic orbitals. These hybrid orbitals are localized on the same atom, are identified by the same label ("*sp*" in the case of BeH$_2$), and are identical except for their orientations.

2. The hybrid orbitals on one atom are combined with atomic orbitals, either hybrid or unhybridized, on a second atom to obtain molecular orbitals localized in the bond between the two atoms.

The number of hybrid atomic orbitals formed must always equal the number of orbitals that are mixed; thus, one 2*s*-orbital and one 2*p*-orbital in Be gives two *sp*-orbitals in BeH$_2$.

Now consider the bonding in the hydrides of the other second-period elements. The ground-state valence-electron configuration of the boron atom is $2s^2 2p$, and from the VSEPR theory, the boron atom in BH$_3$ has steric number 3, corresponding to a trigonal planar structure. Promotion of one of the 2*s*-electrons creates the excited-state configuration $2s^1 2p_x^1 2p_y^1$, and these three atomic orbitals can be mixed to form the three equivalent ***sp*^{2} hybrid atomic orbitals** shown in Figure 18–9. They lie in a plane, with an angle of 120 degrees between them. If each overlaps with a 1*s* atomic orbital on a hydrogen atom, then trigonal planar BH$_3$ forms. Experimentally, a collection of BH$_3$ molecules turns out to be unstable: the molecules react rapidly to form compounds called boranes (B$_2$H$_6$, B$_4$H$_{10}$, and others). However, the BF$_3$ molecule, which is closely related to BH$_3$, has the trigonal planar geometry characteristic of sp^2 hybridization.

In CH$_4$, VSEPR theory assigns a steric number of 4 to the central carbon atom. Promotion of one 2*s*-electron to a 2*p*-orbital gives the electron configuration $2s^1 2p_x^1 2p_y^1 2p_z^1$ on the central carbon atom. The 2*s*-orbital and three 2*p*-orbitals can then be combined to form four equivalent ***sp*^{3} hybrid atomic orbitals,** which point toward the corners of a tetrahedron. These orbitals are illustrated in Figure

* This notation breaks a previous pattern. The superscript "2" in *sp*^{2} gives the number of parent *p*-orbitals contributing to the hybrid, *not* the number of electrons that occupy it. Thus, the symbol "*sp*^{2}" refers to a single orbital, and "(*sp*^{2})^2" would signify the occupancy of that orbital by two electrons. The term "*sp*^{2}" is read "*sp* two," not "*sp* squared."

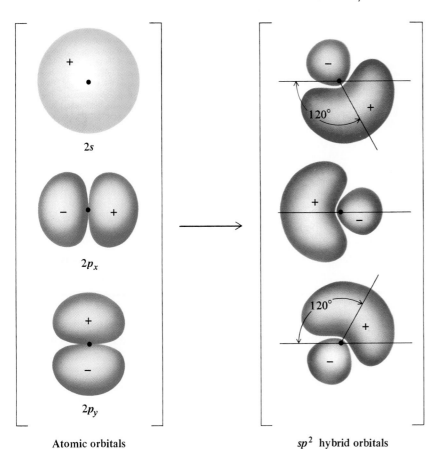

Atomic orbitals sp^2 hybrid orbitals

Figure 18–9 Shapes and relative dispositions of the three sp^2 hybrid orbitals, formed from the $2s$, $2p_x$, and $2p_y$ atomic orbitals on a single atom.

18–10. Each can overlap with a $1s$-orbital on one of the hydrogen atoms to give an overall tetrahedral structure for CH_4.

Lone-pair electrons can occupy hybrid orbitals. The nitrogen atom in NH_3 has a steric number of 4 (like the carbon atom in methane), and its bonding can be explained in terms of sp^3 hybridization. Of the eight valence electrons in NH_3, six are involved in σ-bonds between the nitrogen and the hydrogen atoms. The other two occupy the fourth sp^3 hybrid orbital as a lone pair. Oxygen in H_2O likewise has a steric number of 4 and can be described with sp^3 hybridization, with two lone pairs in sp^3-orbitals. This hybrid-orbital description of NH_3 and H_2O is consistent with bond angles close to the tetrahedral angle (109.5 degrees).

• For the hydrides of the heavier Group V and VI elements, the bond angles are much closer to 90 than 109.5 degrees. Thus, the bond angles in PH_3 are 93.6 degrees, and those in AsH_3 are 91.8 degrees. The bond angle in H_2S is 92.1 degrees, and that in H_2Se is 91 degrees. The approach of these angles to 90 degrees is not accounted for in the VSEPR theory or the hybrid-orbital description.

EXAMPLE 18–3

Predict the structure of ethane (C_2H_6) by writing down the Lewis structure and using VSEPR theory. What is the hybridization of the two carbon atoms?

Solution

Ethane has two carbon atoms, each bonded to the other and to three hydrogen atoms. Its Lewis structure is

$$\begin{array}{ccc} & \text{H} & \text{H} \\ & \overset{..}{} & \overset{..}{} \\ \text{H} : & \text{C} : \text{C} & : \text{H} \\ & \overset{..}{} & \overset{..}{} \\ & \text{H} & \text{H} \end{array}$$

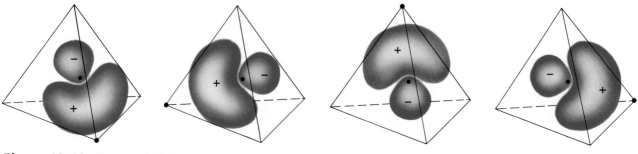

Figure 18-10 Shapes and relative dispositions of the four sp^3 hybrid orbitals, which point to the corners of a tetrahedron.

Because the identical carbon atoms have four bonds and no lone pairs, their steric numbers are 4, and bonds point outward from the carbon atoms toward the corners of a tetrahedron. Figure 18–11 shows the result.

The steric number of 4 implies sp^3 hybridization for each of the two carbon atoms.

Exercise

Predict the structure of hydrazine (H_2NNH_2) by writing down its Lewis structure and using VSEPR theory. What are the hybridizations of the two nitrogen atoms?

Answer:

$$\begin{array}{cc} H & H \\ \ddot{} & \ddot{} \\ :N\!\!-\!\!N: \\ \ddot{} & \ddot{} \\ H & H \end{array}$$

Both of the nitrogen atoms are sp^3 hybridized, with H—H and H—N—N angles of approximately 109.5 degrees. The extent of rotation around the N—N bond cannot be predicted from VSEPR theory or the hybrid-orbital model. The structure of hydrazine appears on page 200.

Our discussion has discovered an intimate relationship between the VSEPR theory of Section 3–6 and valence-bond theory, with steric numbers of 2, 3, and 4 corresponding to sp, sp^2, and sp^3 hybridization, respectively. Both theories are based on the energy savings resulting from reducing electron–electron repulsion. In fact, valence-bond theory can be thought of as an attempt to model the shapes from VSEPR theory using the mathematical formalism of orbitals. Because hybrid orbitals are mathematical entities, it is a mistake to treat them as physically real; they are ultimately no more than a device to rationalize observed bond angles and the observed equivalence of bonds in some molecules.

It is possible to extend the valence-bond picture for steric numbers greater than 4 by mixing selected d-orbitals with s-orbitals and p-orbitals. This creates new hybrid orbitals that help rationalize trigonal bipyramidal and octahedral structures. This is not very useful, however, and there is some question about whether d-orbitals are significantly involved in the bonding of molecules like SF_6 with expanded valence shells. Like the simple Lewis electron-dot model, the hybridization approach has its limits and is most useful in describing the bonding of the lighter elements with steric numbers up to 4. For molecules involving larger atoms situated later in the periodic table, the hybridization method becomes more com-

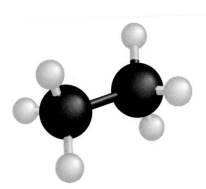

Figure 18-11 The structure of ethane. The four atoms surrounding each carbon lie at the corners of a tetrahedron.

plicated and less useful for qualitative interpretation. The success of the simple VSEPR theory in predicting molecular geometries in the compounds of these heavier atoms is all the more striking.

Double and Triple Bonds

The discussion of polyatomic molecules has to this point been restricted to single bonding. In molecules with double or triple bonds, electrons take part in π-bonds over and above the electrons in hybrid orbitals that form an underlying framework of σ-bonds. As an example, consider ethylene (C_2H_4), which has the Lewis structure

$$\begin{array}{cc} H & H \\ \cdot\cdot & \cdot\cdot \\ C & \colon\colon C \\ \cdot\cdot & \cdot\cdot \\ H & H \end{array}$$

Each carbon atom forms three attachments to other atoms, has no lone pairs, and has a steric number of 3; the hybridization at each carbon atom is sp^2. Let us take the three orbitals that comprise such a hybrid to be the $2s$-, $2p_x$-, and $2p_y$-orbitals, so the three σ-bonds at each carbon lie in the xy-plane. The σ-bond structure then involves a total of five bonds and uses 10 of the 12 valence electrons in ethylene. There are still two electrons left, and there is one atomic orbital (the $2p_z$) left on each carbon atom that has not been used in making hybrid orbitals. The two $2p_z$ orbitals, which project "side by side" perpendicular to the plane of the molecule, can be combined to make a π bonding molecular orbital into which the last two electrons are placed. (A π^* antibonding molecular orbital also results, but it remains empty.) This is the same procedure used in Section 18–1 for diatomic molecules and shown in Figure 18–4. Thus, π-overlap gives a bond order of 2 to the link between the carbon atoms and accounts for the last two valence electrons.

In writing this structure for ethylene, we have assumed that all of the atoms lie in a plane. What happens if we rotate the left CH_2 group around the $C=C$ double bond, so that its two hydrogen atoms depart from the plane of the paper (Fig. 18–12)? During such a rotation, the overlap of the two p_z-orbitals is reduced

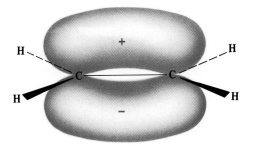

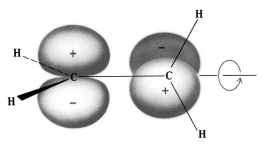

Figure 18–12 If the ethylene molecule is twisted around the $C=C$ axis, the overlap of the two p_z-orbitals decreases, giving a higher energy. The most stable conformation of ethylene is planar.

as their regions of high amplitude move out of alignment. When the rotation reaches 90 degrees, overlap vanishes. The regions in which one p_z-orbital has largest amplitude are in exact conflict with the zero-amplitude nodal plane of the other, and the π-bond between the carbon atoms ceases to exist. The σ-bond is intact, but the energy of the molecule is substantially higher. For this reason, rotation about a double bond rarely occurs at ordinary temperatures. In contrast, rotation about a single σ-bond takes place freely; the strength of the bond stays the same regardless of the orientation of the fragments it connects.

A similar procedure can be applied to the molecule of acetylene (C_2H_2), which contains a triple bond. Here the Lewis structure is

$$H:C \equiv C:H$$

and the steric number of each carbon atom is 2; the molecule is linear. If the axis passing through the two carbon nuclei is labelled the z-axis, then mixing $2s$ atomic orbitals and $2p_z$ atomic orbitals on each carbon atom forms sp hybrid orbitals, as in BeH_2. These overlap with one another and with the two hydrogen $1s$-orbitals to form three σ-bonds, using six of the ten available valence electrons in the molecule. The $2p_x$-orbitals remaining on the two carbon atoms then combine to form a π_{2p_x} molecular orbital, and the $2p_y$-orbitals simultaneously mix to give a π_{2p_y} molecular orbital. Filling both of these bonding orbitals with the remaining four valence electrons gives a triple bond between the carbon atoms. The two π^* antibonding orbitals, which are at higher energy, again remain empty.

• The naming of organic compounds is discussed in Chapter 24.

EXAMPLE 18–4

Consider the molecule propene (CH_2CHCH_3). Its Lewis structure is

$$\begin{array}{ccc} H & H & H \\ \ddot{} & \ddot{}\ddot{} & \ddot{} \\ C & :C:C & :H \\ \ddot{} & & \ddot{} \\ H & & H \end{array}$$

with a double bond between the first and the second carbon atoms. Discuss its bonding and predict its geometry.

Solution

The first step is to determine the hybridization of each carbon atom (identified for reference as C_1, C_2, and C_3, from left to right). The steric number of an atom in a molecule gives the hybridization; according to VSEPR theory, the steric number equals the number of atoms directly attached to an atom plus the number of lone pairs it has. Atoms C_1 and C_2 have steric number 3 and are described by sp^2 hybridization, and atom C_3 has steric number 4 and is described by sp^3 hybridization. The bond geometry at C_1 and C_2 is trigonal planar, and at C_3 is tetrahedral.

Next, electron pairs are placed in each localized (σ) molecular orbital, giving a single bond between each pair of linked atoms. This uses 16 of the 18 valence electrons in propene and creates a total of eight single bonds.

Now, the p-orbitals not involved in hybridization are combined to form π molecular orbitals. Here the p_z-orbitals of C_1 and C_2 can mix to form a π bonding molecular orbital as well as a higher-energy π^* antibonding orbital.

The last step is to put the remaining valence electrons into the π molecular orbitals. In the case of propene, only two valence electrons remain, and both go into the bonding π molecular orbital, giving a double bond between C_1 and C_2.

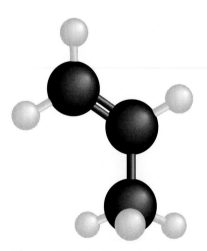

Figure 18–13 The molecular structure of propene. Valence orbitals are sp^2 hybridized on two carbons and sp^3 hybridized on the third.

The final predicted (and observed) three-dimensional structure is shown in Figure 18–13. As in the case of ethylene, the energy is lowest when the lobes of the two p_z-orbitals are parallel to each other. Thus, in the most stable molecular geometry, the hydrogen atoms on C_1 and C_2 should lie in the same plane as the C—C—C carbon skeleton. This prediction is verified by experiment.

Exercise

The Lewis structure of a molecule of propyne (CH_3CCH) is

$$\begin{array}{c} H \\ \cdot\cdot \\ H:C-C\equiv C:H \\ \cdot\cdot \\ H \end{array}$$

Discuss its bonding and predict its geometry.

Answer: One carbon atom in the structure is sp^3 hybridized, and the other two carbon atoms are sp hybridized. The atoms in the molecule are on a single straight line with the exception of the three hydrogen atoms on the sp^3 hybridized carbon atom, which point outward toward three of the corners of a tetrahedron (Fig. 18–14).

Figure 18–14   The molecular structure of propyne. Valence orbitals are sp hybridized on two carbons and sp^3 hybridized on the third.

The compound 2-butene has the formula $CH_3CHCHCH_3$ and the Lewis structure

$$\begin{array}{c} H\ H\ \ \ \ H \\ \cdot\cdot\ \cdot\cdot\ \ \ \cdot\cdot \\ H:C:C::C:C:H \\ \cdot\cdot\ \ \ \ \cdot\cdot\ \cdot\cdot \\ H\ \ \ \ H\ H \end{array}$$

The bonding at the outer two carbon atoms in this structure can be described by sp^3 hybridization and at the inner two by sp^2 hybridization. The bonding molecular orbital formed from the overlap of the p_z-orbitals on the two central atoms has its lowest energy when the lobes of these atomic orbitals are parallel, giving a planar structure for the carbon skeleton of the molecule. However, there are two ways the outer CH_3 groups can be placed relative to the double bond (Fig. 18–15). In the first, called the *cis* form, the two methyl (CH_3) groups lie on the same side of the double bond. In the *trans* form, they lie on opposite sides. Conversion of one form to the other requires breaking the central π-bond (by rotating the two p_z-orbitals 90 degrees with respect to each other, as in Fig. 18–12) and then reforming it in the other configuration. Both *cis* and *trans* forms of 2-butene have been prepared. Because breaking a π-bond costs a significant amount of energy, interconversion between the two is very slow at room temperature. Molecules that have the same chemical formula but different structures are isomers; this type of isomerism is called ***cis–trans* isomerism.**

18-3   THE CONJUGATION OF BONDS AND RESONANCE STRUCTURES

When two or more double or triple bonds occur close to one another in a molecule, the valence electrons tend to mingle together from one bond to another, a phenomenon called conjugation. The molecular orbital picture, with its emphasis on de-localization, explains this effect more readily than the localized valence-bond ap-

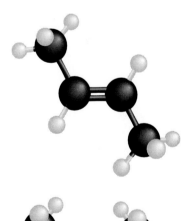

Figure 18–15   The two *cis-trans* isomers of 2-butene.

Figure 18–16 The four π molecular orbitals formed from four $2p_z$ atomic orbitals in 1,3-butadiene, viewed from the side. The locations of planar nodes are indicated by red lines; orbitals with more nodes are higher in energy. Note the similarity in nodal properties to the first four harmonics of a vibrating string (*right*). Only the two lowest-energy orbitals are occupied in the molecule in the ground state (*left*).

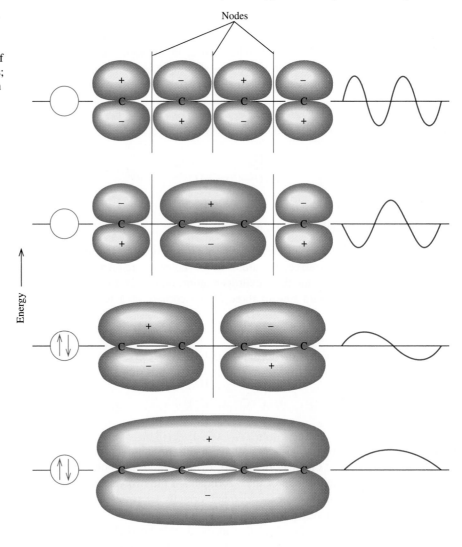

• Recall the analogy of the guitar string in Section 16–4. The higher harmonics have higher energy.

proach. An example is 1,3-butadiene ($CH_2CHCHCH_2$), which has the Lewis structure

$$
\begin{array}{ccc}
H & H & H \\
\cdot\cdot & \cdot\cdot & \cdot\cdot \\
C::C:C::C \\
\cdot\cdot & \cdot\cdot & \cdot\cdot \\
H & & H \quad H
\end{array}
$$

All four carbon atoms have steric numbers of 3, so all are sp^2 hybridized. The remaining p_z-orbitals have maximum overlap when the four carbon atoms lie in the same plane; accordingly, this molecule is predicted to be planar. From these four p_z atomic orbitals, four π molecular orbitals can be constructed by combining phases as shown in Figure 18–16. The four electrons that remain after the σ-orbitals are filled go into the two lowest energy π-orbitals. The first of these is bonding among all four carbon atoms, and the second is bonding between the outer carbon-atom pairs but antibonding between the central pair; 1,3-butadiene therefore has stronger and shorter bonds between the outer carbon pairs than between the two central carbon atoms. It is a **conjugated π-electron system,** a system in which two or more double or triple bonds alternate with single bonds. The molecular orbital approach predicts a lower energy in conjugated systems than localized valence-bond theory does. Observation confirms this lowering in energy.

Resonance Hybrids

Conjugation is also seen in some molecules that, in the language of the Lewis theory of Chapter 3, are "resonance hybrids." Benzene is a ring molecule with the formula C_6H_6. Two Lewis structures can be written for it:

These structures are also called "Kekulé structures," after the 19th-century German chemist August Kekulé, who first formulated the ring structure of benzene. The modern view is that the molecule does not flip-flop from moment to moment between the two structures, but rather has a single time-independent electron distribution in which the π bonding takes place through delocalized molecular orbitals. Each carbon atom has sp^2 hybridization, which uses three of its four valence orbitals. The remaining six p_z-orbitals (one from each carbon atom) combine to give six molecular orbitals delocalized over the entire molecule. The π-orbitals and their energy-level diagram are shown in Figure 18–17. The lowest-energy π-orbital has one nodal plane (the plane of the molecule). The next two have two nodal planes each (the plane of molecule and a plane perpendicular to it). The next two have three nodal planes each, and the highest-energy π-orbital (which is strongly antibonding) has four nodal planes. The C_6H_6 molecule has 30 valence electrons,

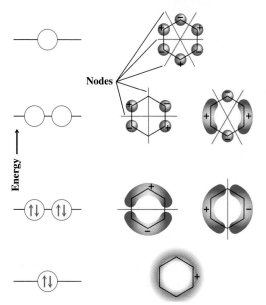

Figure 18–17 The six π molecular orbitals in benzene, viewed from above. The six molecular orbitals arise from overlap among the six $2p_z$ atomic orbitals lying perpendicular to the plane of the molecule. Note the similarity in nodal properties to the standing waves on a loop shown in Figure 16–16. Only the three lowest-energy molecular orbitals are occupied by electrons in benzene in the ground state.

CHEMISTRY IN PROGRESS

Until 1985, just two forms of carbon, diamond and graphite, were known (see Section 3–1). In that year, a new form was discovered: buckminsterfullerene, which consists of ball-like molecules with the formula C_{60}. These molecules have a highly symmetrical structure: 60 carbon atoms arranged in a closed net with 20 hexagonal faces and 12 pentagonal faces. Each pentagonal face touches five hexagonal faces and each hexagonal face touches alternating hexagonal and pentagonal faces (Fig. 18–A). This is exactly the design on the surface of a soccer ball.

C_{60} is named after the architect Buckminster Fuller, who invented the geodesic dome, which the molecular structure of C_{60} also resembles. In C_{60}, every C atom has two single bonds and one double bond for a steric number of 3; all 60 atoms are accordingly sp^2 hybridized. As

in benzene, the π-electrons of the double bonds are delocalized. This means that the 60 p-orbitals (one each from the 60 carbon atoms) do not overlap merely in local double bonds but mix to give 60 molecular orbitals spread over both sides of the entire closed surface of the molecule. The lowest 30 of these molecular orbitals are occupied by the 60 π-electrons. Delocalization makes up for the strain imposed in this structure by the distortion of one of the three bond angles at each carbon atom from the 120 degrees associated with sp^2 hybridization down to 108 degrees. The first conjecture that this structure might be chemically stable came in 1970 and was guided by just such bonding theory.

In 1985, Harold Kroto, Robert Curl, and Richard Smalley were interested in certain long-chain carbon molecules that had been discovered spectroscopically by

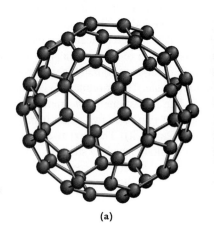

Figure 18–A (a) The structure of C_{60}, buckminsterfullerene. Note the pattern of hexagons and pentagons. (b) The design on the surface of a soccer ball has the same pattern as the structure of C_{60}.

(a)   (b)

of which 24 occupy sp^2 hybrid orbitals and form σ-bonds. When the six remaining valence electrons are placed into the three lowest-energy π-orbitals, the electron distribution that results is the same in each of the six carbon–carbon bonds. As a result, benzene in its ground-state has six carbon–carbon bonds of equal length that are intermediate between single and double bonds in their properties.

Other resonance structures can also be described using delocalized molecular orbitals. For example, as is discussed in Section 3–5, the bent molecule SO_2 has the two resonance Lewis structures

$$\left[\ddot{O}=\ddot{S}-\ddot{O}: \longleftrightarrow :\ddot{O}-\ddot{S}=\ddot{O} \right]$$

The alternative to (or amplification of) the concept of resonance in this case is to say that there is sp^2 hybridization on the central sulfur atom (giving a σ-bond between the S and each O), together with a pair of electrons in a π bonding orbital

The Fullerenes

radioastronomers in the vicinity of red giant stars. They sought to duplicate the conditions near these stars by vaporizing a graphite target with a laser beam. Analysis by mass spectrometry (see Section 1–6) of the debris blasted out of the target serendipitously revealed (along with what they hoped for) the presence of a large proportion of molecules of molar mass 720 g mol^{-1}, which corresponds to the molecular formula C_{60}. Although the amount of C_{60} present in their beam was far too small for isolation and direct determination of structure, they correctly suggested the cage structure shown in Figure 18–A and named it. In 1990, other workers succeeded in synthesizing C_{60} in gram quantities by striking an electric arc between two carbon rods held under an inert atmosphere. The carbon vapors condensed to a soot. Extraction of the soot with an organic solvent and chromatography (see Section 7–7) allowed separation of C_{60} from various impurities. Since 1990, it has been found that C_{60} forms in sooting flames when hydrocarbons are burned. Thus, the new form of carbon was (in the words of one of its discoverers) "under our noses since time immemorial," but it was not seen until a clue came from outer space. Kroto, Curl, and Smalley shared the 1996 Nobel Prize in chemistry for their contribution.

Buckminsterfullerene is not the only new form of carbon that emerges from the chaos of carbon vapor condensing at high temperature. A whole family of closed-cage carbon molecules, the *fullerenes*, has been synthesized. These molecules have formulas like C_{70} or C_{84} or even C_{400}. Proper conditions favor *nanotubes*, which consist of seamless, cylindrical shells of thousands of sp^2 hybridized carbon atoms arranged in hexagons. The ends of the tubes are capped by the introduction of pen-

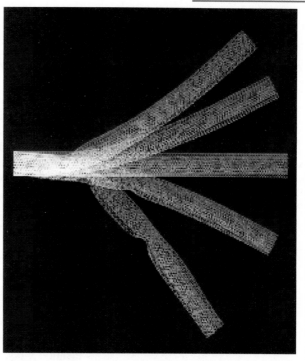

Figure 18–B A simulation of the oscillations experienced by a 300-Å-long multiwalled carbon nanotube after being bent to its point of buckling.

tagons into the hexagonal network. These structures all have a delocalized π-electron system that covers the inner and outer surfaces of the cage or cylinder. Nanotubes and the fullerenes offer exciting prospects for materials science and technical applications. For example, the mechanical properties of nanotubes suggest applications as high-strength fibers (Fig. 18–B).

---

perpendicular to the plane of the molecules and spread out over all three atoms. Delocalizing a pair of bonding electrons over three atoms adds a bond order of $\frac{1}{2}$ to each link. Thus, the bond order of each of the S—O bonds is 1 (from the σ-bond) $+ \frac{1}{2}$ (from the π-bond) $= \frac{3}{2}$. This is intermediate in character between a single and a double bond and is consistent with what is observed experimentally.

18-4 THE INTERACTION OF LIGHT WITH MOLECULES

In Chapter 16, we discussed atomic spectroscopy and showed how the frequency of the light absorbed or emitted by an atom is related to the differences in the energies of its quantum states. In particular, the spectrum of the hydrogen atom was interpreted in terms of its electronic structure. Spectroscopy also has been intensively used to investigate the energy levels of molecules, and careful study of the results leads to useful conclusions about molecular structure, bond lengths, and bond energies. In fact, molecular spectroscopy is the main source of information about chemical bonding.

Molecular Energy Levels

The excited states of atoms arise from the promotion of an electron or electrons into orbitals of higher energy. Excited electronic states in molecules arise similarly and are readily described in the framework of the molecular-orbital theory developed so far. In addition to the excited electronic states, the relative motion of the nuclei must be considered. This was not a factor with atoms, which have only one nucleus. The additional modes of motion that are possible mean that molecules have many more allowed energy levels than atoms. The absorptions and emissions of electromagnetic radiation that accompany transitions between these levels form patterns that are quite complex even in small molecules. This section outlines the changes produced in molecules by radiation of various frequencies. The results are summarized in Table 18–3.

> • Molecules can spin, flex, bend, and stretch in motions that are possible only because they have two or more atoms.

Let us begin with the highest-frequency (highest-energy) photons, those called gamma rays and x-rays. These photons carry enough energy to excite core electrons out of very stable orbitals close to the nuclei up to high-energy outer orbitals or to remove core electrons from the molecule altogether via photoemission. Experimental studies of absorption in the x-ray region have established that the exact frequencies of absorption depend on the environment of the absorbing atom as well as its identity; x-ray absorption thus provides information about bonding.

Light in the ultraviolet and visible region of the spectrum carries somewhat less energy (typically 50 to 500 kJ mol^{-1}), not enough to unseat a tightly bound core electron. Instead, such radiation excites *valence* electrons to higher unoccupied electronic energy levels or frees valence electrons through photoemission. Studies of spectra in this region can reveal the energy ordering of σ and π bonding and antibonding molecular orbitals.

> • Note that the range of energy carried by photons in the ultraviolet and visible region indeed matches the typical energies of chemical bonds (see Section 17-4).

Infrared radiation is less energetic yet. It consists of photons carrying 2 to 40 kJ mol^{-1}, too little energy to perturb the electron clouds of most molecules significantly. This radiation does cause changes in the *vibrational* states of molecules,

Table 18–3
Spectroscopic Experiments

Spectral Region	Frequency (s^{-1})	Energy Levels Involved	Information Obtained
Radio waves	10^7–10^9	Nuclear spin states	Electronic structure near the nucleus
Microwave, far infrared	10^9–10^{12}	Rotational	Bond lengths and bond angles
Near infrared	10^{12}–10^{14}	Vibrational	Stiffness of bonds
Visible, ultraviolet	10^{14}–10^{17}	Valence electrons	Electron configuration
X-ray	10^{17}–10^{19}	Core electrons	Core-electron energies

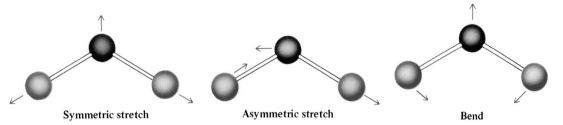

Symmetric stretch   Asymmetric stretch   Bend

Figure 18–18 The three types of vibrational motion possible for a nonlinear triatomic molecule. The displacements of each atom during each type of vibration are shown by arrows.

however. A simple model for the motion of the two nuclei in a diatomic molecule is that they are connected by a spring that allows them to oscillate back and forth. Incoming radiation with a frequency equal to the natural frequency of this oscillation can cause the molecule to absorb energy and reach a higher vibrational state. For polyatomic molecules, several modes of vibrational motion are possible (Fig. 18–18). These different motions have different natural frequencies and so absorb infrared radiation at different wavelengths. Infrared spectroscopy thus gives information about the types of bonds that are present in a molecule, the stiffness of the "springs" connecting the atoms, and the amount of energy needed to dissociate the bond.

Radiation in the far (long wavelength) infrared and microwave regions of the electromagnetic spectrum carries energies that range from 0.001 to 1 kJ mol^{-1}; such radiation excites the low-energy rotational states of a molecule. We think of objects as being able to rotate at any speed, but for an object the size of a molecule, the quantum (discrete) nature of the rotational energy becomes important, and only certain rotational speeds and energies are possible. The rotational energy depends on the masses of the atoms and on the bond lengths and angles, so microwave spectroscopy provides an excellent method of determining molecular structure to high precision.

• A microwave oven operates by exciting the water molecules in the food to higher-energy states—the water molecules gain rotational energy. As they dissipate this energy in collisions, the food is heated.

Finally, **nuclear magnetic resonance (NMR)** spectroscopy employs low-energy (between 0.00002 and 0.00020 kJ mol^{-1}) radio waves to "tickle" the nuclei in a molecule. Some kinds of nuclei have spin (just as electrons have spin), and the energies of different nuclear spin states are split apart by a magnetic field: the stronger the field, the greater the splitting. Nuclei in a magnetic field can change their spin state ("spin-flip") by absorbing or emitting photons of FM radio frequencies. Nuclear spins are sensitive both to the chemical environment of a nucleus (thus to the bonding of the atom) and to the other nuclear spins nearby. NMR spectrometry accordingly offers a way to identify what bonding groups are present in a molecule. For example, ^{13}C nuclei have two spin states: spin up and spin down. When acetic acid (CH_3COOH) is put into a NMR spectrometer, the magnetic field in the device raises the energy of one state and lowers the energy of the other in the ^{13}C atoms that are present. Irradiation with photons of the right frequency delivers energy to spin-flip ^{13}C nuclei from the lower energy spin state to the higher. The exact split in energy between spin states depends on the local chemical environment of the nucleus. Therefore the ^{13}C NMR spectrum of acetic acid has two peaks: a methyl resonance where the ^{13}C nuclei in CH_3 groups absorb energy and, a carboxylic acid resonance where the ^{13}C nuclei in COOH groups absorb energy. The ^{13}C NMR experiment indicates the chemical environment of all the different carbon atoms in even rather complex molecules.

• The ^{13}C NMR experiment succeeds in compounds containing naturally occurring carbon despite the low abundance of ^{13}C nuclei (1.11%).

Figure 18–19 A magnetic resonance imaging (MRI) machine. The patient is placed on the platform and rolled forward into the opening in the magnet.

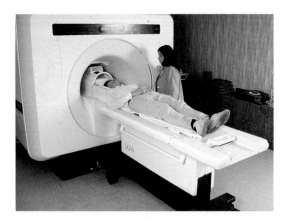

An adaptation of NMR spectrometry that is used in medical diagnosis is called magnetic resonance imaging (MRI). It relies on emission of radio-frequency radiation by the ^1H nuclei in the water contained by the body's organs. The patient is placed in the opening of a large magnet (Fig. 18–19), and a radio transmitter raises the ^1H nuclear spins in the relevant part of the body to their high-energy state. The radio-frequency photons subsequently emitted are detected by a radio receiver coil. The amplitude of the signal indicates the concentration of water present. Thus, MRI can identify tumors by the excess water in their cells. The time delay until emission occurs is related to the type of tissue. Figure 18–20 shows an MRI brain scan.

Although the modes of interaction of radiation with molecules have been described separately, they do not usually occur separately. Most experimental spectroscopic lines correspond to changes in more than one kind of energy. For example, a line in the visible region of the absorption spectrum may connect two energy levels that differ in vibrational and rotational as well as electronic energy (Fig. 18–21).

Absorption Spectra and Color

The absorption of visible or ultraviolet light causes changes in the valence electronic states of molecules. Consider the case of ethylene (C_2H_4). The bonding of the two carbon atoms in ethylene is described as the overlap of sp^2 hybridized atomic orbitals on the two carbon atoms, with the two remaining carbon $2p$-orbitals combined into a π bonding molecular orbital and a π^* antibonding molecular orbital (Fig. 18–22). The lowest-energy state of ethylene has two electrons in the π-orbital. Exciting an electron from the π- to the π^*-orbital gives a higher-energy state of the molecule. An ethylene molecule in this excited state has quite different properties from one in the ground state. The electron in the π^* antibonding orbital cancels the effect of the electron in the π bonding orbital, and the net bond order is reduced to 1 (the σ-bond) from 2. Consequently, the molecule in the excited state is more easily dissociated: it has a lower bond enthalpy and a longer C—C bond length. The wavelength of light required to excite this $\pi \longrightarrow \pi^*$ transition in ethylene is 162 nm. Samples of ethylene strongly absorb light of this wavelength, which is well into the ultraviolet region of the spectrum. Because ethylene does not absorb strongly in the visible region of the spectrum, visible light passes almost unaffected through samples of the substance, which is therefore colorless. As a general rule, substances composed of small molecules with isolated π-bonds (or with σ-bonds only) do not absorb light in the visible region and so are colorless.

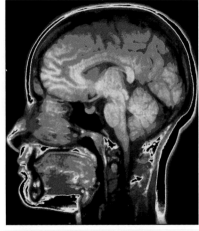

Figure 18–20 A computer-enhanced MRI scan of a normal human brain with the pituitary gland highlighted.

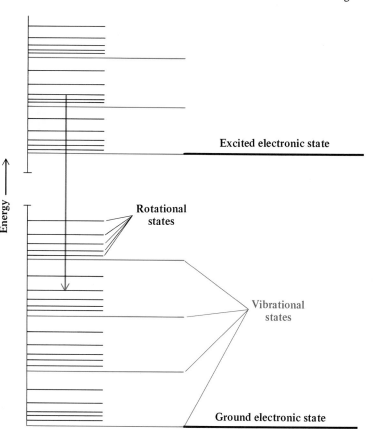

Figure 18-21 An energy-level diagram for a diatomic molecule, showing the electronic, vibrational, and rotational levels. The red arrow indicates one of the many possible transitions; in this particular transition the rotational, vibrational, and electronic states of the molecules all change.

Now, suppose that a molecule has a conjugated π bonding system, that is, bonding in which double bonds alternate with single bonds, as in butadiene or benzene (see Section 18–3). Conjugation lowers the energy of π-orbitals by allowing the orbitals to span several atoms, as in butadiene (see Fig. 18–16), rather than remain localized between a pair of atoms, as in ethylene (see Figure 18–12). Computations of the degree of lowering show (and experiments confirm) that conjugation in general lowers the energy of excited-state orbitals *more* than it does the energy of ground-state π-orbitals. Consequently, the energy differences surmounted in $\pi \longrightarrow \pi^*$ transitions in conjugated molecules are less than those in nonconjugated ones. Increasing the degree of conjugation (adding more alternating single and double bonds) shrinks the difference in energy further. Thus, increased conjugation shifts the energy and the frequency of the absorbed light lower and the wavelength of that light higher (Table 18–4). For a long enough chain, the first

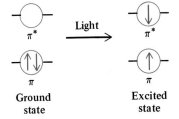

Figure 18-22 Occupancy of the π molecular orbitals in the ground state and first excited state of ethylene (C_2H_4). Ethylene in the excited state is produced by irradiation of ground-state ethylene with ultraviolet light in the appropriate frequency range.

Table 18–4
Absorption of Light by Molecules with π-Electron Systems

Molecule	Number of C=C Bonds	Wavelength of Absorption Maximum (nm)
C_2H_4	1 (nonconjugated)	162
C_4H_6	2 (conjugated)	217
C_6H_8	3 (conjugated)	251
C_8H_{10}	4 (conjugated)	304

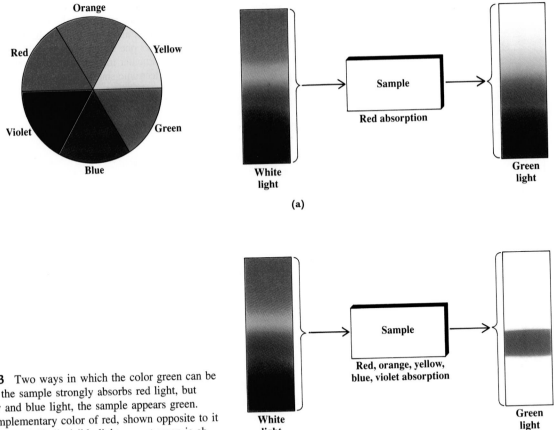

Figure 18-23 Two ways in which the color green can be produced. (a) If the sample strongly absorbs red light, but transmits yellow and blue light, the sample appears green. Green is the complementary color of red, shown opposite to it on the color wheel. (b) If all visible light *except* green is absorbed by the sample, the sample also appears green.

absorption is shifted into the visible region of the spectrum, and the substance takes on color. The color of a material is related to its absorption spectrum in an indirect way. We see the light that is transmitted *through* the material or reflected by the material, *not* the light that is absorbed. In other words, the eye reports the color **complementary** to that which is most strongly absorbed by the sample (Fig. 18–23). An example is the dye beta-carotene (Fig. 18–24), which is responsible for the orange color of carrots, the yellow and orange in certain bird feathers, and the colors of some processed foods. The strength of its absorption at different

Figure 18-24 The molecular structure of the dye beta-carotene. Eleven double bonds alternate with single bonds in this conjugated molecule.

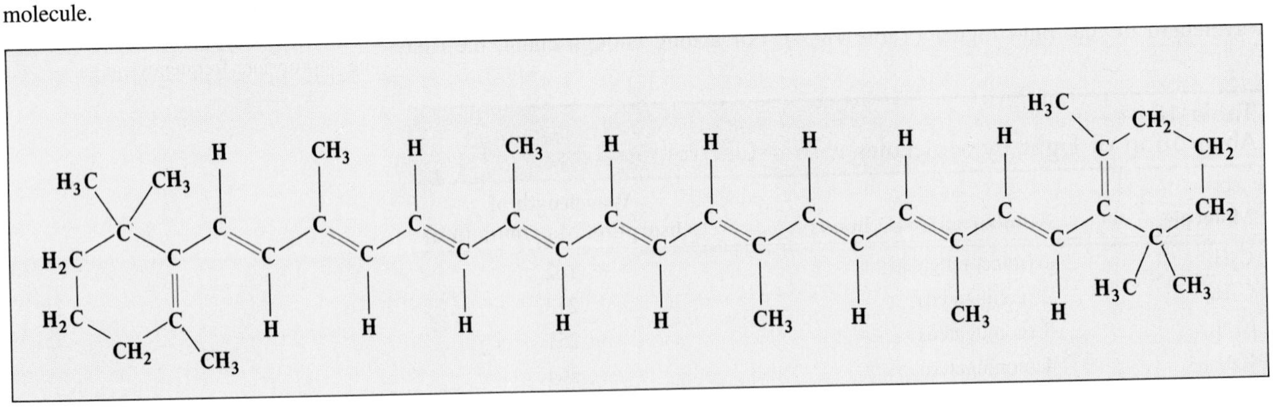

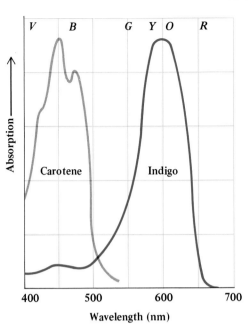

Figure 18-25   The absorption spectra of the two dyes beta-carotene and indigo differ in the visible region, giving them different colors. The letters stand for the color of the light at each wavelength (violet, blue, green, yellow, orange, and red).

wavelengths is graphed in Figure 18–25 in comparison with the absorption of the dye indigo. The beta-carotene absorbs strongly in the blue and violet region of the spectrum, and so it appears orange or yellow (the colors complementary to blue and violet) to an observer. Indigo absorbs at longer wavelengths in the yellow-orange region and appears bluish violet.

EXAMPLE 18-5

Suppose that you set out to design a new green dye. Over what range of wavelengths would you want your trial compound to absorb light?

Solution

A good green dye must transmit green light and absorb other colors of light. This can be achieved by a molecule that absorbs *both* in the violet-blue and in the orange-red regions of the spectrum. Thus, you would look for a dye with strong absorptions in both these regions.

The naturally occurring substance chlorophyll, which is responsible for the green color of grass and leaves, absorbs light over just these wavelength ranges, converting solar energy into chemical energy for the growth of the plant.

Exercise

A translucent layer of red plastic is held over a sheet of white paper so that the sun shines through it, casting a red light onto the paper. The red plastic is then covered with a translucent layer of yellow plastic. (a) What color is seen on the paper? (b) What wavelengths of light does the yellow plastic absorb?

Answer: (a) Orange. (b) Violet (400 to 500 nm, approximately).

Photochemistry

What happens to molecules after they absorb radiation? Some emit one or more photons after a longer or shorter delay and so return to their original states. Others lose all of their added energy in collisions with neighboring molecules that raise the local temperature. Yet others "relax" to ground states by a combination of emissions and collisions. An additional outcome exists if a molecule absorbs enough energy to reach an excited electronic state—it may react chemically. Molecules in excited states often fragment or rearrange to give new molecules. **Photochemistry,** the study of the chemical reactions that follow the excitation of molecules to higher electronic states through absorption of photons, is an active area of modern research.

Some compounds must be stored in the dark because of their great photolytic sensitivity: they break down rapidly when exposed to light. An example is anhydrous hydrogen peroxide, which reacts explosively when illuminated to give water and oxygen:

$$H_2O_2 \longrightarrow 2\ OH \longrightarrow H_2O + \tfrac{1}{2}\ O_2$$

• The common 3% aqueous solution of hydrogen peroxide is completely safe, but it is usually stored in dark brown bottles to slow down its gradual degradation by photochemical reaction.

The rather weak O—O bond in H_2O_2 breaks apart when the energy of a visible photon is added to the molecule in a process called **photodissociation.**

Another type of photochemical reaction involves changes in molecular structure without the complete breaking of bonds. Molecules like those of *trans*-2-butene (shown in Fig. 18–15) absorb ultraviolet light by the excitation of an electron from a π molecular orbital, just as ethylene does (see Fig. 18–22). In the excited electronic state of *trans*-2-butene, the carbon–carbon double bond is effectively reduced to a single bond, and one CH_3 group can rotate relative to the other to form *cis*-2-butene. Absorption of ultraviolet light has led to the formation of an isomer of the original molecule.

Chemiluminescence

In **chemiluminescence,** a chemical reaction produces atoms or molecules in excited energy states that then emit light as they return to their ground states. An example is the highly exothermic reaction of hydrogen peroxide with hypochlorite ion:

$$H_2O_2(aq) + OCl^-(aq) \longrightarrow H_2O(\ell) + Cl^-(aq) + O_2^*$$

The star on the oxygen on the right side of the equation indicates that the atom is in an excited electronic state and can emit light to return to the ground state. Another example is the greenish glow given off by solid white phosphorus in moist air. The mechanism is not fully understood but appears to involve partial oxidation of phosphorus (by the oxygen in the air) in a vapor-rich layer just over the solid to produce excited states of unstable molecules like $(PO)_2$ and HPO than then emit light.

A third example of chemiluminescence is the reaction used in novelty light-sticks (Fig. 18–26). A molecule derived from oxalic acid (by replacement of its two hydrogen atoms with hydrocarbon groups indicated as R—) reacts with hydrogen peroxide according to the equation

$$RO-\overset{\overset{\displaystyle O}{\|}}{C}-\overset{\overset{\displaystyle O}{\|}}{C}-OR + H_2O_2 \longrightarrow \underset{\underset{\displaystyle O-O}{|\quad|}}{\overset{\overset{\displaystyle O\quad O}{\|\quad\|}}{C-C}} + 2\ ROH$$

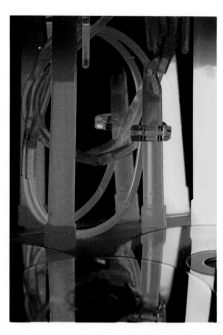

Figure 18–26 These lightsticks and lighttubes emit light through chemiluminescence. The different colors arise from the use of different fluorescers in combination with the same energizing chemical reaction.

The product of this reaction, C_2O_4, is a high-energy species that has not been isolated. It decomposes rapidly and exothermically to form carbon dioxide. It can also pass some of its energy along to a molecule of some fluorescent substance (represented by the symbol "Fl") if such is present:

$$Fl + C_2O_4 \longrightarrow 2\ CO_2 + Fl^*$$

where the star indicates that the fluorescer is in an excited electronic state. As molecules of the fluorescer return to their ground states, they emit visible light. In commercial lightsticks, the fluorescer is mixed with the oxalic-acid derivative, and the hydrogen peroxide and a catalyst are kept separate in an inner sealed tube with thin glass walls. Bending the lightstick breaks the glass tube, the reactants mix, and the reaction begins.

Bioluminescence is the chemical generation of light by living organisms. The most familiar example is the light produced by the firefly, which uses chemical energy from ATP (see Fig. 11–A) in a complex mechanism to excite molecules of substances called "luciferins." The flash of the firefly comes as luciferin in an excited state returns to its ground state by emitting a photon. The process is so efficient that luciferins can be used in biological analysis to measure small concentrations of ATP.

18–5 ATMOSPHERIC CHEMISTRY AND AIR POLLUTION

Nowhere is the ability of light to cause chemical reactions so apparent as in the earth's atmosphere. Although the chemical composition given in Table 5–1 is correct for the average make-up of the portion of the atmosphere closest to the earth's surface, it does not do justice to the strong variations in atmospheric properties with altitude, the dramatic role of local fluctuations in the concentration of trace gases, and the dynamics underlying the average concentrations seen. The atmo-

Figure 18-27 The temperature and pressure of the atmosphere vary with altitude. The temperature variation shows the layered structure of the atmosphere. The pressures show how rapidly the atmosphere thins: thus, 90% of the atmosphere lies within 20 km of the surface and 99.999% within 80 km.

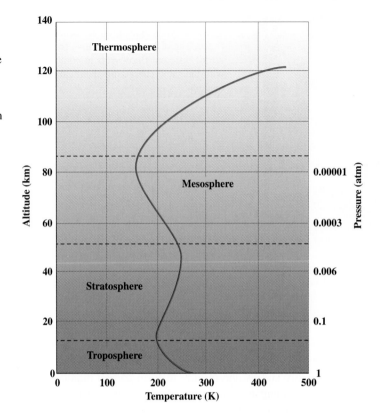

sphere is a complex chemical system that is far from equilibrium. Its properties are determined by an intricate combination of thermodynamic and kinetic factors. It is a multilayered structure (Fig. 18–27), bathed in radiation from the sun and interacting at the bottom with the oceans and land masses. At least four layers are identifiable, each with a characteristic variation of temperature. In the outer two layers (the **thermosphere** and the **mesophere**), atmospheric density is low, and the intense radiation from the sun leads to extensive ionization of the particles that are present. The third layer, the **stratosphere,** is the region between 12 and 50 km (approximately) above the earth's surface. The **troposphere** is the lowest 12 km out from the earth's surface. In the troposphere (like the mesophere), warmer air lies beneath cooler air. This is a dynamically unstable situation because warm air is less dense and tends to rise, so convection takes place, mixing the gases in the troposphere and determining the weather. In the stratosphere (like the thermosphere), the temperature increases with height, and there is little vertical mixing from convection. Mixing across the borders between the four layers is also slow, so most of the chemical processes in the layers can be described separately.

Atmospheric chemistry is not a new subject but dates back to the 18th century. Cavendish, Priestley, Lavoisier, and Ramsay were the first scientists to study the composition of the atmosphere. In recent years, atmospheric chemistry has developed in two different, but related, directions. First, the sensitivity of chemical analysis has greatly improved, and analyses for substances at concentrations below the part-per-billion (ppb) level are now carried out routinely. Airplanes and satellites allow the full distribution of trace substances to be mapped on a global scale.

Second, advances in gas-phase chemical kinetics have led to a better quantitative understanding of the ways in which substances in the atmosphere react. Much of the impetus for these studies of atmospheric chemistry has come from concern over the effect of air pollution on life.

EXAMPLE 18-6

In a sample of polluted air at 1 atm pressure and 300 K, the amount of SO_2 was measured to be 200 ppbv (200 parts per billion by volume). Calculate the partial pressure of the SO_2 and its concentration in moles per liter.

Solution

First, 200 ppb by volume means that if all of the SO_2 could be removed from a billion liters of the polluted air, it would have a volume of 200 liters (under the same conditions of temperature and pressure). By Avogadro's hypothesis (see Section 1–4), there must be 200 molecules of SO_2 for every billion molecules in the sample. The mole fraction of SO_2 is then

$$X_{SO_2} = \frac{n_{SO_2}}{n_{tot}} = \frac{\text{molecules } SO_2}{\text{total molecules}} = \frac{200}{1.0 \times 10^9} = 2.0 \times 10^{-7}$$

The partial pressure of a gas is (from Section 5–6) the product of its mole fraction and the total pressure (1 atm here), so

$$P_{SO_2} = 2.0 \times 10^{-7} \text{ atm}$$

Finally, the concentration in moles per liter (n/V) is, from the ideal gas law,

$$[SO_2] = \frac{n_{SO_2}}{V} = \frac{P_{SO_2}}{RT}$$

$$= \frac{2.0 \times 10^{-7} \text{ atm}}{(0.08206 \text{ L atm mol}^{-1} \text{ K}^{-1})(300 \text{ K})} = 8.1 \times 10^{-9} \text{ mol L}^{-1}$$

Exercise

Cigarette smoke contains 200 to 400 ppmv (parts per million by volume) of carbon monoxide (CO). Find the range of partial pressure (in atmospheres) and the range of concentration (in *molecules* per liter) of CO in a puff of cigarette smoke (at ordinary room conditions).

Answer: $P_{CO} = 2 \times 10^{-4}$ to 4×10^{-4} atm; 5×10^{18} to 1×10^{19} molecules of CO per liter.

Stratospheric Chemistry

As Figure 16–6 shows, the sun emits light over a broad range of wavelengths, with the highest intensity at about 500 nm, in the visible region of the spectrum. The intensity at wavelengths down to 100 nm in the ultraviolet is quite substantial, and because the energy $h\nu$ carried by a photon is inversely proportional to the wavelength λ ($h\nu = hc/\lambda$), the ultraviolet photons carry much more energy than do photons of visible light. If substantial numbers of these photons penetrated to the earth's surface, much harm would be done to living organisms. Fortunately, the outer portions of the atmosphere (especially the thermosphere) absorb ultraviolet radiation through the photodissociation of oxygen molecules:

$$O_2 + h\nu \longrightarrow 2\ O$$

where "$h\nu$" symbolizes a photon. This reaction reduces the number of high-energy photons, especially those with wavelengths less than 200 nm, reaching the lower parts of the atmosphere.

EXAMPLE 18-7

The bond dissociation energy of O_2 is 496 kJ mol^{-1}. Calculate the maximum wavelength of light that can photodissociate an oxygen molecule.

Solution

Because 496 kJ dissociates one mole of O_2 molecules, the energy to dissociate one molecule equals 496 kJ divided by Avogadro's number:

$$\frac{496 \times 10^3 \text{ J mol}^{-1}}{6.022 \times 10^{23} \text{ mol}^{-1}} = 8.24 \times 10^{-19} \text{ J}$$

A photon carrying this energy has a wavelength λ given by

$$8.24 \times 10^{-19} \text{ J} = h\nu = hc/\lambda$$

so that

$$\lambda = \frac{hc}{8.24 \times 10^{-19} \text{ J}} = \frac{(6.626 \times 10^{-34} \text{ J s})(2.998 \times 10^8 \text{ m s}^{-1})}{8.24 \times 10^{-19} \text{ J}}$$

$$= 2.41 \times 10^{-7} \text{ m} = \boxed{241 \text{ nm}}$$

Any photons with wavelengths *shorter* than this are energetic enough to dissociate oxygen molecules. Those with wavelengths shorter than 200 nm are the most efficient in causing photodissociation.

Exercise

The bond dissociation energy of nitrogen is 943 kJ mol^{-1}. Compute the minimum frequency of light that can photodissociate nitrogen. Calculate the minimum wavelength of such light.

Answer: $\nu = 2.36 \times 10^{15}$ s^{-1}; $\lambda = 127$ nm. Nitrogen absorbs only higher-frequency (shorter-wavelength) ultraviolet light. It lets through ultraviolet light that oxygen absorbs.

Rather few photons with wavelengths less than 200 nm are able to penetrate to the stratosphere, but those that do establish a small concentration of oxygen atoms in that layer. These atoms can collide with the much more prevalent oxygen molecules to form excited-state molecules of ozone (O_3):

$$O + O_2 \rightleftharpoons O_3^*$$

The star indicates that the ozone that is formed is in a highly excited state. It has 106 kJ mol^{-1} more energy than does ground-state ozone. The excited-state O_3^* can dissociate in a unimolecular reaction back to O and O_2, as indicated by the reverse arrow in this equilibrium. Alternatively, if another atom or molecule comes along and collides with it soon enough, it can transfer some of its excess energy to that atom or molecule, a process represented as

$$O_3^* + M \longrightarrow O_3 + M$$

where M stands for the atom or molecule with which it collides (the most likely ones are oxygen and nitrogen molecules, because these are the most abundant species in the atmosphere). The net effect of these two reactions is to produce a small concentration of ozone in the stratosphere.

Under laboratory conditions, ozone is an unstable compound. Its conversion to oxygen is thermodynamically favored

$$O_3(g) \longrightarrow \tfrac{3}{2} O_2 \qquad \Delta G^\circ = -163 \text{ kJ}$$

but takes place quite slowly in the absence of light. In the stratosphere, ozone photodissociates readily to O_2 and O in a reaction that requires about 106 kJ per mole of ozone, much less than the dissociation of O_2:

$$O_3 + h\nu \longrightarrow O_2 + O$$

This process occurs most efficiently for wavelengths between 200 and 350 nm. The energy of light of these wavelengths is too small to be absorbed by molecular oxygen but quite large enough to damage organisms at the earth's surface. The ozone layer shields the earth's surface from 200- to 350-nm ultraviolet radiation coming from the sun. The balance between formation and photodissociation leads to a steady-state concentration of more than 10^{15} molecules of ozone per liter in the stratosphere.

Certain types of air pollution give rise to *radicals* that catalyze the depletion of ozone in the stratosphere. Radicals are chemical species containing unpaired electrons. They are usually formed by the breaking of a covalent bond to form a pair of neutral species. One pressing concern involves chlorofluorocarbons, compounds of chlorine, fluorine, and carbon that are used as refrigerants and in some aerosol sprays. These compounds are nonreactive at sea level but are photodissociated in the stratosphere in reactions such as

* Radicals were previously mentioned in connection with reaction mechanisms (see Section 14-5).

$$CCl_2F_2 + h\nu \longrightarrow CClF_2 + Cl$$

The atomic chlorine thus released soon reacts with O atoms or O_3 to give ClO:

$$Cl \longrightarrow ClO$$
$$Cl + O_3 \longrightarrow ClO + O_2$$

The ClO radical is the immediate culprit in the destruction of stratospheric ozone: local increases in ClO concentration are directly correlated with decreases in O_3 concentration. It catalyzes the conversion of O_3 to O_2, probably by the mechanism

$$2 \text{ ClO} + M \longrightarrow ClOOCl + M$$
$$ClOOCl + h\nu \longrightarrow ClOO + Cl$$
$$ClOO + M \longrightarrow Cl + O_2 + M$$
$$\underline{2 \times (Cl + O_3 \longrightarrow ClO + O_2)}$$
$$\text{net reaction: } 2 O_3 \longrightarrow 3 O_2$$

where M stands for N_2 and O_2 molecules. This catalytic cycle is ordinarily disrupted rather quickly as Cl reacts with other stratospheric species to form less reactive "reservoir molecules" such as HCl and $ClONO_2$. The lack of mixing in the stratosphere keeps the reservoir molecules around for long periods, however.

Figure 18–28 The extent of the ozone hole over the South Pole near its maximum in October, 1994. 100 Dobson units of ozone is equivalent to a 1-mm-thick layer of pure ozone gas at STP (273 K and 1 atm pressure).

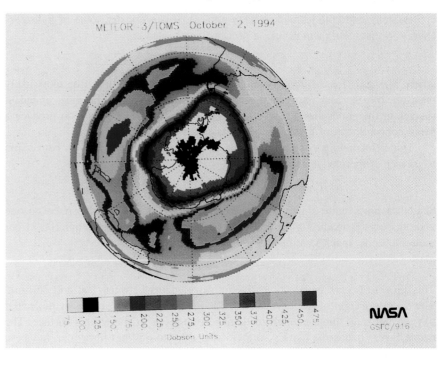

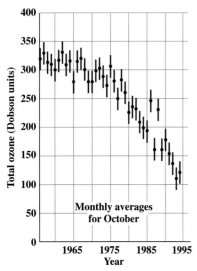

Figure 18–29 Worsening ozone depletion over Halley Bay, Antarctica. These measurements were all taken in the Antarctic spring (October), when depletion is at its worst.

• Exposure to ultraviolet radiation appears to cause certain skin cancers. A predicted consequence of depletion of the stratospheric ozone layer is an increased worldwide incidence of skin cancer.

It is believed that atomic Cl escapes its reservoirs through heterogeneous reactions on the icy stratospheric clouds that form during the intense Antarctic winter. The escaped Cl gives rise to the annual "ozone holes" above the Antarctic (Fig. 18–28). During these episodes more than 70% of the ozone is depleted before the values rise again. The severity of the annual episode of ozone depletion above the Antarctic increases each year (Fig. 18–29), and measurements have shown smaller, but still serious, depletions over other parts of the globe. International agreement (the Montreal protocol) has ended the production of chlorofluorocarbons in the industrial countries, but production continues in developing nations. Even if all production stopped today, stratospheric ozone concentrations would continue to decrease for some time as chlorofluorocarbons already released mix upward into the stratosphere.

Tropospheric Chemistry

The troposphere is the part of the atmosphere in contact with the earth's surface. It is therefore most directly and immediately influenced by human activities, especially by the gases or small particles put into the air by automobiles, power plants, and factories. Some air pollutants have long lifetimes and are spread fairly evenly over the earth's surface; others are more ephemeral and attain large concentrations only around particular cities or industrial areas.

The oxides of nitrogen are major air pollutants. Their persistence in the atmosphere shows the importance of kinetics, as opposed to thermodynamics, in the chemistry of the atmosphere. All of the oxides of nitrogen are thermodynamically unstable with respect to the elements at 25°C, as shown by their positive standard Gibbs functions of formation at that temperature. They form by direct reaction of nitrogen and oxygen whenever air is heated to high enough temperatures, either in

CHEMISTRY IN YOUR LIFE

Sunscreens

Ultraviolet radiation of wavelengths between 290 and 320 nm (sometimes called "UV-B") penetrates to the epidermal layer of the skin, where it causes sunburn. Repeated exposure to UV-B appears to cause certain cancers. Death rates from skin melanomas in the United States correlate positively with southerly latitude: a white male from Florida is about twice as likely to develop this form of cancer as one from Montana. Ozone in the stratosphere absorbs much UV-B, but some penetrates even an undepleted ozone layer. Hence the growing popularity of sunscreen lotions and creams among those working or playing out of doors.

The active ingredients in these products are compounds that absorb ultraviolet photons by undergoing a transition to an excited electronic state. For example, the compound *para*-aminobenzoic acid (see Fig. 18–C) absorbs strongly at a wavelength of 283 to 289 nm. This molecule has numerous electronic energy levels and many transitions connecting them. In one, an electron is promoted from a π molecular orbital mainly associated with the $C=O$ group to a π^* molecular orbital in the same region. The resulting excited-state molecule quickly relaxes to its ground state, either by emitting radiation or by collisions with other molecules that transfer the energy to the surroundings as heat. This step makes it ready to

absorb another UV photon. The re-emitted radiation is at longer wavelengths than the ultraviolet and is not harmful to the skin. Certain sunscreens, however, give the skin a faint bluish cast as their active ingredients re-emit absorbed energy at a visible wavelength.

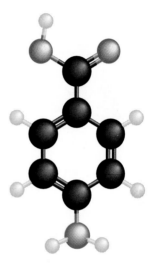

Figure 18–C The structure of *para*-aminobenzoic acid. Other sunscreen compounds have structures that are close variants.

an industrial process or in the engine of a car. They accumulate to much higher than equilibrium concentrations in the atmosphere because they decompose slowly. Interconversions among NO, NO_2, and N_2O_4, the most important nitrogen oxides in air pollution, are rapid and depend strongly on temperature, and so in pollution reports they are generally grouped together as "NO_x." Photochemical smog (Fig. 18–30) is formed by the action of light on nitrogen dioxide, followed by subsequent reaction to produce ozone:

$$NO_2 + h\nu \longrightarrow NO + O$$
$$O + O_2 + M \longrightarrow O_3 + M \qquad (M = N_2 \text{ or } O_2)$$

It may seem paradoxical that the problems with ozone in the stratosphere involve its *depletion*, but in the troposphere its *production*. In fact, the different effects arise from the different concentrations of species in the two layers of the atmosphere. Although ozone is beneficial in reducing the radiation penetrating to the earth's surface, it is quite harmful in direct contact with organisms because of its strong oxidizing power; levels of 10 to 15 ppmv are sufficient to kill small mammals, and a concentration as low as 3 ppmv in urban air triggers an "ozone

Figure 18-30 Photochemical smog casts a pall over Mexico City, which is said to have the highest level of air pollution in the world.

752   CHAPTER 18   Molecular Orbitals, Spectroscopy, and Atmospheric Chemistry

Figure 18–31 The effect of acid rain on a stand of trees and the natural regeneration of young spruce trees in the Karkonoski National Park, southwestern Poland. Here, forests have experienced acid rain in which the pH was as low as 1.7.

alert." In addition, ozone can react with incompletely oxidized organic compounds from gasoline and with nitrogen oxides in the air to produce harmful irritants such as methyl nitrate (CH_3NO_3).

The oxides of sulfur create global pollution problems because they have longer lifetimes in the atmosphere than the oxides of nitrogen. A certain fraction of the SO_2 and SO_3 in the air originates from biological processes and from volcanoes, but most comes from the oxidation of sulfur present in petroleum and in coal that is burned for fuel. If the sulfur is not removed from the fuel or the exhaust gas, SO_2 enters the atmosphere as a stable but reactive pollutant. Further oxidation by radicals leads to sulfur trioxide:

$$SO_2 + OH \longrightarrow SO_2OH$$
$$SO_2OH + O_2 \longrightarrow SO_3 + OOH$$
$$OOH \longrightarrow O + OH$$

The atomic oxygen produced can then react with molecular oxygen to form ozone.

Acid rain results from the reaction of NO_2 and SO_3 with hydroxyl radicals and water vapor in the air to form nitric acid (HNO_3) and sulfuric acid (H_2SO_4), which are soluble in water and return to the earth in the rain. This process occurs naturally to a certain extent and is beneficial in providing the critical nutrients nitrogen and sulfur to soils that are poor in the two elements. Large-scale emission of sulfur oxides from burning coal and from metal smelting, however, has altered the atmospheric balance and increased the acidity of rainfall in much of the northern hemisphere to harmful levels. The effects are to reduce fish populations in lakes, to damage forests severely (Fig. 18–31), to speed the corrosion of metals, and to erode stone and marble structures. To combat acid rain, we must reduce emissions of SO_2. The methods currently used involve powdered limestone, which decomposes at high temperatures to lime and carbon dioxide.

$$CaCO_3(s) \longrightarrow CaO(s) + CO_2(g)$$

The lime can combine with sulfur dioxide (in an acid–base reaction) to produce solid calcium sulfite:

$$CaO(s) + SO_2(g) \longrightarrow CaSO_3(s)$$

The enormous problem of the disposal of the $CaSO_3$ solid waste that is produced by this process has slowed its adoption as an antipollution measure. The best option would be to convert the SO_2 to economically useful sulfuric acid, and some progress has been made in this direction, as discussed in Chapter 22.

The Greenhouse Effect

As we have seen, most of the short-wavelength photons from the sun are absorbed by the outer atmosphere and do not reach the earth's surface. The radiation that does reach the earth maintains a livable temperature, in balance with re-radiation from the earth back into space. Certain gases in the troposphere play a crucial role in this balance because they absorb infrared radiation emitted by the warm surface of the earth rather than letting it pass out to space (Fig. 18–32). Water vapor and water in the form of clouds are the most obvious of these gases. On cloudy winter nights, the temperature does not fall as low as it does on clear nights because clouds provide a thermal blanket that absorbs outgoing radiation with wavelengths near 20,000 nm.

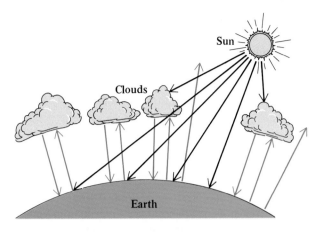

Figure 18–32 Visible light from the sun strikes the earth's surface and heats it. Much energy is radiated back into space in the longer-wavelength infrared region of the spectrum, but molecules in the atmosphere and in clouds can block some of this loss, keeping the lower atmosphere warmer.

Two other gases that absorb infrared radiation to a significant extent are carbon dioxide and methane. Both are uniformly distributed in fairly low concentrations throughout the troposphere. The concentrations of both these gases have increased steadily over the 200 to 300 years since the beginning of the industrial revolution. The tremendous increase in the burning of fossil fuels as energy sources is without a doubt the reason for this trend (Fig. 18–33). It is estimated that by the year 2050, the concentration of CO_2 in the atmosphere will have doubled over premodern values.

Such changes are viewed with alarm because of the **greenhouse effect.** This refers to a global increase in average surface temperature that will occur if heat given off by the earth's surface is prevented from escaping to space by higher

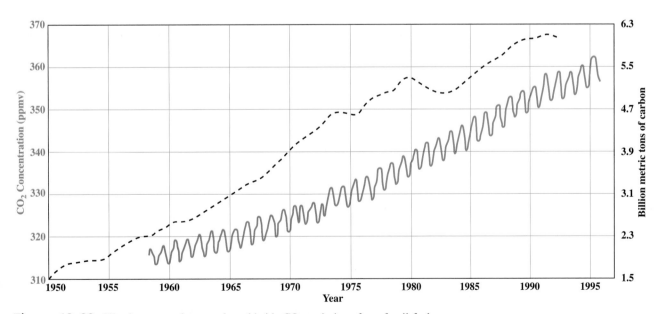

Figure 18–33 Worrisome trends: annual worldwide CO_2 emissions from fossil fuel combustion, 1950–1992 (black); average monthly concentration of atmospheric carbon dioxide at Mauna Loa Hawaii, 1958–1996 (red). The latter varies seasonally.

concentrations of gases that absorb infrared radiation. An increase in average surface temperature of 2 to 5°C over the 21st century would have profound climatic repercussions, including the melting of some polar ice, a consequent increase in the level of the sea, and the conversion of arable land to desert. To prevent these undesirable changes, it will be necessary to develop new energy sources that are not based on the burning of fossil fuels by early in the 21st century.

SUMMARY

18-1 Molecular orbitals of diatomic molecules can be formed from superpositions of atomic orbitals: constructive (in-phase) interference gives **bonding orbitals,** and destructive (out-of-phase) interference gives **antibonding orbitals.** In **correlation diagrams** (energy-level diagrams for molecules), the molecular orbitals are arranged in order of increasing energy and filled with available electrons to determine the ground-state electron structure of the molecule. The molecular orbital description of bonding agrees with the Lewis model in predicting **bond orders** and goes beyond the Lewis model to explain the paramagnetism of O_2 and other species.

18-2 Valence-bond theory focuses on orbitals localized as individual bonds rather than delocalized over the entire molecule. In polyatomic molecules, atomic orbitals on central atoms are first mixed to form new **hybrid atomic orbitals** that contain both s and p character. These hybrid orbitals project from the central atoms in such a way as to minimize electron–electron repulsion, as in the VSEPR model. As the other atoms of the molecule are brought up, electron-pair σ-bonds are formed that are largely localized between pairs of atoms. Multiple bonds result from the occupancy of π **molecular orbitals,** which are formed from p-orbitals that are left over after hybridization is completed. Rotation about double bonds requires breaking the π-bond, which requires energy. This allows the existence of *cis–trans* isomers in compounds with double bonds.

18-3 Isolated π-orbitals are localized between atoms, but **conjugated π-orbitals,** which result from alternating single and double bonds, are better described by delocalized molecular orbitals. The bonding in molecules represented by resonance Lewis structures can be understood in terms of a valence-bond framework of σ-bonds overlaid by π-orbitals that are spread over the entire molecule. The recently discovered fullerenes are examples.

18-4 X-rays are energetic enough to remove core electrons from molecules; ultraviolet and visible radiation can remove valence electrons from molecules or raise molecules to excited electron states. Lower-energy radiation excites vibrations of atoms relative to each other or excites rotations of the whole molecule. In **nuclear magnetic resonance (NMR),** a magnet splits the energies of nuclear spin states in molecules, and transitions between these states are observed at radio frequencies: the exact energy of the split depends on the chemical environment of the nucleus. Measurements of absorptions and emissions of radiation by molecules provide information on bond lengths, bond energies, and bond angles. Light can induce chemical reactions, exciting molecules to higher-energy electronic states from which they break apart or rearrange. The study of such processes is **photochemistry.** In **chemiluminescence,** chemical reactions generate light as they proceed.

18-5 In the outer layers of the atmsophere, **photodissociation** of oxygen molecules absorbs much of the light of highest frequency. In the **stratosphere,** the resulting oxygen atoms collide with oxygen molecules to produce ozone. Ozone in turn absorbs at somewhat longer wavelengths and plays a crucial role in shielding the surface of the earth from harmful ultraviolet light. The destruction of stratospheric ozone by atmospheric pollutants, especially chlorofluorocarbons, is a very serious concern. Other pollution concerns focus on the **troposphere,** the lowest level of the atmosphere. Emissions of oxides of nitrogen and sulfur can lead to irritating and dangerous photochemical smog and to acid rain. Burning of coal, oil, and natural gas has increased the amounts of carbon dioxide in the atmosphere and may cause a serious increase in global temperatures over the next century through the **greenhouse effect.**

PROBLEMS

Note: Answers to blue-numbered problems are given in Appendix F. Problems that are more challenging are indicated with asterisks.

Diatomic Molecules

1. (See Example 18–2.) Write the ground-state valence-electron configurations for the nine diatomic ions X_2^{2+} where X stands for the elements from He to Ne in the periodic table. Give the bond order of each species.

2. (See Example 18–2.) Write the ground-state valence-electron configurations for the seven diatomic ions X_2^- where X stands for the elements from Li to F in the periodic table. Give the bond order of each species.

3. If an electron is removed from a fluorine molecule, an F_2^+ molecular ion is formed.
   (a) Give the ground-state valence-electron configurations for F_2 and F_2^+.
   (b) Give the bond order of each species.
   (c) Predict which species should be paramagnetic.
   (d) Predict which species has the larger bond energy.

4. When one electron is added to an oxygen molecule, a superoxide ion (O_2^-) is formed. The addition of two electrons gives a peroxide ion (O_2^{2-}). Removal of an electron from O_2 leads to O_2^+.
   (a) Construct the correlation diagram for O_2^{2-}.
   (b) Give the ground-state electron configuration for each of the following: O_2^+, O_2, O_2^-, O_2^{2-}.
   (c) Give the bond order of each species.
   (d) Predict which species should be paramagnetic.
   (e) Predict the order of increasing bond energy among the species.

5. Predict the ground-state valence-electron configuration and the total bond order for the molecule S_2, which forms in the gas phase when sulfur is heated to a high temperature. Is S_2 paramagnetic or diamagnetic? (*Hint*: Make use of the group relationship between sulfur and oxygen.)

6. Predict the ground-state valence-electron configuration and the total bond order for the molecule I_2. Is I_2 paramagnetic or diamagnetic? (*Hint*: Make use of the group relationship between iodine and fluorine.)

7. For each of the following valence-electron configurations of a homonuclear diatomic molecule or molecular ion, identify the element X, Q, or Z:
   (a) X_2: $(\sigma_{2s})^2(\sigma_{2s}^*)^2(\sigma_{2p_z})^2(\pi_{2p})^4(\pi_{2p}^*)^4$
   (b) Q_2^+: $(\sigma_{2s})^2(\sigma_{2s}^*)^2(\pi_{2p})^4(\sigma_{2p_z})^1$
   (c) Z_2^-: $(\sigma_{2s})^2(\sigma_{2s}^*)^2(\sigma_{2p_z})^2(\pi_{2p})^4(\pi_{2p}^*)^3$

8. For each of the following valence-electron configurations of a homonuclear diatomic molecule or molecular ion, identify the element X, Q, or Z:
   (a) X_2: $(\sigma_{2s})^2(\sigma_{2s}^*)^2(\sigma_{2p_z})^2(\pi_{2p})^4(\pi_{2p}^*)^2$
   (b) Q_2^-: $(\sigma_{2s})^2(\sigma_{2s}^*)^2(\pi_{2p})^3$
   (c) Z_2^{2+}: $(\sigma_{2s})^2(\sigma_{2s}^*)^2(\sigma_{2p_z})^2(\pi_{2p})^4(\pi_{2p}^*)^2$

9. For each of the electron configurations in problem 7, determine the total bond order of the molecule or molecular ion.

10. For each of the electron configurations in problem 8, determine the total bond order of the molecule or molecular ion.

11. For each of the electron configurations in problem 7, determine whether the molecule or molecular ion is paramagnetic or diamagnetic.

12. For each of the electron configurations in problem 8, determine whether the molecule or molecular ion is paramagnetic or diamagnetic.

13. Following the pattern of Figure 18–7, work out the correlation diagram for the CN molecule, showing the relative energy levels of the atoms and the bonding and antibonding orbitals of the molecule. Indicate the occupation of the molecular orbitals by arrows. State the order of the bond, and comment upon the magnetic properties of CN.

14. Following the pattern of Figure 18–7, work out the correlation diagram for the BeN molecule, showing the relative energy levels of the atoms and the bonding and antibonding orbitals of the molecule. Indicate the occupation of the molecular orbitals by arrows. State the order of the bond, and comment upon the magnetic properties of BeN.

15. The bond length of the transient diatomic molecule CF is 1.291 Å. Explain why the CF bond shortens with the loss of an electron. Refer to the proper molecular-orbital correlation diagram.

16. The compound nitrogen monoxide (NO) forms when the nitrogen and oxygen in air are heated. Predict whether the nitrosyl ion (NO^+) has a shorter or a longer bond than the NO molecule. Is NO^+ paramagnetic like NO, or diamagnetic?

17. What would be the electron configuration for an HeH^- molecular ion? What bond order would you predict? How stable should such a species be?

18. The molecular ion HeH^+ has an equilibrium bond length of 0.774 Å. Draw an electron-correlation diagram for this ion, indicating the occupied molecular orbitals. Is HeH^+ paramagnetic? When the HeH^+ ion dissociates, is a lower-energy state reached by forming $He + H^+$ or $He^+ + H$?

Polyatomic Molecules

19. (See Example 18–3.) Formulate a localized bond picture for the amide ion (NH_2^-). What hybridization do you expect the central nitrogen atom to have, and what geometry do you predict for the molecular ion?

20. Formulate a localized bond picture for the hydronium ion (H_3O^+). What hybridization do you expect the central oxygen atom to have, and what geometry do you predict for the molecular ion?

21. (See Example 18–3.) Write a Lewis electron-dot structure for each of the following molecules and ions. Formulate the hybridization for the central atom in each case, and give the molecular geometry.
    (a) CCl_4 (d) CH_3^-
    (b) CO_2 (e) BeH_2
    (c) OF_2

22. (See Example 18–3). Write a Lewis electron-dot structure for each of the following molecules and ions. Formulate the

hybridization for the central atom in each case, and give the molecular geometry.
    (a) BF_3 (d) CS_2
    (b) BH_4^- (e) CH_3^+
    (c) PH_3

23. Describe the hybrid orbitals used by the chlorine atom in the ClO_3^+ and ClO_2^+ molecular ions. Sketch the expected geometries of these ions.

24. Describe the hybrid orbitals used by the chlorine atom in the ClO_4^- and ClO_3^- molecular ions. Sketch the expected geometries of these ions.

25. The sodium salt of the unfamiliar orthonitrate ion (NO_4^{3-}) has been prepared. What hybridization is expected on the N atom at the center of this ion? Predict the geometry of the NO_4^{3-} ion.

26. Describe the hybrid orbitals used by the carbon atom in $N\equiv C-Cl$. Predict the geometry of the molecule.

27. How many σ-bonds and how many π-bonds does each of the following molecules contain?
    (a) ethylene (C_2H_4)
    (b) fluorine (F_2)
    (c) propane (C_3H_8)
    (d) carbon tetrachloride (CCl_4)

28. How many σ-bonds and how many π-bonds does each of the following molecules contain?
    (a) nitrogen (N_2)
    (b) formaldehyde (CH_2O)
    (c) boron trifluoride (BF_3)
    (d) acetylene (C_2H_2)

29. (See Example 18–4.) The compound acetone has the chemical formula $CH_3(CO)CH_3$. Two of the carbon atoms and the oxygen atom are all bonded to the third carbon atom. Draw the Lewis structure for this molecule, give the hybridization of each carbon atom, and describe the π-orbitals and the number of electrons that occupy each one. Sketch the three-dimensional structure of the molecule, showing all angles.

30. (See Example 18–4.) The compound acetic acid has the chemical formula CH_3COOH. Both oxygen atoms are bonded to the second carbon atom. Draw the Lewis structure for this molecule, give the hybridization of each carbon atom, and describe the π-orbitals and the number of electrons that occupy each one. Draw the three-dimensional structure of the molecule, showing all angles.

31. (See Example 18–4.) Acetic acid can be made by the oxidation of acetaldehyde (CH_3CHO). Molecules of acetaldehyde have a CH_3 group, an oxygen atom, and a hydrogen atom attached to a carbon atom. Draw the Lewis structure for this molecule, give the hybridization of each carbon atom, and describe the π-orbitals and the number of electrons that occupy each one. Draw the three-dimensional structure of the molecule, showing all angles.

32. (See Example 18–4.) Acrylic fibers are polymers made from a starting material called acrylonitrile ($H_2C(CH)CN$). In acrylonitrile a $C\equiv N$ group replaces an H atom on ethylene.

Draw the Lewis structure for this molecule, give the hybridization of each carbon atom, and describe the π-orbitals and the number of electrons that occupy each one. Draw the three-dimensional structure of the molecule, showing all angles.

33. In Section 18–2, it was explained that the compound 2-butene exists in two isomeric forms that can be interconverted only by breaking the central double bond. A third isomer also exists, called 2-methylpropene. In it, both CH_3 groups are attached to the *same* carbon atom, with two H atoms on the other carbon atom. Bearing this fact in mind, how many isomers are possible for each of the following molecular formulas, if each molecule contains a central $C{=}C$ double bond? Remember that a simple rotation of an entire molecule does not give a different isomer.
    (a) $C_2H_2Br_2$
    (b) C_2H_2BrCl

34. Repeat the determination of the number of isomers as discussed in the preceding problem for the following molecular formulas.
    (a) C_2Cl_2BrF
    (b) $C_2HBrClF$

The Conjugation of Bonds and Resonance Structures

35. Discuss the nature of the bonding in the nitrite ion (NO_2^-). Draw the possible Lewis resonance structures for this ion. Use VSEPR theory to determine the steric number and the hybridization of the central N atom, as well as the geometry of the ion. Show how the use of resonance structures can be avoided by introducing a delocalized π molecular orbital. What bond order is predicted by the molecular orbital model for the N—O bonds in the nitrite ion?

36. Discuss the nature of the bonding in the nitrate ion (NO_3^-) in all the particulars listed in the preceding problem.

The Interaction of Light with Molecules

37. Discuss the effect of the absorption of microwave radiation on a molecule. What information is gained from a study of the absorption spectrum in this region?

38. Discuss the effect of the absorption of x-ray radiation on a molecule. What information is gained from a study of the absorption spectrum in this region?

39. Suppose that the ethylene molecule gains an additional electron to give the $C_2H_4^-$ ion. Does the bond order of the carbon–carbon bond increase or decrease? Explain.

40. Suppose that the ethylene molecule is ionized by a photon to give the $C_2H_4^+$ ion. Does the bond order of the carbon–carbon bond increase or decrease? Explain.

41. (See Example 18–5.) The color of the dye indanthrene brilliant orange is evident from its name. In what wavelength range would you expect the maximum in the absorption spectrum of this molecule to lie? Refer to the color spectrum in Figure 16–5.

42. (See Example 18–5.) In what wavelength range would you expect the maximum in the absorption spectrum of the dye crystal violet to lie?

43. The structure of the molecule cyclohexene is shown below:

Does the absorption of ultraviolet light by cyclohexene occur at longer or at shorter wavelengths than in benzene? Explain.

44. The naphthalene molecule has a structure that corresponds to two benzene molecules fused together:

The π-electrons in this molecule are delocalized over the entire molecule. The wavelength of maximum absorption in the UV-visible part of the spectrum in benzene is 255 nm. Is the corresponding wavelength shorter or longer than 255 nm for naphthalene?

Atmospheric Chemistry and Air Pollution

45. (See Example 18–7.) The bond dissociation energy of a typical C—F bond in a chlorofluorocarbon is approximately 440 kJ mol^{-1}. Calculate the maximum wavelength of light that can photodissociate a molecule of CCl_2F_2, breaking such a C—F bond.

46. (See Example 18–7.) The bond dissociation energy of a typical C—Cl bond in a chlorofluorocarbon is approximately 330 kJ mol^{-1}. Calculate the maximum wavelength of light that can photodissociate a molecule of CCl_2F_2, breaking such a C—Cl bond.

47. Draw Lewis structure(s) for the ozone molecule (O_3). Determine the steric number and hybridization of the central oxygen atom, and identify the molecular geometry. Describe the nature of the π-bonds in ozone, and give the bond order of its O—O bonds.

48. The compounds carbon dioxide (CO_2) and sulfur dioxide (SO_2) are formed in the burning of coal. Their similar formulas mask underlying differences in molecular structure. Determine the shapes of these two types of molecules, iden-

tify the hybridization at the central atom, and compare the nature of their π-bonds.

Additional Problems

49. (a) Sketch the occupied molecular orbitals of the valence shell for the N_2 molecule. Label the orbitals as σ-orbitals or π-orbitals, and specify which are bonding and which are antibonding.

(b) If one electron is removed from the highest occupied orbital of N_2, does the equilibrium N—N distance become longer or shorter? Explain briefly.

50. Calcium carbide (CaC_2) is an intermediate in the manufacture of acetylene (C_2H_2). It is the calcium salt of the carbide (also called acetylide) ion (C_2^{2-}). What is the electron configuration of this molecular ion? What is its bond order?

51. Show how the fact that the B_2 molecule is paramagnetic indicates that the energy ordering of the orbitals in this molecule is given by Figure 18–5a rather than 18–5b.

52. The Be_2 molecule has been detected experimentally. It has a bond length of 2.45 Å and a bond dissociation enthalpy of 9.46 kJ mol^{-1}. Write the ground-state electron configuration of Be_2 and predict its bond order, using the theory developed in the text. Compare the experimental bonding data on Be_2 with the data given for B_2, C_2, N_2, and O_2 in Table 18–2. Is the prediction of the simple theory seriously incorrect?

*53. (a) The ionization energy of molecular hydrogen (H_2) is *higher* than that of atomic hydrogen (H), but that of molecular oxygen (O_2) is *lower* than that of atomic oxygen (O). Explain. (*Hint*: Think about the stability of the molecular ion that forms, in relation to bonding and antibonding electrons.)

(b) What prediction would you make for the relative ionization energies of atomic and molecular fluorine (F and F_2)?

54. The stable molecular ion $H_3^+(g)$ is triangular, with H—H distances of 0.87 Å. Sketch the ion, indicating the region of greatest electron density of the lowest-energy molecular orbital.

55. Sketch correlation diagrams for $F_2(g)$, HF(g), and NaF(g). How does molecular orbital theory account for the increasing polarity of the bonds in these molecules?

56. Use Lewis theory, VSEPR theory, and molecular orbital theory to compare the bonding in the species O_3, O_3^- ion and O_3^{2-}. Which is likely to have the shortest O—O bonds? Predict the bond angles in the three species.

*57. According to recent spectroscopic results, the nitramide molecule

is nonplanar. Previously, it had been thought to be planar.

(a) Predict the bond order of the N—N bond in the nonplanar structure.

(b) If the molecule really were planar after all, what would be the bond order of the N—N bond?

58. *Trans*-tetrazene (N_4H_4) consists of a chain of four nitrogen atoms with the two end atoms bonded to two hydrogen atoms each. Use the concepts of steric number and hybridization to predict the overall geometry of the molecule. Give the expected structure of *cis*-tetrazene.

59. Compare the bonding in formic acid (HCOOH) with that in its conjugate base formate ion (HCOO$^-$). Both species have a central carbon atom bonded to the two oxygen atoms and to a hydrogen atom. Draw Lewis structures, determine the steric numbers and hybridization of the central carbon atom, and give the molecular geometries. How do the π-orbitals differ in formic acid and the formate molecular ion? The bond lengths of the C—O bonds in HCOOH are 1.23 Å (for the bond to the lone oxygen) and 1.36 Å (for the bond to the oxygen with a hydrogen atom attached). In what range of lengths do you predict the C—O bond length in the formate ion to lie?

*60. (a) Use data from Appendix D to calculate the enthalpy change when one mole of gaseous benzene is formed from carbon and hydrogen *atoms*, all in the gas phase at 298.15 K.

(b) Compare this result with the enthalpy change that you calculate for forming one of the resonance structures for benzene, using the bond enthalpies for C=C, C—C, and C—H given in Table 10–3.

(c) The additional lowering of the enthalpy found experimentally (the more negative value of $\Delta H°$) is due to *resonance stabilization*: the delocalization of electrons over the whole carbon ring. How large is the enthalpy change due to resonance stabilization in benzene?

61. It has been suggested that a compound of formula $C_{12}B_{24}N_{24}$ might exist and have a structure like that of C_{60} (buckminsterfullerene).

(a) Explain the logic of this suggestion by comparing the number of valence electrons in C_{60} and $C_{12}B_{24}N_{24}$.

(b) Propose the most symmetrical pattern of C, B, and N atoms in $C_{12}B_{24}N_{24}$ to occupy the 60 atom sites in the buckminsterfullerene structure.

62. The standard enthalpy of combustion of $C_{60}(s)$ equals -25891 kJ mol^{-1}. Compute the standard enthalpy of formation of $C_{60}(s)$.

63. Figure 18–23 shows two ways in which green light emerges from an absorbing sample through which white light is passed. How could you use a prism to determine which of these two mechanisms is the origin of the green light seen in a particular case?

64. An electron in the π-orbital of ethylene (C_2H_4) is excited by a photon to the $\pi*$-orbital. Do you expect the length of the bond in the excited ethylene molecule to be greater or less than in ground-state ethylene? Explain your reasoning.

65. One isomer of retinal is converted into a second isomer by the absorption of a photon:

This process is a key step in the chemistry of vision. Although free retinal (in the form shown to the left of the arrow) has an absorption maximum at 376 nm, in the ultraviolet region of the spectrum, this absorption shifts into the visible range when the retinal is bound in a protein, as it is in the eye.

(a) How many of the C=C double bonds are *cis* and how many are *trans* in each of the structures above? (Consider the relative positions of the two largest groups attached at each double bond when assigning labels.) Describe the motion that takes place upon absorption of a photon.

(b) If the ring and the —CHO group in retinal were replaced by —CH_3 groups, would the absorption maximum of the molecule in the UV-visible portion of the spectrum shift to longer or to shorter wavelengths?

* **66.** The ground-state electron configuration of the H_2^+ molecular ion is $(\sigma_{1s})^1$.

(a) An ion of H_2^+ absorbs a photon and is excited to the σ_{1s}^* molecular orbital. Predict what happens to the ion.

(b) Another ion of H_2^+ absorbs even more energy in an interaction with a photon and is excited to the σ_{4s} molecular orbital. Predict what happens to this ion.

67. Compare and contrast the chemical behaviors of ozone (O_3) and nitrogen dioxide (NO_2) in the stratosphere and in the troposphere.

68. Write balanced chemical equations that describe the formation of nitric and sulfuric acid in rain, starting with sulfur in coal, and oxygen, nitrogen, and water vapor in the atmosphere.

69. Describe the greenhouse effect and its mechanism of operation. Give three examples of energy sources that contribute to increased CO_2 in the atmosphere, and three that do not.

CUMULATIVE PROBLEM

Bromine

Elemental bromine is a brownish red liquid that was first isolated in 1826. The current method of production is to oxidize bromide ions in natural brines with elemental chlorine.

(a) Bromine compounds have been known and used for centuries. The deep purple color symbolic of imperial power in Roman times originated from the compound dibromoindigo, which was extracted in tiny quantities from purple snails (about 800 snails per gram of compound). What color and maximum wavelength of *absorbed* light would give this deep purple color?

(b) What is the ground-state electron configuration of the valence electrons of bromine molecules, Br_2? Is bromine paramagnetic or diamagnetic?

(c) What is the electron configuration of the Br_2^+ molecular ion? Is its bond stronger or weaker than that in Br_2? What is its bond order?

(d) What excited electronic state is responsible for the brownish red color of bromine? Refer to Figures 18–5 and 18–22.

(e) One of the most extensively used compounds of bromine is an additive in leaded gasoline colloquially called "ethylene dibromide" (CH_2BrCH_2Br). What is

A chemical plant uses bromine to manufacture brominated flame retardants for plastics and fibers.

the hybridization at the two carbon atoms in this compound? (*Note*: Each C atom is bonded to the other C atom, to two H atoms, and to one Br atom.)

(f) The action of light on bromine compounds released into the air (such as from leaded gasoline) causes the formation of the BrO radical. Give the bond order of this species by comparing it to the related radical OF.

(g) Synthetic bromine-containing compounds in the atmosphere contribute to the destruction of ozone in the stratosphere. The BrO (see part (f)) can take part with ClO in the following catalytic cycle:

$$Cl + O_3 \longrightarrow ClO + O_2$$
$$Br + O_3 \longrightarrow BrO + O_2$$
$$ClO + BrO \longrightarrow Cl + Br + O_2$$

Write the overall equation for this cycle, and explain why it would be important even in the Antarctic winter, when there is not enough sunlight to split apart many O_2 molecules.

Coordination Complexes

The color of these crystals of rhodochrosite ($MnCO_3$) arises from the interaction of Mn^{2+} ions with their environment.

$\mathbf{S}$olid copper(II) sulfate is made by reacting copper and hot concentrated sulfuric acid ("oil of vitriol"); its traditional name, "blue vitriol," recalls this origin and reports the color that is its most obvious property. There is more to this compound than copper and sulfate, however; it contains water as well. The water is important in blue vitriol because when it is driven away by strong heat, the blue color vanishes, leaving greenish white anhydrous copper(II) sulfate (Fig. 19–1). The blue of blue vitriol comes from a **coordination complex** in which H_2O molecules bond directly to Cu^{2+} ions to form composite ions with the formula $[Cu(H_2O)_4]^{2+}$. As a Lewis acid, the Cu^{2+} ion *coordinates* the four water molecules into a group by accepting electron density from the lone pairs on each water molecule. By acting as electron-pair donors (Lewis bases) and sharing electron density with the Cu^{2+} ion, the four water molecules, which in this interaction are called **ligands,** come into the **coordination sphere** of the Cu^{2+} ion. Blue vitriol has the chemical formula $Cu(H_2O)_4SO_4 \cdot H_2O$; the fifth water molecule is not coordinated directly to copper.

The positive ions of every metal in the periodic table accept electron density to some degree and can therefore coordinate surrounding electron donors, even if only weakly. The solvation of the K^+ ion by water molecules in aqueous solution (see Fig. 4–1) is an example of weak coordination. The ability to make fairly strong, *directional* bonds by accepting electron pairs from neighboring molecules or ions is characteristic of the transition-metal elements. Coordination occupies a middle place energetically between the weak intermolecular attractions in solids and liquids (see Section 6–1) and the stronger covalent and ionic bonding (see Chapter 17). Thus, heating blue vitriol disrupts the $Cu—H_2O$ bonds at temperatures well below those required to break the covalent bonds in the SO_4^{2-} group. The enthalpy change in breaking a $+2$ transition-metal ion away from a coordinated water molecule falls in the range of 170 to 210 kJ mol^{-1}. This is far less than the energy required to disrupt the strongest chemical bonds, but it is by no means small. The strengths of metal–ligand interactions do vary, of course, depending on the identities of both the positive ion and the ligand. Metal ions with $+3$ charges, for example, always have stronger coordinate bonds with water than do $+2$ ions.

• This is the bond enthalpy discussed in Sections 10–5 and 17–4.

19–1 THE FORMATION OF COORDINATION COMPLEXES

The total number of metal–ligand bonds in a complex (usually two to six) is the **coordination number** of the metal. The particular atom of a ligand molecule that actually provides the electron density to a central metal atom is a donor, or **ligating,**

Figure 19–1 Hydrated copper(II) sulfate ($CuSO_4 \cdot 5H_2O$) is blue (*left*), but the anhydrous compound ($CuSO_4$) is greenish white (*right*).

Table 19–1
Common Monodentate Ligands and Their Names

Ligand	Formula	Name
Fluoride ion	:F⁻	Fluoro
Chloride ion	:Cl⁻	Chloro
Nitrite ion	:NO$_2^-$	Nitro
	:ONO⁻	Nitrito
Carbonate ion	:OCO$_2^{2-}$	Carbonato
Cyanide ion	:CN⁻	Cyano
Thiocyanate ion	:SCN⁻	Thiocyanato
	:NCS⁻	Isothiocyanato
Hydroxide ion	:OH⁻	Hydroxo
Water	:OH$_2$	Aqua
Ammonia	:NH$_3$	Ammine
Carbon monoxide	:CO	Carbonyl
Nitrogen monoxide	:NO	Nitrosyl

The ligating atom is indicated by a pair of red dots to show a lone pair of electrons. In the CO_3^{2-} ligand, either one or two of the oxygen atoms can donate a lone pair to the metal.

atom. Some common ligands are the halide ions (F^-, Cl^-, Br^-, I^-), ammonia (NH_3), carbon monoxide (CO), and water. These and other ligands are listed in Table 19–1.

The formation of a coordination complex is a Lewis acid–base reaction (see Section 8–8):

central atom	+	ligands	⟶	coordination complex
Lewis acid	+	Lewis bases	⟶	composite ion or molecule
Fe^{2+}	+	6 CN⁻	⟶	$[Fe(CN)_6]^{4-}$
Pt^{4+}	+	6 NH$_3$	⟶	$[Pt(NH_3)_6]^{4+}$
Ni	+	4 CO	⟶	$[Ni(CO)_4]$

As the third example shows, the central atom in a complex need not have a positive charge to coordinate ligands.

Brackets are used in the chemical formulas of coordination complexes to group together the symbols of the central atom and its coordinated ligands. In the formula $[Pt(NH_3)_6]Cl_4$, the portion in brackets represents a positively charged coordination complex in which Pt coordinates six NH_3 ligands. The brackets emphasize that complexes are distinct chemical entities with their own properties. The symbol of the central atom comes first within the brackets.

Coordination modifies the chemical and physical properties of both central atom and ligands. Consider the chemistry of aqueous cyanide (CN^-) and iron(II) (Fe^{2+}) ions. The former reacts immediately with acid to generate gaseous hydrogen cyanide (HCN), a deadly poison. The latter instantly precipitates a gelatinous hydroxide when mixed with aqueous base. Reaction between the two gives the complex ion $[Fe(CN)_6]^{4-}(aq)$, which undergoes neither these two reactions nor any others considered diagnostic of simple CN^- ion and Fe^{2+} ion. Because coordination changes a ligand's chemical behavior, a given ligand may be present in multiple forms in the

• This use of brackets has nothing to do with signifying the concentrations of species in solution, as in $[Cu^{2+}]$. The intended meaning of brackets is usually clear from the context.

same compound. The two Cl^- ions in $[Pt(NH_3)_3Cl]Cl$ differ chemically because one is coordinated and one is not. Treatment of an aqueous solution of this substance with Ag^+ ion immediately precipitates the uncoordinated Cl^- as $AgCl(s)$ but *not* the coordinated Cl^-.

Coordination can also change the acid–base behavior of a ligand. Ammonia is a base—a hydrogen ion acceptor (Brønsted–Lowry definition) or an electron-pair donor (Lewis definition). When NH_3 is coordinated in the $[Pt(NH_3)_6]^{4+}$ ion, the Pt(IV) accepts electrons from ammonia so strongly that the N—H bonds are weakened, and the hydrogen atoms of the ammonia become acidic. The equilibrium constant of the reaction

• This change, which involves the fragment —Pt—NH₂—H, recalls the increase in acidity of oxoacids as the electronegativity of X in the fragment —X—O—H increases. See Section 17-6.

$$[Pt(NH_3)_6]^{4+}(aq) + H_2O(\ell) \Longleftrightarrow H_3O^+(aq) + [Pt(NH_3)_5(NH_2)]^{3+}(aq)$$

is 1.0×10^{-7}. This value of K_a means that the complex ion $[Pt(NH_3)_6]^{4+}$ is a weak acid, comparable to carbonic acid or sulfurous acid in strength. The acidity of aqueous solutions of transition-metal ions is a parallel phenomenon (see Section 9–5).

Types of Ligands

The ligands in Table 19–1 are capable of forming only a single bond to a central metal atom. They are called *monodentate* (from Latin *mono*, meaning "one," plus *dens*, meaning "tooth," indicating that they bind at only one point). Other ligands are capable of forming two or more such bonds are referred to as *bidentate, tridentate*, and so forth.

• The Lewis structure of ethylenediamine is

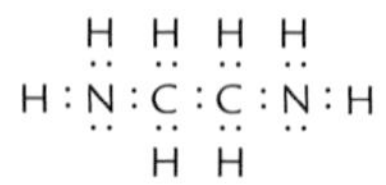

```
    H  H  H  H
    ··    ··
H : N : C : C : N : H
    ··    ··
    H  H
```

Ethylenediamine ($NH_2CH_2CH_2NH_2$), in which two NH_2 groups are held together by a carbon backbone, is a particularly important bidentate ligand. Both nitrogen atoms in ethylenediamine have lone electron pairs to share. If all the nitrogen donors of three ethylenediamine molecules bind to a single ion, say Co^{3+}, then that Co^{3+} ion has a coordination number of 6, and the formula of the resulting complex is $[Co(en)_3]^{3+}$ (where "en" is the accepted abbreviation for ethylenediamine). Complexes in which a ligand coordinates via two or more donors to the same central atom are called **chelates** (from Greek *chela*, meaning "claw," because the ligand bites onto the central atom like a pincers). The structures of some important chelating ligands are given in Figure 19–2.

Charge and Oxidation State in Complexes

The overall electric charge on a coordination complex is the sum of the oxidation number of the metal ion and the charges of the ligands that surround it. Thus, the complex of copper(II) (Cu^{2+}) with four Br^- ions is an anion with a -2 charge, $[CuBr_4]^{2-}$.

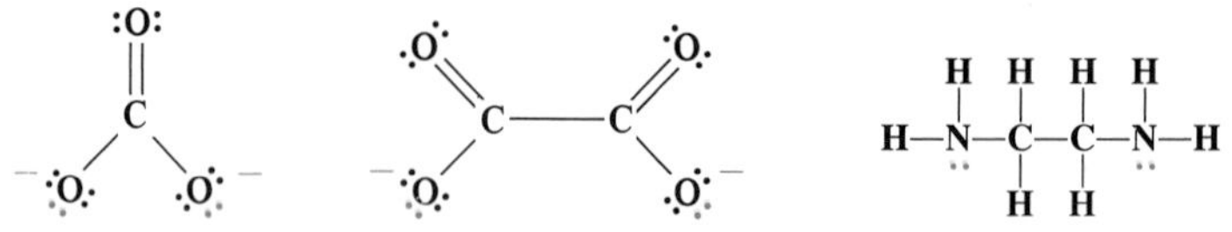

Figure 19–2 The Lewis structures of three bidentate ligands. Each is capable of ligating (donating a pair of electrons to an acceptor) at the sites marked in red.

Carbonate ion, CO_3^{2-}
(a)

Oxalate ion, $C_2O_4^{2-}$
(b)

Ethylenediamine, $NH_2CH_2CH_2NH_2$
(c)

EXAMPLE 19-1

Write formulas, including the overall electric charge, of coordination complexes containing the following metal ions and ligands:
(a) Co^{3+} with six F^- ligands.
(b) Pt^{4+} with two NH_3 ligands, two H_2O ligands, and two Cl^- ligands.
(c) Cr^{3+} with one NH_3 ligand, one ethylenediamine (en) ligand, and three NO_2^- ligands.

Solution

(a) The complex is an anion with the formula $[CoF_6]^{3-}$. The six -1 charges of the ligands combine with the $+3$ of the cobalt to give an overall charge of -3.
(b) The complex is a cation, $[Pt(NH_3)_2(H_2O)_2Cl_2]^{2+}$.
(c) The complex has the formula $[Cr(NH_3)(en)(NO_2)_3]$. There is no net charge because the charges of the three NO_2^- ligands and the Cr^{3+} metal ion add up to zero (and the other ligands are uncharged). Zero charges on complexes are usually not shown explicitly.

Exercise

Write formulas, including the overall electric charge, of coordination complexes containing the following metal ions and ligands: (a) Zn^{2+} with four Cl^- ligands. (b) Cr with six CO ligands. (c) Ni^{2+} with two H_2O ligands and two $C_2O_4^{2-}$ (oxalate) ligands.

Answer: (a) $[ZnCl_4]^{2-}$. (b) $[Cr(CO)_6]$. (c) $[Ni(H_2O)_2(C_2O_4)_2]^{2-}$.

The procedure used in Example 19–1 can be reversed to determine the oxidation state of the central metal atom from the chemical formula of the coordination compound. This is demonstrated in the following example.

EXAMPLE 19-2

Determine the oxidation state of the coordinated metal atom in each of the following compounds:
(a) $K[Co(NH_3)_2(CN)_4]$
(b) $[Os(CO)_5]$
(c) $Na[Co(H_2O)_3(OH)_3]$

Solution

(a) The oxidation state of K is known to be $+1$, so the complex in brackets is an anion with a -1 charge, $[Co(NH_3)_2(CN)_4]^-$. The charge on the two NH_3 ligands is zero, and the charge on each of the four CN^- ligands is -1. The oxidation state of the Co must then be $+3$ because 4×-1 (for the CN^-) $+ 2 \times 0$ (for the NH_3) plus 3 (for Co) equals the required -1.
(b) The ligand CO has zero charge; the complex has zero charge as well. Therefore, the oxidation state of the osmium is zero.
(c) There are three neutral ligands (the water molecules) and three ligands with -1 charges (the hydroxide ions). The Na^+ ion contributes only $+1$, so the oxidation state of the cobalt must be $+2$.

Exercise

Determine the oxidation state of the coordinated metal atom in each of the following compounds: (a) $K_3[Fe(CN)_6]$. (b) $[Co(en)_2(SCN)_2]Cl$. (c) $Na[Rh(NH_3)_3Cl_3]$.

Answer: (a) +3. (b) +3. (c) +2.

Naming Coordination Compounds

Up to now, only chemical formulas have been used to represent coordination compounds, but for many purposes a name is needed. Some of these substances have names that were given to them before their structures were known. Thus, $K_3[Fe(CN)_6]$ was called potassium ferricyanide, and $K_4[Fe(CN)_6]$ was potassium ferrocyanide (these are complexes of Fe^{3+} (ferric) and Fe^{2+} (ferrous) ions, respectively.) The older names still find some use but are gradually being replaced by systematic names based on the following set of rules:

• Note that the NH_3 ligand is named an ammine (with a double *m*), but the NH_2 group in ethylenediamine is called an amine group (with a single *m*).

1. The names of coordination complexes are written as single words, built from the names of the ligands, prefixes to indicate how many ligands are present, and a name for the central metal.

2. A coordination complex may be an ion or a neutral molecule. If it is ionic, the compound in which it is found is named according to the pattern in simple ionic compounds: the positive ion is named first, followed (after a space) by the name of the negative ion, regardless of which is the complex ion.

3. The names of anionic ligands are obtained by replacing the usual ending with the suffix *-o*. The names of neutral ligands are unchanged. Exceptions to the latter rule are *aqua* (for water), *ammine* (for NH_3), and *carbonyl* (for CO). See Table 19–1.

4. Greek prefixes (*di-, tri-, tetra-, penta-, hexa-*) are used to indicate the number of ligands of a given type attached to the central ion, if there is more than one. The prefix *mono-* is not used when there is only one ligand of a given type. If the name of the ligand itself contains the terms *mono-, di-,* and so forth (as in ethylene*di*amine), then the name of the ligand is placed in parentheses and the prefixes *bis-, tris-,* and *tetrakis-* are used instead of *di-, tri-,* and *tetra-*.

5. If more than one type of ligand is present, the ligands are listed in alphabetical order, ignoring the prefixes that tell how often each type of ligand occurs in the coordination sphere.

6. The oxidation state of the central metal atom is given by a Roman numeral enclosed in parentheses immediately following the name of the metal. If the complex ion has a net negative charge, the ending *−ate* is added to the stem of the name of the metal.

The following examples illustrate the systematic naming of several complexes:

$K_3[Fe(CN)_6]$	Potassium hexacyanoferrate(III)
$K_4[Fe(CN)_6]$	Potassium hexacyanoferrate(II)
$[Fe(CO)_5]$	Pentacarbonyliron(0)
$[Co(NH_3)_5CO_3]Cl$	Pentaamminecarbonatocobalt(III) chloride

$K_3[Co(NO_2)_6]$ Potassium hexanitrocobaltate(III)

$[Cr(H_2O)_4Cl_2]Cl$ Tetraaquadichlorochromium(III) chloride

$[Pt(NH_2CH_2CH_2NH_2)_3]Br_4$ Tris(ethylenediamine)platinum(IV) bromide

$K_2[CuCl_4]$ Potassium tetrachlorocuprate(II)

EXAMPLE 19–3

Interpret the names and write the formulas of these coordination compounds:
(a) Sodium tricarbonatocobaltate(III).
(b) Diamminediaquadichloroplatinum(IV) bromide.
(c) Sodium tetranitratoborate(III).

Solution

(a) In the anion, three carbonate ligands (with -2 charges) are coordinated to a cobalt atom in the $+3$ oxidation state. Because the complex ion thus has an overall charge of -3, three sodium cations are required, and the correct formula is $Na_3[Co(CO_3)_3]$. Writing the formula of this compound requires a previous knowledge of the charge on the carbonate group.

(b) The ligands coordinated to one Pt(IV) are two ammonia molecules, two water molecules, and two chloride ions. Ammonia and water are electrically neutral, but the two chloride ions contribute a total charge of $2 \times (-1) = -2$ that adds with the $+4$ of the platinum and gives the complex ion a $+2$ charge. Two bromide anions are required to balance this, so the formula is $[Pt(NH_3)_2(H_2O)_2Cl_2]Br_2$.

(c) The complex anion has four nitrate ligands, each with a -1 charge, coordinated to a central boron(III). This gives a net charge of -1 on the complex ion and requires one sodium ion in the formula, $Na[B(NO_3)_4]$.

Exercise

(a) Name the compound with the chemical formula $[Ni(NH_3)Cl(en)_2]Cl$.
(b) Write the chemical formula for sodium diammineaquatrichlorocobaltate(II).

Answer: (a) Amminechlorobis(ethylenediamine)nickel(II) chloride.
(b) $Na[Co(NH_3)_2(H_2O)Cl_3]$.

Ligand Substitution Reactions

The advantage of the concept of coordination is that it organizes an immense number of chemical compositions as different combinations of ligands linked in various ratios with central metal atoms or ions. In this way, it establishes a basis for comparison among diverse compounds. For example, an otherwise bewildering collection of information on chemical reactivity is easily rationalized in terms of one ligand substituting for another in coordination complexes.

Changing colors provide a way to follow the progress of ligand substitution as well as other reactions involving coordination complexes. Nickel(II) sulfate, for example is a yellow crystalline solid. If exposed to moist air at room temperature, it takes up six water molecules per formula unit. The water molecules coordinate with the nickel ions to form a bright green complex:

$$NiSO_4(s) + 6\ H_2O(g) \longrightarrow [Ni(H_2O)_6]SO_4(s)$$
$$\text{yellow} \qquad \text{colorless} \qquad\qquad \text{green}$$

Figure 19–3 When ammonia is added to the green solution of nickel(II) sulfate on the left (which contains $[Ni(H_2O)_6]^{2+}$ ions), ligand substitution occurs to give the blue-violet solution on the right (which contains $[Ni(NH_3)_6]^{2+}$ ions).

Heating the green hexaaquanickel(II) sulfate sufficiently (a temperature well above the boiling point of water is required) drives off the water and regenerates the yellow $NiSO_4$ in the reverse of the reaction just written. A similar coordination reaction generates a product with a completely different color when yellow $NiSO_4(s)$ is exposed to gaseous ammonia ($NH_3(g)$). This time, the product is a blue-violet complex:

$$NiSO_4(s) + 6\ NH_3(g) \longrightarrow [Ni(NH_3)_6]SO_4(s)$$
$$\text{yellow} \qquad \text{colorless} \qquad\qquad \text{blue-violet}$$

Heating the blue-violet product drives off ammonia, a process that can be followed by watching the color of the solid change back to yellow. Given these facts, it is not hard to explain the observation that a green $[Ni(H_2O)_6]^{2+}(aq)$ solution turns blue-violet when treated with $NH_3(aq)$ (Fig. 19–3). The NH_3 must be displacing the H_2O from the coordination sphere:

$$[Ni(H_2O)_6]^{2+}(aq) + 6\ NH_3(aq) \longrightarrow [Ni(NH_3)_6]^{2+}(aq) + 6\ H_2O(\ell)$$
$$\text{green} \qquad\qquad \text{colorless} \qquad\qquad \text{blue-violet} \qquad\qquad \text{colorless}$$

Complexes that undergo substitution of one ligand for another are **labile,** whereas complexes in which substitution proceeds slowly or not at all are **inert.** In an inert complex, a large energy of activation (see Section 14–6) of ligand substitution prevents rapid reaction even though there may be a thermodynamic tendency to proceed. In the substitution reaction

$$[Co(NH_3)_6]^{3+}(aq) + 6\ H_3O^+(aq) \longrightarrow [Co(H_2O)_6]^{3+}(aq) + 6\ NH_4^+(aq)$$

the products are favored thermodynamically by an enormous amount (the equilibrium constant is about 10^{64}). Yet the inert $[Co(NH_3)_6]^{3+}$ complex ion lasts for weeks in acidic solution because no low-energy path for the reaction exists. The $[Co(NH_3)_6]^{3+}$ ion is thermodynamically unstable relative to $[Co(H_2O)_6]^{3+}$ yet kinetically stable (that is, inert). The closely related cobalt(II) complex $[Co(NH_3)_6]^{2+}$ undergoes a similar substitution reaction:

$$[Co(NH_3)_6]^{2+}(aq) + 6\ H_3O^+(aq) \longrightarrow [Co(H_2O)_6]^{2+}(aq) + 6\ NH_4^+(aq)$$

in a few seconds. This cobalt(II) complex is thermodynamically unstable and also labile.

Substitution of one ligand for another can proceed in stages in the coordination sphere of a metal. By controlling the reaction conditions, substitution can usually be stopped at intermediate stages. For example, all of the possible four-coordinate compositions of Pt(II) with the two ligands NH_3 and Cl^-

$$[Pt(NH_3)_4]^{2+} \quad [Pt(NH_3)_3Cl]^+ \quad [Pt(NH_3)_2Cl_2] \quad [Pt(NH_3)Cl_3]^- \quad [PtCl_4]^{2-}$$

can be prepared and isolated (the ions are isolated as salts with suitable negative or positive ions). Such mixed-ligand complexes add to the richness of coordination chemistry.

19-2 STRUCTURES OF COORDINATION COMPLEXES

The Alsatian-Swiss chemist Alfred Werner pioneered the field of coordination chemistry in the late 19th century. At the time, a number of complexes of cobalt(III) chloride with ammonia were known and had these chemical formulas and colors:

Compound 1:	$CoCl_3 \cdot 6NH_3$	Orange-yellow
Compound 2:	$CoCl_3 \cdot 5NH_3$	Purple
Compound 3:	$CoCl_3 \cdot 4NH_3$	Green
Compound 4:	$CoCl_3 \cdot 3NH_3$	Green

Treating these compounds with aqueous hydrochloric acid did not remove the ammonia, suggesting that it was somehow closely bound with the cobalt ions. Treatment with aqueous silver nitrate at 0°C, on the other hand, gave interesting results. With compound 1, all of the chloride present was precipitated as solid AgCl. With compound 2, however, only two thirds of the chloride was precipitated, and with compound 3, only one third was precipitated. Compound 4 did not react at all with silver nitrate. Werner accounted for these facts by postulating the existence of coordination complexes with six ligands (chloride ions or ammonia molecules or both kinds of ligands) attached to each Co^{3+} ion. Specifically, he wrote the formulas for compounds 1 to 4 as follows:

Compound 1	$[Co(NH_3)_6]^{3+}(Cl^-)_3$
Compound 2	$[Co(NH_3)_5Cl]^{2+}(Cl^-)_2$
Compound 3	$[Co(NH_3)_4Cl_2]^+(Cl^-)$
Compound 4	$[Co(NH_3)_3Cl_3]$

Only those chloride ions that were *not* ligands attached directly to cobalt were precipitated upon the addition of cold aqueous silver nitrate.

Werner realized that his proposal gave predictions about the electrical conductivity of aqueous solutions of salts of these complex ions. Compound 1, for example, should have a molar electrical conductivity close to that of $Al(NO_3)_3$, which also yields one +3 ion and three −1 ions per formula unit dissolved in water. His experiments confirmed this resemblance and also showed that compound 2 resembled $Mg(NO_3)_2$ and that compound 3 resembled $NaNO_3$ in their electrical conductivities. Compound 4 behaved like a nonelectrolyte, with a very low conductivity, as expected from the absence of ions in the formula.

Werner and other chemists studied other coordination complexes of different metals and ligands, using both physical and chemical techniques. This research has shown that the most common coordination number by far is 6, as in the cobalt

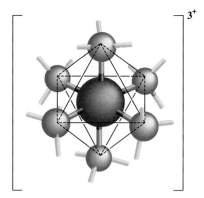

Figure 19-4 The octahedral structure of the $[Co(NH_3)_6^{3+}]$ ion. All six corners of the octahedron are equivalent. The hydrogen atoms are suppressed for clarity.

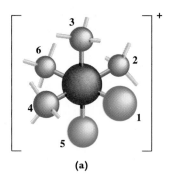

(a)

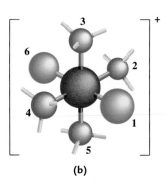

(b)

Figure 19-5 (a) The structure of the *cis*-$[Co(NH_3)_4Cl_2]^+$ ion; (b) the structure of the *trans*-$[Co(NH_3)_4Cl_2]^+$ ion. These isomeric complex ions have distinctly different properties. The *cis* complex is purple in solution, but the *trans* is green, for example. Hydrogen atoms are omitted to simplify the picture.

complexes just discussed. Other coordination numbers ranging from 2 to 12 have been observed, however. Of these, the most important and interesting are 4 (as in $[PtCl_4]^{2-}$), 2 (as in $[Ag(NH_3)_2]^+$), and 5 (as in $[Ni(CN)_5]^{3-}$).

Three-Dimensional Structures

If the central cobalt ion in $[Co(NH_3)_6]^{3+}$ is bonded to six ammonia ligands, what is the geometrical structure of the complex? This question naturally occurred to Werner, who suggested that the arrangement should be the simplest and most symmetrical possible, with ligands located at the six vertices of an octahedron (Fig. 19–4).

Modern methods of x-ray diffraction (see Sections 20–1 and 20–2) allow very precise determinations of atomic positions in crystals and completely confirm Werner's hypothesis of octahedral coordination for this complex. Such techniques were unavailable a century ago, however, and so Werner turned to a study of the properties of substituted complexes to test his hypothesis.

If one ammonia ligand is replaced with a chloride ion, the resulting complex has the formula $[Co(NH_3)_5Cl]^{2+}$, with one vertex of the octahedron occupied by Cl^- and the other five by NH_3. Only one structure of this type is possible, because all six vertices of the original octahedron are equivalent; all ways of drawing singly substituted octahedral structures $[MA_5B]$ (where $A = NH_3$, $B = Cl^-$, $M = Co^{3+}$, for example) become equivalent when the structure is imagined to rotate (tumble) around different axes in space. Now suppose that a second NH_3 ligand is replaced with Cl^-. The second Cl^- can lie in one of the positions closest to the first Cl^- in the plane perpendicular to the $Co-Cl^-$ line (Fig. 19–5a), or it can lie in the sixth position, on the opposite side of the central metal atom (Fig. 19–5b). The former structure, in which the two Cl^- ligands are closer to each other, is called a *cis* structure, *cis*-$[Co(NH_3)_4Cl_2]^+$, and the latter, with the two Cl^- ligands farther apart, is a *trans* structure, *trans*-$[Co(NH_3)_4Cl_2]^+$. The octahedral-structure model predicts that exactly two different ions with the chemical formula $[Co(NH_3)_4Cl_2]^+$ should exist. Such structurally different species are called ***cis–trans* isomers** (and are similar to the *cis–trans* isomers discussed in Section 18–2). When Werner began his work, only the green *trans* form was known, but by 1907, he had prepared the *cis* isomer and showed that it differed from the *trans* isomer in color (it was violet rather than green) and in other physical properties. The isolation of two, and only two, isomers of this atom was good (although not conclusive) evidence that the octahedral structure was correct. Almost all six-coordinated complexes of the transition metals have octahedral structures; hence, many compounds can display similar isomerism. The complex ion $[CoCl_2(en)_2]^+$, for example, also has a purple *cis* form and a green *trans* form (Fig. 19–6).

Figure 19-6 The complex ion $[CoCl_2(en)_2]^+$ is an octahedral complex that has *cis* and *trans* isomers, according to the relative positions of the two Cl^- ligands. Salts of the *cis* isomer are purple, and salts of the *trans* isomer are green.

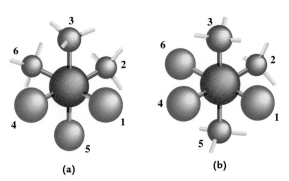

Figure 19-7 Two structural isomers of $[Co(NH_3)_3Cl_3]$.

(a)   (b)

EXAMPLE 19-4

How many structural isomers exist for the octahedral coordination compound $[Co(NH_3)_3Cl_3]$?

Solution

We begin with the two isomers of $[Co(NH_3)_4Cl_2]^+$ and see how many different structures can be made by replacing one more ammonia molecule with Cl^-.

Replacing any of the four NH_3 ligands in the *trans* form (Fig. 19–5b) at sites 2, 3, 4, or 5 gives equivalent structures that can be superimposed by rotation. This isomer is shown in Figure 19–7a. What isomers can be made from the *cis* form of Figure 19–5a? If the ammonia ligand at either site 3 or site 6 is replaced (that is, one of the two that are *trans* to existing Cl^- ligands), the result is simply a rotated version of Figure 19–7a, so these replacements do *not* give another isomer. Replacement of the ligand at site 2 or 4, however, gives a different structure, which is shown in Figure 19–7b. (Note that 1, 2, 5 replacement gives a rotated version of the 1, 4, 5 replacement that is shown.)

We conclude that two, and only two, possible isomers of $[Co(NH_3)_3Cl_3]$ can exist. In fact, only one form of $[Co(NH_3)_3Cl_3]$ has been prepared to date, presumably because the two isomers interconvert rapidly, but exactly two isomers *are* known for the closely related octahedral coordination complex $[Cr(NH_3)_3(NO_2)_3]$.

Exercise

How many different isomers of the octahedral coordination complex $[Co(NH_3)_3(H_2O)Cl_2]^+$ are possible? (*Hint:* Start with Figure 19–7 and replace one Cl^- ligand with an H_2O ligand.)

Answer: 3.

Square-Planar, Tetrahedral, and Linear Structures

Complexes with coordination numbers of 4 are typically either tetrahedral or square planar. The tetrahedral geometry (Fig. 19–8a) predominates for four-coordinate complexes of the early transition metals (those toward the left side of the *d*-block of elements in the periodic table). There is no possibility of *cis–trans* isomerism for tetrahedral complexes of the general form MA_2B_2 because all such structures are superimposable.

The square-planar geometry (Fig. 19–8b, c) is common for four-coordinate complexes of Au^{3+}, Ir^+, and Rh^+ and, most especially, for ions with the d^8 valence-

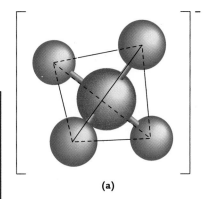

(a)

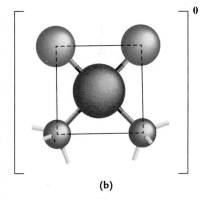

(b)

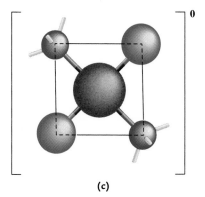

(c)

Figure 19-8 Four-coordinate complexes. (a) Tetrahedral $[FeCl_4]^-$. (b) Square-planar *cis*-$[Pt(NH_3)_2Cl_2]$ (in which the chlorides (green) are *near* to each other). (c) Square-planar *trans*-$[Pt(NH_3)_2Cl_2]$ (in which the chlorides are *far* from each other).

electron configurations: Ni^{2+}, Pd^{2+}, and Pt^{2+}. The Ni^{2+} ion forms a few tetrahedral structures, but four-coordinate Pd^{2+} and Pt^{2+} are nearly exclusively square planar. Square-planar complexes of the type MA_2B_2 can have isomers, as Figures 19–8b and c illustrate for *cis-* and *trans-*$[Pt(NH_3)_2Cl_2]$. The *cis* form of this compound is a potent and widely used anticancer drug called "cisplatin"; the *trans* form has no therapeutic properties.

Finally, linear complexes with coordination numbers of 2 exist, especially for ions with d^{10} configurations such as Cu^+, Ag^+, Au^+, and Hg^{2+}. The central silver atom in a complex such as $[Ag(NH_3)_2]^+$ in aqueous solution strongly attracts several water molecules as well, however, so its actual coordination number under these circumstances may exceed 2.

Chiral Structures

The complex ion $[Co(en)_3]^{3+}$ displays a type of isomerism that differs from the isomerism discussed so far. It is called **chirality** and is illustrated by the two structures shown in Figure 19–9.

These two structures are quite similar, yet not identical. Their ligands are connected in the same relative arrangement, but, taken as a unit, they bear the same relationship to each other that the left hand does to the right. The two are each other's mirror images. No matter how these two structures are tumbled and turned about, they cannot be superimposed. Such mirror-image pairs are referred to as **enantiomers.** The difference between enantiomers is a subtle one, and many of their physical and chemical properties are the same. Members of enantiomeric pairs differ in the way they interact with other objects that have "handedness," just as a right hand, a chiral object, fits well into a right glove but poorly into a left glove. This leads to *stereoselective* reactions, in which one enantiomer reacts preferentially. Many reactions of

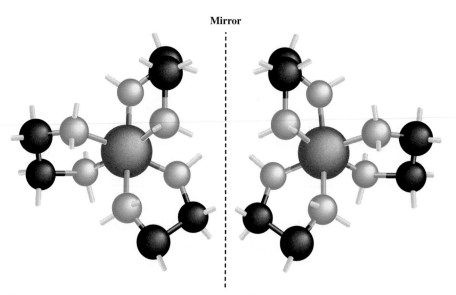

Figure 19–9   The enantiomers of the $[Co(en)_3]^{3+}$ ion. Three coordinate bonds (highlighted in red) are in front of the central cobalt, and three are behind. The ligands wrap clockwise around the cobalt ion from front to back in the isomer on the right but wrap counterclockwise in the isomer on the left. Reflection through the mirror transforms one enantiomer into the other. The two cannot be superimposed by any rotation.

biological importance are stereoselective. Enantiomers can also be distinguished by the way in which their solutions rotate the plane of polarization of plane-polarized light: if one does so in a clockwise sense, the other does so in a counterclockwise sense. For this reason, enantiomers are also called **optical isomers.**

• Figure 16–4 shows a plane wave, which can represent plane-polarized light. When such a wave enters a sample containing one enantiomer and not its mirror image, the plane in which the wave oscillates rotates.

EXAMPLE 19-5

Suppose that the complex ion $[Co(NH_3)_2(H_2O)_2Cl_2]^+$ is synthesized with the two ammine ligands *cis* to each other, the two aqua ligands *cis* to each other, and the two chloro ligands *cis* to each other, as shown in Figure 19–10. Determine whether this complex is optically active—that is, whether it can be superimposed upon its mirror image.

Solution

When all else fails, a structure and its mirror image can be represented by drawing the structure on paper, viewing it in a small mirror, and copying the image. It is faster to indicate a mirror with a line and to create the mirror image by making each point in the structure on the far side of the dotted line lie the same distance from the dotted line as the generating point in the original structure. This is illustrated in Figure 19–10. Compare the original (left) and the mirror image (right) to learn whether they are superimposable. The complex ion *cis*, *cis*-$[Co(NH_3)_2(H_2O)_2Cl_2]^+$ is chiral, because the two structures cannot be superimposed even after turning them. As this example proves, nonchelates can be chiral.

Exercise

Is the square-planar complex $[Pt(NH_3)(H_2O)BrCl]$ (Pt surrounded by four different ligands) chiral? If the answer is yes, draw a representation of a pair of optical isomers of this geometry. If no, explain.

Answer: Forming the mirror image of this square-planar complex is equivalent to simply turning over the plane of the structure; hence, the answer is no.

Approximately enantiomeric seashells. The spiral body of the shell turns to the left in one shell and to the right in the other. The mirror-image relationship is imperfect in other respects, however.

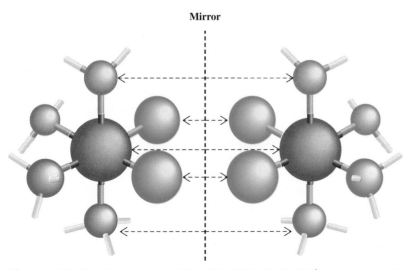

Figure 19–10 The structure of all-*cis*$[Co(NH_3)_2(H_2O)_2Cl_2]^+$ ion, together with its mirror image.

CHEMISTRY IN PROGRESS

Sequestering Agents as Miracle Ingredients

The hexadentate ligand EDTA (ethylenediaminetetraacetate ion) provides some interesting insights into the effects of coordination on the chemistry of central atoms. The structure of this chelating ligand, as coordinated to a Co(III) ion, is shown in Figure 19–A. It is apparent that the central metal ion is literally "enveloped" by the ligand as it coordinates simultaneously at the six corners of the coordination octahedron. Such a chelating agent has a very strong affinity for certain metal ions and can **sequester** them quite effectively in solution. EDTA solubilizes the scummy precipitates that Ca^{2+} ions form with anionic constituents of soap by forming a stable complex

with Ca^{2+}. Thus, it breaks up the main contributor to bathtub rings and is a "miracle ingredient" in some bathtub cleaners. EDTA can be used to remove metal ions that are trace contaminants of water. It has been used as an antidote to lead poisoning because of its great affinity for Pb^{2+} ions. Complexes of EDTA with iron are used in plant foods to permit a slow release of iron to the plant. EDTA also sequesters copper and nickel ions in edible fats and oils. Because these metal ions catalyze the oxidation reactions that make oils rancid, the presence of EDTA prolongs freshness.

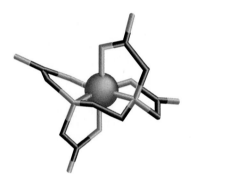

EDTA anion

Figure 19–A The wire-frame structure on the left emphasizes the way in which EDTA enfolds a metal ion as it chelates; hydrogens are omitted for clarity. The diagram on the right gives the full structure of EDTA and shows (in red) the six pairs of electrons that it donates.

19–3 CRYSTAL-FIELD THEORY AND MAGNETIC PROPERTIES

What is the nature of the bonding in coordination complexes of the transition metals? Why does Pt(IV) form only octahedral complexes, whereas Pt(II) forms almost exclusively square-planar ones, and under what circumstances does Ni(II) form octahedral, square-planar, or tetrahedral complexes? Can trends in the lengths and strengths of metal–ligand bonds be understood and predicted? Answering such questions requires a theoretical description of the bonding in coordination complexes.

A simple but useful model for bonding in coordination complexes is **crystal-field theory,** which starts from an ionic description of the metal–ligand bonds, omitting entirely the sharing of electrons from the ligands to the metal. Crystal-field theory considers the response of a central metal to the approach of negatively charged ligands. In an octahedral complex, six negative charges from the lone pairs of six ligands are brought up along the $\pm x$, $\pm y$, and $\pm z$ coordinate axes toward a metal atom or ion located at the origin. In an atom or ion of a transition metal in free

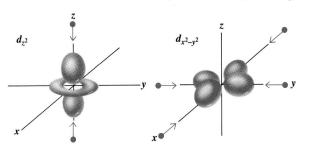

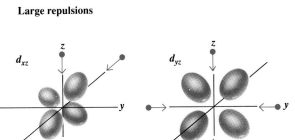

Figure 19–11 Octahedral crystal-field splitting of 3d-orbital energies by ligands. As the external charges approach the five 3d-orbitals, the largest repulsions arise in the d_{z^2}- and $d_{x^2-y^2}$-orbitals, which point directly at two or four of the approaching charges.

space, the energy of an electron is the same in each of the five d-orbitals in a given shell. When the external charges are brought up, however, the energies of the d-orbitals change by different amounts because of differing degrees of repulsion between the electrons in the d-orbitals and the donor electron pairs on the ligands (Fig. 19–11). An electron in a $d_{x^2-y^2}$- or d_{z^2}-orbital is most likely to be found along the coordinate axes, where it comes into close contact with electrons from the ligand in an octahedral complex. This proximity raises the energies of these orbitals. By contrast, an electron in a d_{xy}-, d_{yz}-, or d_{xz}- orbital has its greatest probability density *between* the coordinate axes, so it experiences less repulsion from an octahedral array of approaching charges and has lower energy than the $d_{x^2-y^2}$- and d_{z^2}-orbitals. In the octahedral field created by six ligand lone pairs, the d-orbital energy levels split into two groups (Fig. 19–12). The three lower-energy orbitals, called t_{2g} levels, correspond to the d_{xy}-, d_{yz}-, and d_{xz}-orbitals of the transition-metal atom; the two higher-energy orbitals, called e_g levels, correspond to the $d_{x^2-y^2}$ and d_{z^2}-orbitals of the transition-metal ion. The energy difference between the two sets of levels is Δ_o, the **crystal-field splitting energy** for the octahedral complex that is formed.

Now consider the electron configuration for a transition-metal ion experiencing such an octahedral crystal field. The Cr^{3+} ion, for example, has three d-electrons. According to Hund's rules (see Section 17–1), one electron goes into each of the three t_{2g} levels with parallel spins. For an ion such as Mn^{3+}, which has a fourth d-electron, there are two conceivable ground-state electron configurations. The fourth electron can occupy either in a t_{2g} level, with spin opposite to that of the electron already in that level (Fig. 19–12a), or an e_g level, with spin parallel to those of the three t_{2g} electrons (Fig. 19–12b). The former happens when the splitting Δ_o is large because in that case the energy cost of placing an electron in an e_g orbital is prohibitively high. If the splitting is small, however, the e_g orbital is occupied in order to avoid having two electrons in the same orbital.

• In this model, all the d-orbital energies increase, but the increase is smaller for some orbitals than for others.

• The symbols t_{2g} and e_g come from a mathematical representation of the symmetries of the orbitals that does not concern us here. The subscript "o" on Δ_o stands for "octahedral."

• Recall from Section 17–1 that the Cr atom has the ground-state electronic configuration $[Ar]3d^54s^1$. When they form ions, the fourth-row transition metals lose their $4s$-electrons more easily than their $3d$-electrons, so the valence-electron configuration of Cr^{3+} is $3d^3$.

Figure 19–12 Six point charges in an octahedral array about a central metal ion create an octahedral field about the metal ion. Such a field increases the energies of all five d-orbitals on the metal ion, but the increase is larger for the d_{z^2}- and $d_{x^2-y^2}$-orbitals. As a result, the d-orbitals are split into two sets that differ by the energy Δ_o. The occupancy of these orbitals is shown by the four d-electrons of Mn(III) for (a) the low-spin (large Δ_o) and (b) the high-spin (small Δ_o) case.

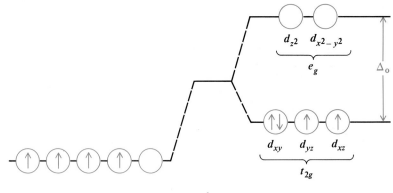

(a) $[Mn(CN)_6]^{3-}$ (low spin)

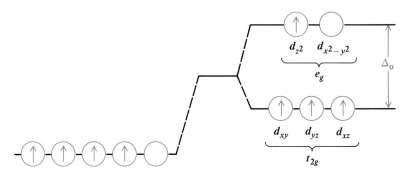

(b) $[Mn(H_2O)_6]^{3+}$ (high spin)

Two types of electron configurations are possible in octahedral complexes of central metal atoms or ions having four to seven d-electrons. When Δ_o is large, **low-spin complexes** are formed in which electrons are paired in the lower-energy t_{2g} orbitals; the e_g orbitals are not occupied until the t_{2g} levels are filled. When Δ_o is small, **high-spin complexes** occur in which electrons are placed singly in both t_{2g} and e_g orbitals and remain unpaired to the fullest extent possible.

There is interesting direct evidence for the reality of the high-spin versus low-spin distributions of d-electrons in coordination complexes. Precise experimental work in which crystals diffract x-rays (see Section 20–1) provides maps of the electron density within crystals. When this technique was recently applied to a compound having Co^{2+} in an octahedral environment of F^- ions, it was possible to "see" the seven $3d$-electrons of the Co^{2+} and confirm that their distribution was that of the configuration $t_{2g}^5 e_g^2$ (high spin) and not $t_{2g}^6 e_g^1$ (low spin).

Magnetic Properties

The existence of high- and low-spin electron configurations accounts for the magnetic properties of many different coordination compounds. As discussed in Section 17–1, substances can be classified as paramagnetic or diamagnetic according to whether they are attracted into a magnetic field. *All* substances are either attracted or repelled by magnets, although most such interactions are so weak that experiments with low-power magnets show nothing. Figure 19–13 depicts an experiment to demonstrate the universal susceptibility of substances to the influence of magnetic fields. A cylindrical sample is suspended so that its bottom is between the poles of a powerful magnet but its top extends out of the field. It is thus in a nonuniform magnetic field, one whose intensity varies from strong to weak. It is weighed very precisely and then reweighed in the absence of the magnet. The net force on the sample changes in the presence of the magnetic field. Substances that are repelled by a nonuniform magnetic field weigh less when dipped into one and are *diamagnetic*. Substances that are attracted by a magnetic field weigh more and are *paramagnetic*. The weighings just described, with due correction and calibration, give numerical values for the **magnetic susceptibility** of a substance, its tendency to be attracted by magnetic fields. The susceptibility of a diamagnet is negative and small (because the substance is repelled by the test magnet). The susceptibility of a paramagnet is positive and can be quite large.

As Section 17–1 explains, paramagnetism is always associated with atoms, ions, or molecules that contain one or more unpaired electrons. Diamagnetic substances have the spins of all of their electrons paired. Thus, measurements of magnetic susceptibility reveal which substances have one or more unpaired electrons and which have completely paired electrons. In many cases, the number of unpaired electrons per molecule in a paramagnet can even be counted, based on the magnitude of its magnetic susceptibility. On a molar basis, a substance with two unpaired electrons per molecule is pulled into a magnetic field more strongly than a substance with only one unpaired electron per molecule.

These facts emerge in connection with coordination complexes because paramagnetism is prevalent among transition-metal complexes; the great majority of other chemical substances are diamagnetic. Among complexes of a given metal ion, the number of unpaired electrons, as observed by magnetic susceptibility, varies with the identity of the ligands. This can be understood through crystal-field theory. Both $[Mn(CN)_6]^{3-}$ and $[Mn(H_2O)_6]^{3+}$ ions, for example, have six ligands surrounding the central Mn^{3+} ion, yet the former has only two unpaired electrons (see Fig. 19–12a) and the latter has four (see Fig. 19–12b). The crystal-field splitting energy is larger for the CN^- ligands than for the H_2O ligands, giving a low-spin complex in the first

• This is one place where the distinction between mass and weight is crucial. The mass of the sample does not change when the magnet is removed, but the weight does, because the net force pulling the sample down is then different.

• Iron and steel weigh much, much more in a magnetic field. They are *ferromagnetic*.

• Indeed, the Lewis-dot representation of bonding (see Chapter 3) calls for displaying all valence electrons in pairs.

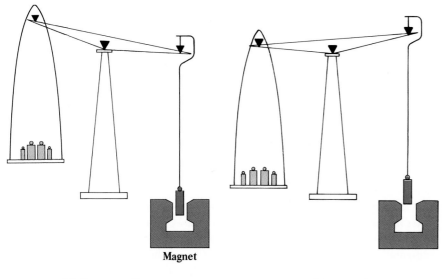

(a) Paramagnetic   **(b) Diamagnetic**

Magnet

Figure 19–13 (a) If a sample of a paramagnetic substance is partially immersed into a magnetic field, it is attracted down into the field. (b) If a diamagnetic substance is put into the same field in the same way, it is buoyed up by the field.

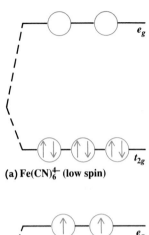

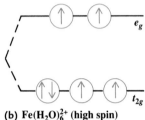

(a) Fe(CN)$_6^{4-}$ (low spin)

(b) Fe(H$_2$O)$_6^{2+}$ (high spin)

Figure 19–14 The occupancy of the 3d-orbitals in a low-spin and a high-spin octahedral complex of Fe(II).

case and a high-spin complex in the second. Similarly, $[Fe(CN)_6]^{4-}$ is diamagnetic, but $[Fe(H_2O)_6]^{2+}$ is paramagnetic to the extent of four unpaired electrons. The six d-electrons on the Fe^{2+} ion are all paired in the lower t_{2g} orbitals in the first case in the configuration $(t_{2g})^6$ (Fig. 19–14a); in the second, two electrons have left the t_{2g} orbital to occupy e_g orbitals in the configuration $(t_{2g})^4(e_g)^2$ (Fig. 19–14b).

EXAMPLE 19–6

The complex ion $[CoF_6]^{3-}$ is paramagnetic, but $[Co(NH_3)_6]^{3+}$ is not. Identify the d-electron configurations in these two octahedral complex ions.

Solution

The Co^{3+} ion, like the Fe^{2+} ion, has six d-electrons. Its high-spin complexes have four unpaired spins $(t_{2g}^4 e_g^2)$ and are paramagnetic (recall Fig. 19–14b); its low-spin complexes have no unpaired spins (t_{2g}^6) and are diamagnetic (Fig. 19–14a). Therefore, $[CoF_6]^{3-}$ must be a high-spin complex, and $[Co(NH_3)_6]^{3+}$ a low-spin complex, with the splitting Δ_o greater for ammonia than for fluoride-ion ligands.

Exercise

The octahedral complex ions $[FeCl_6]^{3-}$ and $[Fe(CN)_6]^{3-}$ are, respectively, high-spin and low-spin complexes. How many unpaired electrons are in each, and in which is the octahedral-field splitting greater?

Answer: Five in $[FeCl_6]^{3-}$ and one in $[Fe(CN)_6]^{3-}$. Δ_o is larger in the cyanide complex.

Square-Planar and Tetrahedral Complexes

Crystal-field theory can be applied to square-planar and tetrahedral complexes as well as to octahedral complexes. Consider a square-planar complex in which four ligands are brought up toward the metal ion along the $\pm x$- and $\pm y$-axes. The $d_{x^2-y^2}$-orbital on the metal has a higher energy than a d_{z^2}-orbital in such a field, because an electron in the $d_{x^2-y^2}$-orbital has maximum probability density along the x- and y-axes, where it is most strongly repelled by ligand electrons. In the same way, the energy of the d_{xy}-orbital lies above that of the d_{xz} and d_{yz} in a square-planar field because the former lies in the plane of the ligands, and the latter wave functions have a node in the xy-plane. Figure 19–15a shows the resulting level structure.

Tetrahedral complexes result from bringing ligands to four of the eight corners of an imaginary cube with the metal ion at its center. The $d_{x^2-y^2}$- and d_{z^2}-oribtals point toward the centers of the cube faces, but the other three orbitals point toward the centers of the cube edges. The cube edges are closer to the ligands, and these three orbitals are therefore more strongly repelled than the $d_{x^2-y^2}$- and d_{z^2}-orbitals. The result is shown in Figure 19–15b and is the reverse of the energy-level ordering found for octahedral complexes. The magnitude of the splitting Δ_t is significantly smaller than Δ_o.

Octahedral complexes are more common than square-planar or tetrahedral complexes because the formation of six bonds to ligands, rather than four, confers greater stability. Square-planar complexes are important primarily for low-spin d^8-electron configurations with large splittings. Forming an octahedral complex in such cases forces two electrons into a high-energy e_g level; in the square-planar complex, they stay in a lower-energy, d_{xy}-orbital. Square-planar complexes predominate for the

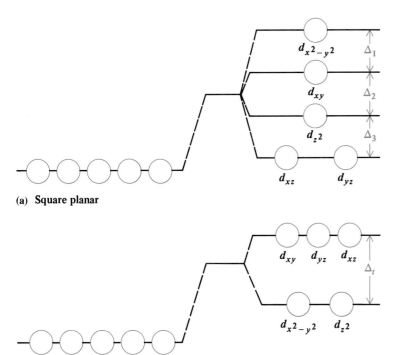

(a) Square planar

(b) Tetrahedral

Figure 19–15   Energy-level structures of the 3d-orbitals in (a) square-planar and (b) tetrahedral crystal fields.

d^8-ions Pt(II) and Pd(II), because the d-orbitals of these ions are strongly split by most ligands. The Ni(II) ion is also a d^8 ion but experiences significantly smaller ligand-field splittings and consequently often forms octahedral complexes. Tetrahedral complexes are less common because of the lower coordination number, the smaller size of ligand-field splitting, and the necessity of placing electrons into a higher-energy level at an earlier stage in the filling process.

19-4   THE COLORS OF COORDINATION COMPLEXES

The most striking properties of the complexes of the transition metals are their colors (Fig. 19–16). Complexes of Co(III), for example, have these colors:

$[Co(NH_3)_6]^{3+}$	Orange
$[Co(NH_3)_4Cl_2]^+$	A green form and a violet form
$[Co(NH_3)_5(H_2O)]^{3+}$	Red

Colors arise when complexes absorb light in some portion of the visible spectrum. As explained in Section 18–4, the color perceived in a sample is the color *complementary* to that which is most strongly absorbed. Suppose that white light passes into an aqueous solution containing the ion $[Co(NH_3)_5Cl]^{2+}$. This ion absorbs strongly near a wavelength of 530 nm, in the yellow-green region of the spectrum. White light consists of photons with a broad distribution of wavelengths, but only the blue and red light is transmitted by the solution, which appears purple. A material that absorbs across all visible wavelengths appears gray or black, and one that absorbs weakly or not at all in the visible range is colorless.

Figure 19–16   Several coordination compounds. From the top left and moving clockwise, they are $[Cr(CO)_6]$ (white), $K_3[Fe(C_2O_4)_3]$ (green), $[Co(en)_3]I_3$ (orange), $[Co(NH_3)_5(H_2O)]Cl_3$ (rose), and $K_3[Fe(CN)_6]$ (red-orange).

The colors of many octahedral transition-metal compounds arise from the excitation of electrons from occupied t_{2g} levels to empty e_g levels. The frequency (ν) of light that is capable of inducing such a transition is related to the energy difference between the two states, which is the crystal-field splitting energy:

$$h\nu = \Delta_o$$

The larger the crystal-field splitting, the higher the frequency of light absorbed most strongly and the shorter its wavelength. In $[Co(NH_3)_6]^{3+}$, an orange compound that absorbs most strongly in the violet region of the spectrum, there is a larger crystal-field splitting Δ_o than in $[Co(NH_3)_5Cl]^{2+}$, a purple compound that absorbs most strongly at lower frequencies (longer wavelengths) in the yellow-green region of the spectrum. A d^{10}-complex (such as those of Zn^{2+} or Cd^{2+}) is colorless because all the d-levels (both t_{2g} and e_g) are filled; consequently, a $t_{2g} \rightarrow e_g$ transition cannot take place, and the absorption in the visible region of the spectrum is very slight. High-spin d^5 complexes such as $[Mn(H_2O)_6]^{2+}$ also show only very weak absorption because processes in which a spin is reversed on excitation by light (required by the Pauli principle in these cases) occur only rarely. For this reason, the Mn^{2+} complex ion with six water ligands is very pale pink (Fig. 19–17).

The simple crystal-field model succeeds in accounting for the strong absorption spectra of transition-metal complexes, just as it accounts for the magnetic properties described in Section 19–3. From the spectra it is possible to rank ligands from those that interact most weakly with the metal ion (and therefore give the smallest crystal-field splitting) to those that interact most strongly and give the largest splitting. Although such an ordering is not invariably followed for every ligand with every central metal atom, it is nonetheless useful. The ordering is called the **spectrochemical series.** For a selection of ligands from weakest to strongest it is:

$$I^- < Br^- < Cl^- < F^-, OH^- < H_2O < NCS^- < NH_3 < en < CO, CN^-$$

| weak-field ligands (high spin) | intermediate-field ligands | strong-field ligands (low spin) |

This ordering cannot be explained within the crystal-field model, and indeed its existence points up an unsatisfactory feature of crystal-field theory. The bonding in coordination compounds *cannot* be fully ionic, as assumed in the crystal-field description. If it were, it would be impossible for neutral ligands such as NH_3 and CO to give larger splittings than small, negatively charged ligands such as F^- ion. The crystal-field theory has been modified to include covalent as well as ionic aspects of coordination. The resulting extended theory, which uses molecular orbital methods of the type described in Chapter 18, successfully explains the spectrochemical series, but a discussion is beyond the scope of this book.

• Aqueous solutions of the transition metals that we have indicated up to now as $Mn^{2+}(aq)$ or $Zn^{2+}(aq)$ can also be represented with the coordinated water molecules shown explicitly: $[Mn(H_2O)_6]^{2+}(aq)$ and $[Zn(H_2O)_4]^{2+}(aq)$. We follow this practice in this chapter to emphasize the nature of the coordination.

Figure 19–17 The colors of the hexaaqua complexes of the metal ions (from left) Mn^{2+}, Fe^{3+}, Co^{2+}, Ni^{2+}, Cu^{2+}, and Zn^{2+} prepared from their nitrate salts. Note that the d^{10} Zn^{2+} complex is colorless. The green color of the Ni^{2+} is due to the absorption of both red and blue from the white light illuminating the solution. The yellow cast of the solution containing $Fe(H_2O)_6^{3+}$ ion is caused by a species produced by hydrolysis of that ion; if this reaction is suppressed, the solution of $[Fe(H_2O)_6^{3+}]$ is pale purple.

CHEMISTRY IN YOUR LIFE

Ruby-Red and Emerald-Green

Crystal-field theory explains the dramatic colors of rubies and emeralds. Rubies result from the substitution of a small amount of chromium for aluminum in aluminum oxide (corundum) crystals. Pure corundum (Al_2O_3) is colorless, but when it is combined with about 1% of the green solid Cr_2O_3, the deep red hue of ruby results. The Cr^{3+} ions replace Al^{3+} ions in slightly distorted octahedral crystal sites, each defined by six oxygen ions. Free Cr^{3+} ions have a $3d^3$ valence electron configuration in their ground states, but at the octahedral sites in the crystal, this becomes $(t_{2g})^3$: three electrons with parallel spins occupy the lower-energy t_{2g} levels. Light of the proper wavelength excites electrons into the higher-energy e_g levels of the Cr^{3+} ions. Two strong absorption bands result, both in the visible region of the spectrum (Fig. 19–B, top). The lower-energy band corresponds to absorption of green and yellow light, and the other to absorption of violet light. The two bands overlap, so that most blue light is also absorbed. The little that passes through combines with the strongly transmitted red light to give a characteristic purple tinge to the red of the ruby.

Emeralds also have Cr^{3+} impurities that substitute for Al^{3+} ions in a distorted octahedral site formed by six oxygen ions. What then causes the dramatic difference between ruby red and emerald green? The answer lies with a small change (about 10%) in the crystal-field splitting parameter Δ_o. The crystal into which chromium substitutes to give emerald is not corundum but beryl, with the formula $Be_3Al_2Si_6O_{18}$. Like corundum, pure beryl is colorless. The presence of beryllium and silicon makes the bonding in beryl weaker than in corundum. The weaker crystal field at the octahedral sites means that both of the Cr^{3+} absorption bands in emerald are shifted to lower energy (longer wavelength, Fig. 19–B, bottom). One absorption now overlaps the red part of the visible spectrum, blocking transmission of red, orange, and yellow light. The other absorption shifts downward from violet toward blue and also changes shape to open a transmission "window" between the two bands. Green light with a trace of blue is transmitted, giving a color that we call "emerald-green." A subtle change in the crystal field completely changes the color of the gem.

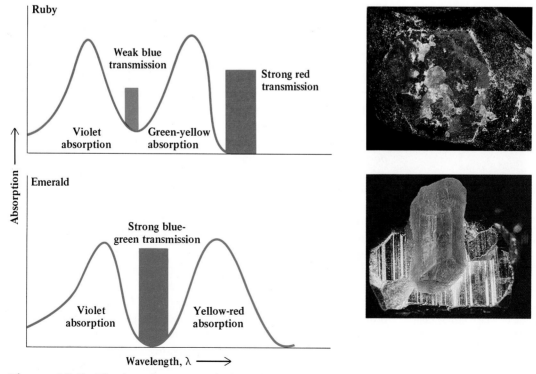

Figure 19–B The absorption spectra of ruby and emerald. Small shifts in the shape of the spectrum have a dramatic effect on the perceived color.

EXAMPLE 19–7

Predict which of the following octahedral coordination complexes has the shortest wavelength of absorption in the visible spectrum: $[FeF_6]^{3-}$, $[Fe(CN)_6]^{3-}$, $[Fe(H_2O)_6]^{3+}$.

Solution

The $[Fe(CN)_6]^{3-}$ ion has the strongest-field ligands of the three complexes, so its energy levels are split by the largest amount. As a result, the frequency of light absorbed is largest for this ion, and its absorption wavelength is the shortest.

The observed color of $[Fe(CN)_6]^{3-}$ solutions is red, indicating absorption of blue and violet light. $[Fe(H_2O)_6]^{3+}$ solutions are (when completely pure) a very pale violet, due to weak absorption of red light, and $[FeF_6]^{3-}$ solutions are colorless, indicating that the absorption lies beyond the long wavelength limit of the visible spectrum.

Exercise

The $[Ti(H_2O)_6]^{3+}$ ion colors its solutions purple and has a single absorption maximum in the visible-frequency range. What is the approximate wavelength of this maximum? What predictions can you make about the wavelengths of maximum absorption in solutions containing $[TiCl_6]^{3-}$ and $[Ti(en)_3]^{3+}$ ion?

Answer: The absorption maximum should be in the yellow-green region (complementary to purple) in the vicinity of 530 nm (observed: 510 nm). The absorption maximum for $[TiCl_6]^{3-}$ should come at a longer wavelength than this, and that for $[Ti(en)_3]^{3+}$ at a shorter wavelength.

19–5   COORDINATION COMPLEXES IN BIOLOGY

Coordination complexes, particularly chelates, play fundamental roles in the biochemistry of both plants and animals. Trace amounts of at least nine transition elements—vanadium, chromium, manganese, iron, cobalt, nickel, copper, zinc, and molybdenum—are essential to life.

Several of the most important complexes are based on the organic compound *porphine*, which has the approximately planar structure shown in Figure 19–18. The π bonding orbitals in these compounds are conjugated. Donation of the two acidic (N-bound) hydrogens from porphine to some base leaves (porphine)$^{2-}$, which has four nitrogen atoms ready to bind to a metal ion M^{2+} and form a chelate structure. This tetradentate ligand, modified by the addition of several side groups, gives a complex with Fe^{2+} ions called *heme* (Fig. 19–18b). The absorption of light by heme is responsible for the red color of blood. In *hemoglobin,* the compound that transports oxygen in the blood, the fifth coordination site of the iron(II) ion binds *globin* (a high-molar-mass protein), and the sixth is occupied by water or molecular oxygen. In cases of carbon monoxide poisoning, CO molecules occupy the sixth coordination site and block the binding and transport of oxygen. Hemoglobin has a complex structure that contains four heme groups (see Fig. 25–11).

Photosynthesis depends on the properties of chlorophyll, which contains a derivative of the porphine molecule with different side groups and with a magnesium ion at its center (Fig. 19–19). The aqueous Mg^{2+} ion does not absorb light in the visible region of the spectrum, but chlorophyll, in which Mg^{2+} is chelated, does.

• The equilibria by which hemoglobin in the blood binds and later releases O_2 are discussed on page 306.

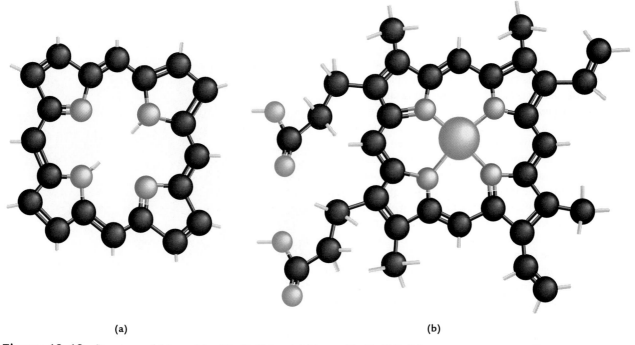

(a)   (b)

Figure 19–18   Structures of (a) porphine ($C_{20}H_{14}N_4$) and (b) heme ($C_{34}H_{32}N_4O_4Fe$), a chelate based on porphine. Note the alternation of double and single bonds in the structures. Porphine is a dark red solid. Heme is isolated as brown needle-like crystals with a violet sheen.

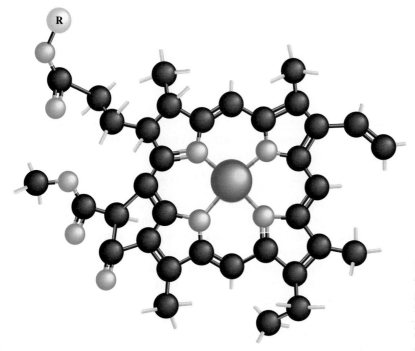

Figure 19–19   The structure of chlorophyll a, $C_{55}H_{72}MgN_4O_5$. The "R" stands for the phytyl group ($C_{20}H_{39}$), a long hydrocarbon "tail" that increases the solubility of the compound in less polar environments. Solid chlorophyll a forms as waxy blue-black (not green) crystals; a solution in ethanol is blue-green.

CHEMISTRY IN YOUR LIFE

Molybdenum, Soil Fertility, and Termites

The fixing of atmospheric nitrogen in compounds suitable for use by plants is essential to life on this planet. Several species of bacteria, including those living in association with such legumes as soybeans, achieve nitrogen fixation by employing a pair of nitrogenase enzymes. One of these compounds has a molar mass of about 60,000 g mol^{-1} and contains 4 iron atoms. The other has a molar mass nearer to 220,000 g mol^{-1} and includes 2 molybdenum atoms and 30 iron atoms. The two work in a complex partnership to provide a low-activation-energy pathway by which atmospheric N_2 is reduced to NH_3, which plants can assimilate. Molybdenum is relatively rare; in soils, it is usually sequestered in the form of highly insoluble Mo_2S_3. Nitrogen-fixing bacteria have evolved efficient mechanisms that involve coordination complexes for accumulating molybdenum. Because nitrogen fixation is essential to the fertility of soils, concern sometimes arises that nitrogen fixation might fail for lack of molybdenum. Farmers at times seek to help nitrogen-fixing bacteria by applying compounds of molybdenum to their legume fields.

Bacteria living in the gut of termites also generate large amounts of the nitrogenases, a fact that makes live termites (Figure 19–C) excellent agents for fixing nitrogen and improving soil fertility. Although this aspect of the termite lifestyle might be beneficial, the damage caused by their attacks on wooden structures throughout the world makes them a pest. It has recently been found that molybdenum compounds kill termites. Baits are dosed with molybdenum at levels that exceed those in the natural environment. Termites feeding on these baits turn

Figure 19–C   Termites.

steel gray after 8 to 10 days, and most die in 30 days. It is thought that nitrogen-fixing bacteria in the termites' guts accumulate so much molybdenum that it spills over and poisons their hosts.

Chemists are now actively trying to design transition-metal–based catalysts that mimic the nitrogen-fixing activity of nitrogenase in bacteria. A good catalyst would greatly reduce the cost of producing nitrogen-based fertilizers, which are now manufactured (see Section 22–4) by direct gas-phase reaction of nitrogen with hydrogen. The high temperatures and pressures that are essential in this method demand elaborate and expensive equipment. Moreover, the reaction relies on dwindling supplies of petroleum or natural gas as sources of hydrogen.

Absorption of light by this complex provides the energy to carry out photosynthesis, producing glucose from water and carbon dioxide:

$$6\ CO_2(g) + 6\ H_2O(\ell) \longrightarrow 6\ O_2(g) + C_6H_{12}O_6(aq) \qquad \Delta G° = 2872\ \text{kJ}$$

The appearance of chlorophyll and related substances in the early biological history of the earth provided the means for the energy of the sun to drive the above reaction to the right. The accumulation of O_2 about two billion years ago profoundly changed the nature of the atmosphere from a reducing environment (dominated by CO_2, CH_4, and NH_3) to an oxidizing environment (dominated by oxygen). The harvesting of the sun's energy using chlorophyll led to life as we know it on earth.

Vitamin B_{12} is useful in the treatment of pernicious anemia and other diseases. Its structure has certain similarities to those of heme and chlorophyll. Again, a metal ion is coordinated by a planar tetradentate ligand, with two other donors completing the coordination octahedron by occupying *trans* positions. In vitamin B_{12}, the

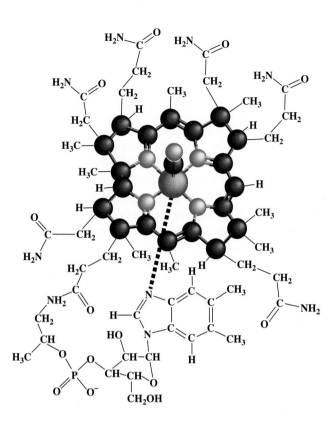

Figure 19–20 The structure of the cobalt complex vitamin B_{12}, $Co(C_{62}H_{88}N_{13}O_{14}P)CN$. For clarity, the structure is "exploded"; atoms that are not part of the corrin ring are shown typographically.

metal ion is cobalt and the planar ring system is not porphine but *corrin* (Fig. 19–20), in which two of the five-membered N-containing rings are joined directly rather than having a carbon atom between them. In the human body, enzymes derived from vitamin B_{12} accelerate a range of important reactions, including those producing red blood cells.

SUMMARY

19–1 A **coordination complex** forms when a metal atom or ion coordinates with the lone-pair electrons of **ligand** molecules in a Lewis acid–base reaction. The number of lone pairs accepted from ligands is the **coordination number** of the metal. One important class of ligands are those that **chelate** (bind to metal atoms or ions at two or more ligating atoms). Complexes differ in chemical and physical properties from their independent metal and ligand components. In aqueous solution they exhibit either thermodynamic or kinetic stability, or both. Complexes that exchange ligands rapidly are said to be **labile,** but those that exchange slowly are **inert.**

19–2 The most common three-dimensional structures of coordination complexes are **octahedral** (for coordination number 6) and **square planar** or **tetrahedral** (for coordination number 4). These structures display two types of *isomerism,* in which the same ligands are attached to the central atom but in arrangements that cannot be superimposed no matter how the complex is rotated. In **optical isomerism,** the two isomers are mirror-image pairs and are called **enantiomers.** In *cis–trans* isomerism, the relative positions of the ligands as they coordinate to the metal are different.

19–3 Crystal-field theory explains many of the properties of coordination complexes. If charges (from ions or lone-pair electrons on molecules) are brought up toward a transition-metal atom or ion, the energies of the five d-orbitals on the metal atom split apart. In the case of an octahedral complex, they form two groups: three t_{2g} orbitals (lower energy) and two e_g orbitals (at higher energy). The energy difference between the two is the **octahedral crystal-field splitting** Δ_o for the complex. When Δ_o is large, electrons pair in the lower levels to form **low-spin complexes.** When Δ_o is small, the same number of electrons spreads across to occupy the upper levels, forming **high-spin complexes.** Low-spin and high-spin complexes of a given central atom have different magnetic properties. The splitting of the levels in tetrahedral and square-planar complexes is different from that in octahedral complexes.

19–4 The frequency of light absorbed as electrons are excited from the lower to the upper level is proportional to the crystal-field splitting, and the observed colors of coordination complexes are complementary to those absorbed most strongly by these transitions. Ligands can be arranged in a **spectrochemical series** from weak-field to strong-field ligands, depending on the magnitude of the crystal-field splitting they induce in metal ions.

19–5 Coordination complexes are significant in living systems. Such biological molecules as hemoglobin, chlorophyll, and vitamin B_{12} contain coordinated transition-metal atoms.

PROBLEMS

Note: Answers to blue-numbered problems are given in Appendix F. Problems that are more challenging are indicated with asterisks.

The Formation of Coordination Complexes

1. (See Example 19–1.) Write formulas, including overall electric charge, of coordination complexes containing the following metal ions and ligands:
   (a) Fe^{2+} with six CN^- ligands.
   (b) Mn^{3+} with one NH_3 ligand, two H_2O ligands, and three Cl^- ligands.
   (c) Pt^{2+} with one H_2O ligand, one Br^- ligand, and one ethylenediamine (en) ligand.

2. (See Example 19–1.) Write formulas, including overall electric charge, of coordination complexes containing the following metal ions and ligands:
   (a) Cr with four CO ligands.
   (b) Co^{3+} with two Cl^- ligands and two $C_2O_4^{2-}$ ligands.
   (c) Rh^{2+} with three NH_3 ligands and three Br^- ligands.

3. Is methylamine (CH_3NH_2) a monodentate or a bidentate ligand? With which of its atoms does it bind to a metal ion?

4. Show how the glycinate ion (H_2N—CH_2—COO^-) can act as a bidentate ligand. (Draw a Lewis structure if necessary.) Which atoms in the glycinate ion bind to a metal ion?

5. Calculate the pH of a 0.124 M solution of $[Pt(en)_3]Cl_4$ at 25°C. The cation is a weak acid with $K_{a1} = 4.5 \times 10^{-6}$ and $K_{a2} = 2.45 \times 10^{-10}$ at this temperature.

6. $[Au(NH_3)_4]PO_4$ has a K_{sp} of 1.0×10^{-11} at 25°C. A simple calculation predicts that the solubility of this substance is 3.2×10^{-6} mol L^{-1}. The observed solubility is greater than this. Use the fact that PO_4^{3-} ion is a base (see Chapter 8) and that $[Au(NH_3)_4]^{3+}$ is a weak acid (with a K_a of 3.3×10^{-8}) to write a reaction that would explain this enhanced solubility.

7. (See Example 19–2.) Determine the oxidation state of the metal of the following coordination complexes: $[V(NH_3)_4Cl_2]$, $[Mo_2Cl_8]^{4-}$, $[Co(H_2O)_2(NH_3)Cl_3]^-$, $[Ni(CO)_4]$.

8. (See Example 19–2.) Determine the oxidation state of the metal in each of the following coordination complexes: $[Mn_2(CO)_{10}]$, $[Re_3Br_{12}]^{3-}$, $[Fe(H_2O)_4(OH)_2]^+$, $[Co(NH_3)_4Cl_2]^+$.

9. (See Example 19–3.) Give the chemical formula of each of the following compounds:
   (a) Sodium tetrahydroxozincate(II).

(b) Dichlorobis(ethylenediamine)cobalt(III) nitrate.

(c) Triaquabromoplatinum(II) chloride.

(d) Tetraamminedinitroplatinum(IV) bromide.

10. (See Example 19–3.) Give the chemical formula of the each of the following compounds:

(a) Silver hexacyanoferrate(II).

(b) Potassium tetraisothiocyanatocobaltate(II).

(c) Sodium hexafluorovanadate(III).

(d) Potassium trioxalatochromate(III).

11. Assign a systematic name to each of the following chemical compounds:

(a) $NH_4[Cr(NH_3)_2(NCS)_4]$

(b) $[Tc(CO)_5]I$

(c) $K[Mn(CN)_5]$

(d) $[Co(NH_3)_4(H_2O)Cl]Br_2$

12. Give the systematic name of each of the following chemical compounds:

(a) $[Ni(H_2O)_4(OH)_2]$

(b) $[HgClI]$

(c) $K_4[Os(CN)_6]$

(d) $[FeBrCl(en)_2]Cl$

13. Describe the behavior of a coordination complex that is both thermodynamically stable and labile.

14. Describe the behavior of a coordination complex that is thermodynamically unstable and kinetically inert.

15. Calculate $\Delta G°$ at 25°C for the dissociation of tetraamminecopper(II) ion to $Cu^{2+}(aq)$ and $NH_3(aq)$ in basic aqueous solution, using data from Appendix D. Is this complex thermodynamically stable with respect to this reaction under standard conditions? Why does it dissociate under acidic conditions? Write a balanced equation for the dissociation reaction at pH = 0, and calculate $\Delta G°$ at 25°C for that reaction.

16. The $\Delta G_f°$ of the tetracyanonickelate(II) anion is +472.1 kJ mol^{-1} in aqueous solution at 25° C. Use this information together with data from Appendix D to calculate the equilibrium constant at this temperature for the dissociation

$$[Ni(CN)_4]^{2-}(aq) \longrightarrow Ni^{2+}(aq) + 4\ CN^-(aq)$$

Isotopically labelled $Na^{13}CN$ is added to a solution of $K_2[Ni(CN)_4]$. Subsequent examination of the $[Ni(CN)_4]^{2-}$ present reveals extensive incorporation of $^{13}CN^-$ into the complex ion. Comment on these observations, and discuss the difference between liability and thermodynamic stability as applied to complexes of nickel(II) with the cyanide ion.

Structures of Coordination Complexes

17. Suppose that 0.010 mol of each of the following compounds is dissolved in separate 1-L portions of water: KNO_3, $[Co(NH_3)_6]Cl_3$, $Na_2[PtCl_6]$, $[Cu(NH_3)_2Cl_2]$. Rank the resulting four solutions in order of their conductivity, from lowest to highest.

18. Suppose that 0.010 mol of each of the following compounds is dissolved (separately) in 1 L of water: $BaCl_2$, $K_4[Fe(CN)_6]$,

$[Cr(NH_3)_4Cl_2]Cl$, $[Fe(NH_3)_3Cl_3]$. Rank the resulting four solutions in order of their conductivity, from lowest to highest.

19. (See Examples 19–4 and 19–5.) Draw the structures of all possible isomers for the following complexes. Indicate which isomers are mirror-image pairs (enantiomers).

(a) Diamminebromochloroplatinum(II) (square planar).

(b) Diaquachlorotricyanocobaltate(III) ion (octahedral).

(c) Trioxalatovanadate(III) ion (octahedral).

20. (See Examples 19–4 and 19–5.) Draw the structures of all possible isomers for the following complexes. Indicate which isomers are enantiometric pairs.

(a) Bromochloro(ethylenediamine)platinum(II) (square planar).

(b) Tetraamminedichloroiron(III) ion (octahedral).

(c) Amminechlorobis(ethylenediamine)iron(III) ion (octahedral).

21. (See Examples 19–4 and 19–5.) Iron(III) forms octahedral complexes. Sketch the structures of all the distinct isomers of $[Fe(en)_2Cl_2]^+$, indicating which pairs of structures are mirror images of each other.

22. (See Examples 19–4 and 19–5.) Platinum(IV) forms octahedral complexes. Sketch the structures of all the distinct isomers of $[Pt(NH_3)_2Cl_2F_2]$, indicating which pairs of structures are mirror images of each other.

Crystal-Field Theory and Magnetic Properties

23. For each of the following ions, draw diagrams like those in Figure 19–14 to show orbital occupancies in both weak and strong octahedral fields. Indicate the total number of unpaired electrons in each case.

(a) Mn^{2+} (d) Mn^{3+}

(b) Zn^{2+} (e) Fe^{2+}

(c) Cr^{3+}

24. Repeat the work of the preceding problem for the following ions:

(a) Cr^{2+} (d) Pt^{4+}

(b) V^{3+} (e) Co^{2+}

(c) Ni^{2+}

25. (See Example 19–6.) Experiments can measure not only whether a compound is paramagnetic, but also the number of unpaired electrons. It is found that the octahedral complex ion $[Fe(CN)_6]^{3-}$ has fewer unpaired electrons than the octahedral complex ion $[Fe(H_2O)_6]^{3+}$. How many unpaired electrons are present in each species? Give the d-electron configuration of each species.

26. (See Example 19–6.) The octahedral complex ion $[MnCl_6]^{3-}$ has more unpaired electrons than does the octahedral complex ion, $[Mn(CN)_6]^{3-}$. How many unpaired electrons are present in each species? Give the d-electron configuration of each species.

27. Explain why octahedral coordination complexes with 3 and 8 d-electrons on the central metal atom are particularly stable. Under what circumstances would you expect complexes

with 5 or 6 *d*-electrons on the central metal atom to be particularly stable?

28. The following standard reduction potentials have been measured in acidic aqueous solution:

$$Mn^{3+} + e^- \longrightarrow Mn^{2+} \qquad \mathscr{E}° = 1.51 \text{ V}$$

$$Fe^{3+} + e^- \longrightarrow Fe^{2+} \qquad \mathscr{E}° = 0.77 \text{ V}$$

$$Co^{3+} + e^- \longrightarrow Co^{2+} \qquad \mathscr{E}° = 1.84 \text{ V}$$

Explain why the reduction potential for Fe^{3+} lies below those for the +3 oxidation states of the elements on either side of it in the periodic table.

The Colors of Coordination Complexes

29. An aqueous solution of zinc nitrate contains the $[Zn(H_2O)_6]^{2+}$ ion and is colorless. What conclusions can be drawn about the absorption spectrum of the $[Zn(H_2O)_6]^{2+}$ complex ion and the configuration of the *d*-electrons of the central metal in this complex?

30. An aqueous solution of sodium hexaiodoplatinate(IV) is black. What conclusions can be drawn about the absorption spectrum of the $[PtI_6]^{2-}$ complex ion?

31. Estimate the wavelength of maximum absorption for the octahedral ion hexacyanoferrate(III) from the fact that light transmitted by a solution of it is red. Estimate the crystal-field splitting energy Δ_o (in kJ mol^{-1}).

32. Estimate the wavelength of maximum absorption for the octahedral ion hexaaquanickel(II) from the fact that its solutions transmit green light. Estimate the crystal-field splitting energy Δ_o (in kJ mol^{-1}).

33. The chromium(III) ion in aqueous solution has a blue-violet color.
    (a) What is the color complementary to blue-violet?
    (b) Estimate the wavelength of maximum absorption for a $Cr(NO_3)_3$ solution.
    (c) Does the wavelength of maximum absorption increase or decrease if cyano ligands are substituted for the coordinated water? Explain.

34. An aqueous solution containing the hexaamminecobalt(III) ion has a yellow color.
    (a) What is the color complementary to yellow?
    (b) Estimate the wavelength of maximum absorption in the visible spectrum by this solution.

35. (a) A solution of $Fe(NO_3)_3$ has only a pale color, but one of $K_3[Fe(CN)_6]$ is bright red. Do you expect a solution of $K_3[FeF_6]$ to be brightly colored or pale? Explain your reasoning.
    (b) Would you predict that a solution of $K_2[HgI_4]$ is colored or colorless? Explain.

36. (a) A solution of $Mn(NO_3)_2$ has a very pale pink color, but one of $K_4[Mn(CN)_6]$ is a deep blue. Explain why the two differ so much in the intensities of their colors.
    (b) Which of the following does crystal-field theory predict

to be colorless in aqueous solution: $K_2[Co(NCS)_4]$, $Zn(NO_3)_2$, $[Cu(NH_3)_4]Cl_2$, $CdSO_4$, $AgClO_3$, $Cr(NO_3)_2$?

37. Why are there fewer tetrahedral complexes than octahedral complexes?

*38. Predict the color of a d^5 tetrahedral complex. Explain the prediction.

Coordination Complexes in Biology

39. Vitamin B_{12s} is 4.43% cobalt by mass. Determine the molar mass of vitamin B_{12s} if each molecule of it contains one atom of cobalt.

40. Hemoglobin, the compound that transports oxygen in the bloodstream, has a molar mass of 6.8×10^4 g mol^{-1}. It is also 0.33% iron by mass. Determine how many iron atoms are in a single hemoglobin molecule.

Additional Problems

41. Explain why ligands are usually negative or neutral in charge and only rarely positive.

42. If *trans*-$[Cr(en)_2(NCS)_2]SCN$ is heated, it reacts to form gaseous ethylenediamine and $[Cr(en)_2(NCS)_2][Cr(en)(NCS)_4]$, a solid. Write a balanced chemical equation for this reaction. What are the oxidation states of chromium in the reactant and in the two complex ions in the product?

43. A coordination complex has the molecular formula $[Ru_2(NH_3)_6Br_3](ClO_4)_2$. Determine the oxidation state of ruthenium in this complex.

44. Write balanced chemical equations for the production of $HCN(g)$ from $CN^-(aq)$ and HCl, for the production of iron(II) hydroxide from aqueous Fe^{2+} and aqueous NaOH, and for the production of hexacyanoferrate(II) ion from $CN^-(aq)$ and aqueous Fe^{2+}.

45. Heating 2.0 mol of a coordination compound gives 1.0 mol NH_3, 2.0 mol H_2O, 1.0 mol HCl, and 1.0 mol $(NH_4)_3[Ir_2Cl_9]$. Write the formula of the original (six-coordinated) coordination compound and name it.

46. Match each compound in the group on the left with the compound on the right most likely to have the same electrical conductivity per mole in aqueous solution.
    (a) $[Fe(H_2O)_5Cl]SO_4$ HCN
    (b) $[Mn(H_2O)_6]Cl_3$ $Fe_2(SO_4)_3$
    (c) $[Zn(H_2O)_3(OH)]ClO_4$ NaCl
    (d) $[Fe(NH_3)_6]_2(SO_4)_3$ $ZnSO_4$
    (e) $[Cr(NH_3)_3Br_3]$ $LaCl_3$

47. Three different compounds are known with the same empirical formula $CrCl_3 \cdot 6H_2O$. When exposed to a dehydrating agent, compound 1 (which is dark green) loses two moles of water per mole of compound, compound 2 (light green) loses one mole of water, and compound 3 (violet) loses no water. What are the probable structures of these compounds? If an excess of silver nitrate solution is added to 100.0 g of each of these compounds, what mass of silver chloride precipitates in each case?

48. The octahedral structure is not the only possible six-coordinate structure. Other possibilities include a planar hexagonal structure and a triangular prism structure. In the latter, the ligands are arranged in two parallel triangles, one lying above the metal atom and other below the metal atom with its corners directly in line with the corners of the first triangle. Show that the existence of two and only two isomers of $[Co(NH_3)_4Cl_2]^+$ provides evidence against both of these possible structures.

49. Cobalt(II) forms more tetrahedral complexes than any other ion except zinc(II). Draw the structure(s) of the tetrahedral complex $[CoCl_2(en)]$. Could this complex exhibit geometric or optical isomerism? If one of the Cl^- ligands is replaced by Br^-, what kinds of isomerism, if any, are possible in the resulting compound?

50. Crystal-field theory treats the energy-level splittings induced in the five d-orbitals. The same procedure could be applied to p-orbitals. Predict the level splittings (if any) induced in the three p-orbitals by octahedral and square-planar crystal fields.

51. The complex ions $[Mn(CN)_6]^{5-}$, $[Mn(CN)_6]^{4-}$, and $[Mn(CN)_6]^{3-}$ have all been synthesized and all are low-spin octahedral complexes. For each complex, determine the oxidation number of Mn, the configuration of the d-electrons (how many t_{2g} and how many e_g), and how many unpaired electrons are present.

52. Is the coordination compound $[Co(NH_3)_6]Cl_2$ diamagnetic or paramagnetic?

*53. A coordination compound has the empirical formula $PtBr(en)(SCN)_2$ and is diamagnetic.
  (a) Examine the d-electron configurations on the metal atoms, and explain why the molecular formula $[Pt(en)_2(SCN)_2][PtBr_2(SCN)_2]$ is preferred for this substance.
  (b) Name this compound.

54. The coordination geometries of $[Mn(NCS)_4]^{2-}$ and $[Mn(NCS)_6]^{4-}$ are tetrahedral and octahedral, respectively. Explain why the two have the same room-temperature molar magnetic susceptibility.

55. The compound $Cs_2[CuF_6]$ is bright orange in color and paramagnetic. Determine the oxidation number of copper in this compound, the most likely geometry of the coordination around the copper, and the possible configurations of the d-electrons of the copper.

56. In the coordination compound $(NH_4)_2[Fe(H_2O)F_5]$, the Fe is octahedrally coordinated.
  (a) Based on the fact that F^- is a weak-field ligand, predict whether this compound is diamagnetic or paramagnetic. If it is paramagnetic, tell how many unpaired electrons it has.
  (b) By comparison to other complexes discussed in the chapter, predict the likely color of this compound.
  (c) Determine the d-electron configuration of the iron in this compound.
  (d) Name this compound.

57. Molecular nitrogen (N_2) can act as a ligand in certain coordination complexes. Predict the structure of $[V(N_2)_6]$, which is isolated by condensing V with N_2 at 25 K. Is this compound diamagnetic or paramagnetic? What is the formula of the carbonyl compound of vanadium that has the same number of electrons?

58. Discuss the role of transition-metal complexes in biology. Consider such aspects as their absorption of light, the existence of many different structures, and the possibility of multiple oxidation states.

CUMULATIVE PROBLEM

Platinum

The precious metal platinum was first used by South American Indians who found impure, native samples in the gold mines of what is now Ecuador and made small items of jewelry. Its high melting point (m.p.) of 1772°C makes platinum harder to work than gold (m.p. 1064°C) and silver (m.p. 962°C), but this same property, and its resistance to chemical attack, make platinum quite suitable as a material for high-temperature crucibles. Although platinum is a noble metal, it forms a wide variety of compounds in the +4 and +2 oxidation states, many of which are coordination complexes. Its coordinating abilities make it an important catalyst for organic and inorganic reactions.

Crystals of potassium hexachloroplatinate(IV) (K_2PtCl_6).

(a) The anticancer drug cisplatin, cis-[Pt(NH$_3$)$_2$Cl$_2$] (see Fig. 19–8b), can be prepared from K$_2$PtCl$_6$ via reduction with N$_2$H$_4$ (hydrazine), giving K$_2$PtCl$_4$, followed by replacement of two chloride-ion ligands with ammonia. Give systematic names to the three platinum complexes referred to in this statement.

(b) The coordination compound diamminetetracyanoplatinum(IV) has been prepared, but salts of the hexacyanoplatinate(IV) ion have not. Write the chemical formulas of these two species.

(c) In a certain compound, Pt(II) is coordinated to two chloride ions and to two molecules of ethylene (C$_2$H$_4$) to give an unstable yellow crystalline solid. Can this complex have more than one isomer? If so, describe the possible isomers.

(d) Platinum(IV) is readily complexed by ethylenediamine. Draw the structures of both enantiomers of the complex ion cis-[Pt(Cl)$_2$(en)$_2$]$^{2+}$. In this ion, the Cl$^-$ ligands are cis to one another.

(e) In platinum(IV) complexes, the octahedral crystal-field splitting Δ_o is relatively large. Is K$_2$PtCl$_6$ diamagnetic or paramagnetic? What is its d-electron configuration?

(f) Is cisplatin diamagnetic or paramagnetic?

(g) The salt K$_2$[PtCl$_4$] is red, but [Pt(NH$_3$)$_4$]Cl$_2 \cdot$H$_2$O is colorless. In what regions of the spectrum do the dominant electronic absorptions lie for these compounds?

(h) When the two salts from part (g) are dissolved in water and the solutions mixed, a green precipitate, called Magnus's green salt, forms. Propose a chemical formula for this salt, and assign the corresponding systematic name.

Structure and Bonding in Solids

Microcrystals of realgar, As_4S_4.

Bonding among atoms gives rise to the structures called molecules. Intermolecular forces, which are weaker than chemical bonds, then act among molecules to organize them into the larger-scale structures of the liquid and solid states that were described in general terms in Chapter 6. In this chapter, the solid state is considered more closely, with emphasis on the diffraction experiments that have revealed so much about its nature.

We subdivide solids into two kinds, depending on the range of microscopic order within them: **crystalline solids** possess long-range order in their structures, whereas **amorphous solids** do not. Although highly ordered, crystalline solids do not contain molecules (or atoms or ions) lined up in perfect infinite rows as if in a chemical cemetery. Defect and disorder play an important role in many technologically important crystalline materials. Other substances form "plastic crystals" and "liquid crystals" that are intermediate in their properties between the solid and liquid states.

20–1 PROBING THE STRUCTURE OF CONDENSED MATTER

In the condensed phases, the organization of the surroundings of the individual molecule, atom, or ion has enormous importance. How is such structure investigated?

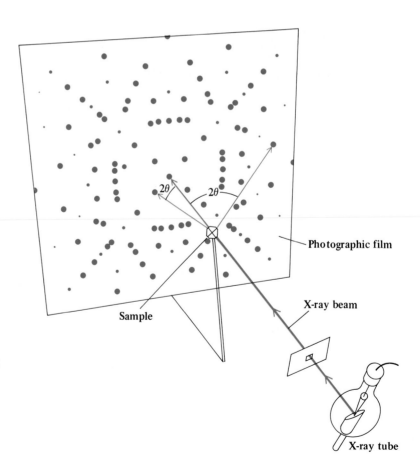

Figure 20–1 A well-defined beam of "white" x-rays (x-rays having a range of wavelengths) strikes a thin sample of a crystalline solid. Some of the x-rays are absorbed; some pass through unchanged; others are scattered at various angles 2θ to the straight-through beam.

Figure 20-2 Scattering of visible light causes the halos that appear around streetlamps in the fog. Light rays exit the lamp headed in all directions. In foggy conditions they strike water droplets suspended in the air and are scattered toward the viewer. These redirected rays appear as a diffuse glow surrounding the "direct beam" that the viewer usually sees.

X-Ray and Neutron Diffraction

The most useful probe for the structure of the condensed states of matter is a beam of x-radiation. In **x-ray diffraction** experiments, x-rays strike samples of solids or liquids (or even gases), interact, and exit. Different arrangements of atoms in the sample affect the x-rays differently. X-rays are electromagnetic waves, similar to visible light but with wavelengths three orders of magnitude shorter. X-rays are not visible, but they register on photographic film and are easily detected by other means as well.

• The x-ray portion of the electromagnetic spectrum includes wavelengths from approximately 0.1 to 100 Å (see Fig. 16–5).

The set-up for a typical x-ray experiment on a crystalline solid is diagrammed in Figure 20–1. Much of the radiation passes straight through a thin sample; however, some is diverted in differing amounts at various angles from the main beam (Fig. 20–2). The intensity of a beam of scattered x-radiation is easily measured by how much it darkens a piece of photographic film; a record of an x-ray diffraction experiment is the **diffraction pattern** of the material under investigation. A complete diffraction record consists of the intensities of all possible diffracted beams and their directions relative to the main beam. The **scattering angle, 2θ,** is defined as the angle between the direction of the diffracted beam and the direction of the main beam (see Fig. 20–1).

X-rays are easy to generate. Electrons are accelerated by high voltage and directed against a metallic target inside an evacuated tube. As they hit, they slow suddenly, emitting *white x-radiation* (x-rays with a range of wavelengths). If energetic enough, the impacts excite atoms in the target to higher energy levels. These atoms subsequently emit nearly *monochromatic* (single wavelength) x-rays as they return to their ground states. The wavelength of this *characteristic radiation,* which was first studied by Moseley (see Section 16–2), depends on the element used in the target. Tubes designed to produce characteristic radiation are used when monochromatic x-rays are required. X-ray diffraction patterns contain a maximum of structural information when the x-rays have wavelengths of about 0.5 to 3 Å because this is roughly the range of distances between neighboring atoms in liquids and solids. The size of the probe is well suited to the size of the objects being probed.

X-rays are not unique as a probe of the structure of condensed matter. **Neutron diffraction,** in which a beam of neutrons replaces the x-rays, is closely related. Like all particles, a neutron possesses wave properties, with an associated wavelength given by the de Broglie relation, $\lambda = h/p$ (see Section 16–4). Applying this equation shows that neutrons moving at a speed of 2.8×10^3 m s^{-1} (a typical speed for neutrons at room temperature) have a wavelength of 1.4 Å, well matched to the distances between atoms in solids and liquids. The nuclei in condensed matter scatter such neutrons easily, and a beam of neutrons makes an excellent structural probe. **Electron diffraction,** based on the same principle, is also possible and is used as a probe of structure.

How X-Rays Interact with Matter

• The instability of the nuclear atom in classical physics (see Section 16–2) arises because orbiting charged particles are oscillating (from one side of the atom to the other) and must emit radiation.

When an x-ray beam strikes any material target, it causes the electrons in the target to move back and forth rapidly—that is, to oscillate as their electric charges are first pushed and then pulled by the changing electric field of the x-radiation. An oscillating electron emits electromagnetic radiation. Thus, each electron in the path of an x-ray beam becomes a minuscule, secondary x-ray source of its own, emitting an expanding sphere of x-radiation. This is the phenomenon called **scattering.**

Some scattered x-rays make fresh encounters with new electrons and are rescattered. Far more important is the *interference* that the scattered waves experience among themselves. Interference occurs with all sorts of waves. It is apparent, for example, among the waves formed when a handful of pebbles is cast into a quiet pool of water. The expanding rings of ripples run into each other, interfering as they do so. The outcome of interference between two traveling waves moving in the same direction with the same wavelength depends on their relative *phase* (just as for standing waves, as discussed in Section 18–1). Interference is constructive if the two waves are exactly in phase, matching crest for crest and valley for valley as they go along; their amplitudes add to give a stronger resultant wave. Destructive interference occurs if two waves are exactly out of phase—that is, if crest matches trough and trough matches crest. Destructive interference leads to the cancellation of both waves. Interference between two waves that are neither exactly in nor exactly out of phase gives an intermediate result.

Suppose that x-radiation strikes two neighboring scattering centers. Each becomes a source of scattered radiation. The expanding spheres of scattered waves soon encounter each other and interfere. In some directions, the waves are in phase and reinforce each other (Fig. 20–3a); in other directions they are out of phase and cancel each other out (Fig. 20–3b). Reinforcement (constructive interference) oc-

Each pebble falling into a still pool starts a widening circle of waves. Where the circles overlap, the waves interfere.

Figure 20–3 A beam of x-rays (not shown) is striking two scattering centers, which are emitting scattered radiation. The difference in the lengths of the paths followed by the scattered waves determines whether they interfere constructively (a) or destructively (b). This path difference depends on the distance between the centers and also on the direction in which the scattered waves are moving.

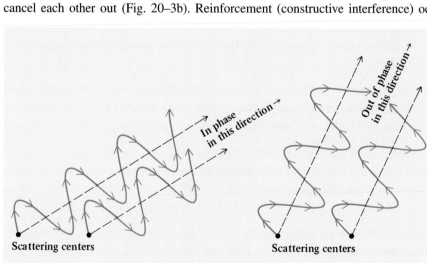

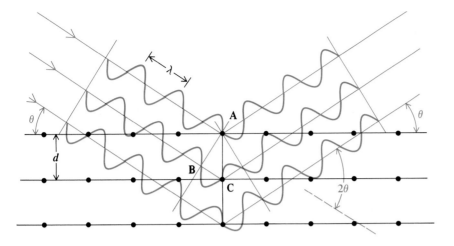

Figure 20–4 The x-rays coming in from the upper left are all in phase, have the same wavelength λ, and make an angle θ with a set of planes of atoms in a crystalline solid. The x-rays scattered by atoms in the top plane must be fully reinforced by those scattered from all the lower planes (only two of which are shown). If this is not the case, then no scattered beam is observed. X-rays scattered by atoms in lower planes travel farther. One half of the difference in path length of the beams scattered by the two topmost planes is equal to the side BC of the triangle ABC. The side BC has length $d \sin \theta$, where d is the distance between the planes. The total difference in path length is then $2d \sin \theta$. For full reinforcement, the difference between the path lengths of the two scattered beams must equal an integral number n of wavelengths. Hence $n\lambda = 2d \sin \theta$ is the condition for Bragg scattering of x-rays by a crystal.

curs when the paths traversed by two waves differ in length by a whole number of wavelengths. Thus, in Figure 20–3a, the scattered waves reinforce because one has traveled exactly one wavelength farther than the other. As Figure 20–3 shows, the difference in path length between the scattered waves depends on the distance between the centers and also on the direction considered. This is the fundamental physical basis of x-ray diffraction as a structural probe.

The Bragg Law

To illustrate the connection between x-ray diffraction patterns and the locations of atoms in a sample, let us consider the scattering by a large group of atoms that are arranged in a regular pattern of parallel planes. This is the way atoms are arranged in crystalline solids. The **Bragg law** sets an essential condition for the allowed angles of x-ray diffraction in such a case. This law was formulated by W. H. Bragg and W. L. Bragg (father and son) in 1914. The scattering of x-rays by crystals is now often called Bragg scattering.

Crystalline solids often possess well-developed planar faces. To understand the Bragg law, suppose first that *visible* light strikes such a smooth crystal face at an angle θ. The face acts as a mirror, reflecting the light. The reflected beam goes off in a different direction but makes the same angle θ to the plane of the face as the incoming beam. The angle between the direction of the incoming beam and that of the reflected beam is 2θ (see Fig. 20–1).

The geometrical relationship between the incoming and outgoing beams is unchanged if single-wavelength (monochromatic) x-rays replace the visible light. Now, however, a flat exterior face is no longer needed because the x-rays penetrate the crystal, and "reflection" (actually scattering followed by interference) comes from evenly spaced stacks of atoms in parallel planes within the crystal. It is clear that x-rays scattered by atoms in planes lower in the stack must travel longer paths than those scattered at the top. The traversal of different paths means different relative phase among "reflected" beams and gives rise to interference. Destructive interference among the beams limits the number of observable x-ray "reflections" from a given stack of planes to a select few values of θ. The Braggs used the analysis shown in Figure 20–4 to quantify this argument. Their conclusion was

$$n\lambda = 2d \sin \theta \qquad \text{where } n \text{ is a positive integer}$$

In this equation d stands for the perpendicular distance between the parallel planes in the stack, λ is the wavelength of the x-rays, θ is as just defined, and n is a whole number. The case $n = 1$ is called first-order Bragg diffraction (or the "first-order reflection"), $n = 2$ is second-order, and so forth.

EXAMPLE 20-1

A crystalline sample scatters a beam of x-rays of wavelength 0.7093 Å at an angle 2θ of 14.66°. If this is a second-order Bragg reflection ($n = 2$), compute the distance between the parallel planes of atoms from which the scattered beam appears to have been reflected.

Solution

The problem requires substituting in the Bragg law and solving for d, the interplanar spacing:

$$d = \frac{n\lambda}{2 \sin \theta} = \frac{2(0.7093 \text{ Å})}{2 \sin\left(\dfrac{14.66°}{2}\right)} = \boxed{5.559 \text{ Å}}$$

Exercise

A beam of x-rays of wavelength 1.936 Å is scattered at an angle 2θ of 88.30°. What is the order of this Bragg reflection if the distance between the parallel planes of atoms involved in the scattering is 5.559 Å?

Answer: Fourth-order.

EXAMPLE 20-2

In a crystal of silver, planes of silver atoms are separated by a distance of 1.446 Å. Compute the θ's for all possible Bragg reflections for these planes if the wavelength of the incident x-radiation is 0.7093 Å.

Solution

Write down the Bragg law:

$$n\lambda = 2d \sin \theta$$

The problem gives the values of λ and d. Rearranging the equation and inserting these values gives

$$\sin \theta = n\frac{\lambda}{2d} = n\frac{0.7093 \text{ Å}}{2(1.4446) \text{ Å}} = n \times 0.2455$$

If $n = 1$, then $\sin \theta$ is 0.2445 and θ is $\sin^{-1} (0.2445) = 14.21°$. Substituting $n = 2$ and $n = 3$ gives θ's of $29.41°$ and $47.43°$. Only one further Bragg reflection is possible: when $n = 4$, θ is $79.11°$. Larger values of n give values of $\sin \theta$ exceeding 1, which corresponds to no possible angle.

Exercise

In a crystal of potassium chloride, parallel planes of ions are separated by a distance of 3.146 Å. Compute the θ's for all possible Bragg reflections from these planes if the wavelength of the incident x-radiation is 1.5405 Å.

Answer: $\theta_{n=1} = 14.17°$; $\theta_{n=2} = 29.32°$; $\theta_{n=3} = 47.27°$; $\theta_{n=4} = 78.33°$.

Experimental Results from X-Ray Diffraction

When a crystalline solid is put in an x-ray beam and rotated, one of two types of diffraction pattern is observed: (1) many sharply defined *spots* appear across the pattern at different angles from the primary beam (Fig. 20–5a), and (2) many sharply defined *rings* encircle the primary beam (Fig. 20–5b).

A pattern such as that shown in Figure 20–5a arises when the specimen is a **single crystal,** so that every set of atomic planes runs straight through the sample from one end to the other. Each spot in the diffraction pattern registers a different diffracted beam of x-rays. The angles that these diffracted beams make with the primary beam are the allowed angles (2θ's) for Bragg reflections from planes of atoms

(a)   (b)

Figure 20–5   Order in x-ray diffraction patterns derives from order in the structure of the material that is doing the scattering. (a) This pattern was obtained by irradiating a crystal of the mineral almandite ($Fe_3Al_2(SiO_4)_3$) with x-rays of a narrow range of wavelengths in a specialized camera. Note the symmetry of the pattern. If the x-radiation had one wavelength only, the streaks would disappear. (b) This pattern is observed by irradiating a microcrystalline powder of albite, $NaAlSi_3O_8$, with x-rays and rotating the sample.

Figure 20-6 A close-up of poly-crystalline sphalerite (ZnS).

in the crystal. Rotation of the sample successively lines up different atomic planes so that the Bragg equation is satisfied, and different diffracted beams shoot out to create the diffraction pattern.

If the x-ray diffraction pattern consists of many sharply defined rings (see Fig. 20–5b), the solid is **polycrystalline.** It consists of randomly oriented, tiny crystalline grains either cemented together as a conglomerate or present as a loose powder (Fig. 20–6). Typical microcrystals have radii as small as 10^{-3} mm. Rings arise because the microcrystals are randomly oriented in the powder or conglomerate. Each tiny crystal gives its own set of sharply defined diffracted beams. Each ring in the diffraction pattern is caused by diffracted beams with the same 2θ coming from many different crystals that are lying tumbled in every possible orientation relative to the beam. Such x-ray diffraction patterns are called "powder patterns."

As the microcrystallines in a polycrystalline solid become smaller in size, the rings in the diffraction pattern become more diffuse, and those at higher angles lose intensity. Extrapolation of this trend gives the diffraction pattern of an amorphous solid, which shows several diffuse, blurred rings encircling the primary x-ray beam. The rings mean that there is some structure, but their diffuseness means the structure is short-range only. The observation of *rings* instead of individual diffracted spots again arises from the random orientation of the individual scattering units. A liquid resembles an amorphous solid in its diffraction pattern and in its microscopic structure. In other ways (such as in the rates of diffusion of atoms), liquids differ from amorphous solids.

20-2 SYMMETRY AND STRUCTURE

Symmetry and order are closely allied. In art and design, symmetry pleases the eye because it satisfies a human desire to detect or impose order (Fig. 20–7).

Symmetry in Crystalline Solids

As suggested by Réné Just Haüy in 1784, the exterior regularities of crystals (Fig. 20–8) are a consequence of interior, microscopic symmetry. This microscopic sym-

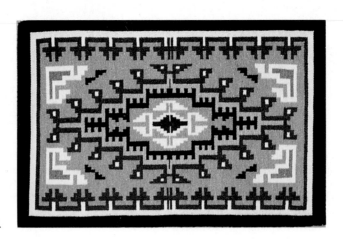

Figure 20-7 The symmetry of the pattern on this Navajo rug is an important aspect of its aesthetic appeal.

Figure 20–8 Crystals often, but not always, have well-developed plane faces. Some of these crystals of sodium chloride have well-developed faces; others have ill-formed exteriors but are still just as crystalline in their interior structure.

metry is at the level of the surroundings of individual atoms and groups of atoms. It is an unalterable part of the crystalline structure. Knowing the microscopic symmetry of a crystal is essential to understanding its structure and its chemical and physical properties. Some crystals have beautifully developed natural faces; others, although they may have just as much interior symmetry, are formless lumps in external appearance. A rough diamond is an excellent example of the latter (Fig. 20–9). Sometimes, the history of a specimen (conditions of synthesis, heat treatment, breakage, erosion, and so forth) robs it of faces. Also, naturally occurring crystals can be deliberately reworked to almost any shape. Such actions do not affect the microscopic interior symmetry of the crystal.

X-rays penetrate crystals and are diffracted by them. The microscopic symmetry inside the crystal leads to observable symmetry in the x-ray diffraction pattern. To learn about the symmetry of a single crystal, even one that has no well-developed faces, we search for symmetry in the diffraction pattern that it produces

Figure 20–9 The facets that give gem-quality diamonds (left) their brilliance and appeal come from the carefully planned cutting and shaping of rough diamonds (above).

when bathed in a beam of x-rays. Figure 20–10 shows views of typical diffraction patterns, one from a crystal of the mineral hydroxyapatite ($Ca_5(PO_4)_3OH$) and the other from a crystal of an organic compound called fluorenone ($C_{13}H_8O$). Crystals give different *projections* of their total diffraction patterns, which are three-dimensional objects, when they are irradiated with x-rays from different angles. Hence, these striking patterns are only parts of the total diffraction patterns of these crystals. The diffraction pattern from hydroxyapatite (Fig. 20–10a) has 6-fold **rotational symmetry.** This means that rotating the pattern by an angle of 60° (one sixth of a full circle) around its center gives back the original pattern. Similarly, a diffraction pattern has 2-fold rotational symmetry if rotation by $360/2 = 180°$ around some axis takes it into itself. It has 3-fold rotational symmetry or 4-fold symmetry if a $360°/3 = 120°$ or a $360°/4 = 90°$ rotation achieves this self-coincidence. The pattern from fluorenone (Fig. 20–10b) has rotational symmetry (a 2-fold axis) and has a second kind of symmetry as well—**mirror symmetry.** Every spot on the left side of a vertical central line has a matching spot on the right side; every spot in the upper half of the pattern has a corresponding spot in the lower half: the fluorenone pattern has two lines of mirror symmetry. Figure 20–11 diagrams how rotational and mirror symmetry work.

• Symmetry disclosed in a diffraction pattern is a property of the crystal as a whole, *not* necessarily of the molecules that compose it.

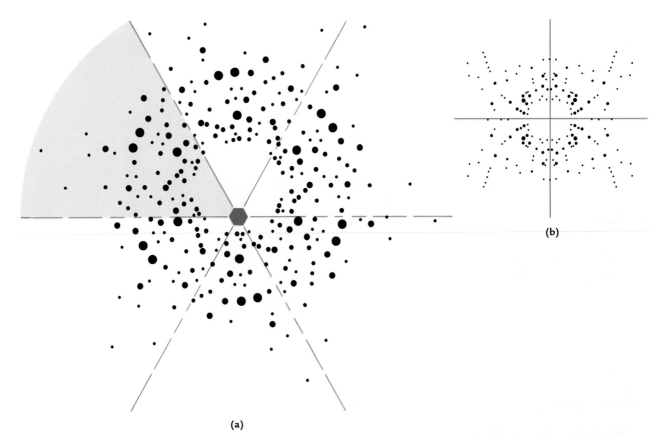

(a)

(b)

Figure 20–10 Views of the diffraction patterns of crystals of (a) hydroxyapatite ($Ca_5(PO_4)_3(OH)$) and (b) fluorenone ($C_{13}H_8O$). The symmetry in these patterns implies the presence of symmetry in the crystals that cast them.

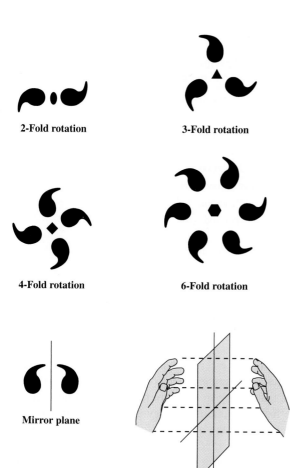

Figure 20-11 Rotational symmetry in an object means that it "goes into itself" upon rotation about an axis by a certain angle. In 2-fold rotation, the angle is 180°. In 3-fold, 4-fold, and 6-fold rotational symmetry, the angles are 120°, 90°, and 60°, respectively. Objects related by mirror symmetry have the same relationship as a left hand and a right hand.

2-Fold rotation

3-Fold rotation

4-Fold rotation

6-Fold rotation

Mirror plane

Mirror plane

EXAMPLE 20-3

List the capital letters of the Roman alphabet that have a vertical mirror line of symmetry.

Solution

A vertical mirror line is present if the right side of the object "goes into" the left side when taken across the line. We carefully print each capital letter and examine it for symmetry. This reveals a vertical mirror line in the letters A, H, I, M, O, T, U, V, W, X, and Y.

Exercise

List the capital letters of the Roman alphabet that have a horizontal mirror line of symmetry.

Answer: B, C, D, E, H, I, K, O, X.

Only a few different kinds of symmetry have been found in the hundreds of thousands of diffraction patterns that have been recorded and studied. These consist of 2-fold, 3-fold, 4-fold, and 6-fold rotational symmetry in various combinations

Table 20–1
The Seven Crystal Systems and the Minimum Essential Symmetry of Their Diffraction Patterns

Crystal System	Minimum Essential Symmetry
Hexagonal	One 6-fold rotation
Cubic	Four independent 3-fold rotations[a]
Tetragonal	One 4-fold rotation
Trigonal	A single 3-fold rotation
Orthorhombic	Three mutually perpendicular 2-fold rotations
Monoclinic	A single 2-fold rotation
Triclinic	No symmetry required

[a]Each of these axes makes a 70.53° angle to the other three.

with each other and with mirror symmetry. Crystals are classified into one of seven **crystal systems** on the basis of the symmetry that is found in their diffraction patterns. The seven crystal systems and their minimum essential symmetry in diffraction are listed in Table 20–1. "Minimum essential symmetry" means the symmetry that *must* be present in the diffraction pattern in order to qualify the crystal for membership in the system. Crystals that belong to a system may and often do have additional symmetry beyond the listed minimum, but they cannot move up the table to join a higher system unless they have the minimum for that system. Also, symmetry is never ignored; crystals are classified as far toward the top of the list in Table 20–1 as possible. Several of the names of the crystal systems describe their minimum essential symmetry. Thus, hexagonal crystals (*hexa* means "six"), tetragonal crystals (*tetra* means "four"), and trigonal crystals (*tri* means "three") have at least 6-fold, 4-fold, and 3-fold interior symmetry, respectively. About 50% of all crystals studied so far have been monoclinic, 25% orthorhombic, and 15% triclinic. The prevalence of the remaining, more symmetric, systems dwindles in the following order: cubic, tetragonal, trigonal, and hexagonal.

The dependence of the pattern of scattering by a crystalline solid on the direction of the x-ray beam is an example of *anisotropy*, the dependence of a property on the direction of its measurement. Many other properties of crystals are directional. If the value of a property changes with direction of measurement, then the crystal is anisotropic with respect to that property. A better-known anisotropy is the tendency of many crystals to cleave preferentially because they have different strengths in different directions. Crystals of mica resemble stacks of sheets of paper. It is easier to peel the sheets apart than to rip across them (Fig. 20–12). All crystalline solids are anisotropic with respect to the property of diffraction. Unlike crystals, amorphous solids are *isotropic* with respect to this and all other properties, as are liquids and gases.

The Crystal Lattice

The numerous sharply defined spots (for single crystals) or sharply defined rings (for polycrystalline solids) in the diffraction patterns of crystals are consequences of the long-range order in their organization at the atomic level. Identical sites within crystals recur regularly, at distances that are on the order of 10^{-10} m. The three-dimensional array made up of all the points within a crystal that have the same envi-

Figure 20–12 The mechanical properties of crystals of mica are quite anisotropic. Thin sheets can be peeled off a crystal of mica easily with a razor blade, but the sheets resist stresses in other directions well.

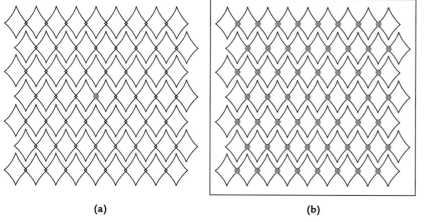

(a)

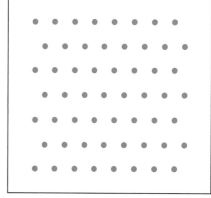

(b)   (c)

Figure 20–13 Generating a lattice from a repeating pattern. (a) Select any point in the pattern and mark it. (b) Move out through the pattern, marking additional points that have the same surroundings in the same orientation. The targeted point recurs regularly, and the collection of marked points quickly grows to resemble a gridwork or lattice. (c) Discard the original pattern and keep the gridwork of points.

ronment in the same orientation is a **crystal lattice.** A crystal lattice is an abstraction lifted away from a real crystal, embodying the scheme of repetition that is at work in that crystal. Figure 20–13 shows (in a two-dimensional example) how a lattice is separated from a repeating pattern. The lattice is imagined to extend infinitely in all directions.

Fundamental differences among lattices are found in their differing symmetries, not their sizes. If we construct a large number of different lattices, we find that the possible symmetries of the surroundings of each lattice point are severely restricted by the requirement that the surroundings of each and every lattice point be identical. It turns out that exactly seven three-dimensional lattices are distinguishable based on the symmetry at their lattice points: the hexagonal, cubic, tetragonal, trigonal, orthorhombic, monoclinic, and triclinic lattices. It is not an accident that the names of these categories of lattices are the same as the names of the seven crystal systems. The correspondence follows from the fact that the symmetry of a particular crystal is always present in its lattice. Thus, a crystal in the tetragonal system always has a lattice of at least tetragonal symmetry.[1]

Unit Cells

Eight points of a crystal lattice define a **unit cell.** The eight points are the corners of the cell and must be chosen so that they define a box having three pairs of parallel faces. Solid figures of this type are **parallelepipeds** (Fig. 20–14). Every unit

[1]The phrase "at least" appears because a lattice may have additional symmetry that the crystal does not have. A lattice contains only points, which are spherically symmetrical, but actual crystals contain atoms in groupings that may have shapes of low symmetry.

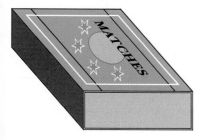 

Figure 20–14 The sliding outer cover of a match box is a parallelepiped and remains one even if it is squashed.

Figure 20–15 Unit cells always have three pairs of mutually parallel faces. Only six pieces of information are required to construct a scale model of a unit cell—the three cell edges (*a*, *b*, and *c*) and the three angles between the edges (α, β, and γ). By convention, the angle between the edges *a* and *b* is labelled "γ," the angle between *b* and *c* is labelled "α," and the angle between *a* and *c* is labelled "β."

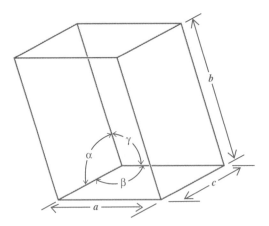

cell in a crystal is packed with atoms of exactly the same kinds, in the same numbers, and in the same relative positions. These contents of the unit cell are the repeating *motif* from which the crystal is constructed. Atoms need not occupy the corners of a unit cell, which are lattice points, although there is nothing to prevent this. The sides and edges of unit cells may cut right through atoms. Only the portions of atoms actually within the walls of a unit cell count as belonging to that cell. Atoms on boundaries count as shared. Unit cells and crystal lattices, which are both products of the human imagination, allow a crystal's structure to be visualized as the result of a tidy stacking up (side by side, front to back, and bottom to top) of little building bricks. Unit cells are the microscopic building bricks in this picture. A single unit cell contains all structural information about its crystal because the crystal could in principle be constructed by making a great many copies of a single original unit cell and stacking them up.

As Figure 20–15 shows, a unit cell has 12 edges in addition to its 8 corners and 6 faces. The size and shape of a unit cell are fully described by three edge lengths (*a*, *b*, and *c*) and by the three angles between these edges (α, β, and γ). These are the *cell parameters* or *cell constants*.

The symmetry that defines the seven different crystal systems imposes conditions on the shape of the unit cell.

Table 20–2
The Conditions Imposed by Symmetry on the Shapes of Unit Cells

Hexagonal	$a = b$; $\alpha = \beta = 90°$, $\gamma = 120°$
Cubic	$a = b = c$; $\alpha = \beta = \gamma = 90°$
Tetragonal	$a = b$; $\alpha = \beta = \gamma = 90°$
Trigonal	$a = b = c$; $\alpha = \beta = \gamma \neq 90° < 120°$
Orthorhombic	$\alpha = \beta = \gamma = 90°$
Monoclinic	$\alpha = \gamma = 90°$
Triclinic	None

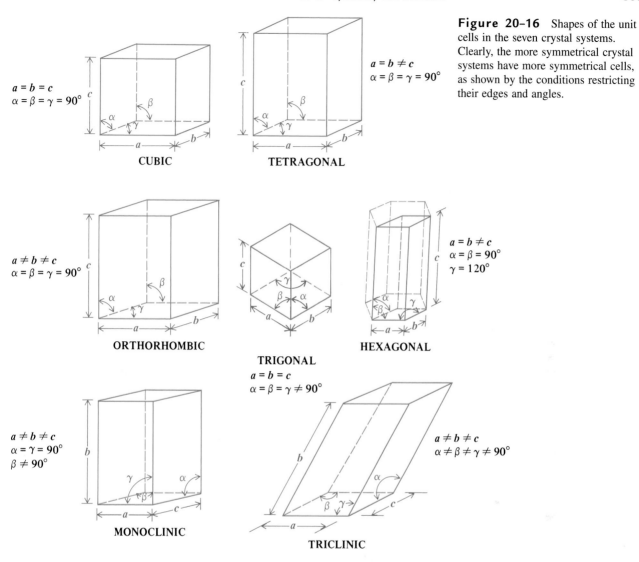

Figure 20-16 Shapes of the unit cells in the seven crystal systems. Clearly, the more symmetrical crystal systems have more symmetrical cells, as shown by the conditions restricting their edges and angles.

These conditions are given in Table 20–2, and the unit cells are shown in Figure 20–16.

There are many possible choices of unit cell in a given crystal lattice. The first rule in selecting a cell is to respect the symmetry of the crystal lattice. Table 20–2 shows how to do this. For example, an acceptable unit cell for a monoclinic crystal *always* has 90° angles between one edge and each of the other two (these are the angles α and γ in Figure 20–16). Other conceivable cells are not used because they ignore the presence of symmetry. The second rule in picking a unit cell is to find the *smallest* cell that still has the symmetry of the crystal lattice; there is no profit in using large cells when small ones will do. A unit cell that contains exactly one lattice point is called a **primitive** unit cell. Such a cell is the smallest possible. A primitive unit cell might seem to contain *eight* lattice points, one at each of its corners, but any corner lattice point is actually shared with seven neighboring cells that just meet at that corner. Only one eighth of each of the eight corner lattice points actually belongs to any unit cell. Because $\frac{1}{8} \times 8 = 1$, the eight shared lattice points at the corners always contribute exactly one lattice point to a unit cell.

• Think of the armrests in a row of theater seats. Each theatergoer has armrests on both sides but must share them. Every person (excluding those at the ends of rows) owns only $2 \times \frac{1}{2} = 1$ armrest.

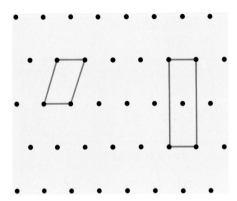

Figure 20–17 In this lattice, every lattice point is at the intersection of a horizontal and vertical mirror line. It is possible to draw a primitive unit cell (red), but the larger centered unit cell (blue) is preferred because as a shape it also has two mirror lines—the full symmetry of the lattice.

Often, a primitive unit cell does not have the full symmetry of the crystal lattice. If so, a larger unit cell that does have the characteristic symmetry is deliberately chosen. Desirable larger cells usually have some combination of 90° or 120° angles between their edges because these angles are built into important symmetry operations (Fig. 20–17). The larger cells are *nonprimitive* and are drawn on a given lattice by selecting unit cell corners in such a way that other lattice points are included either at the center of the cell or at the centers of some or all of its faces. Only three types of nonprimitive cells are commonly used in the description of crystals. If an extra lattice point is at the center of the unit cell, then it is **body centered.** If extra lattice points are at the centers of all six faces of the unit cell (but not at the body center), then that cell is **face centered.** If extra lattice points are at the centers of a single pair of parallel faces of the unit cell (and nowhere else), then that cell is **side centered.** These nonprimitive cells are shown in Figure 20–18.

No matter how a unit cell is selected, it contains a whole number of lattice points. A unit cell with two lattice points automatically has twice the volume of a primitive unit cell drawn on the same lattice. Similarly, if a unit cell contains four lattice points, then it has four times the volume of a primitive unit cell of the same lattice.

• Unit cells on the very boundaries of crystals obviously are exceptions, but they are rare exceptions.

EXAMPLE 20–4

Determine how many lattice points belong to each unit cell in (a) a body-centered lattice and (b) a face-centered lattice.

Solution

(a) In a body-centered lattice, one lattice point lies entirely within the body of each cell and belongs entirely to it. The eight lattice points at the corners are each shared with seven neighboring cells. Hence, each cell owns $1 + (8 \times \frac{1}{8}) = $ 2 lattice points.
(b) In a face-centered lattice, there are lattice points at the centers of the six faces of each cell, as well as its eight corners. The points in the faces are each shared with a single neighboring cell. Therefore, there are $(6 \times \frac{1}{2}) + (8 \times \frac{1}{8}) = $ 4 lattice points per cell.

Exercise
Determine how many lattice points belong to each unit cell in a side-centered lattice.

Answer: 2.

Unit-Cell Volume and Density

The volume of its unit cell tells a great deal about a crystal because it is the size of the repeating motif in a crystal. If the three angles of a unit cell are all 90°, the volume of the unit cell is

$$V = abc$$

If one or more angles is other than 90°, then a formula that takes account of the slanting sides of the nonrectangular box is required:

$$V = \sqrt{abc(1 - \cos^2\alpha - \cos^2\beta - \cos^2\gamma + 2\cos\alpha\cos\beta\cos\gamma)}$$

This formula is easy to use, thanks to the availability of calculators. The quantity under the radical is always less than or equal to 1 but greater than 0.

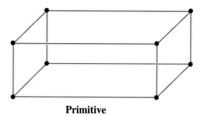

Primitive

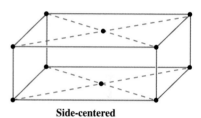

Body-centered

EXAMPLE 20–5

The triclinic unit cell of crystalline malonic acid ($C_3H_4O_4$) has cell edges $a = 5.33$ Å, $b = 8.36$ Å, and $c = 5.14$ Å. The angles are $\alpha = 104°$, $\beta = 94.8°$, and $\gamma = 71.5°$. Compute the volume of this unit cell.

Solution

Use the COS key on a calculator to obtain the cosines of the three angles. They are -0.241922, -0.083678, and $+0.317305$, respectively. Substitute the three cosines into the expression in parentheses under the radical sign in the preceding formula. The result equals 0.846636. The square root of this is 0.920128. The product of the three lengths is 229.032 Å^3, so the actual volume of the unit cell is 0.920128×229.032 Å^3 = 211 Å^3, expressed to three significant figures.

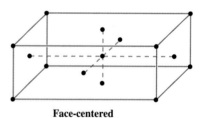

Side-centered

Exercise

One form of the protein myoglobin, extracted from the sperm whale, crystallizes in the monoclinic crystal system. The unit cell has the dimensions $a = 61.6$ Å, $b = 26.9$ Å, and $c = 33.9$ Å. The angles α and γ exactly equal 90° and angle β is 105.5°. Compute the volume of this unit cell.

Answer: 5.41×10^4 Å^3.

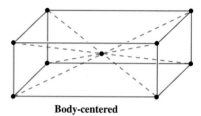

Face-centered

Figure 20–18 Centered unit cells have lattice points in addition to the lattice points at the eight corners. A body-centered lattice has an additional lattice point at the center of the cell; a face-centered lattice has additional lattice points at the centers of the six faces; a side-centered lattice has additional lattice points at the centers of two parallel sides of the unit cell. These lattices in this figure are orthorhombic, but centering can occur in other crystal systems as well. (*Note:* The dots in these diagrams are not atoms, but lattice points.)

The six cell constants—a, b, c, α, β, and γ—of the unit cell are determined from a study of the crystal's x-ray diffraction pattern. If the mass of the unit cell contents is known, the theoretical cell density can be computed. This density in principle equals the measured density of the crystal, a quantity that can be determined by completely independent experiments. Comparing the experimental density of a crystal with the density calculated from x-ray diffraction provides a valuable check on both methods. The following example illustrates this type of calculation.

EXAMPLE 20–6

Crystalline sodium chloride has a cubic unit cell with the cell edge $a = 5.6402$ Å. Each unit cell contains four Na$^+$ and four Cl$^-$ ions. Compute the density of sodium chloride in g cm^{-3}.

Solution

The volume of each unit cell is $(5.6402$ Å$)^3 = 179.43$ Å^3. The molar mass of NaCl is 58.443 g mol^{-1}. A mole of unit cells holds four of these formula units for a total mass of 233.77 g mol^{-1}. A mole of unit cells has a volume of

$$6.0221 \times 10^{23} \, \frac{\text{unit cells}}{\text{mol}} \times \left(\frac{179.43 \, \text{Å}^3}{\text{unit cell}} \right) = 1.0805 \times 10^{26} \, \frac{\text{Å}^3}{\text{mol}}$$

The density of any quantity of matter is its mass divided by its volume or, in this case,

$$d = \frac{m}{V} = \frac{\dfrac{233.77 \text{ g}}{\text{mol}}}{1.0805 \times 10^{26} \, \dfrac{\text{Å}^3}{\text{mol}}} = 2.1635 \times 10^{-24} \, \frac{\text{g}}{\text{Å}^3}$$

Expressing this density in g cm^{-3} requires converting cubic ångströms to cubic centimeters:

$$2.1635 \times 10^{-24} \, \frac{\text{g}}{\text{Å}^3} \times \left(\frac{(10^8)^3 \, \text{Å}^3}{1 \text{ cm}^3} \right) = 2.1635 \, \frac{\text{g}}{\text{cm}^3}$$

• The observed densities of crystals often fall slightly below the theoretical densities due to defects in the crystals (see Section 20-4).

The experimental density of NaCl from conventional measurements is 2.16 g cm^{-3}.

Exercise

Crystals of the compound NiAs$_2$ have an orthorhombic unit cell with edges $a = 5.75$ Å, $b = 5.82$ Å, and $c = 11.43$ Å. Each unit cell contains 8 nickel atoms and 16 arsenic atoms. Compute the density of this compound in g cm^{-3}.

Answer: 7.24 g cm^{-3}.

The same ideas can be applied to determine the molar mass of a compound if it is not known by other means, as shown in the following example.

EXAMPLE 20-7

A hydrate of calcium chloride crystallizes in the hexagonal system and contains one formula unit per unit cell. The cell constants are $a = b = 7.8759$ Å and $c = 3.9545$ Å, and the observed density of the crystal is 1.71 g cm^{-3}.
(a) Compute the molar mass of the hydrate.
(b) Determine the formula of the hydrate.

Solution

(a) First compute the volume of the unit cell from the cell constants according to the general equation for cell volume. A hexagonal unit cell automatically has the angles $\alpha = 90°$, $\beta = 90°$, and $\gamma = 120°$ (see Table 20-2).

$$V = (7.8759 \text{ Å})^2 (3.9545 \text{ Å}) \sqrt{1 - \cos^2 90 - \cos^2 90 - \cos^2 120 + 2 \cos 90 \cos 90 \cos 120}$$
$$= 245.297 \, \text{Å}^3 (0.866025) = 212.43 \, \text{Å}^3$$

If this is the volume occupied per formula unit, then the volume occupied by one mole (the molar volume) must be

$$V_m = \frac{212.43 \, \text{Å}^3}{\text{formula unit}} \times \left(\frac{1 \text{ cm}^3}{1 \times 10^{24} \, \text{Å}^3} \right) \times \left(\frac{6.0221 \times 10^{23} \text{ formula units}}{1 \text{ mol}} \right)$$
$$= 127.93 \, \frac{\text{cm}^3}{\text{mol}}$$

The molar mass of the hydrate equals the density multiplied by this molar volume:

$$\text{molar mass} = 127.93 \; \frac{cm^3}{mol} \times 1.71 \; \frac{g}{cm^3} = 219 \; \frac{g}{mol}$$

(b) The $CaCl_2$ part of the formula contributes $(2 \times 35.45) + 40.08 = 110.98$ g mol^{-1}. The difference of 108 g mol^{-1} must come from the water of hydration. But 108 g mol^{-1} is almost exactly six times the molar mass of H_2O. The formula of the hydrate is therefore $CaCl_2 \cdot 6H_2O$.

Exercise

Corundum is a crystalline material classified in the hexagonal crystal system. It contains six formula units per unit cell. The dimensions of the unit cell are $a = b = 4.7591$ Å and $c = 12.9894$ Å. The density of the crystal is 3.987 g cm^{-3}. Compute the molar mass of the compound that makes up corundum.

Answer: 102.0 g mol^{-1} (the compound is Al_2O_3).

The Intensities of the Diffracted Beams

An x-ray diffraction pattern reveals the interior symmetry of a crystal and allows the unit cell parameters to be determined—that is, the size and shape of the unit cell. This information comes from the relative positions of the diffracted beams in the diffraction pattern. A record of a crystal's diffraction contains more information than this, however. Each diffracted beam has a characteristic *intensity*; some are quite bright, and others are dim. Variations in intensity are visible in the experimental diffraction patterns in Figure 20–5. The intensities of the diffracted beams depend on the distribution throughout the unit cell of the scattering power—that is, of the atoms with their electrons. These intensities are determined by the identities and locations of the atoms in the unit cell.

From an accurate record of the intensities and directions of the many x-ray beams diffracted by a crystal, it is nearly always possible to compute back to the locations and identities of the atoms in the unit cell. This is called "solving the crystal structure." Knowing exact atomic locations is the same as knowing the detailed molecular structure of the substance that makes up a given crystal. This is obviously of tremendous value to chemists, biologists, physicists, and engineers. Solving a crystal structure requires highly accurate intensity data and much computation. The process is now automated, however, and uses specially designed instruments and very efficient computer programs. The determination of molecular structures by analysis of the x-ray diffraction patterns of crystals has become quite routine. High-quality results identify the atoms and give the distances between them to the nearest 0.005 or 0.010 Å.

• Frequently, the most difficult part is growing good crystals of the substance of interest.

20-3 BONDING IN CRYSTALS

The nature of the cohesive forces acting among the particles of substances in condensed phases provides a good basis for classifying crystalline solids (as well as amorphous solids and liquids) and for understanding their properties. Four "bond types" of crystals are usually distinguished. In these types, the principal forces that maintain the structure are van der Waals forces, Coulomb (ionic) forces, metallic bonds, and covalent bonds. We consider each of these types in turn.

Table 20–3
Van der Waals Radii of Several Elements

Element	Radius (Å)
Fluorine	1.47
Chlorine	1.75
Carbon	1.70
Hydrogen	1.20
Nitrogen	1.55
Oxygen	1.52

Molecular Crystals

This class of crystals includes the frozen noble gases, oxygen, nitrogen, the halogens, many covalent and partially covalent compounds such as CS_2, NO_2, Al_2Cl_6, $FeCl_3$, and $BiCl_3$, and the vast majority of organic compounds. In molecular crystals, it is always possible to single out groups of atoms in which the distance from each atom to at least one atom *within* the group is significantly smaller than the distance to any atom that is not within the group. The groups of atoms are, of course, the molecules, within which bonding occurs by means of strong, short covalent bonds (the sharing of electrons). In molecular crystals, the molecules are maintained in their positions in a lattice by van der Waals forces. This is a blanket term introduced in Section 6–1 for all attractive forces that involve complete molecules and neutral atoms. Van der Waals forces include dipole–dipole, ion–dipole, induced dipole, and dispersion forces but do not include attractions between oppositely charged ions or covalent bonds. Van der Waals forces pull neighboring molecules (or atoms) toward each other. Eventually, their effects are balanced by short-range repulsive forces as the electron clouds of closely neighboring particles start to interfere. These repulsions rise so rapidly as distance decreases that a reasonably well-defined radius (called the **van der Waals radius**) can be assigned to every element (see Table 20–3). The van der Waals radius of an atom is effectively the radius it denies to other atoms to which it is not bonded; van der Waals attractions are not strong enough to pull nonbonded atoms closer together than the sum of their van der Waals radii.

The trade-off between attractive and repulsive forces among even small molecules in a molecular crystal is quite complex because so many atoms are involved. A useful simplifying approach is to picture a molecule as a superposition of a set of negatively charged spheres, one centered at each nucleus. The radius of each sphere is the van der Waals radius of the element involved. Figure 20–19 shows such a

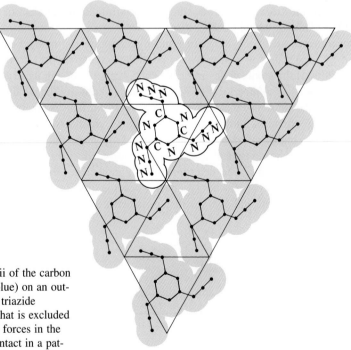

Figure 20–19 The van der Waals radii of the carbon and nitrogen atoms are superimposed (in blue) on an outline of the molecular structure of cyanuric triazide (C_3N_{12}). This shows the volume of space that is excluded by each molecule to others. Van der Waals forces in the molecular crystal hold the molecules in contact in a pattern that minimizes empty space. Note the 3-fold symmetry of the pattern.

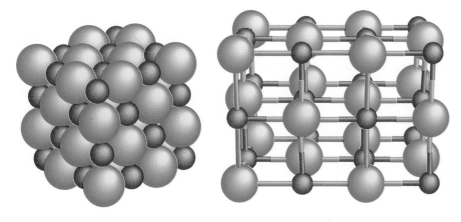

Figure 20–20 The structure of sodium chloride. On the left, the sizes of the Na^+ (gray) and Cl^- (green) ions are shown to scale. On the right, the ions are reduced in size to allow a unit cell (shown by red lines) to be clearly outlined.

"space-filling model" of cyanuric triazide (C_3N_{12}). In molecular crystals, such shapes pack together in such a way that no molecules are suspended in empty space and none overlap. The figure shows the way that nature solves the problem of efficiently packing many copies of the rather complicated molecular shape of C_3N_{12} in a single layer. In the three-dimensional molecular crystal of C_3N_{12}, many such layers stack up with a slight offset that minimizes unfilled space between layers.

Van der Waals forces are much weaker than the cohesive forces in ionic, metallic, and covalent crystals. As a consequence, molecular crystals typically have low melting points and are soft and easily deformed. Although the noble-gas elements crystallize in the highly symmetric face-centered cubic lattice at atmospheric pressure, molecules (especially those with complex geometries) more often form crystals of low symmetry, in the monoclinic or triclinic systems. They do not usually display the striking physical properties of electrical conductivity and magnetism that are characteristic of metals. Their interesting properties are primarily those of their molecular units.

Molecular crystals are of great scientific value. Obtaining proteins and other macromolecules in the crystalline state allows their structures to be determined by x-ray diffraction. Knowing the three-dimensional structures of biological molecules is the starting point for understanding how they work.

• The 1988 Nobel Prize in Chemistry was awarded for the isolation of molecular crystals of a key protein at the photosynthetic reaction center of *Rhodopseudomonas viridis*, a bacterium (see Problem 20–28). X-ray diffraction studies of the crystal structure of this substance, which contains about 630,000 atoms per molecule, have provided important insights into how photosynthesis works.

Ionic Crystals

Compounds formed by elements with significantly different electronegativities are largely ionic, and to a first approximation the ions may be treated as hard, charged spheres that occupy positions on the crystal lattice and interact electrostatically according to Coulomb's law (see Section 17–5). The elements of Groups I and II of the periodic table react with Group VI and VII elements to form ionic compounds, the great majority of which crystallize in the cubic system. The high symmetry of these crystalline solids reflects their structural simplicity. Specifically, the alkali-metal halides (except for the cesium halides), the ammonium halides, and the oxides and sulfides of the alkaline-earth metals all crystallize in the simple and beautiful **rock salt,** or **sodium chloride,** structure, shown in Figure 20–20. It may be viewed as a face-centered cubic lattice with anions at the lattice points, and cations occupying positions exactly between pairs of anions, or, equivalently, as the same lattice with cations at the lattice points and anions occupying positions exactly between. Either way, each ion is surrounded by six equidistant ions of the opposite charge.

Figure 20-21 The structure of cesium chloride. On the left, the sizes of the Cs$^+$ (purple-pink) and Cl$^-$ (green) ions are shown to scale. On the right they are reduced to allow a unit cell (shown by red lines) to be clearly outlined. Note that crystal lattice in this structure is simple cubic—not body-centered cubic, as is sometimes stated.

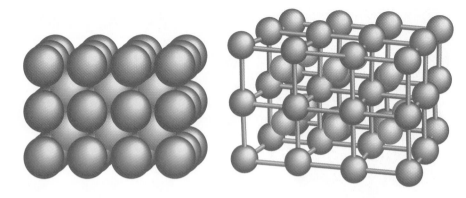

The favored crystal structure for an ionic substance is the one that brings each cation as close as possible to the maximum number of anions (because positive charge attracts negative), keeps it as far away as possible from other cations (because positive repels positive), and simultaneously keeps each anion as far away as possible from other anions (because negative repels negative). Different types of ionic structures emerge as different energetic compromises in the construction of the crystal. The cesium ion, for example, is large enough for eight chloride ions to come up around it without contacting each other. Consequently, in cesium chloride the chloride ions occupy the lattice points of a simple cubic lattice and the cesium ions sit at the centers of the unit cells defined by these eight points (Fig. 20–21). This structure does not work for sodium chloride because the sodium ion is smaller than the cesium ion, and eight chloride ions would interfere seriously with each other if all were brought in contact with a central sodium ion.

The strength and range of Coulomb forces make ionic crystals hard, high-melting, and brittle solids. They are electrical insulators, but melting an ionic crystal disrupts the lattice and sets the ions free to move, so ionic liquids are good electrical conductors.

Metallic Crystals

The most characteristic property of metals is their good ability to conduct electricity and heat. Electrical conduction arises from the flow of electrons from regions of high potential energy to those of low potential energy. Thermal conduction is assisted by the flow of electrons from high-temperature regions (where their kinetic energies are high) to low-temperature regions (where their kinetic energies are low). The clear implication is that the valence electrons in metals move with great ease. Why are the electrons so mobile in a metal but so tightly bound to ions or atoms in an insulating solid such as diamond, sodium chloride, or sulfur?

The answer lies with the type of bonding found in metals, which is quite different from that in other crystals. The valence electrons are delocalized in huge molecular orbitals that extend over the entire crystal. We can regard the metallic crystal as a regular assembly of positive ions surrounded by an "electron sea" that binds them together. Electrons in the sea do not belong to specific metal atoms and move easily throughout the crystal. This explains the high electrical and thermal conductivity of metals. Liquids as well as crystals can be metals; in fact, the electrical conductivity of metals usually drops by only a small amount when they are melted. The electron sea provides very strong binding in some metals, as shown by their high boiling points, but other metals have much lower boiling points. Metals also have

Figure 20–22 Gallium has a low enough melting point that it melts from the heat of the body.

a very large range of melting points. Gallium melts at 29.78°C, below normal body temperature in human beings (Fig. 20–22), and mercury stays liquid at temperatures that freeze water. On the other hand, many metals require temperatures in excess of 1000°C to melt, and tungsten, the highest-melting elemental metal, melts at 3410°C. This high melting point makes it particularly useful for the filaments in incandescent light bulbs.

Most metallic elements have crystal structures of high symmetry and crystallize in body-centered cubic, face-centered cubic, or hexagonal lattices. Why is this? If the attractions between identical atoms in a crystal are nondirectional, as is the case with many metals, lowest energy should favor structures that have the most efficient possible packing together of the atoms. A little experimentation with marbles quickly shows that the most efficient packing of equal-sized balls in a single layer occurs when each one contacts 6 others. Stacking such layers atop one another fills three-dimensional space. The most efficient packing of equal spheres in such a stack occurs when each sphere is surrounded by 12 neighbors—6 in its own layer, 3 from the layer above, and 3 from the layer below. Indeed, many metals do crystallize in exactly such **close-packed** arrangements in which each atom has 12 nearest neighbors. There are two types of close packing, depending on the way in which the layers are arranged relative to one another, as shown in Figure 20–23.

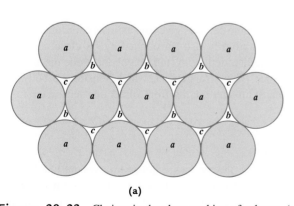

(a)

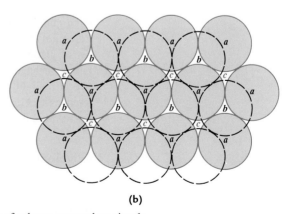

(b)

Figure 20–23 Choices in the close packing of spheres. (a) A layer of spheres centered on sites labelled a. Gaps between spheres are labelled alternately b and c. (b) A second layer of spheres has been added with centers directly over the sites labelled b. Positioning the spheres of a third layer over c-sites gives cubic close packing; positioning them over a-sites gives hexagonal close packing.

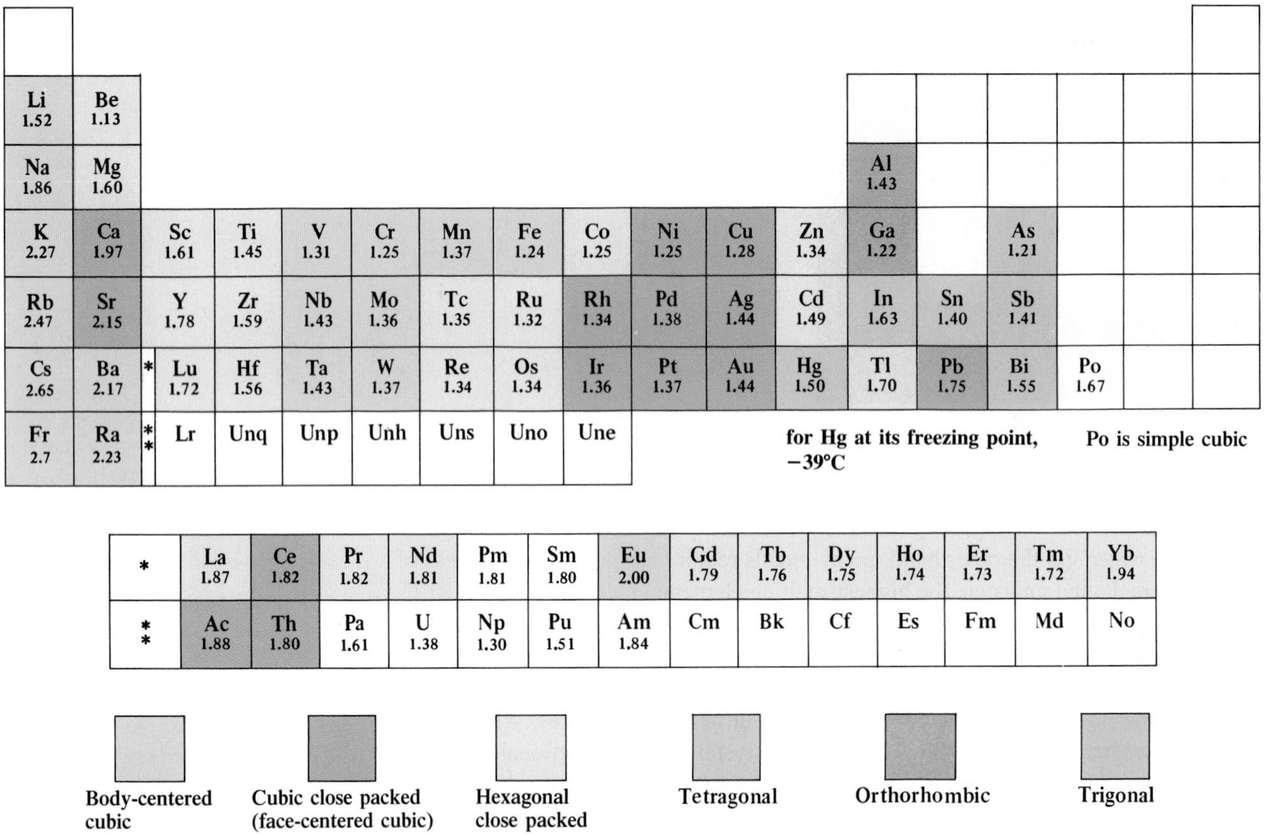

Li 1.52	Be 1.13																
Na 1.86	Mg 1.60											Al 1.43					
K 2.27	Ca 1.97	Sc 1.61	Ti 1.45	V 1.31	Cr 1.25	Mn 1.37	Fe 1.24	Co 1.25	Ni 1.25	Cu 1.28	Zn 1.34	Ga 1.22		As 1.21			
Rb 2.47	Sr 2.15	Y 1.78	Zr 1.59	Nb 1.43	Mo 1.36	Tc 1.35	Ru 1.32	Rh 1.34	Pd 1.38	Ag 1.44	Cd 1.49	In 1.63	Sn 1.40	Sb 1.41			
Cs 2.65	Ba 2.17	* Lu 1.72	Hf 1.56	Ta 1.43	W 1.37	Re 1.34	Os 1.34	Ir 1.36	Pt 1.37	Au 1.44	Hg 1.50	Tl 1.70	Pb 1.75	Bi 1.55	Po 1.67		
Fr 2.7	Ra 2.23	** Lr	Unq	Unp	Unh	Uns	Uno	Une									

for Hg at its freezing point, −39°C   Po is simple cubic

	La 1.87	Ce 1.82	Pr 1.82	Nd 1.81	Pm 1.81	Sm 1.80	Eu 2.00	Gd 1.79	Tb 1.76	Dy 1.75	Ho 1.74	Er 1.73	Tm 1.72	Yb 1.94
*	La 1.87	Ce 1.82	Pr 1.82	Nd 1.81	Pm 1.81	Sm 1.80	Eu 2.00	Gd 1.79	Tb 1.76	Dy 1.75	Ho 1.74	Er 1.73	Tm 1.72	Yb 1.94
**	Ac 1.88	Th 1.80	Pa 1.61	U 1.38	Np 1.30	Pu 1.51	Am 1.84	Cm	Bk	Cf	Es	Fm	Md	No

Body-centered cubic   Cubic close packed (face-centered cubic)   Hexagonal close packed   Tetragonal   Orthorhombic   Trigonal

Figure 20–24 Crystal structures of the metallic elements at 25°C and a pressure of 1 atm. Atomic radii are shown in ångströms.

• Both cubic close packing and hexagonal close packing fill 74.0% of the volume with spherical atoms and leave 26.0% open.

The first generates a face-centered cubic unit cell (like that in Fig. 20–18 but with all cell edges equal) and is called **cubic close packing** because the cell has cubic symmetry. The second type of close packing, **hexagonal close packing,** generates a unit cell with hexagonal symmetry and also achieves a coordination number of 12 for every atom. The two close-packing schemes have the same efficiency but differ fundamentally in their symmetries.

Some crystalline metallic elements do not have close-packed structures, as Figure 20–24 shows. Sodium and the other alkali metals crystallize in body-centered cubic unit cells in which the atoms have a coordination number of only 8. Polonium crystallizes in a simple cubic cell that gives the atoms a coordination number of just 6. In metallic tin each atom is also surrounded by six close neighbors, but the arrangement is somewhat distorted, with four neighbors at one distance and two at a slightly greater distance. This arrangement, which generates a body-centered unit cell with tetragonal symmetry, stems from directional character in the Sn—Sn bonds.

Covalent Network Crystals

In this final type of crystalline solid, the atoms are linked by covalent bonds rather than by the electrostatic attractions of ionic crystals or the valence-electron interactions of metallic crystals. The archetype of the covalent network crystal is diamond, which belongs to the cubic system. The ground-state electron configuration of a carbon atom is $1s^2 2s^2 2p^2$, and, as established in Section 18–2, the four valence orbitals can be replaced by four hybrid sp^3-orbitals directed to the four corners of a regular tetrahedron. If each of the equivalent hybrid orbitals contains one electron, then sp^3-orbitals of neighboring carbon atoms can overlap, with pairing of the spins of the electrons that they contain, to form covalent bonds. Each carbon atom in a large group can link covalently to four others to yield the space-filling network shown in Figure 20–25. In a sense, every atom in a covalent crystal is part of one giant molecule that is the crystal itself. These crystals have very high melting points due to the strong attractions between covalently bound atoms. They are hard and brittle.

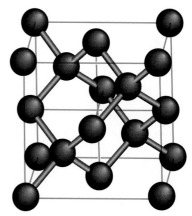

Figure 20–25 The structure of diamond. Each carbon atom has four nearest neighbors covalently bonded to it. They surround it at the corners of a tetrahedron. Silicon has the same organization of atoms, but the unit cell (outlined in red) is larger.

EXAMPLE 20–8

Explain the hardness of diamond in terms of its crystal type.

Solution

Denting a diamond requires the simultaneous bending of many strong directional C—C bonds that support each other in sets of interlocking triangles.

Exercise

Explain the brittleness of diamond in terms of its crystal type.

Answer: The network is strong, but failure of one bond under extreme stress throws extra strain on neighboring bonds closely associated with it in the above-mentioned interlocking triangles. These bonds quickly fail, and the fracture propagates rapidly through the crystal—that is, the crystal shatters.

Bond Lengths and Other Interatomic Distances in Crystals

Interatomic distances in all types of crystals can be computed once the size and shape of the unit cell and the locations of the atoms within it are known. The geometry is straightforward when the unit cell is cubic. We illustrate with three examples.

- In crystalline sodium chloride, Na^+ and Cl^- ions alternate along the edge of the face-centered cubic unit cell, as Figure 20–26 shows. The length of the cell edge is 5.640 Å. This makes the shortest Na^+–Cl^- distance equal to half the cell edge, or 2.801 Å. Chloride ions form rows along the *face diagonals* of the unit cells in sodium chloride. By the Pythagorean theorem, the distance between neighboring Cl^-'s along the face diagonals is $\sqrt{2}(5.640/2) = 3.988$ Å. This is the shortest Cl^-–Cl^- distance in the crystal; others can be calculated by drawing appropriate triangles and doing the geometry. The shortest Na^+–Na^+ distance also equals 3.988 Å, by a similar use of right triangles.

- Iron crystallizes in a body-centered cubic cell having an edge length of 2.866 Å and an Fe atom at every lattice point. Obviously, the Fe–Fe distance along the cell edge equals 2.866 Å, but this is *not* the shortest Fe–Fe distance. Rather, the closest contact is between an Fe at a corner and the Fe at the body center. This distance equals half of the *body diagonal* of the cube. The body diagonals of a cube (there are four) connect its opposite corners and pass through the body center. Their length equals $\sqrt{3}$ times the edge of the unit cube, as shown in Figure 20–27. The shortest Fe–Fe distance in crystalline iron is therefore $\sqrt{3}(2.866)/2 = 2.482$ Å.

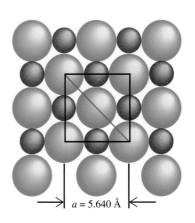

Figure 20–26 Looking down on one face of the face-centered cubic unit cell of NaCl and its immediate surroundings. The blue line is a face diagonal of the unit cell. The distances between the centers of different ions can be computed using geometrical relationships.

$a = 5.640$ Å

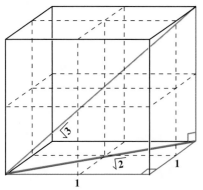

Figure 20-27 Geometrical relationships in a cube. Applying the Pythagorean theorem to the right triangle in blue shows that the face diagonal of a cube equal $\sqrt{2}$ times the edge. Applying the theorem once more, this time to the triangle marked in red, shows that the body diagonal is $\sqrt{3}$ times longer than the edge.

- Diamond has a face-centered cubic unit cell with an edge of 3.567 Å. The distance from one carbon to the next along the cell edge is 3.567 Å (far), and the C–C distance along the face diagonal is $\sqrt{2}(3.567/2) = 2.522$ Å (still rather far). The shortest C–C distance is along a different direction. As Figure 20–25 shows, some carbon atoms in diamond are *not* positioned at lattice points, but one fourth of the way into the unit cell along the body diagonals. The distance from a carbon at a cell corner to one of these atoms equals the length of the body diagonal divided by four, or $\sqrt{3}(3.567)/4 = 1.544$ Å.

Such calculations allow estimation of atomic sizes. The radius of an iron atom, for example, is taken as half the distance between the nearest neighbor atoms in the crystalline element, or 2.48/2 = 1.24 Å. Similarly, the radius of the carbon atom is nominally half the distance between nearest neighbor carbon atoms in diamond, or 0.77 Å. Before the determination of crystal structures by x-ray diffraction became routine, tables of atomic radii derived from a relatively few crystal structure determinations were used to estimate or predict bond lengths in new compounds. Nowadays, new compounds are studied by x-ray diffraction to obtain bond lengths.

The Structures of the Elements

The elements provide examples of three of the four classes of crystalline solids described in this section. Only ionic solids are excluded, because a single element does not provide the atoms of different electronegativities that are needed to form an ionic substance. Some of the structures formed by those metallic elements that readily give up electrons to form the electron sea of metallic bonding have already been discussed. The non-metallic elements are more complex in their structures, reflecting a competition between intermolecular and intramolecular bonding in many cases and producing molecular or covalent solids with quite varied properties.

Each halogen atom has seven valence electrons, and so in the elemental form, each can react with one other halogen atom to form a diatomic molecule. Once this single bond forms, there is no further bonding capacity: the halogen diatomic molecules interact with one another only through relatively weak van der Waals forces to form molecular solids with low melting and boiling points.

The Group VI elements oxygen, sulfur, and selenium display dissimilar structures in the solid state. Each oxygen atom (with six valence electrons) can form one double or two single bonds. Except in ozone (O_3), which is thermodynamically unstable relative to O_2, oxygen prefers to use up all its bonding capacity with an intramolecular double bond, forming a liquid and a solid that are bound only weakly, by van der Waals forces. Diatomic sulfur molecules, S=S, on the other hand, are relatively rare, being encountered experimentally only in high-temperature vapors. The favored forms of sulfur at room temperature involve the bonding of every atom to two other sulfur atoms. This leads to either rings or chains, and both are seen. The most stable form of sulfur at room temperature consists of S_8 molecules, with eight sulfur atoms arranged in a puckered ring (Fig. 20–28). The S_8 molecules in turn interact by van der Waals forces to make elemental sulfur the rather soft molecular solid that it is. Above 160°C, the rings in molten sulfur start to break open and relink to form long chains, producing a highly viscous liquid. An unstable ring form of selenium, Se_8, is known, but the thermodynamically stable form of this element is a gray crystal of metallic appearance that consists of very long covalently bound spiral chains, with weak interchain interaction. Crystalline tellurium has a similar structure. In the Group VI elements, then, there is a trend (moving down the

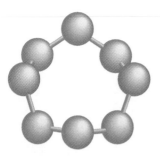

Figure 20-28 The structure of the sulfur molecule. The orthorhombic unit cell of rhombic sulfur, the most stable form of elemental sulfur at room temperature, is large and contains 16 of these S_8 molecules for a total of 128 atoms of sulfur.

Figure 20–29 Structures of elemental phosphorus.

(a) White phosphorus

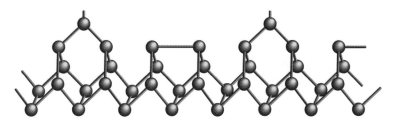

(b) Black phosphorus

(c) Red phosphorus

periodic table) away from the formation of multiple bonds and toward the chains and rings characteristic of collections of atoms that each form two single bonds.

A similar trend is evident among the Group V elements. In this group, only nitrogen forms diatomic molecules with triple bonds, in which all the bonding capacity is used between pairs of atoms. Elemental phosphorus exists in three forms, in all of which the phosphorus atoms form three single bonds rather than one triple bond. White phosphorus (Fig. 20–29a) consists of tetrahedral P_4 molecules that interact with each other through weaker van der Waals forces. Black and red forms of phosphorus (Figs. 20–29b, c), on the other hand, are higher-melting network solids in which the three bonds formed by each atom connect it directly or indirectly with all the other atoms in the solid. Unstable, solid forms of arsenic and antimony that consist of As_4 or Sb_4 tetrahedra like those in white phosphorus can be prepared by rapid cooling of the vapor. The stable forms of these elements have structures related to that of black phosphorus.

The properties of the solids just described are on the borderline between covalent and molecular. Other elements, those of intermediate electronegativity, exist as solids with properties on the borderline between metallic and covalent and are called **semi-metals.** Antimony, for example, has a metallic luster but is a rather poor conductor of electricity and heat. Silicon and germanium are **semiconductors,** with electrical conductivities far lower than those of metals but still significantly higher than those of true insulators like diamond.

Some elements of intermediate electronegativity exist in two crystalline forms with very different properties. White tin has a tetragonal crystal structure and is a metallic conductor. Below 13°C, it crumbles slowly to form a powder of gray tin (with the diamond structure). The latter is a poor conductor and has few, if any, uses. Its formation at low temperature is known as the "tin disease." The thermodynamically stable form of carbon at room conditions is not the insulator diamond, but graphite. Graphite consists of sheets of repeated hexagons, with only rather weak

• The classification of the elements as metals, nonmetals, or semi-metals is discussed in Section 3-1.

• The inherent low conductivity of silicon can be enhanced by adding controlled amounts of impurities such as arsenic or antimony. Such "doped" silicon is used in electronic devices (see Section 21-1).

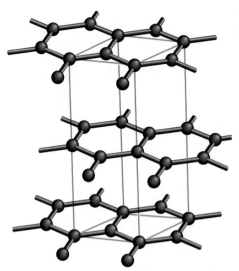

Figure 20–30 The structure of graphite.

interactions between layers (Fig. 20–30). Each carbon atom shows sp^2 hybridization and bonds to three other carbon atoms, with its remaining p-orbital (perpendicular to the graphite layers) taking part in extended π bonding interactions over the whole plane. Graphite can be pictured as a series of interlocked benzene rings, with π-electron delocalization contributing significantly to its stability. The delocalized electrons give graphite a conductivity approaching that of the metallic elements. The conductivity and chemical inertness of graphite make it useful for electrodes in electrochemistry.

20–4 DEFECTS IN SOLIDS

So far, crystalline solids have been discussed as if they consisted of completely orderly arrays of stationary groups of identical atoms extending indefinitely in three dimensions. Such perfect crystals are, of course, unattainable. In the first place, any real crystal has boundaries. At the edges of a crystal, the environment of the atoms obviously differs from that at locations buried in the bulk. Next, the atoms, molecules, or ions in real crystals are not pinned motionless at their assigned positions, but constantly vibrate, vigorously at high temperatures and less vigorously at low. Further, as a practical matter, it is impossible to rid a real crystal of all impurities. Atoms that do not fit the idealized pattern therefore crop up here and there in a real crystal. Even trace-level impurities can exert a large influence on the chemical and physical properties of a crystal. Impurities may substitute for the major constituents in a crystal or, if they are small enough, may fit in the openings (called *interstices*) in the structure of the crystal. One interesting example of the effect of impurities on the physical properties of a crystal concerns the ZnS phosphors used on the inside of television screens to make them glow. These materials contain about one silver atom for every 10,000 formula units of ZnS. Without this impurity, which is deliberately introduced, the ZnS does not fluoresce.

Some further types of lattice imperfections would occur even in an utterly pure crystalline substance. These **point defects** include the simple **vacancy,** in which an atom is missing from a lattice site, and the **interstitial atom,** in which an atom is in-

* The notion of a perfect or ideal crystal is quite analogous to that of a perfect or ideal gas. Neither actually exists. Both provide starting points for understanding the states (crystalline solid and gas, respectively) that they represent.

serted in a site different from its normal site. In real crystals, a small fraction of the normal atom sites remain unoccupied. Such vacancies are called **Schottky defects.** Figure 20–31a illustrates Schottky defects in the crystal structure of an element (metal, noble gas, or nonmetal). Schottky defects also occur in ionic crystals, but with the restriction that the imperfect crystal remain electrically neutral. Thus, in sodium chloride, for every missing Na^+ ion, a Cl^- ion must also be missing (see Fig. 20–31b). Likewise, in $CaCl_2$, for each Ca^{2+} vacancy, there must be two Cl^- vacancies.

In certain kinds of crystals, atoms or ions are displaced from their regular lattice sites to interstitial sites, and the crystal defect consists of the lattice vacancy plus the interstitial atom or ion. Figure 20–31c illustrates this type of lattice imperfection, known as a **Frenkel defect.** The silver halides (AgCl, AgBr, AgI) are examples of crystals in which Frenkel disorder is extreme. The crystal structures of these compounds are established primarily by the anion lattice, and the silver ions occupy highly disordered, almost random sites, much as in a liquid. The rate of diffusion of silver ions in these solids is exceptionally high, as studies using radioactive isotopes of silver have shown. Both Frenkel and Schottky defects in crystals are mobile, jumping from one place to another with frequencies that depend on the temperature and the strength of the atomic forces. Diffusion in crystalline solids is due largely to the presence and mobility of point defects.

Bombarding crystals with subatomic particles or high-energy radiation damages them. The simplest types of damage are the creation of an isolated interstitial atom and an isolated vacancy. The accumulation of defects and their eventual clustering or other interaction can cause marked deterioration in the electrical and mechanical properties of materials. Concerns of this type are fundamental in the engineering of materials for nuclear reactors, which must withstand large fluxes of neutrons. An unexpected loss of strength in a structural part in a nuclear reactor could be catastrophic.

If an alkali halide crystal such as NaCl is irradiated with x-rays, ultraviolet radiation, or high-energy electrons, some Cl^- ions may lose an electron:

$$Cl^- + h\nu \longrightarrow Cl + e^-$$

The resulting Cl atom, being uncharged and much smaller than a Cl^- ion, is no longer strongly bound in the crystal and can diffuse to the surface and escape. The electron can migrate through the crystal quite freely until it encounters an anion vacancy and is trapped in the Coulomb field of the surrounding cations (Fig. 20–32).

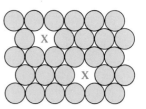

(a) Schottky defects in a metal

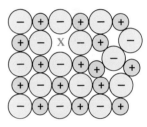

(b) Schottky defect in an ionic crystal

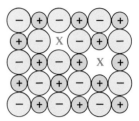

(c) Frenkel defect in an ionic crystal

Figure 20–31 Point-lattice imperfections. The red X's denote vacancies.

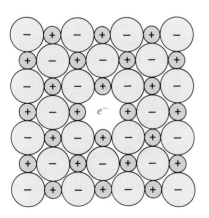

Figure 20–32 A color center in a crystal.

Figure 20–33 Pure calcium fluoride (CaF$_2$) is white, but the natural crystal of calcium fluoride (fluorite) shown here is clouded with purple because of the presence of color centers, lattice sites at which the F$^-$ anion is replaced by an electron only.

This creates a crystal defect called a **color center** (also called an F-center, from the German word *Farbe*, meaning color). It is the simplest of a family of electronic crystal defects and, as the name suggests, color centers impart colors to crystals of the alkali or alkaline earth halides (Fig. 20–33).

Nonstoichiometric Compounds

In Chapter 1, the law of definite proportions was discussed as one of the principal pieces of evidence that led to the acceptance of Dalton's atomic theory. Subsequent careful experiments have shown that in fact the law of definite proportions is violated for a great many solid-state compounds. These materials do *not* have fixed and unvarying compositions after all, but exist over a range of compositions in a single phase. Thus, FeO (wüstite) has the composition range Fe$_{0.85}$O to Fe$_{0.95}$O and is never found with its nominal 1:1 composition. The compounds NiO and Cu$_2$S also deviate considerably from their nominal stoichiometries.

- Nonstoichiometric compounds are sometimes called *berthollides*, in honor of Claude Berthollet, who believed that the proportions of the elements in compounds could vary over a range (see Section 1-3).

The explanation depends on the existence of more than one possible oxidation state for the metal. In wüstite, the iron can exist in both the +2 oxidation state and the +3 state. Consider a sample having the hypothetical composition of Fe$_{1.00}$O$_{1.00}$, with entirely iron(II), into which Fe^{3+} ions are introduced. For every two Fe^{3+} ions put in, three Fe^{2+} ions must be removed, in order to maintain overall charge neutrality. The total number of moles of iron in the sample is then less than that demanded by the ideal FeO stoichiometry. The deviation from the nominal stoichiometry can be far more extreme than that found in wüstite. In "TiO" the composition ranges from Ti$_{0.75}$O to Ti$_{1.45}$O. Nickel oxide varies only from Ni$_{0.97}$O to NiO in composition, but the variation is accompanied by a dramatic change in properties. When the compound is prepared in the 1:1 composition, it is pale green and an electrical insulator. When it is prepared in an excess of oxygen, it is black and conducts electricity fairly well. In the black material, a small number of Ni^{2+} ions are replaced by Ni^{3+} ions, and compensating vacancies occur at some nickel atom sites in the crystal.

- If all the iron atoms were in the +3 oxidation state, the composition would be Fe$_2$O$_3$, which is equivalent to Fe$_{0.667}$O$_{1.00}$. If some are iron(II) and some iron(III), a composition intermediate between Fe$_{0.667}$O$_{1.00}$ and Fe$_{1.00}$O$_{1.00}$ is expected, and is found in wüstite.

The new high-temperature superconductors are, without exception, nonstoichiometric compounds in which the presence of defects is essential to the desired property (see Section 21–5). The formula of the first compound observed to become a superconductor at a temperature exceeding the boiling point of liquid nitrogen (77 K) is YBa$_2$Cu$_3$O$_{9-x}$, in which x somewhat exceeds 2.

EXAMPLE 20-9

The composition of a sample of wüstite is $Fe_{0.930}O_{1.00}$. What percentage of the iron is in the form of iron(III)?

Solution

For every 1.00 mol of oxygen atoms, there is 0.930 mol of iron atoms in this sample. Suppose y moles of the iron are in the $+3$ oxidation state and $0.930 - y$ in the $+2$ oxidation state. Then the total positive charge from the iron, measured in moles of electron charge, is

$$+3y + 2(0.930 - y)$$

This positive charge must exactly balance the 2 mol of negative charge carried by the mole of oxygen atoms (recall that each oxygen atom has oxidation number -2). We conclude that

$$3y + 2(0.930 - y) = +2$$

Solving this equation for y gives

$$y = 0.140$$

The percentage of iron in the form of Fe^{3+} is then the ratio of this to the total number of moles of iron, 0.930, multiplied by 100:

$$\% \text{ iron in form of } Fe^{3+} = \frac{0.140}{0.930} \times 100\% = 15.1\%$$

Exercise
A sample of wüstite, Fe_xO, contains one Fe^{3+} ion for every three Fe^{2+} ions. Calculate the value of x in its formula.

Answer: 0.899.

20-5   LIQUID CRYSTALS

Melting a molecular crystal actually involves *two* effects. The first is what is usually pictured in melting: the thermal energies of the molecules become so large that the lattice breaks down and, as its long-range order is lost, the crystal liquefies. The second effect is the reorientation of the molecules, which begin to tumble and spin in all directions. Usually the two processes occur at the same temperature, and the crystal melts to an ordinary liquid. Sometimes, however, one of the two occurs at a lower temperature than the other. If the molecules are naturally nearly spherical (an example is CCl_4), then they may start to tumble *before* the lattice is disrupted. The substance stays solid and is called a **plastic crystal.** If the molecules are long, stiff, and thin, there may not be enough room for them to tumble after the lattice breaks down. They instead retain some degree of orientational order, and the result is a **liquid crystal.** In both cases, a higher temperature brings on the effect that has been delayed, and the substance becomes an ordinary liquid.

Many organic materials do not show a single solid-to-liquid transition, but rather a cascade of transitions involving new intermediate phases of the types just described. Five percent of all organic compounds exhibit liquid-crystal behavior. Not unexpectedly, the x-ray diffraction patterns of liquid crystals combine characteristics of crystal and liquid diffraction. Thus, an x-ray diffraction pattern from a liq-

Figure 20–34 Different states of structural order for rod-shaped molecules. In a real sample the lining up of the molecules would not be so nearly perfect.

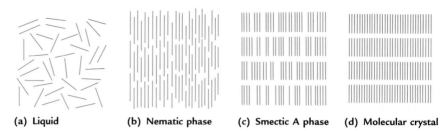

(a) Liquid   (b) Nematic phase   (c) Smectic A phase   (d) Molecular crystal

uid crystal can have both diffuse rings (the signature of a liquid) *and* some fairly sharply defined spots (the signature of a crystal). In recent years, liquid crystals have found a wide variety of practical applications, which range from use as temperature sensors to displays on calculators and other electronic devices.

The Structure of Liquid Crystals

An example of a compound that forms liquid crystals is terephthal-*bis*-(4-*n*-butyl-aniline) (called TBBA), the molecule of which has hydrocarbon groups at each end separated by a relatively rigid backbone of benzene rings and N=C double bonds. Such rod-like molecules have a tendency to line up even in the liquid phase, as shown in Figure 20–34a. Ordering in the liquid phase persists only over small distances, however, and on average a given molecule is equally likely to take any orientation.

$$H_9C_4 - \bigcirc - N{=}C - \bigcirc - C{=}N - \bigcirc - C_4H_9$$

The simplest type of liquid-crystal phase is the **nematic phase** (Fig. 20–34b). In a nematic liquid crystal, the molecules show a preferred orientation in a particular direction, but their centers are distributed at random, as they would be in an ordinary liquid. Although liquid-crystal phases are characterized by a net orientation (or "lining up") of molecules over large distances, it is wrong to think that all the molecules point in exactly the same direction. There are fluctuations in the orientation of each molecule. Only *on average* do the molecules have a greater probability of pointing in one direction.

Some liquid crystals can form one or more **smectic phases.** These show a variety of microscopic structures that are distinguished by the letters A, B, C, and so forth. One of them, the smectic A structure, is shown in Figure 20–34c: the molecules continue to show net orientational ordering but now, unlike the nematic phase, the centers of the molecules also tend to lie in layers. Within each layer, however, these centers are distributed at random, as in an ordinary liquid.

At low enough temperatures, a liquid crystal freezes to a crystalline solid (see Fig. 20–4d) in which the orientations of the molecules are ordered and their centers lie on a regular three-dimensional lattice. The meaning of the term "liquid crystal" can be seen from the progression of structures in Figure 20–34. Liquid crystals are solid-like in showing orientational ordering among their molecules but liquid-like in the random distribution of the centers of those molecules.

A third type of liquid crystal is called **cholesteric.** The name stems from the fact that many of these liquid crystals involve derivatives of the cholesterol molecule. In each layer, the molecules show a nematic type of ordering, but the direction of orientation changes from layer to layer.

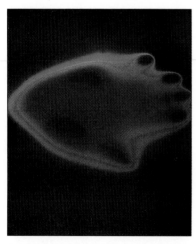

A hand pressed against a cholesteric liquid-crystal film on a black backing. The different colors reveal temperature variations.

CHEMISTRY IN YOUR LIFE

Liquid-Crystal Displays

Digital watches and many calculators use liquid-crystal displays. The operation of such displays relies on the fact that the particular orientation chosen by a liquid crystal is very sensitive to the position and nature of the surfaces with which it is in contact and to small electric or magnetic fields. In a liquid-crystal display, a nematic liquid crystal is contained in a small cell that has interior surfaces treated to specify the orientation of the liquid crystal it holds. Polarizing filters set up on each end of this cell allow only photons with particular polarizations to pass through the cell. In the absence of an electric field, light passes through the liquid crystal and both filters and is reflected by an underlying mirror, so the display appears white (Fig. 20–A, top). If an electric field is applied across some portion of the display, the preferred orientation of the molecules in that area changes, causing a different polarization of light; the "rotated" light is blocked by the second filter, and that part of the display appears black (Fig. 20–A, bottom). When the electric field is turned off, the liquid-crystal molecules relax rapidly to their original orientations, and the display again turns white.

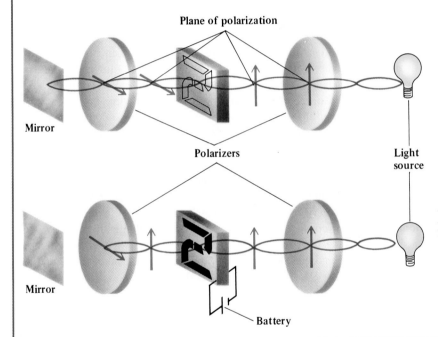

Figure 20–A How a liquid-crystal display device works. The light has a polarization that permits it to pass through the second polarizer and strike the mirror, giving a display that appears bright (top). Imposition of a potential difference (bottom) causes the liquid crystal molecules to reorient, so that the resulting polarized light does not reach the mirror and the display appears dark.

SUMMARY

20–1 In both **crystalline solids** and **amorphous solids,** structure is probed in diffraction experiments using x-rays, neutrons, or electrons. In **x-ray diffraction,** a portion of a beam of x-rays is **scattered** by the electrons of the material being studied. This interaction is followed by interference among the scattered waves. X-ray **diffraction patterns** convey information about the distribution of the scattering centers (atoms) because the distances between atoms control the phase differences of the scattered waves arising from the atoms. The **Bragg law** relates the wavelength of the incoming x-rays, the distance between parallel planes of scattering centers, and the scattering angles at which various "reflected" beams are observed.

20–2 All crystals have a characteristic symmetry that is revealed in x-ray diffraction experiments. Crystals are classified into seven different **crystal systems** that are based on the **symmetry** (including **rotational symmetry** and **mirror symmetry**) of their diffraction patterns. These are the **hexagonal, cubic, tetragonal, trigonal, orthorhombic, monoclinic,** and **triclinic** systems. The three-dimensional array made up of all the points within a crystal that have the same environment in the same orientation is a **crystal lattice;** it represents the scheme of repetition at work in the construction of a crystal. There are seven types of crystal lattice, corresponding to the seven crystal systems. A proper selection of eight points of a crystal lattice defines a **unit cell,** a box that contains all the structural information about a crystal necessary to reproduce it. A crystal is created by stacking up unit cells to fill space. **Primitive** unit cells (the smallest possible unit cells) are the usual choice in discussing crystal structures. Nonprimitive unit cells (including **body-centered, face-centered,** and **side-centered** cells) are used when no primitive cell has the full symmetry of the lattice. The volume of a unit cell can be computed from x-ray measurements of the cell edges (a, b, and c) and the cell angles (α, β, and γ). The density of a crystal is equal to the mass of the contents of a unit cell of that crystal divided by the volume of the cell. From accurate observations of the intensities as well as the directions of the many x-ray beams diffracted by a crystal, it is nearly always possible to compute back to the locations of the atoms in the unit cell.

20–3 Crystals can be classified according to the nature of their bonding interactions as molecular, ionic, metallic, or covalent network. **Molecular crystals** consist of discrete molecules held in positions on a lattice by **van der Waals forces.** These forces cause molecules to approach each other until atoms come into "contact" at their **van der Waals radii. Ionic crystals** are maintained by Coulomb electrostatic attractions between cations and anions. They often have structures of high symmetry, such as the **rock salt** (sodium chloride) structure. **Metallic crystals** are held together by the delocalized valence electrons in huge molecular orbitals that extend across the entire crystal. Elemental metals often use either **cubic close packing** or **hexagonal close packing** to minimize the amount of empty space between the spherical atoms. **Covalent network crystals** contain atoms that are held in a three-dimensional network by covalent chemical bonds. In a sense, an entire covalent crystal is a single molecule. The structures of the chemical elements exemplify three of the four crystal bonding types. Knowledge of the locations of atoms in the unit cell and of the size and shape of the unit cell allows the computation of bond distances and other interatomic distances in crystals.

20–4 Real crystals always contain **lattice imperfections** such as **point defects.** Vacant lattice sites are called **Schottky defects.** The displacement of an atom or ion from its regular lattice site to an interstitial site is a **Frenkel defect.** An **F-center** or **color center** occurs when an anion in an ionic solid is replaced at a lattice site by an electron. A **nonstoichiometric compound** violates the law of definite proportions by having a range of compositions but is readily understood in terms of lattice vacancies.

20–5 In **liquid crystals,** the lattice structure of a crystal vanishes upon melting, but the molecules are still partially constrained by their neighbors and do not move as freely as they do in a true liquid. Liquid crystals form especially with long rod-like molecules. In **nematic phases,** the molecules show a preferred orientation in a particular direction, but their centers are distributed at random, as they would be in an ordinary liquid. In **smectic phases,** the molecules continue to show net orientational ordering, but now the centers of the molecules also lie in layers. In **cholesteric** liquid crystals, the molecules show a nematic type of ordering, but this orientation changes from plane to plane.

PROBLEMS

Note: Answers to blue-numbered problems are given in Appendix F. Problems that are more challenging are indicated with asterisks.

Probing the Structure of Condensed Matter

1. (See Example 20–1.) The second-order Bragg reflection of x-radiation with $\lambda = 1.660$ Å from a set of parallel planes in copper occurs at an angle $2\theta = 54.70°$. Calculate the distance between the scattering planes in the crystal.

2. (See Example 20–1.) The second-order Bragg reflection of x-radiation with $\lambda = 1.237$ Å from a set of parallel planes in aluminum occurs at an angle $2\theta = 35.58°$. Calculate the distance between the scattering planes in the crystal.

3. (See Example 20–2.) The distance between members of a set of equally spaced planes of atoms in crystalline lead is 4.950 Å. If x-rays with $\lambda = 1.936$ Å are diffracted by this set of parallel planes, calculate the angle 2θ at which fourth-order Bragg reflection is observed.

4. (See Example 20–2.) The distance between members of a set of equally spaced planes of atoms in crystalline sodium is 4.28 Å. If x-rays with $\lambda = 1.539$ Å are diffracted by this set of parallel planes, calculate the angle 2θ at which second-order Bragg reflection is observed.

5. (See Example 20–2.) The members of a series of equally spaced parallel planes of ions in crystalline LiCl are separated by 2.570 Å. Calculate all the angles 2θ at which diffracted beams of various orders might be seen, if the x-ray wavelength used is 2.167 Å.

6. (See Example 20–2.) A recent precise re-determination of the structure of vitamin B_{12} used x-rays with a wavelength of 0.6500 Å from a synchrotron source. In the crystal, the spacing between a certain set of planes containing cobalt atoms is 12.55 Å.
   (a) Calculate the smallest five angles 2θ at which reflected beams of various orders involving these planes might be seen.
   (b) Calculate n for the highest-order reflection possible from these planes.

7. Compare the x-ray diffraction pattern of a single crystal with that of a microcrystalline solid. Account for the differences.

8. Compare the x-ray diffraction pattern of an amorphous solid with that of a microcrystalline solid. Account for the differences.

Symmetry and Structure

9. (See Example 20–3.) Which of the following has 3-fold rotational symmetry? Explain.
   (a) an isosceles triangle
   (b) an equilateral triangle
   (c) a tetrahedron
   (d) a cube
   (e) a regular hexagon

10. (See Example 20–3.) Which of the following has 4-fold rotational symmetry? Explain.
    (a) a cereal box (exclusive of the writing on the sides)
    (b) a stop sign (not counting the writing)
    (c) a tetrahedron
    (d) a cube
    (e) a rectangular solid

11. (See Example 20–3.) Which of the following objects has at least one mirror line (or plane)? Explain.
    (a) the number 1881
    (b) a cereal box (exclusive of the writing on the sides)
    (c) a fishhook
    (d) a golf club

12. (See Example 20–3.) Which of the following objects has at least one mirror line (or plane)? Explain.
    (a) a baseball (counting the seam)
    (b) the letters "bood" taken as a group
    (c) the number 6009 printed in block digits
    (d) a closed textbook

13. Identify a symmetry element that must be present in the diffraction pattern of a crystal that belongs to the tetragonal crystal system.

14. Identify a symmetry element that must be present in the diffraction pattern of a crystal that belongs to the orthorhombic crystal system.

15. (See Example 20–5.) An orthorhombic crystal of the sugar D-fructose has a unit cell with edges a, b, and c equal to 8.06 Å, 9.12 Å, and 10.06 Å, respectively. Compute its volume in Å^3 and in m^3.

16. (See Example 20–5.) Compute the volume (in Å^3) of the unit cell of potassium hexacyanoferrate(III) ($K_3Fe(CN)_6$), a substance that crystallizes in the monoclinic system with $a = 8.40$ Å, $b = 10.44$ Å, and $c = 7.04$ Å and with $\beta = 107.5°$.

17. (See Example 20–6.) Diamond has a unit cell that contains eight carbon atoms and has a volume of 45.385 Å^3.
    (a) Compute the volume (in Å^3) per carbon atom in diamond.
    (b) Calculate the density of carbon in diamond.

18. (See Example 20–6.) The unit cell of crystalline gold contains four gold atoms. The cell is cubical and has an edge that is 4.0786 Å long.
    (a) Compute the volume in Å^3 of the unit cell of gold.
    (b) An atom of gold weighs 196.967 u. Compute the density of gold in units of u Å^{-3}.
    (c) Express the density of gold in units of g cm^{-3}. (*Hint:* Recall that 6.0221×10^{23} u $= 1.000$ g and that 1×10^8 Å $= 1$ cm.)

19. At room temperature, the edge length of the cubic unit cell in elemental silicon is 5.431 Å, and the density of silicon at

the same temperature is 2.328 g cm^{-3}. Each cubic unit cell contains eight silicon atoms. Using only these facts:

(a) Calculate the volume of one unit cell (in cubic centimeters).

(b) Calculate the mass (in grams) of silicon present in a unit cell.

(c) Calculate the mass (in grams) of an atom of silicon.

(d) The mass of an atom of silicon is 28.0855 u. Estimate Avogadro's number to four significant figures. (*Note:* The mass of any substance in u divided by its mass in grams is equal to Avogadro's number.)

20. One form of crystalline iron has a body-centered cubic unit cell with an iron atom located at every lattice point. Its density at 25°C is 7.86 g cm^{-3}. The length of the edge of the cubic unit cell is 2.866 Å. Use these facts to estimate Avogadro's number (*Hint:* Follow the procedure suggested in the preceding problem.)

21. The common and important mineral quartz is crystalline SiO_2. There are in fact two distinct forms of quartz: α-quartz (a trigonal crystal) and β-quartz (a hexagonal crystal). In each form, the unit cell contains three SiO_2 units, but the unit cell of α-quartz has a volume of 113.01 Å^3 and the unit cell of β-quartz is slightly larger, with a volume of 118.15 Å^3. The density of α-quartz is 2.648 g cm^{-3}. Compute the density of β-quartz.

22. Compute the density of nickel(II) oxide. This substance crystallizes in the cubic system. The unit cell has $\alpha = 4.177$ Å, and there are four NiO units in each cell.

23. The compound $Pb_4In_3B_{17}S_{18}$ crystallizes in the monoclinic system with a unit cell having $a = 21.021$ Å, $b = 4.014$ Å, and $c = 18.898$ Å, and the only non–90° angle equal to 97.07°. There are two molecules in every unit cell. Compute the density of this substance.

24. Strontium chloride hexahydrate ($SrCl_2 \cdot 6H_2O$) crystallizes in the trigonal system in a unit cell with $a = 8.9649$ Å and $\alpha = 100.576°$. The unit cell contains three formula units. Compute the density of this substance.

25. (See Example 20–7.) Sodium sulfate (Na_2SO_4) crystallizes in the orthorhombic system in a unit cell with $a = 5.863$ Å, $b = 12.304$ Å, and $c = 9.821$ Å. The density of these crystals is 2.663 g cm^{-3}. Determine how many Na_2SO_4 formula units are present in the unit cell.

26. (See Example 20–7.) The density of turquoise ($CuAl_6(PO_4)_4(OH)_8(H_2O)_4$) is 2.927 g cm^{-3}. This gemstone crystallizes in the triclinic system with cell constants $a = 7.424$ Å, $b = 7.629$ Å, and $c = 9.910$ Å, $\alpha = 68.61°$, $\beta = 69.71°$, and $\gamma = 65.08°$. Verify that the volume of the unit cell is 461.40 Å^3, and determine how many copper atoms are present in each unit cell of turquoise.

27. A certain crystal of sodium chloride is a perfect little cube measuring 1.00×10^{-3} m on an edge. The microscopic unit cell of crystalline sodium chloride is a cube that has an edge 5.6402 Å = 5.6402×10^{-10} m in length.

(a) Compute the volume of a unit cell of sodium chloride.

(b) Compute the number of unit cells in this particular crystal.

(c) Compute the number of unit cells that are on the surface of this crystal and the percentage of all the unit cells that are surface unit cells.

28. A crucial protein at the photosynthetic reaction center of the purple bacterium *Rhodopseudomonas viridis* has been separated from the organism, crystallized, and studied by x-ray diffraction. This substance crystallizes with a primitive unit cell in the tetragonal system. The cell dimensions are $a = b = 223.5$ Å and $c = 113.6$ Å.

(a) How many lattice points are in this unit cell?

(b) Determine the volume, in Å^3, of this cell.

(c) One of the crystals in this experiment was box shaped, with dimensions $1 \times 1 \times 3$ mm. Compute how many unit cells were in this crystal.

Bonding in Crystals

29. Classify each of the following crystalline solids as molecular, ionic, metallic, or covalent network:

(a) $BaCl_2$

(b) SiC

(c) CO

(d) Co

30. Classify each of the following crystalline solids as molecular, ionic, metallic, or covalent network:

(a) Rb

(b) C_5H_{12}

(c) B

(d) Na_2HPO_4

31. The melting point of cobalt is 1495°C, and that of barium chloride is 963°C. Rank the four substances in problem 29 from lowest to highest in melting point.

32. The boiling point of pentane (C_5H_{12}) is slightly less than the melting point of rubidium. Rank the four substances in problem 30 from lowest to highest in melting point.

33. Rank the following elements from lowest to highest in electrical conductivity under room conditions: antimony, copper, and oxygen.

34. Rank the following elements from lowest to highest in thermal conductivity under room conditions: chromium, nitrogen, and germanium.

35. By examining Figure 20–21, determine the number of nearest neighbors and second-nearest neighbors of a Cs^+ ion in crystalline CsCl. The nearest neighbors of the Cs^+ ion are Cl^- ions, and the second-nearest neighbors are Cs^+ ions.

36. Repeat the determination of the preceding problem for the NaCl crystal, referring to Figure 20–20.

37. Draw a sketch of a body-centered cubic lattice and use it to determine the number of nearest neighbor and second-nearest neighbor atoms to a given sodium atom in the structure of sodium, which has an atom at each point in this lattice.

38. Draw a sketch of a face-centered cubic lattice and use it to determine the number of nearest neighbor and second-nearest neighbor atoms to a given aluminum atom in the structure of crystalline aluminum, which has an atom at each point in this lattice.

39. White tin crystallizes in a body-centered *tetragonal* unit cell with edges $a = b = 5.8315$ Å and $c = 3.1813$ Å. Tin atoms sit at the lattice points (and at other locations in the unit cell as well). Compute the distance between an Sn atom at a cell corner and an Sn atom at the body center.

40. Thallium bromide crystallizes in the cesium chloride structure. The edge of the unit cell is 3.97 Å. Compute the shortest Tl-to-Br distance in this structure.

Defects in Solids and Liquid Crystals

41. Does the presence of Frenkel defects change the measured density of a crystal?

42. What effect does the (unavoidable) presence of Schottky defects have on the determination of Avogadro's number using the method described in problems 19 and 20?

43. Iron(II) oxide is nonstoichiometric. A particular sample was found to contain 76.55% iron and 23.45% oxygen by mass.
    (a) Calculate the empirical formula of the compound (four significant figures).
    (b) What fraction of the iron in this sample is in the +3 oxidation state?

44. A sample of nickel oxide contains 78.23% Ni by mass.
    (a) What is the empirical formula of the nickel oxide to four significant figures?
    (b) What fraction of the nickel in this sample is in the +3 oxidation state?

45. Compare the nature and extent of order in liquid hydrogen chloride (HCl) with that in the plastic crystalline phase of HCl. Which has the higher entropy? Which has the higher enthalpy?

46. Compare the nature and extent of order in the smectic liquid crystal and isotropic liquid phases of a substance. Which has the higher entropy? Which has the higher enthalpy?

Additional Problems

47. Define the terms *scattering* and *interference* with respect to the effects of a sample on an x-ray beam passing through it.

48. X-rays of wavelength $\lambda = 1.54$ Å are scattered by a crystal at an angle 2θ of 32.15°. Calculate the wavelength of the x-rays in another experiment if this same diffracted beam from the same crystal is observed at an angle of 34.46°.

49. Some water waves with a wavelength of 3.0 m are diffracted by an array of evenly spaced posts in the water. If the rows of posts are separated by a distance of 5.0 m, calculate the angle 2θ at which the first-order "Bragg diffraction" of these water waves is seen.

* 50. If the wavelength λ of the x-rays is too large relative to the spacing of planes in the crystal, no Bragg diffraction is seen because $\sin\theta$ is larger than 1 in the Bragg equation, even for $n = 1$. Calculate the longest wavelength of x-rays that can give Bragg diffraction from a set of planes separated by 4.20 Å.

51. Find all the symmetry operations in the following:
    (a) the number 10801
    (b) the number 86198
    (c) a teacup

52. (a) You receive a valuable golden cube as a gift. To protect this treasure, you construct a tight-fitting wooden case with a plush lining. How many different ways are there for you to place your cube into the case?
    (b) Suppose that in the preceding situation, the valuable bauble had a tetrahedral shape. How many different ways are there to place such a shape into a tight-fitting case?
    (c) How many different ways would there be to place an object with the diagrammed shape into a tight-fitting case?

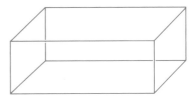

53. Use Figure 20–25 to determine the number of carbon atoms per unit cell of diamond.

* 54. Use Figure 20–30 to determine the number of carbon atoms per unit cell of graphite.

55. Consider a collection of hard spherical atoms arranged at the lattice points of a primitive cubic lattice. A unit cell is constructed so that one eighth of the corner atoms lie within the unit cell. Compute the fraction of the volume of the unit cell that is empty if the atoms are in contact along the edge of the cell.

* 56. Show that hard spherical atoms arranged at the lattice points of a body-centered cubic lattice and in contact along the body diagonal of the conventional unit cell occupy only 68% of the total volume available in the structure.

57. The number of beams diffracted by a single crystal depends on the wavelength λ of the x-rays used and on the volume associated with one lattice point in the crystal—that is, on the volume V_p of a primitive unit cell. An approximate formula is

$$\text{Number of diffracted beams} = \frac{4}{3\pi}\left(\frac{2}{\lambda}\right)^3 V_p$$

(a) Compute the volume of the conventional unit cell of crystalline sodium chloride. This cell is cubical and has an edge of 5.6402 Å.
(b) The NaCl unit cell contains four lattice points. Compute the volume of a primitive unit cell for NaCl.
(c) Use the formula given in this problem to estimate the number of diffracted rays if NaCl is irradiated with x-rays having a wavelength of 2.2896 Å.
(d) Use the formula to estimate the number of diffracted rays that are observed if NaCl is irradiated with x-rays having the shorter wavelength of 0.7093 Å.

58. A compound contains three elements: sodium, oxygen, and chlorine. It crystallizes in a primitive cubic lattice. The oxygen atoms are at the corners of the unit cells; the chlorine atoms are at the centers of the unit cells; and the sodium atoms are at the centers of the faces of the unit cells. What is the formula of the compound?

59. At room temperature monoclinic sulfur has the unit-cell dimensions $a = 11.04$ Å, $b = 10.98$ Å, $c = 10.92$ Å and $\beta = 96.73°$. Each cell contains 48 atoms of sulfur. Compute the density of monoclinic sulfur in units of g cm^{-3}.

60. White tin crystallizes in a body-centered *tetragonal* unit cell with edges $a = b = 5.8315$ Å and $c = 3.1813$ Å. Tin atoms occupy the lattice points and also occupy sites one fourth of the distance along the face diagonals of the *bc* faces of the unit cell. Compute the distance between an Sn atom at a cell corner and one of these Sn atoms.

61. The following graphic appeared on a cardboard box to indicate that the cardboard was recyclable. The graphic does *not* have 3-fold symmetry. What symmetry element or elements does it possess? Redraw the graphic so that it has 3-fold symmetry.

62. Polonium is the only element known to crystallize in the simple cubic lattice.
(a) What is the distance between nearest-neighbor polonium atoms if the first-order reflection of x-radiation with $\lambda = 1.785$ Å from the parallel faces of its unit cells appears at an angle of $2\theta = 30.96°$ from these planes?
(b) What is the density of polonium in this crystal (in g cm^{-3})?

63. Within a recent six-month period, two completely independent reports on the structure of mercury(I) nitrite appeared in the scientific literature. The results were in reasonable agreement, with the following sets of unit-cell constants given:

	Report 1	Report 2
a	4.4145 Å	4.435 Å
b	10.3334 Å	10.344 Å
c	6.2775 Å	6.301 Å
α	Exactly 90°	Exactly 90°
β	108.84°	108.74°
γ	Exactly 90°	Exactly 90°

The two reports agree that the unit cell contains four mercury ions and four nitrite ions.
(a) What is the crystal system of mercury(I) nitrite?
(b) Compute the density of mercury(I) nitrite according to both sets of data.
(c) Explain why there must be equal numbers of Hg atoms and NO$_2^-$ ions in the unit cell.
(d) Comment on the use of significant digits in the two reports.

64. Name two elements that form molecular crystals, two that form metallic crystals, and two that form covalent network crystals. What generalizations can you make about the portions of the periodic table where each type is found?

65. Beryllium chloride is a covalent compound with a rather high melting point (405°C). Draw the Lewis structure of BeCl$_2$, and use it to suggest why the intermolecular attractions in solid BeCl$_2$, a molecular solid, are unusually strong.

66. Solid carbon dioxide (Dry Ice) and solid iodine both sublime at room temperature. What do these solids have in common structurally? Why do many other solids of the same type *not* sublime under the same conditions?

* 67. Sodium hydride (NaH) crystallizes in the rock salt structure, with four formula units of NaH per cubic unit cell. A beam of monoenergetic neutrons, selected to have a velocity of 2.639×10^3 m s^{-1}, is scattered in second order through an angle of $2\theta = 36.26°$ by the parallel faces of the unit cells of a sodium hydride crystal.
(a) Calculate the wavelength of the neutrons.
(b) Calculate the edge length of the cubic unit cell.
(c) Calculate the distance from the center of an Na$^+$ ion to the center of a neighboring H$^-$ ion.
(d) If the radius of an Na$^+$ ion is 0.98 Å, what is the radius of an H$^-$ ion, assuming that the two ions are in contact?

68. A crystal of sodium chloride has a density of 2.165 g cm^{-3} in the absence of defects. Suppose that a crystal of NaCl is grown in which 0.15% of the sodium ions and 0.15% of the chloride ions are missing. What is the density in this case?

* 69. A compound of titanium and oxygen contains 28.31% oxygen by mass.
(a) If its empirical formula is Ti$_x$O, calculate x to four significant figures.
(b) The nonstoichiometric compounds Ti$_x$O can be described as having a Ti^{2+}-O^{2-} lattice in which certain Ti^{2+} ions are missing or are replaced by Ti^{3+} ions. Calculate the fraction of Ti^{2+} sites in the nonstoichiometric compound that are vacant and the fraction that are occupied by Ti^{3+} ions.

70. Classify the bonding in the following amorphous solids as molecular, ionic, metallic, or covalent network:
(a) Amorphous silicon, used in photocells to collect light energy from the sun.
(b) Polyvinyl chloride, a plastic of long-chain molecules composed of —CH$_2$CHCl— repeating units, and used in pipes and siding.
(c) Soda–lime silica glass, used in windows.
(d) Copper–zirconium glass, an alloy of the two elements with

approximate formula Cu_3Zr_2, used for its high strength and good conductivity.

71. The chemical As_2S_3 is common in nature as the crystalline mineral orpiment. Melting a crystal of orpiment and recooling it always gives amorphous As_2S_3. Speculate on why crystals of orpiment cannot be made in the laboratory but are relatively common in nature.

72. Is methane (CH_4) more likely to form plastic crystalline phase or a liquid crystalline phase? Should it be relatively easy or difficult to form an amorphous solid from methane?

CUMULATIVE PROBLEM

Phosphorus

Solid elemental phosphorus appears in a variety of forms, with crystals in all seven crystal systems reported under various conditions of temperature, pressure, and sample preparation.

(a) The thermodynamically most stable form of phosphorus under room conditions is black phosphorus. The unit cell of black phosphorus is orthorhombic with edges of length 3.314, 4.376, and 10.48 Å. Calculate the volume of one unit cell and determine the number of phosphorus atoms per unit cell if the density of black phosphorus is 2.69 g cm^{-3}.

(b) The form of solid phosphorus that is easiest to prepare from the liquid or gaseous state is white phosphorus, which consists of P_4 molecules in a cubic lattice. When x-rays of wavelength 2.29 Å are scattered from the parallel faces of its unit cells, the second-order Bragg diffraction is observed at an angle 2θ of 14.23°. Calculate the length of the edge of the unit cell of white phosphorus. At what angle is fourth-order Bragg diffraction seen?

(c) When liquid white phosphorus is heated to 300°C in the absence of air, a red solid forms that shows diffuse diffraction rings. This material is used in the strip on a matchbox against which a safety match is struck. What is the nature of this red solid?

(d) Red phosphorus (part c) has been reported to convert into monoclinic, triclinic, tetragonal, and cubic red forms with different heat treatments. Identify the changes in the shape of the unit cell as a cubic lattice is converted first to tetragonal, then monoclinic, then triclinic.

(e) A monoclinic form of red phosphorus has been studied that has cell edges of 9.21, 9.15, and 22.60 Å and an angle β of 106.1 degrees. Each unit cell contains 84 atoms of phosphorus. Estimate the density of this form of phosphorus.

(f) Phosphorus forms many compounds with other elements. Describe the nature of the bonding in the following solids: white elemental phosphorus (P_4), black elemental phosphorus, sodium phosphate (Na_3PO_4), and phosphorus trichloride (PCl_3).

Two forms of elemental phosphorus: white and red.

21

Silicon and Solid-State Materials

CHAPTER OUTLINE

Powders of silicon nitride, Si_3N_4, formed in the flame produced by a carbon dioxide laser.

The first 20 chapters of this book treat the underlying principles of chemistry: Chapters 1 through 4 present fundamentals (the atomic theory, mass relationships, the periodic table and classes of reactions); Chapters 5 through 14 consider macroscopic aspects of chemistry (the states of matter, the nature of chemical equilibrium and the thermodynamic laws that underlie it, and the rates of chemical reactions); Chapters 15 to 20 discuss how electrons and nuclei combine to form atoms, which can join together through chemical bonds to make molecules and complexes, which themselves interact further in diverse ways. The remainder of the book takes up some applications of these principles. Recent advances in materials science revolve around the fabrication of engineered materials. These are materials deliberately designed to possess desired properties. For example, chemists and physicists have tailored materials to act as superconductors or semiconductors or to retain strength, flexibility, toughness, and other required properties despite extreme conditions. Materials scientists have embedded sensors in traditional structural materials to signal when a structure is damaged and have developed additives that spring into chemical action to mend structures if corrosion or cracking begins. The elements most prominent in these developments are silicon, which is essential in many ceramics (discussed in this chapter), and carbon, the polymer-builder (discussed in Chapter 25).

• The third important class of materials, besides ceramics and polymers, is the metals. Their preparation and uses are described in Chapter 12.

21–1 SEMICONDUCTORS

As a Group IV element, silicon resembles carbon chemically. Both silicon and carbon exhibit valences of 4 and have the capacity to form extended molecular structures. Both crystallize in the cubic system and possess the same structure. The bonding in solid silicon can be described as the overlap of four sp^3 hybrid orbitals centered on each atom and directed toward the corners of a regular tetrahedron. Each atom is covalently linked to four others. The result is a "giant molecule" network structure like that of diamond (see Fig. 20–25). There are just enough valence electrons in this network to provide a single bond between each silicon atom and its four neighbors.

An alternative to this localized-orbital explanation of the bonding in silicon is to consider that the valence electrons occupy "molecular" orbitals that are spread out, or delocalized, over the entire silicon crystal, Recall how the $1s$-orbitals from two hydrogen atoms mix to form a lower-energy bonding and a higher-energy antibonding molecular orbital in the H_2 molecule (see Fig. 18–2). In the case of silicon, the $4N_0$ valence atomic orbitals from one mole (N_0 atoms) of silicon split into two **bands** in the silicon crystal, each containing $2N_0$ very closely spaced levels (Fig. 21–1). The lower band is called the **valence band,** and the upper one the **conduction band.** Between the top of the valence band and the bottom of the conduction band is an energy region that is forbidden to electrons. The magnitude of the separation in energy between the valence band and the lowest level of the conduction band is called the **band gap (E_g).** For pure silicon the band gap is 1.94×10^{-19} J. This is the amount of energy that an electron must gain to be excited from the top of the valence band to the bottom of the conduction band. If N_0 electrons (1 mol) are excited across the band gap, the energy change is 117 kJ mol^{-1}, which is calculated by multiplying 1.94×10^{-19} J by Avogadro's number N_0.

• Each Si atom contributes one $3s$ and three $3p$ valence orbitals, for a total of four orbitals. Recall that the number of molecular orbitals after mixing is equal to the number of atomic orbitals used.

• Another unit used for band gaps is the electron volt, defined in Section 15-1 as 1.60218×10^{-19} J. The band gap in Si is 1.21 eV.

Each silicon atom in the crystal contributes four valence electrons to the bands of orbitals in Figure 21–1, for a total of $4N_0$ per mole. This is a sufficient number to place two electrons in each level of the valence band (with opposing spins) and

Figure 21–1 The valence orbitals of the silicon atom combine in crystalline silicon to give two bands of very closely spaced levels. The valence band is almost completely filled, and the conduction band is almost empty.

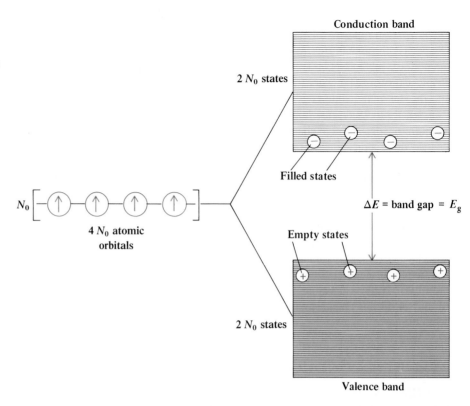

Conduction band

$2 N_0$ states

Filled states

ΔE = band gap = E_g

N_0

$4 N_0$ atomic orbitals

Empty states

$2 N_0$ states

Valence band

leave the conduction band empty. There are no low-lying energy levels for electrons at the top of the valence band to ascend to if given a small boost in energy. In fact, they cannot accept an energy increment at all if it is less than 1.94×10^{-19} J. An electron at the top of the filled valence band must acquire an energy of at least 1.94×10^{-19} J, equivalent to 117 kJ mol^{-1}, to jump to the lowest empty level of the conduction band. This is a large amount of energy. If it were to be supplied by a thermal source, the temperature of the source would have to be on the order of

$$T = \frac{E_g}{R} = \frac{117{,}000 \ \cancel{\text{J mol}^{-1}}}{8.315 \ \cancel{\text{J}} \ \text{K}^{-1} \ \cancel{\text{mol}^{-1}}} = 14{,}000 \ \text{K}$$

At room temperature, only a few electrons per mole in the extreme tail of the Boltzmann distribution have enough energy to jump the gap, so the conduction band in pure silicon is very sparsely populated with electrons. The result is that pure silicon is a poor conductor of electricity. Good electrical conductivity requires a net motion of many electrons under the impetus of a small electrical potential difference. The conductivity of silicon is 11 orders of magnitude smaller than that of a typical metal (copper) at room temperature. Silicon is called a **semiconductor** because its electrical conductivity, although less than that of a metal, is far larger than that of an **insulator** like diamond, which has an even larger band gap. The conductivity of a semiconductor is increased by increasing the temperature, because more electrons are excited into the conduction band. Another way to increase the conductivity of a semiconductor is to irradiate it with a beam of photons that are energetic enough (i.e., of frequency high enough) to excite electrons from the valence band to the conduction band. This process resembles the photoelectric effect described in Section 16–2, with the difference that the electrons are not removed from the material, but only moved into the conduction band, ready to convey a current if a potential difference is imposed.

EXAMPLE 21–1

Calculate the longest wavelength of light that can excite electrons from the valence band to the conduction band in silicon. In what region of the spectrum does this wavelength fall?

Solution

The energy carried by a photon is $h\nu = hc/\lambda$, where h is Planck's constant, ν the photon frequency, c the speed of light, and λ the photon wavelength. For a photon just barely to excite an electron across the band gap in silicon, its energy must equal the band-gap energy, $E_g = 1.94 \times 10^{-19}$ J. We set $h\nu$ equal to E_g, solve for λ, and substitute

$$\frac{hc}{\lambda} = E_g$$

$$\lambda = \frac{hc}{E_g} = \frac{(6.626 \times 10^{-34}\ \mathrm{J\ s}) \times \left(2.998 \times 10^8 \dfrac{\mathrm{m}}{\mathrm{s}}\right)}{1.94 \times 10^{-19}\ \mathrm{J}}$$

$$= 1.02 \times 10^{-6}\ \mathrm{m} = 1020\ \mathrm{nm}$$

This wavelength falls in the infrared region of the spectrum. Photons with shorter wavelengths (for example, visible light) carry more than enough energy to excite electrons to the conduction band in silicon.

Exercise

Calculate the longest wavelength of light that can excite electrons in diamond, in which the band gap is 8.7×10^{-19} J.

Answer: 230 nm, in the ultraviolet region of the spectrum. Visible light does not carry enough energy to excite electrons from the valence band in diamond to the conduction band.

Doped Semiconductors

When certain other elements are added to pure silicon in a process called **doping,** the silicon acquires interesting electronic properties. If, for example, atoms of a Group V element such as arsenic or antimony are diffused into silicon, they substitute for silicon atoms in the network. Such atoms have five valence electrons. Each one therefore introduces one more electron into the silicon crystal than is needed for bonding. The extra electrons occupy energy levels just below the lowest level of the conduction band. Little energy is required to promote electrons from such a **donor-impurity** level into the conduction band, and the electrical conductivity of the silicon crystal is increased. The change is roughly proportional to the amount of dopant that is added. Silicon doped with atoms of a Group V element is called an **n-type semiconductor** to indicate that the charge carrier is negative.

On the other hand, if a Group III element such as gallium is used as a dopant, the valence band contains one fewer electron per dopant atom because Group III elements have only three valence electrons, not four. This situation amounts to creating one **hole** in the valence band, with an effective charge of $+1$, for each Group III atom added. If a voltage difference is impressed across a sample of silicon that is doped in this way, it causes the positively charged holes to move toward the neg-

Figure 21–2 An integrated circuit consists of a series of connected electrical circuits printed onto a tiny silicon wafer. The Pentium chip shown here contains 3.3 million transistors and is about the size of a fingernail.

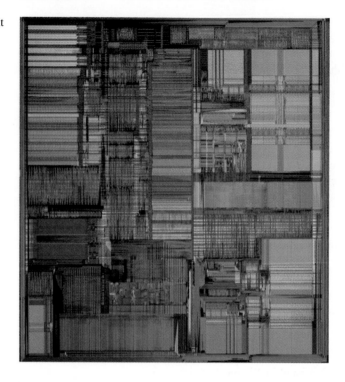

• "Isoelectronic" means that it has the same number of electrons.

ative source of potential. Equivalently, a valence-band electron next to the (positive) hole moves in the *opposite* direction (that is, toward the positive source of potential). Whether one thinks of holes or electrons as the mobile charge carrier in the valence band, the result is the same; the difference is in the words, not the physical phenomenon. Silicon that has been doped with a Group III element is called a **p-type semiconductor** to indicate that the carrier has an effective positive charge.

A different class of semiconductors is based not on silicon but on equimolar compounds of Group III with Group V elements. Gallium arsenide, for example, is isoelectronic to the Group IV semiconductor germanium. Doping GaAs with the Group VI element tellurium produces an *n*-type semiconductor; doping with zinc, which has one *fewer* valence electron than gallium, gives a *p*-type semiconductor. Other III–V combinations have different band gaps and are useful in particular applications. Indium antimonide (InSb), for example, has a small enough band gap that absorption of infrared radiation causes electrons to be excited from the valence to the conduction band; an electric current then flows when a small potential difference is applied. This compound is therefore used as a detector of infrared radiation. Still other compounds formed between elements of the zinc group (zinc, cadmium, and mercury) and Group VI elements such as sulfur also have the same average number of valence electrons per atom as silicon and make useful semiconductors.

Semiconductors perform a wide range of electronic functions, such as amplifying current and converting alternating current into direct current. These functions formerly required the use of vacuum tubes, which occupy much more space, generate larger amounts of heat, and require considerably more energy to operate than **transistors,** their semiconductor counterparts. More importantly, semiconductors can be built into integrated circuits (Fig. 21–2) and made to store information and to process it at great speeds.

The manufacture of integrated circuits requires extraordinary efforts to avoid contamination. Here, metal connections are being deposited on a silicon wafer. The garments protect the work piece from the worker, not the reverse.

Solar cells based on silicon or gallium arsenide provide a way to convert the radiant energy of the sun directly into electrical work by a technology that is virtually nonpolluting. The high capital costs of solar cells make them uncompetitive with conventional fossil-fuel sources of energy at this time. As reserves of fossil fuels dwindle, solar energy will become an important option.

Pigments and Phosphors

The band gap of an insulator or semiconductor has a significant effect on its color. Pure diamond has a large band gap (see Exercise 21–1), so even blue light does not have enough energy to excite electrons from the valence band to the conduction band. As a result, visible light passes through it without being absorbed, and diamond is colorless. Cadmium sulfide has a band gap of 4.2×10^{-19} J, which falls in the visible spectrum (4.2×10^{-19} J corresponds to a wavelength of 470 nm). CdS thus absorbs violet and blue light but strongly transmits yellow, giving it a deep yellow color (Fig. 21–3). Cadmium sulfide is in fact the pigment called cadmium yellow. Cinnabar (HgS) has a smaller band gap of 3.2×10^{-19} J and absorbs all light except red (Fig. 21–4). It has a deep red color and is the pigment vermilion.

Figure 21–4 Crystalline cinnabar, HgS.

Figure 21–3 Mixed crystals of two semiconductors with different band gaps, CdS (yellow) and CdSe (black), show a range of colors, illustrating a decrease in the band-gap energy as the composition of the mixture becomes richer in Se.

Semiconductors with band gaps of less than 2.8×10^{-19} J absorb all wavelengths of visible light and appear black. These include silicon (see Example 21–1), germanium, and gallium arsenide.

Doping silicon creates donor-impurity levels close enough to the conduction band or acceptor levels close enough to the valence band that thermal excitation can cause electrons to move into a conducting level. The corresponding doping of *insulators* or wide-band-gap semiconductors can bring donor or acceptor levels into positions where visible light can be absorbed or emitted. The doping thereby changes the colors and optical properties of the materials. Nitrogen doped into diamond creates a donor-impurity level in the band gap. Transitions to this level can absorb some blue light, giving the diamond a yellowish (undesirable) color. In the other hand, boron doped into diamond gives an acceptor level that absorbs red light most strongly and gives the highly prized and rare "blue diamond."

Phosphors are wide-band-gap materials with dopants that create new levels at energies that favor emission of particular colors of light. Electrons in these materials are excited by light of other wavelengths or by electrons hitting their surfaces, and light is then emitted as they return to lower-energy states. A fluorescent lamp is a mercury-vapor lamp in which the inside of the tube has been coated with phosphors. These phosphors absorb the violet and ultraviolet light emitted by mercury vapor and emit at lower energies and longer wavelengths, giving a nearly white light that is more desirable than the bluish light that comes from a mercury-vapor lamp without the phosphors.

Phosphors are also used in television screens. The picture is formed by scanning a beam of electrons (from an electron gun) over the inside of the screen. The electrons strike the phosphors coating the screen, exciting their electrons and causing them to emit light. In a black-and-white television tube, the phosphors are a mixture of silver doped into ZnS, which gives blue light, and silver doped into $Zn_xCd_{1-x}S$, which gives yellow light. The combination of the two provides a reasonable approximation of white. A color television tube uses three different electron guns, with three corresponding types of phosphor on the screen. Silver doped in ZnS gives blue; manganese doped in Zn_2SiO_4 gives green; and europium doped in YVO_4 gives red light. Masks are used to ensure that each electron beam encounters only the phosphors corresponding to the desired color.

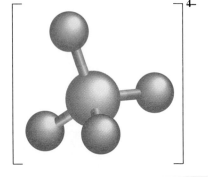

21–2   SILICATE MINERALS

Silicon and oxygen make up most of the earth's crust, with oxygen accounting for 47% and silicon 28% of its mass. The silicon–oxygen bond is partially ionic and strong. It forms the basis for a class of minerals called **silicates,** which make up the bulk of the rocks, clays, sands, and soils in the earth's crust. From time immemorial silicates have provided the ingredients for building materials such as bricks, cement, concrete, ceramics, and glass.

The structure-building properties of silicates originate in the tetrahedral orthosilicate anion, SiO_4^{4-}, in which both elements have their accustomed oxidation numbers: +4 for silicon and −2 for oxygen. The negative charge of the silicate ion is balanced by the compensating charge of one or more cations. The simplest silicates consist of individual SiO_4^{4-} anions (Fig. 21–5), with cations arranged around them on a regular crystalline lattice. Such silicates are properly called **orthosilicates.** Examples are forsterite (Mg_2SiO_4) and fayalite (Fe_2SiO_4), which are at the

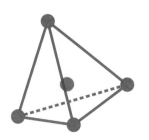

Figure 21–5 The tetrahedral orthosilicate anion SiO_4^{4-}. It is often represented by drawing the tetrahedral outline defined by the oxygen atoms at the corners.

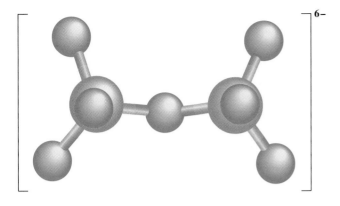

6−

Figure 21-6 The disilicate anion $Si_2O_7^{6-}$. It consists of two silicate tetrahedra linked at a corner. In representations of silicate structures, linked tetrahedra are often projected onto a plane as shown.

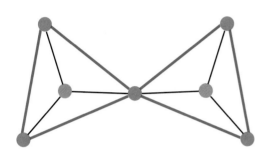

two extremes in a class of minerals called **olivines** ([Mg, Fe]$_2$SiO$_4$). There is a continuous range of proportions of magnesium and iron in the olivines.

Other kinds of silicate structures form when two or more SiO_4^{4-} tetrahedra link together and share oxygen corners. The simplest are the **disilicates,** in which two SiO_4^{4-} tetrahedra are linked (Fig. 21–6), as in thortveitite (Sc$_2$(Si$_2$O$_7$)). Additional linkages of tetrahedra create the ring, chain, sheet, and network structures shown in Figure 21–7 and listed in Table 21–1.

EXAMPLE 21-2

By referring to Table 21–1, predict the structural class into which the mineral Egyptian blue (CaCu(Si$_2$O$_5$)$_2$) belongs. Give the oxidation state of each of its atoms.

Solution

Because the O:Si atom ratio is 5:2, this mineral should have an infinite sheet structure with the repeating unit $Si_2O_5^{2-}$. The oxidation state of Si is +4, that of O is −2, as usual, and that of Ca is +2. In order to make the total oxidation number per formula unit add up to zero, the oxidation state of Cu must be +2.

Exercise

Predict the structural class of the mineral åkermanite (Ca$_2$Mg(Si$_2$O$_7$)), and give the oxidation states of its atoms.

Answer: Disilicate (pairs of tetrahedra). Ca and Mg, +2; Si, +4; O, −2.

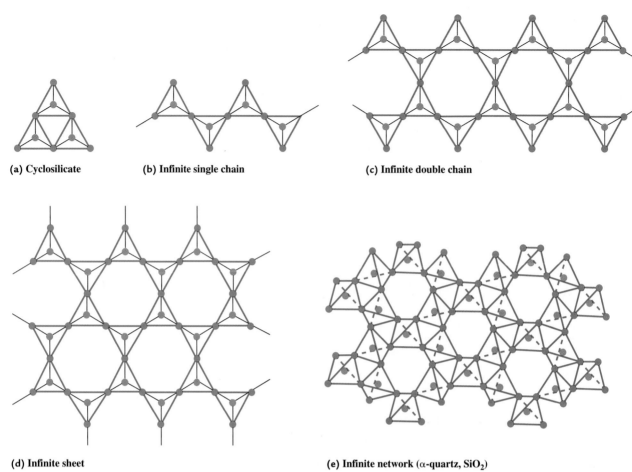

(a) Cyclosilicate   **(b) Infinite single chain**   **(c) Infinite double chain**

(d) Infinite sheet   **(e) Infinite network (α-quartz, SiO$_2$)**

Figure 21–7   Five classes of silicate structures. All of these structures are three-dimensional. In (a) through (d) the SiO$_4$ tetrahedra are projected onto the plane, with the fourth oxygen positioned above the silicon as shown in Figure 21–6. In (e) the tetrahedra are tilted at various angles and linked in 3-fold and 6-fold spirals along the line of viewing. In all, the fundamental structural unit is the same, but the overall O:Si ratio is less than 4 because oxygen atoms are shared between silicon atoms.

Table 21–1
Silicate Structures

Structure	Figure	Corners Shared at Each Si	Repeat Unit	O:Si Ratio	Example
Tetrahedra	21–5	0	SiO_4^{4-}	4:1	Olivines
Pairs of tetrahedra	21–6	1	$Si_2O_7^{6-}$	$3\frac{1}{2}$:1	Thortveitite
Closed rings	21–7a	2	SiO_3^{2-}	3:1	Beryl
Infinite single chains	21–7b	2	SiO_3^{2-}	3:1	Pyroxenes
Infinite double chains	21–7c	$2\frac{1}{2}$	$Si_4O_{11}^{6-}$	$2\frac{3}{4}$:1	Amphiboles
Infinite sheets	21–7d	3	$Si_2O_5^{2-}$	$2\frac{1}{2}$:1	Talc
Infinite network	21–7e	4	SiO_2	2:1	Quartz

The physical properties of the silicates correlate closely with their structures. An example of an infinite sheet structure (Fig. 21–7d) is talc, which has the composition $Mg_3(Si_4O_{10})(OH)_2$. In talc, all of the covalent bonding interactions among the atoms occur in a single layer. Layers of talc sheets are attracted to one another only by van der Waals forces, which, being weak, permit one layer to slip easily across another. This accounts for the slippery feel of talc (called talcum powder). When all four vertices of each tetrahedron are linked to other tetrahedra, three-dimensional network structures result (Fig. 21–7e). The simplest and most abundant example of a three-dimensional network structure is the mineral quartz (Fig. 21–8). Note that the quartz network carries no charge; consequently, there are no cations in its structure. Three-dimensional network silicates such as quartz are much stiffer and harder than the linear and layered silicates.

Asbestos is a generic term for a group of naturally occurring hydrated silicates that can be processed mechanically into long fibers (Fig. 21–9). Some of these show the infinite double-chain structure of Figure 21–7c; an example is tremolite $(Ca_2Mg_5(Si_4O_{11})_2(OH)_2)$. Another kind of asbestos mineral is chrysotile $(Mg_3(Si_2O_5)(OH)_4)$. As the formula indicates, this mineral has a sheet structure (see Fig. 21–7d), but the sheets are rolled up into long tubes. Asbestos minerals are fibrous because the bonds along the tubes are stronger than those that hold different tubes together. Asbestos is an excellent thermal insulator that does not burn, resists acids, and is very strong. For many years, it was used extensively in cement for pipes and ducts, and woven into fabric to make fire-resistant roofing paper and floor tiles. Its use has decreased significantly in recent years because inhalation of its small fibers can cause the lung disease asbestosis. The risk comes with breathing asbestos dust raised during mining and manufacturing or released when asbestos-containing building materials fray, crumble, or are removed.

Figure 21–8 The mineral quartz is one form of silica (SiO_2).

Aluminosilicates

An important class of minerals, called **aluminosilicates,** results from the replacement of some of the silicon atoms in silicates by aluminum atoms. Aluminum is the third most abundant element in the earth's crust (8% by mass), where it occurs largely in the form of aluminosilicates. Aluminum in minerals can be a simple cation (Al^{3+}), or it can replace silicon in tetrahedral coordination. When it replaces silicon, it contributes only three electrons to the bonding framework in place of the four electrons of silicon atoms. The additional required electron is supplied by the ionization of a metal atom such as sodium or potassium; the resulting alkali-metal ions occupy nearby sites in the aluminosilicate structure.

The most abundant aluminosilicate minerals in the earth's surface are the **feldspars,** which result from the substitution of aluminum for silicon in three-dimensional silicate networks like quartz. The aluminum ions must be accompanied by other cations such as sodium, potassium, or calcium to maintain overall charge neutrality. Albite is a feldspar with chemical formula $NaAlSi_3O_8$. In the high-temperature form of this mineral, the aluminum and silicon atoms are distributed at random (in proportions of 1:3) over the tetrahedral sites available to them. At lower temperatures other crystal structures become thermodynamically stable, with partial ordering of Al and Si sites.

If one of the four silicon atoms in the structural unit of talc $(Mg_3(Si_4O_{10})(OH)_2)$ is replaced by an aluminum atom, and a potassium atom is furnished to supply the fourth electron needed for bonding in the tetrahedral silicate framework, the composition $KMg_3(AlSi_3O_{10})(OH)_2$ results. This substance belongs to the family of **micas.** Micas are harder than talc, and their layers slide less readily over one another,

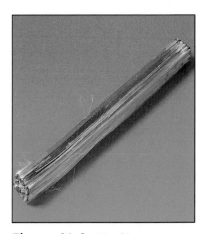

Figure 21–9 The fibrous structure of asbestos is apparent in this sample.

although the crystals still cleave easily into sheets (see Fig. 20–12). The cations occupy sites between the infinite sheets, and the van der Waals attractions that hold adjacent sheets together in talc are augmented by an ionic contribution. The further replacement of the three Mg^{2+} ions in $KMg_3(AlSi_3O_{10})(OH)_2$ by two Al^{3+} ions gives the mineral **muscovite** ($KAl_2(AlSi_3O_{10})(OH)_2$). Writing its formula in this way indicates that there are aluminum atoms in two kinds of sites in the structure: one Al atom per formula unit occupies a tetrahedral site, substituting for an Si atom; the other atom Al atoms are located between the two adjacent layers. The formulas that mineralogists and crystallographers use convey somewhat more information than the usual empirical chemical formula of a compound.

• The importance of x-ray diffraction for determining the structures of silicates cannot be overemphasized. From the empirical formula for muscovite ($KH_2Al_3Si_3O_{12}$), there is no indication of the structural roles of the aluminum atoms in their different sites, nor of the location of the hydrogen atoms in the structure.

Clay Minerals

Clays are minerals produced by the weathering of primary minerals. They have similar structures, but their compositions can vary widely as one element is replaced by another. Invariably they are microcrystalline or powdered in form and are usually hydrated. They are used as supports for catalysts, as fillers in paint, and as ion-exchange vehicles. Those clays that readily absorb water and swell as they do so find use as lubricants and bore-hole sealers in the drilling of oil wells.

The derivation of clays from talcs and micas provides a direct way to understand their structure. Take the infinite sheet mica pyrophyllite ($Al_2(Si_4O_{10})(OH)_2$) as an example. If one out of six Al^{3+} ions in the pyrophyllite structure is replaced by one Mg^{2+} ion and one Na^+ ion (which together carry the same charge), a clay called "montmorillonite" results. Its composition can be represented as $MgNaAl_5(Si_4O_{10})_3(OH)_6$. This clay readily absorbs water, which infiltrates between the infinite sheets and associates by ion-dipole forces with the Mg^{2+} and Na^+ ions located there, causing the montmorillonite to swell (Fig. 21–10).

As noted previously, talc, $Mg_3(Si_4O_{10})_3(OH)_2$, is a layered mineral. If water molecules are allowed to take up positions between the layers, the swelling clay ver-

Figure 21–10 Structure of the clay mineral montmorillonite. Insertion of variable amounts of water causes the distance between layers to swell from 9.6 Å to more than 20 Å. When one Al^{3+} ion is replaced by an Mg^{2+} ion, an additional ion such as Na^+ is introduced into the water layers to maintain overall charge neutrality.

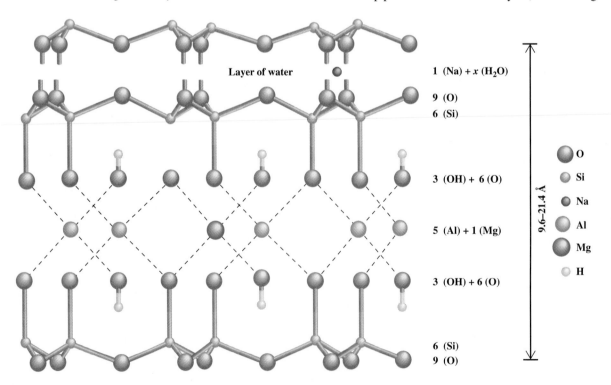

Layer of water

1 (Na) + x (H_2O)

9 (O)

6 (Si)

3 (OH) + 6 (O)

5 (Al) + 1 (Mg)

3 (OH) + 6 (O)

6 (Si)

9 (O)

9.6–21.4 Å

O
Si
Na
Al
Mg
H

miculite results. As is true of all clays, substitution is common, and vermiculite usually contains iron(III) and aluminum, which replace magnesium and silicon in varying proportions. When heated, vermiculite pops like popcorn, as the steam generated by the vaporization of water between the layers puffs the flakes up into a light, fluffy material with air inclusions. Because of its open, porous structure, vermiculite is used for thermal insulation and as an additive to loosen compacted soils.

Zeolites

Zeolites are a class of three-dimensional aluminosilicates. Like the feldspars, they carry a negative charge on the aluminosilicate framework that is compensated by neighboring cations such as the alkali-metal and alkaline-earth ions. Zeolites differ from the feldspars in having much more open structures consisting of polyhedral cavities connected by tunnels (Fig. 21–11). Many zeolites are found in nature, but they can also be synthesized under conditions controlled to favor the formation of cavities of very uniform size and shape. In most zeolites, water molecules are accommodated within these cavities where they provide a mobile phase for the in-and-out migration of the charge-compensating cations. This enables zeolites to serve as ion-exchange materials and is the basis for their use in the softening of water. "Hardness" in water arises from dissolved calcium and magnesium salts such as $Ca(HCO_3)_2$ and $Mg(HCO_3)_2$. Such salts are converted to insoluble carbonates (boiler scale) when the water is heated and also form objectionable precipitates (bathtub ring) with soaps. When hard water is passed through a column packed with a zeolite that has sodium ions in its structure, the calcium and magnesium ions exchange with the sodium ions and are removed from the water phase into the zeolite:

$$Na_2Z(s) + Ca^{2+}(aq) \rightleftharpoons CaZ(s) + 2\,Na^+(aq)$$

When the ion-exchange capacity of the zeolite is exhausted, this reaction can be reversed by passing a concentrated solution of sodium chloride through the zeolite to regenerate it in the sodium form. Ion exchange also works to separate chemically similar cations of the rare-earth elements or the actinides from one another on the basis of their slightly different tendencies to associate with zeolites used as the stationary phase in column chromatography (see Section 7–7).

A second use of zeolites derives from the ease with which they absorb small molecules. Their sponge-like affinity for water makes them useful as drying agents; they are put between the panes of double-pane glass windows to prevent moisture from condensing on the inner surfaces. The pore size of zeolites can be calibrated during their synthesis to allow molecules that are smaller than a certain size to pass through, but to hold back larger molecules. Such zeolites serve as "molecular sieves"; they have been used to capture nitrogen molecules from a gas stream while permitting oxygen molecules to pass through.

Perhaps the most exciting use of zeolites is as catalysts. Molecules of varying sizes and shapes have different rates of diffusion through a zeolite; this enables chemists to enhance the rates and yields of desired reactions and suppress unwanted reactions. The most extensive applications of zeolites at present are in the catalytic cracking of crude oil, a process that involves breaking down long-chain hydrocarbons and re-forming them into branched-chain molecules of lower molecular mass for use in high-octane unleaded gasoline. A relatively new process uses "shape-selective" zeolite catalysts to convert methanol (CH_3OH) to high-quality gasoline. Plants have been built to use this process to make gasoline from methanol derived from coal or from natural gas.

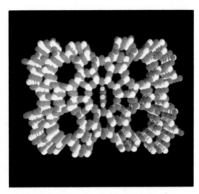

Figure 21–11   The structure of the synthetic zeolite ZSM-5, showing a benzene molecule passing through one of the channels. This material consists of a network of aluminum, silicon, and oxygen atoms, with cations to balance the overall charge. It is used as a catalyst in the conversion of crude oil to petroleum products.

CHEMISTRY IN PROGRESS

Silicones

The chains of alternating silicon and oxygen atoms that are present in silicate minerals form the basis for an important class of synthetic materials—the **silicones**. The simplest of these is dimethyl silicone, a chain molecule that can be derived (structurally) from the silicate chains shown in Figure 21–7b by replacing the two lone oxygen atoms on each silicon atom by methyl groups, —CH_3. It is an example of a polymer molecule, built up from a set of repeating units (in this case, —O—Si(CH_3)$_2$— units) that are connected by chemical bonds. We consider other polymers in Chapter 25.

The starting material for making dimethyl silicone is $(CH_3)_2SiCl_2$, which reacts with water to form a dihydroxo compound:

$$(CH_3)_2SiCl_2(aq) + 2\,H_2O(\ell) \longrightarrow$$
$$(CH_3)_2Si(OH)_2(aq) + 2\,H^+(aq) + 2\,Cl^-(aq)$$

Two molecules of $(CH_3)_2Si(OH)_2$ can then react, with loss of water, to form an Si—O—Si linkage:

Further condensation with additional $(CH_3)_2Si(OH)_2$ molecules leads to long-chain molecules with the overall chemical formula $[(CH_3)_2SiO]_n$. The Si—O bond lengths and Si—O—Si bond angles are much the same as in silicate minerals. Dimethyl silicone fluids are used widely as lubricating greases and in electrical transformers.

Just as silicates form sheets, rings, and networks as well as straight-chain molecules, so also silicones can be synthesized with cross-linkages between chains, leading to three-dimensional networks. One way in which this is done is illustrated by the molecular structure of the silicone sealant present in many bathtub caulks. Acetate and also hydroxide side-groups replace some of the nonreactive methyl side-groups of dimethyl silicone in this material. These groups can condense (Fig. 21–A), with release of acetic acid, to give silicon atoms that have three, instead of two, oxygen linkages to other silicon atoms. The loss of the acetic acid accounts for the characteristic pungent smell of vinegar (acetic acid) after the caulk has been applied. The resulting cross-linked polymer materials are durable and water resistant.

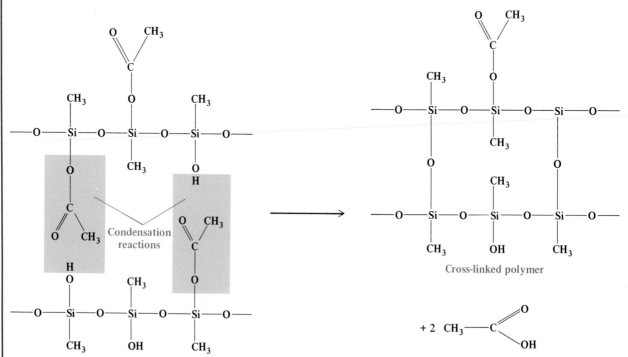

Figure 21–A   Silicone polymers can cross-link, releasing acetic acid.

21-3 THE PROPERTIES OF CERAMICS

The term **ceramics** encompasses synthetic materials that have as their essential components inorganic, non-metallic materials. This broad definition includes cement, concrete, and glass in addition to more traditional fired clay products such as bricks, roof tiles, pottery, and porcelain. The use of ceramics predates recorded history; indeed the emergence of civilization is chronicled in fragments of pottery. No one knows when small vessels were first shaped by human hands from moist clay and left to harden in the heat of the sun. Such containers held nuts, grains, and berries well but lost their shape and slumped into formless mud when water was poured into them. Then someone discovered that if clay were placed in the glowing embers of a fire, it became as hard as rock and withstood water well. Molded figures made in Central Europe 24,000 years ago are the earliest fired ceramic objects discovered so far, and fired clay vessels from the Near East date from 8000 B.C. With the action of fire on clay, the art and science of ceramics began.

Ceramics bear little resemblance to metals and plastics. They offer stiffness, hardness, resistance to wear, and resistance to corrosion (particularly by oxygen and water), even at high temperature. They are less dense than most metals, which makes them desirable metal substitutes when weight is a factor. Although certain highly specialized ceramics become electrical superconductors at low temperatures (as discussed at the end of this chapter), most are good electrical insulators at ordinary temperatures, a property that is exploited in electronics and power transmission (Fig. 21–12). Ceramics retain their strength well at high temperatures. Several important structural metals soften or melt at temperatures a thousand degrees below the melting points of their chemical compounds in ceramics. Aluminum, for example, melts at 660°C, whereas aluminum oxide (Al_2O_3) does not melt until a temperature of 2051°C is reached.

Against these advantages must be listed some serious disadvantages. Ceramics are generally brittle and low in tensile strength. They tend to have high thermal expansion but low thermal conductivity, making them subject to **thermal shock,** in which a sudden local temperature change causes cracking or shattering. Metals and plastics dent or deform under stress, but ceramics cannot dissipate stress in this way. They break instead. A major drawback of ceramics as structural materials is their tendency to fail unpredictably and catastrophically in use. Moreover, some ceramics lose mechanical strength as they age, an insidious and serious problem.

Composition and Structure of Ceramics

Ceramics use a wide variety of chemical compounds, and useful ceramic bodies are nearly always mixtures of several compounds. **Silicate ceramics,** which include the traditional pots, dishes, and bricks, are made from aluminosilicate clay minerals. All contain the tetrahedral SiO_4 grouping discussed in Section 21–2. In **oxide ceramics,** silicon is a minor or nonexistent component. Instead, a number of metals combine with oxygen to give compounds such as alumina (Al_2O_3), magnesia (MgO), or yttria (Y_2O_3). **Nonoxide ceramics** contain as principal components compounds that are free of oxygen. Some important examples are silicon nitride (Si_3N_4), silicon carbide (SiC), and boron carbide (approximate composition B_4C).

A **ceramic phase** is any portion of the whole body that is physically homogeneous and bounded by a surface that separates it from other parts. Distinct phases are visible at a glance in coarse-grained ceramic pieces; in a fine-grained piece, phases can be seen under a microscope (Fig. 21–13). Ceramics are often porous; they have small openings into which air or water can infiltrate. Fully dense ceram-

Figure 21–12 This insulating ceramic, consisting of several interlocking pieces, transmits no electric current despite a potential difference of hundreds of thousands of volts between one end and the other.

• The distinction between ceramics and plastics is not always clear-cut. Glass is a ceramic that can be shaped by plastic deformation at high temperatures, and some silicates behave like plastics upon the addition of water.

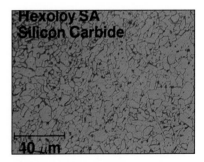

Figure 21–13 A close-up of a silicon carbide ceramic, showing its granular texture.

Figure 21-14 A part freshly fabricated from silicon nitride for use in a gas-turbine engine.

• Some ceramic objects (such as those composed of cement) require no firing in their fabrication. In the hardening of cement, the heat required for chemical and structural transformations is supplied internally by chemical reactions.

ics have no channels of this sort. Two ceramic pieces can have the same chemical composition but quite different densities if the first is porous and the second is not. When examined on a still finer scale, most ceramics, like metals, are microcrystalline. They consist of small crystalline grains cemented together and show this by interacting with x-rays to produce a diffraction pattern comprising numerous well-defined rings (see Section 20–1). The **microstructure** of such objects includes the sizes and shapes of the grains, the sizes and distribution of voids (openings between grains) and cracks, the identity and distribution of impurity grains, and the presence of stresses within the structure. Microstructural variations have enormous importance in ceramics because slight changes at this level strongly influence the properties of individual ceramic pieces. This is less true for plastic and metallic objects.

The microstructure of a ceramic body depends markedly on the details of its fabrication. The techniques of forming and firing a ceramic piece become as important as its chemical composition in determining ultimate behavior because they confer a unique microstructure. This fact explains the biggest problem with ceramics as structural materials: inconsistent quality. Ceramic engineers can produce parts that are stronger than steel, but not reliably because of the difficulties of monitoring and controlling microstructure. Gas-turbine engines fabricated of silicon nitride (Fig. 21–14), for example, run well at 1370°C, which is hot enough to soften or melt most metals. The higher operating temperature increases engine efficiency, and the ceramic turbines weigh less, further boosting fuel economy. Despite these advantages, ceramic gas turbines are not made commercially. Ceramic turbines that work have to be built from selected, pretested components. The testing costs and rejection rates are so high that economical mass production has been impossible so far.

Making Ceramics

The manufacture of most ceramics involves four steps: (a) preparation of the raw material; (b) forming into the desired shape, often achieved by mixing a powder with water or other binder and molding the resulting plastic mass; (c) drying and firing the piece, also called its **densification,** because pores (voids) in the dried ceramic fill in; and (d) finishing the piece by sawing, grooving, grinding, or polishing.

The raw materials for traditional ceramics are natural clays that come from the earth as powders or thick pastes and become plastic enough, after adjustment of their water content, for forming freehand or on a potter's wheel. In contrast, modern specialized ceramics (both oxide and nonoxide) require chemically pure raw materials that are produced synthetically. Close control of the purity of the starting materials for these ceramics is essential to producing finished pieces with the desired properties. In addition to being formed by hand or in open molds, ceramic pieces can be shaped by the squeezing (compacting) of the dry or semi-dry powders in a strong, closed mold of the desired shape, both at ordinary and at elevated temperatures (hot pressing).

Firing a ceramic causes **sintering,** in which the fine particles of the ceramic start to merge together by diffusion at high temperatures. Atoms migrate out of their individual grains, and **necks** grow to connect adjoining particles. The density of the material increases because the voids between grains are partially filled. Sintering occurs below the melting point of the material and shrinks the ceramic body. In addition to the growing together of the grains of the ceramic, firing causes partial melting, chemical reactions among different phases, reactions with gases in the atmosphere of the firing chamber, and recrystallization of compounds with an accompanying increase in crystal size. All of these changes influence the microstructure of a piece and must be understood and controlled. Although firing accelerates physical and chemical changes, thermodynamic equilibrium in a fired ceramic piece is rarely reached. Kinetic factors, including the rate of heating, the length of time at which each temperature is held, and the rate of cooling, influence microstructure. The use of microwave radiation (as in microwave ovens) rather than kilns to fire ceramics is under development in ceramic factories because it promises more exact control of the heating rate and thereby more reliable quality.

A ceramics worker tends a walk-in kiln.

21–4 SILICATE CERAMICS

The silicate ceramics include materials that vary widely in composition, structure, and uses. They range from simple earthenware bricks and pottery to fine porcelain, glass, and cement. Their structural strength is based on the same linking of silicate-ion tetrahedra that gives structure to silicate minerals in nature.

Pottery and Structural Clay Products

Section 21–2 notes that aluminosilicate clays are the products of the weathering of primary minerals. When water is added to such clays in moderate amount, the result is a thick paste that is easily molded into different shapes. Clays expand as water invades the space between adjacent aluminosilicate sheets of the mineral, but they release most of this water to a dry atmosphere and thus shrink. A small fraction of the water or hydroxide ions remains rather tightly bound by ion-dipole forces to cations between the aluminosilicate sheets and is lost only when the clay is heated to a high temperature. The firing of aluminosilicate clays causes irreversible chemical changes to take place. The clay kaolinite ($Al_2Si_2O_5(OH)_4$) undergoes the following reaction:

$$3 \; Al_2Si_2O_5(OH)_4(s) \longrightarrow Al_6Si_2O_{13}(s) + 4 \; SiO_2(s) + 6 \; H_2O(g)$$

kaolinite   mullite   silica   water

The fired ceramic body is a mixture of two phases, mullite and silica. Mullite, a rare mineral in nature, takes the form of needle-like crystals that interpenetrate and confer strength on the ceramic. When the temperature is above 1470°C, the silica phase forms as minute grains of cristobalite, one of the several crystalline forms of SiO_2.

If chemically pure $Al_2Si_2O_5(OH)_4$ is fired, the finished ceramic object is white. Such purified clay minerals are the raw material for fine china. As they occur in nature, clays contain impurities, such as transition-metal oxides, that affect the color of both the unfired clay and the fired ceramic object if they are not removed. The colors of the metal oxides arise from their absorption of light at visible wavelengths, as explained by crystal-field theory (see Section 19–4). Common colors for natural ceramics are yellow or greenish yellow, brown, and red. Bricks are red when the clay used to make them has a high iron content (Fig. 21–15).

Before a clay is fired in a kiln, it must first be freed of moisture by slow heating to about 500°C. A clay body dried at room temperature and placed directly into a hot kiln crumbles or even explodes from the sudden expulsion of water. A fired ceramic shrinks somewhat as it cools, causing cracks to form. Cracks limit the strength of the fired object and are undesirable. Their occurrence can be reduced by coating the surface of a partially fired clay object with a **glaze,** a thin layer that holds the ceramic in a state of tension as it cools. Glazes have no sharp melting tem-

Figure 21–15 Kaolinitic clays with substantial percentages of iron are common. These clays are used to make bricks, and the iron makes the bricks red, which explains the color scheme of many city streets.

Figure 21–16 Glassblowing. When softened by heat, masses of glass can be formed into "balloons" by blowing air into them.

perature; rather, they harden and develop resistance to shear stresses increasingly as the temperature of the high-fired clay object is gradually reduced. Glazes generally are aluminosilicates that have high aluminum content to raise their viscosity and thereby reduce the tendency to run off the surface during firing. They also provide the means of coloring the surfaces of fired clays and imparting decorative designs to them. Again, transition-metal oxides (particularly those of Ti, V, Cr, Mn, Fe, Co, Ni, and Cu) are responsible for the colors. The oxidation state of the transition metal in the glaze is critical in determining the color and is controlled by regulating the composition of the atmosphere in the kiln; atmospheres rich in oxygen give high oxidation states, and those poor or lacking in oxygen give low oxidation states.

Glass

Glassmaking probably originated in the Near East about 3500 years ago. It is one of the oldest domestic arts, and its beginnings, like those of metallurgy, are obscure. Both required high-temperature charcoal-fueled ovens and vessels made of materials that did not easily melt in order to initiate and contain the necessary chemical reactions. A great advance was the invention of glassblowing, which probably occurred in the first century B.C. The glassblower dipped a long iron tube into molten glass and rotated it to cause a rough ball of semisolid material to accumulate on the end. Blowing into the iron tube forced the soft glass to take the form of a hollow ball that could be further shaped into a vessel and severed from the blowing tube with a blade (Fig. 21–16). Artisans in the Roman Empire developed glassblowing to a high degree, but with the decline of that civilization, glassmaking in Europe deteriorated until the Venetians redeveloped the lost techniques a thousand years later.

Glasses are amorphous solids of widely varying composition. Outwardly, glasses often resemble crystalline solids and have mechanical properties similar to those of crystals. On a molecular level, however, glasses resemble liquids in structure, but with the diffusional motions of the molecules brought to a halt. As in crystals, cohesion in glasses may depend on van der Waals forces, Coulomb (ionic) forces, covalent bonds, or metallic bonds. However, in this chapter we use the term "glass" in the restricted and familiar sense to refer to materials formed from silica, usually in combination with metal oxides. The absence of long-range order in glasses makes them isotropic in their physical properties—that is, their properties are the same in all directions. This has certain advantages in technology, including, for ex-

• Glassblowing is possible only because glass has no sharp melting point. Instead, there is a reasonably wide temperature range over which glass is sufficiently plastic to be worked.

Newly formed automobile windows are inspected after careful annealing to eliminate internal stress.

ample, the fact that glasses expand uniformly in all directions with an increase in temperature. The mechanical strength of glass is intrinsically very high, even exceeding the tensile strength of steel provided that the surface is free of scratches and other imperfections. Flaws in the surface provide sites where fractures can start when the glass is stressed. When a glass object of intricate shape and nonuniform thickness is suddenly cooled, internal stresses are locked in. These stresses may be relieved catastrophically when the object is later heated or struck (Fig. 21–17). Heating a strained object to a temperature somewhat below its softening point and holding it there for a while before allowing it to cool slowly, a technique called **annealing,** gives short-range diffusion of atoms a chance to occur and to eliminate internal stresses.

The softening and annealing temperatures of a glass and other properties, such as its density, depend upon its chemical composition (Table 21–2). Silica itself (SiO_2) forms a glass if it is heated above its melting point and then cooled rapidly to avoid crystallization. The resulting **vitreous** (glassy) silica has limited use because the high temperatures required to shape it make it quite expensive. Sodium silicate glasses form in the high-temperature reaction of silica sand with anhydrous sodium carbonate (soda ash, Na_2CO_3):

$$Na_2CO_3(s) + n\ SiO_2(s) \longrightarrow Na_2O{\cdot}nSiO_2(s) + CO_2(g)$$

The melting point of the nonvolatile product is about 900°C, and the glassy state results if cooling through that temperature is rapid. The product, called "water glass," is water soluble and thus unsuitable for making vessels. Its aqueous solutions are used in some detergents and as adhesives for sealing cardboard boxes, however.

Figure 21–17 Internal stresses cause a delicate glass teardrop to shatter when its tail is merely scratched.

Table 21–2 Composition and Properties of Various Glasses				
Silica Glass	**Soda-Lime Glass**	**Borosilicate Glass**	**Aluminosilicate Glass**	**Leaded Glass**
Composition				
SiO_2, 99.9%	SiO_2, 73%	SiO_2, 81%	SiO_2, 63%	SiO_2, 56%
H_2O, 0.1%	Na_2O, 17%	B_2O_3, 13%	Al_2O_3, 17%	PbO, 29%
	CaO, 5%	Na_2O, 4%	CaO, 8%	K_2O, 9%
	MgO, 4%	Al_2O_3, 2%	MgO, 7%	Na_2O, 4%
	Al_2O_3, 1%		B_2O_3, 5%	Al_2O_3, 2%
Coefficient of linear thermal expansion ($°C^{-1} \times 10^7$)[a]				
5.5	93	33	42	89
Softening point (°C)				
1580	695	820	915	630
Annealing point (°C)				
1050	510	565	715	435
Density (g cm^{-3})				
2.20	2.47	2.23	2.52	3.05

[a]The coefficient of linear thermal expansion is defined as the fractional increase in length of a body when its temperature is increased by 1°C.

An insoluble glass with useful structural properties results if lime (CaO) is added to the sodium carbonate–silica starting materials. **Soda-lime glass** is the resulting product, with the approximate composition $Na_2O \cdot CaO \cdot (SiO_2)_6$. Soda-lime glass is easy to melt and shape and is used in many applications, ranging from bottles to window glass. It accounts for over 90% of all the glass manufactured today. The structure of this ionic glass is shown schematically in Figure 21–18. It is a three-dimensional network of the type discussed in Section 21–2 but with random coordination of the silicate tetrahedra. The network is a giant "polyanion," with Na^+ and Ca^{2+} ions distributed in the void spaces to compensate for the negative charge on the network.

Replacement of lime and some of the silica in a glass by other oxides (Al_2O_3, B_2O_3, K_2O, or PbO) modifies its properties noticeably. For example, the thermal conductivity of ordinary (soda-lime) glass is quite low, and its coefficient of thermal expansion is high. This means that internal stresses are set up when its surface is subjected locally to extreme heat or cold; the glass may shatter. The coefficient of thermal expansion is sharply lowered in certain borosolicate glasses, in which many of the silicon sites are occupied by boron. Pyrex, the most familiar of these glasses, has a coefficient of linear expansion about one third that of soda-lime glass. It is the material of choice for laboratory glassware and household ovenware. A

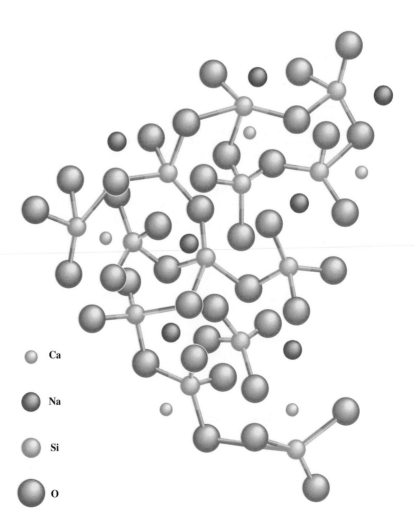

Figure 21–18 The structure of a soda-lime glass. Note the tetrahedral coordination of oxygen atoms around each silicon atom.

Ca

Na

Si

O

Figure 21-19 Fiberoptic lighting from the holes in the pedestals enhances this display of amphoras.

glass called Vycor, which has an even smaller coefficient of expansion (approximating that of fused silica), is made by chemical treatment of borosilicate glass to leach out its sodium. This leaves a porous structure that is densified by raising the temperature and shrinking the glass to its final volume.

A major use of glass is in the transmission of voice messages, television images, and data as light pulses. Tens of thousands of audio messages can be transmitted simultaneously through glass fibers no larger in diameter than a hair by encoding the audio signal into electronic impulses that modulate light from a laser source. The light passes down the glass fiber as though it were a tube. Chemical control of the composition of the glass reduces light loss and permits messages to pass over distances of many kilometers without amplification. Glass fiber also offers greater flexibility in spot illumination. (Fig. 21–19).

Cements

Hydraulic cement was first developed by the Romans, who discovered that a mixture of lime (CaO) and dry volcanic ash reacts slowly with water, even at low temperatures, to form a durable solid. They used this knowledge to build the Roman Pantheon, a circular building in which the concrete dome, spanning 143 feet without internal support, still stands nearly 2000 years after its construction! The knowledge of cementmaking was lost for centuries after the fall of the Roman Empire. It was rediscovered in 1824 by an English bricklayer, Joseph Aspdin, who patented a process for calcining a mixture of limestone and clay. He called the product **portland cement** because, when mixed with water, it hardened to a material that resembled a kind of limestone found on the Isle of Portland. Portland cement is now manufactured in every major country, and world-wide production exceeds that of any other material. Portland cement opened up a new age in constructing highways and buildings. Rock could be crushed and then molded into cement rather than shaped with cutting tools.

Portland cement is a finely ground (powdered) mixture of compounds produced by the high-temperature reaction of lime, silica, alumina, and iron oxide. The lime (CaO) may come from limestone or chalk deposits. The silica (SiO_2) and alumina (Al_2O_3) are often obtained in clays or slags. The blast furnaces of steel mills are a common source of slag, which is a by-product of the smelting of iron ore (see Section 12–1). The composition of slag varies, but it can be represented as a calcium aluminum silicate of approximate formula $CaO \cdot Al_2O_3 \cdot (SiO_2)_2$. This material is crushed and ground to a fine powder, blended with lime in the correct proportion, and burned again in a horizontal rotary kiln at temperatures up to 1500°C to produce "cement clinker." A final stage of grinding and the addition of about 5% by mass of gypsum ($CaSO_4 \cdot 2H_2O$) to lengthen the setting time completes the process of manufacture.

• Cement kilns in the United States are also used for the disposal by burning of wastes such as used oils and solvents.

Table 21–3
Typical Chemical Composition of Portland Cement

Oxide	Percentage by Mass
Lime (CaO)	63.8
Silica (SiO_2)	21.3
Alumina (Al_2O_3)	4.6
Sulfur trioxide (SO_3)	2.9
Magnesium oxide (MgO)	2.8
Iron(III) oxide (Fe_2O_3)	2.6
Minor oxides (Na_2O, K_2O)	0.6
Water and other volatiles	1.3

The composition of a typical portland cement is shown in Table 21–3. That table gives percentages of the separate oxides; these simple materials, which are the "elements" of cementmaking, combine in the cement in more complex compounds such as tricalcium silicate, $(CaO)_3 \cdot SiO_2$, dicalcium silicate, $(CaO)_2 \cdot SiO_2$, tricalcium aluminate, $(CaO)_3 \cdot Al_2O_3$, and tetracalcium aluminum ferrite $(CaO)_4 \cdot Al_2O_3 \cdot Fe_2O_3$.

Cement *sets* when the semiliquid slurry first formed by the addition of water to the powder becomes a solid of low strength. Subsequently, it gains strength in a slower *hardening* process. Setting and hardening involve a complex group of exothermic reactions in which several hydrated compounds form. Portland cement is a **hydraulic cement** because it hardens not by loss of water but by chemical reactions that incorporate water into the final body. The main reaction during setting is the hydration of the tricalcium aluminate, which can be approximated by the equation

$$(CaO)_3 \cdot Al_2O_3(s) + 3\ (CaSO_4 \cdot 2H_2O)(s) + 26\ H_2O(\ell) \longrightarrow$$
$$(CaO)_3 \cdot Al_2O_3 \cdot (CaSO_4)_3 \cdot 32H_2O(s)$$

* "Drying of cement" is thus a somewhat misleading phrase.

The product forms after 5 or 6 hours as a microscopic forest of long crystalline needles that lock together to solidify the cement. Later, the calcium silicates react with water to harden the cement. An example is

$$6\ (CaO)_3 \cdot SiO_2(s) + 18\ H_2O(\ell) \longrightarrow (CaO)_5 \cdot (SiO_2)_6 \cdot 5H_2O(s) + 13\ Ca(OH)_2(s)$$

The hydrated calcium silicates develop as strong tendrils that coat and enclose unreacted grains of cement, each other, and other particles that may be present, binding them in a robust network. Most of the strength of cement comes from these entangled networks, which in turn depend ultimately for strength on chains of —O—Si—O—Si silicate bonds. In the final stages of hardening, the iron-containing compounds react with water. Hardening is slower than setting; it may take as long as a year for a cement to attain its ultimate strength.

Portland cement is rarely used alone. Generally, it is combined with sand, water, and lime to make **mortar,** which is applied with a trowel to bond bricks or stones. When portland cement is mixed with sand and aggregate (crushed stone or pebbles) in the proportions of 1:3.75:5 by volume, the mixture is called **concrete.** Concrete is outstanding in its resistance to compressive forces; it therefore finds primary use for the foundations of buildings and the construction of dams, in which the compressive loads are enormous. The stiffness (resistance to bending) of con-

Architectural forms executed in concrete in a California shopping mall.

crete is high, but its fracture toughness (resistance to impact) is substantially lower, and its tensile strength is relatively poor. For this reason, concrete is usually reinforced with steel rods when it is used in structural elements such as beams that are subject to transverse or tensile stresses.

The water required for chemical combination in the hardening of cement is only about 20% of its mass. Far more water than this is ordinarily added to cement powders to make the slurry easier to work. As the excess water evaporates during hardening, pores form that typically comprise 25 to 30% of the volume of the solid. This porosity weakens concrete, and recent research has shown that the fracture strength is related inversely to the size of the largest pores in the cement. A new material, called "macro-defect-free" (MDF) cement, has pores only a few micrometers across (rather than about a millimeter as in conventional cement). The smaller pores result from the addition of water-soluble polymers that make a dough-like "liquid" cement that is pliable with use of far less water. Unset MDF cement is mechanically kneaded and extruded into the desired shape. The final result possesses substantially increased bending resistance and fracture toughness. MDF cement can even be molded into springs and shaped on a conventional lathe. Reinforcement with fibers further increases its toughness.

21-5 NONSILICATE CERAMICS

Many useful ceramics exist that are *not* based on the Si—O bond and the SiO_4 tetrahedron. They have important uses in electronics, optics, and the chemical industry. Some of these materials are oxides, but others contain neither silicon nor oxygen.

Oxide Ceramics

Oxide ceramics are materials that contain oxygen in combination with any of a number of metals. These materials are named by adding an *-ia* ending to the stem of the name of whatever element is mainly present. Thus, if the main chemical component of an oxide ceramic is BeO, it is a *beryllia* ceramic; if the main component is Y_2O_3, it is an *yttrria* ceramic; and if it is MgO, it is a *magnesia* ceramic. As Table 21–4 shows, the melting points of these and other oxides are substantially higher than the melting points of the elements themselves. Such high temperatures are hard to achieve and maintain, and the molten oxides corrode most container materials. Oxide ceramic bodies are therefore not shaped by melting the oxide and pouring it into a

Table 21–4
Melting Points of Some Metals and Their Oxides

Metal	Melting Point (°C)	Oxide	Melting Point (°C)
Be	1287	BeO	2570
Mg	651	MgO	2800
Al	660	Al_2O_3	2051
Si	1410	SiO_2	1723
Ca	865	CaO	2572
Y	1852	Y_2O_3	2690

mold. Instead, these ceramics are fabricated by sintering, like the silicate ceramics. Oxide ceramic bodies are rarely single pure compounds. When ceramics are fabricated from alumina, for example, high-purity Al_2O_3 is prepared and then deliberately doped with about 1% MgO by mass. The MgO promotes a very fine-grained structure during firing, and the resulting piece is stronger.

Alumina (Al_2O_3) is the most important nonsilicate ceramic material. It melts at a temperature of 2051°C, and it retains strength even at temperatures of 1500 to 1700°C. Alumina has a large electrical resistivity and withstands both thermal shock and corrosion well. These properties make it a good material for spark plug insulators, and most spark plugs now use a ceramic that is 94% alumina.

High-density alumina is fabricated in such a way that open pores between the grains are nearly completely eliminated, and the grains are small, having an average diameter as low as 1.5 μm. Unlike most other ceramics, high-density alumina has good mechanical strength, particularly against impact, which has led to its use in armor plating. The ceramic, suitably backed by a fracture-resistant framing, absorbs the energy of an impacting projectile by breaking, so penetration does not occur. High-density alumina is also used in high-speed cutting tools for machining metals. Such a use is perhaps surprising, but the temperature resistance of the ceramic allows much faster cutting speeds, and a ceramic cutting edge has no tendency to weld to the workpiece, as metallic tools do. High-density alumina is also used in artificial joints (Fig. 21–20).

If Al_2O_3 doped with a small percentage of MgO is fired in a vacuum or in a hydrogen atmosphere (instead of air) at a temperature of 1800 to 1900°C, even very small pores, which scatter light and make the material white, are removed. The resulting ceramic is translucent. The invention of this material has changed the color of streetlights all over the world. It is used to contain the sodium in high-intensity sodium-discharge lamps (Fig. 21–21). With envelopes of high-density alumina, such lamps can be operated at temperatures of 1500°C to give a whiter light and more of it. Old-style sodium-vapor lamps with glass envelopes were limited to a temperature of 600°C because the sodium vapor corroded the glass envelope at higher temperatures. The light from a low-temperature sodium-vapor lamp has an undesirable yellow color.

Magnesia (MgO) is mainly used as a **refractory,** which is a ceramic material that withstands a temperature of more than 1500°C without melting (MgO melts at 2800°C). A major use of magnesia is as insulation in electrical heating devices, because it combines high thermal conductivity with excellent electrical resistance. Magnesia is prepared from magnesite ores, which consist of $MgCO_3$ and a variety of impurities. When purified magnesite is heated to 800 to 900°C, carbon dioxide is driven off to leave fine grains of MgO(s):

$$MgCO_3(s) \longrightarrow MgO(s) + CO_2(g)$$

After cooling, fine-grained MgO reacts avidly with water to form magnesium hydroxide:

$$MgO(s) + H_2O(\ell) \longrightarrow Mg(OH)_2(s)$$

It is therefore called "caustic magnesia." Heating fine-grained MgO(s) to 1700°C causes the MgO grains to sinter. Fully sintered MgO consists of large crystals and does not react with water. It is "dead-burned magnesia." The $\Delta G°$ of the reaction between MgO and H_2O does not change when magnesia is dead burned. The altered microstructure (larger crystals versus small) makes the reaction with water exceedingly slow, however.

* By comparison, pure iron melts at 1535°C, and steel melts at a temperature well below 1500°C.

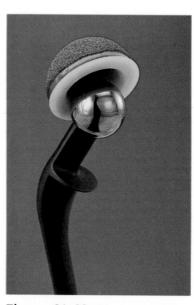

Figure 21–20 The socket in this artificial hip is made of high-density alumina.

Beryllia (BeO) resembles alumina in many of its properties. It is less dense (3.00 g cm^{-3} versus 3.97 g cm^{-3}), however, which makes it attractive in aviation and space. Working with beryllia requires special care because the inhaled dust causes the lung disease berylliosis. Beryllia ceramics also tend to react with water vapor at high temperature, forming poisonous vapors.

Nonoxide Ceramics

In nonoxide ceramics, nitrogen or carbon takes the place of oxygen in combination with silicon or boron. Examples are boron nitride (BN), boron carbide (B_4C), the silicon borides (SiB_4 and SiB_6), silicon nitride (Si_3N_4), and silicon carbide (SiC). All of these compounds possess strong, short covalent bonds and find use as abrasives and cutting tools. They are hard and strong, but brittle. Table 21–5 lists the enthalpies of the chemical bonds in these compounds. Forming these materials into useful finished shapes depends on high-purity starting materials and precise control of the temperature and atmosphere during firing.

Much research has aimed at making gas turbine and other engines out of ceramics. Heat engines run more efficiently at higher temperatures, and above 1000°C certain ceramics have better properties (strength, corrosion resistance, density) for engines than do metals in all respects except their intrinsic brittleness. Of the oxide ceramics, only alumina and zirconia (ZrO_2) are strong enough, but both resist thermal shock too poorly for this application. Attention has therefore turned to the nonoxide **silicon nitride** (Si_3N_4). Silicon nitride is a covalent network solid (Fig. 21–22). In the structure, every silicon atom bonds to four nitrogen atoms that surround it at the corners of a tetrahedron, and each nitrogen atom is surrounded by three silicon atoms. The similarity to the way in which SiO_4 units join together in networks in the silicate minerals (see Section 21–2) is clear.

At first, silicon nitride seems chemically unpromising as a high-temperature structural material. It is unstable in contact with water because the reaction

$$Si_3N_4(s) + 6\ H_2O(\ell) \longrightarrow 3\ SiO_2(s) + 4\ NH_3(g)$$

has a negative $\Delta G°$ at room temperature. In fact, this reaction causes finely ground Si_3N_4 powder to give off a perceptible odor of ammonia in moist air at room temperature. Silicon nitride is also thermodynamically unstable in air, reacting spontaneously with oxygen:

$$Si_3N_4(s) + 3\ O_2(g) \longrightarrow 3\ SiO_2(s) + 2\ N_2(g)$$

(text continued on page 856)

Figure 21–21 The translucent white inner envelope of high-density alumina in this high-intensity sodium-vapor lamp is visible through the outer envelope of glass.

Table 21–5
Bond Enthalpies in Nonoxide Ceramics

Bond	Bond Enthalpy (kJ mol^{-1})
B—N	389
B—C	448
Si—N	439
Si—C	435
B—Si	289
C—C	350

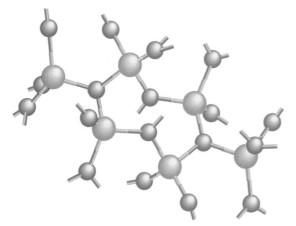

Figure 21–22 The structure of silicon nitride (Si_3N_4). Each Si atom is bonded to four nitrogen atoms, and each N atom is bonded to three Si atoms. The result is a strong network.

CHEMISTRY IN PROGRESS

The oxide ceramics discussed in this section are all oxides of single metals. A natural idea for new ceramics is to make materials containing two (or more) oxides in equal or nearly equal molar amounts. Thus, if $BaCO_3$ and TiO_2 are mixed and heated to high temperature, they react to give the ceramic barium titanate:

$$BaCO_3(s) + TiO_2(s) \longrightarrow BaTiO_3(s) + CO_2(g)$$

Barium titanate is a **mixed-oxide ceramic** (writing the formula as $BaO \cdot TiO_2$ shows this). It has the same structure as the mineral **perovskite** ($CaO \cdot TiO_2$) (Fig. 21–B), except, of course, that Ba replaces Ca. Perovskites typically have two metal atoms for every three oxygen atoms, giving them a general formula of ABO_3, where A stands for a metal atom at the center of the unit cube and B for an atom of a different metal at the cube corners.

Recently, interest in perovskite ceramic compositions has grown explosively following the discovery that some

Figure 21–C   The levitation of a small magnet above a disk of superconducting material. A superconducting material cannot be penetrated by an external magnetic field. At room temperature, the magnet rests on the ceramic disk. When the disk is cooled with liquid nitrogen, it becomes superconducting and excludes the magnet's field, forcing the magnet into the air. Here the magnet floats above a disk of chilled 1-2-3 ceramic.

of them become **superconducting** at relatively high temperatures. A superconducting material offers no resistance whatever to the flow of an electric current. The phenomenon was discovered by the Dutch physicist Kammerlingh-Onnes in 1911 when he cooled mercury below its superconducting transition temperature of 4 K. Such very low temperatures are difficult to achieve and maintain, but if practical superconductors could be made to work at higher temperatures (or even at room temperature!), then power transmission, electronics, transportation, medicine, and many other aspects of human life would be transformed. Over 60 years of research with metallic systems culminated in 1973 in the discovery of a niobium–tin alloy with a world-record superconducting transition temperature of 23.3 K. Progress toward higher-temperature superconductors then stalled until 1986, when K. Alex

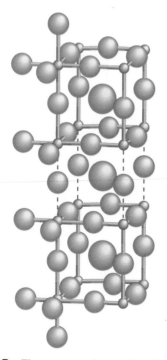

Figure 21–B   The structure of perovskite ($CaTiO_3$). A stack of three unit cells is shown, with some additional O atoms from neighboring cells. Each Ti (blue) is surrounded by 6 O's; each Ca (green) has 12 O's as nearest neighbors.

Superconducting Ceramics

Müller and J. Georg Bednorz, who had had the inspired notion to look for higher transition temperatures among perovskite ceramics, found a Ba-La-Cu-O perovskite phase that had a transition temperature of 35 K. This result motivated other scientists, who soon discovered another rare-earth–containing perovskite ceramic that became a superconductor at 90 K. This result was particularly exciting because 90 K exceeds the boiling point of liquid nitrogen (77 K), a relatively cheap refrigerant (Fig. 21–C). This 1-2-3 compound (so called because its formula $YBa_2Cu_3O_{(9-x)}$ has one yttrium, two

barium, and three copper atoms per formula unit) is not an ideal perovskite. If it were, it would have nine oxygen atoms in combination with its six metal atoms. Instead, a number of the oxygen sites in the solid are vacant. The deficiency makes x in the formula somewhat greater than 2, depending on the exact method of preparation. The 1-2-3 compound is a nonstoichiometric solid whose structure is given in Figure 21–D.

A crucial objective in the application of superconducting ceramics is to fabricate the new materials in desired shapes (tapes, for example, Fig. 21–E).

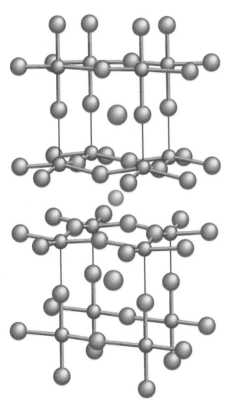

Figure 21–D The structure of the 1-2-3 compound $YBa_2Cu_3O_{(9-x)}$ derives from that of perovskite, as can be seen by comparison to the structure in Figure 21–B. It is a layered perovskite in which one third of the layers contain yttrium (orange) and two thirds contain barium (green). Every third Ba in the hypothetical "$BaCuO_3$" is replaced by a Y, and the O's in the layer occupied by the Y are removed. Note the puckering in the Cu—O layers, as well.

Figure 21–E Winding superconducting tape. This is quite a challenge because these superconductors have the brittleness and fragility typical of ceramics. Superconducting wires and tapes as long as 100 m are now commercially available.

The reaction has a room temperature $\Delta G°$ of -1927 kJ mol^{-1}. Both of these reactions are thermodynamically favored at higher temperature as well. In practice, neither reaction occurs at a perceptible rate when Si_3N_4 is in bulk form. Initial contact of oxygen or water with $Si_3N_4(s)$ forms a surface film of $SiO_2(s)$ that protects the bulk of the $Si_3N_4(s)$ from further attack. When *strongly* heated in air (to about 1900°C), Si_3N_4 does decompose, violently, but below that temperature, it resists attack.

Fully dense silicon nitride parts are stronger than metallic alloys at high temperatures. Ball bearings made of fully dense silicon nitride work well without lubrication at temperatures up to 700°C and last longer than steel ball bearings. Because the strength of silicon nitride shapes increases with the density attained in the production process, the trick is to form a dense piece of silicon nitride in the desired shape. Reaction bonding is one method for making useful shapes of silicon nitride. Powdered silicon is compacted in molds, removed, and then fired under an atmosphere of nitrogen at 1250 to 1450°C. The reaction

$$3 \, Si(s) + 2 \, N_2(g) \longrightarrow Si_3N_4(s)$$

forms the ceramic. The parts neither swell nor shrink significantly during the chemical conversion from Si to Si_3N_4, making it possible to fabricate complex shapes reliably. Unfortunately, reaction-bonded Si_3N_4 is still somewhat porous and is not strong enough for many applications. In the "hot-pressed" forming process, Si_3N_4 powder is prepared in the form of exceedingly small particles by reaction of silicon tetrachloride with ammonia:

$$3 \, SiCl_4(g) + 4 \, NH_3(g) \longrightarrow Si_3N_4(s) + 12 \, HCl(g)$$

The solid Si_3N_4 forms as a smoke. It is captured as a powder, mixed with a carefully controlled amount of MgO additive, placed in an enclosed mold, and sintered at 1850°C under a pressure of 230 atm. The resulting ceramic shrinks to nearly full density (no pores). Because the material does not flow well (to fill a complex mold completely), only simple shapes are possible. Hot-pressed silicon nitride is impressively tough and can be machined only with great difficulty, with diamond tools.

In the silicate minerals discussed in Section 21–2, AlO_4 units routinely substitute for SiO_4 tetrahedra as long as positive ions of some type are present to balance the electric charge. This fact suggests that in silicon nitride some Si^{4+} ions, which lie at the centers of tetrahedra of nitrogen atoms, could be replaced by Al^{3+} if a compensating replacement of O^{2-} for N^{3-} were simultaneously made. Experiments show that ceramic alloying of this type works well, giving many new ceramics with great potential called **sialons.** The discovery of these ceramics illustrates how a structural theme in a naturally occurring material can guide the search for new materials.

Boron has one fewer valence electron than carbon, and nitrogen has one more valence electron. **Boron nitride** (BN) is therefore isoelectronic with carbon, and it is not surprising that it has two structural modifications that resemble the structures of graphite and diamond. In hexagonal boron nitride, the boron and nitrogen atoms take alternate places in extended "chicken-wire" sheets. The sheets stack up in such a way that each boron atom has a nitrogen atom directly above it and directly below it, and vice versa. In graphite (see Fig. 20–30), the carbon atoms lie in similar sheets, but these sheets are offset so that half of the atoms lie on lines between the centers of the hexagons in the sheets above and below. The cubic form of boron nitride has the diamond structure, is comparable in hardness to diamond, and resists oxidation better. Boron nitride is often prepared by **chemical vapor deposition,** a procedure used in the fabrication of several other ceramics as well. In this method,

• This is an excellent example of a kinetic barrier preventing a reaction. It is similar to the well-known corrosion protection that a layer of Al_2O_3 confers on metallic aluminum.

• The name "sialon" comes from the chemical symbols of the four elements involved: Si, Al, O, and N (Si-Al-O-N).

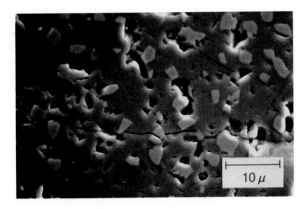

Figure 21–23 Microstructure in a composite ceramic. The light-colored areas are TiB_2, the gray areas are SiC, and the dark areas are voids. The TiB_2 acts to toughen the SiC matrix. Note that the crack shown passing through from left to right is forced to deflect around the TiB_2 particles.

a controlled chemical reaction of gases on a contoured, heated surface gives a solid product of the desired shape. If a cup made of BN is needed, a cup-shaped mold is heated to a temperature exceeding 1000°C, and a mixture of $BCl_3(g)$ and $NH_3(g)$ is passed over its surface. The reaction

$$BCl_3(g) + NH_3(g) \longrightarrow BN(s) + 3\ HCl(g)$$

deposits a cup-shaped layer of BN(s). Boron nitride cups and tubes are used to contain and evaporate molten metals.

Silicon carbide (SiC) is diamond in which half of the C atoms are replaced by Si atoms. Also known by its trade name of Carborundum, silicon carbide was originally developed as an abrasive, but it is now used primarily as a refractory and as an additive in steel manufacture. It is formed and densified by methods similar to those used with silicon nitride. It is a candidate material for high-temperature engines, and much research has been performed in this area. Silicon carbide is often produced in the form of small plates or whiskers. In these forms it works well to reinforce other ceramics. Fired silicon carbide whiskers are quite small (0.5 μm in diameter and 50 μm long) but very strong. They are mixed with a second ceramic material before that material is formed. Firing then gives a **composite ceramic.** Such composites are stronger and tougher than unreinforced bodies of the same primary material. Whiskers serve to reinforce the main material by stopping cracks; they either deflect advancing cracks or soak up their energy, and a widening crack must dislodge them to proceed (Fig. 21–23).

SUMMARY

21-1 Elemental silicon forms a covalently bonded solid in which all the electrons are localized between pairs of atoms, filling the **valence band** and leaving the **conduction band** virtually empty. The result is a low electrical conductivity, which can be increased by heating the solid or illuminating it, thereby exciting electrons across the **band gap.** Other **intrinsic semiconductors** are made by combining elements to give crystals with the same number of valence electrons per atom as silicon (four). An important example is gallium arsenide. An **insulator** such as diamond has a larger band gap than a semiconductor, so its conduction band is unoccupied by electrons at ordinary temperatures. By contrast, the conduction band in a metal is partly filled and electrons can change levels with minimal input of energy. The conductivities of silicon and other intrinsic semiconductors can be increased by **doping,** the deliberate addition of impurity atoms of selected types. If a Group V element is added to pure silicon, the excess electrons occupy levels just below the conduction band, giving an ***n*-type semiconductor.** If atoms of a Group III element substitute for silicon atoms, they make a ***p*-type semiconductor,** in which electrical conduction occurs by the hopping of positive **holes** in the valence band when a voltage is applied.

21–2 Silicates are minerals containing tetrahedral SiO_4 structural units. A variety of silicate structures occur in nature, ranging from single tetrahedra to pairs, rings, chains, sheets, and three-dimensional networks of linked tetrahedra, with cations to balance the total charge in the crystal. In **aluminosilicates,** aluminum atoms substitute for some of the tetrahedrally bonded silicon atoms, and additional cations are present as well to maintain a net charge of zero. In **clay** minerals, water can be absorbed between sheets of bonded atoms and cause swelling. **Zeolites** are network aluminosilicates having pores on a molecular scale. They are useful for exchanging ions, for separating mixtures of gases, and as catalysts.

21–3 Ceramics are synthetic inorganic non-metallic materials. They resist high temperatures and corrosion and are stiff and hard but also brittle, making them liable to sudden failure when stressed. The **microstructures** of these materials are complex, with grains of different phases, cracks, and pores. Ceramics are made by **firing** to a high temperature, which leads to partial melting, to crystal phase changes, and to **sintering,** in which grains grow together and the ceramic body shrinks or **densifies.**

21–4 Many **silicate ceramics** use natural clay minerals such as kaolin as their starting material. A **glaze** is a thin, glassy layer that coats the object, giving it strength and also often decorating it. Different compositions of silicate **glass** are used for different applications. Pure silica (SiO_2) has a high melting point that is lowered by the addition of sodium oxide and calcium oxide to give ordinary **soda-lime glass.** Another silicate ceramic is **portland cement,** a complex mixture of compounds produced from oxides of calcium, aluminum, silicon, and iron, to which calcium sulfate is added.

21–5 Oxide ceramics that do not contain silicon have high melting points and resist corrosion, making them good **refractories. High-density alumina,** a material fabricated with minimal pore size, is useful in cutting tools and armor plating. **Magnesia** combines very high thermal conductivity and heat resistance with very low electrical conductivity. **Nonoxide ceramics** have strong covalently bonded network structures. They include **silicon nitride,** a strong, tough material with uses in the structural components of gas turbines. **Sialons** are ceramic alloys in which some aluminum replaces silicon in silicon nitride with the simultaneous replacement of some nitrogen by oxygen. **Boron nitride** is used for high-temperature ceramic vessels. **Silicon carbide,** or Carborundum, is an abrasive that is now used to reinforce other ceramics. Some oxide ceramics have been found to be **superconducting,** losing all resistance to the flow of an electric current at a certain temperature.

PROBLEMS

Note: Answers to blue-numbered problems are given in Appendix F. Problems that are more challenging are indicated with asterisks.

Semiconductors

1. (See Example 21–1.) Electrons in a semiconductor can be excited from the valence band to the conduction band through the absorption of photons with energies exceeding the band gap. At room temperature, indium phosphide (InP) is a semiconductor that absorbs light only at wavelengths less than 920 nm. Calculate the band gap in InP.

2. (See Example 21–1.) Both GaAs and CdS are semiconductors that are being studied for possible use in solar cells to generate electric current from sunlight. Their band gaps are 2.29×10^{-19} J and 3.88×10^{-19} J, respectively, at room temperature. Calculate the longest wavelength of light that is capable of exciting electrons across the band gap in each of these substances. In which region of the electromagnetic spectrum do these wavelengths fall? Use this result to explain why CdS-based sensors are used in some cameras to estimate the proper exposure conditions.

3. The number of electrons excited to the conduction band by random thermal agitation per cubic centimeter in a semiconductor can be estimated from the equation

$$n_e = (4.8 \times 10^{15} \text{ cm}^{-3} \text{ K}^{-3/2}) \, T^{3/2} e^{-E_g/(2RT)}$$

where T is the temperature in kelvins and E_g the band gap in joules *per mole*. The band gap of diamond at 300 K is 8.7×10^{-19} J. How many electrons are thermally excited to the conduction band at this temperature in a 1.00-cm^3 diamond crystal?

4. The band gap of pure crystalline germanium is 1.1×10^{-19} J at 300 K. How many electrons are excited from the valence band to the conduction band in a 1.00-cm^3 crystal of germanium at 300 K? Use the equation given in the preceding problem.

5. Describe the nature of electrical conduction in (a) Si doped with P and (b) InSb doped with Zn.

6. Describe the nature of electrical conduction in (a) Ge doped with In and (b) CdS doped with As.

7. In a light-emitting diode (LED) of the type is used in displays on electronic equipment, watches, and clocks, a voltage is imposed across an *n-p* semiconductor junction. The electrons on the *n* side combine with the holes on the *p* side and emit light at the frequency of the band gap. This process can also be described as the emission of light as electrons fall from levels in the conduction band to empty levels in the valence band. It is the reverse of the production of electric current by illumination of a semiconductor.

   Many LEDs are made from semiconductors having the general composition GaAs$_{1-x}$P$_x$. When x is varied between 0 and 1, the band gap changes, and with it the color of light emitted by the diode. When $x = 0.4$, the band gap is 2.9×10^{-19} J. Determine the wavelength and color of the light emitted by this LED.

8. When the LED described in the previous problem has the composition GaAs$_{0.14}$P$_{0.86}$ (that is, $x = 0.86$), the band gap has increased to 3.4×10^{-19} J. Determine the wavelength and color of the light emitted by this LED.

9. The pigment zinc white (ZnO) turns bright yellow when heated, but the white color returns when the sample is cooled. Does the band gap increase or decrease when the sample is heated?

10. Mercury(II) sulfide (HgS) exists in two different crystalline forms. In cinnabar, the band gap is 3.2×10^{-19} J; in metacinnabar, it is 2.6×10^{-19} J. In some old paintings with improperly formulated paints, the pigment vermilion (cinnabar) has transformed to metacinnabar upon exposure to light. Describe the color change that results.

Silicate Minerals

11. Draw a Lewis electron-dot structure for the disilicate ion (Si$_2$O$_7^{6-}$). What changes in the structure you have drawn would be necessary to give the structure of the diphosphate ion (P$_2$O$_7^{4-}$) and the disulfate ion (S$_2$O$_7^{2-}$)? What is the analogous compound of chlorine?

12. Draw a Lewis electron-dot structure for the cyclosilicate ion Si$_6$O$_{18}^{12-}$, which forms part of the structure of beryl and emerald.

13. (See Example 21–2.) By referring to Table 21–1, predict the structure of each of the following silicate minerals (network, sheets, double chains, and so forth). Give the oxidation state of each atom.
    (a) Andradite (Ca$_3$Fe$_2$(SiO$_4$)$_3$)
    (b) Vlasovite (Na$_2$ZrSi$_4$O$_{10}$)
    (c) Hardystonite (Ca$_2$ZnSi$_2$O$_7$)
    (d) Chrysotile (Mg$_3$Si$_2$O$_5$(OH)$_4$)

14. (See Example 21–2.) By referring to Table 21–1, predict the structure of each of the following silicate minerals (network, sheets, double chains, and so forth). Give the oxidation state of each atom.
    (a) Tremolite (Ca$_2$Mg$_5$(Si$_4$O$_{11}$)$_2$(OH)$_2$)
    (b) Gillespite (BaFeSi$_4$O$_{10}$)
    (c) Uvarovite (Ca$_3$Cr$_2$(SiO$_4$)$_3$)
    (d) Barysilate (MnPb$_8$(Si$_2$O$_7$)$_3$)

15. By referring to Table 21–1, predict the structure of each of the following aluminosilicate minerals (network, sheets, double chains, and so forth). In each case, the aluminum atoms grouped with the silicon and oxygen in the formula substi-

tute for Si in tetrahedral sites. Give the oxidation state of each atom.

(a) Keatite ($Li(AlSi_2O_6)$)

(b) Muscovite ($KAl_2(AlSi_3O_{10})(OH)_2$)

(c) Cordierite ($Al_3Mg_2(AlSi_5O_{18})$)

16. By referring to Table 21–1, predict the structure of each of the following aluminosilicate minerals (network, sheets, double chains, and so forth). In each case, the aluminum atoms grouped with the silicon and oxygen in the formula substitute for Si in tetrahedral sites. Give the oxidation state of each atom.

(a) Amesite ($Mg_2Al(AlSiO_5)(OH)_4$)

(b) Phlogopite ($KMg_3(AlSi_3O_{10})(OH)_2$)

(c) Thomsonite ($NaCa_2(Al_5Si_5O_{20})\cdot 6H_2O$)

17. Calcite ($CaCO_3$) reacts with quartz to form wollastonite ($CaSiO_3$) and carbon dioxide.

(a) Write a balanced chemical equation for this reaction.

(b) Use the data in Appendix D to calculate $\Delta H°$ and $\Delta S°$ at 25°C for this reaction.

(c) Estimate the temperature at which $\Delta G° = 0$ ($K = 1$) for this reaction.

18. Tremolite ($Ca_2Mg_5Si_8O_{22}(OH)_2$) reacts with calcite ($CaCO_3$) and quartz to give diopside ($CaMgSi_2O_6$), water vapor, and carbon dioxide.

(a) Write a balanced chemical equation for this reaction.

(b) At 25°C, the standard enthalpies of formation of tremolite and diopside are $-12,360$ and -3206.2 kJ mol^{-1}, respectively. Their standard entropies are 548.9 and 142.93 J K^{-1} mol^{-1} at 25°C. Use this information, together with data from Appendix D, to calculate $\Delta H°$ and $\Delta S°$ for this reaction.

(c) Estimate the temperature at which $\Delta G° = 0$ ($K = 1$) for this reaction.

The Properties of Ceramics

19. Ceramics and metals can each have great mechanical strength, but in different ways. Discuss.

20. Compare the effect of slow and rapid heating on a ceramic with the effects on a metal.

21. Compare the ways in which ceramic materials are shaped with the methods used for shaping a metal like steel.

22. Compare the microstructure of a typical ceramic material with that of a metal.

Silicate Ceramics

23. A ceramic that has been much used by artisans and craftsmen for the carving of small figurines is based upon the mineral steatite (commonly known as soapstone). Steatite is a hydrated magnesium silicate that has the composition $Mg_3Si_4O_{10}(OH)_2$. It is remarkably soft—a fingernail can scratch it. When heated in a furnace to about 1000°C, chemical reaction transforms it into a hard two-phase composite of magnesium silicate ($MgSiO_3$) and quartz in much the same way that clay minerals are converted into mullite and cristo-

balite on firing. Write a balanced chemical equation for this reaction.

24. A clay mineral that is frequently used together with or in place of kaolinite is pyrophyllite ($Al_2Si_4O_{10}(OH)_2$). Write a balanced chemical equation for the production of mullite and cristobalite on the firing of pyrophyllite.

25. Calculate the volume of carbon dioxide produced at 0°C and 1 atm pressure when a sheet of ordinary glass of mass 2.50 kg is made from its starting materials—sodium carbonate, calcium carbonate, and silica. Take the composition of the glass to be $Na_2O\cdot CaO\cdot(SiO_2)_6$.

26. Calculate the volume of steam produced when a 4.0-kg brick made from pure kaolinite is completely dehydrated at 600°C and a pressure of 1.00 atm.

27. A sample of soda-lime glass for drinking glasses is analyzed and found to contain the following percentages by mass of oxides: SiO_2, 72.4%; Na_2O, 18.1%; CaO, 8.1%; Al_2O_3, 1.0%; MgO, 0.2%; BaO, 0.2%. (The elements are not actually present as binary oxides, but this is the way compositions are usually given.) Calculate the chemical amounts of silicon, sodium, calcium, aluminum, magnesium, and barium atoms per mole of oxygen atoms in this sample.

28. A sample of portland cement is analyzed and found to contain the following percentages by mass of oxides: CaO, 64.3%; SiO_2, 21.2%, Al_2O_3, 5.9%; Fe_2O_3, 2.9%; MgO, 2.5%; SO_3, 1.8%; Na_2O, 1.4%. Calculate the chemical amounts of calcium, silicon, aluminum, iron, magnesium, sulfur, and sodium atoms per mole of oxygen atoms in this sample.

29. The most important contributor to the strength of hardened portland cement is tricalcium silicate ($(CaO)_3\cdot SiO_2$), for which the measured standard enthalpy of formation at 25°C is -2929.2 kJ mol^{-1}. Calculate the standard enthalpy change for the production of 1.00 mol of tricalcium silicate from quartz and lime.

30. One of the simplest of the heat-generating reactions that takes place when water is added to cement is the production of calcium hydroxide (slaked lime) from lime. Write a balanced chemical equation for this reaction, and use data from Appendix D to calculate the amount of heat generated by the reaction of 1.00 kg of lime with water at room conditions.

Nonsilicate Ceramics

31. Silicon carbide (SiC) is made by the high-temperature reaction of silica sand (quartz) with coke; the by-product is carbon monoxide.

(a) Write a balanced chemical equation for this reaction.

(b) Estimate the standard enthalpy change per mole of $SiC(s)$ produced.

(c) Predict (qualitatively) the following physical properties of silicon carbide: conductivity, melting point, and hardness.

32. Boron nitride (BN) is made by the reaction of gaseous boron trichloride with gaseous ammonia.

(a) Write a balanced chemical equation for this reaction.

(b) Calculate the standard enthalpy change per mole of BN produced, given that the standard molar enthalpy of formation of BN(s) equals -254.4 kJ mol^{-1} at 25°C.

(c) Predict (qualitatively) the following physical properties of boron nitride: conductivity, melting point, and hardness.

33. The standard Gibbs function of formation of cubic silicon carbide (SiC) at 25°C is -62.8 kJ mol^{-1}. Determine $\Delta G°$ when 1.00 mol of SiC(s) reacts with oxygen to form SiO_2 (*quartz*) and $CO_2(g)$. Is silicon carbide thermodynamically stable in air at room conditions?

34. The standard Gibbs function of formation of boron carbide (B_4C) is -71 kJ mol^{-1} at 25°C. Determine $\Delta G°$ when 1.00 mol of $B_4C(s)$ reacts with oxygen to form $B_2O_3(s)$ and $CO_2(g)$. Is boron carbide thermodynamically stable in air at room conditions?

35. Calculate the average oxidation number of the copper in $YBa_2Cu_3O_{(9-x)}$ if $x = 2$. Assume that the rare-earth element yttrium is in its usual $+3$ oxidation state.

36. The mixed oxide ceramic $Tl_2Ca_2Ba_2Cu_3O_{(10+x)}$ has zero electrical resistance at 125 K. Calculate the average oxidation number of the copper in this compound if $x = 0.50$ and thallium is in the $+3$ oxidation state.

Additional Problems

37. Compare the hybridization of silicon atoms in Si(s) with that of carbon atoms in graphite (see Fig. 20–30). If silicon were to adopt the graphite structure, would its electrical conductivity be high or low?

38. Describe how the band gap varies from a metal to a semiconductor to an insulator.

39. Suppose that some people are sitting in a row at a movie theater, with a single empty seat on the left end of the row. Every 5 minutes, a person moves into a seat on his or her left if it is empty. In what direction and with what speed does the empty seat "move" along the row? Comment on the connection with hole motion in p-type semiconductors.

*40. A sample of silicon doped with antimony is an n-type semiconductor. Suppose that a small amount of gallium is added to such a semiconductor. Describe how the conduction properties of the solid vary with the amount of gallium added.

41. Silica (SiO_2) exists in several forms, including quartz (molar volume 22.69 cm^3 mol^{-1}) and cristobalite (molar volume 25.74 cm^3 mol^{-1}).

(a) Use data from Appendix D to calculate $\Delta H°$, $\Delta S°$, and $\Delta G°$ for the conversion of quartz to cristobalite at 25°C.

(b) Which form is thermodynamically stable at 25°C?

(c) Which form is stable at very high temperatures, provided that melting does not take place first?

42. Predict the structure of each of the following silicate minerals (network, sheets, double chains, and so forth). Give the oxidation state of each atom.

(a) Apophyllite ($KCa_4(Si_8O_{20})F\cdot 8H_2O$)

(b) Rhodonite ($CaMn_4(Si_5O_{15})$)

(c) Margarite ($CaAl_2(Al_2Si_2O_{10})(OH)_2$)

43. By referring to Table 21–1, predict the structure of manganpyrosmilite, a silicate mineral with the chemical formula $Mn_{12}FeMg_3(Si_{12}O_{30})(OH)_{10}Cl_{10}$. Give the oxidation state of each atom in this formula unit.

*44. A reference book lists the chemical formula of one form of vermiculite as

$$[(Mg_{2.36}Fe_{0.48}Al_{0.16})(Si_{2.72}Al_{1.28})O_{10}(OH)_2]$$
$$[Mg_{0.32}(H_2O)_{4.32}]$$

Determine the oxidation state of the iron in this mineral.

45. When talc ($Mg_3Si_4O_{10}(OH)_2$) reacts with forsterite (Mg_2SiO_4), enstatite ($MgSiO_3$) and water vapor form.

(a) Write a balanced chemical equation for this reaction.

(b) If these minerals are all present at equilibrium and the pressure is increased, is there net formation of products or net formation of reactants?

46. The most common feldspars are those containing potassium, sodium, and calcium cations. They are called, respectively, orthoclase ($KAlSi_3O_8$), albite ($NaAlSi_3O_8$), and anorthite ($CaAl_2Si_2O_8$). The solid solubility of orthoclase in albite is limited, and its solubility in anorthite is almost negligible. Albite and anorthite, however, are completely miscible at high temperatures and show complete solid solution. Give an explanation for these observations, based on the tabulated radii of the K^+, Na^+, and Ca^{2+} ions, which are 1.33 Å, 0.98 Å, and 0.99 Å, respectively.

47. The clay mineral kaolinite ($Al_2Si_2O_5(OH)_4$) is formed by the weathering action of water containing dissolved carbon dioxide on the feldspar mineral anorthite ($CaAl_2Si_2O_8$). Write a balanced chemical equation for the reaction that occurs. The CO_2 forms H_2CO_3 as it dissolves. As the pH is lowered, does the weathering occur to a greater or a lesser extent?

48. The formulas of certain zeolites can be written as $M_2O\cdot Al_2O_3\cdot ySiO_2\cdot wH_2O$ where M is an alkali metal such as Na or K, y is 2 or more, and w is any integer. Compute the mass percentage of Al in a zeolite that has M = K, $y = 4$, and $w = 6$.

49. Iron oxides are red when the average oxidation state of iron is high and black when it is low. To impart each of these colors to a pot made from clay that contains iron oxides, would you use an air-rich or a smoky atmosphere in the kiln? Explain.

50. In what ways does soda-lime glass resemble and in what ways does it differ from a pot made from the firing of kaolinite? Include the following aspects in your discussion: composition, structure, physical properties, and method of preparation.

51. Leaded glass is used for fine crystal in place of soda-limesilica glass. Table 21–2 shows that it is "heavier" (denser); it also has a higher refractive index. The latter property gives the leaded glass its attractive sparkle. What is the ratio of the number of lead atoms to silicon atoms for leaded glass of the composition listed in Table 21–2?

*52. The cement industry is one of the nation's largest consumers of energy because of the high temperatures needed to form portland cement from its oxide starting materials. One way that has been proposed to reduce energy costs is to incorporate calcium fluoride as a flux in the oxide mixture, in much the same way that limestone is used in steel production (see Section 12–1). What effect would the CaF_2 have, and how would energy consumption be reduced as a result?

53. Burnt dolomite bricks are used in the linings of furnaces in the cement and steel industries. Pure dolomite contains 45.7% $MgCO_3$ and 54.3% $CaCO_3$ by mass. Determine the empirical formula of dolomite.

54. Refractories can be classified as acidic or basic, depending on the properties of the oxides in question. A basic refractory must not be used in contact with acid, and an acidic refractory must not be used in contact with base. Classify magnesia and silica as acidic or basic refractories.

*55. A book states correctly that beryllia (BeO) ceramics have some use but possess only poor resistance to strong acids and bases. It goes on to state that beryllia reacts with water vapor at high temperature to give off toxic vapors. Write likely chemical equations for the reaction of BeO with a strong acid, with a strong base, and with steam, $H_2O(g)$.

56. Some ceramics containing thoria (ThO_2) have been made. Identify a likely problem in the use of such materials.

57. Silicon nitride resists all acids except hydrofluoric, with which it reacts to give silicon tetrafluoride and ammonia. Write a balanced chemical equation for this reaction.

58. Compare oxide ceramics like alumina and magnesia, which have significant ionic character, with covalently bonded nonoxide ceramics like SiC and B_4C (problems 33 and 34), with respect to thermodynamic stability at ordinary conditions.

59. A new inorganic polymer is developed. It is made from plentiful starting materials by simple mixing, followed by heating to high temperature. When hot, it is readily shaped and molded. When cooled, it solidifies to a hard, transparent, brittle solid that can be cut and polished easily. It is completely non-flammable, resists attack by most chemicals, can be sterilized easily for medical uses, and is far more durable than organic polymers ("plastics"). However, its manufacture consumes much energy, and it is not biodegradable. Its brittleness leads to occasional catastrophic loss of structural integrity in use. These episodes create sharp-edged fragments. Sooner or later, everybody using this ceramic will probably suffer injuries from it. As a high government official you must decide whether to license or forbid the production of this material. What do you do? Explain.

Chemical Processes

A plant used in the production of ammonia from nitrogen and hydrogen via the Haber–Bosch process.

At the heart of chemistry lie the chemical processes that transform sets of starting materials into sets of products. We introduce in this chapter a **process-based** approach to chemistry by looking first at the nature of the chemical industry and its major products, emphasizing aspects such as the availability of raw materials and the effect of pollutants on the environment. The rest of the chapter then applies the process-based approach to the large-scale production of compounds of sulfur, phosphorus, and nitrogen and describes some of their uses in the modern world.

22-1 THE CHEMICAL INDUSTRY

A firm grasp of chemical principles is essential to the design of clean and efficient manufacturing processes for chemical products. Industrial processes are under constant development to improve product yield, reduce waste, and minimize expense. Chemical plants are costly, and a new process invariably passes through several **pilot-plant** stages after its development in the research laboratory. In these scaling-up operations, the process is optimized with respect to such variables as temperature, pressure, and time of reaction. Almost all laboratory-scale chemical processes are **batch processes,** in which the reactants are mixed in a container and allowed to react, and the products are removed in a sequence of operations. Many industrial-scale processes, in contrast, are **continuous processes,** designed so that reactants are fed in and products are removed continuously. It is rarely a simple matter to convert a process from batch to continuous operation, but the latter type is generally more economical to run.

Processes in chemical industry can be represented schematically (Fig. 22–1) as

$$\text{starting materials} \longrightarrow \text{desired products} + \text{by-products}$$

Energy, often a great deal of it, plays a part in these transformations. Sometimes energy must be put in to carry out the desired reaction; in other cases it is produced, usually as heat, and must be either recycled or safely discarded. Let us examine each of the terms in this schematic process in light of the basic principles of chemistry.

Starting Materials. In the ideal case, the starting materials for a chemical process are easily obtainable and easily transportable to the site where they will be used. Starting materials tend to have low Gibbs functions simply because substances of high Gibbs function have mostly disappeared, over the long history of the earth, in spontaneous reactions to give substances of low Gibbs function. Thus, aluminum is abundant in the earth's crust but is never found in elemental form. Rather. it is tied up in materials of much lower Gibbs function such as kaolinite, a clay with the chemical formula $Al_2Si_2O_5(OH)_4$ and with a large negative Gibbs function of formation ($\Delta G_f^\circ = -3800$ kJ mol^{-1}). Some compounds of high Gibbs function persist in nature because kinetics is involved as well as thermodynamics; their rates of reaction in their geological deposits are very slow. These compounds are particularly valuable resources for chemistry. Hydrocarbons such as octane occur in petroleum deposits and have high Gibbs functions relative to their combustion products:

$$C_8H_{18}(g) + \tfrac{25}{2} O_2(g) \longrightarrow 9\, H_2O(\ell) + 8\, CO_2(g) \qquad \Delta G^\circ = -5272 \text{ kJ}$$

The list of uses for hydrocarbons is long. It is unfortunate that so much of the world's limited supply of hydrocarbons is being burned as fuel rather than conserved for future use as chemical starting materials for polymers, fibers, detergents, and other goods.

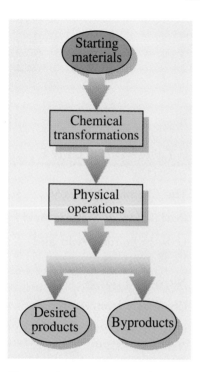

Figure 22–1 A generalized chemical process. This figure employs the color conventions that are followed through the next chapters. Starting materials (red) are converted to products (green) via a series of chemical reactions (blue) and physical operations (orange). Filtration and evaporation are examples of physical operations.

• The earth, taken as a whole, is not at equilibrium. As long as energy in the form of solar radiation continues to arrive, new local build-ups of compounds with high Gibbs function are possible.

• The processing and uses of hydrocarbons are considered in Chapter 24.

Energy. A desired but non-spontaneous reaction (one with $\Delta G > 0$) can in principle be caused to proceed by linking it to another reaction for which $\Delta G < 0$—for instance, the combustion of a fossil fuel such as coal, oil, or natural gas. In many chemical processes, the linkage is completed using electricity. The powerful oxidizing (bleaching) agent sodium chlorate ($NaClO_3$), for example, is made by the overall reaction

$$NaCl(aq) + 3\ H_2O(\ell) \longrightarrow NaClO_3(aq) + 3\ H_2(g) \qquad \Delta G° = +835\ kJ$$

The reaction is driven by an external source of electrical work that is generated by burning fossil fuels or harnessing the energy of a nuclear reaction. Many chemical transformations are practical only if inexpensive energy sources are available. It is very desirable to use the energy from the sun directly to carry out chemical reactions. Nature, of course, does just this in photosynthesis, the process in which light is harvested by chlorophyll in green plants and used to convert water and carbon dioxide (which have low Gibbs functions) to glucose and other sugars (which have high Gibbs functions). Much current research is aimed at understanding photosynthesis and copying its principles to carry out other chemical transformations.

Desired Products. In an ideal chemical process, the desired product would be obtained (1) in 100% yield at room temperature and pressure, (2) at a convenient and controllable rate, and (3) already pure or in a condition that allows easy separation and purification. Obviously, compromises are necessary in real processes. The yield of a chemical reaction is fundamentally limited by thermodynamics, so conditions of pressure and temperature must be adjusted to optimize it, often by installing expensive equipment. The rate of a reaction depends on pressure and temperature but is also sensitive to the presence of catalysts. The control of reaction rates is so important in the chemical industry that researchers search constantly for catalysts that speed the formation of desired products without leading to unwanted by-products. Finally, the last point—separability of the desired product—cannot be overemphasized. A process that yields a product contaminated with large amounts of hard-to-separate (or dangerous) impurities may face abandonment, even if it gives good yields at a convenient rate.

• Good design of a chemical process requires provisions to stop runaway reactions. Unexpected acceleration of the main reaction or sudden onset of an unanticipated side-reaction (caused perhaps by impurities in the starting materials) can occur.

By-products. In some cases, the by-products of chemical processes are relatively harmless, but the question of what to do with them remains. Each year, a small mountain of $CaSO_4$ is produced in the course of synthesizing phosphate fertilizers; although $CaSO_4$ is harmless in itself (it is used in gypsum wallboard), large volumes of it create awkward and expensive disposal problems. Other by-products are toxic to plant and animal life, ranging from slightly irritating to deadly. In the past, some toxic wastes were buried in landfills. This practice risked the contamination of water sources and the direct exposure of nearby residents and is now thoroughly regulated. Strong social, economic, and legal incentives now exist to reduce or eliminate worthless and hazardous by-products; the current concept is that pollution and the cost of its removal are product defects. The United States National Science Foundation and the Council for Chemical Research support a joint program of research in "environmentally benign chemical synthesis and processing," and a recent initiative by the National Science Foundation funds research on "environmentally conscious manufacturing." This means the reduction of environmental pollution at the source by avoiding toxic feedstocks and solvents and by eliminating or combining process steps. Parts of the chemical industry have responded by modifying their processes to minimize the generation of wastes. Other measures that diminish adverse environmental impact are the recycling of materials like glass and paper and the incineration of flammable wastes. When adequate precautions are taken to en-

• Major efforts are being made in the United States and other countries to clean up older toxic waste sites that pose continuing threats to the environment and public health.

• Even complete combustion has its negative side because large-scale emission of CO_2 contributes to the greenhouse effect described in Section 18-5.

Figure 22–2 A process flow diagram for the synthesis of aspirin. The ultimate precursors of aspirin are air, water, salt, sulfur, petroleum, and natural gas. Many of the intermediates are organic chemicals that appear in Chapter 24; their names and chemical formulas need not be learned at this stage.

sure complete combustion, incineration converts even very hazardous hydrocarbons into carbon dioxide and water vapor. One experimental method for the disposal of hazardous wastes injects them into an electrically generated plasma in which the temperature exceeds 10,000°C. Another uses oxygen dissolved in supercritical water to oxygenate pollutants and produce benign smaller molecules (see Chemistry in Your Life: An Exotic (But Useful) State of Matter in Chapter 6). Under these conditions all organic molecules fragment. Wastes that are not combustible can generally be treated chemically to make them nonhazardous.

The ideal way to dispose of by-products is to find a use for them. One example comes from the uranium fuel–processing industry, in which scrap uranium contaminated with dirt and grease is oxidized by nitric acid to uranyl nitrate ($UO_2(NO_3)_2$), which is soluble in water and can therefore be purified. The oxidation reaction gives off the nitrogen oxides NO and NO_2, which are pollutants. Vented

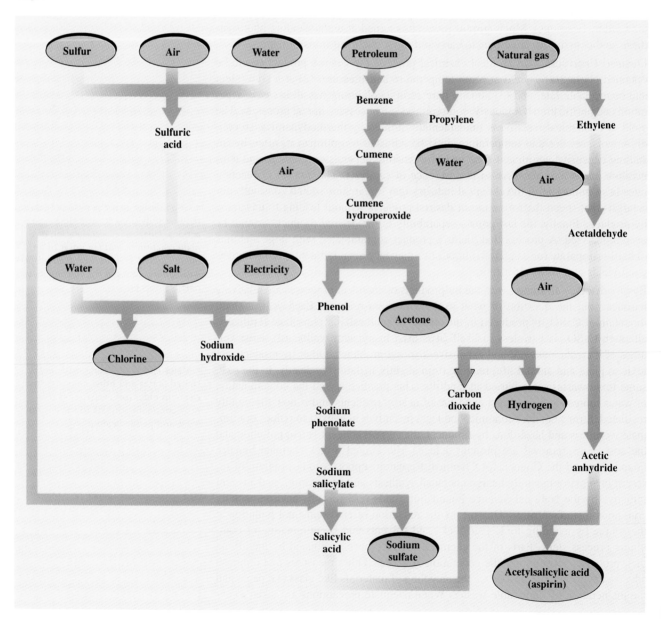

as waste, the gases would contribute to acid rain (see Section 18–5). Instead, they are caused to react first with oxygen and then with water to give nitric acid:

$$NO(g) + \tfrac{1}{2} O_2(g) \longrightarrow NO_2(g)$$

$$3\ NO_2(g) + H_2O(\ell) \longrightarrow 2\ HNO_3(aq) + NO(g)$$

The use of these reactions offers two advantages: reduction of the amount of nitrogen oxides released to the atmosphere and recycling of the nitric acid product for use in the oxidation reaction. A cost saving also results.

Few chemical processes occur in one stage. Most require many steps, with numerous chemicals involved along the way that do not appear in the finished product. Figure 22–2 shows how aspirin is made from primary raw materials: air, water, salt, sulfur, petroleum, and natural gas. Numerous chemicals appear only as intermediates. Aspirin contains no sulfuric acid, but sulfuric acid plays an essential role in several steps in its manufacture. The figure also shows the interdependency built into the chemical industry. An aspirin manufacturer does not purchase salt, sulfur, and the other primary materials but rather buys intermediate chemicals that may have applications in a variety of other fields. Instead of manufacturing sulfuric acid in small quantities, an aspirin company buys it from a supplier who makes far larger quantities—for use by steel mills, for example.

Table 22–1 shows the amounts of the largest-volume chemicals produced in the United States during 1995. Few are used *directly* as consumer products. Household

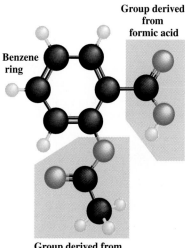

Group derived from formic acid

Benzene ring

Group derived from acetic acid

The structure of acetylsalicylic acid, the active component of aspirin. Note the substructures, which are those of three simpler compounds: benzene, formic acid, and acetic acid.

Table 22–1
Chemicals Produced in Largest Volume in the United States (1995)

Rank	Name	Formula	U.S. Production (billions of kg)	Principal End Use
1	Sulfuric acid	H_2SO_4	43.2	Fertilizers, chemicals, processing
2	Nitrogen	N_2	30.9	Fertilizers
3	Oxygen	O_2	24.3	Steel, welding
4	Ethylene	C_2H_4	21.3	Plastics, antifreeze
5	Lime	CaO	18.7	Paper, chemicals, cement
6	Ammonia	NH_3	16.1	Fertilizers
7	Phosphoric acid	H_3PO_4	11.9	Fertilizers
8	Sodium hydroxide	NaOH	11.9	Chemical processing, aluminum production, soap
9	Propylene	C_3H_6	11.6	Gasoline, plastics
10	Chlorine	Cl_2	11.4	Bleaches, plastics, water purification
11	Sodium carbonate	Na_2CO_3	10.1	Glass
12	Methyl tert-butyl ether	$C_4H_9OCH_3$	7.99	Gasoline additive
13	Ethylene dichloride	$C_2H_4Cl_2$	7.83	Plastics, drycleaning
14	Nitric acid	HNO_3	7.82	Fertilizers, explosives
15	Ammonium nitrate	NH_4NO_3	7.25	Fertilizers, mining

Not included as chemicals are steel and concrete and minerals such as salt (NaCl), gypsum ($CaSO_4$), and sulfur (S). Adapted from *Chemical and Engineering News*, April 8, 1996, p. 16.

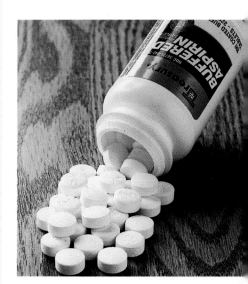

Aspirin tablets.

Figure 22–3 Several of the important raw materials for chemistry. (a) Coal, mined from both open pits and tunnels. (b) Limestone, being quarried here as a source of lime. (c) Water, needed in large quantities and in pure form for many chemical processes.

ammonia, for example, is a negligible fraction of the total ammonia produced, most of which undergoes further chemical transformations for use as fertilizer in forms that include ammonium nitrate, urea, and ammonium phosphate. The following chapters discuss some of the methods by which the large-volume chemicals are produced. Of course, chemicals do not have to be produced in mountainous quantities to be important. For example, pharmaceuticals are vital to health care even though they are produced in relatively small amounts.

Raw Materials

The raw materials of the heavy-chemicals industry (Figure 22–3) are relatively few in number, chiefly air, water, limestone, salt, coal, natural gas and petroleum, sulfur, and phosphate rock, simply because these are abundantly available at low cost. Their sources are listed in Table 22–2. Both geochemical and biochemical transformations are responsible for the generation of economically useful quantities of these raw materials near the earth's surface.

Geochemical transformations are slow. Time scales range from years (for the weathering of the softest rocks) to millions and even billions of years. One reason for this slowness is that geochemical processes are frequently solid-state reactions, and the diffusion of atoms through solids is many orders of magnitude slower than that through liquids and gases. Geochemical transformations begin with the crystallization of a **magma,** a molten silicate fluid forced out from beneath the earth's surface, to form **igneous** rocks and to vent volatile components, such as water, sulfur dioxide, and the oxides of nitrogen, that might be trapped or dissolved in the magma. Although there is some variation in the composition of magmas, the elements oxygen, silicon, and aluminum predominate, with smaller amounts of iron, magnesium, calcium, sodium, and potassium. As a magma cools, it crystallizes in stages, giving minerals with different compositions and structures. Although aluminum, iron, and magnesium are abundant in igneous rocks, they are so tightly

Table 22–2
Sources of Major Raw Materials for Chemistry

Atmosphere	Hydrosphere
Oxygen (O_2)	Water (H_2O)
Nitrogen (N_2)	Sodium bromide (NaBr)
Noble gases	Sodium iodide (NaI)
	Magnesium chloride ($MgCl_2$)
	Sodium chloride (NaCl)

Lithosphere	Biosphere
Silica (SiO_2)	Coal
Limestone ($CaCO_3$)	Petroleum
Sodium chloride (NaCl)	Natural gas
Potassium chloride (KCl)	Sulfur (S)
Sodium carbonate (Na_2CO_3)	Phosphate rock ($Ca_5(PO_4)_3F$)
Sodium sulfate (Na_2SO_4)	Organic natural products
Metal ores	

The atmosphere is the envelope of gases that surrounds the earth. The lithosphere (from the Greek *lithos*, meaning "stone") includes the outer parts of the solid earth. The hydrosphere refers to water in its many forms. The biosphere refers to living organisms, their immediate surroundings, and their products.

bound in minerals that the rocks are not economic sources for these elements. Minerals in igneous rocks do provide some of the less common metals, however, such as molybdenum, tin, and uranium.

Sedimentary rocks form from the weathering of other rocks. Water (especially moving water in streams) breaks rocks into smaller pieces and dissolves them partially. Carbon dioxide, which forms carbonic acid when dissolved, causes further dissolution and chemical reaction. Quartz, a form of silica (SiO_2), is left behind in

A lava flow at the Kilauea rift in Hawaii. Lava is a molten silicate fluid that forms igneous rock as it solidifies.

The foxglove plant produces digitalis, an important drug in the treatment of heart disease.

• Hence the name "wood alcohol."

the weathering process and provides an important source of silicon. Aluminum precipitates in clays and in the mineral bauxite (hydrated Al_2O_3), and iron forms oxide ores. Calcium carbonate ($CaCO_3$) precipitates as limestone. More soluble salts remain in solution and can be found in the oceans, in salt brines in deep wells, and as solid deposits where the water has evaporated. Thus, sylvite (KCl) is primarily obtained as a mineral deposit, and $MgCl_2$ is extracted from brines or from the ocean. Both brines and minerals are important sources of sodium chloride.

The minerals that make up the earth's crust continually undergo structural and chemical transformations. Heat and pressure below the earth's surface cause **metamorphic** rocks to form. The conversion of limestone into marble is a change in texture caused by metamorphism. These two materials have the same chemical composition, but limestone is coarsely crystalline, whereas marble is a compact, coherent mass of small crystals. As metamorphism proceeds, the elemental composition of the metamorphosed rocks becomes more uniform. High-grade metamorphism merges eventually (at high temperatures) with magma formation.

If geochemical processes are characterized by their slowness, biochemical processes are notable for their speed and selectivity. Natural products obtained from plants and animals play a large role in chemistry, being used both directly and as precursors for chemical syntheses of other substances. The terpenes, for example, are a class of compounds found in almost all plants; they are a major component of turpentine, which is obtained from the resin of pine trees. Many terpenes are used directly as artificial scents in the perfume industry; others serve as precursors of making vitamin A (Fig. 22–4).

The distinction between chemical and biological synthesis has become blurred with time. On the one hand, there is increasing use of living agents such as specialized bacteria to carry out the synthesis of complex products that are difficult to produce by conventional chemical means. On the other hand, certain natural products are more easily obtained by synthesizing them from other starting materials than by extracting them from traditional sources. Methanol, for example, was once obtained by the destructive distillation of wood but it is now produced mainly through the reaction of synthesis gas outlined in Section 7–5.

Figure 22–4   (a) The terpene β-ionone ($C_{13}H_{20}O$) is used extensively in perfumes. It is also a precursor in the synthesis of vitamin A. (b) The structure of vitamin A derives from that of β-ionone. It resembles the structures of retinal (crucial for vision, Problem 18–65) and the dye β-carotene (see Fig. 18–24). Carrots in the diet supply vitamin A for metabolic transformation to retinal. This explains why eating carrots can improve night vision.

The fossil fuels are among the major chemical raw materials produced by biological activity. The decay of plant matter mainly through the action of bacterial reducing agents creates first peat, then lignite, and finally bituminous coal. Crude oil (petroleum) is thought to have formed from the decomposition under reducing conditions of the remains of small marine organisms. Hydrocarbons of low molar mass formed in this way are gases rather than liquids. Pockets of natural gas (primarily methane, with smaller amounts of hydrogen, ethane, and propane) therefore are found in association with deposits of petroleum. Although all of these fossil fuels have a biological origin, it is obvious that geological transformations brought them to the form in which we find them today.

A chemical raw material with a somewhat more unexpected biological origin is sulfur. Although sulfur is widely distributed in compounds, its most convenient commercial source is the large deposits of elemental sulfur often associated with salt domes under the surface of the earth. It is thought that sulfur precipitated in these settings as calcium sulfate, in contact with natural gas and carbon dioxide. Reaction between these substances was then catalyzed by enzymes secreted by certain bacteria to give elemental sulfur and calcium carbonate:

$$4\ CaSO_4(s) + 3\ CH_4(g) + CO_2(g) \longrightarrow 4\ CaCO_3(s) + 4\ S(s) + 6\ H_2O(\ell)$$

The origins of phosphate rock, which has the approximate chemical formula $Ca_5(PO_4)_3F$, are less clear. The rock is found in large deposits in certain parts of the world, notably Florida and North Africa. Phosphate ions are present in rather low concentration in seawater, and the solubility of calcium phosphate is comparable to that of calcium carbonate. Therefore, it is surprising that the phosphate is found in such a concentrated form, rather than scattered through the world's limestone deposits. Deposits of phosphate rock may have resulted from extensive dissolving and reprecipitation of phosphate–carbonate mixtures, perhaps in the small, but critical, pH range in which the phosphate precipitates and the carbonate remains in solution. It is more likely, however, that the phosphate rock resulted from accumulation by marine organisms. Phosphate rock is the source of phosphorus and phosphate-based fertilizers. As such, it is the raw material used in greatest volume in the chemical industry, as Section 22–3 develops further.

22-2 PRODUCTION OF SULFURIC ACID

The consumption of sulfuric acid, the chemical produced in the greatest amount in the world today, has come to indicate the health not only of a country's chemical industry but of its entire manufacturing base. The uses of this strongly acidic, extremely reactive compound range from fertilizer manufacture to metal treatment to the production of pharmaceuticals. Its central role in the modern chemical industry was established largely between 1750 and 1900. In 1750, sulfuric acid was produced and used on a very modest scale in metal assaying and treatment, and by some doctors for "cures" that had no scientific basis. New uses for the acid stimulated the search for new ways to produce it; as the price went down, further uses were discovered and exploited. From 1750 to 1900, the price of sulfuric acid declined steadily, and the amount produced grew explosively. This mutual stimulation between new uses and new processes is the hallmark of the chemical industry, and the story of sulfuric acid is a good way to illustrate it.

Sulfuric acid was probably first generated by alchemists who heated crystalline green vitriol, or iron(II) sulfate heptahydrate, in a retort:

$$FeSO_4 \cdot 7H_2O(s) \longrightarrow H_2SO_4(\ell) + FeO(s) + 6\ H_2O(g)$$

- The iron in FeS_2 is in the $+2$, not the $+4$, oxidation state because the sulfur occurs as S_2^{2-} ions, analogous to the peroxide ion O_2^{2-}.

By the 17th century, it was produced on a limited commercial scale from iron pyrite ores. The FeS_2 in these ores was converted to $FeSO_4$ and then oxidized to iron(III) sulfate ($Fe_2(SO_4)_3$). This solid was broken up and strongly heated in small clay vessels, a treatment that decomposed it into iron(III) oxide and sulfur trioxide:

$$Fe_2(SO_4)_3(s) \longrightarrow Fe_2O_3(s) + 3\ SO_3(g)$$

The gas produced was then absorbed in water to make sulfuric acid:

$$SO_3(g) + H_2O(\ell) \longrightarrow H_2SO_4(aq)$$

Production of sulfuric acid was stimulated by the discovery in 1744 that the valued dyestuff indigo, used since antiquity to dye cotton, could also be used as a wool dye after treatment with concentrated sulfuric acid.

- Blue jeans are dyed with indigo.

The Lead-Chamber Process for Sulfuric Acid

The lead-chamber process, which was developed in the second half of the 18th century, probably also had its origins in the laboratories of alchemists, who burned sulfur in earthenware vessels. The small amounts of SO_3 that were produced (alongside the SO_2 that was the main product) were condensed and absorbed into water to make sulfuric acid. An accidental discovery revealed that the addition of sodium or potassium nitrate improved the yield of SO_3. These salts decompose to give nitrogen dioxide, which reacts with SO_2 to give SO_3:

$$SO_2(g) + NO_2(g) \longrightarrow SO_3(g) + NO(g)$$

In 1736, Joshua Ward took the next important step by replacing the earthenware vessels with large glass bottles arranged in series to speed up the process.

The scale of manufacture of sulfuric acid was increased further, and dramatically, by the development of the room-sized **lead chamber,** first used by John Roebuck in 1746. The output of one of the older earthenware vessels was several ounces, and Ward's glass bottles had outputs on the order of several pounds. By contrast, the lead chamber allowed sulfuric acid to be produced in amounts of hundreds of pounds to tons, reducing the price through economy of scale. In the lead-chamber process, a mixture of sulfur and potassium nitrate was placed in a ladle and ignited inside a large chamber lined with lead, the floor of which was covered with water. The gases condensed on the walls and were absorbed in the water. After the process was repeated several times, the dilute sulfuric acid was removed and concentrated by boiling it. Later developments included blowing in steam to speed up the reaction with water and disperse the gases, and separating the burning chamber from the absorption chamber.

Joseph Gay-Lussac took a significant step forward in 1835, when he built a tower to recover the NO that had previously been vented and to convert it back to NO_2 by reaction with oxygen. More precisely, in a Gay-Lussac tower, NO was converted to nitrous acid (HNO_2) dissolved in aqueous sulfuric acid:

$$2\ NO(g) + \tfrac{1}{2}\ O_2(g) + H_2O(\ell) \longrightarrow 2\ HNO_2(aq)$$

This then reacted in a second tower, named after its developer, John Glover, to oxidize sulfur dioxide:

$$2\ HNO_2(aq) + SO_2(g) \longrightarrow H_2SO_4(aq) + 2\ NO(g)$$

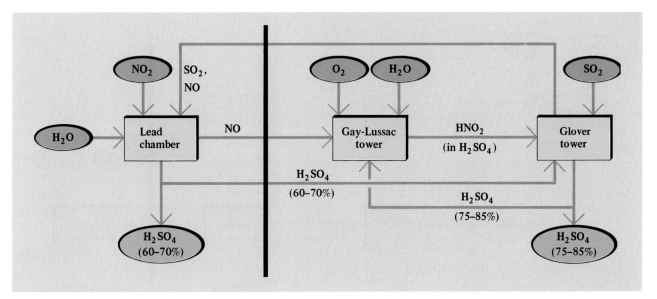

Figure 22–5 The Glover–Gay-Lussac process for making sulfuric acid. The portion of the diagram to the left of the black line shows the original lead-chamber process, which produced a weaker acid and consumed and emitted more nitrogen oxides than the full process as it later developed.

so the overall reaction from the two steps was simply

$$SO_2(g) + \tfrac{1}{2} O_2(g) \longrightarrow SO_3(g)$$

Recycling the oxides of nitrogen greatly reduced the consumption of sodium or potassium nitrate, which was now needed only to make up for losses in the process. In addition, the Glover tower gave a more concentrated form of sulfuric acid, containing 75 to 85% H_2SO_4 by mass, compared with concentrations of 60 to 70% obtained by earlier methods. Figure 22–5 is a flow diagram for the full process after the development of the Glover tower in 1859.

The Contact Process

As early as 1831, the Englishman Peregrine Phillips observed that platinum could catalyze the conversion of SO_2 to SO_3, which lies at the heart of the production of sulfuric acid. Little attention was paid to this discovery until the 1870s, when the growth of the German dye industry stimulated a search for a method of producing sulfuric acid that was more concentrated than that from the Glover tower. Platinum catalysts were rediscovered and patented but initially had limited usefulness because they were poisoned by impurities in the sulfur dioxide feed. As a result, the early applications of this method used, somewhat paradoxically, sulfuric acid from the Glover tower as a raw material. It was decomposed by heating,

$$H_2SO_4(aq) \longrightarrow SO_2(g) + H_2O(\ell) + \tfrac{1}{2} O_2(g)$$

and the relatively pure SO_2 was converted into SO_3 over the catalyst and then back to H_2SO_4. In this way, a highly concentrated acid was obtained, but at great expense. Over the next 30 years, researchers recognized the role of arsenic and other impurities in poisoning the catalyst and found ways to remove the impurities from the SO_2 feed, allowing production of concentrated sulfuric acid without the intermediate lead-chamber step. Different catalysts were studied as well, leading eventually to the selection of an oxide of vanadium (V_2O_5), which is the principal catalyst in use today.

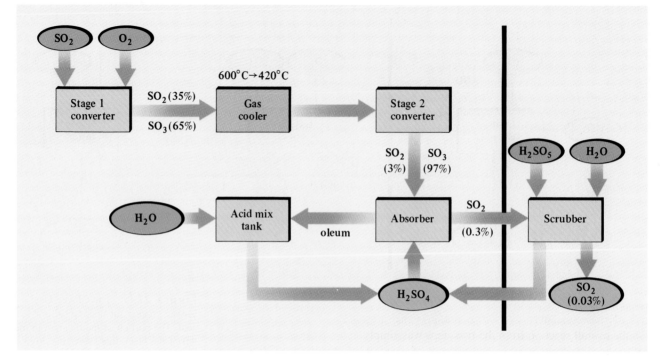

Figure 22–6 A flow chart of the contact process for making sulfuric acid. The portion of the diagram to the right of the black line represents a process that is used in some modern plants to reduce air pollution from residual SO_2.

The catalytic production of sulfuric acid from SO_2 is called the **contact process** and is outlined in Figure 22–6. In the reaction of SO_2 with oxygen,

$$SO_2(g) + \tfrac{1}{2} O_2(g) \longrightarrow SO_3(g)$$

the chemical amount of gas decreases from 1.5 to 1 mol. This suggests running the reaction under high total pressures to increase the yield of product. However, the small advantage gained by using high pressures is more than offset by the greater cost of the equipment that would be needed, so atmospheric pressure is used. A thermodynamic analysis gives

$$\Delta H^{\circ}_{298} = -98.9 \text{ kJ} \qquad \text{and} \qquad \Delta S^{\circ}_{298} = -94.0 \text{ J K}^{-1}$$

Because the reaction is exothermic, the lower the temperature, the higher the equilibrium constant and the greater the degree of conversion to products at equilibrium. The change in Gibbs function is zero (and the equilibrium constant 1) when

$$\Delta G^{\circ} = \Delta H^{\circ} - T\,\Delta S^{\circ} \approx \Delta H^{\circ}_{298} - T\,\Delta S^{\circ}_{298} = 0$$

or

$$T \approx \frac{\Delta H^{\circ}_{298}}{\Delta S^{\circ}_{298}} = \frac{-98,900 \text{ J}}{-94.0 \text{ J K}^{-1}} = 1050 \text{ K}$$

or at approximately 780°C. The temperature must be maintained substantially below this point to achieve a significant equilibrium yield of SO_3.

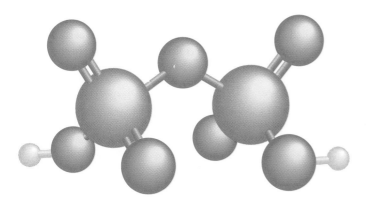

Figure 22-7   The structure of disulfuric acid ($H_2S_2O_7$).

The problem is that at lower temperature the reaction slows, although the catalyst does help to speed it up. A two- to four-stage process is typically used. The entering $SO_2(g)$ reaches the first catalyst at a temperature of 420°C. As the reaction begins, heat is evolved and the temperature of the reacting gas mixture rises. After a few seconds, the mixture has reached equilibrium at about 600°C, with conversion of 60 to 70% of the SO_2. The gases are then cooled back to 420°C and allowed to react one or two more times over the catalyst, using lower temperatures and longer exposure periods. The result is conversion of about 97% of the SO_2 to SO_3. For even greater conversion, the gases are then passed into a tower where SO_3 dissolves in sulfuric acid. This removes the reaction product, so that the equilibrium again shifts to the right when the unreacted SO_2 is passed over the catalyst for a final time. The SO_3 from this last stage is then absorbed, giving an overall yield of about 99.7% of the SO_2 introduced originally. Careful consideration of thermodynamics and kinetics has led to a highly efficient process. Almost all sulfuric acid manufactured today is made by the contact process.

If SO_3 were absorbed in water rather than in sulfuric acid, the product would be more dilute and less SO_3 would be absorbed. In addition, the direct reaction of SO_3 with water produces a fine acidic mist that is difficult to condense. Absorption of SO_3 into sulfuric acid gives fuming sulfuric acid, or **oleum,** which can be used directly or diluted with water to the desired strength. An equimolar amount of SO_3 dissolved in H_2SO_4 gives disulfuric acid ($H_2S_2O_7$), also called "pyrosulfuric acid" (Fig. 22–7).

The amount of SO_2 that escapes from the contact process to pollute the air is very small, but further removal of objectionable SO_2 from the tail gases can be achieved (at additional expense) in another step. Some H_2SO_4 is electrolytically oxidized to peroxodisulfuric acid ($H_2S_2O_8$):

anode: $2\ H_2SO_4(aq) \longrightarrow H_2S_2O_8(aq) + 2\ H^+(aq) + 2\ e^-$ (oxidation)

cathode: $2\ H^+(aq) + 2\ e^- \longrightarrow H_2(g)$ (reduction)

In this compound, a peroxo group (O—O) replaces the bridging oxygen atom in disulfuric acid (Fig. 22–8). It quickly reacts with water to give a mixture of sulfuric acid and peroxosulfuric acid

$$H_2O(\ell) + H_2S_2O_8(aq) \longrightarrow H_2SO_4(aq) + H_2SO_5(aq)$$

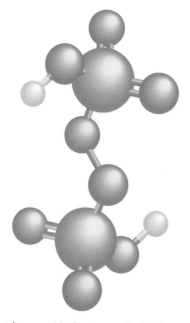

Figure 22-8   Peroxodisulfuric acid ($H_2S_2O_8$). The apparent oxidation number of sulfur in this compound equals +7; this becomes +6 when the two peroxo oxygens that link the sulfurs are counted as having oxidation numbers of −1 (see Section 4–4).

A crystal of sulfur.

In the latter, the peroxo group replaces one of the oxygen atoms in sulfuric acid. This compound is also called "Caro's acid." The exiting gases are passed into a scrubber to mingle with a solution of this powerful oxidizing agent. In the scrubber, the reaction

$$SO_2(g) + H_2SO_5(aq) + H_2O(\ell) \longrightarrow 2\,H_2SO_4(aq)$$

converts $SO_2(g)$ to sulfuric acid. Over 90% of the already small amount of residual $SO_2(g)$ is removed in this way. The product, dilute sulfuric acid, is of course cycled back into the main process.

Sources for Sulfur and Sulfur Dioxide

The primary raw material for making sulfuric acid is sulfur or sulfur dioxide. Sources for these chemicals have changed over the years, driven by considerations of price and the desirability of reducing atmospheric pollution. Centuries ago, the sulfur mines of Sicily were the major source for elemental sulfur. Increases in price, especially after a cartel was established to exploit the mines, led to the search for other sources of sulfur. As the use of metals increased in the 19th century, sulfur was obtained in the form of SO_2 as a by-product of the roasting of sulfide ores of zinc, iron, or copper through reactions such as

$$ZnS(s) + \tfrac{3}{2}\,O_2(g) \longrightarrow ZnO(s) + SO_2(g)$$

Capturing the sulfur dioxide and converting it to sulfuric acid provided a cheap starting material and reduced air pollution from SO_2.

In the late 1890s, a new development shifted interest from sulfide ores back to elemental sulfur as a starting material, at least in the United States. This was the discovery by Herman Frasch of a new method to extract sulfur from underground

Stairs cut into sulfur that is stacked in blocks for shipment.

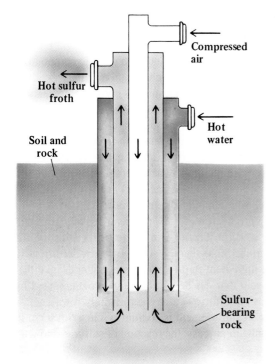

Figure 22-9 The Frasch process uses three concentric pipes driven into the ground. Superheated water (at a temperature of 160°C) is pumped under pressure through the outermost pipe into the sulfur-bearing rock formation. This heats the rock above the melting point of sulfur, 119°C. The molten sulfur is heavier than water and collects in a pool, where heated compressed air pumped through the innermost pipe works it into a froth that rises to the surface through the third pipe.

deposits. The salt domes of Louisiana, Texas, and Mexico contained substantial reserves of sulfur in porous limestone, but extraction had proved difficult. The **Frasch process** (Fig. 22–9) is an ingenious mining method in which sulfur is liquefied by superheated water injected into the deposit and forced to the surface with compressed air. The sulfur is usually shipped as a liquid in heated vessels to sulfuric acid plants.

Since the 1970s, the Frasch process has declined in importance as a source of sulfur for sulfuric acid owing to a competing process that recovers sulfur impurities in natural gas and petroleum. The latter process was developed to reduce sulfur oxide pollution from the burning of oil and gas. Hydrogen sulfide and other sulfide impurities are acidic and so are preferentially extracted into solutions of basic potassium carbonate:

$$H_2S(g) + CO_3^{2-}(aq) \longrightarrow HS^-(aq) + HCO_3^-(aq)$$

After removal from this solution, a portion of the H_2S is burned in air at 1000°C over an aluminum oxide catalyst to yield sulfur dioxide:

$$H_2S(g) + \tfrac{3}{2} O_2(g) \longrightarrow SO_2(g) + H_2O(g)$$

This can be used directly to make sulfuric acid, but an increasing percentage is now being converted back to elemental sulfur by reaction with additional H_2S in the **Claus process.**

$$SO_2(g) + 2 H_2S(g) \longrightarrow 3 S(\ell) + 2 H_2O(g)$$

which uses an Fe_2O_3 catalyst. This method greatly reduces pollution from the burning of natural gas and petroleum. New processes to reduce sulfur pollution from the burning of coal are greatly needed now.

22-3 USES OF SULFURIC ACID AND ITS DERIVATIVES

Pure sulfuric acid is a colorless, viscous liquid that freezes at 10.4°C and boils at 279.6°C. It reacts vigorously with water (Fig. 22–10) and with organic compounds. It may be mixed with water in any proportion, although the mixing must be done cautiously because it generates a good deal of heat. Despite its corrosive power, sulfuric acid is easily handled and transported in steel drums. This fact, together with its acid strength and low cost, has given it a tremendous range of uses. In metals processing, it is used to leach copper, uranium, and vanadium from their ores and to "pickle," or descale, steel. In pickling, layers of oxides on the surfaces of the metal are dissolved away by reaction with the acid. Much sulfuric acid is used as a **dehydrating agent** in the synthesis of organic chemicals and in the processing of petrochemicals. Lesser, but still important, amounts are used in making hydrochloric and hydrofluoric acids and in the production of TiO_2 pigments for paint. As will be discussed, the largest single use of sulfuric acid is in the fertilizer industry. All of these uses are *indirect*, however, in the sense that the sulfur from the sulfuric acid does not become incorporated as an ingredient in products. Rather, it ends up either as sulfate wastes or as **spent acid,** acid that has been diluted and contaminated by its use in processing. The *direct* uses of sulfuric acid are surprisingly few.

• The most familiar use of sulfuric acid is in the lead–acid storage battery (see Section 13-6). This use consumes only a minute fraction of total sulfuric acid production, however.

An important chemical made from sulfuric acid is sodium sulfate (Na_2SO_4). In the past, it was produced in the Mannheim process, the reaction of sulfuric acid with sodium chloride:

$$2 \text{ NaCl} + \text{H}_2\text{SO}_4 \longrightarrow \text{Na}_2\text{SO}_4 + 2 \text{ HCl}$$

The Mannheim process is described in Chapter 23. It requires high temperatures (800 to 900°C) and is now little used because of its high energy cost. A method still in use is the **Hargreaves process,** in which the immediate reactant is SO_2 rather than H_2SO_4:

$$4 \text{ NaCl} + 2 \text{ SO}_2 + 2 \text{ H}_2\text{O} + \text{O}_2 \longrightarrow 2 \text{ Na}_2\text{SO}_4 + 4 \text{ HCl}$$

The SO_2 for this reaction often comes from the decomposition of spent sulfuric acid. Sodium sulfate is also a by-product of the manufacture of rayon. Cellulose from cotton fiber or from wood reacts with carbon disulfide (CS_2) in the presence of NaOH to form a viscous solution, called *viscose*. The solution then reacts with sulfuric acid, regenerating carbon disulfide and cellulose in the form of long strands, suitable for being woven into cloth. The net chemical change that accompanies the production of the rayon strands is a neutralization reaction to produce sodium sulfate and water:

• Rayon and other synthetic fibers are considered further in Chapter 25.

$$2 \text{ NaOH} + \text{H}_2\text{SO}_4 \longrightarrow \text{Na}_2\text{SO}_4 + 2 \text{ H}_2\text{O}$$

Sodium sulfate is used as a phosphate substitute in detergents and in the manufacture of paper products.

Figure 22–10 When diluting sulfuric acid, always add the acid to water rather than the reverse. The dissolution reaction is very exothermic; water can boil and spatter acid out when water is added to acid because heat is generated in a small region where the less dense water collects temporarily atop the more dense acid. Adding acid to water mixes the two much more effectively. Here, methyl orange, an indicator, changes color from yellow to red as acid is added.

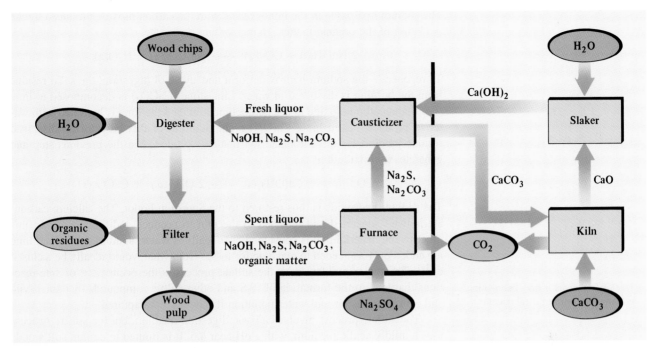

Figure 22–11 The sulfate process for pulping wood. The overall reaction is wood chips + water ⟶ wood pulp + organic residues + carbon dioxide. The entire part of the flow diagram to the right of the black line serves only to make up losses in the process.

Wood Pulp and Paper Processing

One of the most important applications of sulfur chemistry is in processing wood to wood pulp for use in paper and cardboard. Wood consists of three major components: cellulose, lignin, and oils and resins. Cellulose is a fibrous polymeric compound of carbon, oxygen, and hydrogen that provides supporting structure. Lignin, which is also a complex polymeric material, binds the fibers of cellulose together. To produce wood pulp, the cellulose must be separated from the lignin. The other materials in the wood are removed by solvent extraction and in many cases have important uses (for example, turpentine, pine oil, rosin, soaps, and tannin). The chemical reaction involved in digestion, which is the process by which lignin is separated from cellulose, is a hydrolysis. Attack by water in the presence of other chemicals breaks down the lignin polymer into alcohols and organic acids containing long chains of carbon atoms. The primary methods of digestion are the sulfate (Kraft) process and the sulfite process.

The **sulfate process** (Fig. 22–11) is the more widely used of these two methods because it works for almost all kinds of wood and it produces the strongest paper. The digestion liquor is an aqueous solution of $NaOH$ and Na_2S, the active agents in the hydrolysis of lignin. After digestion for several hours at 175°C, the pulp (now primarily cellulose) is separated by filtration, and the liquor is treated to extract any desired organic compounds. The next and crucial step is the removal and recycling of the inorganic components from the spent liquor. First, sodium sulfate is added to make up for the inevitable losses in the process. Then the liquor is concentrated and its organic component is burned, generating heat to run the paper mill. During the burning, the sodium sulfate is reduced by the organic matter in the liquor, a process indicated by the equation

$$Na_2SO_4(s) + 2\ [C] \longrightarrow Na_2S(s) + 2\ CO_2(g)$$

• The name "Kraft" comes from the German word meaning "strength."

• In this equation, the "[C]" stands for the many varieties of carbon-containing matter present.

The sodium hydroxide in the liquor reacts with the carbon dioxide produced by the oxidation of the organic matter, giving sodium carbonate:

$$2 \, NaOH(s) + CO_2(g) \longrightarrow Na_2CO_3(s) + H_2O(g)$$

To regenerate the sodium hydroxide for a new cycle of digestion, the solid residue from the burning is dissolved in water. Limestone ($CaCO_3$) is decomposed to lime (CaO) and carbon dioxide in a separate lime kiln, and the lime is hydrated with water to produce slaked lime, $Ca(OH)_2$. The addition of this base to the dissolved residue precipitates the carbonate ion that was produced in the previous step and generates hydroxide ion:

$$CO_3^{2-}(aq) + Ca(OH)_2(s) \longrightarrow 2 \, OH^-(aq) + CaCO_3(s)$$

The sodium hydroxide is thus restored to the digestion liquor. The calcium carbonate is itself separated and recycled to the kiln to produce additional lime. In this process, the material that is consumed chemically is water; sodium sulfate and lime are added only as needed to make up for losses from what would ideally be a closed cycle. The major problem with the sulfate process is the occurrence of side-reactions that lead to the formation of H_2S and other sulfur compounds that smell vile and cause serious air and water pollution if they are not captured.

* The Kraft process is now banned in Germany for environmental reasons.

The sulfate process involves a basic digestion medium, but the **sulfite process** uses a mildly acidic medium, with a pH near 4.5. It is limited to certain soft woods such as spruce and hemlock, but it produces a lighter pulp that requires less bleaching and is suitable to make high-quality writing paper. In the digestion liquor, gaseous sulfur dioxide is added to a solution containing hydroxide ion (generally prepared by reacting MgO with water) to produce hydrogen sulfite ions:

$$SO_2(g) + OH^-(aq) \longrightarrow HSO_3^-(aq)$$

The action of the hydrogen sulfite ion at 60°C over 6 to 12 hours hydrolyzes the cellulose–lignin complex and dissolves the lignin. After the process is complete, the pulp is recovered by filtration and the liquor is evaporated. The organic residues are again burned to recover heat; during this procedure the $Mg(HSO_3)_2$ decomposes according to the equation

$$Mg(HSO_3)_2(s) \longrightarrow MgO(s) + H_2O(g) + 2 \, SO_2(g)$$

Both the SO_2 and MgO are recovered and recycled through the process.

Phosphorus and Phosphate Fertilizers

Until relatively recently, no single use of sulfuric acid dominated the others. The situation changed with the growing production of phosphate fertilizers, which are processed with sulfuric acid and now account for well over half of the annual consumption of sulfuric acid in the United States. Phosphorus occurs in nature primarily as the PO_4^{3-} ion in phosphate rock, the main component of which is fluorapatite ($Ca_5(PO_4)_3F$). Originally, the rock was simply ground up and applied to fields to make them more fertile. A more active fertilizer is made by treating a slurry of ground-up fluorapatite with sulfuric acid to produce **superphosphate,** a mixture of calcium dihydrogen phosphate and gypsum:

$$2 \, Ca_5(PO_4)_3F(s) + 7 \, H_2SO_4(aq) + 17 \, H_2O(\ell) \longrightarrow$$
$$3 \, Ca(H_2PO_4)_2 \cdot H_2O(s) + 7 \, (CaSO_4 \cdot 2H_2O(s)) + 2 \, HF(g)$$

The world's largest linear disc reclaimer, reclaiming phosphate shale at a phosphorus plant in Idaho.

Superphosphate is more effective because the essential phosphorus nutrient is in a more soluble form. This mixture provides the important secondary nutrients calcium and sulfur as well as phosphorus to growing crops.

A variant of the superphosphate process uses an excess of sulfuric acid to produce phosphoric acid according to the **wet-acid process.** The chemical reaction in this case is

$$Ca_5(PO_4)_3F(s) + 5\ H_2SO_4(aq) + 10\ H_2O(\ell) \longrightarrow$$
$$3\ H_3PO_4(aq) + 5\ CaSO_4 \cdot 2H_2O(s) + HF(g)$$

The hydrogen fluoride that is liberated is carried to an absorption tower that contains SiF_4. In this tower it reacts to give H_2SiF_6, which is used in aqueous solution to fluoridate drinking water. The solid gypsum ($CaSO_4 \cdot 2H_2O$) is filtered out, and the dilute solution of phosphoric acid (H_3PO_4) is concentrated by evaporation.

The major use for wet-process phosphoric acid is again fertilizer manufacture, in which it takes the place of sulfuric acid to produce **triple superphosphate** fertilizer, which does not contain gypsum:

$$Ca_5(PO_4)_3F(s) + 7\ H_3PO_4(aq) + 5\ H_2O(\ell) \longrightarrow 5\ Ca(H_2PO_4)_2 \cdot H_2O(s) + HF(g)$$

This fertilizer contains a greater percentage of phosphorus, a significant advantage in reducing transportation costs. In recent years, the use of triple superphosphate fertilizer has declined. The major phosphorus-containing fertilizer now is ammonium phosphate, which supplies the necessary nutrient nitrogen as well as phosphorus. Ammonium phosphate is obtained by the acid-base reaction of phosphoric acid with ammonia:

$$H_3PO_4(aq) + 3\ NH_3(aq) \longrightarrow (NH_4)_3PO_4(aq)$$

The manufacture of ammonia and other nitrogen-based fertilizers is discussed later in this chapter.

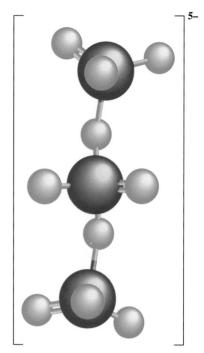

Figure 22-12 The structure of the tripolyphosphate anion $(P_3O_{10})^{5-}$.

• The $P_3O_{10}^{5-}$ ions act as chelating agents; they bind Ca^{2+} and Mg^{2+} ions through their $-O^-$ groups (see Section 19-1) and keep them in solution.

The major nonfertilizer use of wet-process phosphoric acid is in the manufacture of detergents and cleaning agents that are sodium salts of phosphoric acid and its derivatives. Phosphoric acid in excess reacts with sodium carbonate to give sodium dihydrogen phosphate, a weak acid:

$$2\,H_3PO_4(aq) + Na_2CO_3(aq) \longrightarrow 2\,NaH_2PO_4(aq) + CO_2(aq) + H_2O(\ell)$$

When the sodium carbonate is in excess, the product is sodium hydrogen phosphate, a weak base:

$$H_3PO_4(aq) + 2\,Na_2CO_3(aq) \longrightarrow 2\,NaHCO_3(aq) + Na_2HPO_4(aq)$$

Sodium carbonate is too weak a base to remove the last hydrogen ion from HPO_4^{2-}, so addition of the stronger (but more expensive) base NaOH is required:

$$NaOH(aq) + Na_2HPO_4(aq) \longrightarrow Na_3PO_4(aq) + H_2O(\ell)$$

Sodium phosphate (Na_3PO_4), which is also called "trisodium phosphate," or TSP, gives strongly basic solutions that are excellent for washing and cleaning.

Heating solid sodium hydrogen phosphate (Na_2HPO_4) expels one molecule of water for every two formula units:

[chemical structure diagram]

The product, sodium diphosphate (sodium pyrophosphate, $Na_4P_2O_7$), is the sodium salt of diphosphoric (pyrophosphoric) acid $(H_4P_2O_7)$. The diphosphate anion has the same structure as the disilicate anion (see Fig. 21-6): each phosphorus atom is tetrahedrally coordinated to four oxygen atoms (as in the original PO_4^{3-} ion), and the two tetrahedra share a single corner. Heating sodium diphosphate strongly causes the linkage of *three* phosphate tetrahedra

$$2\,Na_2HPO_4(s) + NaH_2PO_4(s) \longrightarrow Na_5P_3O_{10}(s) + 2\,H_2O(g)$$

In this reaction one molecule of water is expelled for each link that forms. The product is called sodium tripolyphosphate $(Na_5P_3O_{10})$, or STPP (Fig. 22-12). It is used as a **builder** in detergents. It helps to sequester Ca^{2+} and Mg^{2+} ions, to maintain the pH in an appropriate range, to keep dirt in suspension, and to increase the efficiency of the detergent itself. The use of detergents that contain phosphates has diminished in recent years because phosphates released into the groundwater fertilize the growth of algae in lakes. Excessive algal growth in a lake leads to its early destruction through **eutrophication.** The algae capture CO_2 from the air photosynthetically. When they die, their bodies fall to the bottom of the lake. The decay of this organic material recycles the phosphates (and other nutrients) to succeeding generations of algae. As the lake becomes more shallow, bottom-rooted plants can take hold. Eventually the lake becomes a marsh and, finally, a meadow. In many communities, concern over eutrophication has led to legislation to limit the use of phosphates in detergents.

The phosphoric acid made by the wet-acid process contains residual gypsum and impurities from the phosphate rock. A second procedure, used for the production of purer phosphoric acid, is called the **furnace process.** Phosphate rock, SiO_2 (in the form of sand), and elemental carbon (in the form of coke) are fed continuously into an electric arc furnace that operates at 2000°C. The overall reaction is approximately

$$12 \ Ca_5(PO_4)_3F(\ell) + 43 \ SiO_2(\ell) + 90 \ C(s) \longrightarrow$$
$$90 \ CO(g) + 20 \ (3 \ CaO \cdot 2 \ SiO_2)(\ell) + 3SiF_4(g) + 9 \ P_4(g)$$

The elemental white phosphorus is burned in air to tetraphosphorus decaoxide (P_4O_{10}) (Fig. 22–13):

$$P_4(g) + 5 \ O_2(g) \longrightarrow P_4O_{10}(s)$$

This is a powerful drying agent because it reacts avidly with water. It dehydrates HNO_3 to N_2O_5, and even H_2SO_4 to SO_3. The ultimate product of its reaction with water is phosphoric acid:

$$P_4O_{10}(s) + 6 \ H_2O(\ell) \longrightarrow 4 \ H_3PO_4(aq)$$

The purer phosphoric acid from the furnace process is used widely in the food industry. Its acidity gives a pleasant tartness to soft drinks (called "phosphates" when they were first formulated). Calcium dihydrogen phosphate monohydrate ($Ca(H_2PO_4)_2 \cdot H_2O$) is mixed with $NaHCO_3$ in some baking powders. When water is added, these ionic compounds dissolve, and the acidic dihydrogen phosphate ions react with the basic hydrogen carbonate ions to generate carbon dioxide:

$$H_2PO_4^- \ (aq) + HCO_3^- (aq) \longrightarrow HPO_4^{2-} \ (aq) + CO_2(g) + H_2O(\ell)$$

The liberated gas makes the bread or other baked goods rise. Another phosphoric acid derivative used in the food industry is sodium hydrogen phosphate (Na_2HPO_4), which is used as an emulsifier to distribute the butterfat uniformly throughout pasteurized processed cheese.

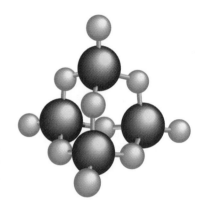

• For historical reasons, P_4O_{10} is often called "phosphorus pentaoxide" after its empirical formula, P_2O_5.

Figure 22-13 The structure of P_4O_{10} is closely related to the tetrahedral P_4 structure in white phosphorus (see Fig. 20–29a).

• Many food labels call Na_2HPO_4 "disodium phosphate."

22-4 NITROGEN FIXATION

Nitrogen at the earth's surface exists almost entirely (99.9%) as gaseous diatomic molecules (N_2), which make up 78% by volume of the atmosphere. The N—N bond is a triple bond, and its enthalpy of dissociation is 945 kJ mol^{-1}, the largest of any diatomic molecule except carbon monoxide. The nitrogen molecule is remarkably unreactive in comparison with other triple-bonded molecules. This is apparent in a simple thermochemical comparison. In each of the two reactions

$$N{\equiv}N(g) + 3 \ H_2(g) \longrightarrow 2 \ NH_3(g) \qquad \Delta H° = -92.2 \ kJ$$
$$HC{\equiv}CH(g) + 3 \ H_2(g) \longrightarrow 2 \ CH_4(g) \qquad \Delta H° = -376.4 \ kJ$$

a triple bond and three H—H bonds are broken and six X—H single bonds are formed (X is N or C). The enthalpy changes in forming an N—H bond and a C—H bond are roughly equal, so the large difference in the $\Delta H°$'s of the reactions must mean that the triple bond in N_2 is substantially stronger than the triple bond in C_2H_2.

The great stability of the N_2 molecule diminishes its reactivity and means that an N_2 atmosphere can be used to prevent air oxidation in metallurgical, chemical, and food-processing operations. For most biological and industrial purposes, however, what is needed is not molecular nitrogen but compounds of nitrogen with other elements. The formation of such compounds is referred to as **nitrogen fixation.** Because fixed nitrogen is present in all amino acids, the building blocks of proteins, it is essential to life and to our daily food supply.

• From Table 10-3, the bond enthalpy of the N—H bond is 391 kJ mol^{-1}, and that of the C—H bond is 413 kJ mol^{-1}, less than 6% greater.

Natural Sources of Fixed Nitrogen

Figure 22–14 illustrates how nitrogen moves through a worldwide cycle involving biological, geological, and atmospheric processes. Herbivores and omnivores use nitrogen from plants and vegetable matter they have consumed in a variety of essential biological functions, including the synthesis of proteins. Eventually, it is excreted and returned to the soil in forms such as urea ($(NH_2)_2CO$). A portion is then reused by other plants as they grow, but some is converted to molecular nitrogen by bacteria and returned to the atmosphere. To continue the cycle, a way is needed to fix it again.

Nature has at least two ways of fixing atmospheric nitrogen. One is the action of lightning on air, in which the high temperature of the electrical discharge causes the reaction of nitrogen with oxygen to form nitrogen monoxide (NO). The nitrogen monoxide is subsequently oxidized to nitrogen dioxide (NO_2), which reacts with hydroxyl radicals (OH) in the atmosphere to form dilute nitric acid. The HNO_3 falls to the earth in the rain and provides a source of fixed nitrogen (in the form of the nitrate ion) for plant growth. Nitrogen fixed in this way is thought to amount to at least 10% of the annual total. A second way in which nitrogen is fixed is by the action of certain bacteria called diazatrophs that live on the roots of legumes such as

• Transport of nitrate to the soil is one of the few benefits of acid rain.

Figure 22–14   The nitrogen cycle.

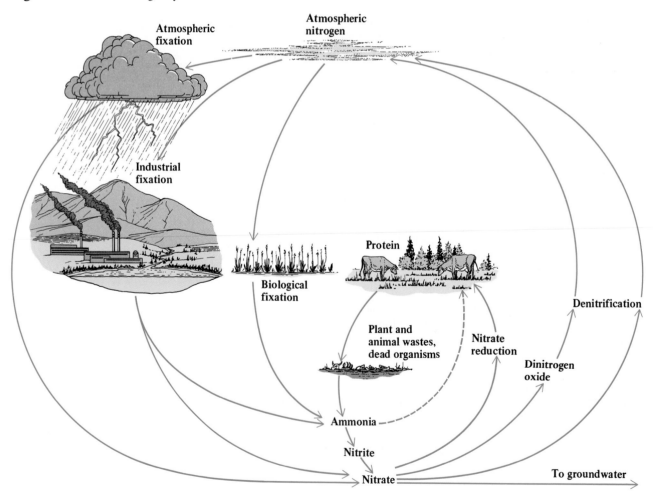

soy beans, clover, and alfalfa. The bacteria reduce N_2 to NH_3, and the plants release a significant amount of the ammonia to the soil in a growing season. Farmers plant legumes every few years to fertilize the soil for other crops. Natural fixation of nitrogen generates the equivalent of 90 billion kg of ammonia annually, according to current estimates.

By the late 18th century, the population of Europe had grown sufficiently that an increased yield of crops was needed. The first step was to reuse animal wastes efficiently. Sailing ships went as far as the coastal regions of South America to bring back guano, the excrement dropped by seagulls, because it was rich in nitrogen and easily mined. Mineral deposits of sodium nitrate in Chile were exploited for fertilizer. By the end of the 19th century, some fixed nitrogen was being recovered from the waste of liquors of natural gas plants, in the form of ammonia. It was generally converted to ammonium sulfate $((NH_4)_2SO_4)$ by reaction with sulfuric acid and applied in that form as a fertilizer.

• Sodium nitrate was called Chile saltpeter. Saltpeter, which was used in the making of gunpowder, is potassium nitrate.

Early Industrial Processes

In 1898, Sir William Crookes addressed the British Association in these words:

> England and all civilised nations stand in deadly peril of not having enough to eat. As mouths multiply, food resources dwindle. . . . It is the chemist who must come to the rescue of the threatened communities. It is through the laboratory that starvation may ultimately be turned to plenty.[1]

He added that "there is a gleam of light amid this darkness and despondency . . . the fixation of atmospheric nitrogen," which would be "one of the great discoveries awaiting the ingenuity of chemists." Within ten years of this prediction, three methods had been developed for fixing atmospheric nitrogen.

The first was the **electric-arc process,** which was based on the observation by Cavendish in the 1780s that a spark causes the combination of nitrogen and oxygen in air. Electrical technology was sufficiently advanced 100 years later to create electric arcs in which the temperature was raised to 2000 to 3000°C. The NO that formed was cooled, oxidized to NO_2, and passed into an absorption tower where it reacted with water to make dilute nitric acid (HNO_3). This process closely resembled the fixing of nitrogen by lightning discharges. The energy costs for running the electric arc were high, and it was difficult to obtain large amounts of fixed nitrogen in a continuous manner. As a result, the electric-arc process was never a significant factor for fixing nitrogen.

More promising was the **cyanamide process,** also called the **Frank–Caro process** after its developers. This method uses calcium carbide (CaC_2) as a starting material. The discovery of this compound has an interesting history. In 1892 T. L. Wilson in North Carolina was trying to make metallic calcium in an electric-arc furnace from lime (CaO) and tar. He obtained a product that was clearly not calcium and threw it into a stream, where he saw to his surprise that it reacted with water to liberate large amounts of a combustible gas. The compound was calcium carbide (CaC_2), and its reaction with water gives acetylene (C_2H_2).

$$CaC_2(s) + 2\ H_2O(\ell) \longrightarrow C_2H_2(g) + Ca(OH)_2(aq)$$

Calcium carbide is now made in an electric-arc furnace at 2000 to 2200°C from lime and coke:

$$CaO(s) + 3\ C(s) \longrightarrow CaC_2(s) + CO(g) \qquad \Delta H° = +465\ kJ$$

---

[1]Quoted in L. F. Haber, *The Chemical Industry, 1900–1930.* Oxford, England: Clarendon Press, 1971, p. 84.

Application of anhydrous ammonia to a soybean field.

In the cyanamide process, calcium carbide then reacts (at $1100°C$ in an electric furnace) with nitrogen obtained from liquefying and distilling air, to give calcium cyanamide:

$$CaC_2(s) + N_2(g) \longrightarrow CaCN_2(s) + C(s) \qquad \Delta H° = -291 \text{ kJ}$$

Steam is added to convert this to ammonia:

$$CaCN_2(s) + 4 H_2O(g) \longrightarrow Ca(OH)_2(s) + CO_2(g) + 2 NH_3(g) \qquad \Delta H° = -154 \text{ kJ}$$

Between 1900 and 1930, calcium cyanamide was used directly as a fertilizer because it furnishes both ammonia and lime to the soil. The major current use of $CaCN_2$ is as a starting material for making plastics.

The Haber–Bosch Process

Both the electric-arc and the cyanamide processes have been superseded by a process that is the predominant source of industrial fixed nitrogen today, the **Haber–Bosch process** for the synthesis of ammonia. The German chemists Fritz Haber and Walther Nernst carried out studies from 1900 to 1910 on the effect of pressure and temperature on the equilibrium between ammonia and its component elements:

$$N_2(g) + 3H_2(g) \rightleftharpoons 2 NH_3(g)$$

The practical ammonia synthesis developed by Haber with chemical engineer Kurt Bosch required both deep chemical insight and equipment capable of operating under conditions of high temperature and pressure. At 298 K, the standard Gibbs function and enthalpy of formation of two moles of ammonia from its elements are

$$\Delta G° = -33.0 \text{ kJ} \qquad \text{and} \qquad \Delta H° = -92.2 \text{ kJ}$$

The equilibrium expression is

$$\frac{(P_{NH3})^2}{(P_{H_2})^3(P_{N_2})} = K$$

and the value for the equilibrium constant, written K_{298}, is

$$\ln K_{298} = \frac{-\Delta G°}{RT} = \frac{33,000 \text{ J mol}^{-1}}{(8.315 \text{ J K}^{-1} \text{ mol}^{-1})(298 \text{ K})} = 13.32$$

$$K_{298} = \text{antiln } (13.32) = e^{13.32} = 6 \times 10^5$$

which is large relative to 1; the formation of the product (ammonia) is favored on thermodynamic grounds. Unfortunately, the reaction is very slow at 298 K, and the temperature must be raised to speed the reaction. Because the process is exothermic, increased temperature lowers the equilibrium constant and therefore decreases product yield. There is thus a competition between thermodynamic factors, which favor a high yield at low temperature, and kinetic factors, which favor the use of a high temperature.

The compromise adopted in industrial production is a temperature between 700 and 900 K. At 800 K, for example, the equilibrium constant can be estimated from the van't Hoff equation (see Section 11–8):

$$\ln \frac{K_{800}}{K_{298}} = \frac{\Delta H°}{R} \left(\frac{1}{T_1} - \frac{1}{T_2} \right)$$

$$= \frac{-92{,}200 \text{ J mol}^{-1}}{8.315 \text{ J K}^{-1} \text{ mol}^{-1}} \left(\frac{1}{298 \text{ K}} - \frac{1}{800 \text{ K}}\right) = -23.35$$

Solving this for the ratio of equilibrium constants gives

$$\frac{K_{800}}{K_{298}} = e^{-23.35} = 7 \times 10^{-11}$$

and the equilibrium constant at 800 K is

$$K_{800} = (7 \times 10^{-11}) K_{298} = 4 \times 10^{-5}$$

This value is only approximate because $\Delta H°$ changes somewhat with temperature between 298 and 800 K. The qualitative result is that the equilibrium yield of ammonia is greatly reduced at the higher temperature. To increase this yield, high total pressures are used. Increased pressure favors the product because two moles of gaseous product are formed from four moles of gaseous reactants. At a total pressure of 200 atm, the reaction yields 15 to 30% of the amount of NH_3 that would be produced by complete consumption of the reactants.

- In light of the enormous economic importance of the Haber–Bosch process, the equilibrium constant of the reaction has been measured and remeasured at elevated temperatures. It is 0.95×10^{-5} at 800 K.

A catalyst increases the rate of this reaction. Haber used osmium, a rare and expensive metal, in his original work. Later, he developed a less expensive catalyst based on partially oxidized iron containing small amounts of aluminum, which has been used ever since. The ammonia produced is liquefied to separate it from the unreacted nitrogen and hydrogen, which are recycled.

Although the principles of the Haber–Bosch process were developed before World War I, its implementation in Germany was greatly speeded by the need for fixed nitrogen to make military explosives when the supplies of nitrates from Chile were cut off. It is an irony of history that the commercialization of this process, which more than any other has expanded the food-producing capacity of the world, began as a military project that probably prolonged World War I by at least a year.

- Worldwide fixation of nitrogen by artificial means amounts to about 90 billion kg of ammonia annually, roughly equal to the amount fixed in natural processes.

In spite of the significant role that it has played since that time, the Haber–Bosch process for the fixing of nitrogen has real problems. Fertilizer factories consume large amounts of energy and require expensive structural materials to operate at high pressure and high temperature, and they depend on hydrogen as a reactant. Because most hydrogen is currently obtained from petroleum and natural gas, both nonrenewable resources, an alternative process using coal or, ultimately, water must eventually be found for the production of hydrogen. Scientists are also trying to design chemical catalysts that imitate the action of nitrogen-fixing bacteria as an economical way of converting atmospheric nitrogen to ammonia or to some other fixed form.

22-5 CHEMICALS FROM AMMONIA

Once molecular nitrogen has been fixed in the form of ammonia, further reactions to produce many other compounds are possible. Reaction with sulfuric, phosphoric, and nitric acids gives ammonium sulfate, ammonium phosphate, and ammonium nitrate, respectively. All of these have been used as chemical fertilizers, with the last two gaining particular popularity in recent years. Some ammonia is also converted to urea $((NH_2)_2CO)$. The modern synthesis of urea uses the reaction of CO_2 with ammonia in hot aqueous solution (180 to 200°C) under pressure to form the intermediate ammonium carbamate, NH_2COONH_4, which is then decomposed to urea by heat:

The active ingredients in the fire retardants being dropped by this airplane are ammonium sulfate and ammonium phosphate. After the fire, these compounds act as fertilizers, speeding the regrowth of vegetation in burned areas. The red color comes from a dye that turns to neutral earth tone after exposure to sunlight for a few days.

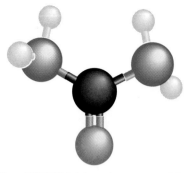

$$2 \, NH_3(aq) + CO_2(aq) \longrightarrow NH_2COONH_4(aq)$$
$$NH_2COONH_4(aq) \longrightarrow (NH_2)_2CO(s) + H_2O(g)$$

Urea is a widely used solid nitrogen fertilizer and an ingredient in other products ranging from glues to skin creams to disinfectants. Urea has historical significance as well, having been synthesized first by Friedrich Wöhler in 1828 from ammonia and cyanic acid (HCNO). Wöhler's work was important in demonstrating that an "organic" compound, formed in human and animal metabolism (and excreted in urine), could be synthesized from strictly inorganic starting materials.

Urea $(CO(NH_2)_2)$ is an important solid component in some fertilizers.

Nitric Acid

One of the most important products made from ammonia is nitric acid, which is generated by the overall reaction

$$NH_3 + 2 \, O_2 \longrightarrow HNO_3 + H_2O$$

This equation is deceptive because direct combination of ammonia with oxygen does *not* produce nitric acid. The most commonly used industrial process (the **Ostwald process,** described in Figure 22–15) involves three steps that, when added together, give the preceding overall reaction. Let us examine these steps.

The first step is the partial oxidation of ammonia in air:

$$4 \, NH_3(g) + 5 \, O_2(g) \longrightarrow 4 \, NO(g) + 6 \, H_2O(g) \qquad \Delta H° = -905.5 \text{ kJ}$$

This is a strongly exothermic reaction, but it occurs only very slowly at room temperature. In addition, once some NO is produced, a competing reaction occurs,

$$4 \, NH_3(g) + 6 \, NO(g) \longrightarrow 5 \, N_2(g) + 6 \, H_2O(g) \qquad \Delta H° = -1808.0 \text{ kJ}$$

which is undesirable because it returns nitrogen to its elemental state. To obtain practical yields of NO, the reaction is carried out at elevated temperatures (800 to 950°C), and a catalyst is used, consisting of a fine gauze made of noble metals such as platinum and rhodium, or gold and palladium (Fig. 22–16). This catalyst promotes the desired reaction in preference to the competing one, and the first step occurs with extraordinary speed: only about 3×10^{-4} seconds of contact time between gases and the gauze is needed to obtain excellent conversion.

The second, and rate-determining, step is the reaction of nitrogen monoxide

Figure 22–15 Flow diagram for the synthesis of nitric acid from ammonia. Both the second step (oxidation of NO to NO_2) and the third step (absorption into water) described in the text take place in the absorber.

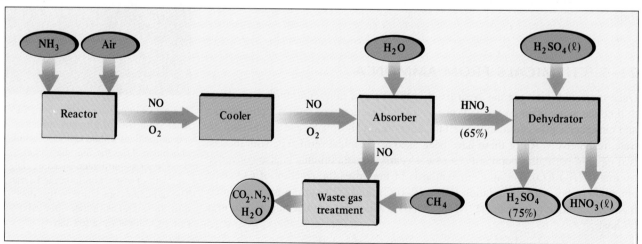

with oxygen to produce nitrogen dioxide:

$$2\ NO(g) + O_2(g) \longrightarrow 2\ NO_2(g) \qquad \Delta H° = -114.1\ kJ$$

The kinetics of this reaction, which requires no catalyst, are discussed in Example 14–4 and Section 14–6. The rate law has the form

$$rate = k[NO]^2[O_2]$$

and the rate constant k *decreases* with increasing temperature. Because the reaction occurs faster at lower temperatures, this step is carried out at the lowest temperature conveniently reached by the cooling water available, 10 to 40°C.

The third step is the absorption of the $NO_2(g)$ into water:

$$3\ NO_2(g) + H_2O(\ell) \longrightarrow 2\ H^+(aq) + 2\ NO_3^-\ (aq) + NO(g) \quad \Delta H° = -113.5\ kJ$$

The $NO(g)$ produced reacts again in the second step to give more $NO_2(g)$. The nitric acid has a concentration in the range of 50 to 65% by mass. Distilling off the water does not increase the concentration beyond 69% HNO_3, which is the "concentrated nitric acid" in normal laboratory use. Adding a powerful dehydrating agent (concentrated sulfuric acid) and then distilling separates more water to give a solution that contains 95 to 98% nitric acid (fuming nitric acid).

The Ostwald process illustrates well the kind of problems involved in practical chemistry. The original overall reaction is quite simple and is strongly favored thermodynamically at room conditions:

$$\frac{[H^+][NO_3^-]}{P_{NH_3}(P_{O_2})^2} = K_{298} = 10^{52}$$

Nevertheless, kinetic effects and the presence of competing equilibria require the careful design of a multistep process to obtain practical yields.

Nitric acid in aqueous solution is an excellent donor of hydrogen ions; it is a strong acid. Its concentrated aqueous solution is a much stronger oxidizing agent than solutions of either sulfuric or phosphoric acid. Thus, in dilute solution it reacts

When a heated platinum wire is placed just above the surface of a solution of concentrated aqueous ammonia, it continues to glow as it catalyzes the continued exothermic oxidation of NH_3 to NO by oxygen from the air.

Figure 22–16 This gold–palladium gauze catalyst is used to make nitric acid from ammonia.

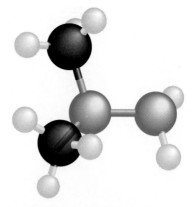

Dimethylhydrazine $(CH_3)_2NNH_2$. This derivative of hydrazine fueled the lunar excursion modules in their departures from the surface of the moon. The oxidant was the nitrogen oxide N_2O_4. Ignition was a certainty because the two burst into flame upon contact.

than solutions of either sulfuric or phosphoric acid. Thus, in dilute solution it reacts with copper and most other metals to give nitrates and NO:

$$8\ HNO_3(aq) + 3\ Cu(s) \longrightarrow 3\ Cu(NO_3)_2(aq) + 2\ NO(g) + 4\ H_2O(\ell)$$

In concentrated HNO_3 the reaction is

$$4\ HNO_3(aq) + Cu(s) \longrightarrow Cu(NO_3)_2(aq) + 2\ NO_2(g) + 2\ H_2O(\ell)$$

Only gold and the platinum metals can resist attack by concentrated nitric acid. The acid also reacts, often violently, with organic compounds. The major use for nitric acid is the manufacture of ammonium nitrate by its reaction with ammonia:

$$HNO_3(g) + NH_3(g) \longrightarrow NH_4NO_3(s)$$

Primarily used as a fertilizer, ammonium nitrate has an important secondary use as an industrial explosive, as explained later.

Hydrazine

Another important product made from ammonia is hydrazine (N_2H_4). In principle direct oxygenation can produce hydrazine from ammonia

$$2\ NH_3 + \tfrac{1}{2}\ O_2 \longrightarrow N_2H_4 + H_2O$$

this fails as a practical method. Competing reactions such as

$$2\ NH_3 + \tfrac{3}{2}\ O_2 \longrightarrow N_2 + 3\ H_2O$$

are thermodynamically favored because the N_2 molecule is so stable. The commercial production of hydrazine relies on the reaction of ammonia with hypochlorite ion (OCl^-) in aqueous solution, known as the **Raschig synthesis:**

$$2\ NH_3(aq) + OCl^-(aq) \longrightarrow N_2H_4(aq) + H_2O(\ell) + Cl^-(aq)$$

Pure liquid hydrazine resembles hydrogen peroxide in many of its physical properties, boiling at 114°C and freezing at 2°C (H_2O_2 boils at 152°C and freezes at -1.7°C). Hydrazine has had some use as a rocket fuel because its reaction with oxygen is extremely exothermic:

$$N_2H_4(\ell) + O_2(g) \longrightarrow N_2(g) + 2\ H_2O(g) \qquad \Delta H^\circ = -534\ kJ$$

A small mass of liquid hydrazine can produce a tremendous thrust as the very hot gaseous products rush out at high speed from the rocket engine.

Nitrogen in hydrazine has an oxidation state of -2, intermediate between those of molecular nitrogen (0) and nitrogen in ammonia (-3). As a result, hydrazine (like hydrogen peroxide) can act either as a reducing agent or as an oxidizing agent in aqueous solution. In basic solution (at pH 14), the standard reduction potential for the half-reaction

$$N_2(g) + 4\ H_2O(\ell) + 4\ e^- \longrightarrow N_2H_4(aq) + 4\ OH^-(aq)$$

is $\mathscr{E}^\circ = -1.16$ V. This means that hydrazine tends to be easily oxidized to nitrogen under these conditions and so should act as a good reducing agent. In a strongly acidic medium, hydrazine is almost completely converted to its conjugate acid, the hydrazinium ion $N_2H_5^+$. The standard reduction potential $\mathscr{E}^\circ = 1.275$ V at a pH of 0 for the half-reaction

$$N_2H_5^+(aq) + 3\ H^+(aq) + 2\ e^- \longrightarrow 2\ NH_4^+(aq)$$

suggests that $N_2H_5^+$ should be easily reduced to NH_4^+ and therefore should serve as an oxidizing agent under these conditions. This reaction is slow, however, and hydrazine is known mainly as a reducing agent.

Hydrazine is used to control corrosion in boilers through the reaction

$$6\ Fe_2O_3(s) + N_2H_4(aq) \longrightarrow 4\ Fe_3O_4(s) + N_2(g) + 2\ H_2O(\ell)$$

The red Fe_2O_3 oxide (ordinary rust) is reduced to Fe_3O_4 that forms a protective black layer to hinder further rusting. Hydrazine can also be used to remove certain metal ions from the wastewaters of chemical plants. Chromate ion (CrO_4^{2-}), for example, is reduced to Cr^{3+} in the reaction

$$20\ H^+(aq) + 4\ CrO_4^{2-}(aq) + 3\ N_2H_4(aq) \longrightarrow 4\ Cr^{3+}(aq) + 3\ N_2(g) + 16\ H_2O(\ell)$$

The Cr^{3+} is then precipitated by adding base:

$$Cr^{3+}(aq) + 3\ OH^-(aq) \longrightarrow Cr(OH)_3(s)$$

Hydrazine also removes halogens (X_2, $X = F$, Cl, Br, I) from industrial wastewater by the reaction

$$N_2H_4(aq) + 2\ X_2(aq) \longrightarrow N_2(g) + 4\ HX(aq)$$

Explosives

An **explosive** is a material that, when subjected to shock, decomposes rapidly and exothermically to produce a large volume of gas. Useful explosives need to be thermodynamically unstable but kinetically stable, so that they can be stored safely for long periods of time before they are detonated. Table 22–3 lists a number of explosives and their decomposition products. It is noteworthy that all of them contain nitrogen. The stability of gaseous nitrogen, a product of the detonation in all of these cases, contributes greatly to the exothermicity of the reactions. The evolution of heat alone, however, is not as important as the *detonation velocity,* a measure of the *rate* at which the heat is released. The combustion of gasoline, for example, releases about 48 kJ g^{-1}, far more than the heat released by the explosives listed in the table. The difference is that the detonation of an explosive is a branching chain reaction in which the release of heat takes place at a tremendous rate.

Table 22–3
Explosives

Name	Formula	Products	ΔH of Explosion (kJ g^{-1})
Gunpowder	$2\ KNO_3 + 3\ C + S$	$N_2 + 3\ CO_2 + K_2S$	-2.1
Nitrocellulose	$C_{24}H_{29}O_9(NO_3)_{11}$	$20.5\ CO + 3.5\ CO_2 + 14.5\ H_2O + 5.5\ N_2$	-4.5
Nitroglycerin	$C_3H_5(NO_3)_3$	$3\ CO_2 + 2.5\ H_2O + 1.5\ N_2 + 0.25\ O_2$	-6.4
Ammonium nitrate	NH_4NO_3	$H_2O + N_2 + 0.5\ O_2$	-1.6
TNT	$C_6H_2CH_3(NO_2)_3$	$3.5\ CO + 3.5\ C + 2.5\ H_2O + 1.5\ N_2$	-4.4
Picric acid	$C_6H_2(OH)(NO_2)_3$	$5.5\ CO + 1.5\ H_2O + 0.5\ C + 1.5\ N_2$	-4.4
Ammonium picrate	$C_6H_2(NO_2)_3OHN_4$	$4\ CO + 2\ C + 3\ H_2O + 2\ N_2$	-2.6
Tetryl	$C_7H_5N_5O_8$	$5.5\ CO + 2.5\ H_2O + 1.5\ C + 2.5\ N_2$	-4.7
Mercury fulminate	$Hg(ONC)_2$	$Hg + 2\ CO + N_2$	-1.5
Lead azide	PbN_6	$Pb + 3\ N_2$	-1.5

Figure 22–17 When it is dry, nitrogen triiodide is so sensitive to shock that it explodes at the gentlest touch.

• One test for sensitivity in explosives is to drop a heavy weight on small samples and measure the average distance of fall required to cause an explosion.

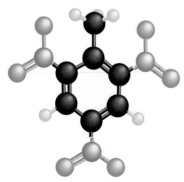

Molecular structure of TNT.

• The great dehydrating power of sulfuric acid drives this reaction (and the previous one) to the right.

Explosives can be characterized as **primary** (or *initiating*) and **secondary.** Primary explosives such as mercury fulminate and lead azide are very sensitive and explode when shocked or heated (Fig. 22–17). Primary explosives are dangerous to handle and are used in small amounts as detonators to start the explosion of larger amounts of secondary explosives. The latter are less sensitive to shock but explode when detonated by a primary explosive.

The earliest explosive was gunpowder, which was developed before A.D. 1000 in China. Early gunpowder (black powder) was a mixture of potassium nitrate, sulfur, and charcoal (carbon) that was moistened and ground up. The dried product was granulated and loaded into cartridges or bombs. This gunpowder was used both for military and civilian applications until the middle of the 19th century. In 1845, nitrocellulose, or guncotton, was discovered as the product of the treatment of cotton with nitric acid. It is far more powerful than early gunpowder and is the major constituent of modern gunpowder. Trinitrotoluene, or TNT, is used for large-scale military purposes. It is manufactured by reaction of toluene ($C_6H_5CH_3$) with a mixture of nitric acid and sulfuric acid, leading to the substitution of three nitro groups (NO_2^-) on the benzene ring:

$$C_6H_5CH_3 + 3\ HNO_3 \text{ (in } H_2SO_4) \longrightarrow C_6H_2CH_3(NO_2)_3 + 3\ H_2O$$

Civilian uses of explosives include mining and road and railroad construction. The development of dynamite in the second half of the 19th century advanced these fields greatly. The major constituent of dynamite is nitroglycerin, made by the addition of high-purity glycerin to a mixture of concentrated nitric and sulfuric acids:

$$C_3H_5(OH)_3 + 3\ HNO_3 \text{ (in } H_2SO_4) \longrightarrow C_3H_5(ONO_2)_3 + 3\ H_2O$$

The Swedish industrialist Alfred Nobel discovered that nitroglycerin could be exploded by shock, and he developed a detonator for it. Nitroglycerin exploded too easily for routine use, however, until Nobel found that absorbing it in a diatomaceous earth (made up of the shells of tiny marine organisms) stabilized it greatly. This mixture was the first dynamite, and Nobel made a vast fortune selling it throughout the world, endowing the Nobel prizes in his will. Modern dynamite still uses nitroglycerin, but mixed with other substances, such as ammonium nitrate or sodium nitrate.

In recent years, ammonium nitrate mixed with small amounts of fuel oil has increasingly supplanted dynamite as an industrial explosive. The advantage of this formulation is its inherent safety. If a misfire occurs and the charge is not set off, the volatile fuel oil evaporates, and the remaining ammonium nitrate is once again safe. Also, the fuel oil can be added after the ammonium nitrate has been placed in the bore hole, avoiding the risks of transporting explosive materials. It is unfortunate that ammonium nitrate and fuel oil are readily obtained by terrorists; such a mixture was used in the 1995 bombing in Oklahoma City.

Ammonium nitrate decomposes nonexplosively above 200°C:

$$NH_4NO_3(s) \longrightarrow N_2O(g) + 2\ H_2O(g) \qquad \Delta G° = -169\ kJ$$

providing a commercial source of dinitrogen oxide (nitrous oxide), which is used as an anesthetic ("laughing gas") and an aerosol propellant. When the mass of ammonium nitrate exceeds a critical value (many tons) or when the ammonium nitrate contains chloride impurities, it may detonate (Fig. 22–18):

$$NH_4NO_3(s) \longrightarrow N_2(g) + 2\ H_2O(g) + \tfrac{1}{2}\ O_2(g) \qquad \Delta G° = -273\ kJ$$

The U.S. government set limits on the quantity of ammonium nitrate that may be stored in one place after a shipload of it blew up in the harbor of Texas City, Texas, in 1947, devastating the town.

• Recall that the major use of ammonium nitrate is as fertilizer; although it is thermodynamically unstable, it is quite safe to handle in small amounts.

Figure 22–18 Ammonium nitrate (NH$_4$NO$_3$) decomposes explosively in the presence of powdered zinc and a catalyst containing the chloride ion.

CHEMISTRY IN PROGRESS

New Explosives

Most "energetic materials" (the term includes both propellants and explosives) in current use have been known for decades; some have been known for centuries. Technology advances slowly in this area because explosives require prolonged and expensive field testing before they can be relied upon. Also, they tend to become military secrets, which prevents widespread participation in their development. Much research on energetic materials, conducted in the former Soviet Union, the United States, Great Britain, and France, became public at the end of the Cold War. This included the discovery of the most powerful single-component chemical explosive yet known—hexanitrohexaazaisowurtzitane (HNIW). HNIW was synthesized in 1986 by U. S. Navy researchers. Its ΔH of explosion equals 5.3 kJ g^{-1}, assuming that its explosive decomposition is fairly closely represented by

$$C_6H_6N_{12}O_{12}(s) \longrightarrow$$
$$3\ CO(g) + 3\ CO_2(g) + 6\ N_2(g) + 3\ H_2O(g)$$

This equation cannot be entirely correct because some oxides of nitrogen do form in the detonation of HNIW. Of more significance for the highly explosive character of HNIW is its density. At 2.05 g cm^{-3}, it is the densest organic compound ever made, and high density in explosives favors high detonation pressures.

The synthesis of HNIW is not easy, and it is expensive (approximately \$600 per pound). Nevertheless, it is a good prospect for commercial as well as military development, and the price should go down as the synthetic method is improved.

The molecular structure of the explosive HNIW $C_6H_6N_{12}O_{12}$. This "wireframe" diagram emphasizes the cage formed by fused 7-, 6-, and 5-membered rings.

Two other new energetic materials with commercial prospects are trinitroazetidine, $(C_3H_3N(NO_2)_3$, called TNAZ), which is less impact sensitive (and therefore safer to transport and use) than many other secondary explosives, and ammonium dinitramide ($NH_4N(NO_2)_2$), which is remarkable in containing no carbon or chlorine. The latter is a possible replacement for ammonium perchlorate (NH_4ClO_4) in the solid-fuel boosters for the space shuttles. A replacement is needed because ammonium perchlorate gives chlorine (see Cumulative Problem, page 220) and also some hydrochloric acid when burned during launches. These substances are expelled as part of the rocket exhaust and harm the environment.

SUMMARY

22–1 An ideal chemical process gives the desired product in 100% yield, at an acceptable and controllable rate, and in acceptable purity or in a form that is easily purified. It causes little pollution and provides for nonpolluting disposal of by-products as an integral part of its design. Chemical processes generally involve many steps and intermediate chemical compounds. Most large-volume chemicals are little used by consumers, but they are essential in making the products that consumers do use. Raw materials for industrial chemistry come from both geochemical and biochemical processes. Geochemical transformations start with the crystallization of a **magma,** a molten silicate fluid, to form **igneous** rocks. **Sedimentary** rocks come from the weathering, dissolution, and precipitation of other minerals. Another type of change is the **metamorphosis** of sedimentary rocks under heat and pressure. The most important chemical raw materials that come from biological activity are coal, petroleum, natural gas, phosphate rock, and sulfur (in some deposits).

22-2 Sulfuric acid was formerly produced in the **lead-chamber** process, in which oxides of nitrogen serve as oxygen carriers for the oxidation of SO_2 to SO_3. The **contact process,** which uses the direct oxidation of SO_2 to SO_3 in the presence of a catalyst, followed by absorption of the SO_3 in H_2SO_4, is currently the main synthetic route to sulfuric acid. Sulfur for sulfuric acid comes from metal sulfides or from deposits of elemental sulfur that are mined by the **Frasch process.** Sulfur present as H_2S in natural gas and petroleum is another source. Most uses of sulfuric acid are indirect in that sulfur from the acid does not become a part of the product but ends up as a sulfate waste or as **spent acid.** Sodium sulfate is synthesized from SO_2, NaCl, and O_2 in the **Hargreaves process** and is a by-product of rayon manufacture.

22-3 The processing of wood to wood pulp for paper and cardboard uses the **sulfate process,** in which the digestion medium is basic, and the **sulfite process,** in which it is weakly acidic. The most important use of sulfuric acid is in the production of phosphate fertilizers. Phosphate rock (principally $Ca_5(PO_4)_3F$) is treated with H_2SO_4 to give **superphosphate** fertilizer. The same type of reaction using an excess of sulfuric acid produces **wet-process** phosphoric acid. A higher grade of phosphoric acid is afforded by the **furnace process,** in which phosphate rock is first reduced to elemental phosphorus, then burned in air to P_4O_{10}, and reacted with water to give H_3PO_4. The reaction of phosphoric acid with ammonia gives ammonium phosphate, the major phosphorus-containing fertilizer in current use. Sodium phosphates from the neutralization of phosphoric acid are used in cleaning products and in detergents.

22-4 The formation of compounds between molecular nitrogen and other elements is called **nitrogen fixation.** In the **cyanamide process,** calcium carbide (CaC_2) was generated by heating lime (CaO) and coke (C) in an electric furnace and then reacted with nitrogen to fix nitrogen in the compound calcium cyanamide ($CaCN_2$). The **Haber–Bosch process** uses the direct combination of N_2 with H_2 to yield NH_3. To be practical, the process requires a compromise between low temperature, which favors high yield of ammonia at equilibrium, and high temperature, which increases the rate of attainment of equilibrium. A catalyst speeds up the reaction, and high pressure favors higher equilibrium yields of ammonia.

22-5 The oxidation of ammonia by air in the **Ostwald process** produces nitric acid (HNO_3). Its major use is in reaction with ammonia to give ammonium nitrate (NH_4NO_3), which is used for fertilizer and as an industrial explosive. The important reducing agent hydrazine (N_2H_4) is made from aqueous ammonia by oxidation with aqueous hypochlorite ion ($ClO^-(aq)$) in the **Raschig synthesis.** Many explosives are nitrogen-containing compounds. **Primary explosives** detonate when shocked or heated; **secondary explosives** are less sensitive and require a primary explosion to induce detonation. Gunpowder, nitrocellulose, TNT, and nitroglycerin are all nitrogen-containing explosives. Dynamite is a mixture of nitroglycerin with diatomaceous earth. Ammonium nitrate mixed with fuel oil has largely supplanted dynamite as an industrial explosive.

PROBLEMS

Note: Answers to blue-numbered problems are given in Appendix F. Problems that are more challenging are indicated with asterisks.

The Chemical Industry

1. Define the terms *batch process* and *continuous process*. How do the designs of the two types of processes differ from each other?

2. Define the terms *by-products* and *waste products*. Is there a distinction between the two?

3. In what ways are geochemical processes similar to and different from processes in the chemical industry? Draw a process diagram for the evolution of marble from calcium and carbonate ions separately present in an igneous rock.

4. In what ways are biochemical processes analogous to and different from processes in the chemical industry? Draw a process diagram for the production of sulfur, beginning with calcium sulfate, carbon dioxide, and marine bacteria.

5. Identify a useful source of each of the following chemical elements as specifically as you can, based on information in the chapter:
   (a) Sulfur   (c) Molybdenum
   (b) Carbon

6. Identify a useful source of each of the following chemical elements as specifically as you can, based on information in the chapter:
   (a) Phosphorus   (c) Magnesium
   (b) Silicon

Production of Sulfuric Acid

7. In the lead-chamber process, the sulfuric acid was contained in a lead container. Explain the great resistance of lead to corrosion by sulfuric acid using the information in Table 4–1.

8. Concentrated sulfuric acid can be stored for years in containers made of mild steel; however, mild steel pipes that are used to carry a stream of the same acid corrode quickly and burst. Suggest an explanation.

9. Propose several ways in which predictions from Le Chatelier's principle could be used to improve the yield of $SO_3(g)$ from $SO_2(g)$ and air.

10. Give a thermodynamic analysis of the Claus process; that is, determine the $\Delta H°$ and $\Delta S°$ of the main reaction and predict conditions that favor a high yield of sulfur. Take S(s) as the physical form of sulfur.

11. Residual SO_2 can be removed from the effluent gases of sulfuric acid plants by oxidation with H_2SO_5, as explained in the text. Suppose that the electrolysis system that supplies the H_2SO_5 fails. Would it be thermodynamically feasible to substitute an aqueous solution of H_2O_2 as the oxidant for the SO_2? Consult Appendix D for necessary data.

12. Amounts of sulfur in a sample can be measured by burning the sample in oxygen and then collecting the oxides of sulfur in a dilute solution of hydrogen peroxide. What is the product formed from this reaction? Write a balanced equation. Suggest a good way to determine quantitatively the amount of the product that is formed.

13. Write a balanced overall chemical equation for the production of sulfuric acid from zinc sulfide ore.

14. Write a balanced overall chemical equation for the production of sulfuric acid from hydrogen sulfide in natural gas.

Uses of Sulfuric Acid and Its Derivatives

15. Hot concentrated sulfuric acid reacts with phosphorus to give sulfur dioxide, phosphoric acid, and water. Write a balanced chemical equation for this reaction.

16. Hot concentrated sulfuric acid reacts with sulfur to give sulfur dioxide and water. Write a balanced chemical equation for this reaction.

17. Write chemical equations to represent the production of HF and of HCl, assuming that supplies of CaF_2 and of NaCl are available.

18. Write an equation to represent the oxidation of copper by hot concentrated sulfuric acid.

19. Describe (via chemical equations) how sodium sulfide for use in the sulfate (Kraft) process can be prepared starting from sulfuric acid and other easily accessible materials.

20. Describe (via chemical equations) how sodium tripolyphosphate for detergents can be prepared starting from phosphoric acid and other easily accessible materials.

21. Predict the product(s) of the reaction of disulfuric acid ($H_2S_2O_7$) with water.

22. Phosphoric acid (H_3PO_4) can be kept pure only in the crystalline state. Above its melting point of 42.4°C, pairs of H_3PO_4 units gradually lose water. What is the other product of this dehydration reaction? What is its structure?

Nitrogen Fixation

23. A minor source of fixed nitrogen is emission from the inside of the earth via volcanoes. One estimate of the total amount of fixed nitrogen generated per year from this source is 2×10^8 kg. How many atoms of fixed nitrogen are emitted per second, on the average, by volcanoes?

24. Recent observations of lightning flashes using a new method suggest that the average lightning stroke produces 10^{27} molecules of NO. The rate of lightning flashes worldwide is about

100 per second. According to these data, how much nitrogen (in metric tons) is fixed annually by lightning flashes?

25. Write an overall equation to represent the production of NH_3 by the cyanamide process, as described in the text. Assume that the starting materials are limestone ($CaCO_3$), coke (C), nitrogen (N_2), and water. Determine the $\Delta H°$ of the process represented by your equation, per mole of NH_3 produced.

26. Suppose that, in the previous problem, the calcium hydroxide and carbon dioxide produced are recycled to give the calcium carbonate starting material. Write the overall equation in this case, and calculate the $\Delta H°$ per mole of NH_3 produced.

27. Some chemical plants manufacture ammonia from the Haber–Bosch process using pressures as high as 700 atm, rather than the 200 atm described in the text. What are the advantages and disadvantages of working at these higher pressures? For a given yield of product, should a higher or a lower temperature be used if the pressure is raised? Is the reaction then faster or slower?

28. Suppose that a catalyst is developed that allows the rapid production of ammonia from the elements at 600 K. Would this be useful in the Haber–Bosch process? Estimate the equilibrium constant for the formation of two moles of NH_3 at this temperature, using the van't Hoff equation.

Chemicals from Ammonia

29. For each of the three steps in the Ostwald process for making nitric acid from ammonia, state whether a high yield of product is favored by low or by high temperature.

30. For each of the three steps in the Ostwald process for making nitric acid from ammonia, state whether a high yield of product is favored by low or by high total pressure.

31. Write balanced equations to represent the reaction of concentrated nitric acid with Cr(s) and with Fe(s). Assume that the nitric acid is reduced to $NO_2(g)$ and that Cr and Fe are oxidized to the +3 oxidation state.

32. Write balanced equations to represent the reaction of concentrated nitric acid with Ni(s) and Tl(s), assuming that the nitric acid is reduced to NO(g) and that Ni and Tl are oxidized to the +2 and +3 oxidation states, respectively.

33. *Trans*-tetrazene has the formula N_4H_4. It consists of a chain of four nitrogen atoms with the two terminal nitrogen atoms bonded to two hydrogen atoms each. Draw a Lewis structure for this compound.

34. Diimine (N_2H_2) cannot be isolated, but there is evidence that it exists in solution.
    (a) Draw a Lewis structure for this compound.
    (b) Would the conversion of hydrazine to diimine in solution require an oxidizing agent or a reducing agent?

35. Two reduction half-reactions involving hydrazine (N_2H_4) are

$$N_2H_4 + 2\,H_2O + 2\,e^- \longrightarrow 2\,NH_3 + 2\,OH^-$$
$$\mathscr{E}° = 0.1\ \text{V}$$
$$N_2 + 4\,H_2O + 4\,e^- \longrightarrow N_2H_4 + 4\,OH^-$$
$$\mathscr{E}° = -1.16\ \text{V}$$

Determine whether hydrazine is thermodynamically stable against disproportionation to N_2 and NH_3 at pH 14.

36. In our discussion of conversions among N_2, N_2H_4, and NH_3, we did not mention a compound of nitrogen named hydroxylamine (NH_2OH).
    (a) What is the oxidation number of nitrogen in NH_2OH?
    (b) Two reduction half-reactions involving NH_2OH are

$$2\,NH_2OH + 2\,e^- \longrightarrow N_2H_4 + 2\,OH^- \qquad \mathscr{E}° = -0.73\ \text{V}$$
$$N_2 + 4\,H_2O + 2\,e^- \longrightarrow 2\,NH_2OH + 2\,OH^-$$
$$\mathscr{E}° = -1.59\ \text{V}$$

Determine whether hydroxylamine is thermodynamically stable against disproportionation to N_2 and N_2H_4 at pH 14.

37. The reaction between liquid hydrazine and liquid nitric acid is suggested to boost a specialized type of rocket because the products, $N_2(g)$ and $H_2O(\ell)$, are completely innocuous in the atmosphere. Write a balanced equation, and compute $\Delta H°$ for this reaction.

38. Write a balanced chemical equation for a reaction between hydrazine and dinitrogen tetraoxide that gives only nitrogen and water. Compute $\Delta H°$ for this reaction.

39. Describe a practical reaction sequence by which $N_2O_4(s)$ could be obtained from the elements.

40. Describe a practical reaction sequence by which $N_2O(g)$ could be obtained from the elements.

41. The decomposition of TNT, according to Table 22–3, proceeds by the reaction

$$2\,C_6H_2(NO_2)_3CH_3 \longrightarrow 3\,N_2 + 7\,CO + 5\,H_2O + 7\,C$$

If the carbon produced could be converted to CO, additional heat would be given off, and the explosion would be still more powerful. Explain why ammonium nitrate (NH_4NO_3) is sometimes mixed with TNT.

42. During World War II, the Allies used to grind up downed Nazi war planes, which contained aluminum, and mix them in with ammonium nitrate to make bombs that were then dropped over Germany. Explain the reason for putting in the aluminum from the airplanes.

43. (a) Write a plausible equation for the explosive decomposition of the energetic material ammonium dinitramide ($NH_4N(NO_2)_2$).
    (b) The $\Delta H_f°$ of ammonium dinitramide is -36 kJ mol^{-1}. Use your equation to estimate a ΔH of explosion of this compound in kJ g^{-1}.

44. Using data given in the chapter and in Appendix D, estimate the enthalpy of formation of the explosive HNIW.

Additional Problems

45. Identify the state of lower Gibbs function in each of the following pairs:
   (a) Aspirin versus (coal, hydrogen, and air).
   (b) Seawater versus (pure water, sodium chloride, and magnesium carbonate).
   (c) (Vitamin A and oxygen) versus (carbon dioxide and water).

46. Define *igneous*, *sedimentary*, and *metamorphic* rocks, based on their different sources. Give an example of a mineral found in each kind of rock.

47. How do plants carry out chemical reactions such as the synthesis of sugars, which correspond to an increase of Gibbs function relative to the starting materials of water and carbon dioxide? How do animals carry out similar processes, such as the building of proteins that have higher Gibbs function than the starting materials?

48. In a recent year, the total consumption of carbonated beverages in the United States corresponded to 63 *billion* 12-oz cans or bottles. If a 12-oz can has a volume of 355 mL, and the concentration of dissolved CO_2 is 0.15 M, calculate the total mass of CO_2 produced and used for making carbonated beverages during that year.

49. Identify four naturally occurring sources of sulfur or sulfur compounds that can be converted into sulfuric acid. How have these changed in importance throughout history?

50. Both SO_2 and oxides of nitrogen (NO and NO_2) are emitted in flue gases from coal-burning power plants. Write chemical equations representing a possible process in which the SO_2 pollutants are removed from smokestack emissions and sulfuric acid is produced.

51. Use thermodynamics to compare the oxidation of SO_2 in the absence of NO_2

$$SO_2(g) + \tfrac{1}{2} O_2(g) \longrightarrow SO_3(g)$$

with the oxidation of SO_2 by NO_2:

$$SO_2(g) + NO_2(g) \longrightarrow SO_3(g) + NO(g)$$

Which has the larger thermodynamic driving force $(-\Delta G°)$ at 25°C? At low temperatures, does the NO_2 act by increasing the yield or by speeding up the reaction?

52. Identify the oxidizing agent in the Glover–Gay-Lussac reaction, which produces SO_3 from SO_2. What change in oxidation state does it undergo? Compare with the corresponding oxidizing agent in the earlier lead-chamber process.

53. The standard reduction potential for the half-reaction

$$S_2O_8^{2-}(aq) + 2\ e^- \longrightarrow 2\ SO_4^{2-}(aq)$$

is 2.0 V. Use the Nernst equation to compute the minimum voltage required to operate an electrolysis unit that converts 1 M SO_4^{2-} to 0.5 M $S_2O_8^{2-}$ and hydrogen at a pressure of 0.10 atm if the pH is 0.

54. The world production of pure H_2SO_4 in a recent year was 5.51×10^{10} kg. Suppose that all locations converted sulfur to H_2SO_4 at 99.97% efficiency but lost the rest of the sulfur as SO_2 to the atmosphere. What then was the total mass of emitted sulfur dioxide (in kilograms)?

55. Write a chemical equation for (a) a precipitation reaction, (b) an acid–base reaction, and (c) an oxidation–reduction, in each of which aqueous sulfuric acid is a reactant.

56. When sulfuric acid is used to remove a layer of oxide from a piece of steel, where do the metallic elements that comprise part of the oxide scale go? Are they removed from the metal and carried away in the acid, or do they remain with the piece of metal?

* 57. Most sulfuric acid (other than that used in making fertilizer) does not form a part of the final product. It lingers inconveniently as spent sulfuric acid that is both diluted and contaminated. Spent acid must not be dumped in the environment. You propose to regenerate H_2SO_4 by heating spent acid to high temperatures (about 1000 K) in the presence of natural gas (CH_4) to make $SO_2(g)$ by the reaction

$$4\ SO_3(g) + CH_4(g) \longrightarrow CO_2(g) + 4\ SO_2(g) + 2\ H_2O(g)$$

and then recycling the $SO_2(g)$ to make more H_2SO_4.
   (a) Estimate $\Delta G°$ at 1000 K for your proposed reaction.
   (b) A critic asserts that the reaction will probably proceed as follows:

$$SO_3(g) + CH_4(g) \longrightarrow CO_2(g) + H_2O(g) + H_2S(g)$$

   Estimate $\Delta G°$ at 1000 K for the critic's proposed reaction.
   (c) How can you tell which reaction will occur? Explain.

58. The Na_2S for the sulfate (Kraft) process can be made by heating Na_2SO_4 with coke (C). Write a balanced chemical equation for this reaction. Can this reaction be used to supply heat for the digestion of the wood, or does it consume heat?

59. The P—O—P linkages in diphosphoric acid are extended by adding other phosphate groups (through reaction with H_3PO_4 and elimination of water). The resulting long-chain phosphate is called "metaphosphoric acid." Determine its empirical formula, assuming that the chain is very long.

60. Suppose that a fertilizer plant produces phosphoric acid through the wet-acid process and uses it to make triple superphosphate fertilizer ($Ca(H_2PO_4)_2 \cdot H_2O$). Write a balanced equation for the overall reaction. How many moles of gypsum are generated per mole of $Ca(H_2PO_4)_2 \cdot H_2O$?

61. In fertilizer manufacture, it is conventional to measure relative amounts of phosphorus in different fertilizers by arbitrarily assuming that the phosphorus is present as P_2O_5, dividing the mass of this P_2O_5 by the total mass, and expressing that mass as a percentage. For example, the formula of triple superphosphate is rewritten as follows:

$$Ca(H_2PO_4)_2 \cdot H_2O = (P_2O_5)(CaH_6O_4)$$

The ratio of the mass of P_2O_5 to the total mass is then the ratio of the molar mass of P_2O_5 to the molar mass of the formula unit, or $141.94/252.07 = 0.563$, so this fertilizer is "56.3% P_2O_5." Carry out the corresponding calculation for ordinary superphosphate fertilizer, which contains seven moles of gypsum for every three moles of $Ca(H_2PO_4)_2 \cdot H_2O$. Account for the origin of the word "triple" in "triple superphosphate."

62. The disposal of by-product gypsum ($CaSO_4 \cdot 2H_2O(s)$) is a significant problem for fertilizer manufacturers.
   (a) A typical fertilizer plant may produce 100 metric tons of phosphoric acid per day by the wet-acid process (1 metric ton = 10^3 kg). What mass of gypsum is produced per *year*?
   (b) The density of crystalline gypsum is 2.32 g cm^{-3}. What volume of gypsum is generated per year? (In fact, the volume is several times *larger* because of the porosity of the gypsum produced.)

63. During the early 1970s, when the price of petroleum and natural gas rose rapidly, the price of nitrogen-based fertilizers increased rapidly as well. Explain why the two commodities are so closely linked.

64. A piece of hot platinum foil is inserted into a container filled with a mixture of ammonia and oxygen. Brown fumes appear near the surface of the foil. Write balanced equations for the reactions taking place.

65. Nitrogen oxides (NO_x) from power stations, steam generators, and other fossil-fuel–burning sources are serious air pollutants. One promising technology (called NOxOUT) for removing NO_x from stack emissions injects an aqueous solution of urea (($NH_2)_2CO$) into various points of the combustion system. Urea reacts with the nitrogen oxides to give N_2, CO_2, and H_2O.
   (a) Write a balanced chemical equation to represent the reaction of urea with NO_2.
   (b) Write a second balanced chemical equation to represent the reaction of urea with NO.

66. The concentration of NO_3^- (aq) in a solution can be determined by adding sulfuric acid and mercury and then collecting the gaseous NO that the reaction generates:

$$2\ NO_3^-(aq) + 8\ H^+(aq) + 3\ SO_4^{2-}(aq) + 6\ Hg(\ell) \longrightarrow$$
$$3\ Hg_2SO_4(s) + 4\ H_2O(\ell) + 2\ NO(g)$$

A 357-mL sample of a solution is treated in this way and generates 11.46 mL of NO(g), measured at a pressure of 0.965 atm and a temperature of 18.8°C. Compute the concentration of nitrate ion in the solution.

67. The formation of hydrazine in the Raschig synthesis is thought to be a two-step process that involves NH_2Cl as an intermediate. Write balanced equations for the two steps in the Raschig synthesis.

68. Molecules of hydrazine (N_2H_4) have a sizable dipole moment. Sketch a structure that is definitely *eliminated* as the structure of the hydrazine molecule by this information.

* 69. (a) Compute the $\Delta G°$ (at 25°C) of the reaction

$$6\ Fe_2O_3(s) + N_2H_4(aq) \longrightarrow$$
$$4\ Fe_3O_4(s) + N_2(g) + 2\ H_2O(\ell)$$

   (b) As explained in the text, this reaction is used to prevent further rusting of the walls of boilers. A boilermaker asks you about a proposal to improve this procedure by making the solution of hydrazine strongly acidic (to "burn off the worst of the rust as the hydrazine simultaneously seals off the rest"). Evaluate this idea.

70. Determine how much (in grams) of each ingredient to mix to prepare 100 g of gunpowder. Use the information in Table 22–3.

71. You have the following chemicals in the laboratory—water, iron(II) sulfide, concentrated sulfuric acid, barium chloride, air. State what procedures you could use to prepare the following: H_2S, $BaSO_4$, SO_2. Include balanced chemical equations.

72. Nitric acid can be prepared in the laboratory by heating a mixture of sodium nitrate and sulfuric acid. A mixture of nitric acid with water distills out of the mixture and may be condensed in a beaker or flask cooled by ice water.
   (a) Write a balanced equation to represent the reaction that generates the nitric acid.
   (b) From the information in the chapter, what is the maximum concentration of HNO_3 in the liquid that distills out?

73. Minor impurities often react with nitric acid prepared by the method described in problem 72 to turn it brown.
   (a) What substance is responsible for this color?
   (b) What kind of reactions are taking place between the nitric acid and the impurities?

23

Chemistry of the Halogens

CHAPTER OUTLINE

The perfluorinated polymer Teflon is used to coat copper wire in computers.

The halogens are reactive elements, so reactive that they occur terrestrially only in combination with other elements. The elemental halogens avidly accept electrons to join in ionic compounds (although they form many covalent compounds as well). Devising conditions to wrest these electrons away and so prepare the pure elements posed a challenge to chemists for many years. The 19th century saw the development of practical processes to obtain elemental chlorine from sodium chloride and, toward its close, the first preparation of elemental fluorine. The halogens form compounds with nearly all the elements in the periodic table. Even the noble gases xenon and krypton are induced into chemical compounds by the oxidizing power of elemental fluorine.

23-1 CHEMICALS FROM SALT

By the second half of the 18th century, several important industries in Europe and North America faced a growing crisis in the supply of raw materials. The glass industry relied on soda ash (sodium carbonate, Na_2CO_3) or potash (potassium carbonate, K_2CO_3) as a flux to lower the viscosity of the molten glass (see Section 21–4). As their common names imply, the two carbonates were derived from the leachings of ashes: soda ash from the ashes of sea plants, and potash from the ashes of trees and shrubs. By 1750, demand from the glass industry was outrunning the supplies from these sources, which varied in quality in any case.

• The other starting materials for glass are lime (CaO) and silica (SiO_2).

Soapmakers faced a supply problem of their own because they needed a cheap base to react with vegetable oils and animal fats to make soap. One base (lime) was readily available. It was produced by high-temperature decomposition of the limestone that was abundantly available throughout the world:

• The soapmaking reaction is discussed in Section 24–2.

$$CaCO_3(s) \xrightarrow{850°C} CaO(s) + CO_2(g)$$

In the 18th century, lime kilns were among the largest industrial structures in Europe. The lime could be **slaked** by the addition of water to give a sparingly soluble hydroxide:

$$CaO(s) + H_2O(\ell) \longrightarrow Ca(OH)_2(s)$$

In this form, it served well to neutralize acids but had no use in the soap industry because the calcium salts that formed from its reaction with oils were insoluble in water. A soap must be water-soluble.

An expanding demand for good clothes greatly increased the need to bleach cotton and linen in the mid-18th century. In early bleaching methods, the cloth or fibers were repeatedly exposed to lactic acid from sour milk, bases from wood ashes, and the bright light of the sun. Cotton took up to three months to bleach, and linen up to six, and the process depended on the weather and the supply of milk. In the closing years of the 18th century, the lead-chamber process (see Section 22–2) began to supply sulfuric acid that could replace the sour milk and greatly hasten that part of the process. The lack of a cheap base and a bleaching agent more effective than sunlight created major bottlenecks.

This section describes how salt (sodium chloride, NaCl) provided an answer to all these problems. Processes were developed to convert salt into sodium carbonate and sodium hydroxide, and the by-products of that conversion included compounds of chlorine that could be made into effective bleaches. Several competing processes rose and fell over the course of the 19th century, with a central role played by the utilization of waste materials to achieve economies. These processes, together with the production of sulfuric acid (discussed in Chapter 22), dominated the 19th-century chemical industry and play an important role to this day.

The Leblanc Process

In 1775, the French Academy of Sciences offered a prize for a satisfactory process to make sodium carbonate from sodium chloride. The problem was solved during the next 15 years by a French doctor and amateur chemist, Nicolas Leblanc. Figure 23–1 outlines the **Leblanc process.**

• This is the Mannheim process mentioned in Section 22–3.

The first step is the production of salt cake, or sodium sulfate (Na_2SO_4). The process Leblanc used was the standard one at the time, adding sulfuric acid to sodium chloride at temperatures of 800 to 900°C:

$$2\ NaCl(s) + H_2SO_4(aq) \longrightarrow Na_2SO_4(aq) + 2\ HCl(g)$$

The sodium sulfate was ground up, mixed with limestone and charcoal or powdered coal, and put in a furnace. The reactions that then took place can be summarized in two equations:

$$Na_2SO_4(s) + 2\ C(s) \longrightarrow Na_2S(s) + 2\ CO_2(g)$$
$$Na_2S(s) + CaCO_3(s) \longrightarrow Na_2CO_3(s) + CaS(s)$$

Figure 23–1 In the original version of the Leblanc process, HCl, CO_2, and CaS were all waste products generated along with the desired Na_2CO_3. Later versions of the process (below black line) converted HCl into useful Cl_2 and recovered sulfur from the CaS to make sulfuric acid.

The result was a porous gray-black mixture called **black ash,** which consisted primarily of calcium sulfide and sodium carbonate, with some unreacted calcium carbonate and other impurities. Filtering systems were developed to leach out the soluble sodium carbonate, leaving behind the solid calcium sulfide as a by-product.

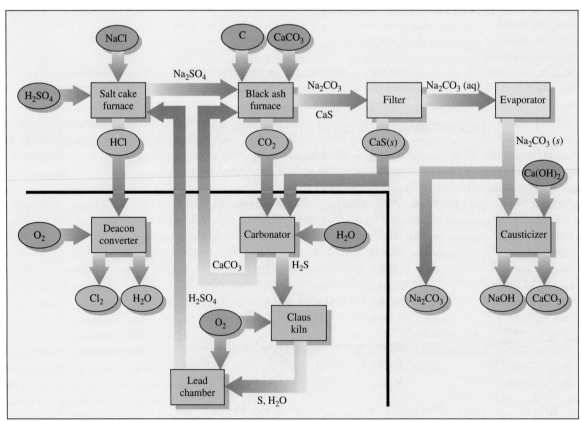

The liquid was then heated in a furnace to remove sulfur and carbon impurities as $SO_2(g)$ and $CO_2(g)$ and leave behind the desired sodium carbonate. If sodium hydroxide was needed, the sodium carbonate was boiled with slaked lime:

$$Na_2CO_3(aq) + Ca(OH)_2(aq) \longrightarrow CaCO_3(s) + 2\ NaOH(aq)$$

The precipitation of the insoluble calcium carbonate left a solution of sodium hydroxide that could be concentrated further or evaporated to dryness.

The use of the Leblanc process spread rapidly, especially in Great Britain, and caused an increased demand for sulfuric acid as a starting material. A look at the overall equation for the process

$$2\ NaCl + H_2SO_4 + 2\ C + CaCO_3 \longrightarrow Na_2CO_3 + 2\ CO_2 + CaS + 2\ HCl$$

shows that, in addition to the primary product, sodium carbonate, and the carbon dioxide, two significant by-products result—calcium sulfide and hydrogen chloride. The former was dumped at sea or left in piles on land, where it weathered to produce noxious H_2S and SO_2 gases. Hydrogen chloride was initially vented as an acidic gas, which laid waste the vegetation for miles around the plants. Tall chimneys (hundreds of feet high) were built, but they only dispersed the HCl farther. In the next remedy attempted, the gaseous hydrogen chloride was passed into towers in which water trickled downward over a packing of coke. The hydrogen chloride dissolved to create a dilute solution of hydrochloric acid that was run off into streams and rivers, substituting water pollution for air pollution. In 1863, the Alkali Act prohibited such pollution in Great Britain. Although some condemned the law as excessive interference by the government in private business, it stimulated the search for ways of using the hydrochloric acid. Efficient processes were soon developed to convert hydrochloric acid to bleaching powder, making a noxious waste into a useful product. Before we consider these processes, let us return to a somewhat earlier period to examine the role of chlorine and its compounds as bleaching agents.

• This law was perhaps the first major piece of environmental legislation.

Chlorine and Bleaches

Chlorine had been discovered in 1774 by the Swedish chemist Carl Wilhelm Scheele, who prepared it in its elemental form by the reaction of hydrochloric acid with pyrolusite, a mineral composed of MnO_2:

$$4\ HCl(aq) + MnO_2(s) \longrightarrow Cl_2(g) + MnCl_2(aq) + 2\ H_2O(\ell)$$

Scheele did not realize that the greenish yellow gas he had produced was an element, and it remained for Humphrey Davy to identify it as such in 1811 and to name it (from the Greek word *chloros*, meaning "green"). In the meantime, Berthollet and de Saussure had already described chlorine's bleaching properties in 1786. These were unsatisfactory in several ways, however. Chlorine disintegrated the cloth unless monitored carefully; it was difficult to handle, and, at the time, could not be transported. The Scottish chemist Charles Tennant made a significant advance in 1799 when he patented a material that he called **bleaching powder,** which was formed by saturating slaked lime with chlorine:

$$Ca(OH)_2(s) + Cl_2(g) \longrightarrow CaCl(OCl)(s) + H_2O(\ell)$$

Together with sulfuric acid, bleaching powder (also called "chlorinated lime") greatly reduced the time needed to bleach cotton and linen, so that by the 1830s, one week (rather than several months) was sufficient to bleach cotton goods.

Bleaching powder acts in aqueous solution by releasing hypochlorite ion (OCl^-):

$$CaCl(OCl)(s) \longrightarrow Ca^{2+}(aq) + Cl^-(aq) + OCl^-(aq)$$

which in turn can decompose, liberating oxygen to carry out the bleaching. One disadvantage of bleaching powder is that it decomposes on standing to calcium chloride and oxygen:

$$CaCl(OCl)(s) \longrightarrow CaCl_2(s) + \tfrac{1}{2} O_2(g)$$

This can be avoided by preparing calcium hypochlorite ($Ca(OCl)_2$), which also produces twice as much hypochlorite ion in solution, making it more effective on the basis of mass. Modern household bleach uses the sodium salt of hypochlorite ion ($NaOCl$) rather than the calcium salt. Sodium chlorite ($NaClO_2$) is an even stronger bleach that is widely used in industrial applications, such as paper bleaching.

For almost a century after its discovery in 1774, the major method for making chlorine for bleach was the original reaction used by Scheele. This is an extremely wasteful method because all of the manganese and much of the chlorine is lost as $MnCl_2$. By the mid-19th century, hydrochloric acid (the noxious by-product of the Leblanc process) was in extensive use for bleach manufacture, and a less wasteful method of oxidizing it was needed. Between 1868 and 1874, the British chemist and industrialist Henry Deacon developed a process to convert gaseous hydrogen chloride into chlorine over a copper-chloride catalyst:

$$2\ HCl(g) + \tfrac{1}{2} O_2(g) \longrightarrow Cl_2(g) + H_2O(g)$$

The **Deacon process** was widely adopted, and hydrochloric acid from the Leblanc process became the major source of chlorine for bleaches.

Decline of the Leblanc Process

As described later in this section, a competitive process for making sodium carbonate came into use in Belgium in the mid-19th century. As time went on, the sodium carbonate from the Leblanc process became a less important product economically than hydrogen chloride, and it was sold at a loss to meet this competition. Profits were still made from the Leblanc process only because of the coproduction of bleaching powder from hydrogen chloride, the former waste product. This turnabout illustrates the way in which the chemical industry changes to meet changing circumstances.

With the successful capture and use of the hydrogen chloride from the Leblanc process, only one major waste product remained—calcium sulfide. This presented a disposal problem, and there was a supply problem with sulfuric acid as well. At the time, sulfur was obtained from mines in Sicily in elemental form and from ores containing pyrite (FeS_2). Supplies and prices were often uncertain. It was clear that the recovery of the sulfur from the calcium sulfide wastes would lead to significant savings in cost. By 1888, a process had been developed in which a slurry of "tank waste" (primarily a water suspension of calcium sulfide) was passed through a series of cylinders where it came into contact with carbon dioxide. The reaction produced hydrogen sulfide:

$$CaS(s) + CO_2(aq) + H_2O(\ell) \longrightarrow H_2S(aq) + CaCO_3(s)$$

Hydrogen sulfide could then be oxidized to sulfur with an iron(III) oxide catalyst:

$$H_2S(g) + \tfrac{1}{2} O_2(g) \longrightarrow S(s) + H_2O(g)$$

Although ingenious, this discovery came too late to save the Leblanc process from the competition of the Solvay process and the electrolytic production of chlorine, the processes we consider next.

The Solvay Process

Figure 23–2 shows a flow diagram for the **Solvay process.** The first and most important reaction is

$$2\,NaCl(aq) + 2\,NH_3(aq) + 2\,CO_2(g) + 2\,H_2O(\ell) \longrightarrow$$
$$2\,NaHCO_3(s) + 2\,NH_4Cl(aq)$$

• The coefficients of 2 appear in this equation to avoid fractional coefficients in the equations for the subsequent steps.

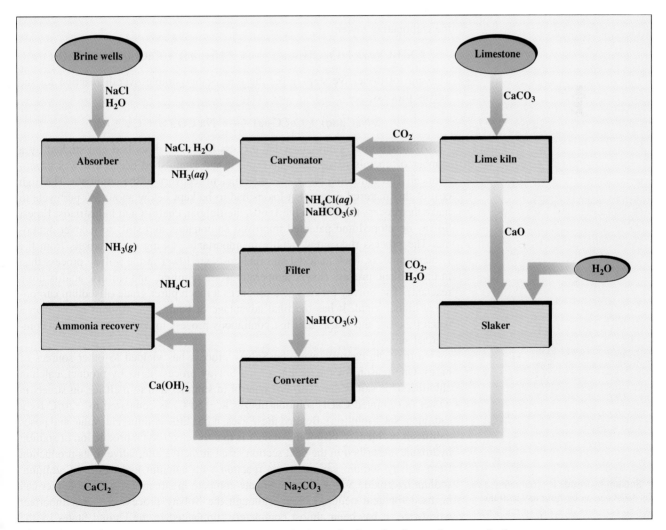

Figure 23–2   The net reaction in the Solvay process is the production of sodium carbonate and calcium chloride from salt and calcium carbonate. Ammonia plays a crucial role in making the process work but is not consumed at all, except for inevitable small losses.

This reaction is carried out in a "carbonating tower," a tall reactor in which brine (concentrated aqueous NaCl) is saturated with ammonia and carbon dioxide. The sodium hydrogen carbonate (sodium bicarbonate, $NaHCO_3$) that forms is relatively insoluble and is removed by filtration on huge rotating drum filters. It is then decomposed thermally in a rotary dryer at over 270°C to give sodium carbonate as the product:

$$2\ NaHCO_3(s) \longrightarrow Na_2CO_3(s) + H_2O(g) + CO_2(g)$$

The carbon dioxide freed in this reaction is returned to the carbonating tower, where it is supplemented by CO_2 generated in a lime kiln:

$$CaCO_3(s) \longrightarrow CaO(s) + CO_2(g)$$

The lime (CaO) formed in this reaction is slaked:

$$CaO(s) + H_2O(\ell) \longrightarrow Ca(OH)_2(aq)$$

The calcium hydroxide is then used in a final step to recover ammonia for reuse in the carbonating tower:

$$2\ NH_4Cl(aq) + Ca(OH)_2(aq) \longrightarrow 2\ NH_3(aq) + CaCl_2(aq) + 2\ H_2O(\ell)$$

The sum of these five steps gives the remarkably simple overall reaction for the Solvay process:

$$2\ NaCl(aq) + CaCO_3(s) \longrightarrow Na_2CO_3(s) + CaCl_2(aq)$$

Although this "ammonia-soda" process had been described and studied by a number of chemists beginning in 1800, it had languished because it was uneconomical. The stumbling block was the efficient recovery of the ammonia. The high cost of this material meant that losses had to be kept below about 5% per cycle to make the process feasible. In the 1860s, the Belgian chemist and industrialist Ernest Solvay developed and patented improved carbonators and stills to recover ammonia. The process then grew rapidly in importance, taking over markets from the Leblanc process. Raw material and fuel costs are lower in the Solvay process than in the Leblanc process (lower temperatures are required, and sulfuric acid is not a reactant). The Solvay process also produces a much purer grade of sodium carbonate at higher yield. It has the final advantage of being a continuous rather than a batch process, and it was the first continuous process to achieve widespread commercial success.

In the 20th century, even the Solvay process has yielded to other sources of bases. One reason for the decline of the Solvay process is its by-product, calcium chloride. This salt is used for dust control in summer and to melt ice on streets in winter (a saturated $CaCl_2$ solution freezes at −50°C, as compared to −20°C for a saturated NaCl solution). Beyond these uses, it has little commercial value and poses a difficult problem of disposal. Another is that the electrolytic production of sodium hydroxide, described in the next section, is an attractive alternative to its production from Solvay sodium carbonate via reaction with calcium hydroxide. In addition, sodium hydroxide and sodium carbonate compete in situations in which either can be used for acid neutralization. Although the Solvay process remains important worldwide, it has been almost completely supplanted in the United States by the mining of a double salt of sodium carbonate and sodium hydrogen carbonate called **trona,** with the approximate composition $(Na_2CO_3) \cdot NaHCO_3 \cdot 2H_2O$. Heating this ore gives sodium carbonate.

• Sodium hydroxide is often referred to as caustic soda because its solutions are so corrosive. It is one of the few soluble inorganic bases (see Section 4-3) and is certainly the most important.

Sodium carbonate is used primarily in the manufacture of glass, especially soda-lime glass (see Section 21–4). Another use is in the production of sodium hydrogen carbonate ($NaHCO_3$), which has the common name bicarbonate of soda. Although $NaHCO_3$ appears as an intermediate in the Solvay process, it is not removed for direct use because it contains too many impurities. Instead, a saturated aqueous solution of sodium carbonate is prepared and carbonated:

$$Na_2CO_3(aq) + CO_2(g) + H_2O(\ell) \longrightarrow 2\ NaHCO_3(s)$$

The suspension of sodium hydrogen carbonate is filtered, washed, and dried, leaving a product of about 99.9% purity. It is used in the kitchen as baking soda because it reacts with acids, such as those present in milk, to generate carbon dioxide for the rising of dough. It is also used in dry chemical fire extinguishers (see problem 56 in Chapter 7).

23–2 ELECTROLYSIS AND THE CHLOR-ALKALI INDUSTRY

In the 1890s, a new process for producing bases and chlorine was developed and grew rapidly in importance: the electrolytic generation of sodium hydroxide and chlorine from salt solutions. The dominant technique in the United States for this process is the **diaphragm cell** (Fig. 23–3). If an electric current is passed through a sufficiently concentrated solution of sodium chloride, the anode reaction is

$$2\ Cl^-(aq) \longrightarrow Cl_2(g) + 2\ e^-$$

and the cathode reaction is

$$2\ H_2O(\ell) + 2\ e^- \longrightarrow H_2(g) + 2\ OH^-(aq)$$

The overall reaction is the sum of these half-reactions:

$$2\ H_2O(\ell) + 2\ Cl^-(aq) \longrightarrow 2\ OH^-(aq) + H_2(g) + Cl_2(g) \qquad \Delta G^\circ = +422\ \text{kJ}$$

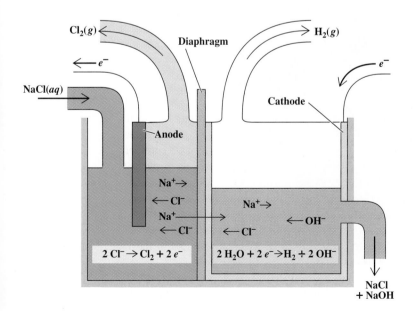

Figure 23–3 In a diaphragm cell, the cathode is a perforated steel box. Note that the level of the solution is deliberately made higher in the anode compartment than in the cathode compartment in order to minimize migration of OH^- through the diaphragm to the anode, where it could react with the chlorine being generated.

Figure 23–4 Chlor-alkali electrolysis cells. These use state-of-the-art membrane technology.

• In a dynamo a magnet rotates inside a wire coil, generating an electric current through the wire.

The gaseous hydrogen and chlorine bubble out and are collected and dried, while the hydroxide ion forms part of a mixed solution of sodium hydroxide and sodium chloride. The less soluble sodium chloride precipitates upon evaporation and can be removed by filtration, leaving a concentrated solution containing 50% by mass of sodium hydroxide in water. This solution may be used directly or evaporated further to make solid sodium hydroxide.

Early diaphragm cells used graphite anodes, but these had to be replaced frequently and have been superseded by anodes made of titanium coated with platinum, ruthenium, or iridium. The cathodes are steel boxes with perforated sides. The diaphragm prevents the H_2 and Cl_2 gases from mixing but allows ions to pass. During electrolysis, the contents of the anode compartment are kept under pressure to minimize the in-migration of $OH^-(aq)$ from the cathode compartment. Hydroxide ion would react with the chlorine to give hypochlorite ions, an undesirable side-reaction:

$$2\ OH^-(aq) + Cl_2(g) \longrightarrow OCl^-(aq) + Cl^-(aq) + H_2O(\ell)$$

Asbestos has been the most widely used material for these diaphragms, but new diaphragms are being developed based on fluorinated polymers. Another approach is to replace the diaphragm with an ion-exchange membrane that permits only sodium ions to pass to the cathode (Fig. 23–4).

The electrolytic production of sodium hydroxide and chlorine was proposed in the mid-19th century, but the only source for electricity at the time was the battery. This made the process so expensive that it was little more than a laboratory curiosity. By the 1890s, however, the development of the dynamo changed the situation dramatically. The electrolysis industry grew up rapidly around sources of cheap electricity from water power—in Norway and near Niagara Falls in the United States. The industry's first effect was to doom the Leblanc process, which had survived so long only due to profits from the production of chlorine. With competing chlorine

The early chlor-alkali industry in the United States and Canada clustered around Niagara Falls to take best advantage of the cheap electricity generated from the waterfall.

produced in greater purity electrolytically, the Leblanc process quickly disappeared. Over a longer period, the electrolytic chlor-alkali process has reduced the role of the Solvay process as well, by yielding a base (NaOH) that competes effectively with sodium carbonate in many applications.

Is the story now complete? Far from it. Consider that a major use of chlorine has long been to bleach wood pulp, giving a pure white cellulose fiber for paper making. Between 50 and 80 kg of chlorine whitens a metric ton of pulp from the sulfate process (see Section 22–3). About 10% of the bleaching chlorine combines with organic compounds present in the wood pulp to generate hundreds of organochlorine compounds. These compounds, some of which are carcinogens or mutagens, appear in the effluent from paper mills that bleach with chlorine and are difficult to remove by waste treatments. As a consequence, the chlorine bleaching of pulp is being phased out by the paper industry.

The loss of a major market for chlorine confronts the chlor-alkali industry with a severe problem of maintaining a balance between stable demand for sodium hydroxide and falling demand for chlorine, because the laws of chemistry require inexorably that every two moles of sodium hydroxide produced by the electrolysis of brine be accompanied by one mole of chlorine. If chlorine were to become an unwanted waste product instead of a valuable by-product, the electrolysis process would suffer heavily. Indeed, the demand for these two products has only rarely been in ideal balance, resulting in price fluctuations in both commodities.

A by-product of the electrolysis of brine is hydrogen. It can be reacted directly with chlorine to give gaseous hydrogen chloride of high purity, although less expensive sources of HCl exist. Alternatively, the chlor-alkali process can be coupled with a fertilizer plant and the hydrogen can be reacted with nitrogen to make ammonia (see Section 22–4).

- Alternative pulp bleaches include hydrogen peroxide (H_2O_2), oxygen (O_2), and ozone (O_3).

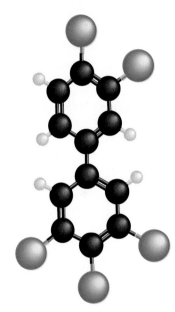

Figure 23–5 One of the many polychlorinated biphenyls (PCBs), which were once used as insulating fluids in electrical transformers. These compounds are being phased out because of their toxic effect on marine life.

23-3   THE CHEMISTRY OF CHLORINE, BROMINE, AND IODINE

Chlorine, bromine, and iodine resemble one another much more closely than they do fluorine, whose special properties are discussed in Sections 23–4 and 23–5.

Chlorine is the most extensively used halogen. Its production and its role in making bleaches have just been described. That role is now overshadowed by chlorine's importance in the manufacture of chlorinated hydrocarbons for use as solvents and pesticides (see Section 24–3) and plastics such as polyvinyl chloride (see Section 25–3). Growing concern over the toxic properties of chlorinated hydrocarbons (Fig. 23–5) has slowed the growth of the industry. Another major use of chlorine is in the purification of titanium dioxide. This substance is mined in impure form. It reacts above 900°C with chlorine and coke to give titanium tetrachloride:

$$TiO_2(s) + 2\ C(s) + 2\ Cl_2(g) \longrightarrow TiCl_4(g) + 2\ CO(g)$$

Because $TiCl_4$ is a volatile liquid (b.p. 136°C) it can easily be purified by distillation. It is then oxidized to pure titanium dioxide, regenerating chlorine:

$$TiCl_4(\ell) + O_2(g) \longrightarrow TiO_2(s) + 2\ Cl_2(g)$$

Titanium dioxide is the major white pigment used in paint. It is more opaque than other pigments, so paints containing it have high covering power.

Bromine is obtained largely from naturally occurring brines, where it occurs in the form of aqueous bromide ion (Br^-). Treatment of these brines with chlorine re-

- Some environmental groups have even called for a ban on all production of chlorine.

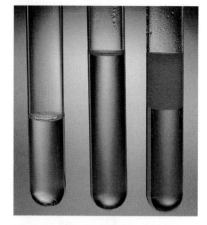

Figure 23–6 Bromide ion in aqueous solution (*left*) is colorless. Oxidation by chlorine yields bromine (*center*), which is colored in aqueous solution. The color is more intense in an organic solvent (upper layer on *right*).

sults in the reaction (Fig. 23–6):

$$2 \, Br^-(aq) + Cl_2(g) \rightleftharpoons 2 \, Cl^-(aq) + Br_2(aq)$$

This equilibrium lies to the right ($\Delta \mathscr{E}° = 0.271$ V and $\Delta G° = -52.4$ kJ): chlorine, a stronger oxidizing agent than bromine, displaces bromide ion from solution. The bromine that is generated is swept out with steam, distilled, and purified. Until recently, the most important use for bromine was in the production of dibromoethane, $C_2H_4Br_2$, which is used in gasoline to scavenge the lead deposited from the breakdown of tetraethyl lead; the use of this "antiknock" additive has diminished sharply, however, with the phase-out of leaded gasoline. A major current use for bromine is in the production of flame-retardant organic compounds. Aqueous solutions of $CaBr_2$ and $ZnBr_2$ are also used as high-density fluids to recover oil from deep wells.

Iodine occurs in seawater to the extent of only $6 \times 10^{-7}\%$, but it is concentrated in certain species of kelp, from whose ash its recovery is commercially feasible. Iodine is present in the metabolism-stimulating hormone thyroxin, secreted by the thyroid gland. Much of the table salt sold has 0.01% NaI added to prevent goiter, the enlargement of the thyroid gland. Silver iodide is used in high-speed photographic film, whereas silver bromide and silver chloride are used in slower-speed film and in photographic printing paper.

The hydrogen halides HCl, HBr, and HI are all gases at room conditions, and all three dissolve in water to form strong acids. Hydrogen chloride is usually prepared by the direct combination of hydrogen and chlorine over a platinum catalyst or as a by-product of organic chemicals processing. Its aqueous solution, hydrochloric acid, is a major industrial acid, used extensively for the cleaning of metal surfaces. Hydrogen bromide and hydrogen iodide have fewer uses, mostly in the area of organic chemical synthesis.

Solution Chemistry of the Halogens

When chlorine, bromine, or iodine is added to water, the halogen disproportionates in part to form its hydrohalic and hypohalous acids according to the equilibrium

$$X_2(aq) + H_2O(\ell) \rightleftharpoons HOX(aq) + X^-(aq) + H^+(aq)$$

The equilibrium constant for this reaction is very small, but the reaction is shifted to the right in basic solution, in which HOX is neutralized and the overall reaction becomes

The structure of thyroxin. Goiters, swellings of the neck caused by enlargement of the thyroid gland, occur in individuals with iodine-poor diets as the gland processes more blood in attempting to obtain sufficient iodine to make thyroxin.

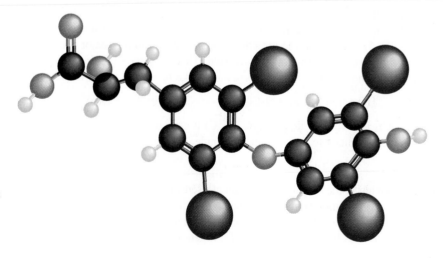

$$X_2(aq) + 2\ OH^-(aq) \rightleftharpoons OX^-(aq) + X^-(aq) + H_2O(\ell)$$

No complications arise with chlorine, and stable solutions of chloride and hypochlorite ions result when the temperature is around 25°C. In this way, a solution of sodium hypochlorite (NaOCl)—used as a laundry bleach—can be prepared. The aqueous hypochlorite ion is thermodynamically unstable with respect to disproportionation into chloride and chlorate ions:

$$3\ ClO^-(aq) \longrightarrow 2\ Cl^-(aq) + ClO_3^-(aq) \qquad \Delta G° = -90.7\ kJ$$

but at ordinary temperatures this reaction is very slow. When the temperature is increased to about 75°C, however, the rate of disproportionation is rapid.

The equilibrium constant for the disproportionation of bromine in water to bromide and hypobromite ions is even smaller than that for the disproportionation of chlorine; if the concentration of base is increased, the further disproportionation of hypobromite to bromide and bromate ions occurs rapidly even at 25°C. The same is true of iodine, but to an even greater extent. The solubility of iodine in water at 25°C is only 1.34×10^{-3} M, but in the presence of iodide ion much more dissolves as triiodide ion forms:

$$I_2(s) + I^-(aq) \rightleftharpoons I_3^-(aq) \qquad K_{298} = 714$$

The I_3^- ion has a linear structure that is symmetric in solution and in certain crystalline salts but unsymmetric (unequal I—I distances) in others.

Among the possible acids of formula HXO_3, only iodic acid exists in anhydrous form, a crystalline solid that is stable up to its melting point. The alkali chlorates, bromates, and iodates have the generic formula MXO_3, and all are well-characterized salts that can be crystallized from aqueous solution. Acidic solutions of the salts are powerful oxidizing agents that react quantitatively with their corresponding halide ion to yield the halogen:

$$IO_3^-(aq) + 5\ I^-(aq) + 6\ H^+(aq) \longrightarrow 3\ I_2(aq) + 3\ H_2O(\ell)$$

When potassium chlorate is carefully heated in the absence of catalysts, it disproportionates to potassium perchlorate and potassium chloride:

$$4\ KClO_3(s) \longrightarrow 3\ KClO_4(s) + KCl(s)$$

Perchloric acid can be formed by heating potassium perchlorate with concentrated sulfuric acid, but this is a hazardous reaction that is prone to explosions. A safer procedure is to oxidize an aqueous solution of an alkali chlorate at the anode of an electrolytic cell:

$$ClO_3^-(aq) + H_2O(\ell) \longrightarrow ClO_4^-(aq) + 2\ H^+(aq) + 2\ e^-$$

Perchloric acid is a colorless liquid that boils at 82°C. In its anhydrous form it is a powerful oxidizing agent, but in aqueous solution at moderate concentration it displays little oxidizing ability. Aqueous perbromic acid, on the other hand, is a powerful oxidizing agent that is also formed electrolytically. Aqueous periodic acid is a more useful oxidizing agent than either perbromic or perchloric acid because its behavior is more controllable.

Halogen Oxides

Chlorine, bromine, and iodine form several oxides, most of which are thermally unstable. Dichlorine monoxide is prepared by reacting chlorine with mercury(II) oxide:

$$2\ Cl_2(g) + 2\ HgO(s) \longrightarrow Cl_2O(g) + HgCl_2 \cdot HgO(s)$$

• Acid and base anhydrides were introduced in Section 4-3.

In basic solution it forms hypochlorites, as it is the anhydride of hypochlorous acid:

$$Cl_2O(g) + 2\ OH^-(aq) \longrightarrow 2\ OCl^-(aq) + H_2O(\ell)$$

Chlorine dioxide is an unstable gas that can be prepared by reacting chlorine with silver chlorate:

$$Cl_2(g) + 2\ AgClO_3(s) \longrightarrow 2\ ClO_2(g) + 2\ AgCl(s) + O_2(g)$$

It is a very powerful oxidizing agent that is used to bleach wheat flour. Like nitrogen dioxide, it consists of "odd-electron" molecules; unlike that molecule, it shows little tendency to dimerize. In basic solution it disproportionates to form chlorite and chlorate ions and can be regarded as a mixed anhydride of chlorous and chloric acids:

$$2\ ClO_2(g) + 2\ OH^-(aq) \longrightarrow ClO_2^-(aq) + ClO_3^-(aq) + H_2O(\ell)$$

The most stable of the chlorine oxides is dichlorine heptaoxide (Cl_2O_7), a colorless liquid that boils without decomposing at 82°C. It is prepared by dehydration of perchloric acid with P_4O_{10}:

$$12\ HClO_4(\ell) + P_4O_{10}(s) \longrightarrow 6\ Cl_2O_7(\ell) + 4\ H_3PO_4(\ell)$$

The oxides of bromine, Br_2O and BrO_2, can be prepared by methods that parallel those for the corresponding oxides of chlorine. The most important of the iodine oxides is diiodine pentaoxide, a white crystalline solid that can be prepared by thermal dehydration of iodic acid:

$$2\ HIO_3(s) \longrightarrow I_2O_5(s) + H_2O(g)$$

Diiodine pentaoxide is useful in analytical chemistry for the detection and quantitative determination of carbon monoxide, which it readily oxidizes:

$$5\ CO(g) + I_2O_5(s) \longrightarrow I_2(s) + 5\ CO_2(g)$$

Figure 23–7   Iodine reacts with starch to form a blue-black complex.

The iodine released in this reaction can be detected visually by means of the blue-black colored complex it forms with starch added as an indicator (Fig. 23–7).

23-4 FLUORINE: PREPARATION AND USES

The chemistry of fluorine is full of surprises. It ranks 13th in abundance among the elements in the rocks of the earth's crust, and its compounds were used as early as 1670 for the decorative etching of glass. Yet its isolation as a free element defied heroic efforts over many years before finally being accomplished in 1886. The element itself and some of its compounds are highly reactive and toxic, but other compounds are at the opposite extreme and are deliberately chosen for uses demanding inert, nontoxic materials.

• Astatine (At) is technically a halogen as well. Little is known of its chemistry because all of its isotopes are unstable and short-lived. It is therefore excluded from this discussion.

The differences between fluorine and the other halogens are extreme enough to place it in a class by itself. Fluorine is far more reactive than chlorine, bromine, and iodine and is indeed the most reactive of all the elements. Elemental fluorine is among the strongest oxidizing agents. The other halogens are moderate to very good at accepting electrons in aqueous media, as shown by the standard reduction potentials of 0.535 V (for iodine), 1.065 V (for bromine), and 1.358 V (for chlorine). Fluorine is *avid* for electrons:

$$F_2(g) + 2\ e^- \longrightarrow 2\ F^-(aq) \qquad \mathscr{E}° = 2.87\ V$$

Because the fluor*ine* atom in F_2 is so good at gaining an electron, fluor*ide* compounds (containing F^-) strongly resist oxidation, the loss of electrons.

The major source of fluorine is the ore fluorspar, which contains the mineral fluorite (CaF_2) (see Fig. 20–33). The cubic structure of CaF_2 is shown in Figure 23–8. Another source of fluorine is fluorapatite ($Ca_5(PO_4)_3F$), which is mined in great quantity to make phosphate fertilizer (see Section 22–3). At one time fluorine-containing by-products (HF, SiF_4, and H_2SiF_6) from fertilizer plants were valueless and created environmental pollution that was comparable to the hydrogen chloride pollution caused by the Leblanc process (see Section 23–1). Now the fluorine from fluorapatite has economic value in making fluorine compounds.

Hydrogen Fluoride

The starting point for the chemistry of fluorine is hydrogen fluoride, which is made by the reaction of fluorspar with concentrated sulfuric acid:

$$CaF_2(s) + H_2SO_4(aq) \longrightarrow 2\,HF(g) + CaSO_4(s)$$

Anhydrous hydrogen fluoride is a colorless liquid that freezes at $-83.37°C$ and boils at $19.54°C$, near room temperature. The boiling point far exceeds the value predicted by the trend among the other hydrogen halides, which is strong evidence for association of the HF molecules (by hydrogen bonds) in the liquid. In this and other physical properties, HF is closer to water and ammonia than to HCl, HBr, or HI. As pure liquids, both water and hydrogen fluoride contain polar molecules but few ions; both are poor electrical conductors. Liquid hydrogen fluoride undergoes autoionization much as water does:

$$2\,HF(\ell) \rightleftharpoons H^+ + FHF^-$$

The reaction has an equilibrium constant (at $19.5°C$) of 2.0×10^{-14}, a value close to the room-temperature autoionization constant of water (1.0×10^{-14}). Hydrogen fluoride in aqueous solution is a weak acid ($K_a = 6.6 \times 10^{-4}$ at $25°C$), quite unlike HCl, HBr, and HI, which are strong acids. Its weakness as an acid is due both to the strength of the H—F bond and to the extensive hydrogen bonding in its aqueous solutions. Fluoride ions interact so strongly with hydronium ions (H_3O^+) that they remain bound in $F^- \cdots H_3O^+$ complexes, rather than separating, as is the case with the chloride, bromide, and iodide ions.

Liquid hydrogen fluoride has excellent solvent properties; it is able to dissolve polar inorganic salts such as NaF and many nonpolar organic compounds as well. Using it as a solvent for chemical reactions requires more caution and preparation than using water because HF is quite poisonous, causes serious burns, and corrodes silicate glass by the reaction

$$4\,HF(\ell) + SiO_2(s) \longrightarrow SiF_4(g) + 2\,H_2O(\ell)$$

This reaction underlies one long-standing use of aqueous hydrogen fluoride: etching decorative patterns in glass (Fig. 23–9). Hydrogen fluoride does not attack Teflon and certain other fluorine-containing plastics. Their availability in recent years has made it easier to handle hydrogen fluoride safely.

A major use for hydrogen fluoride is in the aluminum industry. It is reacted with $NaAlO_2(aq)$, itself prepared from bauxite ore ($Al_2O_3(s)$), and $NaOH(aq)$, to make synthetic cryolite according to the overall equation

$$NaAlO_2(aq) + 2\,NaOH(aq) + 6\,HF(g) \longrightarrow Na_3AlF_6(s) + 4\,H_2O(\ell)$$

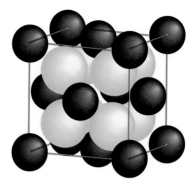

Figure 23–8 Structure of CaF_2. The gray spheres are calcium, and the yellow spheres are fluorine. When heated (but not yet melted), crystals of fluorite become excellent conductors of electricity. The array of Ca^{2+} ions remains crystalline, but the F^- ions become disordered. They are in effect molten and serve to carry the current.

Figure 23–9 Glass etched with hydrofluoric acid. The artist covered areas with wax to create parts of the design and then applied the HF. The wax, which resists HF, was removed later.

Molten cryolite (m.p. 1012°C) dissolves $Al_2O_3(s)$. The very important Hall–Héroult process to produce aluminum (see Fig. 12–15) depends on the electrolysis of such solutions. An essential additive in the molten cryolite–Al_2O_3 solution is aluminum trifluoride (AlF_3), which is synthesized by the acid–base reaction of $HF(g)$ with $Al(OH)_3(s)$.

The Preparation of Fluorine

Elemental fluorine is produced from hydrogen fluoride. The first reproducible preparation was achieved by Henri Moissan in 1886 by the electrolysis of anhydrous hydrogen fluoride at $-50°C$. Today, the element is commercially produced by a similar method, the electrolysis of molten potassium hydrogen fluoride ($KF\cdot 2HF(\ell)$), which can be regarded as a solution of KF in liquid HF. In the solution, F^- ion associates strongly with HF to form FHF^- (the hydrogen difluoride ion), the species that actually loses the electrons:

$$FHF^- \longrightarrow F_2(g) + H^+ + 2\,e^- \quad \text{(anode, oxidation)}$$
$$\underline{2\,H^+ + 2\,e^- \longrightarrow H_2(g) \qquad\qquad \text{(cathode, reduction)}}$$
$$2\,HF \longrightarrow H^+ + FHF^- \longrightarrow F_2(g) + H_2(g) \quad \text{(overall)}$$

This process resembles the electrolytic production of $Cl_2(g)$ (see Section 23–2). As fluorine and hydrogen evolve, $HF(g)$ is fed in continuously to keep the composition of the electrolyte constant at $KF\cdot 2HF$. Commercial fluorine cells use a steel cathode and carbon anode. The cells are carefully engineered to prevent contact between the product gases, which react with catastrophic speed.

Fluorine is a very pale yellow gas. It condenses to a canary-yellow liquid at $-188.14°$ and solidifies at $-219.62°C$. It is an exceedingly toxic substance. Concentrations exceeding 25 ppm in air quickly damage the eyes, nose, lungs, and skin. Fluorine's pungent, irritating odor makes it first detectable in air at a concentration of about 3 ppm, well below the level of acute poisoning but above the recommended safe working level of 0.1 ppm. The ferocious reactivity of fluorine demands the utmost care in its use, and handling procedures that minimize the hazards have been developed. Certain metals (nickel, copper, steel, and the nickel–copper alloy Monel metal) resist attack by fluorine by forming a layer of a fluoride salt on their surfaces. Reactions can be carried out at room temperature in vessels made of these metals. The pure element is regularly packaged and shipped as a compressed gas in special cylinders but is also often generated at the point of use.

The role of elemental fluorine in the chemical industry has grown to major proportions since 1945. Before then, elemental fluorine was mostly a laboratory curiosity. The motive in finding ways to prepare, purify, and handle it was the need to synthesize a volatile uranium compound to separate the ^{235}U and ^{238}U isotopes by gaseous diffusion, as described in Section 5–7. Uranium hexafluoride had the required properties, and gaseous diffusion is still used for preparing ^{235}U-enriched reactor fuel. At present, more than half of the world's production of $F_2(g)$ is used to make UF_6 from uranium(IV) oxide by these reactions:

$$UO_2(g) + 4\,HF(g) \longrightarrow UF_4(s) + 2\,H_2O(g)$$
$$UF_4(s) + F_2(g) \longrightarrow UF_6(g)$$

Most of the remaining fluorine production goes to make sulfur hexafluoride by the reaction

$$S(s) + 3\,F_2(g) \longrightarrow SF_6(g)$$

This unreactive gas is a good electrical insulator. Its presence within high-voltage electrical equipment minimizes arcing or sparking when the switches are thrown. Electrical discharges that do occur under an SF_6 atmosphere decompose the gas to an extent (into sulfur and fluorine), but the dissociation products rapidly recombine. Some triple-pane windows are manufactured with SF_6 sealed in the "dead-air" spaces between the panes because it cuts heat transmission and muffles sound better than air.

For many years, it was believed that no chemical agent could oxidize a fluoride to fluorine. The corollary was that electrochemical methods were essential in the preparation of elemental fluorine. Both ideas were shown to be wrong in 1986, when elemental fluorine was generated by completely chemical means in a reaction at 150°C involving metal fluorides:

$$2 \ K_2MnF_6(s) + 4 \ SbF_5(g) \longrightarrow 4 \ KSbF_6(s) + 2 \ MnF_3(s) + F_2(g)$$

In this reaction, Mn(IV) is reduced to Mn(III) and F is oxidized from the −1 to the 0 oxidation state. Apparently, MnF_4 forms as $SbF_5(s)$, acting as a Lewis acid, removes F^- from $K_2MnF_6(s)$. The MnF_4 is not stable and quickly decomposes to $\frac{1}{2} \ F_2(g)$ and $MnF_3(s)$.

• Neither starting compound in this remarkable reaction requires elemental fluorine for its synthesis, so the process is not an empty circular trick.

23-5 COMPOUNDS OF FLUORINE

In forming compounds, a fluorine atom either shares electrons in a single covalent bond or gains one electron to form the fluoride ion, F^-. In either case, it always attains a noble-gas electron configuration ($1s^2 2s^2 2p^6$). The other halogens tend to behave similarly but tolerate numerous exceptions; fluorine does not. The rules for assigning oxidation numbers (see Section 4–4) recognize the special character of fluorine by setting its oxidation number in compounds always equal to −1. Chlorine, bromine, and iodine may have positive oxidation numbers.

The chemical and physical properties of fluorine (and its compounds) deviate from the values predicted by extrapolating the trends among the other halogens. Figure 23–10 shows that the bond dissociation energy of F_2 is *smaller* than would

Figure 23-10 The measured physical properties of fluorine often do not fall on the trend-lines defined by the behavior of the other halogens. (a) Here the ionization energy of fluorine is unexpectedly high, and its electron affinity and bond dissociation enthalpy are unexpectedly low. (b) The sizes of the fluorine atom and the fluoride ion are both smaller than predicted by comparison with the atoms or ions of the other halogens.

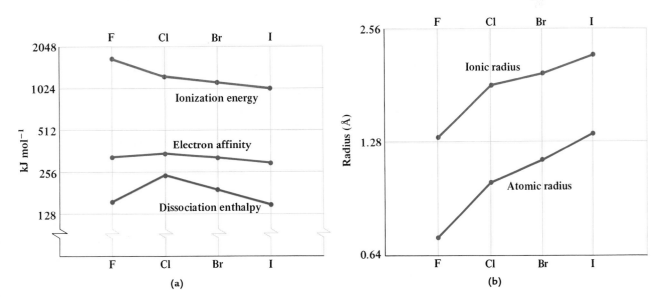

(a)

(b)

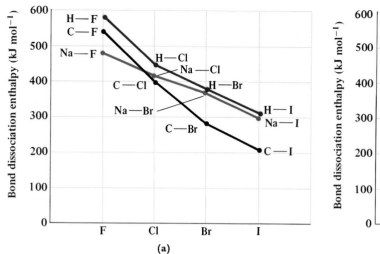

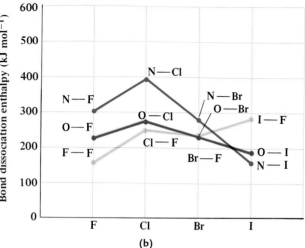

Figure 23-11 (a) Fluorine forms strong bonds with more electropositive elements such as carbon, hydrogen, and sodium. (b) In contrast, fluorine forms unexpectedly *weak* bonds with the electronegative elements nitrogen, oxygen, and fluorine compared with the bonds formed by the other halogens.

be expected from the trend among the other halogens, but the first ionization energy of F is *larger*. Despite its position as the most electronegative of the elements, electron-affinity data show that F(*g*) accepts another electron (to form the gaseous anion) *less* readily than Cl(*g*) and only slightly more readily than Br(*g*). The unexpectedly low electron affinity of F(*g*) is related to the contracted size of the atom. As the graphs of size in Figure 23–10b show, F(*g*) and F$^-$(*g*) are smaller than the extrapolation of the trends among the other halogens would predict, causing a relative crowding and greater mutual repulsion among the electrons in their $n = 2$ quantum levels.

Strong bonds, both covalent and ionic, characterize the interaction of fluorine with most other elements. Figure 23–11a shows that the H—F, C—F, and Na—F bond dissociation enthalpies exceed those of H, C, and Na combined with the other halogens. Fluorine's strong bonds with other elements and weak bond with itself (in the F$_2$ molecule, the bond dissociation enthalpy is only 158 kJ mol^{-1}) together explain its terrific reactivity. Both factors are traceable to the small size of the fluorine atom. The two atoms in the fluorine molecule are only 1.417 Å apart. This puts their lone-pair electrons closer to each other than the lone pairs in the other halogens: repulsion among the lone pairs lowers the bond dissociation enthalpy. Small fluorine makes short bonds to non-fluorine atoms as well. Lone-pair repulsion of the type just described becomes important only with elements that, like fluorine itself, are small and electronegative. Hence, O—F and N—F bonds are, like the F—F bond, weaker than otherwise expected (Fig. 23–11b).

Because fluorine atoms are small, many of them fit around a given central atom before crowding each other. Therefore, atoms often attain a larger coordination number with fluorine than with other elements. The result is the existence of many fluorides, such as K$_2$[NiF$_6$], Cs[AuF$_6$], and PtF$_6$, in which the central element has an unexpectedly high oxidation number. The other halogens offer no comparable compounds.

Reactions of Fluorine

In general, fluorine reacts with other elements and with compounds by oxidizing them. It displaces the other, less reactive halogens from compounds, just as chlorine displaces bromine and iodine, the elements located below it in the periodic table. An example is the reaction

$$2 \text{ NaCl}(s) + \text{F}_2(g) \longrightarrow 2 \text{ NaF}(s) + \text{Cl}_2(g)$$

Fluorine sometimes breaks this pattern by going much further. Thus, if $\text{F}_2(g)$ is heated with potassium chloride, the reaction is

$$\text{KCl}(s) + 2 \text{ F}_2(g) \longrightarrow \text{KClF}_4(s)$$

Here, the chlorine is not displaced from the compound but is oxidized to the $+3$ state and surrounded by fluorine atoms in the "tetrafluorochlorate(III)" ion. Further heating of $\text{KClF}_4(s)$ with F_2 eventually does displace chlorine from potassium, but in an unexpected form, $\text{ClF}_5(g)$:

$$\text{KClF}_4(s) + \text{F}_2(g) \longrightarrow \text{KF}(s) + \text{ClF}_5(g)$$

In a variation, fluorine oxidizes iodine in potassium iodide and also displaces it:

$$\text{KI}(s) + \text{F}_2(s) \longrightarrow \text{KF}(s) + \text{IF}(g)$$

Under different conditions with the same reactants, the theme is carried to an extreme:

$$\text{KI}(s) + 4 \text{ F}_2(g) \longrightarrow \text{KF}(s) + \text{IF}_7(g)$$

Fluorine also can displace oxygen from its compounds:

$$2 \text{ CaO}(s) + 2 \text{ F}_2(g) \longrightarrow 2 \text{ CaF}_2(s) + \text{O}_2(g)$$

With other oxides, only a portion of the oxygen may be displaced, as in the reaction of $\text{SO}_2(g)$ with $\text{F}_2(g)$ to give thionyl fluoride (Fig. 23–12):

$$2 \text{ SO}_2(g) + 2 \text{ F}_2(g) \longrightarrow 2 \text{ SOF}_2(g) + \text{O}_2(g)$$

The most important oxide of all is water. Water vapor at $100°C$ *burns* in fluorine with a weak, luminous flame. The reaction is

$$2 \text{ F}_2(g) + 2 \text{ H}_2\text{O}(g) \longrightarrow 4 \text{ HF}(g) + \text{O}_2(g)$$

As is often the case, different conditions favor different products. An excess of fluorine in contact with ice can give hydrogen fluoride and highly unstable hydrogen oxyfluoride:

$$\text{F}_2(g) + \text{H}_2\text{O}(s) \longrightarrow \text{HF}(\ell) + \text{HOF}(\ell)$$

Oxygen difluoride also forms

$$2 \text{ F}_2(g) + \text{H}_2\text{O}(s) \longrightarrow 2 \text{ HF}(\ell) + \text{OF}_2(g)$$

and the reaction that gives O_2 and HF still takes place to a limited extent. In all three of these reactions, fluorine disrupts stable H_2O, taking oxygen from the -2 to the 0 or $+2$ state. Few materials other than fluorine can oxidize water in this way.

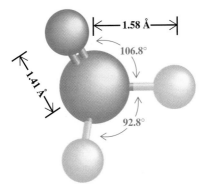

Figure 23-12   In SOF_2, oxygen and fluorine both bond to the central sulfur atom.

• Assigning an oxidation number of -1 to fluorine and $+1$ to hydrogen in hypofluorous acid (HOF) means that the oxidation number of oxygen in this compound is 0.

Metal Fluorides

The alkali and alkaline earth metals form ionic salts with fluorine with the expected stoichiometry. Except for lithium fluoride, the alkali metal fluorides are all quite soluble in water. The differences between the alkaline-earth fluorides (BeF_2, MgF_2, CaF_2, SrF_2, and BaF_2) and the corresponding chlorides, bromides, and iodides are striking: BeF_2 is ionic and soluble in water, but BeCl_2 is covalent enough to sublime; CaF_2 is insoluble in water, but the other calcium halides are quite soluble in water. All of these fluorides are prepared most easily by reaction of the corresponding hydroxide or carbonate with aqueous hydrofluoric acid. In their lower oxidation states, the transition metals and Group II and IV metals also form fluoride salts with a range of solubilities; ZnF_2, SnF_2, and PbF_2 are examples.

• Similar F^- for OH^- replacements occur in many minerals. An example is topaz, a semiprecious mineral with formula $Al_2SiO_4(F, OH)_2$. The fluoride and hydroxide ions are grouped within parentheses because the degree of substitution of one ion for the other varies with the origin of the mineral.

Low concentrations (about 1 mg L^{-1}) of fluoride ion in drinking water help to prevent tooth decay. Fluoride ion substitutes for hydroxide ion in tooth enamel, changing some of the hydroxyapatite ($Ca_5(PO_4)_3OH$) in the enamel to fluorapatite ($Ca_5(PO_4)_3F$). The replacement works because the F^- and OH^- ions have the same charge (-1) and similar radii (1.33 Å for F^- versus approximately 1.2 Å for OH^-). Fluorapatite is less soluble than hydroxyapatite in the acidic oral environment fostered by consumption of sweets. The deliberate fluoridation of civic water supplies to cut down tooth decay is a common public-health measure in the United States, and fluoride salts are often added to toothpaste.

Fluorine forms salts with most metals in their lower oxidation states but is capable of oxidizing many metals to exceptionally high oxidation states as well. The chemistry of cobalt centers on the $+2$ oxidation state, for example, except in complex ions, in which the $+3$ state is not uncommon. Cobalt(III) fluoride, which is obtained by direct fluorination of the metal, is one of the very few simple ionic salts of that element in which the cation has an oxidation number of $+3$. Platinum is directly oxidized to platinum(IV) by fluorine, and (at higher temperatures) to platinum(VI) fluoride. The latter is a dark red solid with a low melting point (56.7°C) and is a covalent molecular compound. Other heavy metals that form volatile hexafluorides include osmium, iridium, molybdenum, tungsten, and uranium. Fluorides of lower oxidation state are known for all of these metals.

• These hexafluorides have significant covalent character and are quite volatile. Fluorine "gives wings to the elements."

In the great majority of its compounds, silver has only a $+1$ oxidation state. Silver in the $+2$ oxidation state, however, can be formed by heating silver(I) fluoride in a stream of fluorine:

$$2\,AgF(s) + F_2(g) \longrightarrow 2\,AgF_2(s)$$

Even the $+3$ state of silver can be formed in a ternary compound when potassium fluoride is mixed with silver fluoride and heated in fluorine:

$$KF(s) + AgF(s) + F_2(g) \longrightarrow KAgF_4(s)$$

In these reactions the filled $4d$-subshell of the silver ion must be opened, at a cost of 2073 and 3360 kJ mol^{-1}, respectively, to form silver in the $+2$ and $+3$ oxidation states.

Covalent Fluorides

Fluorine forms compounds with all the elements in Group III through VII. The Group III elements (boron, aluminum, gallium, indium, and thallium) all form trifluorides, and thallium forms TlF as well. Boron trifluoride and aluminum trifluoride are strong Lewis acids that readily form the complex ions $[BF_4]^-$ and $[AlF_6]^{3-}$ in solutions that contain F^- ion. Boron trifluoride is a gas at room temperature, but the trifluorides of aluminum and the other members of Group III are solids that melt near 1000°C. They are intermediate between ionic and covalent in their bonding, and in the crystalline state their structures consist of extended arrays of metal atoms in sixfold coordination with fluorine atoms. The *trichlorides* of these elements are much more covalent in character; $AlCl_3(s)$ sublimes as $Al_2Cl_6(g)$ when it is heated to only 178°C.

• BF_3 is widely used as a catalyst for polymerization reactions (see Chapter 25).

The transitional character of the bonding is even more evident in the fluorides of Group IV. At room temperature, carbon tetrafluoride and silicon tetrafluoride are gases in which the molecules are covalent; germanium tetrafluoride is a molecular liquid, but the tetrafluorides of tin and lead are solids with considerable ionic character. As mentioned previously, the difluorides of tin and lead are ionic salts. This illustrates a general rule: the halides of the heavier transition and post-transition metals are more covalent when the oxidation number of the metal is high and more ionic when it is low. Carbon tetrafluoride is a kinetically stable compound. Although its reaction with water is a thermodynamically favored process at room conditions, the reaction does not occur at an appreciable rate below 500°C. Silicon and germanium tetrafluorides, however, are extensively hydrolyzed by water but can be successfully distilled from aqueous solutions containing hydrofluoric acid.

The fluorides of the Group V elements are just as varied in composition and behavior. Nitrogen forms fluorides having the composition N_2F_2 (dinitrogen difluoride), N_2F_4 (dinitrogen tetrafluoride), and NF_3 (nitrogen trifluoride). All are gases at room conditions. Nitrogen trifluoride has pyramidal molecules (Fig. 23–13), is rather inert, and has no electron-donor properties. The presence of the electronegative fluorine atoms in fact gives the nitrogen atom in NF_3 a charge of +0.80, according to one calculation. Like CF_4, it is not hydrolyzed by water. It fluorinates certain metals when heated:

$$2\ NF_3(g) + Cu(s) \longrightarrow N_2F_4(g) + CuF_2(s)$$

and is used as an etchant in plasma technology for the fabrication of semiconductors. Dinitrogen tetrafluoride is much more reactive than nitrogen trifluoride, undergoing hydrolysis in water and dissociating when it is heated. The PF_3 molecule has a trigonal pyramidal structure like NH_3 as the VSEPR model predicts. Unlike ammonia, however, it does not donate the lone pair of electrons on the central atom readily, presumably because the highly electronegative fluorine atoms withdraw electron density from the central atom. Like BF_3, PF_5 is a very powerful Lewis acid and is useful as a catalyst for polymerization.

The remaining Group V elements (arsenic, antimony, and bismuth) resemble phosphorus in forming both trifluorides and pentafluorides that readily hydrolyze in aqueous solution. They are powerful fluorinating agents, capable of introducing fluorine into other molecules by substitution reactions.

The highly toxic gas oxygen difluoride (OF_2) can be made by passing fluorine through aqueous alkaline solutions:

$$2\ F_2(g) + 2\ OH^-(aq) \longrightarrow OF_2(g) + 2\ F^-(aq) + H_2O(\ell)$$

It is worthwhile to compare OF_2 with the analogous compound formed between chlorine and oxygen. Both Cl_2O and OF_2 are gases under room conditions, and both have oxygen atoms bonded to two halogens. In both compounds the steric number of the oxygen atom is 4, with two bonds and two lone pairs on the central oxygen atom. Consequently, both have nonlinear molecules (Fig. 23–14). Despite these similarities, the two differ profoundly in their reactivity. Oxygen difluoride contains "positive oxygen" (note the +2 oxidation number for oxygen) because fluorine is highly electronegative. It reacts with water by oxidizing it to liberate O_2:

$$OF_2(g) + H_2O(\ell) \longrightarrow O_2(g) + 2\ HF(aq)$$

Dichlorine oxide contains ordinary "negative oxygen" and behaves quite differently with water. It gives hypochlorous acid without oxidation or reduction:

$$Cl_2O(g) + H_2O(\ell) \longrightarrow 2\ HOCl(aq)$$

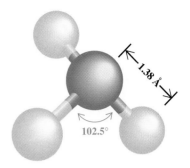

Figure 23–13 The structure of NF_3. The three F—N—F angles are equal; the three N—F bond lengths are equal.

• Recall that "VSEPR" stands for "valence shell electron-pair repulsion." This model for predicting molecular shapes is explained in Section 3-6.

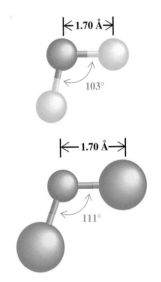

Figure 23–14 The nonlinear structures of Cl_2O (*bottom*) and OF_2. The two are closely related, but the bond distances are substantially longer in Cl_2O and the bond angle is larger.

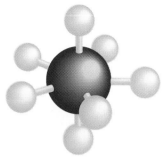

Figure 23-15 The molecular structure of IF_7 is based on the pentagonal bipyramidal geometry. The two axial fluorines are 1.786 Å away from the central iodine; the five equatorial fluorines are farther (1.858 Å) away, presumably to relieve crowding in the equatorial plane.

The chemistries of the compounds diverge along the same lines in many other reactions.

The later Group VI elements also form fluorides. The two most important fluorides of sulfur are quite dissimilar in their properties. Both sulfur tetrafluoride and sulfur hexafluoride are gases at room temperature; the former is highly reactive, and the latter is extremely inert. Their see-saw and octahedral molecular structures are given in Figures 3–20b and 3–17, respectively. Both selenium and tellurium form a tetrafluoride and a hexafluoride with fluorine. These compounds, like the fluorides of their Group V neighbors arsenic and antimony, are excellent fluorinating agents. Their strikingly enhanced reactivity compared with sulfur hexafluoride is not easy to explain, except to note that the larger central atoms are less well shielded by their fluorine atoms.

Fluorine reacts with the other halogens to form several **interhalogen** compounds. Thus, $Br_2(\ell)$ burns in an atmosphere of $F_2(g)$ to give $BrF_3(\ell)$. In these compounds, fluorine atoms, as always in the -1 oxidation state, surround a central chlorine, bromine, or iodine atom as XF, XF_3, and XF_5, where X is Cl, Br, or I. The only XF_7 compound occurs with iodine. The VSEPR model provides accurate predictions of the structures of these compounds. In IF_7, iodine has a steric number of 7, and the structure is based on the pentagonal bipyramid (Fig. 23–15).

Fluorinated Hydrocarbons and Chlorofluorocarbons

Fluorine can substitute for hydrogen in hydrocarbons, giving rise to an important class of compounds called the "fluorinated hydrocarbons." The C—F bond is short and strong, an average of 1.2 times stronger than the C—H bond and 1.4 to 2.2 times stronger than the bonds between carbon and other halogens; it is the strongest single bond. Fluorination of hydrocarbons in which there are no double or triple bonds therefore generally increases thermal and chemical stability. When all the hydrogen atoms in hydrocarbons are replaced by fluorine atoms, compounds called **fluorocarbons** (or **perfluorocarbons**) result (Fig. 23–16). The electron density on the fluorine atoms in fluorocarbons is tightly held and is not easily distorted by approaching atoms or ions. This reduces the reactivity of these compounds and weakens their intermolecular forces. The compounds tend to have low melting and boiling points and low enthalpies of vaporization.

One provocative use of fluorine compounds is in the preparation of blood substitutes. Many fluorocarbons are excellent solvents for oxygen. A milliliter of liquid C_7F_{16}, for example, dissolves 1900 times more oxygen at room conditions than a milliliter of water. Small laboratory animals can survive total immersion in oxy-

• The reverse is true in the reactions of the other halogens with hydrocarbons.

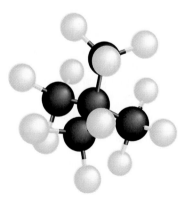

Figure 23-16 The structure of perfluoroneopentane, C_5F_{12}, a perfluorocarbon. The fluorine atoms on the perimeter of the molecule hold their electrons tightly. Perfluorocarbons tend consequently to be unreactive.

Figure 23–17 A Teflon-coated "no-stick" frying pan.

Figure 23–18 The molecular structure of the polymer Teflon. The fluorine atoms shield the carbon chain against attack. In fact, the chain must twist to accommodate their bulk, completing a full spiral every 26 C atoms along the chain.

• The term "halofluorocarbon" includes the possible presence of bromine and iodine.

genated fluorocarbon liquids because their lungs successfully extract the oxygen they need from solution. Fluorocarbons are insoluble in water. For use in the bloodstream, they are prepared as fine-grained emulsions in water. Rats have survived for days without apparent harm after their blood was completely replaced with a concentrated emulsion of $C_8F_{17}Br$ in water. Over a period of time, the composition of the rats' blood reverted to normal as the blood substitute was excreted and new red blood cells grew. Practical fluorocarbon blood substitutes would need no matching of blood types, could be stored for emergency use, and would not transmit disease.

The best-known perfluorocarbon is the solid polytetrafluoroethylene (Teflon), formulated as $-(CF_2CF_2)_n-$ where n is a large number. Chemically, this compound is nearly completely inert, resisting attack from boiling sulfuric acid, molten potassium hydroxide, fluorine gas, and other aggressive chemicals. Physically, it has excellent heat stability, is a very good electrical insulator, and has a low coefficient of friction ("no-stick" Teflon) that makes it useful for bearing surfaces in machines as well as coating frying pans (Fig. 23–17). In Teflon the carbon atoms lie in a long chain that is encased by tightly bound fluorine atoms (Fig. 23–18). Even reactants with a strong innate ability to disrupt C—C bonds (like fluorine itself) fail to attack Teflon at observable rates because there is no route of attack past the surrounding fluorine atoms and their tightly held electrons.

Chlorofluorocarbons (often called CFCs) contain one or more carbon atoms with both fluorine and chlorine atoms attached as side-groups. Chlorofluorocarbons are chemically inert, nontoxic gases or low-boiling liquids such as dichlorodifluoromethane, CF_2Cl_2 (Fig. 23–19). They are relatively cheap to make and have boiling points conveniently near room temperature. Their properties fit them well to serve as propellants in aerosol sprays and as refrigerants. Soon after their discovery, chlorofluorocarbons replaced SO_2 and NH_3 in the latter application. The inertness of chlorofluorocarbons has drawbacks as well because the molecules that escape into the air persist unchanged. In time, they mix into the stratosphere, where ultraviolet radiation from the sun finally breaks them down. Chlorine atoms generated in this decomposition catalyze the destruction of ozone (O_3) in the high reaches of the atmosphere. The phase-out of CFCs under the Montreal Protocol (see Section

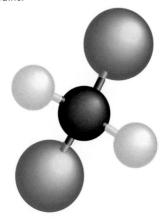

Figure 23–19 The structure of CF_2Cl_2. This compound, also called CFC-12, is being replaced in its major use in air-conditioning units. It harms the stratospheric ozone layer, and production has been ended in the developed nations.

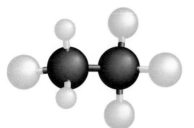

Figure 23–20   The structure of the HFC 1,1,1,2-tetrafluoroethane, or "HFC-134a." This CFC substitute contains no chlorine and so does not harm the ozone layer.

18–5) prompted an intense search for replacements. Substitutes must be reactive enough to break down in the troposphere, yet nontoxic and reasonably nonflammable. They should also have equal (or better) properties for their application, work in existing equipment, and be manufacturable at a reasonable price. Compounds containing at least one hydrogen atom per molecule are generally more reactive than CFCs, and fluorine does not participate in the ozone destruction cycle. Therefore, attention has focused on HCFCs (hydrochlorofluorocarbons) and HFCs (hydrofluorocarbons) as CFC substitutes. The compound "HFC-134a" (Fig. 23–20) works well in many applications, but acceptable substitutes for every current use of CFCs have not been identified.

CHEMISTRY IN YOUR LIFE

Fluorinated Drugs

The compound 5-fluorouracil (Fig. 23–A), an effective anticancer drug, is a planned drug, the result of rational drug design. Cancerous cells use uracil and other essential biochemicals more rapidly than normal cells because they reproduce faster. The idea to substitute fluorine for hydrogen in uracil came from the observation that fluorine-substituted organic compounds are often more toxic than unsubstituted compounds. Converting a C—H bond to a C—F bond in a molecule is, from the point of view of altering molecular shape and size, quite a minimal change. As a consequence, fluorine-substituted compounds often successfully mimic their unsubstituted versions and enter biochemical processes. However, C—F bonds are stronger than C—H bonds (see Table 10–3), strong enough to stop all metabolic steps that require it to break. Thus, 5-fluorouracil enters the cell and assumes the role of uracil. The fluorine atom blocks an essential change at the 5-position and prevents replication of the genetic material of the cell, which no longer reproduces. The drug acts preferentially against tumors because it concentrates in fast-growing cells. Still, 5-fluorouracil is harmful to all cells. This drug is rarely given alone: in combination with other drugs it usually achieves large *synergistic* effects; that is, its ef-

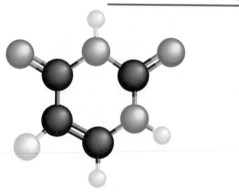

Figure 23–A   The structure of 5-fluorouracil, a drug that is much used in cancer chemotherapy.

fects in combination with those of a second drug exceed the sum of the effects of either drug given separately.

Other fluorine-substituted compounds are effective against viruses such as herpes and against many bacteria. They are also useful in the treatment of the inflammation of rheumatoid arthritis and of malaria.

23-6 COMPOUNDS OF FLUORINE AND THE NOBLE GASES

For many years the noble gases were believed to be chemically inert. They failed to react with the strongest oxidizing agents available, and an early attempt (1895) by Moissan to make argon react with fluorine also failed. In 1933, Linus Pauling suggested on theoretical grounds that xenon and fluorine should form compounds. Soon thereafter, chemists subjected a mixture of the two gases to an electrical discharge, but there was no evidence of reaction. This result, other negative results, and the well-known dangers of working with elemental fluorine discouraged further experiments. It seemed clear that if fluorine could not force a noble gas into combination, then nothing could. Chemists' interest was further stifled because generally accepted simple bonding theory appeared to ratify the complete chemical inertness of the noble gases. Then, in 1962 Neil Bartlett treated gaseous xenon with the powerful fluorinating agent PtF_6, and a yellow-orange solid compound of platinum, fluorine, and xenon resulted. Beyond doubt, xenon had been forced to combine chemically. In 1965, an irony emerged: the same fluorine–xenon mixtures that did not react in the 1933 electrical discharge experiment *did* give $XeF_2(s)$ simply upon exposure to sunlight!

Bartlett's discovery spurred research on the reactions of the noble gases with fluorine, and soon afterward the synthesis of xenon tetrafluoride by direct combination of the elements at high pressure was reported (Fig. 23–21):

$$Xe(g) + 2\,F_2(g) \longrightarrow XeF_4(s)$$

Varying the relative amounts of xenon and fluorine permitted synthesis of $XeF_2(s)$ and $XeF_6(s)$ as well. All three xenon fluorides are colorless crystalline solids at room temperature. VSEPR theory correctly predicts the linear structure of the XeF_2 molecule and the square planar structure of the XeF_4 molecule. Because the steric number of the central Xe is 7 and not 6, XeF_6 molecules are not octahedral; however, the VSEPR theory does not correctly predict their complex structure.

The three xenon fluorides are moderate to powerful fluorinating agents but are thermodynamically stable with respect to decomposition into xenon and fluorine at room temperature. Their behavior toward water is quite diverse. Xenon difluoride hydrolyzes only very slowly in water, and solutions as concentrated as 0.1 M can be prepared at low temperatures. On the other hand, XeF_4 and XeF_6 react vigorously with water to form xenon trioxide (XeO_3):

$$3\,XeF_4(s) + 6\,H_2O(\ell) \longrightarrow XeO_3(aq) + 2\,Xe(g) + 12\,HF(aq) + \tfrac{3}{2}\,O_2(g)$$
$$XeF_6(s) + 3\,H_2O(\ell) \longrightarrow XeO_3(aq) + 6\,HF(aq)$$

Xenon trioxide is soluble in aqueous solution without ionization, but in basic solution it acts as a Lewis acid, accepting OH^-:

$$XeO_3(aq) + OH^-(aq) \longrightarrow HXeO_4^-(aq)$$

This hydrogen xenate ion then disproportionates to form xenon and the perxenate ion, in which the oxidation number of xenon is +8:

$$2\,HXeO_4^-(aq) + 2\,OH^-(aq) \longrightarrow XeO_6^{4-}(aq) + Xe(g) + O_2(g) + 2\,H_2O(\ell)$$

Only one simple krypton halide has been prepared. Irradiating a mixture of krypton and fluorine at low temperature with electrons forms krypton difluoride, a thermodynamically unstable substance. It is a volatile, colorless, crystalline solid. Experiment confirms the linear structure predicted by the VSEPR theory. It forms

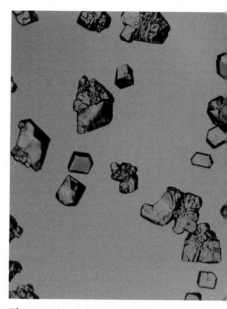

Figure 23–21 Crystals of xenon tetrafluoride formed in making the first binary noble-gas compound.

salt-like compounds with strong fluoride-ion acceptors such as the Lewis acids AsF_5 and SbF_5:

$$KrF_2 + AsF_5 \longrightarrow [KrF]^+[AsF_6]^-$$

Krypton difluoride is a stronger oxidizing agent than fluorine itself. It has a total bond enthalpy (for *both* Kr—F bonds) of only 96 kJ mol^{-1}; this is substantially less than the bond enthalpy of $F_2(g)$ (158 kJ mol^{-1}). It oxidizes $Xe(g)$ to $XeF_6(s)$ at room temperature and oxidizes I_2 directly to IF_7. A solution of KrF_2 in $HF(\ell)$ even attacks gold:

$$7\,KrF_2(g) + 2\,Au(s) \longrightarrow 2\,KrF[AuF_6](s) + 5\,Kr(g)$$

Heating the $KrF[AuF_6]$ to 60°C generates orange AuF_5, in which gold has a +5 oxidation number. By way of comparison, $F_2(g)$ itself requires a temperature of 250°C to act on powdered gold and then generates only AuF_3 (mixed with unreacted gold).

SUMMARY

23–1 Sodium chloride is the source for three important chemicals: sodium carbonate, sodium hydroxide, and chlorine. In the obsolete **Leblanc process,** sodium chloride was heated with sulfuric acid to form sodium sulfate. Heating sodium sulfate with a mixture of carbon (from coal) and limestone ($CaCO_3$) gave a mixture from which the water-soluble sodium carbonate could be extracted. The process had two noxious by-products—CaS and HCl. Oxidation of the HCl from the Leblanc process to Cl_2 either by reaction with the mineral pyrolusite (MnO_2) or by the **Deacon process** changed the HCl from a nuisance to an asset because chlorine, in the form of chlorinated lime, makes an excellent bleach for textiles. The **Solvay process** supplanted the Leblanc process. In it, the net reaction is the conversion of NaCl and $CaCO_3$ to Na_2CO_3 and the by-product $CaCl_2$. Ammonia and ammonium chloride are important intermediates that are recycled through this continuous process.

23–2 Sodium hydroxide and chlorine can also be produced from salt by the electrolysis of concentrated aqueous solutions, with hydrogen a by-product. In the **diaphragm cell** the anode is made of titanium coated with a noble metal, and the cathode is steel. Sodium hydroxide, a strong soluble base, has many uses in chemical and materials processing. Chlorine is used to make bleach, chlorinated hydrocarbons, and plastics, and in the purification of titanium for TiO_2 paint pigments.

23–3 Chlorine, bromine, and iodine all disproportionate with water, producing hydrohalic acid and hypohalous acids. Acidic solutions of chlorates, bromates, and iodates are powerful oxidizing agents. They react with solutions of chlorides, bromides, and iodides, respectively, to produce chlorine, bromine, and iodine.

23–4 The chemistry of fluorine differs sharply from that of the halogens. Fluorine atoms form weak bonds to other fluorine atoms but particularly strong bonds to most other kinds of atoms. These facts explain the high chemical reactivity of F_2. Modern methods for preparing fluorine are based on the original 1886 electrolytic process and use KF dissolved in HF. Most fluorine compounds are made directly or indirectly from hydrogen fluoride, a colorless hydrogen-bonded liquid. Elemental fluorine is used in uranium production and isotope separation and in the manufacture of sulfur hexafluoride.

23-5 Fluorine displaces other halogens from their compounds and oxidizes elements in compounds exposed to it. The compounds of fluorine with metals in their lower oxidation states are salts with strong ionic character. The higher oxidation states of many of the same elements yield volatile fluorides with significant covalent character. Compounds of fluorine with the nonmetals possess a wide range of reactivities and molecular geometries. Some fluorine compounds, such as SF_6, are quite inert toward chemical reactions; others, such as BF_3, are strong Lewis acids that readily accept a share in electron pairs from fluoride ions or other Lewis bases; still others, such as SbF_5, are strong fluorinating agents. **Fluorinated hydrocarbons** arise by the substitution of fluorine for hydrogen in hydrocarbons and often have greater thermal and chemical stability than the parent hydrocarbon. The release of chlorine upon eventual decomposition of **chlorofluorocarbons** in the upper atmosphere leads to depletion of the ozone in that region. The unique properties of fluorine shape the properties of the important fluorinated polymer polytetrafluoroethylene (Teflon).

23-6 Work with PtF_6, a strong fluorinating agent, opened up modern research on the compounds of the noble gases by its reaction with xenon; xenon and krypton fluorides are made by direct reaction with fluorine or with other fluorinating agents.

PROBLEMS

Note: Answers to blue-numbered problems are given in Appendix F. Problems that are more challenging are indicated with asterisks.

Chemicals from Salt

1. Calculate the equilibrium constant at 25°C for the production of sodium hypochlorite bleach from dichlorine oxide and sodium hydroxide:

$$Cl_2O(g) + 2\ OH^-(aq) \rightleftharpoons 2\ OCl^-(aq) + H_2O(\ell)$$

   Relevant data can be found in Appendix D.

2. Calculate the equilibrium constant at 25°C for the preparation of chlorine by the reaction of hydrochloric acid with pyrolusite, used by Scheele in 1774:

$$2\ Cl^-(aq) + 4\ H^+(aq) + MnO_2(s) \longrightarrow$$
$$Cl_2(g) + Mn^{2+}(aq) + 2\ H_2O(\ell)$$

   Relevant data can be found in Appendix D.

3. Compare the Leblanc process to the Solvay process with regard to expense and availability of starting materials, and with regard to purity of the final product.

4. Compare the Leblanc process to the Solvay process with regard to the by-products of the two methods. What uses are made of these by-products?

5. Calculate the standard enthalpy change per mole of Na_2CO_3 produced by the Leblanc process, using data from Appendix D. From this, what do you conclude about the overall heat requirements of that process?

6. Calculate the standard enthalpy change per mole of Na_2CO_3 produced by the Solvay process, using data from Appendix D. From this, what do you conclude about the overall heat requirements of that process?

7. Which is the stronger acid, chlorous acid or hypochlorous acid? Which would be preferred to convert potassium fluoride to hydrofluoric acid? (*Hint:* Refer to Table 8–2.)

8. Which is the stronger base, potassium carbonate or potassium hydroxide? Which would be preferred to convert potassium hydrogen phosphate (K_2HPO_4) to potassium phosphate (K_3PO_4)? (*Hint:* Refer to Table 8–2.)

Electrolysis and the Chlor-Alkali Industry

9. What impurity is likely to be present in NaOH that is made by mixing $Ca(OH)_2$ with Na_2CO_3, but absent in NaOH that is made electrolytically in a diaphragm cell?

10. What impurity is likely to be present in NaOH that is made electrolytically in a diaphragm cell, but absent in NaOH that is made by mixing $Ca(OH)_2$ with Na_2CO_3?

The Chemistry of Chlorine, Bromine, and Iodine

11. Cheap gaseous chlorine can be used to extract bromine from seawater via the reaction

$$2 \, Br^-(aq) + Cl_2(g) \longrightarrow Br_2(g) + 2 \, Cl^-(aq)$$

Calculate the volume of $Cl_2(g)$ at STP that would be needed to extract all the bromine from 1000 m³ of seawater, assuming complete reaction and taking the $Br^-(aq)$ concentration in seawater to be 0.0036 M.

12. Bubbling chlorine through a basic solution containing iodide ions produces chloride ions and iodate (IO_3^-) ions.
    (a) Write a balanced chemical equation for this reaction.
    (b) Suppose that 0.216 L of gaseous chlorine (measured at STP) is needed to react completely with 50.0 mL of the iodide solution in part (a). Determine the initial concentration of iodide ion in that solution.

13. Dichlorine oxide can be prepared by reaction of chlorine with mercury(II) oxide:

$$2 \, Cl_2(g) + 2 \, HgO(s) \longrightarrow Cl_2O(g) + HgCl_2 \cdot HgO(s)$$

    (a) What is the oxidation state of chlorine in Cl_2O?
    (b) What oxoacid results when Cl_2O reacts with water?

14. (a) What is the oxidation state of chlorine in $HClO_4$?
    (b) The compound P_4O_{10} is a powerful enough dehydrating agent to remove water from $HClO_4$, forming phosphoric acid (H_3PO_4) and an oxide of chlorine. What is the chemical formula of this oxide of chlorine, the anhydride of perchloric acid?
    (c) Write a balanced chemical equation for the dehydration of perchloric acid by P_4O_{10}.

15. Given the following reduction potentials,

$$ClO_2 + e^- \longrightarrow ClO_2^- \qquad \mathscr{E}° = 0.954 \text{ V}$$
$$ClO_3^- + H_2O + e^- \longrightarrow ClO_2 + 2 \, OH^- \qquad \mathscr{E} = -0.25 \text{ V}$$

    decide whether ClO_2 is stable with respect to disproportionation in basic aqueous solution under standard conditions.

16. Given the following reduction potentials,

$$ClO^- + H_2O + 2 \, e^- \longrightarrow Cl^- + 2 \, OH^-$$
$$\mathscr{E}° = 0.90 \text{ V}$$
$$ClO_3^- + 2 \, H_2O + 4 \, e^- \longrightarrow ClO^- + 4 \, OH^-$$
$$\mathscr{E}° = 0.48 \text{ V}$$

    decide whether hypochlorite ion (ClO^-) is stable with respect to disproportionation in basic aqueous solution under standard conditions.

Fluorine Chemistry

17. Which bond is stronger: Si—F or Si—Cl? Give reasons for your answer.

18. Which reaction of phosphorus is more exothermic: that with fluorine to give phosphorus trifluoride, or that with chlorine to give phosphorus trichloride? Give reasons for your answer.

19. Write balanced equations for the reaction of elemental fluorine with the following:
    (a) SrO(s)   (c) UF_4(s)
    (b) O_2(g)   (d) NaCl(s)

20. Write balanced equations for the reaction of elemental fluorine with the following:
    (a) SO_2(g)   (c) Na(s)
    (b) MgO(s)   (d) GeO(s)

21. The alkali metals form one or more acid fluorides, with formulas MF·HF, MF·2HF, and MF·3HF (where M stands for the alkali metal). In these compounds, additional HF molecules link with F^- ions by means of hydrogen bonds, in a manner similar to the way water of hydration is incorporated in crystalline hydrates. Nothing similar is observed with the other alkali-metal halides.
    Suppose that a compound is found to contain 2.48 g of fluorine for every gram of sodium. What is the empirical formula of this compound? Assume that any hydrogen present is not detected in this analysis.

22. There are ways to combine carbon and fluorine beyond those mentioned in the text. When carbon in the form of graphite is treated with elemental fluorine, F atoms occupy interstitial spaces between the layers of C atoms, and compounds of empirical formula C_4F, C_2F, and CF result. The bonds between carbon and fluorine in all three of these materials are covalent. Graphite fluorides are used as electrodes in advanced primary batteries and as lubricants.
    Suppose that a graphite fluoride is analyzed and found to contain 2.53 g of carbon for every gram of fluorine. What is the empirical formula of this compound?

23. A gaseous binary compound of chlorine and fluorine has a density at STP of 4.13 g L^{-1} and is 38.35% chlorine by mass. Determine its molecular formula.

24. A gaseous compound has the molecular formula ClO_3F. Determine the volume occupied by 225 g of this compound at STP.

25. Predict the structures of the following fluorine-containing molecules:
    (a) OF_2 (d) BrF_5
    (b) BF_3 (e) IF_7
    (c) BrF_3 (f) SeF_6

26. Predict the structures of the following fluorine-containing molecular ions:
    (a) SiF_6^{2-} (d) AsF_6^-
    (b) ClF_2^+ (e) BF_4^-
    (c) IF_4^+ (f) BrF_6^+

27. Thionyl and selenyl difluoride have the compositions SOF_2 and $SeOF_2$, with the oxygen and fluorine atoms directly linked to the central sulfur or selenium atom. They can be prepared by reaction of a strong fluorinating agent, such as PF_5, with sulfur dioxide or selenium dioxide:

$$SO_2(g) + PF_5(g) \longrightarrow SOF_2(g) + POF_3(g)$$

    (a) Use VSEPR theory to predict the structures of SOF_2 and $SeOF_2$. Draw the structures.
    (b) Thionyl difluoride can coordinate with BF_3 through its lone pair. Does it act as a Lewis acid or a Lewis base in this reaction?

28. When two oxygen atoms take the place of four fluorine atoms in SF_6, the compound has the composition SO_2F_2 and is called "sulfuryl difluoride." Like SF_6, SO_2F_2 is comparatively unreactive and hydrolyzes with difficulty.
   (a) Use VSEPR theory to predict the structure of SO_2F_2. Draw the structure.
   (b) In the conversion of thionyl difluoride (SOF_2) to sulfuryl difluoride,

$$2 \, SOF_2 + O_2 \longrightarrow 2 \, SO_2F_2$$

does the thionyl difluoride act as a Lewis acid or a Lewis base?

29. Although BrF_3 is a covalent compound, $BrF_3(\ell)$ is slightly conductive due to the autoionization of BrF_3 molecules. This is analogous to the small conductivity of water arising from autoionization:

$$2 \, H_2O(\ell) \rightleftharpoons H_3O^+(aq) + OH^-(aq)$$

Write a balanced chemical equation for the autoionization of BrF_3. Can this reaction be described as a Lewis acid–base reaction?

30. The molecules in liquid BrF_3 are linked through "fluorine bonds" that are analogous to the hydrogen bonds in water. Do you expect the entropy of vaporization of $BrF_3(\ell)$ to be larger or smaller than the Trouton's rule value of 88 J K^{-1} mol^{-1} (see Section 11–5)?

31. Rank the following compounds in order of normal boiling point, from lowest to highest: CaF_2, PtF_6, and PtF_4.

32. For each of the following pairs of fluorine compounds, predict which has the higher melting point:
   (a) NaF or PF_3
   (b) AsF_3 or AsF_5
   (c) CF_4 or C_5F_{12}

33. One of the hazards in making Teflon is that the starting material, tetrafluoroethylene ($C_2F_4(g)$), can explode, giving C(*graphite*) and $CF_4(g)$. The standard enthalpy of formation of $C_2F_4(g)$ is -651 kJ mol^{-1}. Use this fact, together with data from Appendix D, to estimate the amount of heat that would be released if a tank containing 1.00 kg of C_2F_4 were to explode.

34. Compute the $\Delta G°$ at 25°C of the reaction

$$CF_4(g) + 2 \, H_2O(\ell) \longrightarrow CO_2(aq) + 4 \, HF(aq)$$

using data from Appendix D. Explain how it is that CF_4 spontaneously decomposes according to the above equation, yet in practice is stable up to 500°C.

Compounds of Fluorine and the Noble Gases

35. At 25°C, the standard enthalpy change $\Delta H°$ for the reaction

$$XeF_6(g) + 3 \, H_2(g) \longrightarrow Xe(g) + 6 \, HF(g)$$

is -1282 kJ. Using this fact and information from Appendix D, calculate the following:

(a) The standard molar enthalpy of formation $\Delta H_f°$ of $XeF_6(g)$.
(b) The average enthalpy of an Xe—F bond in $XeF_6(g)$.

36. At 25°C, the standard enthalpy change $\Delta H°$ for the reaction

$$XeF_4(g) + 2 \, H_2(g) \longrightarrow Xe(g) + 4 \, HF(g)$$

is -887 kJ. Using this fact and information from Appendix D, calculate the following:
(a) The standard molar enthalpy of formation $\Delta H_f°$ of $XeF_4(g)$.
(b) The average enthalpy of an Xe—F bond in $XeF_4(g)$.

37. Xenon oxotetrafluoride ($XeOF_4$) is a liquid at room temperature. It can be prepared by the controlled hydrolysis of xenon hexafluoride:

$$XeF_6(g) + H_2O(\ell) \longrightarrow XeOF_4(\ell) + 2 \, HF(g)$$

At 25°C, the standard enthalpies of formation of the noble-gas compounds in this reaction are

$$\Delta H_f° (XeF_6(g)) = -298 \text{ kJ mol}^{-1}$$
$$\Delta H_f° (XeOF_4(\ell)) = +148 \text{ kJ mol}^{-1}$$

Calculate $\Delta H°$ for the hydrolysis reaction. Consult Appendix D for additional data.

38. The energy of explosion of $XeO_3(s)$ into its gaseous elements was measured in a constant-volume calorimeter:

$$XeO_3(s) \longrightarrow Xe(g) + \tfrac{3}{2} O_2(g)$$

(a) A 2.763×10^{-4} mol sample released 112 J of heat when it was exploded. Calculate the standard *energy* of formation ($\Delta E_f°$) of $XeO_3(s)$.
(b) Is the standard *enthalpy* of formation of $XeO_3(s)$ larger or smaller than your answer to part (a)?

39. The reduction of perxenate ion (XeO_6^{4-}) to xenon is so favored thermodynamically in acidic solution that it even oxidizes Mn^{2+} to MnO_4^-, one of the strongest oxidizing agents ordinarily encountered, with evolution of oxygen. Write balanced half-equations and an equation for the overall reaction in this process.

40. The following standard reduction potentials have been measured for the oxides of xenon in acidic aqueous solution:

$$H_4XeO_6(aq) + 2 \, H^+(aq) + 2 \, e^- \longrightarrow$$
$$XeO_3(aq) + 3 \, H_2O(\ell) \qquad \mathcal{E}° = +2.36 \text{ V}$$
$$XeO_3(aq) + 6 \, H^+(aq) + 6 \, e^- \longrightarrow Xe(g) + 3 \, H_2O(\ell)$$
$$\mathcal{E}° = +2.12 \text{ V}$$

(a) Would you classify perxenic acid (H_4XeO_6) as an oxidizing agent or as a reducing agent?
(b) In the preceding, XeO_3 acts as an oxidizing agent in one half-reaction or as a reducing agent in the other. Is XeO_3 stronger as an oxidizing agent or as a reducing agent at pH 0?
(c) Is XeO_3 stable with respect to disproportionation to Xe(g) and $H_4XeO_6(aq)$ in acidic aqueous solution?

Additional Problems

41. Chlorine-containing bleaches act by oxidizing, but other bleaches, like SO_2, are reducing bleaches. By referring to Appendix E, predict whether bromine is a stronger or a weaker bleach than chlorine.

42. You are stranded in a desert. You have a supply of water, but it is crawling with bacteria that will infect you fatally if you drink it as it is. The standard means of sterilizing water is to add $Cl_2(g)$. You have no chlorine. You do, however, have some $KMnO_4(s)$, $H_2S(aq)$, $NaCl(s)$, $SO_2(g)$, and $Na_2S_2O_3(aq)$. Which chemical (or combination of chemicals) should you test first as a substitute for Cl_2 in sterilizing the water?

43. In the Deacon process, an equilibrium is reached between chlorine, steam, hydrogen chloride, and oxygen at 450°C.
    (a) Use the ΔH_f° and S° values from Appendix D to estimate an equilibrium constant at 450°C for the reaction

$$2\,HCl(g) + \tfrac{1}{2}\,O_2(g) \rightleftharpoons Cl_2(g) + H_2O(g)$$

    (b) Suppose that bromine replaces chlorine in the preceding reaction. After referring to the appropriate bond enthalpies in Table 17–1, state whether the equilibrium constant at 450°C is larger or smaller. Explain.

44. It is estimated that a certain formation at Owens Lake, California, holds about 3.5×10^7 metric tons of trona. Estimate how long this formation could supply the United States with Na_2CO_3 at the current rate of consumption (see Table 22–1).

45. Calculate the theoretical minimum external cell voltage to drive a chlor-alkali cell in which all reactants and products are in their standard states at 25°C.

46. Suppose that a chlor-alkali plant has 250 cells, through each of which a current of 100,000 A passes continuously.
    (a) Calculate the mass of chlorine that this plant can produce per 24-hour day.
    (b) If the cell voltage is 3.5 V (higher than the theoretical minimum voltage from problem 45), calculate the total electrical energy consumed by the plant in one day, and express it both in joules and in kilowatt-hours.
    (c) If electricity costs $0.05 per kilowatt-hour, calculate the cost for the electricity to run the plant for one day.

47. A magazine article about the chlor-alkali process in the United States reports that "production of chlorine [last year] . . . total[ed] about 10.9 million tons. . . . Caustic soda production as usual [was] 4 to 5 percent higher than chlorine production." Explain the use of the phrase "as usual" by computing the theoretical ratio of the yields of these two chemicals in the process.

48. Astatine is a little-studied element whose longest lived isotope, ^{210}At, has a half-life of only 8.3 hours.
    (a) Predict whether HAt is a stronger or a weaker acid than HI.
    (b) The At^- ion is produced when elemental astatine reacts in aqueous solution with zinc, but no reaction is seen with

1 M $Fe^{2+}(aq)$. What range of reduction potentials for conversion of At_2 to At^- is consistent with these observations?
    (c) If $Cl_2(g)$ is bubbled through a solution containing $At^-(aq)$, what reaction results?
    (d) Write a balanced equation for the reaction that should occur when $At_2(s)$ is dissolved in basic aqueous solution.

49. Write a series of reactions that leads to the production of elemental fluorine from calcium fluoride and other readily available chemicals.

50. A commercial fluorine cell generates 3.3 kg of fluorine per hour by electrolysis of $KF \cdot 2HF$. Compute the average current passing through this cell.

51. The reaction

$$K_2[NiF_6] + TiF_4(s) \longrightarrow K_2[TiF_6](s) + NiF_2(s) + F_2(g)$$

generates elemental fluorine. Does $TiF_4(s)$ play the role of a Lewis acid or a Lewis base in this reaction? Explain.

52. (a) Compute the density of $HF(g)$ at 1.00 atm pressure and its normal boiling point, 19.54°C, assuming ideal gas behavior.
    (b) The observed density of $HF(g)$ under the conditions from part (a) is 3.11 g L^{-1}. Explain the large discrepancy with the calculated result of part (a).

*53. The reaction

$$4\,HF(\ell) + SiO_2(s) \longrightarrow SiF_4(g) + 2\,H_2O(\ell)$$

can be used to release gold that is distributed in certain quartz veins of hydrothermal origin. If the quartz contains $1.0 \times 10^{-3}\%$ gold by mass and the gold has a market value of $350 per troy ounce, would the process be economically feasible if commercial (50% by mass) aqueous hydrogen fluoride (density 1.17 g cm^{-3}) costs $0.25 per liter? (1 troy ounce = 31.3 g.)

54. Ingestion of 5 to 10 g of NaF is lethal to a 70-kg man. Poisoning from smaller doses can be treated with calcium therapy, which restores Ca^{2+} levels in the body. Write chemical equation(s) for the effect of NaF on the body.

55. Dioxygen difluoride, O_2F_2 (sometimes written "FOOF"), is a particularly potent fluorinating agent. It is made by irradiating a mixture of O_2 and F_2 at the temperature of liquid nitrogen. The O-to-O distance in O_2F_2 is nearly as short as the distance in O_2, and the O-to-F distances are quite long. Draw the Lewis structure for this molecule, determine the bond order of all bonds, and describe its geometry. What is the name of the analogous compound of oxygen and hydrogen?

56. The fluorinating agent dioxygen difluoride (O_2F_2) (see preceding problem) is important because it converts the plutonium in almost any plutonium-containing material to PuF_6 under mild conditions. The volatility of PuF_6 then allows the separation of radioactive Pu from various impurities. Write balanced chemical equations for the reaction of PuO_2 with FOOF and the reaction of Pu with FOOF.

57. When $SF_4(g)$ reacts with $CsF(s)$, the SF_5^- ion is formed. Use VSEPR theory to predict its geometry.

*58. The compound S_2F_{10} was not mentioned in the chapter but does exist.
    (a) Taking the known bond-forming tendencies of fluorine into account, suggest a probable Lewis structure for this compound.
    (b) Use the VSEPR model to predict the molecular geometry of S_2F_{10}.

59. The result of the accidental first preparation of Teflon was immediately subjected to elemental analysis for both chlorine and fluorine. The report on the new mysterious white powder was that it contained no chlorine and 48.4% fluorine by mass. Compare these first values to the actual percentages of fluorine and chlorine in Teflon.

60. The ionization energy of O_2 is 1180 kJ mol^{-1}. In 1962, Neil Bartlett reported that $O_2(g)$ reacts with $PtF_6(g)$ to form the solid ionic compound $O_2^+ PtF_6^-$. The ionization energy of Xe is 1170.6 kJ mol^{-1}. Explain why it occurred to Bartlett that xenon might also form a compound with PtF_6. Give the formula of the compound.

61. Calculate the equilibrium constant for the reaction

$$Xe(g) + 2\ F_2(g) \rightleftharpoons XeF_4(s)$$

at 25°C. The standard molar Gibbs function of formation of $XeF_4(s)$ is -134 kJ mol^{-1}.

62. Xenon difluoride reacts with arsenic pentafluoride to give an ionic compound:

$$2\ XeF_2 + AsF_5 \longrightarrow [Xe_2F_3]^+[AsF_6]^-$$

Identify which species are Lewis acids in this reaction and which are Lewis bases.

24

From Petroleum to Pharmaceuticals

A single chemical plant can produce more than 1 billion pounds of a product per year. This Texas plant produces ethylene (C_2H_4), a starting material for many plastics.

Carbon is unique among the elements in the large number of compounds it forms and in the variety of their structures. In combination with hydrogen, it forms molecules with single, double, and triple bonds; chains; rings; branched chains; branched and interlocked rings; and cages (Fig. 24–1). The thousands of stable hydrocarbons make a sharp contrast to the mere two stable compounds between oxygen and hydrogen (water and hydrogen peroxide). Even the rather versatile elements nitrogen and oxygen form only six nitrogen oxides.

The unique behavior of carbon relates to its position in the periodic table. As a second-period element, carbon has relatively small atoms, which allow it easily to form the double and triple bonds that are rare in the compounds of related elements, such as silicon. As a Group IV element, carbon can form four bonds, more than the other second-period elements. This gives it wide scope for structural elaboration. Finally, as an element of intermediate electronegativity, carbon can form covalent compounds both with more electronegative elements such as oxygen, nitrogen, and the halogens and with more electropositive elements such as hydrogen and the heavy metals mercury and lead.

The study of the compounds of carbon is the discipline traditionally called **organic chemistry,** although the chemistry of carbon is intimately bound up with the chemistry of inorganic materials and with biochemistry as well.

Figure 24–1 The structure of the hydrocarbon adamantane, $C_{10}H_{16}$. This wire-frame structure emphasizes the interlocking six-membered rings (carbon atoms lie at the black intersections; hydrogen atoms at the white ends). Note the similarity to the structure of diamond (Fig. 20–25).

24–1 PETROLEUM REFINING AND THE HYDROCARBONS

When the first oil well was drilled in 1859 near Titusville, Pennsylvania, the effects that the exploitation of petroleum would have on everyday life in the years to come could not have been anticipated. Today the petroleum and petrochemical industries span the world and touch the most minute details of our daily lives. In the early years of the 20th century the development of the automobile, fueled by low-cost gasoline derived from petroleum, changed many people's lifestyles. The subsequent use of gasoline and fuel oil to power trains and planes, tractors and harvesters, pumps and coolers transformed travel, agriculture, and industry. Finally, the spectacular growth of the petrochemical industry since 1945 has led to the introduction of in-

Much of the oil pumped from wells today comes from beneath the ocean floor. It is extracted using offshore oil rigs, such as this one in the Gulf of Mexico.

numerable new products ranging from pharmaceuticals to plastics and synthetic fibers. Over half of the chemical compounds produced in largest volume stem directly or indirectly from petroleum.

The prospects for the continued enjoyment of cheap petroleum and petrochemicals in the 21st century are clouded. Many wells have been drained, and the remaining petroleum is more costly to extract and often of lower quality. Petroleum appears to have originated from the deposition and decay of organic matter (of animal or vegetable origin) in oxygen-poor marine sediments. Subsequently, this matter migrated to the porous sandstone rocks from which it is extracted today. In the last 100 years, we have consumed a significant fraction of the petroleum accumulated in the earth over many millions of years. The imperative for the future is to save the reserves that remain for uses for which few substitutes are available (such as the manufacture of petrochemicals) while finding other sources for heat and energy.

Petroleum Distillation and the Straight-Chain Alkanes

Although crude petroleum oil contains small amounts of oxygen, nitrogen, and sulfur, its major constituents are **hydrocarbons,** compounds of carbon and hydrogen. The most prevalent hydrocarbons in petroleum are the **straight-chain alkanes** (also called normal alkanes, or n-alkanes), which consist of chains of carbon atoms bonded to one another by single bonds, with enough hydrogen atoms on each carbon atom to bring it to the maximum bonding capacity of four. The simplest alkanes are methane (CH_4), with just one carbon atom, and ethane (C_2H_6). These alkanes have the generic formula C_nH_{2n+2}; Table 24–1 lists the names and formulas of the first few. The ends of the molecules are methyl (CH_3) groups, with methylene (CH_2) groups in between. We could write pentane (C_5H_{12}) as $CH_3CH_2CH_2CH_2CH_3$ to indicate the structure more explicitly, or in abbreviated fashion as $CH_3(CH_2)_3CH_3$ (Fig. 24–2).

Each carbon atom in a straight-chain alkane forms four single covalent bonds that point to the corners of a tetrahedron (exhibiting sp^3 hybridization). Although bond lengths in these molecules do not change very much through vibration, rotation around a C—C single bond occurs quite easily (Fig. 24–3), as explained in Section 18–2. Accordingly, a given hydrocarbon molecule in a gas or liquid constantly changes its conformation as the chain flexes and writhes. The term "straight chain" refers only to the bonding pattern in which each carbon atom is bonded to

• The sulfur content in petroleum is particularly significant because the burning of high-sulfur petroleum products releases quantities of sulfur oxides that seriously pollute the air and cause acid rain.

Table 24–1
Straight-Chain Alkanes

Methane	CH_4
Ethane	C_2H_6
Propane	C_3H_8
Butane	C_4H_{10}
Pentane	C_5H_{12}
Hexane	C_6H_{14}
Heptane	C_7H_{16}
Octane	C_8H_{18}
Nonane	C_9H_{20}
Decane	$C_{10}H_{22}$
Undecane	$C_{11}H_{24}$
Dodecane	$C_{12}H_{26}$
Tridecane	$C_{13}H_{28}$
Tetradecane	$C_{14}H_{30}$
Pentadecane	$C_{15}H_{32}$
Hexadecane	$C_{16}H_{34}$
..	
..	
..	
Triacontane	$C_{30}H_{62}$

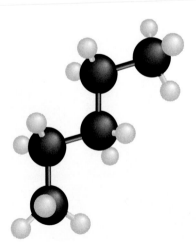

Figure 24–2 The structure of the alkane pentane (C_5H_{12}). It consists of three methylene (—CH_2—) groups in a line capped at each end by methyl (—CH_3) groups.

the next one in a sequence; the carbon atoms are not located along a straight line. An alkane molecule with 15 or 16 carbon atoms looks quite different when it is extended to give a "stretched" molecule and when it turns back on itself (Fig. 24–4). These two conformations (and the many others) interconvert very rapidly at room temperature.

Figure 24–5 shows the melting and boiling points of the straight-chain alkanes; both increase with the number of carbon atoms and thus with molecular mass. This is a consequence of the increasing strength of dispersion forces between heavier molecules, as discussed in Section 6–1. Methane, ethane, propane, and butane are all gases at room temperature, but the first several hydrocarbons that follow them in the alkane series are liquids. Alkanes beyond about $C_{17}H_{36}$ are waxy solids at 20°C, with melting points that increase with the number of carbon atoms present. Paraffin wax, a low-melting solid, is a mixture of alkanes that have 20 to 30 carbon atoms per molecule. Petrolatum (petroleum jelly, or Vaseline) is a different mixture that is semisolid at room temperature.

A mixture such as petroleum does not boil at a single, sharply defined temperature. Instead, as it is heated, the compounds of lower boiling point (the most volatile)

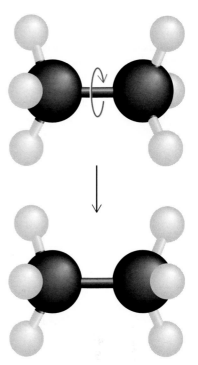

Figure 24–3 The two —CH_3 groups in ethane rotate easily about the bond that joins them.

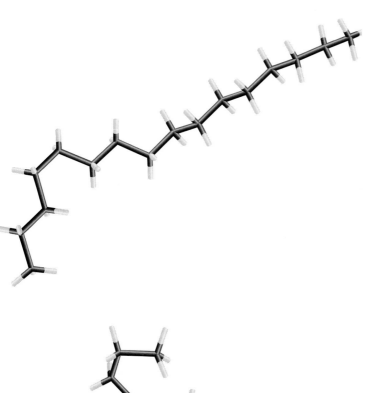

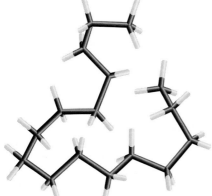

Figure 24–4 Two of the many possible conformations of the straight-chain alkane hexadecane, $C_{16}H_{34}$. Carbons atoms are at the black intersections; hydrogen atoms are at the white ends.

Figure 24–5 The melting and boiling points of the straight-chain alkanes increase with chain length n. Note the alternation in the melting points: alkanes with n odd tend to have lower melting points because it is more difficult to pack them into a crystal lattice.

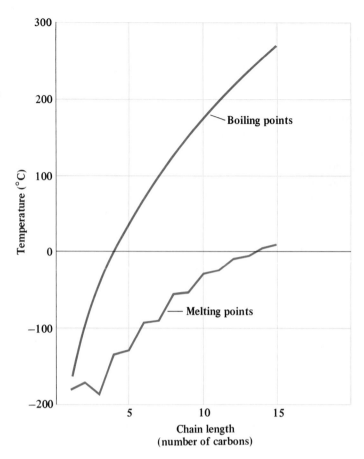

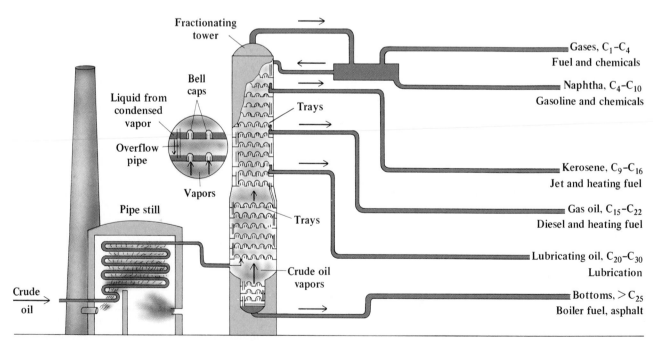

Figure 24–6 In the distillation of petroleum, the lighter, more volatile hydrocarbon fractions are removed from higher up the column, the heavier fractions from lower down.

Figure 24-7 Distillation towers in a modern oil refinery.

boil off first, and as the temperature is raised, more and more of the material vaporizes. The existence of a boiling-point range permits components of a mixture to be separated by distillation, as Section 6–7 discusses. The earliest petroleum distillation was a simple batch process: the crude oil was heated in a still, the volatile fractions were removed at the top and condensed to gasoline, and the still was cleaned for another batch. Modern petroleum refineries use much more sophisticated and efficient distillation methods, in which crude oil is added continuously and fractions of different volatility are tapped off at various points up and down the distillation column (Figs. 24–6 and 24–7). To save on energy costs, heat exchangers capture and recycle the heat released as separated vapors condense to liquids.

Distillation allows hydrocarbons to be separated by boiling point and thus by molecular mass. The gases that emerge from the top of the distillation column resemble the natural gas that collects in rock cavities above petroleum deposits. These gas mixtures can be separated further by redissolving the ethane, propane, and butane in a liquid solvent such as hexane. The methane-rich mixture of gases that remains is used for chemical synthesis or is shipped by pipeline to heat homes. Redistillation of the hexane and its dissolved gases allows their separation and their use as chemical starting materials. Propane and butane are also bottled under pressure as liquefied petroleum gas (LPG), which is used for fuel in rural areas. After the gases, the next fraction to emerge from the petroleum distillation column is **naphtha,** which is used primarily in the manufacture of gasoline. Subsequent fractions of successively higher molecular mass are used for jet and diesel fuel, heating oil, and machine lubricating oil. The thick nonvolatile sludge that remains at the bottom of the distillation unit is pitch or asphalt, which is used to roof buildings and pave roads.

• The boiling range of naphtha is from about 40°C to 180°C. It is a mixture of dozens of hydrocarbons.

Cyclic and Branched-Chain Alkanes

The straight-chain alkanes are not the only hydrocarbons in petroleum. Two other important classes of compound, the cyclic and branched-chain alkanes, are also represented. A **cycloalkane** consists of at least one chain of carbon atoms attached at the ends to form a closed loop. In the formation of this additional C—C bond, two

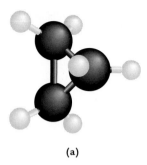

(a)

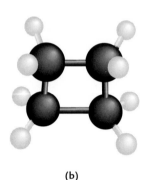

(b)

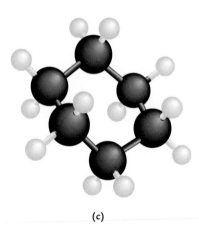

(c)

Figure 24–8 Three cyclic hydrocarbons. (a) Cyclopropane, C_3H_6. (b) Cyclobutane, C_4H_8. (c) Cyclohexane, C_6H_{12}.

hydrogen atoms are eliminated, so the general formula for cycloalkanes with one ring is C_nH_{2n} (Fig. 24–8). The smallest cycloalkanes, cyclopropane and cyclobutane, are very strained compounds because the C—C—C bond angles in both are substantially less than the normal tetrahedral angle of 109.5°. Therefore, they are more reactive than the heavier cycloalkanes or their straight-chain analogs propane and butane.

Branched-chain alkanes are hydrocarbons that contain only C—C and C—H single bonds, but in which the carbon atoms are no longer arranged in a continuous chain. One or more carbon atoms in each molecule is bonded to three or four other carbon atoms, rather than to only one or two as in the straight-chain alkanes or the cycloalkanes. The simplest branched-chain molecule (Fig. 24–9) is that of 2-methylpropane, sometimes referred to as isobutane. This molecule has the same molecular formula as butane (C_4H_{10}), but it has a different structure, in which a central carbon atom is bonded to three CH_3 groups and only one hydrogen atom. The two compounds butane and 2-methylpropane are isomers. Their molecules have the same formula but different three-dimensional structures that can be interconverted only by breaking and reforming chemical bonds.

The number of possible isomers increases rapidly with increasing numbers of carbon atoms in the hydrocarbon molecule. Thus, butane and 2-methylpropane are the only two isomers of molecular formula C_4H_{10}, but there are three isomers of C_5H_{12}, five of C_6H_{14}, nine of C_7H_{16}, and millions of isomers of $C_{30}H_{62}$. A systematic procedure has been developed to name these isomers and has been codified by the International Union of Pure and Applied Chemistry (IUPAC). The following set of rules is a part of that procedure:

1. Find the longest continuous chain of carbon atoms in the molecule. The molecule is named as a derivative of this alkane. Thus, in Figure 24–9b, the longest chain has three carbon atoms, so the molecule is named as a derivative of propane.

2. Determine the names of the hydrocarbon groups attached to the chain selected. These side groups are called **alkyl groups**. Their names are obtained by dropping the ending -*ane* from the corresponding alkane and replacing it with -*yl*. The methyl group (CH_3) is derived from methane (CH_4); the ethyl group (C_2H_5) from ethane (C_2H_6); the propyl group ($CH_2CH_2CH_3$) from propane (C_3H_8), and so on. Note also the *isopropyl* group ($CH(CH_3)_2$), which differs from the propyl group in that it attaches by its middle carbon atom rather than by one at an end.

3. Number the carbon atoms along the chain identified in rule 1. Identify the alkyl groups by the number of the carbon atom at which they are attached to the chain. The methyl group in the molecule in Figure 24–9b is attached to the second carbon atom of the three in the propane chain, so the molecule is called 2-methylpropane. The carbon chain is numbered from the end that gives the lowest number for the position of the first attached group.

4. If more than one alkyl group of the same type is attached to the chain, use the prefixes *di*- (two), *tri*- (three), *tetra*- (four), *penta*- (five), and so forth to specify the total number of such attached groups in the molecule. Thus, 2,2,3-trimethylbutane has two methyl groups attached to the second carbon atom and one to the third carbon atom of the four-atom butane chain. It is an isomer of heptane (C_7H_{16}).

5. If several types of alkyl groups appear, name them in alphabetical order. Ethyl is listed before methyl, which appears before propyl, and so forth.

As an example, let us name the following branched-chain alkane:

$$CH_3 \quad CH_2 \quad CH_2{-}CH_3$$
$$\diagdown C \diagup \diagdown C \diagup$$
$$CH_3 \quad H \; CH_3 \quad CH_2{-}CH_3$$

The longest continuous chain links six carbon atoms, so this is a derivative of hexane. We number the carbon atoms starting from the left:

$$CH_3 \quad \overset{3}{CH_2} \quad \overset{5}{CH_2}{-}\overset{6}{CH_3}$$
$$\diagdown \overset{2}{C} \diagup \diagdown \overset{4}{C} \diagup$$
$$\overset{1}{CH_3} \quad H \; CH_3 \quad CH_2{-}CH_3$$

Methyl groups are attached to carbon atoms 2 and 4, and an ethyl group is attached to atom 4. The name is thus 4-ethyl-2,4-dimethylhexane. Note that if we had started numbering from the right, the higher number 3 would appear for the position of the first methyl group, and so the numbering from the left is preferred.

The fraction of branched-chain alkanes in a gasoline affects how it burns in an engine. Gasoline composed entirely of straight-chain alkanes burns very unevenly, causing a "knocking" that can damage the engine. Blends that are richer in branched-chain and cyclic alkanes burn with much less knocking. Smoothness of combustion is measured quantitatively through the **octane number** of the gasoline, which was defined in 1927 by selecting two compounds present in ordinary gasoline that lie at extremes in the knocking they caused. Pure 2,2,4-trimethylpentane (commonly known as "isooctane") burns very smoothly and was assigned an octane number of 100. Pure heptane gave the most knocking of the compounds examined at the time and was assigned octane number 0. Mixtures of heptane and isooctane cause intermediate amounts of knocking. Standard mixtures of these two compounds define a scale to evaluate the knocking caused by real gasolines, which are complex mixtures of branched-chain and straight-chain hydrocarbons. If a gasoline sample gives the same amount of knocking in a test engine as a mixture of 90% (by volume) 2,2,4-trimethylpentane and 10% heptane, it is assigned the octane number 90.

Certain additives increase the octane rating of gasoline. The cheapest is tetraethyllead ($Pb(C_2H_5)_4$), a compound that has very weak bonds between the central lead atom and the ethyl carbon atoms, so it readily releases ethyl radicals, $\cdot C_2H_5$, into the gasoline during combustion. These reactive species speed and smooth combustion, reduce knocking, and give better fuel performance. Unfortunately, the lead is released into the atmosphere, causing serious long-term health hazards. As a result, the use of lead compounds in gasoline has been almost completely phased out in the United States, and other additives such as "MTBE" (methyl *t*-butyl ether, see Section 24–2) have been introduced to increase octane ratings. Chemical processing to make branched- from straight-chain compounds represents another solution to the problem, although only a partial one.

Alkenes and Alkynes

The hydrocarbons discussed so far are referred to as **saturated** because all the carbon–carbon bonds are single bonds. Hydrocarbons that have carbon–carbon double and triple bonds are **unsaturated** (Fig. 24–10). Ethylene (C_2H_4) has a double bond between its carbon atoms and is the simplest **alkene**. The simplest **alkyne** is acetylene (C_2H_2), which has a triple bond between its carbon atoms. In naming these compounds, the *-ane* ending of the corresponding alkane is replaced by *-ene* when a double bond is present, and by *-yne* when a triple bond is present. Ethene is thus

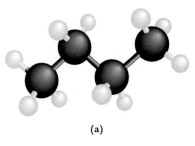

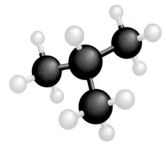

(a)

(b)

Figure 24–9 Two isomeric hydrocarbons with the molecular formula C_4H_{10}. (a) Butane. (b) 2-Methyl propane.

• The generic formula of the alkenes is C_nH_{2n}, and the generic formula of the alkynes is C_nH_{2n-2}.

• Alkenes are also referred to as "olefins."

Figure 24–10 One way to distinguish alkanes from alkenes is by their differing behaviors with aqueous $KMnO_4$. This strong oxidizing agent undergoes no reaction with hexane, retaining its purple color (*left*). When $KMnO_4$ is placed in contact with 1-hexene, however, a redox reaction takes place in which the brown solid MnO_2 is formed (*right*) and OH groups are added to either side of the double bond in the 1-hexene, giving a compound with the formula $CH_3(CH_2)_3CH(OH)CH_2OH$.

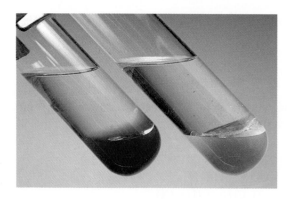

the systematic name for ethylene, and ethyne for acetylene. The nonsystematic names of these two compounds are so well established, however, that we shall use them. For compounds with a carbon backbone of four or more carbon atoms, it is necessary to specify the location of the double or triple bond. This is done by numbering the carbon–carbon bonds and putting the number of the lower-numbered carbon involved in the multiple bond before the name of the alkene or alkyne. Thus, there are two different, isomeric alkynes with the formula C_4H_6:

$$HC\equiv CCH_2CH_3 \qquad \text{(1-butyne)}$$
$$CH_3C\equiv CCH_3 \qquad \text{(2-butyne)}$$

There is yet another subtlety for alkenes. Recall from Section 18–2 that rotation does not take place readily about a carbon–carbon double bond, as it does about a single bond. Many alkenes therefore exist in isomeric forms, depending on whether bonded side-groups are located on the same side or on opposite sides of the double bond. There is only a single isomer of 1-butene, but there are two of 2-butene:

These compounds differ in melting point, boiling point, density, and other physical and chemical properties.

Compounds with two double bonds are called **dienes**, those with three are **trienes**, and so forth. The compound 1,3-pentadiene, for example, is a derivative of pentane with two double bonds:

$$CH_2=CHCH=CHCH_3$$

In such **polyenes,** each double bond may lead to *cis* and *trans* conformations, depending on its neighboring groups, and so there may be several isomers with the same bonding patterns but different molecular geometries and physical properties.

Alkenes are not present to a significant extent in crude petroleum. They are key compounds for the synthesis of valuable organic chemicals and polymers, however, and so ways of producing them from alkanes are of great importance. One way to obtain alkenes is through **cracking** the petroleum by heat or with catalysts. In **catalytic cracking,** the heavier fractions from the distillation column (compounds of C_{12} or higher) are passed over a silica–alumina catalyst at temperatures of 450 to 550°C. Reactions such as

$$CH_3(CH_2)_{12}CH_3 \longrightarrow CH_3(CH_2)_4CH=CH_2 + C_7H_{16}$$

• "Eneynes," which contain both double and triple bonds within the same molecule, also exist.

occur to break the long chain into fragments. This type of reaction accomplishes two purposes. First, the shorter-chain hydrocarbons have lower boiling points and can be added to gasoline. Second, the alkenes that result have higher octane numbers than the corresponding alkanes and perform better in the engine. **Thermal cracking** uses higher temperatures of 850 to 900°C and no catalyst. It produces shorter-chain alkenes such as ethylene and propylene through reactions such as

$$CH_3(CH_2)_{10}CH_3 \longrightarrow CH_3(CH_2)_8CH_3 + CH_2{=}CH_2$$

This is not so useful for gasoline production because the short-chain alkenes are too volatile. It is of great importance for the chemical industry, however, because ethylene and propylene are among the most important starting materials for synthesis.

Aromatic Hydrocarbons

One last group of hydrocarbons present in crude petroleum is the **aromatic hydrocarbons,** of which benzene is the simplest example. The benzene molecule consists of a six-carbon ring in which the delocalization of π-electrons significantly increases the molecular stability (see Section 18–3). Benzene is sometimes represented simply by its chemical formula C_6H_6 and sometimes (to show structure) by a hexagon with a circle inside it:

The six points of the hexagon represent the six carbon atoms, with the hydrogen atoms omitted for simplicity. The circle represents the delocalized π-electrons, which are spread out evenly over the ring. The molecules of other aromatic compounds contain benzene rings with various side-groups, or two (or more) benzene rings linked by alkyl chains or fused side by side, as in naphthalene ($C_{10}H_8$):

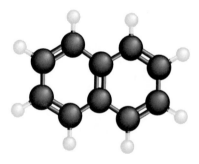

A less abstract representation of the structure and bonding in naphthalene.

Besides benzene, the most prevalent aromatic compounds in petroleum are toluene, in which one hydrogen atom on the benzene ring is replaced by a methyl group, and the xylenes, in which two such replacements are made:

toluene   o-xylene   m-xylene   p-xylene

This set of compounds is referred to as BTX (standing for *benzene-toluene-xylene*). The BTX in petroleum is of great importance to polymer synthesis, as we see in Chapter 25. These components also are very important in gasoline formulations because they significantly increase octane number. In fact, they can be used to make high-performance fuels with octane numbers above 100, as are required in modern aviation.

• The prefixes *o*-, *m*-, and *p*- stand for *ortho*-, *meta*-, and *para*- and are used to name molecules with doubly substituted benzene rings.

A major advance in petroleum refining has been the development of **reforming** reactions, which allow the production of BTX aromatics from straight-chain alkanes that contain the same number of carbon atoms. A fairly narrow distillation fraction containing only C_6 to C_8 alkanes is taken as the starting material. The reforming reactions use high temperatures and transition-metal catalysts such as platinum or rhenium on alumina supports, and their detailed mechanisms are not fully understood. Apparently, a straight-chain alkane such as hexane is cyclized and dehydrogenated to give benzene as the primary product. Heptane yields mostly toluene, and octane yields a mixture of xylenes. Toluene is of increasing importance as a solvent because tests show it to be far less carcinogenic than benzene. Benzene is more important for chemical synthesis, however, so a large fraction of the toluene produced is converted to benzene by **hydrodealkylation:**

• We can write the hydrodealkylation reaction from an interpretation of the name: "hydro" means the addition of hydrogen, and "de-alkyl" means the removal of the alkyl group.

toluene　　benzene

This reaction is carried out at high temperatures (550 to 650°C) and pressures of 40 to 80 atm.

24–2 FUNCTIONAL GROUPS AND ORGANIC SYNTHESIS

The first section of this chapter describes the diverse compounds formed from the two elements carbon and hydrogen, and the extraction of these compounds from petroleum. The hydrocarbons are just the start of organic chemistry, however. Hydrocarbon chains can be extensively modified by attachment or insertion of non-carbon elements such as oxygen, nitrogen, or a halogen, or of combinations of such elements. In considering the effects of these changes, attention is focused away from the structures of entire molecules to the structures of **functional groups,** which consist of the non-carbon atoms plus the portions of the molecule that adjoin them. Functional groups tend to be the reactive sites in organic molecules, and their chemical properties depend only rather weakly on the nature of the hydrocarbon chain to which they are linked. This fact permits us to regard an organic molecule as a hydrocarbon frame, which mainly governs size and shape, to which are attached functional groups that mainly determine the chemistry of the molecule. Table 24–2 shows some of the most important functional groups.

A small number of hydrocarbon building blocks from petroleum and natural gas (methane, ethylene, propylene, benzene, and xylene) are the starting points for the synthesis of most of the high-volume organic chemicals produced today. We discuss the ways in which the largest-volume organic chemicals are synthesized.

Halides

One of the simplest functional groups consists of a single halogen atom. **Alkyl halides** form when mixtures of alkanes and halogens (except iodine) are heated or exposed to light:

• For illustrative purposes, we take the halogen to be chlorine. Similar compounds of the other halogens exist as well.

$$CH_4 \ + \ Cl_2 \ \xrightarrow{\text{250–400° or light}} \ CH_3Cl \ + \ HCl$$

methane　　chorine　　　　　　　　　chloromethane　hydrogen chloride

The mechanism is a chain reaction, as described in Section 14–5. Ultraviolet light initiates the reaction by dissociating a small number of chlorine molecules into highly reactive atoms. These take part in linked reactions of the form

$$\left. \begin{array}{l} Cl\cdot + CH_4 \longrightarrow HCl + \cdot CH_3 \\ \cdot CH_3 + Cl_2 \longrightarrow CH_3Cl + Cl\cdot \end{array} \right\} \text{(propagation)}$$

Chloromethane (also called "methyl chloride") is used in synthesis to add methyl groups to organic molecules. If sufficient chlorine is present, more highly chlorinated methanes form. This provides a method for the industrial synthesis of dichloromethane (CH_2Cl_2, also called "methylene chloride"), trichloromethane ($CHCl_3$, chloroform), and tetrachloromethane (CCl_4, carbon tetrachloride). All three are used as solvents and have vapors with an anesthetic or narcotic effect. Both chloroform and carbon tetrachloride are now known to be carcinogenic. This has eliminated the use of the latter as a household solvent for cleaning soiled clothing and removing grease.

A more important industrial route to alkyl chlorides than the free-radical reactions just described is the addition of chlorine to $C=C$ double bonds. Billions of kilograms of 1,2-dichloroethane (commonly called "ethylene dichloride") are manufactured each year by the addition of chlorine to ethylene over an iron(III) oxide

• Ethylene dichloride is the largest-volume derived organic chemical in the petrochemical industry.

Table 24–2
Some Important Functional Groups in Organic Compounds

Functional Group	Type of Compound	Examples
—F, —Cl, —Br, —I	Alkyl or aryl halide	CH_3CH_2Br (bromoethane)
—OH	Alcohol	CH_3CH_2OH (ethanol)
	Phenol	![phenol structure] —OH HO (1,3-dihydroxybenzene, or resorcinol)
—O—	Ether	$CH_3—O—CH_3$ (dimethyl ether)
$-C{\overset{O}{\underset{H}{\diagdown}}}$	Aldehyde	$CH_3CH_2CH_2—\overset{\overset{O}{\|}}{C}—H$ (butyraldehyde, or butanal)
$\diagdown C=O$	Ketone	$CH_3—\overset{\overset{O}{\|}}{C}—CH_3$ (propanone, or acetone)
$-C{\overset{O}{\underset{OH}{\diagdown}}}$	Carboxylic acid	CH_3COOH (acetic acid, or ethanoic acid)
$-C{\overset{O}{\underset{O—}{\diagdown}}}$	Ester	$CH_3—C{\overset{O}{\underset{O—CH_3}{\diagdown}}}$ (methyl acetate)
—NH_2	Amine	CH_3NH_2 (methylamine)
$-C{\overset{O}{\underset{N}{\diagdown}}}$	Amide	$CH_3—C{\overset{O}{\underset{NH_2}{\diagdown}}}$ (acetamide)

catalyst at moderate temperatures (40 to 50°C) either in the vapor phase or in a solution of 1,2-dibromoethane:

$$CH_2\!=\!CH_2 + Cl_2 \longrightarrow CH_2ClCH_2Cl$$

Almost all the 1,2-dichloroethane produced is used to make chloroethylene (vinyl chloride, $CH_2\!=\!CHCl$). This is accomplished by heating the 1,2-dichloroethane to 500°C over a charcoal catalyst to abstract HCl:

$$CH_2ClCH_2Cl \longrightarrow CH_2\!=\!CHCl + HCl$$

The HCl can be recovered and converted to Cl_2 for further production of 1,2-dichloroethane from ethylene. Vinyl chloride has a much lower boiling point than 1,2-dichloroethane (-13°C compared to 84°C), so the two are easily separated by fractional distillation. The only use of vinyl chloride, but a very important one, is in the production of polyvinyl chloride plastic, as discussed in Section 25–3.

- This conversion of HCl to Cl_2 occurs through the Deacon process (see Section 23-1): $2\ HCl + \frac{1}{2} O_2 \longrightarrow Cl_2 + H_2O$

Alcohols and Phenols

Alcohols have the —OH functional group attached to an alkyl group. The simplest alcohol is methanol (CH_3OH), which is also known as wood alcohol because, until 1926, it was made entirely from the distillation of wood. Now it is made exclusively from synthesis gas, as described in Section 7–5. The next higher alcohol, ethanol (CH_3CH_2OH), can be produced from the fermentation of sugars. Although this is the major source of ethanol for alcoholic beverages and for "gasohol" (automobile fuel made up of 90% gasoline and 10% ethanol), it is not of significance for industrial production, which uses the direct hydration of ethylene:

- *Hydration* means addition of water.

$$CH_2\!=\!CH_2 + H_2O \longrightarrow CH_3CH_2OH$$

Temperatures of 300 to 400°C and pressures of 60 to 70 atm are used, with a phosphoric-acid catalyst. Both methanol and ethanol are widely used as solvents and as intermediates for further chemical synthesis.

Two three-carbon alcohols exist, depending on whether the —OH group is attached to a terminal carbon atom (1-propanol) or the central carbon atom (2-propanol)

$$CH_3CH_2CH_2OH \qquad \underset{\underset{\text{2-propanol}}{\overset{|}{OH}}}{CH_3CHCH_3}$$

1-propanol   2-propanol

- Contrast the names of alcohols with the names of alkyl halides. If the —OH group is replaced by a chlorine atom, the names of the resulting compounds are 1-chloropropane and 2-chloropropane.

The two are frequently referred to as propyl alcohol and isopropyl alcohol, respectively. The systematic names of alcohols are obtained by replacing the *-ane* ending of the corresponding alkane by *-anol* and using a numerical prefix where necessary to identify the carbon atom to which the —OH group is attached. Isopropyl alcohol is made from propylene by a hydration reaction that is catalyzed by sulfuric acid. The reaction sequence is a rather interesting one. The first step is addition of H^+ to the double bond,

$$CH_3CH\!=\!CH_2 + H^+ \longrightarrow CH_3\!-\!CH^+\!-\!CH_3$$

producing a transient charged species in which the positive charge is centered on the central carbon atom. Attack by negative HSO_4^- ions occurs at this positive site and leads to a neutral intermediate that can be isolated.

$$\underset{\underset{OSO_3H}{\overset{|}{}}}{CH_3\!-\!CH\!-\!CH_3}$$

Further reaction with water then causes the replacement of the $-OSO_3H$ group by an $-OH$ group and the regeneration of the sulfuric acid:

$$H_2O + CH_3-\underset{\underset{OSO_3H}{|}}{CH}-CH_3 \longrightarrow CH_3-\underset{\underset{OH}{|}}{CH}-CH_3 + H_2SO_4$$

This mechanism explains why only 2-propanol, and no 1-propanol, is formed.

The compound 1-propanol is a **primary** alcohol: the carbon atom to which the $-OH$ group is bonded has exactly one other carbon atom attached to it. The isomeric compound 2-propanol is a **secondary** alcohol because the carbon atom to which the $-OH$ group is attached has two carbon atoms (in the two methyl groups) attached to it. The simplest **tertiary** alcohol (in which the carbon atom that is attached to the $-OH$ group is at the same time bonded to three other carbon atoms) is 2-methyl-2-propanol (also called "tertiary butyl alcohol"):

$$CH_3-\underset{\underset{CH_3}{|}}{\overset{\overset{OH}{|}}{C}}-CH_3$$

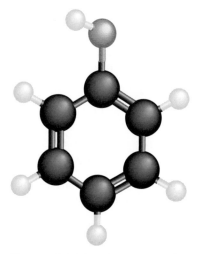

Figure 24-11 The structure of phenol, C_6H_5OH.

This structural classification of alcohols into three groups is made because it effectively divides the alcohols according to their chemical properties, as well.

Phenols are organic compounds in which an $-OH$ group is attached directly to an aromatic ring. The simplest example is phenol itself, C_6H_5OH (Fig. 24–11). Phenols differ substantially from alcohols in both physical and chemical properties. One of the most important differences is in their acidity. Phenol (also called "carbolic acid") has an acidity constant of 1×10^{-10}, much larger than that of typical alcohols, which have K_a's ranging from 10^{-16} to 10^{-18}. The reason for this difference is the greater stability of the conjugate base (the phenolate ion, $C_6H_5O^-$) due to the spreading out of the negative charge over the aromatic ring. Phenol, although not a strong acid, does react readily with sodium hydroxide to form the salt sodium phenolate:

• It might seem that alcohols and phenols would be bases rather than acids because of the presence of $-OH$ groups. Recall from Section 17-6, however, that compounds with $-X-O-H$ bonds act as bases only if X is quite electropositive. They act as oxoacids if X is even somewhat electronegative (as is the case with X = C here).

$$C_6H_5OH + NaOH \longrightarrow H_2O + C_6H_5O^- \; Na^+$$

The corresponding reaction between NaOH and alcohols does not occur to a significant extent.

The manufacture of phenols uses quite different types of reactions from those employed to make alcohols. One method, introduced in 1924 and still used to a small extent today, involves the chlorination of the benzene ring followed by reaction with sodium hydroxide:

$$C_6H_6 + Cl_2 \longrightarrow C_6H_5Cl + HCl$$
$$C_6H_5Cl + 2\,NaOH \longrightarrow C_6H_5O^- \; Na^+ + NaCl + H_2O$$

and then the addition of a hydrogen-ion donor. The first step in this sequence illustrates a characteristic difference between the reactions of aromatics and alkenes. If chlorine reacts with an alkene, it *adds* across the double bond (as we have seen in the production of 1,2-dichloroethane). When an aromatic ring is involved, on the other hand, chlorine *substitutes* for hydrogen instead, and the aromatic π-bonding structure is preserved.

The method used to make almost all phenol today is a different one, however.

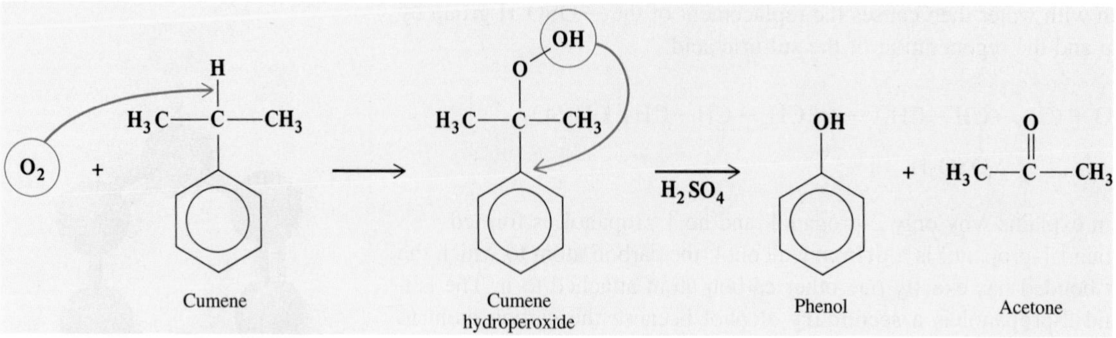

Figure 24–12 The production of phenol and acetone from cumene is a two-step process involving insertion of O_2 to make a peroxide, followed by acid-catalyzed migration of the OH group to form the products.

It starts with the acid-catalyzed reaction of benzene with propylene to give cumene, or isopropyl benzene:

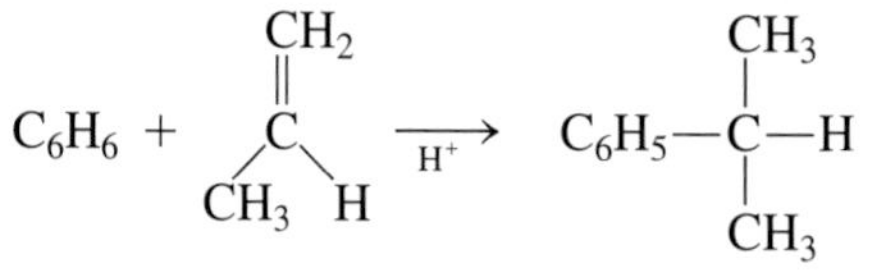

As in the production of 2-propanol, the first step is the addition of H^+ to propylene to give CH_3—CH^+—CH_3. This ion then attaches to the benzene ring to give the cumene and regenerate the H^+ ion. Subsequent reaction of cumene with oxygen (Fig. 24–12) gives phenol and acetone, an important compound discussed later in this section. The ultimate products of the manufacture of phenol are mostly polymers and aspirin (see Fig. 22–2).

Ethers

Ethers are characterized by the —O— functional group, in which an oxygen atom provides a link between two separate alkyl or aromatic groups. One important ether is diethyl ether, often simply called "ether," in which two ethyl groups are linked to the same oxygen atom:

$$C_2H_5—O—C_2H_5$$

This is a useful solvent for organic reactions and was formerly used as an anesthetic. It can be produced by a **condensation reaction** (a reaction in which a small molecule such as water is split out) between two molecules of ethanol in the presence of concentrated sulfuric acid, which serves as a dehydrating agent.

$$CH_3CH_2\boxed{OH + H}OCH_2CH_3 \xrightarrow[H_2SO_4]{} CH_3CH_2—O—CH_2CH_3 + H_2O$$

Another ether of growing importance is methyl t-butyl ether (MTBE):

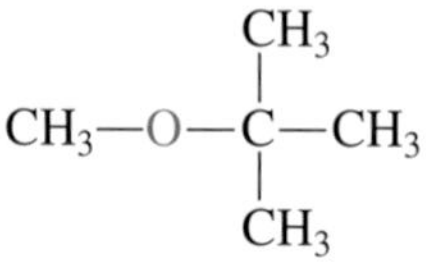

This compound has largely replaced tetraethyllead as an additive to increase octane ratings of gasolines. The bond between the oxygen and the *t*-butyl group is relatively weak. It breaks to form radicals that assist the smooth combustion of the gasoline.

In a cyclic ether, oxygen forms part of a ring with carbon atoms, as in the molecule of tetrahydrofuran (Fig. 24–13), a common solvent. The smallest such ring has two carbon atoms bonded to each other and to the oxygen atom; it occurs in ethylene oxide,

$$\overset{\displaystyle O}{\underset{\displaystyle CH_2-CH_2}{\diagdown\diagup}}$$

which is made by the direct oxidation of ethylene over a silver catalyst:

$$CH_2{=}CH_2 + \tfrac{1}{2}\,O_2 \xrightarrow[\text{Ag}]{} \overset{\displaystyle O}{\underset{\displaystyle CH_2-CH_2}{\diagdown\diagup}}$$

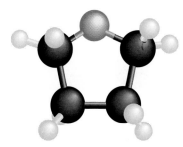

Figure 24–13 The structure of tetrahydrofuran, C_4H_8O. The five atoms forming the ring are not coplanar.

Ethylene oxide and other cyclic ethers with three-membered rings are called **epoxides.** The major use of ethylene oxide is in making ethylene glycol:

$$\overset{\displaystyle O}{\underset{\displaystyle CH_2-CH_2}{\diagdown\diagup}} + H_2O \longrightarrow HO-CH_2-CH_2-OH$$

This reaction is carried out at 195°C under pressure, or at lower temperatures (50 to 70°C) with sulfuric acid as a catalyst. Ethylene glycol is a dialcohol, or **diol,** in which two —OH groups are attached to adjacent carbon atoms. Its primary use is as antifreeze: when it is added to water, it substantially lowers the freezing point and is put in automobile radiators for this reason.

Aldehydes and Ketones

An **aldehyde** contains the characteristic $-\overset{\displaystyle O}{\underset{\displaystyle H}{\overset{\displaystyle \|}{C}}}$ functional group in its molecules.

• The name *aldehyde* arises from *alcohol dehydr*ogenated.

Aldehydes result from the dehydrogenation of (removal of H_2 from) primary alcohols. Formaldehyde, for example, results from the dehydrogenation of methanol at high temperatures over an iron oxide–molybdenum oxide catalyst:

$$CH_3OH \longrightarrow \underset{\displaystyle H}{\overset{\displaystyle H}{\diagdown\diagup}}C{=}O + H_2$$

Another reaction giving the same product is the direct oxidation of the alcohol

$$CH_3OH + \tfrac{1}{2}\,O_2 \longrightarrow \underset{\displaystyle H}{\overset{\displaystyle H}{\diagdown\diagup}}C{=}O + H_2O$$

Formaldehyde is readily soluble in water. A 40% aqueous solution, called formalin, is used to preserve biological specimens. Formaldehyde is a component of wood smoke and helps to preserve smoked meat and fish, probably by reacting with nitrogen-containing functional groups in the proteins of the cells of decay-producing bacteria. Its major use is in making polymeric adhesives and insulating foam.

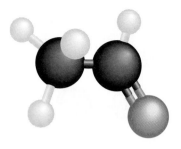

Figure 24–14 The structure of acetaldehyde, CH₃CHO.

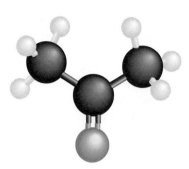

Figure 24–15 The structure of acetone, CH₃COCH₃.

The next aldehyde in the series is acetaldehyde, the dehydrogenation product of ethanol (Fig. 24–14). Acetaldehyde is not actually made industrially from ethanol but rather by the oxidation of ethylene, using a PdCl₂ catalyst.

Ketones have the $\diagdown$C=O functional group in which a carbon atom forms a double bond to an oxygen atom and single bonds to two separate alkyl or aromatic groups. Such compounds can be described as the products of dehydrogenation or oxidation of secondary alcohols, just as aldehydes come from primary alcohols. The simplest ketone, acetone (Fig. 24–15) is in fact made commercially by the dehydrogenation of 2-propanol over a copper oxide or zinc oxide catalyst at 500°C.

$$CH_3-\underset{\underset{OH}{|}}{CH}-CH_3 \longrightarrow CH_3-\underset{\underset{O}{\|}}{C}-CH_3 + H_2$$

Acetone is also produced (in greater volume) as the co-product with phenol of the oxidation of cumene, as discussed earlier. It is a widely used solvent and the starting material for the synthesis of a number of polymers.

Carboxylic Acids and Esters

Carboxylic acids contain the $-C\overset{\displaystyle\nearrow O}{\underset{\displaystyle\searrow OH}{}}$ functional group (also written as —COOH).

They are the products of the oxidation of aldehydes, just as aldehydes are the products of the oxidation of primary alcohols. (The turning of wine to vinegar is a two-step oxidation leading from ethanol through acetaldehyde to acetic acid.) Industrially, acetic acid can be produced by the air oxidation of acetaldehyde over a manganese or cobalt acetate catalyst at 55 to 80°C:

$$CH_3C\overset{\displaystyle\nearrow O}{\underset{\displaystyle\searrow H}{}} + \tfrac{1}{2}\,O_2 \xrightarrow[\text{Mn(CH}_3\text{COO)}_2]{} CH_3C\overset{\displaystyle\nearrow O}{\underset{\displaystyle\searrow OH}{}}$$

The reaction now preferred for acetic acid production for reasons of economy is the combination of methanol with carbon monoxide (both derived from natural gas) over a catalyst containing rhodium and iodine. The overall reaction is

$$CH_3OH(g) + CO(g) \xrightarrow[\text{Rh, I}_2]{} CH_3COOH(g)$$

and can be described as a **carbonylation,** the insertion of CO into a C—O bond.

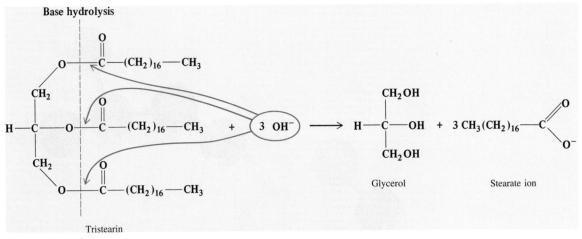

Figure 24–16 Animal fats are largely mixtures of fatty acid triesters of glycerol. Tristearin (*left*) is an example. Treatment of tristearin with sodium hydroxide breaks the three ester bonds to form glycerol and sodium stearate (other fats react similarly). The resulting sodium salt of a fatty acid is a soap. Soaps cut grime by simultaneously interacting with grease particles at the hydrocarbon tail and with water at the carboxylate-ion end-group. This disperses the grease.

Acetic acid is only one of a series of carboxylic acids with formulas $H(CH_2)_nCOOH$. Before acetic acid (with $n = 1$) comes the simplest carboxylic acid, formic acid (HCOOH), in which n equals 0. This compound was first isolated from extracts from the crushed bodies of ants, and its name stems from the Latin word *formica*, meaning "ant." The longer-chain carboxylic acids are called "fatty acids," and their connections to fats and to soap are shown in Figure 24–16.

Carboxylic acids react with alcohols or phenols to give **esters**, forming water as by-product. An example is the reaction of acetic acid with methanol to give methyl acetate:

• The reaction of a carboxylic acid with an alcohol to give an ester plus water is formally similar to the reaction of an inorganic acid with an inorganic base to give a salt plus water.

$$CH_3C\overset{O}{\overset{\|}{-}}\boxed{OH + H}\!-\!OCH_3 \longrightarrow CH_3C\overset{O}{\overset{\|}{-}}O\!-\!CH_3 + H_2O$$

Esters are named by first stating the name of the alkyl group of the alcohol (the methyl group in this case) followed by the name of the carboxylic acid with the ending *-ate* (acetate). One of the most important esters in commercial production is vinyl acetate, with the structure

$$CH_3\!-\!C\!\!\diagup\!\!\overset{O}{\diagdown}\!\!_{O-CH=CH_2}$$

• The $CH=CH_2$ side group is called a vinyl group.

It is prepared by the reaction of acetic acid not with an alcohol but with ethylene and oxygen over a catalyst such as $CuCl_2$ or $PdCl_2$:

$$CH_3C\overset{O}{\overset{\|}{-}}OH + CH_2=CH_2 + \tfrac{1}{2}O_2 \xrightarrow[CuCl_2]{} CH_3C\overset{O}{\overset{\|}{-}}O\!-\!CH=CH_2 + H_2O$$

Figure 24–17 The structures of (a) isoamyl acetate, $CH_3COO(CH_2)_2CH(CH_3)_2$, and (b) benzyl acetate, $C_6H_5CH_2OOCCH_3$.

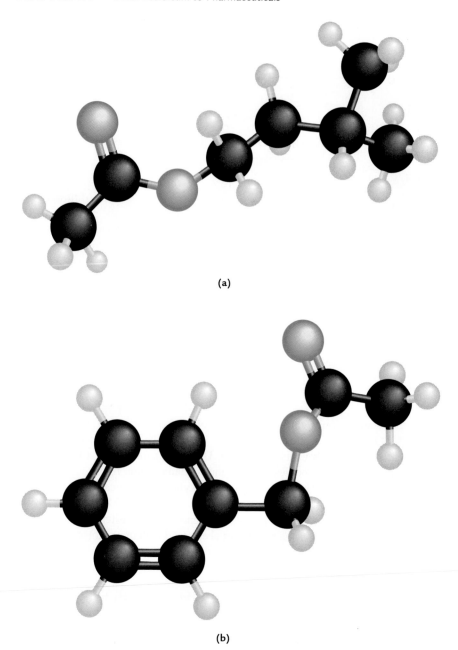

(a)

(b)

Esters are colorless, volatile liquids that frequently have pleasant odors. Many occur naturally in flowers and fruits. Isoamyl acetate (Fig. 24–17a), for example, is generated in apples as they ripen, and contributes to the flavor and odor of the fruit. Benzyl acetate, the ester formed from acetic acid and benzyl alcohol (Fig. 24–17b), is a major component of oil of jasmine and is used in the preparation of perfumes.

• The scents of fine perfumes involve subtle blending of many different compounds, often in very small quantities.

Amines and Amides

The **amines** can be regarded as derivatives of ammonia (NH_3). They have the general formula R_3N, where R can represent a hydrocarbon group or hydrogen. If only one hydrogen atom in the ammonia molecule is replaced by a hydrocarbon group, the result is called a **primary** amine; examples are ethylamine and aniline:

ethylamine   aniline   dimethylamine   trimethylamine

If two hydrocarbon groups replace hydrogen atoms in the ammonia molecule, the compound is a **secondary** amine (such as dimethylamine). Replacement of all three hydrogen atoms gives a **tertiary** amine (such as trimethylamine). Amines are bases because the lone electron pair on the nitrogen can accept a hydrogen ion in the same way that the lone pair on the nitrogen in ammonia does.

A primary or secondary amine (or ammonia itself) can react with a carboxylic acid to form an **amide.** This is another condensation reaction and is analogous to the formation of an ester from the reaction of an alcohol with a carboxylic acid. An example of amide formation is

$$CH_3C \overset{O}{\diagup}\boxed{OH + H} - N(CH_3)_2 \longrightarrow CH_3C \overset{O}{\diagup} N(CH_3)_2 + H_2O$$

If ammonia is the reactant, an —NH_2 group replaces the —OH group in the carboxylic acid as the amide is formed:

$$CH_3C \overset{O}{\diagup} OH + NH_3 \longrightarrow CH_3C \overset{O}{\diagup} NH_2 + H_2O$$
acetamide

Amide linkages are present in every protein molecule and therefore have great importance in biochemistry (see Section 25–2).

24-3   PESTICIDES AND PHARMACEUTICALS

The organic compounds discussed so far all have relatively small molecules; most are produced in large volume. Substances such as these provide starting materials for synthesis of numerous structurally more complex organic compounds that find applications in agriculture, medicine, and consumer products. In this section, we discuss a selection of these compounds, all of which are used as pesticides or as pharmaceuticals. The structures and synthesis of these compounds can be quite intricate. In reading this section, it is not important to memorize complicated structures. A wiser aim is to note the hydrocarbon frames of the molecules, to recognize functional groups, and to begin to appreciate the extremely diverse structures and properties of organic compounds.

• Recall the complexity of the synthesis of the small molecule of acetylsalicylic acid in Figure 22-2.

Because the molecules we shall be discussing are somewhat more complex than those considered so far, we use a shorthand notation for structure, as illustrated in Figure 24–18. In this notation, the symbol "C" for a carbon atom is omitted, and only the C—C bonds are shown. A carbon atom is assumed to lie at each end of the line segments that represent bonds. In addition, hydrogen atoms attached to carbon are omitted. Terminal carbon atoms (those at the end of chains) and their associated hydrogen atoms are shown explicitly, however. To generate the full structure (and the molecular formula) from such a shorthand formula, carbon atoms must be inserted at the end of each bond, and enough hydrogen atoms must be attached to each carbon atom to give it a valence of four.

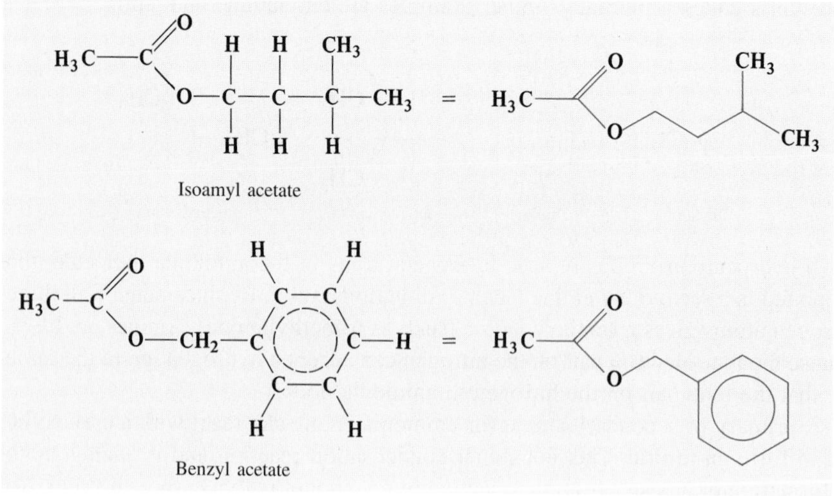

Figure 24–18 In the shorthand notation for the structures of organic compounds illustrated on the right side of this figure, carbon atoms are assumed to lie where the lines indicating bonds intersect. Further, enough hydrogen atoms are assumed to be bonded to each carbon atom to give it a total valence of four. Terminal carbon atoms (those at the ends of chains) and their associated hydrogen atoms are shown explicitly, however.

Isoamyl acetate

Benzyl acetate

Insecticides

The chemical control of insect pests dates back thousands of years. The earliest insecticides were inorganic compounds of copper, lead, and arsenic, as well as some naturally occurring organic compounds such as nicotine (Fig. 24–19a). Very few of these "first-generation" insecticides are in use today because of their adverse side effects on plants, animals, and human beings.

After World War II, controlled organic syntheses gave rise to a second generation of insecticides. The success of these agents fueled rapid growth in the use of chemicals for insect control. The leading insecticide of the 1950s and 1960s was DDT (an abbreviation for *d*ichlor*d*iphenyl*t*richloroethane; see Fig. 24–18b). Although DDT was of worldwide importance in slowing the spread of typhus (transmitted by body lice) and malaria (transmitted by mosquitoes), its use was banned

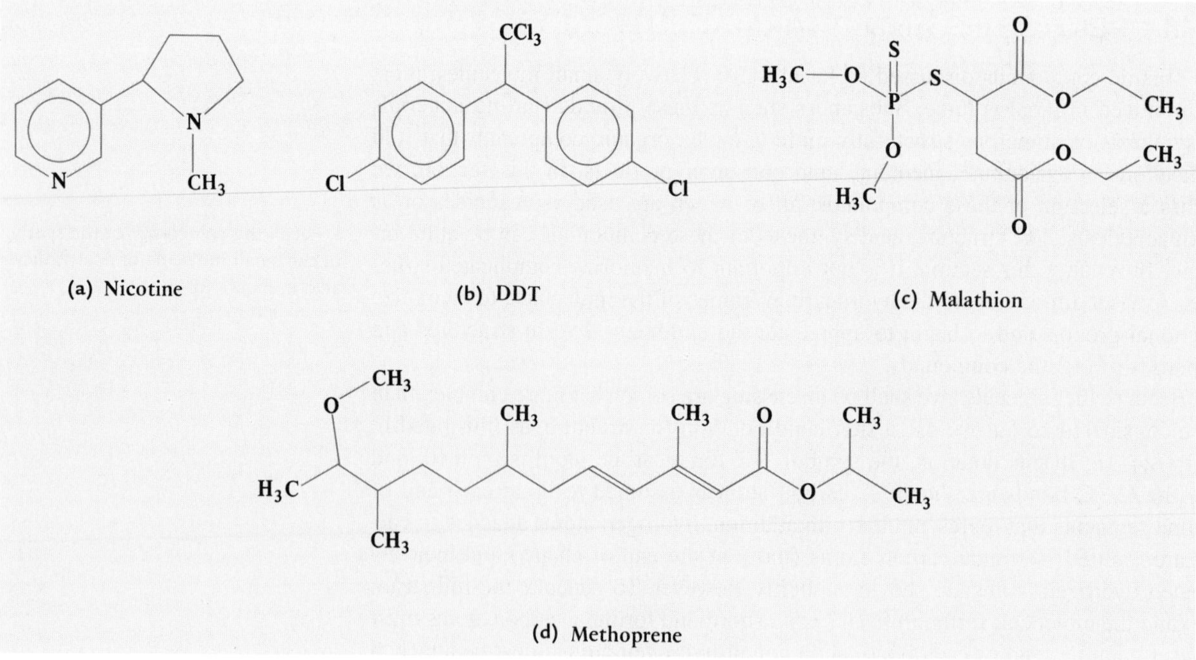

(a) Nicotine   (b) DDT   (c) Malathion

(d) Methoprene

Figure 24–19 The structures of several insecticides.

in the United States in 1972 because of its adverse effects on birds, fish, and other forms of life that can accumulate DDT to high concentrations and because of the increasing resistance of mosquitoes to it. Many other chlorine-substituted hydrocarbons are no longer used as insecticides for the same reason. A class of insecticides widely used in their place are organophosphorus compounds. Figure 24–19c gives the structure of the insecticide malathion, in which the element phosphorus appears together with organic functional groups. Note the two ester groups, the two kinds of sulfur, and the "expanded octet" on the central phosphorus atom that lets it form five bonds.

Unless applied at the right times and in properly controlled doses, second-generation insecticides frequently kill beneficent insects along with the pests. Third-generation insecticides are more subtle. Many are based on sex attractants (to collect insects together in one place before exterminating them) or on juvenile hormones (to prevent insects from maturing to reproduce). The advantages of such compounds are that they are specific against the pests and do little or no harm to other organisms; they can be used in small quantities; and they degrade rapidly in the environment. An example is the juvenile hormone methoprene (Fig. 24–19d), which is used in controlling mosquitoes. It consists of a branched dialkene chain that bears a methyl ether (methoxy) and an isopropyl-ester functional group.

• The use of DDT against malaria-bearing mosquitoes caused the incidence of malaria in India to fall from about 75 *million* cases per year in the early 1950s to 50 *thousand* cases per year in 1961. After spraying stopped, the incidence of the disease increased to 6 million cases in 1976. It has since fallen with the development of other pesticides.

Herbicides

Chemical weed control has contributed (together with fertilizers) to the "green revolution" of the last 45 years, which have seen dramatic increases in agricultural productivity throughout the world. The first herbicide of major importance, introduced in 1945 and still in use today, was 2,4-D (2,4-dichlorophenoxyacetic acid; Fig. 24–20a), a derivative of phenol with chlorine and carboxylic acid functional groups. It kills broadleaf weeds in wheat, corn, and cotton without unduly persisting in the environment, as the chlorinated insecticides discussed earlier do. A related compound is 2,4,5-T (2,4,5-trichlorophenoxyacetic acid), in which a hydrogen atom in 2,4-D is replaced by a chlorine. In recent years, much attention has been given to TCDD ("dioxin," or 2,3,7,8-*tetra*chloro*dibenzo-p-d*ioxin; Fig. 24–20b), which occurs as a trace impurity (10 to 20 ppb by mass) in 2,4,5-T and which, in animal tests, is the most toxic compound of low to moderate molar mass currently known. The use of 2,4,5-T as a defoliant (Agent Orange) in Vietnam led to a lawsuit by veterans who claimed health problems arising from contact with the traces of TCDD present in the 2,4,5-T. Although such a direct connection has never been proved, the use of chlorinated phenoxy herbicides has decreased, and that of other herbicides, such as atrazine (Fig. 24–20c), has grown.

An aircraft applying herbicides and pesticides to a lettuce field in Arizona.

(a) 2,4-D   (b) TCDD   (c) Atrazine

Figure 24–20 The structures of some herbicides.

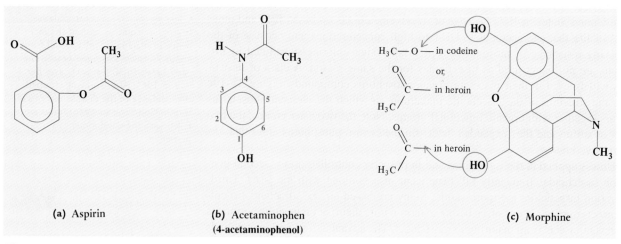

(a) Aspirin

(b) Acetaminophen
(4-acetaminophenol)

(c) Morphine

Figure 24–21   The molecular structures of some analgesics. Note the slight differences between molecules of morphine, codeine, and heroin.

Analgesics

Drugs that relieve pain are called **analgesics.** The oldest and most widely used analgesic is aspirin, which has the chemical name acetylsalicylic acid (Fig. 24–21a). Over 15 million kg of aspirin is synthesized each year, using the synthesis outlined in Figure 22–2. Aspirin acts to reduce fevers as well as to relieve pain. Some recent studies have suggested that regular consumption may reduce the chances of heart disease. As an acid, aspirin can irritate the stomach lining, a side-effect that can be reduced by combining it in a buffer with a weak base such as sodium hydrogen carbonate. Another important pain reliever is 4-acetaminophenol, or acetaminophen (Fig. 24–21b). This compound is sold under many trademarks, most prominently Tylenol. Both acetylsalicylic acid and 4-acetaminophenol are derivatives of phenol, the former being converted to an acetic acid ester with an additional carboxylic acid functional group and the latter having an amide functional group.

A much more powerful pain reliever, which is available only by prescription because it is addictive, is morphine (Fig. 24–21c). Morphine acts on the central nervous system, apparently because its shape fits a receptor site on the nerve cell, and

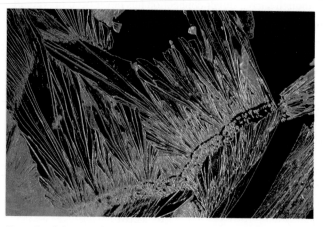

Crystals of 4-acetaminophenol (Tylenol) viewed under polarized light.

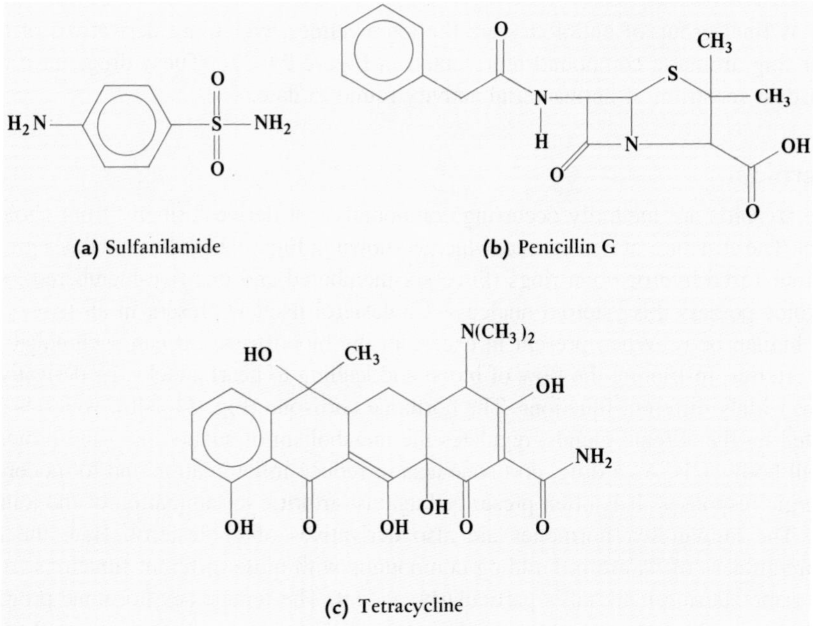

(a) Sulfanilamide

(b) Penicillin G

(c) Tetracycline

Figure 24-22 The molecular structures of some antibiotics.

Figure 24-23 Fermentation tanks used in modern penicillin production.

blocks the transmission of pain signals to the brain. Its structure contains five interconnected rings. A very small change (the replacement of one —OH group by an —OCH$_3$ group, giving a methyl ether) converts morphine into codeine, a prescription drug used as a cough suppressant. Replacing *both* OH groups by acetyl groups (—COCH$_3$) generates the notoriously addictive substance heroin.

Antibacterial Agents

The advent of antibacterial agents changed the treatment of bacterial diseases such as tuberculosis and pneumonia dramatically beginning in the 1930s. The first "wonder drug" was sulfanilamide (Fig. 24–22a), a derivative of aniline. Other "sulfa drugs" are obtained by replacing one of the hydrogen atoms on the sulfonamide group by other functional groups. Bacteria mistake sulfanilamide for *p*-aminobenzoic acid, a molecule with a closely similar shape but a —COOH carboxylic acid group in place of the —SO$_2$NH$_2$ group. The drug then interferes with the organism's synthesis of folic acid, an essential biochemical, and kills it. Mammals do not synthesize folic acid (they obtain it from their diet) and are not affected by sulfanilamide.

• This is the same *p*-aminobenzoic acid used in sun-blocking creams (see Section 18–5).

The molecule of penicillin (Fig. 24–22b) contains an amide linkage connecting a substituted double ring (including S and N atoms) to a benzyl (phenylmethyl) group. It is a natural product formed by certain molds. Although the total synthesis of penicillin was achieved in 1957, that chemical route is not competitive economically with biosynthesis via fermentation. The mold grows for several days in tanks that may hold up to 100,000 L of a fermentation broth (Fig. 24–23). The penicillin is later separated by solvent extraction. Penicillin works by deactivating enzymes in the bacteria that are responsible for building cell walls. Derivatives of natural penicillin have been developed and are also used.

• The name *tetracycline* reflects molecular structure: *tetra*-(four) + *cycl*-(ring).

Small crystals of the antibiotic tetracycline.

A final group of antibiotics are the tetracyclines, which are derivatives of the four-ring aromatic compound represented in Figure 24–22c. These drugs have the broadest spectrum of antibacterial activity found to date.

Steroids

The **steroids** are naturally occurring compounds that derive formally from cholesterol. The structure of cholesterol, which is shown in Figure 24–24a, contains a group of four fused hydrocarbon rings (three six-membered and one five-membered). All steroids possess this "steroid nucleus." Cholesterol itself is present in all tissues of the human body. When present in excess in the bloodstream, it can accumulate in the arteries, restricting the flow of blood and leading to heart attacks. Its derivatives have widely different functions. The hormone cortisone (Fig. 24–24b), which is secreted by the adrenal glands, regulates the metabolism of sugars, fats, and proteins in all body cells. As a drug, cortisone acts to reduce inflammation and to moderate allergic responses. It is often prescribed against arthritic inflammation of the joints.

The human sex hormones are also derivatives of cholesterol. Here the resourcefulness of nature in building compounds with quite different functions from the same starting material is particularly evident. The female sex hormone progesterone (Fig. 24–24c), for example, differs from the male sex hormone testosterone only by the exchange of an acetyl (—$COCH_3$) group for a hydroxyl (—OH) group. Oral contraceptives are synthetic compounds with structures that are closely related to progesterone.

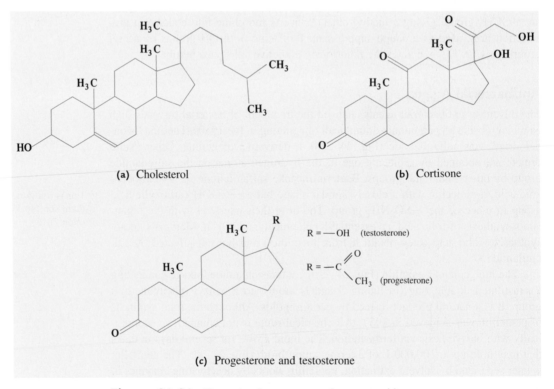

(a) Cholesterol

(b) Cortisone

(c) Progesterone and testosterone

$R = $ —OH   (testosterone)

$R = $ —C$\overset{O}{\underset{CH_3}{}}$ (progesterone)

Figure 24–24   The molecular structures of some steroids.

SUMMARY

24–1 The **straight-chain alkanes** are hydrocarbons with the general formula C_nH_{2n+2} in which carbon atoms are linked one by one in extended chains by single covalent bonds. They are the major constituents of petroleum. The distillation of petroleum separates its constituents according to boiling point. In **branched-chain alkanes,** the chain of carbon atoms has branches so that at least one carbon atom bonds to three other carbon atoms. In **cycloalkanes,** a carbon chain closes a loop upon itself. The general formula for cycloalkanes with one closed ring is C_2H_{2n}. Alkanes are named as derivatives of the alkane that corresponds to the longest continuous chain of carbon atoms in the molecular formula. The naming system then uses the names of the various **alkyl groups** to specify side-chains and numbers to tell their locations. **Alkenes** have at least one carbon–carbon double bond; **alkynes** have at least one carbon–carbon triple bond. Both categories are **unsaturated.** Isomeric alkenes have double bonds at different locations along a chain of carbon atoms of the same length. *Cis–trans* isomerism in alkenes is a consequence of the lack of free rotation around the carbon–carbon double bond. In **polyenes,** two or more double bonds occur along a given chain of carbon atoms. **Catalytic cracking** and **thermal cracking** produce shorter-chain alkenes from the alkanes in the starting material. The **aromatic hydrocarbons** contain rings of carbon atoms in which delocalized π-electrons significantly increase the molecular stability. Petroleum contains useful quantities of benzene, toluene, and the isomeric xylenes. In **reforming** reactions, six-, seven-, and eight-carbon alkanes are converted to the aromatic hydrocarbons, which are used in high-performance fuels and in synthesis.

24–2 Functional groups are sites of greater reactivity in organic molecules. They arise from insertion of non-carbon atoms in a hydrocarbon chain or the attachment of non-carbon atoms to a chain. The replacement of a hydrogen atom by a halogen creates an **alkyl halide.** A similar replacement by an —OH group gives an **alcohol.** When an —OH group attaches to an aromatic hydrocarbon ring, the result is a **phenol. Ethers** are characterized by the —O— functional group, in which an oxygen atom links two alkyl or aromatic groups. Cyclic ethers with various ring sizes exist; those with three members are called **epoxides.** An **aldehyde** has the characteristic $-\overset{\displaystyle O}{\underset{\displaystyle H}{C}}$ functional group and comes from the dehydrogenation of an alcohol. A **ketone** also contains a carbon atom that is double bonded to an oxygen atom but has two other carbon atoms (and therefore no hydrogen atoms) bonded to that carbon atom. **Carboxylic acids** contain the $-\overset{\displaystyle O}{\underset{\displaystyle OH}{C}}$ functional group and condense with alcohols to give **esters.** The **amines** are derivatives of NH_3 in which one, two, or all three hydrogen atoms are replaced by hydrocarbon groups. Such replacements give **primary, secondary,** and **tertiary** amines, respectively. The condensation of a primary or secondary amine with a carboxylic acid gives an **amide.**

24–3 The organic compounds produced from petroleum in large amounts have rather small molecules of simple structure. They are elaborated into materials with more complex molecular structures that serve a range of purposes. These include insecticides; herbicides; pharmaceuticals, including **analgesics** to relieve pain; and antibacterials and antibiotics to control infections. The **steroids** are biologically active compounds that are derivatives of cholesterol.

PROBLEMS

Note: Answers to blue-numbered problems are given in Appendix F. Problems that are more challenging are indicated with asterisks.

Petroleum Refining and the Hydrocarbons

1. Is it possible for a gasoline to have an octane number exceeding 100? Explain.

2. Is it possible for a motor fuel to have a negative octane rating? Explain.

3. (a) Use average bond enthalpies from Table 10–3 and the ΔH_f° of C(g) and H(g) from Appendix D to estimate the standard enthalpy of formation of cyclopropane ($C_3H_6(g)$).
   (b) The standard enthalpy change at 25°C for the combustion of cyclopropane

$$C_3H_6(g) + \tfrac{9}{2}\, O_2(g) \longrightarrow 3\, CO_2(g) + 3\, H_2O(g)$$

   is $\Delta H^\circ = -1959$ kJ. Use this, together with data from Appendix D, to calculate the standard enthalpy of formation of cyclopropane.
   (c) By comparing your answers from (a) and (b), estimate the "strain enthalpy" associated with forming a ring of three carbon atoms (with C—C—C bond angles of only 60 degrees) in cyclopropane.

4. (a) Use average bond enthalpies form Table 10–3 and the ΔH_f° of C(g) and H(g) from Appendix D to estimate the standard enthalpy of formation of cyclobutane ($C_4H_8(g)$).
   (b) The enthalpy change for the combustion of cyclobutane

$$C_4H_8(g) + 6\, O_2(g) \longrightarrow 4\, CO_2(g) + 4\, H_2O(g)$$

   is $\Delta H^\circ = -2568$ kJ. Use this, together with data from Appendix D, to calculate the standard enthalpy of formation of cyclobutane.
   (c) By comparing your answers from (a) and (b), estimate the "strain enthalpy" associated with forming a ring of four carbon atoms (with bond angles less than 109.5 degrees) in cyclobutane.

5. (a) Write a chemical equation involving structural formulas for the catalytic cracking of decane into an alkane and an alkene that contain equal numbers of carbon atoms. Assume that the reactant and both products have a straight chain of carbon atoms.
   (b) How many isomers of the alkene are possible?

6. (a) Write an equation involving structural formulas for the catalytic cracking of 2,2,3,4,5,5-hexamethylhexane. Assume that the cracking occurs between carbon atoms 3 and 4.
   (b) How many isomeric alkenes of formula C_6H_{12} are possible?

7. Write structural formulas for the following:
   (a) 2,3-dimethylpentane
   (b) 3-ethyl-2-pentene
   (c) Methylcyclopropane
   (d) 2,2-dimethylbutane
   (e) 3-propyl-2-hexene
   (f) 3-methyl-1-hexene
   (g) 4-ethyl-2-methylheptane
   (h) 4-ethyl-2-heptyne

8. Write structural formulas for the following:
   (a) 2,3-dimethyl-1-cyclobutene
   (b) 2-methyl-2-butene
   (c) 2-methyl-1,3-butadiene
   (d) 2,3-dimethyl-3-ethylhexane
   (e) 4,5-diethyloctane
   (f) Cyclooctene
   (g) Propadiene
   (h) 2-pentyne

9. Write structural formulas for *trans*-3-heptene and *cis*-3-heptene.

10. Write structural formulas for *cis*-4-octene and *trans*-4-octene.

11. Name the following hydrocarbons:
   (a)

$$CH_2{=}C{=}\overset{\displaystyle H}{\overset{|}{C}}{-}CH_2{-}CH_2{-}CH_3$$

   (b)

$$H_2C{=}\overset{\displaystyle H}{\overset{|}{C}}{-}\overset{\displaystyle H}{\overset{|}{C}}{=}\underset{\overset{|}{H}}{\overset{|}{C}}{-}\underset{\overset{|}{H}}{C}{=}CH_2$$

   (c)

$$H_2C{=}\overset{\displaystyle CH_3}{\overset{|}{C}}{-}CH_2{-}CH_2{-}CH_2{-}CH_3$$

   (d)

$$CH_3{-}CH_2{-}C{\equiv}C{-}CH_2{-}CH_3$$

12. Name the following hydrocarbons:
   (a)

$$CH_2{=}\underset{\overset{|}{CH_3}}{C}{-}\!\!-\!\!\underset{\overset{|}{CH_3}}{C}{=}CH_2$$

   (b)

$$CH_3{-}\underset{\overset{|}{H}}{C}{=}\underset{\overset{|}{H}}{C}{-}\underset{\overset{|}{H}}{C}{=}\underset{\overset{|}{H}}{C}{-}CH_3$$

   (c)

$$CH_3{-}\underset{\overset{|}{CH_3}}{\overset{\overset{\displaystyle CH_3}{|}}{C}}{-}CH_2{-}CH_3$$

   (d)

$$CH_3{-}\underset{\overset{\|}{CH_2}}{C}{-}CH_3$$

13. State the hybridization of each of the carbon atoms in the hydrocarbon structures in problem 11.

14. State the hybridization of each of the carbon atoms in the hydrocarbon structures in problem 12.

Functional Groups and Organic Synthesis

15. Write balanced equations for the following reactions. Use structural formulas to represent the organic compounds.
   (a) The production of butyl acetate from butanol and acetic acid.
   (b) The conversion of ammonium acetate to acetamide and water.
   (c) The dehydrogenation of 1-propanol.
   (d) The complete combustion (to CO_2 and H_2O) of heptane.

16. Write balanced equations for the following reactions. Use structural formulas to represent the organic compounds.
   (a) The complete combustion (to CO_2 and H_2O) of cyclopropanol.
   (b) The reaction of isopropyl acetate with water to give acetic acid and isopropanol.
   (c) The dehydration of ethanol to give ethylene.
   (d) The reaction of 1-butyl iodide with water to give 1-butanol.

17. Outline, using chemical equations, the synthesis of the following from easily available petrochemicals and inorganic starting materials:
   (a) Vinyl bromide ($CH_2{=}CHBr$)
   (b) 2-butanol
   (c) Acetone (CH_3COCH_3)

18. Outline, using chemical equations, the synthesis of the following from easily available petrochemicals and inorganic starting materials:
   (a) Vinyl acetate ($CH_3COOCH{=}CH_2$)
   (b) Formamide ($HCONH_2$)
   (c) 1,2-difluoroethane

19. Write a general equation (using the symbol R to represent a general alkyl group) for the formation of an ester by the condensation of a tertiary alcohol with a carboxylic acid.

20. Explain why it is impossible to form an amide by the condensation of a tertiary amine with a carboxylic acid.

21. In a recent year, the United States produced 6.26×10^9 kg of ethylene dichloride (1,2-dichloroethane) and 15.87×10^9 kg of ethylene. Assuming that all significant quantities of ethylene dichloride were produced from ethylene, what fraction of the ethylene production went to make ethylene dichloride? What mass of chlorine was required for this conversion?

22. In a recent year, the United States produced 6.26×10^9 kg of ethylene dichloride (1,2-dichloroethane) and 3.73×10^9 kg of vinyl chloride. Assuming that all significant quantities of vinyl chloride were produced from ethylene dichloride, what fraction of the ethylene dichloride production went to make vinyl chloride? What mass of hydrogen chloride was generated as a by-product?

Pesticides and Pharmaceuticals

23. (a) The insecticide methoprene (see Fig. 24–19d) is an ester. Write the structural formulas for the alcohol and the carboxylic acid that react to form it. Name the alcohol.
   (b) Suppose that the carboxylic acid from part (a) is chemically changed so that the OCH_3 group is replaced by an H atom and the COOH group is replaced by a CH_3 group. Name the hydrocarbon that would result.

24. (a) The herbicide 2,4-D (see Fig. 24–20a) is an ether. Write the structural formulas of the two alcohol or phenol compounds that, upon condensation, would form this ether. (The usual method of synthesis is different, however.)
   (b) Suppose that the chlorine atoms are replaced by hydrogen atoms and the carboxylic acid group is changed to a CH_3 group in the two compounds in part (a). Name the resulting compounds.

25. (a) Write the molecular formula of acetylsalicylic acid (see Fig. 24–21a).
   (b) An aspirin tablet contains 325 mg of acetylsalicylic acid. Calculate the chemical amount (in moles) of that compound in the tablet.

26. (a) Write the molecular formula of acetaminophen (see Fig. 24–21b).
   (b) A tablet of extra-strength Tylenol contains 500 mg of acetaminophen. Calculate the chemical amount (in moles) of that compound in the tablet.

27. Describe the changes in hydrocarbon structure and functional groups that are needed to make cortisone from cholesterol (see Fig. 24–24).

28. Describe the changes in hydrocarbon structure and functional groups that are needed to make testosterone from cortisone (see Fig. 24–24).

Additional Problems

29. At 25°C, the standard enthalpy of formation of heptane is -187.82 kJ mol^{-1}, and the standard enthalpy of formation of isooctane (2,2,4-trimethylpentane) is -224.13 kJ mol^{-1}.
   (a) Compute the standard enthalpies of combustion of one mole of each of the two if they are burned to $CO_2(g)$ and $H_2O(g)$ in an internal-combustion engine.
   (b) Which compound delivers more heat per gallon? (Isooctane has a density of 5.77 lb per U.S. gallon, and heptane has a density of 5.71 lb per U.S. gallon.)

*30. Ethylene (C_2H_2) is an organic compound of great economic significance that is often transported by pipeline. It has a critical temperature of 282.65 K and a critical pressure of 50.096 atm.
   (a) Express the critical temperature and pressure of ethylene in °F and in pounds per square inch.
   (b) A pipeline contains ethylene at a pressure of 54 atm and a temperature of 15°C. Discuss the physical state of the ethylene within the pipe.
   (c) The pipeline is 100 km long and has an interior diame-

ter of 250 mm. Estimate the mass of ethylene in the pipeline, assuming that it is an ideal gas.

(d) The density of the ethylene within the pipe from part (c) is 0.20 g cm^{-3}. Compute the actual mass of ethylene in the pipe.

(e) If the temperature of the entire pipeline and its contents falls to 0°C on a cold day, suggest what the operators could do to make sure that the ethylene in the pipe does not change its physical state.

31. *Trans*-cyclodecene boils at 193°C, but *cis*-cyclodecene boils at 195.6°C. Write structural formulas for these two compounds.

32. Compare catalytic cracking with thermal cracking of hydrocarbons. What is the purpose of each type of process?

33. Alkenes and aromatic compounds, as well as branched-chain alkanes, increase the octane number of gasoline. Describe the methods used to increase the proportions of these two types of compounds in gasoline.

*34. Consider the following proposed structures for benzene, each of which is consistent with the molecular formula C_6H_6:

(i)

(ii)

(iii)

(iv) $CH_3C{\equiv}CC{\equiv}CCH_3$
(v) $CH_2{=}CHC{\equiv}CCH{=}CH_2$

(a) When benzene reacts with chlorine to give C_6H_5Cl, only one isomer of that compound is formed. Which of the five proposed structures for benzene are consistent with this observation?

(b) When C_6H_5Cl reacts further with chlorine to give $C_6H_4Cl_2$, exactly three isomers of the latter compound are formed. Which of the five proposed structures for benzene are consistent with this observation?

35. How does *dehydration* differ from *dehydrogenation*? Give an example of each.

36. When an ester forms from an alcohol and a carboxylic acid, an oxygen atom links the two parts of each ester molecule. This atom could have come originally from the alcohol, from the carboxylic acid, or randomly from either. Propose an experiment using isotopes to determine which is the case.

37. Glycerol ($C_3H_8O_3$) has a three-carbon chain with one —OH group attached to each carbon atom.
(a) Draw the structure of glycerol.
(b) What kind of reaction does glycerol undergo with stearic acid to form tristearin (see Fig. 24–16)? What type of organic compound is tristearin?

38. (a) It is reported that ethylene is released when pure ethanol is passed over alumina (Al_2O_3) that is heated to 400°C, but that diethyl ether is obtained at a temperature of 230°C. Write balanced equations for both of these dehydration reactions.
(b) If the temperature is raised well above 400°C, an aldehyde forms. Write a chemical equation for this reaction.

39. In what ways do the systematic development of pesticides and of pharmaceuticals resemble each other, and in what ways do they differ? In your answer, consider such aspects as "deceptor" molecules, which are mistaken by living organisms for other molecules, side effects, and the relative advantages of a broad versus a narrow spectrum of activity.

40. The steroid stanolone is an androgenic steroid (a steroid that develops or maintains certain male sexual characteristics). It is derived from testosterone by adding a molecule of hydrogen across the C—C double bond in testosterone.
(a) Using Figure 24–24c as a guide, draw the molecular structure of stanolone.
(b) What is the molecular formula of stanolone?

Polymers: Natural and Synthetic

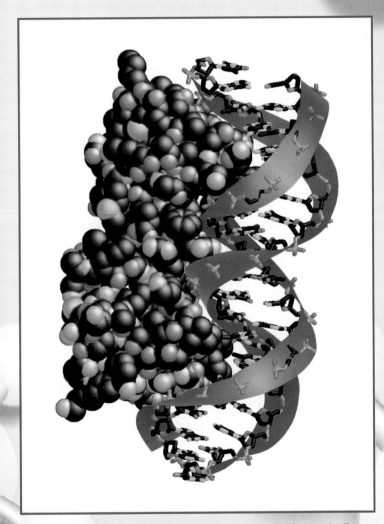

An interaction between biopolymers. Here, the approach of a protein distorts the famous double helix (purple) of DNA.

Hundreds of thousands of organic compounds possess molecules having 20 or fewer carbon atoms combined with hydrogen, oxygen, nitrogen, and the halogens in the functional groups discussed in Chapter 24. Thousands more such compounds remain to be discovered or synthesized. All of this barely touches the full potential of carbon in compound formation, however, because carbon atoms can string together in stable chains of essentially unlimited length that provide the backbone of truly huge molecules, ones that contain hundreds of thousands or even millions of atoms. Such compounds, called **polymers,** are formed by the linking of numerous separate small **monomer units** in strands and webs.

Polymers are ubiquitous both in nature and in modern industrial chemistry. Most foods and fibers are polymers. The diverse, useful materials lumped under the name "plastic" are synthetic (or semi-synthetic) carbon-based polymers. Natural carbon-based polymers include energy-storage and structural materials (such as starch, cellulose, and protein), the enzymes that catalyze essential biochemical reactions, and the genetic substances DNA and RNA, which store and transmit the information required for the replication of life. The prominence of carbon-based polymers in biology and industry sometimes obscures the many polymers that contain no carbon whatever. The chains of alternating Si and O atoms found in some silicate minerals and in synthetic silicones (see Section 21–2) are examples.

25–1 MAKING POLYMERS

To construct a polymer a very large number of monomer units must add to a growing polymer molecule; the reaction must not falter after the first few molecules have reacted. Continued growth is achieved by having the polymer molecule retain highly reactive functional groups at all times during its synthesis. The two major types of polymer growth are addition polymerization and condensation polymerization.

Addition Polymerization

In **addition polymerization,** monomers react to form a polymer chain without net loss of atoms. The most common type of addition polymerization involves free-radical chain reaction of molecules that have C=C double bonds. As an example, consider the polymerization of vinyl chloride (chloroethylene, $CH_2=CHCl$) to polyvinyl chloride:

$$n \; CH_2=\overset{\displaystyle H}{\underset{\displaystyle Cl}{\overset{|}{\underset{|}{C}}}} \longrightarrow \left[CH_2-\overset{\displaystyle H}{\underset{\displaystyle Cl}{\overset{|}{\underset{|}{C}}}} \right]_n$$

As in the chain reactions considered in Section 14–5, the overall process consists of three steps: initiation, propagation (repeated many times to build up a long chain), and termination.

The polymerization of vinyl chloride (Fig. 25–1) can be initiated by a small concentration of molecules that have bonds weak enough to be broken by the action of light or heat, giving radicals. An example of such an **initiator** is a peroxide,

(a)

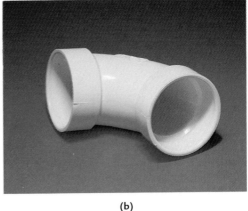

(b)

Figure 25-1 (a) Several billion kilograms of polyvinyl chloride are produced each year in this chemical plant in Texas. (b) A pipe fitting of polyvinyl chloride.

which can be represented as R—O—O—R′, where R and R′ represent alkyl groups. The weak O—O bonds can break

$$R—\overset{..}{\underset{..}{O}}—\overset{..}{\underset{..}{O}}—R' \longrightarrow R—\overset{..}{\underset{..}{O}}· + ·\overset{..}{\underset{..}{O}}—R' \quad \text{(initiation)}$$

to give radicals. In this equation the valence electrons of the oxygen atoms are represented explicitly to show that their valence shells are incomplete after the reaction. The radicals remedy this by reacting avidly with vinyl chloride, accepting electrons from the C=C double bonds to re-establish a closed-shell electron configuration on the oxygen atoms:

$$R—\overset{..}{\underset{..}{O}}· + CH_2{=}CHCl \longrightarrow R—\overset{..}{\underset{..}{O}}—CH_2—\overset{\overset{\displaystyle H}{|}}{\underset{\underset{\displaystyle Cl}{|}}{C}}· \quad \text{(propagation)}$$

One of the two π-electrons in the vinyl chloride double bond has been used to form a single bond with the R—O· radical. The other remains on the second carbon atom, leaving it as a seven-valence-electron atom that reacts with another vinyl chloride molecule:

$$R—O—CH_2—\overset{\overset{\displaystyle H}{|}}{\underset{\underset{\displaystyle Cl}{|}}{C}}· + CH_2{=}\overset{\overset{\displaystyle H}{|}}{\underset{\underset{\displaystyle Cl}{|}}{C}} \longrightarrow R—O—CH_2—\overset{\overset{\displaystyle H}{|}}{\underset{\underset{\displaystyle Cl}{|}}{C}}—CH_2—\overset{\overset{\displaystyle H}{|}}{\underset{\underset{\displaystyle Cl}{|}}{C}}· \text{(propagation)}$$

At each stage, the end-group of the lengthening chain is left one electron short of a valence octet and remains quite reactive. The reaction can continue, building up long-chain molecules of high molecular mass. The vinyl chloride monomers always attach to the growing chain with their CH_2 group, because the odd electron is more stable on a CHCl end-group. This gives the polymer a regular alternation of —CH_2— and —CHCl— groups.

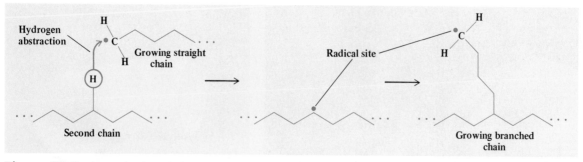

Figure 25-2 A growing branched chain results during free-radical polymerization when a radical at one end of a growing straight chain by chance extracts a hydrogen atom from the middle of a second chain.

Termination occurs when the radical end-groups on two different chains encounter each other and the two chains couple to give a longer chain:

$$R—O—(CH_2—CHCl)_m—CH_2—\overset{\overset{\displaystyle H}{|}}{\underset{\underset{\displaystyle Cl}{|}}{C}} \cdot \; + \; \cdot \overset{\overset{\displaystyle H}{|}}{\underset{\underset{\displaystyle Cl}{|}}{C}}—CH_2—(CHCl—CH_2)_n—O—R' \longrightarrow$$

$$R—O—(CH_2—CHCl)_m—CH_2—\overset{\overset{\displaystyle H}{|}}{\underset{\underset{\displaystyle Cl}{|}}{C}}—\overset{\overset{\displaystyle H}{|}}{\underset{\underset{\displaystyle Cl}{|}}{C}}—CH_2—(CHCl—CH_2)_n—O—R'$$

(termination)

Alternatively, a hydrogen atom may transfer from one end-group to the other:

$$R—O—(CH_2—CHCl)_m—\overset{\overset{\displaystyle \textcircled{H}}{|}}{\underset{\underset{\displaystyle H}{|}}{C}}—\overset{\overset{\displaystyle H}{|}}{\underset{\underset{\displaystyle Cl}{|}}{C}} \cdot \; + \; \cdot \overset{\overset{\displaystyle H}{|}}{\underset{\underset{\displaystyle Cl}{|}}{C}}—CH_2—(CHCl—CH_2)_n—O—R' \longrightarrow$$

$$R—O—(CH_2—CHCl)_m—CH{=}CHCl + CH_2Cl—CH_2—(CHCl—CH_2)_n—O—R'$$

(termination)

The latter termination step leaves a double bond on one chain end and a $—CH_2Cl$ group on the other. When polymer molecules are long, the exact nature of the end-groups has little effect on the physical and chemical properties of the material.

A different type of hydrogen-transfer step often has a much larger effect on the properties of the resulting polymer. Suppose that hydrogen-atom transfer occurs not from the monomer unit on the *end* of a second chain, but from a monomer unit in the *middle* of that chain (Fig. 25–2). Then the first chain stops growing, but the radical site moves to the middle of the second chain, and growth resumes from that point, forming a *branched* polymeric chain with very different properties.

Addition polymerization can be initiated by ions as well as by free radicals. An example is the polymerization of acrylonitrile:

$$n \; CH_2{=}\underset{\underset{\displaystyle C{\equiv}N}{|}}{CH} \longrightarrow \left[CH_2\underset{\underset{\displaystyle C{\equiv}N}{|}}{CH} \right]_n$$

A suitable initiator for this process is butyl lithium $((CH_3CH_2CH_2CH_2)^- Li^+)$. The butyl anion (abbreviated Bu^-) reacts with the end carbon atom in a molecule of acrylonitrile to give a new anion:

$$Bu^{\ominus}Li^{\oplus} + CH_2{=}CH \longrightarrow Bu{-}CH_2{-}CH^{\ominus}Li^{\oplus} \qquad \text{(initiation)}$$
$$\overset{|}{C{\equiv}N} \qquad\qquad\qquad \overset{|}{C{\equiv}N}$$

The new anion then reacts with an additional molecule of acrylonitrile according to

$$Bu{-}CH_2{-}CH^{\ominus}Li^{\oplus} + CH_2{=}CH \longrightarrow Bu{-}CH_2{-}CH{-}CH_2{-}CH^{\ominus}Li^{\oplus}$$
$$\overset{|}{C{\equiv}N} \qquad\qquad \overset{|}{C{\equiv}N} \qquad\qquad\qquad \overset{|}{C{\equiv}N} \qquad \overset{|}{C{\equiv}N}$$

$$\text{(propagation)}$$

The process continues, building up a long-chain polymer.

Ionic polymerization differs from free-radical polymerization because the negatively charged end-groups repel one another, ruling out the coupling of two active chains. The ionic group at the end of the growing polymer is stable at each stage. Once the supply of monomer has been used up, the polymer can exist indefinitely with its ionic end-group, in contrast with the free-radical case, in which some reaction must take place to terminate the process. Ion-initiated polymers are called "living polymers" because when additional monomer is added (even months later), they resume growth and increase in molecular mass. Termination can be achieved by adding water to replace the Li^+ with a hydrogen ion:

$$\overset{}{{\text{---}}(CH_2{-}CH)_{\overline{n}}CH_2{-}CH^{\ominus}Li^{\oplus} + H_2O \longrightarrow}$$
$$\overset{|}{C{\equiv}N} \quad \overset{|}{C{\equiv}N}$$
$$\qquad\qquad\qquad {\text{---}}(CH_2{-}CH)_{\overline{n}}CH_2{-}CH_2 + Li^{\oplus} + OH^{\ominus}$$
$$\qquad\qquad\qquad\qquad \overset{|}{C{\equiv}N} \qquad \overset{|}{C{\equiv}N}$$

$$\text{(termination)}$$

Condensation Polymerization

A second important mechanism for polymerization is **condensation polymerization,** in which a small molecule (frequently water) is split off as each monomer unit is attached to the growing polymer. An example is the polymerization of 6-aminohexanoic acid. The first two molecules react upon heating according to

An amide linkage and water form from the reaction of an amine with a carboxylic acid. The new molecule still has an amine group on one end and a carboxylic acid group on the other end, so it can react with two more molecules of

• Condensation reactions have appeared several times outside the context of polymer synthesis. For example, two molecules of sulfuric acid (H_2SO_4) condense to form disulfuric acid $(H_2S_2O_7)$ (see Fig. 22-7), and a carboxylic acid condenses with an alcohol to form an ester (see Section 24-2).

Figure 25-3 The proportions of long- and short-chain molecules differ in addition and condensation polymerization throughout the course of the reaction, but similarly long chains are formed at completion.

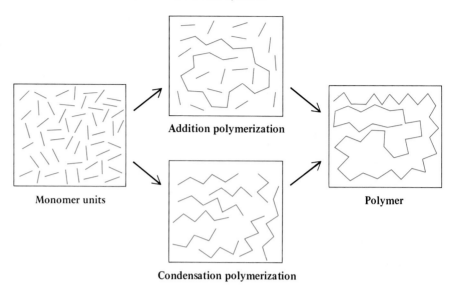

Monomer units

Addition polymerization

Condensation polymerization

Polymer

6-aminohexanoic acid. The process repeats to build up a long-chain molecule. For each monomer unit added, one molecule of water is split off. The final polymer in this case is called "nylon-6"; it is used in fiber-belted radial tires and in carpets.

Addition and condensation polymerization differ in the distribution of chain lengths over the course of the reaction (Fig. 25–3). In addition polymerization, reaction begins only at the initiator sites, so relatively few polymer chains are growing at any one time. If the reaction is interrupted short of completion, the result is a number of polymer molecules mixed with unreacted monomer. In condensation polymerization, every monomer is a potential initiator, and so dimers and trimers begin to form throughout the mixture almost immediately upon heating. If the reaction is stopped short of completion, the mixture contains a large number of "oligomers" (molecules with short polymer chains) and almost no monomer. If the reaction later resumes, the oligomers react with each other to give long-chain polymers.

Copolymers

A **copolymer** is a polymer formed from two or more different monomers (Fig. 25–4). Consider butadiene and styrene (vinyl benzene). Free-radical initiation causes each of these compounds to polymerize. The polymerization of styrene to

Figure 25-4 Lego and Duplo toy building bricks, which can themselves be clicked together in models of polymers, are made from a copolymer of acrylonitrile, butadiene, and styrene.

polystyrene proceeds similarly to the polymerization of vinyl chloride outlined at the beginning of this section. Butadiene has two C=C bonds per monomer, and its polymerization is different and interesting. The π-electrons shift upon initiation and during propagation, causing the systematic relocation of double bonds:

$$R-O\cdot + CH_2=CH-CH=CH_2 \longrightarrow R-O-CH_2-CH=CH-CH_2\cdot$$

(initiation)

One π-electron from each of the two double bonds has moved to the central C—C bond, converting it into a double bond. A third π-electron is used to make the O—C bond to the initiator, and the fourth remains as an unpaired electron on the terminal —CH$_2$ group, ready to react further according to

$$R-O-CH_2-CH=CH-CH_2\cdot + CH_2=CH-CH=CH_2 \longrightarrow$$
$$R-O-CH_2-CH=CH-CH_2-CH_2-CH=CH-CH_2\cdot \quad \text{(propagation)}$$

In the polybutadiene that ultimately forms $-(CH_2-CH=CH-CH_2)_n$, all the C—C single bonds from the monomers have been converted into double bonds, and all the double bonds have become single bonds.

Now let us suppose that polymerization is initiated in a *mixture* of styrene and butadiene. In this case, the growing polymer chain may encounter and react with either type of monomer. The result is a **random copolymer,** with butadiene (B) and styrene (S) units in an irregular sequence along the chain. A segment of this random copolymer might be symbolized as

$$-B-B-S-B-S-S-B-B-B-S-B-S-S-$$

The probability of attachment of a given type of monomer unit depends not only on its concentration in the mixture but also on its inherent reactivity. Thus, the fraction of B monomer units in a chain does not generally equal the fraction of B in the original mixture. Nonetheless, varying the composition of the mixture does change the ratio of the two monomer units in the polymer that results. For example, a 1:6 molar ratio of styrene to butadiene monomers is used to make styrene–butadiene rubber (SBR) for automobile tires, and a 2:1 ratio gives a copolymer used in latex paints.

A second type of copolymer is called a **block copolymer.** It consists of a series of blocks connected together, in which each block is a chain of one type of monomer unit. A typical block copolymer of butadiene and styrene might be represented as

$$-(B-B-B-B-B-B)_m(S-S-S-S-S-S-S)_n$$

in which a large block of butadiene units and a large block of styrene units are linked to form the final polymer chain. More complicated block copolymers might have several alternating blocks. Such copolymers can be made by using the "living-polymer" process described earlier: polymerization in pure butadiene is initiated by butyl lithium and allowed to proceed until long chains are created and the butadiene is used up, after which some styrene is added and the chains grow further with styrene units. Block copolymers have very different properties from random ones. They also differ from simple mixtures of two polymers. A mixture of polystyrene and polybutadiene tends to separate into two phases, each rich in one of the polymers. This cannot occur in a copolymer because the polystyrene and polybutadiene portions of the chain are connected by a chemical bond.

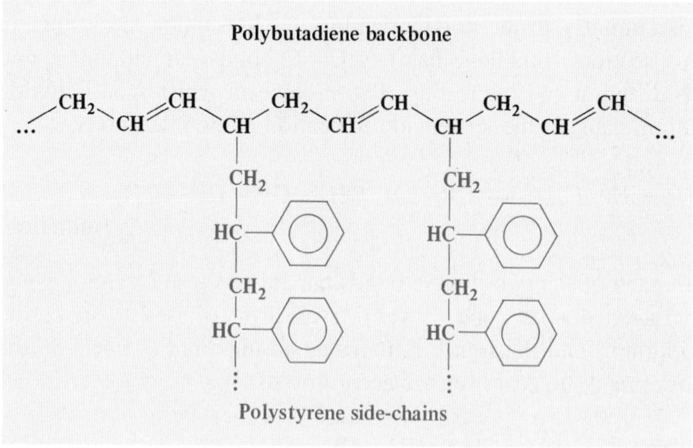

Figure 25-5 The graft copolymer of butadiene with styrene shows excellent resistance to impact.

A third type of copolymer is called a **graft copolymer.** It is made by polymerizing one component and then "grafting on" side-chains of the second to the chains of the first (Fig. 25–5). In high-impact polystyrenes, for example, a small amount of butadiene is polymerized first. Upon addition of styrene and a free-radical initiator, radical sites are formed along the polybutadiene chain and serve as starting points for polystyrene growth. The result is a polybutadiene chain onto which are grafted a series of polystyrene chains, just as branches can be grafted onto tree limbs in an orchard. The resulting plastic is significantly harder and more resistant to impact than ordinary polystyrene.

Cross-Linking

If every monomer forming a polymer has only two reactive sites, then only chains and rings can be made. The 6-aminohexanoic acid used in making nylon-6, for example, has one amine and one carboxylic acid group per molecule. When both functional groups react, one link is forged in the polymer chain, but that link cannot react further. If some or all of the monomers in a polymer have three or more reactive sites, however, then cross-linking is possible to form sheets or networks. This is often desirable in synthetic polymers because it leads to a stronger material.

• Note the analogy to the different solid-phase structures of the elements, among which the number of possible bonds per atom profoundly affects the nature of the solid (see Chapter 20).

One important example of cross-linking involves copolymers of phenol and formaldehyde (Fig. 25–6). When these two compounds are mixed with the phenol in excess (in the presence of an acid catalyst), straight-chain polymers form. The first step is the addition of formaldehyde to phenol to give methylolphenol:

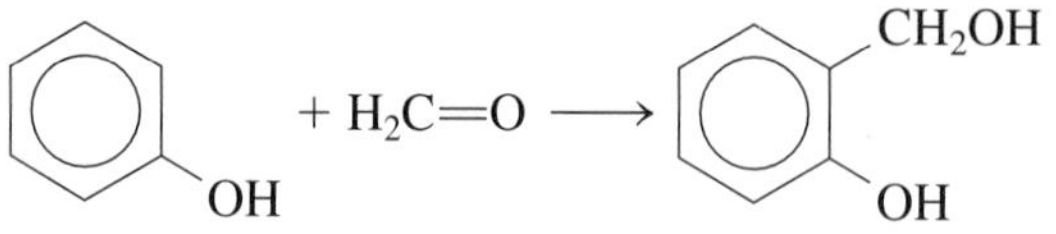

Molecules of methylolphenol can then react through condensation polymerization (releasing water) to form a linear polymer called "novalac":

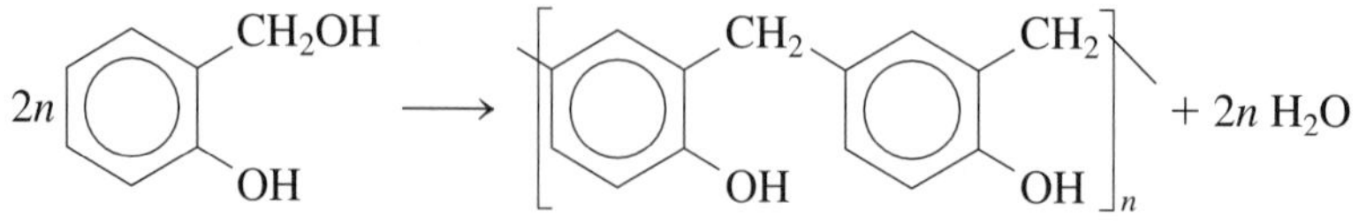

Figure 25-6   When a mixture of phenol (C_6H_5OH) and formaldehyde (CH_2O) dissolved in acetic acid is treated with concentrated hydrochloric acid, a cross-linked phenol–formaldehyde polymer grows.

If, on the other hand, the reaction is carried out with an excess of formaldehyde, then di- and trimethylolphenols form.

Each of these monomers has more than two reactive sites and can react with up to three others to form the type of cross-linked structure shown in Figure 25–7.

Figure 25-7   Cross-linking in the phenol–formaldehyde polymer.

• One of the first uses of Bakelite was in the manufacture of billiard balls.

The cross-linked polymer is much stronger than the linear polymer. The very first synthetic plastic, Bakelite, was made in 1907 from cross-linked phenol and formaldehyde. Modern phenol–formaldehyde polymers are used extensively in the construction industry as adhesives for plywood, with more than 1 billion kg produced per year in the United States.

Cross-linking is often desirable because it leads to a stronger material. Sometimes cross-linking agents are added deliberately to form additional bonds between polymer chains. Polybutadiene, for example, contains double bonds that can be linked upon addition of appropriate oxidizing agents. One especially important kind of cross-linking occurs through sulfur chains in rubber, as discussed in Section 25–3.

25–2 NATURAL POLYMERS

All the products of human ingenuity in the design of polymers pale beside the products of nature. Plants and animals employ a tremendous variety of long-chain molecules with different functions: some for structural strength, others to act as catalysts, and still others to provide instructions for the synthesis of vital components of the cell. In this section, we discuss three important classes of natural polymers: proteins, carbohydrates, and nucleic acids.

Amino Acids

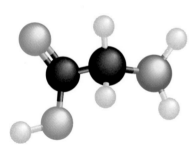

Figure 25–8 The structure of glycine.

The monomeric building blocks of the important biopolymers called proteins are the α-amino acids. The simplest amino acid is glycine, which has the molecular structure shown in Figure 25–8. An amino acid, as indicated by the name, contains an amine group ($-NH_2$) and a carboxylic-acid group ($-COOH$). In α-amino acids, the two groups are bonded to the same carbon atom. Other types of amino acids exist, but we restrict our attention to α-amino acids. In acidic aqueous solution, the amine group of an amino acid gains a hydrogen ion to form $-NH_3^+$; in basic solution, the carboxylic acid group loses a hydrogen ion to form $-COO^-$. At intermediate pH, both reactions occur, that is, the hydrogen ion transfers from the carboxylic acid group to the amine group. The amino acid is then a *zwitterion* (a positive ion on one end and a negative ion on the other). The simple amino acid form shown in Figure 25–8 is rarely present.

Two glycine molecules can condense with loss of water to form an amide:

$$\underset{OH}{\overset{O}{\underset{\|}{C}}}-CH_2-N\overset{H}{\underset{H}{\diagdown}} \ + \ \underset{OH}{\overset{O}{\underset{\|}{C}}}-CH_2-NH_2 \longrightarrow \overset{O}{\underset{\|}{\underset{OH}{C}}}-CH_2-\underset{H}{\overset{}{N}}-\overset{O}{\overset{\|}{C}}-CH_2-NH_2 \ + \ H_2O$$

• The $\Delta G°$ for the formation of the peptide linkage is positive; the biosynthesis of polypeptides requires energy.

The amide functional group connecting two amino acids is referred to as a **peptide linkage,** and the resulting molecule is called a *dipeptide*—in this case, diglycine. Because the two ends of the molecule still have carboxylic acid and amine groups, further condensation reactions to form a **polypeptide,** a polymer composed of many amino-acid groups, are possible. If glycine were the only amino acid available, the result would be polyglycine, a rather uninteresting protein. Of course, nature does not stop with glycine as a monomer unit. Any of 20 different α-amino acids are found in most natural polypeptides. In each of these, one of the two hydrogen atoms on the central carbon atom of glycine is replaced by another side-group.

Alanine is the next simplest α-amino acid after glycine; it has a $-CH_3$ group in place of an $-H$ atom. This substitution has a profound consequence. In alanine, four different groups are attached to a central carbon: $-COOH$, $-NH_2$, $-CH_3$, and $-H$. There are two ways in which four different groups can be arranged in a tetrahedral structure around a central atom:

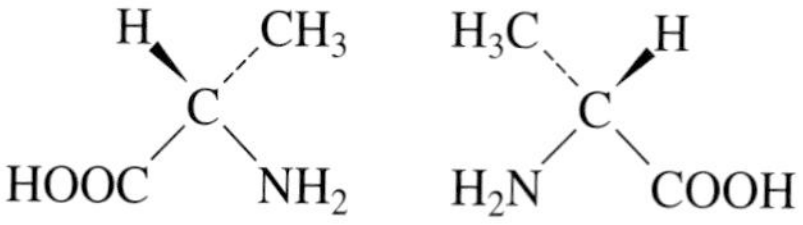

These two structures are mirror images (enantiomers) of each other. They cannot be interconverted without breaking and re-forming bonds, a very slow process. The above two forms of alanine are designated with the prefixes L- and D- for *levo* and *dextro* (from the Latin for "left" and "right," respectively). A carbon atom with four different groups attached to it is called a **chiral center**.

If a mixture of L- and D-alanine were caused to polymerize into chains of any great length, nearly every polymer molecule would have a different structure because the sequences of the D-alanine and L-alanine monomeric units would differ. To create polymers with definite structures for particular roles, there is only one recourse: to build all polypeptides from just one of the optical isomers so that the properties are reproducible from molecule to molecule. Nearly all naturally occurring α-amino acids are the L-form, and most earthly organisms have no use for D-α-amino acids in making polypeptides. Terrestrial life could presumably go on equally well using mainly D-amino acids (all biomolecules would be mirror images of their present forms). The mechanism by which the established preference was initially selected is not known.

The $-H$ group of glycine and the $-CH_3$ group of alanine give just two α-amino-acid building blocks. All 20 important α-amino acids are shown in Table 25–1 by listing their side-groups. Note the variety of their chemical and physical

• Mirror-image isomers, or enantiomers, also occur in coordination complexes, as discussed in Section 19-2.

• The drug cyclosporin involves an exception. First isolated from an obscure fungus, it has saved thousands of lives by preventing the rejection of transplanted organs. It contains a polypeptide ring with 11 amino-acid links, ten of which have the L-form, but one of which, an alanine, has the D-form.

Table 25–1
α-Amino-Acid Side-Groups

	Symbol	Structure of Side-Group
Hydrogen "side-group"		
Glycine	Gly	$-H$
Alkyl side-groups		
Alanine	Ala	$-CH_3$
Valine	Val	$-CH-CH_3$ $\quad\mid$ $\quad CH_3$
Leucine	Leu	$-CH_2-CH-CH_3$ $\qquad\quad\mid$ $\qquad\quad CH_3$
Isoleucine	Ile	$-CH-CH_2-CH_3$ $\quad\mid$ $\quad CH_3$
Proline	Pro	(structure of entire amino acid) 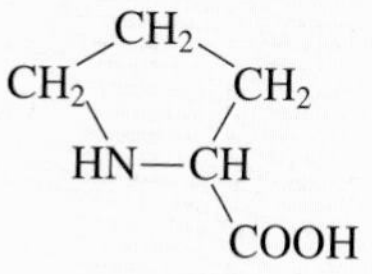

Table 25–1
(*Continued*)

	Symbol	Structure of Side-Group
Aromatic side-groups		
Phenylalanine	Phe	$-CH_2-$ ⬡
Tyrosine	Tyr	$-CH_2-$ ⬡ $-OH$
Tryptophan	Trp	$-CH_2-C$ ... indole ring
Alcohol-containing side-groups		
Serine	Ser	$-CH_2OH$
Threonine	Thr	$-CH$ with OH and CH_3
Basic side-groups		
Lysine	Lys	$-CH_2CH_2CH_2CH_2NH_2$
Arginine	Arg	$-CH_2CH_2CH_2NH-C$ with $=NH$ and NH_2
Histidine	His	$-CH_2-C=C-H$ ... imidazole ring
Acidic side-groups		
Aspartic acid	Asp	$-CH_2COOH$
Glutamic acid	Glu	$-CH_2CH_2COOH$
Amide-containing side-groups		
Asparagine	Asn	$-CH_2\overset{O}{\overset{\|}{C}}-NH_2$
Glutamine	Gln	$-CH_2CH_2\overset{O}{\overset{\|}{C}}-NH_2$
Sulfur-containing side-groups		
Cysteine	Cys	$-CH_2-SH$
Methionine	Met	$-CH_2CH_2-S-CH_3$

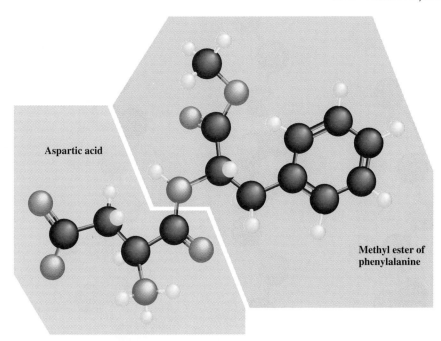

Aspartic acid

Methyl ester of phenylalanine

The structure of the artificial sweetener Aspartame, $C_{14}H_{18}N_2O_5$. In this dipeptide, the peptide linkage unites the amino acid phenylalanine with the methyl ester of the amino acid aspartic acid.

properties. Some side-groups are basic; others are acidic. Some are compact; others are bulky. Some can take part in hydrogen bonds; others can complex readily with metal ions to form coordination complexes. These varied properties lead to even more variety in the polymers derived from the α-amino acids.

Proteins

A **protein** is a long polypeptide made by joining α-amino acids through peptide linkages. The term is usually applied to polymers having more than about 50 amino-acid groups; large proteins may contain many thousand such monomer units. Given the fact that any one of 20 α-amino acids may appear at each point in the chain, the number of possible sequences of amino acids in even small proteins is staggering. Moreover, the amino-acid sequence describes only the first aspect of the molecular structure of a protein. It contains no information about the three-dimensional conformation adopted by the protein. The $\backslash C{=}O$ group and the $\backslash N{-}H$ group in each amino acid along the protein chain are potential sites for hydrogen bonds. Hydrogen bonds may also involve functional groups on the amino-acid side-chains. Also, the cysteine side-groups ($-CH_2-SH$) can react with one another, with loss of hydrogen, to form $-CH_2-S-S-CH_2-$ disulfide bridges between different cysteine groups in a single chain or between neighboring chains (the same kind of cross-linking by sulfur occurs in the vulcanization of rubber). As a result of these strong intrachain interactions, the molecules of a given protein have a rather well-defined conformation even in solution, as compared with the much more varied range of conformations available to a simple alkane chain (see Fig. 24–3). The three-dimensional structures of many proteins have been

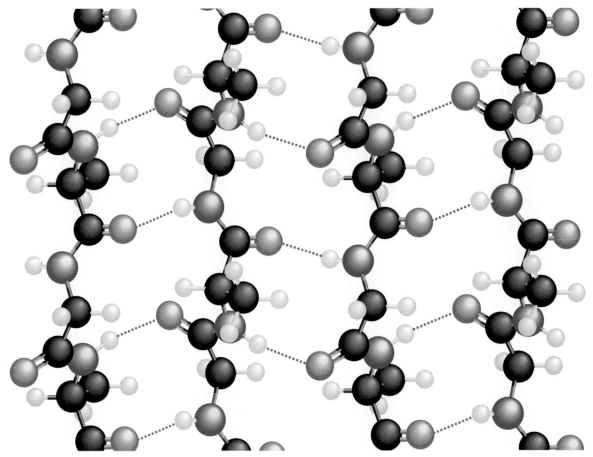

Figure 25–9 The structure of silk. In silk, the amino-acid side-groups fit into a sheet-like structure. Here, all side-groups are shown as methyl (CH$_3$) groups, but H and other small side-groups are possible.

determined by x-ray diffraction (see Section 20–1). This method relies on getting the protein to crystallize, something that is often hard to do. Nonetheless, these determinations are the main source of information about protein structure.

There are two primary categories of proteins: fibrous and globular. **Fibrous proteins** are usually structural materials in nature. They consist of sheets of regularly linked polymer chains or else of long fibers comprising protein chains twisted into spirals. Silk is a fibrous protein in which the monomer units are primarily glycine and alanine, with smaller amounts of serine and tyrosine. The protein chains are cross-linked by hydrogen bonds to form sheet-like structures (Fig. 25–9) that are arranged so that the nonhydrogen side-groups all lie on one side of the sheet; the sheets then stack into layers. The relatively weak forces between sheets give silk its characteristic smooth feel. Wool and hair contain proteins that are composed predominantly of amino acids having side-chains that are larger, bulkier, and less regularly distributed than those in silk, and so sheet structures do not form. Instead, the protein molecules twist into a right-handed coil called an **α-helix** (Fig. 25–10). This is a spiral structure in which each $\diagdown$C$=$O group is hydrogen-bonded to the $\diagdown$N—H group of the fourth amino acid further along the chain; the bulky side-groups jut out from the helix so that they do not overlap and interfere with one another. In hair, three such α-helices are twisted in a *left*-handed coil to form a protofibril that is held together by sulfur bridges and hydrogen bonds. Many of these protofibrils are bound together in a hair cell.

The second type of protein is the **globular protein.** Globular proteins include the carriers of oxygen in the blood (hemoglobin) and in cells (myoglobin). They have irregular folded structures (Fig. 25–11) and typically consist of between 100 and 1000 amino-acid groups in one or more chains. Globular proteins frequently have parts of their structure in α-helices and sheets, with other portions in more disordered forms. Hydrocarbon side-groups tend to cluster in regions that exclude water, whereas charged and polar side-groups tend to remain in close contact with water. The sequence of amino-acid residues (structural units) has been worked out for many such proteins by cleaving them into smaller pieces and analyzing the structure of the fragments. Overlapping sequences are needed to obtain a complete picture, and the required analysis is quite a complex puzzle. It took Frederick Sanger ten years to complete the first such determination of sequence for the 51 amino acids in bovine (cattle-derived) insulin (Fig. 25–12). His persistence earned him the Nobel Prize in chemistry in 1958. Now, automated procedures allow the rapid determination of amino-acid sequences in much longer protein molecules.

Enzymes constitute a very important class of globular proteins. They catalyze nearly every reaction in living cells, such as those synthesizing or breaking down other proteins, transporting substances across cell walls, or recognizing and resisting foreign bodies. Enzymes must have two characteristics: they must be effective in lowering the activation barrier for a reaction, and they must be selective so as to act only on a restricted group of substrates. As an example, consider the enzyme "carboxypeptidase A," whose structure has been determined by x-ray diffraction. It acts to remove one amino acid at a time from the carboxylic-acid end of a polypeptide. Figure 25–13 shows the structure of the active site (with a peptide chain in place, ready to be cleaved). A special feature of this enzyme is the role played by the zinc ion, which is coordinated to two histidine residues in the enzyme and to a carboxylate group on a nearby glutamic-acid residue. The zinc ion, which has a positive charge, acts to remove electrons from the carbonyl group of the peptide linkage, making it more positive and thereby more susceptible to attack by water or by the carboxyl group of a second glutamic-acid residue. The side-chain on the outer amino acid of the peptide being cleaved is located in a hydrophobic cavity, which favors large aromatic or branched side-chains (such as that in tyrosine) over smaller hydrophilic side-chains (such as that in aspartic acid). Carboxypeptidase A thus shows selectivity in the rates with which it cleaves peptide chains.

The "molecular engineering" that lies behind nature's design of carboxypeptidase A and other enzymes is truly remarkable. The amino-acid residues that form the active site and determine its catalytic properties are *not* adjacent to one another in the protein chain. As indicated by the numbers after the residues in Figure 25–13, the two glutamic-acid residues are the 72nd and 270th amino acids along the chain. The enzyme adopts a conformation in which the key residues, distant from one another in terms of chain position, are nonetheless quite close in their positions in three-dimensional space, allowing the enzyme to carry out its specialized function.

text continued on p. 975

• In fingernails, the same types of protein α-helices are much more extensively bonded by sulfur links, giving a much harder material.

• Hydrocarbons are nonpolar and are very poorly soluble in the highly polar solvent water. Thus, hydrocarbon side-groups "fear" water and are said to be hydrophobic. Polar side-groups interact well with water and are said to be hydrophilic (water-loving).

• Figure 14–19 gives a schematic picture of enzyme–substrate binding and subsequent reaction.

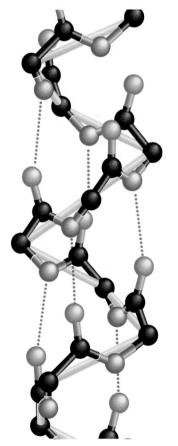

Figure 25–10 The structure of the α-helix for a stretch of linked glycine monomer units. The superimposed yellow line highlights the helical structure, which is maintained by hydrogen bonds (red dotted lines). The hydrogen atoms themselves are omitted for clarity.

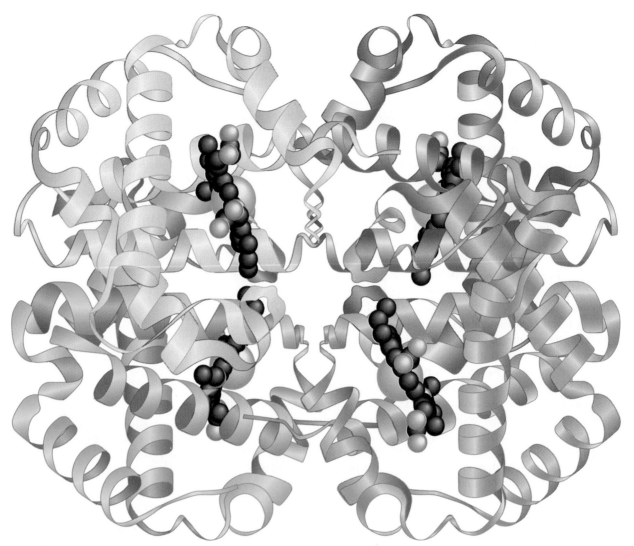

Figure 25–11 A computer-generated model of the structure of hemoglobin. Much of this globular protein, represented by the colored ribbons, is coiled in stretches of α-helix. The four heme groups have their central iron atoms enlarged for visibility. Oxygen binds at these atoms during oxygen transport by hemoglobin.

Crystals of insulin viewed under polarized light.

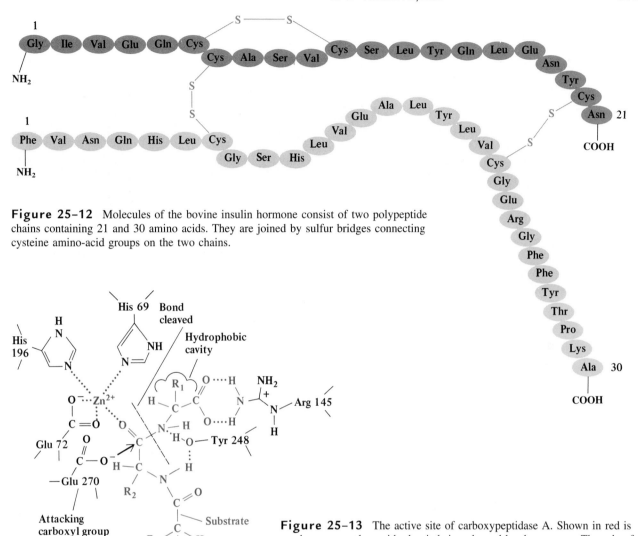

Figure 25-12 Molecules of the bovine insulin hormone consist of two polypeptide chains containing 21 and 30 amino acids. They are joined by sulfur bridges connecting cysteine amino-acid groups on the two chains.

Figure 25-13 The active site of carboxypeptidase A. Shown in red is a substrate, a polypeptide that is being cleaved by the enzyme. The role of Zn^{2+} as a complexing ion is shown in green, and the hydrogen bonds that maintain the geometry near the active site are shown in blue.

Carbohydrates

Glucose ($C_6H_{12}O_6$) is a leading representative of a class of organic compounds called **carbohydrates.** Carbohydrates are widely distributed natural products, most of which are produced photosynthetically from carbon dioxide and water by growing plants. Their name comes from the form of their chemical formulas, which can be written as $C_n(H_2O)_m$, suggesting a "hydrate" of carbon (Fig. 25–14). Simple **sugars,** or **monosaccharides,** are carbohydrates with the chemical formula $C_nH_{2n}O_n$. Those with three, four, five, and six carbon atoms are called trioses, tetroses, pentoses, and hexoses, respectively.

Glucose is a hexose sugar that can exist in three forms in solution (Fig. 25–15). There is a rapid equilibrium between a straight-chain form (a six-carbon

molecule with five —OH groups and one aldehyde —C group), and cyclic

• Carbohydrates are not true hydrates because they do not contain distinct water molecules.

Figure 25–14 Concentrated sulfuric acid is an excellent dehydrating agent; it is able to remove water from a carbohydrate (here, sucrose [$C_{12}H_{22}O_{11}$]), leaving mostly carbon.

• Recall from earlier in this section that there are two distinct ways to attach four different groups to a tetrahedral carbon atom. This is what is meant by a chiral center.

forms, in which the rings are composed of five carbon atoms and one oxygen atom, with four —OH side-groups and one —CH$_2$OH side-group. The formation of six-membered or five-membered rings of this type is quite common among the hexoses and pentoses. Note that in the straight-chain form, four of the carbon atoms of glucose (those numbered 2 through 5) are chiral centers, because each has four different groups bonded to it. Each of these four chiral carbon atoms can exist in two configurations (just as in the D- and L-forms of the amino acids), giving rise to $2^4 = 16$ distinct hexose sugars. The glucose formed in plant photosynthesis always has the chirality shown in Figure 25–15 at each carbon atom. This is D-glucose; the mirror-image sugar L-glucose is not found in nature. Of the 14 other straight-chain hexose sugars, the only ones found in nature are D-galactose and D-mannose (a plant sugar, named for *manna* in the Bible).

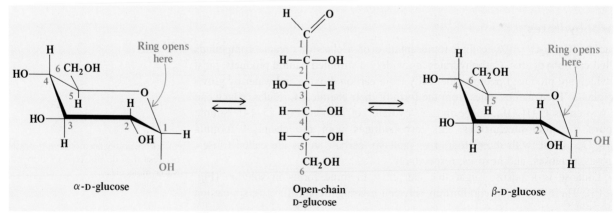

Figure 25–15 In solution, D-glucose exists in two ring forms (a and c), which interconvert via an open-chain form (b). The two rings differ in the placement of the —OH and —H groups of carbon atom 1.

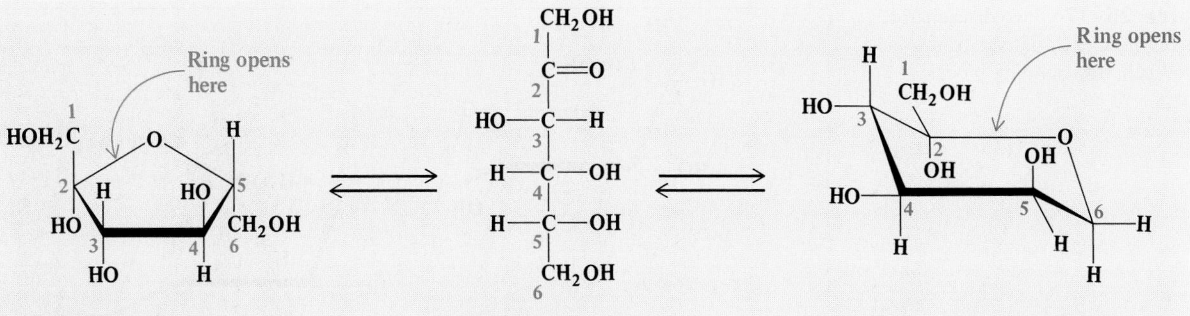

(a) Five-membered ring   (b) Open-chain form   (c) Six-membered ring

Figure 25–16 In aqueous solutions of the sugar D-fructose, an equilibrium exists between a five-atom ring, an open chain, and a six-atom ring. Both ring forms have α isomers, in which the —CH₂OH and —OH on carbon-2 are exchanged, as well as the β isomers shown here.

Figure 25–15 shows that glucose actually has two different ring forms, depending on whether the —OH group created from the aldehyde by the closing of the ring (and shown in blue in the figure) projects down from or out away from the ring. Another way to see this is to note that closing the ring creates a fifth chiral carbon atom in glucose. The two ring forms are called α-D-glucose (Fig. 25–15a) and β-D-glucose (Fig. 25–15c). In aqueous solution, the two ring forms interconvert rapidly via the open-chain glucose form and cannot be separated. They can be isolated separately in crystalline form, however.

D-fructose, a common sugar found in fruit and honey, has the same molecular formula as D-glucose but is a member of a class of hexose sugars that are ketones rather than aldehydes. In their straight-chain forms, these sugars have the C=O double bond at carbon atom 2 rather than carbon atom 1. In solution, the straight-chain form of fructose is in equilibrium with both a five-atom ring and a six-atom ring (Fig. 25–16). Ribose is an aldehyde sugar that is a pentose, with a five-atom ring in solution. In both the chain and the ring form of these and other sugars, the exposed —OH groups interact strongly with water through hydrogen bonds, making the sugars quite soluble in water.

Many plant cells do not stop the process of photosynthesis with simple sugars like glucose but continue by linking sugars together as monomer units to form more complex carbohydrates. **Disaccharides** are composed of two simple sugars linked together in a condensation reaction, with the elimination of water. Examples shown in Figure 25–17 are the milk sugar lactose and the plant sugar sucrose (ordinary table sugar, extracted from sugar cane and sugar beets). The different sugars differ in their sweetness. Fructose is sweeter than sucrose, which is sweeter than glucose. It follows that fructose is a lower-calorie natural sugar because less of it gives the same sweet taste. Sucrose solutions can be treated with an enzyme called *invertase*, which hydrolyzes the bond between the two rings and leaves a mixture of fructose and glucose, called *invert sugar*, which is sweeter than the original sucrose because of its fructose content.

Further linkages of sugar units lead to polymers called **polysaccharides.** The position of the oxygen atom linking the monomer units has a profound effect on the properties and function of the polymers that result. Starch (Fig. 25–18a) is a polymer of α-D-glucose and is metabolized by humans and animals. Cellulose (Fig. 25–18b), a polymer of β-D-glucose can be digested only by certain bacteria that live in the digestive tracts of goats, cows, and other ruminants, and in some insects, such as termites. It forms the structural fiber of trees and plants and is present in linen, cotton, and paper. It is the most abundant organic compound on earth. Despite the great solubility of the monomer D-glucose in water, neither starch nor cellulose is water soluble.

Figure 25–17 Two disaccharides, showing how each derives from monosaccharide building blocks.

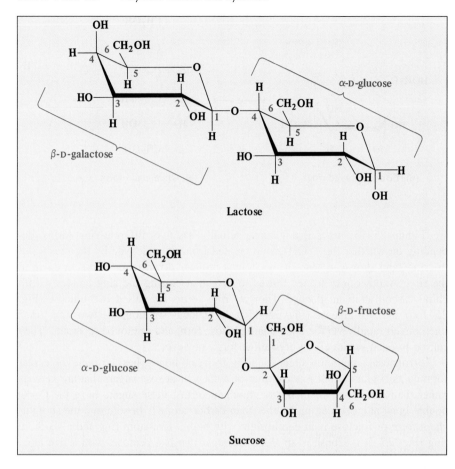

Lactose

Sucrose

Nucleic Acids

As described earlier, proteins are copolymers made up typically of 20 types of monomer unit. Simply mixing the amino acids and dehydrating them to form polymer chains at random would never lead to the particular structures needed by living cells. How does the cell preserve information about the amino-acid sequences that make up its proteins, and how does it transmit this information to daughter cells when it reproduces? These questions lie in the field of molecular genetics, an area in which chemistry is of central importance.

The primary genetic material is deoxyribonucleic acid (DNA). This biopolymer is made up of four types of monomer units called **nucleotides.** Each nucleotide is constructed of three parts:

1. One molecule of a pyrimidine or purine base. The four bases are thymine, cytosine, adenine, and guanine (Fig. 25–19a).
2. One molecule of the pentose sugar D-deoxyribose ($C_5H_{10}O_4$).
3. One molecule of phosphoric acid (H_3PO_4).

The cyclic sugar molecule links the base to the phosphate group, undergoing two condensation reactions with loss of water to form the nucleotide (Fig. 25–19b). The first key to discovering the structure of DNA was the observation that, although the proportions of the four bases in DNA from different organisms are quite variable, the chemical amount of cytosine (C) is always approximately equal to that of guanine (G), and the chemical amount of adenine (A) is always approximately equal to that of thymine (T). This suggested some type of "base-pairing" in DNA

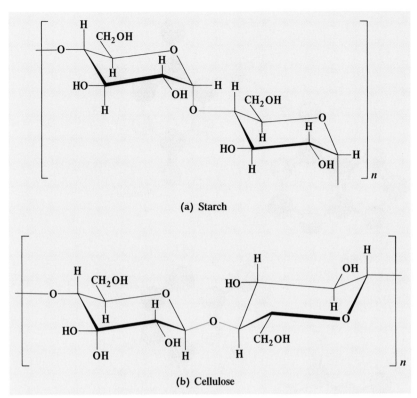

(a) Starch

(b) Cellulose

Figure 25–18 Both starch (a) and cellulose (b) are polymers of glucose. In starch all the cyclic glucose units are α-D-glucose, but in cellulose all the monomer units are β-D-glucose.

that could lead to association of C with G and A with T. The second crucial observation was an x-ray diffraction study by Rosalind Franklin and Maurice Wilkins that suggested the presence of helical structures of more than one chain in DNA.

These two pieces of information were put together by James Watson and Francis Crick in their famous 1953 proposal of a double-helix structure for DNA. They concluded that DNA consists of two interacting helical strands of nucleic-acid polymer (Fig. 25–20), with each cytosine on one strand linked through hydrogen bonds to a guanine on the other, and each adenine to a thymine. This accounted for the observed molar ratios of the bases, and it also provided a model for the replication of the molecule, which is crucial for passing on information during the reproductive process. One DNA strand serves as a template upon which a second DNA strand is synthesized. A DNA molecule replicates by starting to unwind at one end. As it does so, new nucleotides are guided into position opposite the proper bases on each of the two strands. If the nucleotide does not fit the template, it cannot link to the polymeric strand under construction. The result of the polymer synthesis is two double-helix molecules, each containing one strand from the original and one new strand, identical to the original in every respect.

Information is encoded in DNA in the order of the base pairs. Subsequent research has succeeded in breaking this genetic code and establishing the connection between the base sequence in a segment of DNA and the amino-acid sequence of the protein synthesized according to the directions in that segment. The code in the nucleic acids consists of consecutive, non-overlapping triplets of bases, with each triplet standing for a particular amino acid. Thus, a polymeric nucleic-acid strand containing only the side-group cytosine was found to give a polypeptide of pure proline, meaning that the triplet CCC codes for proline. The nucleic-acid

Figure 25-19 (a) The structures of the purine and pyrimidine bases. Hydrogen bonding between pairs of bases is indicated by red dots. (b) The structure of the nucleotide adenosine monophosphate (AMP).

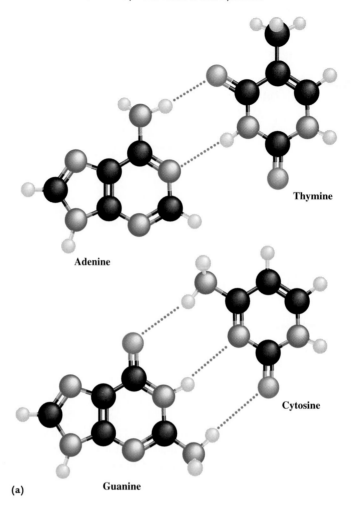

Thymine

Adenine

Cytosine

Guanine

(a)

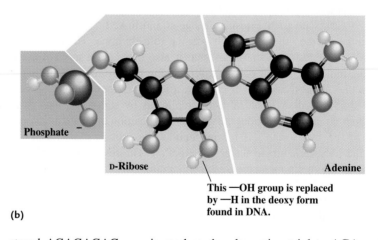

Phosphate −

D-Ribose

Adenine

This —OH group is replaced by —H in the deoxy form found in DNA.

(b)

strand AGAGAGAG . . . is read as the alternating triplets AGA and GAG and gives a polypeptide consisting of alternating arginine (coded by AGA) and glutamic acid (coded by GAG) monomer units. There are 64 (= 4^3) possible triplets, so there is typically more than one code for a particular amino acid. Some triplets serve as signals to terminate a polypeptide chain. Remarkably, the genetic code appears to be universal, independent of the particular species of plant or animal, a finding that suggests a common origin of all terrestrial life.

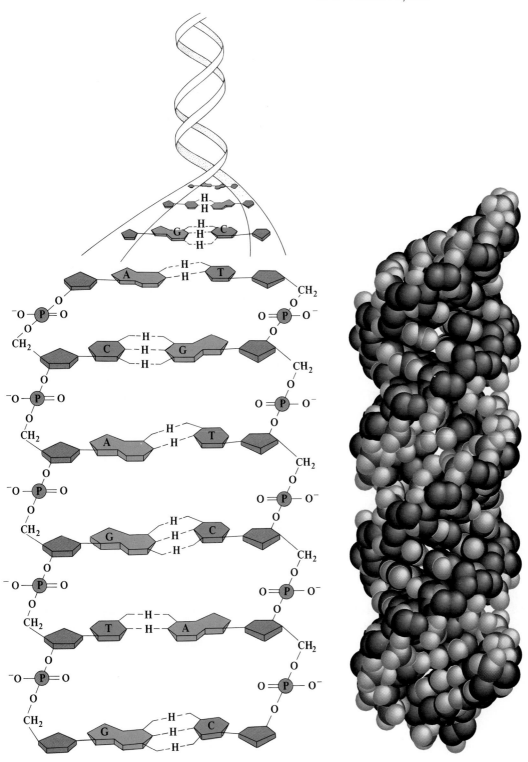

Figure 25-20 The double-helix structure of DNA.

25-3 USES FOR POLYMERS

If food is excluded, the three largest uses for polymers are in fibers, plastics, and elastomers (rubbers). These three types of materials are distinguished on the basis of their physical properties, especially their resistance to stretching. A typical fiber strongly resists stretching and elongates by less than 10% before breaking. Plastics are intermediate in their resistance to stretching and elongate 20 to 100% before breaking. Elastomers stretch readily, with elongations of 100 to 1000% (that is, some rubbers can be stretched by a factor of 10 without breaking). In this section, we examine the major kinds of polymers and their uses.

Fibers

Many important fibers, including cotton and wool, are naturally occurring polymers. The first commercially successful synthetic polymers were not made by polymerization reactions, but through the chemical regeneration of the natural polymer cellulose (Fig. 25–18b). The origins of the modern synthetic fiber industry are said to lie in an accident in the laboratory of the German chemist Christian Schonbein, who in 1845 wiped up a spill of nitric and sulfuric acids with a cotton apron, hung up the apron to dry, and thereby produced the polymer cellulose trinitrate (guncotton). In guncotton, all three —OH groups in the glucose subunits of the cellulose (see Fig. 25–18b) are replaced by —ONO_2 nitrate groups. The compound is an explosive (see Section 22–5); it was subsequently developed into cordite and continues in extensive use today as a propellant in guns and rockets.

By varying the amounts of acid used, one can prepare a second nitrated derivative of cellulose. In nitrocellulose (cellulose dinitrate), only the two —OH groups directly attached to each ring are nitrated. When this compound is dissolved in a mixture of camphor and alcohol and the solvent carefully evaporated, the product is celluloid, a material used for photographic film in the early motion picture industry. The nitrate groups in celluloid make it quite flammable, as might be expected from a polymer closely related to guncotton. This fact, combined with the high price of camphor, led to the use of cellulose acetate instead of celluloid in photographic film. In cellulose acetate all three —OH groups on the glucose rings are esterified by treatment with acetic acid that contains a small amount of sulfuric acid to form —O—C side-groups.

The discovery of nitrocellulose also led to the production of the first synthetic fiber. In 1884, the French chemist Hilaire Bernigaud, Count of Chardonnet, showed how to remove the nitrate groups in nitrocellulose and to spin the resulting reconstituted cellulose into fibers. Modest commercial success for the resulting "Chardonnet silk" synthetic fiber ensued until the development of the alternative and cheaper **viscose rayon** some ten years later. In the viscose-rayon process (Fig. 25–21), which is still used today, cellulose is digested in a concentrated solution of NaOH to convert the —OH groups into —O^-Na^+ ionic groups. Reaction with CS_2 leads to the formation of about one "xanthate" group for every two glucose monomer units:

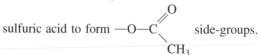

Filter paper (cellulose) dissolves in a concentrated ammonia solution containing $[Cu(NH_3)_4]^{2+}$ ions. When the solution is extruded into aqueous sulfuric acid, a dark-blue thread of rayon (regenerated cellulose) precipitates.

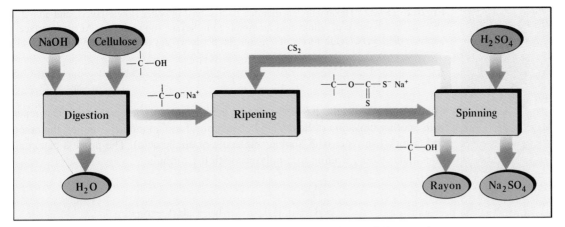

Figure 25–21 The viscose-rayon process. The chemical state of the —OH groups is shown at each stage. Sodium sulfate is a commercially significant by-product of the process.

Such substitutions weaken the hydrogen-bond forces holding polymer chains together. In the ripening step, some of these xanthate groups are removed with regeneration of CS_2 and others migrate to the —CH₂OH groups from the ring —OH groups. After this has taken place, sulfuric acid is added to neutralize the NaOH and to remove the remaining xanthate groups. At the same time, the viscose rayon is spun out to form fibers. The chemical composition of the completed rayon is very close to that of the cellulose that began the process, but the polymer molecules are stretched out and aligned in fibers.

Rayon is a semi-synthetic fiber because it is prepared from a natural polymeric starting material. The first truly synthetic polymeric fiber was nylon, which was developed in the 1930s by Wallace Carothers at the Du Pont Company. He knew of the condensation of an amine with a carboxylic acid to form an amide linkage (see Section 24–2) and noted that if each molecule had *two* amine or carboxylic acid functional groups, then long-chain polymers could be formed. The specific starting materials upon which Carothers settled after a number of attempts were adipic acid and hexamethylenediamine:

$$HO-\overset{\overset{\displaystyle O}{\|}}{C}-(CH_2)_4-\overset{\overset{\displaystyle O}{\|}}{C}-OH \qquad H_2N-(CH_2)_6-NH_2$$

$$\text{adipic acid} \qquad\qquad \text{hexamethylenediamine}$$

The two condense with loss of water according to the equation

$$HO-\overset{\overset{\displaystyle O}{\|}}{C}-(CH_2)_4-\overset{\overset{\displaystyle O}{\|}}{C}-\boxed{OH + H}-\underset{\underset{\displaystyle H}{|}}{N}-(CH_2)_6-NH_2 \longrightarrow$$

$$HO-\overset{\overset{\displaystyle O}{\|}}{C}-(CH_2)_4-\overset{\overset{\displaystyle O}{\|}}{C}-\underset{\underset{\displaystyle H}{|}}{N}-(CH_2)_6-NH_2 + H_2O$$

• If the —OH groups are esterified with acetic acid to make "acetate" fiber, the reduced interchain hydrogen-bond interaction and the bulk of the acetate groups cause the fiber to be softer, but also weaker, than rayon.

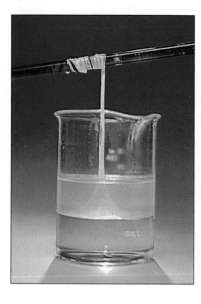

Figure 25–22 Hexamethylenediamine is dissolved in water (lower layer), and adipyl chloride, a derivative of adipic acid, is dissolved in hexane (upper layer). At the interface between the layers, nylon forms; the nylon can be drawn out and wound up on a stirring rod.

The resulting molecule has a carboxylic-acid group on one end (which can react with another molecule of hexamethylenediamine) and an amine group on the other end (which can react with another molecule of adipic acid). The process can continue indefinitely, leading to a polymer with the formula

$$\left[\begin{array}{c} \overset{\displaystyle O}{\overset{\|}{C}}-(CH_2)_4-\overset{\displaystyle O}{\overset{\|}{C}}-\underset{\underset{H}{|}}{N}-(CH_2)_6-\underset{\underset{H}{|}}{N} \end{array}\right]_n$$

called nylon-66 (Fig. 25–22). The nylon is extruded as a thread or spun as a fiber from the melt. The combination of well-aligned polymer molecules and N—H··O hydrogen bonds between chains makes nylon one of the strongest materials known.

The designation "66" indicates that this nylon has six carbon atoms on the starting carboxylic acid and six on the diamine. Other nylons can be made with different numbers of carbon atoms. After nylon-66, the most important is called "nylon-6" and can be made from the polymerization of 6-aminohexanoic acid, as outlined in Section 25–1:

$$n\ HO-\overset{\displaystyle O}{\overset{\|}{C}}-(CH_2)_5-\underset{\underset{H}{|}}{N}-H \longrightarrow \left[\overset{\displaystyle O}{\overset{\|}{C}}-(CH_2)_5-\underset{}{N}-\underset{}{H}\right]_n + n\ H_2O$$

Note that this uses a monomer in which a single molecule has both a carboxylic acid and an amine group. Such a molecule can "bite its own tail" to form a cyclic molecule with loss of water:

$$\underset{HO}{\overset{\displaystyle O}{\overset{\|}{C}}}-(CH_2)_5-NH_2 \longrightarrow \begin{array}{c} \overset{O}{\overset{\|}{C}}-N-H \\ H_2C \qquad CH_2 \\ H_2C \qquad CH_2 \\ CH_2 \end{array} + H_2O$$

This compound, called "caprolactam," is the normal starting material for the production of nylon-6, which is a little cheaper but not quite as strong as nylon-66. The bonding in these nylons is just like the bonding in naturally occurring polypeptides and proteins. Polyglycine could in fact be referred to as "nylon-2," a simple polyamide in which each repeating unit contains two carbon atoms.

Just as a carboxylic acid reacts with an amine to give an amide, so it reacts with an alcohol to give an ester. This suggests the possible reaction of a dicarboxylic acid and a glycol (dialcohol) to form a polymer. The polymer produced most extensively in this way is polyethylene terephthalate, which is built up from terephthalic acid (a benzene ring with —COOH groups on either end) and ethylene glycol. The first two molecules react according to

A room full of nylon bobbins for use in making fire hose.

Strong, lightweight Mylar covered the wings of the Gossamer Albatross, which succeeded in a flight across the English Channel powered solely by a single man.

$$\text{HO}-\underset{\text{terephthalic acid}}{\text{C}}-\text{C}=\text{O} \quad \boxed{\text{OH}+\text{H}}\text{O}-\text{CH}_2-\text{CH}_2-\text{OH} \quad \longrightarrow$$

$$\text{HO}-\text{C}-\bigcirc-\text{C}-\text{O}-\text{CH}_2-\text{CH}_2-\text{OH} \quad + \text{H}_2\text{O}$$

Further reaction then builds up the polymer, which is referred to as "polyester" and sold under trade names such as Dacron. The planar benzene rings in this polymer make it stiffer than nylon, which has no aromatic groups in its backbone, and help to make polyester fabrics crush-resistant. The same polymer, when formed in a thin sheet rather than a fiber, is known as Mylar, a very strong film used for audio and video tapes.

A final class of polymer fibers are the acrylics, which are built up from the free-radical polymerization of acrylonitrile:

$$n\,\text{CH}_2\!=\!\underset{\underset{\text{C}\equiv\text{N}}{|}}{\text{CH}} \quad \longrightarrow \quad \left[\text{CH}_2-\underset{\underset{\text{C}\equiv\text{N}}{|}}{\text{CH}}\right]_n$$

The resulting polymer is then dissolved and spun into fibers as the solvent evaporates. Pure polyacrylonitrile has an inconveniently high melting point and cannot be dyed, so most acrylics are copolymers with vinyl acetate, vinyl chloride, styrene, or other monomers. The presence of chlorine atoms in the copolymer reduces the flammability of the fabric.

Table 25–2 summarizes the structures, properties, and uses of some important fibers.

Table 25–2
Fibers

Name	Structural Units	Properties	Sample Uses
Rayon	Regenerated cellulose	Absorbent, soft, easy to dye, poor wash and wear	Dresses, suits, coats, curtains, blankets
Acetate	Acetylated cellulose	Fast drying, supple, shrink-resistant	Dresses, shirts, draperies, upholstery
Nylon	Polyamide	Strong, lustrous, easy to wash, smooth, resilient	Carpeting, upholstery, tents, sails, hosiery, stretch fabrics, rope
Dacron	Polyester	Strong, easy to dye, shrink-resistant	Permanent-press fabrics, rope, sails, thread
Acrylic (Orlon)	$-(\text{CH}_2-\underset{\underset{\text{C}\equiv\text{N}}{|}}{\text{CH}})_{\overline{n}}$	Warm, lightweight, resilient, quick-drying	Carpeting, sweaters, baby clothes, socks

Adapted from P.J. Chenier, *Survey of Industrial Chemistry.* New York: John Wiley & Sons, 1986, Table 18.4.

CHEMISTRY IN PROGRESS

Paint

Paint contains three major components: pigment, binder, and thinner. The pigment gives opaqueness and color to the paint and is frequently an inorganic solid such as cadmium yellow (CdS) or chrome green (Cr_2O_3) that is ground up into small particles before being dispersed in the paint. The thinner is a volatile solvent that is essential to spread the paint but then mostly evaporates. The binder serves to bring the pigment into the thinner; upon evaporation of the thinner, the binder cross-links to provide mechanical strength to the paint film.

Most binders in current use are polymeric. One broad class of binders are the "alkyd resins," which are polyesters with unsaturated hydrocarbon side-chains. As the thinner evaporates, these side-chains cross-link. Alkyd paints are quite versatile and can be formulated with flat, semigloss, and high-gloss finishes. Other binders are used in latex paints, in which water is the thinner. Copolymers of styrene and butadiene (in a 2:1 ratio) are low in cost and were the first binders to be used in water-thinned paints. Acrylic paints, which are more expensive but tougher and more durable, also use water as the thinner (Fig. 25–A). Binders in acrylic paints are copolymers that contain methylmethacrylate monomer units:

$$-CH_2-\underset{\underset{\underset{OCH_3}{|}}{\overset{\overset{CH_3}{|}}{\underset{\|}{\overset{|}{C}}}}{\overset{}{C}}-$$

Figure 25–A Acrylic paint.

Plastics

* Under this broad definition ordinary glass, discussed in Section 21-4, is a plastic as well. This is quite correct. "Plastic" means "capable of being molded" and is a property possessed by inorganic as well as organic polymers.

Plastics are loosely defined as polymeric materials that can be molded or extruded into desired shapes and that harden upon cooling or solvent evaporation. Rather than being spun into threads in which the molecules are aligned, as in fibers, plastics are cast into three-dimensional forms or spread into films for packaging applications. Although celluloid articles were fabricated by plastic processing by the late 1800s, the first important synthetic plastic was Bakelite, the phenol–formaldehyde resin whose cross-linking was discussed in Section 25–1. Table 25–3 lists some of the most important organic plastics, along with some properties and uses. Also given are the symbols that appear on plastic food and beverage containers to aid in separating the materials according to chemical constitution during the recycling process.

Ethylene ($CH_2{=}CH_2$) forms polyethylene through a free-radical–initiated addition polymerization mechanism at high pressures (1000 to 3000 atm) and temperatures (300 to 500°C), conditions that require elaborate equipment (Fig. 25–23). The product is not the perfect linear chain implied by the simple equation

$$n\ CH_2{=}CH_2 \longrightarrow -\!\!\left[CH_2CH_2\right]_n\!\!-$$

Free radicals frequently abstract hydrogen from the middles of chains in this synthesis. This means that the polyethylene is heavily branched with hydrocarbon sidechains of varying lengths. It is called **low-density polyethylene** (LDPE) because the difficulty of packing the irregular side-chains gives it a lower density (<0.94 g cm^{-3}) than that of perfectly linear polyethylene. This irregularity also

text continued on p. 988

Table 25-3
Plastics

Name	Structural Units	Properties	Sample Uses	Recycling Code
Polyethylene terephthalate	$+(OCH_2CH_2O-\overset{O}{\underset{\|\|}{C}}-\bigcirc-\overset{O}{\underset{\|\|}{C}})_n$	High tensile strength, tear resistance	Clothing fiber, tire cord, plastic film	1 PETE
Polyethylene	$+(CH_2-CH_2)_n$	High density: Hard, strong, stiff	Molded containers, lids, toys, pipe	2 HDPE
		Low density: Soft, flexible, clear	Packaging, trash bags, squeeze bottles	4 LDPE
Polyvinyl chloride	$+(CH_2-\underset{Cl}{CH})_n$	Nonflammable, resistant to chemicals	Water pipes, roofing, credit cards, records	3 PVC
Polypropylene	$+(CH_2-\underset{CH_3}{CH})_n$	Stiffer, harder than high-density polyethylene, higher melting point	Containers, lids, carpeting, luggage, rope	5 PP
Polystyrene	$+(CH_2-\underset{\bigcirc}{CH})_n$	Brittle, flammable, not resistant to chemicals, easy to process and dye	Furniture, toys, refrigerator linings, insulation	6 PS
Phenolics	Phenol-formaldehyde copolymer	Resistant to heat, water, chemicals	Plywood adhesive, fiberglass binder, circuit boards	7 OTHER

Adapted from P. J. Chenier, *Survey of Industrial Chemistry*. New York: John Wiley & Sons, 1986, pp. 252–264.

Figure 25-23 A 720-million-pound-per-year polyethylene plant in Louisiana.

Figure 25-24 The structures of (a) isotactic, (b) syndiotactic, and (c) atactic polypropylene. In these structures the purple spheres represent —CH$_3$ (methyl) side-groups on the long chain.

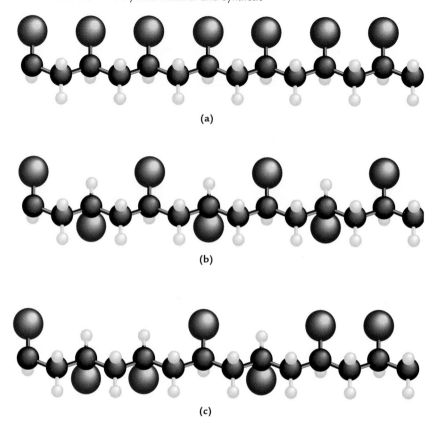

(a)

(b)

(c)

makes it relatively soft, so its primary uses are in coatings, plastic packaging, trash bags, and squeeze bottles in which softness is an advantage, not a drawback.

A major breakthrough occurred in 1954, when the German chemist Karl Ziegler showed that ethylene could also be polymerized with a catalyst consisting of TiCl$_4$ and an organoaluminum compound (for example, Al(C$_2$H$_5$)$_3$). The addition of ethylene takes place at each stage within the coordination sphere of the titanium atom, so that monomer units can add only at the end of the growing chain. The result is linear polyethylene, also called **high-density polyethylene** (HDPE) because of its density (0.96 g cm^{-3}). Because its linear chains are so regular, HDPE contains large crystalline regions at low temperatures, which make it much harder than LDPE and thus suitable for molding into plastic bowls, lids, and toys.

A third kind of polyethylene introduced in the late 1970s is called **linear low-density polyethylene** (LLDPE). This is made by the same metal-catalyzed reactions as HDPE, but it is a deliberate copolymer with other 1-alkenes such as 1-butene. It has some side-groups, which reduce the crystallinity and the density, but these are of a controlled short length, as opposed to the irregular, long side-branches in LDPE. LLDPE is stronger and more rigid than LDPE; it is also cheaper because lower pressures and temperatures are used in its manufacture.

If one of the hydrogen atoms of the ethylene monomer unit is replaced with a different type of atom or functional group, the plastics that form upon polymerization have different properties. Substitution of a methyl group (that is, the use of propylene as monomer) leads to polypropylene:

$$\left[\begin{array}{c} \text{CH}_2 - \text{CH} \\ | \\ \text{CH}_3 \end{array}\right]_n$$

This reaction cannot be carried out successfully by free-radical polymerization, however. It was first performed in 1953–1954 by Ziegler and the Italian chemist Giulio Natta, who used the Ziegler catalyst employed in making HDPE. In polypropylene, the methyl groups attached to the carbon backbone can be arranged in different conformations (Fig. 25–24). In the **isotactic** form, all the methyl groups are arranged on the same side, whereas in the **syndiotactic** form they alternate in a regular fashion. The **atactic** form shows a random positioning of methyl groups. Natta showed that the Ziegler catalyst led to isotactic polypropylene, and he developed another catalyst using VCl_4 that gave the syndiotactic form. Polypropylene plastic is stiffer and harder than HDPE and has a higher melting point, so it is particularly useful in applications in which somewhat higher temperatures are used (such as in the sterilization of medical instruments).

* Ziegler and Natta were awarded the Nobel Prize in 1963 for their development of polymerization catalysts.

Substituting a chlorine atom for one of the hydrogen atoms in ethylene gives vinyl chloride, which polymerizes to polyvinyl chloride (PVC), according to the equation

$$n\ CH_2{=}CHCl \longrightarrow \left[\begin{array}{c} CH_2{-}CH \\ | \\ Cl \end{array}\right]_n$$

The largest uses of PVC are in the construction industry, where a rigid and chemically resistant plastic is needed for making pipes and architectural products such as gutters, downspouts, and siding. **Plasticizers** are frequently added to PVC to make it less rigid for other uses. A plasticizer is an additive that softens a plastic and makes it more flexible. In a rigid amorphous plastic, nonbonded attractions between the tangled molecules of the polymer define a honeycomb-like structure. A plasticizer, which typically has small molecules, intervenes to mask some of these attractions. The result is fewer points of attachment between polymer chains and a reduction in the rigidity of the structure.

In polystyrene, a benzene ring replaces one hydrogen atom in the polyethylene molecule. Because of the great bulk of such a ring, atactic polystyrene does not crystallize to any significant extent. The most familiar use of this polymer is in polystyrene foam, which is used in disposable containers for food and drinks and as insulation. A volatile liquid or a compound that dissociates to gaseous products on heating is added to the molten polystyrene. It forms bubbles that remain as the polymer is cooled and molded. The large number of gas-filled pockets in the final product make it a good thermal insulator. Polyurethane is another polymer that is fabricated as a foam insulation. It is made by the polymerization of a glycol with a diisocyanate, as in the reaction

* Water plasticizes proteins, which are natural polymers, by masking some of the hydrogen bonds that would otherwise form between the long-chain molecules. A moist cake of soybean curd (a vegetable protein) has a jelly-like consistency. Thorough drying shrinks it and makes it hard and tough.

$$n\ HO{-}CH_2{-}CH_2{-}OH + n\ O{=}C{=}N{-}(CH_2)_{\overline{m}}{-}N{=}C{=}O \longrightarrow$$

$$\left[\begin{array}{c} \qquad\qquad\qquad O \qquad\qquad\qquad H \\ \qquad\qquad\qquad \| \qquad\qquad\qquad | \\ O{-}CH_2{-}CH_2{-}O{-}C{-}N{-}(CH_2)_m{-}N{-}C \\ \qquad\qquad\qquad\quad | \qquad\qquad\qquad\quad \| \\ \qquad\qquad\qquad\quad H \qquad\qquad\qquad\quad O \end{array}\right]_n$$

Rubber

An **elastomer** is a plastic that can be deformed to a large extent and still recover its original shape when the deforming stress is removed. The term **rubber** was introduced by Joseph Priestley, who observed that such materials can be used to

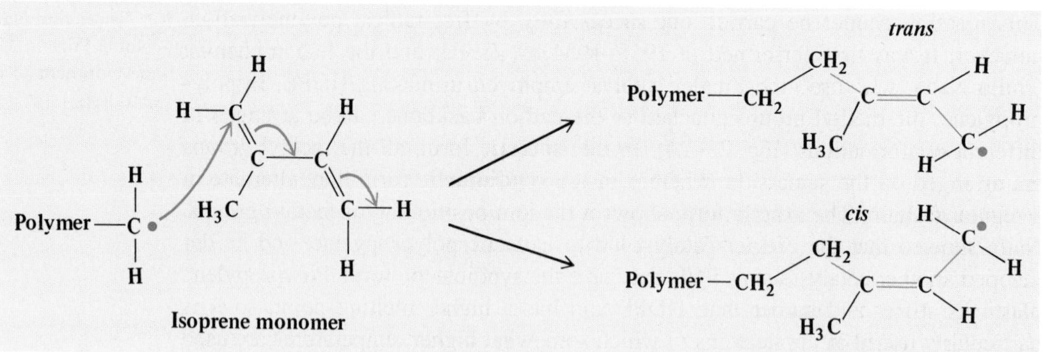

Figure 25-25 In the polymerization of isoprene, *cis* or *trans* configurations can result at each double bond in the polymer. For the monomer conformation shown, a *trans* linkage appears most likely to result. Rotation about the central C—C bond in isoprene favors the *cis* linkage. The blue arrows show the redistribution of the electrons upon bond formation.

• Chewing gum is made from chicle, which exudes from the cut bark of yet another tree. Chicle is a mixture of *cis*- and *trans*-polyisoprene and several other substances.

rub out pencil marks. Natural rubber, the specific elastomer Priestley used, occurs in over 200 plant species (including the common dandelion). The only important source is the tropical rubber tree (*Hevea braziliensis*). When its bark is cut, this tree exudes latex, a milky fluid that contains about 35% natural rubber. Natural rubber is a polymer of **isoprene** (2-methylbutadiene). The isoprene molecule contains two double bonds, and polymerization removes only one of these, so natural rubber is unsaturated, containing one double bond per isoprene unit. In polymeric isoprene, the geometry at each double bond can be either *cis* or *trans* (Fig. 25–25). Natural rubber is all-*cis* polyisoprene. The all-*trans* form also occurs in nature in the sap of certain trees; it is called *gutta-percha*. This material is used to cover golf balls because it is particularly tough. Isoprene can be polymerized by free-radical addition polymerization, but the resulting polymer contains a mixture of *cis* and *trans* double bonds and is useless as an elastomer.

Pure natural rubber has limited usefulness because it melts, is soft, and does not spring back fully to its original form after being stretched. In 1839 the American inventor Charles Goodyear discovered that if sulfur is added to rubber and the mixture is heated, the rubber becomes harder and more resilient and does not melt. This process is referred to as **vulcanization.** It involves the formation of sulfur bridges between the methyl side-groups on different chains. Small amounts of sulfur (<5%) yield an elastic material in which sulfur links between chains remain after stretching and enable the rubber to regain its original form when the stretching force is removed. Large amounts give the very hard, non-elastic material ebonite.

• Neoprene is still produced today as a specialty rubber because of its resistance to chemical attack and heat.

Research on synthetic substitutes for natural rubber began in the United States and Europe before World War II. An accidental discovery at the Du Pont Company led to the production of neoprene, a rubbery polymer formed from monomeric chloroprene (2-chloro-1,3-butadiene), in which the methyl group on isoprene is replaced by chlorine. Further attention focused on copolymers of styrene with butadiene (now called SBR) and copolymers of acrylonitrile with butadiene (NBR). The Japanese occupation of the rubber-producing countries of Southeast Asia during World War II sharply curtailed the supply of natural rubber to the Allied nations, and rapid steps were taken to increase production of synthetic rubbers. The initial production goal was 40,000 tons per year of SBR. By 1945, U.S. production had reached a nearly incredible 600,000 tons per year. During those few years, many advances were made in production techniques, quantitative analysis, and basic understanding of rubber elasticity. Styrene–butadiene rubber production continued after the war, and in 1950 SBR exceeded natural rubber in overall production volume for the first time. More recently, several factors have favored natural rubber: increasing cost of the hydrocarbon feed stock for synthetic rubber, gains in productivity of natural rubber, and the preference for radial-belted tires, which use more natural rubber.

The development of the Ziegler–Natta catalysts had an effect on rubber production as well. First, it facilitated the synthesis of all-*cis* polyisoprene and the demonstration that its properties were nearly identical to those of natural rubber. (A small amount of "synthetic natural rubber" is produced today.) Second, a new kind of synthetic rubber was developed: all-*cis* polybutadiene. It now ranks second in production after styrene–butadiene rubber.

• The silicones, which were discussed at the end of Section 21–2, are another important class of elastomer.

SUMMARY

25–1 Polymers are giant molecules that are formed by stringing together large numbers of separate **monomer units** in repetitive sequences to form chains and webs. They grow by **addition polymerization** and **condensation polymerization.** Addition polymerization proceeds by free-radical chain reaction or by an ionic mechanism. In condensation polymerization, a small molecule, often water, is split out. **Copolymers** form when chemically different monomers are mixed in the polymerization process. Depending on how the differing monomer units join the growing polymer chain, the result is a **random, block,** or **graft copolymer.** Polymer chains growing from monomers with more than two reactive sites can be **cross-linked** into sheets and networks.

25–2 In proteins, the polymer chain is formed of α-amino acid monomer units. The carboxylic-acid group of one amino acid joins with the amine group of a second in a **peptide linkage.** Each of the 20 commonly occurring α-amino acids (except glycine) has a carbon atom that is a **chiral center,** meaning that it can be made in two forms that are mirror-image isomers. Natural proteins contain almost exclusively L-amino acids. The side-groups along a polymer chain determine how the protein is folded, coiled, looped back on itself by hydrogen bonds, or cross-linked to other protein chains. **Fibrous proteins** are structural materials that form sheets or fibers. In a fibrous protein the polymer may be twisted into a right-handed coil called the α-**helix,** which is maintained by hydrogen bonds. **Globular proteins** such as hemoglobin have irregular folded structures. Enzymes are globular proteins in which folding of the polymer chain forms active sites at which reactions are catalyzed. Sugars are **carbohydrates,** compounds of the general formula $C_n(H_2O)_m$. **Monosaccharides,** or simple sugars, can be linked to form long chains, or **polysaccharides.** Cellulose and starch are both polymers of glucose that differ only in the position of the oxygen atoms linking the monomer units. **Nucleic acids** that maintain and transmit information for the synthesis of proteins in cells. Deoxyribonucleic acid (DNA) is made up of monomer units called **nucleotides.** Each nucleotide is formed by condensation of a pyrimidine (thymine or cytosine) or purine (adenine or guanine), a deoxyribose sugar, and a molecule of phosphoric acid. The sequence of the four types of nucleotide down the DNA polymer encodes the information for the sequence of the amino acids in proteins. The DNA molecule has two complementary strands in which base pairs are held together by hydrogen bonds.

25-3 Useful polymers occur in fibers, plastics, and elastomers (rubber). Rayon, a semi-synthetic fiber, is a regenerated form of cellulose. The first true synthetic polymeric fiber was nylon, a polyamide formed by the condensation of a dicarboxylic acid and a diamine. **Plastics** are polymeric materials that can be molded or extruded into appropriate shapes and that harden upon cooling or solvent evaporation. The polymerization of ethylene gives low-density polyethylene (LDPE), high-density polyethylene (HDPE), and linear low-density polyethylene (LLDPE). The difference concerns the number and length of side-chains projecting from the polymer molecules. If a methyl group replaces one of the hydrogen atoms in every monomer unit of polyethylene, the result is polypropylene. The relative positions of the methyl groups attached to the carbon backbone may be **isotactic** (all on the same side), **syndiotactic** (alternating in a regular pattern), or **atactic** (distributed at random). Substitution of a chlorine atom for one of the hydrogen atoms in ethylene gives vinyl chloride, which polymerizes to polyvinyl chloride. The **elastomer** natural rubber is a polymer of **isoprene** (2-methylbutadiene), in which the geometry at every double bond along the polymer chain is *cis*. Natural rubber is cross-linked by S—S bonds in a process called **vulcanization,** giving a material that is used for automobile tires. Synthetic rubbers include neoprene, in which a chlorine atom replaces the methyl group in natural rubber, and copolymers of butadiene with styrene or with acrylonitrile.

PROBLEMS

Note: Answers to blue-numbered problems are given in Appendix F. More challenging problems are indicated with asterisks.

Making Polymers

1. Write a balanced chemical equation to represent the addition polymerization of 1,1-dichloroethylene. The product of this reaction is Saran, used as a plastic wrap.

2. Write a balanced chemical equation to represent the addition polymerization of tetrafluoroethylene. The product of this reaction is Teflon.

3. A polymer produced by addition polymerization consists of —(CH_2—O)— groups joined in a long chain. What was the starting monomer?

4. The polymer polymethyl methacrylate is used to make Plexiglas. It has the formula

$$\left[CH_2 - \underset{\underset{O\diagdown \diagup OCH_3}{\overset{|}{C}}}{\overset{CH_3}{\underset{|}{C}}} \right]_n$$

Draw the structural formula of the starting monomer.

5. The monomer glycine (NH_2—CH_2—COOH) can undergo

condensation polymerization to form polyglycine, in which the structural units are joined by amide linkages.
(a) What is the molecule split off in the formation of polyglycine?
(b) Draw the structure of the repeat unit in polyglycine.

6. The polymer $-(NH-CH(CH_3)-\overset{\overset{\textstyle O}{\|}}{C})_{\overline{n}}$ forms upon condensation polymerization with loss of water. Draw the structure of the starting monomer.

7. Sodium amide, $NaNH_2$, can act as an ionic initiator for addition polymerization, with liquid ammonia as solvent. By making an analogy with the role of butyl lithium described in the text, write a chemical equation for the process by which $NaNH_2$ initiates the polymerization of acrylonitrile.

8. The polymerization initiated by butyl lithium is anionic because the growing polymer retains an anionic functional group during its growth. Cationic polymerization is also possible. Write a chemical equation for the cationic addition polymerization of propylene (CH_2CHCH_3) initiated by molecular iodine, I_2. (*Hint*: I_2 dissociates into I^+ and I^-, and the I^+ attacks the monomer.)

Natural Polymers

9. How many tripeptides could be synthesized using just three different species of α-amino acid?

10. How many dipeptides could be formed using the 20 amino acids listed in Table 25–1?

11. Draw the structure of the pentapeptide alanine–leucine–phenylalanine–glycine–isoleucine. Assume that the free —NH_2 group is at the alanine end of the peptide chain. Would this compound be more likely to dissolve in water or in octane? Explain.

12. Draw the structure of the pentapeptide aspartic acid–serine–lysine–glutamic acid–tyrosine. Assume that the free —NH_2 group is at the aspartic acid end of the peptide chain. Would this compound be more likely to dissolve in water or in octane? Explain.

13. By referring to Figure 25–17a, draw the structure of the ring form of β-D-galactose. How many chiral centers are there in the molecule?

14. By referring to Figure 25–19b, draw the structure of the ring form of D-ribose. How many chiral centers are there in the molecule?

15. Suppose that a long-chain polypeptide is constructed entirely from phenylalanine monomer units. What is its empirical formula? How many amino acids does it contain if its molar mass is 17,500 g mol^{-1}?

16. A typical bacterial DNA has a molar mass of 4×10^9 g mol^{-1}. Approximately how many nucleotides does it contain?

Uses for Polymers

17. (a) What is the empirical formula of starch?
    (b) What is the empirical formula of cellulose?

18. (a) What is the empirical formula of guncotton (cellulose trinitrate)?
    (b) What is the empirical formula of cellulose acetate?

19. Determine the mass of adipic acid and the mass of hexamethylenediamine needed to make 1.00×10^3 kg of nylon-66 fiber.

20. Determine the mass of terephthalic acid and the mass of ethylene glycol needed to make 10.0 kg of polyester fiber.

21. Diamond has been called "the ultimate branched-chain alkane." Explain why (refer to the structural diagram of diamond in Fig. 20–25).

22. High-density polyethylene has been called "the ultimate straight-chain alkane." Explain why.

23. In a recent year, 4.37 billion kg of low-density polyethylene was produced in the United States. What volume of gaseous ethylene at STP would give this amount?

24. In a recent year, 2.84 billion kg of polystyrene was produced in the United States. Polystyrene is the addition polymer formed from the styrene monomer, $C_6H_5CH{=}CH_2$. How many styrene monomer units were incorporated in that 2.84 billion kg of polymer?

Additional Problems

25. In the addition polymerization of acrylonitrile, a very small amount of butyl lithium causes a reaction that can consume hundreds of pounds of the monomer; however, the butyl lithium is called an initiator, not a catalyst. Explain why.

26. Based on the fact that the free-radical polymerization of ethylene is spontaneous and the fact that polymer molecules are less disorganized than the starting monomers, decide whether the polymerization reaction is exothermic or endothermic. Explain.

27. Sketch the structures of random, block, and graft copolymers of vinyl chloride with butadiene.

* 28. Can vinyl chloride and styrene form a graft copolymer? Explain.

* 29. Polypeptides are synthesized from a 50:50 mixture of L-alanine and D-alanine. How many different isomeric molecules containing 22 monomer units are possible?

30. Based on the information in the text, explain the similarity in the odors of burning hair and burning fingernail clippings.

31. An osmotic-pressure measurement on a solution containing hemoglobin shows that the molar mass of that protein is approximately 65,000 g mol^{-1}. A chemical analysis shows it to contain 0.344% of iron by mass. How many iron atoms does each hemoglobin molecule contain?

32. At 25°C, the standard Gibbs function of formation of aqueous glucose is -917.2 kJ mol^{-1}. Use this value, together with data from Appendix D, to calculate $\Delta G°$ at 25°C for the degradation of one mole of glucose to aqueous ethanol and aqueous carbon dioxide, a chemical reaction carried out by certain bacteria.

33. L-sucrose tastes sweet, but it is not metabolized. It has been suggested as a potential nonnutritive sweetener. Draw the molecular structure of L-sucrose, using Figure 25–17b as a starting point.

34. The sequence of bases found in one strand of DNA reads ACTTGACCG. Write the sequence of bases in the complementary strand.

35. Approximately 950 million lb of ethylene dichloride was exported from the United States in a recent year, according to a trade journal. The journal goes on to state that "between 500 million and 550 million pounds of PVC could have been made from that ethylene dichloride." Compute the range of percentage yields of PVC from ethylene dichloride that is implied by these figures.

36. The complete hydrogenation of natural rubber (the addition of H_2 to all double bonds) gives a product that is indistinguishable from the product of the complete hydrogenation of gutta-percha. Explain how this strengthens the conclusion that these two substances are isomers of each other.

* 37. What similarities and what differences are there between the formation of isotactic, syndiotactic, and atactic polypropylene on the one hand, and the formation of all-*cis*, all-*trans*-, and mixed-*cis* and *trans*-polyisoprene on the other hand?

Scientific Notation and Experimental Error

A–1 SCIENTIFIC NOTATION

Very large and very small numbers are common in chemistry. Repeatedly writing down such numbers in the ordinary way (as in 602 213 700 000 000 000 000 000) would be tedious and inevitably engender errors. **Scientific notation** offers a better way.

In scientific notation, a number is expressed as a coefficient between 1 and 10 multiplied by 10 raised to some power. The coefficient is sometimes called the *pre-exponential* and the 10-to-a-power the *exponential* part of the notation. The power must be an integer but may be either negative, positive, or zero. Any number can be represented in scientific notation, as the following examples show:

$$643.8 = 6.438 \times 10^2$$
$$19\ 000\ 000 = 1.9 \times 10^7$$
$$0.0236 = 2.36 \times 10^{-2}$$
$$-0.00297 = -2.97 \times 10^{-3}$$
$$602\ 213\ 700\ 000\ 000\ 000\ 000\ 000 = 6.022137 \times 10^{23}$$

When a number is to be converted to scientific notation, the power to which 10 is raised is $+n$ if the decimal point is moved n places to the *left* in the process of repositioning it and $-n$ if the required movement is n places to the right. A minus sign in an exponent does *not* make a number negative. In scientific notation, negative numbers are signified by placing a minus sign in front of the coefficient, just as in ordinary notation. A negative exponent means that the number is a fraction (lies between 0 and 1 or between 0 and −1 if it has a negative sign in front).

> • The power to which the 10 is raised when a number like 6.42 is expressed in scientific notation equals 0, because no movement of the decimal point is required. Simply writing 6.42 is acceptable scientific notation for this number; few scientists would trouble to write 6.42×10^0.

When two or more numbers written in scientific notation are to be added or subtracted, they must first be re-expressed as multiples of the *same* power of 10:

$$
\begin{aligned}
6.43 \times 10^4 &\longrightarrow 6.43 \times 10^4 \\
+2.1 \times 10^2 &\longrightarrow +0.021 \times 10^4 \\
+3.6 \times 10^3 &\longrightarrow \underline{+0.36 \times 10^4} \\
? & 6.81 \times 10^4
\end{aligned}
$$

Note the rounding off in the final result.

> • The rounding off of numbers is considered in Section A–3.

When two numbers in scientific notation are multiplied, the coefficients are multiplied, and the powers of 10 are multiplied separately by adding the exponents. The product is written as the combination of the new pre-exponential part and the new exponential part:

$$(1.38 \times 10^{-16}) \times (8.80 \times 10^3) = (1.38 \times 8.80) \times 10^{(-16+3)}$$
$$= 12.1 \times 10^{-13}$$

If the new pre-exponential part exceeds 10 or is less than 1, it is usually desirable to rewrite the answer to bring the pre-exponential part between 1 and 10. In this example, the pre-exponential part of the answer exceeds 10, so we write

$$12.1 \times 10^{-13} = 1.21 \times 10^1 \times 10^{-13} = 1.21 \times 10^{-12}$$

One number is divided by a second in scientific notation by separately dividing the coefficient of the first by the coefficient of the second and then multiplying the answer by 10 raised to a power equal to the first exponent minus the second exponent:

$$\frac{6.63 \times 10^{-27}}{2.34 \times 10^{-16}} = \frac{6.63}{2.34} \times \frac{10^{-27}}{10^{-16}}$$
$$= 2.83 \times 10^{[-27-(-16)]}$$
$$= 2.83 \times 10^{-11}$$

Remember that subtracting a negative quantity is the same as adding a quantity of that magnitude ("minus a minus is a plus").

All calculators and computers equipped for scientific and engineering calculations can accept and display numbers in scientific notation. However, they cannot determine whether a key-punch error has occurred or whether the answer makes sense. These are the user's responsibility! Develop the habit of mentally estimating the order of magnitude of the answer as a rough check on the calculator's result.

A-2   EXPERIMENTAL ERROR

As in all other experimental sciences, quantitative measurements in chemistry are subject to some degree of error. Error can be reduced by carrying out additional measurements or by changing or improving the experimental apparatus, but it can never be eliminated altogether. The degree of error in an experiment obviously has a profound effect on the conclusions that can be drawn from it. It is therefore important to assess the error in every measurement. Error appears in two ways: (1) from a lack of precision (random errors), and (2) from a lack of accuracy (systematic errors).

Precision and Random Errors

Precision refers to the degree of agreement in a collection of experimental results obtained by repeating a measurement under conditions as close to identical as possible. If the conditions are truly identical, then the differences among the trials are due to random error. As a specific example, let us consider some actual results of an early, important experiment to measure the charge e on the electron. This experiment was first carried out by the American physicist Robert Millikan in 1909. The experimental method, discussed in Section 1–5, involved a study of the motion of charged oil drops suspended in air in an electric field. Millikan made hundreds of measurements on many different oil drops; one set of results for one particular drop is quoted in Table A–1. Values were found that ranged from 4.894 to 4.941×10^{-10} esu. What

Table A–1

					Measurement Number $e\ (10^{-10}\text{esu})$							
1	2	3	4	5	6	7	8	9	10	11	12	13
4.915	4.920	4.937	4.923	4.931	4.936	4.941	4.902	4.927	4.900	4.904	4.897	4.894

From R. A. Millikan, *Phys. Rev.* 32:349, 1911. [1 esu = 3.3356 × 10^{-10} C.]

should be reported as the best estimate for e? The proper procedure is first to examine the data to see whether any of the results are especially far from the rest (a value above 5×10^{-10} esu would fall into this category). Such values are likely to result from some mistake in carrying out or reporting that particular measurement and therefore are excluded from further consideration. In Millikan's data, no such points appear. The best estimate for e is then obtained by calculating the average or **mean** value. The mean value of any property after a series of N measurements $x_1, x_2, \ldots, x_N$ is

$$\bar{x} = \frac{1}{N}(x_1 + x_2 + \cdots + x_N)$$

where the bar above the x means that this is the average value. In the present case, the average for e is 4.917×10^{-10} esu.

The average by itself does not convey any estimate of uncertainty. If all of the measurements gave results falling between 4.91 and 4.92×10^{-10} esu, the uncertainty would surely be less than if the results were scattered across the interval between 4×10^{-10} and 6×10^{-10} esu. Furthermore, an average of 100 measurements should assuredly have less uncertainty than an average of five measurements. How can these ideas be made quantitative? A statistical measure of the spread of data, called the **standard deviation σ,** is useful in this regard. It is given by the formula

$$\sigma = \sqrt{\frac{(x_1 - \bar{x})^2 + (x_2 - \bar{x})^2 + \cdots + (x_N - \bar{x})^2}{N - 1}}$$

The standard deviation is found by adding up the squares of the deviations of the individual data points from the average value x, dividing by $N - 1$, and taking the square root. The reasons for this particular (and apparently rather complicated) formula are beyond the scope of this presentation. We merely state the conclusion—that the standard deviation σ is a useful measure of the uncertainty in any experimental result. For Millikan's data, $N = 13$ and $\sigma = 0.017 \times 10^{-10}$ esu, so the result of these 13 measurements of the charge on the electron is $e = (4.917 \pm 0.017) \times 10^{-10}$ esu.

Accuracy and Systematic Errors

The charge on the electron, e, has been measured by several different techniques since Millikan's day. The current best estimate for e is

$$e = (4.8032068 \pm 0.0000015) \times 10^{-10} \text{ esu}$$
$$= (1.6021773 \pm 0.0000005) \times 10^{-19} \text{ C}$$

This value lies well outside the range of uncertainty we have estimated from Millikan's original data. In fact, it is distinctly less than the smallest of his 13 measurements of e. Why?

• Cases are on record, however, where just such exceptional results have been real, not errors, and have led to significant breakthroughs.

To understand this discrepancy, we need to remember that any experiment has a second source of error: *systematic* error that causes a shift in the measured values from the true value and reduces the **accuracy** of the result. By making more measurements, a scientist can reduce the uncertainty due to *random* errors and improve the *precision* of a result, but if systematic errors are present, the average value continues, to deviate from the true value. Systematic errors can result from a miscalibration of the experimental apparatus or from fundamental inadequacies in a specific technique for measuring a property. In the case of Millikan's experiment, the then-accepted value for the viscosity of air (used in calculating the charge *e*) was subsequently found to be wrong. This caused Millikan's results to be systematically too high.

Error therefore arises from two sources. Lack of precision (from random errors) can be estimated by a statistical analysis of a series of measurements, but lack of accuracy (from systematic errors) is much more problematical. If a systematic error is known to be present, it should be corrected for before the result is reported (for example, if the apparatus has not been calibrated correctly, it should be recalibrated). The problem is that systematic errors of which we have no knowledge may be present. In this case the experiment should be repeated with different apparatus to eliminate the systematic error caused by one particular piece of equipment; better still, a different and independent way of measuring the property might be devised. Only after independent confirmatory experimental data are available can a scientist be convinced of the accuracy of a result—that is, how closely it approximates the true result.

A–3 SIGNIFICANT FIGURES

The number of **significant figures** (significant digits) in a measured quantity, or in a number calculated from a measured quantity, is the number of digits used to express it, not counting zeros that are present for the sole purpose of positioning the decimal point.

• This estimate of uncertainty might come from our knowledge of the measuring instrument and our experience with similar measurements.

Suppose that the mass of a sample of sodium chloride is measured to be 8.241 g and the uncertainty in this measurement is estimated to be ± 0.001 g. The mass is said to be given to four significant figures because we are confident of the first three digits (8, 2, 4) and the uncertainty appears in the fourth (1), which nevertheless is still meaningful. Because the 1 is subject to uncertainty, digits beyond it would be either meaningless or misleading. Such additional digits should not be written unless the precision of the weighing is improved.

A volume recorded as 22.4 L implies that the uncertainty in the measurement is in the last digit written ($V = 22.4 \pm 0.3$ L, for example). A volume written as 22.48 L, on the other hand, implies that the uncertainty is far less and appears only in the fourth significant figure. Similarly, writing 10.000 m is quite different from writing 10.0 m: the implied uncertainty in "10.000 m" is on the order of a few thousandths of a meter (0.001 m), whereas the implied uncertainty in "10.0 m" is on the order of a tenth of a meter (0.1 m), which is 100 times greater. The second measurement could easily be completed with a common meterstick. The first would require a more complex method.

The following procedure is used to count the number of significant figures in a measured number:

1. All nonzero digits are counted. Thus, 489 g has three significant figures, and 111.1 g has four significant figures.
2. Zeros that precede the first nonzero digit in the number are *not* counted. Such zeros serve only to position the decimal point. Thus, the zeros in "0.00463°C"

are excluded in tallying up significant figures, and "0.00463°C" has three significant figures.

3. Zeros surrounded by nonzero digits *are* counted. The zeros in "1005 g" are significant because omitting them does not merely reposition the decimal point but changes the meaning of the number in the same way that leaving out the 1 or the 5 would change it.

4. Zeros that follow the last nonzero digit are *sometimes* counted. The trailing zeros in the measurements "9.0 g" and "13.620 mL" do nothing to position the decimal point. Because the only possible reason for their presence is to communicate the degree of uncertainty in the measurement, they are definitely significant. In "700 m," however, the two trailing zeros may or may not be significant. They may be present solely to position the decimal point but may also be intended to convey the precision of the measurement. The uncertainty in the measurement is on the order of ± 1 m or ± 10 m or perhaps ± 100 m; it is impossible to tell which without further information. To avoid this ambiguity, such measurements can be written using the scientific notation described in Section A–1. The measurement "700 m" translates in scientific notation to any of the following:

7.00×10^2 m	Three significant figures
7.0×10^2 m	Two significant figures
7×10^2 m	One significant figure

Significant Figures in Calculations

Frequently several different experimental measurements must be combined to obtain a final result. Sometimes these operations involve addition or subtraction, and other times they entail multiplication or division. These operations affect the number of significant figures that should be retained in the calculated result.

Suppose, for example, that a weighed sample of 8.241 g of sodium chloride is dissolved in 160.1 g of water. What is the mass of the resulting solution? It is tempting simply to write 160.1 + 8.241 = 168.341 g, but this is *not* correct. Stating that the mass of water is 160.1 g implies an uncertainty about the number of tenths of a gram measured. This uncertainty must also apply to the sum of the masses, so that the last two digits in the sum are not significant and should be **rounded off,** leaving 168.3 as the final result.

• Significant figures must also be considered in determining logarithms or inverse logarithms. The rules for these operations are best presented in the context of a review of logarithms and are given in Appendix C–3.

> Following addition or subtraction, round off so that the sum or difference has the same number of decimal places as there are in the measurement with the smallest number of decimal places.

Here are some more examples:

94.17 g	11.171 m
+ 0.023 g	− 2.1 m
94.193 g (round off to 94.19 g)	9.071 m (round off to 9.1 m)

Rounding off is a straightforward operation. It consists of first discarding the digits that are not significant and then adjusting the last digit that remains. Follow these rules for rounding off:

1. If the first discarded digit (reading from left to right) is less than 5, the remaining digits are left as they are. For example, when 168.341 is rounded off

to four significant figures, it is rounded down to 168.3, because the first discarded digit, 4, is less than 5.

2. If the first discarded digit is greater than 5, then the last digit is increased by 1. By this rule, 168.374 becomes 168.4 when rounded off to four digits; the first discarded digit, the 7, exceeds 5.

3. If the first digit discarded equals 5 and is followed by one or more nonzero digits, then the last digit is increased by 1. Thus, 168.3503 becomes 168.4 when rounded off to four digits. Similarly, 16.25001 becomes 16.3 when rounded off to three digits.

4. If the first digit discarded is 5 and *all* subsequent discarded digits are zeros, the last digit retained is rounded to the nearest even digit. By this rule, 168.350 becomes 168.4 when rounded off to four digits and 168.450 also becomes 168.4. The change in rounding up numbers of this type exactly equals the change in rounding them down. No reason exists to favor either choice, and a policy of always rounding down (or up) might cause errors to accumulate. The nearest-even-digit rule assures that, over many cases, as many numbers are rounded up as down. Other rules are sometimes used.

In multiplication or division, it is not the number of decimal places that matters (as in addition or subtraction) but the number of significant figures in the least precisely known quantity. Suppose, for example, that the volume of a sample of sodium chloride is 4.34 cm^3. The density is given by the mass divided by the volume. Dividing gives

$$\frac{8.241 \text{ g}}{4.34 \text{ cm}^3} = 1.89884 \ldots \text{ g cm}^{-3}$$

How many significant figures should be reported? Because the volume is the less precisely known quantity (three significant figures rather than four for the mass), it controls the precision that may properly be reported in the answer. In this case, there are only three significant figures, so the result is rounded to 1.90 g cm^{-3}.

In multiplication and division, the proper number of significant figures in the product or quotient equals the smallest number of significant figures in the measured quantities used as input.

The following examples show this rule in action:

$$2.3 \times 19.987 = 45.9701 \qquad \text{(round off to 46)}$$
$$\frac{0.0114}{0.0045} = 2.533333 \qquad \text{(round off to 2.5)}$$
$$\frac{11.187 \times 22.2}{17.76 \times 14.0} = 0.9988393 \quad \text{(round off to 0.999)}$$

The rule also applies to computations in which multiplication and division are mixed, as the last example shows.

It is best to carry out arithmetical operations and *then* to round the final answer to the correct number of significant figures, rather than to round off the input data first. The difference is usually small, but following this recommendation can sometimes prevent wrong answers. For example, the correct way to add the three distances 15 m, 6.6 m, and 12.6 m is

15 m		15 m $\longrightarrow$	15 m
+ 6.6 m	rather than	6.6 m $\longrightarrow$	7 m
+ 12.6 m		12.6 m $\longrightarrow$	13 m
34.2 $\longrightarrow$ 34 m			35 m

For the same reason, extra digits are frequently carried through the intermediate steps of a worked example and discarded only at the final stage. If a calculation is done entirely on a scientific calculator or a computer, several extra digits are usually carried along automatically. Before reporting the final answer, however, it is important to round off to the proper number of significant figures.

● Alert readers will notice that this practice is followed in this book.

Sometimes mathematical constants and **exact numbers,** numbers obtained by counting and not by measurements, appear in expressions. Exact numbers should be considered to have an infinite number of significant figures. The precision of the result is then controlled by the precision of the remaining input. If the mass of an object is 6.142 g, then the mass of three identical such objects is

$$3 \times 6.142 = 18.43 \text{ g}$$

This answer is *not* rounded off to one significant digit because the number 3 comes from counting the objects. It is known exactly and has an infinite number of significant digits, not just one. Similarly, the uncertainty in the volume of a sphere, computed from the formula $V = 4/3 \, \pi r^3$, depends only on the uncertainty in the radius r. This follows because 4 and 3 are exact values (4.000 . . . and 3.000 . . . , respectively) and not the results of measurements, and π, which is known to many millions of digits ($\pi = 3.14159265358979$ . . .), can be used as required. Note that mathematical constants and exact numbers are not necessarily whole numbers.

PROBLEMS

Note: Answers to blue-numbered problems are given in Appendix F.

Scientific Notation

1. Express the following in scientific notation:
   (a) 0.0000582
   (b) 1402
   (c) 7.93
   (d) −6593.00
   (e) 0.002530
   (f) 1.47

2. Express the following in scientific notation:
   (a) 4579
   (b) −0.05020
   (c) 2134.560
   (d) 3.825
   (e) 0.0000450
   (f) 9.814

3. Convert the following from scientific notation to decimal form:
   (a) 5.37×10^{-4}
   (b) 9.390×10^6
   (c) -2.47×10^{-3}
   (d) 6.020×10^{-3}
   (e) 2×10^4

4. Convert the following from scientific notation to decimal form:
   (a) 3.333×10^{-3}
   (b) -1.20×10^7
   (c) 2.79×10^{-5}
   (d) 3×10^1
   (e) 6.700×10^{-2}

5. Determine which of the following numbers is largest algebraically: 1.90×10^2, 9.7×10^{-2}, -4.90×10^2, -4.10×10^{-3}.

6. Arrange the following from largest to smallest in the algebraic sense: 11.7×10^{-7}, -14.8×10^{-4}, -2.17×10^3, 3.19×10^{-2}.

7. A certain chemical plant produces 7.46×10^8 kg of polyethylene in 1 year. Express this amount in decimal form.

8. A microorganism contains 0.0000046 g of vanadium. Express this amount in scientific notation.

Experimental Error

9. A group of students takes turns using a laboratory balance to weigh the water contained in a beaker. The results they report

are 111.42 g, 111.67 g, 111.21 g, 135.64 g, 111.02 g, 111.29 g, and 111.42 g.

(a) Should any of the data be excluded before calculating the average?

(b) Calculate the average value of the mass of the water in the beaker from the valid measurements.

(c) Calculate the standard deviation σ of the valid data.

10. By measuring the sides of a small box, a group of students make the following estimates for its volume: 544 cm^3, 590 cm^3, 523 cm^3, 560 cm^3, 519 cm^3, 570 cm^3, and 578 cm^3.

(a) Should any of the data be excluded before calculating the average?

(b) Calculate the average value of the volume of the box from the valid measurements.

(c) Calculate the standard deviation σ of the valid data.

11. Of the measurements in problems 9 and 10, which is more precise?

12. A more accurate determination of the mass in problem 9 (using a better balance) gives the value 104.67 g, and a more accurate determination of the volume in problem 10 gives the value 553 cm^3. Which of the two measurements in problems 9 and 10 is more accurate, in the sense of having the smaller systematic error relative to the actual value?

Significant Figures

13. State the number of significant figures in each of the following measurements:

(a) 13.604 L

(b) $-0.00345°C$

(c) 340 lb

(d) 3.40×10^2 miles

(e) 6.248×10^{-27} J

14. State the number of significant figures in each of the following measurements:

(a) -0.0025 in.

(b) 7000 g

(c) 143.7902 s

(d) 2.670×10^7 Pa

(e) 2.05×10^{-19} J

15. Round off each of the measurements in problem 13 to two significant figures.

16. Round off each of the measurements in problem 14 to two significant figures.

17. Round off the measured number 2,997,215.548 to nine significant digits.

18. Round off the measured number in problem 17 to eight, seven, six, five, four, three, two, and one significant digits.

19. Express the results of the following additions and subtractions to the proper number of significant figures. All of the numbers are measured quantities and have the same units.

(a) $67.314 + 8.63 - 243.198 =$

(b) $4.31 + 64 + 7.19 =$

(c) $3.2156 \times 10^{15} - 4.631 \times 10^{13} =$

(d) $2.41 \times 10^{-26} - 7.83 \times 10^{-25} =$

20. Express the results of the following additions and subtractions to the proper number of significant figures. All of the numbers are measured quantities and have the same units.

(a) $245.876 + 4.65 + 0.3678 =$

(b) $798.36 - 1005.7 + 129.652 =$

(c) $7.98 \times 10^{17} + 6.472 \times 10^{19} =$

(d) $4.32 \times 10^{-15} + 6.257 \times 10^{-14} - 2.136 \times 10^{-13} =$

21. Express the results of the following multiplications and divisions to the proper number of significant figures. All of the numbers are measured quantities.

(a) $\dfrac{-72.415}{8.62} =$

(b) $52.814 \times 0.00279 =$

(c) $(7.023 \times 10^{14}) \times (4.62 \times 10^{-27}) =$

(d) $\dfrac{4.3 \times 10^{-12}}{9.632 \times 10^{-26}} =$

22. Express the results of the following multiplications and divisions to the proper number of significant figures. All of the numbers are measured quantities.

(a) $129.578 \times 32.33 =$

(b) $\dfrac{4.7791}{3.21 \times 5.793} =$

(c) $\dfrac{10,566.9}{3.584 \times 10^{29}} =$

(d) $(5.247 \times 10^{13}) \times (1.3 \times 10^{-17}) =$

23. Compute the area of a triangle if its base and altitude are measured to equal 42.07 cm and 16.0 cm, respectively. (The area of a triangle is 1/2 its base times its altitude.) Explain your use of significant figures in the answer.

24. An inch is defined as exactly 2.54 cm. The length of a table is measured as 404.16 cm. Compute the length of the table in inches. Explain your use of significant figures in the answer.

SI Units and the Conversion of Units

B

B-1 THE INTERNATIONAL SYSTEM OF UNITS

Most measurements in physics and chemistry require the use of units, which are reference standards to which experimentally observed quantities can be related. These measurements are really *ratios* between the thing being measured and the size of the "measuring stick" (the reference unit). If a zoologist tells us that a snake is 5 long, we still know nothing about the snake until we learn whether inches, feet, meters (Heaven forbid!), or some other unit of length is intended. Whenever the magnitude of an experimental observation depends on the units selected to express it (as in the case of the snake), the observation is said to have **dimensions.** Over the course of history, different countries and regions evolved numerous sets of locally accepted units to express length, mass, and many other physical dimensions of importance in commerce and industry. These diverse units are gradually being replaced by international standards that allow easy comparison of measurements made in different localities and avoid misunderstanding. The unified system of units that is currently recommended by international agreement is a metric system called the SI, which stands for "*Système International d'Unités*," or *International System of Units*. This appendix outlines the use of SI units and discusses interconversions with other systems of units.

> • We say "a" metric system because several variant systems exist that use the metric ideas first adopted in France in 1790.

The SI uses seven **basic units,** which are listed in Table B–1. These units are the defined standards of reference for seven physical quantities that were chosen for their importance and the ease with which they are measured. Of these quantities, only luminous intensity is not used in this book. Length, mass, and time, the first three quantities in Table B–1, are familiar from everyday life. A **length** is a distance

Table B–1
Base Units in the International System of Units

Quantity	Unit	Symbol
Length	meter	m
Mass	kilogram	kg
Time	second	s
Temperature	kelvin	K
Chemical amount (of substance)	mole	mol
Electric current	ampere	A
Luminous intensity	candela	cd

or a spatial separation; the **mass** of an object is the quantity of matter that it contains; and a period of **time** is a temporal separation. Measurements of **temperature,** the degree of hotness or coldness of an object on some defined scale, are of daily importance. The full definition of temperature requires some care (see Section 5–3). Many units of temperature exist in addition to the SI base unit, the kelvin. Of these, the Fahrenheit degree is the most familiar in the United States, and the Celsius degree is the most important in science. The relationships among the most important temperature scales and units of temperature are discussed later. The quantity called amount of substance, or **chemical amount,** is perhaps unfamiliar, but its name reveals its importance in chemistry. Chemical amount is less concerned with the total quantity of matter in a sample, which is given by its mass, than with the number of molecules or other entities in that sample that exhibit the same behavior in a chemical reaction. This point is discussed in full in Section 1–7. Finally, the measurement of **electric current,** the rate of passage of electric charge through a circuit, is important in the study of electrochemistry (see Section 12–3).

The seven base units in the SI all have more or less complex technical definitions that are intended to allow their precise reproduction in independent laboratories. We give none of these definitions here. Knowing the definitions is less important than having a sense of the size of the units. Thus, the second presents no problems; a meter is about 10% longer than 1 yard (a length of 39.37 inches), and a kilogram is very close to 10% heavier than 2 lb (it is 2.205 lb). These comparisons are helpful only up to a point; mastery of the SI system requires learning to "think metric" instead of estimating and planning in terms of other units and then translating. A table of conversion factors between SI units, other metric units, and units in the U.S. customary system is given at the front of this book.

Units other than the seven SI base units also see a great deal of use. Depending on the scientific field, they are indeed often encountered more frequently than the base units themselves. Such units are called **derived units,** and several are listed in Table B–2. All derived units are obtained as some combination (by means of multiplication or division) or one or more of the seven base units. The following are some important physical quantities that are measured in derived units.

Table B–2
Some Derived Units in the International System of Units

Quantity	Definition	Unit	Name/Abbreviation
Area	Length $\times$ length	m^2	square meter
Volume	Length $\times$ length $\times$ length	m^3	cubic meter
Density	Mass/length^3	$kg\ m^{-3}$	kilogram per cubic meter
Velocity	Length/time	$m\ s^{-1}$	meter per second
Acceleration	Length/time^2	$m\ s^{-2}$	meter per second squared
Force	Mass $\times$ acceleration	$kg\ m\ s^{-2}$	newton (N)
Pressure	Force/area	$N\ m^{-2} = kg\ m^{-1}\ s^{-2}$	pascal (Pa)
Energy	Force $\times$ distance	$N\ m = kg\ m^2\ s^{-2}$	joule (J)
Power	Energy/time	$W = kg\ m^2\ s^{-3}$	watt (W)
Electric charge	Electric current $\times$ time	$A\ s$	coulomb (C)
Electric potential difference	Energy/electric charge	$J\ C^{-1} = kg\ m^2\ s^{-3}\ A^{-1}$	volt (V)

Area and Volume. Area is spatial extent in two directions, and volume is spatial extent in three directions. (Recall that length is extent in one direction.) It follows that area has dimensions of length × length, and volume has dimensions of length × length × length. The natural SI unit of area is formed simply by squaring the SI unit of length. It is the meter × meter (m^2), or square meter. The natural SI unit of volume is the meter × meter × meter, or cubic meter (m^3).

Density. It is often important to know how tightly or loosely the matter that comprises a given sample is packed together. This information is given by the **density** of the sample, defined as its mass divided by its volume

$$\text{density} = \text{mass/volume} \quad \text{or} \quad d = m/V$$

From this definition and the discussion of volume in the previous paragraph, it follows that density has dimensions of mass divided by length × length × length. The natural SI unit of density is accordingly the kilogram per cubic meter (abbreviated kg/m^3 or $kg\ m^{-3}$). Density is widely used to aid in the identification of materials because it takes sample size into account and is therefore an inherent physical characteristic of the sample. (Density does depend on sample temperature, especially for liquids and gases.) For example, the room-temperature density of fool's gold (iron pyrites) is always far less than the density of true gold, although a big lump of fool's gold could easily have a larger mass than a small nugget of true gold.

• Note the use of a negative exponent to indicate the "m^3" in the denominator of a fraction. This useful notation and other aspects of exponents are reviewed in Appendix C.

Speed and Velocity. The speed of a moving object is the rate of change of its position with time—that is,

$$\text{speed} = \frac{\text{change in position}}{\text{change in time}}$$

A change in position (the distance between position 1 and position 2) is a length and has the units of length. An elapsed time, of course, has units of time. Speed thus has the units of length divided by time, and the natural unit of speed in the International System is the meter per second, or $m\ s^{-1}$. Velocity has the same units as speed. The difference between speed and velocity is not in their units but in the fact that speed has magnitude only but velocity has both magnitude *and* direction. Driving a car faster than the posted limit often earns the driver a speeding ticket; driving quite slowly the wrong way on a one-way street is a different violation that merits what could be called a "velocity ticket," because it is the direction of travel that is forbidden, not the rate of travel.

Acceleration. Acceleration is the rate of change of velocity with time:

$$\text{acceleration} = \frac{\text{change in velocity}}{\text{change in time}}$$

From this definition, acceleration must have the units of velocity, which are length divided by time, again divided by time. This is equivalent to length divided by time squared. The definition is used in just this way in evaluating automobiles. A powerful car is able to accelerate from a standstill up to a velocity of 60 miles per hour in 6 seconds. It gains speed at the rate of 10 miles per hour every second, for an average acceleration of 10 miles $hr^{-1}\ s^{-1}$. This unit of acceleration uses a mixture of time units. The natural SI unit of acceleration measures time exclusively in seconds. It is the meter per second per second, equivalent to the meter per second squared, or $m\ s^{-2}$.

Force.   A physical force is easily understood as a push or pull. The force exerted by the earth's gravity, which tugs all terrestrial objects downward, is the most familiar force in everyday experience. An unbalanced force acting on an object causes it to change its velocity—that is, to accelerate. This observation is part of Newton's second law of motion, which gives the size of the force as the product of the mass of the object and the observed acceleration:

$$\text{force} = \text{mass} \times \text{acceleration}$$

Newton's second law provides the definition of the SI unit of force. The SI unit of mass (kilogram) is multiplied by the SI unit of acceleration (the meter per second per second) to give the kg m s^{-2}. This derived unit has a special name; 1 kg m s^{-2} is called a *newton* (abbreviated "N"), in honor of Sir Isaac Newton.

• One newton is approximately the gravitational force exerted by the earth on an apple.

Pressure.   A pressure is the force exerted on an object (such as the wall of a container) divided by the area that receives that force:

$$\text{pressure} = \frac{\text{force}}{\text{area}}$$

Substituting the SI units of force and area into this definition establishes that the natural SI unit of pressure is the newton per square meter, or N m^{-2}. This unit is named a *pascal* (abbreviated "Pa"), in honor of the French physicist Blaise Pascal. Combining the fact that a newton equals 1 kg m s^{-2} with the definition of the pascal gives the pascal in terms of SI base units:

$$\text{pascal} = \frac{\text{N}}{\text{m}^2} = \frac{\text{kg m s}^{-2}}{\text{m}^2} = \text{kg m s}^{-2} \text{ m}^{-2} = \text{kg m}^{-1} \text{ s}^{-2}$$

Energy.   Perhaps the most familiar manifestation of energy is mechanical work, and it is through the definition of mechanical work that one arrives at the units of energy most easily. Mechanical work is defined as the product of the external force on an object times the distance through which the force acts:

$$\text{work} = \text{force} \times \text{distance}$$

The natural SI unit of work is therefore the SI unit of force (the newton) multiplied by the SI unit of distance (the meter). It is the *newton-meter* (N m). Because mechanical work is a form of energy, the newton-meter is also the SI unit of energy. A newton-meter is called a *joule* (abbreviated "J"), in honor of the English physicist James Joule. In terms of the SI base units, a joule is

$$\text{joule} = \text{newton} \times \text{meter} = \text{kg m s}^{-2} \times \text{m} = \text{kg m}^2 \text{ s}^{-2}$$

An object in motion has energy by virtue of that motion. This is called its *kinetic energy*. The kinetic energy of an object is defined by the equation

$$\text{kinetic energy} = \tfrac{1}{2} \times (\text{mass of the object}) \times (\text{velocity of the object})^2$$

To verify that kinetic energy is indeed an energy, simply multiply out the SI units of mass and velocity as required by the definition

$$\text{kg} \times (\text{m s}^{-1})^2 = \text{kg m}^2 \text{ s}^{-2} = \text{J}$$

The 1/2 in the definition contributes no units and is omitted from the equation for this reason.

Prefixes in SI Units

Because scientists work on scales ranging from the microscopic to the astronomical, a tremendous range exists in the magnitudes of measured quantities. For this reason a set of **prefixes** has been incorporated into the International System of Units to simplify the description of small and large quantities (Table B–3). The prefixes are used to specify various powers of 10 times the base and derived units. The natural SI unit of volume, the cubic meter (m^3), for example, is inconveniently large for ordinary laboratory work, in which measurements of volume are typically on the order of 10^{-4} to 10^{-6} m^3. Adding the prefix *centi-* to meter defines the centimeter (abbreviated "cm"), a unit of length equal to exactly 10^{-2} m. The *cubic* centimeter (cm^3) is then $10^{-2} \times 10^{-2} \times 10^{-2} = 10^{-6}$ *cubic* meter. Typical laboratory-scale measurements of volume thus come out between 1 and 100 when expressed in cubic centimeters. Similarly, the natural SI unit of density is inconveniently small for many purposes. The room-temperature density of water, for example, is nearly 1000 kg m^{-3}. Removing the prefix on the unit of mass (switching to the gram instead of the kilogram) and inserting a prefix in the unit of volume (going to cubic centimeters from cubic meters) gives a new SI unit of density, the gram per cubic centimeter, that is much larger:

$$1 \text{ g cm}^{-3} = 1000 \text{ kg m}^{-3}$$

Most solids and liquids at ordinary conditions have densities between 0.1 and 20 g cm^{-3}.

Non-SI Units

In addition to base and derived SI units, certain additional units that are not part of the SI are used in this book. These units are summarized in Table B–4. The first is the *liter,* which, like the cm^3, has a convenient size for the measurement of the volumes used in laboratory-scale chemistry. One liter equals 10^{-3} m^3, or 1 dm^3:

$$1 \text{ L} = 1 \text{ dm}^3 = 10^{-3} \text{ m}^3 = 10^3 \text{ cm}^3$$

The SI prefixes can be combined with this unit (and other non-SI units as well). For example, the *milliliter* (1 mL = 10^{-3} L = 1 cm^3) is quite common in chemistry, and the *deciliter* (1 dL = 10^{-1} L = 100 cm^3) is much used in the health sciences.

• It is particularly worth noting that 1 mL = 1 cm^3.

Second, chemists frequently prefer the *ångström* (abbreviated "Å") as a unit of length in discussing sizes of atoms and molecules:

$$1 \text{ Å} = 10^{-10} \text{ m} = 100 \text{ pm} = 0.1 \text{ nm}$$

Table B–3
Prefixes in SI

Fraction	Prefix	Symbol	Factor	Prefix	Symbol
10^{-1}	deci-	d	10	deca	da
10^{-2}	centi-	c	10^2	hecto-	h
10^{-3}	milli-	m	10^3	kilo-	k
10^{-6}	micro-	μ	10^6	mega	M
10^{-9}	nano-	n	10^9	giga-	G
10^{-12}	pico-	p	10^{12}	tera-	T
10^{-15}	femto-	f	10^{15}	pect-	P
10^{-18}	atto-	a	10^{18}	exa-	E

Table B–4
Some Non-Si Units Used in This Book

Physical Quantity	Name of Unit	Symbol for Unit	Value in SI Units
Volume	liter	L	$1 \text{ dm}^3 = 10^{-3} \text{ m}^3$
Length	ångström	Å	10^{-10} m
Mass	atomic mass unit	u	$1.66054 \times 10^{-27} \text{ kg}$
Mass	metric ton (tonne)	t	10^3 kg
Energy	electron volt	eV	$1.60218 \times 10^{-19} \text{ J}$
Pressure	standard atmosphere	atm	$1.01325 \times 10^5 \text{ Pa}$
Time	minute	min	60 s
Time	hour	h	3600 s
Time	day	d	$8.64 \times 10^4 \text{ s}$

Most atomic sizes and chemical bond lengths fall in the range of one to several ångströms, and the use of either picometers or nanometers is slightly awkward. Similarly, the *atomic mass unit* is just right for expressing the masses of individual atoms (which range from 1 to 270 u), the *electron volt* is well-sized to express energy changes on the atomic scale, and the *metric ton* is convenient for talking about the industrial production of chemicals.

Finally, the *standard atmosphere* (abbreviated "atm") is used in this book as a unit of pressure. It is not a simple power of ten times the pascal, which is the natural SI unit of pressure, but is defined as 1.01325×10^5 Pa. This definition makes the standard atmosphere quite close to the observed sea-level atmospheric pressure. It is used because most chemistry is carried out at pressures near atmospheric pressure, for which the pascal is too small to be convenient. A slightly different unit of pressure called the *bar* (defined as exactly 10^5 Pa) has been recommended to replace the standard atmosphere but is not used here.

Many non-Si units continue in common use in the United States. Most are now defined in terms of SI units. Table B–5 lists a few familiar units and their current official definitions in terms of SI units. Note that the number 3.78541 and the others like it in Table B–5 have an infinite number of significant figures because they are exact numbers (as explained in Appendix A–3) and not measurements.

Temperature Scales

The two most important units of temperature in the United States are the Celsius degree (°C) and the Fahrenheit degree (°F). The first is a derived SI unit that is ex-

Table B–5
Some Familiar United Defined in Terms of SI Units

Physical Quantity	Name of Unit	Symbol for Unit	Value in SI Units
Volume	U.S. gallon	gall (US)	3.78541 dm^3 exactly
Length	inch	in.	$2.54 \times 10^{-2} \text{ m}$ exactly
Mass	pound avoirdupois	lb	0.45359237 kg exactly
Energy	calorie	cal	4.184 J exactly

actly equal to the kelvin in magnitude; the second is a non-SI unit equal to 5/9 of a kelvin:

$$1°C = 1\ K \qquad \text{and} \qquad 1°F = 5/9\ K \quad \text{\footnotesize (for size of unit only)}$$

Measurements of temperature, unlike those of most other physical quantities, require a point of reference as well as a unit. This requirement leads to scales of temperature, of which three appear in this book: the absolute scale, which uses the kelvin as its unit of temperature difference, the Celsius scale, which uses the °C, and the Fahrenheit scale, which uses the °F. On the Fahrenheit scale, the freezing point of water is 32°F and the boiling point of water under a pressure of 1 atm is 212°F. The interval between these two temperatures is 180°F. On the Celsius scale, these two points have values of 0°C and 100°C, respectively, making the interval 100°C, and on the absolute scale the two points have values of 273.15 K and 373.15 K, making the interval 100 K. Evidently, the kelvin and the Celsius degree are equal in size, as already stated. The two are larger than the Fahrenheit degree by the factor 180/100 because 100 Celsius degrees (or kelvins) are enough to span the same temperature interval as 180 Fahrenheit degrees (see Fig. 5–9). It follows that the Fahrenheit degree is 100/180 = 5/9 of the Celsius degree (or kelvin).

> • These equalities refer to the sizes of the different kinds of degrees and *not* to temperatures.

To convert a temperature given in degrees Fahrenheit to degrees Celsius, this fact is used in the formula

$$t_{°C} = \frac{5°C}{9°F}\,(t_{°F} - 32°F)$$

Solving the formula for t_F in terms of t_C gives

$$t_{°F} = \frac{9°F}{5°C}\,(t_{°C}) + 32°F$$

> • In this formula, as well as the next two formulas, units are included and displayed to show the way that they cancel out. The utility of this approach is discussed in the next section of this appendix.

The Kelvin scale uses absolute zero as its zero point. It is impossible for a temperature to reach or fall below absolute zero; temperatures on the Kelvin scale are always positive. Temperatures on the Kelvin scale are related to Celsius temperature by the formula

$$T_K = \frac{1\ K}{1°C}\,(t_{°C} + 273.15°C)$$

Note the practice of using an uppercase T for a temperature on an absolute scale and a lowercase t for temperatures that use other scales.

B-2   THE CONVERSION OF UNITS USING THE UNIT-FACTOR METHOD

If all the quantities in a calculation are inserted in units that are either base units of the International System or combinations of base units, the final result automatically comes out in SI units as well. This coherence is the great advantage of the SI. Nevertheless, it is essential to be able to convert at will among units because non-SI units are unavoidable and sometimes even desirable, both in scientific and non-scientific situations. The **unit-factor method** works to solve all problems of unit conversion. The basis of the method is in two facts: first, the measured value of a physical quantity is mathematically equivalent to a pure number multiplied by a unit (example: 64.3 g = 64.3 × grams); second, number × unit combinations may be manipulated under the ordinary rules of algebra. To see how these facts help in the conversion of units, suppose that a mass has been found to equal 64.3 g and must be expressed in kilograms (the SI base unit of mass) for use in a formula. To convert, start with the equation 1 kg = 1000 g and divide both sides by 1000 g:

$$\left(\frac{1 \text{ kg}}{1000 \text{ g}}\right) = 1$$

The starting value is now multiplied by this newly created **unit-factor,** so called because it equals 1 (unity):

$$64.3 \text{ g} \times \left(\frac{1 \text{ kg}}{1000 \text{ g}}\right) = 0.0643 \text{ kg}$$

Canceling the unit of grams in the starting value against the grams in the denominator of the unit-factor gives the unit of the final result. The unit-factor was set up with kilograms in the numerator and grams in the denominator just so that this cancellation would work. If this unit-factor had been inserted upside down, the desired cancellation would have been impossible. On the other hand, it would have been equally correct to write 1 g = 10^{-3} kg and carry out the same conversion as

$$64.3 \text{ g} \times \left(\frac{10^{-3} \text{ kg}}{1 \text{ g}}\right) = 0.0643 \text{ kg}$$

Attention to the cancellation of units in this method eliminates the common error of multiplying by a conversion factor instead of dividing, or vice versa.

Other conversions of units may involve more than just powers of ten, but they are equally easy to set up. For example, to express 16.4 inches in meters, use the fact that 1 inch = 0.0254 m to construct a unit-factor:

$$16.4 \text{ in.} \times \left(\frac{0.0254 \text{ m}}{1 \text{ in.}}\right) = 0.417 \text{ m}$$

More complicated combinations are possible as well. To convert from liter-atmospheres to joules (the SI unit of energy), multiply by two successive unit-factors to get units that are part of the SI:

$$1 \text{ L atm} \times \left(\frac{10^{-3} \text{ m}^3}{1 \text{ L}}\right) \times \left(\frac{101,325 \text{ Pa}}{1 \text{ atm}}\right) = 101.325 \text{ Pa m}^3$$

Manipulation of the SI units on the right side of the equation confirms that a "Pa m³" is truly equivalent to a joule:

$$101.325 \text{ Pa m}^3 = 101.325 \frac{\text{N}}{\text{m}^2} \text{ m}^3 = 101.325 \text{ N m} = 101.325 \text{ J}$$

The new unit-factor

$$\left(\frac{101.325 \text{ J}}{1 \text{ L atm}}\right)$$

or its reciprocal

$$\left(\frac{1 \text{ L atm}}{101.325 \text{ J}}\right)$$

can be used in any subsequent conversions between the liter-atmosphere and the joule.

It is very important to write out units explicitly when doing chemical calculations and to cancel out these units thoughtfully in intermediate steps. Cancellation provides an essential check that units have not been inadvertently changed without proper conversions and that an incorrect formula has not been used. If a result that is supposed to be a temperature comes out with units of $\text{m}^3 \text{ s}^{-2}$, a mistake has clearly been made!

PROBLEMS

Note: Answers to blue-numbered problems are given in Appendix F.

The International System of Units

1. Rewrite the following in scientific notation, using only the base units of Table B–1, without prefixes:
   (a) 65.2 nanograms
   (b) 88 picoseconds
   (c) 5.4 terawatts
   (d) 17 kilovolts

2. Rewrite the following in scientific notation, using only the base units of Table B–1, without prefixes:
   (a) 66 μK
   (b) 15.9 MJ
   (c) 0.13 mg
   (d) 62 GPa

3. A certain quantity is measured in an SI unit called the weber. A weber equals a volt-second (1 Wb = 1 V s). Use the information in Table B–2 to express the weber in terms of base SI units.

4. A certain quantity is measured in the SI system in joules per tesla (J T^{-1}). Use the information in Table B–2 and the fact that 1 T = 1 kg s^{-2} A^{-1} to show that 1 J T^{-1} equals 1 A m^2.

5. Arrange the following volumes in order from smallest to largest: 100 cm^3, 500 mL, 100 dm^3, 150 L, 1.5 gall (US).

6. Arrange the following in order from smallest to largest: 10^6 mm, 10^6 μm, 10^{-4} Mm, 10^{-2} km, 1000 in.

7. Express the following temperatures in °C:
   (a) 9001°F
   (b) 98.6°F (the normal body temperature of human beings)
   (c) 20°F above the boiling point of water at 1 atm pressure
   (d) −40°F

8. Express the following temperatures in °F:
   (a) 5000°C
   (b) 40.0°C
   (c) 212°C
   (d) −40°C

9. Express the temperatures given in problem 7 on the Kelvin scale.

10. Express the temperatures given in problem 8 on the Kelvin scale.

11. Construct an "inverted" version of Table B–5 in which the SI units are given in terms of the more familiar inch, pound, and so forth.

12. The following routines are sometimes recommended as easy-to-remember ways to convert a temperature on the Celsius scale to the Fahrenheit scale and vice versa:

$t_{°C} \longrightarrow t_{°F}$: add 40, multiply the result by 9/5, subtract 40;

$t_{°F} \longrightarrow t_{°C}$: add 40, multiply the result by 5/9, subtract 40.

Show algebraically why these steps give the right answers.

The Conversion of Units Using the Unit-Factor Method

13. Express the following in SI units, either base or derived:
    (a) 55.0 miles per hour (1 mile = 1609.344 m)
    (b) 1.51 g cm^{-3}
    (c) 1.6×10^{-19} C Å
    (d) 0.15 mol L^{-1}
    (e) 5.7×10^3 L atm min^{-1}

14. Express the following in SI units, either base or derived:
    (a) 67.3 atm
    (b) 1.0×10^4 V cm^{-1}
    (c) 7.4 Å hr^{-1}
    (d) 22.4 L mol^{-1}
    (e) 14.7 lb in^{-2}

15. Which of the following is not a possible unit of density: lb in^{-3}, μg L^{-1}, Pa m^2 s^{-2}, J m^{-1} s^{-2}.

16. Which of the following is not a possible unit of speed: mile h^{-1}, inch d^{-1}, N s kg^{-1}, kg J^{-1}.

17. The kilowatt-hour is a common unit in measurements of the consumption of electricity. What is the conversion factor between the kilowatt-hour and the joule? Express 15.3 kilowatt-hours in joules.

18. A car's rate of fuel consumption is often measured in miles per gallon.
    (a) Determine the conversion factor between miles per gallon and the SI unit of meters per cubic decimeter.
    (b) Express 30.0 miles per gallon in SI units.
    (c) Express the meter per cubic decimeter in base SI units (with no use of prefixes).

19. A certain V-8 engine has a displacement of 404 in^3. Express this volume in cubic centimeters (cm^3) and liters.

20. Light travels in a vacuum at a speed of 3.00×10^8 m s^{-1}.
    (a) Convert this speed to miles per second.
    (b) Express this speed in furlongs per fortnight, a little-used unit of speed. (A furlong, a distance used in horse racing, is 660 feet; a fortnight is exactly 2 weeks.)

C

Mathematics for General Chemistry

Mathematics is essential in chemistry. This appendix reviews some of the most important mathematical facts and techniques for general chemistry.

C–1 USING GRAPHS

In many situations in science, it is important to know how one quantity (measured or predicted) depends on a second quantity. The position of a traveling car depends on the time, for example, or the pressure of a gas depends on its volume at a given temperature. One very useful way to show such a relation is through a **graph,** in which one quantity is plotted against another.

The usual convention in drawing graphs is to use positions on the horizontal axis to represent values of the variable over which we have control, and to use positions on the vertical axis to represent values of the measured or calculated quantity. Suppose that the volume of a sample of gas is adjusted to the value 2.60 L. On a pressure gauge the pressure reads 3.22 atm. The volume in this case is the independent variable because it is controlled from outside the experiment, and the pressure is the dependent, or response, variable. To display the result of this experiment on a graph, locate a point by moving along the (horizontal) volume axis to 2.60 L and along the vertical (pressure) axis to 3.22 atm, and make a mark. Further measurements at different volumes would generate further points on the graph. After a series of such measurements, the points on the graph frequently lie along a recognizable curve, which can be drawn in (Fig. C–1). Because any experimental measurement involves some degree of uncertainty, some scatter in the points occurs, so no purpose is served in drawing in a sawtooth curve that passes precisely through every measured point. If a relation exists between the quantities measured, however, the points have a systematic trend and a curve can be drawn that represents this trend as the result of the experiment.

Such a graph can then be used to predict the results of future measurements. Suppose, for example, that a gas pressure of 5.00 atm is needed for an experiment (see Fig. C–1). Read across from 5.00 atm to the curve that shows pressure against volume, and then move down to the horizontal axis to read off the volume (1.67 L). If the pressure gauge reads 5.00 atm, the gas volume should be approximately 1.67 L. Alternatively, one could predict the pressure that results from a given gas volume.

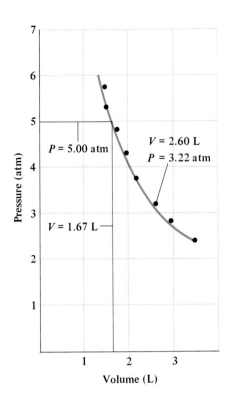

Figure C–1 The dependence of one experimental quantity on another is represented by linking the experimentally determined points with a smooth curve. Note the slight scatter of the points in this case.

The curves plotted on graphs can have many shapes. Of these, the simplest and most important is a straight line. A straight line is a graph of a relation such as

$$y = 4x + 7$$

or, more generally,

$$y = m\,x + b$$

where the dependent variable y is plotted along the vertical axis and the independent variable x along the horizontal axis (Fig. C–2). The quantity m is called the **slope** of the line that is plotted, and b is the **y-intercept.** The intercept is the point on the y-axis at which the line crosses that axis. This can be seen by setting x to zero and noting that y is then equal to b. The slope is a measure of the steepness of the line: the larger the value of m, the steeper the line. If the line goes up and to the right, the slope is positive; if it goes down and to the right, the slope is negative.

The slope of a straight line can be determined from the coordinates of two points on it. Suppose, for example, that when $x = 3$, $y = 5$, and that when $x = 4$, $y = 7$. These two points can be written in shorthand notation as (3,5) and (4,7). The slope of the line can then be defined as "the rise over the run": the change in the y-coordinate divided by the change in the x-coordinate:

• This is true if x increases from left to right on the horizontal axis and y increases from bottom to the top on the vertical axis, which is the accepted practice.

$$\text{slope} = m = \frac{\Delta y}{\Delta x} = \frac{7 - 5}{4 - 3} = \frac{2}{1} = 2$$

Figure C–2 A straight-line or linear relationship between two experimental quantities is a very desirable result because it is easy to graph and easy to represent mathematically. The equation of this straight line ($y = 2x - 1$) fits the general form $y = mx + b$. Its y-intercept is -1, and its slope is 2.

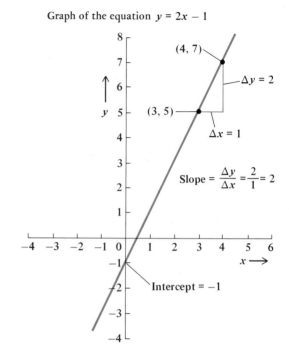

Graph of the equation $y = 2x - 1$

The symbol Δ (capital Greek "delta") indicates the change in a quantity—the final value minus the initial value. If the two quantities being graphed have units, the slope has units as well. If a graph of distance traveled (in meters) against time (in seconds) is a straight line, its slope has units of meters per second ($m\ s^{-1}$).

C–2   SOLUTION OF ALGEBRAIC EQUATIONS

Frequently in chemistry an algebraic equation must be solved for an unknown quantity, such as a concentration or a partial pressure in an equilibrium-constant expression. Let us represent the unknown quantity by the symbol x. If the equation is linear, the method of solution is straightforward:

$$5x + 9 = 0$$
$$5x = -9$$
$$x = -\frac{9}{5}$$

or, more generally, if $ax + b = 0$, then $x = -b/a$.

Nonlinear equations are of many kinds. One of the most common in chemistry is the quadratic equation. Quadratic equations in the unknown x can always be rearranged into the form

$$ax^2 + bx + c = 0$$

in which each of the constants (a, b, and c) may be positive, negative, or zero. Although quadratic equations may be solved in other ways, it is probably best for applications in chemistry simply to memorize the quadratic formula

$$x = \frac{-b \pm \sqrt{b^2 - 4ac}}{2a}$$

• The symbol $\pm$ is a compact way to represent two solutions to this equation (called "roots" of the equation). One corresponds to using the $+$ sign and the other to the $-$ sign.

As an example, suppose the equation

$$x = 3 + \frac{7}{x}$$

arises in a chemistry problem. Multiplying by x and rearranging the terms gives

$$x^2 - 3x - 7 = 0$$

This is plainly a quadratic equation having $a = 1$, $b = -3$, and $c = -7$. Inserting these values into the quadratic formula gives

$$x = \frac{-(-3) \pm \sqrt{(-3)^2 - 4(1)(-7)}}{2} = \frac{3 \pm \sqrt{9 + 28}}{2} = \frac{3 \pm \sqrt{37}}{2}$$

The two roots of the equation are

$$x = 4.5414 \quad \text{and} \quad x = -1.5414$$

The choice of the proper root in a chemistry problem can frequently be made on physical grounds. If x corresponds to a concentration, for example, the negative root is "unphysical" and can be discarded.

The solution of cubic or higher-order algebraic equations (or more complicated equations involving sines, cosines, logarithms, or exponentials) becomes more difficult, and approximate or numerical methods need to be used. As an illustration, consider the equation

• If the quantity $b^2 - 4ac$ under the radical sign is negative, *no* real solution exists because the square root of a negative number is imaginary. If this happens in the course of a chemical calculation, it is almost always an indication that an error has been made.

$$x^2\left(\frac{2.00 + x}{3.00 - x}\right) = 1.00 \times 10^{-6}$$

If x is assumed to be small relative to both 2.00 and 3.00, the simpler approximate equation

$$x^2\left(\frac{2.00}{3.00}\right) \approx 1.00 \times 10^{-6}$$

• The symbol $\approx$ means "is approximately equal to."

results and leads to the roots $x = \pm 0.00122$. We immediately confirm that 0.00122 *is* small compared to 2.00 (and 3.00) and that the approximation is a good one. When a quantity (x, in this case) is added to or subtracted from a larger quantity in a complicated equation, it is usually worth a try to solve the equation by simply neglecting the small quantity. Note that this tactic works only for addition and subtraction, never for multiplication or division.

Suppose now that the equation is changed to

• Whenever an equation is solved by approximation, a formal check should be run to confirm that the approximation is valid.

$$x^2\left(\frac{2.00 + x}{3.00 - x}\right) = 1.00 \times 10^{-2}$$

In this case, the equation that comes from neglecting x compared to 2.00 and 3.00 is

$$x^2\left(\frac{2.00}{3.00}\right) \approx 1.00 \times 10^{-2}$$

which leads to $x \approx \pm 0.122$. The number 0.122 is smaller than 2.00 and 3.00 but not so small that it can be ignored. In this case, more precise results can be obtained by **numerical approximation.** Substituting the approximate root $x = 0.122$ into the left side of the original equation gives:

$$x^2\left(\frac{2.00 + x}{3.00 - x}\right) = (0.122)^2\left(\frac{2.122}{2.878}\right) = 0.0110$$

The result 0.0110 exceeds the correct value, which is 0.0100. We therefore adjust x. Trying $x = 0.120$ gives

$$x^2\left(\frac{2.00 + x}{3.00 - x}\right) = (0.120)^2\left(\frac{2.12}{2.88}\right) = 0.0106$$

This is closer to 0.0100. If our trial had moved the value of the left side of the equation farther away from 0.0100 instead of closer, we would have adjusted x in the other direction. The correct value of the right side of the equation can generally be bracketed by two or three adjustments of x. Fine adjustment to any desired degree of precision can then be carried out. Such computations are readily mechanized on electronic calculators. In this case, continued trials reveal that x equals 0.117 to three significant figures:

$$(0.117)^2\left(\frac{2.117}{2.883}\right) = 0.0101$$

Note that another root of this equation is obtained in a similar fashion if the procedure starts with $x = -0.122$.

Numerical approximation can proceed without any assumptions about the value of x being small or large and can often be guided using mathematical knowledge. Consider, for example, the equation

$$x^2\left(\frac{2.00 + x}{3.00 - x}\right) = 10.0$$

There is no particular reason to believe that x should be small compared to 2.00 or 3.00. However, x cannot exceed 3, because that would force the left side of the equation to be negative. An initial guess of $x = 2.90$ gives 412 for the value of the left side of the equation. Clearly a much lower value is needed. When $x = 2.00$, the left side of the equation equals 16.0. Continued (but less drastic) adjustment of x quickly shows that when $x = 1.79$, the left side of the equation has a value of 10.04; this is a root of the equation to three significant figures.

C–3   POWERS AND LOGARITHMS

Raising a number to a power and the inverse operation of taking a logarithm are important in many chemical problems. Although the ready availability of electronic calculators makes the mechanical execution of these operations quite routine, it remains important to understand what is involved in these "special functions."

Powers

The mathematical expression 10^4 is read as "ten raised to the fourth power." Such an expression implies multiplying 10 by itself 4 times to give 10,000:

$$10^4 = 10 \times 10 \times 10 \times 10 = 10,000$$

Ten, or any other number, when raised to the power 0 gives 1:

$$10^0 = 1$$

Negative powers of 10 give numbers smaller than 1, and are equivalent to raising 10 to the corresponding *positive* power and then taking the reciprocal:

$$10^{-3} = 1/10^3 = 0.001$$

The idea of raising to a power can be extended to include powers that are not whole numbers. Raising to the power $1/2 = 0.5$, for example, is the same as taking the square root:

$$10^{0.5} = \sqrt{10} = 3.1623 \ldots$$

• It is obvious that when the power is not a positive integer, we can no longer talk sensibly about multiplying 10 by itself that number of times. The quantity 10^x can still be calculated, however.

Scientific calculators have a 10^x (or INV LOG) key that can be used for calculating powers of 10 in cases in which the power is not a whole number.

Numbers other than 10 can be raised to powers as well; these numbers are referred to as **bases.** Many calculators have a y^x key that lets any positive number y be raised to any power x. One of the most important bases in scientific problems is the transcendental number called e (2.7182818 . . .). The e^x (or INV LN) key on a calculator is used to raise e to any power x. The quantity e^x is called the **exponential** of x. It is also often represented as exp (x).

A key property of powers is that a base raised to the sum of two powers is equivalent to the product of the base raised separately to each of those powers. Thus,

$$10^{(21+6)} = 10^{21} \times 10^6 = 10^{27}$$

The same type of relationship holds for any base, including the base e.

Logarithms

Logarithms can occur frequently in chemistry problems. The logarithm of a number is the exponent to which some base has to be raised to obtain the number. The base is almost always either 10 or e. Thus,

$$B^a = n \qquad \text{and} \qquad \log_B n = a$$

where a is the logarithm, B is the base, and n is the number.

Common logarithms are base-10 logarithms—that is, they are powers to which 10 has to be raised in order to give the numbers. For example, $10^3 = 1000$, so $\log_{10} 1000 = 3$. We shall generally omit the explicit mention of the base 10 when showing common logarithms and write this equation as $\log 1000 = 3$. On an electronic calculator, a common logarithm is found by putting the number in the display and pressing the "log" key. The answer appears in the display. Only the logarithms of 1, 10, 100, 1000, and so forth are whole numbers. The logarithms of other numbers are decimal numbers. The decimal point in a logarithm divides it into two parts. To the left of the decimal point is the *characteristic*; to the right is the *mantissa*. Thus, the logarithm in the equation

$$\log (7.310 \times 10^3) = 3.8639$$

has a characteristic of 3 and a mantissa of 0.8639. As may be verified with a calculator, the base-10 logarithm of the much larger (but closely related) number 7.310×10^{23} is 23.8639. As this case illustrates, the characteristic is determined solely by the location of the decimal point in the number and not by its precision, so it is *not* included when counting significant figures. When the logarithm is taken of a measured quantity, the mantissa should be written with as many significant figures as are present in that quantity.

A logarithm is truly an exponent; as such, it follows the same rules of multiplication and division as other exponents. In multiplication and division

$$\log (n \times m) = \log n + \log m$$

$$\log \left(\frac{n}{m}\right) = \log n - \log m$$

Thus, the logarithm of 15 is

$$\log 5 + \log 3 = 0.69897 + 0.47712 = 1.17609$$

because 5 times 3 is 15. Also, the logarithm of 5/3 is

$$\log 5 - \log 3 = 0.69897 - 0.47712 = 0.22185$$

Furthermore,

$$\log n^m = m \log n$$

so the logarithm of 3^5 is

$$\log 3^5 = 5 \log 3 = 5 \times 0.47712 = 2.3856$$

There is no such thing as the logarithm of a negative number because there is no power to which 10 (or any other base) can be raised to give a negative number. This point can be understood as follows: the numbers between 0 and 1 have all of the negative numbers, ranging from $-\infty$ (negative infinity) up to 0, as their logarithms; the logarithm of 1 itself is 0; the numbers that exceed 1 have the entire set of positive numbers as their logarithms; thus, no real numbers remain to serve as the logarithms of negative numbers.

Another frequently used base for logarithms is the number e ($e = 2.7182818\ldots$). The logarithm to the base e is called the **natural logarithm** and is indicated by $\log_e$ or ln. The problem "find the logarithm to the base e of 8.23" could be written as "find ln 8.23." On a calculator, use the "ln" key to get natural logarithms. For example,

$$\ln 8.23 = 2.108 \qquad \text{and} \qquad \ln 0.0147 = -4.220$$

- "The characteristic comes free" in determining how many significant digits to write in the logarithm of a measured number. The four significant digits in 7.310×10^{23} give rise to the four digits after the decimal point (in the mantissa) in 23.8639.

- Logarithms can be negative; negative numbers do not have logarithms. Trying to compute the logarithm of a negative number on a calculator gives an error message.

Base-e logarithms are related to base-10 logarithms by the formula

$$\ln n = 2.3025851 \log n$$

As stated previously, calculating logarithms and calculating powers are inverse operations. Thus, to find the number for which 3.8639 is the common logarithm, simply calculate

$$10^{3.8639} = 7.310 \times 10^3$$

To find the number that has a natural logarithm of 2.108, calculate

$$e^{2.108} = 8.23$$

The number of significant digits in the answer should correspond to the number of digits in the *mantissa* of the logarithm.

C–4   RATIOS

One way of comparing two numbers is to form their **ratio.** A ratio is written by using a division sign (as in $7 \div 11$), by separating the two numbers with a colon (as in 7:11), by constructing a fraction (as in 7/11), or by computing a decimal equivalent (7/11 = 0.636363 . . .). Two ratios come from every pair of numbers because it matters which number divides into which. The other ratio involving 7 and 11 is 11:7, or 11/7, which equals 1.57142857 . . . . This ratio is the **reciprocal** of the first ratio. The product of a number and its reciprocal equals 1.

In chemistry, the ratios among sets of numbers are quite often more important than the numbers themselves. These ratios are routinely computed with electronic calculators, but calculators display their results as decimals. This creates a problem because ratios in the form of fractions are required for many purposes in chemistry. The easiest solution is to know and recognize the decimal equivalents of simple ratios such as the following:

1/8 = 0.125	1/7 = 0.142857 . . .	1/6 = 0.166 . . .	1/5 = 0.2
1/4 = 0.25	2/7 = 0.285714 . . .	1/3 = 0.333 . . .	3/8 = 0.375
2/5 = 0.4	3/7 = 0.428571 . . .	1/2 = 0.5	4/7 = 0.571428 . . .
2/3 = 0.666 . . .	3/5 = 0.6	5/8 = 0.625	5/7 = 0.714295 . . .
3/4 = 0.75	4/5 = 0.8	5/6 = 0.833 . . .	6/7 = 0.857142
7/8 = 0.875			

Suppose that a calculation gives 2.375 as a desired ratio. The preceding list allows this to be recognized as equal to $2\frac{3}{8}$ It is then easy to express the ratio as a fraction: $2\frac{3}{8} = 19/8$.

A common application of ratios in chemistry is to convey the relative amounts of different substances making up some mixture. To do this, the amount of each component of the mixture is in turn divided by the total amount of all of the components. When the various amounts are measured as masses, the result is a **mass fraction** for each component.

$$\text{mass fraction of component 1} = \frac{m_1}{\Sigma m_i}$$

where the sigma in the denominator denotes a sum over the components i.

The units in the numerator and denominator of this fraction are the same, so a mass fraction has no units. A **mass percentage** is a mass fraction put on a per-hundred basis. This is done by multiplying the ratio of the mass of the part to the mass of the whole by 100%:

$$\text{mass percentage} = (\text{mass fraction}) \times 100\%$$

When one component is present in a small or very small relative amount, its contribution to the whole is sometimes expressed on a **parts per million (ppm)** or **parts per billion (ppb)** basis:

$$\text{ppmm} = \text{mass fraction} \times 10^6$$
$$\text{ppbm} = \text{mass fraction} \times 10^9$$

where the final m's attached to the abbreviations mean that these are parts per million and parts per billion by mass. Use of parts per million and parts per billion avoids inconveniently small numbers.

If the amounts in a mixture are compared on the basis of their relative volumes, then the **volume fraction** of each component is computed

$$\text{volume fraction of component 1} = \frac{V_1}{\Sigma V_i}$$

where the V's are the volumes of the components prior to mixing. The volume percent, parts per millions by volume (ppmv), and parts per billion by volume (ppbv) of a component are then obtained by multiplying its volume fraction by 100%, 10^6, 10^9, respectively.

C–5   FORMULAS FOR THE AREA AND VOLUME OF SIMPLE SHAPES

Geometry plays an important role in understanding the three-dimensional structures of molecules, as well as in other fields of chemistry. The following *mensuration formulas* are given for reference.

1. Area of a circle having radius r:

$$A = \pi r^2$$

2. Area of a triangle having base b and altitude a:

$$A = \frac{ba}{2}$$

3. Volume of a cube having edge e:

$$V = e^3$$

4. Volume of a rectangular prism having edges a, b, and c:

$$V = abc$$

5. Volume of a sphere having radius r:

$$V = \frac{4}{3} \pi r^3$$

PROBLEMS

Note: Answers to blue-numbered problems are given in Appendix F.

Using Graphs

1. A graph of distance traveled against time elapsed is found in one case to be a straight line. After an elapsed time of 1.5 hours, the distance traveled was 75 miles. After an elapsed time of 3.0 hours, the distance traveled was 150 miles. Calculate the slope of the graph of distance against time, and give its units.

2. The pressure of a sample of gas in a rigid container is measured at several different temperatures, and it is found that a graph of pressure against temperature is a straight line. At 20.0°C, the pressure is 4.30 atm, and at 100.0°C, the pressure is 5.47 atm. Calculate the slope of the graph of pressure against temperature, and give its units.

3. Rewrite each of the following linear equations in the form $y = mx + b$, and give the slope and intercept of the corresponding graph. Then draw the graph.
   (a) $y = 4x - 7$
   (b) $7x - 2y = 5$
   (c) $3y + 6x - 4 = 0$

4. Rewrite each of the following linear equations in the form $y = mx + b$, and give the slope and intercept of the corresponding graph. Then draw the graph.
   (a) $y = -2x - 8$
   (b) $-3x + 4y = 7$
   (c) $7y - 16x + 53 = 0$

5. Graph the equation

$$y = 2x^3 - 3x^2 + 6x - 5$$

   from $x = -2$ to $x = +2$. Is the graph a straight line?

6. Graph the equation

$$y = \frac{8 - 10x - 3x^2}{2 - 3x}$$

   from $x = -3$ to $x = +3$. Is the graph a straight line?

Solution of Algebraic Equations

7. Solve the following linear equations for x:
   (a) $7x + 5 = 0$
   (b) $-4x + 3 = 0$
   (c) $-3x = -2$

8. Solve the following linear equations for x:
   (a) $6 - 8x = 0$
   (b) $-2x - 5 = 0$
   (c) $-4x = -8$

9. Solve the following quadratic equations for x:
   (a) $4x^2 + 7x - 5 = 0$
   (b) $2x^2 = -3 - 6x$
   (c) $2x + \frac{3}{x} = 6$

10. Solve the following quadratic equations for x:
    (a) $6x^2 + 15x + 2 = 0$
    (b) $4x = 5x^2 - 3$
    (c) $\frac{1}{2 - x} + 3x = 4$

11. Solve each of the following equations for x, using the approximation of small x, or numerical approximation, as appropriate:
    (a) $x(2.00 + x)^2 = 2.6 \times 10^{-6}$
    (b) $x(3.00 - 7x)(2.00 + 2x) = 0.230$
    (c) $2x^3 + 3x^2 + 12x = -16$

12. Solve each of the following equations for x, using the approximation of small x, or numerical approximation, as appropriate:
    (a) $x(2.00 + x)(3.00 - x)(5.00 + 2x) = 1.58 \times 10^{-15}$
    (b) $\dfrac{x(3.00 + x)(1.00 - x)}{2.00 - x} = 0.122$
    (c) $12x^3 - 4x^2 + 35x = 10$

Powers and Logarithms

13. Calculate each of the following expressions, giving your answers to the proper number of significant figures:
    (a) $\log (3.56 \times 10^4)$
    (b) $e^{-15.69}$
    (c) $10^{8.41}$
    (d) $\ln (6.893 \times 10^{-22})$

14. Calculate each of the following expressions, giving your answers to the proper number of significant figures:
    (a) $10^{-16.528}$
    (b) $\ln (4.30 \times 10^{13})$
    (c) $\exp (14.21)$
    (d) $\log (4.983 \times 10^{-11})$

15. What number has a common logarithm of 0.4793?

16. What number has a natural logarithm of -15.824?

17. Determine the common logarithm of 3.00×10^{121}. It is quite likely that your calculator does *not* give a correct answer. Explain why.

18. Compute the value of $10^{-107.8}$. It is quite likely that your calculator does *not* give the correct answer. Explain why.

19. The common logarithm of 5.64 is 0.751. Without using a calculator, determine the common logarithm of 5.64×10^7 and of 5.64×10^{-3}.

20. The common logarithm of 2.68 is 0.428. Without using a calculator, determine the common logarithm of 2.68×10^{192} and of 2.68×10^{-289}.

21. Use a calculator to solve the equation

$$\log \ln x = -x$$

for x. Give x to four digits.

22. Use a calculator to find a number that is equal to the reciprocal of its own natural logarithm. Report the answer to four digits.

Ratios

23. A mixture contains 24.60 g of water, 17.2 g of NaCl, and 0.0020 g of mercury. The volumes of these three components are 24.67 mL, 7.944 mL, and 1.47×10^{-4} mL, respectively.
    (a) Compute the mass fraction of all three components in the mixture.
    (b) Compute the volume fraction of all three components in the mixture.
    (c) Compute the parts per million by volume (ppmv) of mercury in the mixture.
    (d) Compute the parts per billion by pass (ppbm) of mercury in the mixture.

24. A gaseous mixture contains 13.6 parts per million by volume of xenon. If the total of the volumes of all the components in the mixture equals 716 L, compute the volume of xenon in the mixture.

Standard Chemical Thermodynamic Properties

This table lists standard enthalpies of formation ΔH_f° standard third-law entropies S°, and standard free energies of formation ΔG_f° for a variety of substances all at 25°C (298.15 K) and 1 atm. The table proceeds from the left to the right side of the periodic table. Binary compounds are listed under the element that occurs to the left in the periodic table, except that binary oxides and hydrides are listed with the other element. Thus, KCl is listed with potassium and its compounds, but ClO_2 is listed with chlorine and its compounds.

Note that the *solution-phase* entropies are not absolute entropies but are measured relative to the arbitrary standard $S^\circ(H^+(aq)) = 0$. It is for this reason that some of them are negative.

Most of the thermodynamic data in these tables were taken from the *NBS Tables of Chemical Thermodynamic Properties* (1982), changed where necessary from a standard pressure of 0.1 MPa to 1 atm. The data for organic compounds C_nH_m ($n > 2$) were taken from the *Handbook of Chemistry and Physics* (1981).

Substance	ΔH_f° (25°C) kJ mol^{-1}	S° (25°C) J K^{-1} mol^{-1}	ΔG_f° (25°C) kJ mol^{-1}
H(g)	217.96	114.60	203.26
H$_2$(g)	0	130.57	0
H$^+$(aq)	0	0	0
H$_3$O$^+$(aq)	−285.83	69.91	−237.18
Li(s)	0	29.12	0
Li(g)	159.37	138.66	126.69
Li$^+$(aq)	−278.49	13.4	−293.31
LiH(s)	−90.54	20.01	−68.37
Li$_2$O(s)	−597.94	37.57	−561.20
LiF(s)	−615.97	35.65	−587.73
LiCl(s)	−408.61	59.33	−384.39
LiBr(s)	−351.21	74.27	−342.00
LiI(s)	−270.41	86.78	−270.29
Na(s)	0	51.21	0
Na(g)	107.32	153.60	76.79
Na$^+$(aq)	−240.12	59.0	−261.90
Na$_2$O(s)	−414.22	75.06	−375.48

(continued)

Substance	ΔH_f° (25°C) kJ mol^{-1}	S° (25°C) J K^{-1} mol^{-1}	ΔG_f° (25°C) kJ mol^{-1}
NaOH(s)	−425.61	64.46	−379.53
NaF(s)	−573.65	51.46	−543.51
NaCl(s)	−411.15	72.13	−384.15
NaBr(s)	−361.06	86.82	−348.98
NaI(s)	−287.78	98.53	−286.06
NaNO$_3$(s)	−467.85	116.52	−367.07
Na$_2$S(s)	−364.8	83.7	−349.8
Na$_2$SO$_4$(s)	−1387.08	149.58	−1270.23
NaHSO$_4$(s)	−1125.5	113.0	−992.9
Na$_2$CO$_3$(s)	−1130.68	134.98	−1044.49
NaHCO$_3$(s)	−950.81	101.7	−851.1
K(s)	0	64.18	0
K(g)	89.24	160.23	60.62
K$^+$(aq)	−252.38	102.5	−283.27
KO$_2$(s)	−284.93	116.7	−239.4
K$_2$O$_2$(s)	−494.1	102.1	−425.1
KOH(s)	−424.76	78.9	−379.11
KF(s)	−567.27	66.57	−537.77
KCl(s)	−436.75	82.59	−409.16
KClO$_3$(s)	−397.73	143.1	−296.25
KBr(s)	−393.80	95.90	−380.66
KI(s)	−327.90	106.32	−324.89
KMnO$_4$(s)	−837.2	171.71	−737.7
K$_2$CrO$_4$(s)	−1403.7	200.12	−1295.8
K$_2$Cr$_2$O$_7$(s)	−2061.5	291.2	−1881.9
Rb(s)	0	76.78	0
Rb(g)	80.88	169.98	53.09
Rb$^+$(aq)	−251.17	121.50	−283.98
RbCl(s)	−435.35	95.90	−407.82
RbBr(s)	−394.59	109.96	−381.79
RbI(s)	−333.80	118.41	−328.86
Cs(s)	0	85.23	0
Cs(g)	76.06	175.49	49.15
Cs$^+$(aq)	−258.28	133.05	−292.02
CsF(s)	−553.5	92.80	−525.5
CsCl(s)	−443.04	101.17	−414.55
CsBr(s)	−405.81	113.05	−391.41
CsI(s)	−346.60	123.05	−340.58

	Substance	ΔH_f° (25°C) kJ mol^{-1}	S° (25°C) J K^{-1} mol^{-1}	ΔG_f° (25°C) kJ mol^{-1}
II	Be(s)	0	9.50	0
	Be(g)	324.3	136.16	286.6
	BeO(s)	−609.6	14.14	−580.3
	Mg(s)	0	32.68	0
	Mg(g)	147.70	148.54	113.13

(continued)

Substance	ΔH_f° (25°C) kJ mol^{-1}	S° (25°C) J K^{-1} mol^{-1}	ΔG_f° (25°C) kJ mol^{-1}
$Mg^{2+}(aq)$	-466.85	-138.1	-454.8
$MgO(s)$	-601.70	26.94	-569.45
$MgCl_2(s)$	-641.32	89.62	-591.82
$MgSO_4(s)$	-1284.9	91.6	-1170.7
$Ca(s)$	0	41.42	0
$Ca(g)$	178.2	154.77	144.33
$Ca^{2+}(aq)$	-542.83	-53.1	-553.58
$CaH_2(s)$	-186.2	42	-147.2
$CaO(s)$	-635.09	39.75	-604.05
$CaS(s)$	-482.4	56.5	-477.4
$Ca(OH)_2(s)$	-986.09	83.39	-898.56
$CaF_2(s)$	-1219.6	68.87	-1167.3
$CaCl_2(s)$	-795.8	104.6	-748.1
$CaBr_2(s)$	-682.8	130	-663.6
$CaI_2(s)$	-533.5	142	-528.9
$Ca(NO_3)_2(s)$	-938.39	193.3	-743.20
$CaC_2(s)$	-59.8	69.96	-64.9
$CaCO_3(s, calcite)$	-1206.92	92.9	-1128.84
$CaCO_3 (s, aragonite)$	-1207.13	88.7	-1127.80
$CaSO_4(s)$	-1434.11	106.9	-1321.86
$CaSiO_3(s)$	-1634.94	81.92	-1549.66
$CaMg(CO_3)_2 (s, dolomite)$	-2326.3	155.18	-2163.4
$Sr(s)$	0	52.3	0
$Sr(g)$	164.4	164.51	130.9
$Sr^{2+}(aq)$	-545.80	-32.6	-559.48
$SrCl_2(s)$	-828.9	114.85	-781.1
$SrCO_3(s)$	-1220.0	97.1	-1140.1
$Ba(s)$	0	62.8	0
$Ba(g)$	180	170.24	146
$Ba^{2+}(aq)$	-537.64	9.6	-560.77
$BaCl_2(s)$	-858.6	123.68	-810.4
$BaCO_3(s)$	-1216.3	112.1	-1137.6
$BaSO_4(s)$	-1473.2	132.2	-1362.3
$Sc(s)$	0	34.64	0
$Sc(g)$	377.8	174.68	336.06
$Sc^{3+}(aq)$	-614.2	-255	-586.6
$Ti(s)$	0	30.63	0
$Ti(g)$	469.9	180.19	425.1
$TiO_2(s, rutile)$	-944.7	50.33	-889.5
$TiCl_4(\ell)$	-804.2	252.3	-737.2
$TiCl_4(g)$	-763.2	354.8	-726.8
$Cr(s)$	0	23.77	0
$Cr(g)$	396.6	174.39	351.8

(continued)

Substance	ΔH_f° (25°C) kJ mol^{-1}	S° (25°C) J K^{-1} mol^{-1}	ΔG_f° (25°C) kJ mol^{-1}
$Cr_2O_3(s)$	−1139.7	81.2	−1058.1
$CrO_4^{2-}(aq)$	−881.15	50.21	−727.75
$Cr_2O_7^{2-}(aq)$	−1490.3	261.9	−1301.1
$W(s)$	0	32.64	0
$W(g)$	849.4	173.84	807.1
$WO_2(s)$	−589.69	50.54	−533.92
$WO_3(s)$	−842.87	75.90	−764.08
$Mn(s)$	0	32.01	0
$Mn(g)$	280.7	238.5	173.59
$Mn^{2+}(aq)$	−220.75	−73.6	−228.1
$MnO(s)$	−385.22	59.71	−362.92
$MnO_2(s)$	−520.03	53.05	−465.17
$MnO_4^-(aq)$	−541.4	191.2	−447.2
$Fe(s)$	0	27.28	0
$Fe(g)$	416.3	180.38	370.7
$Fe^{2+}(aq)$	−89.1	−137.7	−78.9
$Fe^{3+}(aq)$	−48.5	−315.9	−4.7
$Fe_{0.947}O(s, \text{wüstite})$	−266.27	57.49	−245.12
$Fe_2O_3(s, \text{hematite})$	−824.2	87.40	−742.2
$Fe_3O_4(s, \text{magnetite})$	−1118.4	146.4	−1015.5
$Fe(OH)_3(s)$	−569.0	88	−486.6
$FeS(s)$	−100.0	60.29	−100.4
$FeCO_3(s)$	−740.57	93.1	−666.72
$Fe(CN)_6^{3-}(aq)$	561.9	270.3	729.4
$Fe(CN)_6^{4-}(aq)$	455.6	95.0	695.1
$Co(s)$	0	30.04	0
$Co(g)$	424.7	179.41	380.3
$Co^{2+}(aq)$	−58.2	−113	−54.4
$Co^{3+}(aq)$	92	−305	134
$CoO(s)$	−237.94	52.97	−214.22
$CoCl_2(s)$	−312.5	109.16	−269.8
$Ni(s)$	0	29.87	0
$Ni(g)$	429.7	182.08	384.5
$Ni^{2+}(aq)$	−54.0	−128.9	−45.6
$NiO(s)$	−239.7	37.99	−211.7
$Pt(s)$	0	41.63	0
$Pt(g)$	565.3	192.30	520.5
$PtCl_6^{2-}(aq)$	−668.2	219.7	−482.7
$Cu(s)$	0	33.15	0
$Cu(g)$	338.32	166.27	298.61
$Cu^+(aq)$	71.67	40.6	49.98
$Cu^{2+}(aq)$	64.77	−99.6	65.49
$CuO(s)$	−157.3	42.63	−129.7

(continued)

Substance	ΔH_f° (25°C) kJ mol^{-1}	S° (25°C) J K^{-1} mol^{-1}	ΔG_f° (25°C) kJ mol^{-1}
$Cu_2O(s)$	-168.6	93.14	-146.0
$CuCl(s)$	-137.2	86.2	-119.88
$CuCl_2(s)$	-220.1	108.07	-175.7
$CuSO_4(s)$	-771.36	109	-661.9
$Cu(NH_3)_4^{2+}(aq)$	-348.5	273.6	-111.07
$Ag(s)$	0	42.55	0
$Ag(g)$	284.55	172.89	245.68
$Ag^+(aq)$	105.58	72.68	77.11
$AgCl(s)$	-127.07	96.2	-109.81
$AgNO_3(s)$	-124.39	140.92	-33.48
$Ag(NH_3)_2^+(aq)$	-111.29	245.2	-17.12
$Au(s)$	0	47.40	0
$Au(g)$	366.1	180.39	326.3
$Zn(s)$	0	41.63	0
$Zn(g)$	130.73	160.87	95.18
$Zn^{2+}(aq)$	-153.89	-112.1	-147.06
$ZnO(s)$	-348.28	43.64	-318.32
$ZnS(s, sphalerite)$	-205.98	57.7	-201.29
$ZnCl_2(s)$	-415.05	111.46	-369.43
$ZnSO_4(s)$	-982.8	110.5	-871.5
$Zn(NH_3)_4^{2+}(aq)$	-533.5	301	-301.9
$Hg(\ell)$	0	76.02	0
$Hg(g)$	61.32	174.85	31.85
$HgO(s)$	-90.83	70.29	-58.56
$HgCl_2(s)$	-224.3	146.0	-178.6
$Hg_2Cl_2(s)$	-265.22	192.5	-210.78
III $\quad B(s)$	0	5.86	0
$B(g)$	562.7	153.34	518.8
$B_2H_6(g)$	35.6	232.00	86.6
$B_5H_9(g)$	73.2	275.81	174.9
$B_2O_3(s)$	-1272.77	53.97	-1193.70
$H_3BO_3(s)$	-1094.33	88.83	-969.02
$BF_3(g)$	-1137.00	254.01	-1120.35
$BF_4^-(aq)$	-1574.9	180	-1486.9
$BCl_3(g)$	-403.76	289.99	-388.74
$BBr_3(g)$	-205.64	324.13	-232.47
$Al(s)$	0	28.33	0
$Al(g)$	326.4	164.43	285.7
$Al^{3+}(aq)$	-531	-321.7	-485
$Al_2O_3(s)$	-1675.7	50.92	-1582.3
$AlCl_3(s)$	-704.2	110.67	-628.8
$Ga(s)$	0	40.88	0
$Ga(g)$	277.0	168.95	238.9

(continued)

Substance	ΔH_f° (25°C) kJ mol^{-1}	S° (25°C) J K^{-1} mol^{-1}	ΔG_f° (25°C) kJ mol^{-1}
Tl(s)	0	64.18	0
Tl(g)	182.21	180.85	147.44
IV C(s, graphite)	0	5.74	0
C(s, diamond)	1.895	2.377	2.900
C(g)	716.682	157.99	671.29
CH$_4$(g)	−74.81	186.15	−50.75
C$_2$H$_2$(g)	226.73	200.83	209.20
C$_2$H$_4$(g)	52.26	219.45	68.12
C$_2$H$_6$(g)	−84.68	229.49	−32.89
C$_3$H$_8$(g)	−103.85	269.91	−23.49
n-C$_4$H$_{10}$(g)	−124.73	310.03	−15.71
C$_4$H$_{10}$(g, isobutane)	−131.60	294.64	−17.97
n-C$_5$H$_{12}$(g)	−146.44	348.40	−8.20
C$_6$H$_6$(g)	82.93	269.2	129.66
C$_6$H$_6$(ℓ)	49.03	172.8	124.50
CO(g)	−110.52	197.56	−137.15
CO$_2$(g)	−393.51	213.63	−394.36
CO$_2$(aq)	−413.80	117.6	−385.08
CS$_2$(ℓ)	89.70	151.34	65.27
CS$_2$(g)	117.36	237.73	67.15
H$_2$CO$_3$(aq)	−699.65	187.4	−623.08
HCO$_3^-$(aq)	−691.99	91.2	−586.77
CO$_3^{2-}$(aq)	−677.14	−56.9	−527.81
HCOOH(ℓ)	−424.72	128.95	−361.42
HCOOH(aq)	−425.43	163	−372.3
HCOO$^-$(aq)	−425.55	92	−351.0
CH$_2$O(g)	−108.57	218.66	−102.55
CH$_3$OH(ℓ)	−238.66	126.8	−166.35
CH$_3$OH(g)	−200.66	239.70	−162.01
CH$_3$OH(aq)	−245.93	133.1	−175.31
H$_2$C$_2$O$_4$(s)	−827.2	120	−697.9
HC$_2$O$_4^-$(aq)	−818.4	149.4	−698.34
C$_2$O$_4^{2-}$(aq)	−825.1	45.6	−673.9
CH$_3$COOH(ℓ)	−484.5	159.8	−390.0
CH$_3$COOH(g)	−432.25	282.4	−374.1
CH$_3$COOH$^-$(aq)	−485.76	178.7	−396.46
CH$_3$COO$^-$(aq)	−486.01	86.6	−369.31
CH$_3$CHO(ℓ)	−192.30	160.2	−128.12
C$_2$H$_5$OH(ℓ)	−277.69	160.7	−174.89
C$_2$H$_5$OH(g)	−235.10	282.59	−168.57
C$_2$H$_5$OH(aq)	−288.3	148.5	−181.64
CH$_3$OCH$_3$(g)	−184.05	266.27	−112.67
CF$_4$(g)	−925	261.50	−879

(continued)

Substance	ΔH_f° (25°C) kJ mol^{-1}	S° (25°C) J K^{-1} mol^{-1}	ΔG_f° (25°C) kJ mol^{-1}
$CCl_4(\ell)$	-135.44	216.40	-65.28
$CCl_4(g)$	-102.9	309.74	-60.62
$CHCl_3(g)$	-103.14	295.60	-70.37
$COCl_2(g)$	-218.8	283.53	-204.6
$CH_2Cl_2(g)$	-92.47	270.12	-65.90
$CH_3Cl(g)$	-80.83	234.47	-57.40
$CBr_4(s)$	79	357.94	67
$CH_3I(\ell)$	-15.5	163.2	13.4
$HCN(g)$	135.1	201.67	124.7
$HCN(aq)$	107.1	124.7	119.7
$CN^-(aq)$	150.6	94.1	172.4
$CH_3NH_2(g)$	-22.97	243.30	32.09
$CO(NH_2)_2(s)$	-333.51	104.49	-197.44
$Si(s)$	0	18.83	0
$Si(g)$	455.6	167.86	411.3
$SiC(s)$	-65.3	16.61	-62.8
$SiO_2(s, quartz)$	-910.94	41.84	-856.67
$SiO_2(s, cristobalite)$	-909.48	42.68	-855.43
$Ge(s)$	0	31.09	0
$Ge(g)$	376.6	335.9	167.79
$Sn(s, white)$	0	51.55	0
$Sn(s, gray)$	-2.09	44.14	0.13
$Sn(g)$	302.1	168.38	267.3
$SnO(s)$	-285.8	56.5	-256.9
$SnO_2(s)$	-580.7	52.3	-519.6
$Sn(OH)_2(s)$	-561.1	155	-491.7
$Pb(s)$	0	64.81	0
$Pb(g)$	195.0	161.9	175.26
$Pb^{2+}(aq)$	-1.7	10.5	-24.43
$PbO(s, yellow)$	-217.32	68.70	-187.91
$PbO(s, red)$	-218.99	66.5	-188.95
$PbO_2(s)$	-277.4	68.6	-217.36
$PbS(s)$	-100.4	91.2	-98.7
$PbI_2(s)$	-175.48	174.85	-173.64
$PbSO_4(s)$	-919.94	148.57	-813.21
V $N_2(g)$	0	191.50	0
$N(g)$	472.70	153.19	455.58
$NH_3(g)$	-46.11	192.34	-16.48
$NH_3(aq)$	-80.29	111.3	-26.50
$NH_4^+(aq)$	-132.51	113.4	-79.31
$N_2H_4(\ell)$	50.63	121.21	149.24
$N_2H_4(aq)$	34.31	138	128.1
$NO(g)$	90.25	210.65	86.55
$NO_2(g)$	33.18	239.95	51.29

(continued)

Substance	ΔH_f° (25°C) kJ mol^{-1}	S° (25°C) J K^{-1} mol^{-1}	ΔG_f° (25°C) kJ mol^{-1}
NO$_2^-$(aq)	−104.6	123.0	−32.2
NO$_3^-$(aq)	−205.0	146.4	−108.74
N$_2$O(g)	82.05	219.74	104.18
N$_2$O$_4$(g)	9.16	304.18	97.82
N$_2$O$_5$(s)	−43.1	178.2	113.8
HNO$_2$(g)	−79.5	254.0	−46.0
HNO$_3$(ℓ)	−174.10	155.49	−80.76
NH$_4$NO$_3$(s)	−365.56	151.08	−184.02
NH$_4$Cl(s)	−314.43	94.6	−202.97
(NH$_4$)$_2$SO$_4$(s)	−1180.85	220.1	−901.90
P(s, white)	0	41.09	0
P(s, red)	−17.6	22.80	−12.1
P(g)	314.64	163.08	278.28
P$_2$(g)	144.3	218.02	103.7
P$_4$(g)	58.91	279.87	24.47
PH$_3$(g)	5.4	210.12	13.4
H$_3$PO$_4$(s)	−1279.0	110.50	−1119.2
H$_3$PO$_4$(aq)	−1288.34	158.2	−1142.54
H$_2$PO$_4^-$(aq)	−1296.29	90.4	−1130.28
HPO$_4^{2-}$(aq)	−1292.14	−33.5	−1089.15
PO$_4^{3-}$(aq)	−1277.4	−222	−1018.7
PCl$_3$(g)	−287.0	311.67	−267.8
PCl$_5$(g)	−374.9	364.47	−305.0
As(s, gray)	0	35.1	0
As(g)	302.5	174.10	261.0
As$_2$(g)	222.2	239.3	171.9
As$_4$(g)	143.9	314	92.4
AsH$_3$(g)	66.44	222.67	68.91
As$_4$O$_6$(s)	−1313.94	214.2	−1152.53
Sb(s)	0	45.69	0
Sb(g)	262.3	180.16	222.1
Bi(s)	0	56.74	0
Bi(g)	207.1	186.90	168.2

	Substance	ΔH_f° (25°C) kJ mol^{-1}	S° (25°C) J K^{-1} mol^{-1}	ΔG_f° (25°C) kJ mol^{-1}
VI	O$_2$(g)	0	205.03	0
	O(g)	249.17	160.95	231.76
	O$_3$(g)	142.7	238.82	163.2
	OH$^-$(aq)	−229.99	−10.75	−157.24
	H$_2$O(ℓ)	−285.83	69.91	−237.18
	H$_2$O(g)	−241.82	188.72	−228.59
	H$_2$O$_2$(ℓ)	−187.78	109.6	−120.42
	H$_2$O$_2$(aq)	−191.17	143.9	−134.03
	S(s, rhombic)	0	31.80	0
	S(s, monoclinic)	0.30	32.6	0.096
	S(g)	278.80	167.71	238.28

(continued)

Substance	ΔH_f° (25°C) kJ mol^{-1}	S° (25°C) J K^{-1} mol^{-1}	ΔG_f° (25°C) kJ mol^{-1}
$S_8(g)$	102.30	430.87	49.66
$S^{2-}(aq)$	33.1	−14.6	85.8
$H_2S(g)$	−20.63	205.68	−33.56
$H_2S(aq)$	−39.7	121	−27.83
$HS^-(aq)$	−17.6	62.8	12.08
$SO(g)$	6.26	221.84	−19.87
$SO_2(g)$	−296.83	248.11	−300.19
$SO_3(g)$	−395.72	256.65	−371.08
$H_2SO_3(aq)$	−608.81	232.2	−537.81
$HSO_3^-(aq)$	−626.22	139.7	−527.73
$SO_3^{2-}(aq)$	−635.5	−29	−486.5
$H_2SO_4(\ell)$	−813.99	156.90	−690.10
$HSO_4^-(aq)$	−887.34	131.8	−755.91
$SO_4^{2-}(aq)$	−909.27	20.1	−744.53
$SF_6(g)$	−1209	291.71	−1105.4
$Se(s, black)$	0	42.44	0
$Se(g)$	227.07	176.61	187.06
VII $F_2(g)$	0	202.67	0
$F(g)$	78.99	158.64	61.94
$F^-(aq)$	−332.63	−13.8	−278.79
$HF(g)$	−271.1	173.67	−273.2
$HF(aq)$	−320.08	88.7	−296.82
$XeF_4(s)$	−261.5	—	—
$Cl_2(g)$	0	222.96	0
$Cl(g)$	121.68	165.09	105.71
$Cl^-(aq)$	−167.16	56.5	−131.23
$HCl(g)$	−92.31	186.80	−95.30
$ClO^-(aq)$	−107.1	42	−36.8
$ClO_2(g)$	102.5	256.73	120.5
$ClO_2^-(aq)$	−66.5	101.3	17.2
$ClO_3^-(aq)$	−103.97	162.3	−7.95
$ClO_4^-(aq)$	−129.33	182.0	−8.52
$Cl_2O(g)$	80.3	266.10	97.9
$HClO(aq)$	−120.9	142	−79.9
$ClF_3(g)$	−163.2	281.50	−123.0
$Br_2(\ell)$	0	152.23	0
$Br_2(g)$	30.91	245.35	3.14
$Br_2(aq)$	−2.59	130.5	3.93
$Br(g)$	111.88	174.91	82.41
$Br^-(aq)$	−121.55	82.4	−103.96
$HBr(g)$	−36.40	198.59	−53.43
$BrO_3^-(aq)$	−67.07	161.71	18.60
$I_2(s)$	0	116.14	0
$I_2(g)$	62.44	260.58	19.36

(continued)

Substance	ΔH_f° (25°C) kJ mol^{-1}	S° (25°C) J K^{-1} mol^{-1}	ΔG_f° (25°C) kJ mol^{-1}
$I_2(aq)$	22.6	137.2	16.40
$I(g)$	106.84	180.68	70.28
$I^-(aq)$	−55.19	111.3	−51.57
$I_3^-(aq)$	−51.5	239.3	−51.4
$HI(g)$	26.48	206.48	1.72
$ICl(g)$	17.78	247.44	−5.44
$IBr(g)$	40.84	258.66	3.71

	Substance	ΔH_f° (25°C) kJ mol^{-1}	S° (25°C) J K^{-1} mol^{-1}	ΔG_f° (25°C) kJ mol^{-1}
VIII	$He(g)$	0	126.04	0
	$Ne(g)$	0	146.22	0
	$Ar(g)$	0	154.73	0
	$Kr(g)$	0	163.97	0
	$Xe(g)$	0	169.57	0

Standard Half-Cell Reduction Potentials at 25°C

Half-Reaction	$\mathcal{E}°$ (volts)
$F_2(g) + 2e^- \longrightarrow 2\,F^-$	2.87
$H_2O_2 + 2\,H^+ + 2\,e^- \longrightarrow 2\,H_2O$	1.776
$PbO_2(s) + SO_4^{2-} + 4\,H^+ + 2\,e^- \longrightarrow PbSO_4(s) + 2\,H_2O$	1.685
$Au^+ + e^- \longrightarrow Au(s)$	1.68
$MnO_4^- + 4\,H^+ + 3\,e^- \longrightarrow MnO_2(s) + 2\,H_2O$	1.679
$HClO_2 + 2\,H^+ + 2\,e^- \longrightarrow HClO + H_2O$	1.64
$HClO + H^+ + e^- \longrightarrow \frac{1}{2}\,Cl_2(g) + H_2O$	1.63
$Ce^{4+} + e^- \longrightarrow Ce^{3+}$ (1 M HNO_3 solution)	1.61
$2\,NO(g) + 2\,H^+ + 2\,e^- \longrightarrow N_2O(g) + H_2O$	1.59
$BrO_3^- + 6\,H^+ + 5\,e^- \longrightarrow \frac{1}{2}\,Br_2(\ell) + 3\,H_2O$	1.52
$Mn^{3+} + e^- \longrightarrow Mn^{2+}$	1.51
$MnO_4^- + 8\,H^+ + 5\,e^- \longrightarrow Mn^{2+} + 4\,H_2O$	1.491
$ClO_3^- + 6\,H^+ + 5\,e^- \longrightarrow \frac{1}{2}\,Cl_2(g) + 3\,H_2O$	1.47
$PbO_2(s) + 4\,H^+ + 2\,e^- \longrightarrow Pb^{2+} + 2\,H_2O$	1.46
$Au^{3+} + 3\,e^- \longrightarrow Au(s)$	1.42
$Cl_2(g) + 2\,e^- \longrightarrow 2\,Cl^-$	1.3583
$Cr_2O_7^{2-} + 14\,H^+ + 6\,e^- \longrightarrow 2\,Cr^{3+} + 7\,H_2O$	1.33
$O_3(g) + H_2O + 2\,e^- \longrightarrow O_2 + 2\,OH^-$	1.24
$O_2(g) + 4\,H^+ + 4\,e^- \longrightarrow 2\,H_2O$	1.229
$MnO_2(s) + 4\,H^+ + 2\,e^- \longrightarrow Mn^{2+} + 2\,H_2O$	1.208
$ClO_4^- + 2\,H^+ + 2\,e^- \longrightarrow ClO_3^- + H_2O$	1.19
$Br_2(\ell) + 2\,e^- \longrightarrow 2\,Br^-$	1.065
$NO_3^- + 4\,H^+ + 3\,e^- \longrightarrow NO(g) + 2\,H_2O$	0.96
$2\,Hg^{2+} + 2\,e^- \longrightarrow Hg_2^{2+}$	0.905
$Ag^+ + e^- \longrightarrow Ag(s)$	0.7996
$Hg_2^{2+} + 2\,e^- \longrightarrow 2\,Hg(\ell)$	0.7961
$Fe^{3+} + e^- \longrightarrow Fe^{2+}$	0.770
$O_2(g) + 2\,H^+ + 2\,e^- \longrightarrow H_2O_2$	0.682
$BrO_3^- + 3\,H_2O + 6\,e^- \longrightarrow Br^- + 6\,OH^-$	0.61
$MnO_4^- + 2\,H_2O + 3\,e^- \longrightarrow MnO_2(s) + 4\,OH^-$	0.588
$I_2(s) + 2\,e^- \longrightarrow 2\,I^-$	0.535
$Cu^+ + e^- \longrightarrow Cu(s)$	0.522

(continued)

Half-Reaction	$\mathscr{E}°$ (volts)
$O_2(g) + 2\ H_2O + 4\ e^- \longrightarrow 4\ OH^-$	0.401
$Cu^{2+} + 2\ e^- \longrightarrow Cu(s)$	0.3402
$PbO_2(s) + H_2O + 2\ e^- \longrightarrow PbO(s) + 2\ OH^-$	0.28
$Hg_2Cl_2(s) + 2\ e^- \longrightarrow 2\ Hg(\ell) + 2\ Cl^-$	0.2682
$AgCl(s) + e^- \longrightarrow Ag(s) + Cl^-$	0.2223
$SO_4^{2-} + 4\ H^+ + 2\ e^- \longrightarrow H_2SO_3 + H_2O$	0.20
$Cu^{2+} + e^- \longrightarrow Cu^+$	0.158
$S_4O_6^{2-} + 2\ e^- \longrightarrow 2\ S_2O_3^{2-}$	0.0895
$NO_3^- + H_2O + 2\ e^- \longrightarrow NO_2^- + 2\ OH^-$	0.01
$2\ H^+ + 2\ e^- \longrightarrow H_2(g)$	0.000 exactly
$Pb^{2+} + 2\ e^- \longrightarrow Pb(s)$	−0.1263
$Sn^{2+} + 2\ e^- \longrightarrow Sn(s)$	−0.1364
$Ni^{2+} + 2\ e^- \longrightarrow Ni(s)$	−0.23
$Co^{2+} + 2\ e^- \longrightarrow Co(s)$	−0.28
$PbSO_4 + 2\ e^- \longrightarrow Pb(s) + SO_4^{2-}$	−0.356
$Mn(OH)_3(s) + e^- \longrightarrow Mn(OH)_2(s) + OH^-$	−0.40
$Cd^{2+} + 2\ e^- \longrightarrow Cd(s)$	−0.4026
$Fe^{2+} + 2\ e^- \longrightarrow Fe(s)$	−0.409
$Cr^{3+} + e^- \longrightarrow Cr^{2+}$	−0.424
$Fe(OH)_3(s) + e^- \longrightarrow Fe(OH)_2(s) + OH^-$	−0.56
$PbO(s) + H_2O + 2\ e^- \longrightarrow Pb(s) + 2\ OH^-$	−0.576
$2\ SO_3^{2-} + 3\ H_2O + 4\ e^- \longrightarrow S_2O_3^{2-} + 6\ OH^-$	−0.58
$Ni(OH)_2(s) + 2\ e^- \longrightarrow Ni(s) + 2\ OH^-$	−0.66
$Co(OH)_2(s) + 2\ e^- \longrightarrow Co(s) + 2\ OH^-$	−0.73
$Cr^{3+} + 3\ e^- \longrightarrow Cr(s)$	−0.74
$Zn^{2+} + 2\ e^- \longrightarrow Zn(s)$	−0.7628
$2\ H_2O + 2\ e^- \longrightarrow H_2(g) + 2\ OH^-$	−0.8277
$Cr^{2+} + 2\ e^- \longrightarrow Cr(s)$	−0.905
$SO_4^{2-} + H_2O + 2\ e^- \longrightarrow SO_3^{2-} + 2\ OH^-$	−0.92
$Mn^{2+} + 2\ e^- \longrightarrow Mn(s)$	−1.029
$Mn(OH)_2(s) + 2\ e^- \longrightarrow Mn(s) + 2\ OH^-$	−1.47
$Al^{3+} + 3\ e^- \longrightarrow Al(s)$	−1.706
$Sc^{3+} + 3\ e^- \longrightarrow Sc(s)$	−2.08
$Ce^{3+} + 3\ e^- \longrightarrow Ce(s)$	−2.335
$La^{3+} + 3\ e^- \longrightarrow La(s)$	−2.37
$Mg^{2+} + 2\ e^- \longrightarrow Mg(s)$	−2.375
$Mg(OH)_2(s) + 2\ e^- \longrightarrow Mg(s) + 2\ OH^-$	−2.69
$Na^+ + e^- \longrightarrow Na(s)$	−2.7109
$Ca^{2+} + 2\ e^- \longrightarrow Ca(s)$	−2.76
$Ba^{2+} + 2\ e^- \longrightarrow Ba(s)$	−2.90
$K^+ + e^- \longrightarrow K(s)$	−2.925
$Li^+ + e^- \longrightarrow Li(s)$	−3.045

All voltages are standard reduction potentials (relative to the standard hydrogen electrode) at 25°C and 1 atm pressure. All species are in aqueous solution unless otherwise indicated.

Solutions to Selected Odd-Numbered Problems

Chapter 1

1. Table salt: heterogeneous mixture; wood: heterogeneous mixture; mercury: homogeneous substance (element); air: homogeneous mixture; water: homogeneous substance (compound); seawater: homogeneous mixture; sodium chloride: homogeneous substance (compound); mayonnaise: heterogeneous mixture (mayonnaise appears homogeneous to the naked eye, but magnification reveals it to be water droplets suspended in oil).

3. To a chemist, "pure" means "containing only a single substance." To the student, "pure" means "as found in nature" or "not adulterated or made unnatural by commercial processing."

5. Salt at the bottom of a cup of water.

7. Vodka is not a substance, but a (homogeneous) mixture of substances.

9. 0.056 g   11. It has escaped into the air, largely as carbon dioxide.

13. 0.41 g   15. 16.9 g

17. (a) 1/1   (b) 8/3   (c) 4/1   (d) 8/1   (e) 13/5

19. $2.005/1.504 = 4/3$. A multiple of SiN.

21. (a) 1.0, 1.3, and 0.50 g   (b) 1.0, 0.75, and 2.0 g

23. HO (or any multiple such as H_2O_2)   25. 2.0 L N_2O, 3.0 L O_2

27. Mass is conserved; volume is not.

29. 6, 6, 12, 5; $^{32}S^{2-}$, 16; ^{27}Al, 13; 208, 82

31. (d) Manhattan Island is 10,000 to 100,000 times larger than 10 cm.

33. 95 p, 146 n, 95 e 35. 1.54:1, 95   37. 12.0004

39. 12.35 g. This answer is the weighted average; it reflects the fact that green marbles are more numerous than red ones.

41. 11.01   43. (a) 10   (b) 94   45. 2.0×10^3 g

47. (a) 283.89   (b) 115.36   (c) 164.09   (d) 197.13   (e) 132.13

49. The total count of 2.52×10^9 atoms of gold has a mass of only 8.3×10^{-13} g, far too small to detect with a balance.

51. (a) 1.031×10^{23} molecules of SF_2 (b) 0.8440 g   (c) 5.157×10^{22} molecules, 0.8440 g   (d) The S_2F_4 has half as many molecules as SF_2 in any mass of substance, but the same proportion of sulfur to fluorine.

53. 1.09×10^{26} atoms   55. $2.1072997 \times 10^{-22}$ g

57. (a) 0.5375 mol   (b) 3.237×10^{23} molecules

59. (a) 17.5 g (b) 17.5 lb   **61.** 99 kegs

63. 26 ng = 2.6×10^{-8} g   **65.** 11.1 g; 0.0840 mol

67. 0.0911 g cm^{-3} **69.** 2540 cm^3 = 2.54 L

71. 2.95×10^{-23} cm^3 per atom

Chapter 2

1. Both have empirical formula CH_2.   **3.** 996.84 g mol^{-1}, $C_3H_3N_2O$

5. 9.4×10^{23} molecules of H_2O **7.** (a) 504 mg

9. 4.794% H (by mass)   **11.** N_4H_6, H_2O, LiH, $C_{12}H_{26}$

13. 79.12%   **15.** Pt: 47.06%; F: 36.67%; Cl: 8.553%; O: 7.720%

17. 64.11%, 75.84%   **19.** 0.225%   **21.** $Zn_3P_2O_8$

23. $SiCl_2$ **25.** Fe_3Si_7 **27.** BaN, Ba_3N_2

29. 452 atoms   **31.** Na_2SO_4 **33.** C_7H_{10}

35. $COCl_2$ **37.** $C_{30}H_{50}$

39. (a) $N_2 + O_2 \longrightarrow 2\,NO$

    (b) $2\,N_2 + O_2 \longrightarrow 2\,N_2O$

    (c) $K_2SO_3 + 2\,HCl \longrightarrow 2\,KCl + H_2O + SO_2$

    (d) $2\,NH_3 + 3\,O_2 + 2\,CH_4 \longrightarrow 2\,HCN + 6\,H_2O$

    (e) $CaC_2 + 3\,CO \longrightarrow 4\,C + CaCO_3$

41. (a) $3\,H_2 + N_2 \longrightarrow 2\,NH_3$

    (b) $2\,K + O_2 \longrightarrow K_2O_2$

    (c) $PbO_2 + Pb + 2\,H_2SO_4 \longrightarrow 2\,PbSO_4 + 2\,H_2O$

    (d) $2\,BF_3 + 3\,H_2O \longrightarrow B_2O_3 + 6\,HF$

    (e) $2\,KClO_3 \longrightarrow 2\,KCl + 3\,O_2$

    (f) $CH_3COOH + 2\,O_2 \longrightarrow 2\,CO_2 + 2\,H_2O$

    (g) $2\,K_2O_2 + 2\,H_2O \longrightarrow 4\,KOH + O_2$

    (h) $3\,PCl_5 + 5\,AsF_3 \longrightarrow 3\,PF_5 + 5\,AsCl_3$

43. (a) $2\,C_6H_6 + 15\,O_2 \longrightarrow 12\,CO_2 + 6\,H_2O$

    (b) $2\,F_2 + H_2O \longrightarrow OF_2 + 2\,HF$

    (c) $CaC_2 + 2\,H_2O \longrightarrow Ca(OH)_2 + C_2H_2$

    (d) $4\,NH_3 + 7\,O_2 \longrightarrow 4\,NO_2 + 6\,H_2O$

    (e) $2\,Al + 6\,NaOH \longrightarrow 3\,H_2 + 2\,Na_3AlO_3$

45. (a) $Fe_2O_3 + 3\,H_2 \longrightarrow 2\,Fe + 3\,H_2O$

    (b) $3\,Fe_2O_3 + H_2 \longrightarrow 2\,Fe_3O_4 + H_2O$

47. (a) 56.0 g

    (b) 780.3 g

    (c) 239.9 g

49. 9.88 mol

51. (a) 0.2507 g   (b) 0.6262 g   (c) 20.42 g   (d) 6.117 g

53. 1.18×10^3 g   **55.** 418 g KCl, 199 g Cl_2

57. 7.1 L   **59.** 7.83 g of $K_2Zn_3[Fe(CN)_6]$

61. 12   **63.** 0.134 g of SiO_2

65. 250 cheeseburgers; 100 hamburger patties

67. 75 tables   **69.** NH_3 is limiting; 53.9 g of H_2O

71. (a) $BaCl_2$ (b) 7.285 g $BaSO_4$

73. 1.126 g H_2O **75.** 13.3 L CO_2

77. Theoretical yield: 303.0 g, % yield = 83.93%

79. (a) 2609 cookies   (b) 209 cookies

81. 3×10^{-9}%   **83.** 745 metric tons of NaOH

85. 2 kmol CO_2; 4 kmol O; 2.41×10^{27} atoms O

Chapter 3

1. PH_3, HCl, SiH_4, H_2S **3.** K_2Po, $SrCl_2$, BaS

5. Melting points are highest in Group IV and fall off on both sides. Only sulfur is out of line.

7. (a) Alkaline earth metals (Group II elements)   (b) Strontium (Sr)

9. (a) Examples are hydrogen, helium, nitrogen, oxygen, fluorine, neon, chlorine, argon, krypton, and xenon.
(b) Yes. These elements tend to lie in the upper right part of the periodic table.

11. Melting point 1250°C (obs. 1541°C), boiling point 2386°C (obs. 2831°C), density 3.02 g cm^{-3} (obs. 2.99)

13. SbH_3, HBr, SnH_4, H_2Se

15. (a) 10, 5 $: \overset{\cdot}{\underset{\cdot}{P}} \cdot$ (b) 46, 7, $: \overset{\cdot \cdot}{\underset{\cdot \cdot}{I}} \cdot$ (c) 10, 3, $\cdot \overset{\cdot}{Al} \cdot$ (d) 36, 2 $\overset{\cdot}{Sr} \cdot$

17. $: \overset{\cdot \cdot}{\underset{\cdot \cdot}{Kr}} :$ $: \overset{\cdot \cdot}{\underset{\cdot \cdot}{Te}} :^{2-}$ $: \overset{\cdot}{\underset{\cdot \cdot}{S}} \cdot^{+}$ $Mg \cdot^{+}$

19. Kr: 36 electrons (8 valence, 28 core); Te^{2-}: 54 electrons (8 valence, 46 core); S^{+}: 15 electrons (5 valence, 10 core); Mg^{+}: 11 electrons (1 valence, 10 core)

21. (a) Strontium selenide, SrSe   (b) Rubidium iodide, RbI   (c) Sodium sulfide, Na_2S (d) Gallium chloride, $GaCl_3$ (e) Francium oxide, Fr_2O

23. (a) Aluminum oxide   (b) Rubidium selenide   (c) Ammonium sulfide
(d) Calcium nitrate   (e) Cesium sulfate   (f) Potassium hydrogen carbonate

25. (a) AgCN   (b) $Ca(OCl)_2$ (c) K_2CrO_4 (d) Ga_2O_3 (e) KO_2
(f) $Ba(HCO_3)_2$

27. Na_3PO_4, sodium phosphate   **29.** Potassium dihydrogen phosphate

31. (a) One electron missing   (b) One extra electron   (c) Two extra electrons

33. (a)

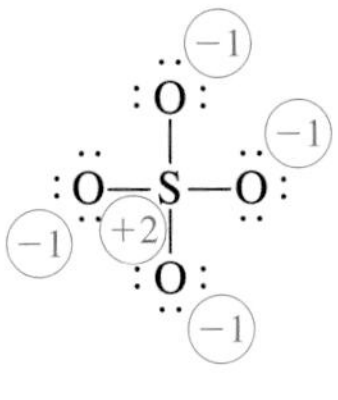

(b)

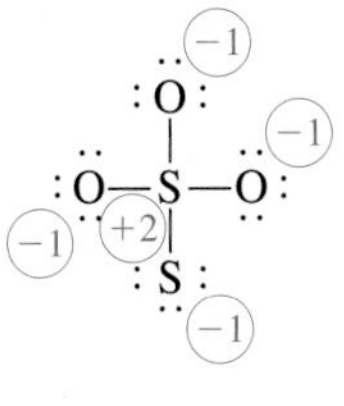

(c)

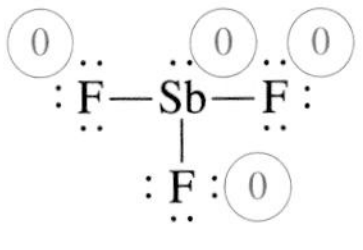

(d)

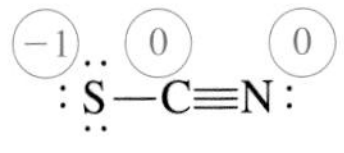

35.

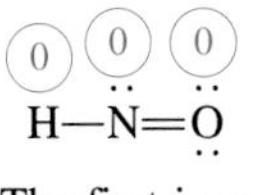

H—N=O   and   H—O=N

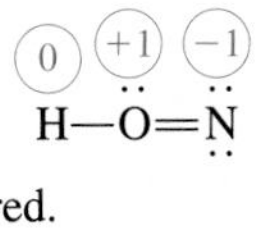

The first is preferred.

37. (a) Group IV, CO_2 (b) Group VII, Cl_2O_7 (c) Group V, NO_3^-
    (d) Group VI, HSO_4^-

39. 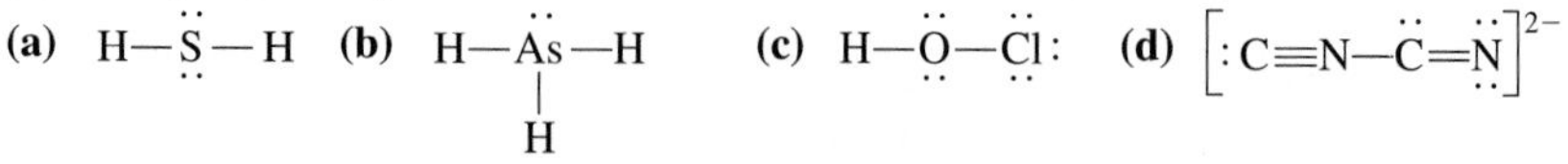 (a) H—S̈—H (b) H—Äs—H   (c) H—Ö—C̈l: (d) $\left[:C \equiv N - C = \ddot{N} \right]^{2-}$
               |
               H

41. 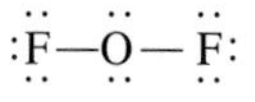 :F̈—Ö—F̈:   **43.**

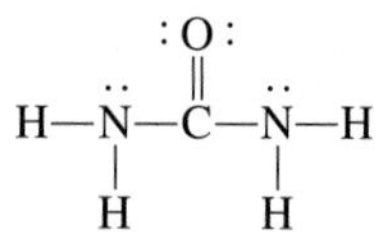

                            :O:
                            ‖
                 H—N̈—C—N̈—H
                    | |
                    H   H

45. 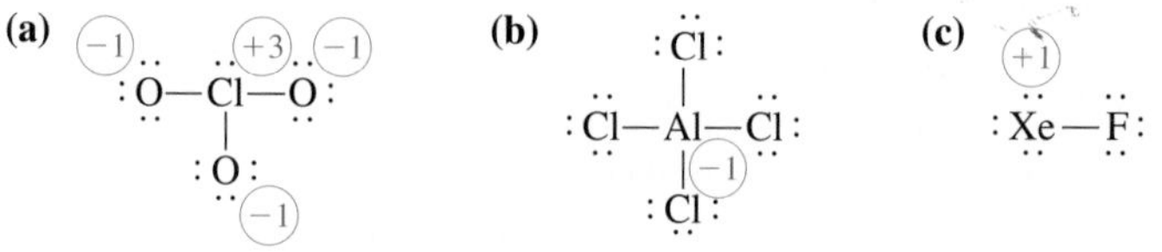 (a) -1, $+3$, -1 charges on :Ö—C̈l—Ö:
                        |
                       :Ö: (-1)

(b)   :C̈l:
        |
    :C̈l—Al—C̈l:
        ‖ (-1)
        :C̈l:

(c) $(+1)$:Xe—F̈:

47.  (0) (0) (-1) on Ö=N—Ö: ⟷ (-1) (0) (0) on :Ö—N=O
    Range of O-to-N bonds is 1.18 to 1.43×10^{-10} m

49. 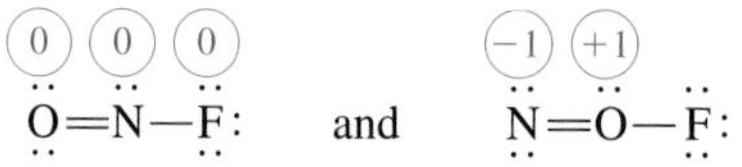 (0) (0) (0) on Ö=N—F̈:   and (-1) $(+1)$ on N=Ö—F̈:

51. 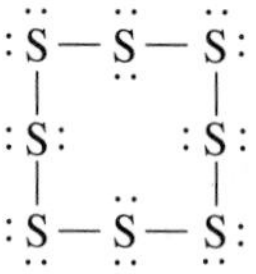 :S—S̈—S:
      | |
      :S̈:   :S̈:
      | |
      :S—S̈—S:

53. 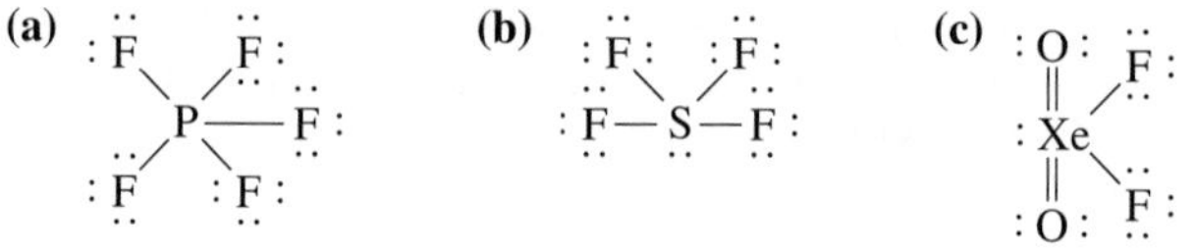 (a) :F̈, :F̈, P—F̈:, :F̈, :F̈ around P
    (b) :F̈, :F̈, F̈—S—F̈ around S
    (c) :O, F̈, Xe, F̈, :O (O double bonded) around Xe

55. (a) 4   (b) 3   (c) 6   (d) 4   (e) 5

57. (a) tetrahedral   (b) trigonal planar   (c) octahedral   (d) pyramidal
    (e) distorted T-shape

59. (a) $SN = 6$, square planar   (b) $SN = 4$, bent, angle $< 109.5°$ (c) $SN = 4$,
    pyramidal, angle $< 109.5°$ (d) $SN = 2$, linear

61. (a) SO_3 (b) NF_3 (c) NO_2^- (d) CO_3^{2-}

63. Only (d) and (e) are polar.

65. No, because VSEPR theory predicts a steric number of 3 and a bent molecule in both cases.

67. (a) linear   (b) the N end

69. (a) N_2O_5 (b) HI   (c) BN   (d) PH_3

71. (a) Hydrogen telluride   (b) Phosphorus trichloride   (c) Ammonia
    (d) Tetrasulfur tetranitride

73. $P_4O_{10} + 6\ PCl_5 \longrightarrow 10\ POCl_3$

75. (a) Iron(III) oxide   (b) Titanium(IV) bromide   (c) Tungsten(VI) oxide
    (d) Lead(IV) chloride   (e) Manganese(III) fluoride

77. (a) SnF_2 (b) Re_2O_7 (c) CoF_3 (d) WCl_5 (e) $Cu(NO_3)_2$

79. $Re_2S_7 + H_2 \longrightarrow 2\ ReS_3 + H_2S$

Chapter 4

1. Electrolytes give solutions that conduct electricity. Strong electrolytes dissociate essentially completely into ions in aqueous solution; weak electrolytes are slightly dissociated; non-electrolytes dissolve without dissociating.

3. $Mg(ClO_4)_2(s) \longrightarrow Mg^{2+}(aq) + 2\,ClO_4^-(aq)$

5. Deoxyribose and ethylene glycol should be freely soluble in water; potassium perchlorate is only slightly soluble in water.

7. Sodium chloride, potassium bromide, sodium bromide, and potassium chloride

9. Add ethanol to the aqueous solution to force the rubidium chloride out of solution.

11. (a) $Ag^+(aq) + Cl^-(aq) \longrightarrow AgCl(s)$
    (b) $K_2CO_3(s) + 2\,H^+(aq) \longrightarrow 2\,K^+(aq) + CO_2(g) + H_2O(\ell)$
    (c) $2\,Cs(s) + 2\,H_2O(\ell) \longrightarrow 2\,Cs^+(aq) + 2\,OH^-(aq) + H_2(g)$
    (d) $2\,MnO_4^-(aq) + 16\,H^+(aq) + 10\,Cl^-(aq) \longrightarrow 5\,Cl_2(g) + 2\,Mn^{2+}(aq) + 8\,H_2O(\ell)$

13. $NaBr(aq) + AgNO_3(aq) \longrightarrow AgBr(s) + NaNO_3(aq)$
    $Ag^+(aq) + Br^-(aq) \longrightarrow AgBr(s)$

15. The second solution contains sodium and sulfate ions, and the first solution does not. Also, in the first solution, the concentrations of the barium ions and sulfate ions are equal. In the second solution, they probably differ.

17. (a) $Zn(NO_3)_2(aq) + K_2S(aq) \longrightarrow ZnS(s) + 2\,KNO_3(aq)$
    $Zn^{2+}(aq) + S^{2-}(aq) \longrightarrow ZnS(s)$
    (b) $2\,AgClO_4(aq) + CaCl_2(aq) \longrightarrow 2\,AgCl(s) + Ca(ClO_4)_2(aq)$
    $Ag^+(aq) + Cl^-(aq) \longrightarrow AgCl(s)$
    (c) $3\,NaOH(aq) + Fe(NO_3)_3(aq) \longrightarrow Fe(OH)_3(s) + 3\,NaNO_3(aq)$
    $Fe^{3+}(aq) + 3\,OH^-(aq) \longrightarrow Fe(OH)_3(s)$
    (d) $3\,Ba(CH_3COO)_2(aq) + 2\,Na_3PO_4(aq) \longrightarrow Ba_3(PO_4)_2(s) + 6\,NaCH_3COO(aq)$
    $3\,Ba^{2+}(aq) + 2\,PO_4^{3-}(aq) \longrightarrow Ba_3(PO_4)_2(s)$

19. (a) $Be^{2+}(aq) + 2\,CH_3COO^-(aq) \longrightarrow Be(CH_3COO)_2(s)$
    (b) $Ba^{2+}(aq) + SO_4^{2-}(aq) \longrightarrow BaSO_4(s)$
    (c) $Ca^{2+}(aq) + 2\,OH^-(aq) \longrightarrow Ca(OH)_2(s)$ (d) N.R.

21. NaCl and KNO_3 are both soluble salts.

23. A soluble sulfate salt such as Na_2SO_4

25. (a) Hydrosulfuric acid   (b) Periodic acid
    (c) Carbonic acid   (d) Hydrobromic acid

27. (a) I_2O_5, diiodine pentaoxide, acid anhydride
    (b) BaO, barium oxide, base anhydride
    (c) CrO_3, chromium(VI) oxide, acid anhydride
    (d) N_2O, dinitrogen oxide, acid anhydride

29. (a) base anhydride, $Mg(OH)_2$, magnesium hydroxide
    (b) acid anhydride, HOCl, hypochlorous acid
    (c) acid anhydride, H_2SO_4, sulfuric acid
    (d) base anhydride, CsOH, cesium hydroxide

31. (a) $2\,HNO_3(aq) + K_2CO_3(s) \longrightarrow 2\,K^+(aq) + 2\,NO_3^-(aq) + H_2O(\ell) + CO_2(g)$
    (b) $2\,HBr(aq) + Zn(s) \longrightarrow Zn^{2+}(aq) + 2\,Br^-(aq) + H_2(g)$
    (c) $H_2SO_4(aq) + Zn(OH)_2(s) \longrightarrow Zn^{2+}(aq) + SO_4^{2-}(aq) + 2\,H_2O(\ell)$

33. (a) $HClO_3(aq) + KOH(aq) \longrightarrow K^+(aq) + ClO_3^-(aq) + H_2O(\ell)$
    (b) $N_2O_5(g) + 2\,NaOH(aq) \longrightarrow 2\,Na^+(aq) + 2\,NO_3^-(aq) + H_2O(\ell)$
    (c) $(NH_4)_2SO_4(aq) + Ba(OH)_2(aq) \longrightarrow 2\,NH_3(g) + BaSO_4(s) + 2\,H_2O(\ell)$

35. **(a)** $2\ HBr(aq) + Ca(OH)_2(s) \longrightarrow CaBr_2(aq) + 2\ H_2O(\ell)$

(b) $NH_3(aq) + H_2SO_4(aq) \longrightarrow NH_4^+(aq) + HSO_4^-(aq)$

(c) $LiOH(s) + HNO_3(aq) \longrightarrow LiNO_3(aq) + H_2O(\ell)$

37. **(a)** $Ca(OH)_2(aq) + 2\ HF(aq) \longrightarrow CaF_2(s) + 2\ H_2O(\ell)$
hydrofluoric acid, calcium hydroxide, calcium fluoride

(b) $2\ RbOH(aq) + H_2SO_4(aq) \longrightarrow Rb_2SO_4(aq) + 2\ H_2O(\ell)$
sulfuric acid, rubidium hydroxide, rubidium sulfate

(c) $Zn(OH)_2(s) + 2\ HNO_3(aq) \longrightarrow Zn(NO_3)_2(aq) + 2\ H_2O(\ell)$
nitric acid, zinc hydroxide, zinc nitrate

(d) $KOH(aq) + CH_3COOH(aq) \longrightarrow KCH_3COO(aq) + H_2O(\ell)$
acetic acid, potassium hydroxide, potassium acetate

39. **(a)** Phosphorus trifluoride, phosphorous acid, hydrofluoric acid (hydrogen fluoride)

(b) $PF_3 + 3\ H_2O \longrightarrow H_3PO_3 + 3\ HF$

41. Sodium sulfide

43. $2\ Al(s) + 6\ H^+(aq) \longrightarrow 2\ Al^{3+}(aq) + 3\ H_2(g)$

45. **(a)** $\overset{0\ +1-2}{CH_2O}$ **(b)** $\overset{+1+4-2}{H_2CO_3}$

(c) $\overset{+1\ -1}{RbH}$ **(d)** $\overset{+5\ -2}{N_2O_5}$

47. $\overset{+3}{As_2O_3}(aq) + 2\ \overset{0}{I_2}(s) + 2\ H_2O(\ell) \longrightarrow 5\ \overset{+5}{As_2O_5}(aq) + 4\ \overset{-1}{HI}(aq)$

Iodine is reduced ($2 \times 2 \times 1\ e^- = 4e^-$ gained); arsenic is oxidized ($1 \times 2 \times 2\ e^- = 4\ e^-$ lost).

49. **(a)** $3\ \overset{+3}{P}F_2\overset{0}{I}(\ell) + 2\ \overset{+2}{Hg}(\ell) \longrightarrow P_2F_4(g) + \overset{+1}{Hg_2I_2}(s)$

(b) $2\ \overset{+5\ -2}{KClO_3}(s) \longrightarrow 2\ \overset{-1}{KCl}(s) + \overset{0}{O_2}(g)$

(c) $4\ \overset{-3}{NH_3}(g) + 5\ \overset{0}{O_2}(g) \longrightarrow 4\ \overset{+2\ -2}{NO}(g) + 6\ \overset{-2}{H_2O}(g)$

(d) $2\ \overset{0}{As}(s) + 6\ \overset{+1}{Na}OH(\ell) \longrightarrow 2\ Na_3\overset{+3}{As}O_3(s) + 3\ \overset{0}{H_2}(g)$

51. **(a)** P reduced, Hg oxidized:
$2 \times 1 \times 1\ e^-$ gained by P $= 2 \times 1 \times 1\ e^-$ lost by Hg

(b) Cl reduced, O oxidized;
$2 \times 1 \times 6\ e^-$ gained by Cl $= 2 \times 3 \times 2\ e^-$ lost by O

(c) O reduced, N oxidized;
$5 \times 2 \times 2\ e^-$ gained by O $= 4 \times 1 \times 5\ e^-$ lost by N

(d) H reduced, As oxidized;
$6 \times 1 \times 1\ e^-$ gained by H $= 2 \times 1 \times 3\ e^-$ lost by As

53. $2\overset{0}{Au}(s) + 6\ \overset{0\ +6-2}{H_2SeO_4}(aq) \longrightarrow + 3\ \overset{+3\ -6\ -2}{Au_2(SeO_4)_3}(aq) + \overset{+1+4\ -2}{H_2SeO_3}(aq) + 3\ \overset{+1\ -2}{H_2O}(\ell)$

Gold is oxidized, and selenium (half of it) is reduced.

55. $2\ Al_2O_3 + 3\ C \longrightarrow 4\ Al + 3\ CO_2$

C is oxidized, and Al is reduced.

57. **(a)** $4\ HCl(g) + O_2(g) \longrightarrow 2\ H_2O(\ell) + 2\ Cl_2(g)$ oxygenation

(b) $H_2C{=}O(g) + H_2(g) \longrightarrow H_3COH(\ell)$ hydrogenation (across a double bond)

(c) $Mg(s) + 2\ HCl(aq) \longrightarrow MgCl_2(aq) + H_2(g)$ displacement

(d) $N_2(g) + 3\ H_2(g) \longrightarrow 2\ NH_3(g)$ combination, hydrogenation

59. (a) $\overset{+1+5-2}{4\ HClO_3^-} \longrightarrow \overset{+4\ -2\quad 0}{4\ ClO_2 + O_2} + \overset{+1-2}{H_2O}$

   $\overset{+1\ +5-2}{2\ HClO_3^-} \longrightarrow \overset{+1\ -1}{2\ HCl} + \overset{0}{3\ O_2}$

   $\overset{+1+5-2}{4\ HClO_3^-} \longrightarrow \overset{+1+7-2}{3\ HClO_4} + \overset{+1-1}{HCl}$

   (b) The last reaction is a disproportionation.

61. (a) $NH_4NO_2(s) \longrightarrow N_2(g) + H_2O(g)$

   (b) The oxidation number of N is -3 in ammonium ion, $+3$ in nitrite ion. The "overall" oxidation state of N in ammonium nitrite equals 0. On the basis of the overall oxidation number, the reaction is a simple dehydration. It is still usually thought of as a redox reaction: the N in the ammonium ion is oxidized and the N in the nitrite ion is reduced.

63. $Ba(s) + H_2(g) \longrightarrow BaH_2(s)$

   $BaH_2(s) + 2\ H_2O(\ell) \longrightarrow Ba^{2+}(aq) + 2\ OH^-(aq) + 2\ H_2(g)$

   $BaH_2(s) + 2\ HCl(aq) \longrightarrow Ba^{2+}(aq) + 2\ Cl^-(aq) + 2\ H_2(g)$

   $BaH_2(s) + 2\ H_2O(\ell) + Zn^{2+}(aq) + SO_4^{2-}(aq) \longrightarrow BaSO_4(s) + Zn(OH)_2(s) + 2\ H_2(g)$

65. (a) $H_2Te(aq) + H_2O(\ell) \longrightarrow H_3O^+(aq) + HTe^-(aq)$ acid–base

   (b) $SrO(s) + CO_2(g) \longrightarrow SrCO_3(s)$ does not fit any of these categories (but note that SrO is a base anhydride and CO_2 is an acid anhydride)

   (c) $2\ HI(aq) + CaCO_3(s) \longrightarrow Ca^{2+}(aq) + 2\ I^-(aq) + CO_2(g) + H_2O(\ell)$ a characteristic reaction of an acid

   (d) $Na_2O(s) + 2\ NH_4Br(aq) \longrightarrow 2\ Na^+(aq) + 2\ Br^-(aq) + 2\ NH_3(g) + H_2O(\ell)$ acid–base (assuming hydration of NaO to NaOH)

67. 56 mmol, or 0.056 mol   69. 83 mL of 12 M HCl

71. 84.90%   73. 1.07 kg of N_2H_4

75. 16.8 mL of 7.91 M HNO_3 77. 11.2 kg of $CO_2(g)$

79. $[H_3PO_3] = 0.0010$ M; $[HF] = 0.0030$ M   81. 0.04841 M HNO_3

83. 7.175×10^{-3} M $Fe^{2+}(aq)$

Chapter 5

1. $H_2S(g)$

3. $NH_4HS(s) \longrightarrow NH_3(g) + H_2S(g)$

5. $NH_4Br(s) + NaOH(aq) \longrightarrow NH_3(g) + H_2O(\ell) + NaBr(aq)$

7. Oxygen, $2\ BaO_2(s) \longrightarrow 2\ BaO(s) + O_2(g)$

9. 0.0431 atm   11. 2.59 m   13. 76 cm

15. 1697.5 atm, 1.7200×10^3 bar   17. 1.09 atm   19. 0.857 atm

21. (a) 0.0992 L   (b) 1.08×10^3 mL   (c) 1.48×10^3 torr   (d) 4.00 atm

23. 37.0°C, 310.2 K   25. 8.00 L   27. 14.3 gills   29. 134 L

31. (a) 253 L   (b) 9.0 quarts   (c) 373 mL   (d) 1.26×10^3°C

33. 3.38 atm   35. (a) 19.8 atm   (b) 23.0 atm

37. 741°F   39. 262 K   41. -20°C

43. 4.1×10^{14} atoms of Kr   45. 5.82 g L^{-1} 47. C_4F_8

49. 3.0×10^6 L of $H_2 = 3.0 \times 10^3$ m^3

51. (a) $2\ Na(s) + 2\ HCl(g) \longrightarrow 2\ NaCl(s) + H_2(g)$ (b) 4.23 L

53. 17.5 g   55. (a) 932 L H_2S (b) 1.33 kg, 466 L SO_2

57. C_3H_8 59. $P_{N_2} = 1.41$ atm, $X_{N_2} = 0.580$, $X_{O_2} = 0.420$

61. $X_{SO_3} = 0.0688$; $P_{SO_3} = 0.0654$ atm

63. $X_{N_2} = 0.027$, $P_{N_2} = 1.6 \times 10^{-4}$ atm

65. **(a)** $X_{CO} = 0.444$ **(b)** $X_{CO} = 0.33$

67. **(a)** 1.93×10^3 m s^{-1} = 1.93 km s^{-1} **(b)** 226 m s^{-1}

69. 6100 m s^{-1} (6000 K), 790 m s^{-1} (100 K)

71. Greater   **73.** 92.3 g mol^{-1} **75.** 162 atm = 2.39×10^3 psi

77. **(a)** 14.77 atm   **(b)** 13.24 atm   **(c)** Both equations of state are only approximate.

Chapter 6

1. **(a)** 2.0×10^{-10} m; 2.5×10^{-10} m
(b) The Cl-to-Cl bond is shorter than the K$^+$-to-Cl$^-$ bond but not as strong (only about 240 kJ mol^{-1} compared to about 480 kJ mol^{-1}).

3. **(a)** *Ion-ion*, induced dipole, dispersion
(b) *Dipole-dipole*, dispersion
(c) *Hydrogen bonding* (dipole-dipole), dispersion
(d) *Dispersion*
(e) *Dispersion*

5. Bromide ion

7.

$$
\begin{array}{c}
\qquad\qquad\text{CH}_3 \\
\qquad\qquad | \\
\text{CH}_3\text{—O—H}\cdots\text{O} \\
\qquad\ \ \vdots\qquad\quad | \\
\qquad\ \ \text{H}\qquad\ \ \text{H} \\
\qquad\ \ |\qquad\quad \vdots \\
\qquad\ \ \text{O}\cdots\text{H—O—CH}_3 \\
\qquad\ \ | \\
\qquad\ \ \text{CH}_3
\end{array}
$$

9. In all three phases, the diffusion constant should decrease as its density is increased. At higher densities, molecules are closer to each other. In gases, they collide more often and travel shorter distances between collisions. In liquids and solids, less space is available for molecules to move around each other.

11. The mean free path of molecules in the gas phase should be longer than the mean free path of molecules in the liquid phase.

13. 6.16 L mol^{-1}, several times smaller than the volume of 22.4 L mol^{-1} at STP

15. 0.9345 g L^{-1} **17.** $Br_2 < SnBr_4 < SnBr_2$

19. 0.69 atm; 31% lies below   **21.** Reduce the pressure on the sample.

23. No phase change occurs.

25. **(a)** Liquid   **(b)** Gas   **(c)** Solid   **(d)** Gas

27. **(a)** Above. If gas and solid coexist at $-84.0°C$, their coexistence must extend upward in temperature to the triple point.
(b) The solid sublimes at some temperature below $-84°C$.

29. The meniscus between gas and liquid phases disappears at the critical temperature, 126.19 K.

31. 3.05% H_2O_2 by mass; $X_{H_2O_2} = 0.0164$; molality $H_2O_2 = 0.924$ mol kg^{-1}

33. molality$_{HCl}$ = 16.81 mol kg^{-1}; molarity$_{HCl}$ = 12.38 M

35. 90.7 mol kg^{-1} **37.** 0.3395 atm

39. 0.059   **41.** 3.4×10^2 g mol^{-1} **43.** N_2H_4

45. $-2.8°C$. As the solution becomes more concentrated, its freezing point decreases further.

47. Acetone does not dissociate. The calculated freezing point (assuming no dissociation) is $-0.993°C$, which lies within 2.4% of the experimental value.

49. (a) 2.7 particles (complete dissociation of Na_2SO_4 gives three particles)
(b) 0.85

51. 7.46×10^3 g mol^{-1}

53. $T_{freez} = -0.028°C$; $T_b = 100.0078°C$; $\Pi = 0.37$ atm. The computation of osmotic pressure assumes that the molarity of the solution equals its molality.

55. (a) 0.17 mol CO_2 (b) Because the partial pressure of CO_2 in the atmosphere is much less than 1 atm, the excess CO_2 bubbles out from the solution and escapes when the cap is removed.

57. 4.14×10^2 atm   **59.** 0.774

61. (a) 0.491   (b) 0.250 atm   (c) 0.575

63. A solution (such as ethanol in water) is homogeneous down to the molecular scale, but a colloidal suspension (such as gold particles in water) contains particles of one phase in a second. Even in solutions, clustering of particles of a component can occur, so the distinction is not a sharp one.

65. Steam and water vapor are invisible. The white clouds are a colloidal suspension of droplets of liquid water that have condensed in the cool air above the beaker.

Chapter 7

1. (a) $\dfrac{P_{H_2O}^2}{P_{H_2}^2 P_{O_2}} = K$

(b) $\dfrac{P_{XeF_6}}{P_{Xe} P_{F_2}^3} = K$

(c) $\dfrac{P_{CO_2}^{12} P_{H_2O}^6}{P_{C_6H_6}^2 P_{O_2}^{15}} = K$

3. $P_4(g) + 6\ Cl_2(g) + 2\ O_2(g) \rightleftharpoons 4\ POCl_3(g)$

$$\dfrac{P_{POCl_3}^4}{P_{P_4} P_{Cl_2}^6 P_{O_2}^2} = K$$

5. 1.04×10^{-4} **7.** 14.6   **9.** 53.5

11. (a) $P_{SO_2} = P_{Cl_2} = 0.58$ atm; $P_{SO_2Cl_2} = 0.14$ atm
(b) 2.4

13. $K_1 = (K_2)^3$ **15.** $\dfrac{K_2}{K_1}$

17. (a) $Q = 213$; right to left
(b) $Q = 55.9$; left to right
(c) $Q = 107$; left to right

19. (a) 9.83×10^{-4} (b) Net consumption

21. $K > 5.1$ **23.** 1.8×10^{-5} atm

25. (a) $Q = 0$; P increases
(b) $P_{SO_2Cl_2} = 0.455$ atm; $P_{Cl_2} = P_{SO_2} = 1.045$ atm

27. (a) 0.180 atm   (b) 0.244

29. $P_{PCl_5} = 0.078$ atm; $P_{PCl_3} = P_{Cl_2} = 0.409$ atm

31. $P_{Br_2} = 0.0116$ atm; $P_{Cl_2} = 0.0016$ atm; $P_{IBr} = 0.0768$ atm

33. $P_{Al_2Cl_6} = 0.020$ atm; $P_{Al_3Cl_9} = 0.020$ atm

35. (a) [NO] = 4.29×10^{-3} M; [Br$_2$] = 1.63×10^{-2} M; [NOBr] = 3.96×10^{-2} M
   (b) [NO] = 1.63×10^{-2} M; [Br$_2$] = 4.29×10^{-3} M; [NOBr] = 3.96×10^{-2} M
   (c) [NO] = 8.58×10^{-3} M; [Br$_2$] = 3.27×10^{-2} M; [NOBr] = 7.93×10^{-2} M

37. 5.6×10^{-5} mol L^{-1}

39. 7.43 atm

41. (a) Shifts left   (b) Shifts right   (c) Shifts right   (d) The volume must
   have been increased to keep the total pressure constant. Shifts left.   (e) No effect

43. First step: use low temperature, high pressure. Second step: use high temperature, low
   pressure.

45. (a) Exothermic   (b) Decrease

47. (a) $\dfrac{(P_{H_2S})^8}{(P_{H_2})^8} = K$ (b) $\dfrac{(P_{COCl_2})(P_{H_2})}{P_{Cl_2}} = K$

   (c) $P_{CO_2} = K$ (d) $\dfrac{1}{(P_{C_2H_2})^3} = K$

49. (a) $\dfrac{[Zn^{2+}]}{[Ag^+]^2} = K$

   (b) $\dfrac{[VO_3(OH)^{2-}][OH^-]}{[VO_4^{3-}]} = K$

   (c) $\dfrac{[HCO_3^-]^6}{[As(OH)_6^{3-}]^2\, P_{CO_2}^6} = K$

51. (a) The graph is a straight line passing through the origin.
   (b) The experimental K's range from 3.42×10^{-2} to 4.24×10^{-2}, with a mean of
   3.83×10^{-2}.

53. (a) $Q = 2.05$; reaction shifts to right.   (b) $Q = 3.27$; reaction shifts to left.

55. (a) 8.46×10^{-5} (b) 0.00336 atm   **57.** 17.6 atm   **59.** 76

61. (a) $K_1 = 1.64 \times 10^{-2}$; $K_2 = 5.4$ (b) 330

Chapter 8

1. (a) Cl$^-$ cannot act as a Brønsted–Lowry acid.
   (b) SO$_4^{2-}$ (c) NH$_3$ (d) NH$_2^-$ (e) OH$^-$

3. (a) NH$_4$Br(aq) + H$_2$O(ℓ) $\longrightarrow$ NH$_3$(aq) + H$_3$O$^+$(aq) + Br$^-$(aq)
   (b) H$_2$S(aq) + H$_2$O(ℓ) $\longrightarrow$ H$_3$O$^+$(aq) + HS$^-$(aq)
   (c) (NH$_4$)$_2$SO$_4$(aq) + 2 H$_2$O(ℓ) $\longrightarrow$ 2 NH$_3$(aq) + 2 H$_3$O$^+$(aq) + SO$_4^{2-}$(aq)

5. pH = 3.70; pOH = 10.30

7. 3×10^{-7} M < [H$_3$O$^+$] < 3×10^{-6} M

   3×10^{-9} M < [OH$^-$] < 3×10^{-8} M

9. pH = 3.22; [OH$^-$] = 1.7×10^{-11} M   **11.** pH = 1.63

13. Ammonium ion

15. (a) C$_{10}$H$_{15}$ON(aq) + H$_2$O(ℓ) $\rightleftharpoons$ C$_{10}$H$_{15}$ONH$^+$(aq) + OH$^-$(aq)
   (b) 7.1×10^{-11}
   (c) stronger

17. (a) ClO$^-$(aq) + H$_2$O(ℓ) $\rightleftharpoons$ HClO(aq) + OH$^-$(aq)
   (b) 3.3×10^{-7}

19. (a) $C_6H_5O^-$ (b) A weak base
    (c) $NaC_6H_5O(s) \longrightarrow Na^+(aq) + C_6H_5O^-(aq)$ (dissolution)
    $C_6H_5O^-(aq) + H_2O(\ell) \rightleftharpoons C_6H_5OH(aq) + OH^-(aq)$ (acid-base reaction)
    The second reaction occurs only to a small extent.

21. $2.4 \times 10^1 = 24$

23. (a) methyl orange   (b) 3.8 to 4.4

25. (a) 2.22; 0.030
    (b) 2.44; 0.018
    (c) 4.98; 5.3×10^{-7}
    (d) The pH increases in going from strongest to weakest acid; the fraction dissociated decreases in going from the strongest to the weakest acid.

27. (a) pH = 2.35   (b) pH = 2.50
    (c) The pH is higher in the more dilute solution; the fraction ionized is also higher in the more dilute solution.

29. (a) 2.67   (b) 2.7×10^{-2} mol per liter

31. 1.16   33. 1.2×10^{-6} 35. 10.4   37. $[HNO_2] = 0.21$ M

39. (a) Basic   (b) Acidic   (c) Neutral   (d) Neutral   41. 0.90

43. The reaction gives a moderately weak base, the acetate ion, in solution so the pH > 7.

45. (a) $[H_3O^+] = 2.4 \times 10^{-4}$; pH = 3.62   (b) pH = 3.79   47. 8.08

49. *m*-chlorobenzoic acid because its pK_a is closest to the required pH.

51. 13.88, 11.23, 7.00, 2.77

53. pH = 2.86; 4.72; 8.71; 11.00

55. 0.357 g, pH = 4.97, bromocresol green

57. 0.1164 M, $K_a = 6.5 \times 10^{-7}$

59. (a) RbOH by HBr   (b) pH~7, bromothymol blue   (c) 0.020 M NaOH

61. $[H_3AsO_4] = 8.0 \times 10^{-2}$ M, $[H_2AsO_4^-] = [H_3O^+] = 2.0 \times 10^{-2}$ M,
    $[HAsO_4^{2-}] = 9.3 \times 10^{-8}$ M, $[AsO_4^{3-}] = 1.4 \times 10^{-17}$ M

63. $[PO_4^{3-}] = 0.020$ M, $[HPO_4^{2-}] = [OH^-] = 0.030$ M, $[H_2PO_4^-] = 1.6 \times 10^{-7}$ M,
    $[H_3PO_4] = 7.1 \times 10^{-18}$ M

65. $[H_2CO_3] = 8.5 \times 10^{-6}$ M, $[HCO_3^-] = 1.5 \times 10^{-6}$ M, $[CO_3^{2-}] = 2.8 \times 10^{-11}$ M

67. (a) Lewis acid: Ag^+; Lewis base: NH_3 (b) Lewis acid: BF_3; Lewis base: $N(CH_3)_3$ (c) Lewis acid: $B(OH)_3$; Lewis base: OH^-

69. (a) $CaO(s) + H_2O(\ell) \longrightarrow Ca(OH)_2(s)$
    (b) The CaO acts as a Lewis base, donating a pair of electrons (located on the oxide ion) to one of the hydrogen ions (the Lewis acid) on the water molecule.

71. (a) Fluoride acceptor   (b) Acids: BF_3, TiF_4; bases: ClF_3O_2, KF

Chapter 9

1. 45.1 g of solid   3. $CoSO_4 \cdot 7\,H_2O$

5. About 48°C   7. Solubility increases with temperature.

9. (a) 0.0376 M $SrBr_2$ (b) 0.0752 M Br^-

11. (a) $SrI_2(s) \rightleftharpoons Sr^{2+}(aq) + 2\,I^-(aq)$ $[Sr^{2+}][I^-]^2 = K_{sp}$
    (b) $I_2(s) \rightleftharpoons I_2(aq)$ $[I_2] = K$
    (c) $Ca_3(PO_4)_2(s) \rightleftharpoons 3\,Ca^{2+}(aq) + 2\,PO_4^{3-}(aq)$ $[Ca^{2+}]^3[PO_4^{3-}]^2 = K_{sp}$
    (d) $BaCrO_4(s) \rightleftharpoons Ba^{2+}(aq) + CrO_4^{2-}(aq)$ $[Ba^{2+}][CrO_4^{2-}] = K_{sp}$

13. $Fe_2(SO_4)_3(s) \rightleftharpoons 2\ Fe^{3+}(aq) + 3\ SO_4^{2-}(aq)$ $[Fe^{3+}]^2[SO_4^{2-}]^3 = K_{sp}$

15. $[I^-] = 6.2 \times 10^{-10}$ M, $[Hg_2^{2+}] = 3.1 \times 10^{-10}$ M

17. 0.0665 g per 100 mL water

19. (a) $[Pb^{2+}] = 0.016$ M; $[Cl^-] = 0.032$ M

    (b) No. The estimated concentration is 33,000 ppm; the actual concentration certainly exceeds this somewhat.

21. 1.9×10^{-12} **23.** 1.6×10^{-8}

25. Yes. The initial reaction quotient is $6.2 \times 10^{-10} > K_{sp}$.

27. No   **29.** No. $Q = 2.6 \times 10^{-9} < K_{sp}$

31. $[Pb^{2+}] = 2.3 \times 10^{-10}$ M; $[IO^-_3] = 0.033$ M

33. $[Ag^+] = 1.8 \times 10^{-2}$ M; $[CrO_4^{2-}] = 6.2 \times 10^{-9}$ M

35. 2.4×10^{-8} mol L^{-1}

37. (a) 3.4×10^{-6} mol L^{-1} (b) 1.6×10^{-14} mol L^{-1}

39. In pure water: 1.2×10^{-4} mol L^{-1}; at pH 7: 0.15 mol L^{-1}

41. 8.6   **43.** (a) 8.6×10^{-4} M, Pb^{2+} in solid   (b) 3.1×10^{-7}

45. 2×10^{-13} M

47. pH = 2.4; $[Pb^{2+}] = 6 \times 10^{-11}$ M

49. $[Cu(NH_3)_4^{2+}] = 0.10$ M; $[Cu^{2+}] = 6 \times 10^{-14}$ M

51. $Cu^{2+}(aq) + 2\ H_2O(\ell) \rightleftharpoons CuOH^+(aq) + H_3O^+(aq)$ or

    $Cu(H_2O)_4^{2+}(aq) + H_2O(\ell) \rightleftharpoons Cu(H_2O)_3OH^+(aq) + H_3O^+(aq)$

53. 5.3   **55.** $K_a = 9.6 \times 10^{-10}$ **57.** $[I^-] = 5.3 \times 10^{-4}$ M

59. First: dissolve in water, adjust the pH to 9.5, and add ammonium carbonate. If a precipitate forms, the sample is barium nitrate; otherwise, it is cesium nitrate. Second: flame test. Blue is cesium, and green is barium.

61. Acidify to pH 0.5 and saturate with H_2S. If ppt (precipitate), then Hg^{2+} is present. Adjust the pH of supernatant liquid to 9 with NH_4^+/NH_3 buffer. If ppt, then Ni^{2+} is present. Remove H_2S and ammonium ion from supernatant; bring pH back to 9.5 and add ammonium carbonate. If ppt, then Sr^{2+} is present.

63. Group 1 is not present. At least one ion from group 2 *or* 3 is present (note that the sulfides of the ions in group 2 are less soluble than those of the ions in group 3 and precipitate in basic solution as well as acidic). Group 4 or 5 may be present.

Chapter 10

1. Electrical energy to turn the drill has been converted to heat by friction.

3. 1 kg iron < 2 kg iron < 1 kg water < 2 kg water

5. 24.8, 28.3, 29.6, 31.0, 32.2 J K^{-1} mol^{-1}. Extrapolating gives about 33.5 J K^{-1} mol^{-1} for Fr.

7. 26.1, 25.4, 25.0, 24.3, 25.4, 27.6 J K^{-1} mol^{-1} **9.** 93.8 J K^{-1} mol^{-1}

11. 97.8°C   **13.** (a) 2.46×10^3 J K^{-1} (b) 10.8 kJ   **15.** ΔH is negative.

17. (a) -6.68 kJ   (b) $+7.49$ kJ   (c) $+0.594$ kJ

19. 41.3 kJ mol^{-1}

21. (a) $\Delta H > 0$ (b) $\Delta H < 0$ (c) $\Delta H < 0$

23. 45.07 kJ   **25.** $\Delta H = +513$ J   **27.** -2.7 kJ   **29.** -623.5 kJ

31. Diamond. This is recommended as a source of heat only as a last resort, however!

33. -555.93 kJ   35. (a) -878.26 kJ   (b) -1.35×10^7 kJ of heat absorbed

37. (a) -81.4 kJ   (b) $55.1°C$

39. $H_f^\circ = -152.3$ kJ mol^{-1} 41. -78.5 kJ   43. $+1425$ kJ

45. (a) $+1304$ kJ   (b) $+614$ kJ   (c) $+22$ kJ

47. (a)

(b) -1.58×10^3 kJ mol^{-1}

49. Each side of the equation has 3 B—Br and 3 B—Cl bonds.

51. -2.16×10^4 L atm $= -2.19 \times 10^6$ J

53. 4.10 L atm $= 416$ J   55. w is zero.

57. (a) w is zero, q is positive, and ΔE is positive.
    (b) w is zero, q is negative, and ΔE is negative.
    (c) $(w_1 + w_2)$ is zero. $(\Delta E_1 + \Delta E_2) = (q_1 + q_2)$. The latter two sums could be any of three possibilities: both positive, both negative, or both zero.

59. 6.65 kJ

61. (a) $C_{10}H_8(s) + 12\ O_2(g) \longrightarrow 10\ CO_2(g) + 4\ H_2O(\ell)$ (b) -5157 kJ
    (c) -5162 kJ   (d) $+84$ kJ mol^{-1}

Chapter 11

1. (a) $2\ Al(s) + Fe_2O_3(s) \longrightarrow Al_2O_3(s) + 2\ Fe(s)$
   (b) $H_2O(\ell) \longrightarrow H_2O(g)$ (above $100°C$) [many other answers are correct]

3. Boiling water above $100°C$, melting ice above $0°C$, and dissolving NH_4Cl in water

5. (a) $6 \times 6 = 36$ (b) 1 in 36

7. (a) $\Delta S > 0$ (b) $\Delta S > 0$ (c) $\Delta S < 0$

9. (a) $\Delta S < 0$ (b) $\Delta S < 0$ (c) $\Delta S > 0$ (d) $\Delta S > 0$

11. (a) -116.58 J K^{-1} (b) Lower (more negative)

13. $\Delta S^\circ = -162.54, -181.12, -186.14, -184.72,$ and -191.08 J K^{-1}. The entropy changes in these reactions become increasingly negative with increasing atomic mass, except that the rubidium reaction is out of line.

15. 1.1 J K^{-1} 17. The tendency for entropy to increase

19. Matter has no place in the second law. Both the energy and the entropy of a system can increase or decrease in a spontaneous process.

21. ΔS_{surr} must be positive, and greater in magnitude than 44.7 J K^{-1}.

23. 9.61 J K^{-1} mol^{-1}

25. (a) $+740$ J   (b) 2.65 kJ   (c) No   (d) 196 K

27. 29 kJ mol^{-1} 29. (a) $+33.0$ kJ   (b) -840.1 kJ

31. $\Delta H^\circ = -87.9$ kJ, $\Delta G^\circ = -32.3$ kJ, $\Delta S^\circ = -186$ J K^{-1}

33. (a), (b) $\Delta G < 0$;   (c), (d) $\Delta G > 0$

35. (a) $0 < T < 3000$ K   (b) $0 < T < 1050$ K
    (c) Spontaneous at all temperatures

37. $WO_3(s) + 3\ H_2(g) \longrightarrow W(s) + 3\ H_2O(g)$; $\Delta H^\circ = 117.41$ kJ; $\Delta S^\circ = 131.19$ J K^{-1}; $\Delta G < 0$ for $T > \Delta H^\circ/\Delta S^\circ = 895$ K

39. $\Delta H° = 62.1$ kJ mol^{-1}; $\Delta G° = 25.7$ kJ mol^{-1}; $\Delta S° = 122$ J K^{-1} mol^{-1}

41. $\Delta G° = -550.23$ kJ; $K = 2.5 \times 10^{96}$

43. **(a)** 2.6×10^{12}, $\dfrac{P_{SO_3}}{(P_{SO_2})(P_{O_2})^{1/2}} = K$

(b) 5.4×10^{-35}, $(P_{O_2})^{1/2} = K$
(c) 5.3×10^3, $[Cu^{2+}][Cl^-]^2 = K$

45. 6.2×10^{-8}

47. The ΔS of the process is evidently negative.

49. 2.5×10^{-3} **51.** ΔH is negative.

53. Low temperature and high pressure

55. 4.6×10^{-3} **57.** -58 kJ

59. $\Delta H° = -172$ kJ; $\Delta S° = -506$ J K^{-1}

61. **(a)** -56.9 kJ   **(b)** $\Delta H° = -55.61$ kJ, $\Delta S° = +4.2$ J K^{-1}

63. **(a)** $\Delta H°_{vap} = 23.8$ kJ   **(b)** $T_b = 240$ K

65. $+34$ kJ mol^{-1} **67.** $+2810$ kJ

Chapter 12

1. $MnO_2(s) + \frac{4}{3} Al(s) \longrightarrow Mn(s) + \frac{2}{3} Al_2O_3(s)$;
$\Delta H° = -597.1$ kJ, $\Delta G° = -589.7$ kJ

3. 839 K $= 566°$C   **5.** 9.8 kg

7. **(a)** $Fe_2O_3(s) + 3 CO(g) \longrightarrow 2 Fe(s) + 3 CO_2(g)$

$Fe_3O_4(s) + 4 CO(g) \longrightarrow 3 Fe(s) + 4 CO_2(g)$

$FeCO_3(s) + CO(g) \longrightarrow Fe(s) + 2 CO_2(g)$

(b) From Fe_2O_3, -14.7 kJ per mole of Fe; from Fe_3O_4, -4.4 kJ per mole of Fe; from $FeCO_3$, $+15.1$ kJ per mole of Fe. Fe_2O_3 is thermodynamically easiest to reduce per mole of Fe produced.

9. CoO, Co, 2, -107, smelting

11. **(a)** $Fe^{2+}(aq) \longrightarrow Fe^{3+}(aq) + e^-$ oxidation
$H_2O_2(aq) + 2 H^+(aq) + 2 e^- \longrightarrow 2 H_2O(\ell)$ reduction

(b) $2 H_2O(\ell) + SO_2(aq) \longrightarrow HSO_4^-(aq) + 3 H^+(aq) + 2 e^-$ oxidation
$MnO_4^-(aq) + 8 H^+(aq) + 5e^- \longrightarrow Mn^{2+}(aq) + 4 H_2O(\ell)$ reduction

(c) $ClO_2^-(aq) \longrightarrow ClO_2(g) + e^-$ oxidation
$ClO_2^-(aq) + 4 H^+(aq) + 4 e^- \longrightarrow Cl^-(aq) + 2 H_2O(\ell)$ reduction

13. **(a)** $2 VO_2^+(aq) + SO_2(g) \longrightarrow 2 VO^{2+}(aq) + SO_4^{2-}(aq)$
(b) $Br_2(\ell) + SO_2(g) + 2 H_2O(\ell) \longrightarrow 2 Br^-(aq) + SO_4^{2-}(aq) + 4 H^+(aq)$
(c) $Cr_2O_7^{2-}(aq) + 3 Np^{4+}(aq) + 2 H^+(aq) \longrightarrow$
$2 Cr^{3+}(aq) + 3 NpO_2^{2+}(aq) + H_2O(\ell)$
(d) $5 HCOOH(aq) + 2 MnO_4^-(aq) + 6 H^+(aq) \longrightarrow$
$5 CO_2(g) + 2 Mn^{2+}(aq) + 8 H_2O(\ell)$
(e) $Pb_3O_4(s) + 4 H^+(aq) \longrightarrow 2 Pb^{2+}(aq) + PbO_2(s) + 2 H_2O(\ell)$
(f) $3 Hg_2HPO_4(s) + 2 Au(s) + 8 Cl^-(aq) + 3 H^+(aq) \longrightarrow$
$6 Hg(\ell) + 3 H_2PO_4^-(aq) + 2 AuCl_4^-(aq)$

15. **(a)** $2 Cr(OH)_3(s) + 3 Br_2(aq) + 10 OH^-(aq) \longrightarrow$
$2 CrO_4^{2-}(aq) + 6 Br^-(aq) + 8 H_2O(\ell)$
(b) $ZrO(OH)_2(s) + 2 SO_3^{2-}(aq) \longrightarrow Zr(s) + 2 SO_4^{2-}(aq) + H_2O(\ell)$

(c) $7 \text{ HPbO}_2^-(aq) + 2 \text{ Re}(s) \longrightarrow 7 \text{ Pb}(s) + 2 \text{ ReO}_4^-(aq) + \text{H}_2\text{O}(\ell) + 5 \text{ OH}^-(aq)$

(d) $4 \text{ HXeO}_4^-(aq) + 8 \text{ OH}^-(aq) \longrightarrow 3 \text{ XeO}_6^{4-}(aq) + \text{Xe}(g) + 6 \text{ H}_2\text{O}(\ell)$

(e) $3 \text{ Ag}_2\text{S}(s) + 2 \text{ Cr(OH)}_3(s) + 7 \text{ OH}^-(aq) \longrightarrow$
$$6 \text{ Ag}(s) + 3 \text{ HS}^-(aq) + 2 \text{ CrO}_4^{2-}(aq) + 5 \text{ H}_2\text{O}(\ell)$$

(f) $\text{N}_2\text{H}_4(aq) + 2 \text{ CO}_3^{2-}(aq) \longrightarrow \text{N}_2(g) + 2 \text{ CO}(g) + 4 \text{ OH}^-(aq)$

17. $\text{PbS}(s) + 2 \text{ O}_2(g) + 4 \text{ OH}^-(aq) \longrightarrow \text{PbO}_2^{2-}(aq) + \text{SO}_4^{2-}(aq) + 2 \text{ H}_2\text{O}(\ell)$

19. (a) $\text{Ca}(s) + \text{Cl}_2(g) \longrightarrow \text{CaCl}_2(s)$

(b) $2 \text{ Fe}^{3+}(aq) + \text{Sn}^{2+}(aq) \longrightarrow 2 \text{ Fe}^{2+}(aq) + \text{Sn}^{4+}(aq)$

(c) $\text{Fe}_2\text{O}_3(s) + 3 \text{ H}_2(g) \longrightarrow 2 \text{ Fe}(s) + 3 \text{ H}_2\text{O}(g)$

(d) $2 \text{ K}(s) + \text{H}_2\text{O}_2(aq) \longrightarrow 2 \text{ KOH}(aq)$

21. $\text{Cu}_5\text{FeS}_4(s) + 7 \text{ FeCl}_3(aq) \longrightarrow 5 \text{ CuCl}(s) + 8 \text{ FeCl}_2(aq) + 4 \text{ S}(s)$

23. $3 \text{ HNO}_2(aq) \longrightarrow \text{NO}_3^-(aq) + 2 \text{ NO}(g) + \text{H}^+(aq) + \text{H}_2\text{O}(\ell)$

25. 54 C   **27.** 6.0 A

29. Anode: $\text{H}_2\text{O}(\ell) \longrightarrow \frac{1}{2} \text{ O}_2(g) + 2 \text{ H}^+(aq) + 2 \ e^-$

Cathode: $2 \text{ H}^+(aq) + 2 \ e^- \longrightarrow \text{H}_2(g)$

Total: $\text{H}_2\text{O}(\ell) \longrightarrow \frac{1}{2} \text{ O}_2(g) + \text{H}_2(g)$

33. $2 \text{ Cr}^{2+}(aq) + \text{Cu}^{2+}(aq) \longrightarrow 2 \text{ Cr}^{3+}(aq) + \text{Cu}(s)$

35. (a) $5 \text{ Fe}^{2+}(aq) + 8 \text{ H}^+(aq) + \text{MnO}_4^-(aq) \longrightarrow 5 \text{ Fe}^{3+}(aq) + 4\text{H}_2\text{O}(\ell)$

(b) $2 \text{ Ag}(s) + \text{I}_2(aq) \longrightarrow 2 \text{ Ag}^+(aq) + 2 \text{ I}^-(aq)$

37. 9.36×10^{20} electrons

39. (a) $9.5 \times 10^4 \text{ C}$ (b) $3.86 \times 10^5 \text{ C}$ (c) $1.45 \times 10^5 \text{ C}$ (d) $2.32 \times 10^6 \text{ C}$

41. 0.180 mol, 21.4 g

43. (a) $\text{Zn}(s) + \text{Cl}_2(g) \longrightarrow \text{Zn}^{2+}(aq) + 2 \text{ Cl}^-(aq)$

(b) $1.20 \times 10^3 \text{ C}; 1.24 \times 10^{-2} \text{ mol } e^-$

(c) Decrease by 0.407 g   (d) 0.152 L Cl_2 consumed

45. 0.248 g   **47.** +2

49. $2 \text{ Cl}^-(melt) \longrightarrow \text{Cl}_2(g) + 2 \ e^-$ anode $\text{Na}^+(melt) + e^- \longrightarrow \text{Na}(\ell)$ cathode

51. 4.4×10^4 kg   **53.** $2 \text{ Mg} + \text{TiCl}_4 \longrightarrow \text{Ti} + 2 \text{ MgCl}_2$; 101 kg

55. 42.4 minutes

Chapter 13

1. 15 cents   **3.** 0.933 seconds

5. (a) 0.4665 V, −90.02 kJ   (b) 0.721 V, −348 kJ per mole of MnO_4^-

(c) 0.265 V, −51.1 kJ per mole of I_2

7. $\Delta G° = -921 \text{ J}; w_{max} = +921 \text{ J}$

9. (a) Anode: $\text{Co}(s) \longrightarrow \text{Co}^{2+}(aq) + 2 \ e^-$; cathode: $\text{Br}_2(\ell) + 2 \ e^-(aq) \longrightarrow 2 \text{ Br}^-(aq)$;

total: $\text{Co}(s) + \text{Br}_2(\ell) \longrightarrow \text{Co}^{2+}(aq) + 2 \text{ Br}^-(aq)$ (b) 1.34 V

11. (a) Anode: $\text{Zn}(s) \longrightarrow \text{Zn}^{2+}(aq) + 2 \ e^-$; cathode: $\text{In}^{3+}(aq) + 3 \ e^- \longrightarrow \text{In}(s)$

(b) −0.338 V

13. It will not, because the potential difference $\Delta\mathscr{E}° = -1.029 - 1.51 = -2.54 \text{ V} < 0$

15. $2 \text{ Fe}^{3+}(aq) + \text{Fe}(s) \longrightarrow 3 \text{ Fe}^{2+}(aq)$; 1.179 V

17. Reducing agent   **19.** Br_2 less effective than Cl_2

21. (a) No   (b) Yes   (c) Yes   (d) No

23. (a) Yes   (b) P_4 **25.** 0.381 V

27. **(a)** 0.513 V   **(b)** 0.629 V   **(c)** 0.146 V

29. -0.30 V   **31.** The $Cd^{2+} | Cd$ half-cell   **33.** 5.17

35. **(a)** 0.31 V   **(b)** 1×10^{-8} M

37. **(a)** Anode: $Ni(s) \longrightarrow Ni^{2+}(aq) + 2\,e^-$
     Cathode: $Ag^+(aq) + e^- \longrightarrow Ag(s)$
     Overall: $Ni(s) + 2\,Ag^+(aq) \longrightarrow Ni^{2+}(aq) + 2\,Ag(s)$
     (b) 1.00 V   **(c)** 6×10^{34}

39. $K = 3 \times 10^{31}$, orange   **41.** 3×10^6 **43.** pH $= 2.53$; $K_a = 0.0029$

45. **(a)** 1.065 V   **(b)** 1.3×10^{-11} M   **(c)** 7.6×10^{-13}

47. 2.041 V; 12.25 V   **49.** 9.3×10^6 C   **51.** 1.9×10^7 J

53. No, because $H_2SO_4(aq)$ is not the only substance the amount of which changes during discharge. The accumulated $PbSO_4$ must also be removed and replaced by Pb and PbO_2.

55. 7900 J g^{-1}

57. $\Delta\mathscr{E}° = -0.42$ V, not spontaneous under standard conditions (pH $= 14$). If $[OH^-]$ is small enough, the equilibrium shifts to the right and the reaction becomes spontaneous.

59. According to its reduction potential, yes. In practice, however, the sodium would react instantly and explosively with the water and would therefore be useless for this purpose.

Chapter 14

1. 1.18 miles per minute; 70.9 mph

3. 50 mph; 46 mp h; 42 mph; 46 mph

5. 5.3×10^{-5} mol L^{-1} s^{-1}

7. rate $= -\dfrac{\Delta[N_2]}{\Delta t} = -\dfrac{1}{3}\dfrac{\Delta[H_2]}{\Delta t} = \dfrac{1}{2}\dfrac{\Delta[NH_3]}{\Delta t}$

9. **(a)** 0.225 mol L^{-1} s^{-1} **(b)** 0.075 mol L^{-1} s^{-1}

11. **(a)** second order   **(b)** first order   **13.** L^2 mol^{-2} s^{-1}

15. **(a)** Rate $= k[NOBr]^2$ **(b)** $k = 0.800$ L mol^{-1} s^{-1}

17. **(a)** Increase by a factor of 10   **(b)** Decrease by a factor of 4

19. **(a)** Rate $= k[NO]^2[H_2]$; k has the units L^2 mol^{-2} s^{-1}
     (b) An increase by a factor of 18

21. **(a)** Rate $= k[C_5H_5N][CH_3I]$ **(b)** $k = 75$ L mol^{-1} s^{-1}
     (c) 7.5×10^{-8} mol L^{-1} s^{-1}

23. **(a)** Rate $= k[Cr(H_2O)_6^{3+}][SCN^-]$ **(b)** $k = 2.0 \times 10^{-6}$ L mol^{-1} s^{-1}

25. **(a)** 1.5×10^{-5} s^{-1} **(b)** 0.41   **27.** 5.3×10^{-3} s^{-1}

29. Only (a) and (c) are true.   **31.** 2.34×10^{-5} mol L^{-1}

35. **(a)** Bimolecular, rate $= k[HCO][O_2]$
     (b) Termolecular, rate $= k[CH_3][O_2][N_2]$
     (c) Unimolecular, rate $= k[HO_2NO_2]$

37. **(a)** The first step is unimolecular; the others are bimolecular.
     (b) $H_2O_2 + O_3 \longrightarrow H_2O + 2\,O_2$ **(b)** O, ClO, CF_2Cl, Cl

39. 0.26 L mol^{-1} s^{-1}

41. **(a)** $A + B + E \longrightarrow D + F$ rate $= \dfrac{k_1 k_2}{k_{-1}} \dfrac{[A][B][E]}{[D]}$

(b) $A + D \longrightarrow B + F$ rate $= \dfrac{k_1 k_2 k_3}{k_{-1} k_{-2}} \dfrac{[A][D]}{[B]}$

43. Only mechanism (b)   **45.** Only mechanism (a)

47. **(a)** 4.25×10^5 J mol^{-1} **(b)** 1.54×10^{11} L mol^{-1} s^{-1}

49. **(a)** 1.49×10^{-3} L mol^{-1} s^{-1} **(b)** 1.30×10^3 s

51. **(a)** 4.3×10^{13} s^{-1} **(b)** 1.7×10^5 s^{-1}

53. 70.3 kJ mol^{-1} **55.** CF_2Cl_2 is a catalyst.

Chapter 15

1. **(a)** $2 \, {}^{12}_{6}C \longrightarrow {}^{23}_{12}Mg + {}^{1}_{0}n$ **(b)** ${}^{15}_{7}N + {}^{1}_{1}H \longrightarrow {}^{12}_{6}C + {}^{4}_{2}He$
(c) $2 \, {}^{3}_{2}He \longrightarrow {}^{4}_{2}He + 2 \, {}^{1}_{1}H$

3. **(a)** 3.300×10^{10} kJ mol^{-1} = 342.1 MeV total; 8.551 MeV per nucleon
(b) 7.312×10^{10} kJ mol^{-1} = 757.8 MeV total; 8.711 MeV per nucleon
(c) 1.738×10^{11} kJ mol^{-1} = 1801.7 MeV total; 7.570 MeV per nucleon

5. 4.01251 u

7. The two ^{4}He atoms are more stable, with a mass lower than that of one ^{8}Be atom by 9.87×10^{-5} u.

9. **(a)** ${}^{39}_{17}Cl \longrightarrow {}^{39}_{18}Ar^+ + {}^{0}_{-1}e^- + \tilde{\nu}$ **(b)** ${}^{22}_{11}Na \longrightarrow {}^{22}_{10}Ne^- + {}^{0}_{1}e^+ + \nu$
(c) ${}^{224}_{88}Ra \longrightarrow {}^{220}_{86}Rn + {}^{4}_{2}He$ **(d)** $({}^{82}_{38}Sr^+ + {}^{0}_{-1}e^-) \longrightarrow {}^{82}_{37}Rb + \nu$

11. ${}^{30}_{14}Si + {}^{0}_{1}n \longrightarrow {}^{31}_{14}Si \longrightarrow {}^{31}_{15}P^+ + {}^{0}_{-1}e^- + \tilde{\nu}$

13. 1.312 MeV   **15.** 2.2164 MeV

17. -0.39 MeV   **19.** 3.7×10^{10} min^{-1}

21. **(a)** 1.0×10^6 **(b)** 5.9×10^4

23. 30.3 yr   **25.** Old: 167 s; New: 133 s   **27.** 1.5×10^{18} Bq

29. **(a)** 0.250   **(b)** 1.85×10^9 yr

31. 4220 yr   **33.** 5×10^4 yr

35. ${}^{11}_{6}C \longrightarrow {}^{11}_{5}B^- + {}^{0}_{1}e^+ + \nu$; ${}^{15}_{8}O \longrightarrow {}^{15}_{7}N^- + {}^{0}_{1}e^+ + \nu$

37. Exposure from ^{15}O is greater by a factor of $1.72/0.99 = 1.74$.

39. **(a)** 2.3×10^{10} Bq   **(b)** 2.5 mrad
(c) Yes. A dose of 500 rad has a 50% chance of being lethal; this dose is greater than 850 rad in the first 8 days.

41. **(a)** ${}^{90}_{38}Sr \longrightarrow {}^{90}_{40}Zr^{2+} + 2 \, {}^{0}_{-1}e^- + 2 \, \tilde{\nu}$
(b) 2.8 MeV   **(c)** 5.24×10^{12} Bq
(d) 4.45×10^{11} Bq

43. Increase (assuming no light isotopes of U are products of the decay)

45. 7.59×10^7 kJ g^{-1}

Chapter 16

1. 0.66 m s^{-1} **3.** 3.04 m   **5.** **(a)** 5.00×10^5 s^{-1} **(b)** 4.4 minutes

7. $\lambda = 1.313$ m, time $= 0.0873$ s   **9.** Blue

11. Red   **13.** 550 nm, green

15. **(a)** 3.371×10^{-19} J   **(b)** 203.0 kJ mol^{-1} **(c)** 4.926×10^{-3} mol s^{-1}

17. Part of the yellow light, together with green and blue light, ejects electrons from Cs. No visible light ejects electrons from Se; ultraviolet light is required.

19. (a) 7.4×10^{-20} J   (b) 4.0×10^5 m s^{-1}

21. (a) $n = 3$: -0.111 Ry, -2.42×10^{-19} J;
  $n = 6$: -0.0278 Ry, -6.06×10^{-20} J;
  $n = 8$: -0.0156 Ry, -3.41×10^{-20} J;
  $n = 8$ is highest.
  (b) $n = 3$: $9a_0$, 4.76×10^{-10} m;
  $n = 6$: $36a_0$, 1.90×10^{-9} m;
  $n = 8$: $64a_0$, 3.39×10^{-9} m;
  $n = 8$ is largest.

23. (a) 4.21×10^{-21} Ry, 9.18×10^{-20} J
  (b) 4.86×10^{-2} Ry, 1.06×10^{-19} J
  (c) -4.86×10^{-2} Ry, -1.06×10^{-19} J

25. (a) 2.16×10^{-6} m;   (b), (c) 1.87×10^{-6} m

27. $r_3 = 0.0952$ nm; $E_3 = -6.06 \times 10^{-18}$ J; energy per mole $= 3.65 \times 10^3$ kJ mol^{-1};
  $v = 1.14 \times 10^{16}$ s^{-1}; $\lambda = 2.63 \times 10^{-8}$ m $= 26.3$ nm

29. 72.90 nm, ultraviolet

31. (a) 100 cm, 33 cm   (b) 2 nodes

33. (a) 7.27×10^{-7} m   (b) 3.96×10^{-10} m   (c) 1.48×10^{-34} m

35. (a) 5.8×10^4 m s^{-1} (b) 7.9 m s^{-1}

37. (a) $4p$ (b) $2s$ (c) $6f$

39. (a) 2 radial, 1 angular   (b) 1 radial, 0 angular   (c) 2 radial, 3 angular

41. It is most likely to be found along the y-axis. The probability of finding it vanishes in the xz plane, and on the sphere where it has a radial node.

43. 3 quantum numbers in this one-electron ion Cl^{16+}

45. 2   1 $2p$ 3
  3   2 $3d$ 5
  4   3 $4f$ 7

47. Only (b) is allowed.

Chapter 17

1. (a) [Ne] $3s^2 3p^2$ (b) [Ne]$3s^2 3p^4$ (c) [Ar] $3d^7 4s^2$

3. Be$^+$: $1s^2 2s^1$; Cl$^-$: $1s^2 2s^2 2p^3$; Ne^{2+}: $1s^2 2s^2 2p^4$; Mg$^+$: [Ne]$3s^1$; P^{2+}: [Ne]$3s^2 3p^1$; Cl$^-$: [Ne]$3s^2 3p^6$; As$^+$: [Ar]$3d^{10} 4s^2 4p^2$; I$^-$: [Kr]$4d^{10} 5s^2 5p^6$. All except Cl$^-$ and I$^-$ are paramagnetic.

5. (a) In   (b) P^{3-} (c) V^{2+} **7.** 117

9. First "noble gases" at $Z = 1, 5, 9$

11. (a) Sr   (b) Rn   (c) Xe   (d) Sr

13. (a) Cs   (b) F   (c) K   (d) At

17. 318.4 nm, near ultraviolet

19. The number of electrons increases, and they cannot be placed in the same regions of space as the core electrons.

21. (a) Rb   (b) K   (c) Kr   (d) K   (e) O^{2-}

23. (a) S^{2-} (b) Ti^{2+} (c) Mn^{2+} (d) Sr^{2+}

25. Bond length tends to increase and bond enthalpy to decrease from top to bottom in the periodic table. Exceptions occur: for the bond enthalpy, the dissociation enthalpy of F$_2$ is smaller than that of Cl$_2$, for instance.

27. The As—H bond length lies between 1.42 and 1.71 Å (observed: 1.52 Å). SbH_3 has the weakest bond.

29. (a) CI_4 (b) OF_2 (c) SiH_4

31. (a) $+70$ and $+1203$ kJ mol^{-1} (b) $+147$ and $+1198$ kJ mol^{-1}

33. HF, 40%; HCl, 19%; HBr, 14%; HI, 8%; CsF, 87%

35. (a) N—P > C—N > N—O > N—N
(b) N—O : N, N—P : P, C—N : C

37. $+5, +5, +7, +2$

39. The $5d$ and $4f$ electrons do not shield the $6s$ electrons well, so the $6s$ electons are held very tightly in Pb^{2+}.

41. Metal ions in high oxidation states have high electronegativities. The compounds are better described as covalent than as ionic.

43. K_{a1} should lie between about 5×10^{-4} and 8×10^{-2} (experimental: 2×10^{-4}). At pH 14, HCOO$^-$ should dominate (monoprotic acid).

45. $H_3BO_3 < H_5B_3O_7 < H_6B_4O_9 < H_2B_4O_7$

47.

H : O—S—O : H   H : O—S—O : H

H_2SO_4 is stronger because it has more lone oxygen atoms.

Chapter 18

1. He_2^{2+}: $(\sigma_{1s})^2$, bond order 1.
Li_2^{2+}: $(\sigma_{1s})^2(\sigma_{1s}^*)^2$; bond order 0.
Be_2^{2+}: $(\sigma_{1s})^2(\sigma_{1s}^*)^2(\sigma_{2s})^2$; bond order 1.
B_2^{2+}: $(\sigma_{1s})^2(\sigma_{1s}^*)^2(\sigma_{2s})^2(\sigma_{2s}^*)^2$; bond order 0.
C_2^{2+}: $(\sigma_{1s})^2(\sigma_{1s}^*)^2(\sigma_{2s})^2(\sigma_{2s}^*)^2(\pi_{2p})^2$; bond order 1.
N_2^{2+}: $(\sigma_{1s})^2(\sigma_{1s}^*)^2(\sigma_{2s})^2(\sigma_{2s}^*)^2(\pi_{2p})^4$; bond order 2.
O_2^{2+}: $(\sigma_{1s})^2(\sigma_{1s}^*)^2(\sigma_{2s})^2(\sigma_{2s}^*)^2(\sigma_{2p})^2(\pi_{2p})^4$; bond order 3.
F_2^{2+}: $(\sigma_{1s})^2(\sigma_{1s}^*)^2(\sigma_{2s})^2(\sigma_{2s}^*)^2(\sigma_{2p})^2(\pi_{2p})^4(\pi_{2p}^*)^2$; bond order 2.
Ne_2^{2+}: $(\sigma_{1s})^2(\sigma_{1s}^*)^2(\sigma_{2s})^2(\sigma_{2s}^*)^2(\sigma_{2p})^2(\pi_{2p})^4(\pi_{2p}^*)^4$; bond order 1.

3. (a) F_2: $(\sigma_{2s})^2(\sigma_{2s}^*)^2(\sigma_{2p})^2(\pi_{2p})^4(\pi_{2p}^*)^4$
F_2^+: $(\sigma_{2s})^2(\sigma_{2s}^*)^2(\sigma_{2p})^2(\pi_{2p})^4(\pi_{2p}^*)^3$
(b) F_2: 1, F_2^+: 3/2   (c) F_2^+ should be paramagnetic   (d) $F_2 < F_2^+$

5. $(\sigma_{3s})^2(\sigma_{3s}^*)^2(\sigma_{3p_z})^2(\pi_{3p})^4(\pi_{3p}^*)^2$; bond order = 2; paramagnetic

7. (a) F   (b) N   (c) O

9. (a) 1   (b) 5/2   (c) 3/2

11. (a) is diamagnetic; (b) and (c) are paramagnetic

13. Bond order: 2 1/2; paramagnetic

15. The outermost electron in CF is in a π_{2p}^* molecular orbital. Removing it gives a stronger bond.

17. $(\sigma_{1s})^2(\sigma_{1s}^*)^2$, bond order 0, should be unstable

19. sp^3 hybridization of central N$^-$, bent molecular ion

21. (a) sp^3 on C, tetrahedral   **(b)** sp on C, linear   **(c)** sp^3 on O, bent
   (d) sp^3 on C, pyramidal   **(e)** sp on Be, linear

23. sp^2 hybrid orbitals. ClO_3^+ is trigonal planar, ClO_2^+ is bent.

25. sp^3 hybrid orbitals, tetrahedral

27. (a) 5 σ-bonds, 1 π-bond   **(b)** 1 σ-bond   **(c)** 10 σ-bonds   **(d)** 4 σ-bonds

29. Central carbon atom sp^2, other carbon atoms: sp^3

One π orbital involving $2p_z$ orbitals on C and O; doubly occupied

H—C—C, H—C—H angle $\approx 109.5°$

C—C—O, C—C—C angles $\approx 120°$

31. —CH_3 carbon sp^3, other carbon sp^2

A π-orbital with two electrons bonds the second C atom with the O atom. The three groups around the second carbon form an approximately trigonal planar structure, with bond angles near 120 degrees. The geometry around the first carbon atom is approximately tetrahedral, with angles near 109.5 degrees.

33. (a) 3   **(b)** 3

35.

$$\left\{ :\!\ddot{O}::\!\ddot{N}\!:\!\overset{-1}{\ddot{O}}: \longleftrightarrow :\!\overset{-1}{\ddot{O}}\!:\!\ddot{N}::\!\ddot{O}: \right\}^-$$

$SN = 3$, sp^2 hybridization, bent molecule. The $2p_z$-orbitals perpendicular to the plane of the molecule can be combined into a π molecular orbital containing one pair of electrons. Complete occupancy of this orbital (and vacancy in the π^* orbital) adds bond order 1/2 to each N—O bond, for a total bond order of 3/2 per bond.

37. Absorption of microwave radiation causes molecules to rotate faster. Information is gained about bond lengths and angles and molecular geometry.

39. Decrease, because the electron enters a π^* antibonding orbital.

41. In the blue range, around 450 nm

43. Shorter; the π bonding is localized in cyclohexene.

45. 270 nm

47. $\{ \ddot{O}\!=\!\ddot{O}\!-\!\ddot{O}: \longleftrightarrow :\!\ddot{O}\!-\!\ddot{O}\!=\!\ddot{O} \}$
   $SN = 3$, sp^2 hybridization, bent molecule.

Two electrons in a π-orbital formed from the three $2p_z$-orbitals perpendicular to the molecular plane, total bond order of 3/2 for each O—O bond.

Chapter 19

1. (a) $[Fe(CN)_6]^{3-}$ **(b)** $[Mn(NH_3)(H_2O)_2(Cl)_3]$ **(c)** $[Pt(H_2O)Br(en)]^+$

3. Monodentate, at the N-atom lone pair

5. 3.13   **7.** +2, +2, +2, 0

9. (a) $Na_2[Zn(OH)_4]$ **(b)** $[CoCl_2(en)_2]NO_3$
   (c) $[Pt(H_2O)_3Br]Cl$ **(d)** $[Pt(NH_3)_4(NO_2)_2]Br_2$

11. (a) Ammonium diamminetetraisothiocyanatochromate (IV)
   (b) Pentacarbonyliodotechnetium(I)
   (c) Potassium pentacyanomanganate(IV)
   (d) Tetraammineaquachlorocobalt(III) bromide

13. Such a complex persists in solution but exchanges ligands with other complexes, as can be verified by isotopic labeling experiments.

15. $\Delta G° = +70.56$ kJ, stable against dissociation. In acid, $[Cu(NH_3)_4]^{2+}(aq) + 4 H_3O^+(aq) \longrightarrow Cu^{2+}(aq) + 4 NH_4^+(aq) + 4 H_2O(\ell)$
$\Delta G° = -140.68$ kJ, spontaneous

17. $[Cu(NH_3)_2Cl_2] < KNO_3 < Na_2[PtCl_6] < [Co(NH_3)_6]Cl_3$

23. **(a)** Strong: 1, weak: 5   **(b)** Strong or weak: 0   **(c)** Strong or weak: 3
    (d) Strong: 2, weak: 4   **(e)** Strong: 0, weak: 4

25. $[Fe(CN)_6]^{3-}$: 1 unpaired, (t_{2g}^5)
    $[Fe(H_2O)_6]^{3+}$: 5 unpaired, $(t_{2g}^3 e_g^2)$

27. d^3 always gives half-filled and d^{10} filled subshells for metal ions in an octahedral field. d^5 is half-filled and stable for high-spin (small Δ_0), and d^6 is a filled subshell and stable for low-spin (large Δ_0).

29. This ion does not absorb any significant amount of light in the visible range; it has a $t_{2g}^6 e_g^4$ configuration.

31. $\lambda \approx 480$ nm; $\Delta_0 \approx 250$ kJ mol^{-1}

33. **(a)** Orange-yellow   **(b)** Approximately 600 nm (actual: 575 nm)
    (c) Decrease because CN^- is a strong-field ligand and increases Δ_0.

35. **(a)** Pale, because F^- is an even weaker field ligand than H_2O, so it should be high-spin d^5. This is a half-filled subshell and the complex absorbs only weakly.
    (b) Colorless, because Hg(II) is a d^{10} filled-subshell species.

37. The formation of fewer bonds means that tetrahedral complexes are less stable than octahedral complexes.

39. 1330 g mol^{-1}

Chapter 20

1. 3.613 Å   3. 102.9°   5. 49.87° and 115.0°

7. A single crystal gives a diffraction pattern with spots, and a microcrystalline solid gives one with rings. The difference comes from the many random orientations of the small crystals in the second case.

9. (b), (c), (d), and (e)   11. (a), (b), (c), and (d)

13. One fourfold rotation.   15. 739 Å^3 = 7.39×10^{-28} m^3

17. **(a)** 5.6731 Å^3 per C atom   **(b)** 3.5157 g cm^{-3}

19. **(a)** 1.602×10^{-22} cm^3 **(b)** 3.729×10^{-22} g   **(c)** 4.662×10^{-23} g
    (d) 6.025×10^{23} (in error by 0.05%)

21. 2.533 g cm^{-3} 23. 4.059 g cm^{-3} 25. 8

27. **(a)** 1.7943×10^{-28} m^3 **(b)** 5.57×10^{18} **(c)** 1.89×10^{13}, 3.38×10^{-4}%

29. **(a)** Ionic   **(b)** Covalent network   **(c)** Molecular   **(d)** Metallic

31. $CO < BaCl_2 < Co < SiC$ 33. $O_2 < Sb < Cu$ 35. 8, 6

37. 8 nearest neighbors, 6 second-nearest neighbors   39. 4.4197 Å

41. Not by a significant amount, because each vacancy is accompanied by an interstitial. In large numbers, such defects could cause a small bulging of the crystal and lower its density.

43. **(a)** $Fe_{0.9352}O$ **(b)** 0.1386 of the iron

45. The liquid is less ordered than the plastic crystal because the molecular centers are free to move. It has a higher entropy and a higher enthalpy.

Chapter 21

1. 2.16×10^{-19} J

3. 6.1×10^{-27} electrons; that is, no electrons

5. **(a)** Electron movement (n-type)   **(b)** Hole movement (p-type)

7. 680 nm, red   **9.** Decrease

11.

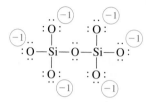

The structures of $P_2O_7^{4-}$ and $S_2O_7^{2-}$ have the same number of electrons, and are found by writing P or S in place of Si. The formal charges on the P's are $+1$ in the resulting structure, and those on the S's are $+2$. The chlorine compound is Cl_2O_7, dichlorine heptaoxide.

13. **(a)** Tetrahedra. Ca: $+2$, Fe: $+3$, Si: $+4$, O: -2
    (b) Infinite sheets: Na: $+1$, Zr: $+2$, Si: $+4$, O: -2
    (c) Pairs of tetrahedra: Ca: $+2$, Zn: $+2$, Si: $+4$, O: -2
    (d) Infinite sheets. Mg: $+2$, Si: $+4$, O: -2, H: $+1$

15. **(a)** Infinite network. Li: $+1$, Al: $+3$ Si: $+4$, O: -2
    (b) Infinite sheets. K: $+1$, Al: $+3$, Si: $+4$, O: -2, H: $+1$
    (e) Closed rings or infinite single chains: Al: $+3$, Mg: $+2$, Si: $+4$, O: -2

17. **(a)** $CaCO_3(s) + SiO_2(s) \longrightarrow CaSiO_3(s) + CO_2(g)$
    (b) $\Delta H° = +89.41$ kJ, $\Delta S° = +160.8$ J K^{-1} **(c)** 556 K

19. The strength of typical ceramics is in their ability to resist compression (related to their hardness). Typical metals are stronger in their ability to resist tensile forces (stretching).

21. If the ceramic is a product of firing of clay, the starting clay material can be shaped by hand; other ceramic compounds are shaped in a mold by compression at high temperature. Metals like steel can be melted and poured into molds.

23. $Mg_3Si_4O_{10}(OH)_2(s) \longrightarrow 3\, MgSiO_3(s) + SiO_2(s) + H_2O(g)$ **25.** 234 L

27. 0.418 mol Si, 0.203 mol Na, 0.050 mol Ca, 0.0068 mol Al, 0.002 mol Mg, 0.0005 mol Ba

29. -113.0 kJ

31. **(a)** $SiO_2(s) + 3\, C(s) \longrightarrow SiC(s) + 2\, CO(g)$ **(b)** $+624.6$ kJ mol^{-1}
    (c) Low conductivity, high melting point, very hard

33. -1188.2 kJ, unstable   **35.** 7/3

Chapter 22

1. A batch process is carried out by mixing reactants, letting them react, and then extracting products in a series of separate steps. In a continuous process, all the steps run simultaneously. A batch process can take place in a single container with one opening, but a continuous process requires either flow through several containers or at least different places into which reactants can be introduced and from which products are removed.

3. Both use high temperatures and pressures for certain processes, and both exploit solubility differences to separate products, but geochemical processes are much slower.

$$Ca^{2+}, CO_3^{2-} (\textit{in igneous rocks}) \xrightarrow[\text{water}]{} CaCO_3(aq) \xrightarrow[\text{evaporation}]{}$$

$$CaCO_3 \ (\textit{limestone}) \xrightarrow[\text{heat and pressure}]{} CaCO_3(\textit{marble})$$

5. **(a)** In salt domes, produced by bacteria from $CaSO_4$, CH_4, and CO_2
   (b) In coal, produced from plant matter by bacterial reduction, heat, and pressure
   (c) In igneous rocks

7. The lead is protected (passivated) by a layer of lead(II) sulfate.

9. Operate at high total pressures and low temperatures; continuously remove SO_3 as it is produced.

11. Yes. The reaction that would occur, $H_2O_2(aq) + SO_2(g) \longrightarrow H_2SO_4(aq)$, has $\Delta G° = -321.7$ kJ < 0, assuming that $H_2SO_4(aq)$ is entirely $HSO_4^-(aq)$ and $H^+(aq)$. Thus, the process is thermodynamically feasible.

13. $ZnS(s) + 2\ O_2(g) + H_2O(\ell) \longrightarrow ZnO(s) + H_2SO_4(\ell)$

15. $5\ H_2SO_4(aq) + 2\ P(s) \longrightarrow 5\ SO_2(g) + 2\ H_3PO_4(aq) + 2\ H_2O(\ell)$

17. $CaF_2(s) + H_2SO_4(\ell) \longrightarrow 2\ HF(g) + CaSO_4(s)$

    $2\ NaCl(s) + H_2SO_4(\ell) \longrightarrow 2\ HCl(g) + Na_2SO_4(s)$

19. $H_2SO_4(\ell) + 2\ NaOH(s) \longrightarrow Na_2SO_4(s) + 2\ H_2O(\ell)$

    $Na_2SO_4(s) + 2\ C(s) \longrightarrow Na_2S(s) + 2\ CO_2(g)$

21. $H_2S_2O_7(s) + H_2O(\ell) \longrightarrow 2\ H_2SO_4(\ell)$

23. 3×10^{26} atoms per second.

25. $CaCO_3(s) + 2\ C(s) + N_2(g) + 4\ H_2O(\ell) \longrightarrow$
    $$2\ NH_3(g) + 2\ CO_2(g) + CO(g) + Ca(OH)_2(s);\ \Delta H° = +374.4\ kJ.$$

    Dividing by 2 gives $+187.2$ kJ per mole of $NH_3(g)$ produced.

27. Higher pressures mean higher yields but also more expensive equipment. At fixed yield, a higher temperature can be used at high pressure, giving a faster reaction.

29. High yield is favored by low temperature for all three steps.

31. $3\ HNO_3(aq) + Cr(s) + 3\ H^+(aq) \longrightarrow 3\ NO_2(g) + Cr^{3+}(aq) + 3\ H_2O(\ell)$

    $3\ HNO_3(aq) + Fe(s) + 3\ H^+(aq) \longrightarrow 3\ NO_2(g) + Fe^{3+}(aq) + 3\ H_2O(\ell)$

33.

35. Unstable

37. $5\ N_2H_4(\ell) + 4\ HNO_3(\ell) \longrightarrow 7\ N_2(g) + 12\ H_2O(\ell);\ \Delta H° = -2986.71$ kJ

39. $N_2(g) + 3\ H_2(g) \longrightarrow 2\ NH_3(g)$ (Haber–Bosch)

    $4\ NH_3(g) + 5\ O_2(g) \longrightarrow 4\ NO(g) + 6\ H_2O(g)$ (Pt catalyst, first step of Ostwald process)

    $2\ NO(g) + O_2(g) \longrightarrow 2\ NO_2(g)$ (second step of Ostwald process)

    $2\ NO_2(g) \longrightarrow N_2O_4(s)$ (low temperature)

41. The decomposition of NH_4NO_3 liberates some O_2, which can react with the C from the TNT to give $CO_2(g)$.

43. $NH_4N(NO_2)_2(g) \longrightarrow 2\ N_2(g) + 2\ H_2O(g) + O_2(g)$

Chapter 23

1. 3.2×10^{16}

3. The Leblanc process requires carbon (coke) and sulfuric acid, in addition to the salt and calcium carbonate required by the Solvay process. The sulfuric acid in particular adds

to the cost. The product, Na_2CO_3, is purer from the Solvay process because it does not contain CaS impurities.

5. $\Delta H° = +259$ kJ for the reaction: 2 NaCl(s) + H_2SO_4(ℓ) + 2 C(s) + $CaCO_3$(s) $\longrightarrow$ Na_2CO_3(s) + 2 CO_2(g) + CaS(s) + 2 HCl(g). Endothermic, so heat must be supplied to make it run.

7. Chlorous acid is stronger, and should be used.   **9.** Calcium hydroxide

11. 4.0×10^4 L = 40 m^3 **13. (a)** +1   **(b)** HClO, hypochlorous acid

15. Unstable

17. Si—F, because F atom is more electronegative and smaller

19. (a) 2 SrO(s) + 2 F_2(g) $\longrightarrow$ 2 SrF_2(s) + O_2(g)   **(b)** O_2(g) + 2 F_2(g) $\longrightarrow$ 2 OF_2(g)
(c) UF_4(s) + F_2(g) $\longrightarrow$ UF_6(s)   **(d)** 2 NaCl(s) + F_2(g) $\longrightarrow$ 2 NaF(s) + Cl_2(g)

21. NaF·2 HF   **23.** ClF_3

25. (a) Bent   **(b)** Trigonal planar   **(c)** Distorted T   **(d)** Square pyramidal
(e) Pentagonal bipyramidal   **(f)** Octahedral

27. (a) Pyramidal   **(b)** Lewis base

29. 2 BrF_3 $\rightleftharpoons$ BrF_4^- + BrF_2^+. Here, BrF_2^+ and BrF_3 are both Lewis acids (electron pair acceptors), whereas F^- is the Lewis base (electron pair donor).

31. $PtF_6 < PtF_4 < CaF_2$ **33.** 2740 kJ

35. (a) -345 kJ mol^{-1} **(b)** 136 kJ mol^{-1}

37. $+190$ kJ mol^{-1}

39. XeO_6^{4-}(aq) + 12 H^+(aq) + 8 e^- $\longrightarrow$ Xe(g) + 6 H_2O(ℓ)
Mn^{2+}(aq) + 4 H_2O(ℓ) $\longrightarrow$ MnO_4^-(aq) + 8 H^+(aq) + 5 e^-
5 XeO_6^{4-}(aq) + 8 Mn^{2+}(aq) + 2 H_2O(ℓ) $\longrightarrow$ 5 Xe(g) + 8 MnO_4^-(aq) + 4 H^+(aq)

Chapter 24

1. Yes, if the amount of knocking is less than that of isooctane. Examples are the BTX compounds.

3. (a) -64 kJ mol^{-1} **(b)** $+53$ kJ mol^{-1} **(c)** $+117$ kJ mol^{-1}

5. (a) $C_{10}H_{22}$ $\longrightarrow$ C_5H_{10} + C_5H_{12} **(b)** Two: 1-pentene and 2-pentene

7. (a)

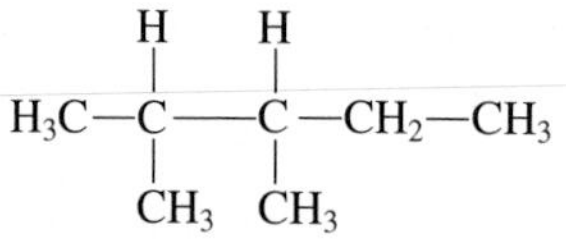

(b)

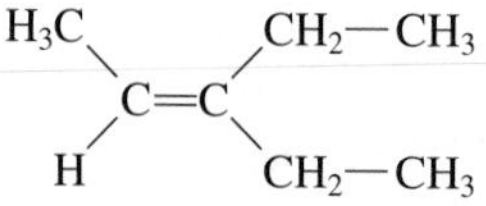

(c)

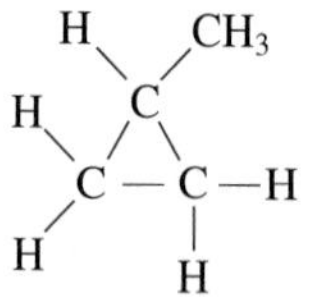

(d)

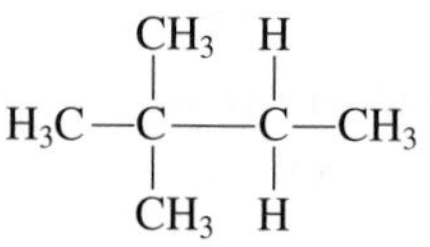

(e)

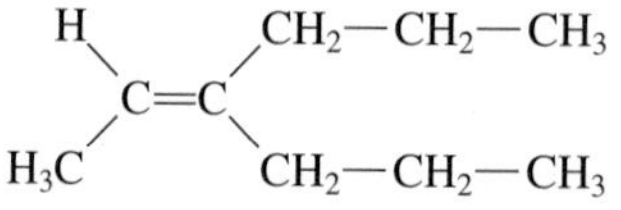

(f)

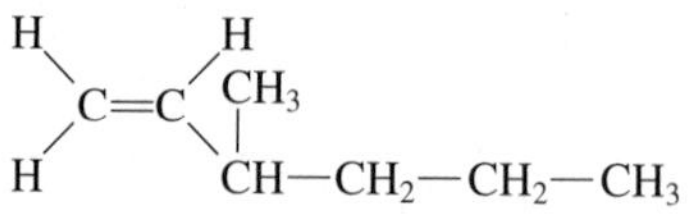

(g)

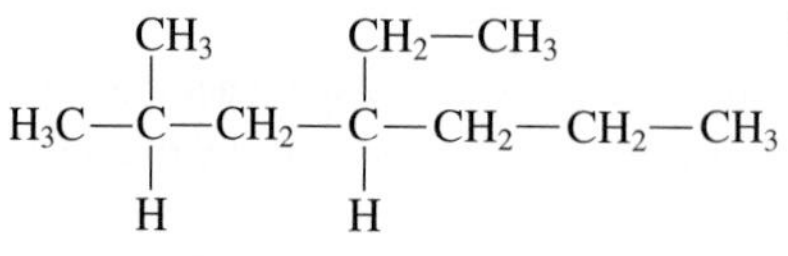

(h)

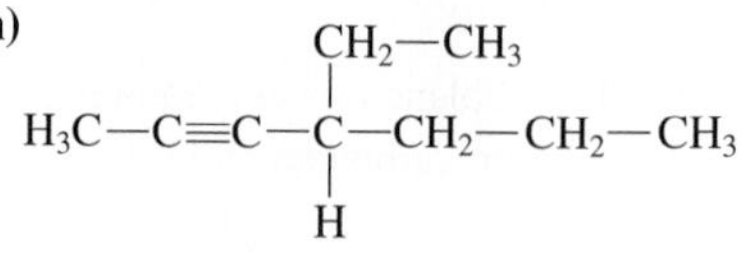

9.

$$\begin{array}{ccc} H & & CH_2-CH_2-CH_3 \\ & C=C & \\ H_3C-CH_2 & & H \end{array}$$

$$\begin{array}{ccc} H_3C-CH_2 & & CH_2-CH_2-CH_3 \\ & C=C & \\ H & & H \end{array}$$

11. **(a)** 1,2-hexadiene   **(b)** 1,3,5-hexatriene   **(c)** 2-methyl-1-hexene

   (d) 3-hexyne

13. **(a)** The first three are sp^2, and the last three are sp^3.

   (b) All sp^2

   (c) The first two on the main chain are sp^2; all others are sp^3

   (d) The third and fourth are sp; all others are sp^3.

15. **(a)** $CH_3CH_2CH_2CH_2OH + CH_3COOH \longrightarrow CH_3COOCH_2CH_2CH_2CH_3 + H_2O$

   (b) $NH_4CH_3COO \longrightarrow CH_3CONH_2 + H_2O$

   (c) $CH_3CH_2CH_2OH \longrightarrow CH_3CH_2CHO$ (propionaldehyde)

   (d) $CH_3(CH_2)_5CH_3 + 11\ O_2 \longrightarrow 7\ CO_2 + 8\ H_2O$

17. **(a)** $CH_2{=}CH_2 + Br_2 \longrightarrow CH_2BrCH_2Br$; $CH_2BrCH_2Br \longrightarrow CH_2{=}CHBr + HBr$

   (b) $CH_3CH_2CH{=}CH_2 + H_2O \longrightarrow CH_3CH_2CH(OH)CH_3$ (using H_2SO_4)

   (c) $CH_3CH{=}CH_2 + H_2O \longrightarrow CH_3CH(OH)CH_3$ (using H_2SO_4);

   $CH_3CH(OH)CH_3 \longrightarrow CH_3COCH_3 + H_2$ (copper oxide or zinc oxide catalyst)

19.

$$-C{\overset{O}{\Big\backslash}}OH + (R')_3C-OH \longrightarrow R-C{\overset{O}{\Big\backslash}}O-C(R')_3$$

21. 11.2%; 4.49×10^9 kg

23. **(a)** Alcohol: $CH_3CH(OH)CH_3$, isopropyl alcohol

   Carboxylic acid:

   $CH_3C(CH_3)(COCH_3)(CH_2)_3CH(CH_3)\ CH_2CH{=}CHC(CH_3){=}CHCOOH$

   (b) 3,7,10-trimethyl-2,4-dodecadiene

25. **(a)** $C_9H_8O_4$ **(b)** 1.80×10^{-3} mol

27. Dehydrogenate to make C=O double bond from C—OH bond on first ring; move C=C bond from first to second ring; insert C=O group on third ring; remove hydrocarbon side chain on fourth ring together with H atom and replace it with an —OH group and a —COCH$_2$OH group.

Chapter 25

1. $n\ CCl_2{=}CH_2 \longrightarrow \left(CCl_2-CH_2\right)_n$ **3.** Formaldehyde, $CH_2{=}O$

5. **(a)** H_2O **(b)** $\left(NH-CH_2-CO\right)$

7. $NH_2^{\ominus}Na^{\oplus} + CH_2{=}CH \longrightarrow NH_2-CH_2-CH^{\ominus}Na^{\oplus}$

$$\qquad\qquad\qquad\ \ \underset{C{\equiv}N}{|} \qquad\qquad\quad \underset{C{\equiv}N}{|}$$

9. 27   **11.** In octane, because the side groups are all hydrocarbon.

13. 5 chiral centers   **15.** C_9H_9NO; 119 units

17. **(a), (b)** $C_6H_{10}O_5$

19. 646 kg adipic acid and 513 kg of hexamethylenediamine

21. Every C atom has the maximum number of bonds to other C atoms (4).

23. 3.49×10^{12} L $= 3.49 \times 10^9$ m$^3 = 3.49$ cubic kilometers

Appendix A

1. (a) 5.82×10^{-5} (b) 1.402×10^3 (c) 7.93 (d) -6.59300×10^3
   (e) 2.530×10^{-3} (f) 1.47

3. (a) 0.000537 (b) $9,390,000$ (c) -0.00247 (d) 0.006020
   (e) $20,000$

5. 1.90×10^2 7. $746,000,000$ kg

9. (a) The value 135.6 (b) 111.34 g   (c) 0.22 g

11. That of the mass in problem 9

13. (a) 5 (b) 3 (c) Either two or three   (d) 3 (e) 4

15. (a) 14 L   (b) $-0.0034°C$
    (c) 340 lb $= 3.4 \times 10^2$ lb   (d) 3.4×10^2 miles   (e) 6.2×10^{-27} J

17. $2,997,215.55$

19. (a) -167.25 (b) 76 (c) 3.1693×10^{15} (d) -7.59×10^{-25}

21. (a) -8.40 (b) 0.147 (c) 3.24×10^{-12} (d) 4.5×10^{13}

23. $A = 337$ cm^2

Appendix B

1. (a) 6.52×10^{-11} kg   (b) 8.8×10^{-11}s   (c) 5.4×10^{12} kg m^2 s^{-3}
   (d) 1.7×10^4 kg m^2 s^{-3} A^{-1}

3. 1 Wb $= 1$ kg m^2 s^{-2} A^{-1}

5. 100 cm^3 < 500 mL < 1.5 gall (US) < 100 dm^3 < 150 L

7. (a) $4983°C$ (b) $37.0°C$ (c) $111°C$ (d) $-40°C$

9. (a) 5256 K   (b) 310.2 K   (c) 384 K   (d) 233 K

11. 1 m^3 = 264.1722 . . . gall (US); 1 m $= 39.37008$ . . . in.;
    1 kg $= 2.204623$ . . . lb; 1 J $= 0.2390057$ . . . cal.

13. (a) 24.6 m s^{-1} (b) 1.51×10^3 kg m^{-3} (c) 1.6×10^{-29} A s m
    (d) 1.5×10^2 mol m^{-3} (e) 9.6×10^3 kg m^2 s^{-3} = 9.6×10^3 W

15. Pa m^2 s^{-2} and J m^{-1} s^{-1} are not units of density.

17. 1 kW-hr $= 3.6 \times 10^6$ J; 15.3 kW-hr $= 5.51 \times 10^7$ J

19. 6620 cm^3 or 6.62 L

Appendix C

1. 50 miles per hour

3. (a) Slope $= 4$, intercept $= -7$
   (b) Slope $= 7/2 = 3.5$, intercept $= -5/2 = -2.5$
   (c) Slope $= -2$, intercept $= 4/3 = 1.3333$

5. Graph is not a straight line.

7. (a) $-5/7 = -0.7142857$ . . .   (b) $3/4 = 0.75$ (c) $2/3 = 0.6666$ . . .

9. (a) 0.5447 and -2.295 (b) -0.6340 and -2.366 (c) 2.366 and 0.6340

11. (a) 6.5×10^{-7} (also two complex roots)   (b) 4.07×10^{-2}, 0.399 and -1.011
    (c) -1.3732 (also two complex roots)

13. (a) 4.551 (b) 1.53×10^{-7} (c) 2.6×10^8 (d) -48.7264

15. 3.015 17. 121.477 19. 7.751 and -2.249 21. $x = 1.086$

23. **(a)** Mass fraction of water = 0.588;
Mass fraction of NaCl = 0.411;
Mass fraction of mercury = 4.8×10^{-5}.

(b) Volume fraction of water = 0.7564
Volume fraction of NaCl = 0.2436
Volume fraction of mercury = 4.51×10^{-6}

(c) ppmv of mercury = 4.51

(d) ppbm of mercury = 4.8×10^4

ANSWERS TO CUMULATIVE PROBLEMS

Chapter 1

(a) The masses of gallium present per 1.000 g of sulfur in the five samples are: 1.450 g Ga in sample 1; 2.175 g Ga in sample 2; 4.349 g Ga in sample 3; 1.450 g Ga in sample 4; and 2.175 g Ga in sample 5. The small whole number ratios (2 to 3 to 6 to 2 to 3) among these masses are consistent with the law of multiple proportions, which applies to compounds, not mixtures.

(b) There are at least three different compounds. Samples 1 and 4 have the same proportions of gallium and sulfur. They may be the same compound, but this is not certain because it is possible for different compounds to have the same elements in the same definite proportions. A similar statement applies to samples 2 and 5. Samples 1 and 2 must be different compounds, despite their similar appearance.

(c) If sample 1 is GaS, then sample 2 is Ga_3S_2 (or a multiple), sample 3 is Ga_3S (or a multiple), sample 4 is GaS (or a multiple), and sample 5 is Ga_3S_2 (or a multiple).

(d) If sample 1 is Ga_2S_3, then sample 2 is GaS (or a multiple), sample 3 is Ga_2S (or a multiple), sample 4 is Ga_2S_3 (or a multiple), and sample 5 is GaS (or a multiple).

(e) The relative atomic mass of gallium would be 46.49. (Note that this is two thirds of the accepted value.)

(f) Ga_2S_3, GaS, Ga_2S, Ga_2S_3, and GaS.

(g) ^{32}S and ^{69}Ga. The only way for the chemical relative atomic mass of S to come out so close to 32 is for ^{32}S, the lightest isotope, to dominate in abundance; the argument is the same in the case of gallium.

(h) There are 1.167×10^{22} gallium atoms and 1.750×10^{22} sulfur atoms. The ratio is 1.500:1 or 3:2.

(i) 221 cm^3 of SO_2.

(j) The densities of the five samples are 3.65, 3.86, 4.18, 3.65, and 3.86 g cm^{-3}, respectively. The equality of the densities of samples 1 and 4 supports but does not prove the conclusion that they are the same compound. The same is true for samples 2 and 5.

(k) Each atom occupies an average of 2.14×10^{-23} cm^3 (samples 1 and 4), 2.19×10^{-23} cm^3 (samples 2 and 5), and 2.26×10^{-23} cm^3 (sample 3).

Chapter 2

(a) $TiCl_4$

(b) $TiO_2 + C + 2 Cl_2 \longrightarrow TiCl_4 + CO_2$ **(c)** 59.2 g

(d) $TiCl_4 + 2 Mg \longrightarrow Ti + 2 MgCl_2$

(e) Mg; 62.3 g   **(f)** 1.7×10^5 metric tons

Chapter 3

(a) The ion has a total of ten electrons, of which-eight are valence electrons (an octet) and two are core electrons. TiO_2 is titanium(IV) oxide.

(b) The ion in BaO_2 must be the peroxide ion O_2^{2-}, and the compound is barium peroxide. A Lewis structure, with formal charges, for the peroxide ion is

$$\overset{(-1)}{:}\overset{..}{O}:\overset{..}{O}\overset{(-1)}{:}$$

The O—O bond is a single bond.

(c) The ion in KO_2 is the superoxide ion O_2^-; the compound is named potassium superoxide. The superoxide ion has 17 total electrons, of which 4 are core electrons. The best Lewis structures one can draw are the resonance structures

$$\left\{ :\overset{..}{O}:\overset{..}{O}\overset{(-1)}{:} \quad \longleftrightarrow \quad \overset{(-1)}{:}\overset{..}{O}:\overset{..}{O}: \right\}$$

in which only one of the oxygen atoms attains an octet electron configuration.

(d) The Lewis structure of dihydrogen trioxide is

$$H:\overset{..}{O}:\overset{..}{O}:\overset{..}{O}:H$$

(e) Table 3–3 gives the average length of O—O single bonds as 1.48×10^{-10} m; take this as the predicted length. (The experimental lengths both equal 1.445×10^{-10} m.) Table 3–3 gives the average length of O—H single bonds as 0.96×10^{-10} m; take this as the predicted length. (The experimental O—H lengths in H_2O_3 both equal 0.982×10^{-10} m.)

(f) The steric numbers of all three oxygen atoms equal 4, with two bonded pairs and two lone pairs. The H—O—O bond angles and the O—O—O bond angle should all be somewhat less than the tetrahedral angle of 109.5°. (Experiment gives values of 100.8° for both the H—O—O angles and 106.4° for the O—O—O angle.)

Chapter 4

(a) CO_2 is the anhydride of H_2CO_3.

(b) The second step is a precipitation reaction:

$$2\,Na^+(aq) + 2\,HCO_3^-(aq) \longrightarrow 2\,NaHCO_3(s)$$

(c) This decomposition reaction is not an acid–base reaction in the Arrhenius sense.

(d) CaO is the anhydride of $Ca(OH)_2$, which is prepared in step 5.

(e) $2\,NH_4^+(aq) + 2\,OH^-(aq) \longrightarrow 2\,NH_3(aq) + 2\,H_2O(\ell)$

(f) None of the steps is a redox reaction because oxidation numbers do not change.

(g) $2\,NaCl(aq) + CaCO_3(s) \longrightarrow Na_2CO_3(s) + CaCl_2(aq)$. When calcium chloride and sodium carbonate are mixed in aqueous solution, solid calcium carbonate precipitates and sodium chloride remains in solution. These compounds are the net starting materials in the Solvay process.

Chapter 5

(a) $2\,NH_4ClO_4(s) \longrightarrow N_2(g) + Cl_2(g) + 2\,O_2(g) + 4\,H_2O(g)$

(b) 328 atm   (c) $X_{Cl_2} = 1/8 = 0.125$ (exactly); $P_{Cl_2} = 41.0$ atm

(d) 318 atm

(e) $u_{rms}(H_2O) = 1220$ m s^{-1}; $u_{rms}(Cl_2) = 614$ m s^{-1}

(f) 2.89×10^5 m^3

Chapter 6

(a) Dispersion forces

(b) On the order of 1 to 2 kJ mol^{-1}; a distance of around 4×10^{-10} m (estimated by comparison to the Ar + Ar curve in Figure 6–1).

(c) The high pressure has partially liquefied the chlorine in the tank. Evaporation of the liquid keeps the interior pressure constant as gaseous chlorine is removed. The pressure drops only when the last of the liquid evaporates.

(d) The chlorine is heated and then compressed. During these operations the original gas becomes, by gradual degrees, a supercritical fluid. The supercritical fluid is then cooled at high pressure. It becomes, by gradual degrees, a liquid. The net change is from gas to liquid, but a sharp transition during which two phases are both present is never observed.

(e) Mass percentage = 0.637%; mole fraction = 0.00163; molality = 0.0904 mol kg^{-1}; molarity = 0.0924 mol L^{-1}.

(f) Henry's law constant k_H = 613 atm^{-1}.

(g) $v = 3$. The reaction of 1 particle gives $(1 - 0.326) + 3(0.326) = 1.652$ particles; $i = 1.652$. In this case, water serves as true reactant as well as a dispersing medium.

(h) $\pi = 3.73$ atm.   **(i)** $-0.67°$

Chapter 7

(a)
$$\frac{P_{SO_2}}{P_{O_2}} = K_1$$

$$\frac{P_{SO_3}}{P_{SO_2}P_{O_2}^{1/2}} = K_2$$

$$\frac{1}{P_{SO_3}} = K_3$$

(b) $S(s) + \frac{3}{2} O_2(g) + H_2O(\ell) \rightleftharpoons H_2SO_4(\ell)$; $K = K_1K_2K_3 = 2.2 \times 10^{79}$

(c) $K_2 = 2.63$ **(d)** exothermic   **(e)** $P_{O_2} = 0.029$ atm.

(f) Left to right; the pressure decreases.

(g) An increase in temperature shifts both equilibria to the left. A decrease in volume does not affect equilibrium 1 and shifts equilibrium 3 to the right.

Chapter 8

(a) 4.4 to 4.8.   **(b)** $[H_3O^+] = 5.0 \times 10^{-4}$ M; $[OH^-] = 2.0 \times 10^{-11}$ M.

(c) $H_2SO_3(aq) + H_2O(\ell) \longrightarrow HSO_3^-(aq) + H_3O^+(aq)$. The stronger acid is H_3O^+. The stronger base is HSO_3^-.

(d) pH = 3.41   **(e)** pH = 3.11

(f) $H_3O^+(aq) + CO_3^{2-} \longrightarrow HCO_3^-(aq) + H_2O(\ell)$. The HSO_4^- in the sulfuric acid reacts according to $HSO_4^-(aq) + CO_3^{2-}(aq) \longrightarrow SO_4^{2-}(aq) + HCO_3^-(aq)$. This gives rise to an HCO_3^-/CO_3^{2-} buffer that resists further changes in pH. An excess of acid rain overwhelms the buffer and leads to the formation of H_2CO_3.

(g) 3.16×10^{-4} M   **(h)** The pH is 3.81.

Chapter 9

(a) $MgCO_3 \cdot 3 H_2O$

(b) $CaMg(CO_3)_2(s) \longrightarrow Ca^{2+}(aq) + Mg^{2+}(aq) + 2 CO_3^{2-}(aq)$

$$[Ca^{2+}][Mg^{2+}][CO_3^{2-}]^2 = K_{sp}$$

(c) 9.3×10^{-3} g L^{-1}

(d) 8.7×10^{-6} g L^{-1}, smaller than in part (c) because of the common-ion effect.

(e) $[CO_3^{2-}] = 5.8 \times 10^{-6}$ M; $Q = 2.2 \times 10^{-9} < K_{sp} = 8.7 \times 10^{-9}$, so the lake is slightly less than saturated.

(f) Increase   **(g)** Increase

(h) Dissolve the carbonate sample in acid and heat to remove carbonate ion in the form of carbon dioxide. Saturate with H_2S and adjust the pH to 9 with an NH_4^+/NH_3 buffer. A sulfide precipitate indicates the presence of Group 3 cations (here, Fe^{2+} or Mn^{2+}). Further analysis (not discussed in this chapter) would be required to decide whether both cations are present, or only one.

Chapter 10

(a) 102 kJ   **(b)** 1.2×10^3 kJ   **(c)** -507 kJ

(d) -676.49 kJ   **(e)** 2.11×10^4 kJ

(f) $\Delta H - \Delta E = 43.9$ kJ   **(g)** -0.60 L atm $= -61$ J

Chapter 11

(a) Negative   **(b)** Positive

(c) $\Delta H° = -160.8$ kJ; $\Delta S° = -409.5$ J K^{-1} **(d)** 393 K $= 120°C$

(e) 4.0×10^4 **(f)** 2.0×10^{-5}

(g) 92.0 J K^{-1} mol^{-1}, close to the Trouton's rule value of 88 J K^{-1} mol^{-1}

(h) 16.7 atm

Chapter 12

(a) $PbS(s) + 2\ O_2(g) \longrightarrow PbSO_4(s)$

$PbS(s) + 2\ PbO(s) \longrightarrow 3\ Pb(s) + SO_2(g)$

$PbS(s) + PbSO_4(s) \longrightarrow 2\ Pb(s) + 2\ SO_2(g)$

(b) $PbS(s) + Fe(s) \longrightarrow Pb(s) + FeS(s)$; $\Delta G° = -1.7$ kJ, (barely) spontaneous

(c) Anode: $Pb(s, crude) \longrightarrow Pb^{2+}(aq) + 2\ e^-$

Cathode: $Pb^{2+}(aq) + 2\ e^- \longrightarrow Pb(s, pure)$

(d) An electric current still flows, but it deposits pure lead on the crude lead electrode, and lead from the pure electrode is oxidized and passes into solution.

(e) Between PbO and CuO.

(f) $Pb(s) + PbO_2(s) + 2\ SO_4^{2-}(aq) + 4\ H^+(aq) \longrightarrow 2\ PbSO_4(s) + 2\ H_2O(aq)$

(g) 2.31 g

Chapter 13

(a) Manganese appears at the cathode and MnO_2 at the anode.

(b) Oxidation: $Mn(OH)_2(s) + OH^-(aq) \longrightarrow Mn(OH)_3(s) + e^-$

Reduction: $O_2(aq) + 2\ H_2O(aq) + 4\ e^- \longrightarrow 4\ OH^-(aq)$

Total: $4\ Mn(OH)_2(s) + O_2(aq) + 2\ H_2O(aq) \longrightarrow 4\ Mn(OH)_3(s)$

$\Delta \mathcal{E}° = 0.401 - (-0.40) = 0.80$ V

(c) $K = 1 \times 10^{54}$

(d) Mn^{3+} spontaneously oxidizes I^-

$2\ Mn^{3+}(aq) + 2\ I^-(aq) \longrightarrow 2\ Mn^{2+}(aq) + I_2(s)$ $K = 9 \times 10^{32}$

(e) 3.8×10^3 J = 1.0×10^{-3} kW h

(f)   Permanganate ion   (g)   2.12 V

Chapter 14

(a) 3.40×10^{-4} mol L^{-1} s^{-1} (b)   2.7 mol L^{-1} s^{-1}

(c)   All of the oxygen is consumed, so the first step of the process is over.

(d) 5.43×10^{-5} mol L^{-1} s^{-1} (e) 1.09×10^{-4} mol L^{-1} s^{-1}

(f) 2.45×10^{-5} mol L^{-1} s^{-1} (g)   First order, $k = 0.0133$ s^{-1}

(h)   52.1 s   (i)   It happens too fast for study in this experiment.

(j)   34 kJ mol^{-1} (k) Fe^{2+} acts as a catalyst.

(l) $2\ SO_2(aq) + O_2(aq) + 2\ H_2O(\ell) \longrightarrow 2\ SO_4^{2-}(aq) + 4\ H^+(aq)$

Chapter 15

(a)   An atom of ^{222}Rn has 86 electrons outside the nucleus. Inside the nucleus are 86 protons and $222 - 86 = 136$ neutrons. An atom of ^{220}Rn has the same number of electrons and protons, but only 134 neutrons in the nucleus.

(b)   Four alpha particles are produced in making ^{222}Rn from ^{238}U; three are produced in making ^{220}Rn from ^{232}Th.

(c)   If $^{238}_{92}$U were used to lose four alpha particles, $^{222}_{84}$Po would form instead of $^{222}_{86}$Rn. Somewhere along the way, two beta particles ($_{-1}^{0}e^{-}$) must be ejected from the nucleus to raise Z to 86. The same is true of the production of ^{220}Rn from ^{232}Th.

(d) $^{222}_{86}$Rn $\longrightarrow$ $^{218}_{84}$Po + $^{4}_{2}$He; $\Delta m = -0.0060$ u < 0; allowed

     $^{220}_{86}$Rn $\longrightarrow$ $^{216}_{84}$Po + $^{4}_{2}$He; $\Delta m = -0.00688$ u < 0; allowed

(e) $\Delta E = -6.41$ MeV = -1.03×10^{-12} J

(f) $A = 1.14 \times 10^8$ Bq

(g) $A = 9.0 \times 10^6$ Bq

(h)   Greater

Chapter 16

(a)   91.1 nm   (b) 8.8×10^{-19} J

(c)   0.52 nm = 5.2 Å   (d) 6.40×10^{-7} m = 6400 Å

(e) 5.98×10^{-2} m = 5.98 cm, in the microwave region

(f)   No, because the lowest energy absorption for a ground-state hydrogen atom is in the ultraviolet region of the spectrum.

(g)   4 angular nodes and 104 radial nodes

(h) 9.37×10^{-25} J; 0.564 J mol^{-1} (i)   Too low

Chapter 17

(a) $[Kr]4d^{10}5s^25p^5$

(b)   The first ionization energy is higher for I than for Te, but lower for I than for Xe. The electron affinity is higher for I than for Te or Xe.

(c)   Electronegativity difference = 1.73. The compound (NaI) is ionic.

(d)   Electronegativity difference = 0.11. The C—I bond is polar covalent.

(e)   The +7 oxidation state corresponds to participation of all the $5s$- and $5p$-electrons in bonding. By the end of the fifth period, the $4d$-electrons are so low in energy and so close to the nucleus that they cannot participate in bonding (see Fig. 17–6).

(f) strongest $HIO_4 > HIO_3 > H_5IO_6$ weakest. Note that the strength of the acid with iodine in the $+5$ oxidation state falls between the strengths of the two acids in which iodine is in the $+7$ oxidation state.

Chapter 18

(a) Yellow light, near 585 nm

(b) $(\sigma_{4s})^2(\sigma_{4s}^*)^2(\sigma_{4p_z})^2(\pi_{4p})^4(\pi_{4p}^*)^4$, diamagnetic

(c) $(\sigma_{4s})^2(\sigma_{4s}^*)^2(\sigma_{4p_z})^2(\pi_{4p})^4(\pi_{4p}^*)^3$ Stronger. Bond order is 3/2 versus 1.

(d) The lowest-energy excited state arises from the excitation of an electron from the filled π_{4p}^* orbital to the unfilled $\sigma_{4s}^{*2}\sigma_{4p_z}$-orbital.

(e) sp^3 hybridization   **(f)** 3/2 order

(g) Overall: $2\,O_3 \longrightarrow 3\,O_2$. This mechanism does not require formation of atomic oxygen, as does the mechanism in Section 18–5 that is based only on chlorine.

Chapter 19

(a) *cis*-diamminedichloroplatinum(II), potassium hexachloroplatinate(IV), and potassium tetrachloroplatinate(II)

(b) $[Pt(NH_3)_2(CN)_4]$ and $[Pt(CN)_6]^{2-}$

(c) Pt(II) forms square-planar complexes. Consequently, two forms are possible, arising from *cis* or *trans* placement of the ethylene molecules.

(d)

The two structures derive from those shown in Figure 19–9 by replacement of an en ligand with a pair of Cl^- ligands.

(e) The six *d*-electrons in Pt(IV) are all in the lower t_{2g} level in a low-spin, large Δ_0 complex. All are paired, so the compound is diamagnetic.

(f) Diamagnetic, with the four lowest energy levels in the square-planar configuration (all but the $d_{x^2-y^2}$) fully occupied.

(g) Red transmission corresponds to absorption of green light by $K_2[PtCl_4]$. A colorless solution has absorptions at either higher or lower frequency than visible. Because Cl^- is a weaker field ligand than NH_3, the absorption frequency should be higher for the $[Pt(NH_3)_4]^{2+}$ complex, putting it in the ultraviolet region of the spectrum.

(h) $[Pt(NH_3)_4][PtCl_4]$, tetraammineplatinum(II), tetrachloroplatinate(II)

Chapter 20

(a) Volume is 152.0 Å^3. Eight atoms per unit cell.

(b) 18.5 Å. Angle $2\theta = 28.69°$ **(c)** It is amorphous.

(d) Cubic to tetragonal: one cell edge is stretched or shrunk. Tetragonal to monoclinic: the angles between two adjacent faces (and their opposite faces) are changed from $90°$. Monoclinic to triclinic: the remaining two angles between faces are deformed from $90°$.

(e) The density is 2.36 g cm^{-3}.

(f) P_4 (white) and PCl_3 are molecular solids; P (black) is a covalent network solid; Na_3PO_4 is an ionic solid.

Credits

Chapter 20

Chapter opener: Brian Parker/Tom Stack & Associates; **20–2:** Ray Nelson/Phototake; **20–5a,b:** Stephen J. Guggenheim/UIC Department of Geological Sciences; **20–6, 20–7, 20–9a, 20–12, 20–33, unnum. figure p. 829:** Charles D. Winters; **20–8:** Runk Schoenberger/Grant Heilman Photography, Inc.; **20–9b:** Thomas Hunn Co./Visuals Unlimited; **20–22:** Leon Lewandowski; **unnum. figure p. 794:** Yoav Levy/Phototake; **unnum. figure p. 822:** Kurt Nassau/Phototake.

Chapter 21

Chapter opener: *The MIT Report;* **21–2:** Courtesy of Intel Corporation; **21–3:** Kurt Nassau/Phototake; **21–4:** Julius Weber; **21–8, 21–9, 21–16, 21–21:** Leon Lewandowski; **21–11:** Mobil Corporation; **21–12:** Niagara Mohawk Power; **21–13, 21–23:** Carborundum Company; **21–14:** GTE Labs, Inc.; **21–15:** Herb Pressman/International Stock Photo; **21–17:** James L. Amos/Peter Arnold, Inc.; **21–19:** Courtesy of Schott Fiber Optics; **21–20:** SIU/Photo Researchers, Inc.; **unnum. figure p. 835:** Allied Signal Aerospace Company; **unnum. figure p. 845:** Jet Set—Sygma; **unnum. figure p. 846:** Courtesy of PPG Industries, Inc.; **unnum. figure p. 850:** Coco McCoy/Rainbow; **21–C:** Yoav Levy/Phototake; **21–E:** American Superconductor Company.

Chapter 22

Chapter opener: Dresser Industries/Kellogg Company; **22–3a:** Frank Grant/International Stock Photo; **22–3b:** Ireco, Inc.; **22–3c:** Arturo A. Wesley/Black Star; **22–10, unnum. figure p. 876, unnum. figure p. 889:** Leon Lewandowski; **22–16:** Johnston-Matthey; **22–17, 22–18:** Charles D. Winters; **unnum. figure p. 867:** George Semple; **unnum. figure p. 869:** Soames Summerhays/Science Source/Photo Researchers, Inc.; **unnum. figure p. 870:** Pamela Harper; **unnum. figure p. 876:** C. B. Jones/Taurus Photos; **unnum. figure p. 881:** Manley Prim/FMC Corporation; **unnum. figure p. 886:** Thomas Hovland/Grant Heilman Photography, Inc.; **unnum. figure p. 887:** Phototake.

Chapter 23

Chapter opener: Tom Carroll; **23–4, 23–6:** Charles D. Winters; **23–7:** Leon Lewandowski; **23–9:** The Corning Museum of Glass; **23–21:** Argonne National Laboratory; **unnum. figure p. 908:** International Stock Photos.

Chapter 24

Chapter opener: Occidental Petroleum; Robin Smith/Tony Stone Images; **24–10:** Charles Steele; **unnum. figure p. 931:** Tom Tracy/Black Star; **unnum. figure p. 951:** Jack Fields/Photo Researchers, Inc.; **unnum. figure p. 952:** Phillip A. Harrington/Fran Heyl Associates; **unnum. figure p. 954:** Ulof Bjorg Christianson/Fran Heyl Associates.

Chapter 25

25–1a: Occidental Petroleum; **25–b, 25–4, 25–22, unnum. figure p. 982:** Leon Lewandowski; **25–6, 25–14:** Charles D. Winters; **25–23:** Union Carbide; **unnum. figure p. 974:** Phillip A. Harrington/Fran Heyl Associates; **unnum. figure p. 984:** Tom Carroll; **unnum. figure p. 985:** Corbis-Bettmann; **25–A:** Comstock.

Glossary/Index